# Alluvial Aquifer Processes

# Alluvial Aquifer Processes

Edited by

Milan Dimkić

PUBLISHING

**Published by** **IWA Publishing**
**Unit 104-105, Export Building**
**1 Clove Crescent**
**London E14 2BA**
Telephone: +44 (0)20 7654 5500
Fax: +44 (0)20 7654 5555
Email: publications@iwap.co.uk
Web: www.iwapublishing.com

First published 2021


*British Library Cataloguing in Publication Data*
A CIP catalogue record for this book is available from the British Library

ISBN: 9781789060898 (Paperback)
ISBN: 9781789060904 (eBook)
ISBN: 9781789060911 (ePUB)

# Contents

**Chapter 7**
***Removal of pesticides from groundwater by filtration*** ..... ***397***
*Nevena Živančev and Milan Dimkić*



**Chapter 8**
***Transformation of nitrogen compounds in groundwater*** ..... ***437***
*Marija Perović and Milan Dimkić*

## Chapter 10

# The Editor

**Professor Dr. Milan A. Dimkić** was born in 1953 in Belgrade, Serbia. He attended the Belgrade Gymnasium of Mathematics and received his bachelor's and doctoral degrees from the University of Belgrade, Faculty of Civil Engineering. His scientific and professional career is closely tied with the Jaroslav Černi Water Institute in Belgrade, where he founded and developed the Department of Groundwater Protection. He was Director General of that institute from 1999 to 2018 and then scientific advisor until 2020. He has been an associate professor since 2008 and full professor since 2013 at the University of Novi Sad, Faculty of Technical Sciences, Department of Environmental Protection. He teaches courses that focus on surface water and groundwater management and processes that take place in aquifers.

During his professional career, Prof. Dimkić has been involved in a large number of groundwater protection and post-contamination rehabilitation projects. He has also participated in the design and implementation of a number of groundwater sources, planning and development of hydraulic engineering systems, and the development of water management and water sector governance.

Professor Dimkić is particularly recognized for his scientific and professional contributions in the following areas:

- Groundwater flow in confined aquifers, taking into account the compressible properties of the formation;
    - Purification processes during groundwater flow;
- Technical issues associated with groundwater abstraction;
- Conservation, protection and remediation of groundwater resources; and
- Management of water resources.

Professor Dimkić has published more than 200 papers and managed or participated in a considerable number of scientific projects. He has supervised numerous bachelor's theses and seven doctoral dissertations.

Professor Dimkić has been President of the Serbian Water Pollution Control Society since 2006, a member of the Serbian Academy of Engineering Sciences since 2015, and IWA Fellow since 2016, to mention just a few of many memberships and related pursuits.

Professor Dimkić has authored or edited many publications and books. He is Editor-in-Chief of the journal 'Water Research and Management'. He has chaired or co-chaired numerous international conferences, including three IWA groundwater conferences in Belgrade (2007, 2011 and 2016).

The predecessor of this book, titled 'Water Management in Large River Basins' and printed by IWA Publishing, London in 2008, was co-edited by Prof. Dimkić, Prof. Heinz-Jürgen Brauch and Dr. Michael Kavanaugh.

# Preface

Back in 2008, IWA Publishing released the book 'Water Management in Large River Basins', edited by Milan A. Dimkić, Heinz-Jürgen Brauch and Michael Kavanaugh. The book was well received in a large number of countries.

However, there have been many changes on our planet in the past 12 years. Global population continues to grow, but while it is virtually exploding in some parts of the world, it is declining elsewhere. This is having a major impact on the global economy and, consequently, on macro geopolitics. Stability appears to be on the downslope. People are migrating, largely because of wars but also due to major economic and political disparities. Climate and climate change are part of the counterpoint. While my associates and I are finalizing this book, the COVID-19 pandemic is a threat to the global population and economy.

Pressures on the sustainable use of water resources are increasing. Groundwater accounts for about half of the water used by humankind. Half of that water is of alluvial origin. Shallow groundwater is a source of water for people, animals, rivers, habitats… People tend to concentrate in alluvial plains, where human activity exerts pressures on waters and other resources. Sustainable use and conservation of groundwater hold a prominent place among the complex issues of our planet and civilization. Shallow groundwater is increasingly on the radar of the water profession and science.

Sustainable use and conservation of alluvial groundwater resources are becoming imperative.

Since the previous book, my associates and I have gained additional insights and experience in examining various natural, technical and social processes.

I will mention a few conferences we have organised or managed, whose outcomes are reflected in this book:

- International Conference on Planning and Management of Water Resource Systems, 2008, Novi Sad, Serbia


Editor: Milan Dimkić
doi: 10.2166/9781789060904_xvii

- IWA Specialist Groundwater Conference, 2011, Belgrade, Serbia
- UNESCO Conference on Emerging Pollutants in Water, 2013, Belgrade, Serbia
- International Conference on Climate Change Impact on Water Resources, 2013, SASA, Belgrade, Serbia
- Milankovitch Anniversary UNESCO Symposium, 2014, SASA, Belgrade, Serbia
- IWA Specialist Groundwater Conference, 2016, Belgrade, Serbia
- International Conference: Transition Countries and Water as a Factor of Stability, 2016, Belgrade, Serbia

I have personally supervised the doctoral theses of some excellent students, including Dr. Srđan Kovačević, Dr. Nevena Živančev, Dr. Marija Perović, Dr. Jelena Zarić, Msc David Mitrinović, Dr. Brankica Majkić-Dursun and Msc Aleksandar Čalenić. The outcomes of their work are included in this book.

I decided to compile this book in response to persistent requests from Mr. Mark Hammond of IWA, but also feeling an obligation of sorts as I approach the final stages of my professional career. In the interim, I ceased to manage the Jaroslav Černi Water Institute and some of my students began assuming important roles in scientific, educational and professional circles.

The book comprises 11 chapters and 4 appendices, which encompass current topics associated with the state, processes and problems of alluvial aquifers, where my associates and I felt we could make valid contributions. The book was written in a holistic manner, but care was taken to ensure that each chapter can be perused separately.

I hope that the book will achieve its objective – to be interesting and useful to a wide circle of experts, university professors, planners, students in advanced stages of learning, and other parties concerned.

A considerable number of highly experienced engineers, listed below, have also participated in the preparation of the book and enriched its content.

I would like to highlight the input of Mr. Ivan Zavadsky, Executive Secretary of the International Commission for the Protection of the Danube River (ICPDR), and his young associate Miss Jelena Krstajić.

Of course, I have to mention my esteemed colleagues – active professors, professors emeritus and experts – who have participated in the effort: Prof. Dr. Slobodan Vujasinović (hydrogeology), Prof. Emeritus Mirjana Vojinović-Miloradov (chemical and environmental engineering), Assoc. Prof. Ivana Mihajlović (physical chemistry and environmental engineering), Dr. Ivana Matić-Bujagić, Scientific Associate (analytical chemistry and environmental protection technologist), Assoc. Prof. Dr. Svetlana Grujić (analytical chemistry and environmental protection technologist), Prof. Tatjana Đurkić (analytical chemistry and environmental protection technologist), Prof. Dr. Mila Laušević (analytical chemistry and environmental protection technologist), Dejan Dimkić (hydraulic engineer), Prof. Dr. Milenko Pušić (hydrogeology), and Dr. Milan Dotlić (applied mathematics).

The chapters have been reviewed by leading experts.

Chapter 2: Slobodan Petković, PhD in hydraulic engineering; Prof. Veselin Dragišić, PhD in hydrogeology; Prof. Milan Šojić, PhD in economics; and Milivoje Ilić, MBA in IT.

Chapter 3: Academician Prof. Miljko Satarić, PhD in physics; Prof. Slobodan Knežević, PhD in palaeontology; and Prof. Draženko Nenadić, PhD in hydrogeology.

Chapter 4: Dr.h.c. prof. Ing Ľubomír Šooš, PhD, Dean of the Faculty of Mechanical Engineering at Slovak University of Technology and Prof. Ing Josef Čáslavský, Csc, Faculty of Chemistry at Brno University of Technology, Prof. Aleksandar Đorđević, PhD in chemistry.

Chapter 5: Assoc. Prof. Dragoljub Bajić, PhD in hydrogeology.

Chapter 6: Prof. Jelena Radonjić, PhD in environmental engineering.

Chapter 7: Prof. Ivana Ivančev Tumbas, PhD in environmental protection.

Chapter 8: Academician Prof. Miljko Satarić, PhD in physics, and Assoc. Prof. Dragan Adamović, PhD in environmental engineering.

Chapter 9: Prof. Slavica Ražić, PhD in chemistry.

Chapter 10: Prof. Milenko Pušić, PhD in hydrogeology.

Chapter 11: Assoc. Prof. Aleksandar Đukić, PhD in hydraulic engineering.

Members of the Serbian Academy of Sciences and Arts (SASA), Prof. Dr. Miljko Satarić and Prof. Dr. Zoran Đurić gave an additional impetus, as did Prof. Dr. Prvoslav Marjanović and Prof. Dr. Slobodan Knežević.

Associate Prof. Dr. Srđan Kovačević and Prof. Dr. Slobodan Vujasinović provided major assistance and support.

Jaroslav Černi Water Institute and Geo Margi-Engineering Ltd helped fund the translation and print-ready design of the book. Their backing is much appreciated.

Sincere gratitude is also due to the translators, Mrs. Vida Marjanović and in part Mrs. Dubravka Miladinov.

Miss Snežana Ivanović provided outstanding clerical services and special thanks go to Mr. Aleksandar Čalenić, Mr. Aleksandar Anđelković and Miss Dunja Gajinović for most of the layout.

Mr. Milan Peca Nikolić and his revision studio played a substantial role in the book cover design, as in the case of the book printed in 2008.

My associates and I are extremely grateful to the above-mentioned contributors and supporters, as well as all those who provided assistance but have not been mentioned.

I particularly wish to thank my wife for her patience and support. Gratitude is also due to the families of all those who actively participated in the preparation of the book.

**Milan A. Dimkić**

# Chapter 1
# Introduction

The development of water management can be divided into three distinct stages. *Stage 1* – abundance, in which water use and water pollution are low compared to the quantity and quality of available water resources. *Stage 2* – overexploitation, in which water use is considerable and water quality substantially disturbed. *Stage 3* – aiming at sustainable development, which requires water management to ensure good status of water resources and safeguarding of water quantity and quality for the future.

Global population is expected to grow from more than seven billion at present to nine billion by the year 2050. As a result, resource issues are becoming increasingly severe. Estimates suggest the water demand could grow from the current 3000–4000 $km^3$/year to about 5000 $km^3$/year by 2050.

The questions raised in connection with natural resources are practical, but also philosophical and political. The use and management of resources certainly need to be local, regional and global. We are witnessing tumultuous social, economic and political shifts and changes. The demographic growth is increasing the global drinking water demand. At the time of writing, the COVID-19 pandemic has already taken hundreds of thousands of lives and is likely to have grave economic and uncertain political consequences.

Climate change occurs periodically on our planet and lasts for a long time. Ice ages and interglacial periods alternated in the Pleistocene and Holocene. However, changes have been quite rapid over the past decades. The Intergovernmental Panel on Climate Change indicated in its AR5 Synthesis Report that each of the past three decades has been consecutively warmer than any other decade since 1850. Further warming will cause certain wet regions to become wetter and some presently arid regions drier. This could have a dramatic effect on a number of regions, if not the entire world.

It is becoming increasingly difficult to conserve our planet's nature. Water, air and soil are mostly threatened by humankind. By way of illustration, an expedition noted plastic waste on the bottom of the Mariana Trench (at a depth of some 10,000 m, the lowest point on Earth).

Optimal water governance in variable social and climate conditions is a topical issue worldwide, which largely remains unresolved.


Editor: Milan Dimkić
doi: 10.2166/9781789060904_0001

There is an abundance of water on our planet. However, 97.5% is saline water of the seas and oceans. The remaining 2.5% is fresh water, of which slightly more than three-quarters is ice. The liquid phase accounts for 23% of fresh water, of which 22% is groundwater. Rivers and lakes make up about 0.3% of our planet's fresh water resources. The Food and Agriculture Organisation (FAO) estimates that the volume of fresh groundwater is 10.53 million $km^3$ (FAO, 2016). However, global water resources are not evenly distributed.

How should water scarcity be addressed? Viewed in broad terms, by meeting the water demand and conserving our planet's water resources. However, the water-food-energy counterpoint needs to be taken into account in an appropriate fashion.

The sustainability and conservation of water resources are highly complex. In addition to social, economic and political factors, there are significant natural constraints – local, regional and global.

The response to these various challenges is certainly multifaceted. Above all, the answers should be founded upon a sound understanding of natural and social processes, as well as the condition of water resources. In turn, this requires sufficient knowledge about the water sector and the various concerns in the field of water. To that end, our society needs to ensure the availability of required scientific, professional, institutional and operational capacities. Ever present in this regard is the matter of designing appropriate policies, which are efficient and implementable.

Dispersion of the use of water resources can facilitate adaptation to change. This includes the use of groundwater and surface water, seawater desalination, wastewater recycling, and water storage (primarily in aquifers but also in surface reservoirs).

Agriculture needs to focus on crops compatible with water availability in a given community. Climate change will improve conditions (e.g. two harvests per year) in some places and aggravate them in others. Farmers in certain regions will have to opt for other crops. Where irrigation is applied, water use efficiency will need to be maximised.

Alluvial groundwater is becoming increasingly important for drinking water supply, but also for agriculture and other human needs. Estimates show that groundwater meets about half of the municipal water demand. Half of that groundwater comes from alluviums. The advantages of alluvial aquifers are that they are close to consumers and that the extracted groundwater is renewable.

Human civilisations developed in alluviums during the Holocene. Armies, roads, trade and agglomerations were largely concentrated in river valleys. Alluviums are resource treasuries (water, forests, farmland, etc.), but at the same time they are exposed to pressures from the population and economy. These threats to alluvial plains and alluvial aquifers require very careful planning and persistent implementation of protection measures.

Sustainable management and conservation of alluvial aquifers requires the understanding and identification of alluvium formation and sedimentation conditions (input of materials and mechanisms of erosion, transport and accumulation), as well as the connection of a given aquifer with other aquifers, surface water and other geologic formations. All of these factors are determined by their origin. They differ both globally and locally, so the origin and characteristics of shallow aquifers developed in Quaternary sediments are relevant for the protection, conservation and sustainable use of groundwater resources. Even though Quaternary sediments are rather fragmented, in addition to their significance for the population and water supply, they help us understand and reconstruct paleo climate conditions during the Quaternary period.

Sudden social and natural changes increasingly lead the world into a tangle of global events and problems whose outcomes are not easy to perceive, let alone influence. One of the factors that can help is knowledge or, in other words, expertise focussed on a single problem or area. The ability to synthesise information and master comprehensive models is also extremely important.

This book, a monograph, is a modest attempt to contribute to the understanding of planetary water management, groundwater and certain natural processes in alluviums.

The book is a continuation of sorts of another book released in 2008 by IWA Publishing – 'Groundwater Management in Large River Basins', edited by Milan A. Dimkić, Heinz-Jürgen Brauch and Michael Kavanaugh.

The book comprises 11 chapters and 4 appendices. In essence, each chapter was written in a stand-alone manner but care was taken to ensure that they are complementary and in holistic harmony. The chapters following this introductory one can be divided into three groups.

## GROUP I

This group comprises a single, rather voluminous chapter – Chapter 2. It does not purport to offer prescriptions for water governance or management of groundwater or alluvial water resources. Still, the intent is to bring the reader closer to the main natural, administrative (at different levels), economic, financial and institutional features and processes associated with water management, in particular alluvial groundwater management. Chapter 2 describes general positions in respect of water governance and analyses water availability and the water-food-energy nexus. It discusses three levels of water management – global, regional and local, from administrative, economic, financial, scientific, technical and planning perspectives. It was an honour to incorporate a section on the efforts of the International Commission for the Protection of the Danube River (ICPDR), written by its Executive Secretary Mr. Ivan Zavadsky and his associate Miss Jelena Krstajić.

Chapter 2 emphasises the main concepts and processes associated with the transformation of groundwater quality and contains an overview of the processes discussed in other chapters of the book.

The final section of Chapter 2 includes a very good example of how the expansion of human activity and settling juxtaposes the conservation of alluvial groundwater resources.

## GROUP II – GENERAL CHAPTERS

This group comprises Chapters 3, 4 and 11, which discuss the natural background of processes that take place in alluvial aquifers, which are described in the remainder of the book.

Chapter 3: *Alluvial Aquifers – The Legacy of the Quaternary Period*

This chapter provides insight into the formation of alluvial aquifers, particularly at moderate latitudes. Special attention is devoted to alternating ice ages and interglacial periods of the Pleistocene and the consequences of erosion, alluvial sedimentation, and mechanical and chemical properties (e.g. the oxic state of an aquifer). In essence, the chapter comprises three parts.

- The first part is general and contains an overview of known geologic and cosmic conditions that have determined the mega processes on our planet in the Quaternary period. Phenomena and occurrences relevant to the origin and formation of alluvial sediments are also covered.
- The second part illustrates (we hope) how the age of alluvial deposits is determined.
- The third part reports original investigations aimed at connecting high mountains in the upper basin of the Sava River (a tributary of the Danube) with polycyclic sedimentation in the Sava alluvium, particularly at Belgrade's water supply source that relies on groundwater. The studied processes have had a dominant effect on the creation of conditions for the development of a groundwater source, particularly its yield, the aquifer's purification potential and the oxic state of the groundwater.

Chapter 4: *Oxic State of Alluvial Aquifers*

Chapter 4 discusses the effect of the oxic state of an alluvial aquifer on the direction and rate of the processes discussed in the book. An aquifer's oxygen balance depends on the oxidation-reduction of organic and inorganic substances, groundwater flow rate, and hydrogeologic, hydrologic and similar conditions. It is evident that fine sediment tends to be deposited along the lower courses of large rivers and that, almost as a rule, the prevailing conditions are less oxic. The chapter briefly analyses the transformation of nitrogen compounds into other substances in groundwater, as well as water well clogging.

Chapter 11: *Droughts and Alluvial Groundwater Resources*

Apart from other major pressures on water use and water management in general, population growth, climate and climate change are of great importance. The global population is expected to increase from the current more than seven billion to nine billion in 2050. Consequently, resource issues are becoming increasingly acute.

In addition to the cosmic, geologic and hydrologic changes in the Quaternary period, over the past hundred years there has been an upward average temperature trend, which is largely attributed to human activity. It appears that the trend has accelerated in the past several decades and will likely continue to pick up pace in the future. Continued warming is expected to render certain wet regions wetter and relatively arid regions drier. Water scarcity will likely spread in arid, sub-arid and other low-precipitation regions.

The chapter describes the types and indexing of droughts. Worldwide examples are provided. An interesting method developed in Serbia is described, which correlates average temperatures with average river discharges.

General components of river basin water budgets are discussed, along with the vertical balance of overlying semi-permeable strata. Parameters and relationships relevant to infiltration, evapotranspiration and alluvial aquifer capacities are analysed.

## GROUP III – CHAPTERS 5, 6, 7 AND 8

These chapters describe the types of processes or states that facilitate the dissemination and transformation of a certain substance in nature, particularly alluvial groundwater.

Chapter 5: *Iron Incrustation of Water Wells*

Groundwater extraction from alluvial sediments along large rivers faces a number of challenges caused by declining yields of individual wells and the entire well field.

Several causes potentially lead to declining yields of alluvial groundwater sources, regardless of the condition of the pumping wells. In the case of well fields along rivers, which rely on bank filtration, riverbed colmation frequently leads to yield decline.

The usual causes of water well clogging are

- mechanical clogging (colmation) of the well screen and near-screen zone;
- chemical clogging (incrustation), which is either purely chemical (e.g. carbonate incrustation) or requires a biochemical catalyst (e.g. trivalent iron incrustation); and
- biological clogging (biofouling), where the biological agent is dominant.

This chapter provides an overview of the causes of well clogging and then focusses on trivalent iron. Clogging with trivalent iron occurs under anoxic conditions in the aquifer. We had an opportunity to study the yields of 150 mostly radial collector wells at Belgrade Groundwater Source and other sources of water supply, from 2005 to 2018. We plotted critical well entrance velocities versus screen clogging.

As far as we are aware, that was the first attempt to correlate flow resistance, entrance velocity and biochemical parameters of the groundwater. The chapter describes case studies and contains graphics and calculations. The effects of well clogging by iron under anoxic conditions can be a considerable operational and financial challenge in terms of well field maintenance. Basic information about the rehabilitation of clogged well screens is also provided.

Chapter 6: *Removal of Pharmaceuticals from Groundwater by Filtration*

According to the World Health Organisation (WHO), pharmaceuticals are any substances or pharmaceutical products administered to treat people and animals, as well as those used to modify or investigate physiological conditions or pathologies in favour of the recipient.

The chapter first presents the classification of pharmaceuticals and global experiences in connection with their use, management and occurrence in waters. Sorption and degradation of pharmaceuticals in groundwater are analysed. Then the chapter describes the results of detailed, targeted monitoring of pharmaceuticals in surface water and alluvial groundwater in Serbia from 2009 to 2015. Four rather common pharmaceuticals were selected: trimethoprim, carbamazepine, diclofenac, and N-acetyl-4-aminoantipyrine. A large number of surface water samples from the Sava, Danube, Morava and Tisa rivers and groundwater samples of their riparian aquifers were analysed. In order to gain some insight on the ground, in addition to in the laboratory, solutions of pharmaceuticals and tracers were injected into aquifers, at a certain distance from pumping wells. The data collected in the field and lab, as well as from long-term testing, were used for mathematical modelling. The results are interesting as they show the concentrations of pharmaceuticals in surface water and alluvial groundwater in Serbia. Some insight was also gained into the fate of the pharmaceuticals during bank filtration.

Chapter 7: *Removal of Pesticides from Groundwater by Filtration*

Agriculture has been relying on plant protection agents for 70 years. Previously, undesirable insects and weeds had been controlled by crop rotation, scheduling of sowing and planting to avoid periods of abundance of pests, mechanical weeding, and the like. Modern farming is unthinkable without pesticides because crop yield losses would be too high.

Roughly 90% of pesticides are used in agriculture. On a global scale, this amounts to four million tons per year. According to FAO, 53% of this quantity is applied in Asia, 30% in the US, and 14% in Europe (FAO, 2018).

This chapter comprises three parts. The first contains general information about pesticide classification, application and occurrence in the environment. The second part describes the most likely fate of pesticides. The third part shows the results of investigations conducted in rivers and shallow aquifers in Serbia from 2009 to 2015, along with the results of laboratory analyses and tracing tests in the field. Numerical analyses provide indications of the rate of pesticide sorption to aquifer material and cap rock.

Chapter 8: *Transformation of Nitrogen Compounds in Groundwater*

Population growth has increased nitrate emissions from agricultural and other sources. Fertilising is one of the most common measures for boosting yields and achieving good crop quality. However, excessive and inadequate application can have a direct adverse effect and result in high levels of environmental contamination. Leakage of nitrogen compounds from farmland affects groundwater quality, but also the global nitrogen cycle due to slow groundwater flow and the accumulation and biological transformation of nitrogen compounds.

Nitrogen sources, such as municipal wastewater and inadequate waste disposal sites, are a special threat to the water quality of shallow, aerobic aquifers, where nitrification conditions are favourable. It has been established that elevated concentrations of nitrates in drinking water cause methemoglobinemia, a blood disorder when red blood cells contain more than 1% of methemoglobin.

Nitrogen circulation in the biosphere through biological and chemical processes constitutes the nitrogen cycle and shows the connection between the various forms of nitrogen compounds and the environments in which they occur. Oxidation-reduction processes govern the creation, occurrence, transformation and transport of nitrogen compounds in nature, including water. Moreover, hydrogeologic, hydrologic and hydrodynamic circumstances affect the transport and state of nitrogen compounds. Consequently, the study of nitrogen compounds in water requires a comprehensive approach.

Nitrogen compounds that end up in groundwater are exposed to complex dynamics of diverse biogeochemical processes of nitrogen cycle transformation, such as denitrification (DN), dissimilatory and/or assimilatory nitrogen reduction to ammonia (DNRA, ANRA), nitrification, anaerobic ammonium oxidation (Anammox, Feammox), nitrifying denitrification, sorption, and mineralisation of organic substances. Under anoxic/reduction conditions, nitrogen will be subject to reductive transformation, including assimilation into biomass, respiratory denitrification, Anammox, Feammox and dissimilatory/assimilatory nitrate reduction, generally associated with ammonium cation production.

The chapter contains an overview of the occurrence of nitrogen in the environment and the application of nitrogen fertilisers in agriculture. A special section is devoted to the description and analysis of the transformation pathways of nitrogen compounds. Another section discusses the isotope-based tracing of the origin of nitrogen. Concentrations of nitrogen compounds in surface water and groundwater are shown based on monitoring of rivers and shallow aquifers in Serbia (2009–2015). Also included are the results of analyses of nitrogen compound transformation in two large agricultural regions, one anoxic and the other oxic.

The remaining two chapters (9 and 10) cannot be classified under any of the above groups.

Chapter 9: *Methods for Trace Analysis of Pharmaceuticals and Pesticides in Water and Sediment Samples*

This chapter describes how pharmaceuticals and pesticides in water and sediment are sampled and analysed. Liquid and gas chromatography – mass spectrometry are discussed. A view of analytical trends in this field is also proffered.

A novel analytical method is described, stemming from participation in a monitoring and analysis programme focussing on surface water, groundwater and sediment quality analysis, as well as quality transformation in alluvial aquifers. The programme was implemented in 2009–2015 and the results are discussed in detail in Chapters 6 and 7.

Chapter 10: *Supplemental Recharge of Aquifers and the Compression of Layers of a Hydrogeological Complex due to Groundwater Abstraction*

Increasing use of groundwater frequently leads to overexploitation. This is more likely to happen with deep, confined aquifers than alluvial aquifers. In such cases, excessive groundwater abstraction exceeds the recharge potential of a given aquifer (hydrogeologic complex). This phenomenon is associated with the response of overlying, underlying and intercapillary, compressible aquitards to declining piezometric head. The compressibility of such strata can be several times (to a couple of orders) higher than that of the aquifer. This leads to geomechanical effects – subsidence, as well as considerable hydrodynamic changes due to the inflow of sizeable amounts of water from the compressible aquitards into the aquifer.

The chapter introduces an approach that addresses the compressibility of overlying and underlying aquitards in water budgeting and assessments of the effects of pumping on the aquifer/hydrogeologic complex. It is a straightforward way of explaining certain specific hydrodynamic effects, which helps clarify and solve engineering problems in practice.

The size of the affected area can be much larger than the thickness of the captured hydrogeological complex (by two, three or more orders of magnitude). In such cases, it might turn out that the use of 3D models to assess the hydrodynamics and subsidence is, likely, inefficient.

A quasi-3D approach is proposed instead. This method is applicable under certain assumptions, as discussed in the chapter. Among other things, groundwater flow and expulsion from the aquitard is specified as a one-dimensional process in the direction of the vertical axis ($z$). The flow within the aquifer(s) is set to be horizontal ($x$ and $y$ axes).

The book ends with four appendices. Their purpose is to assist the reader and provide additional insight into some aspects discussed in the chapters. The appendices contain specific data on certain parameters, relations and processes.

Appendix 1: *Surface Tension, Capillarity and Viscosity*

Appendix 2: *Determining the Coefficient of Filtration Based on Data on the Granulometric Composition of Soil*

Appendix 3: *Chemical Structure and Selected Properties that Influence the Fate of Pesticides in the Environment*

Appendix 4: *The Chronology of the Quaternary Period*

That was basically a synopsis of the chapters. The plan was to write them in such a way that they could be read separately. However, they are complementary and the entire book treats the subject matter in a holistic fashion.

Considering the scale of the subject matter, some of the processes discussed in the book are of very short duration, even to a human observer. However, the Quaternary period, on which the book focusses, is very long from a human perspective. It has witnessed enormous changes in the flora, fauna and human species. Large civilisations appeared and disappeared during the Holocene. Still, it was relatively short in Geologic Time and an infinitesimal blink of an eye in Total Time.

Nevertheless, it is on us to strive to conserve streams, rivers and groundwater.

## REFERENCES

FAO. (2016). FAOSTAT. Database. Food and Agriculture Organization of the United Nations (FAO), Rome. Available at http://faostat3.fao.org/browse/R/RP/E (accessed July 2016).

FAO. (2018). FAOSTAT. http://www.fao.org/faostat/en/#data/RL

# Chapter 2

# Water management in large alluvial plains

*Milan Dimkić, Srđan Kovačević, Aleksandar Čalenić and Ivan Zavadsky*

## 2.1 INTRODUCTION

Water management has a very long history, beginning with an attempt to respond to seasonal changes in water availability. Water management was of crucial importance during the transition from a hunting-gathering society to an agricultural society, and it became even more important when the first cities, and industrial and administrative centers emerged. The need for water management occurs, initially, in young civilizations in areas with relative water scarcity. Today, we look with sympathy at the beginnings of water management. An example of the initial management of water distribution is the so-called water clock (Figure 2.1) that regulated the use of a particular water resource in subarid areas – oases.

The whole development of water management can be divided into three basic phases. The *first phase* – the phase of abundance, when water use and water pollution is low relative to the total quantity and quality of water resources. The *second phase* – the over-exploitation phase, when water use and pollution are significant in relation to the total quantity of water resources, and when the water quality is significantly impaired; and the *third phase* – namely the phase of sustainable development, when water management is essential so as to maintain a good status of water resources and not impair the quantity and quality of water resources at the expense of the future.

The water sector, as an extremely complex and specific sector, affects all economic and non-economic activities.

It is necessary to provide a long-term supply of water, ensure water protection and protect against the harmful effects of water, all in order not to make water a limiting factor in the development of society. The adequate development of water management activities which will ensure a smooth social and


Editor: Milan Dimkić
doi: 10.2166/9781789060904_0009

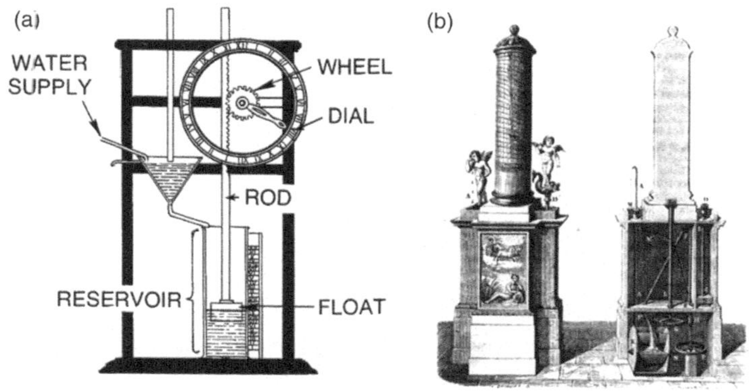


**Figure 2.1** Water clock – the beginning of water management – Ctesibius water clock, visualized by the 17th-century French architect (Extrait de 'Astronomie populaire' Tome 1 de François ARAGO [1854]).

economic development, while not endangering the environment, is a very difficult and delicate task (Dimkić & Milovanović, 2014).

The challenges of water management are a function of the large populations on Earth and the insufficient ability of societies to ensure adequate water security, let alone integrated water management under climate change conditions and in circumstances of major social issues and changes.

Self-awareness, improvement of individual traits and the development of individual resources are needed for man to be independent. Similarly, countries, nations and river basins need to develop water management mechanisms. This requires an ethical mindset and due awareness of water, as well as maintenance of water systems for which appropriate resources are needed (in terms of organization, capacity, funding, etc.).

The preferred approach to water management will largely depend on the economic power of the country in question and the availability of water resources (Figure 2.2).

Countries with a similar economic performance and water availability tend to follow similar approaches to water management.

When we speak of integrated water management, we have sustainability in mind, and then adaptability to various changes, above all climate change and social change. Differences in cultural, social, economic and other circumstances of different nations and changes within, as well as characteristics of river basins and other natural basins, tend to be put forward but with some reluctance (Dimkić & Milovanović, 2009).

There is plenty of water on the planet. About 97.5% of it is saltwater in seas and oceans, and the remaining 2.5% is what can be considered to be freshwater. Of this, surface and other freshwater make about 1.3%, groundwater 30.1% and glaciers and ice about 68.6% (Figure 2.3). Surface and other freshwater is distributed between rivers (0.46%), swamps and marshes (2.5%), lakes (20.1%), soil moisture (3.5%), water in the atmosphere (0.22%), bound water in living things (0.22%) and water in its solid state (snow and ice, 73.1%).

The planet is becoming more and more populated and it is becoming increasingly difficult to preserve the environment on the planet. Water, air and soil are most prominent in terms of invasion by mankind. As an unpleasant matter of interest, we will mention that plastic debris was spotted at the bottom of the Mariana Trench (about 10,000 m deep – the deepest part of the ocean on the planet) when landing a research submarine.

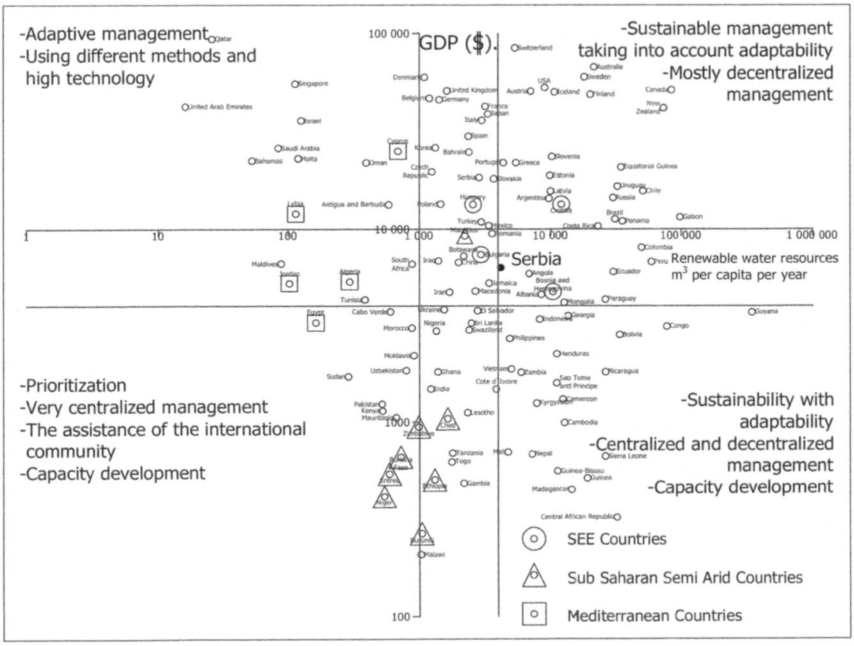


**Figure 2.2** Economic power and availability of water resources as indicators of a water management status (Dimkić and Milovanović, 2014. Note: GDP – gross domestic product, SEE – southeastern Europe.

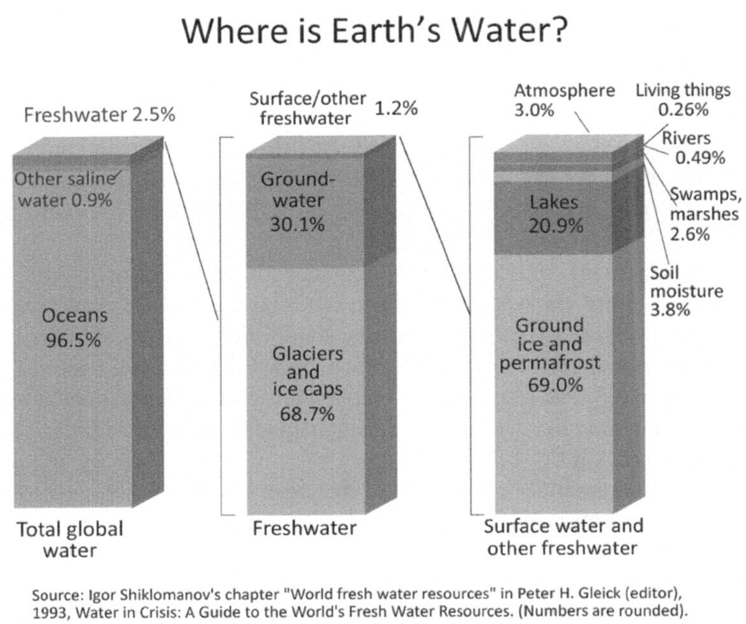


**Figure 2.3** Distribution of the worlds water after Gleick (1993).

Water management is a complex task and must consider:

- Essential natural and social aspects;
- Water availability;
- Demographics;
- Supporting economy and finance;
- Climate;
- Water management capacities;
- Water governance;
- Specific aspects related to the stakeholders claiming the right to use the resources;
- Participation in management taking global, regional and local components into consideration and the complexity, integrity and sustainability of water management;
- The issue of management in terms of the degree of centralization and decentralization of governance;
- The location of groundwaters in the water system, the basic differences between surface and ground waters as well as their management;
- The specific characteristics of alluvial aquifers and processes within;
- The basic technical elements pertaining to finding a solution to the problem of water abstraction, groundwater protection and the interaction with the alluvial groundwater environment.

Alluvial plains have always been areas where roads intersect, where the construction of large agglomerations takes place, where trade and migrations occur and in general, where there is intense human activity and settlement. The use of groundwater from alluvial aquifers is popular because of its proximity to places of consumption, but also because of the general good quality of the groundwater. Groundwater is used by over 60% of the population in Europe for water supply. Of this 60%, more than half use alluvial groundwater. This is the main reason for this book's themes which are: different physical, chemical and biological processes in groundwater, which are mainly specific and significantly expressed in alluvial aquifers.

## 2.2 DEFINITIONS

**Water management** consists of a set of measures and activities aimed at maintaining and improving the water regime, providing the necessary amounts of water of a prescribed quality for various purposes, protecting the water from pollution as well as ensuring protection against the harmful effects of water (Figure 2.4).

**Integrated water management** consists of a system of many interdependent activities and measures that affect water use and water quality, the water regime and water protection. This makes it very complex as it involves many segments of human activities and thereby conditions the survival of the human community.

**Sustainable water management** refers to management aimed at protecting water as a natural resource, as well as at the conservation of water resources for both the present and future generations. It also refers to the provision of sufficient quantities of water of an adequate quality for various uses, and for the economic and social development of mankind.

By **adaptive water management** we mean water management that enables us to adapt as easily as possible to the challenges and changes that occur in natural or social circumstances. It is achieved by diversifying the use of both resources and methods necessary to achieve the set management goals.

The term '**groundwater**' is usually reserved for the subsurface water that occurs in the soil and geological formations that are completely saturated with water. There is a similar approach in the new European Union (EU) legislation Water Framework Directive (WFD): 'groundwater refers to all the

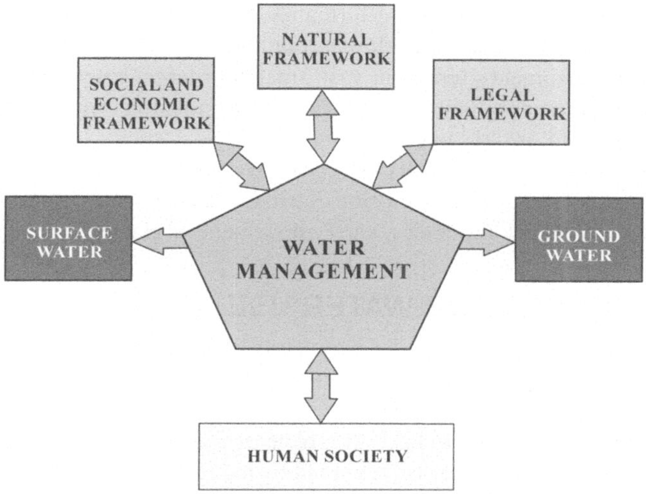


**Figure 2.4** Schematic diagram of frameworks and impacts on water management and mankind (Dimkić *et al.*, 2008b).

water below the surface of the terrain in the saturation zone, which is in direct contact with the soil or subsoil'.

'**Groundwater management**' consists of a subset of activities we refer to as '**water management**'. Groundwater is a very important resource because it represents a significant source of water supply for the population, and an important element of the natural water cycle and natural habitat.

In general groundwater movements are much slower than surface water flows, while the extensive self-purification processes that take place in groundwater are extremely specific. This is why the water quality problems related to groundwaters are more often than not of a local character, regardless of their great global importance (Table 2.1).

**An aquifer** consists of a subsurface layer or of layers of rock mass or other geological formations of sufficient porosity and permeability to allow for a quantitatively significant flow of groundwater or the abstraction of significant quantities of groundwater.

**An aquiclude** is any geological formation that absorbs and retains water, but does not transfer it at a sufficient rate to supply springs and/or wells, so that no economically significant amount of water can be extracted from the aquiclude.

**An aquitard** is a geological formation or layer located near an aquifer that allows only a small amount of groundwater to flow. It is a saturated but poorly permeable layer, which can however store water and transfer significant amounts of water to and from surrounding groundwater aquifers.

**Table 2.1** Groundwater specificity with respect to surface water.

| Parameter | Groundwater | The Rivers |
|---|---|---|
| Velocity of flow | cca 1 m/day | cca 1 m/s |
| Exposure to natural processes | Months, years, hundreds of years | Days |
| Self-purification processes | Water/aquifer skeleton/air | Mostly water/sedimentation/air |
| Baseline quality | More different | Less different |

**An Aquifuge** is a non-permeable geological formation that is neither porous nor permeable - which means that it cannot store water or have water flow through it.

**Alluvions** (lat. Alluvio = sediment), represent vast, smaller or larger reservoirs of river material, which fill the river valley. In other words, they are deposits of considerable widths and lengths of different materials, primarily sand, gravel or finer material.

**Alluvions** are the result of various processes that have, over the course of their history, influenced their formation. They can primarily be the result of erosion due to the flow of water itself, and can also result from geological factors: tectonic, volcanic processes, karst, movement of glaciers, etc.

## 2.3 WORLD RESOURCES AND WATER USE

The shape of the Earth does not fit any regular geometric body and perhaps best resembles a two-axis ellipsoid (spheroid). The true shape of the Earth deviates from the shape of the rotating ellipsoid due to the bulges (continents) and recesses (ocean basins).

Today, our planet Earth is considered to have the shape of a geoid (a term introduced by Listing in the second half of the 19th century), which is a shape limited by the surface occupied by the calm surface of the oceans and seas, extended by channels across continents, and governed at every point in the direction of gravity (Figure 2.5).

The density of the Earth was first determined by Newton in 1736, ranging between 5000 and 6000 kg/m$^3$. Subsequent determinations showed that Earth's mean density equaled 5527 kg/m$^3$. The values of the mean density of rocks in the Earth are (Sučić-Protić & Miletić-Spajić, 1969):

- Upper, 'granite' layer of the Earth's crust – 2700 kg/m$^3$;
- Lower, 'basalt' layer of the Earth's crust – 2900 kg/m$^3$;

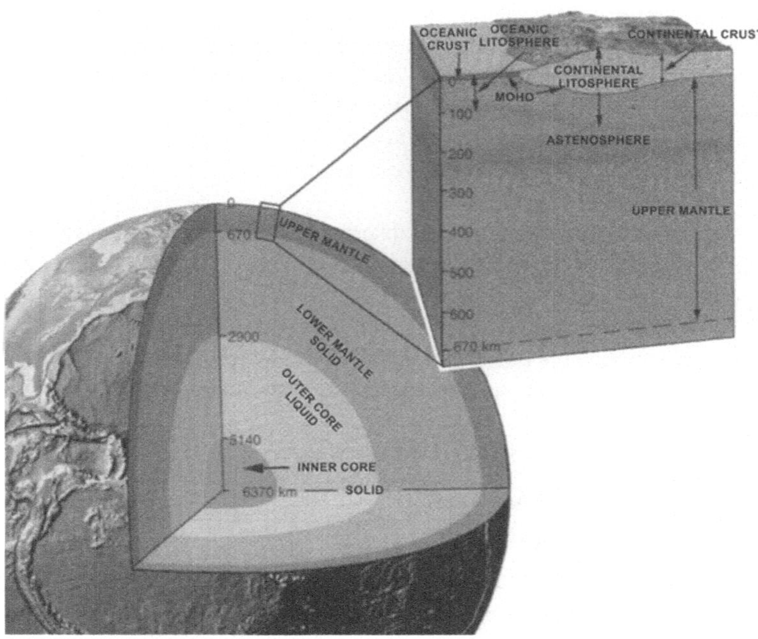


**Figure 2.5** Earth spheres (Milovanovic & Boev, 2001).

- Top layer of the oceanic crust – 3000–3100 kg/m$^3$;
- Sub-layer at a depth of about 100 km – 3400 kg/m$^3$.

At a depth of about 2900 km from the surface the Earth's density equals 5300 kg/m$^3$. Just below that depth, which marks the border of the planetary core, the density reaches 9900 to 10,000 kg/m$^3$ and continues to increase evenly to the center of the core where it reaches a value of 11,000 kg/m$^3$, and even higher (according to Fersman, 1923).

*Hydrosphere*

The hydrosphere does not form a continuous sheath and is made up of the waters of the sea and oceans, the waters on land, and glaciers. Groundwaters, although closely related to the waters of the hydrosphere, have a special place.

Almost all chemical elements are present in the waters of the sea and oceans. However, only the following four: oxygen, hydrogen, chlorine and sodium, are predominant and they make up 99.5% by weight of water. The mean chemical composition of salt water is given in Table 2.2. The salinity varies especially in seawater, while in ocean waters the variations are smaller.

### 2.3.1 Water use

The potential water scarcity due to ever-increasing water needs has had various consequences. It has resulted in an increased demand for energy, food production, water supply and industry, and the additional negative impacts of climate change. In recent decades, the increased global water use expressed in terms of percentage has exceeded twice the population growth. As a result, more and larger regions of the world have been exposed to water stress and the current limited rates of water use and consumption, not to mention the desired rates, have become unsustainable. The supplies and needs for water are changing. What will happen to these in the future is uncertain, but what is certain is that they will continue to change.

Water is increasingly becoming a policy priority at the international level. Both economic development and security are threatened by poor water management. That is why concerns regarding the global energy crisis have recently begun to be accompanied by concerns about the global water crisis. The relationship between energy and water, both expressed by the effects of water use on energy consumption and the effects of energy production on water consumption, is receiving increasing attention.

**Table 2.2** The mean chemical composition of saltwater (Sučić-Protic & Miletic Spajić, 1969).

| Name of the Salt | Content mg/l | % |
|---|---|---|
| NaCl | 27.2 | 77.71 |
| $MgCl_2$ | 3.8 | 10.86 |
| $MgSO_4$ | 1.7 | 4.86 |
| $CaSO_4$ | 1.2 | 3.43 |
| $K_2SO_4$ | 0.9 | 2.57 |
| $CaCO_3$ | 0.1 | 0.29 |
| $MgBr_2$ | 0.1 | 0.29 |
| In total | 35 | 100.00 |

The FAO and the World Water Council (FAO & WWC, 2015) made data projections with appropriate investment and policy interventions, according to which food production will be sufficient to support a global population of 9–10 billion in 2050, although the food and nutrition insecurity will continue in many regions. The pressure for more water for food production can be expected to increase as large segments of the population in developing countries will tend to raise their standard of living.

Today half the world's population lives in urban areas. According to data (UN Habitat, 2012), it is predicted that virtually all of the world's demographic growth over the next 30 years will occur in urban areas, mainly in low-income countries.

According to the standards which have up to now been defined by the right to water (the human right to water entitles everyone to sufficient, safe, acceptable, physically accessible and affordable water for personal and domestic uses), the number of people without access to safe and reliable drinking water in households amounts to between 3 and 4 billion. In addition, 884 million people use inadequate drinking water while approximately 2.6 billion people do not have access to adequate sanitation (World Health Organization [WHO], 2010).

Water consumption per capita varies greatly across the globe. In developed regions an average value of 200 l per person per day can be assumed. The internationally accepted value for basic human water needs is about 50 l per person per day (Gleick, 1996).

From today up to 2050, water demand is expected to increase by 400% for production purposes and by 130% for domestic use (Guppy *et al.*, 2017). As water availability decreases, competition for access to this scarce resource will increase. International shared river basins account for 60% of all freshwater, and it is estimated that there are 592 transboundary aquifers (Petersen-Perlman *et al.*, 2017). The continuous cooperation and coordination between countries is key to ensuring the availability of water for human, economic and environmental needs. Although hundreds of international agreements on water have been signed over time, there is no certainty as to what the response will be to the future continued growing pressure on water resources.

The global demand for water is estimated to amount to 3000–4000 $km^3$ per year and this already represents a difficulty in terms of managing a sustainable system of water resources. An increase of 20–30%, which is expected by 2050, will lead to a demand of 5500–6000 $km^3$ per year (Burek *et al.*, 2016).

In developing countries, about 45 million cubic meters per day is lost due to leakages in the water infrastructure – which is enough to meet the needs of almost 200 million people (Kingdom *et al.*, 2006). This problem will only get worse if the water infrastructure is not properly maintained, even for high-income countries; for example, the capital investment needed to sustain the ageing US water infrastructure will reach an estimated US $ 195 billion in 2040, but if the current funding trends continue, the needs will be underfunded by US $ 144 billion.

Water security is essential for poverty reduction, and water resource management affects almost all aspects of economic activity, including food production and security, industry, energy production and transportation. However, human activities often degrade water resources.

How the available quantities of water resources will be used for different purposes depends on how developed a society is and on the water regime in a particular area. In countries with lower levels of economic development, the participation of individual users expressed as a percentage (water supply, agriculture, industry) differs significantly from that in countries with high levels of economic development.

The quantity, availability and safety of water supply remain key concerns at national, regional and global levels. The link between climate, population increase, food production, energy and water circulation seems fundamental, and it is therefore important to establish a high level of synergy between the above factors. Possible errors in the management of water management systems are most often due to the absence of an all-inclusive approach, e.g. water management at the basin level, while meteorological information is generally available at a wider level and at different spatial scales.

The population affected by the decline in water availability increased from 0.24 billion (14% of the world population) in the 1900s to 3.8 billion (58%) in the 2000s (Kummu *et al.*, 2016).

Water resources show a significant spatial variation and many different coefficients can be used for the purpose of estimation (Cambridge University, 2012; Rijsberman, 2006; UNDP, 2006; United Nations Environment Programme [UNEP], 2008). Some of them are:

- The surface water resource coefficient: measured as the surface water capacity (sum of the annual river runoff) per capita;
- The exploitation coefficient (water intake coefficient): defined as the ratio between the total water intake and the total sum of surface and groundwater resources.

The water poverty index is measured using five indices (Brown, 2002; Chapagain & Hoekstra, 2004):

- Water resources (quantity and quality);
- Availability (distance from the water source, exploitation time, existing conflicts);
- Ability to obtain water (ability to cover the costs of water treatment, type of water management);
- Use of water (water management rules);
- Environment (assessment of the dependence between water and the environment, ecosystem status and erosion).

Just like other forms of shortages, the physical lack of water can be fundamentally divided into two aspects: lack and stress. Water scarcity refers to the impact of low per capita water availability. Given the per capita resource requirements, water scarcity can therefore be seen as a stress on the population. Water stress refers to the impact that increased water use has on water availability.

The use of a large proportion of resources (Falkenmark, 2013; Rockström *et al.*, 2009) can lead to difficulties in accessing a resource, as well as to side effects (Falkenmark, 1997) e.g. a social and environmental impact.

There are, however, multiple ways in which each of the terms can be defined, giving different sets of indicators for lack and stress. For example, use may refer to consumption or exploitation. Availability may refer to water from different sources, of different quality, or to decadal, annual, or seasonal timescales. Despite their high level of abstraction, and the multiple ways they can be used, concepts of deficiency and stress and their indicators are key to understanding the development of water scarcity over time.

North America is the largest consumer of water per capita with a consumption 2.5 times that of Europe. The reason for this is that water is not expensive in North America compared to other industrial countries. In North America and Europe, a resident consumes about 3 $m^3$/day (includes the water consumed in the production and delivery of food), in Asia 1.4 $m^3$/day, and in Africa 1 $m^3$/day, while about 120 million people lack access to safe drinking water in Europe (Kummu *et al.*, 2016).

Globally speaking, the largest amount of water is used for irrigation. The largest share of water use for irrigation purposes is in south Asia (96–98%) due to the extensive cultivation of rice, as well as in the Middle East (97–99%) due to the dry conditions (Siebert *et al.*, 2015). In Western Europe, the share of irrigation in total water consumption is the smallest (62–74%) because it includes areas where irrigation is not used extensively for food production. Moreover, the economy is more industrialized than, for example, Asia.

Household water consumption is approaching industrial water consumption (Kummu *et al.*, 2016). Another significant global trend is water consumption for the production of thermal electricity (~1% share). In some areas (e.g. Western Europe, Australia/Pacific), the share of water used for irrigation has increased and the share of household consumption has decreased. The opposite has happened in other areas (e.g. North America).

Areas with a high water scarcity and high water stress are most pronounced in southern and eastern Asia, North Africa and the Middle East, with 61–89% of the population lacking water. The regions with the lowest percentage of the population lacking water are Australia-Pacific, South America, North America and Southeast Asia (7–29%). Sub-Saharan Africa uses barely 5% of renewable drinking water per year, while access to adequate drinking water remains the lowest. The rapid increase in population has led to an increased demand for water and the accelerated degradation of the water resources in many African countries. At the regional level, only 20% of the total rainfall represents renewable freshwater, representing about 9% globally (FAO, 2005). In sub-Saharan Africa, the amount of water per capita has decreased from 15,000 m$^3$ in 1960 to 5500 m$^3$ in 2005. About 15% of the renewable water resources are groundwaters, about 75% of which are used as the main source of drinking water (UNEP, 2011). Africa has one-third of the world's large international water basins (more than 100,000 km$^2$), between 63 and 80 transboundary rivers and lakes and over 90 international agreements on water (UNEP, 2011).

The Asia-Pacific region is home to 60% of the world's population, with only 36% of the total global water resources (Asia Pacific Water Forum [APWF], 2009). Yet, it represents the largest share of renewable freshwater in the world with an annual average of 21,135 billion m$^3$. Given the population and economic growth, the rate of water abstraction is very high. More than 480 million people have no access to adequate drinking water, and 1.9 billion lack adequate sanitary conditions (APWF, 2009). Agriculture is the largest consumer of renewable water resources in this region, using about 80% of the renewable water (APWF, 2009). The sanitary conditions in Asia are not the same in all parts. Good conditions are found in the north-east and south-east, while the south-west and south are of considerable concern (Guha-Sapir, 2011).

More than 8% of the world's population live in southern America and the Caribbean, about 581 million people – half of whom live in Brazil and Mexico (UNEP, 2011). Population growth, expanded industrial activity, especially mining in the Andean countries, and the high demand for irrigation led to a tenfold increase in the total water exploitation in the region during the 20th century. Between 1990 and 2004, water exploitation increased by 76% (UNEP, 2011). The withdrawal of glaciers affects the water supply of about 30 million people in the region. Droughts are occurring regularly and between 2000 and 2005 they caused serious economic losses and affected 1.23 million people (UNEP, 2011).

There are two main ways to address water shortages:

- Increasing the available water, or
- Reducing the demand for water.

Water availability can be increased by using desalination (Wada *et al.*, 2014), as well as by introducing additional physical water transfer (Gohari *et al.*, 2013; Gupta & van der Zaag, 2008) and/or by reducing evaporation (Rockström, 2003).

Methods to reduce the demand for water would need to include more efficient irrigation (Jägermeyr *et al.*, 2016), a reduction in the amount of food that is wasted (Kummu *et al.*, 2012), a reduction in the use of products that use large quantities of water in production (Jalava, 2014, 2016) and the reduction of losses in water supply systems (European Environment Agency [EEA], 2012).

In many countries, groundwater is under significant pressure due to its various uses including:

- Providing drinking water for people and animals;
- Irrigation;
- Water supply to industry;
- Therapeutic purposes;
- As a raw material for the production of useful substances (such as iodine and bromine);

- Heating purposes (thermal water), etc.

The global amount of groundwater is unknown. It is estimated to be around 15.3 to 60 million $km^3$, of which 8–10 million $km^3$ is freshwater while the rest is saltwater (Marga, 2008). If groundwater is polluted, it is very difficult and expensive to recover. Furthermore, access to freshwater sources is not evenly distributed among the population. Access to these sources is far greater in urban areas, about 94%, while in rural areas it amounts to about 76% (WHO, 2010).

Improving access to safe water and to basic sanitary conditions could have a huge economic impact. World Bank studies in five Southeast Asian countries estimate that ~2% of their total GDP is lost due to poor sanitation and water scarcity, and in the worst case (Cambodia) this figure rises to over 7% (World Bank, 2008). This indicates that water scarcity and water sustainability are closely linked to other socio-economic and political issues.

Based on 2018 data (United Nations World Water Development Report [UN-WWDR], 2018), globally, about 70% of water resources are used for agricultural production, 22% are used by industry and 8% by households. These percentages are different for countries with low levels of economic development and those with high levels of economic development, as shown in Figure 2.6.

When it comes to the use of water, priority is always given to supplying settlements and the food industry with water of adequate quality and quantity, making water an indispensable life-supporting substance. Since ancient times, great efforts have been made and financial resources invested to provide sufficient quantities of water for human consumption that meet the prescribed quality standards.

Modern water supply to the population means that sufficient quantities of water of an appropriate quality should be provided for:

- Households;
- Those industries that require high quality water;
- Public use in the services sector (institutions, schools, hospitals, organizations, banks, restaurants, street washing, fountains, watering green spaces, and more);
- Irrigation and agricultural activities;
- Navigation.

Water resources are also used for:

- Hydropower;
- Tourism;
- Medical treatment;
- Recreation.

Supplying water to settlements is of key importance and is the primary activity of the water sector. When balancing water resources within an area, priority is given to ensuring that the supplied amount of water

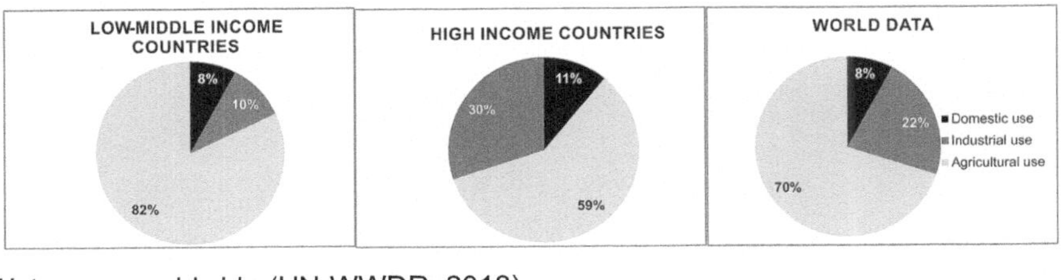


**Figure 2.6** Water use worldwide (UN-WWDR, 2018).

is of satisfactory quality for the purpose for which it is supplied, as it directly affects the health of the population.

Water supply systems (conduit lines) consist of different structures: water intake structures, water treatment plants, pipelines and the necessary structures for the transport and storage of water.

### 2.3.2 Groundwater use

Groundwater is most commonly used for water supply, if possible, because of its good quality, good temperature, and because it is usually clear and relatively clean. The type of groundwater intake depends on the depth at which the groundwater is located, the thickness of the aquifer and the yield. The main types of groundwater withdrawal structures are:

- Dug wells, and drilled wells (vertical water intakes);
- Wells with horizontal drains;
- Galleries;
- Catchment structures at wells and springs.

If, when opening a source, water under pressure appears above the surface of the terrain, it is called an arterial source, whereas if it rises to below the level of the terrain, it is known as subarterial water.

The type of groundwater abstraction that will be used depends on the type of source (Figure 2.7), the abundance, the topographic and geological characteristics of the terrain and on the amount of water to be captured.

Groundwater represents one of the key resources for supplying the population with drinking water. Since water is an irreplaceable substance, this segment of water management, namely providing sufficient quantities of water of appropriate quality to the population, has priority over other uses of water.

Using groundwater has significant advantages over the use of surface water. Generally speaking, the quality of groundwater is better and it is more protected against pollution. In addition, groundwater yields

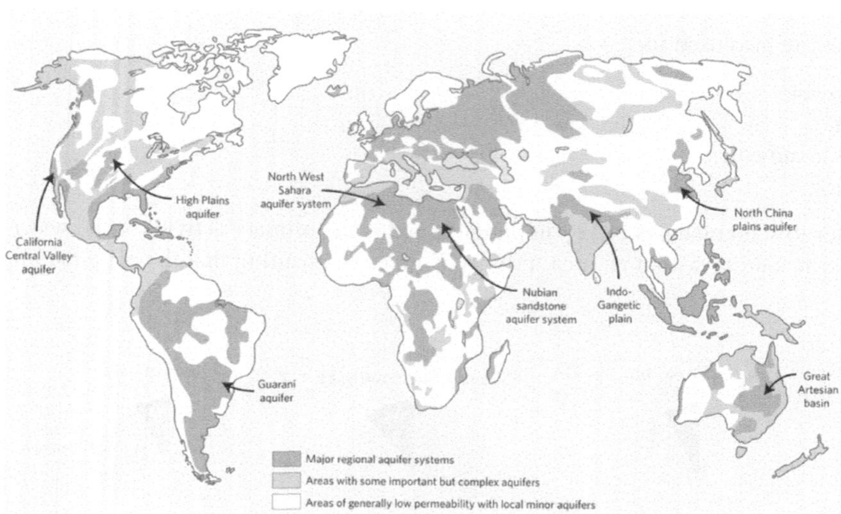


**Figure 2.7** Simplified version of global groundwater resources (Taylor *et al.*, 2013).

are much less dependent on annual and perennial oscillations. In northern areas, when surface water is frozen, and in arid regions, when surface flows dry up, groundwater remains the sole source of water supply. Unlike surface water sources, groundwater sources can be made bigger by drilling additional wells.

The role of groundwater in the overall supply of water has changed over time. Although groundwater has been used in dry regions since the dawn of civilization (remains of dug wells that are about 8000 years old have been found in Israel), the increased demand for water has resulted in the development of settlements alongside waterways, primarily because of the easy access to drinking water.

However, the increased pollution of surface waters during the 19th century and the numerous epidemics of deadly diseases forced large cities to redirect their water systems towards groundwater and the effects of self-purification of aquifers. Groundwater is often transported over long distances, in some cases up to 150 km (e.g. Vienna and Paris).

Today, a large number of countries use groundwater as their primary source of drinking water supply. A significant number of authors (Gleeson *et al.*, 2016; Margate & van der Gun, 2013; Siebert *et al.*, 2015) suggest the following percentages of use of groundwaters, including alluvial sources which use groundwater obtained after coastal filtration: over 95% in Austria, Denmark, Iceland, Hungary, Montenegro, Croatia, Libya, Saudi Arabia, the United Arab Emirates, Oman, Qatar, Indonesia, Mongolia, etc.; over 80% in Italy, Germany, Portugal, Slovakia, Switzerland, Poland, Russia, etc.; and more than 50%, in the Netherlands, Bulgaria, Finland, France, Greece, Serbia, Turkey, Thailand and so on. About 75% of the EU population uses groundwater for water supply. Many European countries mainly use groundwater for water supply (Serbia, Lithuania, Armenia, Belarus, Georgia, etc.).

Groundwater is a particularly significant source of water supply for smaller settlements. Even in countries where most of the water supply comes from surface water (such as in the USA), in residential areas with up to 10,000 inhabitants the percentage of groundwater used for water supply is significantly larger than that of surface water.

The use of groundwater for irrigation for food production purposes is another priority. This type of groundwater use is particularly significant in developing countries that cannot otherwise meet their food needs.

According to Zekster and Everett (2004), groundwater is used – in some cases to a considerable extent - for irrigation purposes in arid regions (Libyan Arab Jamahiriya – 100%, Saudi Arabia – 86%), partly arid areas (Argentina – 70%, Algeria – 56%, Australia – 46%, United States and Mexico – 38%, India – 35%), and in the Mediterranean (Spain, France, Greece and Italy – over 26%). In arid and partly arid areas, groundwaters are often the basic resource for cattle breeding on a larger scale.

When large quantities of water are used for irrigation in areas with limited groundwater resources, it is important to take into account the possible negative effects the processes of over-pumping, salinity, soil subsidence and others can have on the environment.

In many industrialized countries, groundwaters meet a significant proportion of the needs of industry (Figure 2.8) with their own sources: 71% in Greece, 65% in Denmark, 40% in Japan, 27% in France and India and 26% in Germany and the United States (UN FAO, 2015).

This includes water for the cooling of thermoelectric and nuclear power plants, but it does not include hydropower. Water withdrawn by industries that are connected to the public supply network is generally included in municipal water withdrawals.

In many cases, it is more appropriate to use groundwater rather than surface water for the industry's water supply. However, since groundwater is very important for the supply of households, it is necessary to establish priorities in certain cases.

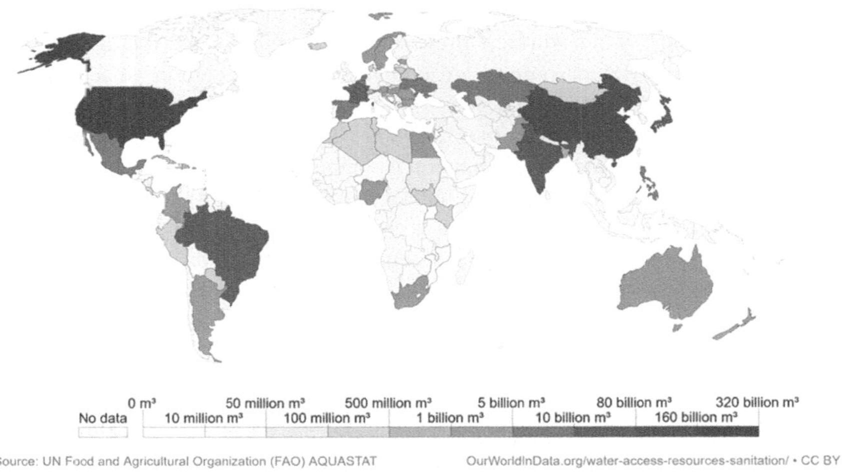


**Figure 2.8** The annual (2015) quantity of self-supplied water withdrawn for industrial uses, in cubic meters ($m^3$) per year (FAO, 2015).

## 2.4 GENERAL WATER MANAGEMENT SCHEME

Figure 2.2 in sub-section 2.1 simultaneously shows the GDP in one coordinate system (national income per capita) and the quantity of renewable fresh water per capita per year. It can be seen that the national income and the amount of fresh water available per capita vary from country to country by several, several dozens and even hundreds of times. This points to the need for different approaches to water management, which will be discussed.

The economic status and water abundance, climate and climate change, as well as the general capacity of society (capacities and their connectivity and functionality) all appear as indicators of water management.

Issues related to water management units (areas) are also significant. The first division can consist of global, regional, local and even micro-local management units. The reasons for this division may be technical, economic or political (Dimkić, 2010).

Another type of division of administrative jurisdiction consists of the division into catchment areas, administrative areas or a combination of the two. Technically speaking, the water catchment area should be an integral water management unit. However, it is absolutely essential that administrative units (states, parts of the state, municipalities) are also involved in water management.

Mixed management is usually present, especially in large catchments. An example of this is the International Commission for the Protection of the Danube River (ICPDR), which covers the entire Danube area, which houses 19 national delegations.

Figure 2.9 provides an outline of a water management scheme that may apply to a basin, an administrative unit and also to mixed management.

Water management frameworks have here been divided into global-general as well as particular-local frameworks.

### 2.4.1 General water management frameworks

General frameworks are those that apply to the planet as a whole i.e. the natural, socio-economic and legal frameworks.

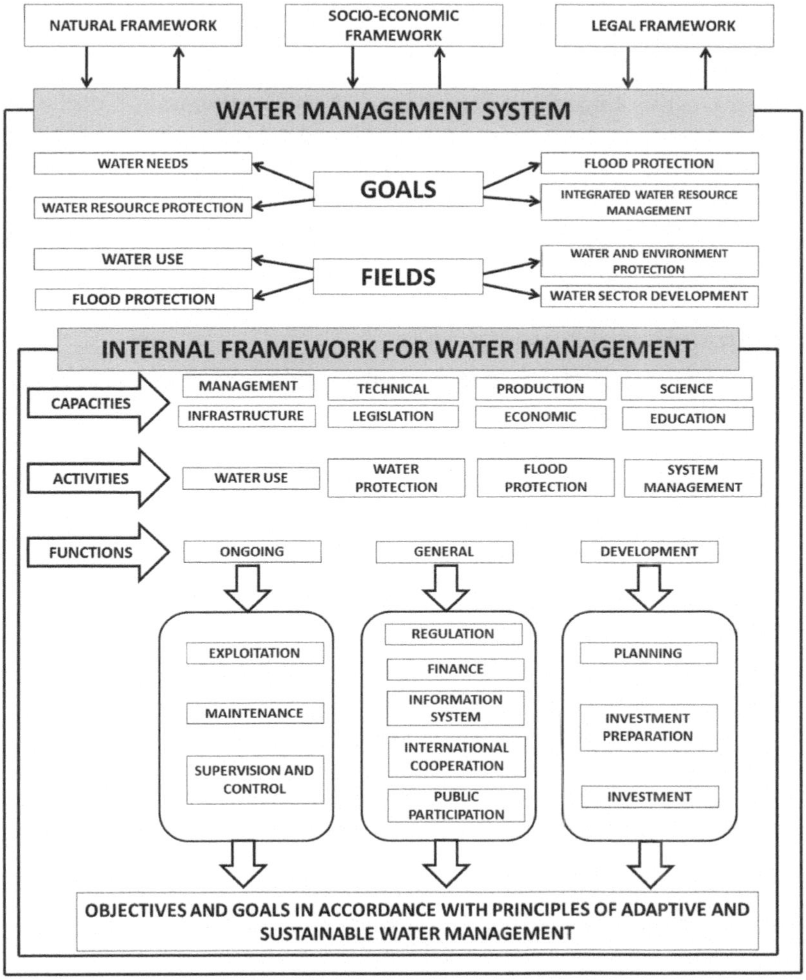


**Figure 2.9** Schematic presentation of the general structure of a water management (Modified from Dimkić *et al.*, 2008a, b).

The natural framework includes:

- The climate on the planet and the changes that occur;
- The state of water resources, their use and protection;
- The total water balance, the capacity of the resources on the planet and their distribution by area.

The socio-economic framework consists of the very complex social conditions and trends in the world and its individual parts. It includes the state of the planet's economy, as well as the economies of individual countries, and of course the manner and efficiency of governance.

At the world level, the UN and its various organizations, funds and agencies, such as, for example, UNESCO, FAO, United Nations Development Programme (UNDP), UNEP and others, deal with water related issues.

The financial institutions that are involved include the International Monetary Fund (IMF), World Bank (WB), African Development Bank (ADB), BRIKS Bank, European Bank, Asian Development Bank (ADB) and others.

The legal framework consists mainly of Declarations, Agendas, Protocols, Conclusions, etc. given at major international meetings/conferences/summits held under the auspices of the UN. Examples of these are the Stockholm Conference, Rio Summit, Kyoto Protocol, Millennium Summit – New York, Rio +10, Rio +20, etc.

It is also important to bear in mind the work of international organizations and associations, such as, among others, the International Water Organization. These global organizations will be discussed in a separate chapter.

### 2.4.2 Water management objectives

Water management objectives are considered in the context of the overall water management system. The basic objectives of water management are:

- Meeting the water demand;
- Conservation of water resources;
- Protection against the harmful effects of water;
- Integrated and sustainable water management.

It is generally believed that water management, including that of groundwaters, should be based on the principles of integrated water resources management:

- Surface and groundwaters are considered to be an integral part of a single water cycle;
- Water is considered both an economic and social category;
- The principles 'user pays' and 'polluter pays' apply;
- The water law is well established;
- There is a relationship between water resources management and socio-economic development;
- Public participation is present.

In order to implement the above, it is necessary to:

- Apply appropriate legal, socio-economic and natural preconditions;
- Have adequate institutional organization and the necessary capacities;
- Have good planning and management;
- Implement measures;
- Implement appropriate monitoring.

### 2.4.3 Basic water management activities

#### *2.4.3.1 Water protection*

The long-term goal in the field of water protection is to protect the quality of surface and groundwaters, to preserve and improve the quality of water to the extent that it can be used for the needs of users with the highest quality requirements (Dimkić *et al.*, 2013).

This task is accomplished by means of integrated, rational and unique water management within a management system which includes:

- The rational withdrawal of water quantities and necessary measures within technological production units;

- The implementation of rational technical and technological measures for wastewater and other waste materials collection, treatment and disposal;
- Measures to improve water regime in the receiving water bodies (other technical measures when water flows into recipients);
- Technical measures in watercourses and control of non-point sources of water pollution, as well as control of potential sources of water pollution;
- Control of movement and use of dangerous substances and other appropriate non-investment measures.

All this is accomplished within the complex and integrated management of the water regime in accordance with the rational use of other resources (especially land), taking into account socio-economic as well as other constraints (climate, climate change and others).

Wastewaters are regarded as a mixture of different waters, where each component has its own composition and where the original chemical composition and physical properties of the water have been changed.

Depending on its origin and characteristics, wastewater can be divided into three categories:

- Used household water (sanitary wastewater);
- Used industrial water (industrial wastewater);
- Atmospheric waters (polluted urban runoff).

In wastewater, substances appear in an undissolved and dissolved state. The composition of wastewater will vary depending on the nature, origin and type of wastewater. The analysis of wastewaters is very complex and usually includes physical, chemical and biochemical parameters.

When designing wastewater treatment plants (determination of appropriate process and recipient loads) the term 'population equivalent' (PE) is commonly used. For example, when the total biochemical oxygen demand ($BOD_5$) of a wastewater from industry is known, the PE is obtained by dividing that value by the $BOD_5$ generated by one average resident.

The quality of municipal wastewater is generally known with some variations depending on the size of the settlement, degree of urbanization and connection of industry to municipal sewers. The main categories of matter in urban-municipal wastewaters are of organic and inorganic origin, and they may be suspended or dissolved. On average wastewater loads can be estimated to be about 70 g of suspended substances and about 60 g of $BOD_5$ per capita per day.

Municipal wastewaters also contain dissolved gases (oxygen, carbon dioxide, nitrogen, sulfur-hydrogen, etc.), as well as bacteria and microorganisms that play a significant role in the wastewater treatment process.

Unlike municipal wastewaters, industrial wastewaters contain a wider range of pollutants, and seasonal and daily variations in quality are possible. If there is a high content of toxic materials (phenols, pesticides, cyanides, etc.), or heavy metals (lead, cadmium, mercury, nickel, etc.), the biological wastewater treatment process could be affected due to negative effects of the mentioned substances on the active biomass responsible for the removal of pollutants from the wastewater treated. In addition, the organic wastewater load from some industries (e.g. machine industry) may be too low or too high (e.g. food industry), which in both cases represents a significant problem in the treatment of these waters.

If wastewater is reused or recycled, often the case in industry and less so in the case of municipal wastewater, the wastewater must be treated to some degree, or even completely, depending on the specific requirements of the water reuse or recycling site. Some forms of reuse or recirculation in the water industry have been routine practice for decades, for example: recirculation of cooling water and condensates or reuse of cooling waters in the heating of greenhouses.

**Table 2.3** Millennium Water Targets (2015).

| | Set Goal | Accomplished |
|---|---|---|
| Accessibility to drinking water | 88% | 91% |
| Accessibility to advanced sanitation | 75% | 67% |

Unlike the very mild targets set by the UN for drinking water for 2015 (the so-called Millennium Goals) which have been met (Table 2.3), the Millennium Goals set for wastewater have not been reached.

**Protecting the quality of the groundwater**

Although groundwater in some parts of the world can be an abundant resource, if the quality of the groundwater is degraded relative to its natural state by the introduction of pollutants, the aquifer may be unusable for water supply.

Pollutants reach aquifers in various ways:

- Infiltration of surface water through the soil, sediments and rocks;
- Direct infiltration of surface waters (such as rivers, streams, lakes, etc.);
- Direct flow through improperly drilled or sealed wells (representing a conduit for pollutants of anthropogenic origin), underground reservoirs, mines, etc.

Pollutants in groundwater can be present due to various anthropogenic activities, which accidentally or systematically cause the introduction of hazardous substances. Three basic groups of activities – industry, population (households) and agriculture – are considered to be the most serious sources of pollution (Figure 2.10).

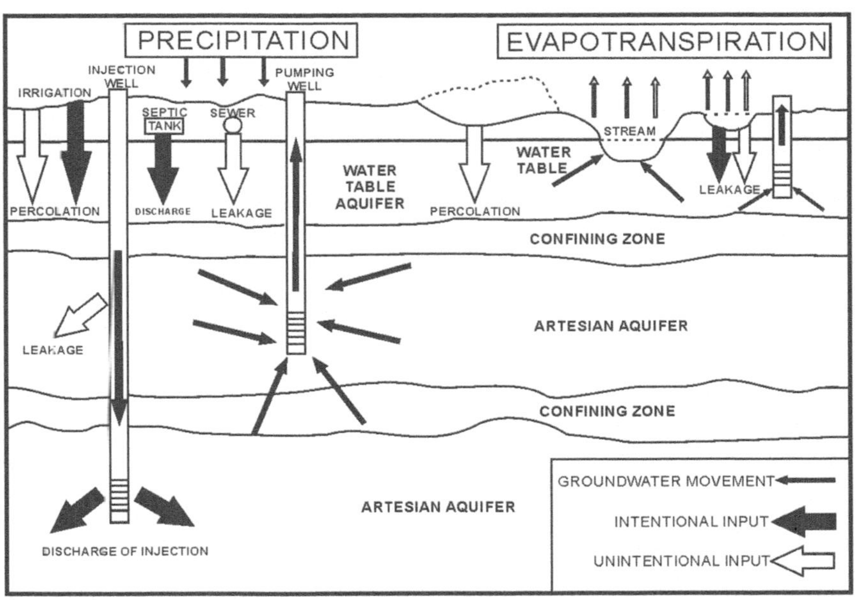


**Figure 2.10** Potential groundwater pollution (Stellman, 1998).

Depending on the physical space they occupy, sources of pollution are usually divided into two basic groups:

- Point sources (various ways of depositing petroleum products, diluents, chemicals and leachates from landfills) and
- Non-point sources (diffuse sources), for example, nitrates originating from agriculture, pesticides, urban pollution due to leakage from sewage collectors, drains, etc.

The list of potential sources of groundwater pollution is not limited to the above. The increased number and types of human activities over time, especially in connection with new technologies, materials and energy sources, are often in conflict with the objectives of protection of the environment.

Many aquifers have a natural self-purifying potential and thus mitigate the effects of pollution, and this explains why groundwater quality remains good in many parts of the world despite the pressure from potentially polluting activities.

The protection of groundwater resources largely depends on a number of factors, such as the type of pollution (behavior in the underground), the type of pollution source (point or diffuse), and on how pollution is introduced into the underground environment. However, how protection will be actively implemented primarily depends on the scale of the problem, i.e. on whether protection is enforced at the regional (multi-state), basin (parts of the basin) or local level (primarily due to the typically low flow velocities characteristic of groundwater flow in aquifers).

Due to underground pollution having propagated over a long period of time, the problems regarding the protection of groundwater quality are mostly of a local character. Basin-level problems occur when existing pressures have similar impacts in different countries and in this case integrated approaches to problem-solving are needed.

**Protecting quantitative status of groundwater**

The uncontrolled and in some cases excessive exploitation of groundwater has called into question its sustainability. The basic question asked is to what extent can groundwater be exploited without compromising the principles of sustainable development. The most commonly cited definition of sustainable development is given in the Bruntland Commission's report on Our Common Future, released during the UN World Commission on the Environment and Development, and it says that sustainable development is: 'development that meets the needs of the present without compromising the ability of future generations to meet their own needs'.

Due to the complexity of the issues, the sustainability of groundwater exploitation must be an interdisciplinary issue, with hydrology, environmental protection, geomorphology and climatology playing significant roles. It is necessary to distinguish shallow groundwater systems from deep ones, since shallow ones come into contact with surface waters and deep ones do not.

In general, groundwater does not recover at the same rate as surface water, with the dynamics of the groundwater renewal period varying from several years to several thousand years, depending on the location, type, depth, characteristics and connectivity of the aquifers.

According to the map (Figure 2.11), the level of exploitation is currently high in many countries in the Middle East, in northern and southern Africa, Asia, and in several European countries and Cuba. In parts of China, India, Mexico, Pakistan and the United States, over-exploitation is also occurring in arid, densely populated areas.

Excessive pumping can result in the over-exploitation of groundwater resources if the catchment dynamics are greater than the potential for recharge. The excessive pumping of resources is usually the result of uncontrolled groundwater exploitation, i.e. where there is no management of exploitation

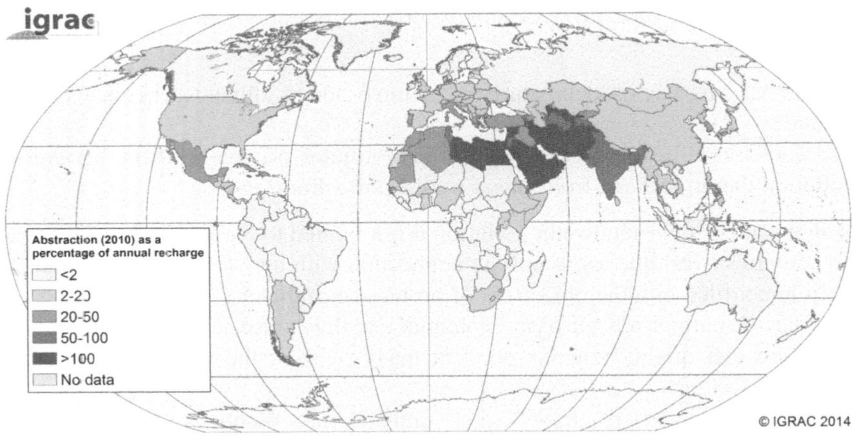


**Figure 2.11** Groundwater development stress (IGRAC, 2014).

based on the water balance (calculation of all inflows, runoff and changes in the specific ecological abundance).

Intensive groundwater exploitation is becoming common in many parts of the world, especially in partially arid and arid areas, on smaller islands and in coastal zones. Abstraction from aquifers has increased worldwide, especially so in the second half of the 20th century. This has had a negative impact resulting in declining groundwater levels, the depletion of groundwater reservoirs, the disturbance of springs, surface waters and groundwater dependent ecosystems, and in the potential subsidence of terrain, etc.

### *2.4.3.2 Protection from water and the regulation of watercourses*

In order to protect people and reduce damage and the negative social and economic consequences of the harmful effects of water, a number of investment (construction) and non-investment (administrative, regulatory and institutional) activities can be carried out. The following are measures that can be taken to provide protection against the harmful effects of water:

- Flood protection from external and internal waters;
- Protection against erosion and torrents;
- Management of watercourses.

The main objective of flood protection is to protect people and reduce potential damage, which is achieved by implementing various measures. How severe the damage caused by flooding will be depends on the size and value of the floodplain. Based on this, it is possible to determine the return period for which flood protection is provided. Human settlements have always developed along the banks of rivers, as discussed in sub-chapter 2.1.

Along with the development of settlements, there has been growing awareness of the fact that a reduction in the risk of floods increases the benefits from water and reduces any damage to people and to socio-economic development. Measures and activities aimed at a reasonable reduction of major flood damage can be divided into investment and non-investment measures.

Investment measures implemented for the purpose of flood protection:

- Construction of defensive and other embankments, such as quays, for the purpose of protection from any direct spillage of water and its penetration into protected areas;

- Improving the discharge capacity on sections of a river (roughness reduction, increase of the flow profile, shortening of the river, etc.);
- Retention of part of the flood wave in reserved spaces in single or multipurpose reservoirs or retention basins;
- Reduction of the downstream flow by diverting the flood wave into discharge channels.

**Non-investment flood protection measures**

Preventive and operational measures represent the overall activity aimed at reducing the negative effects of floods at all stages of flood defense. The most important preventive measure consists of making and implementing flood protection plans and regulations that define the obligations and rights of all participants in flood defense.

Operational flood protection measures include the forecast of a flood wave, the transfer of information to the field, and the obligation to notify and alert the relevant authorities and population in accordance with a prepared plan. This plan must also cover the possible dislocation of the population and movable property in the event of extremely large floodwaters. In the case of rivers where flood protection reservoirs are used, special management instructions for their operation are compiled, which is especially important for multipurpose reservoirs.

Regulatory and institutional measures consist of a set of measures defined by law, regulations or otherwise, which establish the use of relevant measures in flood threatened areas, administrative responsibilities, the organization of expert services, supervision of activities and the implementation of relevant measures to comply with legal regulations. These measures are of great importance when planning development in a flood-threatened area, and they can also affect the population and existing economy in the affected area.

In addition to the operational, regulatory and institutional measures, non-investment measures include solidarity measures, information and rapid response measures, and also education.

In recent times, the prevailing view has been that protection against the harmful effects of water can only be ensured by combining investment and non-investment measures. There is a tendency towards an integrated approach to the 'management of floodplains', which can be defined as the ongoing process of deciding whether and how potential floodplains should be used with specific (allowed) degrees of risk. This approach aims to protect human lives and goods and preserve or restore natural functions and resources within the floodplain. The integrated river basin flood risk management strategy is shown schematically in Figure 2.12.

**Flood protection – drainage (inland water protection)**

Overly moist soil and high groundwater levels require the drainage of excess water and the control of groundwater levels. This is important in order to create favorable sanitary and hygienic conditions for settlements and for the productive use of land for agricultural purposes, as well as to reduce damage in the area exposed to the adverse effects of inland waters. It is also very important to determine the origin of excess water in order to be able to decide on an optimal solution for protection against the harmful effects of inland waters.

Inland waterway protection can also include protection against erosion and torrents. The absence of vegetation, especially on sloping terrains, leads to erosion and, consequently, more intense torrents.

Measures to prevent and control erosion and torrents are very important in terms of saving lives, property and arable land, as well as in terms of preventing the undesirable deposition of suspended sediment at downstream sections of watercourses.

**Regulating watercourses**

Watercourses are regulated in order to increase the flow capacity, to provide more suitable conditions for their use (e.g. navigation) and to reduce the risk of flooding.

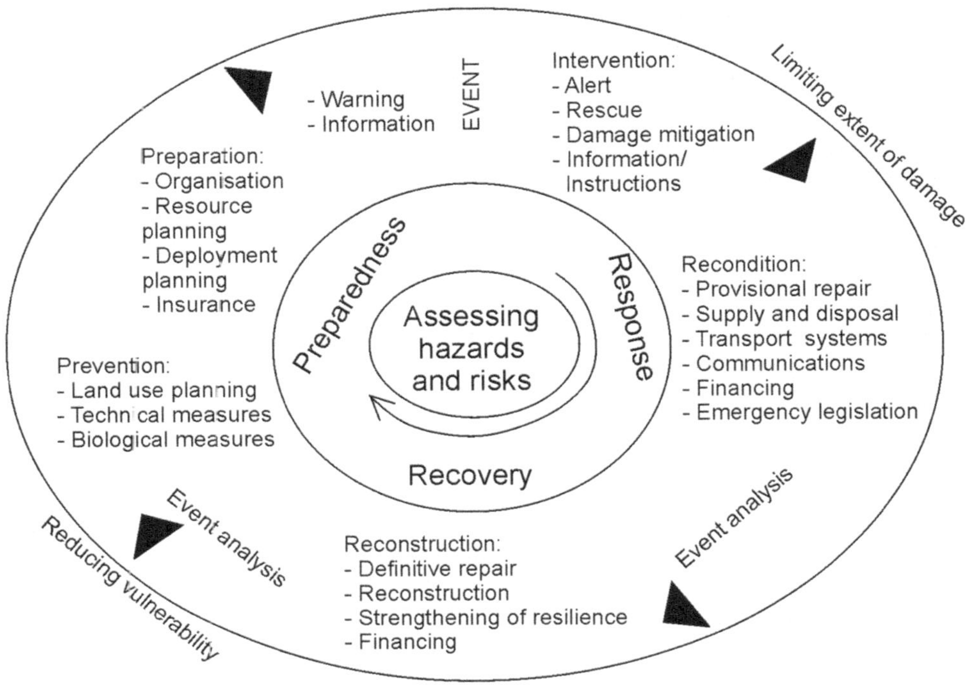


**Figure 2.12** Integrated river basin flood risk management strategy (Federal Office for Civil Protection [FOCP], 2009).

In addition to artificial regulation measures, there is an increasing insistence on regulating the flow of rivers as naturally as possible (for example, the discharge of flood waters into natural cassettes, allowing the flooding of certain areas, etc.).

### 2.4.4 Internal water management framework

The internal framework of a water management system refers to a system of legislative regulations, institutions and organizations in the area of management, the economic situation in the area being managed, as well as the scope, efficiency and quality of financial flows.

Internal legal frameworks rely on external legal frameworks (binding global and regional acts). Internal legal frameworks in the general sense relate to: constitutional provisions, water laws, environmental laws, laws related to water management activities and other relevant laws/groups of laws with accompanying by-laws. Interstate agreements are also important here.

The institutional and organizational framework presupposes the existence of an institutional system of the state/catchment in terms of the governing and operational structures, as well as structures at the level of smaller catchments and/or municipalities, districts, etc. This implies, at the level of the political and administrative entity, existence of appropriate administrative structure, as well as an organization at the level of functions that should be achieved (current business functions, and joint administrative and development functions).

The financial framework covers all the elements pertaining to the financing of the necessary capacities and functions of the water system. This includes the sources and circulation of assets, and their placement and renewal.

### 2.4.5 Capacities and functions

Capacity refers to the ability to perform a particular job – a function. Within the framework of water management, the functions are: current – business, joint – administrative and development functions. Water management functions consist of a set of activities that ensure the achievement of specific water management objectives. The success of the entire water management system is a function of the efficient implementation of these functions. Capacities include human capacities and institutional and economic capacities. Human capacities include scientific, expert, technical and educational capacities. We can distinguish between the capacities of administrative institutions, research and development institutions and companies, and design and contracting capacities, as well as operators.

The functions of day-to-day business include the operational function, exploitation, maintenance (regular and investment), supervision and control (inspection and monitoring). Development functions include planning, investment preparation and investment. Common (general) functions consist of regulations, financing, information systems, international cooperation, public participation, etc.

### 2.4.6 Available water and the national income

The following two major factors largely determine the ability to achieve sustainable water resource management in different countries and regions:

- Natural water resources;
- Economic power.

Based on the above two parameters, countries can be classified into four general groups (Table 2.4), each of which requires a separate approach to water management (Dimkić *et al.*, 2011).

At the regional level, the ultimate goal is to establish a legal and institutional framework for sustainable water management. In some regions, especially those that are economically powerful, the establishment of sustainable water management is already underway.

For example, the EU Water Directive promotes improvement in terms of water protection and the sustainable management of water. Many European countries have invested heavily in water protection. A

**Table 2.4** Four basic groups of countries in terms of the level of economic development and quantity of available water resources (Dimkić *et al.*, 2011a).

| Groups of Countries | Quantity of Available Water Resources | Level of Economic Development | Preferred Way to Manage Water |
|---|---|---|---|
| 1 | High | High | Sustainable and responsive |
| 2 | High | Low | Adaptive that strives to be sustainable – help is essential |
| 3 | Low | High | Adaptive that strives to be sustainable |
| 4 | Low | Low | Adaptive – requires significant assistance |

good example is the Danube river basin, where a large number of countries are cooperating successfully within the framework of the International Commission for the Protection of the Danube River (ICPDR).

However, there are significant regional differences even within the European Union. For example, water management in southern Europe (Spain, Greece, Italy, etc.) is different from the water management systems in northern Europe, because there are significant differences in the volume of water which needs to be stored, water consumption per capita in the public water supply, agriculture etc.

Water supply associations (for example, the International Water Supply Association in the Danube River Basin – IAVD) operating within the same basin, also contribute to setting of unique criteria for sustainable water management. On the other hand, in certain regions (Saharan countries), water resource management frameworks are being put in place for resources that are either slowly-renewable or non-renewable.

In economically more powerful countries, there is a more pronounced possibility of decentralization of water management, due to the greater ability of smaller units to make financial investments.

In poorer countries, an adequate centralized water management is needed, this is especially important in very poor countries (Figure 2.2).

At the local level, which is the basic level of water management, it is important to provide appropriate legal, social, economic and other mechanisms to achieve the highest possible sustainability of water management.

Active water managing at the local level is challenging and it is crucial to ensure adequate funding is available at the local level.

In many countries, especially in arid and semi-arid regions, when assessing the sustainability of water resources, it is important to take into account that the users (for example farmers) are responsible for the intensive use of groundwater, a seemingly unstoppable phenomenon under the current institutional, economic, social and political conditions. In such cases, it is necessary to put in place appropriate mechanisms to prevent the uncontrolled use of a source (over-exploitation) through a better interconnection between the state institutions and users.

One of the causes of inadequate water management is the low cost of water in many countries to ensure social peace.

As a result, groundwater resources are often over-exploited and depleted, and there is a deterioration in the quality of surface and groundwater due to pollutants from industrial, agricultural and other activities.

Therefore, it is important to adopt frameworks that recognize the need to take preventative action, ensure the effective management of social risks and individual needs, and harmonize the way natural resources are managed when bodies of water extend across the territory of several countries. Political priorities and the economic stability of countries must be taken into account.

Ensuring adequate funding for water resource management is a very important factor as it contributes to the sustainable management of those resources. There are significant differences between water prices in different countries and they reflect the economic power of individual countries, but also to some extent the level of awareness of the importance of this resource (Figure 2.13).

Among the listed cities, each with a population of over one million, with the exception of Minsk, Sofia, Belgrade, Kiev and Odessa, the total cost of water and sewerage per cubic meter is greater than 1 euro. On the other hand, in 7 out of the 24 cities shown, the price is more than 3 $EUR/m^3$.

The situation is similar for the slightly smaller cities with 0.5 to 1 million inhabitants shown in Figure 2.14, where only in the cities of Kisnev (Moldova) and Lviv (Ukraine) is the price of water and sewerage less than 1 euro per cubic meter. For all others the price is higher than that amount.

These differences in the price of water can be attributed to the different conditions associated with the provision of this resource, but they might, perhaps to an even greater degree, reflect the difference in the

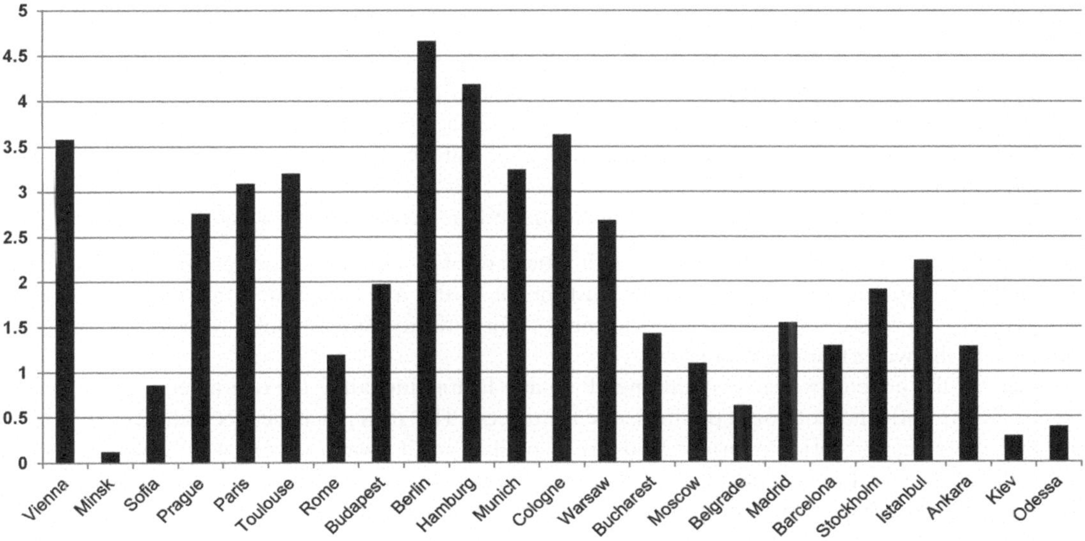


**Figure 2.13** The price of water (EUR/m$^3$) in cities and towns with populations of over one million (inner city) – Global Water Intelligence magazine (GWI, 2017).

approach to ensuring the sustainable management of this resource. The price of water is higher in economically developed countries where there is probably a well-developed awareness and financial basis for establishing the operation and sustainable management of this resource (Germany, France, Poland and Sweden, etc.).

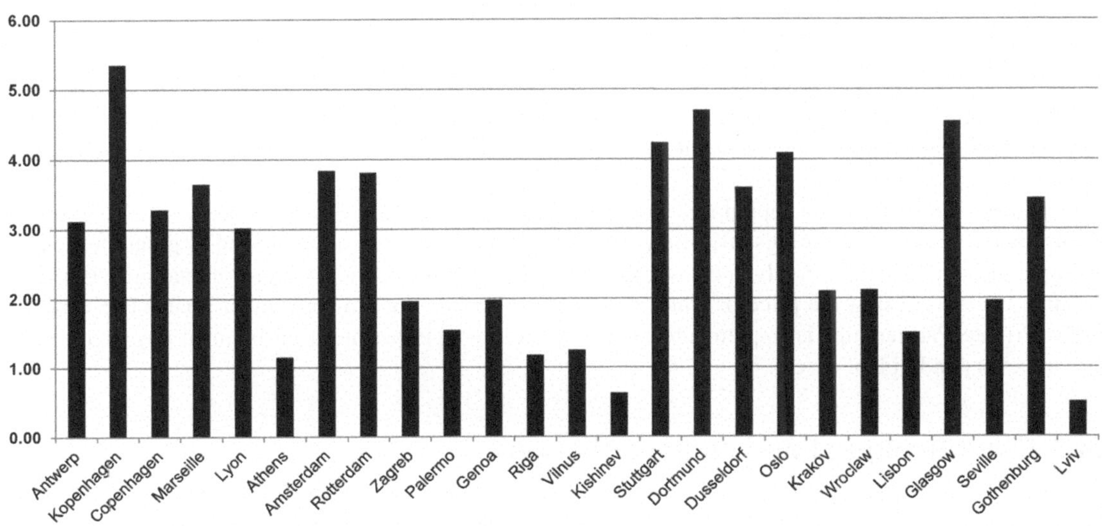


**Figure 2.14** Price of water (EUR/m$^3$) in individual cities with 0.5 to 1 million inhabitants (inner city) – Global Water Intelligence magazine (GWI, 2017).

It can generally be said that the economic cost of water should consist of the following elements:

- Operating and maintenance costs (costs of the daily operation of the water supply system, electricity used for pumping water, chemicals for water treatment, labor costs, etc.);
- Investment maintenance costs (costs related to the replacement of fixed assets);
- Economic costs (costs related to the impact the use of water has on other consumers, whether negative or positive);
- Environmental costs (costs related to maintaining public health and ecosystems).

In conditions where countries and the local authorities are of weaker economic power, local authorities should have an appropriate role (through the instrument of the real economic price of water), but it is also necessary to ensure possible state incentives for priority investments, which are of crucial importance to the water management sector.

In practice, the price of water is conditioned by many factors including the overall economic power of society and the influence of local political circumstances. The difference between the actual price of water and the economic price of water leads to disparities in the water sector at all levels - from the national level, where this is reflected in the insufficient global protection of water resources, and inadequate organizational structure in the water sector, etc., to the local level, where the consequences are most often reflected in the insufficient quantities of water supplied to the population, in the poor quality of drinking water, in the inadequate protection of source water and similar.

Another socio-economic problem that is often encountered is institutional disorganization. Questions and problems that first appear on the economic level become problems at the institutional level.

This is especially true for groundwaters because their natural features are more 'local' than those of surface water. Of course, in both cases, it is necessary to apply the advantages of both approaches.

It is clear that with an ever-growing population, in addition to their local character, resources also have wider, global and regional components, so we speak of:

- Basin management;
- Management of inter-state basins (regional governance);
- Regional solving of water related problems;
- The global role in addressing the issue of protection and use of resources (in this case water).

In richer countries, local government units, as a rule, have greater financing options, including a guarantee function. In poorer countries, local governments tend to have less financial means (Šojić *et al.*, 2014).

The participation of private capital in water management is necessary but it is a very sensitive issue. On the one hand, private equity brings efficiency in terms of greater functionality, as well as greater efficiency per employee, and the wider involvement of capital. On the other hand, it is interested in making a profit and this often leads to raising the price of water services and setting water management goals. The way PPP (Public-Private Partnership) is implemented is a very sensitive issue and a problem in any country, for each and every individual case.

### 2.4.7 Climate impact and climate change

A number of global and regional climate and hydrological models have been developed to estimate future temperatures, precipitation and runoff levels for different climate scenarios (IPCC 2007a).

Not enough light has yet been shed on the mechanisms leading to climate change, as global and regional forecast methods have not been sufficiently tested. Still, there are indicators that precipitation will decrease in a significant number of mid-latitude regions (the Mediterranean, southern Africa, Australia, etc.). The frequency and severity of extreme events – floods and droughts – are also expected to increase in many regions (Figure 2.15).

Climate is generally defined as the regime of multiple parameters in a given area: air temperature, precipitation, humidity, etc. These parameters are subject to pronounced periodic changes: daily, seasonal, annual, long-term, etc. Over the past century, there have been some trends in the change of certain parameters that condition climate change. Some have caused a greater fluctuation in certain climatic parameters, while others have reduced fluctuations.

These changes demonstrated the need for a systematic and multi-state approach to the issue - at the United Nations level. In 1988 the United Nations Environment Programme (UNEP) and the World Meteorological Organization (WMO) formed an Intergovernmental Panel on Climate Change (IPCC), which has three main objectives:

- To determine the current level of scientific knowledge on climate change;
- To explore the effects of climate change on the environment, economy and society;
- To formulate a series of strategies for responding to climate change.

The IPCC has played a key role in providing climate change research data to decision makers. The foundations of the United Nations Framework Convention on Climate Change were laid down in Rio de Janeiro in 1992. Its ultimate goal is to 'stabilize the concentration of greenhouse gases in the atmosphere at a level that will not threaten the global climate. This should be done within a timeframe that allows

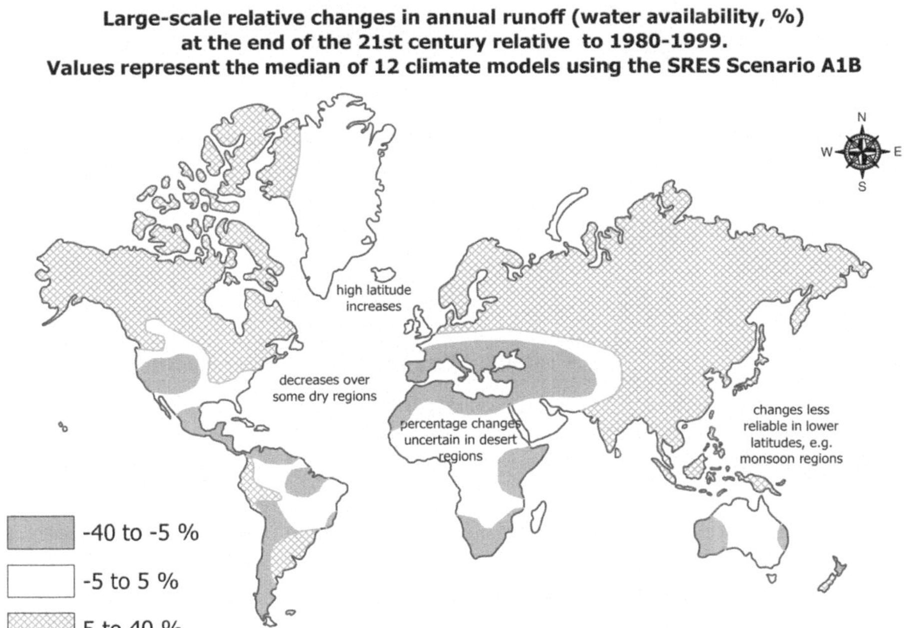


**Figure 2.15** Large-scale relative changes in annual runoff (water availability %) at the end of the 21st century relative to 1980–1999. Values represent the median of 12 climate models using the SRES Scenario A1B (IPCC, 2007b).

ecosystems to adapt to climate change naturally, without compromising food production and thus enabling economic development in accordance with the principles of sustainable development.'

The UN Framework Convention on Climate Change, which entered into force on 21 March 1994, contains general frameworks for interstate cooperation and coordination in response to the challenges posed by climate change to human society as a whole. It treats climate as a common resource whose stability can be affected by emissions of carbon dioxide and other greenhouse gases.

According to the IPCC report from 2007, the following occurrences and impact on water resources due to climate change can be expected (Table 2.5):

Water management targets were set by the UN in 2000 and 2015, respectively:

- 2000 – Millennium Goals - Seven water and natural resource goals: related to forest land cover, $CO_2$ emissions, fish species and natural fish stocks, aquatic resources, land and marine protected areas, protection of species threatened with extinction and availability of drinking water and sanitation;
- The UN Rio +10 and Rio +20 conferences looked at the implementation of the measures prescribed at the Rio World Summit in 1992 (Agenda 21);
- At the UNEP Climate Change Conference in Paris in 2015, declarations were adopted to maintain the rise of the planet's temperature below +2°C for the long term;
- At the UN Conference in New York in September 2015, the so-called Agenda 2030 was adopted. The goals of this agenda are aimed at complementing and delivering goals not covered and pursued by the Millennium Goals. Among other things, the Agenda anticipates taking urgent action on climate change. Agenda item 7 relates to clean water and sanitation. This point seeks to ensure access to clean water and sanitation for everyone by 2030.

The IPCC in its AR5 Synthesis Report has shown that each of the last three decades has been successively warmer at the Earth's surface than any preceding decade since 1850. The globally averaged combined land and ocean surface temperatures, as calculated by a linear trend, show a warming of 0.85 (0.65 to 1.06)°C over the period 1880 to 2012.

Impacts from recent climate-related extremes such as heat waves, droughts, floods, cyclones, and wildfires, reveal the significant vulnerability and exposure of some ecosystems and many human systems to current climate variability. Direct and insured losses from weather-related disasters have increased substantially in recent decades, both globally and regionally.

**Table 2.5** Predicted occurrences and impact on water resources (IPCC, 2007a).

| Occurrence | Predicted Key Impact on Water Resources |
|---|---|
| Warmer and hotter summer days in most land areas | Impact on water resources that depend on snowmelt, impact on specific water supply systems |
| Warmer periods, hotter waves in most land areas | Increased water demand, water quality problems – algal blooms |
| Higher incidence of heavy rainfall | Negative impact on surface and groundwater quality, pollution of water sources |
| Increase in drought-affected areas and regions | Increasing pressure on water resources |
| Increased incidence of tropical cyclones | Disruption of water supply due to power failure |
| More frequent occurrence of high sea levels | Reduction of available freshwater due to saltwater intrusion |

Questions regarding the impact of human beings on the planet's climate remain largely open. The reasons for climate change can be long-term and short-term. The long-term reasons may be related to cosmic parameters (Earth's tilt, Earth's trajectory around the sun, shifting ice ages and interglacials), the medium to short-term reasons to volcanic eruptions or radiation intensity, and the short-term reasons to human impact.

Given the importance of climate and climate change, it is necessary to continue monitoring hydrological-climatological parameters and to carry out hydrological-climatological, geological, biological and other studies, improving prognostic methods and defining mitigation and adaptation measures to climate change as much as possible.

### 2.4.8 Development of scientific, technical and operational capacities

The following capacities are necessary for a water management system to function smoothly:

- Management and planning;
- Science and education;
- Production and services;
- Construction.

One of the main frameworks for water management is science and education (Figure 2.16). Even if there are other water management capacities, there are no human scientific and professional capacities that, as products of education and professional-scientific work, can implement measures and activities for efficient water management without developed scientific knowledge and a developed education system.

Investment in science, education and research is a precondition for providing an adequate response to the increasingly dynamic changes in the socio-economic system, as well as to changes in the natural environment due to climate change.

How much is invested in science and education depends on how developed a country is (Figure 2.17). Ultimately, this affects a country's capacity and ability to implement regional water management frameworks as well as the country's capacities that depend on science and research such as production and services, construction, management and planning.

The issue of professional and operational capacities is also important for the policy and method of financing projects. It is necessary to take into account the professional development of staff and the development of other capacities through participation in projects.

For the successful functioning and development of the water sector, an adequate legislative and institutional framework must be provided to allow all activities to be carried out, respecting all the functions essential for the achievement of water management objectives in a given territory.

## 2.5 WATER-FOOD-ENERGY CONNECTION

Today, we face many challenges, such as climate change (both today and in the future), a rapidly growing population (affecting social and economic development), globalization and urbanization, and political and economic confrontations. In order to respond to these challenges, efficient and adaptive water management is required.

The growing population on our planet is one of the most evident pressures on water resources. This pressure can be divided into two groups. The first group consists of direct pressures, namely the required amount of water that meets the sanitary requirements of the entire population. The second group consists of indirect pressures reflected in the increasing demands and use of water resources for production of food and energy.

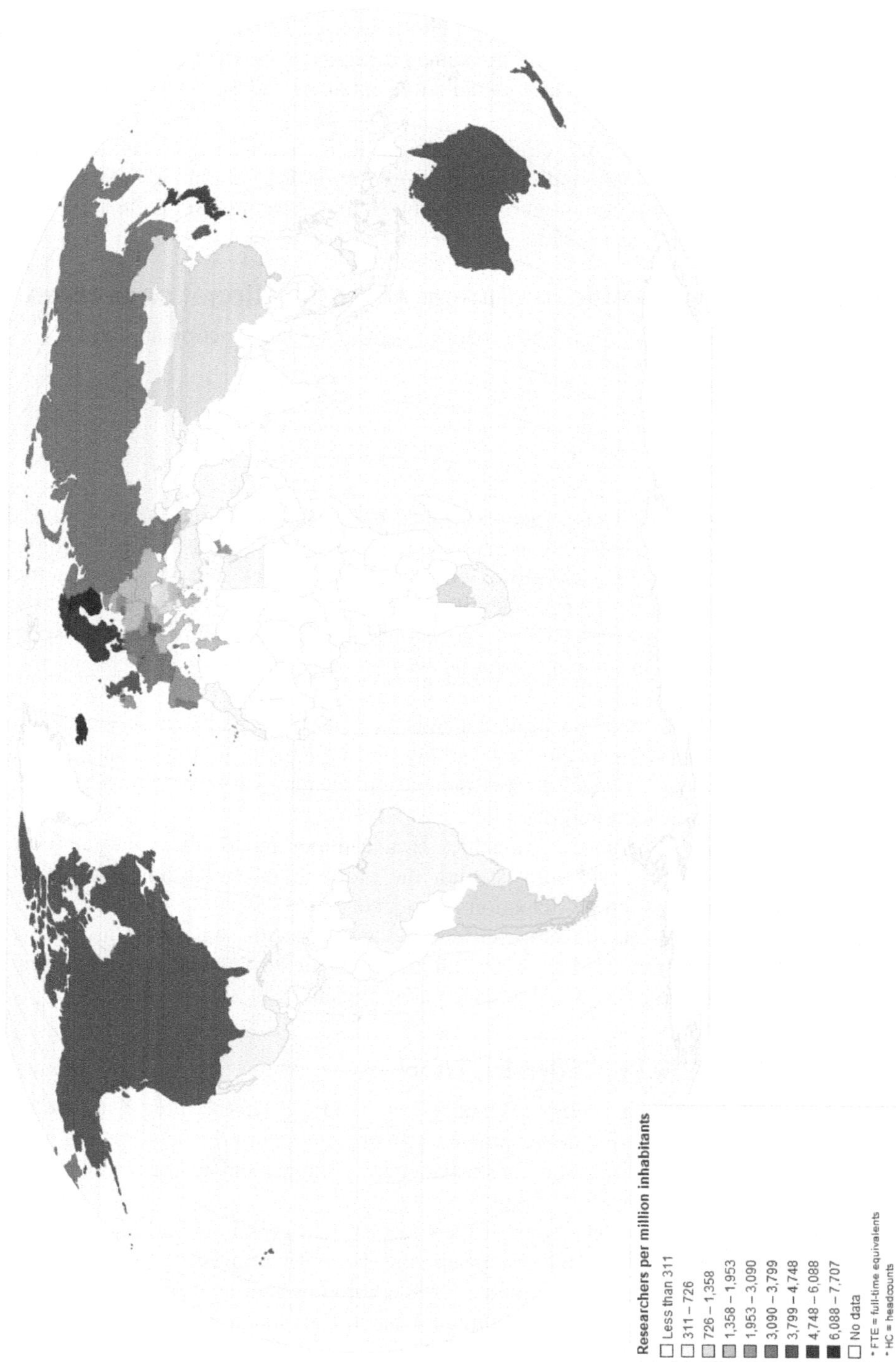


**Figure 2.16** Number of researchers per million inhabitants by country (UNESCO, 2011).

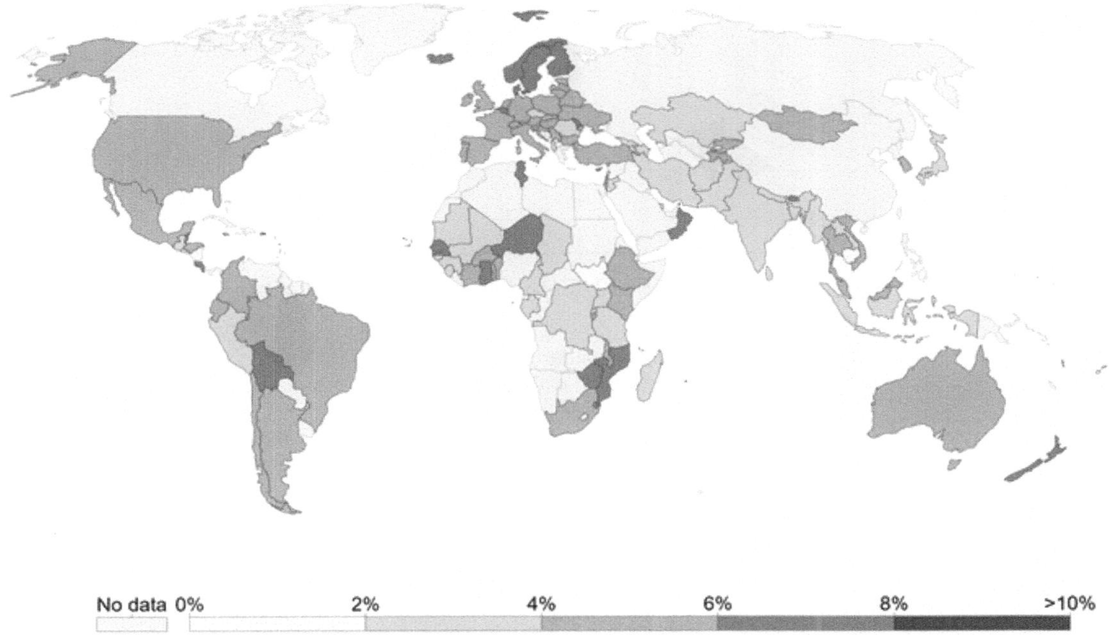


**Figure 2.17** Allocation of funds for education in different regions and countries % GDP (UNESCO, 2018).

Along with the expected population growth, a 50% increase in the demand for food is expected by the year 2030 and an up to 70% increase by 2050 (Bruinsma, 2009), while the demand for energy from hydropower and other renewable energy sources will increase by 60% (World Water Assessment Programme [WWAP], 2009). Future water demand due to population growth is presented in more detail in sub-section 2.3.

In addition to the increase in population, population migrations will also have a significant impact on water resources. Population migrations are caused by wars but they are also the result of people looking for better economic conditions. This leads to increased demands for water in urban areas on the one hand, while on the other hand, the lack of competent staff in rural areas also causes problems. Such social changes lead to changes in the management system as well as to institutional changes.

Climate change also significantly affects the water cycle. Climate systems are not stationary and climate change is increasingly visible. The magnitudes, and spatial and temporal distribution of precipitation is not uniform and this, together with temperature changes, leads not only to significant changes in low, average and high flows, but also to changes in evaporation as well as the runoff in rivers.

While some countries face record high droughts, others are struggling with record floods. Extreme events such as droughts and floods lead to problems of water supply, and thus to problems in food and energy production. It is evident that certain aquifers have been over-exploited, and that bodies of surface and groundwater as well as seas and oceans have been degraded which in turn has resulted in the degradation of the overall quality of the environment.

This degradation has taken place in order to meet short-term economic objectives, which often do not consider long-term related problems, such as the very economic and environmental sustainability of the region or basin.

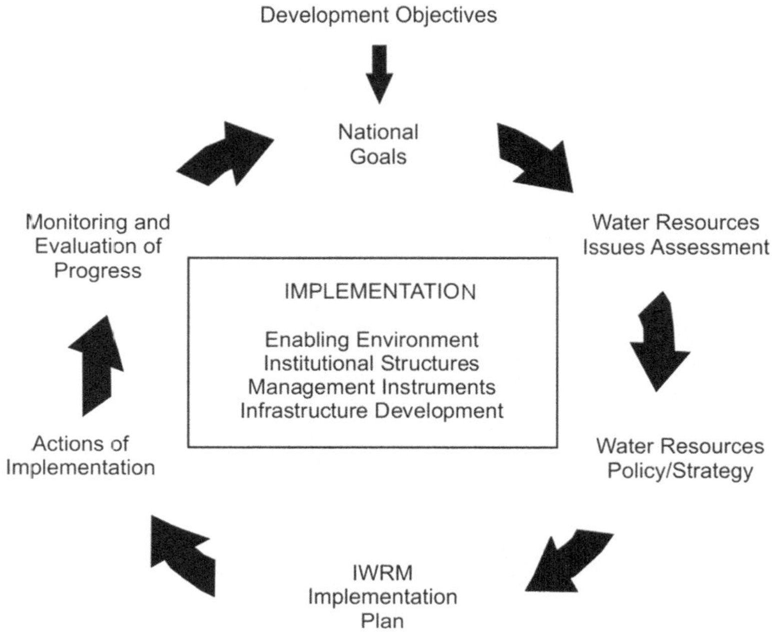


**Figure 2.18** Phases of planning and implementation of an integrated water management system (WMO, 2014).

Good policies at the national level are of crucial importance for the development of a country. Most countries that have successfully found a solution to the problem of poverty have done so primarily by building up their basic production capacity, especially in the fields of agriculture, energy, transport and water management systems. Figure 2.18 shows the key stages of the planning and implementation of integrated water resources management (IWRM).

### 2.5.1 Water required for food production

As there is a strong link between food, water, energy, the environment and human well-being, the water/food/energy link needs to be very sustainable as it can affect both natural resources and the environment. In an era of ever-increasing water scarcity, it is important to identify ways of using our technical capacity and knowledge to reduce inefficiency and the increasing demand for agricultural production.

Poor and unsustainable land use and management leads to desertification and land degradation worldwide and increasing pressure on water resources, which ultimately leads to water scarcity. Soil degradation is on the increase and is most prominent in arid parts of the world, but it has recently also been observed in humid areas (Bai *et al.*, 2008). It is estimated that nearly 2 billion ha of land worldwide (twice the size of the Republic of China) have already been severely degraded and some irreversibly (FAO, 2008a, b).

It is anticipated that climate change will have significant effects on the amount of food available, especially in areas of Africa and Asia where millions of farmers rely solely on atmospheric precipitation for crop growth. This implies a further increase in water use in the agricultural sector, which will cumulatively lead to more pressure on the already overburdened water resources.

Irrigation as a hydro-technical measure enables the optimal moisture of soil during the vegetation period of plants. The water that is used predominantly comes from watercourses, artificial lakes or groundwater, while a very small portion comes from treated wastewater. Globally, crop yields where irrigation is used are 2.7 times higher than those where only rainwater is used.

The area of land with irrigation systems has increased. In 1970, 170 million ha of land were irrigated. By 2010 (Wild *et al.*, 2010) the surface area with irrigation systems had grown to about 300 million ha. There is still space and potential for increasing the surface areas that can be irrigated in parts of Africa and South America where there are available sufficient water resources. It has been estimated that there will be an 11% increase in water consumption between 2008 and 2050, due to increased agricultural areas with irrigation systems (FAO, 2011a, b).

Although an increase in agricultural land is possible, about 5–7 million ha (about 0.6%) of agricultural land is lost annually due to degradation and urbanization. In contrast to the decrease of agricultural land due to degradation, an increase in population is evident, leading to a decrease in the total area of agricultural land per capita from 0.4 ha/capita in 1961 to 0.2 ha/capita in 2005 (FAO, 2011a, b).

The increased amount of freshwater used for irrigation and food production has become a major challenge. Of the total amount of freshwater consumed globally, as much as 70% is used for irrigation and food production. In some countries with a strong economic growth, such as the BRIC S Member States (Brazil, Russia, India and China), approximately 74% of the total amount of extracted water is for agricultural activities, while in India it is around 87%. In some developing countries, the amount of water used for agricultural activities is up to 90% (FAO, 2011a, b).

In many countries, the availability of water for agriculture is already limited and threatened, with a tendency to increase the pressure on water supply.

In order to properly manage agricultural land and to supply sufficient water, it is necessary to know the properties (physical and chemical) of the land that we want to irrigate. Based on these land characteristics, we can determine the retention time of water in the soil, the soil moisture at a specific depth, whether the water is free or physically or chemically bound, and how fast the water will infiltrate into the deeper layers or evaporate. Figure 2.19 shows a vertical profile of soil with the water content in it.

The water capacity of the soil defines the total amount of water stored in the soil within the zone of a plant's root. The texture (predominantly porosity) of the soil and the root of the crop that is planted on

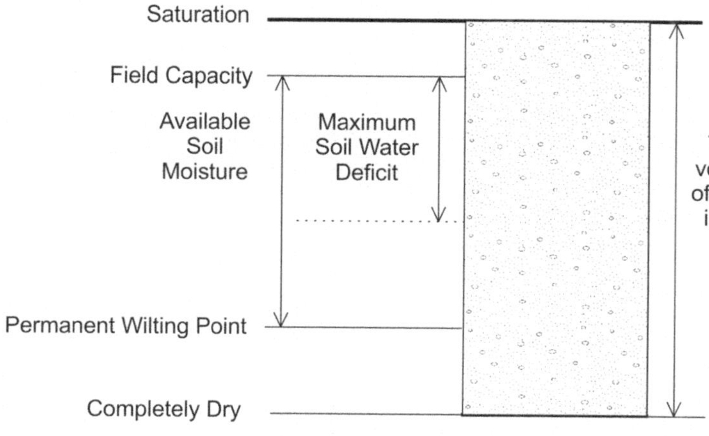


**Figure 2.19** Soil humidity Conditions (Ministry of Agriculture, British Columbia, 2015).

**Table 2.6** Effective rooting depth of mature crops (Ministry of Agriculture, British Columbia, 2015).

| Shallow (0.45 m) | Medium Shallow (0.6 m) | Medium Deep (0.9 m) | Deep (1.2 m) |
|---|---|---|---|
| Cabbages | Beans | Brussels sprouts | Asparagus |
| Cucumbers | Beets | Corn (sweet) | Blackberries |
| Cauliflower | Blueberries | Eggplant | Grapes |
| Onions | Broccoli | Kiwifruit | Loganberries |
| Lettuce | Carrots | Peppers | Raspberries |
| Radishes | Celery | Squash | Sugar beets |
| Turnips | Potatoes | Saskatoons | Tree fruits (spacing 4 m × 6 m) |
| | Peas | Tree fruits (spacing 2 m × 4 m) | |
| | Strawberries | | |
| | Tomatoes | | |
| | Tree fruits (spacing 1 m × 3 m) | | |

the given soil (Tables 2.6 and 2.7) have the greatest influence on the water retention capacity of the soil. Knowledge of the capacity to store water in the soil makes it possible to define the amount of water that should be used for irrigation at all times, the intensity of irrigation, and how long it is necessary to wait between each irrigation period.

For example, the amount of water applied at the same time on sandy soil, which has a low capacity to accumulate water in the soil, would be greater than that for loam, which has a higher water retention capacity. This assumes the rooting depth is the same for both soils. Applying more water to the soil than can be stored leads to the loss of water and its transport to deeper layers of the soil as well as to the leaching of nutrients outside the root zone. This can also lead to the salinization of the soil and thus impair the original quality of the soil. Table 2.7 shows the average root depth of agricultural plants and the potential water storage capacity of different soil types.

Another very significant consequence of the over-irrigation of agricultural land is the pollution of groundwater by various products used in agriculture (e.g. pesticides, fungicides, etc.) and nitrates, which are one of the biggest chemical contaminants of groundwater resources in the world. The USA uses the

**Table 2.7** Guide to available water storage capacities (Ministry of Agriculture, British Columbia, 2015).

| Textural Class | Available Water Storage Capacity (AWSC) (mm) water/(m) soil |
|---|---|
| Clay | 200 |
| Clay loam | 200 |
| Silt loam | 208 |
| Loam | 175 |
| Fine sandy loam | 142 |
| Sandy loam | 125 |
| Loamy sand | 100 |
| Sand | 83 |

largest amount of pesticides, followed by countries in Europe, especially countries in western Europe. In terms of pesticide use per unit area of agricultural land, Japan is first on the list.

Africa's economy relies heavily on agriculture which in turn depends on the amount of atmospheric precipitation. About 87% of pumped water is for agricultural purposes (FAO, 2008a, b). The average amount of annual rainfall in Africa is not small, about 815 mm (FAO, 2008a, b), but it is not evenly distributed spatially and temporally. Most of the rainfall occurs in the equatorial parts of Africa, in the vicinity of the Niger Delta to the Congo Basin. The rainfall is mainly seasonal, which results in alternating droughts and floods. In the region of Saudi Arabia, agriculture is the primary source of water stress. In most countries, agriculture accounts for more than 70% of the total water demand. For example, in Iraq, Oman, the Syrian Arab Republic and Yemen, agriculture accounts for more than 90% of the water demand. However, the region is unable to produce a sufficient amount of food for the entire population, and therefore imports 40–50% of its total cereal consumption (Cline, 2007). Saudi Arabia, which is one of the largest cereal growers in the Middle East, has adopted a program of measures that will reduce cereal production by 12% in order to reduce unsustainable groundwater use. The government of Saudi Arabia issues incentives to corporations to lease large tracts of land in Africa for agricultural production, in order to protect its own available water and secure food production. By investing in Africa to grow its crops, Saudi Arabia saves the equivalent of hundreds of millions of gallons of water per year. India grows corn, sugar cane, lentils and rice in Ethiopia, Kenya, Madagascar, Senegal and Mozambique to meet the demands of the domestic market.

It has been estimated that future global water consumption for agriculture will increase by 19% by 2050, and in the absence of any technological advancement or political intervention there might be an even greater increase in water consumption (UNEP, 2007a, b, c).

Average water demand of crops per region is presented in Table 2.8.

Droughts have the greatest negative economic impact on agriculture. Adapting to drought requires major investment and institutional change, which can be achieved by introducing completely new crop varieties, a new and technologically advanced irrigation infrastructure, as well as new processing plants and the development of education institutions for the transfer of technologies. Water demand of crop production in selected countries and selected river basins is presented in Table 2.9.

**Table 2.8** Average water demand of crops per region (1996–2005).

| Region | Water Demand ($m^3/t$) | Region | Water Demand ($m^3/t$) |
|---|---|---|---|
| Central Africa | 12.5 | Caribbean | 162.5 |
| Eastern Africa | 59 | Central America | 74.5 |
| Northern Africa | 336 | Northern America | 42.5 |
| Southern Africa | 40 | Southern America | 48 |
| Western Africa | 20 | Eastern Europe | 19 |
| Central Asia | 144.5 | Northern Europe | 0.5 |
| Eastern Asia | 119 | Southern Europe | 70 |
| Middle East | 271.5 | Western Europe | 7 |
| Southeastern Asia | 90 | Oceania | 33 |
| Southern Asia | 339 | | |
| Western Asia | 206.5 | | |

**Table 2.9** Water demand of crop production in selected countries and selected river basins (1996–2005).

| River Basin | Water Demand ($Gm^3$/yr) | Country | Water Demand ($Gm^3$/yr) |
|---|---|---|---|
| Mississippi | 20 | India | 115.7 |
| Ganges | 54 | China | 59.45 |
| Yangtze | 9 | USA | 47.95 |
| Indus | 58.5 | Brazil | 4.45 |
| Parana | 1.6 | Russia | 5.2 |
| Niger | 0.85 | Indonesia | 5.75 |
| Nile | 14.5 | Nigeria | 0.55 |
| Huang He | 10.5 | Argentina | 2.15 |
| Nelson | 0.75 | Canada | 0.8 |
| Danube | 0.9 | Pakistan | 37.15 |
| Krishna | 10.5 | **World** | 449.5 |
| Volga | 1.7 | | |
| Ob | 0.9 | | |
| **World** | 449.5 | | |

The Water Footprint is one of the most common and applicable methods for categorizing water consumption (Hoekstra *et al.*, 2011; Hoekstra & Mekonnen, 2011; Mekonnen & Hoekstra, 2011, 2014).

The Water Footprint method is based on the assumption that all the water consumed for the production of a product can be divided into three categories:

- Blue water footprint;
- Green water footprint;
- Gray water footprint.

Table 2.10 shows the average water demand of crops and products footprints.

### 2.5.2 Energy required for water production

Water resources and energy are of crucial importance for human well-being and for sustainable socio-economic development. Globally, the demand for energy will increase over the coming decades to meet the needs of the population, growing economies and changing lifestyles. This will intensify the pressure on the limited natural resources and the ecosystem.

The challenge will be the greatest in countries undergoing a rapid transition and a rapid economic growth, especially where water resources are scarce or where the water-related infrastructure and services are inadequate.

A water management policy should be able to provide sufficient water to produce the energy needed by society. Energy, such as electricity and various fuels, requires water for cooling and production.

If the current consumption trend and the way energy is used continues, there will be an increase in water demand of about 11.2% by 2050 (UNEP, 2011). Already, the water used to cool power plants in the United States accounts for 40% of the country's industrial water use, while it will reach 30% in China by 2030 (UNEP, 2011). Energy is necessary to provide the large amount of water that is needed and that we use,

**Table 2.10** Average water demand of crops and products footprints (Hoekstra *et al.*, 2011; Hoekstra and Mekonnen, 2011; Mekonnen & Hoekstra, 2011, 2014).

| Crop | Water Demand ($m^3$/t) | Crop | Water Demand ($m^3$/t) |
|---|---|---|---|
| Wheat | 171 | Spinach | 7 |
| Wheat flour | 173.5 | Tomatoes | 31.5 |
| Wheat bread | 150.5 | Tomato paste | 126.5 |
| Dry pasta | 173.5 | Tomato ketchup | 79 |
| Rice, paddy | 170.5 | Cauliflowers and broccoli | 10.5 |
| Rice flour | 267.5 | Pumpkins and squash | 12 |
| Maize (corn) | 40.5 | Chilies and peppers, green | 21 |
| Maize (corn) flour | 41.5 | Garlic | 40.5 |
| Maize (corn) oil | 85.5 | Garlic powder | 156.5 |
| Rye | 12.5 | Beans, green | 27 |
| Rye flour | 16 | Peas, green | 31.5 |
| Potatoes | 16.5 | Bananas | 48.5 |
| Sugar cane | 28.5 | Oranges | 55 |
| Refined sugar | 243.5 | Orange juice | 99.5 |
| Beans, dry | 62.5 | Lemons and limes | 76 |
| Peas, dry | 16.5 | Grapefruit | 42.5 |
| Chickpeas | 112 | Apples, fresh | 66.5 |
| Cashew nuts | 460.5 | Strawberries | 54.5 |
| Chestnuts | 87 | Raspberries | 26.5 |
| Almonds, with shell | 954 | Blueberries | 167 |
| Walnuts, with shell | 649.5 | Grapes | 48.5 |
| Pistachios | 3801 | Pineapples | 4.5 |
| Hazelnuts, with shell | 545 | Coffee, green | 58 |
| Soya beans | 35 | Coffee, roasted | 69.5 |
| Coconuts | 1 | Cocoa beans | 2 |
| Olives | 249.5 | Green and black tea | 449 |
| Sunflower seed oil, refined | 149.5 | Vanilla beans | 19524 |
| Cabbages and other brassicas | 13 | Cinnamon | 20.5 |
| Ginger | 20 | Watermelons | 12.5 |

and vice versa, water is needed to provide most of the energy that is required and used. Furthermore, population growth will lead to the construction of new cities and the expansion of the existing ones, which will result in higher energy and water consumption. This can in turn affect future economic and social development as well as human health and the environment.

Of the total global energy consumption, 7–8% is spent on groundwater pumping, piping and the treatment of both groundwater and wastewater (Hoffman, 2011), while in developing countries, this increases to close to 40% (World Economic Forum [WEF], 2011). Large quantities of wastewater continue to be discharged into the final receiving bodies without prior treatment, especially in developing countries with less economic power.

On the other hand, the treatment of wastewater requires large amounts of energy and it can be expected that the need for energy for this purpose will increase by 44% by 2030 (International Energy Agency [IEA], 2009).

Appropriate measures need to be taken to reduce the stress on activities related to water use, including water conservation and the improvement of infrastructure. However, even though the measures that need to be taken are very important, they only lead to improving the quality of the existing freshwater resources, and not to increased quantities.

One possible way to increase the amount of water available for water supply purposes is to use the available amount of freshwater from the hydrological cycle (recirculation) of fresh water, as well as by means of desalination and water reuse (Shannon *et al.*, 2008).

The number and size of desalination plants worldwide has been increasing at a rate of 5 to 6% annually since 2010 (Figure 2.20), corresponding to an additional 3 to 4 million $m^3$/day of fresh water each year from the newly installed plants (Voutchkov, 2017). Between June 2015 and July 2016, the capacity of desalination plants globally amounted to 3.7 million $m^3$/day and there were 512 new plants during this period (GWI, 2016), of which 266 (52%) were of a high or medium capacity. By June 30, 2016, the total number of desalination plants in the world amounted to 18,983, with a cumulative capacity for the production of fresh water amounting to 95.6 million $m^3$/day (GWI, 2016).

Along with the development and improvement of desalination technology, the costs of building desalination plants are declining. Cost reduction is one of the primary factors that influences the wider adoption, growth and success of desalination technology. Since the 1960s, costs for water desalination have decreased by about 90%, from an approximate unit cost of US \$ 10.00/$m^3$ in the 1960s to less than US \$ 1.00/$m^3$. In 2017, the costs were reduced by 20% compared to 2010, mainly due to technological developments, such as better membrane design and the integration of systems (Ghaffour *et al.*, 2012). The Table 2.11 shows the energy required per $m^3$, for different types of water supply.

The data in Table 2.11 indicate that the energy required to desalinate seawater is approximately 8 to 10 times greater than that needed to provide fresh water from conventional sources such as rivers, lakes and freshwater aquifers. However, it should be noted that such resources are limited to less than 2.5% of the water available on the planet.

The most common forms of desalination can be divided into two types of technology:

(1) Thermal desalination (using thermal energy to separate distillates from high salinity water). Mechanical vapor compression (MVC) is primarily used for the desalination of high total

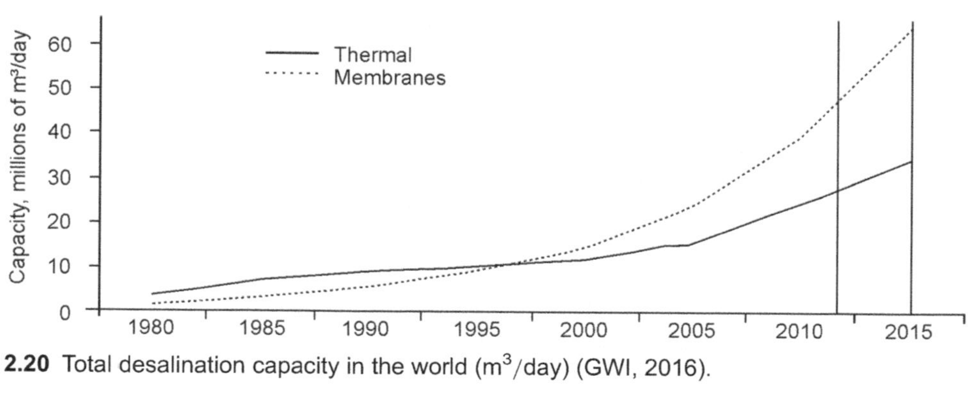


**Figure 2.20** Total desalination capacity in the world ($m^3$/day) (GWI, 2016).

**Table 2.11** Energy used for different types of water supply (Voutchkov, 2017).

| Water Supply Alternatives | Energy Use (kWh/m$^3$) |
|---|---|
| Conventional treatment of surface water | 0.2–0.4 |
| Water reclamation | 0.5–1.0 |
| Indirect potable reuse | 1.5–2.0 |
| Brackish water desalination | 1.0–1.5 |
| Desalination of Pacific Ocean water | 2.5–4.0 |

dissolved solids (TDS) (>45.0 mg/l) and/or industrial wastewater for reuse and not necessarily for drinking water.

(2) Reverse osmosis (RO) membrane separators, which use a membrane barrier and the energy of pumping to separate salts from high salinity water (typically <45,000 mg/l). RO systems use high-pressure pumps to overcome the osmotic pressure of raw feed water. For example, some seawater reverse osmosis (SWRO) factories may require a pressure of up to 70 bar.

The first desalination plants, produced mainly in arid countries, were based on thermal desalination. The plants applying thermal desalination use a large amount of thermal and electrical energy, which results in high emissions of gases that cause the greenhouse effect (Semiat, 2008), in contrast to the plants using reverse osmosis. Currently, reverse osmosis, in terms of energy consumption, represents the most effective technology for the desalination of seawater and is a reference point for the comparison of any new technology for desalination (Menachem & William, 2011).

Almost 95% of the installed capacity for desalination today is either thermal (35%) or membrane (60%) technology (Ghaffour *et al.*, 2012). The largest production capacity is in the Middle East, while the biggest users of desalination according to capacity are the Kingdom of Saudi Arabia, followed by the United States, the United Arab Emirates, Australia, China, Kuwait and Israel (Figure 2.21).

Factors that have a direct and major impact on the costs of desalination include the desalination technology, the quality of the abstracted and produced water, the type of abstraction and discharge, the location of the plant, the type of energy used, the price of electricity, the need for post-treatment, storage and distribution, the costs of the local infrastructure and the regulations pertaining to the environment.

Table 2.12 gives the typical range of costs for freshwater production and energy use by reverse osmosis membrane systems of medium and large seawater desalination plants (i.e. plants with a freshwater production capacity of 40,000 m$^3$/day or more). As can be seen from the SWRO, systems of the best seawater desalination plants use between 2.5–2.8 kWh of electricity to produce 1 m$^3$ of freshwater, while the average energy consumption in industry is about 3.1 kWh/m$^3$.

The cost of producing freshwater from seawater is currently around US $ 1.1/m$^3$. Energy costs typically account for 25 to 40% of this price, depending on the unit of power and design of the SWRO plant, as well as the efficiency of the equipment.

It should be noted that the energy consumption shown in Table 2.12 only covers the operations of the SWRO system and not the energy consumption of the entire seawater desalination plant. Typically, SWRO systems account for between 65% and 80% of the total energy demand of a desalination plant. The figure shows an analysis of the energy consumption of a typical seawater desalination plant using water from the Pacific Ocean with the total concentration of dissolved solids (TDS) equaling 33.5 mg/l as the source of seawater.

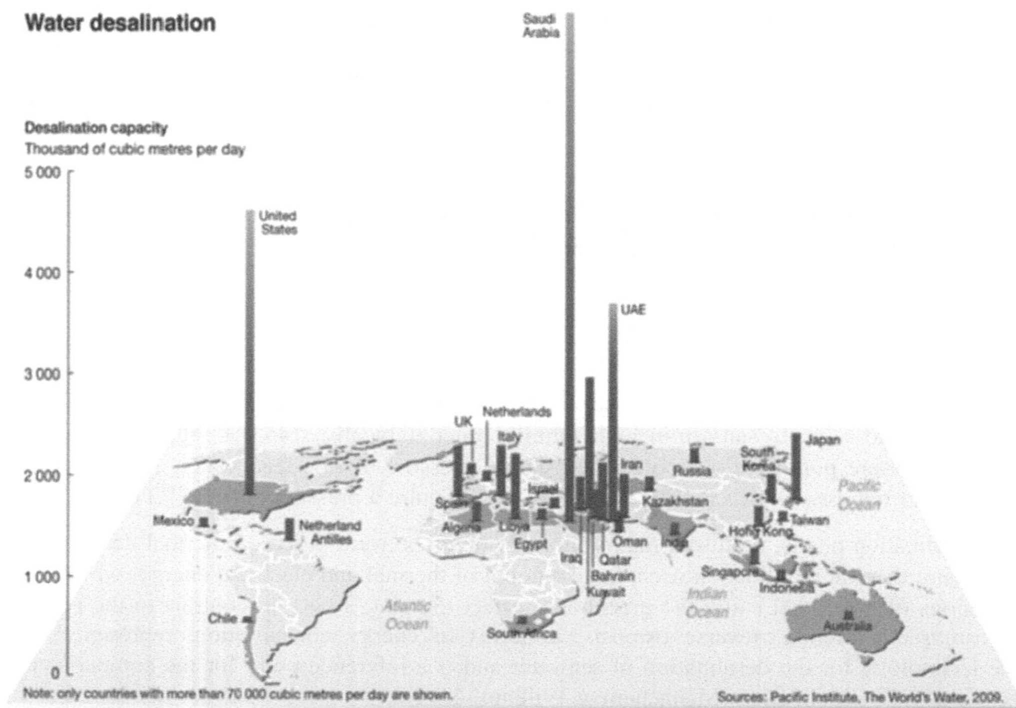


**Figure 2.21** Desalination capacity by country (Gleick *et al.*, 2009).

**Table 2.12** Typical cost and energy consumption for medium and large plants using SWRO systems (Voutchkov, 2017).

| Classification | Cost of Water (US$/m$^3$) | SWRO System Energy Use (kWh/m$^3$) |
|---|---|---|
| Low-end bracket | 0.5–0.8 | 2.5–2.8 |
| Medium range | 0.9–1.5 | 2.9–3.2 |
| High-end bracket | 1.6–3.0 | 3.3–4.0 |
| Average | 1.1 | 3.1 |

## 2.6 WATER MANAGEMENT AT THE GLOBAL LEVEL

Because of the extreme importance of water as a resource due to the increase in water demand caused by the global population growth, as well as because of the rate of pollution of water resources, it is essential to find ways to conserve the natural resources of water. The conservation of water resources ensures the sustainable supply of drinking water and helps future generations to have access to water to be used for the survival and development of human society.

There are many governments, non-profit and business organizations around the world involved in raising awareness about the importance of water conservation. The following are just some of the many organizations that have made a significant contribution towards the conservation and protection of water, be it about the raising of awareness or about setting real new standards for the efficient use of water: Biofuelwatch, 350.org, Center for Development and Strategy, Citizens' Climate Lobby, Climate Action Network, Confederation of European Environmental Engineering Societies, Conservation International, Environmental Defense Fund, Foundation for Environmental Education, Greenpeace, International Union for Conservation of Nature, The Surfrider Foundation USA, World Resources Institute, Sierra Club, Intergovernmental Panel on Climate Change, Pacific Institute, World Water Council, WET Project, Clean Water Action, Water Aid, Charity: Water, Clean Water Fund, Global Water Challenge, Global Water Leaders, The United Nations, The Water Project, and World Resources Institute.

In the world, more than 276 major river basins cross the political borders of two or more countries and are home to about 40% of the world's population. However, approximately two-thirds of river basins do not have a common framework for river basin water management and cooperation. Although there are many international agreements on cooperation in the field of water management, disagreements and disputes still occur. It is therefore important for societies to establish domestic, bilateral and multilateral mechanisms to support peaceful and effective mediation. Cooperation is especially important in water-scarce regions, where the upstream and downstream impacts of consumption and pollution are on the increase. Effective, sustainable, global, regional, legal and institutional frameworks at the basin level can contribute to stable and reliable cooperation, increasing security for competitive water use thereby preventing potential conflicts.

It is necessary to manage water globally in order to manage it adequately. This management relies on the cooperation of international institutions that propose and adopt mechanisms.

## 2.6.1 The United Nations system

The UN is an international organization founded in 1945 after World War II. There are currently 193 Member States involved in the following activities:

- Maintaining international peace and security;
- Protection of human rights;
- Delivery of humanitarian assistance;
- Promoting sustainable development;
- Adoption of international law.

The UN has many programs, funds and special agencies, each with their own separate membership, leadership and budget. The financial means for the programs and funds come from voluntary contributions, such as:

- The United Nations Development Programme (UNDP);
- The United Nations Environment Programme (UNEP);
- The United Nations Human Settlements Programme (UN-Habitat);
- The United Nations Population Fund (UNFPA);
- The World Food Programme (WFP);
- The United Nations International Children's Emergency Fund (UNICEF).

The special agencies are independent international organizations funded by both voluntary contributions and estimated contributions. UN Special Agencies are autonomous organizations working with the United Nations through agreed treaties. Some were created before World War I, some were affiliated with the League of Nations, and others were created almost simultaneously with the creation of the UN.

**Table 2.13** A review of important meetings, declarations and agendas relating to the global water management system.

| Year | Name | Organization |
|---|---|---|
| 1968 | Global Biosphere Conservation | UNESCO with the UN, FAO and WHO |
| 1969 | National Conference | UNESCO |
| 1972 | Conference on the Human Environment or Stockholm Conference | UN |
| 1987 | Brundtland Commission | UN |
| 1992 | World Summit, Rio Conference | UN |
| 1992 | Agenda 21 adopted at the World Summit, Rio Conference | UN |
| 1994 | United Nations Convention to Combat Desertification in Countries Experiencing Serious Drought and/or Desertification, especially in Africa | UN |
| 1997 | Kyoto Protocol | UN |
| 1997 | Convention on the Law of the Non-navigational Uses of International Watercourses | UN |
| 2000 | Millennium Development Goals | UN |
| 2002 | United Nations Conference on Sustainable Development, Rio +10 | UN |
| 2012 | United Nations Conference on Sustainable Development, Rio +20 | UN |
| 2015 | Climate Change Conference | UN |
| 2015 | COP21, Paris Climate Change Conference | UNEP |
| 2015 | Agenda 2030 | UN |

There are also special agencies created by the UN to meet emerging needs:

- The Food and Agriculture Organization (FAO);
- The International Monetary Fund (IMF);
- The United Nations Educational, Scientific and Cultural Organization (UNESCO);
- The United Nations Industrial Development Organization (UNIDO);
- The World Health Organization (WHO);
- The World Meteorological Organization (WMO);
- The World Bank.

Table 2.13 provides an overview of the more important meetings, declarations and agendas related to the global water management system.

## 2.6.2 International financial institutions

The MDB (Multilateral Development Bank) brings together information pertaining to the financing of Climate Change and Water Resources Projects, namely information from the African Development Bank, Asian Development Bank, European Bank for Reconstruction and Development, European Investment Bank, US Inter-Group Development bank and the World Bank group. All these banks are involved in financing a vast majority of multilateral areas of development in different countries. In October 2017, the Islamic Development Bank joined a group dedicated to financing the monitoring of climate change and possible solutions.

**Table 2.14** A review of important international financial institutions relating to the global water management system.

| Name | Financing Level |
|---|---|
| International Monetary Fund | Global, Regional |
| World Bank | Global, Regional |
| African Development Bank | Regional |
| Asian Development Bank | Regional |
| European Bank for Reconstruction and Development | Regional |
| European Investment Bank | Regional |
| Inter-American Development Bank | Regional |
| Islamic Development Bank | Regional |
| New Development Bank of the BRICS countries | Global, Regional |

In the report submitted by the multilateral development banks on financing climate change, data shows that US $ 27.9 billion, or 79% of the total for 2017, was intended for projects to mitigate the effects of climate change and aimed at reducing the emission of harmful greenhouse gases as well as slowing the global warming of the planet. The remaining 21% or US $ 7.4 billion was intended for developing countries, where investment was planned into projects related to adaptation to changing climate.

Table 2.14 provides an overview of the more important international financial institutions related to the global water management system.

## 2.7 WATER MANAGEMENT AT THE REGIONAL LEVEL

**Review of significant watershed commissions**

Table 2.15 shows a list of transboundary water management organizations and partner countries.

**European Commission decisions on water and the environment**

Table 2.16 shows the European Commission directives on water management.

### 2.7.1 Water Framework Directive (Directive 2000/60/EC)

The Water Framework Directive sets out a long-term policy and defines the EU's basic objectives in the water sector based on the following principles:

- Good status of all waters by 2015;
- Determination of hydrological, biological and chemical standards;
- Implementation of the program of measures defined in management plans by 2015;
- Water monitoring based on established standards;
- Groundwater exploitation in accordance with the principles of sustainable use;
- Integrated management of river basin districts based on natural hydrological and geographical units rather than on administrative and political boundaries;
- Facilitating economic development in accordance with the principles of sustainable development and the protection of water resources and the environment;
- A water management plan is a major tool for presenting the water status and for analyzing pressures and impacts on water resources within a river basin district;
- Inclusion of interdepartmental and cross-sector cooperation.

**Table 2.15** List of transboundary water management organizations and partner countries.

| Transboundary Water Management Organizations | Partner Countries |
|---|---|
| Amazon Cooperation Treaty Organization | Bolivia, Brazil, Colombia, Ecuador, Guyana, Peru, Suriname, Venezuela |
| Lake Titicaca Autonomous Binational Basin Authority | Bolivia, Peru |
| Transboundary River Commission between Finland and Sweden | Finland, Sweden |
| Finnish Norwegian Transboundary Water Commission | Finland, Norway |
| Genevese Aquifer Management Commission | France, Switzerland |
| Great Lakes Commission | Canada, United States |
| Guadiana River Commission | Spain, Portugal |
| Intergovernmental Coordinating Committee of the River Plate Basin Countries | Argentina, Bolivia, Brazil, Paraguay, Uruguay |
| International Boundary and Waters Commission | United States and Mexico |
| Lake Geneva International Commission for the Protection of Lake Geneva | France, Switzerland |
| International Commission for the Protection of the Danube River | Austria, Bosnia and Herzegovina, Bulgaria, Croatia, Czech Republic, Germany, Hungary, Moldova, Montenegro, Romania, Serbia, Slovakia, Slovenia, Ukraine |
| International Commission for the Protection of the Elbe | Germany, Czech Republic |
| International Commission for the Protection of the Rhine | Germany, France, Luxembourg, the Netherlands, Switzerland |
| International Joint Commission | United States and Canada |
| International Meuse Commission | Belgium, France |
| International Sava River Basin Commission | Bosnia and Herzegovina, Croatia, Slovenia, Serbia |
| Interstate Commission for Water Coordination of Central Asia | Republic of Kazakhstan, Kyrgyz Republic, Republic of Tajikistan, Turkmenistan, and Republic of Uzbekistan |
| Joint Rivers Commission | Bangladesh and India |
| Lake Chad Basin Commission | Chad, Cameroon, Niger, Nigeria |
| Lesotho Highlands Water Project | Lesotho, South Africa |
| Mekong River Commission | Cambodia, Laos, Thailand, Vietnam |
| Niger Basin Authority | Benin, Burkina Faso, Cameroon, Ivory Coast |
| Nile Basin Initiative | Burundi, DR Congo, Egypt, Ethiopia, Rwanda Sudan, Tanzania, Uganda, Kenya |
| Peipsi Center for Transboundary Cooperation | Estonia, Russia |
| Permanent Indus Commission | India, Pakistan |

**Table 2.16** List of European Commission directives.

| Year of Adoption | Name | Commission |
|---|---|---|
| 1991 | Nitrate directive | EU |
| 1991 | Urban Wastewater Treatment Directive | EU |
| 1991 | Plant Protection Products Directive 91/414/EEC | EU |
| 1998 | Amendments to the Urban Wastewater Treatment Directive | EU |
| 1998 | Drinking Water Directive | EU |
| 1998 | Directive on Biocidal Products (98/8/EC) | EU |
| 1999 | Landfill Directive 99/31/EC | EU |
| 2000 | Directive 2000/60/EC | EU |
| 2006 | Groundwater Directive 2006/118/EC | EU |
| 2007 | Directive 2007/60/EC, Directive on the assessment and management of flood risks | EU |
| 2008 | Directive setting environmental quality standards in the field of water policy | EU |
| 2009 | Technical specifications for chemical analyses and the monitoring the water status | EU |
| 2013 | Priority substances in the field of water policy | EU |

A combined approach to point and non-point sources of pollution will ensure that discharges into recipients are controlled via:

- Emission control measures based on the best available techniques;
- Relevant limit values.

For non-point sources control measures based on the best available practices established by the Integrated Pollution Prevention and Control Directive, the Waste Water Treatment Directive, the Agricultural Pollution Directive, and other relevant community legislation are to be applied.

Proper pricing should be in place which will reflect the true costs applying the following principles:

- The user pays;
- The polluter pays;
- Full cost reimbursement;
- Public involvement;
- Informing, consulting and transparency;
- Bringing into accordance the interests of different interested groups;
- Public participation on a larger scale.

The WFD Implementation Strategy is set out in a separate document: 'The WFD Common Implementation Strategy (May 2001) and contains a series of guidelines and recommendations for operational methods and other supporting documents to be used in the implementation process in EU countries.

The priorities imposed during the initial period of the WFD implementation can be classified into three groups: the legal aspect (implementation of WFD provisions in the legal regulations of the Member States, designation of river basin districts and competent institutions), the organizational aspect, that is, the establishment and operation of a strategic group for coordination and technical working groups for the project activities of the joint EU Strategy at the community level and in the Member States, and urgent technical tasks involving the initial definition of water bodies and the analysis of the characteristics of river basin districts (Ninković *et al.*, 2010).

## 2.7.2 International Commission for Protection of Danube River

### 2.7.2.1 General information

Firstly it is worth mentioning a few facts about the world's most international river basin, the Danube River Basin (DRB). It includes the territories of 19 countries, and is home to 81 million people with a variety of languages and historical backgrounds. The River Danube itself, stretches over 2780 km between its spring in Germany's Black Forest and the Danube Delta. Floodplain forests, marshlands, deltas, floodplain corridors, lakeshores and other wetlands are essential components in the basin's biodiversity and hydrology. Many are transboundary in nature and represent valuable drinking water reserves for millions of people.

The International Commission for the Protection of the Danube River (ICPDR) comprises 15 Contracting Parties who have committed themselves to implement the Danube River Protection Convention (DRPC). All Danube countries with territories of 2000 km$^2$ or more in the Danube River Basin are Contracting Parties to the DRPC. Since its establishment in 1998, the ICPDR has grown into one of the largest and most active international bodies engaged in river basin management in Europe. Its activities relate not only to the Danube River, but also to the tributaries and groundwater resources which will be the focus of this chapter.

### 2.7.2.2 Goals and legal basis

The ultimate goal of the ICPDR is to implement the DRPC and thus to cooperate on fundamental water management issues and to take all appropriate legal, administrative and technical measures to maintain and improve the quality of the Danube River and its environment. The DRPC mission is to promote and coordinate sustainable and equitable water management, including conservation, and improvement and rational use of waters for the benefit of the Danube River Basin countries and their people. The ICPDR pursues its mission by making recommendations for the improvement of water quality, developing mechanisms for flood and accident control, agreeing on standards for emissions, and by assuring that these measures are reflected in the Contracting Parties' national legislations and are applied in their policies.

It is important to mention that the daily work of the ICPDR is also based around the most relevant legislation at European level for river ecosystem management – the Water Framework Directive (WFD). Adopted in the year 2000, it establishes a governance framework for all waters in the European Union, which prevents further deterioration and protects and enhances the status of aquatic ecosystems, promotes sustainable water use, aims at enhanced protection and improvement of the aquatic environment, ensures the progressive reduction of pollution of groundwater and prevents its further pollution, and contributes to mitigating the effects of floods and droughts by applying a river basin approach.

The Directive is legally binding for all EU Member States and serves as an umbrella for measures related to drinking and bathing water, urban wastewater treatment, groundwater protection, floods, and protection of waters against the impacts of agricultural pressures, such as from nitrates in organic and chemical fertilizers and/or industrial emissions. The Contracting Parties of the ICPDR which are not the EU Member States also committed themselves to implement the WFD within the frame of the DRPC.

In addition, the ICPDR serves as a coordination platform for the basin-wide implementation of the EU Floods Directive (EFD, formally Directive 2007/60/EC).

Modern water legislation strives to build the institutional foundation for sustainable water resource management and protection of important habitats and species. At the international level, the most prominent legal acts for improving freshwater governance and fostering the equitable and sustainable sharing of transboundary watercourses are the UNECE (United Nations Economic Commission for Europe) Convention on the Protection and Use of Transboundary Watercourses and International Lakes as well as the UN Convention on the Law of the Non-navigational Uses of International Watercourses. The Sustainable Development Goals (SDGs), launched in 2015, include a target to ensure everyone has access to safe water by 2030, making water a key issue in the fight to eradicate extreme poverty, where goal 6 covers the entire water cycle. The ICPDR, as one of few freshwater River Basin Organizations (RBOs) in the world, sees its role as a direct contributor to achieving the SDGs which are part of a global agenda for the survival of the planet.

#### *2.7.2.3 Modus operandi*

The ICPDR meets twice a year, once at the Ordinary Meeting which is held in Vienna in December, and once for the meeting of the Heads of Delegations which is held in June in the country that currently holds the Presidency. The meetings comprise of delegations of Contracting Parties and observer organizations. Every Contracting Party has one Head of Delegation representing the country. For every decision, the primary goal is to reach some form of consensus between the members while the current ICPDR President chairs each meeting. The ICPDR Presidency is passed on from one country to another in alphabetical order every year.

In addition, much of the work of the ICPDR is done by Expert Groups (EGs), which are formed by national experts from the Contracting Parties and representatives of ICPDR observer organizations. Eight Expert Groups deal with a variety of issues and make recommendations to the ICPDR (see Figure 2.22 below).

The technical work within the ICPDR relies very heavily on the work of these EGs. These panels of specialists are not only essential to the operation of the ICPDR they also link the work of the commission to the national administration of its contracting parties. Indeed, they comprise national experts and, although with no formal vote, delegates from observers.

Currently, the ICPDR has seven regular Expert Groups (River Basin Management, Flood Protection, Pressures and Measures, Accident Prevention and Control, Monitoring and Assessment, Information Management and Geographic Information Systems, Public Participation) and one ad-hoc Expert Group for legal and administrative questions. The seven Expert Groups report to the ICPDR and propose decisions to the commission at its plenary meetings. The EGs typically meet twice a year at changing locations and are chaired by Chairpersons elected by the ICPDR upon nomination of the EG. The Expert Groups have proven vital for allowing the ICPDR to work in a decentralized manner as the ICPDR deals with a broad range of issues, despite its small number of staff. This is only possible because most of the technical decisions are prepared through the Expert Group bodies, which draw from over 200 people who are specialists in their respective countries or organizations. In case of the Groundwater Task Group the responsible EG is the Monitoring and Assessment Expert Group.

For efficiency and proportionality purposes, and also to be in line with one of the governing principles of the ICPDR (the principle of subsidiarity), the management of the DRBD is based on three levels of coordination:

Part A: International, basin-wide level – the Roof level;
Part B: National level (managed through the competent authorities) and/or the international coordinated sub-basin level for selected sub-basins (Tisza, Sava, Prut, and Danube Delta);
Part C: Sub-unit level, defined as management units within the national territory (see Figure 2.23).

One of the main benefits provided by the ICPDR is the ability to help national decision-makers balance the competing needs and uses of the Danube River, such as hydropower, agriculture and the management of

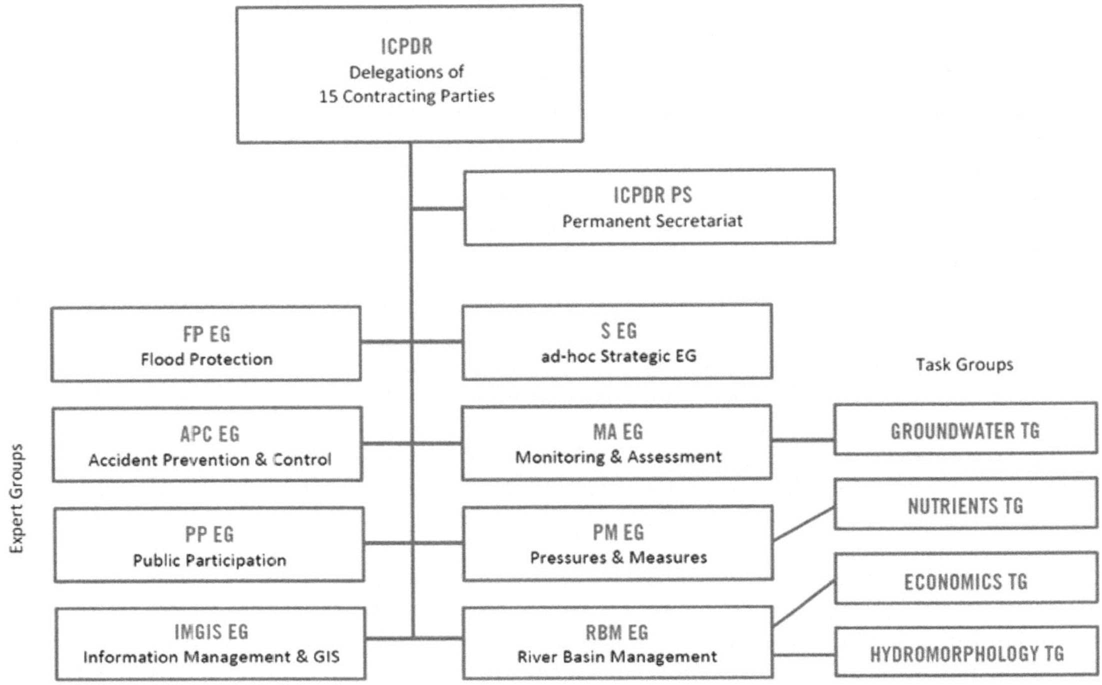


**Figure 2.22** Expert groups in ICPDR organization (ICPDR).

climate risks. The ICPDR serves its member countries as a fully-functioning platform for consultation, coordination and the integrated management (IRBM) of the Danube River Basin.

Given the complexity of the Danube River Basin (the many countries, differences in economic performance, biological assets, past damages and continued threats from human impacts) it was clear from the start that one overall framework or mechanism was required to sustainably manage the basin environment. 'Integrated river basin management', otherwise known as IRBM, became the primary mechanism to address the issues and their impacts. For a set of selected hazardous substances called

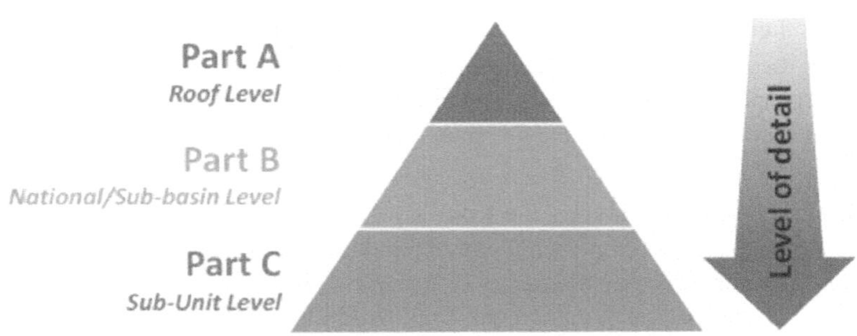


**Figure 2.23** Three levels of management for WED implementation in the DRBD showing the increase of the level of detail from Part A to Parts B and C (ICPDR).

priority substances, limit values were set on the European level which define so-called good chemical status. Clean water, not polluted by organic substances, nutrients and dangerous substances is essential. However, that alone is not enough, in cases where the natural ecosystem including its flora and fauna is significantly damaged or dysfunctional. To meet these objectives, the ICPDR developed its first Danube River Basin Management Plan (DRBM Plan) in 2009, including assessments and measures towards the achievement of good status by 2015. Aware of the fact that not all waters would reach the target values in six years, the WFD foresaw updates to the DRBM Plan in 2015 and 2021.

The measures foreseen in the ICPDR's management plan for the Danube River Basin improve the purity of rivers, lakes and groundwater, and make rivers and lakes thriving ecosystems for animals and plants. The 2015 update of the Danube River Basin Management Plan builds on a previous plan for the period from 2009 to 2015, which resulted in significant improvements to the Danube River and its tributaries. As an example, a 30% reduction of phosphorus emissions was registered when compared to the 2009 figures, a highly positive effect on ecosystems throughout the basin and in the Black Sea.

#### *2.7.2.4 Groundwater and its regional importance*

Despite the fact that there are many publications dealing with groundwater processes and groundwater treatment, their role in overall large river basin management remains a challenge. As groundwater is recognized as a dominant source of drinking water supply, the groundwater related issues are an integral part of the River Basin Management Plan. Self-purification processes which improve the water quality, or at least stabilize its quality, make groundwater sources suitable for water supply. The self-purification ability is especially characteristic for aquifers formed in an intergranular medium. It is important to mention that groundwater is often an essential factor during dry periods when they are the major contributor to river flows. Therefore it is also important for creation and conservation of wetlands and also oases in arid regions.

Despite the fact that the groundwater resources are managed within the scope of the overall natural water cycle and overall human activity in an integrated manner, due to their specific behavior and conditions, related issues have to be addressed in a different manner than that applied in addressing surface water issues. In many regions during the previous decades groundwater used to be an inexhaustible and accessible source of high quality water supply. However, the expansion of various human activities (e.g. agriculture, production) and growth of the human population in general has led to increased rates of abstractions and thus constraints on water management including the use and protection of groundwater.

The Danube region is an example of a noticeable gap in the economic power between countries which makes the implementation of all the required elements of sustainable water management challenging. Besides this, an additional factor which should be considered, as one of the modern agents influencing the nature of water management, is the factor of climate change. The fact that there is a re-distribution of atmospheric precipitation with regard to area, intensity and time is undeniable and therefore water management in general has to be adaptive. Here, it is important to bear in mind that the velocities of groundwater flows are much lower in comparison with the velocities of surface waters and therefore the physical and biochemical processes present in the groundwater systems tend to have a more local nature. The regional management, such as ICPDR, deals with the transboundary aquifers of basin-wide importance and the constant effort on the common approach in the region should be enhanced. One of the main goals at the regional level is to enhance the cooperation between countries in this respect via established legal and institutional frameworks for such water management.

#### *2.7.2.5 Groundwater monitoring network under TNMN*

The ICPDR Transnational Monitoring Network (TNMN) has been in operation since 1996 and its major objective is to provide an overview of the overall status and long-term changes of surface water and, where necessary, groundwater status in a basin-wide context (with particular attention paid to the transboundary pollution load). The transnational groundwater management activities in the DRB were initiated in 2002 and were triggered by the implementation of the WFD. Monitoring of the selected transboundary groundwater bodies (GWBs) of basin-wide importance has been integrated into the TNMN of the ICPDR.

For groundwater monitoring under the TNMN (GW TNMN) a six-year reporting cycle has been set, which is in line with reporting requirements under the WFD.

GW TNMN includes both quantitative and chemical (quality) monitoring. It provides the necessary information to assess groundwater status, identify trends in pollutant concentrations, support GWB characterization and the validation of risk assessment, assess whether drinking water protected area objectives are achieved, and support the establishment and assessment of the programs of measures and the effective targeting of economic resources. To select the monitoring sites, a set of criteria has been applied by the countries, such as aquifer type and characteristics (porous, karst and fissured, confined and unconfined groundwater) and depth of the GWB (for deep GWBs, the flexibility in the design of the monitoring network is very limited). The flow direction has also been taken into consideration by some countries, as well as the existence of associated drinking water protected areas or ecosystems (aquatic and/or terrestrial).

The qualitative monitoring determinants of GW TNMN, which are set as mandatory by the WFD, include dissolved oxygen, pH-value, electrical conductivity, nitrates and ammonium. The measurement of temperature and set of major (trace) ions is recommended as this can be helpful to validate the Article 5 risk assessment and conceptual models. Selective determinants (e.g. heavy metals and relevant basic radionuclides) would be needed for assessing natural background concentrations. It is also recommended to monitor the water level at all chemical monitoring points in order to describe (and interpret) the physical status of the site and to help in interpreting (seasonal) variations or trends in chemical composition of groundwater. In addition to the core parameters, selective determinants will need to be monitored at specific locations, or across GWBs, where the risk assessments indicate a risk of failing to achieve WFD objectives. Transboundary water bodies are also monitored for those parameters that are relevant for the protection of all uses supported by groundwater.

As regards quantitative monitoring, WFD requires only the measurement of groundwater levels but the ICPDR has also recommended monitoring of spring flows, flow characteristics and/or stage levels of surface water courses during drought periods, stage levels in significant groundwater dependent wetlands and lakes, and water abstraction as optional parameters.

The results of the status assessment of the selected transboundary GWBs of basin-wide importance are provided for the whole national part of a particular ICPDR GWB (so called: aggregated GWB). If a national part of an ICPDR GWB consists of several individual national-level GWBs, then poor status in one national-level GWB is decisive in characterizing the whole national part of an ICPDR GWB as having poor status. To indicate the diversity of different status results of individual GWBs within aggregated groundwater bodies the concept of the aggregation confidence levels was developed by the ICPDR. The reason for introducing these specific confidence levels for the DRBM Plan was the need to distinguish between the cases when all individual GWBs in an aggregated GWB have the same status (high confidence) or do not (medium confidence) or the assessment is based on the risk assessment data (low confidence).

A more detailed description of the technicalities of the GW TNMN and groundwater status assessment are given in the ICPDR Groundwater Guidance.

### *2.7.2.6 Groundwater in 2009 vs 2015 Danube River Basin Management Plan*

The analysis and review of GWBs in the DRBD, as required under Article 5 and Annex II of the WFD, was updated in 2013.

The pressures on groundwater bodies of basin-wide importance have not changed since 2009. Pollution by nitrates from diffuse sources is the key factor posing significant pressure on the chemical status while over-abstraction is the key pressure affecting the quantitative status of groundwater bodies. Since the DRBM Plan in 2009 for the GWBs of basin-wide importance failing to achieve good chemical status, new sewer systems were constructed and new legislation (e.g. extension of nitrate vulnerable zones or application of whole territory approach) was developed. GWBs failing good quantitative status were addressed by new projects and new legislation as well.

Nutrient pollution is caused by significant releases of nitrogen (N) and phosphorus (P) into the aquatic environment. Nutrient emissions can originate from both point and diffuse sources. Point sources of nutrient discharges are highly interlinked to those of organic pollution. Municipal wastewater treatment plants with inappropriate technology, untreated wastewater, industrial enterprises and animal husbandry can discharge considerable amounts of nutrients into the surface waters besides organic matter. Diffuse pathways, however, have higher importance when considering nutrients. Direct atmospheric deposition, overland flow, sediment transport, tile drainage flow and groundwater flow can remarkably contribute to the emissions into rivers, conveying nutrients from agriculture, urban areas, atmosphere and even from naturally covered areas.

The importance of the different pathways for diffuse pollution varies for N and P. For N, groundwater flow and urban run-off are the most relevant diffuse pathways. In the case of P, groundwater is usually replaced by sediment transport generated by soil erosion.

A remarkable decrease has been observed regarding the nutrient point source emissions in the Danube basin. For the reference year of the 1st DRBM Plan (2006), 130,000 tons per year TN and 22,000 tons per year TP pollution was reported via direct urban wastewater discharges. The reported emissions from urban wastewater treatment plants (WWTPs) used in the 2nd DRBM Plan are significantly lower in comparison to those of the 1st DRBM Plan, the TN and TP discharges having declined by 32% and 45%, respectively.

The measures under implementation have been substantially contributing to the reduction of nutrient inputs into surface waters and groundwater but further efforts are still needed. The continued implementation of measures in urban wastewater, industrial, market production and agricultural sectors is necessary.

Diffuse pathways have a dominant share of the total nutrient emissions, therefore implementation of measures addressing land management has high importance. A key set of best agricultural practices to reduce nutrient inputs and losses related to farming and land management has been identified as an appropriate management tool to be applied in agricultural areas. In the EU Member States a number of these measures have to be implemented under the action programs required by the Nitrates Directive. Moreover, application of good agricultural practices is either ensured by regulatory actions (cross-compliance) or encouraged by economic incentives (rural development programs) of the financial mechanism of the Common Agricultural Policy. Non-EU Member States are also highly encouraged to implement similar best agricultural practices.

Regions with high agricultural surplus and shorter groundwater residence time, and/or bedrock layers with lower denitrification capacity, produce the highest area-specific emissions. Urban areas with significant point sources and urban runoff generate remarkable local fluxes as well.

Groundwater flow (base flow and interflow) dominates the distribution of the pathways in most of the countries, particularly in the upper and middle basin. Point sources and urban runoff show significant relative contributions in the downstream countries. Regarding the sources, agricultural activities have a principal role in nitrogen emission generation, whereas atmospheric deposition is an equally important nitrogen input as fertilizers in many countries.

When it comes to hazardous substances diffuse emission pathways are substance-specific. In general surface runoff, sediment transport and groundwater flow are the main contributing routes.

The importance of groundwater can also be seen in the case of the decreasing water tables of the Danube and of the associated seriously affected groundwater which endangers the ecology of the floodplains in the 'Donau-Auen National Park'. In addition, inadequate and seasonally strongly fluctuating fairway depths in this section of the river substantially affect navigation. The Integrated River Engineering Project on the Danube to the east of Vienna was launched to improve the hydromorphology of the river and ecology of the floodplains (in line with equivalent levels of flood protection) as well as to improve the fairway conditions in this section of the Danube. One of the main measures is the granulometric improvement of the riverbed to provide long-term stabilization of the riverbed and of groundwater conditions.

Sound quantitative management of water resources is a prerequisite for addressing water scarcity and drought events but also for the achievement of WFD objectives, including the need to ensure the quantitative status of groundwater bodies and to achieve good ecological surface water status (including in terms of supporting river flows).

The basic principles and assessment of pollution sources for surface waters also provide relevant background information for groundwater due to the very close interrelation between the two water categories. Specifically, synergies between groundwater and the three significant water management issues (SWMIs) of organic, nutrient and hazardous substance pollution are of importance.

### *2.7.2.7 Groundwater and Joint Danube Surveys*

The findings of Joint Danube Surveys (JDS) are supportive to the implementation of WFD as they provide an extensive homogeneous dataset, production of which was mainly based on WFD compliant methods commonly used by the Danube experts. Even though these data are not intended to replace the national data used for the assessment of the ecological and chemical status they are an excellent reference database serving for future efforts of method harmonization in the DRB, especially concerning the development of a concerted type-specific approach to the status assessment of large rivers.

During the JDS2 the main hydrological objectives were to use the isotope methods to better understand the impacts of tributary inputs and mixing in the Danube including the distribution of groundwater discharge and to provide a more complete baseline of isotope data for assessing future impacts within the Danube Basin. Isotopes can be good indicators of change because they are often sensitive to land use or major hydro-climatic factors. In addition to being a hydrological tracer, 3H can also be viewed as a water quality parameter related to, for example, nuclear power plant discharges. The radon isotope investigations showed higher radon concentrations in the Upper Danube, which suggests that this is the area where groundwater contributions to the river are the largest (although tributaries may still be major inputs as well).

The isotope data reported in JDS2 also provided an important environmental baseline needed for the implementation of the EU Groundwater Directive, more specifically in monitoring and conceptual assessment of river/groundwater interactions in the major transboundary aquifers in the Danube Basin.

During JDS3, altogether 68 sites were sampled along a 2581 km stretch of the Danube, 15 of which were located in the mouths of tributaries or side arms. Sampling at the JDS3 stations included five different sample types – surface water, biological quality elements, sediment, suspended particulate matter (SPM) and biota for chemical analysis (fish and mussels) – each with a different determinant list.

For the first time the link between contamination of surface water and groundwater was explored. A number of emerging substances were detected during JDS3 in the abstraction wells at bank filtration sites. This phenomenon can be expected for substances like amidotrizoic acid, iopamidol, acesulfame, benzotriazole or carbamazepine which are known to be quite persistent in the aquatic environment and which are mostly not completely retained by bank filtration. However, due to the relatively low concentration levels in the Danube, concentrations in the abstraction wells were mostly below 0.1 μg/l for most substances. An exception was the artificial sweetener acesulfame which occurred in concentrations up to 1.1 μg/l in the Danube and was detected in most of the abstraction wells with a maximum concentration of 0.45 μg/l.

Acesulfame is used as a food additive and the observed concentrations are not considered to be harmful for humans. However, acesulfame can act as an example for a more or less persistent and very mobile substance which is consumed in large quantities.

#### *2.7.2.8 ICPDR Climate Change Adaptation Strategy and groundwater*

Climate Change (CC) is recognized by the ICPDR as one of the burning issues of global importance and therefore the first ICPDR Climate Change Adaptation Strategy was produced in 2012 and later updated in 2018. The issues related to the impact of CC on groundwater resources form an integral part of both publications. The concern regarding the effect of CC on the groundwater regime in the DRB remains the same in both documents. An assumption given in the first Adaptation Strategy (2012) of a general decline in groundwater storage and recharge for Central and Eastern Europe, especially in summer, is still correct. Besides shortages in water availability, a decline in groundwater recharge could also lead to negative effects for groundwater quality. The exact numbers and changes in water temperatures differ regionally and locally, but values of 1–2°C are often mentioned, particularly for surface waters and groundwater in summer. After long droughts, preferential flow paths are of particular relevance in groundwater protection zones, given the fact that pollutants can pass rapidly along them and almost unimpeded into groundwater. That is why the water quality issues are considered as one of the most relevant climate change impacts.

Additionally, a possible increase in irrigation using groundwater resources could intensify the decline. An increase may deteriorate the ecological and chemical balance of freshwater bodies and could lead to an increase of contaminated surface and groundwater bodies after enhanced agricultural use.

An increased risk of conflicts over water use can occur in the event that no adequate adaptation measures are taken. Due to having less precipitation in summer, the extreme events will occur more frequently in summer than in winter. In some parts of the DRB, the drought risk is expected to increase drastically in the future, leading to possible economic loss, potential for water conflicts and water use restrictions. Possible consequences are difficulties in water supply with an increased risk of water shortages and an over-exploitation of aquifers in the future. Besides climate change impacts, future water demand is also triggered by anthropogenic impacts, political regulations and restrictions, and new technologies.

Regarding regional differences, a pronounced decline is particularly indicated for the Pannonian Plain Area, which has already been monitored in the past. For some alpine regions, however, a local increase in groundwater storage is likely to occur.

Some additional, quantitative assessments for smaller regions were carried out within INTERREG projects and other European research projects. Other national projects are covering only local areas or catchments or certain water-related issues like groundwater. The necessity for a basin-wide vulnerability assessment is also addressed in the recently published outline of the Climate Adaptation Strategy of the Sava River Basin.

As a recommendation and further steps, the Adaptation Strategy points out the importance of monitoring and status assessment. Maintaining both surface and groundwater surveillance monitoring sites over long time series is suggested as an important and continuous procedure.

Furthermore, it is necessary to set up an investigative monitoring program for climate change and for monitoring climate change 'hot spots', and to combine these as much as possible with the results from the operational monitoring program.

Integral approaches are needed and aim to enhance synergy effects sought between different sectors. An example of a synergy effect is an increase in water retention areas which can lead to a higher groundwater recharge, a reduction of flood peaks and positive effects for biodiversity.

The River Basin Management Plans are tools to make full use of the Water Framework Directive environmental objectives, e.g. the requirement to achieve good groundwater quantitative status helps to ensure a robust water system, which is more resilient to climate change impacts.

Further positive synergies which can be achieved by such adaptation measures can include the improved linkage between surface and groundwater, leading to increased robustness with regard to potential periods of water scarcity and droughts.

These steps should enhance sustainable groundwater management.

#### *2.7.2.9 Groundwater – the river's invisible twin*

As an illustrative overview for the wider public, the ICPDR has produced a groundwater brochure under the name 'Groundwater – the river's invisible twin'. Nearly 72% of all drinking water consumed in the DRB is produced from groundwater sources, serving an overall population of some 59 million people. Therefore, the brochure describes the groundwater importance and vulnerability, and differences between the media where the groundwater has been stored. This brochure briefly informs the public about the ICPDR's work related to the groundwater issues and describes the role of the relevant EU legislation. The brochure is available online on the ICPDR's website and in hard copy.

#### *2.7.2.10 ICPDR on groundwater related events*

Over the past 20 years the ICPDR has become a globally recognized role model for transboundary management of shared river basins and water resources therein, including groundwater in transboundary aquifers.

The institutional set-up, working methods and established practices in transboundary cooperation on shared groundwater systems have become a subject of interest of many governments and their institutions, inter-governmental organizations, scientists and international non-governmental organizations (NGOs). The ICPDR, therefore, has been given numerous opportunities at international conferences and high-level events to share its experiences and lessons learned on cooperation and management of shared groundwater systems. Several examples follow:

In March 2015 the ICPDR Executive Secretary (ES), Mr Ivan Zavadsky, served on the panel at the European Water Association (EWA) Spring Days in Budapest discussing the importance of sound management and protection of groundwater resources, which make up a large part of the water resources using for drinking water supply in the Danube countries. He delivered similar messages at the 4th EU

Water Conference in Brussel in 2015. In November 2015, he facilitated the side event on Focus on Groundwater in the 2030 Agenda at the 7th Meeting of Parties (MOP) of the UN ECE Water Conference in Budapest, co-organized by UNESCO Intergovernmental Hydrological Programme (IHP).

The Ministers of Contracted Parties to DRPC adopted a Danube Declaration at their meeting in Vienna in February 2016. The Declaration emphasized the importance of implementation of measures addressing the pollution of groundwater by nitrates from diffuse sources – in addition to those addressing various sources of pollution of surface water – and agreed upon the need for appropriate controls regarding over abstraction of groundwater in the DRB. In June 2016, the ES was given an opportunity to address the IWA Specialists Groundwater Conference in Belgrade at the opening high-level session.

The 2017 ICPDR President, Peter Gammeltoft, addressed, among other things, the importance of sound groundwater management and protection of the drinking water supply at the Danube Water Conference, organized by the IAWD and the World Bank in Vienna; he also delivered a speech at the Water and Climate Summit in Rome in October 2017, underlying the role of groundwater for adaptation to climate change.

Another opportunity for the 2018 ICPDR President, Helge Wendenburg, and the ES to address the global freshwater community, including political leaders, scientists and representatives of civil society and youth was the 8th World Water Forum in Brasilia in March 2018. The Global Environment Facility (GEF) Biannual International Waters Conference in Marrakech in November 2018 invited the ICPDR to share its experience and working frameworks from the river basin management (RBM) perspective; the ES, at this conference, moderated a session on RBM planning 'from formulation through implementation', including aspects of groundwater management.

Numerous governmental departments, institutions, universities and international water-focused organizations are organizing study tours to ICPDR premises to learn and extract experiences and practices suitable for them to implement, e.g. a delegation from China led by the Deputy Minister of water resources, several high-level delegations from India, the Korean water agency, and senior water managers for Central Asia countries, facilitated by the International Union for Conservation of Nature (IUCN), etc.

## 2.8 WATER MANAGEMENT FUNCTIONS

### 2.8.1 Functions of water management at the global level

Water management functions are sets of activities aimed at achieving a specific goal within a water management system. The objectives of global water management can be identified as:

- Achieving targets to reduce greenhouse gas emissions and reduce their impact on the rise of global temperature;
- Promoting activities that help reduce the imbalance in the water/food/energy chain, due to the global increase in population which causes a greater need for food and energy;
- Participation in the financing of development and other projects related to water resources;
- Raising the general level of knowledge and promoting activities, methods and knowledge related to water resources.

The global climate and climate change, global legal frameworks and water management goals, as well as global socio-economic changes affect the functioning of the water management system worldwide.

It is estimated that there has been an increase in the average global temperature of 0.8–1.2°C in the aftermath of the industrial revolution, and an increase of about 0.4°C since 1970 (Figure 2.24; NASA, 2019). This increasingly affects water management, due to a decrease in the specific surface runoff and

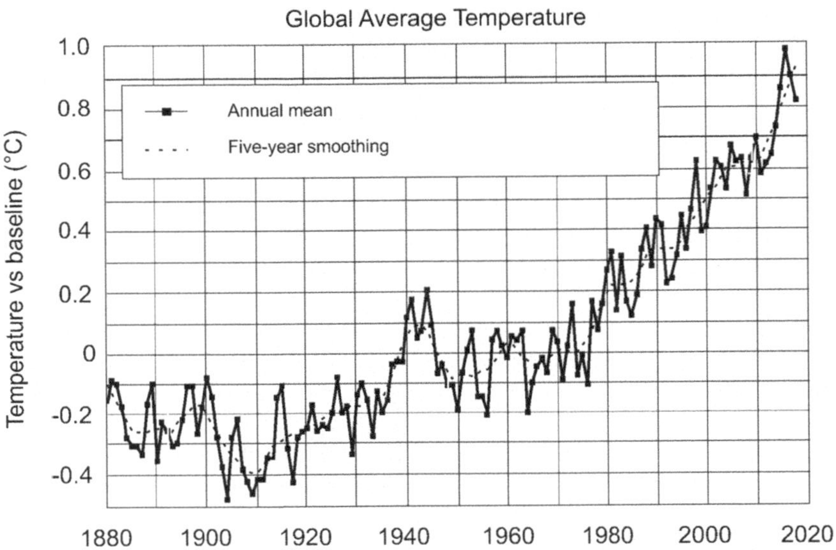


**Figure 2.24** Global mean surface temperature changes since 1880 (NASA Goddard Institute for Space Studies [GISS], 2019).

the reduced available amount of surface and groundwater, especially if an increase in stress (the ratio between the amount of water used for water supply purposes and the abstracted water) that is caused by the increased pumping of water for different needs is also expected.

Figure 2.25 shows an estimate and a perspective pertaining to the global temperature change, with a time period of 500 million years up to the present (the year 2015 is regarded as year zero). The figure shows five special linear segments of temperature, and the left part of the graph shows the average temperature values that are very similar and are only considered as qualitative indicators.

The breaks in the paleo periods are not evenly distributed but are positioned at geologically relevant times, e.g. at the border line between the Mesozoic-Cenozoic period which happened around 65 million years ago. This is the 'KT' boundary line (today called the 'Cretaceous-Paleogene'), which is when the

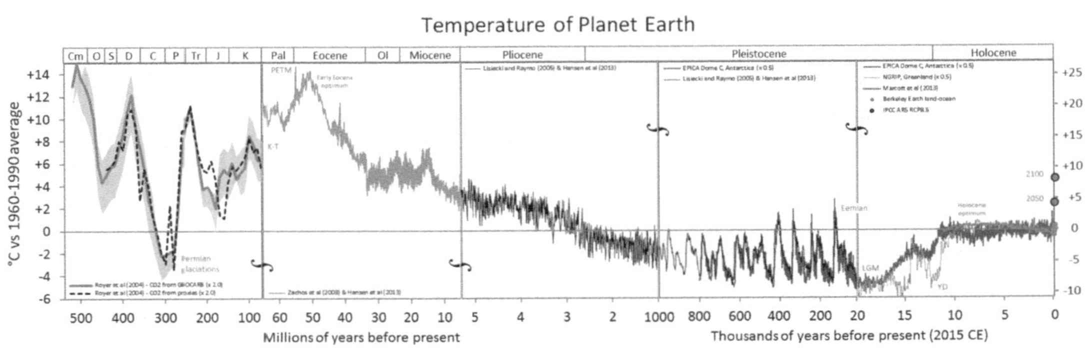


**Figure 2.25** The temperature on Planet Earth during the Paleo Period (Royer, 2014).

dinosaurs became extinct. The next break occurred at the Miocene-Pliocene border line and lasted approximately 5.3 million years. It happened millions of years ago, near the beginning of the current, 100,000-year-old glacial cycle (the previous glaciation was shorter near the last glacial maximum, 20,000 years ago).

Using climate models and socio-economic scenarios, the World Resources Institute (WRI) assessed and ranked the future stress on water resources, namely the competition for water and the reduction in the amount of surface water in 167 countries expected by 2020, 2030 and 2040. Based on the research results, it was concluded that 33 countries would have to deal with an exceptionally high stress on water resources by 2040 (Figure 2.26). In addition, Chile, Estonia, Namibia and Botswana are expected to face significant increases in water stress by 2040. This means that industry, agriculture and the social community in these countries may be exposed to significantly more water stress than is the case today.

Interpretations of data from systematic observations within the World Meteorological Organization's Global Climate Observation System, show that much of the observed global warming of the atmosphere over the last 50 years can be attributed to human influence. According to the data obtained for the 15 hottest years in the last 150 years, 13 were registered at the end of the 20th and the beginning of the 21st century.

This corresponds to the period when a sudden increase in anthropogenic greenhouse gas emissions was registered, i.e. when there was a sudden increase in atmospheric concentrations of these gases. Globally, the 20th century, especially the last decade of the 20th century, and 1998 were the warmest century and year during the last millennium.

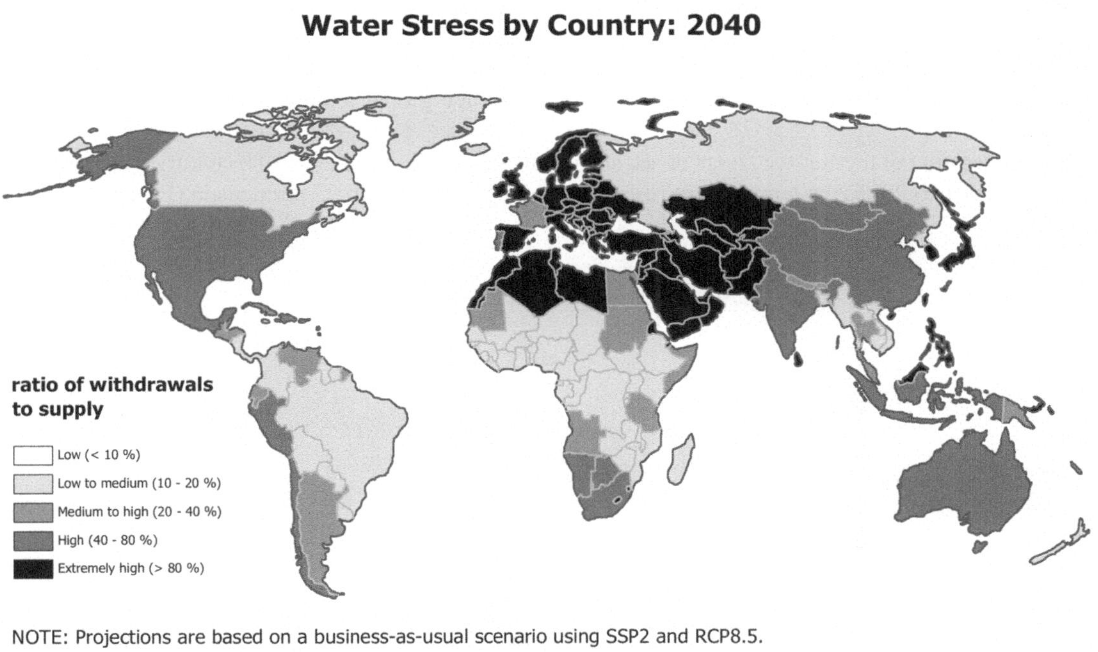


**Figure 2.26** Change in the ratio between the amount of water used for water supply and the water that is abstracted in % (World Resource Institute, 2015).

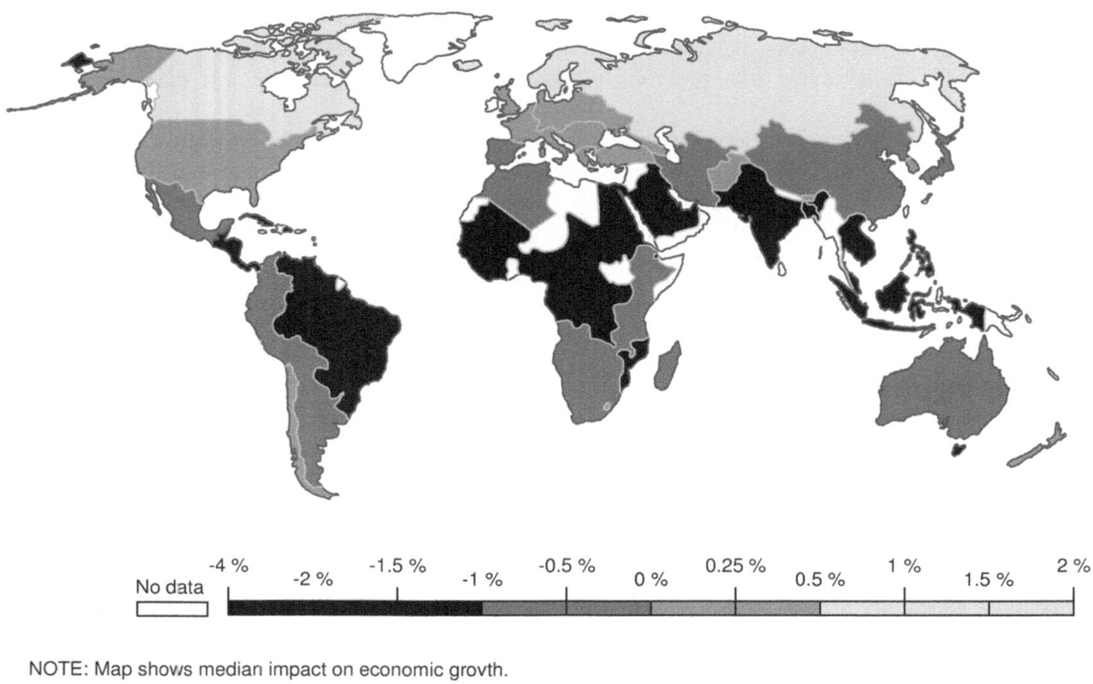


**Figure 2.27** Assessment of the impact of increased average global temperature on GDP (Pretis *et al.*, 2018).

Climate change can also have a potential negative impact on a country's economy, as can be seen in the Figure 2.27 showing the impact of an increase in the average global temperature of 1.5°C on the GDP per capita.

The temperature of the surface layers of sea and ocean water rose during the 20th century and continue to rise. From 1901 to 2015, the temperature rose at an average rate of 0.07°C per decade (Figure 2.28). The

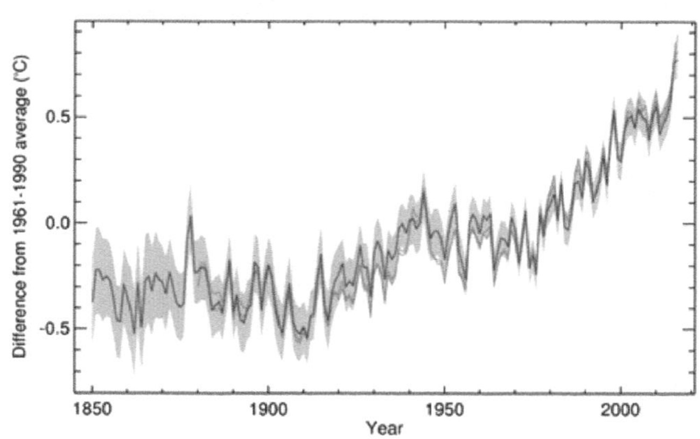


**Figure 2.28** Changes in the temperature of the surface layers of the sea and oceans between 1850 and 2016 (NOAA – National Oceanic and Atmospheric Administration, NASA, Met Office Hadley Centre, 2014).

temperature of the surface layers of seas and oceans has been constantly higher over the last three decades than at any other time since accurate measurements began in 1880. Based on historical records, the rise in sea surface and ocean surface temperatures mainly occurred during two key periods: between 1910 and 1940, and from around 1970 to the present. It is noted that there was a slight decrease in the sea and ocean temperatures between 1880 and 1910.

The changing temperatures of the surface layers of the sea and oceans vary in different regions of the world. While temperatures have been rising in most parts of the world's seas and oceans, some areas are accompanied by a cooling trend - for example, parts of the North Atlantic.

One possible influence on climate change is the increase in the concentration of $CO_2$ in the atmosphere (Figure 2.29). Since carbon dioxide has a very important role in climate change, researchers are focusing much more attention on its impact than on that of the other gases that cause the greenhouse effect, or aerosols. The concentration of $CO_2$ in the atmosphere has increased by about 30% since the 1800s, and continues to increase at a rate of about 1.53 ppm per year, as mankind continues to burn fossil fuels and change the amount of biomass on the surface of the earth (Figure 2.30).

Climate change has a significant impact on water resources and some of the effects are already visible. Almost all of the countries that are part of the United Nations Economic Commission for Europe (UNECE) and others are expected to be negatively affected by climate change. This will be reflected in the increased frequency and intensity of floods and droughts, deterioration of water quality, increased erosion and sedimentation, reduced glacier volume and snow cover, and rising sea levels.

Moreover, the impact of climate change on water resources will have cascading effects on human health and on many parts of the economy and society, as different sectors, such as agriculture, energy production, navigation, health, tourism and also the environment, directly depend on water.

Adapting to climate change is an economic and social imperative and action and measures need to be taken to manage water and adapt the water management strategy to the effects of climate change. Non-adaptive water management and a reduced capacity to adapt to change can jeopardize the principles of sustainable water management. On the other hand, improved prevention, preparedness for catastrophic

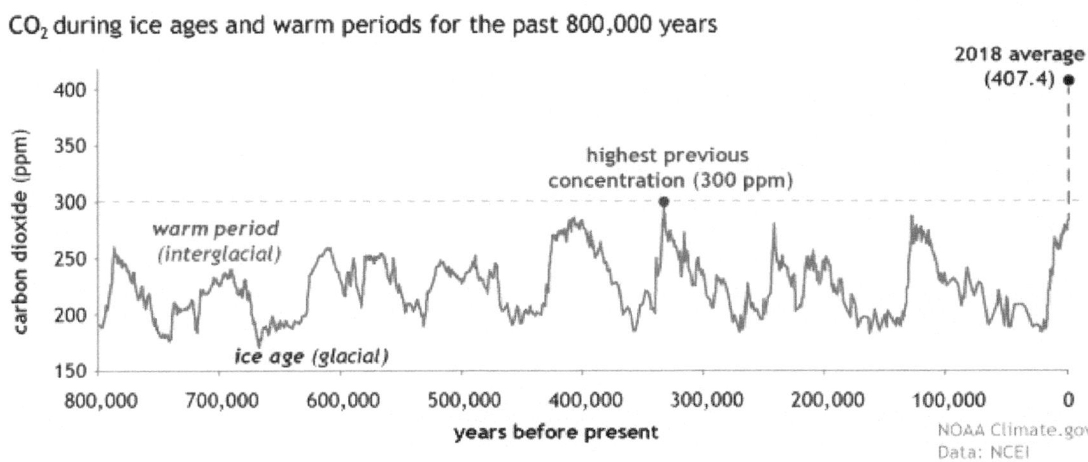


**Figure 2.29** Historical changes in the concentration of $CO_2$ in the atmosphere (NOAA, based on data from Lüthi, *et al.*, 2008).

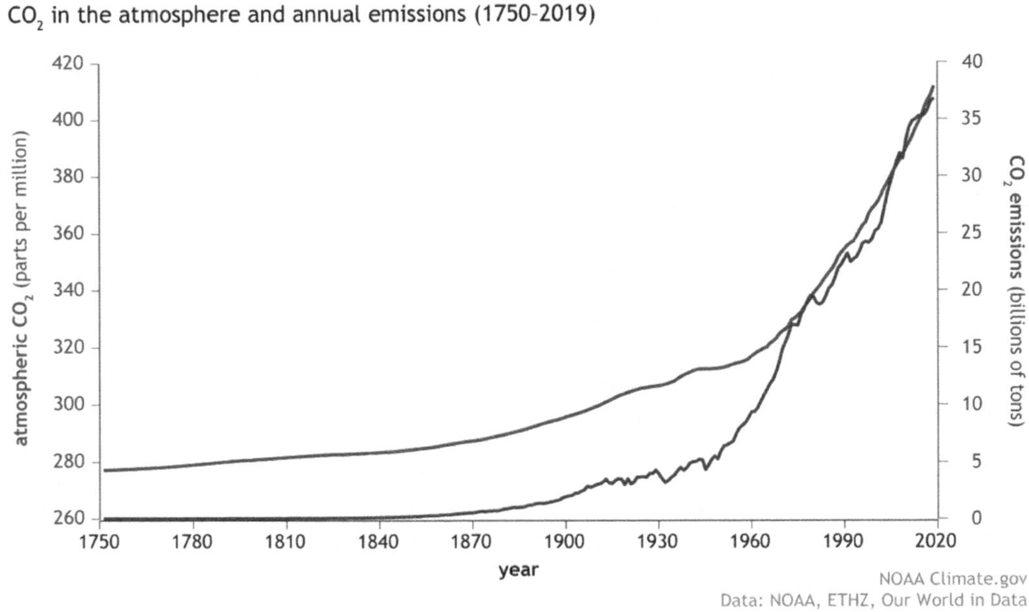


**Figure 2.30** Trend showing the changing $CO_2$ concentration in the 21st century (NOAA Climate.gov graph, adapted from original by Dr. Howard Diamond (NOAA ARL). Atmospheric $CO_2$ data from NOAA and ETHZ. $CO_2$ emissions data from Our World in Data and the Global Carbon Project).

events and other adaptation measures, as well as lifestyle adjustments, can greatly reduce the negative impacts of climate change.

In addition to climate change, the increase in population also has a large impact on water management and the total amount of water. According to Shiklomanov (2000), there has been an increasing trend in the use of water for the water supply of cities, industry and agriculture related to the increase in the world population (Figure 2.31).

The obvious increase in population implies a growing need for food, as well as the growing consumption of energy, the production of which requires significant amounts of water. This directly affects the management of water resources globally, regionally and locally.

### Climate Impact, Mitigation and Adaptation and UN Activities

Numerous international agreements include provisions and tools that can support the development of strategies to adapt to climate change and the implementation of adaptive water management.

A particular challenge for water resource management is the fact that almost half of the total surface water resources are transboundary. In addition, numerous groundwater resources are also transboundary. Transboundary waters create hydrological, social and economic interdependence between countries. As water and climate change do not respect borders, the international dimension of adapting to climate change is becoming increasingly emphasized (Dimkić *et al.*, 2013).

The legal framework, which implies strategies and plans to reduce greenhouse gas emissions, as well as global targets pertaining to water use, water protection and food scarcity, relates to the global legal aspect of water management.

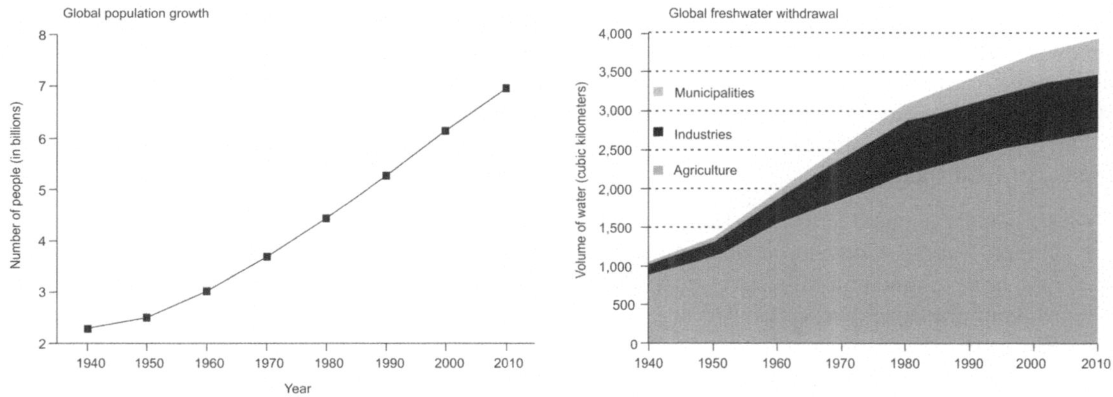


**Figure 2.31** Water use in relation to world population growth (Shiklomanov, 2000).

The following UN institutions and agencies are responsible for managing water resources on a global level:

- UN-Water – UN agencies, programs and other entities dealing with water-related issues;
- UNEP – United Nations Environment Programme;
- UNESCO – UN branch dealing with issues related to education, science and culture;
- FAO – a special UN agency whose primary mission is to eradicate hunger globally;
- WHO – the coordinating body for health related issues in the system;
- WMO – a specialized agency for meteorology;
- IPCC – The Intergovernmental Panel on Climate Change.

In addition to the United Nations system and the institutions and organizations that monitor the implementation of the Global Water Management Goals, conferences and commissions are also very important as they place special emphasis on climate and climate change. These include:

- The Rio World Summit, United Nations Framework Convention on Climate Change (UNFCCC);
- The Kyoto Protocol;
- Rio +10;
- The United Nations Conference on Sustainable Development, Rio +20;
- The UN Conference on Climate Change.

The UNFCCC requires communities to develop, implement and regularly update national and, where appropriate, regional programs of measures to facilitate an adequate adaptation to climate change.

The latest climate and climate change goals were adopted at the UN Conference on Climate Change in Paris in 2015 (COP21), where Member States pledged to limit greenhouse gas emissions to levels that would prevent global warming over 2°C (3.6°F) from pre-industrial levels.

The following are conferences and committees specifically related to water resource management:

- Rio World Summit, Agenda 21;
- Millennium Development Goals;
- Agenda 2030.

The 2030 Agenda, which was presented and adopted in New York in 2015 as the most important global legal framework affecting the water management system, predicts that by 2030, all the people on Planet Earth will have access to drinking water (currently, 89% of the people on the planet have access to drinking water), and to advanced sanitation (63% of people have access to advanced sanitation to date).

The following are conferences and commissions that specifically deal with finding a solution to the problem of food shortages:

- Agenda 21 adopted at the World Summit, Rio Conference;
- United Nations Convention to Combat Desertification in countries experiencing serious drought and/or desertification, especially in Africa;
- Millennium Development Goals;
- Rio +10;
- United Nations Conference on Sustainable Development, Rio +20;
- Agenda 2030.

The FAO, as a special UN agency, formed, within the United Nations Conference on Sustainable Development or Rio +20, the partnership initiative 'Globally Important Agricultural Heritage Systems' (GIAHS) based on the objectives of Agenda 21. It aims to implement and achieve the goals set in Chapter 14, Agenda 21. Also, the Millennium Development Goals from the year 2000, aim at reducing the percentage of hungry people in the world by 50% by 2015, while the latest global legal framework, Agenda 2030, aims at the complete eradication of hunger.

Economic growth, social development and poverty eradication are the first and foremost priorities in developing countries and are key to meeting national and global sustainability goals. In order to achieve the global goals of water management and for the system to function, it is necessary to provide assistance, that is, financial support and the transfer of technology to developing countries, without which it is very difficult or impossible to fully implement the global water management goals.

The goals are as follows:

- Define measures related to financial resources and mechanisms for the implementation of global water management goals;
- Secure new and additional funding that is both adequate and predictable;
- Seek to continuously and qualitatively improve the funding mechanisms that will be used to implement the global water management objectives.

However, in order to successfully achieve the goals of integrated water management, the most important thing is to achieve efficient water management in each country, basin and region. This means that it is necessary to manage the water system in a way that maximizes the internal ability to manage water (Figure 2.9). It is necessary to achieve sufficient capacities and achieve their harmony in the areas of administration (state, local) infrastructures, knowledge, science and education.

In addition, it is necessary to have a developed financial and legal system. Revenues in the water sector should be such that the development and maintenance of the water sector is constant. The question of the character of the administration should be based on relevant parameters such as wealth of water resources, economic power, climate and climate change, etc.

The assistance and participation of the international community in the development of the water sector is desirable and sometimes necessary. However, water management usually takes place locally, in the basin or at the level of the state or region, so building local capacities is extremely important for the development of fundamental water resources in the direction of sustainable and integrated water management. It is recommended that developing countries develop their own funding system. Sources of financing are primarily the adequate price of water and services with an adequate system of accumulation and use of capital.

In addition to internal sources and financing mechanisms, external financing mechanisms are also important. The financing mechanisms for global water management should be implemented and provided in a way that maximizes the availability of new and additional resources and utilizes all available funding sources and mechanisms. Financing mechanisms include, but are not limited to:

(1) Multilateral development banks and other funds;
   - The International Development Association (IDA). Regarding the various issues and options that will be considered in the future under the Global Water Management Objectives, the World Bank represents the major funding mechanism which focuses on the poorest countries to provide loans for projects related to sustainable water management.
   - Regional and sub-regional development banks. Regional and sub-regional development banks and funds play the largest role in providing financial resources, under concession or other favorable conditions that are required for the implementation of water management objectives.
   - The global instrument for financing projects in the field of environmental protection which is jointly managed by the World Bank, UNDP and UNEP, whose additional loans, grants and concessionary funds are designed to achieve global goals pertaining to the environment, should cover the additional costs related to relevant activities set by Agenda 2030, especially for developing countries.

(2) Relevant specialized agencies, other United Nations bodies and other international organizations, which have a role to play in supporting national governments in the implementation of global water management goals (such as the IMF and other bodies);

(3) Multilateral capacity building and technical cooperation institutions. The necessary funding should be provided to UNDP for it to use its network of field offices and its own budget to enable capacity building at the level of a country, taking full advantage of the expertise of specialized agencies and other United Nations bodies within their respective competencies, especially UNEP and multilateral and regional development banks;

(4) Bilateral assistance programs. Programs need to be improved (stepped up and strengthened) in order to promote sustainable development;

(5) Release from debt. It is important to achieve lasting solutions to debt problems in low- and middle-income developing countries in order to provide the necessary resources for sustainable development. All creditors at the Paris Club should apply the December 1991 agreement to provide debt relief for the poorest countries with a high debt that seek structural adjustment;

(6) Private financing. Voluntary contributions through non-governmental channels, which currently amount to some 10% of ODA (Official Development Assistance), may increase.

The major problem in poor developing countries is the financial system, which, due to the low GDP, cannot provide sufficient funding to finance and implement measures and activities related to the use of water and the development of water resources, water protection and river regulation, and protection against the adverse effects of water.

The five basic factors that influence the development of the water management sector are:

- The strength of the national economy;
- The availability of water resources;

- Future (present) climate change;
- Knowledge of technology and knowledge of procedures - both of which are the results of education and professional-scientific work and services;
- Efficient water management.

All the factors that affect a water management system are interrelated and each factor is very important for the functioning of the water management system.

The strength of the national economy in transition countries is increasing slightly (Figure 2.32), but it is lagging behind developing countries due to the process of transition. Strong economic measures and investments into the economy and industry of transition countries are needed for the economies to recover in the shortest possible time.

Government borrowing also occurs during the transition process which significantly affects the ability of states to invest in the water management sector, and the GDP is not great enough to sufficiently allocate funds from the state's own revenues. In addition, the price of water in transition countries is most often a social category, so it is not possible to provide the necessary funds to invest in the water sector.

The problems specific to developing countries are also becoming more pronounced due to the global movement of capital.

According to Piketty (2014), the six decades between 1914 and 1973 saw a sudden change in capital as the rate of economic growth exceeded the rate of return on capital after tax (Figure 2.33). Since then, the growth rate of the economy has been on the decline while the world capital has been increasing to pre-World War I levels.

When there is a large difference between the rate of return on capital and total capital, there is an increase in the risk of an economic crisis, which poses an additional challenge for credit-laden developing countries.

Figure 2.34 shows national debt in EU countries in the 1st quarter of 2019 in relation to gross domestic product.

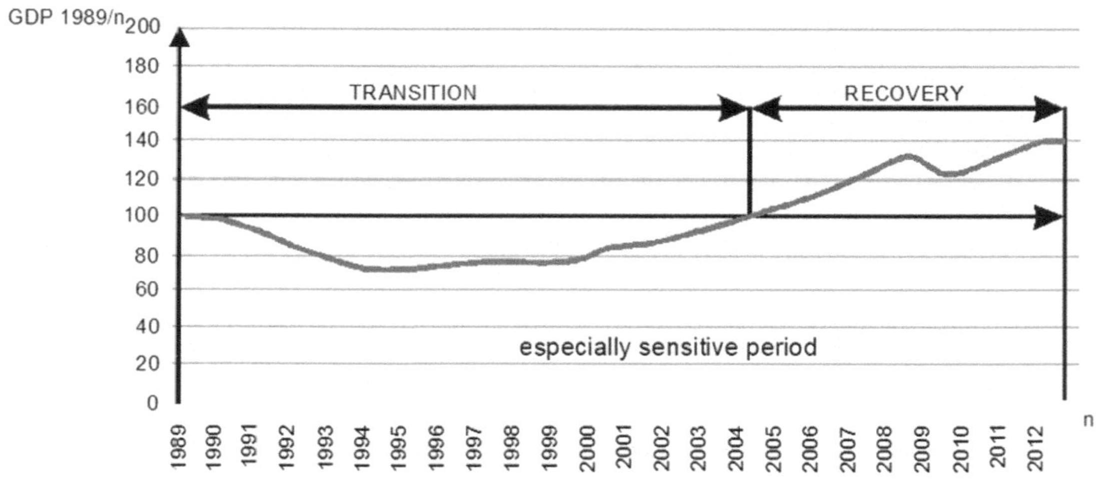


**Figure 2.32** GDP during periods of transition (World development indicators, WB-WDI, May 2016 WEO extent IMF, OECD, IMF website, IFS-IMF, ECB, EUROSTAT, BIS, NBS website, MF-Serbia website).

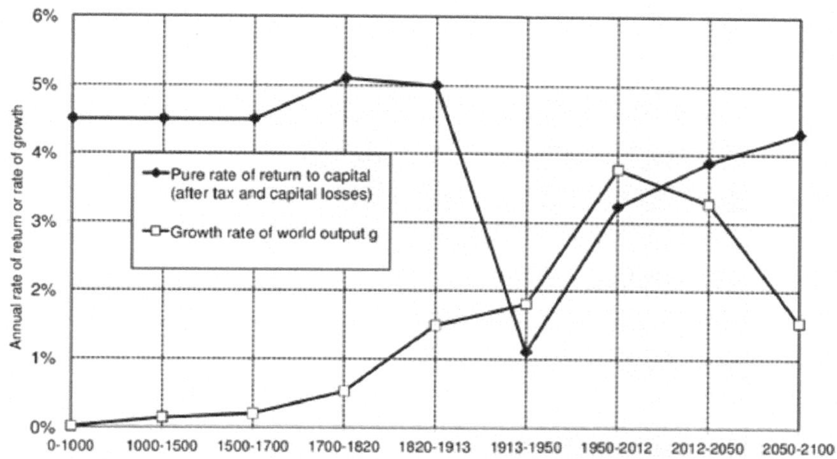


**Figure 2.33** The rate of return on capital (after taxes and losses) relative to the total growth of the world economy (Piketty, 2014).

## 2.8.2 Water management functions at the regional level

Although groundwaters and surface waters are most often a local problem, there are significant transboundary aquifers and rivers where it is of the utmost importance to establish the adequate management of water resources to protect them and ensure optimal use. Special attention should be paid to monitoring transboundary surface and groundwaters, given the sensitivity of transboundary water management related issues, as well as to the formation of good bases for water management in large catchments or wider areas. Bilateral agreements between two countries also play an important role. Unlike multilateral agreements, the problems in bilateral agreements are often of a much more concrete nature.

The Convention on the Protection and Use of Transboundary Watercourses and International Lakes (Water Convention) – does not mention climate explicitly, but it does represent one of the most important legal frameworks in the UNECE region pertaining to cooperation regarding the transboundary aspects of climate change and the development of adaptive water management strategies. The Convention obliges states to prevent, control and reduce transboundary impacts, including those related to adapting to climate change or to mitigating the impact of climate change.

The EU Water Framework Directive (WFD) is based on the principles of IWRM (Integrated Water Resources Management) and on how countries should cooperate when managing water. Together with its climate change adaptation guidelines, the WFD concept supports the development of an adaptive water management strategy.

The most important regional documents include the European Union directives pertaining to water management which are listed here chronologically: the Nitrate directive from 1991, the Directive on the treatment of urban wastewater from 1991, the Directive on drinking water from 1998, the Framework Directive from the year 2000, the Directive on groundwater from 2006, the Directive on the assessment and management of flood risks from 2007, the Directive setting environmental quality standards in the field of water management from 2008, the Directive defining the technical specifications for chemical analyses and for monitoring the water status from 2009, and the Directive on priority substances in the field of water management from 2013.

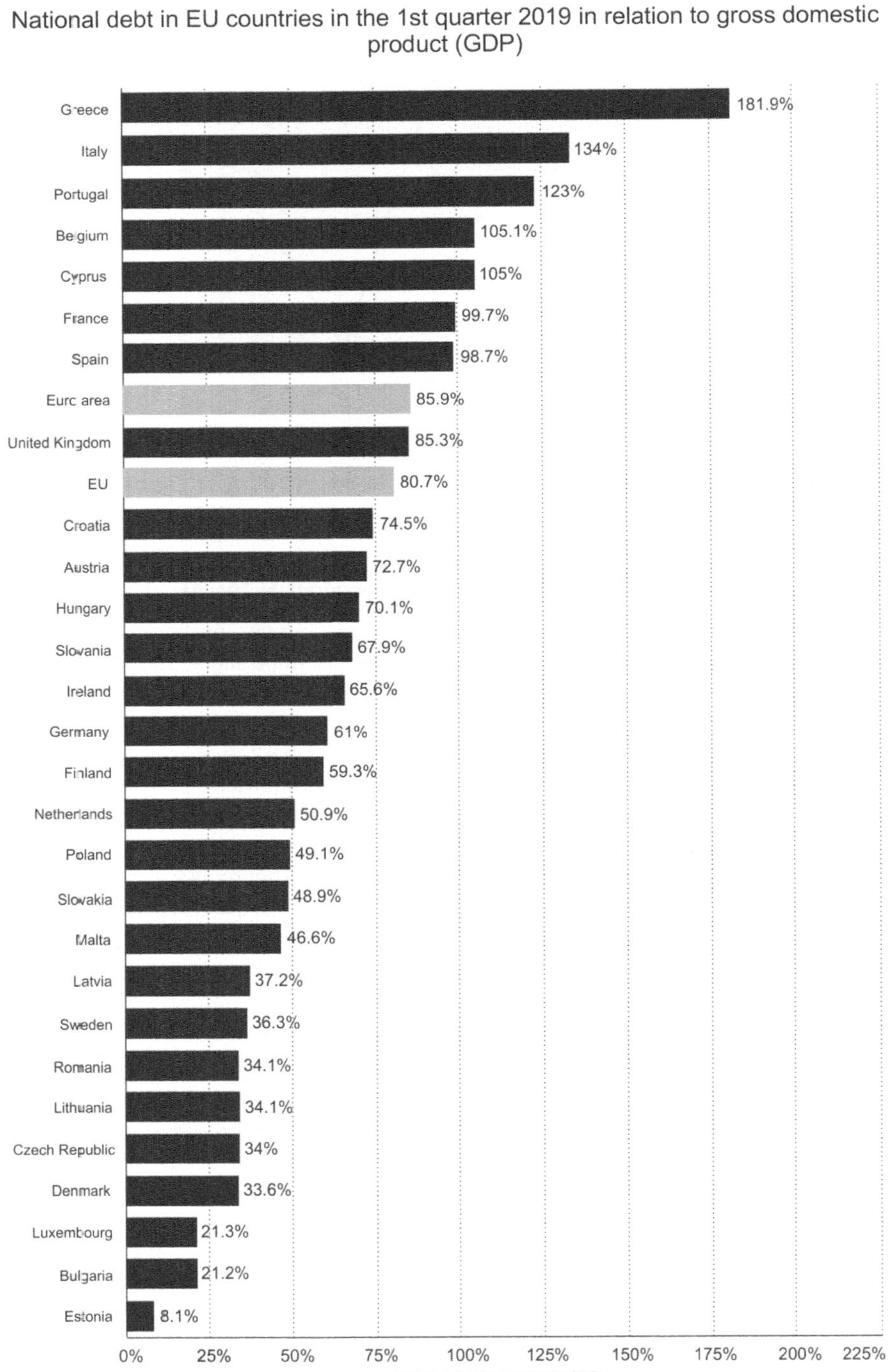


**Figure 2.34** External debts for specific European countries (EUROSTAT, 2019).

The WFD is a key document prescribing the establishment of a framework for EU action in the field of water policy, as well as the specific activities and deadlines necessary to implement an integrated approach to water resources management in EU Member States, which is based on regional water resources management plans (river basin management plans – RBMPs).

In addition to this Directive, there are a number of other directives governing specific areas such as: the Urban Wastewater Treatment Directive, the Groundwater Directive, the Drinking Water Directive, the Flood Directive, etc. (Figure 2.35).

Some of the EU's basic principles in the field of water are:

- High level of protection;
- Precautionary principle;
- Preventive action;
- Addressing pollution at the place of origin;
- The polluter pays principle;
- Integrating environmental conditions into the definition and implementation of other community policies, e.g. in the field of industry, agriculture, transport and energy;
- Promoting sustainable development.

Most EU directives governing the water sector contain:

- A proposal of the basic objectives and provisions;
- Key obligations of Member States;

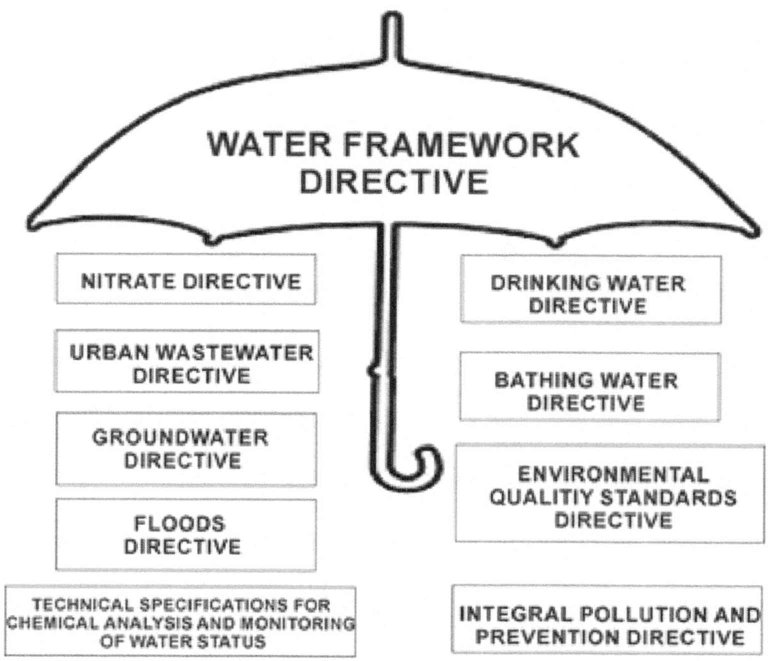


**Figure 2.35** Schematic illustration of the Water Framework Directive and other EU directives relevant to the water sector (author picture – no source).

- Planning;
- Management;
- Monitoring;
- Data sharing and reporting.

Additional legal acts closely related to a given Directive implementation include:

- Defining key tasks;
- Consideration of implementation by stages and implementation guidance relating to planning;
- Expert advice and guidelines;
- Planning within basins;
- Monitoring and management;
- Consultation and reporting;
- Expenditure.

All international documents create a very important legal, organizational and management framework that needs to be incorporated into both the water management strategy and the water-related legislation.

International cooperation with neighboring countries and with the wider international community, which is essential and very significant for the water sector, is governed by international treaties, conventions and agreements, which are or must become an integral part of the legislative framework for water management.

The most important regional organizations in Europe are presented in Table 2.17.

**Table 2.17** List of European Transboundary Water Management Organizations and Partner Countries.

| Transboundary Water Management Organizations | Partner Countries |
|---|---|
| Border River Commission between Finland and Sweden | Finland, Sweden |
| Finnish Norwegian Transboundary Water Commission | Finland, Norway |
| Genevese Aquifer Management Commission | France, Switzerland |
| Guadiana River Commission | Spain, Portugal |
| Lake Geneva International Commission for the Protection of Lake Geneva | France, Switzerland |
| International Commission for the Protection of the Danube River | Austria, Bosnia and Herzegovina, Bulgaria, Croatia, Czech Republic, Germany, Hungary, Moldova, Montenegro, Romania, Serbia, Slovakia, Slovenia, Ukraine |
| International Commission for the Protection of the Elbe | Germany, Czech Republic |
| International Commission for the Protection of the Rhine | Germany, France, Luxembourg, the Netherlands, Switzerland |
| International Meuse Commission | Belgium, France |
| International Sava River Basin Commission | Bosnia and Herzegovina, Croatia, Slovenia, Serbia |
| Peipsi Center for Transboundary Cooperation | Estonia, Russia |
| Permanent Indus Commission | India, Pakistan |

Cooperation in the UNECE region is based on the Convention on the Protection and Use of Transboundary Watercourses and International Lakes, Helsinki, 1992, which is a binding framework for the protection of international surface and groundwaters through prevention, control and environmentally sound water management.

The Commissions operating in Europe were formed as the result of bilateral and multilateral treaties between signatory states in order to adopt water management plans, i.e. to conduct regular monitoring and to monitor the implementation of the measures and activities foreseen by the joint water management plan within a designated transboundary catchment.

## 2.8.3 Water management functions at the local level

The availability of water resources represents a precondition for adequate water management, and in the case of transboundary water resources, the establishment of international transboundary cooperation is required to meet the basic objectives of water management:

- Use and development of water resources;
- Water protection;
- River regulation and protection against the adverse effects of water;
- Integrated water management.

All the segments are interconnected and depend on the measures and activities implemented at the regional, cross-border level.

Socio-economic conditions are one of the most important frameworks for water management at the local level. In Europe, the rate of connection to public water supply systems in developed countries is close to or 100%, but there are still countries in the European Union where the rate of connection to water supply systems is less than 80% (Figure 2.36).

The pressures on the quality and quantity of water resources which are a result of the socio-economic framework in a given country are often significant. The quality of surface and groundwater depends on how well the activities and measures related to protecting the quality and quantity of surface and groundwaters are implemented. In transition countries, a small percentage of wastewater is treated or it is fully untreated, but because of the under-developed industry and agriculture, there is usually no

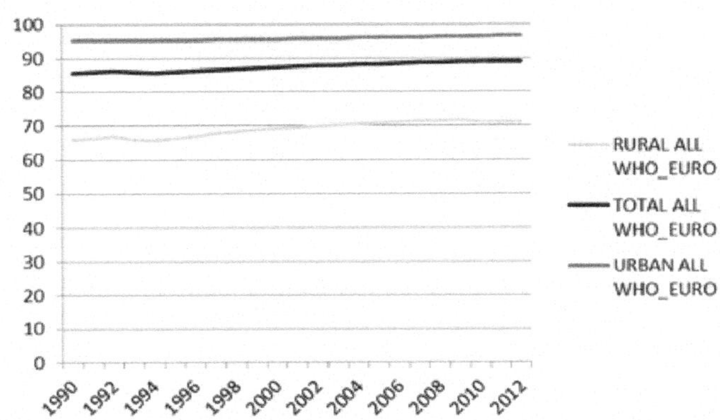


**Figure 2.36** Percentage of the population with house connections to improved water sources in urban and rural areas (WHO European Region, 2015).

significant negative impact on changes in water quantity and quality (Dimkić *et al.*, 2013). In the future, if the development of agriculture and industry is to be expected, there will certainly be increasing pressures on water quantities and quality requiring significant investment to reach the objectives defined on the basis of regional (Water Framework Directive) or local frameworks.

The most important prerequisite for the implementation of the Drinking Water Directive and the Municipal Wastewater Treatment Directive is the restructuring of companies, the improved full cost recovery and the dynamic harmonization of tariffs for financing the operation of a developed infrastructure. Also, in view of the anticipated climate change and water reductions, water needs to be managed adaptively and it is necessary to strive towards sustainable water management, with particular regard to transboundary cooperation and the implementation of measures at regional and local water management levels.

At the local level, the most important internal frameworks that influence the development of the water sector are:

- Legal;
- Institutional and organizational;
- Financial.

In countries where there is a developed legal framework, as well as the institutions and organization necessary for the basic functioning of the water management system, in the case of insufficient financial resources, efficient water management is not possible. In Europe, most countries have implemented external legal frameworks (Water Directives), and there are institutions and bodies involved in water management (Ninković *et al.*, 2010). However, due to insufficient knowledge of the quantities and quality of water resources, as well as insufficient capacity to implement measures and activities, and insufficient financial resources, the water management systems do not function effectively.

The capacities necessary for a water management system to function smoothly are:

- Management and planning;
- Science and education;
- Production and services;
- Construction.

One of the main influencing factors for water management is science and education (this can be of external and internal nature). If there are other water management capacities, without a sufficiently developed science and education system, there are no human scientific and professional capacities that, as products of education and professional-scientific work, can implement measures and activities for efficient water management. Figure 2.37 shows the global distribution of researchers per million inhabitants.

Without investment in science and research in the coming period, it will not be possible to adequately respond to the increasingly dynamic changes in the socio-economic system, as well as to changes in nature due to climate change.

The degree of investment in science and education differs depending on the development of a country. This ultimately affects the country's capacity and ability to implement regional water management frameworks, as well as the country's capacities that depend on science and research such as production and services, construction, management and planning. All these factors affect the ability of the water management system to function. Water management can only be realized if there is a developed awareness of the need to conserve resources, both in terms of quality and quantity (Dimkić & Milovanović, 2014).

Very often, especially in developing countries, the individual users of groundwater and surface water can be very influential from the point of view of both over-exploitation and the pollution of resources, so their

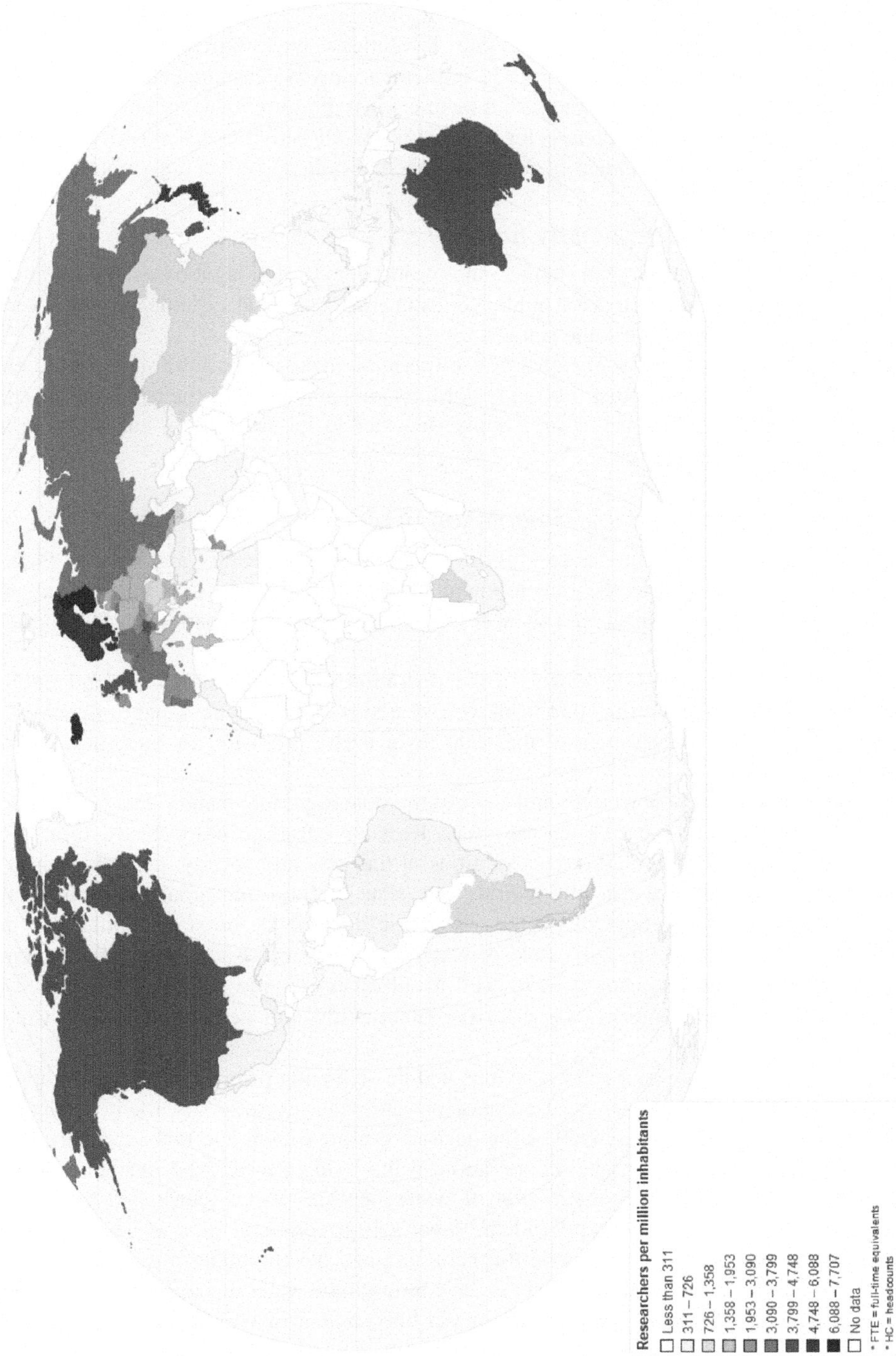


**Figure 2.37** World distribution of researchers per million inhabitants (UNESCO, 2011).

education is of the utmost importance. Also, it is very important to increase the capacity of the state both in terms of the number of people involved and in terms of their education in the field of integrated water management. It is also necessary to better equip institutions in technical terms (introduction of computers, information technology, etc.) in order to implement new water management models.

An adequate legislative and institutional framework must be provided to ensure the successful functioning and development of the water sector. It must allow all activities to be carried out respecting the essential functions for the achievement of water management objectives on a given territory.

#### 2.8.3.1 Water management regulatory function

Water management must be based on well-defined and clearly demarcated legal bases, based on the public ownership of water. Just as surface waters are public goods in almost all legal systems, groundwaters, as part of a single water cycle, must have the same status.

Through a system for the allocation of rights of use (permits, licenses, etc.) the state should ensure the adequate use of groundwater, its fair distribution (priorities), and should meet the needs of different users (supply of water to the population, irrigation, supply of water to industry, etc.), ensuring the adequate protection of resources and sources, etc.

#### 2.8.3.2 Planning functions and implementation of plans

Planning is of great importance for the successful development of any country and society. Planning is a particularly important component in the development of the water sector. It is carried out at several levels: the state level, the river basin district (if necessary, the sub-basin, as part of river basin area), and the local government unit.

Planning must be a mandatory component in the operation of public companies and public utility companies, as well as other legal entities that want to have a successful business and to develop.

The primary objective of management at the state level is the provision of integrated and uniform water management.

Strategic documents set out the objectives and tasks of the state regarding water management. Bearing in mind the situation in the field of water (water resources, legislative framework, infrastructure, human and other capacities), the need for water and the possibilities of the state and society, and taking into account contemporary trends and practices in the world, the state defines the goals pertaining to water management and sustainable development. In these strategic documents the state should also define its strategy for international cooperation in the field of water as well as ways and means of financing water infrastructure maintenance and reconstruction as well as all other activities of interest to state and local administration which are a precondition for effective functioning and development of a water sector in general.

Strategic documents define the necessary activities and those having priority in terms of improving the state of things, the necessary resources and deadlines for achieving the set goals, the possible sources of funds and the dynamics of financing as well as the necessary professional and other capacities. The state should define the strategy for international cooperation in the field of water, the manner and conditions for financing the construction and reconstruction of water facilities and systems and the execution of other tasks of general interest for the state and the local self-government unit as well as all the other elements that are a prerequisite for the successful functioning and development of the water sector.

Planning at the river basin district (RBD) level is done through the water management plan for an RBD - the so called RBMP. This is a new type of planning act, the content of which is in accordance with the requirements of the WFD (EC Directive 2000/60 EC).

The water management plan contains all the necessary elements to ensure the rational use and protection of water in the area under consideration, as well as the protection against water, which includes an assessment of the necessary resources for the implementation of the planned activities, the dynamics and the manner in which they will be provided. It is adopted for a period of six years, after which the plan is updated. If significant differences are found when compared with the proposed solutions, additional measures may be adopted for the achievement of the objectives set out in the plan, i.e. amendments to the previous solutions may be adopted. It is important to note that the preparation of the plan takes place with the active participation of the public. In addition to the management plan, a separate plan, in accordance with an appropriate plan for the river basin district, can be adopted for individual water management issues or for individual sub-basins as parts of the river basin. The active involvement of all stakeholders is required, especially in the process of developing and updating river basin management plans. It is also mandatory that all public comments, including users of water resources, be published and made available to each river basin district.

The third level of planning is planning at the level of the local self-government unit (municipality, city). At this level, investment activities are planned for facilities of importance to that local government unit (LGU), or, in association with other LGUs of significance to the wider area. This includes facilities for municipal infrastructure, for the public water supply and sewerage (water services) the planning of which will require participation by the public utility company.

#### *2.8.3.3 Financing function*

There are two basic goals to be achieved related to financing the water sector:

- Stable financing;
- The principle of self-financing in the water sector.

Stable financing means securing long-term permanent sources of financing, which include an appropriate legal and institutional framework that determines the sources of funds, the amount of funds needed and how they are secured, including revenue collection mechanisms, as well as precisely defined payers of fees for the use and protection resources.

Within this framework, it is also necessary to engage the state in creating a favorable environment for the use of local and international funds to finance development projects, as well as to create the right conditions for the significant investment of private capital.

The operation of the water sector on the principle of self-financing means providing the highest quantities of funds needed for the functioning of the water sector by ensuring the real price of water is considered, i.e. the application of the principle 'user pays', as well as the principle 'polluter pays'. Namely, the price of water must include the total real costs of preparing drinking water and bringing it to the user (the principle of 'full recovery cost'), but also the necessary amount to be used for development. The charge set for the use and protection of resources also affects the price of water. It is therefore advisable that the amount charged for the discharge of wastewater be such that the polluter would be more interested in building a treatment plant as soon as possible, rather than paying the amount charged over a longer period (Šojić *et al.*, 2014).

#### *2.8.3.4 Investment preparation function and investment*

The planning and the realization of investments in the water sector consist of three basic phases: the preparatory or pre-investment, investment and operational phases. The preparatory or pre-investment phase, contrary to widespread belief, is the most important phase in the entire investment cycle. It is of crucial importance for overcoming problems related to the lack of infrastructure, for identifying development priorities and the priority facilities within the priorities, for efficient and successful financial management, etc. It begins with the identification of problems and the selection of ideas and ends with the analyses of the financial or socio-environmental justification of the project, provision of means for the realization of the project and by obtaining the necessary permits for construction.

The investment phase begins when the investment is well prepared. It begins by announcing a tender for the selection of contractors followed by their selection, and then comes the most important part - the construction of the facility. It ends with a technical inspection and obtaining of the necessary permits for the constructed facility.

The operational phase includes the operation of the facility and its maintenance in a functional state.

*The preparatory phase*

The preparatory phase includes:

- Definition and selection of ideas;
- Previous analyses and papers (research, background);
- Feasibility study and the preliminary design;
- Obtaining the necessary permits for construction and the main project design.

*Definition and selection of ideas*

This is the first step in the preparatory phase for the implementation of an investment project. It begins with the idea for development by a potential investor, followed by an analysis of the possibility of building the structure, taking into consideration its compliance with the relevant strategic and planning documents of varying significance (at the national, i.e. river basin, or local community level). Selection of the initial ideas is made based on the results of the analysis, and the priority project elements are then selected, following which further technical and other documentation is developed.

*Baseline studies (research, background)*

Baseline studies provide the basis for the preparation of technical documentation for the facilities requiring a construction permit that is issued by the competent ministry.

The type and scope of research at this stage depend on the investment facility, and include:

- Research and analysis of projects and other professional material;
- Obtaining data needed to analyze and set engineering and geological, geotechnical, geodetic, hydrological, meteorological, urban, technical, technological, economic, energy, seismic, water management and traffic conditions that the design and project must meet;
- Environmental conditions;
- Fire protection conditions (if applicable);
- Other conditions affecting the construction and use of the facility under consideration.

*The investment phase*

The investment phase includes:

- The tender and selection of contractors;
- Registration of works and preparation for construction;
- The construction of a facility and technical inspection;
- Commissioning (if necessary) and the acquisition of the necessary permits to use the facility.

*The operational phase*

The operational phase, i.e. the use of the facility, occurs when the necessary operational permits for the facility are obtained. At this stage, it is necessary to evaluate the results of the investment, i.e. to check that it meets the set parameters and indicators, and in the case of utility infrastructure facilities (water supply and sewerage) whether it also meets the expectations of the population. After assessing the investment that has been realized, it is possible to draw conclusions for the future.

Running costs (business, maintenance, depreciation), which are often considerable, must be taken into account during operation of the facility. In the case of water and sewerage facilities, these costs must be covered by revenue (water and sewerage tariffs).

Problems are sometimes experienced in operation as a result of 'mistakes' made during the selection of investment options considered. Often, cheap investment (low investment) is associated with high operating costs, which outweigh the savings on the investment.

In addition to the operating expenses, the owner of the facility ensures the execution of works on capital maintenance of the facility, and, if necessary, regular, extraordinary and specialist inspections of the facility, in accordance with special regulations.

#### *2.8.3.5 Monitoring and control*

Monitoring is an extremely important segment of water management, and it involves monitoring of certain phenomena. In the area of water management, the quantities and quality of water can be monitored. Systematic monitoring of resources (flows, levels, surface water quality within the basin, as well as groundwater levels and quality within the aquifer) is usually carried out by authorized institutions. Surface and groundwater monitoring data provide us with an insight into the state of resources, as well as trends in change.

In addition to systematic monitoring, there is also operational monitoring, which should be in the service of monitoring changes in the regime of water resources and carried out as needed.

Inspection monitoring includes the work of services that monitor the use, protection and quality of water and possible impacts on the environment.

Considering the complexity of water management, inspection is carried out by:

- An inspector who controls the water management sector, working as part of the ministry responsible for water management;
- A sanitary inspector, as part of the ministry responsible for health and sanitation;
- An environmental inspector as part of the ministry responsible for the environment.

The inspector who controls the water management sector inspects and controls the implementation of the provisions of the law relating to the field of water management, the implementation of regulations adopted on the basis of this law, as well as other regulations and general acts relating to the construction of new and reconstruction of existing facilities and other construction work that may affect changes in the water regime.

The sanitary inspector performs health and sanitary supervision related to the public supply of hygienically safe drinking water as well as to water needed for bathing.

The environmental inspector inspects and controls the sustainable use of natural resources and monitors environmental protection, including the quality of wastewater discharged into a recipient. The inspector also supervises municipal infrastructure facilities and the performance of public utilities.

## 2.9 GROUNDWATER RESOURCE SPECIFICS

The term 'groundwater' is usually reserved for groundwater occurring in soil and in geological formations that are completely saturated with water. There is a similar approach in the new EU legislation (Water Framework Directive (WFD)):

'Groundwater is all the water below the surface of the terrain in the saturation zone, which is in direct contact with the soil or subsoil.' In their classic textbook, Freeze and Cherry (1979) say that their 'approach will be in line with the traditional emphasis on the groundwater flow in shallow, saturated environments, but will also cover the soil moisture regime in the unsaturated zone at the surface, which plays a significant role in the hydrological cycle, as well as the regimes in the much deeper, more saturated environments that significantly affect many geological processes.'

The self-purifying potential of an aquifer is its ability to improve the initial water quality by filtration through the aquifer, which is a function of the time and length of filtration, the conditions in the aquifer and the water movement dynamics (Dimkić & Kečkarević, 1990).

When filtration through the aquifer is long enough, it changes the water quality and most often improves it due to the physical, chemical and biochemical processes that occur. In a lesser number of cases the water quality may be impaired (e.g. due to the dissolving of iron and manganese in the aquifer matrix material).

In the design, development and protection of source areas, an aquifer should be regarded as a physicochemical-biological reactor, whose purification potential should be adequately analyzed and utilized.

All waters found in the lithosphere are called groundwater. They are different in appearance and composition. Groundwaters occur in the form of vapor, in the liquid or solid state and in the supercritical state. They also occur as physically and chemically bound water. The distribution of groundwater in the vertical profile, in terms of character and volume, form and composition, is very uneven.

It depends on the one hand on the character of the porosity and the degree of cracking of the rock masses, and on the other on changes in the thermodynamic and physicochemical conditions that change with depth.

### 2.9.1 Groundwater resource specifics

Groundwater velocity, especially through intergranular aquifers, is much slower than river velocity. The processes that lead to the transformation of the quality of groundwater occur in water, the aquifer skeleton and the skeleton/water interaction. Groundwater quality transformation processes are often slow and take a long time thus leading to a more pronounced specificity of the groundwater quality. The content of the dissolved substances in groundwater can be quite different from place to place, even within a single aquifer. Also, there are numerous processes in groundwater that lead to the self-purification of groundwater, which most often improves the initial quality of groundwater. Due to their specific features, groundwaters are a hidden resource suitable for use, but due to the increasing anthropogenic pressure, they need to be adequately managed (Dimkić *et al.*, 2009). Table 2.18 represents groundwater properties and differences in relation to surface water.

**Table 2.18** Specificity of groundwater properties in relation to surface water.

| | Groundwaters | Rivers |
|---|---|---|
| Flow velocity | 1 m/day | Order of magnitude 1 m/s |
| Exposure to natural processes | Month, year, hundreds of years | Days |
| Self-purification process | Water/aquifer skeleton/air | Mostly water/sedimentation/air |
| Nature | Mostly local | Minor differences |
| Baseline quality | Highly variable | Less variable |

## 2.9.2 Baseline quality

When considering the quality or composition of natural water resources, we must bear in mind the fact that there is no pure water in nature and that natural water contains various impurities, to a greater or lesser extent.

Therefore, the word 'water' has the meaning: chemically pure water, that is, a unique chemical compound which (provided we do not consider the presence of minimal amounts of individual isotopes) has a defined chemical composition, $(H_2O)$ n, as well as well-defined and specific physicochemical properties, the nature and concentrations of which determine the quality of a particular water resource and, based on those standards, determine the intended use. If the concentration of one or more impurities is greater than the limit value for a specific purpose, that concentration must be reduced by water treatment until the conditions for the specific purpose are met.

In certain natural environments and flow regimes water acquires certain characteristic quality attributes (see previous section). This is especially true for groundwater, which is generally slow moving and exposed to long and intense environmental influences.

Basic water quality is the quality of surface and groundwater without any anthropogenic influences.

It can be said that as surface and groundwaters flow they come into contact with the riverbed and are exposed to different processes in the ecosystem, thus acquiring natural characteristics which depend on the length and time of the water flow. Water flows are measured for days and weeks from a source to the estuary. The natural mineralization of water in seas and oceans is around 35 g/l, in surface waters it is about 2500 mg/l, while for groundwaters, mineralization is generally higher than that of surface waters, but much less than the mineralization of seas and oceans, which is a consequence of long water retention times. Basic water quality occurs as a result of numerous water/rock/gas interactions.

Table 2.19 lists the basic characteristic differences in the quality of surface and groundwater.

## 2.9.3 Water quality transformation processes

There are numerous processes that affect the self-purifying potential of alluvial aquifers in removing pharmaceuticals from groundwater. In addition to the convective transport of dissolved matter, self-purification processes include: dispersion processes, sediment/water matrix exchange processes, and degradation processes that result in the loss of the dissolved matter in the sediment/water matrix (Figure 2.38).

In nature, dispersion is a more pronounced process than diffusion, and the way it is expressed (the diffusion law) sometimes does not reflect what really goes on in nature.

Sorption and desorption involve several possible processes:

- Physical sorption;
- Chemisorption;
- Ion exchange;
- Dissolution (so-called winding) of minerals.

Degradation or the 'decomposition' of matter can be the result of several processes including:

- Biodegradation;
- Radioactive decay;
- Chemical oxidation processes;
- Evaporation.

**Table 2.19** Basic characteristic differences in surface and groundwater quality (Knežić, 2007).

| Parameter | Surface Water | Groundwaters | |
|---|---|---|---|
| | | Aerobic | Anaerobic |
| (1) | (2) | (3) | (4) |
| Temperature, $t$ °C | Varies seasonally; the most common range is 4–25°C | Mostly slightly variable; the most common range is 8–15°C | |
| Turbidity suspended matter (real or colloidal) | Variable, occasionally high | Low or zero (excluding karst areas) | |
| Coloring | Primarily tied to suspended material (clay, algae, etc.), except in very soft or acidic water (humic acids) | Less common than in surface water and when it occurs binds to dissolved solids | |
| Total organic matter (COD, TOC, BOD and consumption $KMnO_4$) | Occurrence with frequent anthropogenic contribution | Usually lower concentrations than (2); occurs as a natural, basic concentration (background); in the case of additives due to human activity, by means of self-purifying mechanisms tends to reach the basic concentration. Generally, the decomposition of organic pollutants is better in aerobic rather than in anaerobic conditions. | |
| Total content mineral matter | Variable, depending on precipitation, soil composition and other factors | Usually constant, much higher than in (2) | |
| Iron and manganese | Usually very low concentration except in eutrophic waters | Possible occurrence in lower concentrations $C_{Fe} < 1$ mg/l | Frequent occurrence in higher, and in significantly higher concentrations than in (2) or (3) |
| Aggressive carbon dioxide | Usually not present | More commonly present in smaller or larger quantities | |
| Dissolved oxygen | Often close to saturation (10 mg/l) | Present in smaller concentrations than saturation concentrations | Practically absent |
| Hydrogen sulfide | Usually completely absent | Usually completely absent | Often present in smaller or greater concentration |
| Ammonium | Absent; present only in polluted waters | Absent | Often present in the basic composition of water and/or due to human influence |

(*Continued*)

**Table 2.19** Basic characteristic differences in surface and groundwater quality (Knežić, 2007) (*Continued*).

| Parameter | Surface Water | Groundwaters | |
|---|---|---|---|
| | | Aerobic | Anaerobic |
| (1) | (2) | (3) | (4) |
| Nitrates | Usually low concentration (5–10 mg/l) except in polluted waters | Basically occurs most often in low concentrations; occurs in larger concentrations due to anthropogenic influence | Usually does not occur |
| Silicon dioxide | Moderate content | Often higher content than in (2) | |
| Mineral or organic micropollutants | Present in all developed countries, but prone to rapid disappearance upon the disappearance of the source of pollution | Basic and as a rule absent. However, in case of incidental, anthropogenic pollution, remediation is, as a rule, more difficult than with (2) | |
| Microorganisms | Bacteria (including pathogens), viruses and plankton | Non-pathogenic bacteria, fungi, yeasts, pathogenic organisms basically absent | |

COD – chemical oxygen demand; TOC – total organic carbon; BOD – biological oxygen demand.

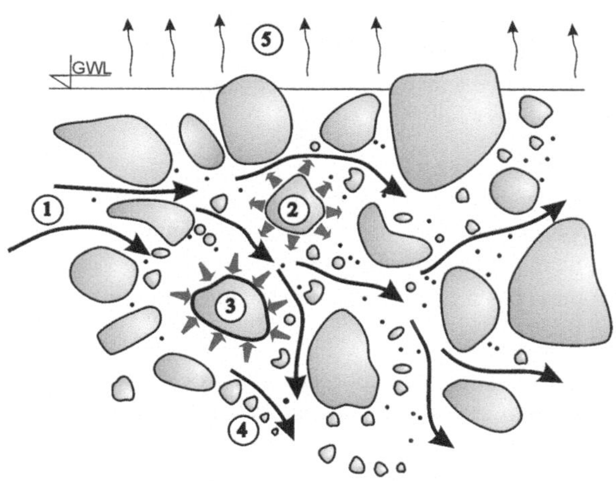


**Figure 2.38** Schematic presentation of various processes of substance transfer and degradation: 1. Transfer of dissolved matter by means of convection, dispersion, diffusion; 2. Transfer of substance from the solid phase to a solution – processes of desorption, dissolution of aquifer material, transition of insoluble the substance into the solution; 3. Processes of transition of the substance from the liquid phase to the aquifer material – physical sorption, chemical sorption, settling; 4. Substance degradation processes – biodegradation, radioactive decomposition, reduction-oxidation processes beyond the biological degradation process; 5. Evaporation (Dimkić *et al.*, 2008a).

The above processes depend on a variety of circumstances and conditions that determine the flow, the speed and the purifying effects of filtration through the underground including:

- The content of the substance in the solution and the nature of the sorbent;
- The content of the substance precipitated or sorbed on the material;
- The conditions in the solution (dissolved oxygen, pH, temperature, etc.);
- Microbiological agents and their activity;
- The mineralogical nature of the sorbent (aquifer material).

The parameters commonly considered to be important for the process of water filtration through the aquifer are:

- In the liquid phase: the pH, the dissolved oxygen content, the redox potential, the degree of saturation of the solution by a certain substance, etc.;
- In the aquifer: the mineral composition, the granulometric composition, porosity and others.

In addition to these parameters, temperature is of great importance, as are some other parameters. Groundwater toxicity is also very important for the different processes that occur in groundwater.

## 2.9.4 Types of aquifers and groundwater sources

There are two endpoints in the spectrum of aquifer types: the free-surface aquifer and the aquifer under pressure (and between them is the aquifer with a partially free surface).

- Aquifers with a free surface are also called phreatic aquifers because their upper boundary is a free water surface or phreatic surface. Typically (although not always), an aquifer at a minimum depth in a certain location is an aquifer with a free surface, which means that between it and the surface of the terrain there is no impermeable layer. Aquifers with a free surface typically feed directly from the surface of the terrain, from precipitation, or from surface waters which have a hydraulic connection with the underlying aquifer.
- For pressurized aquifers, the free water surface is above their upper limit and they usually occur below the free surface aquifer.

**Categorization of groundwater sources**

Groundwater sources can be categorized on the basis of various criteria, such as the type of rock porosity, hydrodynamic conditions, type of flow (free or pressurized), manner of feeding, the size of the hydraulic connection and the source from which the feeding comes, the type of water intake facility, etc. The proposed categorization of sources is a combination of several different categorizations:

*Groundwater sources in crevices*

The abundance of aquifers formed in crack porosity type rocks is usually small and these sources are mainly used for the water supply of smaller settlements. Water is usually exploited through water intakes in the case of karstic springs (on an erosion base) or through wells, water intake galleries and similar.

The quality of water obtained from crack porosity rocks is usually good, except if mineral ore or anthropogenic influences are present. Oscillations in the amount of water that can be exploited are relatively small. The variation in water quality is not pronounced, except in the case of a direct anthropogenic influence or an indirect influence through sources of feeding (rainfall, rivers, etc.).

*Groundwater sources in karst environments*

Karst aquifers are formed in cavernous or fissured porosity rocks (Figure 2.39) and in most cases their abundance varies a great deal. Depending on the volume of the porous media and the source of feeding, the yield of the karst aquifer can range from several liters (or less) to several cubic meters of water per

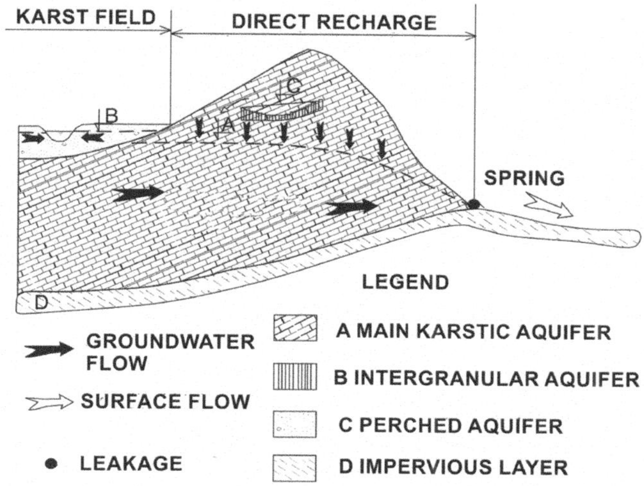


**Figure 2.39** Recharge and emptying of a karst aquifer (Dimkić *et al.*, 2008b).

second. The water derived from karst aquifers is usually of excellent quality, provided the aquifers are not exposed to adverse anthropogenic influences. Assuming that there are no such impacts, there is no significant deviation in the water quality, except in the case of aquifers where turbidity occurs during heavy rainfall. Groundwater from these aquifers is usually captured by water inlets at the point of natural discharge or by water intake facilities.

Because karst formations of extremely cavernous porosity can occur in zones deeper than the natural discharge point, such zones can potentially be used as underground reservoirs to provide additional quantities of water for water supply, using appropriate technical solutions. Saltwater penetration is a fundamental problem facing sources in coastal aquifers, especially if groundwater exploitation is intense.

The protection of these sources must first and foremost ensure the maximum protection of the recharge zone because, due to the presence of privileged groundwater movement routes, any pollutants possibly entering these routes are not treated and reach the point of discharge very quickly.

*Groundwater sources in alluvial environments*

Many groundwater sources are found in the alluvions of larger or smaller rivers (Figure 2.40). These springs are usually located near settlements so the cost of transporting water is low. If underground water is exploited from porous alluvial (water-permeable) river deposits, such a source is called an alluvial source.

With this type of source the aquifer is largely fed from a river. The reason for this is that the exploitation of the springs results in the formation of a local erosion base. A river may have a better or poorer connection to the aquifer, but it always has a primary influence on the abundance of a spring and on the water quality. The manner in which water is abstracted depends on the hydrogeological and hydrodynamic characteristics of the aquifer, although in most cases the water is captured in wells (dug, tubular, with horizontal drains), galleries, water intakes, etc., whose receiving part is placed in more permeable (sandy, gravel) deposits.

The capacity of such sources ranges from small to medium, depending on the thickness and filtration characteristics of the aquifer that the river formed during a geological period. The most common problem with such springs is the decline in abundance over time (especially in the case of intensive exploitation).

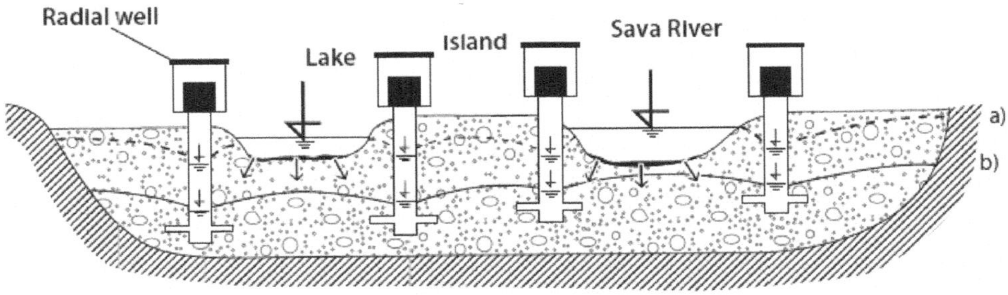


**Figure 2.40** Alluvial type of source (Dimkić *et al.*, 2008a).

During an intensive flow from an open stream to water intake structures and as a consequence of the precipitation of river sediment and of the chemical processes that cause incrustation, there are certain negative influences on the filtration characteristics, primarily:

- At the point of contact between the river or lake and the aquifer at the bottom of the river or lake (clogging the riverbed);
- In the immediate area of the water intake structure (filter area).

The proximity of settlements, industry, and agriculture to the alluvial source represents the most significant anthropogenic influence. The water quality of an alluvial source can significantly depend on the quality of the surface water, which is often the most significant source of alluvial groundwater recharge.

In addition, man's aspiration to create new agglomerations and infrastructure results in the occupation of larger areas within the alluvium, and this requires extensive experience and knowledge in the spatial planning and design of groundwater resource protection zones.

*Groundwater sources in pressurized aquifers*

An artesian-subartesian well is formed when groundwater is exploited from deeper geological environments, where the flow of water is pressurized (Figure 2.41). The piezometric level of the subartesian aquifer is below the surface of the terrain (in the unsaturated zone), while the piezometric level of the artesian aquifer is above the surface of the terrain.

The groundwater quality of these springs is often directly related to the mineral composition of the lithology. Since recharge is slow and since the water in these sources is sometimes over 10,000 years old, interactions with the material over time can lead to a high mineralization of the groundwater.

Water sources in pressurized aquifers may have a large capacity to store groundwater, but their exploitation capacity is limited by the amount of water infiltrating the source. If the exploitation of the aquifer is greater than the feeding, the piezometer levels in the aquifer decrease, resulting also in the compression of the lithological layers and the settling of the terrain.

Since such aquifers form in semi-enclosed geological formations, the negative anthropogenic impacts are much less pronounced and oscillations in the quality of groundwater are small.

*Groundwater sources in coastal aquifers*

The penetration of saltwater is the main problem at sources where water is drawn from coastal aquifers, especially when the exploitation is intense. Groundwater is mainly exploited using water intake structures at a point of natural discharge (usually a karst aquifer), or some other type of structure. The yield of such

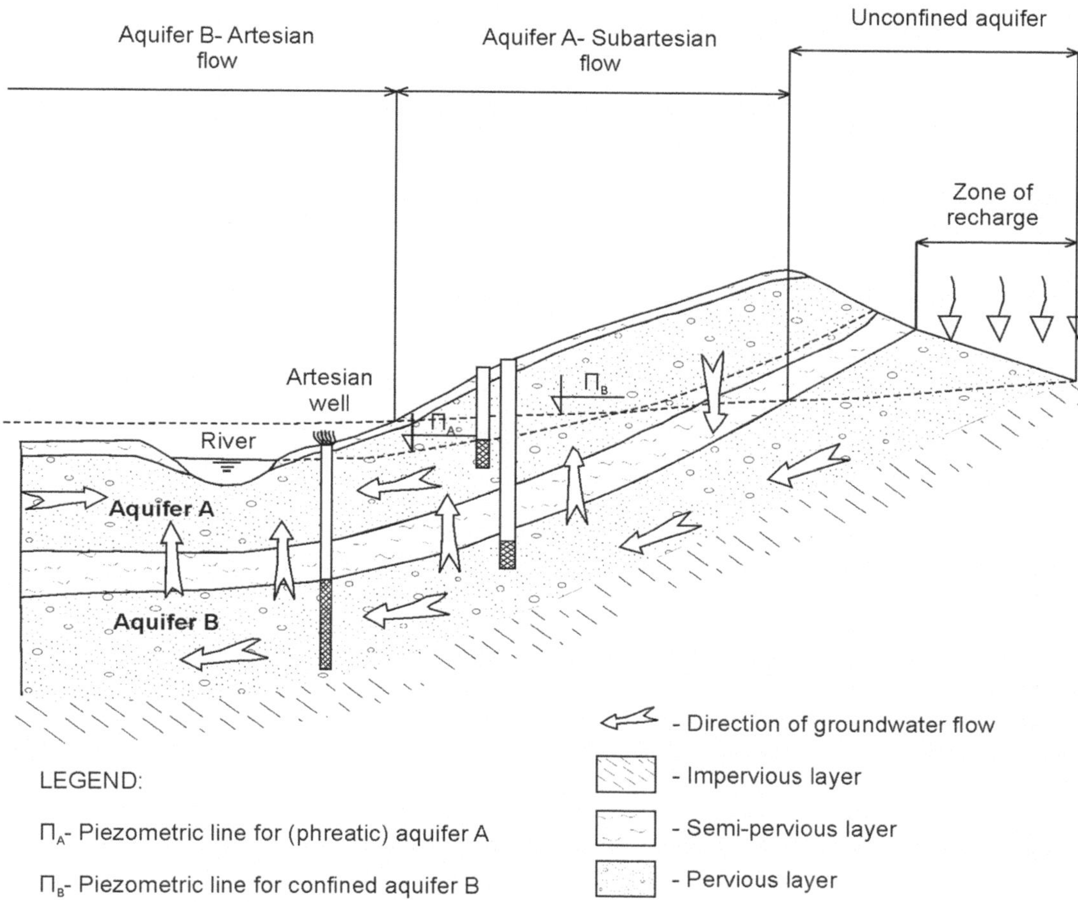


**Figure 2.41** Artesian and subartesian source (Dimkić *et al.*, 2008a).

sources can be considerable, if formed within the karst formation with a large basin area. However, the mixing of freshwater and saltwater is a frequent problem.

The contact zone between saltwater and freshwater is formed and depends on the hydrodynamic conditions and the presence of saltwater in the aquifer beneath the seabed (Figure 2.42).

As a rule, fresh water, whose density is lower, appears above the layer of saltwater. Hydrodynamic dispersion creates a border layer between the saltwater and freshwater. In this case too, the protection of a source primarily consists of the greatest possible protection of the feeding zone, especially of the karst springs, where potential pollutants are quickly transported to the zone of discharge via privileged pathways and are not subject to treatment.

*Groundwater infiltration sources*

Artificial infiltration is most often carried out to achieve a better exploitation capacity of groundwater from an aquifer. Artificial infiltration can be used for both intergranular aquifers and crack porosity aquifers, while the infiltration structures can be lakes, infiltration basins and similar. Water intake

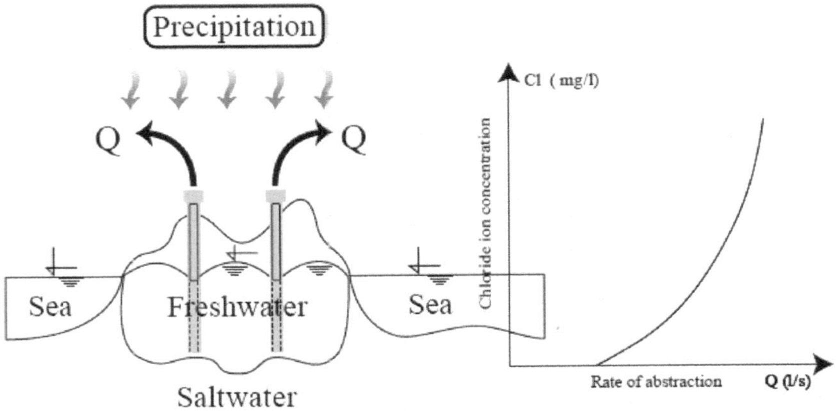


**Figure 2.42** A source in a coastal aquifer, (Dimkić *et al.*, 2008a).

structures are mostly wells and drainage galleries. From the point of view of infiltration itself, systems that have a lower infiltration rate, such as infiltration lakes, are simpler. If a higher rate of infiltration is required, it is necessary to increase the efficiency of the surface water pretreatment system before infiltration, in order to reduce the collimation effect of the infiltration structures.

The formation of an infiltration source (Figure 2.43) typically makes it possible to draw greater and more stable quantities of water per unit of available area. This method can provide significant strategic reserves of groundwater and improve the coexistence of sources and the environment. In addition, such systems are less sensitive to the incidental contamination of resources of surface water. A well-designed and well-built system which enables the adequate duration of filtration in the underground ensures efficient filtration treatment through a porous environment. The quality of the abstracted groundwater is better and requires less treatment than would otherwise be necessary to reach drinking water quality. On the other hand, such a system in improved aerobic conditions greatly accelerates the self-purification potential of an aquifer and has significant advantages. Achieving an aerobic environment will also slow down the ageing of water intake objects.

Table 2.20 represents different types and properties of groundwater sources.

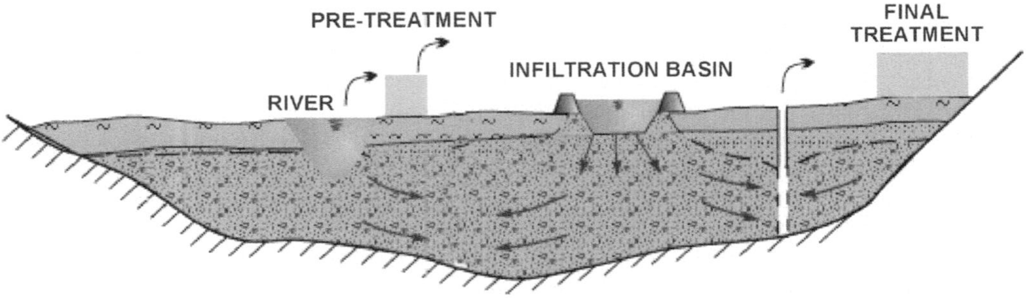


**Figure 2.43** An infiltration source, (Dimkić *et al.*, 2008a).

**Table 2.20** Types of groundwater sources (Dimkić *et al.*, 2007).

| Source Types | Abundance | Oscillations of Abundance | Quality | Quality Oscillations | Degree of Treatment Required | Required Level of Protection | Maintenance Problems |
|---|---|---|---|---|---|---|---|
| Springs in a cracked rock | Small | Low | Usually good to excellent | Insignificant | Usually just densification | Relatively low to high, in a wide area | Water intake |
| Source in a karst environment | Small to Large | Big, seasonal | Usually good to excellent | Insignificant, possible turbidity | Usually only disinfection, in some cases clarification and filtration | Low to high, in the wider zone protection zones are sparsely populated | Sensitivity of the water intake |
| Sources in an alluvial environment | Small to large | Small to moderate | Usually good, sometimes poor, to excellent | Small to significant, depends on the resources surface rainwater | Disinfection, aeration, filtration, active charcoal, ozonization | Medium to high, depends on the environment and the protection of the river | Ageing of water intake structures, collimation of the contact between a river and the porous environment |
| Sources in a pressurized aquifer | Small to medium | Low | Usually good, sometimes excellent, to poor | Usually insignificant | Usually only disinfection, rarely more complex treatment | Low | Ageing of intake structures, over-exploitation |
| Sources in a coastal aquifer | Small to large | Significant, seasonal | Excellent to poor | Insignificant, possible salting and turbidity | Usually only disinfection | Low to high in a wide area | Cleaning of the basin, ageing of water intake structures |
| Infiltration source | Medium to high | Stable, does not depend on the season | Very good | Insignificant | Pre-treatment, usually just disinfection, rarely complex treatment | Medium to high, depending on the environment and the protection of the river | Cleaning of pits, ageing of water intake structures |

## 2.10 ALLUVIAL AQUIFERS

### 2.10.1 Definition of alluvial aquifer, basic specificity

Alluvions are the result of various processes that have, over the course of their history, influenced their formation. This may primarily be due to erosion caused by the flow of water itself, as well as being due to geological factors: tectonic, volcanic processes, karst, the movement of glaciers and similar.

Figure 2.44 shows a typical river alluvion profile from the source to the mouth of the river, where three material deposition zones in the alluvial river valley are emphasized.

The first zone represents the upper part of the river where the flow of the river is faster and there is significant erosion of coarse and drawn sediment (coarse gravel and sand). In the upper part of the alluvial aquifer, oxic groundwater conditions are present. The middle flow of the river is quieter and is marked by the beginning of deposition of sand and gravel of medium granulation and the beginning of the deposition of clay material. In the middle course of the river, the alluvial aquifer still has oxic environmental conditions, but the dissolved oxygen concentration decreases due to the organic matter and minerals that consume oxygen during the oxidation process, during infiltration into the alluvial aquifer. In the lower part of the river, the flow of the river is slow and, because of this, the deposition of mainly fine sand, fine material and clay material is most pronounced. Here, there is also the greatest amount of organic matter and minerals which consume all the oxygen from the river water during the oxidation process which occurs during infiltration into the alluvial aquifer, so that the groundwater is predominantly anoxic in the lower river.

*Sources that use river water use bank filtration and/or artificial filtration methods*

The advantages of both methods are, in the main, the proximity to the places of consumption, as well as the numerous processes/mechanisms of self-purification which, naturally, generally improve the quality of

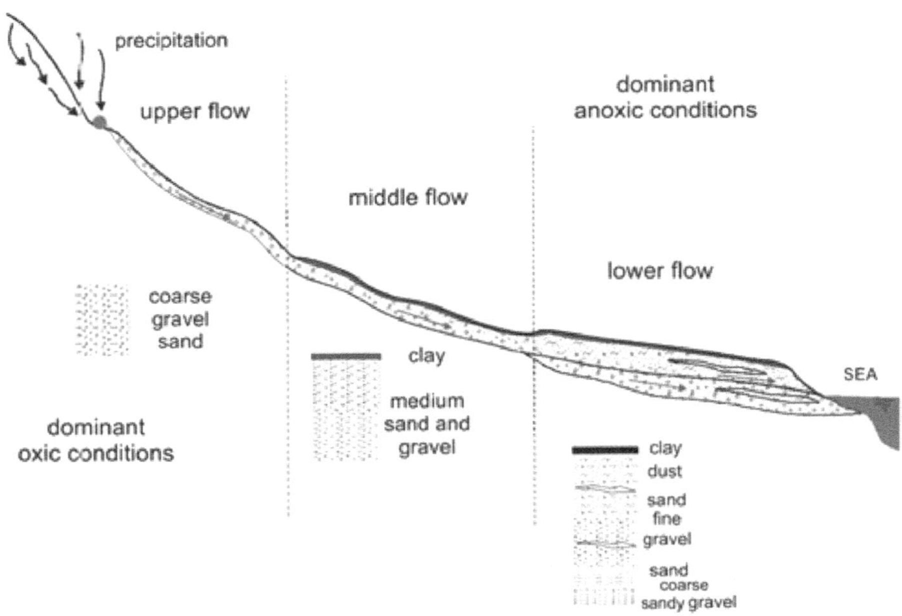


**Figure 2.44** Changes in granulation and the oxic state of alluvial aquifers along the river flow (Dimkić *et al.*, 2011a).

river water very significantly. This water, to a greater or lesser degree, acquires the characteristics of groundwater.

The advantage of bank filtration over the artificial infiltration method is the lower cost of forming and maintaining a source.

Both types of sources are most often under pressure due to human activity:

- Expansion of agglomerations (hinterland);
- Agriculture (hinterland);
- Industry (hinterland and river course);
- Deterioration of the water quality from rivers due to human activity in the river basin, etc.;

The advantages of an infiltration source over the application of the bank filtration method are:

- Reduced impact on the environment and reduced environmental impact on the source;
- Greater amount of water supply;
- Better ability to control quality;
- Specific amount of retention (storage) time.

The excessive pumping of water using the coastal filtration method can lead to a significant decrease in groundwater levels. The separation of groundwater in the aquifer from the clay and sediment layers can lead to the absence of significant capillary climbing of water and to problems pertaining to the drying of crops.

This section further outlines some conditions, elements and processes important for the management of groundwater in alluvial aquifers.

- Encrustation of wells (water wells) with iron;
- Self-purification by means of filtration within an alluvial source;
- Elimination of pharmaceuticals and pesticides by filtration within an alluvial source;
- Transformation of nitrogen components in groundwater;
- Droughts and groundwater alluvial resources.

Groundwater oxicity is one of the most important parameters for most of these processes. The problems of groundwater oxicity are analyzed separately in Chapter 4 which looks at the reasons why oxic environmental conditions occur, as well as the impact of these conditions on the kinetics and way in which certain processes occur in alluvial groundwater.

## 2.10.2 Basic types of groundwater sources in alluvial aquifers

### *2.10.2.1 Riverbank filtration and artificial groundwater recharge*

Riverbank filtration (Figure 2.45) is a method of water treatment that is effectively used in many European countries. The following are examples of the use of riverbank filtration: 50% of drinking water in the Slovak Republic, 45% of drinking water in Hungary, 17% of drinking water in Germany, 5% of drinking water in the Netherlands (Hiscock & Grischek, 2002), and more than 50% of drinking water in the Republic of Serbia (Dimkić *et al.*, 2011a). The artificial water infiltration method is less common and significantly more expensive than riverbank filtration.

There is a relationship between the river water and the groundwater of aquifers. In the new parts of the river flow, the river water infiltrates into the alluvial aquifer, while in certain parts of the flow, the river has a drainage role. There are also situations where the river current and that in the alluvial aquifer are parallel, or the currents may be at a certain angle. The construction of wells by the river, in cases of increased

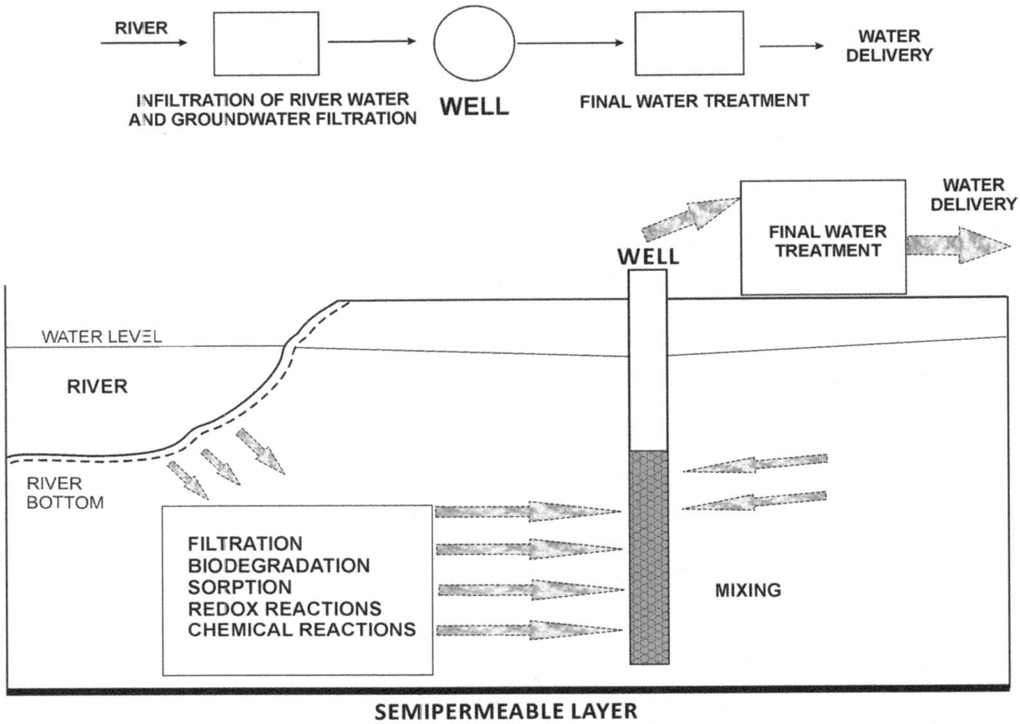


**Figure 2.45** Schematic presentation of the water treatment line using riverbank filtration (Revised: Dimkić and Keckarević, 1990; Dimkić *et al.*, 2008a).

exploitation, results in a greater amount of water in the well being from the river and not from the hinterland. The quality of the water that is abstracted depends on the quality of the river water, and of the groundwater in the hinterland, as well as on the self-purifying mechanisms of groundwater filtration in the aquifer, implying that the protection of water throughout the river basin also involves the protection of the source using the riverbank filtration method.

The advantages of the riverbank filtration method are reflected in its simplicity and the relatively low investment, operating and maintenance costs. Self-purification can have a significant effect on the quality of the abstracted water.

By filtration through the granular part of the aquifer, water may achieve the basic groundwater quality of the aquifer, thereby reducing or eliminating the concentration of substances of an anthropogenic origin from the river or hinterland of the source. However, if the fine-grained structure of the aquifer skeleton contains soluble matter, the groundwater quality may also deteriorate.

There are numerous benefits to using riverbank filtration and artificial groundwater infiltration including improvement of the quality of water and reduction of the drinking water preparation cost. The results of research (Bradley *et al.*, 2014; Kovacevic *et al.*, 2017a; Mazahirali *et al.*, 2015; Storck *et al.*, 2012; Sudhakaran *et al.*, 2013) indicate micropollutants can be removed using the self-purifying potential of the aquifer. In addition, the authors of this chapter conducted research regarding the transformation and transport of selected pollutants, and, during field experiments, analyzed in detail the self-purification processes of selected pollutants (phenols, mineral oils, heavy metals, pharmaceuticals, pesticides and nitrates). Experience and knowledge regarding the effects of groundwater self-purification were also gained from less serious, more difficult and very complicated cases of pollution, protection and recovery of wells (Dimkić, 2007).

The oxic state of the alluvial environment is of great importance when looking at the transformation of water quality, self-purification of groundwater and maintenance of wells (the effects of collimation).

In addition to the sorption and biodegradation of micropollutants and pollutants in general, another very important factor that affects the self-purification capacity is the groundwater filtration time from the river to the wells which depends on the specific hydrogeological conditions of the site. This filtration time can vary from several days to several years (Table 2.21).

Riverbank filtration is characterized by an interaction with the hinterland, such that groundwater flows to the source and anthropogenic influence can compromise the water quality at the source and hence the protection of the source must be well planned. On the other hand, groundwater pumping during coastal filtration can cause a decrease in alluvial groundwater levels. If the groundwater level falls below the level of the preferential, less permeable layer, capillary lifting may be interrupted and vegetation could be threatened. Figure 2.46 provides a simplified scheme showing artificial infiltration.

The three main benefits of an infiltration source are:

- Increased water from aquifers;
- Reduced environmental impact of the source and that of the environment on the source, reducing the required surface area of the protection zones of a source;
- Better water quality at the source.

However, the cost of an infiltration source is considerably higher than for a source by the river.

Riverbank filtration and the artificial infiltration of groundwater represent sustainable techniques and methods for the treatment of drinking water. However, it is important that the system of riverbank filtration and artificial infiltration are designed and set up on an appropriate location in hydrogeological terms, taking into account the characteristics of the aquifer and the oxicity of the groundwater, as well as the time required for the infiltration of water from the river to the water intake well.

**Table 2.21** Characteristics and parameters of riverbank filtration and artificial groundwater recharge (Dimkić *et al.*, 2008a).

| Parameter | Characteristics |
|---|---|
| Typical infiltration time from river to well | day < t < year |
| Typical river distance from wells | 10 m < L < 100 m |
| Pre-treatment | Does not have |
| Redox conditions | Short term high oxygen saturation (high oxicity) zone after which oxygen is reduced |
| Temperature change | Small variations (mostly 12–15°C) |
| The effect of temperature | Low impact on changing redox conditions |
| Share of river water | Variable with respect to the river water level and the hydrogeological conditions of the environment, as well as the intensity of exploitation of springs |

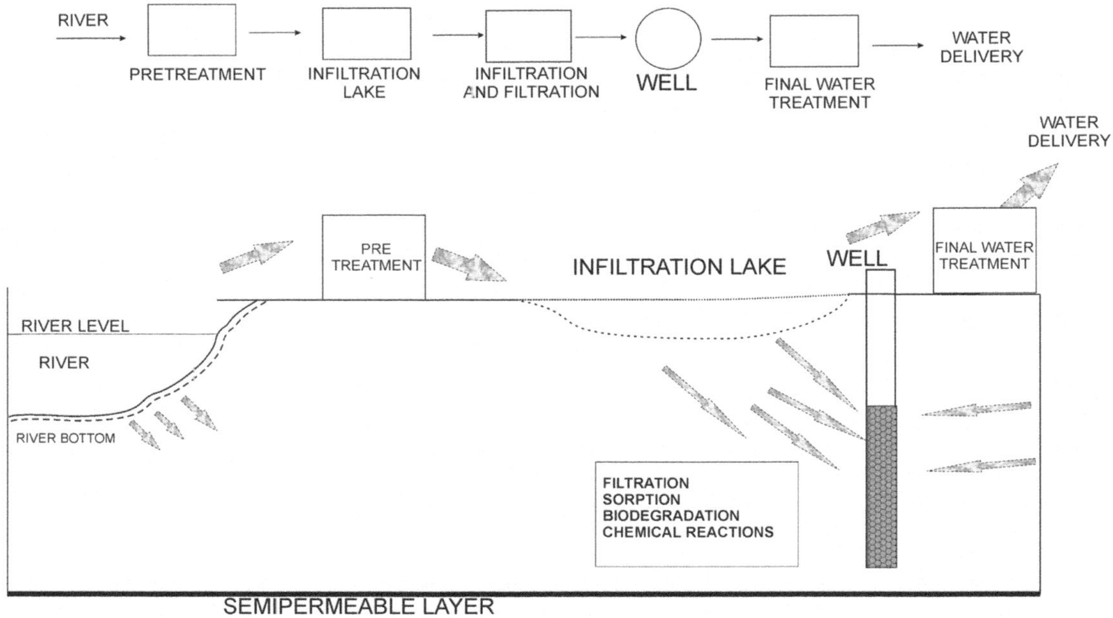


**Figure 2.46** Schematic presentation of the water treatment line using artificial filtration (Revised: Dimkić and Keckarević, 1990; Dimkić *et al.*, 2008a).

### *2.10.2.2 Effects of riverbank filtration and artificial groundwater infiltration*

In comparison with the abstraction and use of surface drinking water, riverbank filtration and the artificial recharge of groundwater have many advantages, such as: the partial or complete elimination of suspended matter and particles of biodegradable substances, bacteria or viruses; the partial elimination of substances that are absorbed; the water temperature is made the same and the impact of shock concentrations in accident situations is reduced; less disinfection by-products are formed; the taste and odor are improved; and biologically more stable water is obtained (Dimkić & Keckarevic, 1990; Dimkić *et al.*, 2011c; Hiscock & Grischeck, 2002; Kühn & Müller, 2000, 2006; Maeng *et al.*, 2010; Ray *et al.*, 2002; Schmidt, 1988).

Some side effects may also occur if minerals dissolve within the aquifer skeleton (iron and manganese), resulting in deterioration of the groundwater quality.

The positive effects of coastal filtration and artificial infiltration can most often be attributed to the processes of sorption and to the degradation of the dissolved substances in groundwater, as well as to the interaction with other self-purification processes.

The main two zones where the most intense self-purification processes take place are the river bottom, where the most intense processes of sorption and degradation occur over a short period of time, as well as the zone between the river bottom and the water intake structure, where sorption and degradation are less pronounced but where the exposure time to these processes is considerably longer (Dimkić *et al.*, 2011c).

During riverbank filtration, there are four phases of filtration from the river to the wells (Figure 2.47). Phase I consists of filtration through the collimated river bottom and is characterized by fine sediments and oxic conditions, while the sorption potential is very high as is biochemical activity. Phase II consists

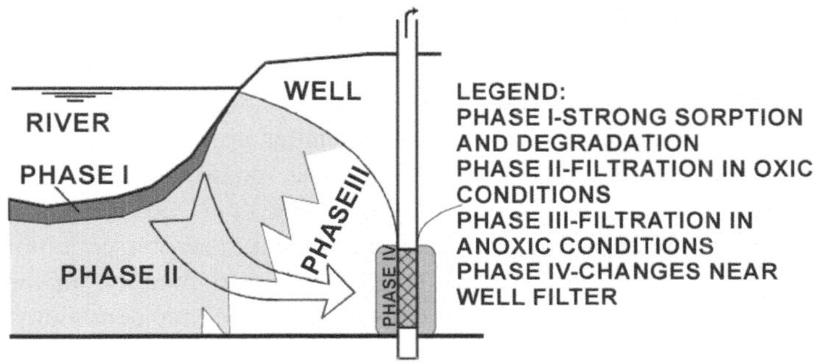


**Figure 2.47** Phases of filtration from river to well (Dimkić *et al.*, 2011c).

of filtration through the aquifer under oxic conditions, where the oxygen dissolved in water is consumed to oxidize organic matter as well as minerals. Phase III consists of filtration under anoxic conditions. Here, transformation occurs due to redox reactions (e.g. $Fe^{3+}$ becomes $Fe^{2+}$), and the total content of dissolved iron, manganese and ammonia increases. Phase IV consists of filtration in the vicinity of the well's filter. There are intense mechanical, chemical and biochemical changes in the aquifer due to the change of redox conditions in the filter zone. The absence of phase III is possible, but this does not mean that groundwater quality transformation does not occur as anoxic and oxic water quality transformation conditions are possible in phase III.

The main advantage of the application of the riverbank filtration method and of artificial groundwater infiltration is the degradation of organic matter. For example, at the Belgrade source (99 radial wells along the Sava River) the organic carbon content decreased (Figure 2.48) from 2.5 mg/l in the Sava River to about 1 mg/l in the groundwater (Dimkić *et al.*, 2011c).

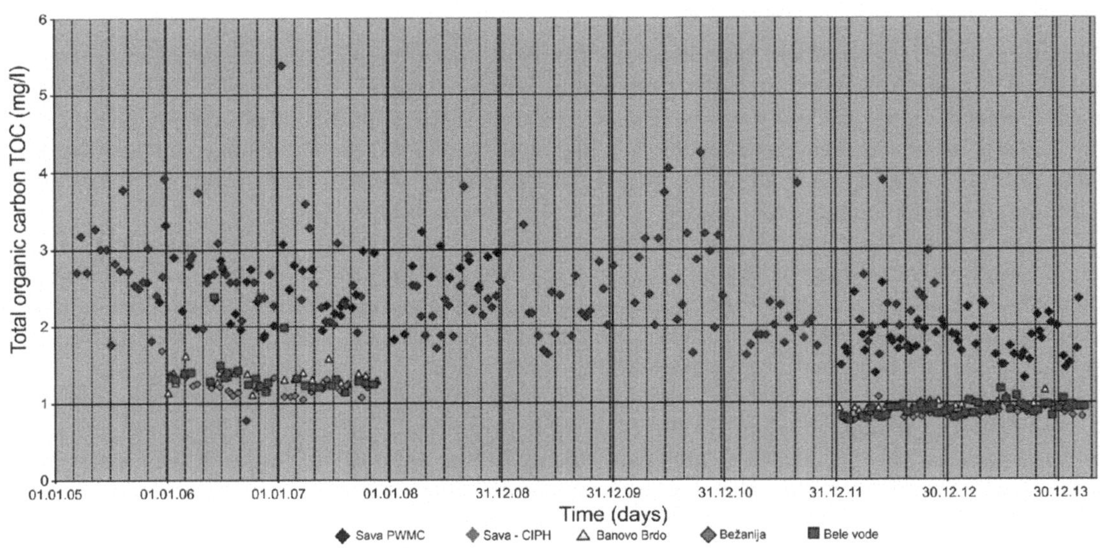


**Figure 2.48** Total organic carbon (TOC) content results in the Sava River and corresponding wells of the Belgrade spring source (Dimkić & Milovanović, 2014).

### 2.10.3 Experiments for defining phenol degradation in an oxic alluvial environment

**Phenol biooxidation pathways under aerobic conditions**

The rapid degradation of phenols in surface waters, under appropriate conditions, is well known (Borighem and Vereecken, 1979; Kostyaev, 1975). The possibility of their decomposition in groundwater has also been reported (Bekins *et al.*, 1993; Dimkić *et al.*, 2003; Thornton *et al.*, 2001).

The processes of hydrocarbon degradation in water are subject to aerobic conditions. The processes of biooxidation and degradation, as a rule, tend to result in the end products in the process, $CO_2$ and $H_2O$. The basic prerequisites for such degradation of phenols are aerobic conditions and the presence of phenol-active microorganisms.

Prof. Dr. Milan Dimkić and his associates conducted a series of experiments and monitored groundwater quality between 1983 and 1994. The results showed an intense degradation of phenols in groundwater in the oxic groundwater environment. It was also found that phenols were not sorbed on the sand and gravel in the alluvial aquifer during the experiments, which were carried out at the following locations:

- Location 1 (Groundwater source in the city of Požega) – Six years of monitoring of the transformation of the phenol concentration in groundwater during filtration through a porous environment. The monitoring carried out was an inspection and showed that phenol degradation occurs in the shallow groundwater source;
- Location 2 (The 'Žičko polje' groundwater source, Kraljevo) – The shallow oxic source was monitored to see the distribution of phenol and of the chloride ion in groundwater. Approximately the same rate of transport of phenol and chloride in groundwater was observed;
- Location 3 (the 'Mediana' groundwater source, Niš) – A dipole with circular groundwater movement was formed as part of the shallow oxic water study. The comparative monitoring of the concentration of tracer (Cl-) and the concentration of the phenol (<100 μg/l) clearly showed a decrease in the concentration of phenol in relation to the concentration of the tracer. The time it took to double the decrease in the concentration of dissolved phenol was 2–3 days.

The surveys were conducted at several locations. Location 1 is represented by a protection system against the spilled D-2 diesel oil, right next to the water supply source of the town of Požega. The D-2 oil was spilled when a train overturned, and part of the protection system consisted of infiltration pits into which water from the river was infiltrated. This water, acting as a hydraulic barrier, did not allow the contamination of the source with dissolved oil, and one of the quality parameters that was monitored was phenol. The effects of this system were monitored for a period of six years (1984 to 1989).

Location 3 is the 'Mediana' infiltration source of the city of Niš. At this source, the experiment consisted of injecting phenols and tracers into groundwater served to evaluate the possibility of the degradation of this potential pollutant between the infiltration lake and the water intake well. This was done for the purpose of designing the source system, i.e. to determine the filtration distance and filtration time between the infiltration lake and the water intake wells (Dimkić *et al.*, 1997).

Phenol degradation in groundwater was monitored at several locations and over a relatively long period (1984–1995). The aquifers in which the field tests were performed were aerobic, with water temperatures generally between 10–15°C and with high permeability (of the order of $10^{-3}$ m/s). It was observed that the sorption of phenol on the relatively coarse-grained aquifer material was practically insignificant (Figure 2.49), while the biodegradation of phenol was evident and relatively rapid (Figure 2.50).

During a complex, long-term experiment at the source for the city of Kraljevo (Location 2), it was found that the half-life period of phenol at concentrations of 0–100 μg/l was generally $t = 2$ (or 3) and up to 5 days.

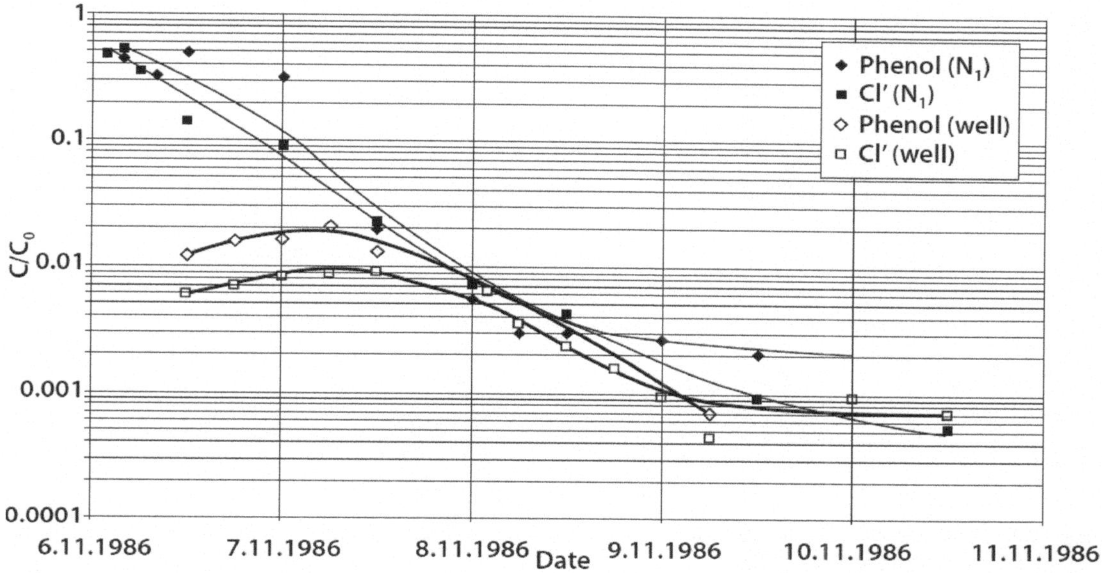


**Figure 2.49** Realistic concentrations of phenol and chloride ions over time, data recorded during a short-term experiment at the 'Žičko Polje' source in 1986 (Dimkić *et al.*, 1986).

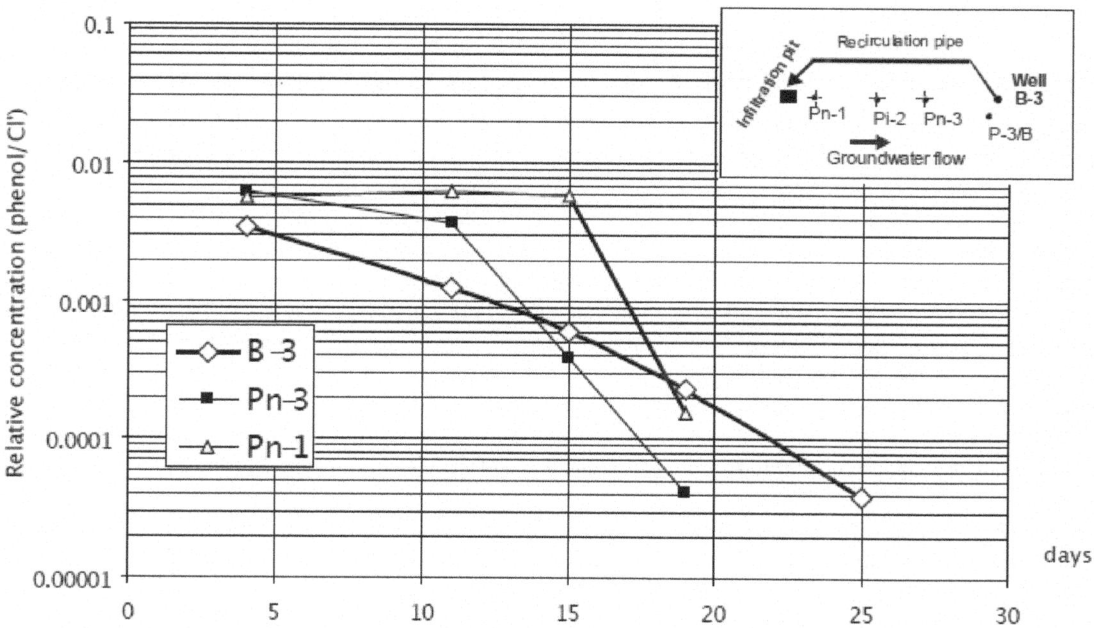


**Figure 2.50** Changes in the relative concentration of phenol (relative to the chloride ion) during the experiment at 'Mediana' infiltration source (Dimkić *et al.*, 2007; Dimkić *et al.*, 1986–1993, 1991).

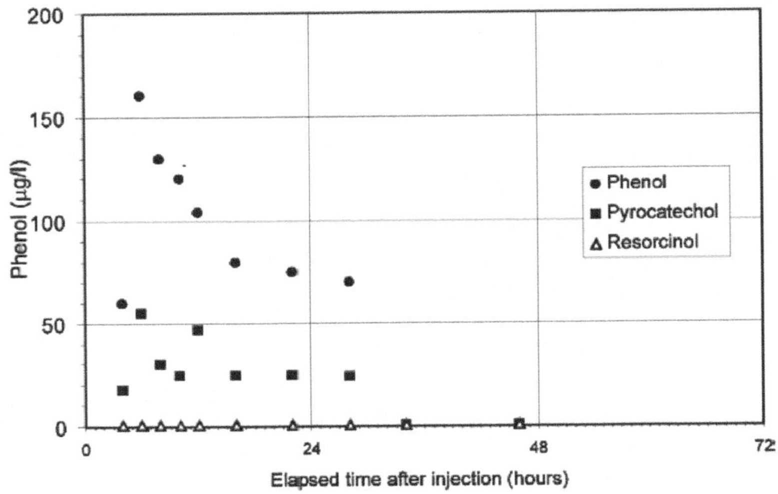


**Figure 2.51** Phenolic matter concentrations in groundwater at the observation site close to the location of phenol infiltration (Dimkić *et al.*, 1991).

In this case, the half-life was longer for higher concentrations of phenol in groundwater. During the experiment at the Mediana source (Location 3), a significant presence of the phenol degradation intermediary (pyrocatechol) was observed just a few hours after the infiltration of phenols into the groundwater (Figure 2.51).

## 2.10.4 Ageing and yield reduction processes in water intake structures

The ageing and deterioration of wells and other catchment structures has been a problem for a long time in the exploitation and regulation of groundwater regimes. Mechanical, chemical and biochemical collimation are the main causes of well decay.

The old masters (Abramov, 1952; Cistin, 1965; Gavrilko & Alekseev, 1985; Johnson, 1975; Kovacs & Ujfaludi, 1983; Pietraru, 1982; Sichardt, 1928) and others analyzed and determined criteria pertaining to mechanical collimation due to excessive input speeds in the filters.

Vukovic and Soro (1990) and Vuković and Pušić (1990) presented an overview of past results. Johnson (1975) and Gavrilko and Alekseev (1985) studied criteria which led to the occurrence of corrosion and incrustation. Gavrilko & Alekseev (1985) proposed a classification of water with regard to type of well ageing, based on the LSA (Langelier Saturation Index) and the RSI (Ryznar Stability Index).

More recently, a number of authors have examined well screen clogging (Cullimore, 2008; Cullimore & Mansuy, 1987; Houben, 2003; Houben & Treskatis, 2007; Mansuy, 1999; McLaughlan, 2002; Pyne, 1995).

Mansuy (1999) and Cullimore (1999) studied the effect of microbial growth within the aquifer on well ageing. This research generally addressed vertical, tube wells. Dimkić *et al.* (2007) and Dimkić and Pušić (2008) included design recommendations in their well ageing studies, combining hydraulic and hydrochemical criteria and using a case study of the Belgrade Groundwater Source (BGWS).

In alluvial environments, chemical and biochemical well clogging is manifested by the formation of sediments, mostly compounds of trivalent iron, manganese and carbonates.

The destruction of the filter of a well can be due to: chemical corrosion, which causes even destruction along the entire surface of the filter structure; selective corrosion, which occurs on the well and filter structure when consisting of alloys of several materials; and electrochemical corrosion, which is due to the activity of micro-galvanic pairs. At temperatures and pressures that exist in the groundwater of alluvial aquifers, the speed at which corrosive processes occur is not great, but the effects can still be very pronounced.

Well ageing due to the collimation of the filter by iron occurs when, under conditions of a relatively reduced redox potential, a sufficient amount of dissolved iron reaches the well of the filter. It is necessary for the well to have slightly more favorable redox conditions (more oxygen and/or a slightly higher redox potential [*Eh*] than in groundwater) to produce a reaction of the following type:

$$Fe^{2+} - e^{-} \rightarrow Fe^{3+} \tag{2.1}$$

These reactions can occur with the catalytic activity of bacteria which use the energy of the reaction, and/or its products for their needs. Iron bacteria can be primarily anaerobic, or facultatively anaerobic and, rarely, slightly aerobic. The processes of transition of $Fe^{2+}$ to $Fe^{3+}$ almost never occur in extremely oxic conditions ($Eh > 300$ mV), they are of a very low intensity in poorly oxic environments (150 mV $< Eh <$ 300 mV) and are present in anoxic environments (Dimkić *et al.*, 2012).

Biochemical collimation takes place due to the processes that occur in conditions of anoxic groundwaters in the vicinity of wells. The issue of anoxicity and the processes leading to the ageing of wells have been studied by a number of authors worldwide. However, extensive research carried out over many years has aimed at establishing a link between some of the most important biochemical parameters of groundwater (*Eh*, Fe), and mechanical parameters (groundwater inflow velocities into well filters, and kinetics, i.e. the collimation rates). To our knowledge, this has been one of the first attempts to address the issue of well ageing in this way and therefore represents pioneering work.

The research was done from 2005 to 2017 at the Belgrade source and at a number of other alluvial sources in Serbia under the guidance of the author, Dimkić, and is also the result of collaboration between Dimkić and Pušić.

The interrelation between the increase of mechanical resistance and the biochemical indicators was analyzed at over one hundred wells with horizontal drains and tubular wells (Dimkić *et al.*, 2011b, 2011d, 2011e).

It was observed that the kinetics of the increase in the local hydraulic resistance (*KLHR*) on the filters of wells depended on the speed of the water entering the filter of the well ($v$), on the concentration of iron in the water ($Fe$), the redox potential ($Eh$), the biochemical indicators ($B$), and the granulometry of the material (grain size distribution) around the well.

$$KLHR = KLHR(v, Fe, Eh, B, \Gamma) \tag{2.2}$$

Figure 2.52 shows experimentally obtained type diagrams. Probably the most interesting diagrams for practical purposes are those showing the permitted groundwater inflow velocities, the concentration of iron and the redox of water:

These diagrams should be considered possible indicators. They primarily point out the character of the process. Well-ageing indicators, with special reference to the Belgrade Groundwater Source, are the issues discussed (Dimkić *et al.*, 2008a).

However, because of the importance of the issue, certain parts of the material have been reinterpreted here, and new findings acquired between 2008–2018 have been added. Relevant material on the ageing of wells due to iron encrustation is given in Chapter 5 of this book.

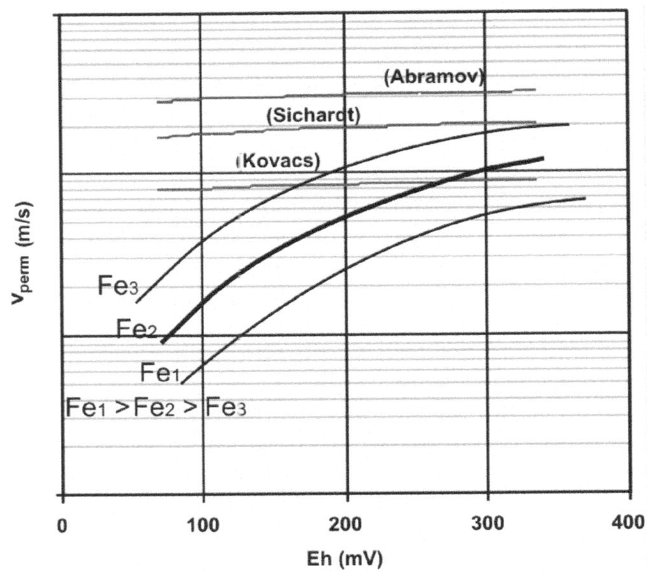


**Figure 2.52** Allowable entrance velocities of groundwater to the well regarding Fe (Dimkić *et al.*, 2011e).

### 2.10.5 Removal of pharmaceuticals from alluvial groundwater by self-purification by means of filtration in an alluvial aquifer

Selected groups of pharmaceuticals, and metabolites, or products of degradation represent indicators of the human influence on the environment and have been separately analyzed as part of research activities. Due to their widespread use, a large number of pharmaceuticals reach aquatic ecosystems and this represents a potential negative environmental impact. These micropollutants appear in very low concentrations of the order of magnitude $10^{-9}$ to $10^{-6}$ (ppb to ppm), registered in river watercourses, and in even lower concentrations in groundwater.

Awareness of the large number of pharmaceuticals detected in the aquatic environment, with a probable increase in their number due to the ever greater sensitivity of new analytical techniques and methods which allow for the more accurate determination of the concentrations of pharmaceuticals, is a serious concern (Pearson, 2002). However, the specific structure of non-organic matter (NOM), which changes in time and space and the great variety of combinations that can occur in the treatment of wastewater (which refers to the mix of substances and the way wastewater is treated) represent a great challenge, when considering the optimization of technology, treatment and sustainable water management.

The fact remains that many pharmaceuticals are not completely removed in wastewater treatment plants, while in some countries most wastewater may not be treated at all.

Most of the existing wastewater treatment plants are not specifically designed to eliminate pharmaceuticals, with the result that many pharmaceuticals pass through the treatment process due to their persistence as well as due to the continuous anthropogenic input into the sewage stream. In addition, the adequate monitoring of the presence of pharmaceuticals in most wastewater treatment plants is not carried out due to the sensitivity of the analytical methods and the high cost of testing.

After treating surface drinking water and drinking water obtained from groundwater, some pharmaceuticals can still be found in drinking water (Tumbas, 2014). Most pharmaceuticals can remain in drinking water when conventional methods of drinking water treatment are used, while most pharmaceuticals are removed when advanced drinking water treatment techniques are applied. However, there are some pharmaceuticals that are not eliminated even with the advanced treatment methods (Tumbas, 2014). Over 100 individual pharmaceuticals have been identified in aquatic and drinking water samples, and the number of newly discovered pharmaceuticals is increasing (Pearson, 2002). On the other hand, there are no standards to date pertaining to the release of pharmaceuticals into the environment.

Alluvial aquifers are under the direct influence of surface water resources, which are loaded with untreated wastewater. As a result, pharmaceuticals occur in surface water and groundwater.

Riverbank filtration (RBF) can be exploited to remove pharmaceuticals through self-purification processes, among which sorption and biodegradation are the most pronounced. Among other factors and processes, the first important one for the self-purification capacity of the subsurface is the groundwater residence time (i.e. the travel time of water from the river to a well) which is dependent on site-specific hydrogeology and can vary from a few days to several years. The second important factor is the ratio of the infiltrated river water to alluvial groundwater, because pharmaceuticals originate from the anthropogenic emissions into the surface water. Therefore, the dilution from the hinterland of the well represents a very important factor for RBF systems.

The aim of the study was to obtain sufficient information about the occurrence of selected pharmaceuticals in surface water and groundwater, as well as about the removal of these pharmaceuticals through RBF, which is generally the method of choice for drinking water supply in Serbia. This was the first time that a long-term study (7-year survey) of pharmaceutical occurrence in surface water and groundwater was conducted in the Republic of Serbia (Kovačević *et al.*, 2011a, 2011b, 2012, 2013a, 2013b, 2013c, 2014, 2017a, 2017b; Radović *et al.*, 2012, 2015).

Preliminary research included a systematic and comprehensive review of basic definitions, the selection and division of pharmaceuticals and a summary of the literature review regarding wastewater treatment efficiency referring to the removal of pharmaceuticals as well as to the efficiency of the purification of drinking water in treatment plant processes. Novel results were obtained based on the complex hydrogeological, hydraulic and analytical testing and monitoring of pharmaceuticals in surface water and groundwater, and based on a complex mathematical model of the transport of pharmaceuticals in groundwater.

Research sites include large rivers in the Republic of Serbia (the river Danube, the Sava River, the Tisa River, and the Velika Morava River) and corresponding wells in the immediate proximity of surface water sampling points. During the investigation period three specific locations (the Danube River alluvion, the Velika Morava River alluvion and Sava River alluvion) were analyzed in more detail.

Between 2009 and 2015, a total of 25 pharmaceuticals and metabolites were studied and analyzed in 184 samples. Ten were detected in surface water, and eight in groundwater. One of the main goals of the research was to identify the pharmaceuticals that occur frequently in surface water and groundwater (Kovačević *et al.*, 2017a).

Research data presented the occurrence frequency, the average concentrations, and the minimum and maximum concentrations of the detected pharmaceuticals (Kovačević *et al.*, 2017a). Scientific and research activities were carried out for the first time in the Republic of Serbia over a wide area of the Danube River Basin, and new and original results were obtained. The occurrence of the selected pharmaceuticals was analyzed based on the characteristic hydrological, hydrodynamic and oxic/anoxic conditions of alluvial groundwater. A systematization of the results was carried out in order to calculate the removal percentage during riverbank filtration.

As a part of the experimental part of the research activities (Kovačević *et al.*, 2017b; Radović *et al.*, 2016), field and experimental research was conducted in order to quantify the sorption potential for the removal of pharmaceuticals during riverbank filtration. The results of field experiments were used as a basis for defining the hydrodynamic groundwater model and the pharmaceutical groundwater transport model (Dimkić *et al.*, 2016). In the last phase of the research, a hydrodynamic model of the groundwater flow was defined as was a pharmaceutical groundwater transport model. Based on the model calibration, the pharmaceutical maximum half-life degradation was calculated. The results were compared with experimental and literature data (Dimkić *et al.*, 2017).

Based on the adopted and optimized sorption and degradation degree, new criteria for the pharmaceuticals transport model in alluvial groundwater were defined and the results obtained could serve as a basis for design of future groundwater sources.

## 2.10.6 Removal of pesticides from alluvial groundwater by self-purification by means of filtration through an alluvial aquifer

Unlike pharmaceuticals, which are the result of concentrations in river water, pesticides are most commonly used on agricultural lands in the hinterland of rivers. It is therefore very important to know the vertical flow of pesticides: pesticide – plant – arable soil – unsaturated layer and aquifer saturated layer.

The need for the use of chemicals in modern agriculture emerged as a result of the increased demand for higher quantities of better quality food, as the use of chemicals such as fertilizers and pesticides makes it possible to provide sufficient food for the ever-increasing global population. Not using pesticides in agricultural production would lead to up to 40% global loss of major crops (Matthews, 2015).

According to 2017 data, a total of about 4 million tons of pesticides are used worldwide, most of which are used in Asia (about 53% of the total quantity) and in North and South America (about 30% of the total quantity), while around 14% of the total pesticide consumption is used in Europe (FAO, 2018).

Pesticides are chemical compounds of different chemical structures and with different effects, but what is common is that they are used on large agricultural areas as well as in urban areas, and therefore represent an important source of diffuse anthropogenic pollution that is difficult to control. After the use of pesticides for their intended purpose, part of the applied quantities remains in the environment and spreads through the environment by various mechanisms. Some are washed away from the surface of the soil, some seep through the soil profile to groundwater, some evaporate and some transform and degrade (Hamilton & Crossley, 2004). The use of pesticides is often associated with various types of adverse health effects in humans and animal organisms, such as dermatological, gastrointestinal, neurological, carcinogenic, respiratory, reproductive and endocrine effects (Nicolopoulou-Stamati *et al.*, 2016). Human exposure to pesticides mainly occurs through the food chain, air, water, soil, flora and fauna (Kim *et al.*, 2017).

A large number of laws worldwide pertain to regulations regarding the standard values for pesticides on land surfaces, in the air in residential areas, in drinking water, surface water, groundwater and food, and they generally state the maximum acceptable concentrations (Li & Jennings, 2017). The maximum allowable concentrations serve as a benchmark to determine the quantities of pesticides that are considered safe for human health. The basic document for water protection adopted at the EU level is the Water Framework Directive 2000/60/EC (WFD). The aim of the Directive is to establish frameworks for the protection of surface and groundwater by implementing various measures, one of which is to gradually reduce the discharge of dangerous substances on the list of priority substances to achieve 'good status' waters. The list of priority substances is given in Directive 2008/105/EC and contains 33 substances that pose a significant risk to water bodies and whose emissions must therefore be controlled. Almost a third of the list of priority substances are pesticides. The WFD and Directive 2008/105/EC have been supplemented by the adoption of the European Commission Decision 2018/840, which in the Annex contains a 'watch list' of substances that must be monitored in surface waters, and it includes a group of pesticides known as neonicotinoids. The substances on the list have been selected based on exposure, hazard and risk assessment, but have also been chosen because of a lack of monitoring data throughout the European Union.

The goal of the 'watch list' is to identify future substances that will be on the priority list. Directive 2006/118/EC on the protection of groundwater against pollution and the deterioration of quality under Annex I lays down the maximum permissible concentration of 0.1 μg/l for individual pesticides and 0.5 μg/l for total pesticides, where the indicated concentrations refer to active substances of pesticides and their significant metabolites and degradation products in groundwater.

The use of pesticides poses a risk to water quality in agricultural areas due to potential seepage through the soil layer leading to groundwater pollution, although groundwater transport can also lead to a continuous uptake of small quantities of pesticides into surface water. In general, the pollution of water resources by pesticides is influenced by the physicochemical properties of the substances that have been mentioned, such as solubility in water, the potential for their retention in soil and degradation, but also by the properties of the areas where they are applied, such as climate, geology, morphology and hydrology. Sorption is the process that most influences the further fate of pesticides in the environment, because it has a strong influence on other processes of distribution: transport, degradation, evaporation and bioaccumulation. Sorption is a process that slows down the movement of pesticides by holding on to soil particles and, if there is no containment, it is more likely the pesticides will leach into groundwater. In the unsaturated surface layer of soil, because of incomplete water saturation, the kinetics of competitive processes between pesticide molecules and water molecules during sorption is lower, and this contributes to the larger amount of pesticide molecules that are sorbed and the smaller amount available for further distribution in the environment. However, for the overall fate of pesticides, it is also necessary to observe the sorption behavior in the saturated zone of soil, in which pesticides move along with groundwater. The relatively slow movement of water in the aquifer makes the slowed down movement of the pollutants very important, providing an opportunity for degradation processes to act for longer. Understanding the complex sorption behavior of pesticides in aquifer materials is of great importance for predicting the distribution of pesticides through the aquifer and for assessing the amount of pesticides in water bodies (Zivancev *et al.*, 2019).

Since the removal of pesticides is not sufficient during the process of preparing drinking water, it is important that the groundwater and surface water used for water supply have low levels of pesticides, which are acceptable according to drinking water standards.

### 2.10.7 Pathways and transformations of nitrogen compounds in the alluvial environment

Nitrogen sources and the chemical processes of nitrogen compounds are very complex. Knowledge of the process of transport and of the transformation of nitrogen compounds in alluvial plains as well as of its vertical transport and transformation to groundwater in general is very important. Chapter 9 is devoted to the above processes, which are very important both for agriculture and for the quality of alluvial groundwater.

Organic waste, wastewater, inorganic and organic fertilizers, processes that naturally occur in the soil (mineralization), animal manure, septic tanks, illegal dumps and atmospheric precipitation are all sources of nitrogen compounds. The characteristics of an area such as climate, geology, morphology and hydrology have a significant impact on the fate of nitrogen compounds. If the application of fertilizers is inappropriate (quantities, application time, inadequate form of nitrogen or the type of soil to which the fertilizer is applied), nitrogen can be infiltrated through the unsaturated layer and can reach groundwater.

The constant increase of nitrogen loading pressures on aquifers and on hydraulically connected water bodies often exceeds the capacity of natural ecosystem services – the self-purifying potential of water. Nitrate anions are directly bioavailable but also highly mobile. Nitrate mobility has been studied in detail and most commonly defined as tracer motion, due to the stability of nitrate anions in the oxic environment because the sorption of negatively charged ions does not occur on sediment and soil particles (except in acidic soils, which can result in significant nitrate sorption (Cahn *et al.*, 1992; Yao *et al.*, 2012)). Inappropriate, as well as heavy rainfall, can cause the seepage of mobile nitrogen compounds and of nitrogen compounds that are persistent in an oxic environment, nitrates, to below the root system. By switching to a lower oxidation state, the amount of nitrogen in groundwater is permanently reduced (emission of gaseous nitrogen oxides) or the ability of the aquifer to conserve nitrogen by transferring nitrate to an ammonium cation becomes obvious. The ammonium cation has a pronounced affinity for sorption and tends to sorb on the negatively charged aquifer skeleton and is typically considered immobile. Nitrogen compounds that reach groundwater are subject to the complex dynamics of various biogeochemical processes of the N cycle transformation: denitrification (DN), dissimilatory and/or assimilatory reduction to ammonium ions (DNRA, ANRA), nitrification, anaerobic ammonium oxidation (Anammox, Feammox), nitrification, denitrification, sorption and mineralization of organic matter (Nikolenko *et al.*, 2018; Weng *et al.*, 2017). Under anoxic/reducing conditions, nitrogen will become the subject of reductive transformations: assimilation into the biomass, respiratory denitrification, Annamox, Feammox, and dissimilatory/assimilatory nitrate reduction, usually associated with the production of ammonium cations (Perovic *et al.*, 2017).

Modern biogeochemical research is focused on defining the cause and effect correlation between hydrogeological, physicochemical and microbiological parameters of soil and water in order to define the dominant influence on the final transformation process. The oxygen concentration and redox potential, the pH value, quantity and quality of available organic carbon, the C:N ratio (DOC:NO3-), and the concentration of other electron donors – cations of iron, manganese and anions of sulfides, will determine the type and form of nitrogen species in groundwater (Burgin & Hamilton, 2007; Dimkić *et al.*, 2008a, b; Nikolenko *et al.*, 2018; Rivett *et al.*, 2008). The ratio between the stable isotopes of nitrogen and oxygen has in recent years been a widely used method for identifying nitrate sources. However, as the isotopic signatures of different sources can overlap but also change under the influence of various complex biogeochemical mechanisms of transformation, such as volatilization, nitrification, and denitrification, it cannot be exclusive in the identification of a source (Pasten-Zapata *et al.*, 2014). By combining the study of stable isotopes of nitrogen and oxygen in nitrates and by monitoring of the movement patterns of conservative tracers, it is possible to contribute to a more accurate determination of the nitrate source and the transformation process in groundwater.

The seriousness of the problem of elevated nitrate concentrations in water bodies in the world can be recognized by the fact that the European Union has adopted a separate directive concerning the protection of waters against nitrate pollution originating from agricultural sources, clearly confirming that agriculture is one of the main sources of nitrates in surface water and groundwater. The Nitrate Directive (Council Directive 91/676/EEC) is one of the most important EU legal bases for environmental protection within the framework of European water policy and represents the basic support for the WFD.

Since 2005, there has been a declining trend of nitrate concentrations in the groundwaters in Europe. In 2011 there was a decrease in nitrate concentrations to the 1992 level. In rivers in the European Union this declining trend in nitrate concentrations led to an average annual decrease of 0.03 mgN/l, or 0.8%, annually, in the 1992–2012 period (EEA, 2018). The reduction of the nitric nitrogen concentration is the result of measures taken to reduce pressures on water bodies originating from agricultural sources (Nitrate

Directive – Directive 91/676/ EEC), and of improved wastewater treatment (Urban Waste Water Directive – Directive 91/271/EEC), but it is also a result of the decline in industrial production in Eastern Europe.

When considering the nitrogen balance as the difference between inputs and total nitrogen loss, the declining trend of the N balance is evident for developed countries: Austria, Belgium, Denmark, France, the Netherlands, Japan, Germany and Sweden. Although less pronounced, but also declining, the nitrogen balance trend is also noticeable in Romania, Slovenia, Croatia and Bulgaria. An increased nitrogen loading trend relative to nitrogen losses has been observed in the Czech Republic, Poland and Hungary. Compared to 1960, the worldwide use of mineral fertilizers has increased tenfold (FAO, 2016). The countries with the highest production and use of fertilizers are China, America and Northern Europe.

North America, Europe and parts of South and East Asia are regions with the most pronounced impact of elevated nutrient concentrations on the quality of water bodies (FAO, 2017). In most high-income and emerging economies, agricultural water pollution significantly outweighs the impact of wastewater from settlements and industry and is a major factor in the degradation of inland and coastal water quality (eutrophication) (FAO, 2017). In low-income countries and developing economies, untreated municipal and industrial wastewaters have the greatest impact on the quality of water bodies (FAO, 2017). On average, high-income countries treat about 70% of the wastewater that is generated, with this share falling to 38% in middle-income countries and 28% in lower-middle-income countries (WWAP, 2017a, b). In low-income countries, only 8% of industrial and municipal wastewater is subject to some type of treatment. These estimates support the fact that globally, over 80% of wastewater is discharged into the environment without adequate treatment (WWAP, 2013, 2015).

The complex chemistry of nitrogen and the wide range of possible, often overlapping, oxidation-reduction transformation reactions, conditioned by the type, amount, and bioavailability of electron donors and acceptors, under different environmental conditions, indicate that in order to give a proper definition of groundwater vulnerability and sensitivity to nitrogen pollution, it is necessary to carry out a complementary and simultaneous analysis of hydrogeological, physicochemical and microbiological parameters. In order to protect groundwater from pollution by nitrogen compounds, it is necessary to define and quantify the anthropogenic sources of nitrogen compounds and the pressures from agriculture (method of land cultivation, type and amount of fertilizers applied) on the quality of water bodies, as well as the locations and degree of influence of potential pollutants (sewage discharges, non-sanitary sewage systems, livestock).

### 2.10.8 Sanitary protection zones and water production line

#### *2.10.8.1 Groundwater self-purification processes and their effects*

The self-purifying potential of an aquifer is the potential for improving the initial water quality by filtration through an alluvial aquifer, as a function of the time and the length of filtration, the conditions in the aquifer, and the dynamics of water movement (Dimkić & Keckarević, 1990).

In any case, in addition to the convective processes, we reduce the mathematical processes to those that are represented by the dispersion term, by sorption, and by degradation.

The combined effect of these groups of processes is important. Depending on the type of event (filtration through a porous medium), certain processes may be more dominant or less dominant.

In the case of a 'batch' intake (Figure 2.53) and the filtration of dissolved matter, dispersion becomes a relatively important process as the intake of dissolved matter is shorter.

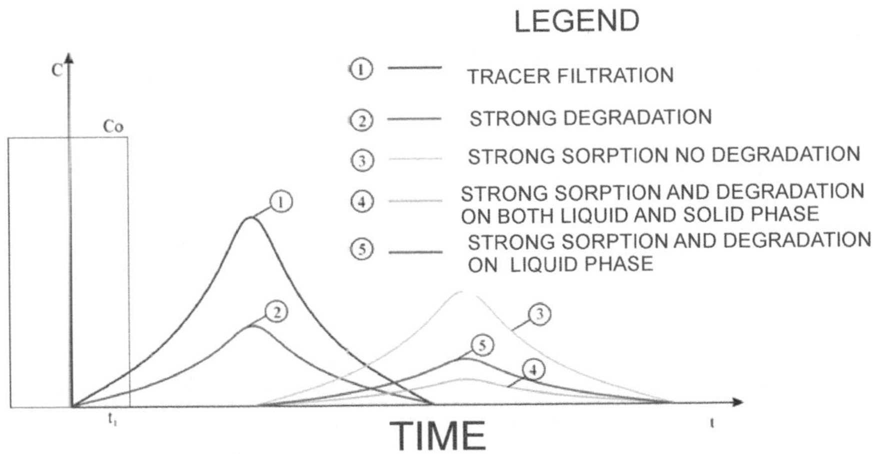


**Figure 2.53** Interaction of different processes - typical output concentration curves (at a water intake structure) in the case of a 'batch' (accidental) input of matter into the aquifer (infiltration source system) (Dimkić, 2007).

The initial conditions for the infiltration of matter are:

$$\begin{aligned} t < t_1 &\Rightarrow C_{IN} = 0 \\ t_1 \le t \le t_2 &\Rightarrow C_{IN} = C_0 \\ t > t_2 &\Rightarrow C_{IN} = 0 \end{aligned} \tag{2.3}$$

where:

$C$ – Concentration in the liquid phase at the exit of the well;

$C_{IN}$ – Input concentration of the substance into the aquifer;

$t$ – Time.

In the case of the 'continuous' intake of dissolved matter, the process of degradation, or better still a combination of the sorption and degradation processes (Figure 2.54), is of greatest importance, while dispersion loses its significance. The initial conditions for the infiltration of matter are:

$$\begin{aligned} t < t_1 &\Rightarrow C_{ul} = 0 \\ t \ge t_1 &\Rightarrow C_{ul} = C_0 \end{aligned} \tag{2.4}$$

There are many factors that influence the type and kinetics of the process. The processes are often interconnected. The aquifer is in fact a physical biochemical reactor. Infiltration springs use the effects of natural purification processes. It is the job of the engineer to optimally support and utilize this.

It is the job of researchers and designers of groundwater sources to define an aquifer's purification potential and to incorporate it properly into the concept of forming or expanding a groundwater source, while meeting certain criteria and standards.

This is not at all simple given the huge number of parameters and factors that are relevant to the water treatment process. It is often not possible to carry out all the necessary research to define all the process parameters that we consider relevant. Even if we were to define all of them, forecasting the final effects can remain a rather delicate task.

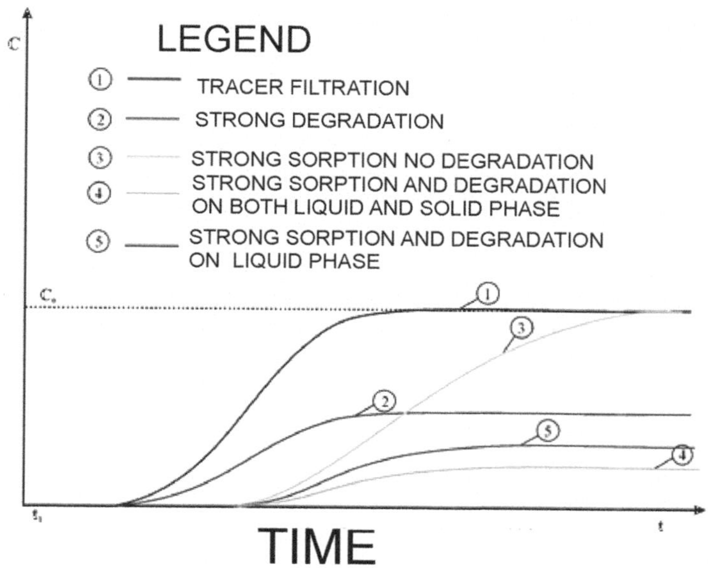


**Figure 2.54** Interaction of different processes – typical output concentration curves (at a water intake structure) in the case of 'continuous' (accidental) inflow of matter into groundwater (infiltration source system) (Dimkić, 2007).

### *2.10.8.2 Design of a filtration line and a sanitary protection zone for wells in alluvial aquifers*

When researching and defining the purification potential, as well as when forecasting the purification effects of a particular system, we need to keep various things in mind:

- Laboratory research;
- Punctual field research;
- Synthesis of field tests, analysis of existing exploitation;
- Development of macro-models and the verification of parameters of the system, based on the already observed effects:
- Forecast, by extrapolation on the adopted model.

By elaborating the stages of research, we can develop the concept for determining the production line of the source or the protection of groundwater sources in alluvial aquifers.

We need to optimize research, based on global knowledge of the process. It is necessary to select the characteristic and critical parameters for water quality and to determine the size of a source (allowable yield) accordingly. The text that follows in this section looks at the most important aspects of the purification process, namely the hydrogeological frameworks of aquifer purification by filtration, and the processes of sorption, biodegradation and dispersion.

When defining the concept of the technological line and the protection of a source, it is necessary to properly determine the location of the source, as well as the technical elements within the technological line of the source. The most important element of the technological line of a well with regard to water quality treatment is, in addition to pre-treatment and/or water treatment, the phase of water filtration through the aquifer.

The goal and task of filtration through the aquifer is to provide, together with the other parts of the technological line of the well, the required water quality and ensure its safety. All the stages of the technological line in the preparation of drinking water need to be properly conceptualized and brought into accordance in order to provide sufficient water of the required water quality, bearing in mind the economic aspect and the relation of the source to the environment.

In order to achieve this, it is necessary to define the present and future quality of raw water resources, after defining the critical (characteristic) general and individual parameters.

It is preferable to set up the well in an aerobic environment, as the aerobic conditions of a groundwater well contribute to better purifying effects of filtration through the aquifer. One of the activities consists of oxidizing the substances in raw water by means of pretreatment to an appropriate level.

The basic elements of water filtration through an alluvial aquifer are: the duration of filtration, and the time and rate of groundwater filtration from the infiltration site to the water intake structure (Dimkić, 2007).

It is necessary to define the typical substances that are relevant for conceptualizing the source. The relevant features can be:

- Mobility in a porous environment;
- (In) sorption;
- (Non) degradability;
- (Non) detrimental;
- Possibility of occurrence in high concentrations;
- The possibility of an accident, etc.

An analysis of the properties and behavior of characteristic substances during an accidental and or continuous inflow of a substance into the aquifer should result in the definition and verification of the elements of the technological line of the source, both for the filtration phase through the aquifer and for the water treatment phase (Figure 2.55).

Source protection zones are defined in more or less the same way as the elements of a filtration line within the aquifer (Figure 2.56). Instead of the elements of the filtration line, the following represent the required output values that characterize the protection zones: $T_{\min}$ – minimum filtration time in the protection zone of the source, $L_{\min}$ – minimum filtration length in the protection zone of the source, $F_{\min}$ – minimum surface area of the protection zone of the source. The elements of the system of protection zones $L_{\min}$, $t_{\min}$, and $F_{\min}$ depend on the degree of restriction when determining the sanitary protection zones of the source.

#### *2.10.8.3 Groundwater source protection zones*

The protection of a groundwater source is essential when developing any type of source. For alluvial sources, this is certainly very important because of:

- Relative aquifer vulnerabilities;
- Pressures from agglomerations, industry and agriculture;
- The importance of a source as a resource for the human community as well as for the environment.

The recharge area of the entire alluvial source must certainly be considered as one of the priorities in establishing a sustainable human-environment relationship.

Administrative studies on the protection of alluvial sources usually envisage three zones:

- Zone I: the zone of immediate protection (zone I is very narrow and protects the well itself from any adverse effects);

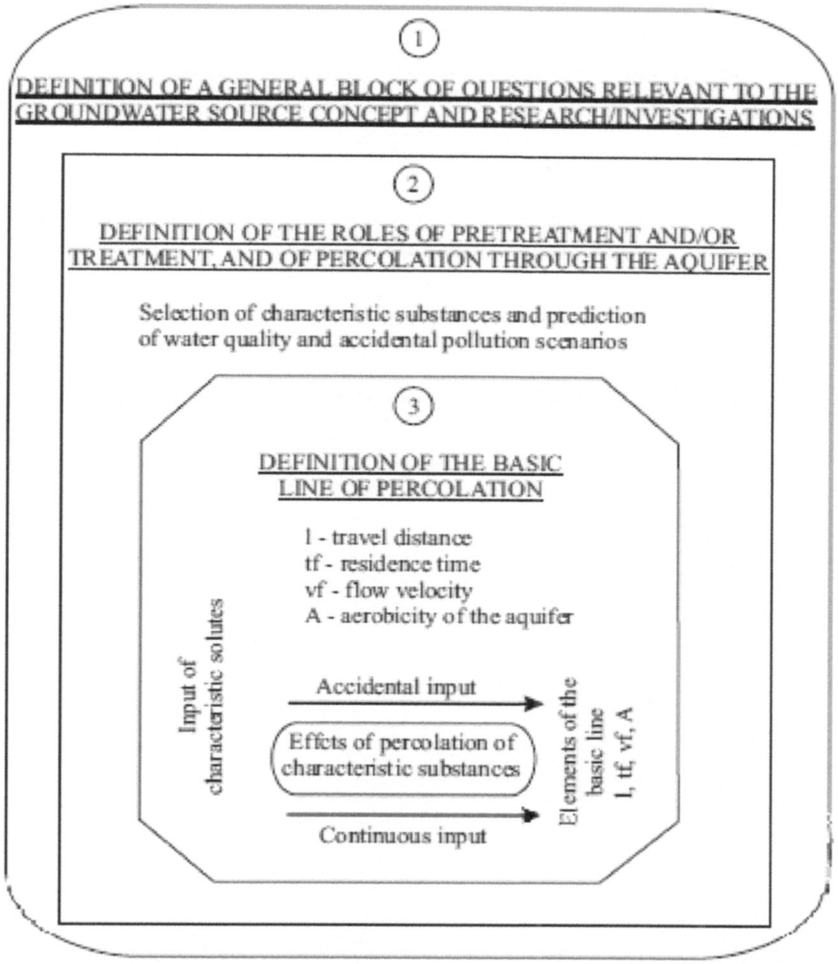


**Figure 2.55** Determination of the basic elements of a filtration line through an aquifer (Dimkić, 2007).

- Zone II: a narrower protection zone (zone II has strict restrictions and is often formed in accordance with the requirement that the filtration of water through this zone should eliminate almost all pathogenic bacteria and viruses). It is believed that 30 days of water filtration through an intergranular source should be enough to eliminate almost all pathogenic bacteria, and 50 days should suffice to eliminate viruses. Also, the rate of travel during water filtration varies along the vertical line of the sedimentary layers. It is therefore necessary to have the right approach when determining the travel time of the water in order to determine the width of the protection zone;
- Zone III: wider zone of protection a source (appropriate monitoring is done in zone III, while the contents/activities in the wider zone of protection are defined by corresponding and necessary constraints).

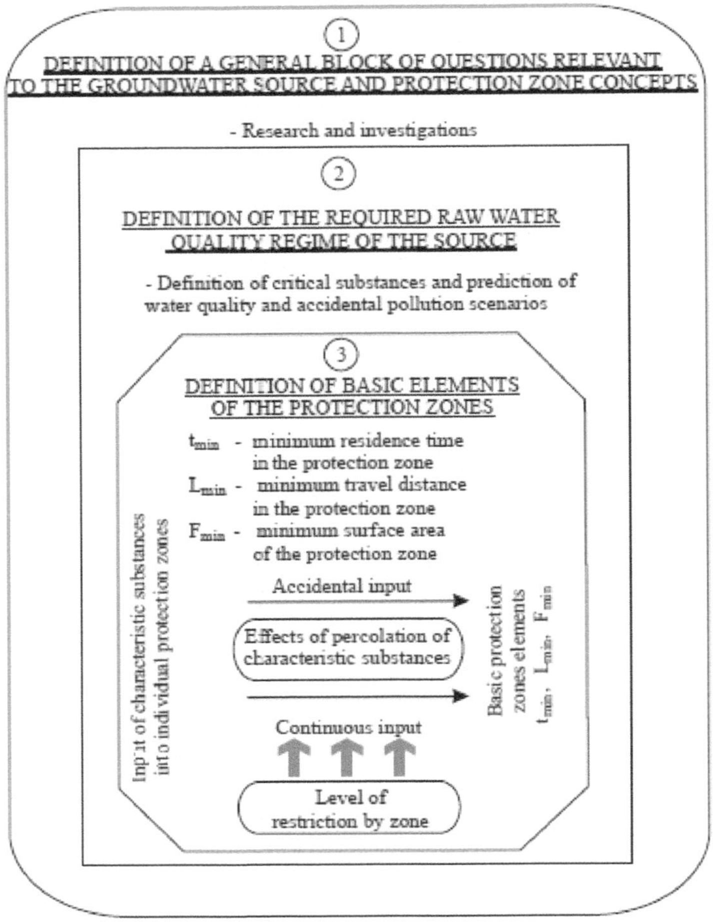


**Figure 2.56** Determination of the basic elements of the protection zones of a groundwater source (Dimkić, 2007).

The use of certain characteristic substances is considered in much the same way as is done when considering the production line of the source. The behavior of characteristic substances is important in determining constraints, especially in the wider zone of sources, and, if appropriate, in the planning of procedures in cases of incidental situations (see Figure 2.56). An overview of protection zones in several European countries is represented in Table 2.22.

This book does not address the definition of source protection zones in other aquifer types.

## 2.10.9 Limit values for groundwater quality

Groundwater resources, although 'hidden', protected by surface layers of sediments, are still exposed to anthropogenic influence. Shallow aquifers, especially those in densely populated areas, are to a greater or lesser degree exposed to anthropogenic influence. Human activity threatens the ecological and utilization status of groundwater resources. The limit values for groundwater quality parameters represent a

**Table 2.22** Overview of protection zones in several European countries (Krešić *et al.*, 2006).

| Country | Wells/Well | Internal Protection Zone | External Protection Zone | The Catchment Area |
|---|---|---|---|---|
| Austria | Internal protection zone | Protection zone, 60 days | Preventive protection zone I, the feeding zone | Preventive protection zone I, observation zone |
| Belgium (Walloon Region) | I: zone of water supply structures, 10 m | IIa: 24 h or more, > 25 days in karst: includes all priority infiltration points associated with the source<br>IIb: 50 days in pressurized aquifers, free surface aquifers: Sand: IIa + 100 m gravel: IIa +500 m Karst: IIa +100 m | III: Observation zone or catchment area | |
| Croatia | Ia: immediate source<br>Ib: immediate catchment area | IIa: 24 h strictly restricted zone<br>III: 1–10 days restriction and control zone | IV: 10–50 days – restricted protection zone, special protection zone, zone of water reserves | |
| Ireland | IA: 0–10 m | IB: 10–300 m | IC: 300–1000 m | |
| Italy | Each local government is responsible for drafting regulations | | | |
| Malta | Expropriated land in the immediate vicinity of a to the well | 500 m or more | | |
| The Netherlands | Basin of one or more wells: 60 days or >30 m for porous or karst aquifers (restriction: only water supply activities allowed | | Protected area: (a) 10 years (b) 25 years karst aquifer; (a) 2 km; (b) recharge zone; (restrictions: unacceptable-transport and storage of hazardous materials, industry, waste disposal, construction of military facilities, intensive agricultural production, mining wastewater discharge | Remote recharge zone: porous aquifer: recharge zone |
| Portugal | | High permeability medium: 20–25 m | Poorly permeable medium: 100–200 m, zones with an | |

(*Continued*)

**Table 2.22** Overview of protection zones in several European countries (Krešić *et al.*, 2006) (*Continued*).

| Country | Wells/Well | Internal Protection Zone | External Protection Zone | The Catchment Area |
|---|---|---|---|---|
| | | Poorly permeable medium: 10–20 m, zones with an impermeable to poorly permeable unsaturated layer of a thickness ≥50 m: 5–10 m | impermeable to poorly permeable unsaturated layer of a thickness ≥50 m: 20 m | |
| Slovakia | 10–50 m | 50 days or >50 m | The catchment area | |
| Spain | Immediate protection zone - 24 h | II: 50 days (porous medium); 100 days (karst) | 1 year | Remote protection zone |
| Turkey | I: 50 m (porous medium) 100 m (karst) | II: 50–250 (porous medium); 100–500 m (karst) | III: Recharge zone | The catchment area |

compromise between protecting and preserving the chemical status of groundwater resources, and the boundary between unacceptable restrictive measures for human society.

A relationship is common between the resource and the groundwater source. One resource is sometimes exploited at multiple locations – sources.

A groundwater resource is characterized by an appropriate degree of protection, which is most closely defined through appropriate limit values pertaining to the quality of the resource water. The limits for a groundwater source are in most cases more stringent than those for groundwater resources. The limit value pertaining to the quality of a groundwater source does not have to be in accordance with the quality standards pertaining to water that are to be used. It is necessary to have an understanding of the definitions and basic terms. A groundwater resource (GWR) is the broadest term and may include an aquifer, a group of groundwater aquifers, or a family of aquifers forming part of a natural framework. It is the resource that meets certain water needs.

A groundwater source (GS) encompasses a limited space and part of an aquifer used to capture groundwater to meet specific needs.

The protection zones (PZs) of springs are zones where restrictive measures are implemented to protect water quality. These zones are usually divided into a narrower protection zone, where restrictive measures are very strict, and into wider protection zones, where the restrictive measures are less pronounced.

An aquifer (Aq) is an underground layer or layers of rocks or other geological deposits of sufficient porosity and permeability for a significant flow of groundwater or for the withdrawal of significant quantities of groundwater.

A groundwater body (GB) is a defined volume of groundwater within one or more aquifers. The limit value for a single groundwater body (LGB) is the upper limit of a given parameter expressed as the concentration of a pollutant, a group of pollutants or an indicator of the pollution of a groundwater body which should not be exceeded in order to protect human health and the environment.

The limit value for one groundwater source (LGS) is the limit value (LV) of one parameter of the quality of raw water abstracted from a source, expressed as the concentration of a pollutant, a group of pollutants or an indicator of the pollution of a source.

The maximum allowable value (MAV) is the highest (most unfavorable) value of a given water quality parameter for a particular use.

Figure 2.57 shows the relationship between the limit values for a groundwater body and a groundwater source.

If the concentration of one or more admixtures is greater than the limit value for a specific purpose, that concentration must be reduced by water treatment until the conditions for the intended purpose are met. For certain natural conditions and modes of movement, water acquires certain characteristic attributes of quality. This fact is especially true for groundwater, which is, for the most part, slow moving and exposed to long and intense environmental influences.

Surface and groundwaters, along their course and during surface contact with the riverbed as well as with the processes in the ecosystem, take on natural characteristics which depend on the length and duration of the water flow. Water flows are measured for days and weeks from source to estuary. The natural mineralization of water in seas and oceans is about 35 g/l, in surface waters about 2500 mg/l, while groundwater mineralization is generally higher than that for surface waters, and considerably lower than the mineralization of the sea and oceans, which is a consequence of the long water retention time. Basic water quality occurs as a result of numerous water/rock/gas interactions (Dimkić, 2007).

The limit value of a groundwater body (LGB) should be close to the natural value of a particular parameter for a particular aquifer/groundwater body – the baseline quality (BQ). The BQ value of certain parameters or types of groundwater bodies may be higher or lower depending on the natural conditions (matrix of composition of a source and the hydrodynamic regime of groundwater).

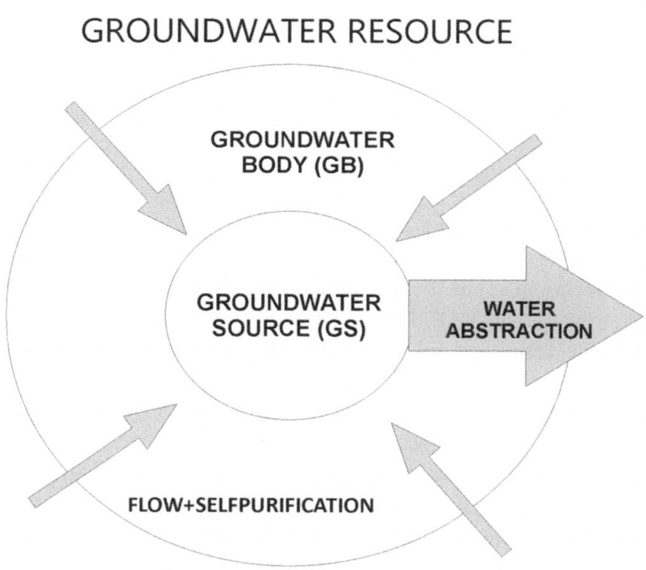


**Figure 2.57** Ratio of limit values: GBL (groundwater body limit) and GSL (groundwater source limit) (Dimkić *et al.*, 2008a, b).

Therefore, the first step when determining groundwater (GW) status is to assess the basic groundwater quality (Dimkić *et al.*, 2008b). This principle is emphasized in the EU Groundwater Directive (Annex IIA.1.d), which contains guidelines for the establishment of limit values.

According to the same directive (Annex II.A), the determination of limit values should be based on:

- The degree of interaction between groundwater and associated aquatic and water dependent terrestrial ecosystems;
- The disturbance of the real or potential justified uses or functions of groundwater;
- The hydrogeological characteristics, including data on natural concentrations.

Based on the mentioned elements, two types of criteria should be considered when establishing limit values (Working Group C – Groundwater, Activity 'Fulfillment of Status Requirements and Trends'),

The chemical status of groundwaters and limit values:

- Environmental criteria (associated aquatic ecosystems and water dependent terrestrial ecosystems);
- Criteria related to use (drinking water in the protection zones of drinking water sources and other justified uses: crop irrigation, industry, etc.).

In order to define the LV, a new term is introduced – the criteria value (CV), i.e. relevant standards for various purposes (drinking water standards, irrigation standards), or relevant environmental protection standards when considering the protection of connected surface waters or groundwater-dependent terrestrial ecosystems (Dimkić *et al.*, 2008b) need to be established.

Two types of limit values (ecological limit values and occupational limit values) can be determined based on the above criteria, although both will be based on a comparison of the CV with the BQ.

In a general sense, there are two options:

- BQ is greater than CV ($CV < BQ$); in this case the usable LV should be equal to the BQ ($LV = BQ$)
- BQ is less than CV ($CV > BQ$); in this case $BQ < LV \approx CV$.

Where:

- BQ – baseline quality;
- CV – criterion value;
- LV – limit value;
- LGB – limit value of a groundwater body

The relation $LGB = BQ + \Delta m$ pertains to a groundwater body and it shows that the limit value for the groundwater body is equal to the natural value for a given parameter to which an increase ($\Delta m$) is added due to anthropogenic influence that can be tolerated. For socio-economic and practical reasons, it is acceptable for LV to be greater than BQ. Tolerance should be acceptable from both perspectives: the degree of restriction on human activities and the impact on groundwater (Dimkić *et al.*, 2008b).

The limit value for a groundwater source may be more stringent than the limit value for a groundwater body (i.e. lower or equal to): $LGS \leq LGB$. The transition from LGB to LGS is due to the effects of self-purification of groundwater during filtration.

The GSL aspires to have the natural values of a given parameter for a given groundwater body: $LGS = BQ$. The LGS (or BQ) may be higher than the prescribed maximum value of a specific parameter for a particular use (maximum allowed value – MAV): $LGS \geq MAV$. If there is any difference between the LGS and MAV, it is eliminated through the technological process of water preparation (Dimkić *et al.*, 2008b).

## 2.11 EXAMPLE OF THE BELGRADE ALLUVIAL GROUNDWATER SOURCE

Water, food and energy are closely interlinked. Population growth on our planet leads to increasing disputes pertaining to water demand, the availability of water resources, and the cost of providing water. Alluviums are water resources from which mankind obtains about one-third of its drinking water supply. Their proximity to users is often an advantage. River water, filtering through water wells, strives to achieve a quality akin to the baseline quality of a given groundwater resource (unaffected by man).

Large cities require large amounts of water. Their water demand often exceeds the available water resources, resulting in the use of resources that are of lower quality, more distant and/or costlier. Since cities often develop in river valleys, urban expansion often collides with, occupies and devastates alluvial resources. The Belgrade Groundwater Source is a 50 km-long alluvial drinking water source on the banks of the Sava River. The alluvial aquifer sediments are polycyclic and generally reflect alternating ice ages. The Belgrade Groundwater Source contains about one hundred horizontal collector wells and several dozen tube wells. Its total capacity is currently less than 3 $m^3/s$. On the one hand, its capacity is limited by the composition of the hydrogeological strata, especially the river-aquifer contact, with the highest achievable capacity of the source being approximately 5 $m^3/s$. On the other hand, corrosion and iron incrustation destroy radial well laterals. Because of insufficient maintenance and replacement of the laterals, the groundwater abstraction rate has dropped from 5 $m^3/s$ to less than 3 $m^3/s$ at present. This reduced capacity is a major short-fall, given that the quality of the groundwater is very high due to the effect of bank filtration and the fact that the aquifer contains polycyclic deposits of clastic material.

There is also a conflict between groundwater source protection and the city's attempt to occupy valuable protective zones. The stripping of undesirable substances dissolved in water, the destruction and maintenance of radial well laterals, and source protection zoning are crucial issues and challenges for both agglomerations and water resource conservation.

Today an average of 2.5–3 $m^3/s$ of groundwater is tapped from the alluvial aquifer of the Sava River. So far, 99 lateral drain wells have been constructed and are located in 2 wells along the Sava River (Figure 2.58). Some 50 tubular wells have also been constructed.

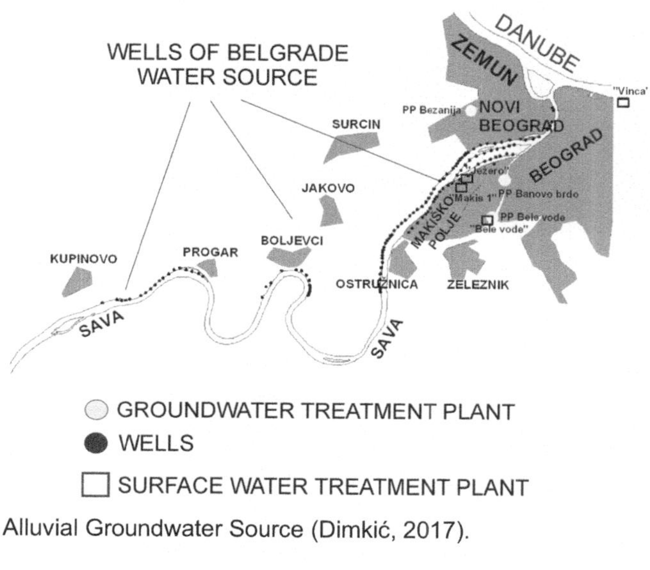


**Figure 2.58** Belgrade Alluvial Groundwater Source (Dimkić, 2017).

The Belgrade source wells are located in the alluvial plain of the Sava River and they exploit groundwater from its sediments. These sediments, in the geological sense, belong to Quaternary formations. In the base of the Quaternary sediment there are mostly sandy marl clays and marls. Although in some smaller areas, the aquifer is made of limestone and flysch, the aquifer consists of gravel, sandy gravels, gravelly sand and sporadically tiny less permeable layers (Dimkić, 2017).

The gravel, sandy gravels and gravelly sand sporadically alternate towards the terrain surface and have intergranular porosity. The water-bearing sediment is up to 25 m thick. Well laterals are located at a depth of about 20 meters (Figure 2.59).

The hydraulic conductivity values of the water-bearing layer generally range from $10^{-3}$ to $10^{-5}$ m/s. This value increases with depth, as the lowest sediments (sandy gravels and gravels) have the highest permeability.

Gravel-sandy series sediments rarely break out onto the surface of the terrain. They are mainly covered with clay, clayish, dusty and humus deposits, whose thickness ranges from 1 to 9 m, and is mostly between 3 and 5 m. Their permeability is poor but they cannot be considered as isolators. In zones with larger thicknesses, they provide good aquifer protection from pollution from the surface of the terrain.

## 2.11.1 Groundwater quality

The waters from the River Sava, after their riverbed and bank filtration, make up 80–90% of the water that is tapped in the Belgrade waterworks wells. The quality parameters are monitored by the Hydrometeorological Service of Serbia at hydraulic stations located along the river flow. The analysis includes a total of 64 quality parameters, among which are heavy metals, organic micropollutants, mineral oils and the total beta activity. The amount of data collected varies for different groups of parameters that are examined. Most samples are

**Figure 2.59** Schematic of a typical longitudinal profile with two lateral wells (Dimkić, 2019).

analyzed for the basic parameters while a smaller number are analyzed for volatile phenols and mineral oils, heavy metals and pesticides.

The Belgrade waterworks' physicochemical laboratory also checks the quality of raw and delivered water. Due to the self-purification processes in an alluvial aquifer and the natural polycyclic sedimentation in the aquifer skeleton media, as well as because of the way the wells have been built, the quality of the trapped water is excellent. Dissolved carbon concentration from the Sava River decreases by a factor of two by the time it reaches the abstraction well. Reduction in the concentration of micropollutants is also large (pharmaceuticals, pesticides, different hydrocarbons etc.).

## 2.11.2 Water source capacity

We can distinguish three periods in the development of a source pertaining to well capacities (see Figure 2.60).

- Period 1: Water saturation of the aquifer and soil. High capacity lateral wells extracting static groundwater reserves.
- Period 2: The dominant process of riverbed clogging and the decreasing capacity along the entire 50 km of the Belgrade alluvial water source.
- Period 3: After the stabilization of the riverbed clogging, the overall hypothetical natural source capacity is about 5 $m^3/s$.

However, the extraction and delivery of water amounts to less than 3 $m^3/s$. This is mostly due to the destruction of old well laterals that corrode and to the clogging of well laterals.

The wells are not sufficiently maintained. Usually 10–15 wells are renovated per year. The installation of driven drains is relatively rare (on average, drains are installed on less than one well in two years).

The overall effect of wells ageing, accompanied by factors which are a result inadequate maintenance, is the loss of the ability to produce over 100 l/s of water on a continuous basis. Therefore, there is a drastic

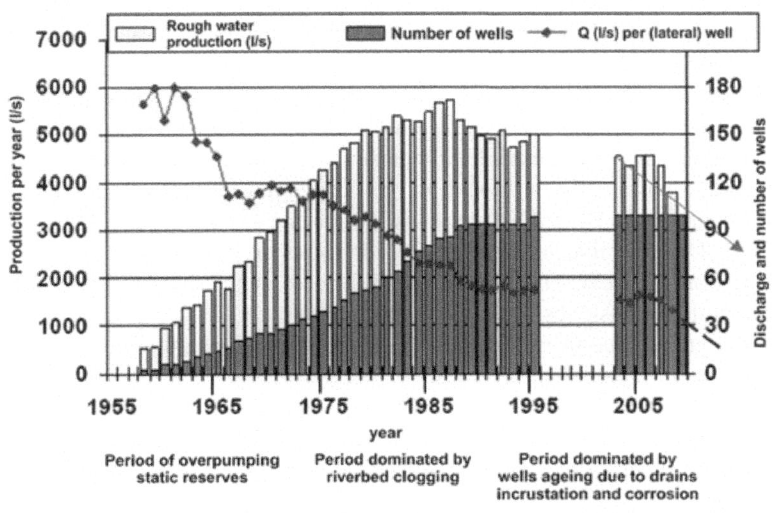


**Figure 2.60** Changing groundwater source exploitation, number of wells and average well yields (Dimkić *et al.*, 2011d).

decrease in the possibility of producing water and sustainability of the source. The maintenance of wells is not sufficient.

## 2.11.3 The city and its groundwater source

Groundwater sources formed in alluvial environments are traditional drinking water sources in Serbia. The benefits of using groundwater resources through bank filtration are multifold, both from the perspective of the generally sound aquifer recharge available from the river and because of the generally favorable water quality (i.e. utilization of the aquifer's purification potential). However, throughout history, many human settlements and cities have tended to occupy the vicinity of rivers or even the very riverbanks, benefiting from the convenient locations (water supply, trade, transportation etc.). As riverside cities expanded, riparian zones and existing water supply sources (e.g. in Belgrade and Novi Sad) came under increasing pressure to ensure the best possible protection of their water wells. In such conditions the importance of optimal and effective protection of water source continues to grow.

The size, position and significance of the Belgrade Groundwater Source, its surroundings and the users of the area that have been potentially endangering it for several decades, as well as the plans pertaining to future investments in the source protection zones, make the issues regarding the future development of the city and relevant source interests very current.

We have a growing demand for water due to population growth; immigration; the connecting of new municipalities (Mladenovac, Eastern Srem, etc.); growing industry demand; available reserves etc.

In view of this, it is important to bear in mind the high quality of the water abstracted from the Belgrade source area, as well as the increased safety of the drinking water supply for the needs of the city.

The general assessment is that 11–12 $m^3/s$ of water from the Belgrade waterworks as a regional source is needed for a long-term stable water supply solution.

The current capacity of the structures for capturing, preparing and distributing the water is:

- Groundwater source $<3\ m^3/s$;
- Surface water of the Sava River $<5\ m^3/s$.

The total capacity in critical periods when the need for water is the greatest, is at the edge of sufficiency, and scarcity occurs in some parts of the system.

This points to the need to restore the capacity of the existing groundwater source to its natural capacity (5 $m^3/s$). It is estimated that by replacing the drains on 5 wells and regenerating about 15 wells annually, it would be possible to return the spring exploitation capacity to 5 $m^3/s$ in about 15 years.

The location of Zidine has also been looked into as a potential infiltration source as well as locations upstream of the left bank of the River Sava (Klenak – Jarak), and the Kovin – Dubovac location on the left bank of the river Danube, that have been considered as sources that could be formed using the method of bank filtration. However, the formation of sources at these locations could be incompatible with the coal exploitation in the slopes of the alluvial segments.

City expansion, the great value of riparian land and good communication possibilities has put a lot of pressure on groundwater source protection zones.

There are various installations in the narrow protection zone and some unsuitable activities. For example petroleum products are stored and transferred, there are chemical plants, textile factories, and metalworking machinery production, and there is some food industry, as well as municipal wastewater and agricultural production (Figure 2.61).

In addition to the existing pollutants, there are significant other potential pressures on the source such as traffic, construction work, recreation, etc.

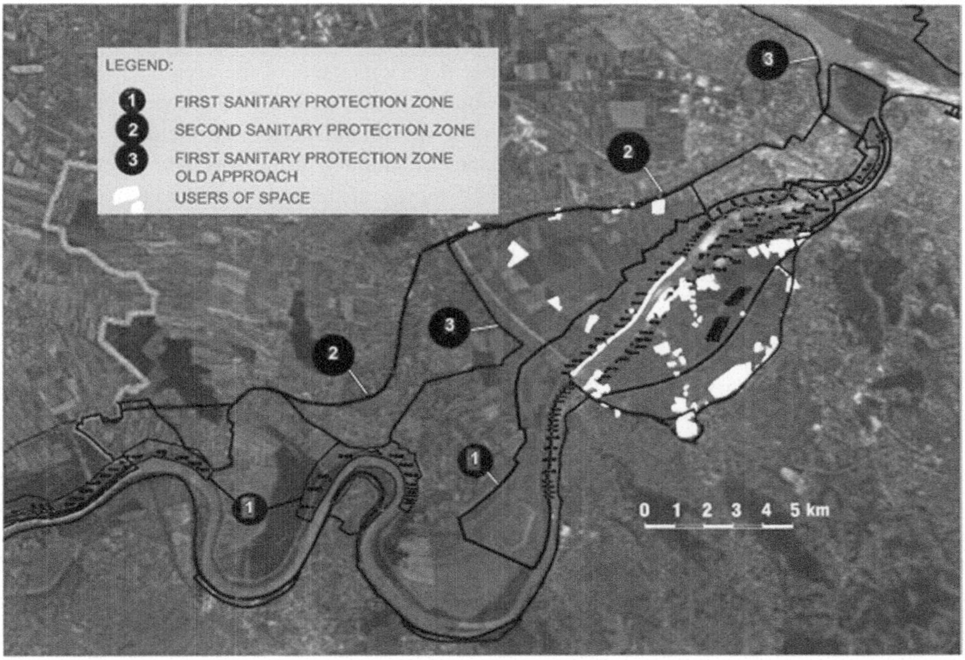


**Figure 2.61** Water pollutant pressure on the Belgrade Groundwater Source (Dimkić, 2019).

However, the very important ability of a source to self-protect (overlaying layer, self-purifying effects of filtration, strong polycyclicity of alluvial sedimentation) should be kept in mind. As a result, the groundwater drawn from the wells is still of a very good quality. The water quality is worsened in certain locations but remains within allowed values. Nevertheless, it is necessary to periodically carry out a detailed review of the quality of raw water in all water intake structures.

### 2.11.4 Source protection zones

The legal basis for the protection of groundwater resources is primarily provided in the Water Law of the Republic of Serbia (WL).

The previously valid sanitary protection zones of the Belgrade Groundwater Source were defined almost 30 years ago (decisions from 1986). They consisted of the 50 km-long Sava riverbank, upstream from the mouth of the river and covered an area of over 40,000 ha. The sanitary protection zones of that time had included the planned but never implemented structures of the source (infiltration sources 'Makis' and the 'lower field') and the existing and planned wells. They were, in theory, very restrictive in terms of how the space could be used but that became very uncertain in actual practice. Over time, the practice proved to be very limiting and made the legal planning of the city's development very difficult.

It became extremely difficult, on such a large area, to control and monitor the content and illegal construction that took place which was incompatible with the protection of the source and the environment. However, legal regulations on the protection of groundwater sources worldwide and in our country, as well as a better understanding of the importance of this protection, has resulted in more attention being paid to the coexistence of a source and the environment. The development of urban units often disregarded the interests of a source and its survival. Nowadays, approaches to preserving sources and the criteria for defining sanitary protection zones have been adapted to modern knowledge both in terms of preserving the sources and in terms of the coexistence of various users of the area.

New source protection zones of groundwater sources were defined to provide protection against intentional or accidental pollution and other influences that could affect the yield of a source and the natural composition of water at the source. Three zones of sanitary protection have been defined: zone I is the area right next to the wells, and zones II and III represent the area 'from which water reaches the water intake structures' and is defined by the water retention time, or the length of the underground water path to wells. The source is in the alluvial environment, with intergranular aquifer porosity, and the minimum water travel time depends on whether the 'underground water is under pressure' or of 'free level': in zone II it is 50 m or 50 days, and in zone III it is 500 m, or 200 days. An analysis is also carried out on any possible impact from the surrounding terrain (hinterland) and existing infrastructure (main roads, industrial zones etc.). The maintenance of the source protection zones is to be achieved through restrictive measures regarding construction work and other activities in order not to 'endanger the quality of water at the source'. This will be accompanied by proposed measures to be implemented in order to establish appropriate water quality monitoring.

This is especially important for future infrastructural projects such as the future Underground and Tesla City, as well as for large sports facilities.

Effort has been made to reduce the clash between the city's tendency to spread to the banks of the Sava River and the need to protect sources of water. Regulations on the Protection Zones of Water Sources of the City of Belgrade were adopted in 2014. These regulations reduced the conflict between the city and the source in two ways:

- They significantly reduced the II and III source protection zone in accordance with appropriate professional and scientific principles, fully respecting the need to protect the quality of the water captured by wells. This increased the prospects for reducing conflicts and respecting protection zones;
- The II protection zone is no longer continuous through the entire length of the Sava and its source. In places where it was feasible, passages (interruptions) were left for the possible communication between the riverbanks and remote riparian areas. The intention was to enable communication between the riverbanks and the inner-city area.

In the future, issues related to the construction of large infrastructural facilities in the riparian area and their impact on the source will certainly have to be considered with the utmost care and responsibility.

Nowadays, the amount of water obtained from the groundwater source is less than 3 $m^3$/s. The total natural (former exploitation) capacity is about 5 $m^3$/s. In order to recover this capacity, it is necessary to permanently invest in the replacement of well drains and well regeneration. It was estimated that by replacing drains of about 5 wells annually and by regeneration of 15 wells in about 15 years the desired working capacity of the source of around 5 $m^3$/s could be reached.

The needs of the city in terms of water supply and water supply security must be achieved by:

- Maintaining the surface water source capacity at about 5 $m^3$/s;
- Maintaining and increasing the groundwater source capacity to about 5 $m^3$/s;
- Forming new groundwater source capacities of about 1.5–2 $m^3$/s (infiltration source or remote sources using the riparian filtration method).

It is also necessary to clearly define the monitoring of the quality, quantity and operation of water intake structures as well as of the infrastructure at the source. Periodically, it is necessary to analyze the impact of the city on the source and that of the source on the city.

To solve the problem of the Belgrade water supply, it is necessary to do some serious work in the coming period. Among other measures (see Chapter 3), it is necessary to increase the current operating capacity of the groundwater sources from less than 3 to 5 $m^3/s$. This should be achieved by efficiently regenerating the wells and replacing the used-up drains. In addition, other very important measures such as the monitoring, study, design, maintenance and formation of new source capacities are also needed.

Adaptive groundwater resources management should ensure the ability to respond effectively to different challenges: social, economic, climate, political, etc. This implies that in addition to the legally mandatory study of the reserves and monitoring of a particular water body, complex studies should be completed and solutions found to the very important groundwater resources issues and water in general.

## REFERENCES

Abramov C. K. (1952). Method of Calculation and Selection of Filters for Drilled Wells (in Russian), Moscow.

Bai Z. G., Dent D. L., Olsson L. and Schaepman M. E. (2008). Proxy global assessment of land degradation. *Soil Use and Management*, **24**(3), 223–234.

Bekins B. A., Godsy E. M. and Goerlitz D. F. (1993). Modeling steady-state methanogenic degradation of phenols in groundwater. *Journal of Contaminant Hydrology*, **14**(3–4), 279–294.

Borighem G. and Vereecken J. (1979). Study of the Biodegradation of Phenol in River Water. *Ecological Modelling*, **4**(1), 51–59.

Bradley P. M., Barber L. B., Duris J. W., Foreman W. T., Furlong E. T., Hubbard L. E., Hutchinson K. J., Keefe S. H. and Kolpin D. W. (2014). Riverbank filtration potential of pharmaceuticals in a wastewater-impacted stream. *Environmental Pollution*, **193**, 173–180, doi: 10.1016/j.envpol.2014.06.028

Brown L. R. (2002). Water deficits growing in many countries. Water shortage may cause food shortages. www.greatlakesdirectory.org (accessed February 2013).

Bruinsma J. (2009). The Resource Outlook to 2050: By How Much do Land, Water and Crop Yields Need to Increase by 2050? Prepared for the FAO Expert Meeting on 'How to Feed the World in 2050', 24–26 June 2009, Rome.

Burek P., Satoh Y., Fischer G., Kahil M. T., Scherzer A., Tramberend S., Nava L. F., and Wada Y. (2016). Water Futures and Solution: Fast Track Initiative (Final Report). IIASA Working Paper (International Institute for Applied Systems Analysis (IIASA), Laxenburg, Austria).

Burgin A. J. and Hamilton S. K. (2007). Have we overemphasized the role of denitrification in aquatic ecosystems? A review of nitrate removal pathways. *Frontiers in Ecology and the Environment*, **5**(2), 89–96.

Cahn M. D., Bouldin D. R. and Cravo M. S. (1992). Nitrate sorption in acid soil. *Plant and Soil*, **143**, 179–183.

Cambridge University. (2012). Managing the risk of extreme events and disasters to advance climate change adaptation. Special Report of the Intergovernmental Panel on climate change. Cambridge.

Chapagain A. K. and Hoekstra A. Y. (2004). Water footprints of nations. Main Report. Vol. 1. UNESCO-IHE, Delft, the Netherlands.

Cistin J. (1965). Some Problems of Deformation of Mechanical Filters (in Slovakian). Vodohospod, Casopis 2.

Cline W. (2007). Global Warming and Agriculture: Impact Estimates by Country. Center for Global Development and Peterson Institute for International Economics, Washington DC.

Cullimore R. (1999). Microbiology of Well Biofouling, CRC Press, Boca Raton.

Cullimore R. (2008). Practical Manual of Groundwater Microbiology, 2nd edn. Taylor & Francis, Boca Raton.

Cullimore R. and Mansuy N. (1987). A screen arc model well to simulate iron bacterial fouling. *Journal of Microbiological Methods*, **7**, 225–232.

Dimkić M. (2007). Selfpurification effects of groundwater filtration 'Zadužbina Andrejević', ISBN:978 86 7244 629 6, Beograd, Srbija, 252.

Dimkić M. (2010). Water management and risks in Serbia. IAWD/IAWR– Symposium Water Resources Protection and Risk Managment in River Basins, 3–4 November 2010, Vienna, Austria, Proceedings, pp. 28–29.

Dimkić M. (2017). Several aspects of the influence of ice ages in Serbia on the formation of the Sava alluvion and water source processes. Conference organized by Serbian Academy of Sciences and Arts, Earth's Climate Change: Science and Impacts, 11–13 October 2017, Belgrade.

Dimkić M. (2019). Urban Development vs Alluvial Groundwater Sources – Belgrade Example. e-Future of Cities - Between Exponential Growth of New Technologies and the Concept of a Human City, October 2019, Belgrade.

Dimkić M. and Keckarević D. (1990). Quantification of the Self-purifying Potential of an Aquifer as a Physical and Biochemical Reactor. 10th Symposium of the Yugoslav Hydraulic Investigations Society, Sarajevo, pp. 467–479.

Dimkić M. and Pušić M. (2008). Recommendations for water well design taking into account iron clogging, based on experience gained at the Belgrade Groundwater Source (in Serbian). *Civil Engineering Calendar*, **40**, 430–496.

Dimkić M. and Milovanović M. (2009). Elements of the Global Approach to Groundwater Management, Monograph 'Management of Serbia's Water Resources **'09'**. Jaroslav Černi Water Institute, Belgrade, 1–21, ISBN 978-86-82565-24-6.

Dimkić M. and Milovanović M. (2014). Water Management in Transition Countries. Milankovitch Anniversary UNESCO Symposium 'Water Management in Transition Countries as Impacted by Climate Change and Other Global Changes, Lessons from Paleoclimate, and Regional Issues', 3–5 September 2014, Belgrade, Serbia, Publishers: Jaroslav Černi Water Institute and Serbian Academy of Sciences and Arts, M. Dimkić, F. Mesinger and A. Berger (eds), Book of Abstracts, ISBN 978- 86-82565-42-0, pp. 16–29.

Dimkić M. and Pušić M. (2014). Correlation between entrance velocities, increase in local hydraulic resistances and redox potential of alluvial groundwater source. *Water Research and Management*, **4**(1), 3–23.

Dimkić M., Pavlović D., Soro A., Radenković Z. and Filipović R. (1997). Revitalization and Upgrading of the Mediana Groundwater Source in the City of Niš. Conference on Groundwater as a Source of Water Supply, Association for Water Technology and Sanitary Engineering, Belgrade, Serbia, pp. 493–515.

Dimkić M., Vrvić M., Keckarević D. and Brčeski I. (1986–1989). Istraživanja degradacije fenola u oksičnim akviferima u periodu 1986–1993. Fondovska dokumentacija.

Dimkić M., Vrvić M., Keckarević D. and Brčeski I. (1991). Definisanje prečišćavajućih potencijala izdani na izvorištu podzemnih voda 'Žičko Polje' – Kraljevo, Konferencija o aktuelnim problemima zaštite voda 'Zaštita voda '91', 20.-22.05.1991, Neum, str. 30–36.

Dimkić M., Pušić M. and Ranković V. (2003). Behavior of Phenols and Discussion of Aquifer Dispersivity Based on In Situ Tests. Jaroslav Černi Institute, Belgrade, Serbia.

Dimkić M., Taušanović V., Pušić M., Boreli-Zdravković Đ., Durić D., Slimak T., Petković A., Obradović V. and Babić R. (2007). Belgrade groundwater source, condition and possible development directions. *Water Practice and Technology*, **2**(3), wpt2007059. https://iwaponline.com/wpt/article-abstract/2/3/wpt2007059/20868/BELGRADE-GROUNDWATER-SOURCE-CONDITION-AND-POSSIBLE?redirectedFrom=fulltext

Dimkić M., Kavanaugh M. and Brauch H. J. (2008a) Groundwater Management in Large River Basins. IWA Publishing, London, UK, ISBN 1843391902.

Dimkić M., Đurić D. and Milovanović M. (2008b). Relation of limit values of groundwater quality parameters to protected zones for intergranular aquifers. *Water and Sanitation Techniques*, **38**(3), 19–36, ISSN 0350-5049.

Dimkić M., Milovanović M. and Dimkić D. (2011). Sustainable and adaptive water management: case study of water management in Serbia. *Water Research and Management*, **1**(4), 9–19, ISSN 2217-5237.

Dimkić M., Pusić M., Vidovic D., Petković A. and Boreli-Zdravković D. (2011a). Several natural indicators of radial well ageing at the Belgrade groundwater source, part 1. *Water Science and Technology*, **63**(11), 2560–2566, doi: 10.2166/wst.2011.590

Dimkić M., Pušić M., Vidović D., Filipović N., Isailović V. and Majkić B. (2011b). Numerical model assessment of radial-well aging. *ASCE Journal of Computing in Civil Engineering*, **25**(1), 43–49, doi: 10.1061/(ASCE)CP.1943-5487.0000063

Dimkić M., Đurić D., Milovanović M., Laušević M., Jevtić G. and Petković A. (2011c). Natural attenuation of emerging pharmaceuticals by bank filtration in addressing regional groundwater management issues. *Water Research and Management*, **1**(1), 29–45, ISSN 2217-5237.

Dimkić M., Pušić M., Petković A. and Boreli-Zdravković Đ. J. (2011d). Several natural indicators of radial well ageing at the Belgrade groundwater source, part 1. *Water Science and Technology*, **63**(11), 2560–2566.

Dimkić M., Pušić M., Obradović V. and Đurić D. (2011e). Several natural indicators of radial well ageing at the Belgrade Groundwater Source. Part **2**. *Water Science and Technology*, **63**(11), 2567–2574.

Dimkić M., Pušić M., Obradović V. and Kovačević S. (2012). The effect of certain biochemical factors on well clogging under suboxic and mildly anoxic conditions. *Water Science and Technology*, **65**(12), 2206–2212, doi: 10. 2166/wst.2012.129, ISSN 0273-1223

Dimkić M., Milovanovic M. and Pavlovic R. (2013). Water for sustainable development and adaptation to climate change. In: 'Free Flow – Reaching Water Security Through Cooperation', J. Griffiths and R. Lambert (eds.), UNESCO, Paris, pp. 319–321, ISBN 978-92-3-104256-0.

Dimkić M., Kovačević S., Pušić M., Dotlić M. and Radišić M. (2016). Mathematical Modeling of Sorption and Degradation of pharmaceuticals: Case study of Belgrade Groundwater Source well Rb 16. IWA Specialist Groundwater Conference, Institute for Water Management 'Jaroslav Černi', 09–11 June, Republic of Serbia, Belgrade, ISBN 978-86-82565-46-8.

Dimkić M., Kovačević S., Pušić M. and Dotlić M. (2017). Modeling of sorption and degradation of selected pharmaceuticals: case study of Belgrade Groundwater Source. *Geološki Anali Balkanskog Poluostrva*, **78**(1), 47–59, doi: 10.2298/GABP1778047D, http://gabp.rgf.bg.ac.rs/index.php/gabp/article/view/13

European Environment Agency. (2012). Towards efficient use of water resources in Europe.

European Environment Agency, EEA. (2018). Nutrients in freshwater in Europe. https://www.eea.europa.eu/data-and-maps/indicators/nutrients-in-freshwater/nutrients-in-freshwater-assessment-published-6 (accessed March 2019).

EUROSTAT. (2019). Eurostat, the statistical office of the European Union, Government finance statistics - quarterly data, National debt in EU countries in the 1st quarter 2019 in relation to gross domestic product (GDP).

Falkenmark M. (1997). Meeting water requirements of an expanding world population. *Philosophical Transactions of the Royal Society of London. Series B, Biological Sciences*, **352**, 929–936.

Falkenmark M. (2013). Growing water scarcity in agriculture: future challenge to global water security. *Philosophical Transactions of the Royal Society A: Mathematical, Physical and Engineering Sciences*, **371**, 20120410. http://dx. doi.org/10.1098/rsta.2012.041

FAO (Food and Agriculture Organization of the United Nations). (2005). Irrigation in Africa in Figures: AQUASTAT Survey – 2005. FAO Water Report No. 29. Rome, FAO. ftp://ftp.fao.org/agl/aglw/docs/wr29_eng.pdf

FAO (Food and Agriculture Organization of the United Nations). (2008a). Sustainable Land Management. FAO, Rome. http://www.un.org/esa/sustdev/csd/csd16/documents/fao_factsheet/land.pdf

FAO (Food and Agriculture Organization of the United Nations). (2008b). Mapping poverty, water and agriculture in sub-Saharan Africa. In: Water and the Rural Poor: Interventions for Improving Livelihoods in sub-Saharan Africa, J.-M. Faurès, G. Santini (eds.), FAO, Rome. ftp://ftp.fao.org/docrep/fao/010/i0132e/i0132e03a.pdf

FAO (Food and Agriculture Organization of the United Nations). (2011a). The State of the World's Land and Water Resources: Managing Systems at Risk. Earthscan, London.

FAO (Food and Agriculture Organization of the United Nations). (2011b). AQUASTAT Online Database. FAO, Rome. http://www.fao.org/nr/water/aquastat/data/query/index.html

FAO. (2016). FAOSTAT. Database. Food and Agriculture Organization of the United Nations (FAO), Rome. Available at http://faostat3.fao.org/browse/R/RP/E (accessed July 2016).

FAO. (2017). Water Pollution From Agriculture: A Global Review. Food and Agriculture Organization of the United Nations (FAO), Rome. Available at http://www.fao.org/3/a-i7754e.pdf (accessed January 2020).

FAO and WWC. (2015). Towards a Water and Food Secure Future: Critical Perspectives for Policy-Makers, Rome.

Fersman A. E. (1923). Chemical Elements of the Earth and the Cosmos, Zemlya Vselennaya, No. 3, 43–44.

FOCP. (2009). Federal Office for Civil Protection.

François ARAGO. (1854) Astronomie populaire, L. Guérin et Cie (Paris), Première édition de 1854 à, 1 vol. (XII-559p.)

Freeze R. A. and Cherry J. C. (1979). Groundwater. Prentice Hall, New York.

Gavrilko V. M. and Alekseev V. S. S. (1985). Water wells screen. Publishing House Nedra, Moscow, pp.300–304

Ghaffour N., Missimer T. M. and Amy G. L. (2012). 'Technical review and evaluation of the economics of water desalination: current and future challenges for better water supply sustainability'. Water Desalination and Reuse Center KAUST.

Gleick P. H. (1993). Water in crisis: a guide to the world's fresh water resource. Oxford University Press, New York, Oxford, p. 473. ISBN: 0-19-507627-3.

Gleick P. H. (1996). Basic water requirements for human activities: meeting basic needs. *Water International*, **21**(2), 83–92.

Gleick P., Palaniappan M., Morikawa M. and Morrison J. (2009). The World's Water 2008-2009: The Biennial Report on Freshwater Resources, Bibliovault OAI Repository, the University of Chicago Press.

Gleeson T., Jasechko K., Luijendijk S. and Cardenas E. (2016). The global volume and distribution of modern groundwater. *Nature Geoscience*, **9**, 161–167. doi: 10.1038/ngeo2590.

Global Water Intelligence Magazine. (2017). The Global Value of Water and The Global Water Tariff Survey.

Gohari A., Eslamian S., Mirchi A., Abedi-Koupaei J., Massah Bavani A. and Madani K. (2013) Water transfer as a solution to water shortage: a fix that can Backfire. *Journal of Hydrology*, **491**, 23–39.

Guha-Sapir D. (2011). Annual Disaster Statistical Review 2010: The Numbers and Trends. CRED, Brussels. http://www.undp.org.cu/crmi/docs/cred-annualdisstats2010-rt-2011-en.pdf

Gupta J. and van der Zaag P. (2008). Interbasin water transfers and integrated water resources management: where engineering, science and politics interlock. *Physics and Chemistry of the Earth, Parts A/B/C*, **33**, 28–40.

GWI, IDA, Desalination Yearbook 2016–2017. (2016) Water Desalination Report, Global Water Intelligence.

Hamilton D. and Crossley S. (2004). Pesticide Residues in Food and Drinking Water: Human Exposure and Risks. John Wiley & Sons, Wiley.

Hiscock K. M. and Grischeck T. (2002). Attenuation of groundwater pollution by bank filtration. *Journal of Hydrology*, **266**, 139–144. doi: 10.1016/S0022-1694(02)00158-0

Hoekstra A. Y. and Mekonnen M. M. (2011). The water footprint of humanity. *PNAS Direct Submission*, **109**(9), 3232–3237.

Hoekstra A. Y., Chapagain A. K., Aldaya M. M. and Mekonnen M. M. (2011). The Water Footprint Assessment Manual: Setting the Global Standard. Earthscan, London and Washington, DC. ISBN: 978-1-84971-279-8.

Hoffman A. R. (2011). The Connection: Water Supply and Energy Reserves. US Department of Energy, Washington DC. http://waterindustry.org/Water-Facts/world-water-6.htm (accessed 10 May 2011).

Houben G. J. (2003). Iron oxide incrustations in wells. Part 1: Genesis, mineralogy and geochemistry. *Applied Geochemistry*, **18**(6), 927–939.

Houben G. and Treskatis C. H. (2007). Water Well Rehabilitation and Reconstruction. McGraw-Hill, New York.

IEA (International Energy Agency). (2009). World Energy Outlook 2009. IEA, Paris.

IGRAC. (2014). International Groundwater Resource Assesment Centre, Groundwater Development Stress Map, https://www.un-igrac.org/resource/groundwater-development-stress-map

IPCC. (2007a). Climate Change 2007: Synthesis Report. Contribution of Working Groups I, II and III to the Fourth Assessment Report of the Intergovernmental Panel on Climate Change. In: Core Writing Team, R. K. Pachauri, A. Reisinger (eds.), IPCC, Geneva, Switzerland.

IPCC. (2007b). Assessment Reports Working Group II: Impacts, Adaptation and Vulnerability. Intergovernmental Panel on Climate Change (IPCC). (2014). AR5 Synthesis Report, https://www.ipcc.ch/site/assets/uploads/2018/05/SYR_AR5_FINAL_full_wcover.pdf

Jägermeyr J., Gerten D., Schaphoff S., Heinke J., Lucht W. and Rockström J. (2016). Integrated crop water management might sustainably halve the global food gap. *Environmental Research Letters*, **11**, 025002.

Jalava M., Kummu M., Porkka M., Siebert S. and Varis O. (2014). Diet change – a solution to reduce water use? *Environmental Research Letters*, **9**, 074016.

Jalava M., Guillaume J. H. A., Kummu M., Porkka M., Siebert S. and Varis O. (2016). Diet change and food loss reduction: what is their combined impact on global water use and scarcity? *Earth's Future*, **4**, 62–78.

Johnson E. E. (1975). Ground Water and Wells: A Reference Book for the Water Well Industry. Edward E. Johnson, St. Paul, Minnesota.

Kim K. H., Kabir E. and Jahan S. A. (2017). Exposure to pesticides and the associated human health effects. *Science of the Total Environment*, **575**, 525–535.

Kingdom B., Liemberger R. and Marin P. (2006). The Challenge of Reducing Non-revenue Water (NRW) in Developing Countries – How the Private Sector can Help: A Look at Performance-based Service Contracting. Water Supply and Sanitation Sector Board Discussion Paper Series No. 8. The World Bank, Washington DC. http://documents.worldbank.org/curated/en/385761468330326484/The-challenge-of-reducing-non-revenuewater-NRW-in-developing-countries-how-the-private-sector-can-help-a-look-at-performance-based-ervice-contract ing

Knežić L. V. (2007). Water Technology (in Serbian), Inedita.

Kostyaev V. Y. (1975). Biological Factors in the Decomposition of Phenol, Tr., Inst. Biol. Vnutr. Vod. Aka. Nauk, 30, 85-88 SSSR.

Kovačević S., Dimkić M., Vojinović – Miloradov M., Mirosavljević Z., Stanisavljević N. and Batinić B. (2011a). 'Identifcation of water quality and distribution of pollutants in the River Danube on a city level, by MFA method'. Annals of Faculty Engineering Hunedoara – International Journal of Engineering, Tome IX, Fascicule 2, pp. 173–176, ISSN: 1584–2665.

Kovačević S., Vojinović-Miloradov M., Milovanović D., Đukić Z., Mirosavljević Z. and Stupavski M. (2011b). Emerging substances and chemical reactions in aquatic phase of landfill system. ISWA Beacon Conference, Novi Sad, Serbia, pp. 235–243, ISBN 978-86-7892-361-6.

Kovačević S., Turk Sekulić M., Radonić J., Špánik I., Vojinović Miloradov M., Milovanović D. and Čepić Z. (2012). Meat industry wastewater as potential source of emerging pollutants in the Vojvodina Province. International Conference on Inovative Technologies-INTECH 2012, Rijeka, Croatia, pp. 377–379. ISBN: 978-953-6326-77-8.

Kovačević S., Dimkić M., Vojinović-Miloradov M., Majkić B., Đogo M., Radonić J. and Turk Sekulić M. (2013a). Occurrence and behavior of selected pharmaceuticals during bank filtration in Danube river – a three year survey in Serbia. International Conference 'Engineering for Environment Protection' TOP 2013, Bratislava, Slovak Republic, pp. 151–155, ISBN: 978-80-227-3955-9.

Kovačević S., Petković A., Pušić M. and Dimkić M. (2013b). Fate of emerging pollutants in alluvial groundwater of big rivers in Serbia. UNESCO Conference on Emerging Pollutants in Water, Belgrade, Republic of Serbia, pp. 59–63, ISBN:978-86-82565-39-0.

Kovačević S., Vojinović Miloradov M., Sremački M., Radonić J., Petković A. and Dimkić M. (2013c). Natural removal of selected emerging polutants due to the river bank filtration. III International Conference 'Ecology of Urban Areas', Kaštel Ečka-Zrenjanin, Republika Srbija, pp. 238–242, ISBN: 978-86-7672-210-5.

Kovačević S., Sremački M., Vojinović Miloradov M. and Dimkić M. (2014). Selected pharmaceuticals load in urban wastewater for the city of Novi Sad. IV International Conference 'Ecology of Urban Areas 2014', Zrenjanin, Serbia, pp. 311–317, ISBN: 978-86-7672-237-2.

Kovačević S., Radišić M., Laušević M. and Dimkić M. (2017a). Occurrence and behaviour of selected pharmaceuticals during riverbank filtration in the republic of Serbia. *Environmental Science and Pollution Research*, **24**(2), 2075–2088, ISSN: 0944-1344, doi: 10.1007/s11356-016-7959-4 https://link.springer.com/article/10.1007/s11356-016-7959-4.

Kovačević S., Mitrinović D., Radišić M., Živančev (Šenk) N., Perović M., Vojt P. and Dimkić M. (2017b). Tracer test and behavior of selected pharmaceuticals. *Water Science and Technology – Water Supply*, **18**(2), 1043–1052, doi: 10.2166/ws.2016.202, ISSN: 1606-9749.

Kovacs G. and Ujfaludi L. (1983). Movement of Fine Grains in the Vicinity of Well Screens. *Hydrological Sciences – Journal des Sciences Hydrologiques*, **28**(2), 6/1983.

Krešić N., Vujasinović S. and Matić I. (2006). Remediation of groundwater and geologic media, 385 pp., Spec. ed. Dpt. Hydrogeol. Fac. Min. Geol., Univ. Belgrade, Belgrade, (in Serbian).

Kühn W. and Müller U. (2000). Riverbank filtration – an overview. *Journal AWWA*, **92**(12), 60–69.

Kühn W. and Müller U. (2006). Exportorientierte F&E auf dem Gebiet der Wasserver- und entsorgung. Teil I: Trinkwasser, Band 2, Leitfaden. DVGW-Technologiezentrum Wasser (TZW) Karlsruhe, Eigenverlag.

Kummu M., de Moel H., Porkka M., Siebert S., Varis O. and Ward P. J. (2012). Lost food, wasted resources: global food supply chain losses and their impacts on freshwater, cropland, and fertiliser use. *Science of the Total Environment*, **438**, 477–489.

Kummu M. J., Guillaume H. A., de Moel H., Eisner S., Flörke M., Porkka M., Siebert S., Veldkamp T. I. E. and Ward P. J. (2016). The world's road to water scarcity: shortage and stress in the 20th century and pathways towards sustainability. *Natureresearch Journal, Scientific Reports*, **6**, 38495, doi: 10.1038/srep38495

Li Z. and Jennings A. (2017). Worldwide regulations of standard values of pesticides for human health risk control: a review. *International Journal of Environmental Research and Public Health*, **14**(7), 826.

Lüthi D., Le Floch M., Bereiter B., Blunier T., Barnola J.-M., Siegenthaler U., Raynaud D., Jouzel J., Fischer H., Kawamura K. and Stocker T. F. (2008). High-resolution carbon dioxide concentration record 650 000-800 000 years before present. *Nature*, **453**, 379–382, doi: 10.1038/nature06949

Maeng S. K., Ameda E. A., Sharma S. K., Grützmacher G. and Amy G. L. (2010). Organic micropollutant removal from wastewater effluent-impacted drinking water sources during bank filtration and artificial recharge. *Water Research*, **44**(11), 4003–4014, doi: 10.1016/j.watres.2010.03.035

Mansuy N. (1999). Water Well Rehabilitation: A Practical Guide to Understanding Well Problems and Solutions. Lewis Publishers, Boca Raton, 174.

Marga J. (2008). Les eaux souterraines dans le monde. BGRM/Editions UNESCO, Paris.

Margat J. and van der Gun J. (2013). Groundwater Around the World: A Geographic Synopsis, CRC Press, Taylor & Francis Group, Boca Raton, Florida.

Massmann G., Dünnbier U., Heberer T. and Taute T. (2008). Behaviour and redox sensitivity of pharmaceutical residues during bank filtration – investigation of residues of phenazone-type analgesics. *Chemosphere*, **71**, 1476–1485, doi: 10.1016/j.chemosphere.2007.12.017

Matthews G. (2015). Pesticides: Health, Safety and the Environment. John Wiley & Sons, Chichester, UK.

Mazahirali A., Shewchuk J. and Drewes J. (2015). Effect of temperature on removal of trace organic chemicals in managed aquifer recharge systems. *Chemosphere*, **122**, 23–31, doi: 10.1016/j.chemosphere.2014.10.064

McLaughlan R. (2002). Managing Water Well Deterioration. Taylor & Francis, New York, 128.

Mekonnen M. M. and Hoekstra A. Y. (2011). The green, blue and grey water footprint of crops and derived crop products. *Hydrology and Earth System Science*, **15**, 1577–1600, doi: 10.5194

Mekonnen M. M. and Hoekstra A. Y. (2014). Water footprint benchmarks for crop production: a first global assessment. *Ecological Indicators*, **46**, 214–223.

Menachem E. and William A. P. (2011). The future of seawater desalination: energy, technology, and the environment. *Science*, **333**(712), 712–717, doi: 10.1126/science.1200488

Milovanović D. and Boev B. (2001). The Earth, Plate Tectonics and Magnetism (in Serbian). Nauka, Belgrade.

Ministry of Agriculture, British Columbia. (2015). Soil Water Storage Capacity and Available Soil Moisture.

NASA Goddard Institute for Space Studies. (2019). GISS Surface Temperature Analysis, GISTEMP Team, version 4. Dataset accessed 2019-10-20 at https://data.giss.nasa.gov/gistemp/.

Nicolopoulou-Stamati P., Maipas S., Kontampasi C., Stamatis P. and Hens L. (2016). Chemical pesticides and human health: the urgent need for a new concept in agriculture. *Frontiers in Public Health*, **4**, 148.

Nikolenko O., Jurado A. V., Borges A. V., Knöller K. and Brouyere S. (2018). Isotopic composition of nitrogen species in groundwater under agricultural areas: a review. *Science of the Total Environment*, **621**, 1415–1432.

Ninković D., Babić Mladenović M., Dimkić M., Milovanović M., Milovanović D., Vranković J. and Paunović M. (2010). Implementation of the EU Water Framework Directive in Serbia, Chapter 3 of 'The Danube in Serbia' – Results of the National Program of the Second Joint Danube Survey 2007. In: The Danube in Serbia' – Results of the National Program of the Second Joint Danube Survey 2007, P. Simonović, V. Simić, S. Simić and M. Paunović (eds.), Republic of Serbia, Ministry of Agriculture, Forestry and Water Management – National Water Directorate; University of Belgrade, Siniša Stanković Institute of Biological Research; University of Kragujevac, School of Natural Sciences, Belgrade, pp. 39–58, ISBN 978-86-81829-96-7. Number of Researchers per million inhabitants by country, United Nations Educational, Scientific and Cultural Organisation (UNESCO) Institute for Statistics, NOAA Climate, https://www.climate.gov/

Pastén-Zapata E., Ledesma R., Harter T., Ramirez A. I. and Mahlkencht J. (2014). Assessment of sources and fate of nitrate in shallow groundwater of an agricultural area by using a multi-tracer approach. *Science of the Total Environment*, **470–471**, 855–864.

Pearson H. (2002). 'Superbug' hurdles key drug barrier. *Nature*, **418**(6897), 469–469, http://dx.doi.org/10.1038/418469b

Perović M., Obradović V., Kovačević S., Mitrinović D., Živančev N. and Nenin T. (2017). Indicators of groundwater potential for nitrate transformation in a reductive environment. *Water Environment Research*, **89**(4), 4–16.

Petersen-Perlman J. D., Veilleux J. C. and Wolf A. T. (2017). International water conflict and cooperation: challenges and opportunities. *Water International*, **42**(2):105–120. http://www.tandfonline.com/doi/full/10.1080/02508060.2017.1276041

Pietraru V. (1982). Rapport general sur les crepines et filtres des puits. International Symposium on Hydraulic Problems of Ground Water Drainage, Gornji Milanovac.

Piketty T. (2014). Capital in the Twenty First Century. Harvard University Press, Cambridge, Massachusetts London, England, ISBN 978-0674430006.

Pyne D. (1995). Groundwater Recharge and Wells. CRC Press LLC, Boca Raton, 376.

Pretis F., Schwarz M., Tang K., Haustein K. and Allen M. (2018). Uncertain impacts on economic growth when stabilizing global temperatures at 1.5°C or 2°C warming. *Philosophical Transactions of The Royal Society A Mathematical Physical and Engineering Sciences A*, **376**, 20160460, doi: 10.1098/rsta.2016.0460.

Radović T., Grujić S., Dujaković N., Radišić M., Vasiljević T., Petković A., Boreli-Zdravković Đ., Dimkić M. and Laušević M. (2012). Pharmaceutical residues in the Danube River Basin in Serbia – a two-year survey. *Water Science and Technology*, **66**(3), 659–665, https://doi.org/10.2166/wst.2012.225

Radovic T., Grujic S., Petkovic A., Dimkic M. and Lausevic M. (2015). Determination of pharmaceuticals and pesticides in river sediments and corresponding surface and ground water in the Danube River and tributaries in Serbia. *Environmental Monitoring and Assessment*, **187**(1), 25232–25244. doi: 10.1007/s10661-014-4092-z

Radović T., Grujić S., Kovačević S., Laušević M. and Dimkić M. (2016). Sorption of selected pharmaceuticals and pesticides on different river sediments. *Environmental Science and Pollution Research*, **23**(24), 25232–25244, doi: 10.1007/s11356-016-7752-4, ISSN: 0944-1344.

Ray C., Melin G. and Linsky R. B. (2002). Riverbank Filtration. Kluwer Academic Publishers, Dordrecht.

Rijsberman F. R. (2006). Water scarcity: fact or fiction. *Agricultural Water Management*, **80**(1–3), 5–22.

Rivett M. O., Buss S. R., Morgan P., Smith J. W. N. and Bemment C. D. (2008). Nitrate attenuation in groundwater: a review of biogeochemical controlling processes. *Water Research*, **42**, 4215–4232.

Rockström J. (2003). Water for food and nature in drought–prone tropics: vapour shift in rain–fed agriculture. *Philosophical Transactions of the Royal Society B: Biological Sciences*, **358**, 1997–2009.

Rockström J., Falkenmark M., Karlberg L., Hoff H., Rost S. and Gerten D. (2009). Future water availability for global food production: the potential of green water for increasing resilience to global change. *Water Resources Research*, **45**, W00A12.

Royer D. (2014). 'Dana Royer comment at RealClimate'. RealClimate. Retrieved 26 March 2014.

Schmidt K. (1988). Pufferwirkung der Untergrundpassage – Leistungsfähigkeit und Grenzen der Uferfiltration und künstlichen Grundwasseranreicherung bei der Eliminierung von Schadstoffen, Gewässerschutz Wasser Abwasser, **109**, 181–194.

Semiat R. (2008). Energy issues in desalination processes. *Environmental Science & Technology*, **42**, 8193–8201. doi: 10.1021/es801330u

Shannon M., Bohn P. and Elimelech M. A., (2008). Science and technology for water purification in the coming decades. Nature, **452**, 301–310, doi: 10.1038/nature06599

Shiklomanov A. I. (2000). Appraisal and assessment of world water resources. *Water International*, **25**(1), 11–32, doi: 10.1080/02508060008686794

Sichardt W. (1928). Method of Stabilization of Drilled Wells (in German). Springer, Berlin.

Siebert S., Kummu M., Porkka M., Döll P., Ramankutty N. and Scanlon B. R. (2015). A global data set of the extent of irrigated land from 1900 to 2005. *Hydrology and Earth System Sciences*, **19**, 1521–1545.

Šojić M., Dimkić M., Milovanović M. and Ilić M. (2014). Financial Aspect of Water Management in Serbia 2014–2023. Milankovitch Anniversary UNESCO Symposium 'Water Management in Transition Countries as Impacted by Climate Change and Other Global Changes, Lessons from Paleoclimate, and Regional Issues', 3–5 September 2014, Belgrade, Serbia, Publishers: Jaroslav Černi Institute for the Development of Water Resources and

Serbian Academy of Sciences and Arts, M. Dimkić, F. Mesinger and A. Berger (eds.), Book of Abstracts, pp. 51–55, ISBN 978-86-82565-42-0.

Stellman J. M. (1998). Encyclopaedia of Occupational Health and Safety, 4th edn. International Labor Office, Geneva, Chapter 55, ISBN 9221092038, 9789221092032.

Storck F. R., Schmidt C. K., Lange F. T., Henson J. W. and Hahn K. (2012). Factors controlling micropollutant removal during riverbank filtration. *Journal of American Water Works Association*, **4**(12), 643–652, doi: 10.5942/jawwa.2012.104.0147

Sučić-Protić Z. and Miletić-Spajić O. (1969). Geology: a Basic Course (in Serbian). Zavod za izdavanje udžbenika Socijalističke Republike Srbije, Belgrade.

Sudhakaran S., Lattemann S. and Amy G. L. (2013). Appropriate drinking water treatment processes for organic micropollutants removal based on experimental and model studies – a multi-criteria analysis study. *Science of Total Environment*, **442**, 478–88, doi: 10.1016/j.scitotenv.2012.09.076

Taylor R., Scanlon B. and Döll P., *et al.* (2013) Ground water and climate change. *Nature Climate Change*, **3**, 322–329, doi: 10.1038/nclimate1744

Thornton S. F., Quigley S., Spence M. J., Banwart S. A., Bottrell S. and Lerner D. N. (2001). Processes controlling the distribution and natural attenuation of dissolved phenolic compounds in a deep sandstone aquifer. *Journal of Contaminant Hydrology*, **53**(3–4), 233–267.

Tumbas I. I. (2014). The fate and importance of organics in drinking water treatment: a review. *Environmental Science and Pollution Research*, **21**, 11794–11810, doi: 10.1007/s11356-014-2894-8

UN Habitat. (2012). Sustainable Urbanization, Thematic Think Piece, Dep. of Econ. and Soc. Affairs and the U. N. Dev. Programme.

UNEP (United Nations Environment Programme). (2007a). Global Environment Outlook (GEO-4). Environment for Development Home Page. http://www.unep.org/geo/geo4.asp (accessed 10 October 2011) – 2010a. Global Envvironment Outlook: Latin America and the Caribbean (GEO LAC) 3. Panama City, UNEP, Regional Office for Latin America and the Caribbean. http://www.unep.org/pdf/GEOLAC_3_ENGLISH.pdf

UNEP (United Nations Environment Programme). (2007b). Global Environment Outlook (GEO-4). Environment for Development Home Page. http://www.unep.org/geo/geo4.asp (accessed 10 October 2011).

United Nations Environment Programme (UNEP). (2007c). Global Environment Outlook 4: Environment for Development. http://www.unep.org/geo/geo/assessments/global-assessments/global-environment-outlook-4

UNEP. (2008). Implementation of the water policy and strategy of the United Nations Environment Programme. UNEP/GC.25/9.

UNEP (United Nations Environment Programme). (2011). Towkards a Green Economy: Pathways to Sustainable Development and Poverty Eradication. UNEP, Nairobi. http://www.unep.org/GreenEconomy/Portals/93/documents/Full_GER_screen.pdf

UNDP (2006). Human development report 2006: Beyond scarcity – Power, poverty and the global water crisis. Palgrave Macmillan Houndmills, Basingstoke, New York, http://hdr.undp.org/en/content/human-development-report-2006

UNESCO. (2011). Number of Researchers per million inhabitants by Country, United Nations Educational, Scientific and Cultural Organisation (UNESCO) Institute for Statistics, viewed 17th December, 2019, http://chartsbin.com/view/1124

United Nations Food and Agricultural Organization (UN FAO). (2015). AQUASTAT Database, http://www.fao.org/nr/water/aquastat/data/query/results.html

United Nations Educational, Scientific, and Cultural Organization (UNESCO). (2018). Institute for Statistics, Data published by World Bank – World Development Indicators, http://data.worldbank.org/data-catalog/world-development-indicators

Voutchkov N. (2017). Energy use for membrane seawater desalination – current status and trends. *Desalination*, **431**, 2–14, http://dx.doi.org/10.1016/j.desal.2017.10.033

Vukovic M. and Pušic M. (1990). Soil Stability and Deformation Due To Seepage. Water Resources Publications, Littleton, Colorado, USA, 80.

Vukovic M. and Sorc A. (1990). Hydraulics of Water Wells. Water Resources Publications, Littleton, Colorado, USA, 354.

Wada Y., Gleeson T. and Esnault L. (2014). Wedge approach to water stress. *Nature Geoscience*, **7**, 615–617.

WEF (World Economic Forum). (2011). Water Security: The Water-Food-Energy-Climate Nexus: The World Economic Forum Initiative. Island Press, Washington DC.

Weng T. N., Liu C. W., Kao Y. H. and Hsiao S. S. Y. (2017). Isotopic evidence of nitrogen sources and nitrogen transformation in arsenic-contaminated groundwater. *Science of Total Environment*, **578**, 167–185.

WHO. (2015). Data on the total population with access to improved water sources and safe drinking-water are collected via questionnaire by WHO/UNICEF country representatives as part of the JMP, https://gateway.euro.who.int/en/indicators/enhis_3-population-with-house-connections-to-piped-water/

WHO (World Health Organization)/UNICEF (United Nations Children's Fund). (2010). Progress on Sanitation and Drinking-Water: 2010 Update. Joint Monitoring Programme for Water Supply and Sanitation. WHO/UNICEF, Geneva/New York.

Wild D., Francke C.-J., Menzli P. and Schön U. (2010). Water: A Market of the Future. A Sustainable Asset Management (SAM) Study. SAM, Switzerland.

World Bank. (2008). Natural Disasters: Counting the Costs. The World Bank, Washington DC. Economic Impact of Sanitation in Indonesia: A Five-country Study Conducted in Cambodia, Indonesia, Lao PDR, the Philippines, and Vietnam Under the Economics of Sanitation Initiative. Water and Sanitation Program. The World Bank, Washington DC.

World Meteorological Organization. (2014). Water Exemplar to the User Interface Platform of the Global Framework for Climate Services: Water exemplar to the user interface platform of the Global Framework for Climate Services. Chair, Publications Board World Meteorological Organization (WMO), Geneva, Switzerland.

World Resource Institute. (2015). Ranking the World's Most Water-Stressed Countries in 2040, https://www.wri.org/resources/data-sets/aqueduct-projected-water-stress-country-rankings

WWAP. (2009). (World Water Assessment Programme) United Nations World Water Development Report 3: Water in a Changing World. UNESCO Publishing/Earthscan, Paris/London.

WWAP. (2013). United Nations World Water Development Report 2013. United Nations World Water Assessment Programme (WWAP). United Nations Educational, Scientific and Cultural Organization, Paris.

WWAP. (2015). United Nations World Water Development Report 2015: Water for a sustainable world. United Nations World Water Assessment Programme (WWAP). United Nations Educational, Scientific and Cultural Organization, Paris.

WWAP. (2017a). (World Water Assessment Programme) United Nations World Water Development Report 2017: Wastewater the Untapped Resource. United Nations Educational, Scientific and Cultural Organization (UNESCO), Paris. http://unesdoc.unesco.org/images/0024/002471/247153e.pdf

WWAP. (2017b). United Nations World Water Development Report 2017: Wastewater, the untapped resource. United Nations World Water Assessment Programme (WWAP). Available at https://unesdoc.unesco.org/ark:/48223/pf0000247553/PDF/247553eng.pdf.multi (accessed January 2020). Paris, United Nations Educational, Scientific and Cultural Organization.

WWAP (United Nations World Water Assessment Programme)/UN-Water. 2018. The United Nations World Water Development Report 2018: Nature-Based Solutions for Water. Paris, UNESCO.

Yao Y., Gao B., Zhang M., Inyang M. and Zimmerman A. R. (2012). Effect of biochar amendment on sorption and leaching of nitrate, ammonium, and phosphate in a sandy soil. *Chemosphere*, **89**(11), 1467–1471. doi: 10.1016/j.chemosphere.2012.06.002

Zektser I. S. and Everett L. G. (2004). Groundwater Resources of the World and Their Use, Published in 2004 by the United Nations Educational, Scientific and Cultural Organization, **7**, place de Fontenoy, 75352 Paris 07 SP.

Živančev N., Kovačević S., Radović T., Radišić M. and Dimkić M. (2019). Mobility and sorption assessment of selected pesticides in alluvial aquifer. *Environmental Science and Pollution Research*, **26**(28), 28725–28736.

# Chapter 3

# Alluvial aquifer – The legacy of the quaternary period

*Jelena Zarić, David Mitrinović and Milan Dimkić*

## 3.1 GROUNDWATER RESOURCES OF ALLUVIAL AQUIFERS

The term 'groundwater' usually refers to subsurface water that occurs and moves in soil and geological formations, called aquifers, which are completely saturated with water.

Alluviums are spacious, smaller or larger accumulations of loose, unconsolidated material (alluvium), eroded, transported and redeposited by surface watercourses, which fill a river valley, consisting above all of gravel, sand, siltstone and clay. The formation of rivers and alluvial valleys is the result of the influence and action of endogenous and exogenous factors and forces – relief conditions, orogenic and epeirogenic, tectonic movements and climate. The Pleistocene period is the coldest part of the Cenozoic. Within it, we can distinguish the Lower, Middle and Upper Pleistocene while the time from the last cold period until today, the last warm (interglacial) period, is called the Holocene. The entire Pleistocene period is characterized by alternating cold and warm periods – glacial and interglacial. With the cyclic change of hot and cold phases, the ice cover periodically expanded and retracted. The creation and melting of glaciers resulted in great changes on the surface of the planet. Glacial erosion, as well as the formation of very powerful watercourses during the warm phases (interglacials) strongly influenced the formation of alluvial valleys and alluvial aquifers.

It is believed that only about 2.5% of the overall water on the planet is made up of freshwater resources. Of these, two thirds are glaciers, and less than one third are groundwaters.

According to the Food and Agriculture Organization (FAO), the estimated volume of fresh groundwater is 10.53 million km$^3$, which is almost 99% of all the liquid freshwater on Earth (FAO, 2016).


Editor: Milan Dimkić
doi: 10.2166/9781789060904_0135

Groundwater plays a key role in the domestic water sector as a source of drinking water for half of the world's population. The share of groundwater resources in the water supply of the population varies significantly from country to country:

- 71% in South Asia;
- 70% in Europe (including Russia);
- 66% in Australia;
- 55% in the Middle East;
- 33% in the USA;
- 32% in North Africa;
- 29% in China.

Thirty-seven global mega aquifers, cover about 26% of the world's land mass and contain about two thirds of the world reserves of groundwater, but receive less than 10% of the water used for the recharge of groundwaters. Numerous alluvial aquifers receive the bulk of the world's groundwater recharge. This makes them a very important resource, and creates the need for their better governance.

The total global withdrawal rates of groundwater are currently estimated at 982 $km^3$/year, which is equivalent to 8% of the mean global groundwater renewal. There are huge differences regarding the exploitation and recharge in different parts of the world, with India, the USA and China abstracting about half of this annual volume.

It can be roughly estimated that about half of the groundwater that is used can be attributed to alluvial waters.

For the sustainable management and preservation of alluvial aquifers, it is very important to understand and define the conditions of sedimentation and alluvial aquifer formation (material yield and the mechanisms of erosion, transport and accumulation), and their connection with other aquifers, surface waters and other geological formations. All these factors are determined by genesis – the conditions of the formation and existence of alluvial aquifers and they vary greatly from area to area. The study and understanding of the characteristics and genesis of aquifers formed in Quaternary sediments is very important for the protection, preservation and sustainable use of groundwater resources. In addition, quaternary sediments are essential for the understanding and reconstruction of paleo-climatic conditions of the Quaternary period.

In this chapter, we have limited ourselves to some aspects and factors regarding the formation of alluvial deposits and the determination of their age. Special attention has been paid to the appearance and influence that the changes of cold and warmer periods had on the formation and processes in alluvial aquifers. The final subchapter refers to research conducted at the Belgrade Groundwater Source and in the Sava River Basin (the largest tributary of the Danube River), and related to the significant role the glacial period had on the formation and characteristics of alluvial deposits.

## 3.2 CLIMATE DURING THE QUATERNARY PERIOD

The Quaternary is the youngest geological period that began 2.59 million years ago and continues today, characterized by the variability and dynamic aspect of physical and biological processes. Periods of warmer and colder climates change, sea and lake shores change, glaciers occupy larger areas and recede, and flora and fauna migrate.

The Quaternary covers two epochs: the Pleistocene, which began 2.59 million years ago and ended about 10,000 years ago, characterized by alternating cold and warm intervals, and the Holocene, which has so far lasted 11,700 years and is characterized by a relatively stable and warmer climate.

### 3.2.1 Global climate change over the course of geological history with an emphasis on the climate during the Quaternary period

For most of its existence, the Earth has been warmer than it is today, without surfaces covered with a permanent ice sheet. However, large climate fluctuations which caused changes in the formation of glaciers left traces in sediments from the geological past. The Huron ice age occurred between 2,700 million years and 2,300 million years ago – in the Proterozoic. The most extreme ice age occurred 850 to 630 million years ago (Cryogenic age). It is believed that during the Cryogenic period, the world's oceans were frozen. The more moderate Andean-Saharan ice age lasted from 460 to 430 million years ago (Ordovician, Silurian). From 350 to 260 million years ago, during the Carboniferous and Permian periods, polar caps periodically expanded (the Kara ice age).

The last great cold period – ice age, began 40 million years ago with a growth of the ice cover in Antarctica. It intensified during the late Pliocene, about 3 million years ago when ice sheets spread in the northern hemisphere. The ice age then continued during the Pleistocene (sometimes the term 'ice age' is used for the Pleistocene). Since then, cycles of ice cover expansion, i.e. glaciation, have taken place and are characterized by cycles of the order of magnitude of 40,000, up to approximately 1,050,000 years ago, and 100,000 years onwards.

Stages of ice ages, i.e. of glacial periods, drier and cooler than the periods between them (interglacials), are characterized by lower elevations at which mountain glaciers appear, thicker glaciers and by a wider spread of the polar ice caps, and have been divided and named according to the geographical areas where they were found and the times when they occurred. Within the glacial periods there may be stages, i.e. cold interglacials separated by warmer interstadials. In the classical Alpine stratigraphic division, glacials are named after the rivers that spring in the Alps. Using morphostratigraphic methods, i.e. by extracting from the relief of moraines, gravel terraces formed from material washed from the glacier zone, valleys deepened by the work of glaciers, stretch marks on rocks, etc., the following four glacials were initially defined – Günz, Mindel, Riss and Würm, to which the following periods correspond: periods 866,000 to 533,000 (424,000), 476,000 to 424,000 (378,000 to 337,000), 300,000 to 126,000 and 115,000 to 11,700 BC, respectively. These periods are generally accepted for Riss and Würm, while for Günz and Mindel there are several views (the main alternative view is given in parentheses), due to the less reliable absolute dating of morphological structures, and significantly more than two cycles based on a pre-Riss climate change indicator (Figure 3.1). Consecutive glaciations tend to erase the evidence from previous stages which makes it difficult to separate and date glacials and interglacials, especially if some of the subsequent glaciations caused a greater ice cover expansion.

The movement of continents, volcanic activity and large changes in the relief and size and position of the oceans were the main factors that caused the beginning and end of most ice ages. The Pleistocene epoch was not marked by volcanic and tectonic activity of such intensity that it could have been a major factor in climate change.

According to the currently widely accepted 'cosmic' theory of the Serbian scientist Milutin Milanković (Milanković, 1941), changes in the insolation due to changes in the tilt and orientation of the Earth's axis and the shape of its orbit are the main causes of cyclical climate change and glacial and interglacial changes during the Pleistocene and Holocene. Milanković's insolation calculations were performed using a comprehensive calculation of the movement of our planet carried out primarily by the French astronomer, Le Verrier, and the German mathematician Pilgrim who, using Le Verrier's calculations, added changes to the inclination of the Earth's axis of rotation. The calculations were polished at the Astronomical Observatory in Belgrade by students doing part-time work, who later became renowned

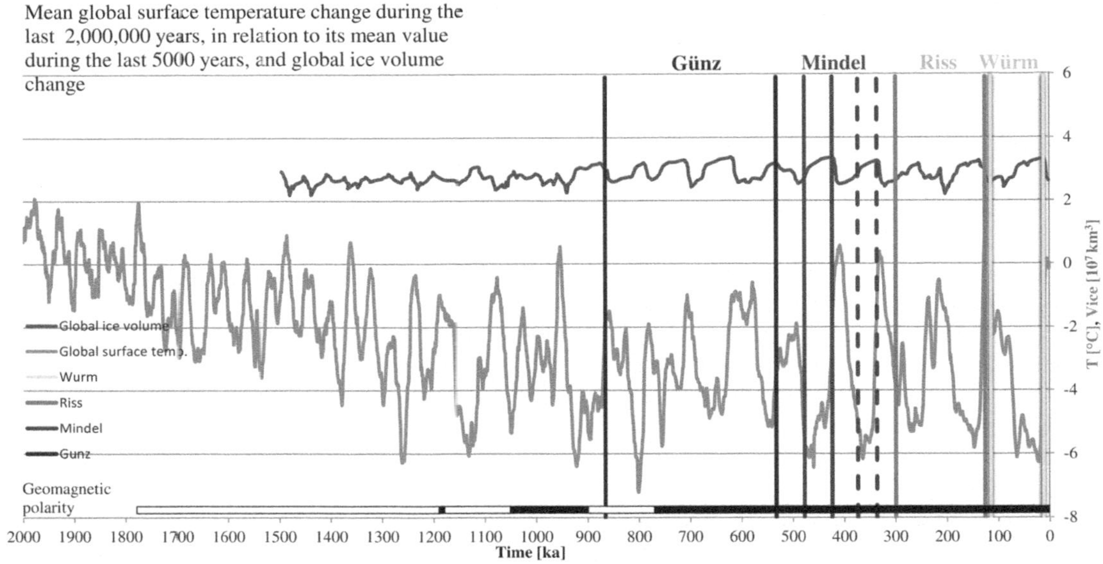


**Figure 3.1** Change in global mean surface temperature over the last 2,000,000 years, relative to its mean in the last 5,000 years (Snyder, 2016), and classical glacial periods of Alpine Quaternary stratigraphy (Deutsche Stratigraphische Kommission, 2016).

Serbian mathematicians, Dragoslav Mitrinović and Stanimir Fempl who carried out extensive calculations that lasted almost three years (Mesinger, 2010).

Insolation is defined as the energy of radiation from the Sun that falls on a certain horizontal surface on Earth in a unit of time.

Insolation depends on:

- solar constants such as a flux of the incoming radiation energy through a perpendicular unit area at the top of the atmosphere, whose current value is 1368,W/m;
- distances from the Sun;
- changes on the Sun (spots);
- the height of the Sun (angle) above the horizon;
- the percentage of incident radiation that returns from the top of the atmosphere back into space;
- the percentage of absorbed radiation in the atmosphere;
- the percentage of reflected radiation from the Earth's surface (albedo).

Insolation, like the solar constant, is most often expressed in watts per square meter. At medium latitudes, the average summer value of insolation is 340 $W/m^2$, while when it is cloudy, it is only 120 $W/m^2$. Changes in insolation are the main astronomical factor of climate change. At latitudes greater than 65°, an increase in the tilt of the Earth's axis relative to the plane in which it orbits (ecliptic) above a certain value, provided that the Earth is approximately the closest to the Sun at the time when the summer solstice is in the northern hemisphere, causes winter snow and part of the ice and snow previously accumulated to melt completely during the summer due to the higher irradiated energy, resulting also in less precipitation in slightly colder winters. The cumulative effect is that the boundary of constant snow on the mountains slowly rises and the polar caps shrink, which reduces the albedo (reflectivity) of the Earth's surface, which leads to an increase in the total annual irradiated energy of the Sun, so over time, the snow and ice practically recede from today's temperate climate zone. The phenomenon is complicated by the simultaneous precession of the Earth's axis, i.e. by the change in its orientation, and the changing of the shape of the orbit, in such a way that at the same inclination of the axis the distance from the Sun is different, as is the shape of the orbit.

The summer insolation at 65° north latitude is highest when the axis is tilted the most and the distance from the Sun is the smallest, while the winter insolation is then the smallest, which most likely causes less precipitation during the winter. Due to this complex interaction of different cycles and other factors, the periodicity of the occurrence of glacial periods is not so simply determined. The glacials and stadials end in a sudden increase in global temperature which coincides with a maximum insolation at 65° north latitude, but almost always provided that the difference in insolation between the North Pole (90° north latitude) and 65° north latitude is also close to the maximum (Figure 3.2). It is most likely that a larger difference in insolation intensifies the movement of warm air masses from lower latitudes towards the poles. In general, the stage can be said to last as long as the insolation difference between the North Pole and 65° north latitude is close to the minimum, and as long as the insolation maximum does not coincide with the maximum insolation difference between the North Pole and 65° north latitude (Best, 2016).

Changes in the concentration of greenhouse gases, the albedo of the Earth's surface, the intensity of evaporation, the path and flow of ocean currents, vegetation and winds are factors that affect the Earth's climate and can increase or decrease the effects of astronomical factors.

The share of greenhouse gases in the atmosphere, primarily carbon dioxide, increases with the withdrawal of ice covers and an increase in temperatures, and decreases with more ice formation and a decrease in temperatures. Carbon dioxide is directly related to glacial periods. The rise and fall in the concentration of carbon dioxide in the atmosphere coincides well with the rise and fall of the average temperatures through the glacial periods. As a greenhouse gas, it causes an increase in temperature, thereby reducing the capacity of the ocean to absorb carbon dioxide, which results in its increased share in the atmosphere.

Snow and ice increase the Earth's albedo, which is why the Earth reflects more and absorbs less energy from the Sun's radiation. The ratio of ocean and ice albedo is 20% to 80%. Decreasing the forest area also increases albedo.

The degree of evaporation is higher over oceans without an ice cover and from there precipitation can form, which will form snow and ice over the polar regions.

Ocean currents change their course during the ice ages and glacial periods so that thermal energy is transported to another latitude or to a completely different area on the planet.

During glaciation, a long-term field of high atmospheric pressure, an anticyclone, is formed over large, ice-covered continental areas, and the paths of humid ocean air masses are pushed to the south. This caused extremely wet periods – pluvials, in areas that were not under the direct influence of the northern ice sheets.

### 3.2.2 Quaternary cold stages

The Pleistocene is known for the large climatic oscillations, shifts of cold and warm cycles, which caused significant changes within ecosystems. These changes are a direct consequence of the changes in terms of insolation and radiation, the amount of heat the Earth receives and the heat that is released. An important role in this process was played by the tectonic movements from earlier stages, which influenced the formation of high mountain ranges.

In 2009, the International Geological Survey (IUGS) decided that the Quaternary began 2.588 million years ago, and not 1.8 million years ago as previously thought. The lower limit is

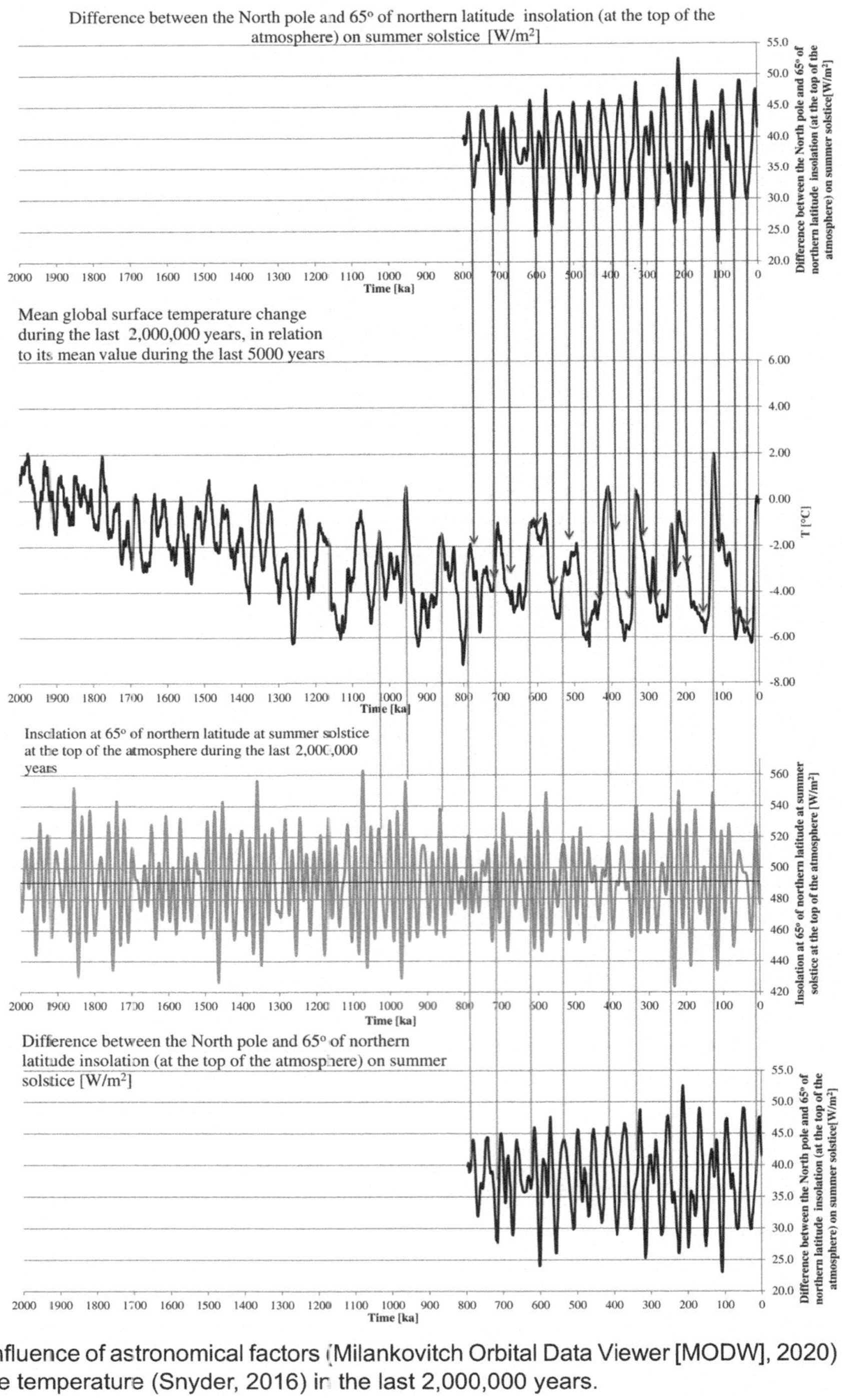


**Figure 3.2** Influence of astronomical factors (Milankovitch Orbital Data Viewer [MODW], 2020) on changes in global surface temperature (Snyder, 2016) in the last 2,000,000 years.

magnetostratigraphically defined as the basis of the Matuyama (C2r) chronozone or marine isotopic stage (MIS) zone 103.

The boundary between the Lower and Middle Pleistocene has been magnetostratigraphically defined as the Brines-Matuyama shift of magnetic epochs (since 1996). The age of this border is estimated at 781,000 years. The boundary between the Middle and Upper Pleistocene coincides with the beginning of MIS 5. The beginning of the Upper Pleistocene corresponds to the beginning of the last interglacial (which in northern Europe is called the Ems; that is, the Riss-Würm according to the Alpine division). The age of this boundary is estimated at 127,000 years. The boundary between the Pleistocene and the Holocene is set at the end of the last glaciation and according to the calculation of M. Milanković is about 11,500 years before modern times.

There were few cycles when the glaciers would advance and retreat as they thawed and refroze. On a global scale, scientists have identified Gelasian, Calabrian, Ionian and Tarantian as the Pleistocene epoch's four key stages, or ages.

Due to the different local conditions, several regional divisions have been adopted, all correlated with MIS phases.

The recurrence of hot and cold cycles in the younger Quaternary varied between 80 and 120 thousand years, with an average repetition interval of about 100,000 years. The typical Quaternary cycle implied a long period of cooling, during which ice masses were formed and during which they expanded, this was followed by rapid and intense melting. These periods of rapid warming are called terminations. There were seven terminations in the Pleistocene. The most recent one, 'I' Termination, is the last glacial termination at the end of the Marine Isotope Stage 2; Termination II is at the end of Marine Isotope Stage 6, Termination III is at the end of Marine Isotope Stage 8; Termination IV is at the end of Marine Isotope Stage 10, and so forth. During the last termination huge northern hemisphere ice sheets melted away, the sea level rose by about 120,m and the warm interglacial climate spread across the planet in just a few thousand years.

The cycles of hot and cold phases during the Pleistocene have been defined based on the results obtained by dating marine sediments. Marine sediments are present over wide areas and can be correlated globally, in contrast to terrestrial sediments which are fragmented and highly dependent on local conditions. This is why the division of the Pleistocene was made according to the 'marine isotope stages' (MIS). Unlike deep-sea sediments, continental evidence of sedimentation during these periods is very incomplete (due to the hiatus in sedimentation or subsequent erosion) and regionally very different, hence terrestrial sediments must be correlated with oceanic evidence of hot and cold phase shifts in order to globally correlate terrestrial sediments.

### 3.2.3 Pleistocene glaciations in different parts of the world

During maximum glaciation, sheets of ice covered all of Antarctica, large parts of Europe and South America, and small areas of Asia. In North America they stretched over Greenland and Canada and parts of the northern United States (Figure 3.3).

Climate oscillations had a significant impact on the paleogeographic conditions and this was reflected in the type of sediments deposited in glacial and periglacial areas, on the fluctuations in the levels and temperatures of the world's seas and oceans, and on the shift and migration of flora and fauna.

Glaciers were formed in the humid climate zone and at higher latitudes, while in dry areas, permanently frozen soil or permafrost was formed. Strong winds blew out dust from moraine deposits which was deposited in periglacial areas in the form of loess. In tropical and subtropical areas, climatic oscillations led to the alternation of wet (pluvial) and arid phases.

**Figure 3.3** Global glacial coverage during the last glacial maximum (approximate extent of the ice cover, presented on Digital Elevation - Shuttle Radar Topography Mission (SRTM) file format).

In North America, Pleistocene glaciers formed at higher altitudes in the Rocky Mountains and at high latitudes in the Greenland area and in central parts of northern Canada. From these areas, the ice spread in all directions, following the topography of the terrain. Glaciers from the Rocky Mountains and central Canada merged in North America to form a thick ice sheet that stretched from the Pacific to the Atlantic Ocean. In the south of this continent, ice masses stretched up to 40° north latitude.

In the territory of Eurasia, thick ice masses covered the British Isles, Scandinavia and Siberia in the north, while significant amounts of ice were also found in the Alps, as well as on other high mountain massifs. The ice masses of Eurasia were not formed from one center but were formed by the merging of thick ice deposits from different glaciation centers. Their southernmost border reached 45° north latitude.

In the glaciated areas, the ice caps at higher altitudes and the superficial deposits mainly date back to the late Quaternary period because glaciers and melt-waters eroded and redistributed older deposits.

In tropical and equatorial regions, Pleistocene climate oscillations had less influence on the erosion and deposition of Quaternary sediments.

South America and Africa are two continents that transect the equator with a large latitude range.

South America has a very small land area at high altitudes and a large land area close to the sea level. The Andes were formed as a result of the collision of part of the mainland of the former Gondwana and the Pacific plate, which resulted in a tectonic uplift and the formation of the Andean fold mountain chain which stretches along the entire length of the west coast of South America.

Africa has a large area at an intermediate elevation, no areas above 6,000 m a.s.l., and very little area close to sea level (Summerfield, 1996), it lacks continuous mountain chains owing to its passive tectonic margins, and features a broadly elevated plateau (cf 650 m a.s.l.). In the African continent, evidence of former glaciation occurs on relatively isolated highlands, such as the Atlas Mountains bordering the Mediterranean in the north, elevated plateaus and volcanoes in the East African Rift zone and the escarpment of South Africa. The highest mountains on the continent are volcanic cones which follow two parallel systems of faults.

Some of the main geographical changes in the Pleistocene included the formation of the Bosphorus Straits and the Skagerrak Straits. Freshwater aquatoriums such as the Black and Baltic Seas returned to the group of salt waters after the melting and rising sea levels.

The periodic melting and repointing of the English Channel formed the morphology of the soil and seabed between Britain and mainland Europe, (the English Channel used to be a land connection between Britain and France at the time of glaciation when the sea level decreased up to 120 m) while the periodic closure of the Bering Straits formed a land connection between Asia and North America.

During the Pleistocene epoch coastal lines changed due to glaciation and melting, and many lakes that exist today in the plateau areas in glacier melting zones (Scablands) are the result of repeated glaciation and melting and of the retreat of glaciers.

Today, ice sheets cover about 10% of the Earth's surface, while during most of the Pleistocene about 30%, was covered with ice. Today, ice masses are mainly concentrated in Antarctica, Greenland and the Arctic, while smaller glacier-type deposits are associated with higher altitudes, both in areas of higher and lower latitudes.

## 3.3 FORMING OF ALLUVIAL AQUIFERS

Water appeared on the planet Earth in a liquid state during the Archaic period. The most common theory is that the hydrosphere was formed as a result of gravity differentiation and the degassing of the mantle substance, although a certain amount was brought by comets that fell in large numbers on the planet's surface during that period. The amount of water in the hydrosphere changed over the geological past, as did the level of the world's seas. In cold periods, large amounts of water were trapped in glaciers and ice caps. In addition, the contours of water and land surfaces changed due to tectonic activity. Large bodies of water in the form of lake lagoons and shallow seas established or lost their connection with the ocean during certain stages thanks to these changes. In depressions and basin structures which completely lost their connection with the ocean due to the tectonic uplift of the peripheral terrains, sedimentation continued with the material coming from the peripheral terrains (denudation, runoff and surface flows), in shallow marine environments, over time sweetened by rivers and precipitation. In conditions of intensive sedimentation, such forms were quickly filled with thick sediment deposits.

Most of the answers about the development of river flows are based on paleoclimatic reconstructions, especially for the Pleistocene epoch, because glaciation played an important role in the formation of rivers, lakes and swamps.

The complex processes of water circulation, and the development of soil erosion, as well as climate change during the Earth's geological history played an important role in the formation of rivers. The development and morphology of flows were influenced by changes in water surfaces (seas and lakes), in water and land distribution, in the level of the world sea and erosion base, and by large glaciations that occurred several times in the geological past. All these processes influenced the erosion, transport and deposition of sediments and the formation of the relief. River sedimentation formed alluvial aquifers which are valuable alluvial groundwater resources used for water supply.

In general, the formation of alluvial aquifers through geological time is determined by tectonics and climate (determined by astronomical factors). These factors indirectly determine many other conditions that in turn determine the lithological composition, and the mineralogical and granulometric composition of sediments, and the form of occurrence and existence of groundwater in alluvial aquifers. Together they all determine the local and regional conditions under which groundwaters occur and exist, and under which groundwaters from an alluvial aquifer in a particular area can be exploited.

## 3.3.1 Factors pertaining to the formation of alluvial aquifers

Plate tectonics had a very significant influence on the processes causing the onset of ice ages. The lithosphere is divided into tectonic plates that 'float' on the asthenosphere. There are seven main and many smaller plates. The boundaries between the plates can be convergent (subduction zones), divergent (separation zones, such as the East African trench) and transform (plate slip zones along the faults – San Andreas Fault between the Pacific and North American plates) (Figure 3.4).

The lateral movement of the plates ranges from 0.66 to 8.50 cm per year. In the geological past, these processes influenced the movement of continents, and the distribution of land and water surfaces, as well as the spread of glaciers. Earthquakes, volcanic activity, the rise of mountain ranges and the formation of ocean trenches occur along the boundaries of the plates (Figure 3.5) and they represent a number of additional factors that have influenced glacial processes during the geological past.

The position of the continents is very important because it controls long periods of complex glacial processes. Today, large ice sheets are formed in the high latitudes of the northern and southern hemispheres, where large continental masses are located, because larger ice masses cannot spread across the ocean. These are Antarctica in the south, and large parts of North America, the Arctic and Eurasia in

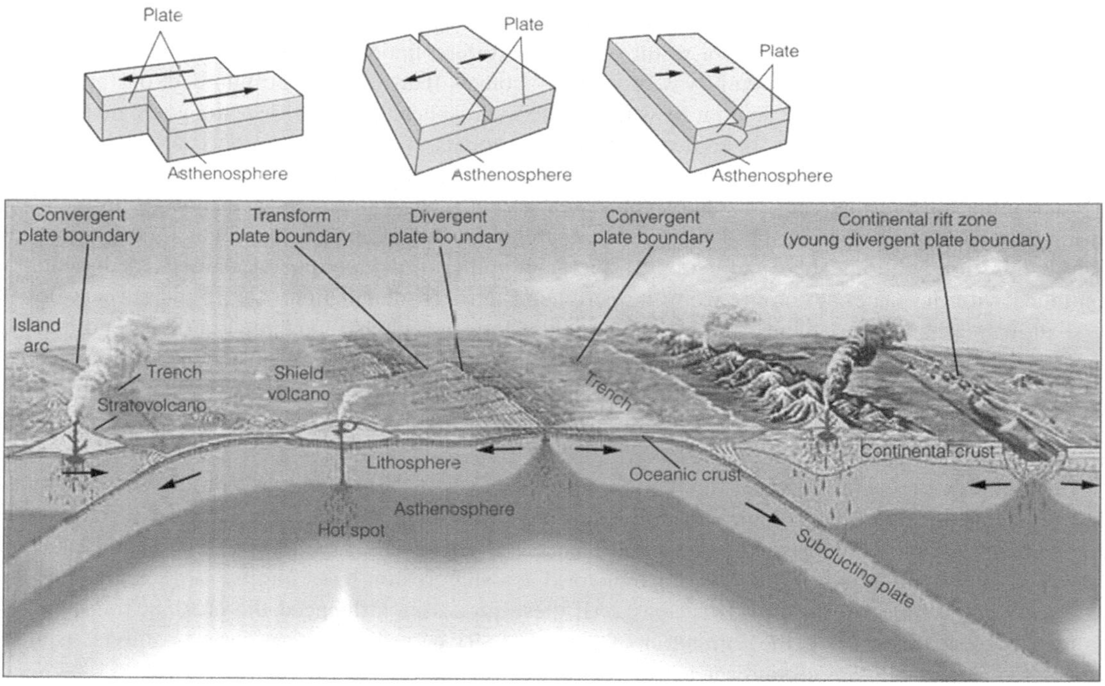


**Figure 3.4** Slab boundaries: (1) asthenosphere, (2) lithosphere, (3) hot spot, (4) oceanic crust, (5) subduction plate, (6) continental crust, (7) continental rift zone (young plate boundary), (8) converging plate boundary, (9) divergent boundary, (10) transform plate boundary, (11) volcanic shield, (12) expanding ocean ridge, (13) convergent plate boundary, (14) stratovolcano, (15) island arch, (16) plate, (17) asthenosphere trench (Vigil F. Jose, USGS/USGov. (1997), modified by Zimbres E., 1997).

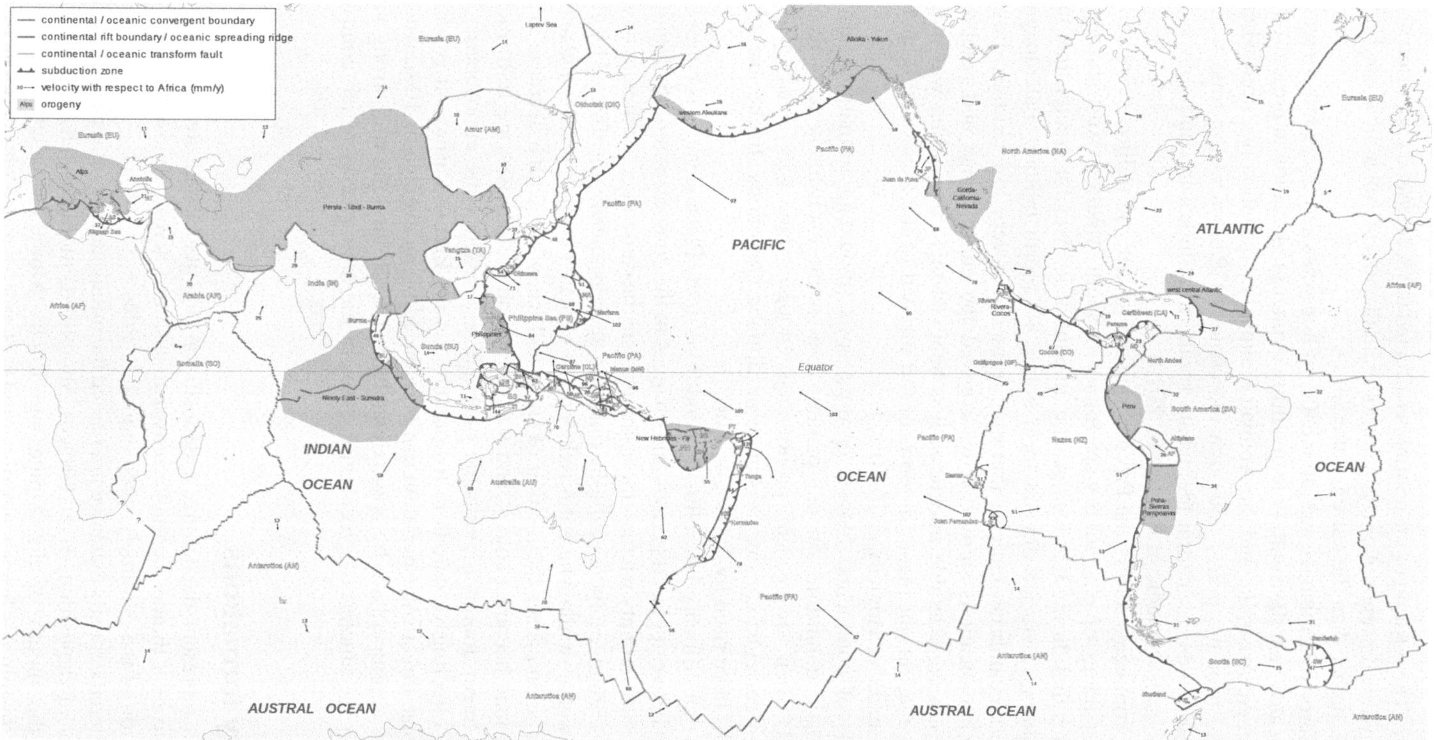


**Figure 3.5** Map of tectonic plates, continental rift boundaries, oceanic reefs, convergent boundaries, oceanic and continental transform faults and subduction zones (Gaba, 2015).

the northern hemisphere. The position of the continents affected the accelerated expansion of the ice cover in the north of our planet, while the amount of ice in the south, in the area of Antarctica, was limited by the land area and was only 10% higher than the current situation.

During the Carboniferous-Permian ice age (about 300 million years ago), the southern parts of the Pangea supercontinent were located near the South Pole, resulting in intense glaciation in present-day Africa, South America, India, Antarctica, Australia, and the Arabian Peninsula. Glaciation occurred when large parts of the land on Earth were near the poles. Although the center of Pangea at the time was located at the equator, its southern parts were located at the distant southern latitudes, in conditions where ice masses could form and stay as such (Nenadić & Bogićević, 2010).

The position of the continents during the Precambrian era (about 700 million years ago), has not been clearly determined, but traces of glaciation were found in South America, Australia and Africa. More recent research indicates that during that ice age, the amount of ice was far greater than in any later cold stage. During the Quaternary era ice did not form below 40° north latitude. Paleomagnetic studies of glacier rocks from the Precambrian indicate that they formed at a very low latitude, only a few degrees from the equator. Such indications show that ice covered the entire planet and reached as far as the equator where it was of a considerable thickness. It is believed that this cooling was one of the biggest catastrophes that have ever hit our planet in terms of the survival of the living world, and it is known as the Earth - ice ball (Nenadić & Bogićević, 2010).

Vertical movements of continental blocks, or orogenesis (mountain formation), also have a great influence on the climatic conditions on our planet. They occur exclusively as a result of the plate tectonic process, and when strong and pronounced enough, they will cause global changes in atmospheric and ocean currents, causing a change in climatic conditions. In the last 15 million years, some continental blocks have risen by 600 m, or more.

As for other factors that influenced the climatic conditions in the Pleistocene, it is believed that there is a significant relationship between glacial periods and changes in the composition of the atmosphere, and it is not yet possible to say with certainty whether these changes are the cause of cold climates or whether cooling affects large atmospheric changes. Increased solar radiation in areas that are not affected by glaciation causes the climate to become wetter, while where it is reduced, the climate becomes continental and dry. At higher latitudes, increased radiation causes the melting of snow and ice, while with a decrease in the intensity of solar radiation, the climate becomes colder, and vast snow and ice deposits are re-formed. After strong volcanic eruptions, a huge amount of volcanic dust and ash reaches the atmosphere. These particles reflect solar radiation, effecting variations in temperature on the Earth's surface. It is believed that after larger volcanic eruptions, direct solar radiation can be reduced by an average of 10%, which is enough to lower the average annual temperature by approximately 2–3°C. In the long run, strong volcanic activity leads to an increase in the amount of greenhouse gases in the atmosphere, which contributes to global warming.

### 3.3.2 Quaternary formations

Quaternary formations represent the youngest deposits on our planet and are characterized by a wide distribution in the form of a thinner or thicker cover that spread across almost the entire surface of the Earth, with the exception of the extremely steep mountain sides.

These are often scattered deposits that have not passed the initial stage of lithogenesis, forming landforms representing processes of deposition (river terraces, shorelines, moraines, and drumlins, for example). Glacial moraines at the mouth of the Alpine and Rocky Mountain valleys are preserved ancient examples of moraines forming at higher elevations today. Large marine delta deposits of the Quaternary age are often associated with the same major rivers still actively building their modern deltas, such as the Mississippi River and the Ganges River.

Tectonically, they are mostly undisturbed, so that folding and fault deformations are very rare and are often the result of glacier movements or mass slips. These formations generally have a small thickness, usually up to several tens, rarely hundreds, of meters, although in some depressions it is up to 2,000 m.

The type of Quaternary deposits that will be formed depends on the interaction of several factors, among which tectonic movements, climate and relief are the most prominent.

- Tectonic movements are the main factor in relief modeling, and therefore the key influence on the appearance, distribution and thickness of Quaternary formations. The constant antagonism between the endogenous and exogenous forces, has caused the creation of different forms of relief that indirectly affect the character of denudation-accumulation processes and the creation of different genetic categories of Quaternary sediments.
- Climate determines the regularity in the distribution of different genetic types of deposits, and the intensity of exogenous processes: erosion, transport and accumulation, strongly influencing the appearance and type of newly formed sediments.
- The formation of Quaternary deposits is closely related to the relief, which indirectly affects the creation of different types of Quaternary formations, their composition and distribution. The modern relief to a great extent gained its appearance during the Quaternary due to the combined effect of endogenous and exogenous processes.

The influence of the endogenous factor is reflected in the formation of prominent morphostructures or anomalies. Using geological and geophysical methods, it is possible to investigate the tectonic deformations of different morphostructural levels: the presence of folding and faulting elements, and the influence of plutonic and volcanic bodies on the appearance and development of relief.

Traces of the exogenous formation of the relief have left a clear imprint on the Quaternary deposits and are largely preserved, be they glacial, aeolian, alluvial or slope formations.

The climatic factor has a great influence on exodynamic shapes, and monitoring of modern exogenous processes can explain the climatic changes that prevailed in the period when the morphostructures and sediments were formed.

Quaternary age deposits mostly consist of unbound and loosely bound formations such as gravel, sand, siltstone and clay.

- Debris and other material that are the result of the action of glaciers are called the till. This is deposited on the top, at the sides, base layer and on the surface of glaciers forming various types of moraines. The material that is washed out as a result of the melting of glaciers is deposited in valleys at the base of glaciers, such as karst poljes. Fluvioglacial processes first transported the material by streams formed by melted glaciers down to large river systems that transported it further and shaped and redeposited it in the form of alluvial sediments.
- A classic example of Quaternary sediments are loess deposits. These are very characteristic formations of the Quaternary related to the cold stages of the Pleistocene, and deposited in wide periglacial areas.
- The material deposited on slopes can be moved by the action of rivers which is how alluvial deposits of gravel, sand, siltstone and clay are formed. Such deposits are very common in terrestrial conditions: in the mountains and in plains, at the mouths of large rivers etc. Alluvial deposits are among the most widespread continental formations of the Quaternary period and are found along all major rivers around the world, where they build river terraces and alluvial plains.

#### 3.3.2.1 Glacial forms and glacial erosion

Glacial erosion is a major feature of high mountains such as the Alps, the Himalayas, the Andes and the Rocky Mountains. Glacial erosion formed glacial valleys in the Alps, while peaks such as the Matterhorn on the border between Switzerland and Italy remained. Valley glaciers near the coastline of oceans formed the deep fjords of Norway, Greenland, Canada, Alaska, Chile and Antarctica. Many glacial formations and striking glacial landforms are today submerged beneath the sea as a result of rising of the world sea after the last glaciation.

The till on the sides of glaciers formed lateral moraines, while at the top, as far as the glacier reached, frontal moraines were formed. Due to multiple glaciations and the changes in the size of glaciers, there are several generations of moraines from different glacial phases. Many of them were completely or partially destroyed by subsequent glaciations which were more intense. Such forms of glacial erosion and accumulation are found in Europe, Scandinavia and South America. The importance of moraines in the formation of modern relief can be seen in numerous examples around the world, as in the area of Long Island (New York) where the entire island is formed of the two large terminal moraines, which are created as part of the Laurentide Ice Sheet. These moraines were formed about 22,000 to 18,000 years ago, this was followed by glacial sediments shaped by coastal processes and stream processes into the sandy barrier islands of Long Island's south shore. In the southern hemisphere, south of 40–50° south latitude, various forms of glacial formations that participate in the formation of the modern relief are widespread and have influenced the formation and appearance of today's coastlines.

#### 3.3.2.2 Loess and aeolian sands

Loess deposits represent very widespread Quaternary formations in the northern hemisphere. They were deposited during the Pleistocene in the unglaciated areas of Eurasia and America, south of the former glacial sheet.

Significant deposits of sand and loess were formed in these dry areas. Loess formations were formed in cold and dry climates when strong winds blew over glaciers and the periphery of glaciers, carrying fine siltstone material and transporting it for miles across areas of grassy steppes and cold dry tundra. During the interglacial, there was a partial destruction and decomposition of the loess formations at their highest horizons, resulting in the formation of the surface crust of land or paleo-land (buried land). Due to the large presence of relatively lush vegetation, these horizons are rich in humus, and the loess itself has characteristic vertical cavities which indicate dust filled grassy vegetation.

Thick loess deposits are found in Europe and Asia, and also on extensive sections along the Yellow River west of Beijing. In the valley of the Mississippi River the accumulated deposits are over 8 m thick but their thickness decreases towards the east under the influence of winds from the west.

Two types of sediment can be distinguished: primary (periglacial) loess created as a periglacial formation by dust that came from the moraine area of Central Europe, and secondary (extraglacial) loess created by the redeposition of local material (Nenadić & Bogićević, 2010).

Loess deposits are located in areas between 55° and 24° north latitude, while in the southern hemisphere there are loess-like sediments formed between 24° and 45° south latitude (Figure 3.6). The thickness of these formations varies being 30 m in the Danube and Rhine valleys, 80 m in Ukraine, and up to 300 m in China (Yellow River area). In North America, the thickness of loess formations is somewhat lower ranging between 5 m and 15 m.

Loess is essentially a massive, loose and porous sedimentary rock of a siltstone-clay composition, mostly yellow-brown in color. It is characterized by the absence of large rock fragments and macroscopically visible stratification. Its color largely depends on the presence of the sand component, so that its

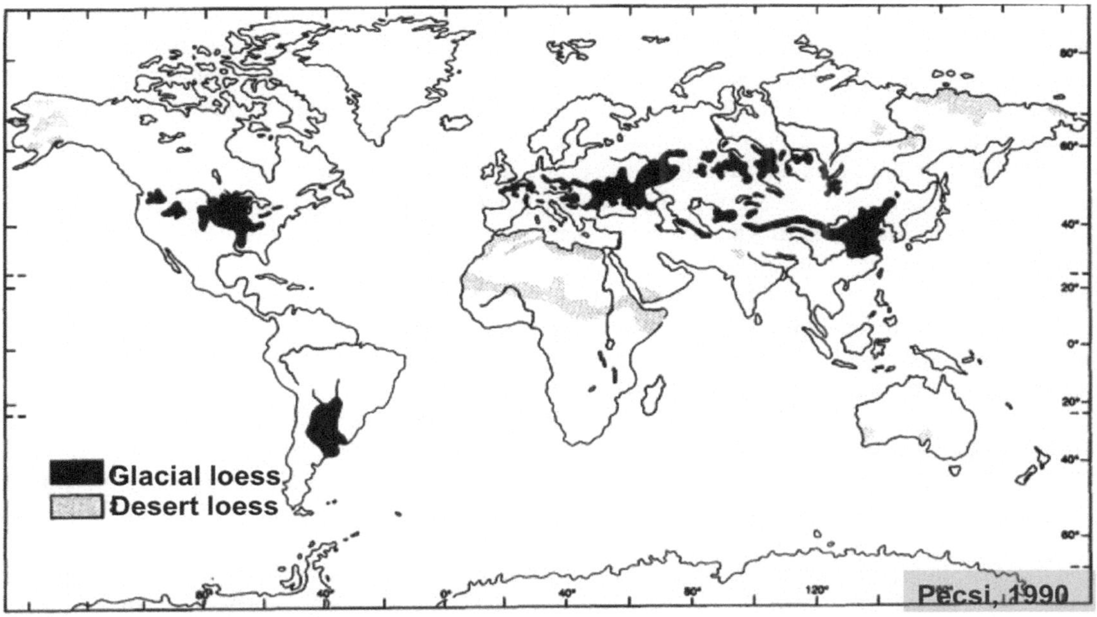


**Figure 3.6** Loess distribution map of the World (from Pecsi, 1990).

increased presence causes the yellow color to become more pronounced. The vertical round cavities allow it to last for a relatively long time in steep sections. Due to the manner in which sedimentation occurred, the position in the relief, the alternation with buried soil and because of certain paleontological characteristics, this sediment is considered a reliable paleoclimatological marker.

The trapping of dust (loess) is mostly related to vegetation cover (e.g. Danin & Ganor, 1991; Hatté *et al.*, 1998, 2013; Svirčev *et al.*, 2013; Tsoar & Pye, 1987). Therefore, the northern border of the loess distribution in Euroasia probably coincides with the northern fringe of past vegetation systems. Almost no loess accumulated south of the northern timberline during the last glacial maximum, as the accumulation of loess was mostly related to both tundra and steppe environments.

Large amounts of sand deposits are fossilized in Eurasia and North America. They were formed during the Pleistocene by the deflationary blowing of fine-grained material from moraine sediments, and were deposited not far away, while the fine-grained siltstone fraction was carried by the wind much further distances and accumulated in the form of loess (Nenadić & Bogićević, 2010). The thickness of the Aeolian sands ranges from a few tens to hundreds of meters.

### *3.3.2.3 Alluvial sediments*

Fluvial processes are associated with rivers and streams and the deposits and landforms created by them. The term fluvioglacial process is used when a stream or rivers are associated with glaciers, ice sheets, or ice caps.

The base level of erosion is an imaginary surface of an irregular shape, inclined toward the lower end of the principal stream of a basin, below which the stream and its tributaries are presumed not to be able to erode. The 'ultimate base level' is the plane that results from the projection of the sea level under landmasses (Goudie, 2004). The level of the world's seas has changed drastically during the geological past. In the Pleistocene during the cold phases, when a large amount of water was trapped in glaciers, the level of the world sea was drastically lower than today. The global sea level has fluctuated through multiple cycles, ranging from a few meters above the current level to 139 m below. The record of ice-volume changes in oxygen isotopes, calibrated to sites on oceanic islands, implies that at the peak of the most recent glacial stage 18,000 calendar years ago, the sea level was some 120 m lower than it is today. About 130,000 years ago due to the rise in temperature, the global ice volume declined, and because of that, eustatic sea levels were approximately 5–6 m higher than today (Lambeck & Chappell, 2001). Base levels may be local when large landmasses are far from the sea or disconnected from it, like for example the Danube upstream of the Iron Gate.

The higher the river flow, and the steeper the gradient, the faster the flow and the higher the energy with which it performs erosion and the transport of materials. The river current is usually strongest near the source. A swift current can move even large boulders. These break apart, and the pieces that are carried in the moving water scrape and dig into the river bottom, or bed.

Little by little, a river tears away rocks and soil along its bed and carries them downstream. It carves a narrow, V-shaped valley. Rapids and waterfalls are common to rivers, particularly near their sources.

Eventually, the river flows to lower land. As the slope of its course flattens, the river cuts less deeply into its bed. Instead, it begins to wind from side to side in looping bends called meanders. This action widens the river valley.

At the same time, the river begins to leave behind sediments, the rocks, sand, and other solid material it collected upstream. Once the sediment is deposited, it is called alluvium. Alluvium may contain a great deal of eroded topsoil from upstream and from the banks of its meanders.

Near the end of its journey, the river slows and may appear to move sluggishly. It has less energy to cut into the land, and it can no longer carry a heavy load of sediment. Where the river meets the ocean or a lake, it may deposit so much sediment that new land, a delta, is formed. Some rivers do not form deltas at their mouths because this is not possible due to terrain and relief conditions. The Amazon does not have a true delta, as the strength of the tides and currents of the Atlantic Ocean prevent the build-up of sediment.

Alluvial systems, according to the morphology of the flow and the river network, can generally be divided into two types between which there is a stepped transition. Alluvial fans[1] are formed by short-term torrential flows. Such flows deposit unbound alluvium[2], forming mainly braided channels. Alluvial fans often grade into, or occasionally flank alluvial floodplains. The alluvial plain of a river usually consists of one main river channel (which meanders along its course) and which has a continuous flow.

Braided flows are characterized by large gradients, short-term and strong currents, and coarse-grained unsorted material. In systems of meandering flows, the gradient is less steep, the flow is continuous, and the sediments are fine-grained.

Watercourses, or specifically riverbeds, are most often predisposed by the tectonics and relief and later sculpted by erosion-accumulative processes. That is why river flows often drastically changed their directions, meandered, broke through gorges and deepened canyon valleys, all depending on the relief, soil composition and tectonic activity of the given area. The level of the world sea played a major role in

---

[1]Alluvial fans typically form in elevated or even mountainous regions where there is a rapid change in slope from a high to low gradient. When the slope decreases rapidly into a relatively planar area or plateau, the flow loses strength and sedimentation of the material occurs.

[2]Alluvium is loose, unconsolidated soil or sediments, which has been eroded, reshaped by water in some form, and redeposited in a non-marine setting, made up of a variety of materials, including fine particles of silt and clay and larger particles of sand and gravel.

the formation of flows and the relief since the changes in the erosion base caused sedimentation and erosion in the given terrain.

Climate had a great influence because it conditioned the intensity of glaciation and thus the amount of water that formed flows in the interglacial periods. The faster and more intense the temperature fluctuations between the glacial and interglacial phases, the stronger the torrents formed by the melting of glaciers and the greater the power of erosion and transport of matter. Valleys having a 'U' profile indicate a glacial process, while those having a 'V' point to a fluvioglacial and fluvial process.

V.V. Lamakin (1948) singled out three dynamic phases of alluvial accumulation: instrative, perstrative and constrative (Nenadić & Bogićević, 2010).

The instrative phase is characterized by an intensive cutting of a river or stream into the river valley, where in the initial phase, as a rule, there are no alluvial deposits. At this stage, narrow valleys of a gorge or canyon are formed. The riverbed is more or less straight, and sharp meanders, if they exist, are due to tectonic causes. Due to the high velocity of a flow, only erosion takes place in this phase, and the bottom of the riverbed is only covered with large and unrounded material in exceptional cases. In the later stages, when there is a decrease in the slope of the longitudinal river profile and when there is further strengthening of lateral erosion, clearly defined gravelly-sandy riverbeds are formed.

The perstrative phase is characterized by weak tectonic activity or by conditions of a dynamic balance between endogenous and exogenous processes. Lateral erosion widens the river valley, so that in addition to riverbed formations (gravels and sands), downstream deposits of (siltstones and clays) are also present.

The constrative phase is characterized by the multiple alternation of riverbed sediments and flood sediments formed in wide alluvial plains where there is a synchronous lowering of the terrain and an intensive accumulation of river sediments. This phase is characterized by the reduction or cessation of deep erosion when the river acquires an equilibrium profile, and meanders, oxbows, ponds, backwaters, etc. are formed in the relief of the valley. These shapes, due to the easy movement of the riverbed, very often change their position, which results in the complex structure of the relief of the valley. Such a ratio of deposits indicates rhythmic polycyclic sedimentation with formations whose thickness can be tens or even hundreds of meters.

Delta deposits are a special type of alluvial formations located at the transition between continental and marine sediments. Within a delta, it is possible to distinguish the upper – above-water part built mainly of alluvial deposits and the underwater part made up of both alluvial and marine formations. The upper or above-water part of the delta is lithologically represented by sands, siltstones and clays often enriched with organic matter. The stratification is clearly expressed and mostly sloped. The largest delta in the world is jointly formed by the rivers Ganges and Brahmaputra in India at their confluence at the Bay of Bengal. It covers an area of about 80,000 km$^2$. The Nile delta in Egypt also covers a large area, as do the Danube delta in Romania, the Amazon delta in Brazil, the Po river delta in Italy, etc.

### 3.3.3 Rivers and the Pleistocene relief

Rivers and river valleys were heavily influenced by climate changes in the Pleistocene. The morphology of river systems, as well as the characteristics of sediments that are transported and deposited by river flows, depend largely on the amount and type of sediment that enters a river from the areas surrounding the river basin. In addition to amount of material, the river flow and its strength also affect the morphology of the canal network and sediment deposition. Both these factors are closely related to climate. Climate influences precipitation and evaporation, the seasonal fluctuations of temperature and rainfall, the propagation and type of vegetation cover and the manner and intensity of the process of rock decomposition (weathering processes). Due to fluctuations in the level of the world sea during the Pleistocene, there was often a change in the erosion of the base level of rivers in coastal regions. All this resulted in the very dynamic and varied development of rivers and river basins.

Flows, resulting from the melting of glaciers formed braided river flows (braided-channel patterns) due to the large amount of material that was introduced as a result of glacial erosion and that was carried by strong torrential flows of the dissolved glaciers. The intensity and strength of the flows within a period depended on the daily and seasonal temperature fluctuations which determined the intensity of melting. Each glacial cycle resulted in the formation of thick layers of gravel and sand that were deposited in the river valleys. With each subsequent cycle, new amounts of material were created and deposited over the older layers of gravel and sand, while some layers were partially eroded or re-deposited under the influence of new torrential flows. In such dynamic environments, sediments were formed in the valleys of large rivers such as the Thames, the Elbe, the Rhine, and the Seine in Europe, as well as in the valleys of the Hudson, Mississippi, and Ohio rivers in the United States.

The gravelly-sandy layers of the older glacial cycles have been particularly preserved in areas where tectonic activity resulted in the lowering of the land which prevented the erosion of these layers. Thick deposits of sand and gravel or lacustrine silt have also been preserved in those parts of river valleys where glacial lakes were formed as a result of glacial damming.

During the transition from glacial to interglacial climatic conditions, river networks evolved from a system of intertwined river flows to a system of meandering rivers. The main reason for this was the reduced amount of sediments as a result of less water and the smaller flow of watercourses. In conditions of reduced flows, significantly smaller river flows continued to meander in the wide river valleys eroding the older alluvial deposits. This is how river terrace systems were formed. The position of the erosion base, local tectonic conditions and the intensity and strength of a flow influenced the number and morphology of river terrace systems.

Present day flows are significantly smaller than the wide valleys formed during the Pleistocene thaws and within which several generations of river terraces from different glacial periods can often be observed.

The deserts of the subtropical regions during the Pleistocene were affected by intense aeolian activity. While the southern hemisphere saw the peak of glaciation between around 20,000 to 12,000 years ago, the climate in Australia reached a peak in aridity and the aeolian processes were then the most dynamic. Similar conditions prevailed in the Sahara and in the deserts in Africa, India and the Middle East. It is assumed that tropical and arid areas covered about a five-times greater surface area during glacial periods. The level of the world sea was lower, water masses were colder and tropical cyclones were less widespread, which resulted in less precipitation. Intense aeolian activity left traces not only in the form of terrestrial sediments. Oceanic sediments in areas that were reached by winds also contain a proportion of aeolian sediments deposited during the period of maximum aeolian activity. Observed also in the ice wells in Greenland and Antarctica were microparticles that were deposited at the time of maximum glaciation and which were blown away and carried by the wind from areas of extremely dry climates which then prevailed in the subtropical zone. During periods of milder and warmer climates in glacial areas, the climate also changed in the subtropical zone. The climate became wetter and desert areas became smaller, dunes were covered with vegetation and were thus stabilized, and aeolian activity was reduced. For example, in the period before about 12,000 up to 5,500 years ago the Sahara was fertile soil – fertile savannah, and this fertility was primarily due to monsoons.

Continental plates continued to move during the Pleistocene, but the position of continents and the distribution of land masses at the beginning of the Pleistocene essentially correspond to their modern day position. Tertiary tectonics had a much greater influence on the changes in climatic conditions, and on the events and phenomena that marked the Quaternary. The rising of land that occurred at the end of the Tertiary affected ocean currents and atmospheric circulation, which resulted in climate change.

The vertical rising of the Earth's crust in the Quaternary, in addition to the reactivation of existing faults, was also caused by the reduced loading due to the melting of Pleistocene glaciers. While parts of land masses sank during the glacial period because of the pressure of ice masses from the surface of the land, some land masses rose during the interglacial phases. The ascents caused by the unloading of the land during the melting of glaciers were initially very fast but slowed down over time.

About 44 million years ago, strong pressure from the south formed the Pyrenees and then the Alps. This pressure led to the folding and rise of deep layers of rocks deposited in the Tethys Sea, which with a crystalline core rose to a height that reached the height of today's Himalayas. These movements lasted until 9 million years ago. During the Quaternary, the newly formed orogenic massifs were reduced substantially by erosion, and the relief gained the outlines of its present day appearance.

### 3.3.4 Modern rivers

Rivers, as large natural flows of surface waters, can be found on all continents and on almost all terrains, regardless of their climate, latitude, lithological structure and other characteristics.

The present day appearance of river courses was formed in the Holocene (Figure 3.7), as a consequence of previous global climate changes and regional and local tectonic conditions. Even today, their appearance

**Figure 3.7** Global river network of major rivers (Lehner and Grill 2013).

is constantly changing under the influence of present day processes of vertical displacement, erosion, transport and accumulation.

Today's river flows can be formed by water outflow from the lakes. One such example is Lake Itasca which is the source of the Mississippi River, the largest river in North America (3,766 km). The Mississippi starts from Lake Itasca, and then along its course it feeds on the waters of other lakes and river tributaries on its way to the Gulf of Mexico.

Lake Toma in the Swiss canton of Graubünden is generally regarded as the source of the Rhine. Its outflow is called Rein da Tuma and after a few kilometers, it forms the Anterior Rhine.

The Nile and the Congo are the two largest rivers in Africa. The White Nile originates from numerous streams on Mount Burundi, and then flows through Lake Victoria, the largest lake in Africa. The Blue Nile is formed by the outflow of water from Lake Tana in Ethiopia and meets the White Nile on the territory of Sudan, forming the Nile River. The Nile then flows through the Sahara Desert and discharges into the Mediterranean Sea.

The springs of the Congo River lie in the mountains around the Great East African Trench. It is formed there by the confluence of the rivers Luapula and Lualaba. The Congo with its tributaries flows through tropical forests and is as a result very rich in water. The Congo is the largest river in Africa according to the amount of water it carries and the surface of its catchment area, and it is the second largest globally (after the Amazon). It is also the deepest river in the world (over 220 m of depth).

The source of a river may be a melting glacier, such as the Gangotri Glacier which is the source of the Ganges River in Asia.

A river flow can also be fed by melted snow as in the snow from the Andes that feeds the Amazon River.

The Amazon begins as an icy stream high in the Andes mountains of Peru. It flows through Brazil and empties into the Atlantic Ocean.

A source, in the form of a groundwater outflow can also be the headwaters of a river. The source of the Danube River is a spring in the Black Forest of Germany. This karst spring discharges at 15 to 70 l/s.

The waters of this karst spring sink and after 90 m of underground flow spring out forming the surface stream Brigach. The surface streams of Brigach and Breg form the course of the Danube after 1.5 km.

Much of Australia is arid, but rivers still run through it. Australia's principal rivers are the Murray and the Darling, both in the south-eastern part of the continent. The Murray is about 2,590 km long and flows from the Snowy Mountains to a lagoon in the Indian Ocean. The Murray is joined by the Darling, a 2,739 km long river that flows from the highlands on the eastern coast.

From its source, a river flows downhill as a stream which is fed by water from atmospheric precipitation, groundwater, melted snow and ice from the surrounding terrain in the catchment area of the stream, and by water from other surface streams (tributaries) or lakes.

Along its course, the Amazon River is fed by more than 1,000 tributaries. Part of the river basin of the Congo River lies around the equator so the water level in the river is very stable due to the warm and humid climate of the equatorial belt.

Some rivers are only occasionally active, while in dry periods they dry up. Wadis, or dry river valleys were formed during the Pleistocene glaciation when streams of melted snow and ice flowed during the interglacial phases. These rivers formed today's riverbeds which were once extremely rich in water and surrounding vegetation.

At the end of the Pleistocene and at the beginning of the Holocene, the climate in these areas became drier and warmer becoming a desert climate. The riverbeds were left without water, except during occasional sudden showers or during the colder winter period of the year when snow and ice melted from the surrounding higher terrain. Wadis are polygenetic in origin as fluvial and aeolian processes participated in their formation. Most wadis are in the Sahara and in the Arabian desert.

The flowing water of a river has great power to carve and shape the landscape. Many landforms, like the Grand Canyon in the US state of Arizona, were sculpted by rivers over time by a process of weathering or erosion.

## 3.4 METHODS USED FOR THE RECONSTRUCTION OF CLIMATE CHANGE IN THE PAST AND THE ABSOLUTE DATING OF QUATERNARY SEDIMENTS

It is easy in some areas to map terminal moraines as they trace across the landscapes of Indiana and Ohio, or from Denmark through northern Germany into Poland and Lithuania. However, correlation of glacial and interglacial events from continent to continent and between hemispheres requires extensive radiometric dating of fossils. The advent of marine oxygen isotopic records gave fresh impetus to attempts to correlate the long-held 'four glaciation' models from various parts of the northern hemisphere. Isotopic records have shown that there were actually as many as 30 distinct glaciations and many shorter events within the Pleistocene alone.

Studies of river terraces in the Alps, resulted in a chronology that suggested the Pleistocene consisted of four or five major glacial stages which were separated by interglacial stages with climates generally similar to today's. Beginning with studies in the 1950s, a much better chronology and record of Pleistocene climatic events have evolved through analyses of deep-sea sediments, particularly from the oxygen isotope record of the shells of microorganisms that lived in the oceans.

### 3.4.1 Isotopic methods (stable isotopes) – oxygen isotope ratio

An oxygen atom has a 'light' and two 'heavy' isotopes. 'Light' oxygen-16 has 8 protons and 8 neutrons, and is by far the most represented isotope, while there is much less 'heavy' oxygen-18, which is stable like oxygen-16, with 8 protons and 10 neutrons. The quantitative relationship between them in water globally varies depending on the climate. Differences in this relationship in marine sediments, ice cores, or fossils relative to the convention-specific standard (ocean water between 200 and 500 m depth), indicate climate change in the past (Riebeek, 2005). Evaporation increases the proportion of oxygen-18 because water molecules with oxygen-16 more easily pass into the gaseous phase. Condensation affects the spatial changes in the ratio of oxygen isotopes in water because water molecules condense faster with oxygen-18, so it is richer in rain at lower latitudes, while the snow in the polar regions is even poorer in this isotope than the water that originally evaporated (Scot & Cary, 2013).

During the glacial period, water poorer in oxygen-18 is permanently retained in the form of ice and snow in polar ice sheets, glaciers and permafrost, so the water in the ocean becomes increasingly rich in this isotope. Based on a comparison with the current state and standard, it is possible to determine the volume of ice that corresponds to an oxygen isotope ratio in ocean water which is directly related to the global temperature of the sea, land and atmosphere, (Scot & Cary, 2013). In 1999, it was shown that the variation in the isotopic composition of oxygen in sediments on the ocean floor confirms Milanković's predictions about the changes of cycles (Rial, 1999).

Based on changes in the relationship between oxygen isotopes in seawater, the Quaternary has been divided into marine isotope stages (MIS), denoted by numbers 1 to 103, where 1 corresponds to the

**Table 3.1** Chronology of marine isotopic stage and alpine chronology of glaciation (Deutsche Stratigraphische Kommission, 2016).

| M. Isotopic Stage | The Beginning of MIS in ka | M. Isotopic Stage | The Beginning of MIS in ka | M. Isotopic Stage | The Beginning of MIS in ka |
|---|---|---|---|---|---|
| MIS 1 | 14 | MIS 8 | 300 | MIS 15 | 621 |
| MIS 2 | 29 | MIS 9 | 337 | MIS 16 | 676 |
| MIS 3 | 57 | MIS 10 | 374 | MIS 17 | 712 |
| MIS 4 | 71 | MIS 11 | 424 | MIS 18 | 761 |
| MIS 5d | 115 | MIS 12 | 478 | MIS 19 | 790 |
| MIS 6 | 191 | MIS 13 | 533 | MIS 20 | 814 |
| MIS 7 | 243 | MIS 14 | 563 | MIS 21 | 866 |
| Würm | Riss | | Mindel | Günz | |

Holocene and 103 to the beginning of the Pleistocene. Odd numbers indicate stages with a reduced isotope ratio compared to the standard (warmer climate), and even numbers indicate an increased isotope ratio compared to the standard (colder climate), see Table 3.1.

### 3.4.2 Isotopic methods (radioactive isotopes) – U/Th dating

U/Th series – natural isotopes of uranium and thorium $^{238}U$, $^{235}U$ and $^{232}Th$ (i.e. isotopes that have a sufficiently long half-life not to disappear from the natural environment) are the initial parts of radioactive decay chains ending in stable isotopes of lead $^{206}$ Pb, $^{207}Pb$ and $^{208}Pb$ (Scot & Cary, 2013). Because thorium is practically insoluble in water, unlike uranium, only uranium is included in the precipitation of calcium carbonate from water (abiotic or biotic), so theoretically all thorium registered in carbonates originates from the decay of uranium that coprecipitated with calcium carbonate (Noller *et al.*, 2000). Impurities of $^{232}Th$ found in clay and dust particles during the diagenesis, e.g. of gravel into the conglomerate incorporated into a carbonate filling, can affect the test results (Noller *et al.*, 2000). Because of this, the $^{238}U/^{232}Th$ ratio in nearby silty and clayey sediments is often determined before the analysis in order to make corrections to the calculation used to determine the aging of calcium carbonate using a series with an initial $^{238}U$ value. The easiest way to make a correction is to, assuming an equilibrium between $^{232}Th$ and $^{230}Th$ in the allochthonous material present inside the carbonate, calculate the concentration of $^{230}Th$ in the impurity based on the concentration of $^{232}Th$, and to then based on the content of the impurity in the carbonate mineral concentration calculate the 'excess' $^{230}Th$ (Hughes *et al.*, 2011). After $^{238}U$ whose half-life is 4,500,000 years, the next two members in this series are $^{234}U$ and $^{230}Th$, whose half-lives are 245,000 and 76,000 years, respectively. The decay of $^{238}U$ creates an increasing amount of products, until the moment when the intensity of their decay (which is directly proportional to their quantity) reaches the intensity of their creation by the decay of the previous member of the series (Scot & Cary, 2013). The equilibrium between two consecutive members of a series is established after approximately six half-lives. Therefore, the upper age limit that can be determined using the $^{230}Th$ can be taken to be a time of about 500,000 years.

This method of dating can be applied to alluvial, glacial and related sediments in all situations where calcium carbonate is formed – the carbonate cement in moraines, conglomerates or sandstones formed within alluvial deposits etc. U/Th dating of carbonates is one of the most accurate and precise methods in geochronology (Scot & Cary, 2013).

### 3.4.3 Luminescence due to decay of $^{238}U$, $^{235}U$, $^{232}Th$ and $^{40}K$

#### 3.4.3.1 Basics

Radiation resulting from radioactive decay leads to the ejection of electrons from the orbits of atoms in minerals such as quartz and orthoclase, whereby they remain in a metastable state in parts of the crystal lattice (Noller *et al.*, 2000). For each specific content of these radioactive elements in relation to some of these minerals, there is a certain time after which these electrons can no longer accumulate, i.e. there is a certain dose of radiation (which is equal to the product of that time and the intensity of radiation, which is a function of the content of radioactive elements). Heat or light can cause these electrons to be expelled from the metastable state, resulting in the emission of photons – stimulated luminescence (Duller, 2008). The excitation of quartz and orthoclase with visible light (blue or green for optically stimulated luminescence – OSL; infrared for infrared stimulated luminescence – IRSL), or heat (heat-stimulated luminescence TL), results in luminescence, and the total energy thus expressed corresponds to the dose that was received (equivalent dose $D_e$). The time that was required for this energy to be accumulated is calculated from the content of the radioactive elements from which, by calculations, we get the intensity of radiation (dose rate $D_r$), or from the directly measured intensity of radiation (Duller, 2008). During transport by water, wind or gravity, sediment is usually exposed to solar radiation which causes luminescence and the resetting of minerals. About two minutes are sufficient for the complete reset of quartz, while resetting takes several hours for orthoclase (Duller, 2008). Incomplete resetting gives too much age. Typically, 95% of the light perceived as OSL originates from less than 5% of the grains (Duller, 2008). The reason for this variability in the luster of individual mineral grains, although collected from the same sample, is not sufficiently clear, but it is thought to relate to the initial conditions of quartz grain formation and its history of redeposition and the repetition of resetting and recharging cycles. General observations show that quartz derived from granite and hydrothermal wanes is dark (Duller, 2008). When the grains have gone through many processing cycles and have formed e.g. sandstone, the percentage of light grains is usually higher.

The penetration length of alpha, beta, and gamma radiation in sediment is approximately 10 to 20 μm, 2 mm, and 30 cm, respectively (Noller *et al.*, 2000), and therefore mineral inhomogeneities may affect how accurately radiation intensity is determined if found within a radius of 30 cm from the sampling point. Fine-grained mineral particles (4–11 μm) are generally used for analysis because their whole volume was exposed to all forms of radiation and they are typically monomineralic (Noller *et al.*, 2000). The second often-used size ranges between 100 to 150 μm, where the outer zone exposed to alpha radiation and about 20 μm thick, is removed by dissolving in hydrofluoric acid (Noller *et al.*, 2000). Most of the irradiated dose for larger quartz grains is from external beta and gamma radiation, while orthoclase receives a significant internal dose from the radioactive isotope of potassium $^{40}K$ (Noller *et al.*, 2000). The calculated radiation intensity in the case of external radiation is corrected for the amount of absorption in water if the sediment was water-saturated continuously since deposition, and if not, it is necessary to estimate how much of that period was water-saturated.

The age of quartz determined by OSL can go up to between 100,000–250,000 years, while the age of orthoclase determined by IRSL can go up to 600,000 years. Very old ages can be determined if the radiation is of low intensity.

#### 3.4.3.2 Analytical methods

$D_e$ can be determined by using the regression function based on the energies of the luminescence signal of one part of the sample (kept out of any light), caused by stepwise values of additive beta or gamma doses from a calibrated radioactive source to extrapolate back to the basic luminescence signal energy of the other part of the sample exposed in a laboratory to light without the irradiation from other sources (Duller, 2008).

An alternative method is to, after the reduction of the luminescence signal by laboratory light exposure, incrementally add dosage until the original luminescence level, equivalent to the measured level of luminescence accumulated during the burial period, is reached (Duller, 2008). This method, the single aliquot regenerative dose (SAR) procedure is usually performed in such a way as to make corrections for changes in sensitivity of luminescence signal energy to increasing irradiation dosages. In each cycle, after irradiation and OSL signal measurement and resetting of the carrier mineral, the sample is irradiated by a fixed test dose (smaller than other doses) and luminescence signal energy resulting from each dose ($L_X$) is divided by the luminescence signal's energy caused by the test dose ($T_X$). In Figure 3.8 the first cycle is the measurement of the natural signal ($L_N$) and of the first response to the test dose ($T_N$), enabling calculation of the sensitivity-corrected natural OSL luminescence signal ($L_N/T_N$). In this step we still do not know the $D_e$ corresponding to the natural OSL signal and to the associated sensitivity-corrected natural OSL value. Next, regeneration doses of 16, 32, 64, 96, 128 and 0 Gy (Gray – unit of radiation dose, equal to absorbed radiation energy in joules per kilogram) give sensitivity-corrected values $L_1/T_1$, $L_2/T_2$, $L_3/T_3$, $L_4/T_4$, $L_5/T_5$ and $L_6/T_6$ (these data define the dose response curve for this aliquot). In this way, if the sample aliquot becomes more or less sensitive to irradiation during the increasing irradiation in the sense of produced luminescence signal, the response to the test dose will change, and division of the luminescence signal value for each step by the response to the test dose will correct this (Duller, 2008). By interpolating the value of $L_N/T_N$ onto the dose response curve, $D_e$ is obtained (in this case $25.6 \pm 1.3$ Gy); The final cycle repeats the measurement for a regeneration dose of 32 Gy ($L_7/T_7$). The ratio of the two measurements for 32 Gy ($L_7/T_7$ divided by $L_2/T_2$) is known as the recycling ratio (Duller, 2008), and for this aliquot has a value of $0.99 \pm 0.05$ – close to the ideal value of 1.00, and well within the accepted range of 0.9 to 1.1. All of the measurements are preceded by preheating of the sample to a fixed temperature (normally between 160°C and 300°C) and holding it at this level e.g. for 10 s. This procedure removes unstable electrons from

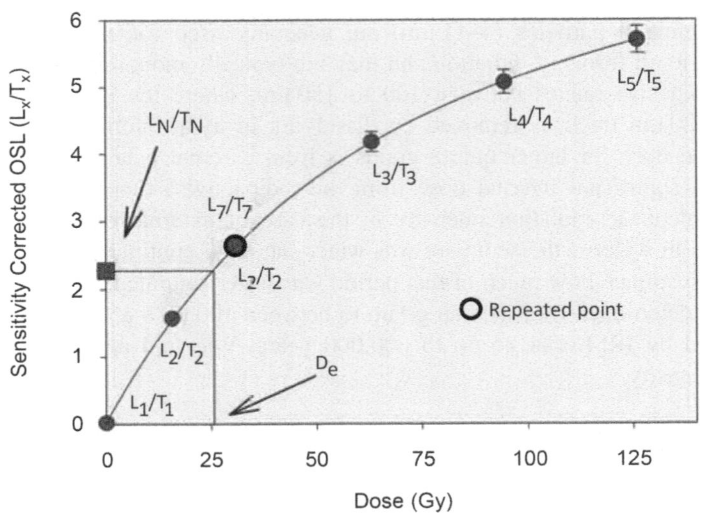


**Figure 3.8** Single aliquot regenerative dose (SAR) procedure (Duller, 2008).

shallow traps so that the OSL signal originates only from electrons that would have been stored safely through the burial period (Duller, 2008).

The dose recovery test is commonly undertaken with SAR. For sediments this involves exposing an aliquot of naturally irradiated grains to daylight or to an artificial light source, irradiating the aliquot with a known dose (e.g. 10 Gy), and then treating this dose as an unknown. The described SAR procedure is applied. The value of the calculated dose should match the known laboratory dose.

Single aliquot measurements reduce the time involved in determining $D_e$ by automating much of the procedure, thus making it feasible to make replicate measurements of $D_e$ for each sample, i.e. to perform measurements for multiple subsamples so it is routinely possible to assess the reproducibility of $D_e$ within each sample (Duller, 2008). Replicate determinations of $D_e$ would be expected to form a normal or log-normal distribution, and in the case that the distribution is normal and not too wide, the mean $D_e$ is taken as the dose acquired since the event of interest (Central Age Model – CAM; Duller, 2008).

If replicate measurements of $D_e$ yield a wide range of values this implies that the sample is complex, and requires further investigation. The most common cause of large scatter is that a large portion of mineral grains in the sample was not completely reset during the event being dated – this is known as incomplete bleaching. In such cases there are different levels of bleaching in the grain population. The luminescence signal measured in the laboratory will then be the sum of the dose acquired since the sample was deposited, plus the residual signal not bleached at the time of deposition (Duller, 2008). In such cases, the best estimate of the dose acquired since the event of interest will be the one based on aliquots with the lowest $D_e$ values (Minimum Age Model – MAM). The second potential cause of $D_e$ variation is the differences between the dose rates in different parts of the sample – small portions of the sample are highly radioactive grains, or gypsum with low radioactivity is present together with wood ash with relatively high radioactivity due to a high content of potassium (Duller, 2008). The third common cause is mixing of material from two adjacent layers of differing age.

The results of single aliquot analyses will also depend on how many grains are contained in each aliquot. In typical luminescence measurements, sand-sized mineral grains around 0.2 mm in diameter are measured. In such cases it is possible to make single aliquots containing a thousand grains, hundreds of grains, tens of grains or even a single grain. In Figure 3.9 the results of the replicate $D_e$ measurements for c 1000, c 200 and a single-grain aliquot sizes are shown. For the first aliquot size the distribution is normal, for the second there is some tailing towards the higher $D_e$ values, while for the single-grain aliquot size the distribution is clearly skewed towards the higher $D_e$ values. If a sample is likely to be affected by incomplete bleaching or mixing, replicate measurements on aliquots with relatively few grains should be used (Duller, 2008).

Aliquots may contain several hundred grains, but the OSL signal observed is likely to originate from a much smaller number of grains because, as mentioned before, 5% of grains dominate with their brightness. Thus, measurements using small aliquots, consisting of tens of grains each, may yield results that are similar to those from single grains. This can reduce the need for time consuming single-grain analyses and the requirement for the laser stimulation system necessary for that. The large variation in the brightness of individual grains also has an impact on the precision with which $D_e$ can be determined which is another downside of single-grain measurements (Duller, 2008). The precision of the measurements may be limited for a grain that does not emit a bright luminescence signal, while brighter grains give more precise $D_e$ estimates.

The range of ages over which luminescence dating can be used varies with the site, depending on the nature of the luminescence signal and the environmental dose rate. The upper age limit is usually controlled by the saturation of the luminescence signal. As samples approach saturation, errors in the $D_e$ value increase, because as the dose response curve flattens, small uncertainties in the value of the natural

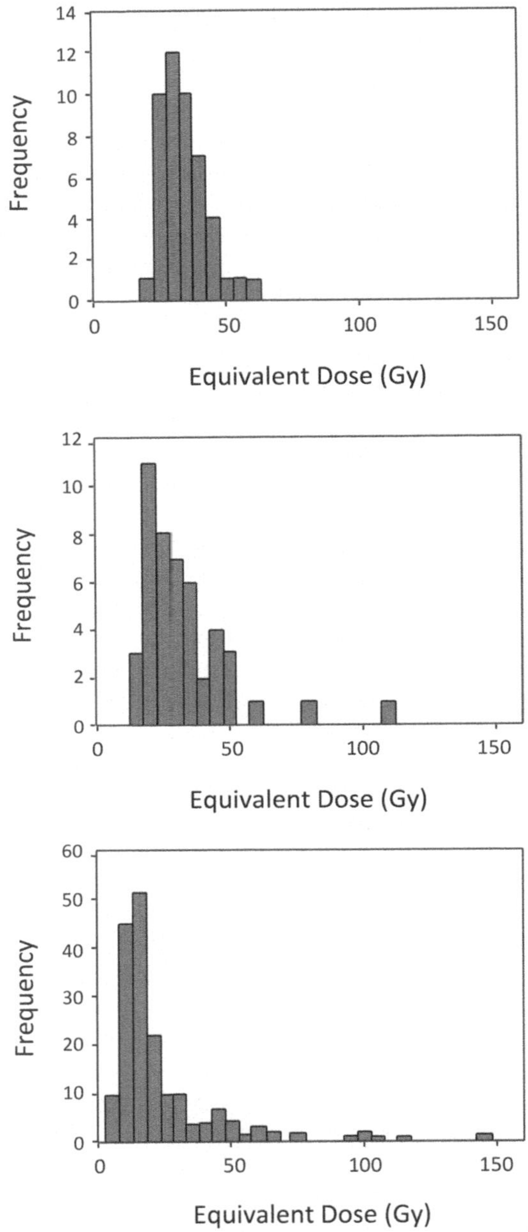


**Figure 3.9** Replicate $D_e$ values obtained for different aliquot sizes (a) c 1000 grains; (b) c 200 grains; (c) single grain (Duller, 2008).

luminescence signal have a big impact on $D_e$ values read from the dose response curve – thus ages obtained also have a high uncertainty level. The usual upper limit of a usable luminescence signal is 85–90% of the maximum saturation value (Duller, 2008).

Many types of sediment do not contain quartz and for their dating the exposure to light of infrared stimulated luminescence is used. Also, the IRSL of potassium-rich feldspars has the potential to extend the datable age range because dose-response curves of K-rich feldspars or the luminescence emissions of polymineral feldspar dominated grains show a much higher saturation dose compared to quartz. The IRSL signals are often much brighter than the quartz OSL signal which enables high precision luminescence measurements to be made (Li & Li, 2011). The downsides of the IRSL are that resetting takes much more time than for OSL, and that the IRSL signal can show anomalous fading, which is a signal loss over burial time causing significant age underestimation (Li & Li, 2011). Recent studies have presented a measuring protocol, in which a high temperature IRSL signal is measured at an increased temperature after a first IR illumination at low temperature (Li & Li, 2011). This pIR-IRSL (post infrared IRSL) signal appears to be less prone to fading and is more stable than the low-temperature IRSL measured at 50°C (IRSL50) and has the same high saturation dose limits as the low-temperature IRSL.

### 3.4.3.3 Dose rate determination

The most important of the radiation sources are radioactive isotopes of uranium (U), thorium (Th) and potassium (K) (Duller, 2008). The radioactive isotope of potassium is $^{40}K$, which decays to $^{40}Ca$ (calcium) or $^{40}Ar$ (argon) by emitting beta particles and gamma rays. Both $^{40}Ca$ and $^{40}Ar$ are stable. Uranium and thorium are more complex: $^{238}U$ decays to $^{234}Th$, which further decays to form protactinium ($^{234}Pa$), and further on until reaching a stable isotope of lead ($^{206}Pb$). Similar decay series exist for $^{235}U$ and for $^{232}Th$ (Duller, 2008). The U and Th chains emit alpha, beta and gamma radiation. The radiation dose from the radioactive elements surrounding the minerals can be calculated from measured concentrations of U, Th and K, or by directly measuring the emission of radiation (Duller, 2008). K occurs in the greatest concentrations, typically 0.5% to 3% by mass, and is measured using atomic absorption spectrophotometry (AAS), X-ray fluorescence and inductively coupled plasma mass spectrometry (ICP-MS) with a precision of 5%. U and Th occur in much lower concentrations which are measured using ICP-MS and neutron activation analysis (NAA). Once the concentrations of these three elements are known, conversion factors enable the calculation of the radiation dose rate (Duller, 2008). For long decay chains, and parallel ones such as those of $^{235}U$ and $^{238}U$, the crucial assumption behind this approach is that the isotopes in the decay chain are present in such mutual ratios that the rate of decay of the previous isotope in the chain (equal to the rate of production of products of this decay) is equal to the rate of decay of product isotopes. This is a state of dynamic equilibrium in which the radiation intensity is constant for a certain concentration of radioactive element like uranium or thorium (Duller, 2008). This is usually the case if there was no frequent percolation of fresh atmospheric waters which could dissolve uranium which would increase the portion of thorium, causing the disequilibrium and different ratio between the uranium concentration and radiation intensity. Precipitation of minerals like calcite from the percolating water can increase the uranium portion, as thorium is practically insoluble in water, and therefore not coprecipitated as uranium is (as mentioned earlier, this disequilibrium is the basis for the U/Th dating method).

The potential problem of inaccuracy in calculating the dose rate from concentrations of radioactive isotopes can be avoided by using in situ or laboratory direct measurement of the radiation dose rate. Owing to the distances they travel, laboratory measurements of the alpha and beta particles flux on a small (c 10 g) sub-sample are very probably an accurate measure of the natural flux of the sample. Due to the much bigger penetrating distances of gamma radiation, the measurement of the gamma dose rate from the sample material could introduce a significant error in the case of inheterogeneous distribution of gamma radiation sources in sediment.

Alpha particles can be counted by using either a thick source alpha counter (TSAC) which is able to measure the total number of alpha particles, or alpha spectroscopy which requires the sample to be dissolved, but can measure the energy of each alpha particle, which helps identify the isotopes undergoing decay (Duller, 2008).

Beta particles, unlike alpha particles and gamma rays, are emitted over a range of energies, so there is no point in using spectroscopy. Beta particles can be simply counted, for example by using Geiger-Müller based equipment, and then the beta dose rate can be directly assessed (Duller, 2008).

The gamma radiation dose rate can be determined in situ using a portable device based on sodium-iodide (NaI) crystals, even though concentrations of individual isotopes cannot be accurately determined (Duller, 2008). Usually a hole more than 50 mm wide must be drilled, and measurement usually takes at least an hour.

This field method is hard to use in deep boreholes, especially in unconsolidated sediments due to the presence of the drilling column keeping the borehole from collapsing, so in these cases laboratory measurements are performed.

There is another factor that affects the effective dose rate received by the minerals – water content of the sediment. The higher the content of water, the less radiation is absorbed by the minerals, the effect of which can be compensated for by calculations, which require an estimate of the water content throughout the burial period. Typically, a 1% increase in water content will increase the calculated age by approximately 1% (Duller, 2008).

All the uncertainties in the measurement of the dose rate – issue of water content often dominating, and in $D_e$ determination, produce errors in luminescence ages normally in a 5 to 10% range (Duller, 2008).

### *3.4.3.4 Sampling basics*

The basis for good sampling is to use geological and sedimentologic knowledge of the deposit to sample well bleached sediments (exposed sufficiently to sunlight before deposition), which are homogenous and of the same age. Well sorted rippled sands are usually well bleached as they are transported slowly and are exposed to daylight, while debris flow deposits are transported quickly and buried and consequently are probably not bleached well. Eolian deposits usually work best, but fluvial, lacustrine and hillslope deposits may also be used (Nelson *et al.*, 2015; Utah State University Luminescence Laboratory [USULL], 2007).

It is best to sample from thick deposits (at least 30 cm thick). Dose rate will vary with heterogeneity of the deposit. Sampling widely different grain-size variations (i.e. clay layer in sand lens) should be avoided. Sediments immediately above an unconformity may contain material eroded from the underlying deposit that was deposited without exposure to sunlight (USULL, 2007).

Water content must be separately determined or preserved in a sample. The knowledge of a site's geological history should help estimate if the moisture content is representative of the burial history of the sample.

For the surficial, or samples close to surface, note the depth of burial to gage the cosmic contribution. In the case of recent erosion or excavation, pre-erosion burial depth should be estimated. If the sample was collected from an accumulating land surface and there are stable land-surface horizons in the stratigraphy (soils), these should also be noted along with the burial depth below those past land surfaces – especially if they were long-lived. The latitude, longitude and elevation of the sample should be noted and used to calculate the cosmic contribution.

The size of the sample depends on the nature of the sediment, although a good value is 500 g. If the material is consolidated it can be carved out. The block must be wrapped in opaque material, like a black plastic or aluminum foil, and packed in such way to prevent it from disintegrating during transport to the luminescence laboratory. The outer surface of the block, which will have been exposed to daylight during its excavation, has to be removed. Thus the block must be large enough – typically 100 mm in each direction – to allow for this. When it is especially difficult to exclude daylight during sampling or a block sample is impracticable, the sample can be performed at night, using a red light (USULL, 2007). In the case of deeper boreholes, the sample can be extracted using a coring tube, bailer tool etc. The content should immediately after reaching the surface be emptied into an opaque container with minimal exposure to daylight. The container (waterproof) should then be sealed and wrapped in a reflective foil.

### 3.4.4 Magnetic methods – sediment magnetization

Magnetism basically arises from two sources: electric current and magnetic spin moments of elementary particles (the spin magnetic moment is the magnetic moment induced by the spin of elementary particles). A magnetic dipole, a small magnet of microscopic to subatomic dimensions, is equivalent to a magnetic field that generates the flow of electric charges into a circuit. Electrons orbiting the atomic nucleus, rotating electrons, and rotating positively charged atomic nuclei are all magnetic dipoles. The magnetic moments of atomic nuclei are usually thousands of times smaller than the magnetic moments of electrons, so they are negligible in the context of material magnetization. Taken together, these effects can be canceled out mutually so that certain atoms cannot be magnetic dipoles. If these effects are not canceled out mutually, the atom is a permanent magnetic dipole, as are iron atoms. Many millions of iron atoms that are spontaneously locked in the same alignment to form a ferromagnetic system also represent a magnetic dipole. Magnetic compass needles are an example of macroscopic magnetic dipoles.

The *strength of a magnetic dipole*, also called the *moment of a magnetic dipole* ***m***, can be considered a measure of the dipole's ability to orient itself towards an external magnetic field. In a homogeneous magnetic field, the strength of the magnetic dipole is proportional to the maximum torque on the dipole, i.e. the one that occurs when the dipole is at a right angle to the magnetic field. The *moment of a magnetic dipole*, often called the *magnetic moment*, can then be defined as the maximum torque $\boldsymbol{\tau}$ caused by the action of a magnetic force on the dipole, per unit force of the surrounding magnetic field in vacuum. The magnetic moment is also defined as a vector that connects the torque caused by an external magnetic field and the vector of that field:

$$\tau = m \times B \tag{3.1}$$

When the magnetic dipole is conceptualized as a circular flow of electric charge, the intensity of the dipole moment ***m*** is proportional to the power of the current multiplied by the area inside the circle, and is expressed as $Am^2$.

Magnetic induction (magnetic flux density) ***B*** induces a magnetic moment in any material within the field. The unit used to express quantity ***B*** is Tesla. The *magnetic induction* ***B*** is a field that is the vector sum of the *magnetic field* ***H*** and the *magnetization* ***M*** of the material (Tauxe *et al.*, 2018) (Figure 3.10):

$$B = \mu_o(H + M) \tag{3.2}$$

Where $\mu_o$ is the constant of magnetic permeability of vacuum ($4\pi\cdot10^{-7}$ H/m).

The intensity of magnetization, ***M*** or ***J*** [A/m], is the intensity of the magnetic moment per unit volume, ***V*** – Tauxe *et al.* (2018), or the density of the net magnetic moment:

$$M = m/V\,[A/m] \tag{3.3}$$

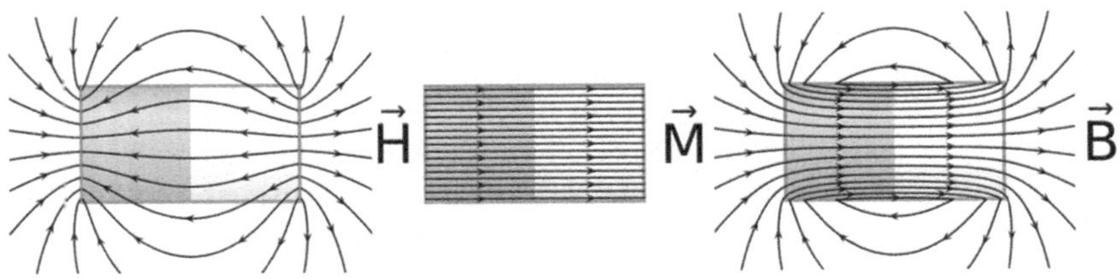


**Figure 3.10** Relationship between the magnetic field, magnetization and magnetic induction (© 2020 Geek3 /https://creativecommons.org/licenses/by-sa/4.0/deed.en/https://commons.wikimedia.org/wiki/File: VFPt_magnets_BHM.svg, with modifications)

Magnetic susceptibility $\boldsymbol{\kappa}$ is the ratio of the intensities of magnetization and the magnetic field – Tauxe *et al.* (2018):

$$\kappa = M/H\,[-] \tag{3.4}$$

while the mass magnetic susceptibility $\chi$ is equal to:

$$\chi = \kappa/\rho\,[m^3/kg] \tag{3.5}$$

When generated magnetic fields pass through magnetic materials that contribute to internal magnetic fields, ambiguities may arise about which part of the field originates from external fields and what comes from the material itself. That is why the magnetic field $\boldsymbol{H}$ is defined as the magnetic influence of external fields in the material, independent of the magnetic response of the material itself.

In terms of magnetization, materials are divided into:

- Diamagnetic, in which $\boldsymbol{M}$ is of very low intensity and is of an opposite direction from $\boldsymbol{B}$ (Vočadlo, 2014). They do not retain magnetization after the cessation of the external magnetic field. Temperature does not affect their magnetic properties so that the susceptibility for a fixed value of the magnetic field intensity is the same regardless of temperature;
- Paramagnetic, in which $\boldsymbol{M}$ is of very low intensity and grows from zero with intensity of $\boldsymbol{B}$, has the same direction as $\boldsymbol{B}$, and they do not retain magnetization after the cessation of the external magnetic field. In these materials, susceptibility decreases with increasing temperature;
- Ferromagnetic, in which $\boldsymbol{M}$ is of high intensity (hundreds of times higher than that of paramagnetic materials) and grows from some initial value (which can be zero) with intensity of $\boldsymbol{B}$, and has the same direction as $\boldsymbol{B}$. They retain part of the magnetization after the action of an external magnetic field ceases. When heated, when the Curie temperature is reached, the magnetization is lost. These are magnetic metals such as iron and nickel. The magnetic moments of whole blocks of atoms, the so-called domains, about 10 μm in size, acquire and retain the same orientation due to the action of an external magnetic field;
- Ferrimagnetic, in which $\boldsymbol{M}$ is of high intensity (hundreds of times higher than that of paramagnetic materials) and grows from some initial value with intensity of $\boldsymbol{B}$, and has the same direction as B. They retain part of the magnetization after the external magnetic field ceases. When heated, when the Curie temperature is reached, the magnetization is lost. In these materials, the atoms are arranged so that the adjacent crystal sublattices have antiparallel magnetic moments of varying intensity, and consequently have a natural permanent magnetization. Their domains are about 1 μm in size, and each has a characteristic identifiable Curie temperature. Certain magnetic minerals are very important for paleomagnetism and mineralogical analyses – magnetite ($Fe_3O_4$) with a

Curie temperature of 580°C, ilmenite ($FeTiO_3$) and titanomagnetite ($Fe^{2+}(Fe3+, Ti)_2O_4$) with a Curie temperature ranging from 200°C to 580°C;

- Antiferromagnetic, in which ***M*** is of low intensity. In these materials, the atoms are arranged so that the adjacent crystal sublattices have antiparallel magnetic moments of the same intensity, and consequently do not have a natural permanent magnetization. Magnetization occurs due to the presence of impurities and defects in the crystal lattice. This group includes the mineral hematite ($Fe_2O_3$) and the metal chromium.

Declination is the azimuth of the magnetic field line and is given by the angle in relation to the geographical north in a clockwise direction and can have a value that ranges from 0° to 359°. The inclination is the slope of the magnetic field lines and is given by an angle that can have values from −90° (up) to 90° (down) – Figure 3.11. In the northern hemisphere, the field is directed downwards. It is perpendicular at the North Magnetic Pole and rotates upward as the magnetic latitude decreases until it becomes horizontal (0°) at the magnetic equator – Figure 3.11. It continues to rotate upwards until it is perpendicular upwards at the South Magnetic Pole.

Looked at in more detail, the Earth's magnetic field is not just a product of a geocentric magnetic dipole (which is generally not axiparallel when observing individual time sections). Multiple fields are superimposed on each other and interact through inductive processes. The most important of these geomagnetic fields are (National Centers for Environmental Information [NCEI], 2020):

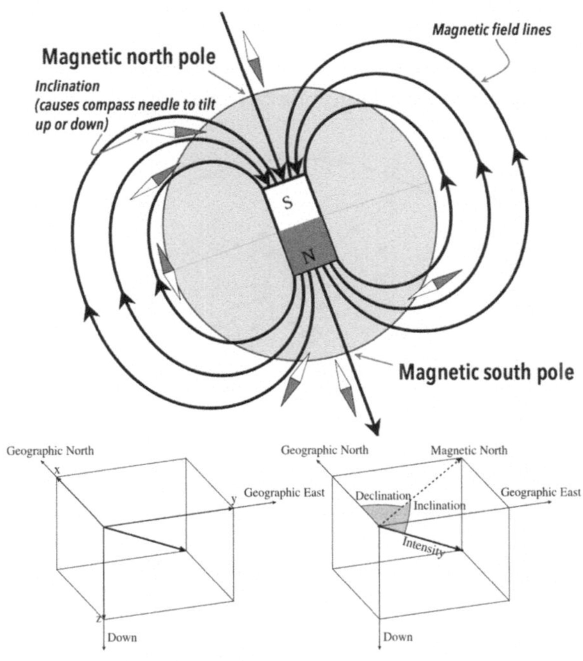


**Figure 3.11** Geomagnetic field lines (modified from Panchuk K. (2019)/https://creativecommons.org/licenses/by-nc-sa/4.0/) and magnetic field vector components (© 2012 Chymæra/ https://creativecommons.org/licenses/by-sa/3.0/deed.en/https://commons.wikimedia.org/wiki/File:XYZ-DIS_magnetic_field_coordinates.svg, with modifications).

- The Earth's main magnetic field generated in the liquid outer core;
- The field formed in the Earth's crust and upper mantle;
- The combined field originating from electric currents in the upper atmosphere and magnetosphere which induce electric currents in the sea and land.

The most suitable dipole for today's field is inclined in relation to the axis of rotation of the Earth. The poles of this best fit dipole, i.e. the points at which the dipole axis passes through the Earth's surface, better known as the 'geomagnetic poles', are located (2019) at 80.6°N, 73.1°W, and at 80.6°S, 106.9°E (World Data Center for Geomagnetism [WDCG], 2020). This dipole field represents about 90% of the total field. The coordinates of virtual geomagnetic poles (VGPs) (Vočadlo, 2014) are calculated based on measurements of declination and inclination at various points on Earth, assuming that the only source of the magnetic field is the geocentric dipole. The axis of the geocentric dipole that at various points on Earth gives the inclinations and declinations that deviate the least from the actual values, intersects the Earth's surface at the geomagnetic poles. The remaining 10%, which represents the non-dipole field, explains the irregularities and anomalies observed on the global maps of declination, inclination and intensity of the Earth's magnetic field. For example, magnetic poles, points at which the field lines are vertical (tilt 90°) and at which the free magnetic needle would face perpendicular to the Earth, do not coincide with geomagnetic poles. The differences are different in the two hemispheres, so that the magnetic poles are not antipodal. The north magnetic pole is at 86.4°N, 175.3°W, while the south magnetic pole is at 64.1°N, 135.9°E (2019) (WDCG, 2020).

According to the hypothesis of a geocentric axial dipole, the geomagnetic field averaged over a period of several thousand years is a dipole field with the center of the dipole in the center of the Earth parallel to the axis of rotation (Figure 3.12).

Based on this assumption, it can be assumed that paleogeomagnetic poles, obtained by averaging the results of paleomagnetic measurements of a large number of samples from an outcrop or from a sample

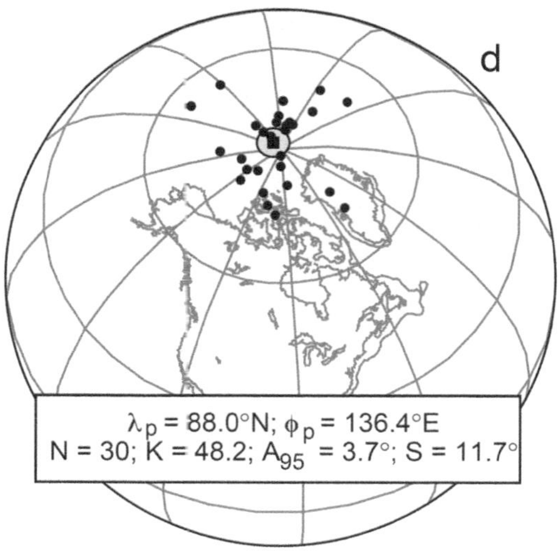


**Figure 3.12** VGPs of rocks with remanent magnetization for 30 sites belonging to Holocene lavas of the western United States. The solid circles show the site-mean VGPs, averaged to determine the 'paleomagnetic pole' shown by the solid square; the stippled circle about the paleomagnetic pole is the region of 95% confidence (Butler, 1998). The rocks are too young to be significantly affected by plate motion.

taken from a drilling core (temporally close but distant enough to include sufficient variations in the dipole geocentric field orientation for the mean inclination and declination to correspond to a geocentric axial dipole), or samples from a number of different boreholes or outcrops of at least a similar age, coincide with the geographic poles. The position of geomagnetic poles determined in this way deviates from the geographical poles more and more the older the samples are (Figure 3.13), which indicates that the sampling point is shifted geographically in relation to the position during the acquisition of remanent magnetization.

The movement of continents and tectonic plates through geological history was reconstructed based on such analyses – Figure 3.14.

During geological time, a large number of rotations of the polarity of the Earth's magnetic field, i.e. a substitution of the location of geomagnetic poles, took place. Figure 3.15 shows the succession of the geomagnetic polarity during the last 5,250,000 years, with the names of the periods of normal and reverse polarity (chron) and of the subperiods of reverse polarity (subchron) from the prevailing polarity of the Earth's magnetic field during the chron. Originally, this scale was formed based on the results of dating geologically young basalt rocks. The application of newer methods, such as dating by measuring the ratio of argon isotopes of $^{40}Ar/^{39}Ar$, has made it possible to determine the age of older effusive rocks in which information on the orientation of the magnetic field at the time of their solidification has been preserved. Based on the results obtained by measuring the magnetic moment in the rocks on the ocean floor, the timescale of geomagnetic polarity for the last 160 million years has been determined. A new crust is being formed on the ocean ridges and the ocean floor is widening (Figure 3.13 and Figure 3.14). Magnetization measurements have shown that it consists of alternating 'stripes' of rocks parallel to the ridge, with a remanent magnetization of opposite polarity.

Ferromagnetic minerals orient themselves in accordance with the Earth's magnetic field if the crystals are small enough and if the other forces acting on them are weaker than the magnetic one. They are ferrimagnetic as long as their temperature is lower than their Curie temperature. When sediment compaction prevents the further movement of mineral grains within them, or when minerals grow to a size that prevents them from moving within the sediment matrix, or molten rock hardens and minerals that crystallize during the process are fixed, they remain fixed in a position aligned with the magnetic the field of the Earth at that moment. It is a *natural remanent magnetization* (NRM), whose main forms are:

- *Depositional remanent magnetization* (DRM) – the phenomenon whereby the magnetic minerals that are deposited as small particles (clay and dust) in a still environment are oriented towards the current magnetic field, especially at the sediment surface. Based on the measurement of the specific direction of the magnetic field towards which the minerals were oriented, it is possible to determine the age of the sediment (Figure 3.15). By combining the determined relationship between oxygen isotopes in seawater and by determining the isotopic ratio in calcium carbonate taken from marine shells deposited on the seabed in fine-grained slowly deposited sediments, the paleomagnetic analysis of their depositional remanent magnetization provided the age of each of the marine isotope stages. The current direction is taken as the normal direction of the magnetic field;
- *Thermo-remanent magnetization* (TRM) – magnetization acquired when a rock cools below the Curie temperature of its magnetic minerals, fixing magnetic domains along positions aligned with the Earth's magnetic field and creating a remanent magnetization that can be stable at room temperature for billions of years;
- *Chemical remanent magnetization* (CRM) – magnetization acquired when a ferrimagnetic mineral grows within the matrix to a critical blocking grain volume at which the current orientation is aligned with the magnetic field and continues to be fixed. The acquired remanent magnetization can be stable over a very long period.

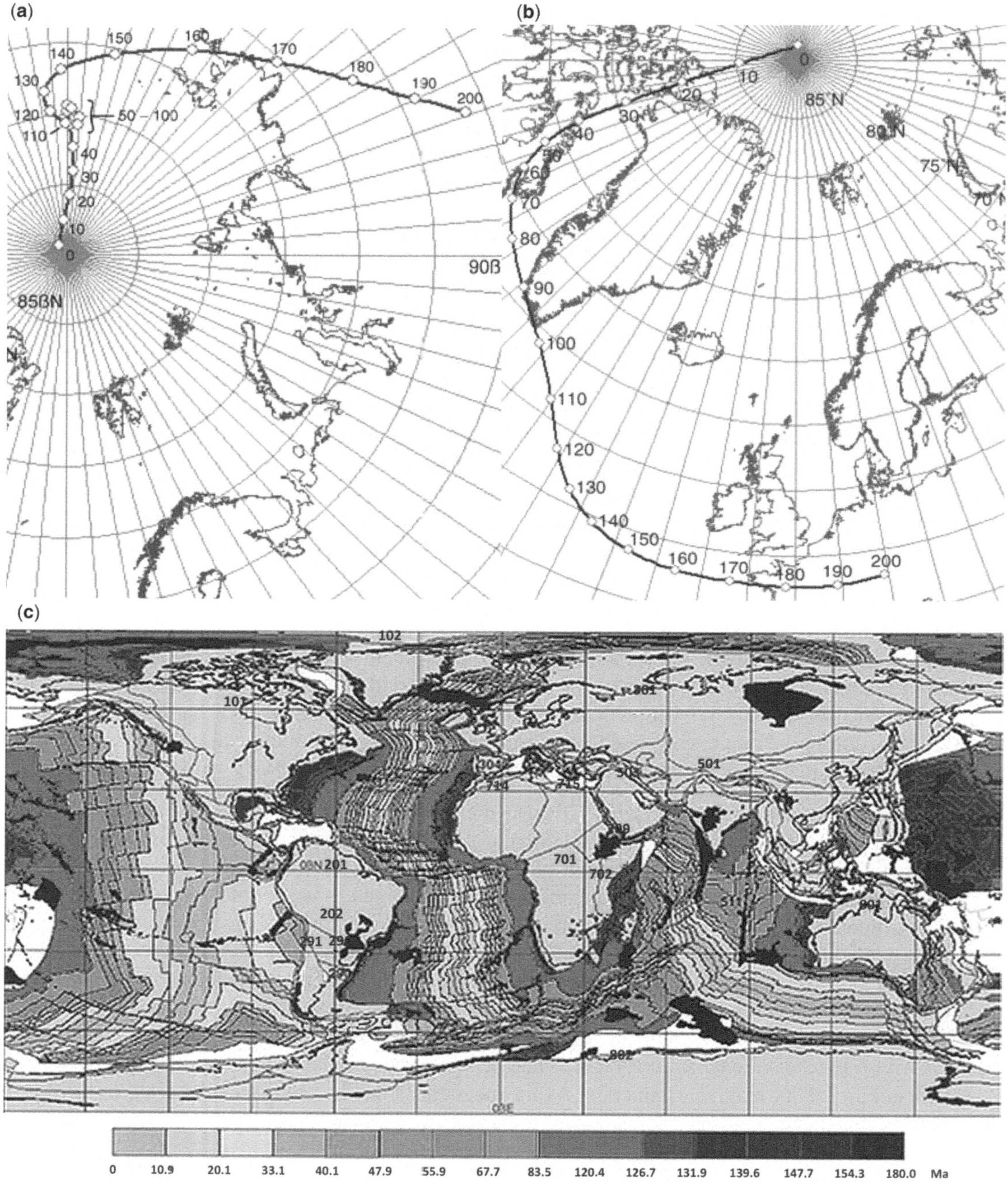


**Figure 3.13** Displacement of the north magnetic pole under the assumption of stillness in geological history of the places where the samples were taken – Eurasia (top left) and Australia (top right); today's age of the ocean floor in millions of years – Ma (bottom) (Schettino & Scotese, 2005).

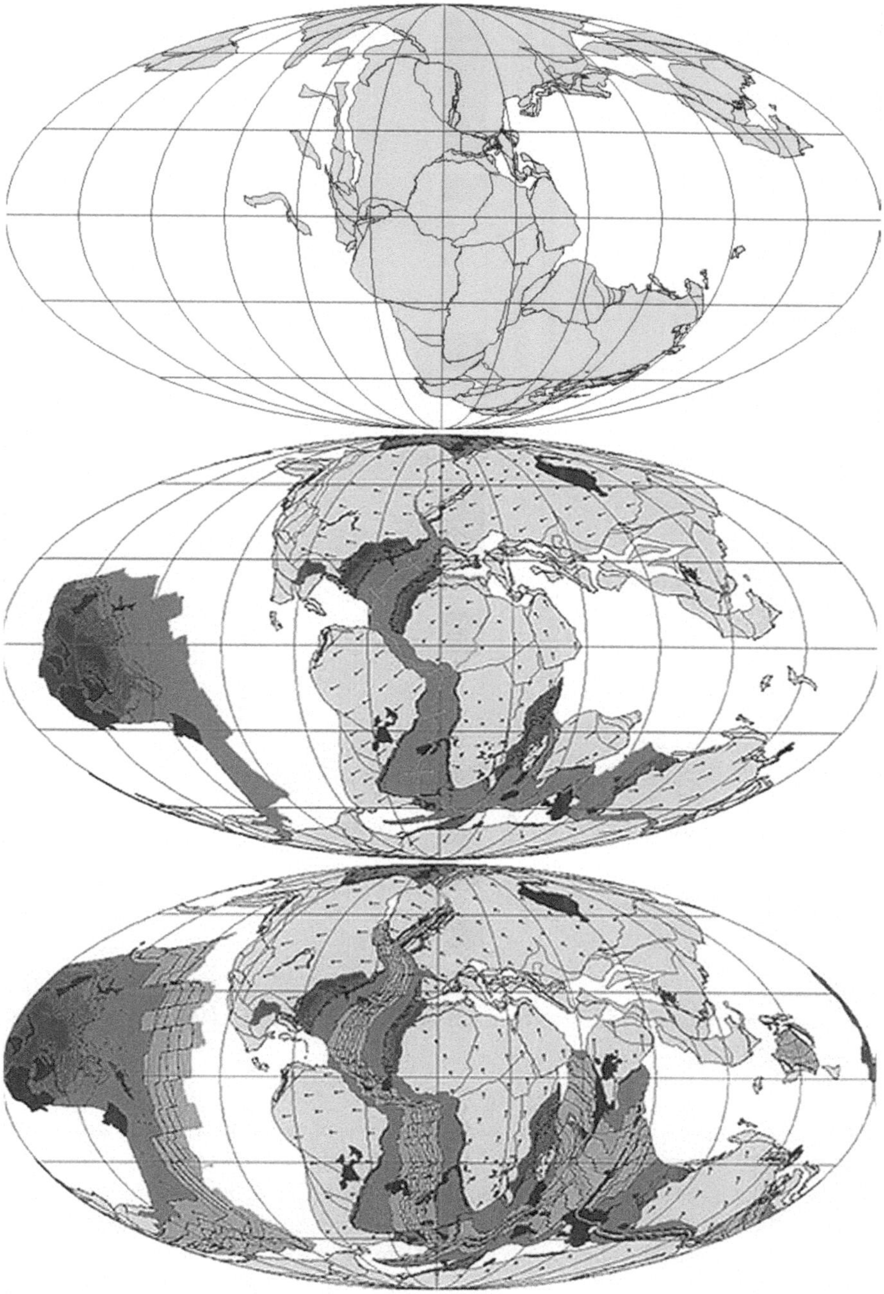

**Figure 3.14** Reconstruction of the position of tectonic plates before 200, 120.4 and 40.1 Ma with the assumption of a geocentric axial dipole. The arrows represent the direction and speed of the plates relative to the geomagnetic reference frame. The shades of the ocean floor correspond to the scale of shades of gray in Figure 3.13 (Schettino & Scotese, 2005).

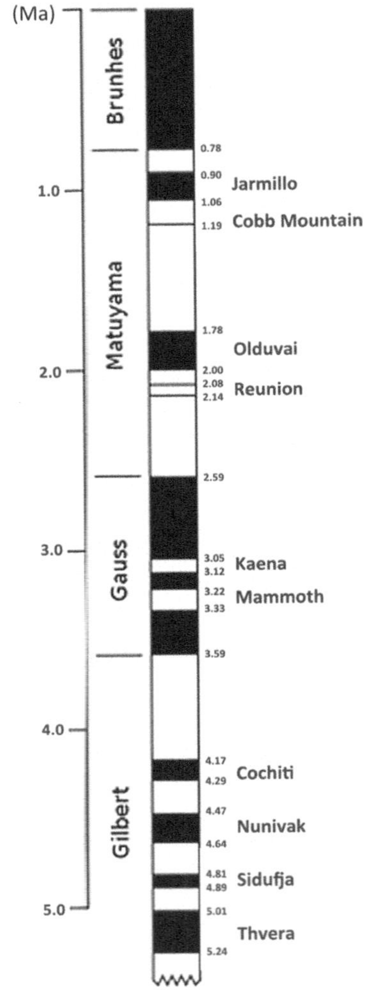


**Figure 3.15** Geomagnetic polarity timescale over the last 5,250,000 years (Mankinen & Wentworth, 2003).

Often, the total natural remanent magnetization is a vector sum of different magnetic components, because the primary NRM, i.e. the magnetization formed during rock formation, can subsequently be masked by the magnetic components acquired later in geological history through reactions of mineral formation at low temperatures, or via thermochemical reactions associated with tectonic movements and the processes of sinking and backfilling of layers, lightning strikes, etc. Components that have been subsequently acquired can be removed by 'magnetic cleaning' techniques, i.e. by gradual demagnetization if the primary magnetization is stronger. Thermal demagnetization and demagnetization techniques in a magnetic field of variable direction are used for his purpose. The first technique is based on the gradual heating of the sample at certain temperature increments, whereby the vector of the magnetic moment of the sample is determined after each demagnetization step. The second technique is based on gradually increasing the strength of the alternating magnetic field in certain increments of strength, whereby the vector of the magnetic moment of the sample is determined after each step of demagnetization.

### 3.4.4.1 Sampling

In order for the orientation (determined by declination and inclination) and intensity of the paleomagnetic field to be accurately determined, the orientation of the sample taken from the rock outcrop must be precisely determined. The angle of deviation of the z axis of the sample (normal to the outgrowth plane) from the vertical, and angle x of the axis (intersection of the horizontal plane and the plane containing the z axis of the sample and the vertical) in relation to the geographical north clockwise are measured – Figure 3.16. When taking a sample from a deep well, it is not possible to establish the exact horizontal orientation, but only the inclination of the sample in relation to the vertical, therefore only the inclination of the paleomagnetic field can be established based on the measurement of paleomagnetism.

When the cylinder is pressed or cut into the rock, the x-axis and its orientation should be incised on the sample before removing the cylinder.

Smaller samples of a standard size corresponding to the sample carriers in instruments for determining the magnetic moment or magnetic susceptibility are taken from a single core removed from a rock outcrop or from a wellbore. Their orientation is determined in relation to the sample, and through its orientation in relation to the geographical north and vertical, the orientation of the sample in relation to the geographical north and vertical is obtained. The magnetization vector of the sample can then be transformed into a vector of the paleomagnetic field.

### 3.4.4.2 Measurement of paleomagnetism and analysis of results

The vector of the magnetic moment of samples is determined by using magnetometers which can be astatic, rotational or cryogenic.

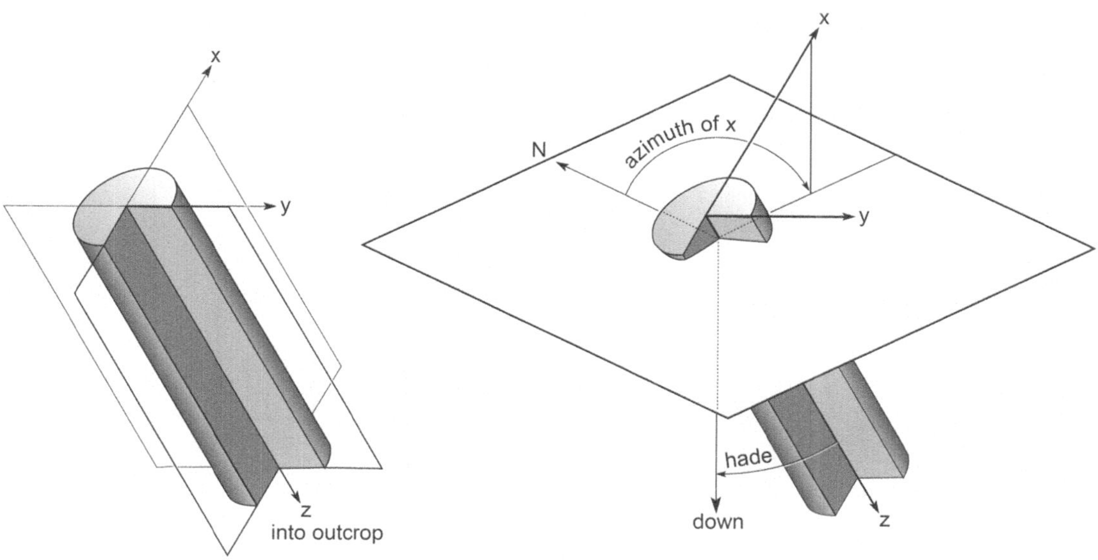


**Figure 3.16** Orientation of a sample from a rock outcrop (Butler, 1998).

In an *astatic magnetometer*, two identical magnets are mounted antiparallel to the magnetic shafts inside a vertical cylinder suspended from a wire – Figure 3.17. In this magnet system there is no net torque in a homogeneous magnetic field. The rock sample placed under the cylinder creates a stronger torque moment on the lower magnet than on the upper one (Vočadlo, 2014). The horizontal component of the magnetic moment of the sample is proportional to the angular deviation of the cylinder. The measurement of the magnetic moment for three mutually normal positions of the sample makes it possible to determine the direction of the magnetization vector.

With a *rotary magnetometer*, the sample is rotated in the center of the system of coils. This induces an electromotive force (creates a voltage) in the coils that is proportional to the component of the magnetic moment normal to the axis of rotation. The direction of this component is obtained from the voltage phase. The highest voltage occurs when the component is aligned with the axis of the coil. Comparing the voltage with the voltage of a standard sample placed in the sample holder or with the voltage that is synchronously caused by a reference magnet within a second system of coils makes it possible to determine the phase (Vočadlo, 2014).

*Cryogenic magnetometers* are particularly suitable for measurements on samples of weakly magnetized sedimentary rocks. The sample is lowered into a series of superconducting coils inside a chamber that is cooled to a very low temperature with liquid nitrogen or helium. The chamber is protected from the Earth's magnetic field. By lowering the sample through the coils, a direct current is induced which is proportional to the component of the magnetic moment normal to the axis of the coils (Vočadlo, 2014).

The results of determining the magnetic moment vector are most often shown on a modified Zijderveld diagram and on an equal-area Lambert stereographic projection of the lower or upper hemisphere (Figure 3.18b, Figure 3.19 bottom and Figure 3.21 right). In addition to the declination and inclination,

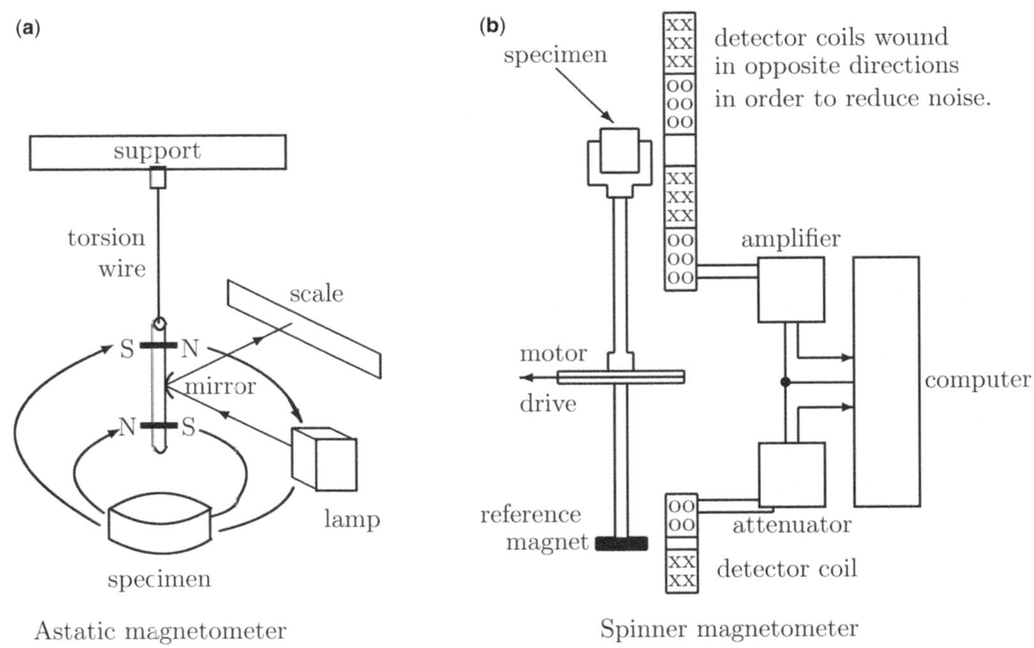


**Figure 3.17** Astatic (a) and rotary (b) magnetometers (Vočadlo, 2014).

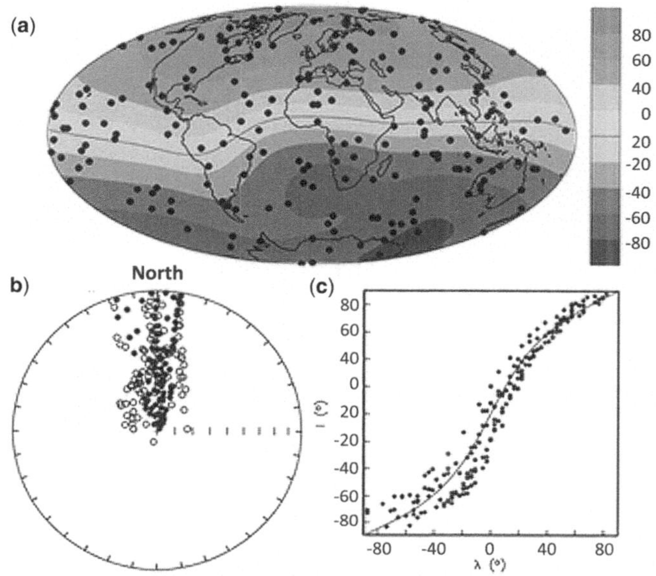


**Figure 3.18** Declination and inclination of the magnetic field in the Lambert projection (b), based on the IGRF (International Geomagnetic Reference Field) for 200 selected locations worldwide (a), the ratio of inclination and latitude (c) where the grey curve is the theoretical function for a dipole geocentric field (Tauxe *et al.*, 2018).

this data representation also shows the intensities of the magnetic moment and the temperature or strength of the alternating magnetic field for each step of demagnetization.

In the modified Zijderveld diagram the inclination is given through the ratio of the vertical and horizontal components of the magnetic moment vector in the vertical plane containing this vector, while in the Zijderveld diagram each vector is projected onto one vertical plane, so the heights and horizontal distances refer to the vector projection on that plane.

As for the inclination, the vertical axis is often marked as Up in the upper part, where the positive values indicate the upward direction (therefore the inclination of magnetization corresponding to the reverse polarized field for locations closer to the north geographic pole, i.e. the north pole of the geocentric axial dipole, is in the upper part of the Zijderveld and modified Zijderveld diagrams). The part of the horizontal axis to the right of the vertical is usually marked with H (H - Horizontal) it shows the intensity of the horizontal component of the magnetization vector. That is why the inclinations are shown only to the right of the vertical axis.

As for declination, the horizontal axis is marked with E and W, and the vertical with N and S, and it is read directly from the diagram as an angle with respect to the north in a clockwise direction.

The intensity of the magnetization vector is given by the distance of the point showing the inclination from the coordinate system origin, while the intensity of its horizontal component is given by the distance of the point showing the declination from the coordinate system origin.

In the Lambert projection, the lower hemisphere or upper hemisphere is projected on the horizontal plane (in the northern hemisphere the direction of the magnetic field vector is mainly downwards) (in the southern hemisphere the direction is mainly upwards). When referring to the projection of the upper hemisphere, i.e. to the magnetic moment vector pointing upward, this is usually indicated by an empty symbol, while the opposite orientation is indicated by a full symbol (Figure 3.18b and Figure 3.21 right). The outer circle

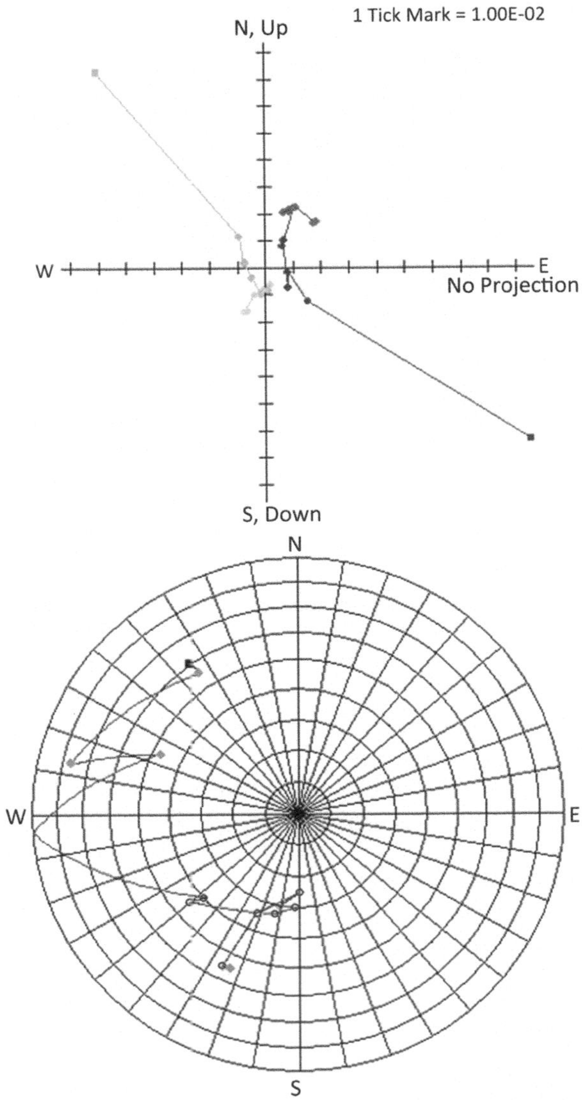


**Figure 3.19** Presentation of the results of the demagnetization process on a modified Zijderveld diagram and Lambert projection (Chunfu, 2004).

represents an inclination of 0°, the inner circles represent inclinations from 10° to 90° (Figure 3.18b, Figure 3.19 bottom and Figure 3.21 right). The declination is read from the diagram as an angle with respect to the north in the clockwise direction.

The procedure for analyzing standard size samples taken from core samples is shown graphically in Figures 3.19, 3.21 and 3.22. The first measurement of remanent magnetization is often the furthest from the coordinate origin because the intensity of the magnetic moment is usually the highest before the demagnetization process begins. The intensity is read on a scale given in mA/m.

The magnetization vector at the beginning of the analysis process is the vector sum of two or more magnetization components (Lowrie, 2007). If, for example, there are two of them and the angle between them is less than 90° (Figure 3.20(a)) then the demagnetization of the weaker one (BC in the first three steps of demagnetization in Figure 3.20(a)) will cause the intensity of the vector sum (AC in Figure 3.20 (a)) to be smaller than before (Figure 3.20(c) curve (a)). If the angle is greater than 90° (Figure 3.20(b)), decreasing the intensity of the weaker component will lead to an increase in the intensity of magnetization due to the higher intensity of the vector sum (Figure 3.20(c) curve (b)).

Figure 3.19 shows an example in which the declination (bright grey curve) and the inclination (dark gray curve) change significantly during the process of demagnetization all the way to its end, and do not tend to go towards the coordinate origin, which is the most desirable behavior of the sample. This desirable characteristic indicates that the primary magnetization has been detected, and that because there are no other superimposed remanent magnetizations, further demagnetization does not change the declination and inclination, but only weakens the intensity of magnetization. The starting moment is in the northwest direction (the light grey square on the modified Zijderveld diagram) and at a slope of 30° downwards

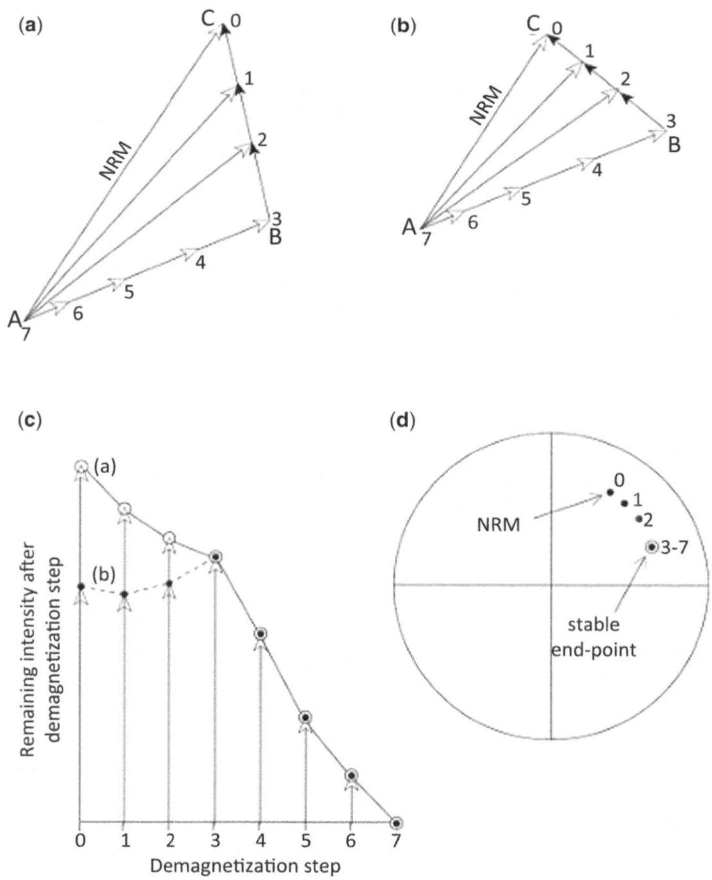


**Figure 3.20** Demonstration of the influence of the mutual orientation of two superimposing magnetizations on the results of the demagnetization process (Lowrie, 2007).

(the dark grey square on the modified Zijderveld diagram). In Lambert's projection, the starting point is shown as a dark grey square. After three demagnetization steps shown on the Lambert projection with full light grey circles, the inclination takes an upward direction and remains so oriented until the last demagnetization step, which is shown on the Lambert projection with empty circles.

Figure 3.21 shows the demagnetization of a sample in which the primary and subsequent magnetization overlap only partly (i.e. after reaching a certain maximum intensity of alternating fields, in addition to the demagnetization of the carriers of secondary magnetization, the carriers of primary magnetization also begin to demagnetize), which is reflected in the relatively rapid changes in the declination (the black dots and curve on the Zijderveld diagram) after the second demagnetization step (at an alternating field intensity of 150 Oersteds, which is a unit for the strength of the magnetic field that does not belong to the SI system). In this Zijderveld diagram, the magnetization vector is projected onto the north-south-vertical and the horizontal plane, so that the horizontal inclination component is projected on the north-south-vertical plane. After the fifth demagnetization step, both declination and inclination stabilize. The standard procedure in this case is to perform a linear regression for several points (demagnetization steps) after stabilization, and before the last few demagnetization steps in which the magnetic moment of the sample is lost, so that the linear function passes through the coordinate origin (interrupted dark grey and light grey lines on the Zijderveld diagram). This gives the so-called primary component of magnetization, which is why this analysis is called the primary component analysis (PCA). On the Lambert projection these points are observed as a densely grouped set of bigger light grey circles. In the Zijderveld projection, the points selected for the mentioned analysis are shown as larger dark grey and light grey circles. The same presentation uses black dots to show the inclination and declination of the magnetization vector of very low intensity that remained after the demagnetization of the primary magnetization which had a stable declination and a decreasing intensity during demagnetization.

Figure 3.22 shows an overview of the results of the demagnetization process on the modified Zijderveld diagram. It shows the change in the relative intensity of magnetization relative to the initial intensity as a function of the maximum strength of the field for demagnetization, and as a function of the demagnetization temperature. In both cases, the intensity of magnetization decreases during the

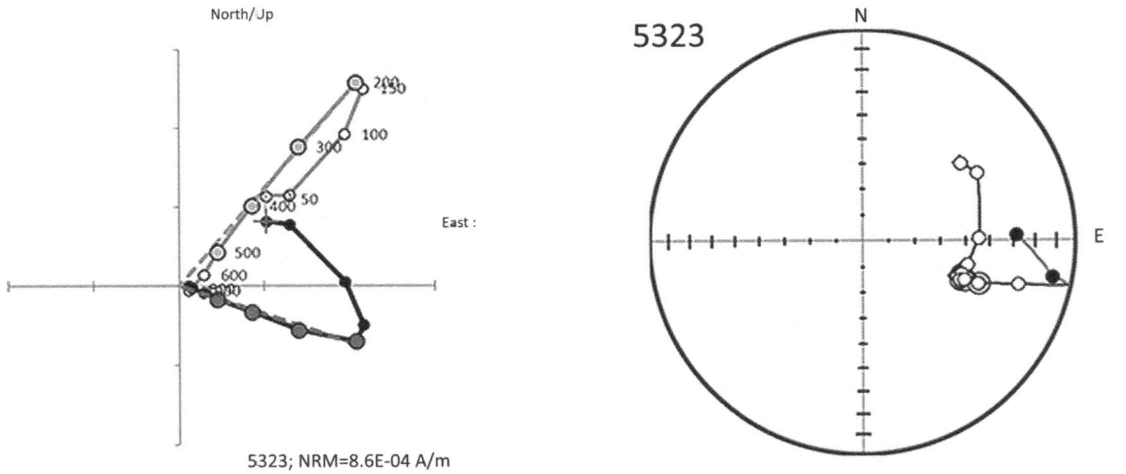


**Figure 3.21** Presentation of the results of the demagnetization process on the Zijderveld diagram and the Lambert projection, and the method of determining the primary component of magnetization (Sagnotti, 2012).

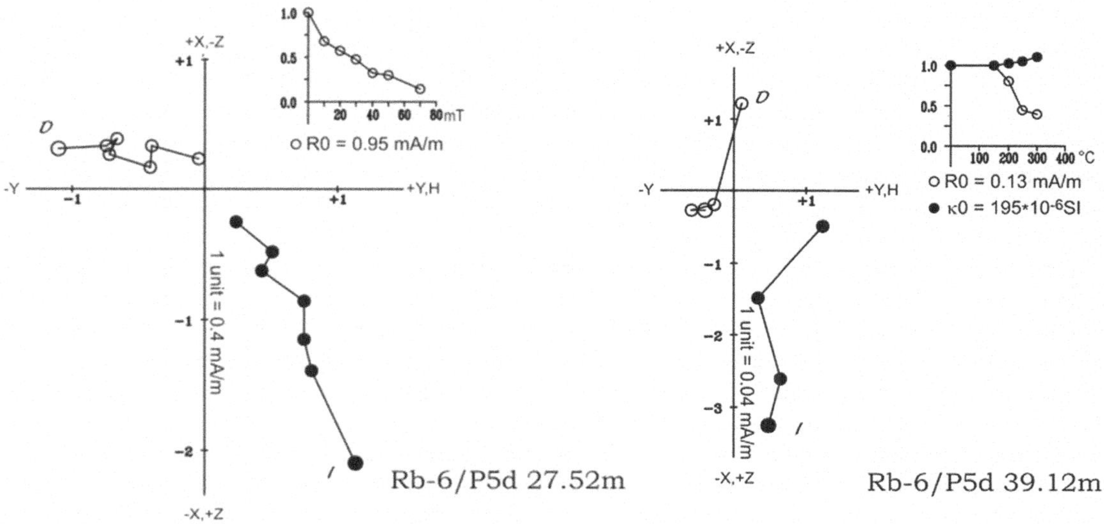


**Figure 3.22** Presentation of the results of the demagnetization process on the modified Zijderveld diagram, showing the change of the relative magnetization intensity (empty circles top right) in relation to the initial intensity (R0), as a function of the maximum demagnetization field strength (left), and the change of relative magnetization intensity (empty circles top right) and the magnetic susceptibility (black circles top right), as a function of the demagnetization temperature (right) (Marton, 2017).

demagnetization process. Besides determining the intensity of the remanent magnetization during thermal demagnetization, the magnetic susceptibility was also determined after each step as its more significant change would indicate a change in the mineral composition due to heating.

When a similar declination and inclination of the primary component vector is determined on several samples in one core, it indicates that the conditions of rock formation, its previous history and the sampling process itself were appropriate to preserve information about the paleomagnetic field at the time when the ferrimagnetic minerals that are the carriers of remanent magnetization were aligned with it. If both normal and reverse polarizations that alternate along the core are found, a series of such cores can indicate the age of a rock based on a comparison with the timescale of successions of geomagnetic polarity (Figure 3.15) and thus dating is performed.

In terms of using alluvial sediments for dating, it should be noted that lacustrine (lake) or palustrine (wetland) sediments formed between, before or after the alluvial sedimentation period may contain particles that have retained remanent magnetization. This can occur either in a very calm aquatic environment where fine mineral particles in the surface layer of dusty or clayey sediment at the bottom can be oriented in accordance with the magnetic field, or more likely, in geochemical conditions within such sediment in which ferrimagnetic minerals are formed that assume an orientation in accordance with the magnetic field until they exceed the blocking size.

### 3.4.5 Amino acid dating

Amino acids have at least one asymmetric carbon atom or a center of symmetry (except for simple glycine). Such molecules are called chiral molecules (from the Greek word for 'hand'), which means that their mirror images are different from them, i.e. are non-superimposable after the rotation (Scot & Cary, 2013). Stereoisomers of chiral molecules are molecules with identical molecular formula and sequence of bonded atoms, chemical bonds and functional groups, but different three-dimensional shape, and can be enantiomers which are mirror images of each other, and diastereomers which are not. Stereoisomers of the same amino acid differ for example by the direction in which they rotate plane-polarized light which passes through a solution containing only one configuration of particular amino acid. Stereoisomers rotating the plane-polarized light to the left are called L-amino acids, whereas those rotating it to the right are D-amino acids (Scot & Cary, 2013).

Amino acids in living organisms are almost exclusively L-amino acids, but after the death of the organism the amino acids, released by the breakdown of proteins or still incorporated inside them, randomly at some rate convert to D-amino acids, which also undergo this process called racemization in the opposite direction. The equilibrium is reached when the shares of both forms are equal in the case of enantiomers, for diastereomers equilibrium is reached at some other ratio (Scot & Cary, 2013). The rates at which amino acids are transformed, depend primarily on the particular amino acid, temperature and the factors unique to each taxon. Without microbe induced breakdown, the simplest amino acids do not break down at usual surface temperatures for hundreds of millions of years. Conditions for the preservation of amino acids are favorable in the carbonate skeleton of mollusks and Foraminifera, eggshells etc. Organic material is incorporated either between the mineral crystals (intercrystalline), usually determining the form of the mineral phase during its forming, or within crystals (intracrystalline), providing durability to the mineral phase.

Not all amino acids are suitable for dating purposes – glycine, as mentioned, is not a chiral molecule, alanine and some other simple amino acids are created by the decomposition of more complex amino acids, and thermodynamically unstable serine and threonine have too short half-lives. The slow-racemizing amino acids such as isoleucine, leucine, valine, and glutamic acid are widely used, together with fast-racemizing aspartic acid. In addition to the listed constraints, the number of stereoisomers is of concern too. The amino acids commonly utilized for geochronological applications (except for isoleucine) have a single chiral carbon atom and exist only in their D- or L-forms (Scot & Cary, 2013), enantiomers having identical physical and chemical characteristics (when interacting chemically with non-chiral molecules). Isoleucine has two chiral carbon atoms (marked as alpha and beta) and can racemize about either or both of them, although in nature L-isoleucine racemizes almost exclusively about the alpha carbon, producing a D-alloisoleucine molecule (Figure 3.23), which is not a mirror image of L-isoleucine and has different chemical and physical characteristics, making it easier to separate in the laboratory. Racemization including two diastereomers is called epimerization (Scot & Cary, 2013). The rates of interconversion by racemization (more strictly epimerization) are different for opposite directions of the reaction, so in the state of equilibrium there is 57% D-alloisoleucine.

The rate constants for racemization reactions are usually derived from $D$ to $L$ stereoisomers quantity ratios of radiocarbon-dated samples at sites of near-constant temperature (Scot & Cary, 2013), with $D$ and $L$ being the concentrations of different stereoisomers rotating the plane-polarized light right and left, respectively. If the rate of reaction is determined at different temperatures (for example due to different burial depths of samples, different altitudes or latitudes) for a particular amino acid stereoisomer in a shell of particular mollusk taxon for example, the activation energy $EA$ and pre-exponential factor $A$ (i.e. Arrhenius factor) in the Arrhenius equation (Equation (3.6)) can be determined for that particular case.

$$k_1 = Ae^{-EA/RT} \tag{3.6}$$

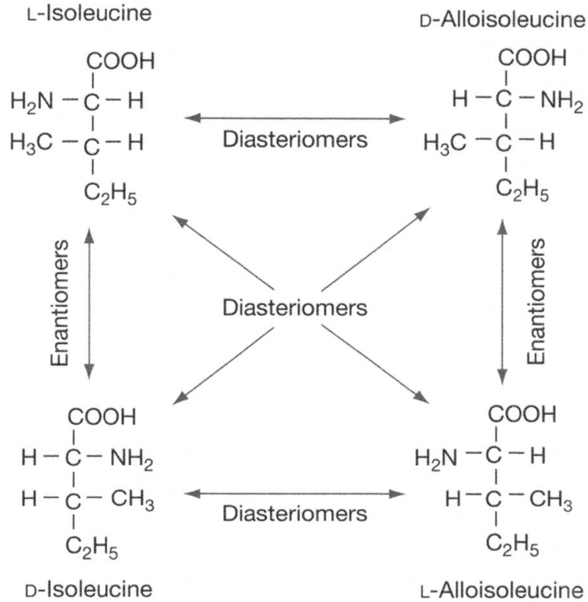


**Figure 3.23** Stereoisomers of amino acid isoleucine, and racemization and epimerization pathways (Scot & Cary, 2013).

Rate constant $k_1$ for the racemization reaction in direction from $L$ to $D$ stereoisomer can be determined for a particular temperature (under the condition of approximately constant temperature of the sample since the racemization started) from the D to L ratio and time elapsed since the death of the organism (determined by independent method) using Equation (3.7) (Engel & Macko, 1993):

$$\left(\ln\frac{1+\frac{D}{L}/}{1-\frac{k_2}{k_1}\frac{D}{L}} - \ln\frac{1+\frac{D_0}{L_0}}{1-\frac{k_2}{k_1}\frac{D_0}{L_0}}\right)\left(\left(1+\frac{k_2}{k_1}\right)t\right)^{-1} = k_1 \qquad (3.7)$$

where $k_2/k_1$ is the ratio of rate constants in both directions, and in the case of enantiomers is equal to 1, $D/L$ is the ratio of quantities of two stereoisomers, $t$ is the elapsed time since the start of racemization, and $D_0/L_0$ is the ratio of two stereoisomers in the modern, live specimen (since the racemization takes place outside the living cells) (Scot & Cary, 2013). For L-isoleucine and D-alloisoleucine the value of $k_2/k_1$ is 0.77 instead of the usual 1. When $k_1$ and $k_2$ are determined for different temperatures (using matching ages and $D/L$ ratios), *EA* and *A* factors in the Arrhenius equation can be determined. An illustrated example of the set of needed data is provided in Figure 3.24.

In this way, when the $D/L$ ratio of a specimen of a particular mollusk's shell is determined, together with the $D_0/L_0$ ratio of a modern specimen of the same or similar species, and temperature history is assumed, Equation (3.7) can be used to derive the age of the specimen $t$. The correlation between the $D/L$ ratios of two amino acids and the ages of ostracode *Candona* at the temperatures characterizing the bottom of deep lakes in the western USA is shown in Figure 3.25.

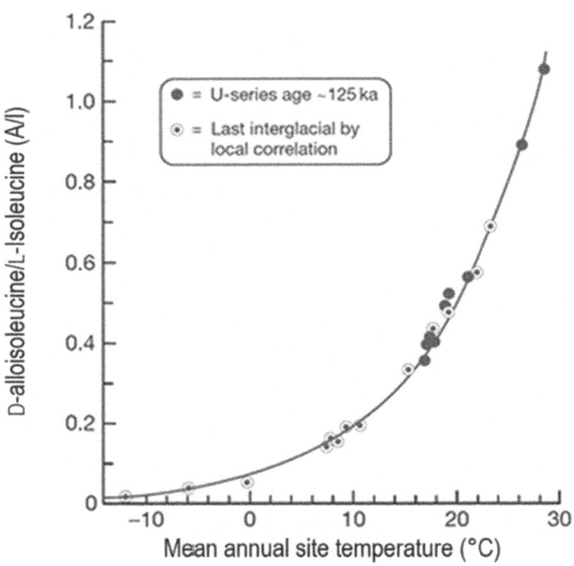


**Figure 3.24** Correlation between the D-alloisoleucine/L-isoleucine ratio in moderate racemization rate mollusks approximately 125 ka old from sites across Europe (ranging from Arctic to the Mediterranean, temperatures ranging accordingly), and the current mean annual site temperature. Sites were dated by the U/Th method used on corals or correlated to the last interglacial (Scot & Cary, 2013).

The temperature history is a very important factor because of the exponential dependence of rate constants on temperature. In practice there are two kinds of temperature variations of significance:

- seasonal cycle of temperatures in continental climates that can extend to a depth of several meters, which increases the effective mean annual temperature (the mean temperature that would yield a

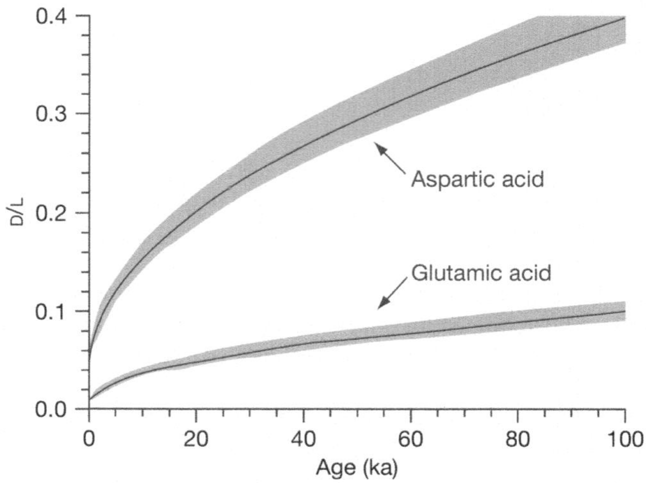


**Figure 3.25** The correlation between the *D/L* ratios of aspartic and glutamic amino acids and the ages of ostracode Candona. Shaded envelopes denote one standard deviation uncertainty (Scot & Cary, 2013).

proper age) because the higher temperatures due to the exponential dependence mean disproportionally higher rate constants and vice versa. Mean annual surface temperature and the temperature at 10 m of depth which is constant, are equal, while the surface effective temperature can be higher by 10 degrees or more, than both the mean surface annual temperature and the effective temperature at 10 m of depth (which is equal to the mean annual temperature at this depth) (Scot & Cary, 2013);
- glacial cycle of temperatures that can cause the mean annual temperature to differ by up to 10 degrees Kelvin between the different periods of time within the age of the specimen.

Because of this, the time that the burial process took, and the depth of the burial are of great importance. The period some time before, and since the death of an organism or laying of an egg can be divided into periods with different temperature conditions as assumed based on, for example, glacial history of the site. This timescale should be longer than the supposed age. Then Equation (3.7) can be used to determine the $D/L$ ratio at the end of each period, which at the same time is the $D_0/L_0$ ratio for the next period. If the final calculated ratio is for example larger than the measured one, the first assumed time period should be shortened or eliminated, and so on towards the present, until the final ratio matches the measured one. This can be implemented manually or through automated calculation. The other approach is to calculate the effective temperature based on the exponential dependence and the length of periods with particular temperatures.

Besides the inherent characteristics of each amino acid and the influence of temperature, taxonomic factors significantly influence the racemization rates.

The first reason for this is tied to differences in proteins in the exoskeleton or eggshells of different genera. Laboratory experiments indicate that amino acids located at interior positions within peptide chains have low rates of racemization, whereas amino acids located at terminal positions racemize most rapidly. The $D/L$ ratio of free amino acids is usually high even though the reaction rate is lower than when they are a part of protein chain. This is explained by the preferential release of relatively highly racemized amino acids from terminal positions. Larger polypeptides have a higher ratio of interior to terminal amino acids than smaller polypeptides, and consequently feature a lower extent of racemization (Scot & Cary, 2013).

The second reason is that mollusk and eggshells have several or more structural layers each usually containing a different array of amino acids. The racemization rate of an amino acid also depends to a certain extent on its bonds with adjacent amino acids in the peptide chain, thus racemization rates commonly vary in different structural layers for this reason too. Analyses of mollusk stereoisomers ratio variability indicate differences of up to 40% between different parts of even simple shells (Scot & Cary, 2013). In bivalve shells, protein residues are mostly present along intercrystalline growth, where they serve to nucleate the development of the next incremental segment of shell growth. Leaching of amino acids and small peptide chains by water is more present at the intersections of these growth bands within the shell surface (Scot & Cary, 2013). Molecules most susceptible to leaching are the low-molecular-weight solutes like free amino acids and small peptides. As they tend to be more racemized, leaching decreases the overall level of racemization. High temperatures and a succession of wet and dry periods are most conducive to leaching. Protein residues in intracrystalline sites are much less susceptible to leaching, so isolating this fraction of amino acids can improve the accuracy of amino acid dating (Scot & Cary, 2013).

Peptide bonds are hydrolyzed prior to amino acid analysis by heating the decalcified residue (typically 22 h at 110°C) in a strong acid (6 mol/dm$^3$ HCl) which breaks the proteins down to amino acid and the final result is the total acid hydrolysate on which analyses are usually performed for determination of amino acid stereoisomers (Scot & Cary, 2013). Gas or high-pressure liquid chromatography (HPLC) are used to separate, detect, and quantify the D- and L-amino acid enantiomers and diastereomers. L-isoleucine can be separated from its D-alloisoleucine pair by conventional ion-exchange HPLC because of differing physical and chemical characteristics. Most other D-/L-amino acid pairs are separated by gas chromatography or reverse-phase (non-polar sorbent) liquid chromatography using a chiral stationary or mobile phase to separate the enantiomers – Figure 3.26.

The residue which is not hydrolyzed, or free fraction, is composed of naturally hydrolyzed amino acids, and the level of racemization in this pool is more advanced.

Some studies have shown that after 'bleaching', i.e. treating the finely ground mollusk shells with concentrated sodium hypochlorite for 48 h effectively reducing the amino acid content to the intracrystalline fraction, variability between the determined $D/L$ ratios in different same species shells from the same sample was significantly reduced, both total hydrolysate and free fraction (Scot & Cary, 2013).

The time span over which amino acid racemization dating is effective is determined by the reaction rate, so it is much longer in cold than in tropical climates (high rate). At room temperature conversion from pure

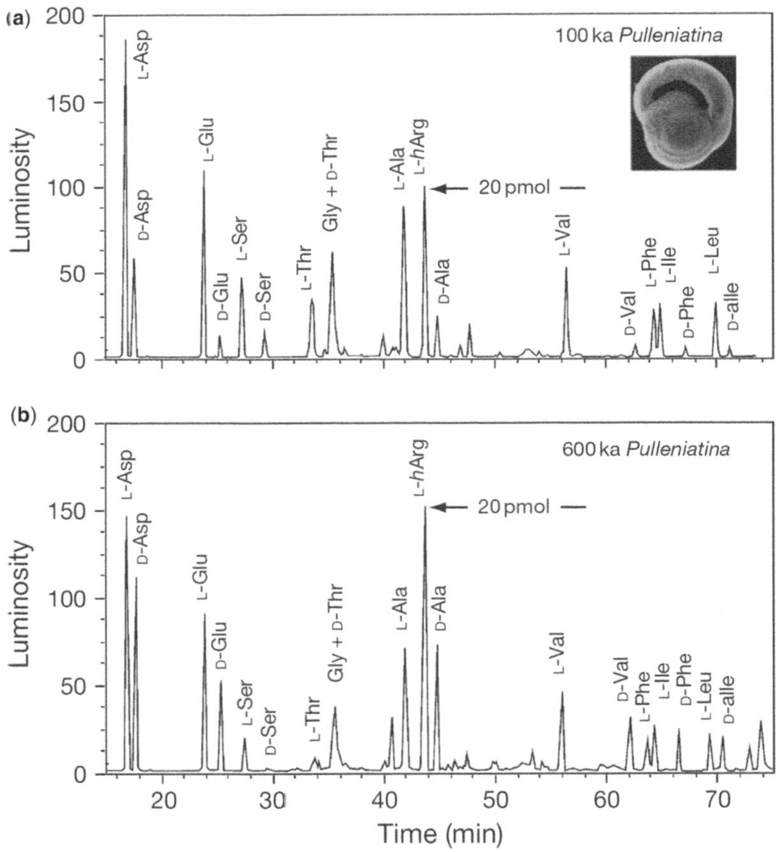


**Figure 3.26** Reverse-phase HPLC chromatograms of planktonic foraminifer Pulleniatina, approximate ages of 100 ka (a) and 600 ka (b) (Scot & Cary, 2013).

L-amino acid to a racemic mixture equilibrium takes several tens of thousands of years for most amino acids, whereas in Arctic regions reaching racemic equilibrium takes 1–2 million years. Aspartic acid racemizes approximately 10 times faster than other amino acids and is used for dating very young materials or samples from polar, or high-altitude sites (Scot & Cary, 2013).

Alternatively, climate change record can be derived from $D/L$ ratios if independently derived ages are available for samples from a particular region, as the evolution of temperature in the region can be calculated using Equation (3.7).

### 3.4.6 Cosmogenic nuclides dating and geomorphologic processes quantification

The Earth is constantly exposed to cosmic radiation, i.e. primary cosmic rays, most of which are of galactic origin, and are composed of high energy particles (0.1–10 GeV) such as protons (87%), α-particles (12%) and heavier nuclei (1%) (Darvill, 2013). The rays are affected by the Earth's magnetic field and partly by solar activity. Only the incoming rays with the right inbound angle and high enough kinetic energy are able to penetrate the Earth's magnetic field, this getting easier with increasing latitude. Primary rays that are not deflected by the Earth's magnetic field react with nuclei of atmospheric gases which produces an array of secondary cosmic rays – high and low energy nucleons (e.g. protons and neutrons) and mesons (e.g. kaons, pions and muons), their intensity decreasing down the atmosphere (roughly exponential attenuation) (Darvill, 2013). Significant long-term variations in primary ray influx are caused by changes in the Earth's magnetic field. Upon encountering the Earth's surface, by hitting certain minerals these particles start a number of nuclear reactions that can result in the production of cosmogenic nuclides:

- Spallation reactions involve high energy neutrons and produce most cosmogenic nuclides ($^{3}$He, $^{21}$Ne, $^{10}$Be, $^{26}$Al, $^{14}$C, $^{36}$Cl), almost exclusively in the first 3 m or so as the neutron flux diminishes to less than 1% at this depth (Darvill, 2013);
- Capture of thermal neutrons which are slowed down during the atmospheric cascading, which can be important for the production of mainly $^{36}$Cl. As these low energy neutrons can leak back out of a rock, thermal neutron flux reaches a maximum at around 20 cm of depth (Darvill, 2013);
- Muonic reactions are triggered by muons which have lower mass than neutrons and are much less reactive and penetrate deeper into the rock, accounting for some 2% of cosmogenic reactions at the surface, and 100% of reactions below around 3 m (Darvill, 2013).

Stable nuclides accumulate at the rate of their production (Figure 3.27) and will equilibrate with the erosion rate of the rock surface when enough of them are accumulated per unit volume of rock per unit of time that the nuclides lost, with this rock volume per unit of time equal to the production rate. Radioactive nuclides are lost by radioactive decay in addition to loss by erosion. When the intensity of radioactive decay, which is directly proportional to the quantity of nuclide per unit volume, equalizes with the rate of their production caused by secondary cosmic radiation, plus the erosion loss, the concentration of nuclides reaches equilibrium, constant value. During the first part of the period before reaching the equilibrium, the radioactive nuclide concentration increases approximately linearly with age, as the rate of decay is still much lower than the rate of production – Figure 3.27, and this is the time span over which the dating can be performed. This saturation limits the timescales for dating.

There are several issues inherent to this dating technique that can at the same time be instrumental in quantifying erosion rates or duration of burial periods:

- Inherited nuclides – pre-exposure to the radiation (mostly reworking of clasts) and insufficient depth of erosion of the rock surface in the erosion event to be dated. The inherited nuclide component will

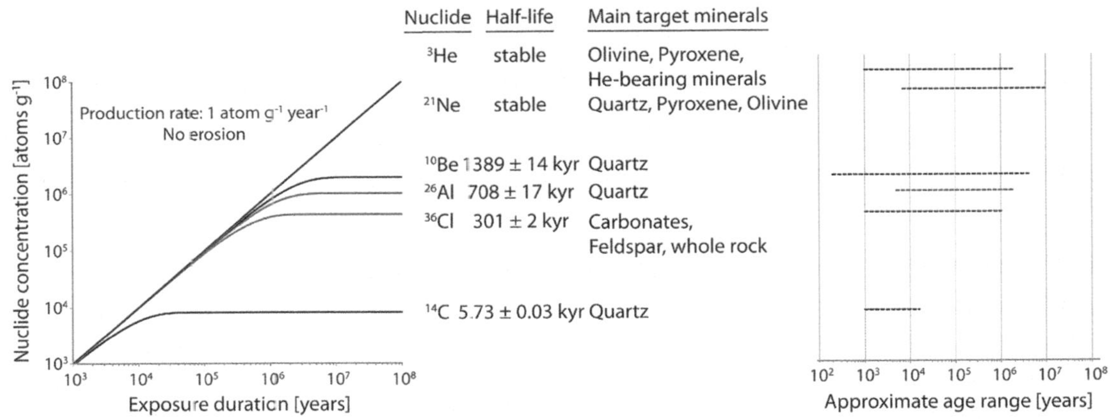


**Figure 3.27** Six most commonly-used cosmogenic nuclides, the change of their concentrations with time if the production rate is 1 atom/g/a and there is no erosion (left) and approximate age ranges over which they are effective for dating (Darvill, 2013).

lead to an overestimated age. A nuclide concentration in a deep sample, roughly deeper than 3 m, can only be a product of inheritance, and this concentration can be used as a base level for shallower samples (Darvill, 2013);

- Changing depth – due to erosion the exposure of a sample to cosmic radiation can increase in time as its depth decreases. The loss of the finer matrix from moraines leading to greater cosmic radiation exposure of boulders (Noller *et al.*, 2000), or slope erosion are examples of possible causes of depth reduction. The increasing production rate causes the nuclide concentration to increase, and age, calculated upon measuring the concentration and determining the contemporary secondary cosmic radiation intensity, is underestimated. The underestimation is higher when erosion took place recently and suddenly than when it proceeded at a steady rate during a long time period. In the case of dating events, like erosion episode or exposure due to tectonic movement, the erosion subsequent to these events can expose material less influenced by secondary cosmic radiation and therefore erroneously younger. To avoid the described effects, use horizontal surfaces with no debris, preferably with some sign of surface preservation like striations or glacial polish from past glacier movements (Noller *et al.*, 2000);
- Erosion loss – nuclides formed just under the surface can be depleted steadily which reduces the concentration of nuclides for a given age, and causes equilibrium to be reached earlier – Figure 3.28. If not taken into account, this can reduce the calculated age. The rate of erosion must be estimated based on, for example, differences between depth of erosion between the minerals of different weathering resilience (Noller *et al.*, 2000), or the depth of surface grooves and pits (Darvill, 2013) etc. Also, the nature of different nuclide's production and decay rates can yield significant insight on erosion rate – Figure 3.29. For example, as mentioned before, due to thermal neutrons flux reaching a maximum at some 20 cm of depth, the concentration of $^{36}$Cl reaches a maximum at the same depth, so if the concentration profile peaks at the significantly lower depth that is a sign of erosion, or if at lower depth, burial,
- Subsequent burial – when at some moment after the exposure to cosmic radiation has begun, the rock or soil is covered by debris, landslide, ice or snow cap, glacier etc., then the nuclide production stops, and consequently the concentration of radioactive ones diminishes while the concentration of stable ones remains at the pre-burial level.

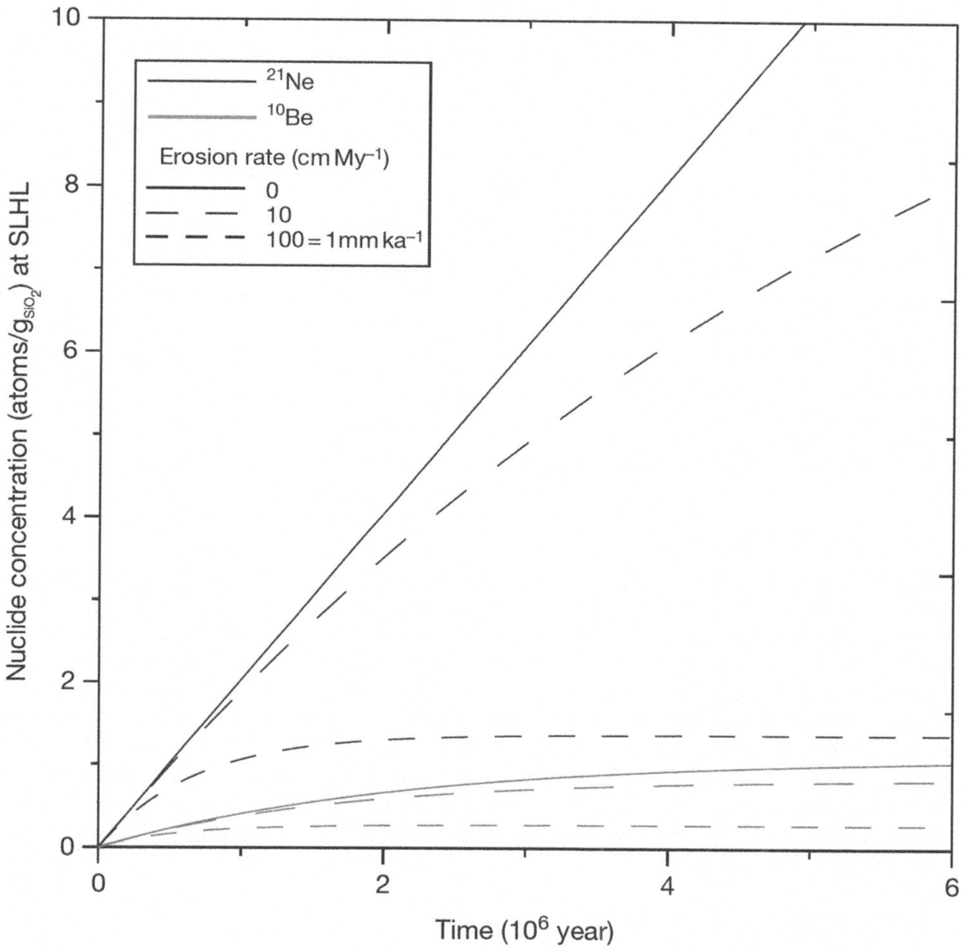


**Figure 3.28** Increase in $^{10}Be$ and $^{21}Ne$ concentrations with time at sea level at high latitudes (SLHL). With increasing erosion, saturation is reached increasingly earlier and at lower concentrations (Scott & Cary, 2013).

Erosion and burial effects on concentrations of two radio nuclides with different half-lives are graphically shown in Figure 3.29. When exposed to radiation, the concentration of both nuclides increases until the concentration of the nuclide with the shorter half-life reaches equilibrium, from which point on only the ratio of the other nuclide to this one continues to increase until the decay rate of the nuclide with the longer half-life is great enough to, added to the erosion rate, equalize the production rate. If at some point the surface is buried to the depth that precludes the radiation to reach the surface, the concentrations of both radioactive nuclides start to diminish, and as the one with longer half-life decays slower, the ratio between it and the other nuclide continues to increase (Scott & Cary, 2013). In Figure 3.29 the burial period follows the production/erosion that lasted long enough for equilibrium concentrations to be reached – point *C*.

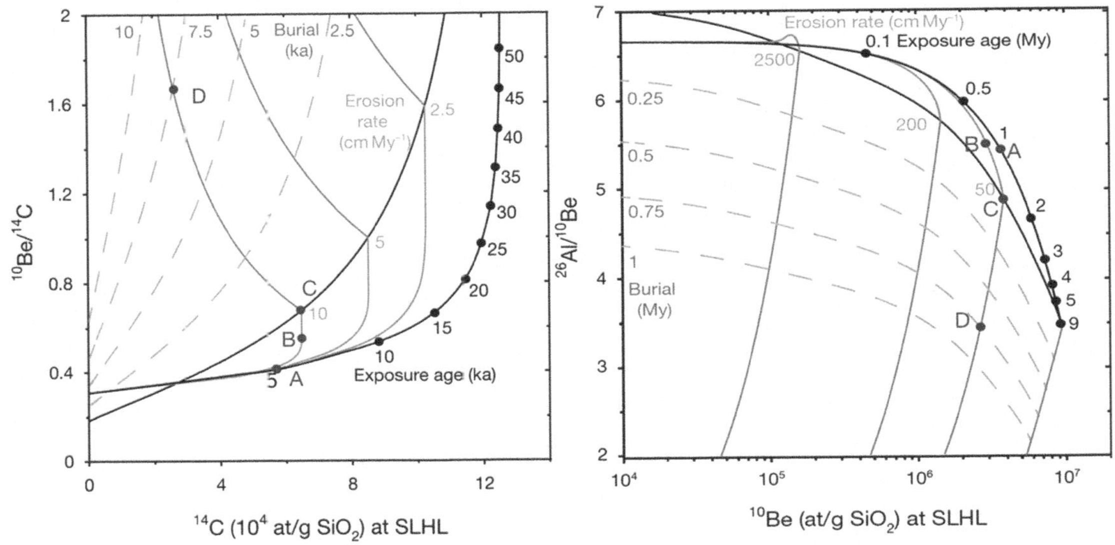


**Figure 3.29** Nuclide concentrations increase and, as they decay at different rates, the nuclide ratio decreases ($^{26}$Al/$^{10}$Be) or increases ($^{10}$Be/$^{14}$C) with time. Data points pertaining to continuously exposed rock surface and no erosion and no cover fall on the 'zero erosion' outer black line (A), with length of exposure shown as black numbered points. The lighter grey lines are trajectories for continuously exposed rock surface undergoing continuous erosion (B), points showing length of exposure, equivalent to ones on the 'zero erosion' line are omitted. Inner black line indicates equilibrium of both nuclides with respect to a specific erosion rate (C) – after the exposure time corresponding to this point the concentrations of both nuclides are constant. The area between the two lines is called the erosion island. Darker grey lines represent trajectories after burial (in this case trajectories starting at points of equilibrium of both nuclides with respect to three different erosion rates), with broken lines crossing them connecting the dots with the same burial time (D) (Scott & Cary, 2013).

### 3.4.6.1 *Fields of use*

Cosmogenic nuclides analyses are used for example for the determination of rates and for the dating of events such as:

- Glacier moraine stabilization – mountain glaciers erode bedrock and transport the debris down the valley, and deposit part of the material as moraines. The key assumption is that the glacial erosion is deep enough to excavate material not affected by cosmic radiation (Noller *et al.*, 2000), which is most likely on the surface of the largest boulders on the crest of the moraine;
- Lava flows, which are practically ideal for surface exposure dating as they form instantaneously and have surface structures that reveal their morphological stability (Noller *et al.*, 2000);
- Fault activity, as during a faulting event, rocks buried until then are exposed to cosmic rays at the fault face (Noller *et al.*, 2000), which usually forms at a steep angle, reducing the portion of the sky the surface is exposed to as well as changing incidence angles compared to horizontal surfaces, hence the need to use suitable correction calculations;
- A landslide event over a surface previously exposed to cosmic radiation long enough for several radioactive nuclides to reach equilibrium, so the starting concentrations can be assumed with satisfactory accuracy, or determined in the open part of this surface;

- Erosion rate of the exposed surface, especially in the case of constant concentrations of pairs of radioactive nuclides due to the equilibrium between production, decay and erosion rates – Figure 3.29.

### 3.4.6.2 Production rates determination

The production of cosmogenic nuclides varies with latitude, altitude and exposure to the sky (regarding the geographical and geometrical factors), the thickness and density of a sample, and its mineral composition. The rate at which the nuclide is produced at the sampling site must be estimated, so the production rates of a particular nuclide in a particular mineral at different locations are usually established using events firmly dated using an alternative technique such as radiocarbon dating, stimulated luminescence, U/Th series etc. The production rates are usually published in a normalized form to sea-level and high latitude (SLHL) in units of number of atoms per gram of the particular mineral per annum (Noller *et al.*, 2000). Either the most appropriate regional production rate can be applied or a reference production rate can be used, which consists of an average of several different regional studies. Models for production rate temporal change calculation mainly include the atmospheric effects, the geomagnetic field variability and solar variability in different ways. Spatial scaling factors must be carefully determined to recalculate the production rate to the cosmic radiation intensity on a particular site, taking into account latitude and altitude. Usually, the scaling factor is calculated by a polynomial (e.g. cubic) in which the variable is altitude and coefficients are given in a table as a function of latitude. The value of the scaling factor rises with latitude and altitude (Noller *et al.*, 2000). The data on change of a particular nuclide production rate (in a particular mineral) with depth in typical lithologies are also available (Darvill, 2013).

### 3.4.6.3 Sampling procedure

Geomorphological processes at the site should be well understood and the temporal sequence of events and states assumed as best as possible. Uncertainties should be avoided as much as possible, for example the boulder chosen for analysis should not show signs of rolling or prior burial, surfaces chosen for dating should not show signs of intense erosion etc. In general, all other processes or events, apart from the ones the rate or date of which are being determined, should not be present, or should have prepared a known base level of cosmogenic nuclides, e.g. exposition of the surface long enough to reach equilibrium concentrations prior to burial event, with no subsequent re-exposure.

Practical recommendations are:

- A minimum of three samples should be collected for one spot, they should not be collected from edges and corners of boulders and rock outcrops, should be roughly 5 cm deep, and deeper for $^{36}Cl$ (Darvill, 2013);
- Photos of the sample site should be taken from many angles, from near and afar, of the rock or depth profile before and after sampling, all with scale included, with 360° panorama around the sample site;
- All the relevant characteristics of the sample and the sites from which it is taken should be documented, like shape, lithology, color, grain size, vegetation cover, glacial polish and striations, cracks with aperture width, depth and frequency of occurrence, weathering pits and grooves with their depths, and any other relevant characteristics (Darvill, 2013);
- The angular elevation of the horizon should be recorded at regular intervals for a full 360° (Darvill, 2013);
- The strike and dip of the fault face or other surface from which the samples ae collected should be recorded.

#### 3.4.6.4 Nuclide analysis

Rocks containing the required mineral for a particular set of nuclides are sampled in the field, crushed and the target mineral is physically (magnetic separation, density separation etc.) and/or chemically separated (dissolution in different acids for example). Then the target nuclide within that mineral is isolated (for example by precipitation from the acid solution of mineral, by addition of appropriate ion), which is a complex, lengthy and costly process. Nuclide concentration is then measured, by determining the ratio between the isotopes (e.g. $^{9}Be/^{10}Be$) in an accelerator mass spectrometer (Noller *et al.*, 2000).

### 3.4.7 Time ranges of dating methods on the background of Pleistocene timeline

The overview of the temporal change of mean annual global surface temperature, ice volume and $^{18}O$ isotope content (with matching MIS chronology) during the last 2,000,000 years of the Pleistocene is shown in Figure 3.30. Chronologies based on the polarity of the paleogeomagnetic field, and glacio-geomorphological features in the Alps partially supported by absolute dating are also presented, as is the temporal change of insolation at the top of the atmosphere at key 65° northern latitude, and the difference between this and the insolation at the North pole.

## 3.5 NEW RESEARCH OF ALLUVIAL SEDIMENTS OF THE LOWER COURSE OF THE SAVA RIVER – FORMING CHARACTERISTICS AND THE SUBSTANTIAL EXPLOITATION OF ALLUVIAL AQUIFERS

### 3.5.1 General introductory remarks

The Sava River is the biggest tributary of the second largest river in Europe – the Danube (Figure 3.31).

Its course and alluvial deposits developed in the Pleistocene. Tectonic activity and climate were especially important in this case. The way the results of tectonic activities developed – namely faulting and the lowering of the Sava trench, as well as climate and climate change enabled the creation of a very significant alluvial aquifer (Dimkić, 2017; Jaroslav Cerni Water Institute [JCI], 2010; Mitrinović *et al.*, 2019).

The cooling that occurred during the Pleistocene (roughly in the last million years), enabled the formation of sufficiently massive glaciers whose melting and consequent erosion made possible the transport of large amounts of clastic (gravel, sand), and weakly bound material (clay, mud), as well as their polycyclic deposition. The alluvial groundwater source of Belgrade is located at the confluence of the Sava and the Danube Rivers – Figure 3.31. The groundwater source stretches on the left and right banks of the Sava River for about 50 km. It has 99 wells with horizontal drains (radial wells) and dozens of tube wells – Figure 3.32. Owing to the pronouncedly layered structure of the aquifer (Figure 3.33), the quality of water extracted from the wells with horizontal drains is very good.

Figure 3.31 The river Sava with its tributaries and the river Danube (cartographic basis from ESRI ArcGIS Online World Topographic Map service). The white frame represents the boundaries of the map from Figure 3.32.

Raw water extracted from radial wells typically has a small amount of organic matter, around or below 1 mg/l, decreased from the initial value of about 2.5 mg/l in the river Sava (Dimkić *et al.*, 2019). The other characteristics of the raw water are very good. The pharmaceuticals content is extremely low ($C_f \sim$ up to 5 ng/l) (Dimkić *et al.*, 2019). The pesticide content is of the order of up to a few nanograms $C_p \sim (0–3)$ ng/l.

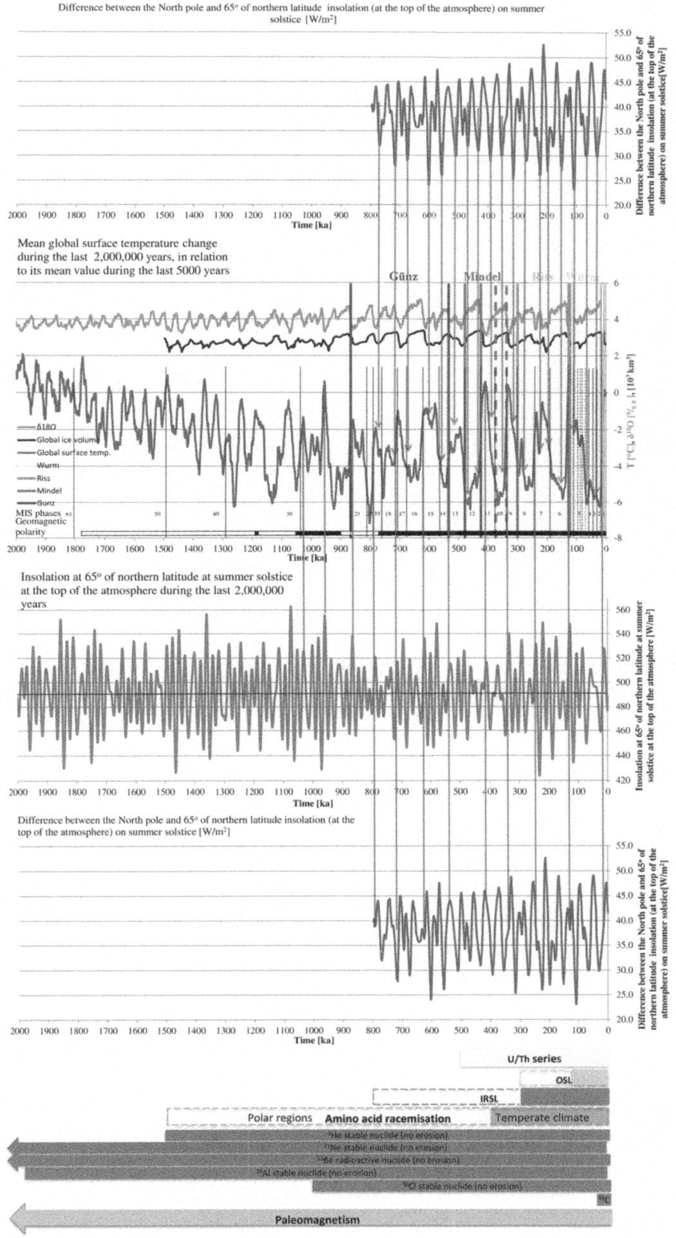


**Figure 3.30** Influence of astronomical factors (MODW, 2020) on the mean global surface temperature (Snyder, 2016) and global ice volume (Pollard & DeConto, 2009) during the last 2,000,000 years, MIS phases and 18O isotope content (Snyder, 2016), Alpine chronology (Deutsche Stratigraphische Kommission, 2016), geomagnetic timescale with chrones and subchrones, and reach of dating methods (IRSL – Infra Red Stimulated Luminescence, OSL – Optically Stimulated Luminescence, cosmogenic nuclides, paleomagnetism, amino acid racemization, U/Th series).

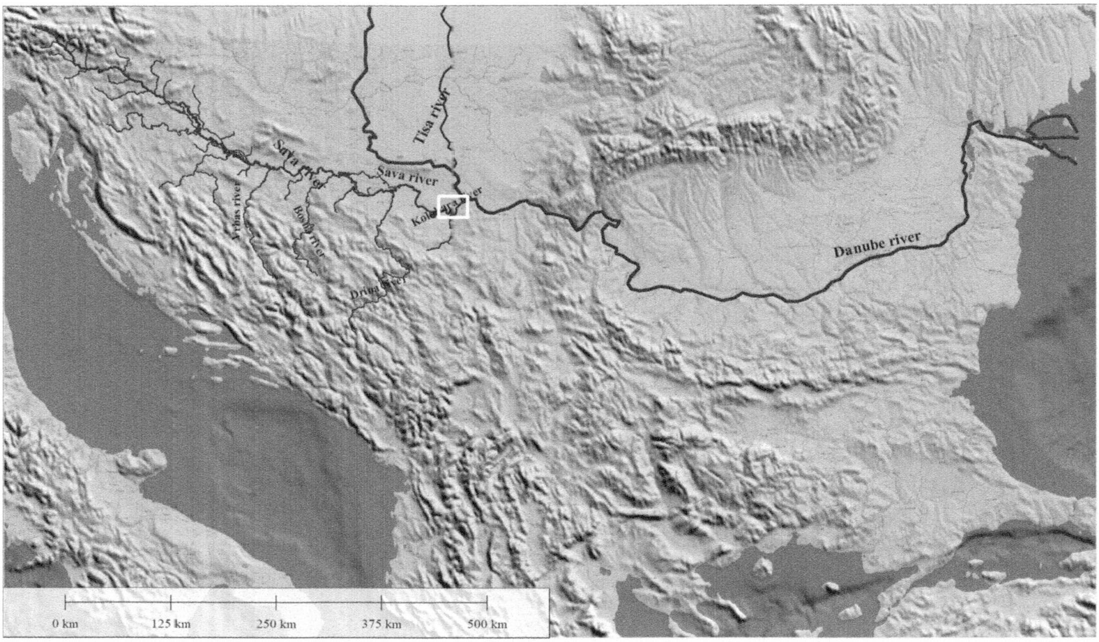


**Figure 3.31** The river Sava with its tributaries and the river Danube (cartographic basis from ESRI ArcGIS Online World Topographic Map service). The white frame represents the boundaries of the map from Figure 3.32.

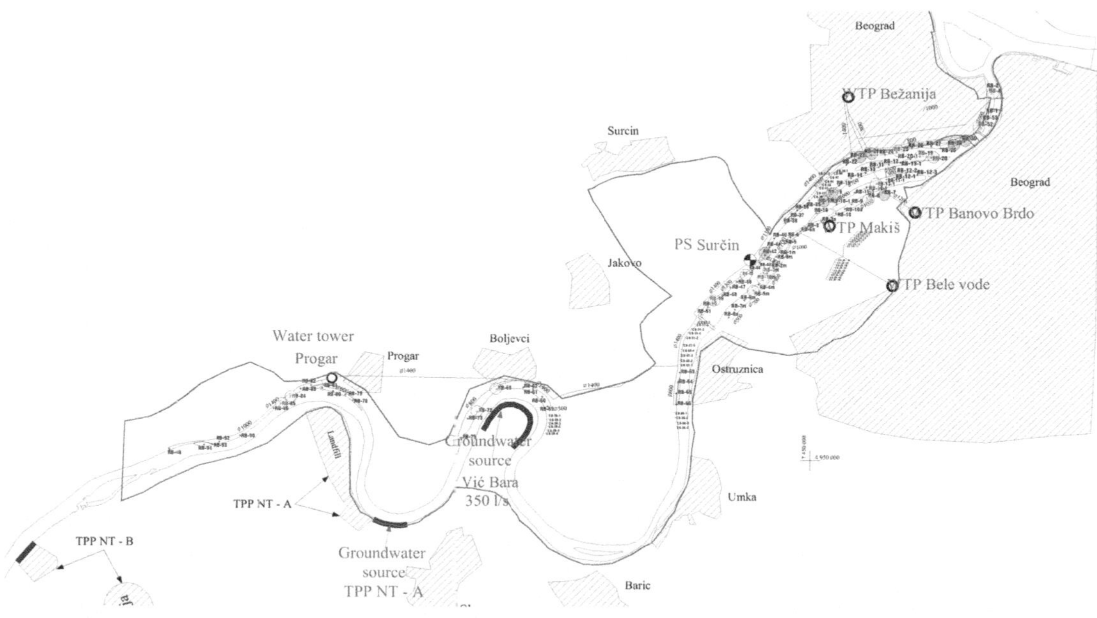


**Figure 3.32** Disposition of radial (designation RB) and tubular wells (designation CB) of the Belgrade Groundwater Sou‑ce (taken with modifications from JCI, 2010).

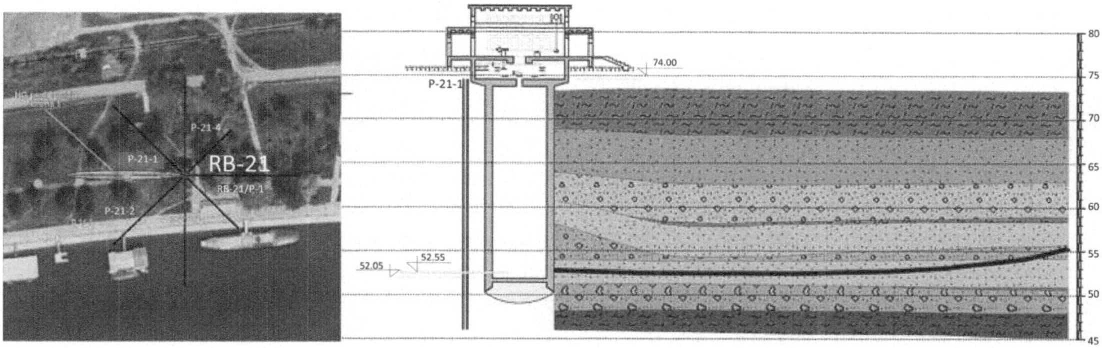


**Figure 3.33** Characteristic profile of one segment of the Belgrade Groundwater Source – at the well with horizontal drains Rb-21 (taken from JCI, 2010). The black curve is the well drain.

The average oxygen content is around 0.15 mg/l, and iron content in the raw water ranges mostly between 0 and 2.5 mg/l. The colmation of the wells by iron minerals results in an increase in the hydraulic resistance at the filter drains and a significant decline in the water yield of the well.

In 2017 about 40 new observation wells were drilled at the locations of radial wells planned for the new radials installation and further away back from the river, to determine the hydrogeological and lithological characteristics of sediments, and the hydraulic losses at the radials of wells (JCI, 2017). Two deeper observation wells (51 m and 65 m) were drilled. In addition to the frequent granulometric (at every one meter of depth) paleontological and mineralogical analyses that were carried out, analyses were also completed to determine the age of the samples using the luminescence method as well as by determining the orientation of the paleomagnetic field at the time of deposition of the lithologic layers (Marton, 2017; Mitrinović *et al.*, 2019). A comparison was made with the results of studies related to the extent and age of glaciation (Adamson *et al.*, 2015; Hughes *et al.*, 2011) in the Orjen and Durmitor mountains (Durmitor is located in the Sava River Basin). It turned out that the age of the moraines at Mt. Durmitor was very similar to the age of the polycyclic sediments at the Belgrade Groundwater Source. Certain mineralogical tests, as well as the shape of the sediments at the source, in addition to the determined age of the sediments, clearly fit into the thesis about the polycyclic formation of the aquifer complex at the Belgrade source. The older layers of sediments are predominantly represented by very fine-grained, weakly bound material (so-called 'motley clays') and, occasionally, with a small amount of gravelly material. This can be largely attributed to the warmer climate in the older Pleistocene (Figure 3.30) and the lack of mountains that were high enough for glaciers to form in warmer climates.

The way the fluvial-glacial polycyclic sediments were formed caused the hydro-geologic layering of the Belgrade Groundwater Source aquifer, which resulted in a longer path for the river water to the wells and this in turn resulted in the excellent self-purification effects of the natural filtration of water from the river to the radial wells (Dimkić, 2017).

The anoxicity of groundwater in the aquifer of the Belgrade Groundwater Source is due to the length of transport of sediment from areas of erosion and consequently to the sufficient percentage of fine material with oxidizable minerals. The appearance of iron in water, along with relatively low redoxes ($70 < Eh < 200$ mV) has caused the colmation of well filters and certain problems related to maintaining the capacity of the source (Dimkić *et al.*, 2011a; Dimkić *et al.*, 2011b; Dimkić, 2012).

### 3.5.2 The genesis and formation of the river Sava and the Sava River Basin

The older and middle Miocene sediments of the Pannonian Basin indicate a transition from continental, river-lake conditions to a completely marine environment (Ter Borgh *et al.*, 2013). Most extensive tectonic activity in this area happened during the Badenian-Sarmatian rift phase, approximately 15–11 million years ago (Fodor *et al.*, 1999).

The Pannonian lake was formed at the end of the Middle and the beginning of the Upper Miocene about 11.6 million years ago by the rise of the Carpathians which separated the lake from the sea (Magyar *et al.*, 1999; Sztanó *et al.*, 2015; Ter Borgh *et al.*, 2013). By breaking its contact with the Mediterranean, the sea became shallower and sweeter, so that in the Sarmatian, the Pannonian Sea became a salty lake and the fauna that is characteristic of a marine environment became extinct (Figure 3.34). The sedimentation and filling of the Pannonian basin went on and lasted for more than 7 million years. It was filled with lacustrine and fluvial sediments (from rivers that flowed into the lake) and with the material that was eroded and was brought in from the orogenic formations surrounding the lake.

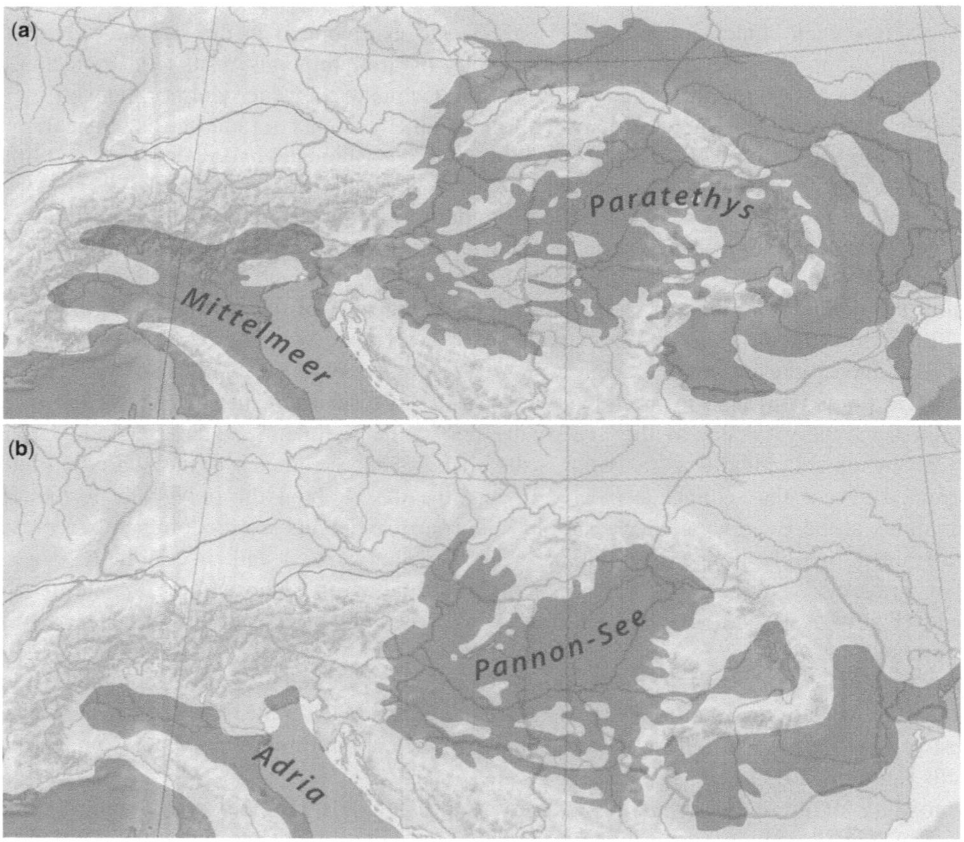


**Figure 3.34** (a) Connection between the Pannonian and the Mediterranean Sea in Middle Miocene (Badenian Stage) and (b) spread of the Pannonian Sea in the Upper Miocene (Magyar *et al.*, 1999).

At the beginning of the Pliocene, the brackish great Caspian-Pannonian lake was reduced to a significantly smaller, freshwater Paludina lake. During the Miocene and Pliocene pre-river flows transported and deposited chemically and mechanically resistant minerals and sediments such as quartzite. The rivers flowed in shallow channels, along cuts and meanders. Most of the older Pleistocene (from 2.58 million years ago to 1.20 million years ago) was characterized by a slightly warmer climate than that of the Middle and Late Pleistocene. The differences in temperature between the colder and warmer were smaller in the older than in the middle and younger Pleistocene (Figure 3.1).

During the older Pleistocene, the gravel and sands of Pliocene were being eroded and redeposited by inter-twined river flows inherited from the Pliocene. Great Neogene lakes disappeared from the area of the Pannonian Plain leaving behind smaller lakes, ponds and wetlands which together with a network of wide braided river flows represented a large area of sedimentation. The entire peripheral part of the Pannonian Plain, including the Sava River trench at the very periphery continued to settle during the Pleistocene, while the mountainous parts at the periphery began to rise, forming a large sedimentation area in which the deposited sediments were similar to those in the lakes. Even though these sediments had a lot of horizontal distribution, the lateral facial variability, finger laying facies and the layering point to river sedimentation.

In the older Pleistocene, glaciation also occurred in the Alps (the so-called Preatiglian phase lasts from 2.588 million years ago up to 2.4 million years ago and includes MIS 100 to MIS 96) but these glaciations did not have a great impact on the terrain of the Balkan Peninsula or on the glaciation in the Dinaric Alps.

In the middle Pleistocene, the post-Neogenic tectonics led to the rising of mountain massifs immediately prior to the great glaciation. This caused the formation of larger glacial covers and more intensive melting in the interglacial periods that alternated with colder glacial phases. Many Balkan mountains, i.e. a considerable part of their massifs, would not have surpassed the permanent snow line – Equilibrium Line Altitude (ELA), if there had not been tectonic movements just before the glacial epoch.

Intense and rapid thawing, more precisely, the sudden changes from completely glacial to fully interglacial conditions (Broecker & van Donk, 1970) were very pronounced in the Middle Pleistocene. In the last 620,000 years, seven terminations have been identified, limited by six glacial cycles. They have been marked using Roman numbers, from I, from the transition from MIS 6 to 5, up to VII at the border between MIS 16 and 15. The intensive melting of glaciers formed strong river currents and resulted in cutting of the pre-riverbeds and intense erosion and sedimentation in river basin areas. This period left a clear mark in terms of the characteristics of sediments.

The schematic in Figure 3.35 under (b) shows that, according to some authors, the course of the river Sava did not yet exist in the older Pleistocene. The same schematic presentation shows that the river Danube was formed as a longitudinal basin at the foot of the Alps, and linearly flowed eastwards following the former coastal line of the Paratethys which evolved into the Black Sea. The river Sava course was formed much later and was predetermined by the Sava fault, while the formation of the river flow itself was caused by glacial melting from the orogenic formations that surround today's river course.

We can say that the Sava trench, as the southern rim of the Pannonian Basin, had different sedimentation conditions during the Pleistocene compared to those in the central part of the Pannonian Basin. The material from the Alps and the Carpathians was brought by the tributaries of the pre-flows of large rivers such as the Tisza and the Danube. Glaciation in the Carpathian and Alpine areas began much earlier, in the older Pleistocene given their geographical position and proximity to the boundary with the northern ice sheet, as well as higher altitude in relation to the Dinaric Mountains. Such conditions during the Pleistocene caused different conditions of

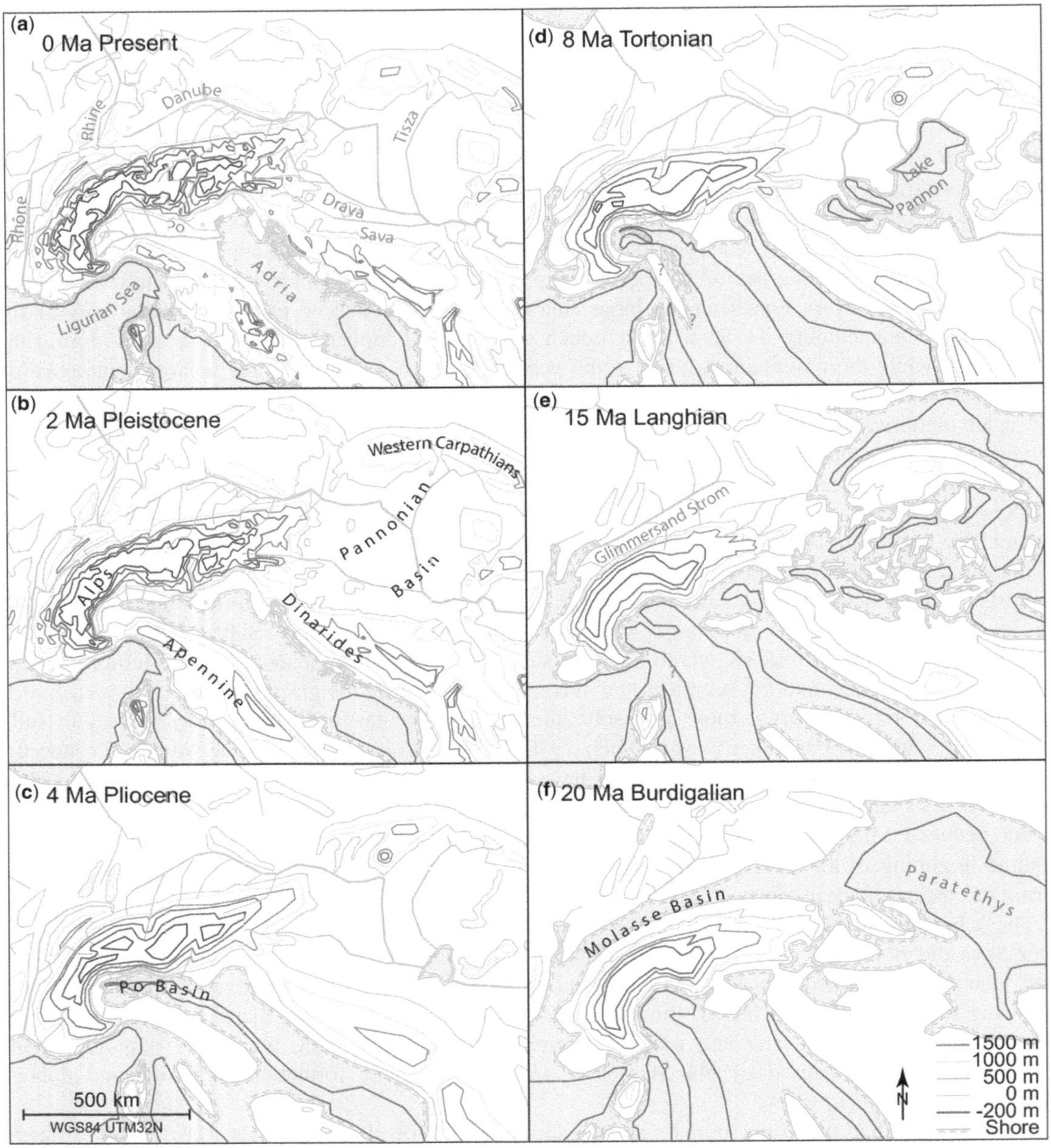


**Figure 3.35** Paleo-geographic scheme of the evolution of the Alps and the Pannonian Plain (Winterberg & Willett, 2019).

sedimentation in the area of the Sava trench, which is located on the southern edge of the Pannonian Basin, where sedimentation under the influence of glacier melting began much later and was dominated by flows from the Dinarides.

### 3.5.3 Investigations of alluvial sediments in the lower course of the Sava River – Case study of the Belgrade Groundwater Source

#### *3.5.3.1 Starting hypotheses*

According to the modern climate-stratigraphic division, the Quaternary is divided according to marine isotope stages (Cita, 2008; Emilliani, 1955; Shackleton & Opdyke, 1973). This model has already been successfully applied in the correlation of loess series in the Pannonian Basin with loess – paleo – land profiles of the Eurasian mainland, and deep-sea sediments and ice cores of Antarctica and Greenland (Antoine *et al.*, 2009; Buggle *et al.*, 2009; Hambach *et al.*, 2008; Marković *et al.*, 2005, 2007, 2008, 2009).

The results of dedicated research at the Belgrade source during 2016 and 2017 (JCI, 2017) served as the basis for defining the alluvial sediments in the lower course of the Sava according to marine isotope stages (MIS), as well as for determining their correlation with the age of corresponding generations of moraines on Mt. Durmitor and with the washed out material in the karst poljes at the foot of Mt. Orjen, as well as for the superposition with the loess Quaternary formations at the peripheral part of the Pannonian Basin. The idea was to, by examining the age of sediments and their mineralogical and granulometric composition, and other characteristics, contribute to knowledge regarding the genesis of alluvial sediments in the Sava aquifer. The correlation of the age and of the mineralogical and petrological conditions with the high massifs in the Sava basin (Mt. Durmitor and others) was meant to significantly help understand the influence the Pleistocene glacial periods had on the genesis of alluvial layers of aquifers in the Sava trench (Dimkić, 2017). This idea was put into practice by the authors of this chapter through various projects. The results have so far been published in one doctoral thesis (Zarić, 2020) and several articles (Mitrinović *et al.*, 2019; Zarić *et al.*, 2019a, 2019b). In addition to this, the connection between the climate and the tectonics during the Pleistocene was also analyzed, as factors that affected the sedimentation of polycyclic alluvial sediments and their characteristics (distribution of the grain sizes in the different layers, presence of certain minerals, oxidation-reduction conditions and other characteristics), which determine the natural capacity of an aquifer (Dimkić *et al.*, 2011a; Dimkić *et al.*, 2011b) and the processes which follow the exploitation of groundwaters (Dimkić, 2012).

In the Balkan Peninsula, the remains of glaciers in the form of moraines, circuses and erratic blocks have been registered throughout the Dinaric Alps (e.g. Mt. Durmitor, Mt. Orjen, Prokletije), the Pindus Mountains, at Rila, the Pirin Mountains on the Peloponnese, etc. The lowest confirmed boundary of a glacier is at about 500 m above sea level on Mt. Orjen, with the mean altitude of the most extensive glaciation at 1,250 m, while it descends to about 900 m on Mt. Durmitor, with the mean altitude of the boundary of the most extensive glaciation at 1,600 m. According to the results of research in the last 20 years, using the U/Th method of dating the carbonate binding material of moraines and of the washed glacier material in the karst poljes below the glacier, it was found that the most extensive glaciation occurred in the period before 478,000 to 424,000 years (according to the Alpine stratigraphy in Mindel) and that the next period of glaciation was between 190,000 and 130,000 years (Riss), while the latest occurred between 115,000 and 11,700 years ago (Würm). Based on some results obtained by dating the moraines on Mt. Durmitor and the washed out material at the foot of Mt. Orjen, it is possible that a trace was found of glaciation from before between 300,000 and 243,000 years ago. The age of all the moraines from which no samples were taken was determined by comparing the hypsometric positions and the development of the pedological layer that formed on them. The distribution of glaciers in different periods of glaciation on Mt. Durmitor and on the surrounding mountains is shown in Figure 3.36.

Based on data regarding the average annual global temperatures of the Earth's surface and the volume of global ice, it is known that there were periods that were equally as cold periods previously, especially MIS

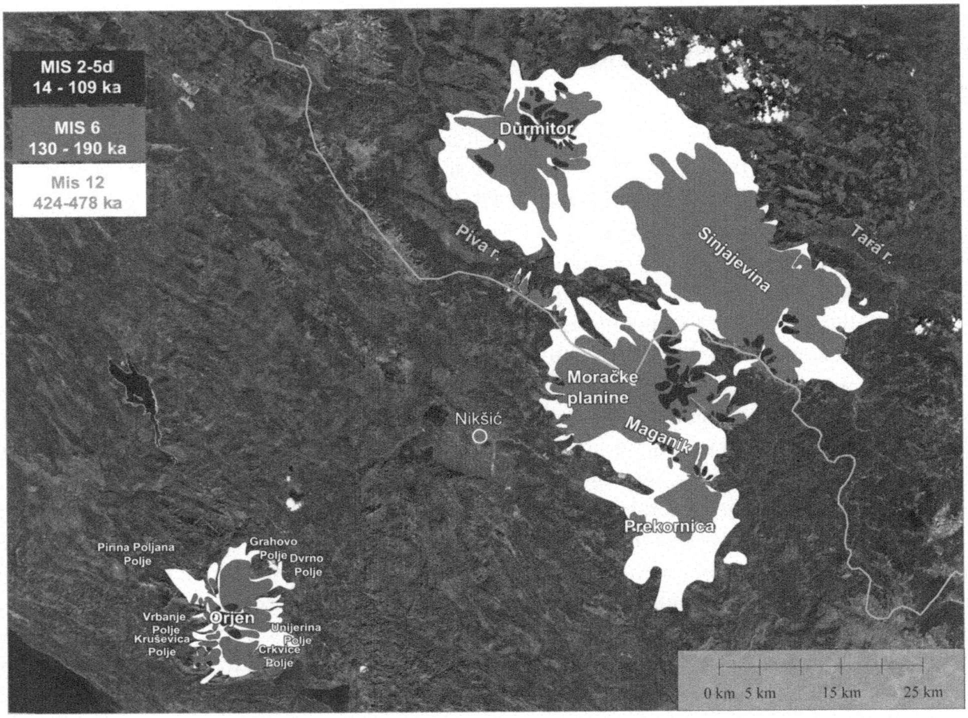


**Figure 3.36** Distribution of glaciers in the last half million years on Mt. Durmitor during three, so far isolated, glaciation periods, with the border of the Sava River Basin – the light gray line (prepared according to Hughes *et al.*, 2011; Adamson *et al.*, 2015).

16 (according to Alpine geochronology – Günz). It is possible that the remains of these glaciations have disappeared and it is also possible that the obtained ages that are greater than 350,000, for several samples taken from Mt. Durmitor and Mt. Orjen using the U/Th method, point to a greater age than the estimated 350,000 years.

The basic assumption is that the Dinaric Alps and the outer Dinaric ranges belonging to the Sava River Basin are the main erosive belt and source of materials from which Holocene and Pleistocene alluvial sediments were formed in the Sava River Basin and at the Belgrade source, and that this gravelly and sandy material was transported by rivers whose flows were much stronger than the current flows of the Sava and Danube, in conditions of the rapid melting of glaciers and ice caps at the beginning of the interglacial and interstadial periods. In that case, the age of the moraines left after several glacial cycles in the Dinaric Alps and the peripheral Dinaric massifs should correspond to cycles of warm phases with alluvial sedimentation preceded by cold phases of glaciation in the Dinarides. During these very short periods (in geological terms) of the rapid melting of glaciers and ice caps, strong river flows eroded to the local erosion base fine-grained sediments deposited during the previous interglacial and glacial periods (or interstadial and stadial), and in parallel to this process first deposited gravelly material and then the finer sandy material. The end of this period is followed by an interglacial (or interstadial) and then a glacial (or stadial) period during which mainly fine grained floodplain sediment was deposited at the rate of tectonic subsidence. Their older, deeper parts, which due to tectonic subsidence descended to below the local erosion base were preserved during the next transition with intensive melting, forming the shortened cycles that form the Pleistocene polycyclic layers (Mitrinović *et al.*, 2019).

Under the same assumption, the age of the material washed from moraines in the karst poljes at the foot of the mountains should correspond to the interglacial phases in which the polycyclic river sediments were deposited in the Sava River Basin.

Also, the basic assumption is that the age of the sediments of the alluvial aquifer is less than 780,000 years, i.e. that the alluvial aquifer is a product of processes that occurred during the Middle and Late Pleistocene. The age and genesis of the underlying layer of the alluvial aquifer are also interesting. In the past it was believed that the underlying layer of the alluvial aquifer was a product of the Pliocene (at that time, the nomenclature of the time division of the Pleistocene was different). Most of the underlying layers consist of thick motley clays and dust, interspersed here and there with sparse layering of sand and fine gravel. This can potentially be related to the different climate of the older Pleistocene and, by all accounts, to the somewhat lower massif of mountains in the Sava River Basin.

## 3.5.4 Research locations and methodology

At Belgrade Groundwater Source, 40 observation wells were drilled in 2016 and 2017 on the locations of 6 radial collector wells and in the hinterland. Two deeper, special-purpose boreholes (Rb-6/p-5d and Rb-36/p-4d) were drilled deep into the aquifer base clays, in order to perform a stratigraphic analysis (Dimkić, 2017, Zarić *et al.*, 2019a), and conduct additional investigations (like the analysis of paleomagnetic properties of base clays) to investigate the characteristics, origin and absolute age of the strata. Stimulated luminescence was used to date the mineral's last exposure to sunlight. The same set of analyses apart from paleomagnetic analysis was also performed on the material from two boreholes drilled just to the bottom of the aquifer (Rb-1 m/p-3d and Rb-44/p-4d) (Zarić *et al.*, 2019b). The locations of the boreholes Rb-6/p-5d, Rb-36/p-4d and Rb-1 m/p-3d are shown in Figure 3.37. Along

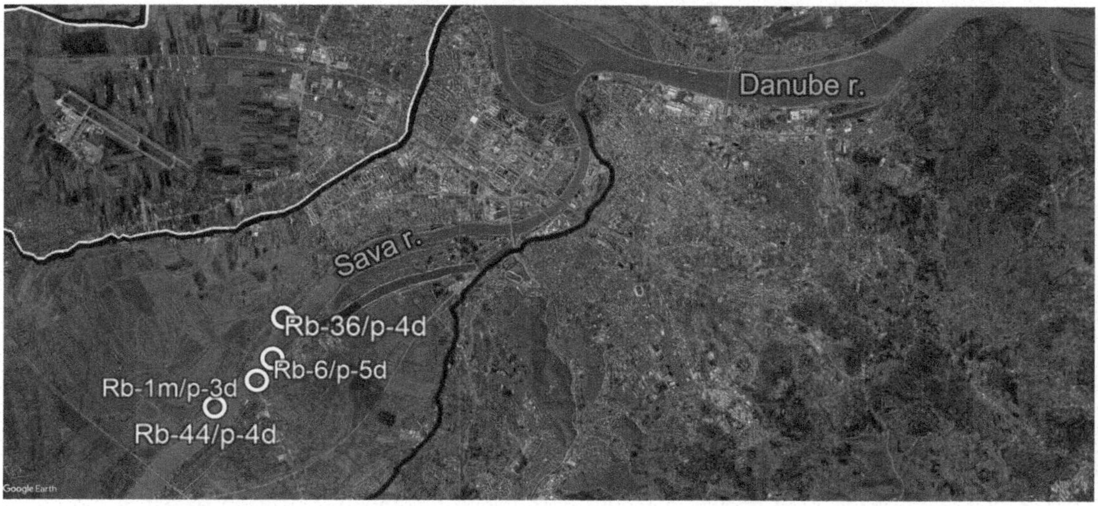


**Figure 3.37** Location of the Rb-6/p-5d, Rb-1 m/p-3d and Rb-36 m/p-4d boreholes, areas between the black lines are the alluvial plane and terraces, to the west and north of the black line (inside white lines) is the loess plateau with alluvial sediments underneath (Mitrinović *et al.*, 2019 – modified).

with investigation methods and results of absolute dating of the alluvial sediments from the two boreholes, indications and hypotheses pertaining to the connection of the genesis and characteristics of the strata (age, grain-size distribution, sorting, etc.) with climate changes in the Pleistocene and dated glaciations in the Dinaric mountains are also presented (Mitrinović *et al.*, 2019; Zarić, *et al.*, 2019a).

Samples were also taken for granulometric, geochemical, mineralogical, petrological and sedimentological analysis. Mineralogical and geochemical analyses were performed and they included determining the elemental composition of the entire sample by means of X-ray fluorescence, the separation of heavy minerals by using heavy liquid and the determination of their mass share in the fraction selected for mineralogical tests, the magnetic separation of heavy minerals and the determination of their mass share in the heavy fraction, the determination of the mineralogical composition of all magnetic fractions within the heavy fraction and the determination of the petrographic and mineralogical composition of the light fraction, the testing performed by using optic microscopic methods, by scanning using electronic microscopy – energy-dispersion spectroscopy, by the X-ray diffraction of powder and by calcimetry. The results of the laboratory mineral and geochemical analyses were processed in detail.

The paleontological material was also identified, with an emphasis on the index fossils.

The glaciation extent in the central Montenegrin mountains and Mt. Orjen (Adamson *et al.*, 2015; Hughes *et al.*, 2011) is shown in Figure 3.38. Studies performed at these locations by Hughes *et al.* (2011) and Adamson *et al.* (2015), which include the up to date most detailed dating of glacial features of the Dinaric Alps, are indicative for the glaciations in the rest of Dinaric mountains. Hughes *et al.* (2011) reported dating results for glacial landforms on several central Montenegrin mountains and Mount Orjen,

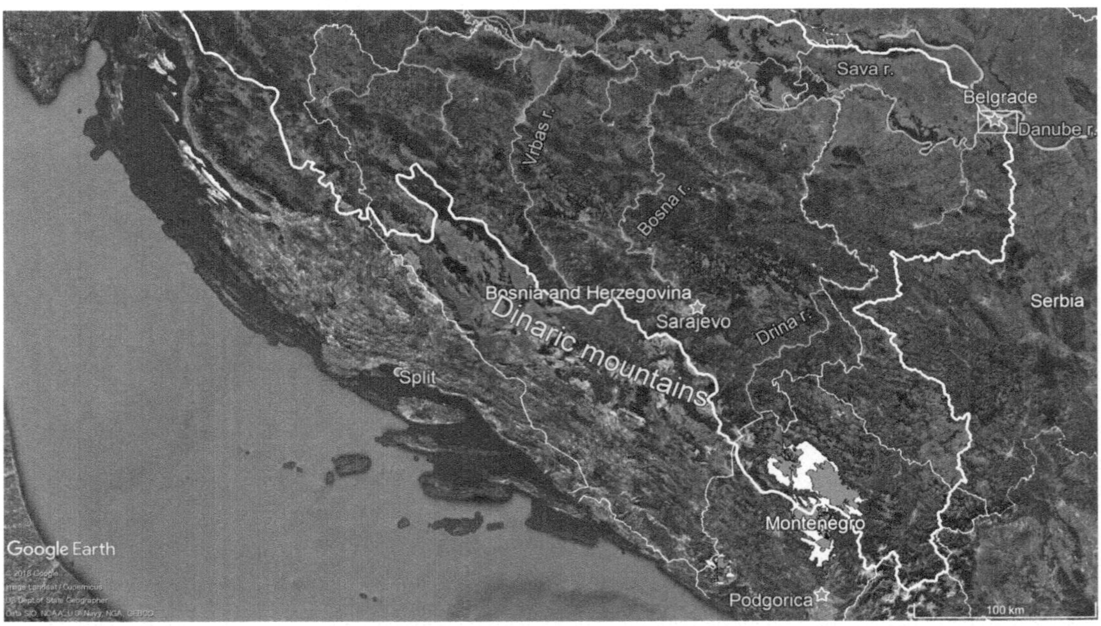


**Figure 3.38** Extent of the area shown in Figure 3.37 (white rectangle), glaciation extent in the central Montenegrin mountains and Mt. Orjen according to Hughes *et al.*, 2011 and Adamson *et al.*, 2015 (white, light and dark grey areas), Sava River catchment (outlined in white), Danube River, Sava River and its tributaries (light grey).

using hypsometric, lithological and pedological classification and the geochronological method for dating the calcite cement by uranium/thorium series, while Adamson *et al.* (2015) provided the results of dating of the glacial outwash in karst poljes at the foothills of Mt. Orjen. Mount Orjen belongs to the Adriatic drainage basin, but the changes in the spread and thickness of the ice and snow cover of the glaciers occurred at the same time as those on Mt. Durmitor and other central Montenegrin mountains, due to geographic proximity. The same would be true for the entire Dinaric Alps.

The melting of glacial cover on Dinaric mountains within the Sava River catchment (outlined in white in Figure 3.38) at the beginning of interglacials and some interstadials will have released very large quantities of water in a short time period, resulting in very high flows in the Sava River and its southern tributaries, resulting in intensive erosion in the upper catchment area, and combined erosion of fine-grained sediment and deposition of coarser sediments in the downstream part of the Sava River valley (Dimkić, 2017; Mitrinović *et al.*, 2019).

### 3.5.4.1 Grain size analysis

During the course of drilling of observation well Rb-6/p-5d and Rb-36/p-4d using a rotary drill and a coring tube, and Rb-1 m/p-3d, using an auger and a bailer tool, samples for detailed grain-size analysis to identify the sedimentation cycles were collected. The grain size distribution curves were calculated from the results of sieving and sedimentation procedures, and characteristic grain sizes determined. There were 67 samples, collected from the loose aquifer matrix and semi-permeable, partially consolidated interbeds, in general at each major lithologic change but in parts also at shorter intervals.

### 3.5.4.2 Mineralogical-petrological analysis

By analyzing the results of the mineralogical analysis of alluvial sediments in the Sava valleys near Belgrade, it was noticed that the highest content of mineral-petrological components in the gravel of the Pleistocene alluvial layers have grains of quartz (over 75%) and red and black chert (from 5 to 7%). Also noted was the presence of Miocene marl and limestone occurrences, while from the Mesozoic, most often noted were Cretaceous sandstones and clays, and less often serpentinites and igneous rock cobbles (diabase, dacito-andesite, granite, etc.).

Since quartz and chert are composed of silicon dioxide, their general property is that they are very resistant to physical and chemical decomposition. Due to their resilience, they were able to endure transport from remote areas, and form, in significant quantities, the material that formed the alluvial, primarily Pleistocene, deposits of the Belgrade Groundwater Source. The high share of rounded chert suggests that the parent rocks from which the coarse grains of this mineral component originate were located in remote areas. If we look at the regional distribution of geological formations that contain chert, we can see that they are also found in the ranges of the high mountains of central and eastern Bosnia, northern Montenegro and south-western Serbia (Zarić *et al.*, 2019a, 2019b).

### 3.5.4.3 Biostratigraphic - paleontological analysis

Biostratigraphic-paleontological studies were performed to determine the relative age and genesis of sediments. An examination of the entire core in the boreholes was performed, and samples were taken from fossil-bearing sediment. This was followed by laboratory biostratigraphic analysis of fossils. During drilling and the preliminary mapping of boreholes, the occurrence of characteristic fossils and their number and physical condition were recorded, and fossils were collected from part of the drilled material that was not stacked in a box and it was put into small plastic Ziploc bags. During the specialist paleontological examination of the cores, the results of the preliminary mapping and sampling were checked. During the examination of the borehole column, close-up photographs of characteristic fossils were taken and the depth from where they were taken was recorded. The samples were taken with a clean steel spoon and they were packed in plastic bags. The fossils that were found during the granulometric analysis were also identified. The laboratory part of the research consisted of preparing the fossils (making oriented cross-sections of the fossils, washing or dissolving of the sediment in order to extract the fossils), their identification and the paleontological estimate of their minimum age.

### 3.5.4.4 OSL analysis

Sampling for dating by optically stimulated luminescence and infrared stimulated luminescence (OSL, IRSL) was conducted by collecting the sediment from a certain depth, (which represented a sedimentation cycle), and emptying the bailer tool or coring tube directly into a short opaque 120 mm diameter PVC tube sealed at one end, which was then immediately plugged and secured at the other end. The closed tube containing the sample was wrapped in a reflective foil, to prevent even minimal exposure to light – Figure 3.39.

The samples were analyzed at the University of Szeged (Hungary), Department of Physical Geography and Geophysics, Laboratory for Optically Stimulated Luminescence. Both quartz and feldspar fractions were extracted and subjected to dating for the samples from Rb-6/p-5d, and as the results of feldspar dating were assessed as unreliable, only quartz dating was performed on the sample from Rb-1 m/p-3d.

Samples were dried and sieved to obtain the 150–220 μm sand fraction. Subsequent to acid treatment (HCl and $H_2O_2$) heavy liquid was applied first to isolate quartz grains, by setting densities to 2.68 and 2.62 g/cm$^3$. Subsequently, from the remaining light fraction, by using a liquid with 2.58 g/cm$^3$ density, a K-feldspar rich extract was produced. Quartz samples were etched with 40% hydrofluoric acid (HF) for

**Figure 3.39** Sediment sampling for dating by optically stimulated and infrared stimulated luminescence (Mitrinović *et al.*, 2019).

45 min, K-feldspar samples with 10% HF for 20 min. Samples were mounted on stainless steel (quartz) and aluminum (feldspar) discs using silicone oil and a 2 mm mask to prepare small aliquots.

Both fractions of each sample were subjected to luminescence dating using a RISO TL/OSL DA-20 luminescence reader equipped with a $^{90}Sr/^{90}Y$ source calibrated to both stainless steel and aluminum discs. Quartz and feldspar stimulation was carried out by using blue (470 nm) and IR (870 nm) LEDs, respectively (Bøtter Jensen *et al.*, 2010). Detection was made through either a U-340 filter (quartz) or the combination of BG39 and CN-7–59 filters (feldspar).

Quartz measurements were made using the SAR protocol (Murray & Wintle, 2000; Murray & Wintle, 2003; Wintle & Murray, 2006) without using a 280°C hot bleach at the end of the SAR cycles (Tóth *et al.*, 2017). Based on combined preheat and dose recovery tests, the applied preheat temperature was set either to 200°C or 220°C. Concerning feldspar samples the pIRIR290 protocol was applied. When determining quartz and K-feldspar *De*, 78 and 12 aliquots were subjected to SAR and pIRIR290 measurements, respectively. Rejection criteria were set to usual values (recycling ratio: $1.00 \pm 0.10$; maximum dose error: 10%; maximum recuperation: 5%). In the case of quartz samples *De* was calculated using the central age model (CAM) (Galbraith *et al.*, 1999).

Environmental dose rate $D^*$ was determined using high resolution, extended range gamma spectrometry (Canberra XtRa Coaxial Ge detector), using 500 $cm^3$ marinelli beakers. Dry dose rates were calculated using the conversion factors of Liritzis *et al.* (2013).

#### *3.5.4.5 Paleomagnetic analysis*

Samples for paleomagnetic testing were collected by means of coring tubes, making sure that there were no discontinuities along the core length (about 1.5 m). Upon extraction from the coring tube, a groove was made in the sample with a downward arrow and the sample cut into segments of selected finest grain sizes. They were wrapped in several layers of cellophane and a piece of paper placed in-between, with a down arrow, borehole designation, date, and depth interval. The remainder of the core was placed in a core crate. The samples were immediately refrigerated at 5°C, awaiting analysis. From Rb-6/p-5d a total of 10, and from Rb-36/p-4d 20 core segments, assessed to be the most fine grained, were sent for paleomagnetic testing to the Hungarian Geophysical Institute in Budapest, Department of Geophysics, Paleomagnetism Laboratory. To measure magnetic susceptibility a KLY-2 Kappabrides (AGICO, Brno, Czech Republic) instrument was used, for measuring natural remanent magnetization JR-4 magnetometers were used (AGICO, Brno, Czech Republic), thermal demagnetization was performed using a TSD-1 thermal demagnetizer (Schonstedt Instruments Company, Reston, USA), variable alternating magnetic field demagnetization was performed using LDA3A and DEMAG0179 demagnetizers (AGICO, Brno, Czech Republic and Technical University of Budapest, Hungary, respectively), and to determine the Curie points a CS-3 instrument (AGICO, Brno, Czech Republic) was used.

### 3.5.5 Dating results for the alluvial sediments in the lower course of the Sava River

#### *3.5.5.1 Grain-size analysis*

The results of the grain-size analysis are shown in Figures 3.43, 3.44 and 3.45, in the form of depth profiles of typical grain sizes – $d_{10}$, $d_{15}$, $d_{20}$, $d_{50}$, $d_{60}$ and $d_{85}$. The abrupt and significant increase in the typical grain diameters size was interpreted as the start of a new sedimentation cycle. The boundaries of the cycles are marked on lithological columns.

### 3.5.5.2 Biostratigraphic - paleontological analysis

The paleontological material in the sediments of the wells contained a macrofauna of mollusks, and in some places also a microfauna of ostracods.

Based on the analysis of the collected fossil fauna and of the lithological columns, it was determined that Quaternary sediments were discovered in borehole Rb-36/p-4d, while the Quaternary and Upper Miocene (Pannonian sediments) were discovered in boreholes Rb-6/p-5d and Rb-1 m/p-3d:

#### Upper Miocene – Pannonian

Layers of the Upper Miocene-Pannonian were discovered in the deepest part of well Rb-6/p-5d, below a depth of about 40.6 m to its bottom. They are made of grey massive marly clays with a frequent microfauna of ostracods and rarely of fossil mollusks. The following species were identified – *Congeria banatica R.Hoernes, Congeria czizeki M.Hoernes, Paradacna syrmiense R.Hoernes, Limnocardium sp.*

#### Quaternary

Quaternary sediments were discovered in most of the borehole's profile Rb-6/p-5d (from 0 to 40.6 m) and Rb-1 m/p-3d (from 0 to 30 m), as well as along the entire profile of the well Rb-36/p-4d.

#### Pleistocene

The sediments of the Pleistocene epoch have the largest thickness among the drilled sediments in the Belgrade Groundwater Source. According to their superposition, sedimentological and paleontological characteristics, two geological formations can be distinguished: 1) the older formation of polygenetic deposits and 2) the younger formation of polycyclic river sediments.

(1) The formation of the older, polygenetic Pleistocene sediments was deposited over Neogene Pannonian marly clays and lies under the water-bearing polycyclic river deposits of the Pleistocene deposits. It is made of motley clays (grey-brown with frequent oolites and lenses of manganese and reddish rust-like iron oxides, gray-blue clays etc.), with some layers and lenses of sand and sandy gravel. No fossils have been found, so it is not possible to determine the age by paleontological methods. In borehole Rb-6/p-5d, this formation has a thickness of about 13 m, and on the left bank of the Sava in borehole Rb-36 the thickness is greater than 30 m. We believe that these are polygenetic (marsh-lake terrestrial) formations, with coarser sediments that were occasionally deposited by torrential flows due to climate change.

(2) The formation represented by polycyclic river Pleistocene sediments, lies over the base made of older Pleistocene deposits in boreholes Rb-1 m/p-3d, Rb-6/p-5d and Rb-36/p-4d. It consists mostly of sandy gravels in alternation with gravelly sands, and layers of silt and clay are also occasionally present, with cyclic repetition sets of coarser and finer grained clastics.

Fossil river fauna appears in these layers and is similar to the one that today inhabits the Danube and its larger tributaries, including the Sava River. The species Corbicula fluminalis and Viviparus boechi, which are index fossils by which Pleistocene polycyclic river deposits are distinguishable from modern alluvial deposits, were present in all boreholes. Fossil fauna is very unevenly represented, so that it rarely occurs in some boreholes, while at other localities fossil shells occur en masse.

It was found that the species Viviparus boechi occurs more often in the older layers of the formation, while the species Corbicula fluminalis is much more common in the younger layers (for example, in well Rb-1 m/p-3d).

This formation has been known as 'layers with Corbicula fluminalis' (Laskarev, 1938; Stevanović, 1977) according to the characteristic species of the same name. These authors believed that they occurred due to sedimentation in the river-lake environment during the Mindel/Riss interglacial. Rakić (1977) points out that these are polycyclic river sediments of the older Pleistocene created in conditions of climate change and the rhythmic tectonic lowering of the river valley.

#### Holocene

The modern river sediments of the Sava River belong to the Holocene. It is possible to distinguish the older package of sediments of the riverbed facies built of sands, and the younger package of sediments of floodplain facies, built of grey-brown silt and sandy clays. In the contemporary alluvial sediments, fauna appears in layers that are very unevenly distributed. Most modern species are inherited from the Pleistocene, such as: *Litoglyphus naticoides, Litoglyphus fuscus, Microcolpia daudebartii aciculris, Esperiana esperi, Theodoxus danubialis, Theodoxus transversalis, Amphimelania holandrei* and others. The species *Dreissena polymorpha* has been found quite frequently, and it was not found in the Pleistocene layers.

### *3.5.5.3 OSL and IRSL analysis*

The results of dating by optically stimulated luminescence are shown in Table 3.2. The age model decision process between the different statistical approaches has suggested the CAM age model is more suitable for the determination of the sample ages.

Feldspar dating of the younger samples provided very different ages from the quartz dating, which was considered more reliable and accurate, as quartz grains take much less time to bleach during transport. Also, the sample from the 9.3–9.6 m interval in Rb-6/p-5d (Figure 3.44) belongs to the last sedimentation cycle ending with recent sediments and almost certainly belonging to the Holocene, which is also reflected by the quartz dating results, while the feldspar result shows significantly higher age (Table 3.2). Quartz and feldspar dating results also differ very much for the 16.0–16.5 m and 22.2–22.6 m intervals in Rb-6/p-5d, and 16.5–16.7 m and 21.5–21.7 m intervals in Rb-36/p-4d, with feldspar showing again much higher ages (Table 3.2). Feldspar results are likely to have been influenced by impartial bleaching and consequently were not used for dating.

### *3.5.5.4 Paleomagnetic analysis*

After preliminary screening, four core segments from Rb-6/p-5d, from the upper and lower parts of the so-called 'motley clays', were selected for testing. Twenty-eight samples in total were collected from core segments, to measure magnetic susceptibility, determine the Curie point, and, after thermal demagnetization and/or demagnetization by variable magnetic field, determine the direction and intensity of natural remanent magnetization (NRM). NRM could not be determined for two segments. For the remaining two core segments (27.35–27.65 m and 38.80–39.20 m intervals shown in Figure 3.44, as dark grey rectangles denoted by PM, for paleomagnetism), the polarization of Earth's magnetic field was normal when the orientation of the mineral crystals, the carriers of NRM (most likely magnetite), was fixed in these two intervals. The reliability of the determined NRM direction and intensity during demagnetization of examined samples from those two core segments was satisfactory, as during the course of alternating field demagnetization magnetic inclination and declination remained moderately stable. The results of the demagnetization procedure on two of the samples are presented as modified Zijderveld diagrams in Figure 3.40.

**Table 3.2** Results of dating of samples from Rb-6/p-5d, Rb-36/p-4d and Rb-1 m/p-3d boreholes by optically stimulated luminescence of quartz (Q) and feldspar (F). $D_e$ is equivalent dose, $D^*$ is the environmental dose rate (from Mitrinović *et al.*, 2019).

| Lab Code | Field ID | Depth Interval (m) | w (%) | Mineral | Age Model | U (ppm) | Th (ppm) | K (mas.%) | $D^*$ (Gy/ka) | $D_e$ (Gy) | Age (ka) |
|---|---|---|---|---|---|---|---|---|---|---|---|
| | | | | Q | CAM | | | | 1.05 ± 0.07 | 13.80 ± 0.43 | 13.20 ± 0.93 |
| 1530 | RB-6/P-5d | 9.3–9.6 | 12.87 | | MAM | 1.06 ± 0.01 | 3.11 ± 0.06 | 0.68 ± 0.03 | | 11.55 ± 0.75 | 11.05 ± 1.0 |
| | | | | F | Avg. | | | | 1.668 ± 0.13 | 50.75 ± 2.00 | 30.21 ± 2.61 |
| | | | | Q | CAM | | | | 0.88 ± 0.07 | 70.94 ± 3.97 | 80.9 ± 7.8 |
| 1531 | RB-6/P-5d | 16–16.5 | 11.35 | | MAM | 0.69 ± 0.01 | 2.27 ± 0.04 | 0.65 ± 0.03 | | 43.74 ± 3.97 | 49.9 ± 6.0 |
| | | | | F | Avg. | | | | 1.51 ± 0.12 | 414.84 ± 22 | 274.3 ± 27.38 |
| | | | | Q | CAM | | | | 0.80 ± 0.07 | 109.48 ± 6.26 | 137.5 ± 14.1 |
| 1532 | RB-6/P-5d | 22.2–22.6 | 11.23 | | MAM | 0.72 ± 0.01 | 2.3 ± 0.05 | 0.56 ± 0.02 | | 81.49 ± 6.8 | 102.3 ± 12.2 |
| | | | | F | Avg. | | | | 1.42 ± 0.12 | 672.91 ± 53.00 | 470.8 ± 55.70 |
| 1533 | RB-6/P-5d | 27–27.3 | 8.16 | F | Avg. | 0.63 ± 0.01 | 2.06 ± 0.04 | 0.53 ± 0.02 | 1.38 ± 0.12 | >962.47 ± 48 | >693.5 ± 71.90 |
| 1534 | RB-6/P-5d | 35.9–36.3 | 10.03 | F | Avg. | 0.79 ± 0.01 | 2.22 ± 0.05 | 0.49 ± 0.02 | 1.37 ± 0.12 | >577.96 ± 23 | >420.7 ± 42.05 |
| 1625 | RB-1 m/P-3d | 27.8–28.0 | 6.06 | Q | CAM | 0.59 ± 0.01 | 1.85 ± 0.5 | 0.47 ± 0.02 | 0.67 ± 0.08 | 131.19 ± 6.74 | 193.38 ± 23.62 |
| | | | | Q | CAM | 0.89 ± 0.01 | 2.42 ± 0.05 | 0.54 ± 0.02 | 0.87 ± 0.03 | 48.99 ± 2.27 | 56.0 ± 3.3 |
| 1527 | RB-36/P-4d | 16.5–16.7 | 6.68 | | MAM | | | | | 48.99 ± 4.29 | 56.0 ± 5.4 |
| | | | | F | Avg. | | | | 1.59 ± 0.13 | 405.18 ± 27.85 | 254.6 ± 26.8 |
| | | | | Q | CAM | 0.84 ± 0.01 | 2.71 ± 0.06 | 0.68 ± 0.03 | 0.97 ± 0.04 | 86.58 ± 6.31 | 88.8 ± 7.2 |
| 1528 | RB-36/P-4d | 21.5–21.7 | 8.97 | | MAM | | | | | 67.99 ± 7.54 | 69.8 ± 8.1 |
| | | | | F | Avg. | | | | 1.71 ± 0.13 | 756.0 ± 54.6 | 442.1 ± 46.2 |

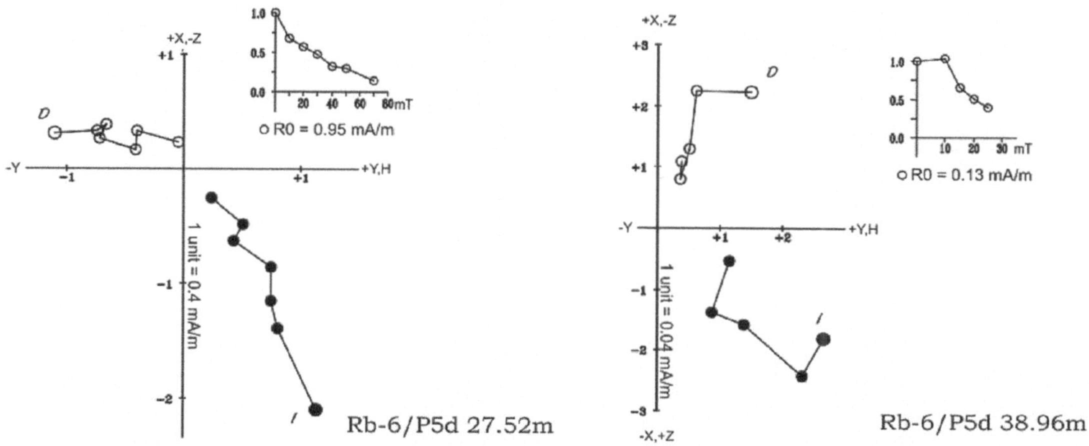


**Figure 3.40** Modified Zijderveld diagrams of demagnetization procedure on two of the samples. D – declination, I – inclination, intensity of the remanence is plotted as a function of the demagnetization field (Mitrinović *et al.*, 2019).

Twenty core segments from Rb-36/p-4d, assessed to be the most fine grained, were sent for paleomagnetic analysis. After preliminary screening, 10 core segments, 4 from the upper and 6 from the lower parts of the so-called 'motley clays' at the base of polycyclic sediments were selected for testing. In total, 24 samples were collected from the core segments, to measure magnetic susceptibility, determine the Curie point (2 samples), and, after thermal demagnetization and/or demagnetization by variable magnetic field, determine the direction and intensity of natural remanent magnetization (NRM). During the preparation larger or smaller pieces of rock fragments or pebbles were observed in all the core segments, except for the segment 53.40–53.65 m. Some samples disintegrated during laboratory processing due to their internal inhomogeneity. The results of the demagnetization procedure on nine of the samples from five core segments (43.40–43.65, 46.75–47.00, 49.60–49.85, 53.40–53.65 and 60.80–61.00 m) are presented as modified Zijderveld diagrams in Figure 3.41. Stability of the paleomagnetic signal with clear tendency to decay towards the origin was not a general feature (five samples out of nine in Figure 3.41 exhibited such behavior). For the 46.93 and 60.93 m samples it was not even possible to define the polarity. The paleomagnetic directions from the same core segments were far from consistent, as exemplified by data in Figure 3.41. Finally, the samples with normal and reversed polarities were in some instances intermixed within the same core segment (Figure 3.41 – modified Zijderveld diagrams No. 2, 3, 4, 8, core segment 53.40–53.65 m). It seems that the material was basically re-deposited in the form of 'mudcakes', i.e. that the orientation of the mineral crystals, the carriers of NRM (most likely magnetite), was fixed in clayey lumps which were deposited and then agglutinated in random orientations. The Curie-point measurements (Figure 3.41c) also document extreme variation in magnetic mineralogy. While the core segment 53.40–53.65 m, more clayey than the rest and devoid of small pebbles, contains magnetite (Figure 3.41c, up), another one contains an insignificant amount of magnetic mineral, probably iron sulfide, which converts to magnetite on heating above 400°C (Figure 3.41c, bottom). The conclusion is that the results of paleomagnetic analysis could not provide any age assessment as core material proved to be very inhomogeneous, which demonstrated that the sedimentation of even the fine grained part of the material took place in significantly more turbulent conditions than those present during the deposition of sediment from borehole Rb-6/p-5d (Mitrinović *et al.*, 2019).

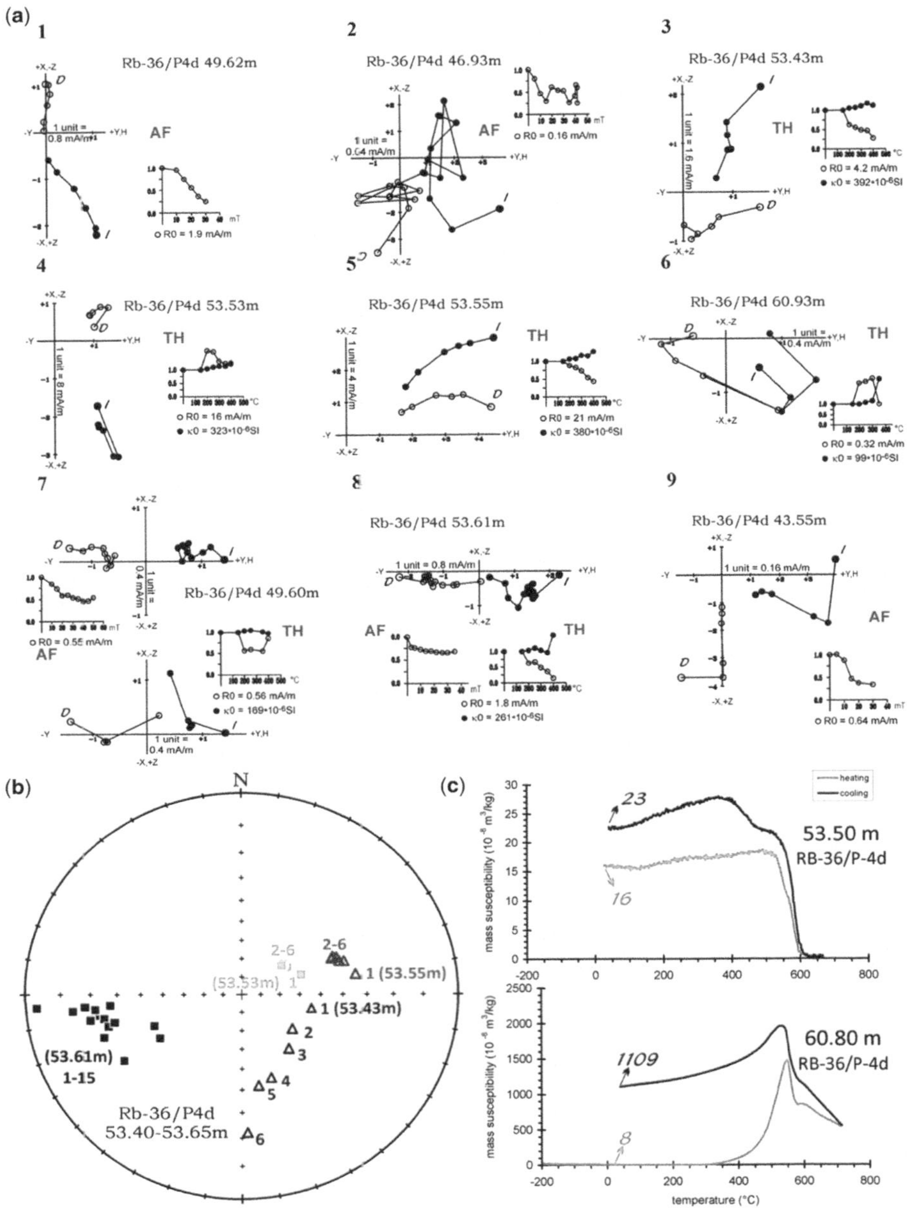


**Figure 3.41** (a) Modified Zijderveld diagrams of demagnetization procedure on nine samples. D – declination, I – inclination, o – intensity of the remanence plotted as a function of the demagnetization field/temperature, ● – susceptibility as a function of the temperature, TH – thermal demagnetization, AF – alternate field demagnetization. (b) The directions of the remanent magnetizations for samples from core segment 53.40–53.65 m on stereonet. (c) Curie-point measurements – the magnetic susceptibility as a function of the temperature (grey curve: heating, black curve: cooling). Initial and final mass susceptibilities of the powdered samples are shown as grey and black numbers (Marton, 2019).

### 3.5.6 Correlation of lower Sava alluvial sediments with the results of moraine dating at the Dinaric Alps

The Dinaric Alps and outer Dinaric ranges that belong to the Sava River Basin are assumed to be the main erosion zone and source of the materials from which the Quaternary fluvial sediments in the Belgrade Groundwater Source (BGS) area of the Sava Valley were formed. These sandy and gravelly materials were transported by rivers that were much more powerful than the present day Sava and Danube, under conditions of rapid melting of glaciers and ice caps at the beginning of the interglacial and interstadial periods, hence the age of the moraines remaining after several glacial cycles in the Dinaric Alps and outer Dinaric ranges should correspond to the glacial and stadial periods previous to those interglacials and interstadials. The age of the moraine outwash in the karst poljes at the foot of the mountains should then correspond to the age of the cycles in the Quaternary fluvial sediments at the BGS, as well as the age of the underlying polygenetic sediments (Mitrinović *et al.*, 2019).

#### *3.5.6.1 Glacial studies in Montenegro – dating of moraines*

Six stratigraphic units were identified and their ages determined during the course of investigations on Mt. Durmitor, Mt. Sinjajevina, the Morača mountains, Mt. Maganik, Mt. Prekornica, Mt. Vojnik and Mt. Orjen (Hughes *et al.*, 2011). The results of absolute dating using U/Th series are shown in Table 3.3 (source: Hughes *et al.*, 2011).

In the final reconstruction of the spread of the ice cover that formed glacial landforms on the central Montenegrin mountains, and the age of these maxima, the absolute dating results were taken as an indication of the minimal age of the calcite cement, where the mutual hypsometric positions of the moraines and the degree of formation of a soil layer (quantified by means of the Profile Development Index – PDI), were pivotal for the assessment of the relative age and temporal and spatial correlation of the geochronologically dated and non-dated moraines (Hughes *et al.*, 2011). Five ice masses and their ages were identified. The reconstruction results are shown in Table 3.4 and Figure 3.36.

There are strong indications that the glaciers that existed after the largest ones from MIS 12, and before those from MIS 6 (with a similar extent to the latter), existed in MIS 8 (and possibly MIS 10), so an unknown unit was entered in Table 3.4. In one place on Durmitor, the age of the sample was 231.9 ± 17.5 ka, whereas the ages of two samples of glacial material from Orjen, deposited in karst poljes at the foothills, were found to be 248.6 ± 16.7 and 213.5 ± 11.3 ka (Adamson *et al.*, 2015). The highest intensity glaciation with the most extensive ice cover in MIS 12 removed or disturbed the traces of prior glaciations, which made it difficult to determine their spread accurately and even confirm their existence (Mitrinović *et al.*, 2019).

#### *3.5.6.2 Glacial studies in Montenegro – dating of glacial outwash in karst poljes*

According to Adamson *et al.* (2015), the carbonate cement, formed within the glacial outwash in karst poljes on the foothills of Mt. Orjen, dated using U/Th series could have occurred much later than the deposition of sediments, at a time of favorable hydrologic and climate conditions. For this reason, as for moraines, the results of cement dating were taken as an indication of minimum age. In this case as well, the PDI values of the soil that covers the poljes were correlated with the age of the underlying sediments, and the oldest age for a certain PDI range was the age that corresponded to that PDI range. If any of the dating results indicated a much younger age of the cement than that which corresponded to the PDI, they were deemed irrelevant. If the PDI indicated an age younger than the one determined for the cement sampled at a certain depth below the surface soil, it was assumed that the material between the depth of the sample and the ground surface was deposited later (Mitrinović, *et al.*, 2019).

**Table 3.3** Results of absolute dating and identification of stratigraphic units of central Montenegrin mountains and Mt. Orjen (from Hughes *et al.*, 2011).

| Unit | Corrected U/Th Series Ages (ka) | Soil PDI | Moraine Age | Characteristics |
|---|---|---|---|---|
| 6 Debeli Namet Member (Durmitor + Central Montenegro) | | | AD 1850 to present | Modern glacier |
| 5 Gornji do Member (Orjen) | 9.6 ± 0.8<br>8.0 ± 0.2 | | Younger Dryas | Cirque glaciers |
| 4 Karlica Member (Durmitor + Central Montenegro) | 10.9 ± 0.3<br>10.5 ± 0.3<br>9.8 ± 0.3<br>9.1 ± 0.3<br>7.9 ± 0.3<br>2.2 ± 0.09 | 14.3 | Younger Dryas | Valley and cirque glaciers |
| 4 Reovci Member (Orjen) | 17.3 ± 0.6<br>13.9 ± 0.4<br>12.5 ± 0.4 | | MIS 2 | Valley and cirque glaciers |
| 4 Međed Member (Durmitor + Central Montenegro) | 13.4 ± 0.4 | 16.8 | MIS 2 | Valley glaciers |
| 3 Crkvice Member (Orjen) | 124.6 ± 5.7<br>124.5 ± 3.5<br>124.0 ± 3.4<br>102.4 ± 3.1 | | MIS 6 | Ice caps and valley glaciers |
| 2 Žabljak Member (Durmitor + Central Montenegro) | 120.2 ± 6.4<br>105.4 ± 5.2<br>104.1 ± 2.9<br>88.1 ± 3. 4 | 31.5 | MIS 6 | Ice caps and valley glaciers |
| 2 Žabljak Member (Durmitor + Central Montenegro, absent from Orjen) | 231.9 ± 17.5<br>162.7 ± 7.9 | 32.7 | MIS 8 or 10 | Ice caps and valley glaciers |
| 1 Knežlaz Member (Orjen) | >350<br>>350<br>309.3 ± 42.4 | | MIS 12 | Ice caps |
| 1 Ninkovići Member (Durmitor + Central Montenegro) | >350<br>>350<br>396.6 ± 87.2 | 43.1 | MIS 12 | Ice caps |

*Note*: PDI – Profile Development Index.

**Table 3.4** Results of dating of various glaciation maxima on the central Montenegrin mountains (according to Hughes *et al.*, 2011).

| Unit | Time Range [ka] | MIS Stage |
|---|---|---|
| Ninkovići Member ice masses | 478–424 | MIS 12 |
| ? | 374–243 | MIS 10 or MIS 8 |
| Žabljak Member ice masses | 190–130 | MIS 6 |
| Međed Member ice masses | 109–11.7 | MIS 5d-2 |
| Karlica Member, Debeli Namet Member ice masses | 12.9 - present | Younger Dryas, late Holocene, respectively |

The ages of the eroded moraine material from the karst poljes below Mt. Orjen are shown in Table 3.5 (according to Adamson *et al.*, 2015):

### *3.5.6.3 Study of samples from boreholes Rb-6/p-5d, Rb-36/p-4d and Rb-1 m/p-3d – grain size distribution, OSL and paleomagnetic analysis*

Based on grain-size analysis and lithological mapping, seven cycles were identified within the polygenetic strata (three), and polycyclic, water-bearing strata of BGS and Holocene sediments (four) in Rb-6/p-5d and Rb-36/p-4d boreholes (Figure 3.44 and Figure 3.45), while four cycles were identified within polycyclic and Holocene alluvial sediments from the Rb-1 m/p-3d borehole (Figure 3.43). A layer of fine material that separates two sand or gravel strata or a sudden increase in the grain size of sand accompanied by a package of finer overlying material, was taken as the boundary between two cycles. Six cycles in Rb-6/p-5d were identified in this manner. The last three, according to OSL dating of quartz (Table 3.2), belong to the beginning of MIS 1 (Holocene), MIS 5b (according to the Alpine classification, short Würmian stadial) and the end of MIS 6 (end of Rissian glaciation) (Figure 3.44; Mitrinović *et al.*, 2019). The results of absolute dating and grain-size analysis of the samples collected from Rb-6/p-5d confirmed the connection between the moraines and fluvioglacial material from the karst poljes in the Dinaric Alps, and the three youngest sedimentation cycles at the BGS (Figures 3.42, 3.43 and 3.44). Based on the comparison of the results of dating quartz from samples 1532 and 1533 (Table 3.2) by OSL, another cycle was identified within the first cycle of polycyclic sediments that was determined on the basis of

**Table 3.5** Absolute ages and PDIs (according to Adamson *et al.*, 2015).

| Karst Polje | PDI [m] | Profile Height [m] | Age [ka] |
|---|---|---|---|
| Dvrsno | 11.80, 7.35 | 10 | – |
| Grahovo | 9.91 | 3 | >350 |
| Pirina Poljana | 9.12 | 4 | 213.5 ± 11.3 |
| Kruševice | 13.75 | 9 | – |
| Vrbanje | 4.88 | 7 | 126.6 ± 4.5 |
| Unijerina | 1.43 | 11 | >350, 248.6 ± 16.7, 80.3 ± 5.9,16.6 ± 0.4 |
| Crkvice | 3.92 | 2 | 144.2 ± 5.1 |

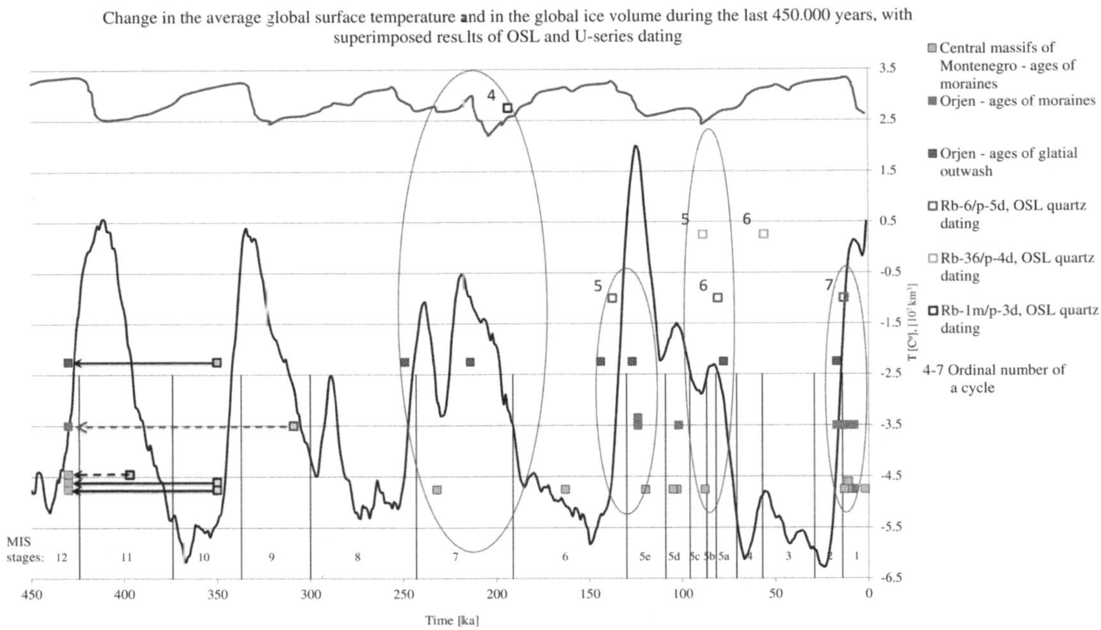


**Figure 3.42** Change in mean global temperature and global ice volume during the past 450 ka, with superimposed results of OSL dating (CAM) of BGS sediment samples and results of glaciological investigations of central Montenegrin mountains and Mt. Orjen (Mitrinović *et al.*, 2019).

grain-size analysis. The age of this cycle was greater than 150 ka, the approximate saturation age of the sample. It was dated according to the correlation with the sample 1625 from Rb-1 m/p-3d, which was collected at almost the same depth, and in the same sedimentation cycle based on grain size analysis (Mitrinović *et al.*, 2019), and with fluvioglacial material from the karst poljes on the foothills of Mt. Orjen, as MIS 7 (Figures 3.43 and 3.44). As regards Rb-36/p-4d, the seventh cycle, the most recent one, almost certainly belongs to MIS 1, the sixth and fifth cycles according to the OSL results belong to the end of MIS 4 and beginning of MIS 3, and to MIS 5b, respectively, while the age of the fourth cycle (first within polycyclic sediments), inferred from the succession of glaciations, would be the end of MIS 6 or MIS 5e.

Within the polygenetic sediments that underlie the polycyclic Pleistocene strata, three cycles were identified both in Rb-6/p-5d and Rb-36/p-4d, based on the results of the grain-size analysis and lithological mapping. The ages of Rb-6/p-5d cycles assumed on the basis of correlation with moraine deposits and karst polje materials and succession of MIS phases are MIS 9, MIS 11 and MIS 13 (Figure 3.44). The age of the oldest underlying polygenetic strata would then be 533 ka (Mitrinović *et al.*, 2019). The basic assumption is that the surface traces of the glaciations in MIS 8 and MIS 10 in Dinaric mountains, which were almost erased by the glaciation in MIS 6 were preserved in the cycles of deposition of the sediments in the lower course of the Sava River at the BGS. Using the same assumptions on Rb-36/p-4d, the inferred ages of cycles before polycyclic strata would be MIS 7, MIS 9 and MIS 11. The cycles defined by the results of grain-size distribution analysis and OSL dating in Rb-1 m/p-3d, Rb-6/p-5d and Rb-36/p-4d boreholes are separated by red lines in Figures 3.43, 3.44 and 3.45, and are listed in Table 3.6.

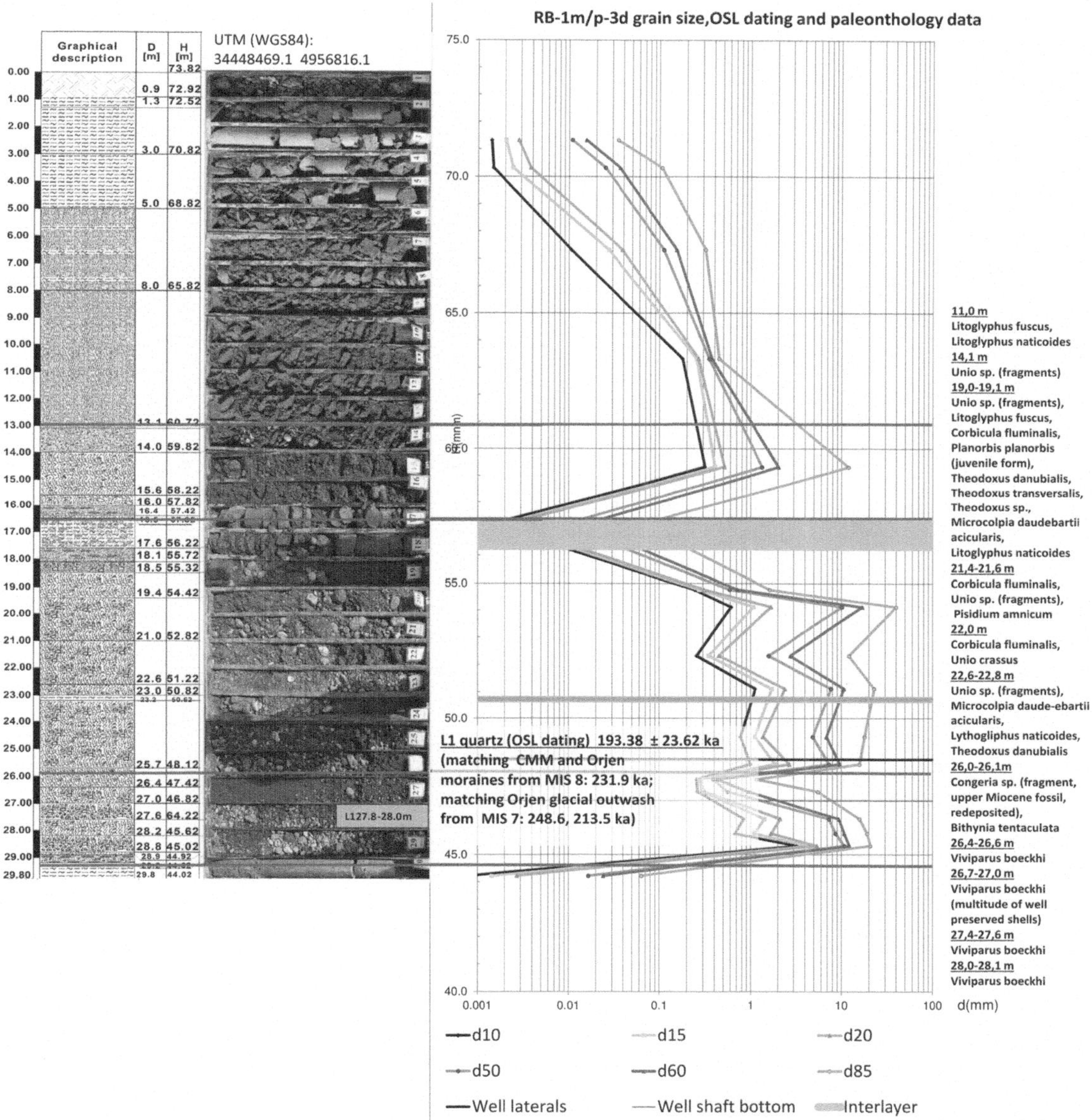


**Figure 3.43** Lithologic section of borehole Rb-1 m/p-3d, showing sampling interval for OSL dating (light grey rectangle), results of grain-size analysis, and OSL-determined age of sample (JCI, 2017; Mitrinović *et al.*, 2019; Zarić, 2020).

Ages greater than 400 ka cannot be determined by OSL, and rarely by IRSL and U-series methods, so paleomagnetic analyses are used if possible for older sediments. This requires the sediment sample to be fine-grained and homogeneous, and the material to have been deposited in a very calm environment. Also, samples need to be carefully collected and spatial orientation information retained. It is then possible to identify changes in orientation and magnetic field intensity, and through correlation with

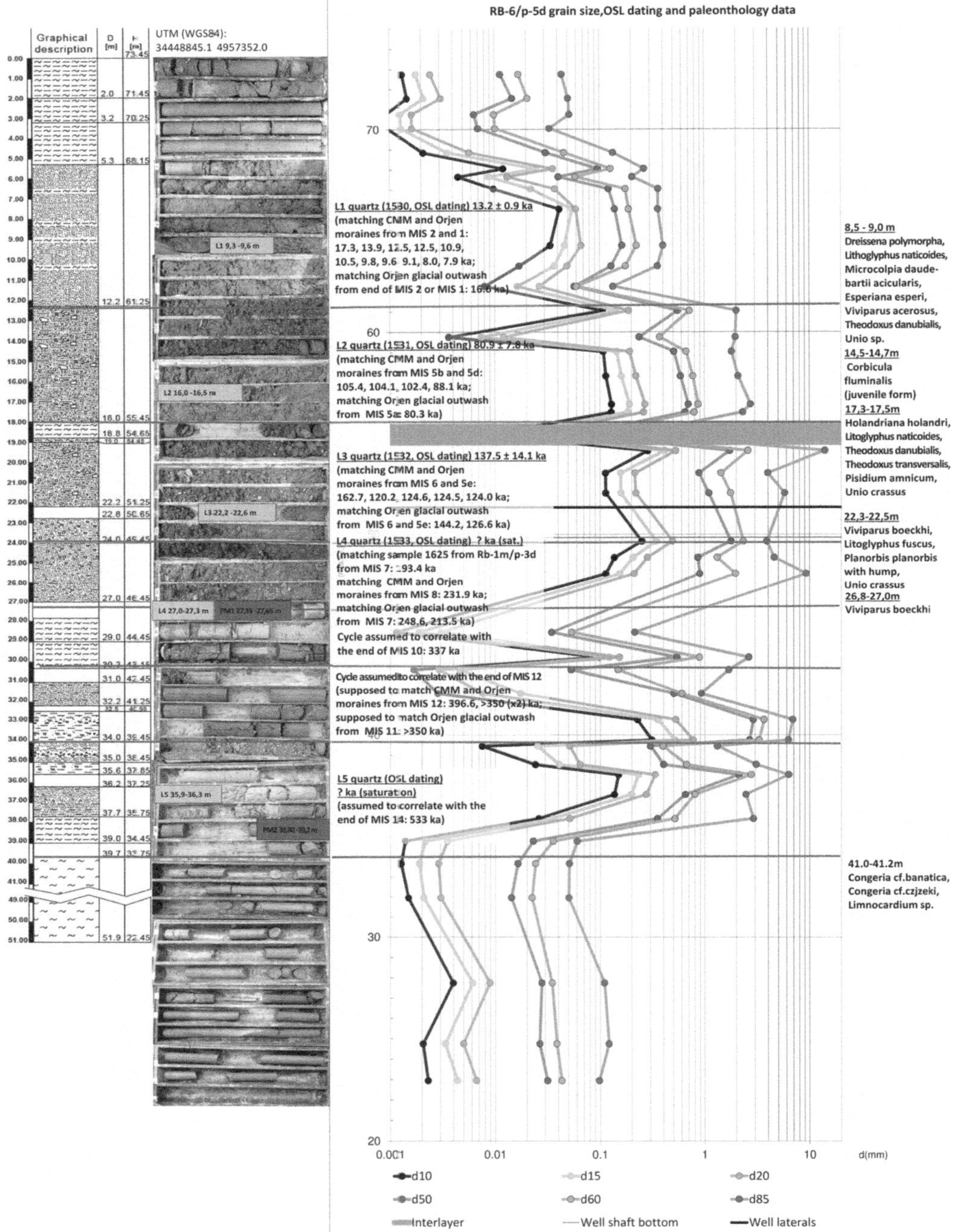


**Figure 3.44** Lithological section of borehole Rb-6/p-5d, showing sampling intervals for OSL and IRSL dating (light grey rectangles) and paleomagnetic testing (dark grey rectangles), results of grain-size analysis, and OSL-determined ages of samples (JCI, 2017; Mitrinović *et al.*, 2019; Zarić, 2020).

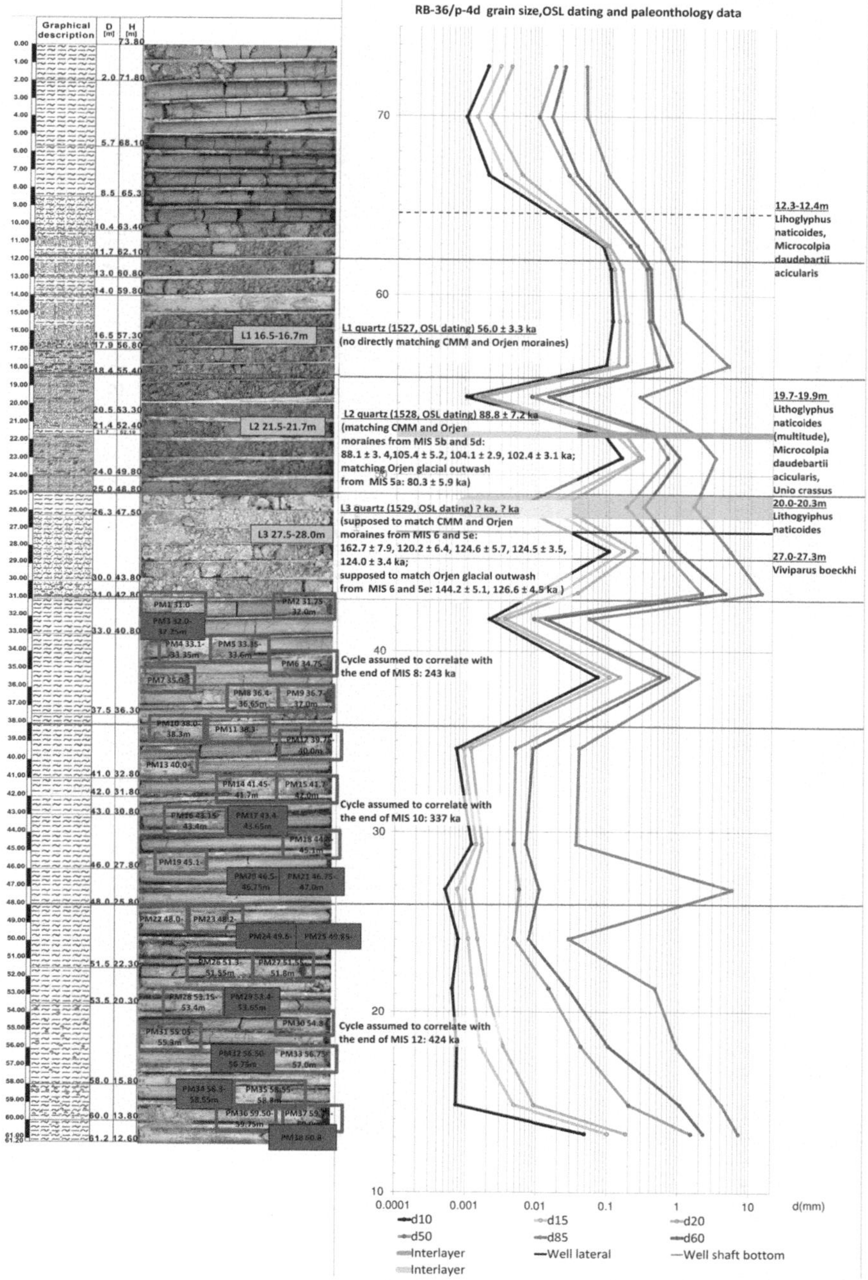


**Figure 3.45** Lithological section of borehole Rb-36/p-4d, showing sampling intervals for OSL and IRSL dating (light grey rectangles) and paleomagnetic testing (dark grey rectangles), results of grain-size analysis, and OSL-determined ages of samples (JCI, 2017).

**Table 3.6** Results of dating of samples from Rb-6/p-5d, Rb-36/p-4d and Rb-1 m/p-3d boreholes by OSL of quartz (Q), correlation with ages of CMM and Mt. Orjen moraines and Mt. Orjen glacial outwash, and sedimentation cycles demarcation.

| Cycle No. | Lab Code | Field ID | Depth Interval of Sample (m) | Age (ka) | Age (MIS) | Inferred Age from Succession of Cycles (MIS) | Related CMM and Mt. Orjen Moraines | Related Mt. Orjen Glacial Outwash |
|---|---|---|---|---|---|---|---|---|
| 7 | 1530 | RB-6/P-5d | 9.3 9.6 | 13.20 ± 0.93 | beginning of MIS 1 | | end of MIS 2 and MIS 1 (17.3, 13.9, 12.5, 12.5, 10.9, 10.5, 9.8, 9.6, 9.1, 8.0, 7.9 ka) | MIS 2 (16.6 ka) |
| 6 | 1531 | RB-6/P-5d | 16–16.5 | 80.9 ± 7.8 | MIS 5b | | MIS 5d and 5b (105.4, 104.1, 102.4, 88.1 ka) | MIS 5b (80.3 ka) |
| 5 | 1532 | RB-6/P-5d | 22.2–22.6 | 137.5 ± 14.1 | end of MIS 6 | | MIS 6 and the beginning of 5e (162.7, 120.2, 124.6, 124.5, 124.0 ka) | MIS 6 and the beginning of MIS 5e (144.2, 126.6 ka) |
| 4 | 1533 | RB-6/P-5d | 27.0–27.3 | | | MIS 7 | MIS 8 (231.9 ka) | MIS 7 (248.6, 213.5 ka) |
| 3 | | RB-6/P-5d | | | | end of MIS 10 | | |
| 2 | | RB-6/P-5d | | | | end of MIS 12 | MIS 12 (396.6, >350 (x2) ka) | MIS 11 (>350 ka) |
| 1 | 1534 | RB-6/P-5d | 35.9–36.3 | | | end of MIS 14 | | |
| 7 | | RB-1 m/P-3d | | | beginning of MIS 1 | | end of MIS 2 and MIS 1 (17.3, 13.9, 12.5, 12.5, 10.9, 10.5, 9.8, 9.6, 9.1, 8.0, 7.9 ka) | MIS 2 (16.6 ka) |
| 6 | | RB-1 m/P-3d | | | MIS 5b | | MIS 5d and 5b (105.4, 104.1, 102.4, 88.1 ka) | MIS 5b (80.3 ka) |
| 5 | | RB-1 m/P-3d | | | end of MIS 6 | | MIS 6 and the beginning of 5e (162.7, 120.2, 124.6, 124.5, 124.0 ka) | MIS 6 and the beginning of MIS 5e (144.2, 126.6 ka) |
| 4 | 1625 | RB-1 m/P-3d | 27.8–28.0 | 193.38 ± 23.62 | MIS 7 | | MIS 8 (231.9 ka) | MIS 7 (248.6, 213.5 ka) |

| | | | | | | | |
|---|---|---|---|---|---|---|---|
| 7 | | RB-36/P-4d | | | MIS 1 | | end of MIS 2 and MIS 1 (17.3, 13.9, 12.5, 12.5, 10.9, 10.5, 9.8, 9.6, 9.1, 8.0, 7.9 ka) | MIS 2 (16.6 ka) |
| 6 | 1527 | RB-36/P-4d | 16.5–16.7 | 56.0 ± 3.3 | beginning of MIS 3 and the end of MIS 4 | | | |
| 5 | 1528 | RB-36/P-4d | 21.5–21.7 | 88.8 ± 7.2 | end of MIS 5b or MIS 5d | | MIS 5d and 5b (105.4 ± 5.2, 104.1 ± 2.9, 102.4 ± 3.1, 88.1 ± 3.4 ka) | MIS 5b (80.3 ± 5.9 ka) |
| 4 | 1529 | RB-36/P-4d | 27.5–28.0 | | | end of MIS 6 | MIS 6 and 5e (162.7 ± 7.9, 120.2 ± 6.4, 124.6 ± 5.7, 124.5 ± 3.5, 124.0 ± 3.4 ka) | MIS 6 and 5e (144.2 ± 5.1, 126.6 ± 4.5 ka) |
| 3 | | RB-36/P-4d | | | | end of MIS 8 | | |
| 2 | | RB-36/P-4d | | | | end of MIS 10 | | |
| 1 | | RB-36/P-4d | | | | end of MIS 12 | MIS 12 (396.6, >350 (x2) ka) | MIS 11 (>350 ka) |

paleomagnetic records from reference boreholes determine the age. In the case of polygenetic sediments underlying fluvial polycyclic sediments at the BGS, deposition did not take place in a truly calm environment, as evidenced by the heterogeneity of the sediments, so NRM that recorded the geomagnetic field orientation from the time close to deposition was most likely the chemical remanent magnetization (CRM), carried by post-depositional mineral crystals formed within the sediments. Assuming that this occurred within a period of several thousand years after the deposition, the results of paleomagnetic testing of material from Rb-6/p-5d indicated that the age of polygenetic sediments was less than 780 ka, as the deeper segment with normal polarization lies immediately at the bottom of these sediments (Mitrinović, *et al.*, 2019). In the case of Rb-36/p-4d, the sedimentation conditions were more turbulent than for borehole Rb-6/p-5d, hence paleomagnetic analysis could not provide any age assessment.

Figures 3.43, 3.44 and 3.45 show the lithological sections of boreholes Rb-1 m/p-3d, Rb-6/p-5d and Rb-36/p-4d, along with sampling intervals for OSL and IRSL dating and paleomagnetic testing, the results of grain-size analysis, and the results of glaciological and geochronological investigations undertaken on the central Montenegrin mountains and Mt. Orjen with its accompanying karst poljes (Adamson *et al.*, 2015; Hughes *et al.*, 2011).

The results of OSL dating of samples collected from boreholes Rb-1 m/p-3d, Rb-6/p-5d, and Rb-36/p-4d at the BGS, as well as the results of glaciological and geochronological investigations undertaken on the central Montenegrin mountains and Mt. Orjen with its accompanying karst poljes (Adamson *et al.*, 2015; Hughes *et al.*, 2011) are presented in Figure 3.42, on the background of change of global average surface temperature (Snyder, 2016) and the global ice volume change (Pollard & DeConto, 2009). Hughes *et al.*, based on their insight into the glaciations in northwestern Greece and Albania, assigned to MIS 12 all samples older than 350 ka (i.e. in which isotope equilibrium was reached after 350 ka), as well as two samples whose ages were found to be 309 and 397 ka (Hughes *et al.*, 2011), as shown in Figure 3.42 by empty symbols connected by arrows with full symbols. The broken line indicates that cementation was assumed to have occurred later.

The ages of the alluvial sediments from boreholes Rb-6/p-5d and Rb-1 m/p-3d based on OSL dating of quartz correlate very well with the moraines on the central Montenegro massifs (CMM) and Mt. Orjen, and glacial outwash in karst poljes around Mt. Orjen. Rb-36/p-4d dating results show younger ages, approximately translated by two MIS stages compared to the other two boreholes (Figures 3.42 and 3.45).

The major difference between the Rb-36/p-4d and Rb-6/p-5d is the thickness of polygenetic strata, as it is more than 30 m, and 11.3 m, respectively. Based on this, and other data (Zarić *et al.*, 2019a; Zarić, 2020) the boreholes are positioned on two tectonic blocks divided by the major fault that follows, in part, the Sava riverbed. The block on the left bank of the Sava was descending faster during polygenetic strata deposition, hence their greater thickness in Rb-36/p-4d. The comparison between the Rb-1 m/p-3d and Rb-6/p-5d dating (which correlates very well with the moraines on the central Montenegro massifs and Mt. Orjen, as well as with glacial outwash deposited in karst poljes around Mt. Orjen), and Rb-36/p-4d dating resulting in younger ages, shows translation by two MIS stages in Rb-36/p-4d compared to the first two boreholes, and the age of the sixth cycle stands out from all the other results (Table 3.6).

Alternating cold and warm stages during the Pleistocene led to the cyclic deposition of sediments of varying grain sizes and sorting (Knežević *et al.*, 2018; Knežević *et al.*, 2019). The incoming directions and the formations that provided the material deposited at the BGS resulted in different mineralogical/petrological compositions of the sediments. Figure 3.44 shows the typical grain diameters; going downward from the ground surface, there is a distinct transition from Holocene to Pleistocene strata. Early Holocene is characterized by finer grains with better sorting than the older strata, lack of gravel, and elevation of its base of approximately 60 m a.s.l. In the case of Pleistocene polycyclic sediments, there are alternating coarse sand and sand-and-gravel sediments eroded and transported by torrential streams at the beginnings of interglacials, with clay-and-silt interbeds and strata deposited during periods of much lower flows (Mitrinović *et al.*, 2019).

As can be observed in Figure 3.44, the grain size (particularly of the largest particles) increases with age, especially when the sediments at the lower part of the youngest (Holocene) deposits are compared to the upper part of the polycyclic deposits (Pleistocene). When comparing the extent of glaciation during MIS 2 to 5d (Würm) (Figure 3.36), at the end of which the lower part of the Holocene deposits were formed (Figure 3.43 and Figure 3.44), to the extent of glaciation during MIS 6 (Riss), at the end of which the youngest Pleistocene deposits were formed (Figure 3.44), the grain size correlates with the previous glaciation extent and the amount of water released at the beginning of the interglacial. Another increase in grain size takes place under the distinct one meter thick interlayer of clay with dust at 18 m of depth. Comparison with the inferred ages shows that the size of the largest grains for the deposits from the ends of MIS 8, MIS 12 and MIS 14 were the largest and of similar sizes, and larger than the size of the largest grains for the deposits from the end of MIS 10 (Mitrinović *et al.*, 2019).

The presented results of OSL dating and grain size distribution analysis, and glacial studies of Hughes *et al.* and Adamson *et al.*, show that every cycle of sedimentation that was to be expected is present in the lithological column, at least as it pertains to polycyclic alluvial sediments. This is at odds with the commonly accepted notion that the cycles are missing between the Lower Pleistocene polycyclic sediments and Holocene alluvial sediments in the zone of exposed alluvial sediments, for which the strongest argument is paleontological data (Nenadić and Gaudeny 2013; Nenadić *et al.*, 2009, 2016). In addition to these new findings, in the conditions of constant subsidence of the surface and filling in with sediment, it may be expected that a few cycles of sedimentations would be totally eroded at some point.

According to earlier works, the older part of the polycyclic formation is comprised of fluvial sediments of the Late Pleistocene, also referred to as 'strata with Corbicula fluminalis' (Laskarev, 1938; Stevanović, 1977). These authors believe that the youngest formation, represented by contemporary Holocene alluvial sediments of the Sava River, discordantly overlies the polycyclic strata. The contemporary alluvial sediments of the Sava are believed to have been deposited after the final cold stage (i.e. generally during the Holocene), while opinions about the underlying younger part of Pleistocene polycyclic sediments differ. According to Laskarev (1938) and Stevanović (1977), they originate from the Mindelian-Rissian interglacial in the Middle Pleistocene. Rakić (1977) believes that they are older and formed in the Günzian-Mindelian interglacial. Knežević *et al.* (1998) are of the opinion that these strata belong to the Early Eopleistocene, while Nenadić *et al.* (2016) identify them as 'younger stages of the Lower-Middle Pleistocene'.

As shown, the results of OSL and IRSL dating and paleomagnetic testing of samples collected from two boreholes (Rb-6/p-5d and Rb-1 m/p-3d) at the BGS correlate very well with the results of recent dating and glaciological investigations undertaken on the central Montenegrin mountains and Mount Orjen, as well as with the results of grain-size analysis. The results for Rb-36/p-4d differ to a certain extent which is yet to be explained. This supports the hypothesis that the coarse grained material is mostly deposited during the glacial-interglacial transitions characterized by fast glacier melting and extremely high surface and river flows, that the fine, mostly floodplain material was deposited during the course of interglacials at the rate of tectonic lowering, and that the constant tectonic subsidence preserved, although shortened, all the cycles of sedimentation. There is a general correlation between the size of the largest grains and the extent of the previous glaciation.

However, there are also certain inconsistencies between earlier literature sources, that describe the age and sedimentation and erosion conditions of polycyclic sediments at the BGS, and presented more recent dating results.

This disagreement is largely a consequence of successive changes in perceptions and the official view about the duration of the Pleistocene. At the time of Laskarev, Rakić and Stevanović, the term Pleistocene meant what we now call the Middle and Upper Pleistocene. The International Community of Geological Sciences (IUGS) made a decision in 2009 according to which the Quaternary is considered to have started 2.588 million years ago, and the research presented here has certainly shown that the age of the polycyclic alluvial aquifer sediments from the Belgrade source is less than 780,000 years. This largely corresponds to the results of dating of the moraine materials on Mt. Durmitor and Mt. Orjen. In addition to age, the mineralogical and petrological analyses also strongly indicate that the formation of polycyclic layers in the Sava aquifer was strongly induced by the melting of glaciers on high mountains in the Sava River Basin.

The existence of glaciers on mountains in the Sava River Basin has not been established and it is not realistic to expect any significant existence of glaciers in the mountains in this basin in the period older than 1.2 million years due to the warmer climate and the slightly lower altitude of the mountains.

## 3.6 CONCLUDING REMARKS

Groundwater represents over 99% of fresh water in liquid form on the planet and over half of the water used for supplying the population with drinking water comes from groundwater. Of this amount, a large part comes from alluvial groundwater. Human settlements, agriculture, industry, traffic, etc., put pressure on alluvial resources. Climate change and global warming are also endangering resources and primarily so shallow aquifers. It is necessary to overcome these contradictions in order to use alluvial aquifers in a sustainable manner. In this chapter, we mainly dwell on the genesis of aquifers, especially taking into account the climate factor and the influence of the alternating glacial and warm periods.

Alluvial aquifers are, above all, quaternary formations. Their formation was significantly influenced by various factors and forces – endogenous and exogenous. There are relief conditions, and orogenic and epeirogenic tectonic movements. Climate is also an extremely important factor here, especially in the middle latitudes. Most of the older Pleistocene (from before 2.58 Ma to 1.2 Ma) is characterized by slightly higher temperatures. The periodic oscillations in the temperature regime between colder and warmer temperatures were moderate during this period. The youngest part of the older Pleistocene, the middle Pleistocene and the younger Pleistocene generally differ with slightly lower temperatures and larger differences in the temperatures between the cold periods and periods of thawing. This had a great influence on the formation of alluvial aquifers, especially within the middle latitudes.

This chapter is divided into four subchapters. The subchapter on the climate of the Pleistocene, the processes of genesis in the Pleistocene and the methods for determining the age of alluvial deposits is the part of the chapter prepared for the reader to get an insight into the already dealt with topics.

Subchapter 3.5 is dedicated to research carried out at the Belgrade Groundwater Source that captures the alluvial aquifer of the Sava River along the last 50 km upstream from the confluence with the Danube River. The investigations were conducted as an important extensive research study (JCI, 2005–2010), and through research work (JCI, 2017) within the scientific project of the Government of Serbia TP 37014 entitled 'Methodology for the assessment, design and maintenance of wells in alluvial environments depending on oxygenation environment,' as well as through the preparation of the doctoral thesis of Zarić (2020). This project was partially presented or shown in Dimkić (2017), Zarić *et al.* (2019a, 2019b), and Mitrinović *et al.*, (2019).

The Belgrade Groundwater Source is very important. The research that ran in parallel with its planning, formation and monitoring, and that preceded the above, was quite extensive. Here we list just some of the significant and published or archived results: Laskarev (1938), Rakić (1977), Stevanović (1977), Dimkić *et al.* (2007, 2010, 2011a, 2011b), JCI (2010), etc.

The position of the Belgrade Groundwater Source (the most downstream part of the alluvial aquifer of the Sava) influenced the formation of mainly anoxic conditions in the aquifer ($O_2$ content in water ~ 0.15 mg/l, and the Fe content mainly $0.5 \leq Fe^{2+} \leq 2$ mg/l). This affects the colmation of wells with $Fe^{3+}$ compounds and decreases their yield.

The profiles showing the results of the granulometric analysis of the samples from examined deep boreholes: Rb-6/p-5d (52 m), Rb-36 /p-4d (61 m) point to the polycyclic nature of sedimentation. The alternation of periods that resulted in deposition of layers of gravelly-sandy material with layers of finer or clayey material, led to the extraordinary filtering, purifying capabilities of the alluvial aquifer. This is all the more so because the water intake facilities at the source are radial wells, with drains in positions mostly parallel to the deposited polycyclic layers.

The significant faults that occurred during the Pleistocene formed a network of faults where the main directions of the faults were more or less parallel to the flow of the Sava River. As a rule, the left, north side of the fault subsided and enabled the permanent storage of a certain amount of material. The processes of erosion and the re-deposition of the already deposited material was also significantly present.

Among other things, specific research was carried out at the source on two deep boreholes: Rb-36/p-4 d (62 m) and Rb-6/p-5d (52 m) (JCI, 2017).

In addition to extracting and mapping the entire core of these boreholes, grain size analysis and mineralogical, petrological and paleontological analyses were also performed. Age analysis was performed by luminescence and paleomagnetism methods. The following was identified and/or indicated:

- The formation of polycyclic, gravelly-sandy layers of aquifers occurred in the last 780 000 years;
- Granulometric analysis and age tests indicate about 6 or 7 cycles of the deposition of aquifer skeletal material;
- It has been indicated that the underlying layer of the aquifer, which consists mainly of colorful siltstone clays with rare interlayers of fine gravel and sand, was also formed in the Pleistocene, and is of an age, apparently, greater than 780,000 years.

Research that was conducted (Adamson *et al.*, 2015; Hughes *et al.*, 2011) on high mountains in the Sava River Basin (Mt. Durmitor) and the nearby mountain (Orjen) pointed to traces of glaciers from three glacial cycles (Figure 3.46). All traces of glaciers are younger than 500,000 years.

A comparison of the ages of sediments in the moraines of Mt. Durmitor and Mt. Orjen with the ages of sediments at the Belgrade source indicate a very interesting correlation. This is also indicated by the cyclic sedimentation, which can be noticed from the profile (log) of granulometric analysis. The mineralogical and petrological analysis of the hardest grains and chemically most stable minerals in the aquifer (quartz, chert) their roundness and mineralogical composition indicate the length and location of the erosion from which they were transported.

The age of the polycyclic sediments of the Belgrade source as well as traces of glaciers on Mt. Durmitor and Mt. Orjen indicate that glacial erosion and powerful streams that deposited coarser material of the alluvial aquifer of the Sava River Basin were dominant in the late middle period and during the Early Pleistocene. The warmer climate during the Late Pleistocene and, probably, the somewhat lower height of the mountain massifs in the Sava River Basin, probably did not enable the significant formation of glaciers in that period, and certainly not in the period older than 1.2 million years.

The Sava River formed its alluvial deposits with eroded material from the area south of the Pannonian Plain. Among other complex factors and processes, the role of erosion in the melting and lowering of glaciers from high mountains in the basin is underlined here.

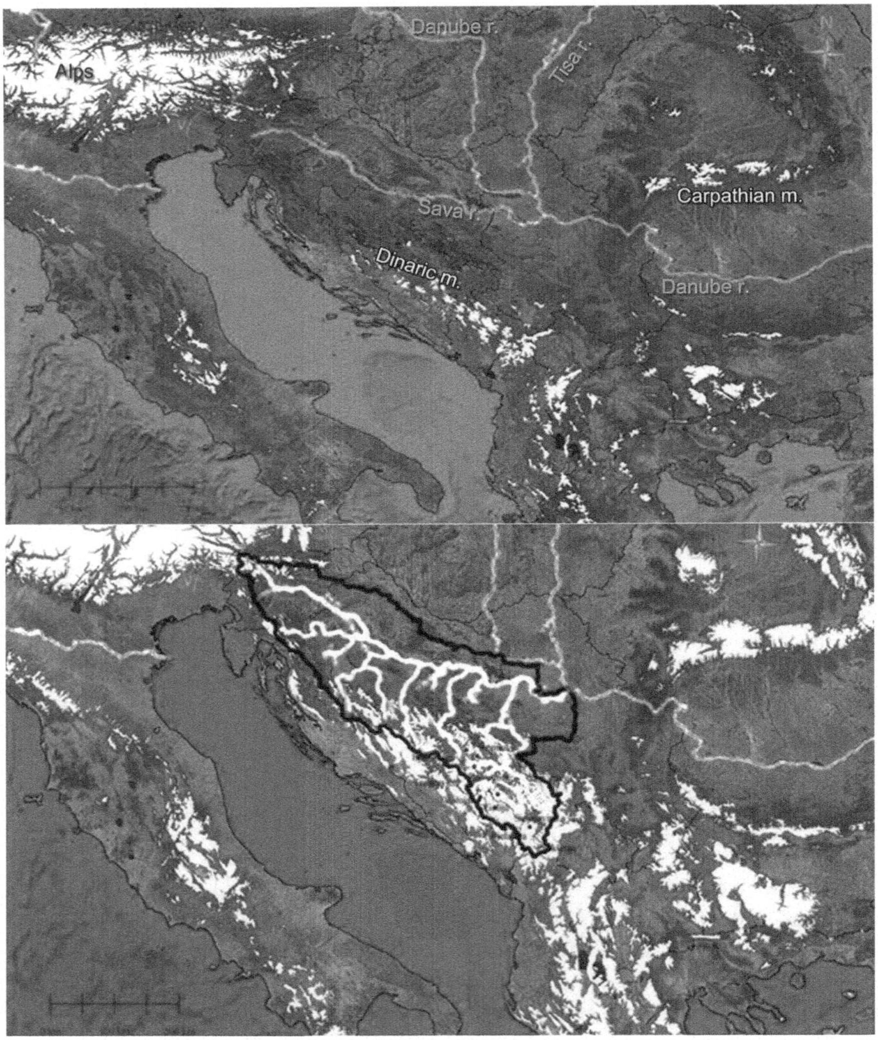


**Figure 3.46** Areas under glaciers in the Dinarides during peak glaciation during the Pleistocene, in the case of the glacial boundary (ELA) above 1,500 m a.s.l. and 1,000 m a.s.l. The contour of the Sava River Basin is marked with a black line, the white line shows the rivers from the Sava River Basin (modified from Dimkić, 2017).

North of the Sava trench in the area of the former Pannonian Sea, layers of the basic water-bearing complex were created in the past one million years. These layers, whose origin was mostly the eroded material formed dominantly due to glacial erosion, was brought in by the river Danube, the Tisza and others. Due to the higher altitude and latitude of the Alps, the formation of glaciers and the fluvioglacial formation of aquifers could have happened earlier than the genesis of the Sava aquifer. The polycyclic aspect of the formation of the Sava aquifer provided additional purification possibilities. It is evident that today very significant self-purifying processes are occurring when filtering water from the Sava into the water intake facilities of the Belgrade Groundwater Source.

Issues regarding the oxicity of groundwater, the purification potential of aquifers and the aging of water intake facilities are very important and also depend on the genesis of aquifers. Some of these processes have been covered in much more detail in the various chapters of this book.

## REFERENCES

Adamson K. R., Woodward J. C and Hughes P. D. (2015). Middle Pleistocene glacial outwash in poljes of the Dinaric karst. In: Caves and Karst Across Time: Geological Society of America Special Paper, J. Feinberg, Y. Gao and E. C. Alexander Jr. (eds.), **516**, doi: 10.1130/2015.2516(20)

Antoine P., Rousseau D. -D., Fuchs M., Hatté H., Gauthier C., Marković S. B., Jovanović M., Gaudenyi T., Moine O. and Rossignol J. (2009). High-resolution record of the last climatic cycle in the southern Carpathian Basin (Surduk, Vojvodina, Serbia). *Quaternary International*, **198**, 19–36.

Best C. (2016). Milankovitch Insolation study. http://clivebest.com/blog/?p=7368 (accessed on 25 March 2020)

Bøtter-Jensen L., Thomsen K. J. and Jain M. (2010). Review of optically stimulated luminescence (OSL) instrumental developments for retrospective dosimetry. In: Radiation Measurements. Vol. **45**(3–6), pp. 253–257, ISSN 1350-4487. doi: 10.1016/j.radmeas.2009.11.030

Broecker W. S. and Van Donk J. (1970). *Rev. Geophys. Space Phys.*, **8**, 169–197.

Buggle B., Hambach U., Glaser B., Gerasimenko N., Marković S. B., Glaser I. and Zoeller L. (2009). Stratigraphy, and spatial and temporal paleoclimatic trends in Southeastern/Eastern European loess paleosol sequences. *Quaternary International*, **196**, 86–106.

Butler R. F. (1998). PALEOMAGNETISM: Magnetic Domains to Geologic Terranes. Electronic Edition, by Robert F. Butler, Department of Geosciences University of Arizona Tucson, Arizona (Originally published by Blackwell Scientific Publications, 1992)

Chunfu Z. (2004). Windows-based paleomagnetic analysis program that integrates on-screen visualization and analysis of paleomagnetic data and printing graphic outputs. https://www.eaps.purdue.edu/paleomag/ (accessed on 18 May 2020)

Cita M. B. (2008) Summary of Italian marine stages of the Quaternary. *Episodes* **31**, 251–254.

Danin A. and Ganor E. (1991). Trapping of airborne dust by mosses in the Negev Desert, Israel. *Earth Surface Processes and Landforms*, **16**(2), https://doi.org/10.1002/esp.3290160206/

Darvill C. M. (2013). Section 4.2.10: Cosmogenic nuclide analysis. In: Geomorphological Techniques (Online Edition), S. J. Cook, L. E. Clarke and J. M. Nield (eds.), British Society for Geomorphology, London, UK. ISSN: 2047–0371

Deutsche Stratigraphische Kommission. (2016). Stratigraphische Tabelle von Deutschland 2016 (Stratigraphic table for Germany 2016). http://www.stratigraphie.de/std/Bilder/5_2.pdf (accessed on 25 March 2020)

Dimkić M. (2012). Importance of the Aerobic State of Alluvial Aquifers for Groundwater Use. Nakdong River International Water Week, 7th World Water Forum, Andong, Republic of Korea, pp. 69–96.

Dimkić M. (2017). Several aspects of the influence of ice ages in Serbia on the formation of the Sava alluvion and water source processes. Conference organized by Serbian Academy of Sciences and Arts, Earth's Climate Change: Science and Impacts, Belgrade, 11–13 October 2017.

Dimkić M., Taušanović V., Pušić M., Boreli-Zdravković Đ., Đurić D., Slimak T., Petković A., Obradović V. and Babić R. (2007). Belgrade Groundwater Source Condition and Possible Development Directions. *Water Practice and Technology*, **2**(3): WPT 2007059, pp.

Dimkić M., Pušić M., Vidović D., Isailović V., Majkić B. and Filipović N. (2010). Numerical Model Assessment of Radial-Well Ageing. *Journal of Computing in Civil Engineering*, **25**(1), 43–49. doi: 10.1061/(ASCE)CP.1943-5487.0000063

Dimkić M., Pušić M., Vidović D., Petković A. and Boreli-Zdravković Đ. (2011a). Several natural indicators of radial well ageing at the Belgrade Groundwater Source, Part 1. In: Water Science & Technology, IWA Publishing, London, ISSN 0273-1223, Vol. **63**(11), pp. 2560–2566.

Dimkić M., Pušić M., Obradović V. and Djurić D. (2011b). Several natural indicators of radial well ageing at the Belgrade Groundwater Source, Part 2. In Water Science & Technology, IWA Publishing, London, ISSN 0273-1223, **63**(11), pp. 2567–2574.

Dimkić M., Mitrinović D., Slimak T., Zarić J., Kovačević S., Čalenić A. and Jelovac M. (2019). Urban development vs. alluvial groundwater sources – Belgrade example. The e-Future of Cities, 'e-FoC'-proceedings, Belgrade, 24–25 October 2019, pp. 1–11.

Duller G. A. T. (2008). Luminescence Dating: Guidelines on Using Luminescence Dating in Archaeology. English Heritage, Swindon, United Kingdon.

Emilliani C. (1955). Pleistocene temperatures. *Journal of Geology*, **63**, 538–575.

M. H. Engel and S. A. Macko (ed.) (1993). Organic Geochemistry: Principles and Applications. Springer Science+Business, New York, USA. ISBN 978-1-4615-2890. doi: 10.1007/978-1-4615-2890

FAO. (2016). Global Diagnostic on Groundwater Governance, ISBN 978–92-5-109259-0

Fodor L., Csontos L., Bada G., Györfi I. and Benkovics L. (1999). Tertiary tectonic evolution of the Pannonian basin system and neighbouring orogens: a new synthesis of paleostress data. In: The Mediterranean Basins: Tertiary Extension within the Alpine Orogen, B. Durand, L. Jolivet, F. Horvath and M. Seranne (eds.), Geological Society of London, Special Publication, **156**, pp. 295–334.

Gaba E., Background map: NGDC World Coast Line data, Data: Prof. Bird P. (2006, updated 2015), Detailed world map in English showing the tectonic plates with their movement vectors. https://commons.wikimedia.org/wiki/File:Tectonic_plates_boundaries_detailed-en.svg (accessed on 23 October 2020)

Galbraith R. F., Roberts R. G., Laslett G. M., Yoshida H. and Olley J. M. (1999). Optical dating of single and multiple grains of quartz from Jinmium rock shelter, northern Australia: Part I, experimental design and statistical models. *Archaeometry*, **41**, 339–364. doi: 10.1111/j.1475-4754.1999.tb00987.x

Goudie A. S. (2004). 'Base level'. In: Encyclopedia of Geomorphology, A. S. Goudie (ed.), Routledge, p. 62.

Hambach U., Rolf C. and Schnepp E. (2008). Magnetic dating of Quaternary sediments, volcanites and archaeologicl materials: an overview. *Quaternary Science Journal*, **57**(1–2), 25–51.

Hatté C., Fontugne M., Rousseau D. D., Antoine P., Zöller L., Tisnérat-Laborde N. and Bentaleb I. (1998). δ13C variations of loess organic matter as a record of the vegetation response to climatic changes during the Weichselian. *Geology*, **26**, 583–586.

Hatté C., Gauthier C.-D., Rousseau D., Antoine P., Fuchs M., Lagroix F., Markovic S. B., Moine O. and Sima A. (2013). Excursions to C-4 vegetation recorded in the Upper Pleistocene loess of Surduk (Northern Serbia): an organic isotope geochemistry study. *Climate of the Past, European Geosciences Union (EGU)*, **9**(3), 1001–1014. doi: 10.5194/cp-9-1001-2013

Hughes P. D., Woodward J. C., van Calsteren P. C. and Thomas L. A. (2011). The glacial history of the Dinaric Alps, Montenegro. *Quaternary Science Reviews*, **30**(23–24).

Jaroslav Cerni Institute (JCI) (2005 – 2010). Studija "Beogradsko izvorište - stanje i pravci razvoja" (Study of Belgrade Groundwater Source – current state and avenues of development). Beograd.

Jaroslav Cerni Water Institute (JCI). (2010). Study on Belgrade Groundwater Source – Present Status and Development Directions (in Serbian), Belgrade

Jaroslav Cerni Water Institute (JCI ). (2017). Hydrogeological Report on the Results of Installation of 30 Observation Wells near Exisiting Radial Collector Wells (in Serbian).

Knežević S., Nenadić D. and Stejić P. (1998). Prelesni kvartarni i pliocenski sedimenti Zemuna i Novog Beograda. *Geološki anali Balkanskoga poluostrva*, **62**, 57–73.

Knežević S., Zarić J., Nenadić D., Radulović B. and Bogićević K. (2018). Influence of climate change in Quaternary on formation of sediments of the Belgrade groundwater source (Uticaj klimatskih promena u kvartaru na stvaranje sedimenata beogradskog izvorišta). 17. Kongres geologa Srbije 17–20th of May, pp. 149–154, Vrnjačka Banja

Knežević S., Nenadić D. and Zarić J. (2019). Influence of climate change in Quaternary on formation of alluvial deposits of the Sava Valley near Belgrade. International Conference 'The Life and Work of Milutin Milanković: Past, Present, Future', Belgrade, 19–21 July, 2019.

Lambeck K. and Chappell J. (2001). Sea Level Change Through the Last Glacial Cycle. *Paleoclimate*, **292**(5517), 679–686

Laskarev V. (1938). Troisiěme note le Quaternaire des environs de Belgrade. (Third note of the Quaternary of theritory of Belgrade). *Geological annals of the Balkan Peninsula*, **15**, 1–35. (in Serbian, French summary)

Lehner B. and Grill G. (2013). Global river hydrography and network routing: Baseline data and new approaches to study the world's large river systems. *Hydrological Processes*, **27**(15), 2171–2186. https://www.hydrosheds.org

Li B. and Li S. (2011). Luminescence dating of K-feldspar from sediments: a protocol without anomalous fading correction. *Quaternary Geochronology*, **6**(5), 468–479.

Liritzis I., Vafiadou A., Zacharias N., Polymeris G. S. and Bednarik R. G. (2013). Advances in surface luminescence dating: new data from selected monuments. *Mediterranean Archaeology and Archaeometry*, **13**(3), 105–115.

Lowrie W. (2007). Fundamentals of Geophysics, 2nd edn. Cambridge University Press, New York, US.

Magyar I., Geary D. H. and Müller P. (1999). Paleogeographic evolution of the Late Miocene Lake Pannon in Central Europe. *Palaeogeogr. Palaeoclimatol. Palaeoecol.*, **147**, 151–167.

Mankinen E. A. and Wentworth C. M. (2003). Preliminary paleomagnetic results from the Coyote Creek outdoor classroom drill hole, Santa Clara Valley, California, Open-File Report 03-187. US Geological Survey, Menlo Park, CA, US.

Marković S. B., McCoy W. D., Oches E. A., Savić S., Gaudenyi T., Jovanović M., Stevens T., Walther R., Ivanišević P. and Galović Z. (2005). Paleoclimate record in the Late Pleistocene loess-paleosol sequence at Petrovaradin Brickyard (Vojvodina, Srbija). *GEOLOGICA Carpathica*, **56**, 545–552.

Marković S. B., Oches E. A., McCoy W. D., Frechen M. and Gaudenyi T. (2007). Malacological and sedimentological evidence for „warm 'glacial climate from the Irig loess sequence (Vojvodina, Serbia). *Geochemistry Geophysics Geosystems*, **8**, Q09008. doi: 10.1029/2006GC001565.

Marković S. B., Bokhorst M. P., Vandenberghe J., McCoy W. D., Oches E. A., Hambach U., Gaudenyi T., Jovanović M., Stevens T., Zöller L. and Machalett B. (2008). Late Pleistocene loess-paleosol sequences in the Vojvodina region, North Serbia. *Journal of Quaternary Science*, **23**, 73–84.

Marković S. B., Hambach U., Catto N., Jovanović M., Buggle B., Mashalett B., Zöller L., Glaser B. and Frechen M. (2009) The middle and late Pleistocene loess-paleosol sequences at Batajanica, Vojvodina, Serbia. *Quaternary International*, **198**, 255–266.

Marton E. (2017). Report - Experiments to test the suitability of the RB-6/P-5d drill core for magnetostratigraphy correlation by checking the paleomagnetic signals of selected core segments. Terradat and Geological and Geophysical Institute of Hungary, Department of Earth Physics, Paleomagnetic Laboratory, Budapest, Hungary.

Marton E. (2019). Report - Experiments to test the suitability of the RB-36/P-4d drill core for magnetostratigraphy correlation by checking the paleomagnetic signals of selected core segments. Terradat and Geological and Geophysical Institute of Hungary, Department of Earth Physics, Paleomagnetic Laboratory, Budapest, Hungary.

Mesinger F. (2010). Srbin među džinovima nauke (A Serb Among the Giants of Science), newspaper article. Politika, Belgrade, Serbia. http://www.politika.rs/sr/clanak/119622/Srbin-medu-dzinovima-nauke (accessed on 24 March 2020)

Milanković M. (1941). Kanon Der Erdbestrahlung und Seine Anwendung Auf Das Eiszeitenproblem (Canon of Insolation and the Ice-Age Problem). Royal Serbian Academy, Belgrade, Kingdom of Yugoslavia

Milankovitch Orbital Data Viewer (MODW). (2020). Milankovitch orbital parameters over time based on the calculations of: J. Laskar et al. (2004). A long term numerical solution for the insolation quantities of the Earth. *Astronomy and Astrophysics*, **428**, 261–285, doi: 10.1051/0004-6361:20041335. https://biocycle.atmos.colostate.edu/shiny/Milankovitch/ (accessed on 22. May 2020)

Mitrinović D., Zarić J, Anđelković O., Sipos G., Polomčić D. and Dimkić M. (2019). Dating of alluvial sediments from borehole at the lower course of the Sava river and indications of the connection between their genesis and climate changes in the Pleistocene, article in press. doi: 10.1016/j.quaint.2019.09.028

Murray A. and Wintle A. G. (2000). Luminescence dating of quartz using an improved single-aliquot regenerativedose protocol. *Radiation Measurements*, **32**, 57–73.

Murray A. and Wintle A. G. (2003). The single aliquot regenerative dose protocol: Potential for improvements in reliability. *Radiation Measurements*, **37**(4), 377–381, doi: 10.1016/S1350-4487(03)00053-2

National Centers for Environmental Information (NCEI). (2020). The World Magnetic Model – Background. https://www.ngdc.noaa.gov/geomag/WMM/back.shtml (accessed 18 May 2020)

Nelson M. S., Gray H. J., Johnson J. A., Rittenour T. M., Feathers J. K. and Mahan S. A. (2015). User Guide for Luminescence Sampling in Archaeological and Geological Contexts. *Advances in Archaeological Practice*, **3** (2), 2015, 166–177. The Society for American Archaeology. doi: 10.7183/2326-3768.3.2.166

Nenadić D. and Bogićević K. (2010). Geologija kvartara. (Geology of Quaternary) Rew. Knežević, S., Eremija, M., University of Belgrade, faculty of Mining and Geology, 281 pp, ISBN 978-86-7352-208-1

Nenadić D. and Gaudeny T. (2013). *Stratigraphical Position of Pleistocene Fluvial Polycyclic Deposits of Serbia. Neogene and Quaternary Stratigraphy, Actual Terminology and Nomenclature*. Serbian Geological Society, Beograd.

Nenadić D., Knežević S. and Bogićević S. (2009). Stratigraphical and palaeogeographical characteristics of Pleistocene series in the Sava riparian area at Belgrade (Serbia). *Bulletin of the Natural History Museum*, **2**, 63–80.

Nenadić D., Gaudenyi T., Bogićević K. and Tošović R. (2016). Stratigraphic and Lithologic characteristics of Pleistocene Fluvial Deposits in the Danube and Sava Riparian area near Belgrade (Serbia). *Stratigraphy and Geological Correlation*, **24**(4), 427–437.

J. S. Noller, J. M. Sowers and W. R. Lettis eds. (2000). Quaternary Geochronology: Methods and Applications. American Geophysical Union, Washington DC, USA.

Panchuk K. (2019). Physical Geology, First University of Saskatchewan Edition. University of Saskatchewan, Saskatoon, Saskatchewan, Canada.

Pécsi M. (1990) Loess is not just the accumulation of dust. *Quaternary International*, **7/8**, 1–21.

Pollard D. and DeConto R. M. (2009). Modelling West Antarctic ice sheet growth and collapse through the past five million years. *In: Nature*, **458**, 329–332.

Rakić M. (1977). Pleistocen Srema i Banata - Geologija Srbije II-3, Stratigrafija (Kenozoik). (Pleistocene of Srem and Banat - Geology of Serbia II-3, Stratigraphy (Cenozoic)). University of Belgrade, pp. 399–405.

Rial J. A. (1999). Pacemaking the Ice Ages by Frequency Modulation of Earth's Orbital Eccentricity. *Science*, **285** (5427), 564–568. doi: 10.1126/science.285.5427.564

Riebeek H. (2005). Paleoclimatology: the Oxygen Balance. https://earthobservatory.nasa.gov/features/Paleoclimatology_OxygenBalance (accessed on 28 March 2020)

Sagnotti L. (2012). Demagnetization Analysis in Excel (DAIE) User Manual (Version 1.0). Istituto Nazionale di Geofisica e Vulcanologia, Roma, Italy

Schettino A. and Scotese C. R. (2005). Apparent polar wander paths for the major continents (200 Ma to the present day): a palaeomagnetic reference frame for global plate tectonic reconstructions. *Geophysical Journal International*, **163**, 727–759.

A. E. Scot and J. M. Cary (Ass.ed.) (2013). Encyclopedia of Quaternary Science, 2nd edition. Elsevier, Amsterdam, Netherlands.

Shackleton N. J. and Opdyke N. D. (1973). Oxygen isotope and palaeomagnetic stratigraphy of Pacific Core V28-238: oxygen isotope temperatures and ice volumes on 105 and 106 year scale. *Quaternary Research*, **3**, 39–55.

Snyder C. W. (2016). Evolution of global temperature over the past two million years. Nature, **538**.

Stevanović P. (1977). Kvartar (antropogen). Opšti pregled facija i njihovo rasprostranjene s posebnim osvrtom na prelesne, lesne i antropogene naslage severne Srbije, u: Geologija Srbije, II-3, Stratigrafija (Kenozoik). Quaternary (Anthropogene). General overview of facies and their distribution with special reference to pre-loess, loess and anthropogenic deposits of northern Serbia, in: Geology of Serbia, II-3, Stratigraphy (Cenozoic). University of Belgrade

Summerfield M. A. (1996). Tectonics, geology and long-term landscape development. In: The Physical Geography of Africa, W. M. Adams, A. S. Goudie and O. R. Orme (eds.), Oxford University Press, Oxford, pp. 1–77.

Svirčev Z., Marković S. B., Stevens T., Codd G., Smalley I., Simeunović J., Obreht I., Dulić T., Pantelić D. and Hambach U. (2013). Importance of biological loess crusts for loess formation insemi-arid environments. *Quaternary International*, **296**, 206–215.

Sztanó O., Sebe K., Csillag G. and Magyar I. (2015) Turbidites as indicators of paleotopography, Upper Miocene Lake Pannon, Western Mecsek Mountains (Hungary). *Geologica Carpathica*, **66**, 331–344.

Tauxe L., Banerjee S. K., Butler R. F. and van der Voo R. (2018). Essentials of Paleomagnetism, 5th Web Edition, University of California Press, CA, USA.

Ter Borgh M., Vasiliev I., Stoica M., Knežević S., Matenco L., Krijgsman W., Rundić L. and Cloetingh S. (2013). The isolation of the Pannonian basin (Central Paratethys): new constraints from magneto - and biostratigraphy. *Global and Planetary Change*, **103**, 99–118.

Tóth O., Sipos G., Kiss T. and Bartyik T. (2017). Variation of OSL residual doses in terms of coarse and fine grain modern sediments along the Hungarian section of the Danube. *Geochronometria*, **44/1**, 319–330.

Tsoar H. and Pye K. (1987). Dust transport and the question of desert loess formation. *Sedimentology*, **34**(1), 139–153.

Utah State University Luminescence Laboratory (USULL). (2007). OSL Sampling, Do's and Don't's. http://www.usu.edu/geo/luminlab/OSL%20sampling%20-%20Do's%20and%20Dont's-1.pdf (accessed on 25 May 2020)

Vigil F. Jose, USGS/USGov. (1997). modified by Zimbres E. A cross section illustrating the main types of plate boundaries. http://commons.wikimedia.org/wiki/File:Tectonic_plate_boundaries2.png (accessed on 23 October 2020)

Vočadlo L. (2014). Geophysics Course Lecture 10 – Paleomagnetism. University College London, London, UK.

Winterberg S. and Willett D. S. (2019) Greater Alpine river network evolution, interpretations based on novel drainage analysis. *Swiss Journal of Geosciences*, **112**(1), 3–22.

Wintle A. G. and Murray A. S. (2006). A review of quartz optically stimulated luminescence characteristic and their relevance in single aliquot regeneration dating protocols. *Radiation Measurement*, **41**, 369–391. doi: 10.1016/j.radmeas.2005.11.001)

World Data Center for Geomagnetism (WDCG), Kyoto. (2020). Magnetic North, Geomagnetic and Magnetic Poles. http://wdc.kugi.kyoto-u.ac.jp/poles/polesexp.html (accessed on 18 May 2020)

Zarić J. (2020). Formiranje i održivo korišćenje izvorišta podzemnih voda u funkciji geneze kvartarnih sedimenata donjeg toka reke Save (Formation and sustainable exploitation of groundwater sources in as a function of the genesis of quaternay aquifers in the lower sava river basin). PhD thesis, University of Belgrade, Faculty of mining and geology, Engineering geology, Hydrogeology, 375 pages, UDK: 556.34:551.79(043.3); 552.5:551.79(043.3)

Zarić J., Mitrinović D., Anđelković O., Perović M. and Dimkić M. (2019a). Polyciclic strata forming along the cross section from Rb-1M to Rb-36 well at the Belgrade water source, with retrospect on local tectonics. 48. konferencija o aktuelnim temama korišćenja i zaštite voda "VODA 2019", 04.-06. jun 2019., Zlatibor. Srpsko društvo za zaštitu voda, Beograd, (2019), pp. 231–242, ISBN 978-86-916753-5-6.

Zarić J., Nenadić D., Mitrinović D., Knežević S. and Dimkić M. (2019b). Genesis Conditions and Characteristics of Pleistocene Alluvial Deposits in the Lower Sava River Basin. *Comptes rendus de l'Acadéemie bulgare des Sciences*, **72**(7), 924–930, ISSN 1310-1331. Akademichno Izdatelstvo Prof. Marin Drinov (Prof. Marin Drinov Academic Publishing House)

# Chapter 4

# Oxic conditions of alluvial aquifers

*Milan Dimkić, Mirjana Vojinović Miloradov and Ivana Mihajlović*

## 4.1 INTRODUCTION

The degree of oxicity in groundwater is of great importance. The oxicity pattern affects the directions and kinetics of physicochemical and biochemical processes. In relatively clear surface waters, the oxygen concentration moves around the saturation concentration. This is not the case in groundwater. Unlike surface waters, the groundwater velocity is generally smaller by several orders of magnitude. In alluvial groundwater, in addition to the oxidation of organic matter, the reduction of groundwater oxicity is also affected by the oxidation of dissolved minerals in water and minerals from the aquifer skeleton.

The diffusion of air through the overlying layer of the aquifer can also be significant when referring to the total oxygen in the alluvial aquifer. It is noticeable that in river alluvions increased groundwater oxicity occurs mainly in the upper part of the basin. In the downstream part of the basin, groundwater oxicity is generally reduced and/or the water goes into anoxic state. This can be due to the energy of the flow of the river and the deposition of sediment. Namely, there is increased erosion in the upstream part. The velocity of the river flow is also higher, so a larger deposit precipitates, with fewer finer particles carrying oxidizing minerals. In the downstream part of the watercourse the energy of the river is generally smaller so that proportionately there are more deposits of fine particles and an increased amount of oxidable minerals. Consequently, in this chapter, attention is given to the processes of oxidation and reduction of Fe(II) and Fe(III), Mn(II) and Mn(IV), sulphur and nitrogen.

The processes arising from the degree of oxicity of alluvial aquifers are very important for the sustainable use and protection of groundwater. They point to the processes of transformation of the quality and protection of groundwater, as well as to the aging and reduction of the capacity of water intake facilities due to biochemical colmation, as well as to other significant processes.


Editor: Milan Dimkić
doi: 10.2166/9781789060904_0227

The genesis and the occurrence of multilayer sediment deposition are discussed in Chapter 3. The effects of the transformation of groundwater quality contaminated by pharmaceuticals, pesticides, and nitrogen compounds are discussed in Chapters 5, 6 and 7.

The monograph uses both worldwide experience on groundwater oxicity as well as our experience acquired over many years along with the results of monitoring and testing of a large number (over 100) of wells in the alluvium of Serbia.

## 4.2 SURFACE WATER OXICITY

Water is a liquid medium with an extremely complex icosahedral cluster structure, with hydration clouds around each chemical species (molecules, ions, radicals, organic or inorganic substances) dissolved in an aquatic system (Marrin, 2002).

Dissolved oxygen is essential for the life of organisms, and for the development of algae and all macrophytes, as well as for the many oxic chemical and biochemical reactions that are essential and necessary for the functioning of the living world in water. The aeration and reaeration of water, as well as the process of self-purification is one of the most important processes in natural watercourses. The arrival of detritus in the aquatic environment catalyses a series of enzymatic biological activities of bacteria and microfungi that in order to function, and for their metabolism, consume dissolved oxygen. Detritus is a biological mass consisting of very fine particles of organic matter formed as a result of the decomposition of biological tissue, cells and other biosystems. Oxygen is an essential element of life present in all human, animal and plant proteins. In nature it occurs in three allotropic modifications: as diatomic $O_2$, triatomic $O_3$, and as a third new allotropic modification – tetra metastable atomic O. This in turn results in the conversion of detritus into the final mineral products. The dissolved oxygen deficiency is compensated for by the reaeration of the oxygen in the air. How efficient the reaeration is depends on how turbulent the movement of the gas phase is at the contact surfaces of the water-atmosphere heterogeneous system and from the deficit of oxygen in the water. Reaeration and biological oxygen consumption as the two basic chemical and biochemical reactions that occur during the self-purification process are the basis for determining the spatial and temporal distribution of dissolved oxygen, namely the oxygen balance. At 25°C water has a maximum dissolved oxygen concentration of 8.3 mg/L.

Table 4.1 shows the maximum dissolved oxygen concentration values in surface waters at certain water temperatures.

As the temperature increases, the concentration of dissolved oxygen in the water decreases, and the gaseous oxygen molecules leave the aquatic environment and move into the gaseous phase, which is in accordance with Henry's law pertaining to the solubility of gas as a function of temperature. Nutrients in water, such as nitrogen compounds and phosphates, stimulate the growth of aquatic vegetation and

**Table 4.1** Maximum concentration of dissolved oxygen (DO) in surface water at different temperatures.

| Temperature (°C) | Dissolved Oxygen (mg/L) |
|---|---|
| 0 | 14.62 |
| 5 | 12.80 |
| 10 | 11.33 |
| 15 | 10.15 |
| 25 | 8.38 |

disturb the oxygen balance, i.e. causing a decrease in oxygen concentration and accelerating eutrophication processes.

## 4.3 REDOX PROCESSES

There is a conceptual analogy between acid-base and redox chemical reactions. Acids and bases can be interpreted in a similar manner as proton donors and proton acceptors, as reducing and oxidizing agents are defined as electron donors and electron acceptors according to Brønsted-Lowry acid-base theory. Lewis theory is based on electron pair exchange which has even closer similarity with redox reactions.

Oxidation is a process where the atom of a certain chemical species (ion, radical, molecule – organic or inorganic) releases one or more electrons. Reduction is a parallel simultaneous chemical process to oxidation in which an atom from another chemical species welcomes one or more electrons. As there are no free electrons in the system, any oxidation must be accompanied by a reduction and vice versa. An oxidizing agent accepts electron(s) from an oxidized substance and is reduced, the reducing agent transfers electron(s) to the reduced compound and is oxidized.

From the thermodynamic aspects of chemical processes, chemical energy can be transformed into electrical as well as electrical energy into chemical. Chemical reactions in which electricity is produced are typical redox reactions in which electrons move from an atom of one substance to an atom of another substance resulting in the process of electron loss or electron gain.

The physicochemical processes that take place in natural aquatic systems basically belong to the electrochemical phenomena that occur in a theoretical electrochemical cell.

The transfer of electrons from a metal or other chemical species into an electrolyte, such as in natural systems of river water, lakes, seawater or pore water in the interspaces of sandy or silt sediments, will only take place if the electrolytic system contains molecules or ions that can easily accept electrons, through the phase boundary of the metal-electrolyte contact interface. This type of chemical species is called an oxidized species. However, the opposite process of oxidation occurs at the same time as the reduction process at the second metal-electrolyte boundary phase. The chemical component that is oxidized serves as a generally rich source of electrons, and represents a reduced chemical species. In complex natural conditions, elements having multiple oxidation states from a minus to a plus oxidation number have a dual role and serve as oxidizing or reducing agents, simultaneously reducing or oxidizing, depending on the pH of the medium, the degree of oxicity and the chemical species present.

Redox chemical reactions with an emphasis on their simultaneous occurrence are redox mechanisms whereby a transfer (donor-acceptor transformation processes) occurs through the metal/chemical type electrolyte phase boundaries. In redox processes, the cation of the higher oxidation number (state) $Me^{n+}$ is involved as the oxidized species, while the cation of the same element but of a lower oxidation state $Me^{(n-1)+}$ is involved as the reduced type.

A redox reaction in which a simultaneous oxidation/reduction of the same type of atom occurs is known as a disproportionation reaction. Chemical substances in which an element has a greater number of different oxidation states/numbers (with a mean value of oxidation number) exhibit the properties of both an oxidizing and reducing agent. Sometimes the mean oxidation number of an element is not the most stable oxidation state, and therefore new chemical species of the same element are formed in the redox system but with a higher or lower oxidation number. For example, all the iron present is characterized by three different oxidation numbers 0, +2, and +3. The oxidation state of Fe(II) is assumed to be a medium oxidation state.

Such chemical species, in which the observed chemical element has a medium oxidation state, can act both as oxidizing and reducing agents.

One of the vital and key redox reactions in the aquatic environment as well as in the pore waters of soil and sediment is the oxidation-reduction reaction disproportioning ammonium nitrate, $NH_4NO_2$ to elemental nitrogen $N_2$, wherein the nitrogen from the $NH_4^+$ cation of oxidation state −3 oxidizes to elemental nitrogen $N_2$, and at the same time the nitrogen from the nitric anion of oxidation state +3 is reduced to elemental nitrogen, with an exchange of six electrons.

In the field of redox reactions, it is accepted that chemical redox reactions can be represented as two different half-reactions (the half-reaction of oxidation and the half-reaction of reduction), whereby a conjugated redox couple (reduced and oxidized species) is formed. The conjugated pair is written as Ox/Red, and the corresponding half-reaction as:

$$\mathrm{Ox} + \mathrm{ne} \rightleftarrows \mathrm{Red} \quad (4.1)$$

The number of electrons that are exchanged, released and received must be equal in one redox reaction.

The key electrochemical magnitude in redox reactions is the redox potential (Petkovic *et al.*, 2008). The redox potential is a physicochemical term that describes the relationship between the oxidized and reduced forms of metabolites in a substrate. The redox potential is the ability of a substance to receive electrons (reduction) or to lose electrons (oxidation) and it is expressed as an *Eh* value. The reason for the transition of electrons from one compound to another is a difference in the electrical potential between the oxidant and the reductant, and it is measured in *V* or mV. The redox potential is an electromotive force in solution between the indication inert electrode and reference electrode whose value is a function of the ratio of concentrations of substances in the reduced form (reductants) and in the oxidized form (oxidants). A highly reducing agent that readily releases electrons has a high negative redox potential, while a highly oxidizing agent that readily accepts electrons has a high positive redox potential. The value *Eh* represents the measured potential at the typically platinum electrode, although a gold or graphite electrode could be used as well, corrected in relation to the standard hydrogen electrode (SHE).

In biotic processes it is correct to use the term substrate instead of matter. The substance (substrate) which is dominated by oxidation processes has positive *Eh* values, while the substrate where reduction processes are dominant has negative *Eh* values. Oxidizing substances increase the redox potential while reducing substances lower the redox potential. Selected important redox pairs and their standard redox potentials are presented in Table 4.2 (Atkins & de Paula, 2006).

**Table 4.2** Some important redox pairs in the (a) abiotic and in (b) biotic water matrix, and corresponding standard redox potentials.

| (a) Abiotic matrix | | |
|---|---|---|
| Couple | Redox half-reaction[1] | $E_o$ (mV) |
| ½ $O_2/H_2O$ | $O_2 + 4H^+ - 4e^- \rightarrow 2H_2O$ | +1230 |
| $Fe^{+3}/Fe^{+2}$ | $Fe^{3+} + e^- \rightarrow Fe^{2+}$ | +770 |
| $Mn^{+4}/Mn^{+2}$ | $MnO_2 + 4H^+ + 2e^- \rightarrow Mn^{2+} + 2e^- + 2H_2O$ | +1230 |
| $S^0/S^{2-}$ | $S + 2e^- \rightarrow S^{2-}$ | −480 |
| (b) Biotic matrix | | |
| Cytochrome C Ox/red | | +250 |
| Dehydroascorbic vinegar/ascorbic vinegar. | | +80 |
| Pyruvate/lactate | | −190 |
| Nicotinamide adenine dinucleotide $NAD^+$/NADH | | −320 |

[1]According to convention, oxidizing-reduction reactions are always written as half-reactions of reduction.

## 4.4 REDOX POTENTIAL *Eh* IN THE BIOLOGICAL OXIDATION SYSTEM

Experimentally determined *Eh* defines the state of the environment and the presence of oxidized or reduced biological substances. In each system during the process of biological oxidation, the ratio of the reductant to the oxidant content changes continuously. At the beginning of the process, the system is in a state of low redox potential. As the process of biological oxidation progresses, the concentration of the reductants decreases, which causes an increase in the *Eh* values, that is of the potential required to transfer electrons from the oxidant to the reductant.

In biological systems, much more complex, reversible equilibrium, interconnected biochemical redox reactions take place. The *Eh* value is a measure of the general state of each aquatic environment in the oxidation-reduction system – and of groundwater. The measured *Eh* defines the oxidation-reduction state of the tested groundwater samples.

The optimum results of *Eh* in oxygenated oxidation systems are positive *Eh* values between +200 and +600 mV, while anoxic processes are characterized by redox potentials ranging between −100 and 200 mV.

Most microorganisms develop and replicate under standard conditions such as when the pressure is 101.325 kPa, the pH 6.5–8.5, the temperature 8–45°C, the redox potential (*Eh*) −50 to +150 mV and the salt concentration lower than 6% NaCl.

The most commonly found microbial organisms in aquatic systems are specific biological species of bacteria that exhibit activities such as biodegradation and biofouling (iron bacteria, sulphate reducing and 'slime' bacteria) and can develop and replicate in both oxic and anoxic conditions.

During biochemical activity, bacteria transform nutrients into forms suitable for cell reproduction, and they receive energy through redox electron transfer reactions from electron donors to electron acceptors. Good electron donors are natural organic substances, that is, carbon atoms, while electron acceptors are atoms of elements with high oxidation numbers such as nitrogen(V) – nitrate ($NO_3^-$), $Fe^{3+}$, sulphur(VI) – $SO_4^{2-}$ and carbon dioxide.

The activities of the consortium of microorganisms are present not only in the shallow but also in the deeper parts of an aquifer and play a key role in various biochemical processes in groundwater (Dimkić and Knezic, 2012).

## 4.5 THERMODYNAMIC EQUILIBRIUM PROCESSES IN WATER, pH-*pE* DIAGRAMS

Water is an amphiprotic substance of acid-base properties which, in the process of autoprotolysis, forms a hydrated hydronium cation $H_3O^+$ ($H^+ \cdot H_2O$) and a hydrated hydroxide anion $H_3O_2^-$ ($OH^- \cdot H_2O$). In chemical equations, hydrated forms of the hydrogen cation and hydroxide anion, for simplicity purposes, use the simplified ionic formulas $H^+$ and $OH^-$. The equilibrium constant of the autoprotolysis process ($K_w$ or $K_{auto}$) depends on temperature and has a value of $K_w = 1.01 \cdot 10^{-14}$ at 25°C. Each chemical species (molecule, cation, anion, radical) in an aqueous solution is hydrated and is located in a cloud, a layered layer of icosahedral water molecules which is referred to as aq. As autoprotolysis is an endothermic process, the concentration of $H^+$ and $OH^+$ ions depends on temperature. At a temperature of 0°C the constant $K_w$ has a value of $1.148 \cdot 10^{-15}$ and pH = 7.47. As an amphiprotic substance, water has the ability to receive and give protons from or to other substances and to act as a typical Brønsted base or acid in the presence of a proton donor or proton acceptor, respectively.

Oxidation and reduction processes, redox processes, play an extremely important role in aquatic chemistry. In the sphere of redox reactions water has dual characteristics, both oxidizing and reducing.

Water oxidation results in the evolutionary formation of oxygen molecules:

$$2H_2O \rightarrow O_{2(g)} + 4H^+_{(aq)} + 4e^- \tag{4.2}$$

Based on the Nernst equation, the redox potential for this molecular electron ion half-reaction is pH dependent and is defined by the following equation at 25°C:

$$Eh = 1.23 - 0.0591\ \text{pH} \tag{4.3}$$

Under extremely reducing conditions, water is reduced and decomposed into gaseous molecular hydrogen and aquatic hydroxide anion:

$$2H_2O + 2e^- \rightarrow H_{2(g)} - 2OH^-_{(aq)} \tag{4.4}$$

The ratio of the redox potential to the pH value for the above reaction at 25°C is:

$$Eh = -0.0591\ \text{pH} \tag{4.5}$$

When the dependence of redox potential on pH value is shown graphically in volts on the y-axis (Figure 4.1), according to the pH value on the x-axis, the linear ratio of the redox potential equations defines the graphical surface of water stability.

The stability of the water corresponds to the value of the redox potential of 0 to 1.229 at pH = 0, and from −0.829 to 0.401 at pH = 14.

The chemical reactions that occur spontaneously have a negative value of Gibbs free energy change $\Delta G$, but therefore a positive value of $\Delta Eh$.

$$\Delta G = -nF\Delta Eh \tag{4.6}$$

In the dynamic equilibrium state of the chemical system, $\Delta G = 0$. It follows that the electrode potential $Eh$ is in direct proportion to $G$, that is, $G = -n \cdot F \cdot Eh$, where $F$ is the Faraday constant. The value of

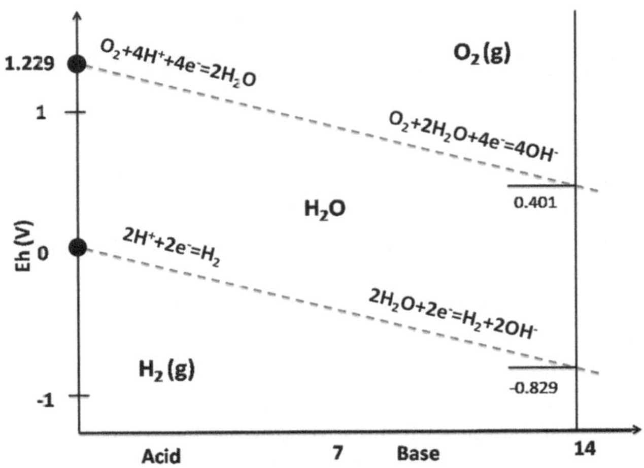


**Figure 4.1** The areas of water stability are shown in the diagram $Eh = f$ (pH) (modified according to Pourbaix, 1974).

the Faraday constant is $F = 96\ 485.3329\ \text{C/mol} = N_A \cdot e^-$, where $N_A$ is the Avogadro number ($6.02214076 \cdot 10^{23}$) and $e^-$ is electron charge ($1.602176634 \cdot 10^{-19}$ C).

The measured electrode potential, analogously to pH, is defined by the negative logarithm of electron activity as:

$$pE = -\log a_e \tag{4.7}$$

A very widely used diagram of two *pE*/pH variables by which the behaviour of the aquatic hydrosphere can be defined is known as the Pourbaix diagram. For environmental processes, pH and *pE* are two key unique physicochemical properties that define acidic or basic environment, oxicity and anoxicity in the aquatic system.

By analysing the Pourbaix two variables water diagram of the pH dependence on *pE* shown in Figure 4.2, the negative value of *pE* means a high value for the activity of electrons in solution, implying that reducing (anoxic) conditions have been achieved in water bodies. *pE*/pH diagrams graphically represent thermodynamic data as well, and it is important to know the most stable forms of an element under certain conditions. The practical application of the *pE*-pH diagram is in a relatively simple graphical representation of thermodynamic data. *pE*-pH diagrams do not provide information on the kinetics and the rate of chemical redox reaction, but may provide oxidation reduction conditions in natural aquatic environments and defined by the correlation between the oxic/anoxic conditions, i.e. pH/pOH, *pE*/pH and other natural phenomena.

Conversely, a high positive value of *pE* means low electron activity $a_{e-}$ and highly oxic conditions, such as in the case of well aerated surface waters. The *pE* value in an aqueous medium approximately ranges from −12 to 25.

However, the activity of electrons is not easily measured, such as when determining the concentration of $H^+$ ions and the correspondent pH values. Electron activity defines the tendency of a solution to receive or release electrons.

Water has a limited range of *pE* and pH values at which it is stable. Under high reducing conditions (low *pE*), water is reduced

$$2H_2O + 2e^- \rightleftarrows H_{2(g)} + 2OH^-_{(aq)} \tag{4.8}$$

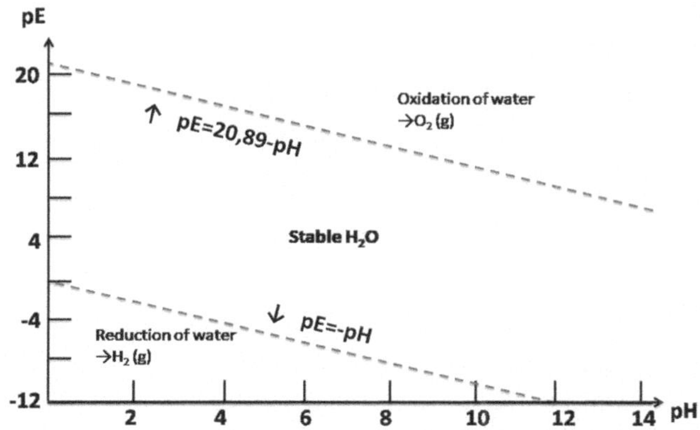


**Figure 4.2** The Pourbaix two variables diagram of water *pE*/pH (modified according to Davis, 2000).

For this reaction $E^0 = -0.828\,\mathrm{V}$, respectively, $pE^0 = -14.0$, wherein van Loon and Duffy (2005) developed a functional dependency:

$$pE = pE^0 + \frac{1}{2}\log\left(\frac{p_{H_2}}{p^\circ} \cdot (a(OH^-))^2\right) \tag{4.9}$$

For the boundary with the gas phase of the atmosphere, conditions for the partial pressure of hydrogen were selected $p_{H_2} = P^0 = 101325\,\mathrm{Pa}$. The equation developed for $pE$ takes the form:

$$pE = -14.0 - \log(a_{OH^-}) = -14.0 + \mathrm{pOH} \tag{4.10}$$

Since $\mathrm{pH} + \mathrm{pOH} = 14$ then $pE = -\mathrm{pH}$.

The limits of $pE$ and pH that define the stability of water are defined in diagram 4.2. When the $pE$ value is less than the pH value, water is 'unstable' and typical water reduction reactions are present:

$$2H_3O +_{(aq)} + 2e^- \rightarrow H_{2(g)} + 2H_2O \tag{4.11}$$

In highly oxic conditions, water is also 'unstable' and degradation occurs:

$$6H_2O \rightleftarrows 4H_3O^+{}_{(aq)} + O_{2(g)} + 4e^- \tag{4.12}$$

For reaction $E^\circ = 1.229\,\mathrm{V}$, that is $pE^\circ = E^\circ/0.0591 = 20.80$.

Water in aquatic systems can be both an oxidizing and reducing agent whereby Pourbaix diagrams of redox reactions in the aquatic environment define boundary oxidation and reduction conditions.

### 4.5.1 Water as an oxidizing agent

The following reactions represent the reduction of water:

$$H^+ + e^- \rightleftarrows 1/2\ H2 \qquad \mathrm{pH} = 7 \qquad Eh = -0.414 \tag{4.13}$$

$$H_2O + e^- \rightleftarrows 1/2H_2 - OH^- \qquad \mathrm{pH} = 14 \qquad Eh = -0.829 \tag{4.14}$$

### 4.5.2 Water as a reducing agent

The following molecular electron-ion reactions show the oxidation of water with corresponding values of the electrode potential in $V$:

$$1/2O_2 + 2H^+ + 2e^- \rightleftarrows H_2O \qquad \mathrm{pH} = 0 \qquad Eh = 1.229 \tag{4.15}$$

$$1/2O_2 + 2H^+ + 2e^- \rightleftarrows H_2O \qquad \mathrm{pH} = 7 \qquad Eh = 0.82 \tag{4.16}$$

$$1/2O_2 + H_2O + 2e^- \rightleftarrows 2OH^- \qquad \mathrm{pH} = 14 \qquad Eh = 0.401 \tag{4.17}$$

On the basis of the calculated electrode potential for water (as oxidizing and reducing agent), the Pourbaix diagram for water with the electrode potential and the pH is shown in Figure 4.3. For better understanding and interpretation of the Pourbaix diagrams, two similar Pourbaix diagrams are shown in the $Eh$/pH functional relationship (Figures 4.1 and 4.3).

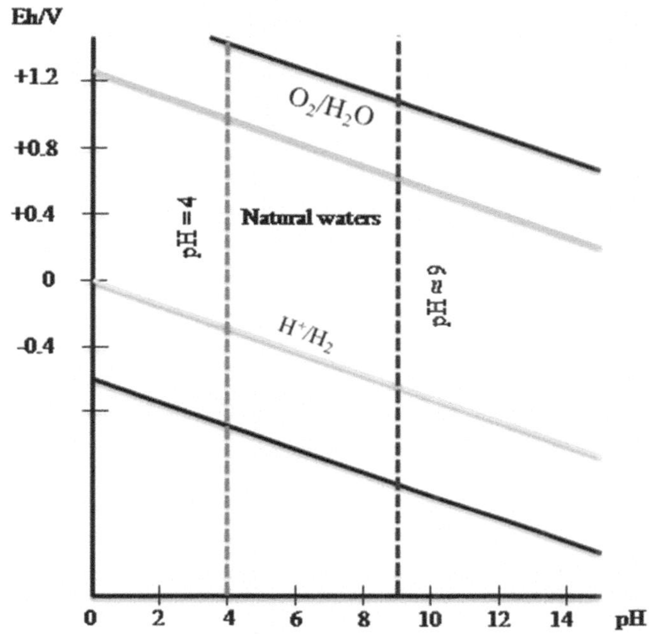


**Figure 4.3** The Pourbaix diagram for water with indicated mean values of the redox potentials and pH values of natural waters (modified according to Van Loon & Duffy, 2005).

### 4.5.3 Thermodynamic equilibrium of iron oxidation in water

Iron has three stable oxidation numbers 0, +2, and +3 with a mean oxidation number +2 and coordination number 6 in complex compounds. The compounds of oxidation number +2 are the most stable, which is why the form of iron (II) is most commonly found in nature, in soil, and in surface and groundwater. Iron compounds are a basic structural component of many minerals such as oxides, silicates, sulphides and others (Nikolić-Bujanović *et al.*, 2012). The most common iron minerals in nature are hematite (iron(III)oxide, $Fe_2O_3$), magnetite (complex iron(II) oxide and iron(III) oxide, $Fe_3O_4$), pyrite (iron(II) disulphide, $FeS_2$), limonite (hydrated iron(III) oxide – hydroxide in varying composition, $Fe_2O_3 \cdot H_2O$), goethite (α crystalline form of hydrated hematite – FeO(OH), component of yellow ochre) and others.

The concentration of iron in natural waters is higher if the iron is present as +2 than as the +3 oxidation number. Groundwater, which is in contact with iron minerals, will dissolve iron as an $Fe^{2+}$ ion. When groundwater containing $Fe^{2+}$ is in contact with oxygen from the air, $Fe^{2+}$ is oxidized to $Fe^{3+}$ and precipitates as an insoluble precipitate of $Fe(OH)_3\downarrow$. At pH 7, the solubility of $Fe^{2+}$ is 15 orders of magnitude greater than $Fe^{3+}$. The solubility product of $Fe(OH)_3$ at 25°C is: $Ks = 4 \cdot 10^{-38}$ $mol^4/dm^{12}$ indicating that the iron(III) hydroxide is a poorly soluble compound that is formed at very low $Fe^{3+}$ concentrations and low pH. In groundwater, with a change in the pH and an increase in the oxic environment, several parallel processes occur, namely oxidation, reduction, hydrolysis and precipitation (Table 4.3 and Figure 4.4).

The concentration of iron in aquatic systems depends on various dynamic equilibrium chemical processes that primarily include the processes of oxidation and reduction, precipitation and the

**Table 4.3** Redox reactions of Fe(II) and Fe(III) ionic and molecular chemical species with thermodynamic parameter values.

| Redox Reaction | *E*° (V) | Δ*G*° (kJ) | Log $K_{red}$ | *pE* |
|---|---|---|---|---|
| $Fe^{2+} + 2e^- \rightarrow Fe_{(s)}$ | -0.44 | 84.91 | −14.8 | −7.4 |
| $H^+ + e^- \rightarrow ½H_{2(g)}$ | 0 | 0 | 0 | 0 |
| $Fe^{3+} + e^- \rightarrow Fe^{2+}$ | +0.77 | 25.09 | 13.0 | 13.0 |
| $Fe(OH)_{3(s)} + 3H^+ + e^- \rightarrow Fe^{2+} + 3H_2O$ | +1.06 | −102.3 | 17.9 | 17.9 |
| $O_{2(g)} + 4H^+ + 4e^- \rightarrow 2H_2O$ | +1.23 | −474.7 | 83.10 | 20.8 |
| $NO_3^- + 6H^+ + 5e^- \rightarrow ½N_{2(g)} + 3H_2O$ | +1.24 | −598.2 | 105.2 | 21 |

dissolution of hydroxides, carbonates and sulphides, and the formation of complexes with organic ligands and biochemical enzymatic reactions (Mrazovac & Vojinovic-Miloradov, 2011; Mrazovac *et al.*, 2013).

The physicochemical parameters that affect the solubility of iron compounds in an aqueous media are the pH, the redox potential *Eh*, and the solubility of sulphur and carbon dioxide compounds. In groundwater, iron can be present in two oxidation states +2 and +3 with a shift of the oxidation-reduction balance towards $Fe^{2+}$ (Čavić *et al.*, 2018).

The form and concentration in which the iron will be present in $Fe^{2+}$ will depend mostly on the pH of the medium. If the pH is greater than 4.8, the solubility of Fe(II) is lower and ranges around 0.01 mg/L. As the oxic environment increases, Fe(II) is oxidized to Fe(III) which (in the low concentration of hydroxide anions), promptly builds up insoluble $Fe(OH)_3$, which is a direct consequence of the low value of *Ks*.

The simple half-reaction of the Fe(III) iron reduction process is represented by the equation:

$$Fe^{3+}_{(aq)} + e^- \rightleftarrows Fe^{2+}_{(aq)} \tag{4.18}$$

The equilibrium constant expressed through activity is given as:

$$K_{eq} = \frac{a_{Fe^{2+}}}{a_{Fe^{3+}} + a_{e^-}} \tag{4.19}$$

The relation between pH and the corresponding equation for *pE* was derived:

$$pE = -\log a_e = \log K_{eq} + \log \frac{a_{Fe^{3+}}}{a_{Fe^{2+}}} \tag{4.20}$$

A thermodynamic approach with the free Gibbs energy transforms the equation:

$$\Delta G^\circ = -2.303RT\log K_{eq} = -nFE^\circ \tag{4.21}$$

In these mathematical and thermodynamic equations *n* has the usual electrochemical meaning i.e. the number of electrons transferred in the redox chemical half-reactions.

With further solving of the equations, which include thermodynamic extensive variables of the thermodynamic temperature 298 K and the universal gas constant $R = 8.314$ J/(K · mol), the equation

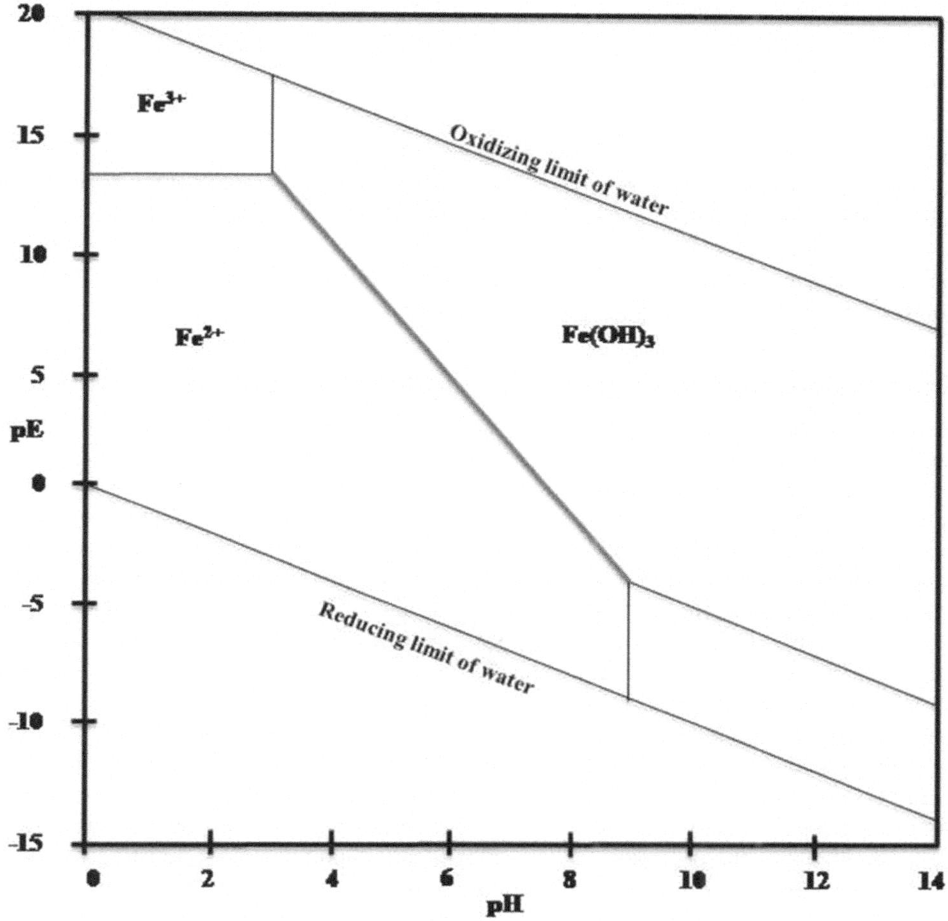


**Figure 4.4** Graphic presentation of *pE*-pH for iron (modified according to Hem, 2005).

under standard conditions takes the form:

$$\log K_{\mathrm{eq}} = \frac{(nFE^{\circ})}{2.303RT} 2.303RT = \frac{(nE^{\circ})}{0.0591} \tag{4.22}$$

For $n = 1$ (single electron transfer) $pE$ is:

$$pE = \frac{E^{\circ}}{0.0591} + \log \frac{a_{\mathrm{Fe}^{3+}}}{a_{\mathrm{Fe}^{2+}}} \tag{4.23}$$

Under standard conditions:

$$a_{\mathrm{Fe}^{3+}} = a_{\mathrm{Fe}^{2+}} = 1, \text{ i.e. } \log \frac{a_{\mathrm{Fe}^{3+}}}{a_{\mathrm{Fe}^{2+}}} = 0 \quad \text{and} \quad pE = pE^{\circ} = \frac{E^{\circ}}{0.0591} \tag{4.24}$$

While for non-standard conditions:

$$pE = pE^{\circ} + log\frac{a_{Fe^{3+}}}{a_{Fe^{2+}}} \tag{4.25}$$

When the standard $pE^{\circ}$ value and activity are approximated to the concentration of cations $Fe^{3+}$ and $Fe^{2+}$ and substituted in these equations, it is relatively simple to calculate the $pE$ value of the aquatic ecosystem.

The values of the thermodynamic parameters for the redox reactions of Fe(II) and for the Fe(III) cationic iron species are shown in Table 4.3.

A graphical presentation of the relation of the logarithm of $E$, and the concentration of hydrogen ions, $pE$-pH (Pourbaix diagram) of iron is shown in Figure 4.4.

### 4.5.4 Thermodynamic equilibrium of manganese oxidation in water

The most common manganese minerals in nature are: pyrolusite $MnO_2$, manganite MnO(OH), hausmanite, thermo luminescent manganese oxide $Mn_3O_4$, braunite $Mn_2O_3$, rhodocrosite $MnCO_3$ and rhodonite $MnSiO_3$. Manganese has six different oxidation states, from zero-state, +2, +3, +4, +6 to +7. The solubility, mobility and species of manganese depend on the redox state of the system. Manganese is less mobile under oxidative and alkaline conditions. In (2005) Hem stated that the rate of oxidation of $Mn^{2+}$ and the precipitation of the oxidized form into $MnO_2$ is accelerated by increasing the pH value and that of the surface of the mineral. The bacteria can affect the rate of the oxidation of manganese similar to how bacteria affect the oxidation of the cations of iron $Fe^{2+}$. Under reducing conditions, the concentrations of manganese in water may reach several mg/L, but the average concentration in continental water is 7 μg/L (Sremacki *et al.*, 2019).

Manganese is most commonly found together with iron in water, but the concentration of iron ($Fe^{2+}$) is higher due to the abundance of iron in the Earth's crust. The concentrations of manganese in groundwater depend on the pH and redox conditions. The mobility of manganese increases with a decrease of the pH, when manganese is predominantly present as $Mn^{2+}$. In pH-neutral conditions, the mobility is determined by the redox conditions of the environment. Under oxidizing conditions characteristic of many shallow aquifers and surface waters, manganese is present in the oxidized form $MnO_2$, Mn(IV) which is insoluble. Much higher concentrations of $Mn^{2+}$ manganese are present under reductive (anoxic) groundwater conditions. As the depth increases, the concentration of dissolved oxygen decreases and therefore higher concentrations of manganese will occur at greater depths. Under reductive conditions and in the presence of dissolved sulphide, manganese may be immobilized by the formation of insoluble manganese sulphide (MnS), if the pH is greater than 8.

At neutral pH values, dissolved $Mn^{2+}$ ions are dominant (Figure 4.5). A small increase in the pH will change the oxidation state to the stability zone of manganese oxide where the $Eh$ and pH affect solubility. The kinetic rate of the oxidation and reduction of manganese are much lower than the kinetic rate of the oxidation and reduction of iron (Hem, 2005).

$$MnO_2 + H_2O + e^- \rightarrow MnO(OH) + OH^- \tag{4.26}$$

The necessary and useful information about the redox reactions that occur in aqueous solutions and about the values of the oxidizing states of manganese and iron can, in a simplified manner, be shown in the following three diagrams: Latimer, Frost and the aforementioned Pourbaix diagram.

The Latimer diagram shows the standard reduction potentials for the corresponding oxidation states of manganese compounds in the basic and acidic environment (Figure 4.6).

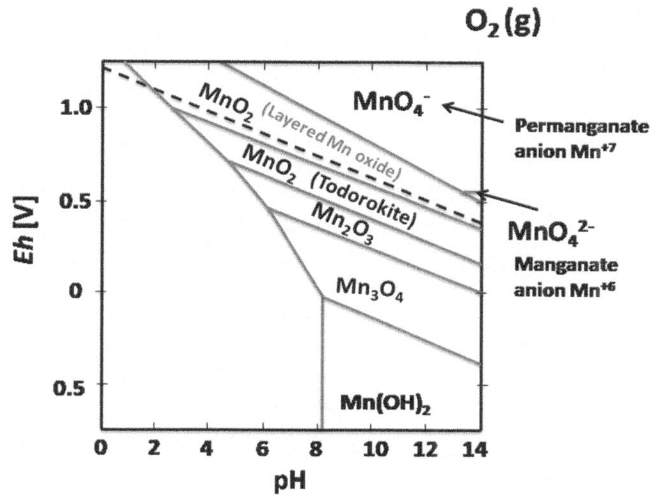


**Figure 4.5** Graphical presentation of *Eh*–pH for different oxidation states of manganese (modified according to Hem, 2005).

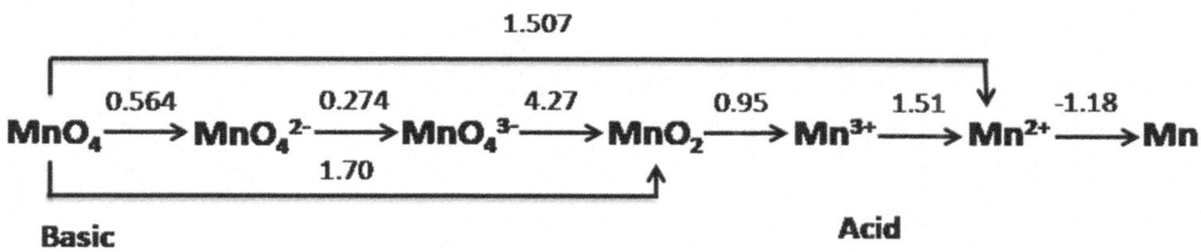


**Figure 4.6** Latimer diagram for the oxidation states of manganese from zero through 2+, 3+, 4+, 6+ and 7+ (modified according to Rayner-Canham, 1996).

The Frost diagram graphically defines the change in Gibbs free energy as a function of different oxidation numbers of manganese from $Mn^{2+}$ to $Mn^{7+}$ (Figure 4.7).

The interpretation and analysis of the two diagrams define the key characteristics of the stable oxidation states of manganese. Elemental manganese as a metal is a medium-strong reducing agent. Mn(II) is the most stable oxidation state of manganese, especially in acidic or highly oxygenated environments. Almost all $Mn^{2+}$ salts are well soluble in water. The reduction of permanganate to Mn(II) ions is thermodynamically favoured, but the redox reaction is slow, except in the presence of a suitable catalytic agent. As a direct consequence of Gibbs free energy values, the aqueous permanganate solutions, $MnO_4^-$ are stable (Figure 4.7). The $Mn^{2+}$ cation is not present in highly basic conditions, which is in accordance with the solubility product of the corresponding hydroxide, $Mn(OH)_2$ ($4.6 \cdot 10^{-14}$). Based on the Frost diagram, the permanganate anion, $MnO_4^-$ (Mn(VII)), is a strong oxidizing agent. The manganate ions $MnO_4^{2-}$ (Mn(VI)) and $Mn_2O_3$ (Mn(III)) belong to redox disproportionation reactions.

The third diagram is the Pourbaix diagram that displays oxidation and reduction of manganese in the relation *Eh* to pH (Figure 4.8).

The interpretation and analysis of each point in the Pourbaix diagram corresponds to the thermodynamically stable and dominant chemical species, i.e. the oxidation or reduction state of

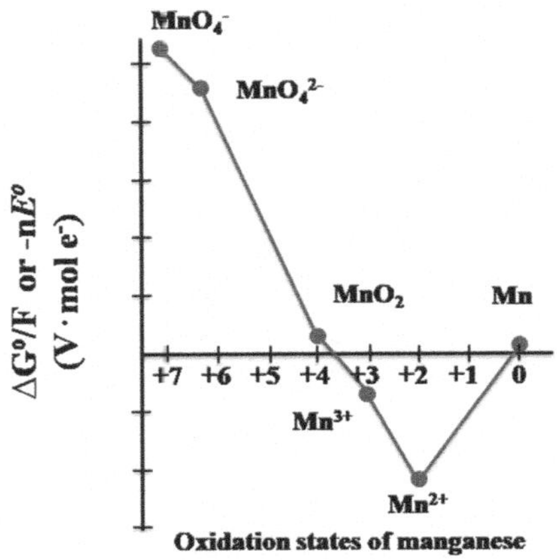


**Figure 4.7** The Frost diagram for manganese (modified according to Rayner-Canham, 1996).

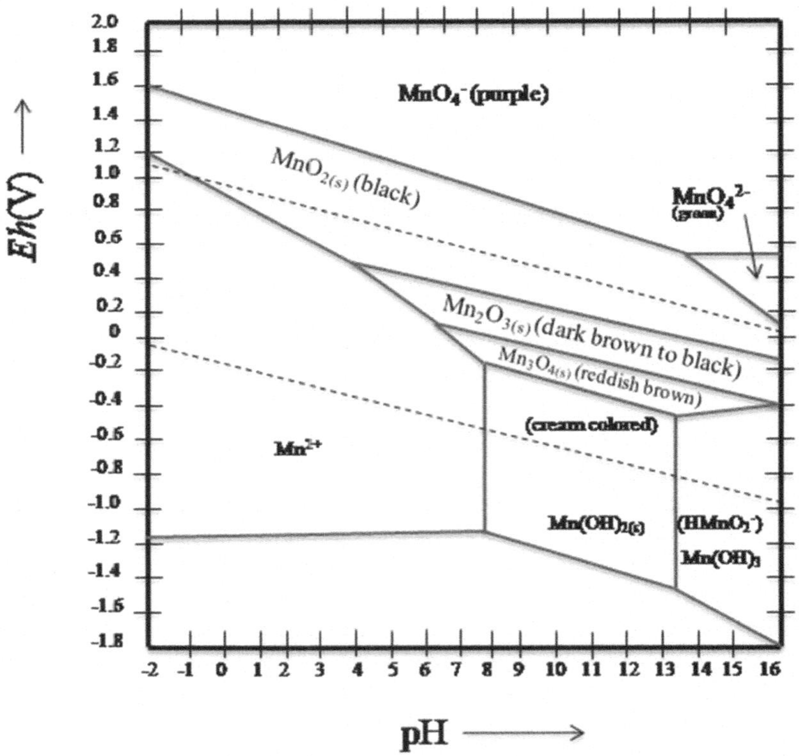


**Figure 4.8** The Pourbaix diagram of *Eh*–pH for manganese (modified according to Hem, 2005).

manganese for the *Eh* and pH values. The generalized conclusion is that strong oxidation agents are always at the very top of the Pourbaix diagram. Thus, for example, the permanganate ion (Mn(VII)) is one of the most powerful oxidizing agents in a wide range of pH values, as well as being a good oxidizing agent at low pH values. Figure 4.8 also shows possible theoretical pH values over and above 14.

The Pourbaix diagram shows that the reducing agents are at the bottom of the diagram. Based on the Pourbaix diagram for manganese, it can be concluded that manganese is, in the elemental form, just like all metals, a reducing agent in a wide range of pH values.

## 4.6 KINETIC REACTIONS OF IRON OXIDATION IN WATER

Redox conditions, and especially the amount of dissolved oxygen, DO, and the oxygen flux in the groundwater represent key factors pertaining to the quality of drinking water, as well as regarding the establishment of the source in the overall water supply cycle.

The coexistence of Fe(II) and DO in groundwater leads to a homogeneous oxidation reaction (Fe(II) in Fe (III)) and to the precipitation of iron(III) hydroxide (Majkić-Dursun *et al.*, 2014; Stumm & Morgan, 1996).

In the first phase, Fe(II) is oxidized to Fe(III) according to the molecular ion half-reaction:

$$4Fe^{2+} + O_2 + 4H^+ \rightarrow 4Fe^{3+} + 2H_2O \tag{4.27}$$

In the second phase, Fe(III) hydrolyses to form the iron(III) hydroxide precipitate due to an increase in the pH of the medium, the presence of $OH^-$ ions and the low value of the solubility product of the order of $10^{-38}$, as shown by the following hydrolysis reaction:

$$4Fe^{3+} + 4H_2O \rightarrow 4Fe(OH)_3 \downarrow + 4H^+ \tag{4.28}$$

The oxidation kinetics of Fe(II) oxide is an extremely important phenomenological process in redox reactions. The reaction rate is linearly dependent on the concentration of iron and its activity, on the partial pressure of the DO and on the square of the hydroxide ion concentration (Menz, 2016).

The half-life of the iron oxidation reaction can be calculated from the equation:

$$t_{1/2} = \frac{\ln 2[H+]^2}{[O_2\ ]} \tag{4.29}$$

The oxidation of Fe(II) to Fe(III) is catalytically accelerated by sorption and hydrolysis of both Fe(II) and Fe(III), and exceptionally in the formation of the coloured iron(III) hydroxide precipitate. When iron(III) hydroxide is present in a water system, and particularly in springs, the process of aging and clogging increases (Houben, 2006).

The kinetic heuristic approach explains and calculates the oxidation of Fe(II) depending on the pH value. The spontaneous chemical oxidation of Fe(II) to Fe(III) is a very complex process that takes place and involves the metastable partially oxidized intermediate chemical species for the ultimate transformation of stable Fe(III) compounds such as hematite, magnetite, goethite, and lepidocrocite. In practical situations, the kinetics pertaining to the formation of the final oxidation products are still unknown, although the basic principles of the kinetics of the oxidation reactions of Fe(II) to Fe(III) are known. One of the few research papers in the literature in this field uses and describes the theory of the chemical equilibrium in aquatic systems as a framework by which a heuristic model can be used to define the kinetics of the oxidation of Fe(II) species to Fe(III) in an oxic environment according to the following relation:

$$\frac{-d[Fe^{2+}]}{dt} = k[Fe^{2+}]\ [OH^-]^2\ p_{O_2} \tag{4.30}$$

According to the equation, the rate of oxidation is a first order reaction with respect to the partial pressure of oxygen $p_{O_2}$ and $Fe^{2+}$ and also a second order one with respect to the hydroxide ion concentration. The rate of the chemical reaction of $Fe^{2+}$ oxidation can be described by the intermediate dissociable forms of

$Fe^{2+}$, $FeOH^{+}$, $Fe(OH)_2^0$, $Fe(OH)_3^-$ in the equation below with units in mol Fe(II)/(L · min), where all the shown Fe(II) species are important for understanding the pH-dependent oxidation rate:

$$\frac{-d[Fe^{2+}]}{dt} = (k_0\,[Fe^{2+}] + k_1\,[Fe(OH)^+] - k_2\,[Fe(OH)^0_{2(aq)}] + k_3[Fe(OH)_3^-])DO \tag{4.31}$$

wherein $k_0$, $k_1$, $k_2$ and $k_3$ are the constants of the rate of the oxidation process.

The graphical presentation of the equation is a sigmoid curve as a function of the pH value. Intriguing, at pH values of approximately 4, the concentration of Fe(II) is dominant and the rate of the oxidation reaction is practically independent of the pH. At pH of about 5 the dominated form is $Fe(OH)_2$, although the cationic forms of $Fe^{2+}$ and $FeOH^+$ are oxidized which is reflected in the rate of the oxidation constant. Between the pH values of 5 and 8, $Fe(OH)_2$ concentrations and the rate of transformational oxidation processes increase with an increase of the pH. At pH values greater than 8, $Fe(OH)_3^-$ is dominant with the sorbed hydroxide anions (negatively charged particles) and the theoretical concentration no longer varies with the pH value, so oxidation is independent of the pH value. Research has shown that a heuristic review of the dependence of the oxidation kinetics of $Fe^{2+}$ on the pH and on the partial pressure of dissolved oxygen is very complex but theoretically important for understanding the processes that occur in groundwater (Morgan & Lahav, 2007) (Figure 4.9).

In a saturated solution of Fe(II) and in the presence of $OH^-$ ions, $Fe(OH)_2$ is a solid substance whose concentration is constant and the previous equation can be simplified to:

$$\frac{-d[Fe^{2+}]}{dt} = k\,DO \tag{4.32}$$

The mathematical form of the equation shows that at higher pH values, the limiting factor of the reaction rate is only the partial concentration of dissolved oxygen, DO.

The rate of the chemical reaction of oxidation is simultaneously thermodynamically and kinetically supported by the adsorption of dissolved chemical species of hydroxide anions onto the surface of solid iron(II) hydroxide, $Fe(OH)_2$.

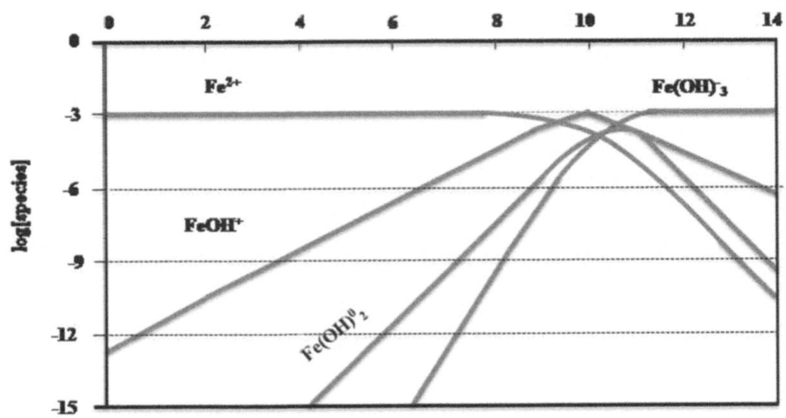


**Figure 4.9** Fe(II) species important for understanding the pH-dependent oxidation rate (modified according to Morgan & Lahav, 2007).

The rate constants for the oxidation reactions of soluble Fe(II) species ($Fe^{2+}$, $FeOH^+$, $Fe(OH)_2$) differ from each other by five orders of magnitude ($6 \cdot 10^{-5}$, 1.7, $4.3 \cdot 10^5$ L/min). In practical terms, this means that the overall rate of the oxidation reaction depends dramatically on the distribution of three mentioned species in an aqueous solution which primarily depends on the pH and much less so on the temperature and ionic concentration.

## 4.7 KINETIC REACTIONS OF MANGANESE OXIDATION IN WATER

Another research endeavour explains the kinetic reactions between the oxygen concentration and the dissolved manganese(II) with an interpretation of the rate of the $Mn^{2+}$ chemical reaction in natural waters (Morgan, 2005).

The oxidation of Mn(II) to the more oxidized state of Mn(III) and Mn(IV) by dissolved oxygen is one of the major processes that regulate the kinetics of the manganese cycle in aquatic geochemistry. For example, in the overall reaction:

$$4Mn^{2+}_{(aq)} + O_{2(aq)} + 6H_2O = 4MnOOH_{(s)} + 8H^+ \tag{4.33}$$

the dissolved Mn(II) ions transition to a partially soluble Mn(III) oxy hydroxy form that binds to soil or sediment particles or is further oxidized to insoluble, solid $MnO_2$. The vertical and horizontal profiles of the dissolved Mn(II) ion in water are determined by the dynamics of these oxidation reduction processes.

The elementary stages of the model predicting the kinetic flow of the manganese oxidation reaction and the corresponding reaction rate constants are shown below (Rosso & Morgan, 2002):

$$Mn^{2+} \rightarrow Mn^{3+} + e^- \qquad \log k = -17.82 \tag{4.34}$$

$$MnOH^+ \rightarrow MnOH^{2+} + e^- \qquad \log k = -9.20 \tag{4.35}$$

$$Mn(OH)_2 \rightarrow Mn(OH)_2^+ + e^- \qquad \log k = -3.12 \tag{4.36}$$

The total rate of oxidation of $Mn^{2+}$, is shown in a simplified manner as $\Sigma k_i\, a_{ij}$, wherein $k_i$ is the oxidation reaction rate constant for the chemical species $i$, and $a_{ij}$ is the partial concentration that refers to the products of the hydrolysis of $Mn^{2+}$, carbonate, ammonium, halide and sulphate complexes, including the free aquatic ions, and the starting kinetic equation is defined as:

$$\frac{-d[Mn^{2+}]}{dt} = k_1[Mn^{2+}] + k_2\,[MnO_x][Mn^{2+}] \tag{4.37}$$

The slower oxidation of Mn(II) compared to the oxidation of Fe(II) can be explained by the different thermodynamic constants and electronic structures of iron and manganese, as well as by the different values of the bond strengths between the aquatic complexes $Fe(H_2O)_6^{3+}$ and $Mn(OH)_6^{3+}$ as illustrated in Table 4.4.

The change in the standard free energy of oxidation of Mn(II) to Mn(III), compared to the analogue Fe(II)/Fe(III) is shown in Table 4.5.

**Table 4.4** Rate constants by single-electron oxygen oxidation for Mn(II) and Fe(II) ions 1/(M · s) at 25°C (King, 1998; Morgan, 2005; Rosso & Morgan, 2002).

| Species *i* | $k_{i,\ axis}$ [a] | $k_{i,\ fit}$ [b] |
|---|---|---|
| $Mn^{2+}$ | $1.6 \cdot 10^{-18}$ | $<1 \cdot 10^{-10}$ |
| $Fe^{2+}$ | $1.6 \cdot 10^{-6}$ | $1.0 \cdot 10^{-6}$ |
| $MnOH^{+}$ | $6.3 \cdot 10^{-10}$ | $1.66 \cdot 10^{-2}$ |
| $FeOH^{+}$ | $2.0 \cdot 10^{-2}$ | 6.3 |
| $Mn(OH)_2$ | $7.9 \cdot 10^{-4}$ | 20.9 |
| $Fe(OH)_2$ | $1.6 \cdot 10^{2}$ | $7.9 \cdot 10^{5}$ |

[a]Predicted concentrations based on the Marcus theory of electron transfer rate (Rosso & Morgan, 2002)
[b]King (1998); Morgan (2005).

**Table 4.5** Standard free energies of $\Delta G$ oxidation of Mn(II) and Fe(II) chemical species in kJ/mol.

| Chemical Species | $\Delta G$ (kJ/mol) |
|---|---|
| $Mn^{2+}/Mn^{3+}$ | 164 |
| $MnOH^{+}/MnOH^{2+}$ | 100 |
| $Mn(OH)_2/Mn(OH)_2^{+}$ | 34 |
| $Fe^{2+}/Fe^{3+}$ | 90 |
| $FeOH^{+}/FeOH^{2+}$ | 48 |
| $Fe(OH)_2/Fe(OH)_2^{+}$ | 5 |

## 4.8 THERMODYNAMIC AND KINETIC REACTIONS OF THE OXIDATION OF SULPHIDE BY OXYGEN IN BIOTIC AND ABIOTIC MEDIA

For the abiotic medium, the following equation was used to show the oxidation rate of sulphide by oxygen:

$$\frac{d[H_2S]}{dt} = k_2[H_2S][O_2] \tag{4.38}$$

$k_2$ is the second order reaction rate constant for the range pH = 4–8, log $k_2$ is calculated as:

$$\log k_2 = 10.5 + 0.16\text{pH} - (3 \cdot 10^3)\,T^{-1} + 0.49I^{1/2} \tag{4.39}$$

From the thermodynamic point of view, half-reactions of reduction and half-reactions of oxidation are used to calculate the values of $\Delta G$ or log $K$

$$1/2O_{2(aq)} + H^+ + e^- \rightarrow 1/2H_2O \qquad pE = pE^0 - \text{pH} = 13.18 - \text{pH} \tag{4.40}$$

$$1/2S + H^+ + e^- \rightarrow 1/2H_2S \qquad pE = pE^0 - \text{pH} = 2.44 - \text{pH} \tag{4.41}$$

The application of the thermodynamic and kinetic approach shows that the sulphide flux in the oxic environment will be determined by the size of the microbial population and by the activity of the consortium of microorganisms in the anoxic-oxic transition zones, rather than by the chemical oxidation processes themselves (Luther *et al.*, 2011). This statement is explained by the thermodynamic and kinetic limitations of the transition of one electron relative to the transition of two electrons in reactions with $HS^-$ to $O_2$.

## 4.9 OXIC CONDITIONS AND THE BASIC QUALITY OF ALLUVIAL UNDERGROUND WATER

Groundwater, as part of a complex water cycle in nature, is extremely specific and unique because of its various kinetically thermodynamic and redox characteristics. Groundwater movement velocity is several (generally 4–6) orders of magnitude slower than the velocity of rivers. The dissolved substances and other material in groundwater are, over long periods of time, exposed to a variety of physical, biochemical and redox processes in the groundwater, in the skeleton of the aquifer, or at the boundary contact surfaces between the different phases of the heterogeneous system (water, aquifer, air).

The rate of the flux and the extensiveness of hydrological and redox processes as the specific characteristics of groundwater are of a more local character than is the case with surface water. This is reflected in the characteristic differences in the physical and biochemical parameters of the solution of surface and groundwater and species in the oxic and anoxic aqueous environments shown in Table 4.6.

**Table 4.6** Key differences in the basic quality of surface and groundwater.

| The Parameter | Surface Water | Ground Water | |
|---|---|---|---|
| | | Oxic | Anoxic |
| Total organic matter (COD, TOC, BOD and consumption $KMnO_4$)[1] | Occurrence with frequent anthropogenic contribution | Usually lower concentrations; occurs as a natural concentration (background) – basic quality; due to human activity, the activity of self-purifying mechanisms aims to reach a basic concentration. The degradation of organic pollutants is better in oxic rather than in anoxic conditions. | |
| Iron ($Fe^{2+}$) and manganese ($Mn^{2+}$) | Usually very low concentration except in eutrophic waters | Possible occurrence at lower concentrations <1 mg/L | Frequent occurrence of higher concentrations than in surface water or in groundwater in the oxic environment. Iron occurs as a soluble cationic form of $Fe^{2+}$ as well as $Mn^{2+}$ |
| Dissolved oxygen (DO) | Often close to saturation (8–9 mg/L) | Present at lower concentrations than the saturation concentration | Practically absent DO <0.2–0.5 mg/L |
| Nitrates ($NO_3^-$) | Usually low concentration (5–10 mg/L) except in polluted waters | Most often occurs at low concentrations; at higher concentrations occurs due to anthropogenic influence | It usually does not occur |
| $H_2S_{(g)}$ | Usually does not occur | It usually does not occur | Occasionally present |

[1]COD – chemical oxygen demand, BOD – biochemical oxygen demand, TOC – total organic carbon.

Due to disturbed water quality as a result of increased anthropogenic activities, purification processes (physical, chemical and biochemical) are used to restore the water quality to the basic natural quality. The natural basic quality depends on the mineralogical composition of the aquifer skeleton, the oxic conditions and the dynamics of groundwater movement.

The oxic conditions of alluvial aquifers are an extremely important feature of groundwater (Dimkić, 2012a, b). The first set of reasons for this relates to the self-purifying potential of aquifers, which is more intense in oxic environments. Another set of causative agents are the processes of the biochemical colmation of wells in conditions of very reduced oxic – anoxic aquifers.

The oxic conditions in groundwater greatly depend on the consumption of oxygen which cannot simply be explained by the oxidation of organic matter in water, since where the mineralogical oxygen consumption is occurring, such as the oxidation of iron $Fe^{2+}$ of mineralogical origin to a higher degree of $Fe^{3+}$, also plays an important role.

The basic indicators that stand out and that determine the level of oxic conditions in alluvial coastal groundwater are illustrated in (Figures 4.10 and 4.11) (Dimkić *et al.*, 2011a, b, 2012). They are:

- The oxygen content of river water,
- The organic matter content in a river and aquifer,
- The genesis of alluvial sediments (geochemical characteristics of rock deposits, mineralogical composition, granulometric composition, lithofacial sorting and other factors of genesis),
- The content of oxidizable minerals in an aquifer,
- The degree of aquifer renewal with oxygen,
- The hydrodynamic character of the groundwater flow from a river to wells.

The thickness of alluvial deposits from a source to the mouth of a river increases and the number and structure of layers is compounded as one goes downstream, and smaller grain fractions gradually predominate (Figure 4.10).

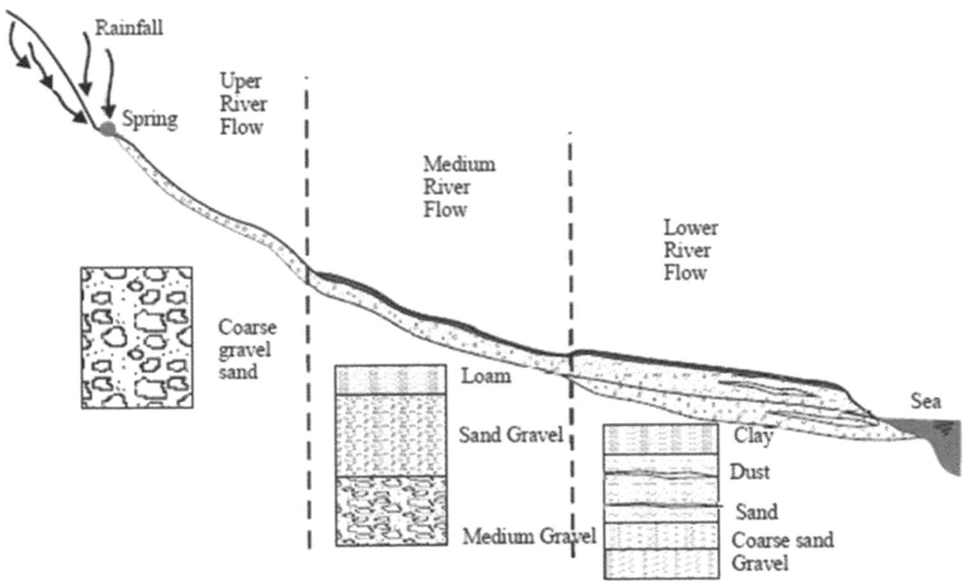


**Figure 4.10** Changes in the grain size of alluvial aquifer material along the flow of a river (Dimkić *et al.*, 2012).

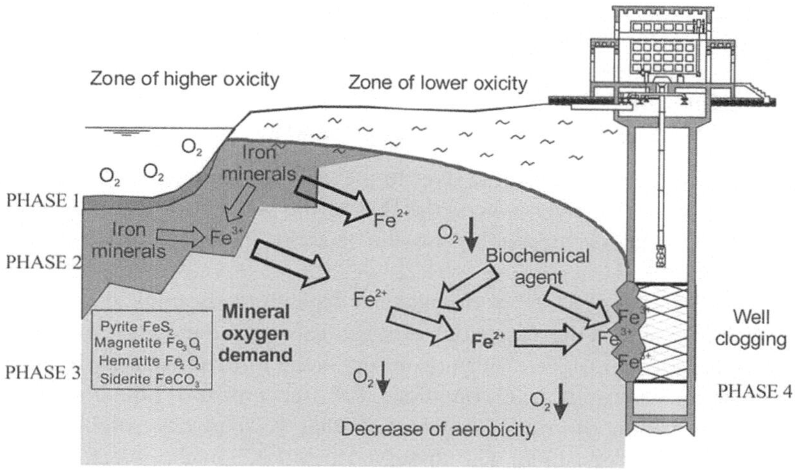


**Figure 4.11** Phases of river water filtration from the riverbed to the well with zones of oxicity (Dimkić *et al.*, 2012).

The schematic profile in Figure 4.10 shows the existing sedimentation segments, in the upper, middle and lower river flows, with different consequences in relation to oxic conditions of groundwater.

In the upper part, formed from coarse-grained material (gravel and sand), the oxic conditions are good with a considerable amount of dissolved oxygen. In the middle section of the flow, coarse sediment is gradually replaced by fine grained sediments, with a high content of clay minerals. In the lower parts of the alluvion of large rivers, fine-grained aluminosilicate sediment is prevalent, often with an increased content of iron and manganese cations.

When groundwater moves from the river to the wells, it loses some oxygen that is used for the oxidation of organic matter at the river-aquifer contact – while part is lost during transport through the aquifer environment. Oxygen consumption also occurs due to the oxidation of inorganic substances. Of particular importance is the oxidation of minerals in which iron and manganese occur in the lowest oxidation states of $Fe^{2+}_{(aq)}$ and $Mn^{2+}_{(aq)}$.

A high level of oxic conditions is present in the upper parts of river flows compared to the lower sections of the alluvions of large rivers. In the alluvions of riverbanks there are four zones/phases of filtration between the river flow and the water intake structures (Figure 4.11).

(1) Phase 1 – The first phase refers to the layer of deposit on the riverbed. Phase 1 is characterized by fine grain deposits, by highly oxygenated conditions, by the high sorption of organic and other substances dissolved in water, and by high biochemical activity.
(2) Phase 2 – The filtration of water in oxic conditions is dominant in Phase 2, and the dissolved oxygen in water is used for the oxidation of organic matter and other redox processes. The consumption of dissolved oxygen for the oxidation of iron minerals into the insoluble iron(III) form (hydroxides, carbonates, oxides, complex compounds) is very important indeed.
(3) Phase 3 – Phase 3 consists of filtration under low oxic conditions. Because of the different oxic conditions of the redox processes in water, *Eh* and $\Delta G$ are different in comparison to phase 2. In the redox reactions $Fe^{3+}$ is reduced into $Fe^{2+}$, which increases the concentration of Fe(II) cations in aqueous systems.

(4) Phase 4 – Phase 4 is characterized by filtration in the immediate vicinity of a well. The mechanical, chemical, redox and biochemical changes occur in the area around the water well filter, due to the intensification of turbulence and the mixing of different oxic layers in contact. The redox systems favour the oxidation of $Fe^{2+}$ to $Fe^{3+}$ ($Fe^{2+}_{(aq)} \leftrightharpoons Fe^{3+}_{(aq)} + e^-$), which causes the rapid aging of wells and significantly increases the cost of maintaining a system of wells.

The redox transformations that occur from the river to the wells depend on the *Eh*, $\Delta G$ and the pH in the aquifer. For highly aerated surface river waters, the flux of the water flow causes a transition from oxic to semi-oxic or anoxic conditions, depending on the reducing potential of the iron minerals in the sediment that can be oxidized.

The presence of cations $Fe^{2+}$ and $Mn^{2+}$ in groundwater depends on the mineralogical composition of the components in the sediment and on the degree of the redox reaction and the value of pH. Iron salts of $Fe^{2+}$ are more soluble than the salts of iron $Fe^{3+}$ that are present in the form of insoluble hydroxides, carbonates and other salts. With an increase in the oxic conditions, $Fe^{2+}$, according to Equation (4.27), is oxidized to $Fe^{3+}$ along with the formation of iron(III) hydroxide that is sparingly soluble, accompanied by a sedimentation process and by deposition. The dissolved iron $Fe^{2+}$ in the water binds to aquifers with plenty of fine-grained sediment in conditions of low oxygen concentrations and with less intense microbial activity.

However, in a biotic system, the oxidation of iron from $Fe^{2+}$ to $Fe^{3+}$ is much faster in conditions of catalytic enzyme reactions, where catalytic biochemical processes dominate.

Minerals with the $Fe^{2+}$ ion, distributed in a crystalline structural grid, are of key importance for geochemical and biochemical redox reactions in alluvial sediments as well as for the aging of wells (e.g. Pyrite $FeS_2$, magnetite FeO $Fe_2O_3$, hematite $Fe_2O_3$, siderite $FeCO_3$, etc.).

For the overall approach, the knowledge regarding the complex processes of oxicity and the mineralogical composition of alluvial sediments is of crucial importance, as metal cations are found in the mineral structures and/or are dissolved in water as, $Me^{n+/n+1}_{(aq)}$: Fe, Mn, Al, Cu and Zn.

## 4.9.1 Natural processes that determine the basic quality of groundwater

The basic quality of groundwater is a balanced quality that is formed as a result of the dissolving of minerals in deposited material through which groundwater flows, and as a result of redox processes that occur, and inflow quality of the water of an aquifer without human influence. The basic quality of groundwater is relatively stable for a given microlocation. However, the basic groundwater quality in different aquifers and for short distances in the same aquifer may be different.

Research results show that there are three main factors that affect the solubility of minerals in groundwater: the type and kind of structure constituents of minerals in the watertight layer, the length of the contact interface between water and minerals, and the characteristic chemical properties of wastewater.

Atmospheric processes dissolve certain minerals in the watertight layer. With a greater contact surface with the mineral, the possibility of a reaction is logically higher, resulting in an increased content of the constituent components of the minerals in the groundwater. The self-purifying capacity of the aquifer is only a property of purification by the natural filtration of the aquifer, which depends on the time and length of the contact surfaces during filtration, and the physicochemical parameters of the aquifer itself, as well as on the dynamics of flux and flow. An extremely important phenomenon is that in the case of an unbalanced water quality, the self-purifying processes have the capacity to return the contaminated water to the basic quality (Dimkić *et al.*, 2008).

**Table 4.7** Processes of the transport of pollutants in aquifers (Dimkić, 2007).

| Processes of Transport | Physicochemical and Mechanical Processes | Specific Processes |
|---|---|---|
| Transport of chemical species (solution) *via* the liquid phase | Convective solution transport<br>Solution dispersion/diffusion | Convection<br>Diffusion<br>Dispersion |
| Transfer of chemical species from the liquid to the solid phase | Transfer from the solid phase to the solution | Desorption<br>Dissolving of rocks<br>Partition of a chemical substance from a solid phase into a solution |
| | Transfer from the solution into the solid phase | Adsorption – physisorption<br>Chemisorption<br>Sedimentation |
| Loss of chemical substance from aquifers | Degradation processes | Biodegradation<br>Chemical redox processes |
| | Evaporation<br>Radioactive processes | Evaporation<br>Radioactive decay |

The processes pertaining to the formation and attainment of the basic groundwater quality are also the processes of selective transport, as well as the various transformational redox processes of the substances present in the groundwater.

### 4.9.2 Transformation processes in groundwater

Two basic cases of the transformation processes are registered in groundwater: in the first case the basic quality of the water remains the same, whereas in the second form of transformation there are rapid and significant changes in the basic quality (accidental contamination and others) (Dimkić *et al.*, 2008). Advective, convective, dispersion and diffusion transport processes that occur during self-purification processes by coastal filtration are the basic natural ways of maintaining the quality of both surface and groundwater (Dimkić, 2013). Described in Table 4.7 are the sequential phases of different dispersion and diffusion processes of pollutant transport in self-purification processes that cause changes in basic water quality.

The river coastal filtration is a very efficient natural process in groundwater purification. Various pollutants of both organic and inorganic origin are removed by self-purification during the flow of water from the river to the source.

The sorption of organic matter in an aquifer has a twofold role relative to the total biochemical oxidation of matter. Sorption can significantly (in the order of tens or hundreds) slow the transport of organic matter along the water flow, exposing the organic matter to microbial degradation for a long period of time.

## 4.10 OXIC/ANOXIC CONDITIONS IN GROUNDWATER

### 4.10.1 Oxic conditions in the biotic matrix

Redox processes depend on microbiological enzymatic activity which consumes the released energy that allows transfer of electrons from the donor (most commonly, the carbon atom of organic matter – Corg) as a reducing agent, to the acceptor atom (in most cases atoms in inorganic compounds). In redox processes, electron acceptors that release more energy than other chemical substances are the first to participate in redox reactions, while the types of electron acceptors that release less energy are involved later in the redox processes in a decreasing order with respect to the energy that is released. The redox process will proceed according to the following scheme of compounds strength and energy (Jurgens *et al.*, 2010).

$$O_{2(g)} \rightarrow NO^{-}_{3(aq)} \rightarrow Mn^{4+}_{(aq)} \rightarrow Fe^{3+}_{(aq)} \rightarrow SO^{2-}_{4(aq)} \rightarrow CO_{2(q)} \quad (4.42)$$

The highest energy content is released and obtained by reducing oxygen, while the nitrate reduction results in slightly lower energy. In the case where iron bacteria use $Fe^{3+}$ energy from $Fe(OH)_3$ to grow the iron mass, it is necessary for microorganisms to reduce high amounts of Fe(III) cations (because the reduction reaction results in energy yields below 20%), while during sedimentation high quantities of harmful undissolved iron(III) hydroxide occur as a side product. Sulphate reduction has similar energy contents.

When evaluating redox processes in groundwater that depend on the type and kind of environment, four types can be differentiated: oxic, sub-oxic, mixed and anoxic (Table 4.8).

The identification of the dominant redox processes in groundwater as well as the redox categories of the appropriate chemical or the cationic and anionic species is described in the following table methodology recommended by the US Geological Survey (USGS) (Jurgens *et al.*, 2010).

The electron-ion half-reactions of the redox are balanced by scaling and electrodeionizing/electronizing the two half-reactions with the participation and exchange of the same number of electrons between the reducing and oxidizing agents.

Zobrist *et al.* (2000) showed the scheme of redox process (Figure 4.12) as a function of the value of the redox potential, which corresponded very well with Jurgens' classification (Jurgens *et al.*, 2010). In the classification Zobrist focuses on the significance of redox indicators in water without specifically analysing the sub-oxic environment.

Molecules are oxidized and reduced at different redox potentials. By observing the functional relationship of the redox potential and of the redox reaction, it can be concluded that, with a decrease of

**Table 4.8** Basic redox processes in groundwater as well as types of chemical substances, cations and anions according to the USGS methodology shown as electron-ion half-reactions (Jurgens *et al.*, 2010).

| Type and Type of Medium | Substances that Participate in Redox Processes | Redox Reactions |
|---|---|---|
| Oxic | $O_2$ | $O_2 + 4H^+ + 4e^- \rightarrow 2H_2O$ |
| Sub-oxic | $O_2$ – Mn(II)<br>$O_2$ – Fe(III) | Low $O_2$; additional data are needed to define the redox process |
| Anoxic | $NO_3^-$ | $2NO_3^- + 12H^+ + 10e^- \rightarrow N_{2(g)} + 6H_2O$;<br>$NO_3^- + 10H^+ + 8e^- \rightarrow NH_4^+ + 3H_2O$ |
| | Mn(IV) | $MnO_{2(s)} + 4H^+ + 2e^- \rightarrow Mn^{2+} + 2H_2O$ |
| | Fe(III)/$SO_4^{2-}$ | Fe(III) and (or) $SO_4^{2-}$ |
| | Fe(III) | $Fe(OH)_{3(s)} + H^+ + e^- \rightarrow Fe^{2+} + H_2O$;<br>$FeOOH_{(s)} + 3H^+ + e^- \rightarrow Fe^{2+} + 2H_2O$ |
| Mixed (anoxic) | Fe(III) – $SO_4^{2-}$ | Fe(III) and $SO_4^{2-}$ |
| Anoxic | $SO_4^{2-}$ | $SO_4^{2-} + 9H^+ + 8e^- \rightarrow HS^- + 4H_2O$ |
| | $CH_4$ genesis – methanogenesis | $CO_{2(g)} + 8H^+ + 8e^- \rightarrow CH_{4(g)} + 2H_2O$ |

All ionic species, cationic and anionic, in an aqueous solution are dissolved, $K^{n+}_{(aq)}$ and $A^{n-}_{(aq)}$ – aquatized ions.
*s* is the solid state of a substance, almost insoluble.

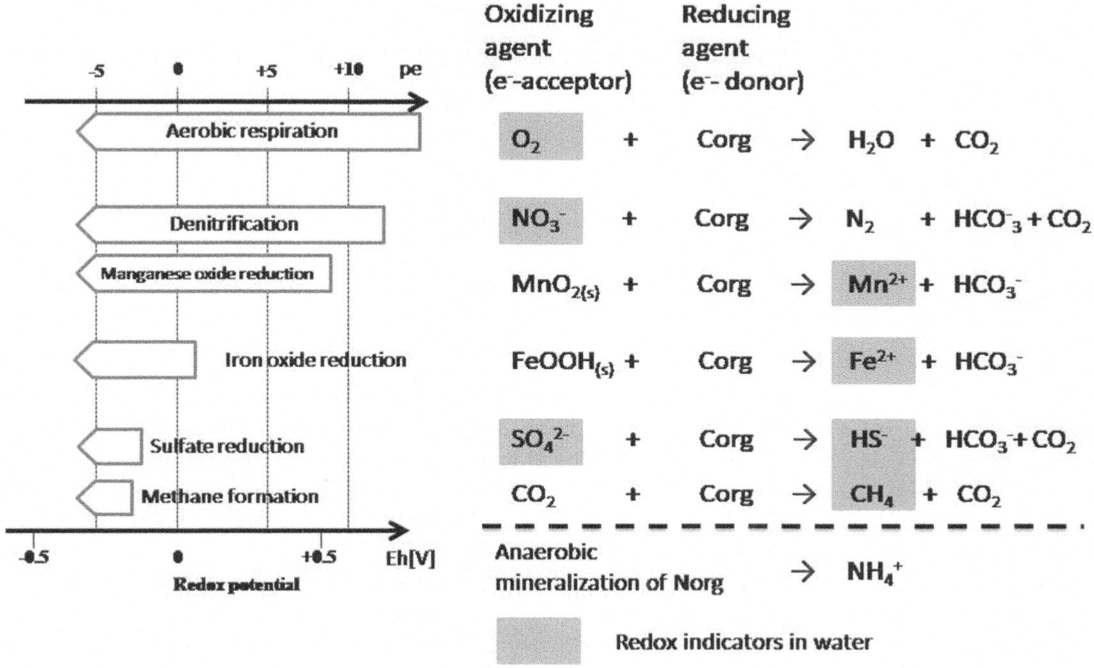


**Figure 4.12** Sequential key stages of the redox process in aquatic systems under the influence of certain microbial species (Zobrist *et al.*, 2000).

the redox potential, the following reductive reactions will occur under anoxic conditions: denitrification-reduction of nitrate anions to elemental nitrogen, the reduction of sparingly soluble cations of Mn(IV) and Fe(III) into soluble Mn(II) and Fe(II) cations, the reduction of the sulphate anion to the sulphide anion, with the formation of methane from carbon dioxide.

### 4.10.2 Oxidation states of certain significant bioelements in groundwater

Groundwater is a highly complex heterogeneous electrolyte system with a diverse geological environment and origin, and it contains numerous inorganic and organic chemical specimens in liquid or solid aggregation.

The solubility of metals in natural waters depends on the different oxidation numbers in different mineral components (Table 4.9). The dynamic interactions of certain contact phase surface processes of solvent-solution-solid matter are determined by the transfer and transport of metal cations in the liquid and solid phases (Obrovski *et al.*, 2016; Pap *et al.*, 2016).

The concentration of iron and manganese cations as well as of other heavy metal cations decreases during natural groundwater filtration processes, predominantly as a result of sorption processes. Under oxic conditions in groundwater, the decrease in the cationic concentration is the result of ionic exchange on the negatively charged surfaces of clay materials of amorphous $Fe_2O_3$ and $Al_2O_3$ oxides as well as of the presence of solid organic matter. In anoxic conditions, the removal of metal ion cations is predominantly followed by the formation of corresponding sparingly soluble salts of metal sulphides, carbonates and hydroxides.

**Table 4.9** Oxidation numbers of certain significant bioelements in groundwater.

| Chemical Formula | Oxidation Number | Chemical Formula | Oxidation Number |
|---|---|---|---|
| $MnO_{2(s)}$ | +4 | $Fe_2O_{3(s)}$ | +3 |
| $Mn^{2+}_{(aq)}$ | +2 | $NH_{3(g)}$ | −3 |
| $FeOOH_{(s)}$ | +3 | $NH^+_{4(aq)}$ | −3 |
| $Fe(OH)_{3(s)}$ | +3 | $N_2$ | 0 |
| $FePO_{4(aq)}$ | +3 | $NO^-_{2(aq)}$ | +3 |
| $FeS_{2(s)}$ | +2 | $NO^-_{3(aq)}$ | +5 |
| $FeCO_{3(s)}$ | +2 | $SO^{2-}_{4(aq)}$ | +6 |

The $Fe^{2+}$ cation is more easily soluble in the reduced form than in the oxidized form of $Fe^{3+}$ when it most often occurs as an oxide, $Fe_2O_3$, as a hematite mineral or as a precipitate of iron(III) hydroxide, $Fe(OH)_3$. In sub-oxic environmental conditions in which the iron is present in the $Fe^{2+}$ cation form, groundwater also contains high concentrations of iron $Fe^{3+}$. The most common case of reduction reactions is due to the presence of organic matter in a solid state, in the form of dissolved organic carbon. The reactions of manganese cations ($Mn^{2+}$) are very similar to those of $Fe^{2+}$.

In groundwater sulphur most commonly appears as a sulphate ($SO_4^{2-}$). If sulphur appears in the reduced form of gaseous hydrogen sulphide ($H_2S_{(g)}$) (or the dissolved sulphide anion $S^{2-}_{(aq)}$, $HS^-_{(aq)}$) the unpleasant smell of rotten eggs is strongly present even at concentrations below 10 mg/L. Redox reactions in anoxic conditions cause the reduction of sulphate to hydrogen sulphide in the presence of organic matter.

### 4.10.3 Biochemical processes of nitrification and denitrification

Nitrogen is a biogenic element with nine different oxidation states, from a maximum of +5 to a minimum −3 in many chemical compounds, nitrates ($NO_3^-$), nitrite ($NO_2^-$), the ammonium ion ($NH_4^+$), in the zero state of gaseous nitrogen ($N_2$) and amino compounds of organic nitrogen. As a rule, all nitrate and ammonia salts dissolve well in water. When the nitrate anion is present in groundwater, $NO_3^-$ moves with the groundwater flow without further attenuation and dispersion.

The ever present natural nitrogen cycle in which nitrogen is transformed, oxidized or reduced from one to another different compound with different oxidation numbers of $N_2$ is extremely important. The nitrogen cycle is catalysed by microbial enzymatic systems, either as a process that generates energy needed for metabolism or as an assimilation process of elemental nitrogen by which nitrate and ammonium ions are incorporated into biota (Figure 4.13). The opposite process of assimilation is the degradation of organic matter by heterotrophic bacteria which releases $NH_4^+$ in the process of mineralization (ammonification).

Denitrification is the process of reducing nitrate and nitrite anions as terminal electron acceptors, wherein the following gases are formed as reaction products: nitric oxide (NO), nitrous oxide ($N_2O$) and at the end the diatomic molecular gas ($N_2$). Denitrification begins in the presence of an organic substance having an oxygen content of at least 0.5 mg/L as a result of the biological activities of nitrate reducing and denitrifying microorganisms. In accordance with the electrode potential and the stability of the nitrogen compounds of the oxidation state +3, the nitrite anion $NO_2^-$ is the first product of a series of reduction processes, where denitrifying bacteria use $NO_3^-$ as a good electron acceptor reducing $NO^-_{2(aq)}$, $N_2O_{(g)}$ to elemental nitrogen, $N_{2(g)}$. (by reducing the oxidation state of nitrogen by +5 through sequential intermediate reactions of nitrogen +3 in $NO^-_{2(aq)}$, nitrogen +2 in $NO_{(g)}$ to +1 in $N_2O_{(g)}$ and finally to the $N_{2(g)}$). Elemental nitrogen as a gas is located in the pores of the soil. Since the denitrification bacteria

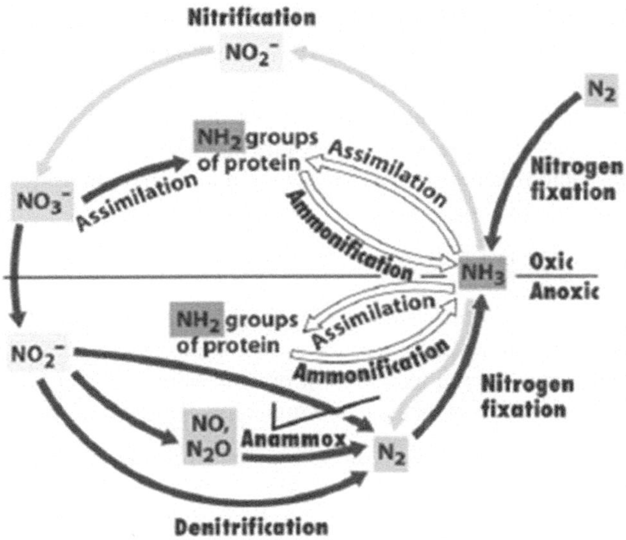


**Figure 4.13** Biochemical processes of nitrification and denitrification (Madigan *et al.*, 2005).

need organic carbon as an energy source for growth and metabolism, the process of denitrification primarily takes place in the soil where organic material is easily available. In accordance with the basic principles of redox reactions, under oxygenated conditions the ammonia anion, $NH_4^+$ is oxidized via elemental nitrogen to nitrite and nitrate by nitrifying bacteria.

Simplified nitrification and denitrification processes, without specifying the electronic component $e^-$, with the nitrification biochemical reactions equalized only by charge, while the fixation equation is scaled by the number of nitrogen and hydrogen atoms, are shown in Table 4.10 and in Figure 4.13 with examples of biological bacterial species.

**Table 4.10** Key processes and prokaryotes participating in the nitrogen cycle.

| Processes | Examples of Bacteria |
|---|---|
| Nitrification ($NH_4^+ \leftrightharpoons NO_3^-$) | *Nitrosomonas* |
| $NH_4^+ \leftrightharpoons NO_2^-$ | *Nitrobacter* |
| $NO_2^- \leftrightharpoons NO_3^-$ | |
| Denitrification ($NO_3^- \leftrightharpoons N_2$) | *Bacillus, Paracoccus, Pseudomonas* |
| $N_2$ fixation ($N_2 + 8H^+ \leftrightharpoons 2NH_3 + H_2$) | |
| Oxic | *Azotobacter, Cyanobacteria* |
| Anoxic | *Clostridium, blue and green bacteria* |
| Symbiotic | *Rhizobium, Bradyrhizobium, Frankia* |
| Amoninitrification (organic N $\leftrightharpoons NH_4^+$) | *Many microorganisms* |
| Anammox ($NO_2^- + NH_3 \leftrightharpoons 2N_2$) | *Brocadia* |

### 4.10.4 Biochemical redox processes of iron and manganese

Different species of reducing microorganisms reduce the iron cation $Fe^{3+}$ to the $Fe^{2+}$ oxidation state. As a typical simultaneous reduction $Fe^{3+}$ is, via the redox system, coupled with a wide-range and a variety of organic and inorganic reduction agents, electron donors, because the standard electrode potential of the conjugated redox pair $Fe^{3+}/Fe^{2+}$ has a high positive value (+770 mV).

The enzymatic processes of the microbiological reduction of iron(III) and manganese(IV) in groundwater that are supported by microbiological reducers are:

- Enzymatic Fe(III) and Mn(IV) reducers – *bacillius, Ferribacterium, Clostridium, Desulfovibrio, Escherichia, Geobacter*
- Sulphur oxidizing Fe(III) and Mn(IV) reducers – *Thiobacillus spp., Sulfolobus spp.*
- Hydrogen oxidizing Fe(III) and Mn(IV) reducers – *Pseudomonas spp., Shewanella spp.*
- Fe(III) and Mn(IV) reducers *via* the oxidation of organic acids
- Fe(III) reducer *via* the oxidation of aromatic compounds – *Geobacter spp.*

The standard electrode potential of $Mn^{4+}/Mn^{2+}$ of the conjugated pair is very high (+1230 mV) and many compounds easily donate electrons in the reduction of $Mn^{4+}$.

A small amount of energy is required to oxidize $Fe^{2+}$ to $Fe^{3+}$, therefore ferrous bacteria need to oxidize a large amount of iron from the minerals to develop and grow. The end product of the oxidation-reduction reaction is the formation of the insoluble precipitate of Fe(III) oxide. The largest numbers of iron oxidizing bacteria are acidophiles, so that at a neutral pH, the cation of iron $Fe^{2+}$ quickly and strongly oxidizes to the $Fe^{3+}$ state, forming a foamy voluminous brown mass precipitate which is very stable under anoxic conditions. *Thiobacillus ferrooxidans* is one of the most famous iron bacteria that can grow autotrophically using either $Fe^{2+}$ iron or reduced anion species of sulphur ($S^{2-}$) as a donor electron.

The net summary redox reaction of iron $Fe^{2+}$ is described by the following aggregated ion-molecular reaction, without showing the exchanged electrons by applying the least stoichiometric numbers:

$$Fe^{2+}_{(aq)} + 1/4O_{2(g)} + H^{+}_{(aq)} \rightarrow Fe^{3+}_{(aq)} + 1/2H_2O_{(aq)} \quad (4.43)$$

$$Fe^{3+}_{(aq)} + 3H_2O_{(aq)} \rightarrow Fe(OH)_{2(aq)} + 3H^{+}_{(aq)} \quad (4.44)$$

Other iron bacteria that develop in near-neutral conditions are always present with the occurrence of the $Fe^{2+}$ cation, and change state from anoxic to oxic in a dynamic equilibrium of the redox system. This type of bacteria includes the following microbial specimens: *Gallionella ferruginea, Sphaerotilus natans*, and *Leptothrix ochracea* (Figure 4.14).

*Leptothrix ochracea* and other bacteria can oxidize $Mn^{2+}$. The biological oxidation of $Mn^{2+}$ cations is possible at pH values below pH = 8, but it does not occur easily and spontaneously due to the electronic configuration of $Mn^{2+}$. There are no registered data that *Leptothrix ochracea* releases energy from the oxidation process, and so the benefit for microorganisms is not known, probably because the cation of manganese is an important coenzyme in biochemical processes.

### 4.10.5 Biochemical redox processes of sulphur compounds

The sulphate-reducing bacteria (SRB) belong to a highly specific physiological group of mostly obligate anaerobes, morphologically very diverse in shape, which are able to perform a dissimilatory reduction of sulphate to hydrogen sulphide $HS^{-}$. SRB have a wide range of metabolic possibilities, which explains their wide distribution and important role in processes of anaerobic decomposition of organic matter in water.

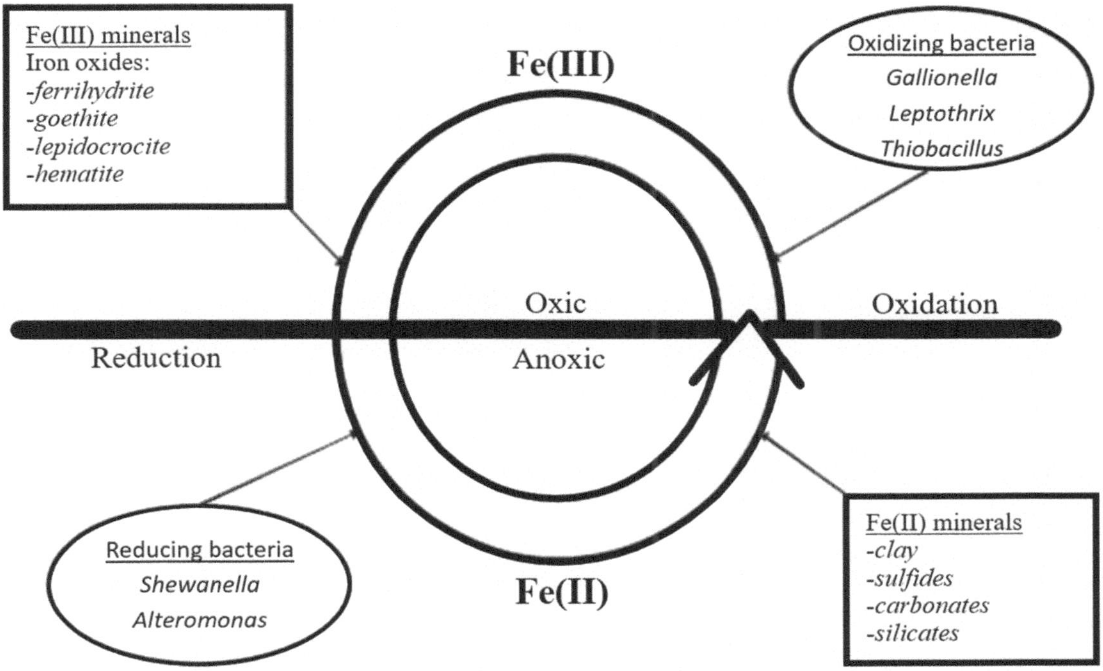


**Figure 4.14** The iron cycle (Houben, 2006).

Although SRB are strictly anaerobes, some of them remain stable in the presence of oxygen in the environment. About 50 types of SRB are known and they are substantially different according to their morphological characteristics and metabolic reactions. SRB create a biofilm within the reduced environment and produce $H_2S_{(g)}$ (the characteristic smell of water like rotten eggs) and cause harmful corrosion which is often named 'under-deposit corrosion'. Subsurface hydrogen sulphide, sulphur, sulphides, and other sulphur compounds such as thiosulphate, terrathionate, and sulphites can be oxidized by colourless bacteria of the genera *Thiobacillus, Beggiatoa, Thiotrix*, and *Thoploca*, which as autotrophic organisms obtain their energy for carbon assimilation by chemosynthesis. $H_2S$ is first converted to free sulphur (S°), which can be temporarily stored inside or outside the cell before it oxidizes to sulphuric acid. Many types oxidize only the hydrogen sulphide, to elemental sulphur, which is then oxidized by other species (Table 4.11).

## 4.11 CASE STUDIES OF ANOXIC ALLUVIAL AQUIFERS (BELGRADE GROUNDWATER SOURCE – SAVA, KOVIN-DUBOVAC – DANUBE) AND OXIC AQUIFERS (KLJUč-POžAREVAC SOURCE – VELIKA MORAVA)

The total water chemistry, solubility and transport, distribution and partitioning of chemical species of organic and inorganic origin, the kinetics of transformation processes and degradation all depend on the oxic conditions in an aquifer, as well as on the redox potential and thermodynamic values. In research studies, the relationship between the degree of oxic conditions and the oxidation state of iron, manganese, sulphur and the nitrogen in groundwater is particularly important and characteristic

**Table 4.11** Key processes and microorganisms that participate in the sulphur cycle.

| Processes | Examples of Organisms |
|---|---|
| Sulphide/sulphur oxidation ($H_2S_{(g)} \leftrightharpoons S^0 \leftrightharpoons SO^{2-}_{4(aq)}$) | |
| Oxic | Sulphur Chemolithotrophs (*Thiobacillus, Beggiatoa*, many microorganisms) |
| Anoxic | Blue and green phototrophic bacteria, some Chemolithotrophs |
| Sulphate reduction (anoxic) ($SO^{2-}_{4(aq)} \leftrightharpoons H_2S_{(g)}$) | *Desulfovibrio, Desulfobacter* |
| Sulphur reduction (anoxic) ($S^0 \leftrightharpoons H_2S_{(g)}$) | *Desulfuromonas*, many hyperthermophilic *Archaea* |
| Sulphur disproportionation ($S_2O_3^{2-} \leftrightharpoons H_2S_{(g)} + SO^{2-}_{4(aq)}$) | *Desulfovibrio* and others |
| Oxidation and reduction of organic sulphur compounds ($CH_3SH \leftrightharpoons CO_2 + H_2S$) (DMSO $\leftrightharpoons$ DMS) dimethylsulfoxide $\leftrightharpoons$ dimethylsulfide | Many microorganisms |
| Desulphurization (organic S $\leftrightharpoons H_2S$) | Numerous microorganisms |

(Figure 4.15). Environmentally and economically strategic processes are highly dependent on the *oxic conditions* in the aquifers of river alluvions in Serbia.

### 4.11.1 Belgrade Groundwater Source (alluvion of the Sava River)

Significant national projects and extensive studies in Serbia, include:

- Defining the impact of the degree of oxic conditions on the protection and use of aquifers in intergranular porous environments in Serbia, 2008–2011,
- Methodology for the assessment, design and maintenance of groundwater sources in alluvial environments depending on the degree of oxicity, period 2011–2014,
- Study of the Belgrade Water Supply Source (Jaroslav Černi Institute for Water Management, Belgrade 2005–2010),
- Study of the Aging of Wells and the Maintenance of Facilities (Jaroslav Černi Institute of Water Management, Belgrade 2010) under the guidance of the first author of this chapter. This resulted in valuable and completely new insights into the phenomenon of aging of wells and its functional dependence on oxic conditions in the alluvial aquifers.

On the basis of numerous results collected within the framework of scientific research and studies, it is possible to, in a complex and integral manner, view the various parameters directly related to the oxic conditions in the Belgrade source aquifer formed in the alluvion of the lower course of the Sava River.

The results are based on the monitoring and interpretation of the correlation between the most important parameters such as the well capacity, the hydraulic character of the groundwater flow (well inflow velocity), the granulometric composition, the dissolved oxygen $O_2$ values, the pH of water, the redox potential (*Eh*),

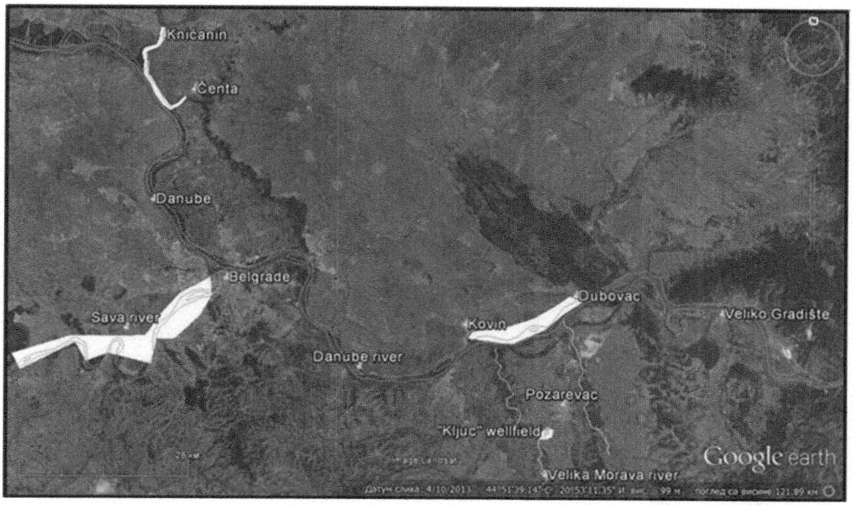


**Figure 4.15** Locations of the Belgrade source, the Kovin-Dubovac source and the Ključ source on the Sava, the Danube and the Sava (www.Google Earth).

the iron concentration ($Fe^{2+}$) and other ions and substances in water, and the biochemical and other parameters.

By considering all these processes, regional and local results on the degree of oxicity in the alluvial aquifers of the Belgrade source were analysed in detail.

The majority of Belgrade's groundwater source consists of wells with horizontal drains which are placed along the banks of the river Sava, from the mouth of the river Sava into the Danube, and upstream for a distance of about 50 km. In total there are 99 wells with horizontal drains and 49 tubular wells, 20 of which are located in Makiš Polje while the others are on the left bank of the Sava River (Figure 4.16).

The wells extract water from alluvial Quaternary (Pleistocene and Holocene) sediments, which are also the main groundwater reservoirs in the area. Below them are waterproof older sandy marl clay and marls. The water in the wells is mostly formed from the inflow from the river (about 80%), and to a lesser extent it originates from the hinterland.

The alluvial aquifer was formed over several cycles of sedimentation. The cycles, sequences of sedimentation, are: sandy gravel, sands of different granulation, silt and clayey sediments. The thickness of these Quaternary sediments is up to 25 m. Depending on the size of the sediments, two zones can generally be distinguished in the profile (Figure 4.17):

- The lower zone, sediments of courser granulometric composition, within which are the drains of the radial wells of the source. In these sediments there are occasional interlayers and lenses of clay, sandy clays and silt.
- The upper zone, finer-grained sediments, poorer filtration properties.

The distribution of the granulometric composition along the depth of the vertical profile of the sediments shows several cycles of the process of deposition (three cycles, some places four). The sizes of the grain

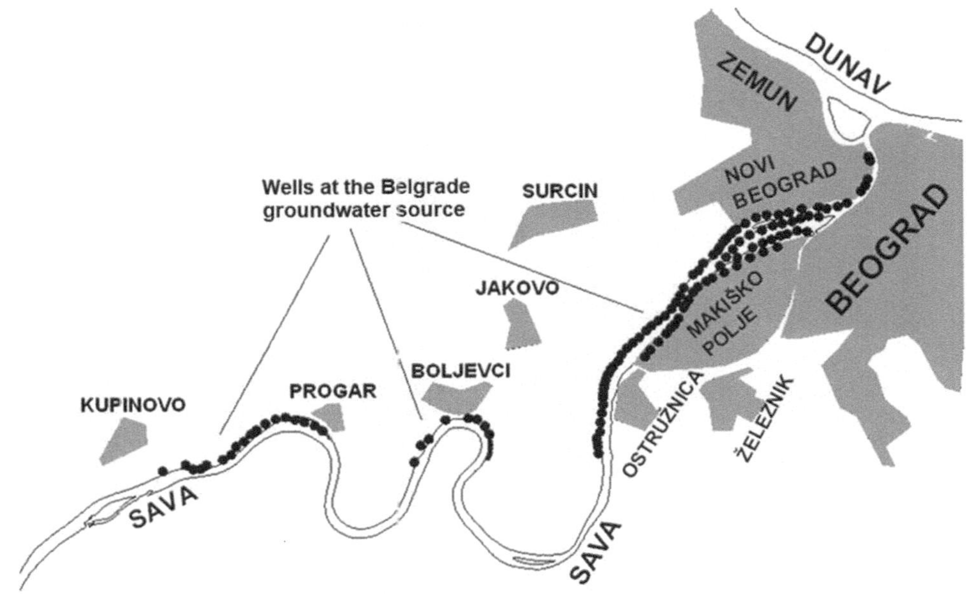


**Figure 4.16** Arrangement of wells at the Belgrade Groundwater Source (Jaroslav Černi Institute for Water Management).

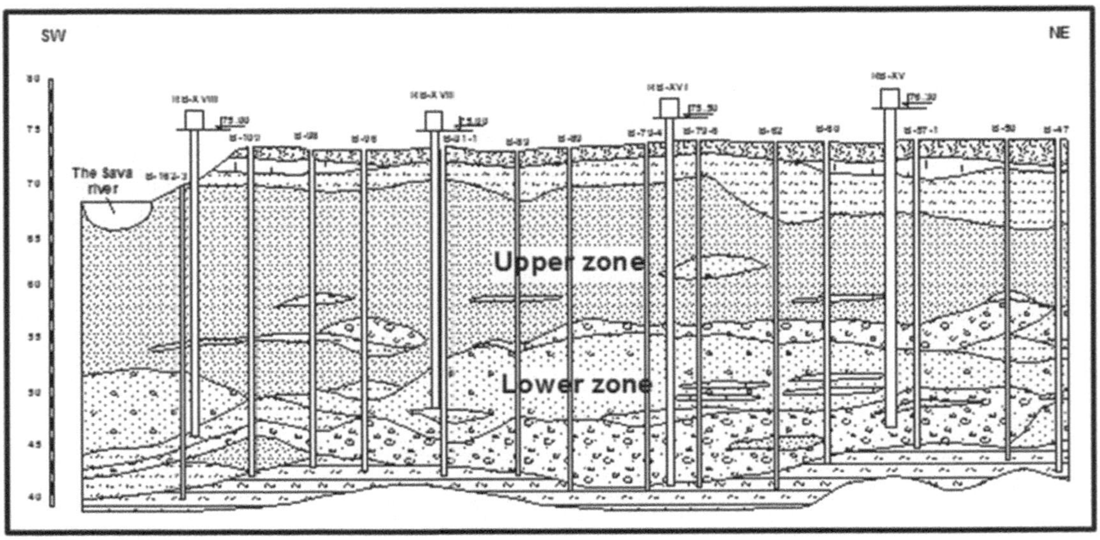


**Figure 4.17** The Belgrade water supply source: A typical lithological profile along the banks of the Sava River (Jaroslav Černi Institute for Water Management).

particles as well as the value of the filtration coefficient increase with depth, so that the lowest sediments (sandy gravel and gravels) have the highest water permeability.

The water-bearing layers, from which wells exploit groundwater for the city, consist predominantly of gravel, sandy gravel, gravel sands, sands, and siltstone clay. The layers alternate, from the underlying older sediments to the surface of the terrain.

The filtration characteristics of this complex can be monitored by analysing the granulometric composition of the sediments. The grain sizes in the lower zone in which the wells' filters are located range from medium-grained sand to fine-grained gravel (Figure 4.18).

It can be seen that these are gravelly sandy sediments, although in the upstream sector of the source the gravel component is missing. Based on the analysed granulometric composition, it can be estimated that the filtration coefficients range between $10^{-5}$ and $10^{-3}$ m/s, although analyses on regional hydrodynamic models showed that the filtration coefficient is of the order of $10^{-3}$ m/s.

The initial results of the operation of the wells were very favourable: the capacity of most wells exceeded 150, and at the smaller ones it was over 200 L/s. However, declining yields on almost all wells were soon observed. The decline in yields was not given due attention for a long time. For decades, the overall decline of the capacity of the wells and the city's increased need for water were compensated for by the accelerated construction of more and more wells. By 1990, virtually all suitable riverfront sites were occupied by wells that were built.

Meanwhile, the average well yield, from about 170 L/s in the 1960s dropped to about 45 L/s (2000–2007) (Figure 4.19). The periodic regeneration of wells usually provides only a temporary increase in their yields.

In the first period of the operation of the spring (1953–1964), the wells used the dynamic water reserves. The groundwater levels were high and the piezometer level sufficient to ensure that hydraulic losses at the drainage and at the river-aquifer contact did not jeopardize the wells too much. This period is characterized by rather high groundwater levels, but with an initial tendency to decline (Figure 4.20).

In the second period (1965–1986), the colmation of the riverbed, caused by the continuous operation of the wells near capacity, as well as by the beginning of a significant aging of the wells, resulted in a decrease in the abundance of wells. During this period, due to the insufficient inflow of water from the river, the wells used static reserves. The groundwater levels at the source declined significantly in the mid-1980s.

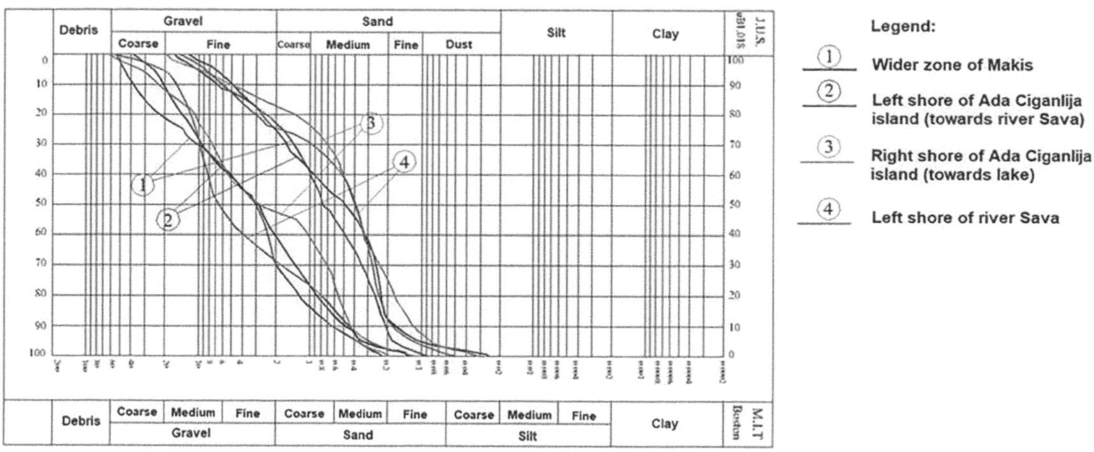


**Figure 4.18** The envelopes of the curves of the granulometric composition of the Belgrade water supply source: the curves refer to the medium in which the drains were placed (Dimkić, 2010).

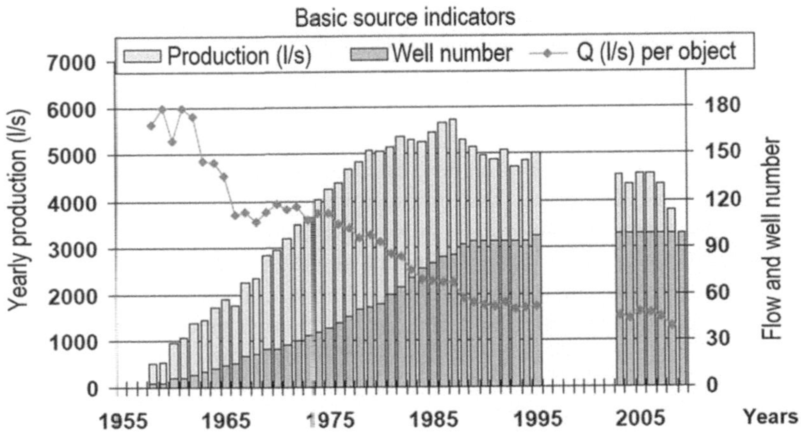


**Figure 4.19** Basic indicators of groundwater exploitation at the Belgrade water supply source (Dimkić, 2010).

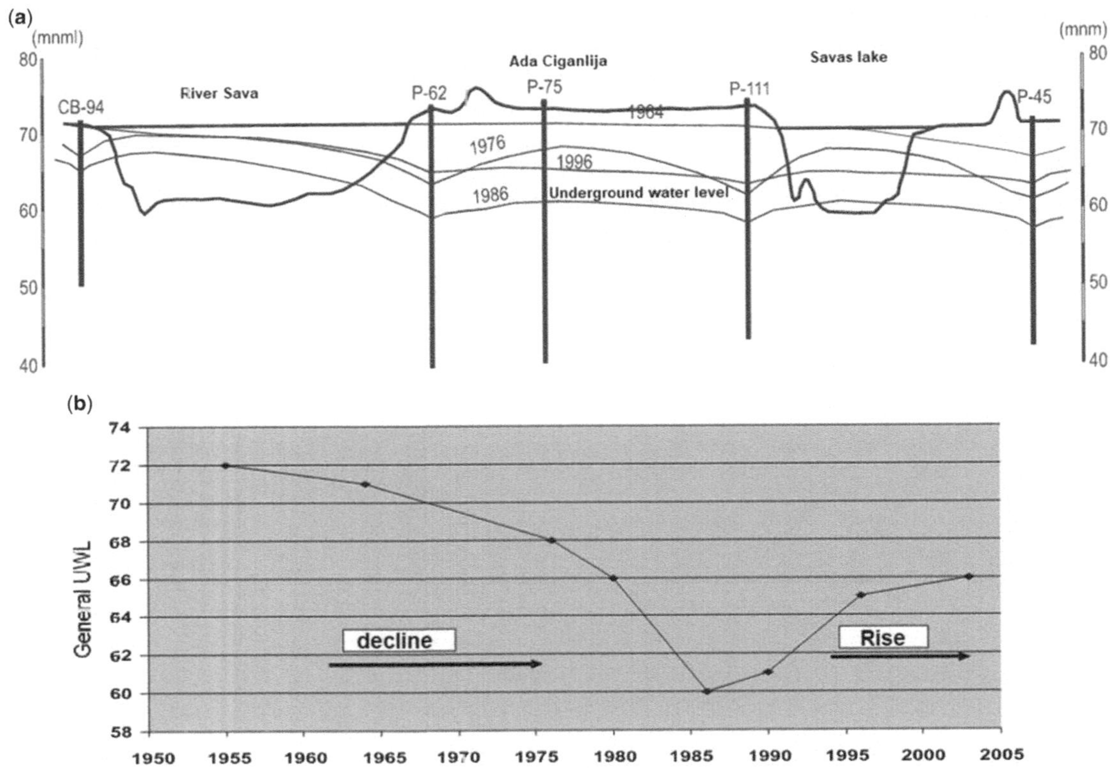


**Figure 4.20** (a) Groundwater levels at several characteristic time sections (1964, 1976, 1986 and 1996) at the Belgrade water supply source, cross-section across the river (site of Ada Ciganlija); (b) General change of the groundwater level at the Belgrade water supply source (Dimkić & Pušić, 2007).

### 4.11.2 Relationship between groundwater exploitation, oxic conditions and some physicochemical parameters

It is remarkable to see that some wells in the Belgrade source have retained a greater abundance than others even after decades of work. The reasons for this are multiple, including the combination of local hydrogeological, hydro-chemical and microbiological conditions of an aquifer observed in light of the different degrees of oxic conditions in individual parts of the aquifer. Among others, parameters considered included:

- The granulometric composition of aquifers – the natural filtration characteristic of an aquifer,
- The electrochemical characteristics of the water (electric conductivity and redox potential) – the natural characteristics of the overall quality of the groundwater,
- Chemical parameters (total iron content, mineralization, carbonate content) – natural characteristics
- Microbiological activity in the water, analysed by reaction tests of biological activity – BART index.

The characteristic values of chemical and other parameters that pertain to the whole source for the 2005–2013 period, and that were derived from the results from all of the wells are shown in Table 4.12.

Figure 4.21 shows the results of interpretation of some groundwater parameters, the average inlet velocities into wells and the size $d_{50}$ (50%) of the granulometric composition with respect to the selected redox potential intervals. In addition to the *Eh* interval, the horizontal axis also shows the number of samples used at a given interval. The basic characteristics of the Belgrade spring are relatively low *Eh* values, generally below 150 mV.

Based on the obtained results it can be concluded that the Belgrade spring is characterized by a relatively low oxygen concentration level and a low redox potential and is characterized by a varying total concentration of $Fe^{2+}$ cations. The groundwater is at a border between oxic and anoxic conditions, it is a relatively inert medium for corrosive activity and the mechanical clogging of the well filter.

### 4.11.3 The drainage system and the potential Kovin-Dubovac source (the Danube alluvion)

The Kovin-Dubovac area occupies the southern part of the Banat depression and includes the alluvial plain of the Danube between these two settlements (Figure 4.22). The construction of the Iron Gate 1 hydropower

**Table 4.12** Groundwater source in Belgrade: values of some of the analysed parameters (Dimkić, 2010, updated and edited).

| Parameter | Minimum | Average Value | Maximum | Number of Measurements |
|---|---|---|---|---|
| *Eh* (mV) | 48 | 107 | 261 | 350 |
| $O_2$ (mg/L) | 0.00 | 0.17 | 1.8 | 358 |
| $NO_3^-$ (N) (mg/L) | 0.03 | 0.28 | 3.60 | 385 |
| $NH_4^+$ (mg/L) | 0.00 | 0.55 | 3.10 | 389 |
| $Fe^{2+}$ (mg/L) | 0.11 | 1.51 | 5.69 | 407 |
| $Mn^{2+}$ (mg/L) | 0.02 | 0.33 | 0.82 | 370 |
| $SO_4^{2-}$ (mg/L) | 8.30 | 32.69 | 94 | 399 |
| Electrical conductivity (25°C) (μS/cm) | 344 | 519 | 867 | 390 |
| pH | 7.1 | 7.3 | 7.6 | 600 |
| $d_{50}$ (mm) | 0.37 | 1.75 | 10.3 | 2500 |

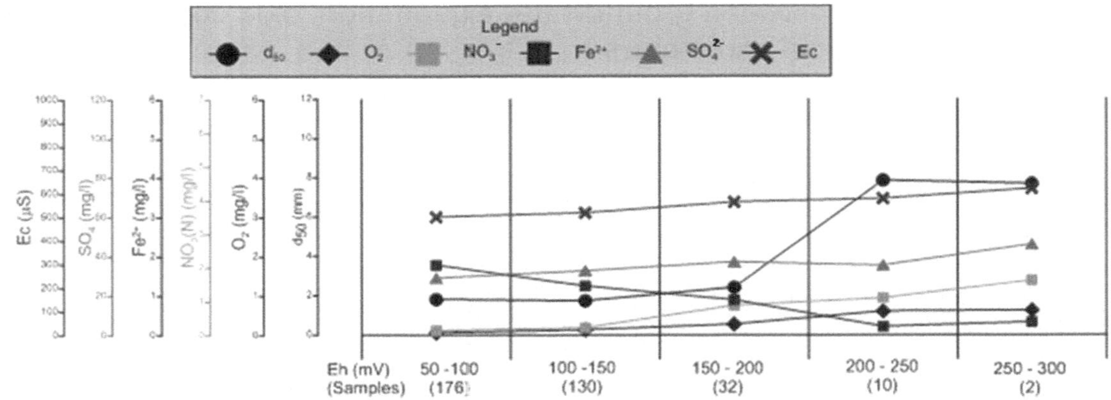


**Figure 4.21** Average values of selected Belgrade source parameters, by segments of redox potential change (Legend: Eh – redox potential (mV), $d_{50}$ – aquifer grain size, taken as 50% of the granulometric composition curve (mm), $O_2$ – oxygen concentration (mg/L), $NO_3^-$ – nitrate concentration (mg/L), $Fe^{2+}$ – iron(II) cation concentration (mg/L), $SO_4^{2-}$ – sulphate anion concentration (mg/L), Ec – electroconductivity of water (µS/cm)) (Jaroslav Černi Institute for Water Management).

plant (HPP) downstream of this area and the formation of a reservoir in 1972 permanently increased groundwater levels in the littoral area. For this reason, a defence system was constructed consisting of embankments, drainage canals and a line of drainage wells. The role of this system is to maintain

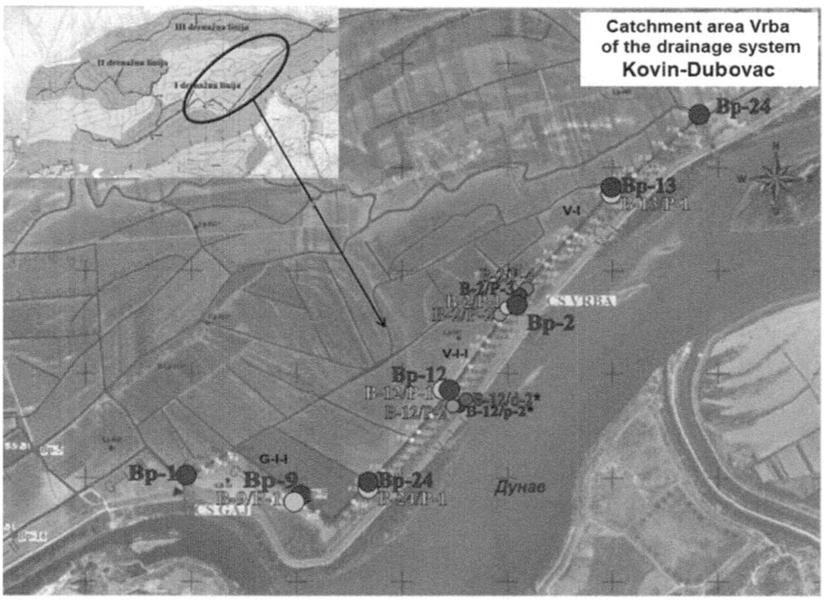


**Figure 4.22** Series of drainage wells along the left bank of the Danube between Kovin and Dubovac (Jaroslav Černi Institute for Water Management).

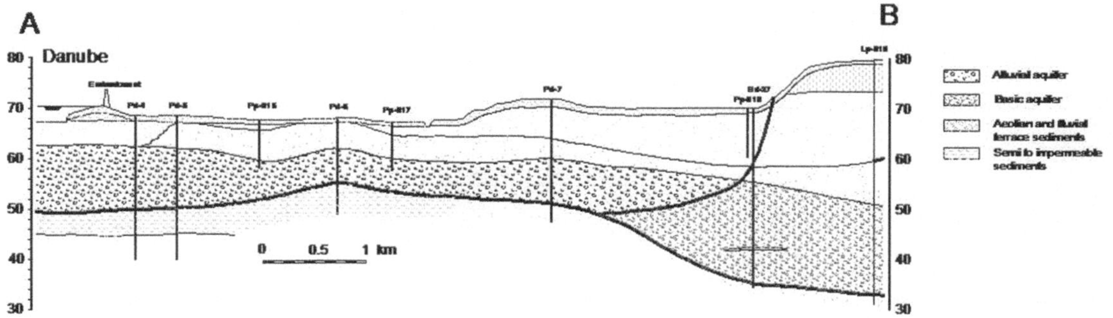


**Figure 4.23** Typical profiles of alluvial sediments between Kovin and Dubovac (Jaroslav Černi Institute for Water Management).

groundwater levels at specified depths and prevent the Iron Gate HPP reservoir impacting deeper into the area.

In the lower part of the aquifer, there are younger Quaternary – upper Pleistocene and Holocene – gravels and sandy gravels, with extremely uneven granulations. Above them are medium-grained gravel, sandy gravel and gravelly sands. The sediment series ends with semipermeable siltstone sands, siltstones, and bar clays. The average thickness of the aquifer sediments is about 20 m (Figure 4.23).

The size of $d_{50}$ grains of aquifer gravel sediments is generally in the range of about 0.5 to 20 mm (Figure 4.24). The sandy sediments above them have $d_{50}$ grains between 0.08 and 0.3 mm in size, while in the overlying layer the poorly permeable sediments $d_{50}$ values are between 0.007 and 0.02 mm.

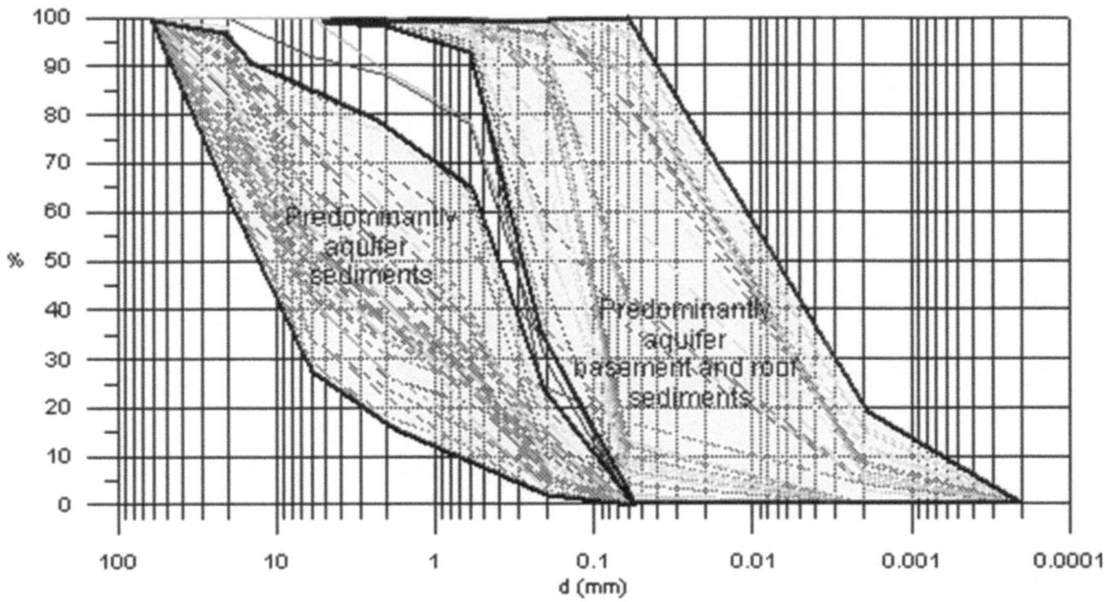


**Figure 4.24** Curves showing the granulometric composition of the alluvial aquifer of the left bank of the Danube from Kovin to Dubovac (Jaroslav Černi Institute for Water Management).

**Table 4.13** Groundwater source in Kovin-Dubovac: values of some of the analysed parameters.

| The parameter | Minimum | Average | Maximum | Number of Measurements |
|---|---|---|---|---|
| *Eh* (mV) | 20 | 96 | 279 | 61 |
| $O_2$ (mg/L) | 0 | 0.06 | 0.80 | 79 |
| $NO_3^-$(mgN/L) | 0.05 | 0.25 | 1.31 | 62 |
| $NH_4^+$(mgN/L) | 0.02 | 0.75 | 2.12 | 76 |
| $Fe^{2+}$(mg/L) | 0.10 | 2.06 | 5.30 | 75 |
| $Mn^{2+}$(mg/L) | 0.04 | 0.38 | 1.08 | 81 |
| $SO_4^{2-}$(mg/L) | 1.00 | 19.17 | 147.80 | 70 |
| Electroconductivity (20°C) (μS/cm) | 367 | 627 | 1078 | 80 |
| $d_{10}$ (mm) | 0.01 | 0.21 | 0.7 | 41 |
| $d_{50}$ (mm) | 0.14 | 4.38 | 13 | 41 |

Water samples were collected during the 2010–2013 period. Table 4.13 provides results on the selected wells of the drainage system, which can be seen in Figure 4.25.

The average *Eh* values are in the range of 50 to 150 mV. A very small number of samples have values over 200 mV (Figure 4.25). The low values of the well inflow velocities into the wells ($10^{-3}$ m/s) are a result of the use of the wells as well as of the self-discharge phenomena.

## 4.11.4 The Ključ source at the town of Požarevac

The Ključ source is used for the water supply of the town of Požarevac. It is located along the river Velika Morava and includes part of the alluvial plain (Figure 4.26).

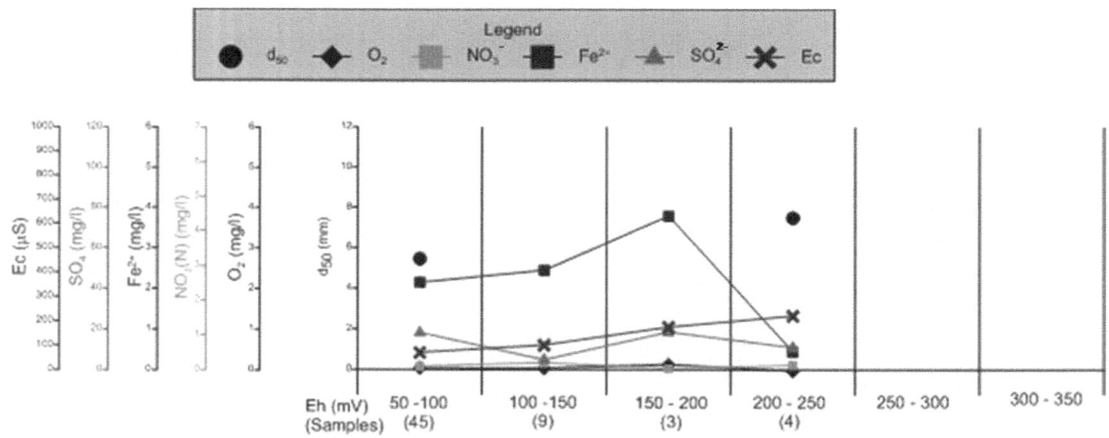


**Figure 4.25** The average values of selected well parameters on the Kovin-Dubovac section, by segments of redox potential (Legend: *Eh* – redox potential (mV), $d_{50}$ – aquifer grain size, taken as 50% of the particle size distribution curve (mm), $O_2$ – oxygen concentration (mg/L), $NO_3^-$ – nitrate concentration (mg/L), $Fe^{2+}$ – iron(II) cation concentration (mg/L), $SO_4^{2-}$ – sulphate anion concentration (mg/L), Ec – water electroconductivity (μS/cm)) (Jaroslav Černi Institute for Water Management).

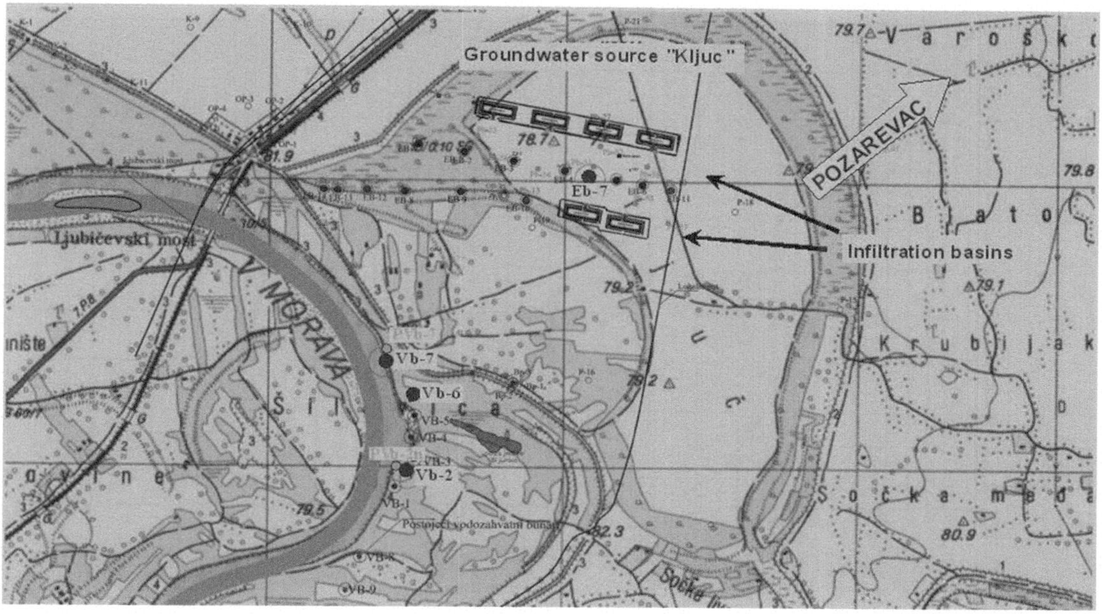


**Figure 4.26** Map of the location of the Ključ source (Jaroslav Černi Institute for Water Management).

The aquifer of the Ključ source consists of river and lacustrine and 'polycyclic' alluvial sediments; with a total thickness of 15 to 20 m. Sandy gravels are predominant in the part of the aquifer from where groundwater is captured. The thickness of the gravel in the zone of the Ključ source ranges between 6 and 11 m, and along the river Velika Morava it is between 8 and 12 m. The series of gravels is at some locations precipitated by sands that are between 2 and 6 m thick. Above the sand are poorly permeable mainly loess and clay sediments that are between 3 and 5 m thick. A typical lithological profile of source sediments is given in Figure 4.27. The figure also shows the riverbed of the Velika Morava River, in which a good hydraulic connection of the groundwater and the river can be seen, as is confirmed by the

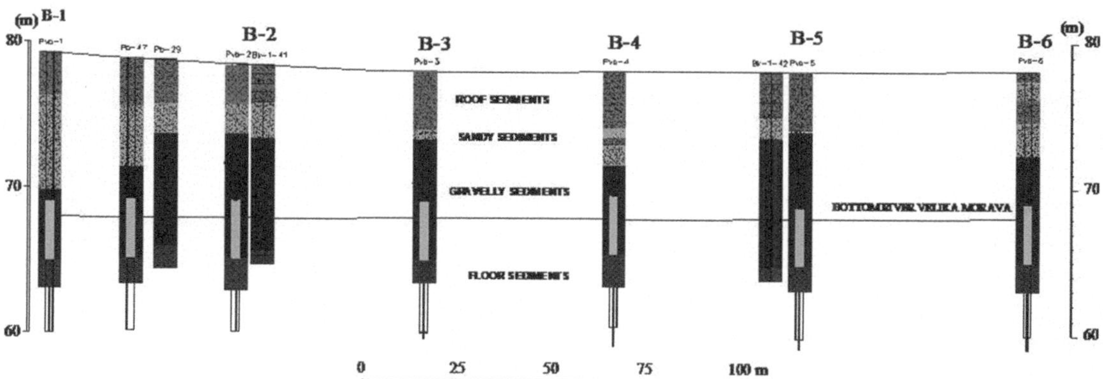


**Figure 4.27** Lithological profile of the Ključ source (Jaroslav Černi Institute for Water Management).

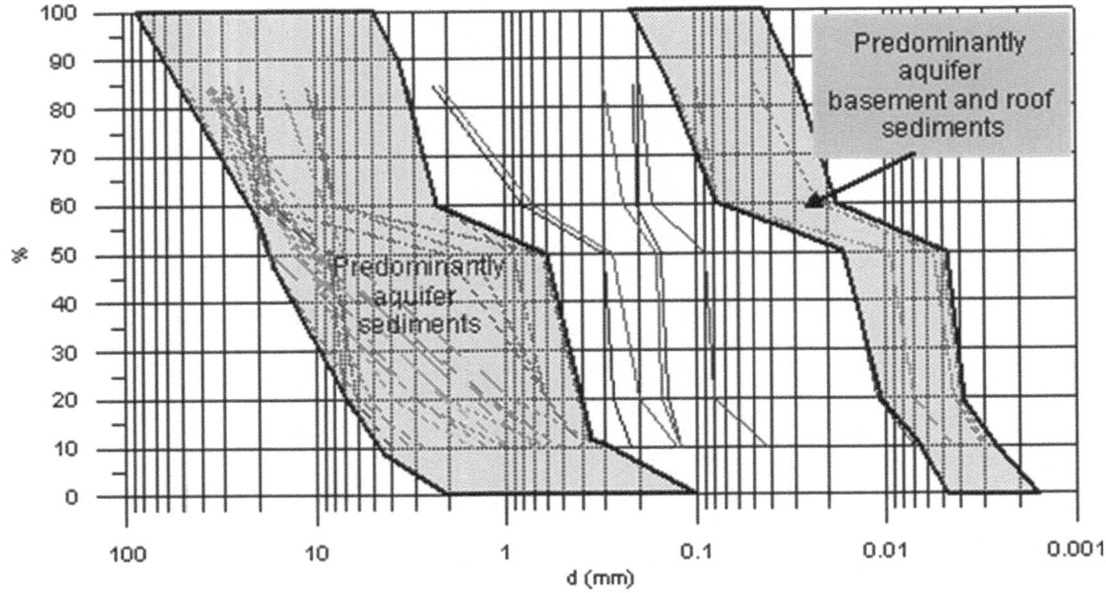


**Figure 4.28** Granulometric compositions of alluvial sediments in the zone of the Ključ source (Jaroslav Černi Institute for Water Management).

results on the granulometric composition of the sediments in the source zone (Figure 4.28). The sediment grains of the overlying and underlying layers of the profile are ten to a thousand times smaller than the aquifer grains.

The groundwater of the Ključ source on the bank of the Velika Morava River is characterized by a high redox potential, above 300 mV, with relatively high dissolved oxygen content (Table 4.14, Figure 4.29). There is practically no $Fe^{2+}_{(aq)}$, but in accordance with the relatively high dissolved oxygen concentration and the oxic conditions, the nitric and sulphate concentrations are increased, which may indicate external pollution from the surface of the terrain as well.

**Table 4.14** The Ključ source, values of measured parameters.

| The parameter | Minimum | Average | Maximum | Number of Measurements |
|---|---|---|---|---|
| *Eh* (mV) | 312 | 371 | 420 | 15 |
| $O_2$ (mg/L) | 4.4 | 5.77 | 9.1 | 18 |
| $NO_3^-$(mgN/L) | 2.1 | 5.5 | 8.12 | 18 |
| $NH_4^+$(mgN/L) | 0.01 | 0.02 | 0.05 | 18 |
| $Fe^{2+}$(mg/L) | <0.1 | <0.1 | 0.1 | 5 |
| $Mn^{2+}$(mg/L) | 0.0 | 0.01 | 0.06 | 9 |
| $SO_4^{2-}$(mg/L) | 75 | 104 | 148 | 18 |
| Electroconductivity (20°C) (μS/cm) | 620 | 843 | 1124 | 19 |

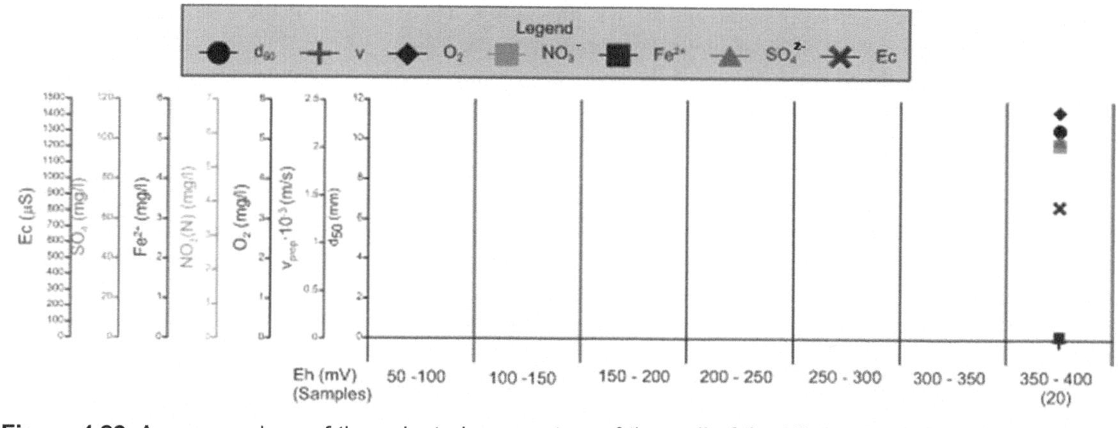


**Figure 4.29** Average values of the selected parameters of the well of the Ključ source near Požarevac, by segments of the redox potential (Legend: *Eh* – redox potential (mV), $d_{50}$ – size of aquifer grains taken as 50% of the curve of granulometric composition (mm), $O_2$ – oxygen concentration (mg/L), $v_{prop}$ – recommended inflow velocity into the well (m/s), $NO_3^-$ – nitrate concentration (mg/L), $Fe^{2+}$ – iron(II) cation concentration (mg/L), $SO_4^{2-}$ – sulphate anion concentration (mg/L), Ec – water electroconductivity (µS/cm)) (Jaroslav Černi Institute for Water Management).

## 4.11.5 Comments on the oxic conditions at the studied locations

A summary of the parameters for all the analysed locations is given in Figure 4.30. The results were previously interpreted as representative of each individual well, and then a statistical analysis was done of the average values at given *Eh* intervals (intervals at 50 mV each). The wells of the Belgrade sources

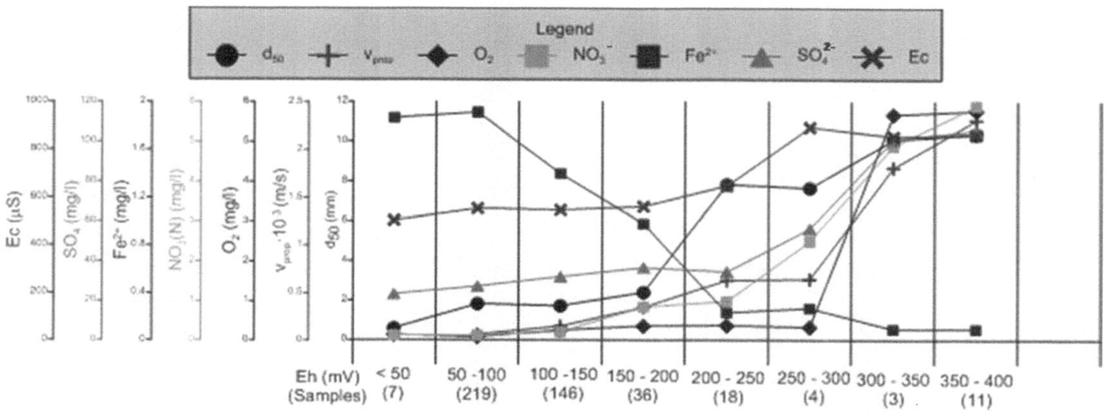


**Figure 4.30** Average values of selected well parameters of all analysed source locations by segments of the redox potential (Legend: *Eh* – redox potential (mV), $d_{50}$ – aquifer grain size taken as 50% of the granulometric composition curve (mm), $O_2$ – oxygen concentration (mg/L), $v_{prop}$ – recommended inflow velocity into the well (m/s), $NO_3^-$ – nitrate concentration (mg/L), $Fe^{2+}$ – iron(II) cation concentration (mg/L), $SO_4^{2-}$ – concentration of sulphate anions (mg/L), Ec – electroconductivity of water (µS/cm)) (Jaroslav Černi Institute for Water Management).

are dominant with the redox potential of 150 mV prevailing. The decrease of the $Fe^{2+}$ concentration is significant, as is the increase in the $d_{50}$, with an increase in the *Eh*.

Figure 4.31 shows the average values of the analysed parameters of the three most typical localities: the source of Belgrade (river Sava), the drainage system of Kovin-Dubovac (river Danube) and the source of the town of Požarevac, Ključ (river Velika Morava).

Figure 4.32 shows the summary results of the oxic conditions at three alluvial sources in Serbia.

The results of the physicochemical parameters of water quality at five wells reflect an oxygen deficient hydrogeological environment, three wells were dominantly in a mildly anoxic environment and two were in the middle of the transition between oxic and anoxic environments. The monitoring of riverbank filtration in an alluvial aquifer over a period of several years (Dimkić & Kečkarević, 1990) indicated the important decrease in the concentration of toxic compound phenols under oxic conditions in the aquifer (Dimkić, 2007).

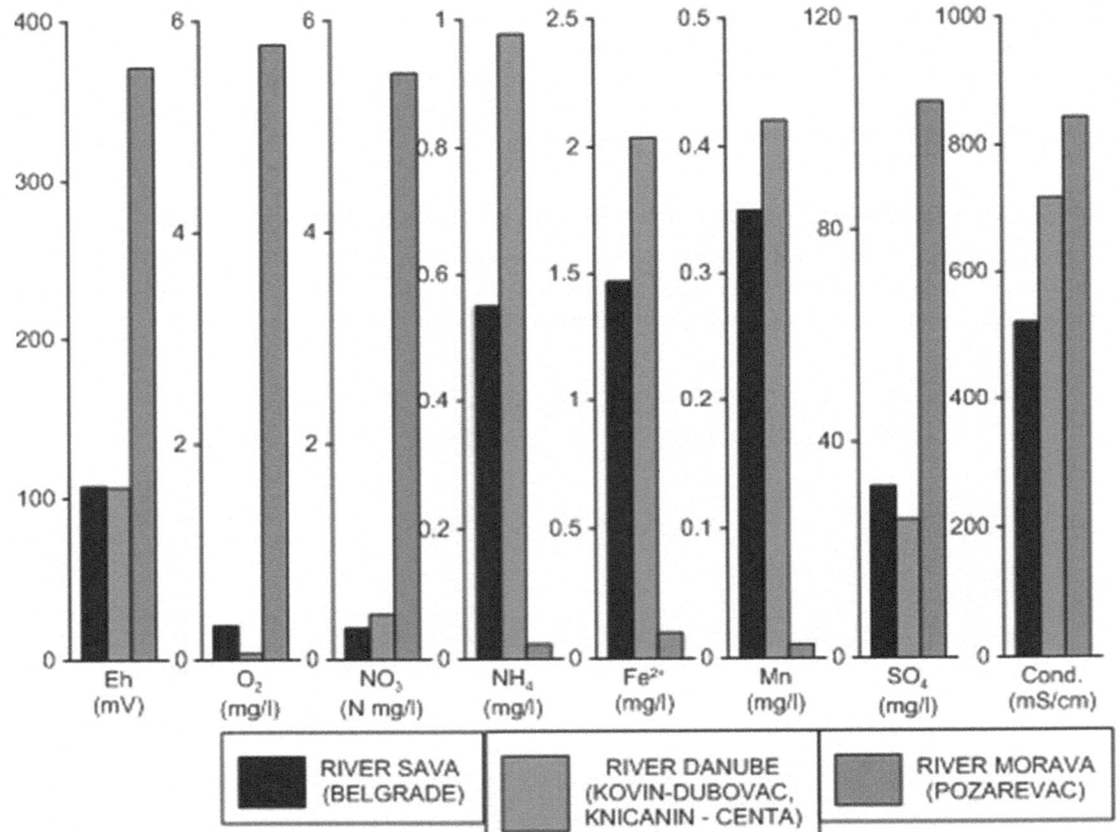


**Figure 4.31** Comparison of the average values of selected groundwater parameters (*Eh*, $O_2$, $NO_3^-$, $Fe^{2+}$, $Mn^{2+}$, $SO_4^{2-}$, Cond. – electroconductivity) and the environment in the alluvion of the Sava River (Belgrade), the Danube River (Kovin-Dubovac, Knićanin-Čenta) and the Velika Morava River (Požarevac) (Jaroslav Černi Institute for Water Management).

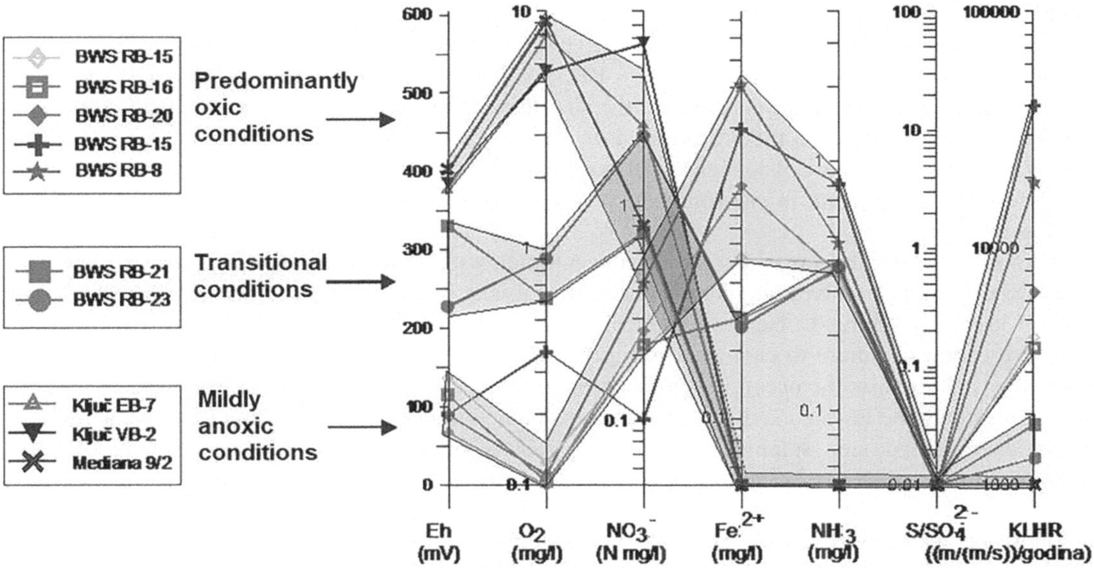


**Figure 4.32** Graphical presentation of oxic condition indicators (*Eh*, $O_2$, $NO_3^-$, $Fe^{2+}$, $NH_4^+$, $S/SO_4^{2-}$, KLHR – Permitted Rate of Well Sealing) in wells at three alluvial sources in Serbia (alluviums of the river Sava, the river Morava and the river Nišava) (Dimkić *et al.*, 2012).

## 4.12 CONCLUDING REMARKS

The alluviums of groundwater are essential and vital sources for drinking water supply, development of society, and strategic and economic progress, as well as for the whole natural environment.

Oxic conditions and dissolved oxygen are of key importance in alluvial aquifers for defining the background groundwater quality, as well as for explaining the various complex redox processes, degradation, metabolic transformations, and transport and diffusion of pollutants, ions and other chemical species that occur in groundwater.

Redox processes affect the physicochemical quality of groundwater in the alluvium. The evaluation of the redox processes, redox potential and ecological and thermodynamic conditions in groundwater depends on many factors such as the category and class of sources and distribution of electron donors and electron acceptors in the aquifer, the kinetics of redox reactions, the flow rate and energy of rivers and the position of the aquifer across the flowing river systems. The biogeochemical processes in groundwater are predominantly based on interaction between the consortium of microorganisms that are present and oxidizing and reducing agents. Variations in the concentrations of key electron donors and electron acceptors: $NO_3^-$, $N_2O$, $NH_4^+$, $SO_4^{2-}$, $CH_4$ and dissolved organic carbon could be the consequence of thermodynamic constants, temperature and Gibbs free energy values.

A higher free energy yield corresponds to a higher value of the redox potential, *Eh*. In oxic systems, *Eh* ranges between +200 and +600 mV, while anoxic processes are characterized by redox potentials between −100 and 200 mV. A low *Eh* value defines reduction processes in anoxic environments, analogously to a higher *Eh* value defining oxic conditions, since low pH values relate to oxidation processes. At lower *Eh*, between −100 and 200 mV, the decrease in energy yield $\Delta G$ of electron acceptors, resulting in decreasing *Eh*, is the result of the following chemical reactions: decrease in the oxygen concentration, the denitrification of the nitrate ion to nitrogen $N_2$, the reduction of manganese from $Mn^{4+}$ to $Mn^{2+}$, iron from $Fe^{3+}$ to $Fe^{2+}$, sulphate to hydrogen sulphide and methanogenesis followed by the declining values of $\Delta G$, −501, −476, −385, −102, −93, and −66 kJ/mol, respectively. The presence of $Fe^{2+}$ and $Mn^{2+}$ in groundwater depends on the structural components of the mineralogical composition of the sediments as well as on the pH values. Low pH values favour oxic conditions and lower oxidation states of iron and manganese. The significant occurrence of $Fe^{2+}$ in water also relates to aquifers with a lot of fine-grained sediment, in environments of lower oxygen concentrations and with less intensity of microbial activities.

Due to the reaction of decreasing oxygen concentration with a highly negative free energy yield value, $\Delta G = -501$ kJ/mol, dissolved oxygen is practically absent in anoxic groundwater, with concentration values of less than 0.5 mg/L. Next in order of spontaneity, with a high value of $\Delta G = -476$ kJ/mol, is the denitrification of nitrate to elemental nitrogen causing a decrease in nitrate concentrations such that nitrate ions do not usually occur in anoxic groundwater. As the reduction of sulphate to hydrogen sulphide is among the last in a series of spontaneous reduction reactions, according to the yield values of the free energy, hydrogen sulphide, $S^{2-}_{(aq)}$ is occasionally present in groundwater, only in an anoxic environment.

Contrary to these redox reactions, as the *Eh* in groundwater increases and reaches values above 200 mV, the following chemical reactions take place: methane oxidation, sulphate formation, nitrification, an increase in the dissolved oxygen concentration, the oxidation of $Mn^{2+}$ to $Mn^{4+}$ and $Fe^{2+}$ to $Fe^{3+}$, followed by a decrease in the energy yield, i.e. $\Delta G$, −408, −399, −181, −179, −164, and −90 kJ/mol, respectively. The natural nitrification process with a relatively low energy $\Delta G = -181$ kJ/mol causes the occurrence of nitrates in low concentrations, while higher concentrations are only possible due to external anthropogenic pollution introduced from the surface of the terrain. The spontaneous formation of oxygen ($\Delta G = -179$ kJ/mol) from the carbon dioxide and water at higher redox potentials above 200 mV results in the presence of dissolved oxygen, most often at concentrations of 4–9 mg/L.

Due to the oxidation of $Fe^{2+}$ and $Mn^{2+}$ into the poorly soluble compounds of iron hydroxide and manganese oxide with oxidation states +3 and +4, $Fe^{2+}$ and $Mn^{2+}$ may be present but at lower concentrations of <1 mg/L. Formation of stable hydroxide, $Fe(OH)_3$, with a very low solubility product ($Ks = 10^{-38}$ $mol^4/dm^{12}$), causes the aging of wells and clogging.

The rate constant of the examined chemical reaction, *k*, depends on the chemical species involved in the reaction and reaction activation energy which in turn depends on temperature per the Arrhenius equation and Gibbs free energy.

The rate of the Fe(II) oxidation reaction depends on the presence of the specific chemical forms $Fe^{2+}$, $FeOH^+$, and $Fe(OH)_2$ in an aqueous solution, which is highly pH dependent but much less dependent on the temperature and ion concentration. The oxidation of Fe(II) to Fe(III) is catalytically accelerated as a consequence of the sorption and hydrolysis processes of both Fe(II) and Fe(III). The kinetics constant of the oxidation of manganese ($k = 1.6 \cdot 10^{-13}$) is significantly lower than the constant of the oxidation of iron ($1.6 \cdot 10^{-6}$) as a consequence of the different electronic structures of iron and manganese.

The conceptual analogy for defining a groundwater quality status and oxidation processes in relation to the *pE* and pH is an extremely important axiom. Low *pE* defines the reduction processes in anoxic environments, analogous to the low pH that defines the acidic and oxic environment. The *Eh* in addition to the dissolved oxygen concentration level (oxic conditions) is the determining factor for the fate and behaviour of iron and manganese cations, sulphur anions, and nitrogen ions in groundwater. The redox conditions and *Eh* potential and the redox sensitivity of the chemical species: $O_2$, $NO_3^-$, $N_2$, $NH_4^+$, $Mn^{2+}$, $Mn^{4+}$, $Fe^{3+}$, $SO_4^{2-}$, $HS^-$, $S^{2-}$ and $CO_2$ will define the redox kinetics, solubility, degradation, transport, mobility, diffusion, distribution and the final groundwater quality.

Higher oxic conditions are more common in the aquifers of the upper river basin than in the mid-sections, and logically less so in the lower parts of the river basin. In addition to the amount of organic load, the mineralogical oxygen consumption and $Fe^{2+}$ are particularly important for the reduction processes in groundwater.

Under highly oxic conditions, oxidation occurs predominantly via dissolved oxygen in water. In an anoxic environment, biochemical redox processes are based on $NO_3^-$ anions, $Fe^{3+}$ and $Mn^{4+}$ cations and on sulphate anions, $SO_4^{2-}$.

The purification of river water via bank filtration is a natural process which is important for the production of drinking water. An oxic aquatic environment is generally suitable for bank filtration processes.

In aquifers where oxic conditions prevail and where the oxygen content is high, intensive biodegradation processes take place. Lipophilic organic substances generally do not dissolve in water, and this, together with the chemical activity of microorganisms in groundwater, is the cause of relatively low concentrations of organic matter.

The monitoring of riverbank filtration in an alluvial aquifer over a period of several years indicated a decrease and degradation of toxic phenols in groundwater under oxic conditions which was determined for the first time in Serbia.

The aging and clogging of wells with abundant iron (Fe(III) hydroxide) sediments is, from ecological, economic and strategic points of view, very challenging in terms of preserving the natural sources of Serbia.

The problem of anoxicity and the processes leading to the aging of wells has been investigated in a small number of scientific research papers. However, the extensive research done by Dimkić *et al.*, has aimed at establishing a relation between some of the most important chemical, biochemical, microbiological, hydrological, mechanical and kinetic parameters of groundwater. During the research period the following parameters were investigated: *Eh*, conductivity, oxidizing states of iron and manganese, nitrogen and sulphur, $\Delta G$, mineralization, ferrous sulphate reduction and 'slime' bacteria, the inflow velocity of groundwater into a well, the kinetics of colmation, the granulometric composition of an aquifer and the filtration properties of a water-bearing environment.

These key mentioned processes have been investigated and studied at several rivers and locations (the Sava River: the Belgrade source, the Danube River: sections of the riverbank: Kovin-Dubovac, the Velika Morava River: Ključ, the source of the town of Požarevac). Studies have shown that biochemical colmation practically does not occur in oxic alluvial groundwater when the redox potential is above 250 mV.

Theoretical knowledge and experimental experiences of the impact of oxic environment levels on the basic natural quality of groundwater as well as on self-purification processes through river filtration are fundamental determinants for the design, use and protection of groundwater sources. This chapter presents a detailed discussion of research undertaken which is relevant to these important groundwater issues.

## REFERENCES

Atkins P. and de Paula J. (2006). ATKINs Physical Chemistry, 8th edn. Oxford University Press, Oxford.

Čavić A., Sremački M., Petrović M., Obrovski B., Šunjević M., Mihajlović I., Bošković A. and Vojinović Miloradov M. (2018). Valorisation of the concentration levels of toxic metal cations in River Danube surface water via multicriteria analysis. *Fresenius Environmental Bulletin*, **27**, 7787–7797.

Davis J. R. (2000). Corrosion: Understanding the Basics. ASM International, Metals Park, OH, 63.

Dimkić M. (2007). The self–purifying effects of groundwater filtration, published by Foundation Andrejevic, Belgrade, 254.

Dimkić M. (2010). Management of Groudwater Sources: Importance of Aerobicity – Belgrade Example. International Congress and Exibition: Environmental Technology & Renewable Energy – ENVIETECH 2010 'Environmental Innovations and Climate Change', Track 3: Waste and Water Management – Abstract, 4–5 February, Vienna, Austria, 44–45.

Dimkić M. (2012a). Importance of the Aerobic State of Alluvial Aquifers for Groundwater Use. Conference 'Contemporary Issues of Adaptive Water Management', 31 October, Belgrade, 29–37.

Dimkić M. (2012b). Processes in alluvial groundwater and their significance. 14th Serbian Symposium on Hydrogeology with International Participation, 17–20 May, Zlatibor, 5–10.

Dimkić M. (2013). Self–Purification Processes in Alluvial Groundwater and Their Significance for Eliminating the Emerging Micropolutants. UNESCO Conference on Emerging Pollutants in Water, 9–11 July, Belgrade, Book of Abstracts, 27–34.

Dimkić M. and Keckarević D. (1990). Quanitification of the purifying potentials of groundwater released as a physical biochemical reactor – in Serbian language, published by the Institute for Water Jaroslav Czerny, Belgrade.

Dimkić M. and Knezic L. (2012). Biochemical Processes. In Groundwater Management in Large River Basins. IWA Regional Conference on Groundwater Management in the Dunav River Basin and Other Large Basins, Beograd, 169–187.

Dimkić M. and Pušić M. (2007). Relationship input speed, increase local hydraulic resistance wells and redox potential in the underground waters of alluvial sources, published by the Institute for Water Jaroslav Czerny, Belgrade.

Dimkić M., Brauch H. and Kavanaugh M. (2008). The Self–Purifying Potential of an Aquifer. In Groundwater Management in Large River Basins. Chapter 3, IWA Publishing Alliance House, London, 129–215.

Dimkić M., Pušić M., Dursun-Majkić B. and Obradović V. (2011a). Implications of Oxic Conditions in Alluvial Groundwater. IWA Specialist Groundwater Conference, Proceedings, 08–10 September, Belgrade, 15–30.

Dimkić M., Pušić M., Majkić-Dursun B. and Obradović V. (2011b). Certain Implications of Oxic Conditions in Alluvial Groundwater. *Water Research and Management*, **1**(2), 27–43.

Dimkić M., Pušić M., Obradović V. and Kovačević S. (2012). The effect of certain biochemical factors on well clogging under suboxic and mildly anoxic conditions *Water Science and Technology*, **65**(12), 2206–2212.

Hem J. D. (2005). Study and Interpretation of Chemical Characteristics of Natural Water, USGS Water-Supply Papero, 2254.

Houben G. (2006). Modelling the buildup of Iron Oxide Encrustations in Wells. *Ground Water*, **42**(1), 78–82. Jaroslav Černi Institute for Water Management, https://www.jcerni.rs/

Jurgens B., Fram M., Belitz K., Burow K. and Landon M. (2010). Effects of Groundwater Development on Uranium. NGWA, Central Valley, California, USA.

King D. W. (1998). Role of carbonate speciation on the oxidation rate of Fe(II) in aqueous systems. *Environmental Science & Technology*, **32**, 2997–3003.

Luther G., Findlay A., MacDonald D., Owings S., Hanson T., Beinart R. and Girguis P. (2011). Thermodynamics and kinetics of sulfide oxidation by oxygen: a look at inorganically controlled reactions and biologically mediated processes in the environment. *Frontiers in Microbiology*, **2**(62), 1–9.

Madigan M.T., Jung D.O., Karr E.A., Sattley W.M., Achenbach L.A. and van der Meer J. (2005). Diversity of anoxygenic phototrophs in contrasting extreme environments. In: Geothermal Biology and Geochemistry in Yellowstone National Park, W.P. Inskeep and T.R. McDermott (eds.), Thermal Biology Institute, Montana State University, Bozeman, pp. 203–219.

Majkić-Dursun B., Petković A. and Dimkić M. (2014). The kinetics of chemical oxidation of iron in groundwater under exploitation conditions. *Journal of the Serbian Chemical Society*, **79**(11), 1–16.

Marrin D. L. (2002). Universal Water: The Ancient Wisdom and Scientific Theory of Water. Inner Ocean Publishing, Maui, HI.

Menz C. (2016). Oxygen delivering processes in groundwater and their relevance for iron-related well clogging processes – a case study on the quaternary aquifers of Berlin. PhD thesis, Freie Universität Berlin, Germany.

Morgan J. (2005). Kinetics of reaction between $O_2$ and Mn(II) species in aqueous solutions. *Geochimica et Cosmochimica Acta*, **69**(1), 35–48.

Morgan B. and Lahav O. (2007). The effect of pH on the kinetics of spontaneous Fe(II) oxidation by $O_2$ in aqueous solution – basic principles and a simple heuristic description. *Chemosphere*, **68**, 2080–2084.

Mrazovac S. and Vojinovic-Miloradov M. (2011). Correlation of main physicochemical parameters of some groundwater in northern Serbia. *Journal of Geochemical Exploration*, **108**(3), 176–182.

Mrazovac S., Vojinovic-Miloradov M., Matic I. and Maric N. (2013). Multivariate statistical analyzing of chemical parameters of groundwater in Vojvodina. *Chemie der Erde-Geochemistry*, **73**(2), 217–225.

Nikolić-Bujanović L. J., Čekerevac M., Vojinović-Miloradov M., Jokić A. and Simičić M. (2012). A comparative study of iron-containing anodes and their influence on electrochemical synthesis of ferrate(VI). *Journal of Industrial and Engineering Chemistry*, **18**, 1931–1936.

Obrovski B., Bajić J., Mihajlović I., Vojinović-Miloradov M., Batinić B. and Živanov M. (2016). Colorimetric fiber optic probe for measurement of chemical parameters in surface water. *Sensors and Actuators B: Chemical*, **228**, 168–173.

Pap S., Radonic J., Trifunovic S., Adamovic D., Mihajlovic I., Vojinovic Miloradov M. and Turk Sekulic M. (2016). Evaluation of the adsorption potential of eco-friendly activated carbon prepared from cherry kernels for the removal of $Pb^{2+}$, $Cd^{2+}$ and $Ni^{2+}$ from aqueous wastes. *Journal of Environmental Management*, **184**, 297–306.

Petkovic A., Dimkic M. and Boreli-Zdravkovic Đ. (2008). Measurement of Redox potential in groundwater: significance, concept and problem. 37th Conference on Current Problems of Water Use and Protection 'WATER 2008', 03–06 June, Mataruška Banja, 221–224.

Pourbaix M. (1974). Atlas of Electrochemical Equilibria in Aqueous Solutions. NACE International, Houston TX.

Rayner-Canham (1996). Descriptive Inorganic Chemistry, 2nd edn., W.H. Freeman, New York.

Rosso K. M. and Morgan J. J. (2002). Outer-sphere electron transfer kinetics of metal ion oxidation by molecular oxygen. *Geochimica et Cosmochimica Acta*, **66**, 4223–4233.

Sremacki M., Dmitrasinovic S., Petrovic M., Mihajlovic I., Obrovski B., Sunjevic M. and Vojinovic Miloradov M. (2019). Evaluation of metal cations in surface water via decision analysis by ranking techniques (Dart). *Journal of Environmental Protection and Ecology*, **20**(2), 579–588.

Stumm W. and Morgan J. J. (1996). Aquatic Chemistry, Chemical Equilibria and Rates in Natural Waters, 3rd edn. John Wiley & Sons, Inc., New York.

Van Loon G. and Duffy S. (2005). Environmental Chemistry, a Global Perspective. Oxford University Press, Oxford.

Zobrist J., Muller S. R., Anmann A., Bucheli T. D. and Boller M. (2000). Quality of roof runoff for groundwater infiltration. *Water Research*, **34**(5), 1455–1462.

# Chapter 5

# Iron incrustation of water wells

*Milan Dimkić, Milenko Pušić, David Mitrinović and Slobodan Vujasinović*

## 5.1 INTRODUCTION

Groundwater abstraction from alluvial sediments along large rivers is accompanied by a series of problems associated with a gradual decline in the capacity of individual wells and the entire wellfield. There are several potential causes, independent of the condition of the wells. In the case of wellfields along rivers, which are based on bank filtration, riverbed colmation often leads to reduced capacity.

Water wells become plugged due to:

- Mechanical clogging (colmation) of the screen and near-screen zone;
- Chemical clogging (incrustation), either purely chemical (e.g. carbonate clogging) or aided by a biochemical catalyst (e.g. clogging by trivalent iron);
- Biological clogging (biofouling), dominated by a biological agent.

This chapter includes an overview of the causes of well clogging and then focusses on clogging by trivalent iron, which occurs under anoxic aquifer conditions. The reason for this is that the authors have had an opportunity to investigate the capacity of ~150 mostly radial collector wells (RCWs) at Belgrade Groundwater Source (BGWS) and other water supply sources that rely on groundwater. The investigations lasted from 2005 to 2018.

Plots of entrance velocities that are critical in terms of well screen clogging are presented:

$$V_{cr} = V_{cr}(Fe, Eh) \tag{5.1}$$


Editor: Milan Dimkić
doi: 10.2166/9781789060904_0275

The oxic (anoxic) state is expressed based on the order of the oxidation agents that dominate the oxidation-reduction process:

$$O_2 \rightarrow NO_3^- \rightarrow Mn^{4+} \rightarrow Fe^{3+} \rightarrow SO_4^{2-} \rightarrow CO_2 \tag{5.2}$$

Under anoxic conditions, $Fe^{2+}$ remains dissolved in groundwater. Where the redox potential changes as groundwater enters the near-screen zone or the screen itself, a process takes place whose half-reaction is of the following type:

$$Fe^{2+} - e^- \rightarrow Fe^{3+} \tag{5.3}$$

Iron bacteria often cause iron incrustation of well screens. These bacteria utilise nutrients and iron to grow and reproduce. However, they do not use bioaccumulates for their metabolism, but incorporate them into the screen and near-screen zone. Bioaccumulates are compounds that do not biodegrade and are composed of metallic ions, such as Fe, Mn, Al, Cu and Zn.

Iron incrustation of water wells occurs in alluviums where the oxic state is low and there is enough bivalent iron in the groundwater. Based on our experience, favourable conditions for pronounced iron incrustation are

- $O_2 < 0.2$ (0.5) mg $O_2$/l of water;
- $Fe^{2+} > 0.5$ (1) mg $Fe^{2+}$/l; and
- Eh < 200 (150) mV/l.

Incrustations are found on screen surfaces and in the near-screen zone, if there are evident differences in redox potential.

## 5.2 PROCESSES THAT REDUCE WELL CAPACITY

The decrease in groundwater abstraction at a groundwater source over time is a result of two primary phenomena or their synergy:

- Ageing of individual wells; and
- Ageing of the entire groundwater source.

Three basic parameters generally characterise the longevity of an operating well:

- Well discharge capacity;
- Hydrogeological and hydrochemical conditions; and
- Well screen.

As a rule, given identical natural conditions and type of wells, it has been noted that the life span of a high-capacity well is shorter than that of a low-capacity well. On the other hand, based on long-term monitoring of wells with the same type of screen in different hydrogeological and hydrochemical environments, the life span varies from 2–3 to 25–30 years. The longer life is associated with relatively soft water (up to 10″ dH), low iron concentrations and no aggressive components. The life span is shorter if wells pump groundwater whose $CaHCO_3$ and iron concentrations are elevated.

As part of a study on well ageing processes, a comparative analysis was conducted of wells with eight different screens in the same hydrogeological and hydrochemical environment. It transpired that the technical characteristics of the screens play a significant role in well yield reduction over time. Screens with a filter pack and screens whose skeleton was wrapped with stainless steel wire were the top performers. As a rule, the reduction in groundwater withdrawal capacity is a result of one of the following two processes (or their combination):

- Corrosion; and
- Incrustation.

Corrosion is a phenomenon associated with physical degradation of the well (primarily screen) material. It is a set of chemical, electrochemical and biochemical processes that damage and ultimately destroy the screen or well casing. Screen destruction in such cases can be due to:

- Chemical corrosion, where the structure is ruined evenly across the surface;
- Selective corrosion, if the structure is composed of alloys of multiple materials; and
- Electrochemical corrosion, which is a result of the action of microgalvanic pairs.

In recent times, an increasing number of authors have come to believe that significant corrosion problems (more than 30%) are caused by microorganisms, especially sulphate-reducing bacteria.

The process has also been noted in non-operating wells, albeit much slower.

According to the Johnson's and Gavrilko's criteria, well casings tend to corrode when groundwater parameters are as follows:

- Low pH (pH $<7$);
- Oxygen concentration in groundwater greater than 2 mg/l;
- Hydrogen sulphide concentration greater than 1 mg/l;
- Carbon dioxide concentration greater than 50 mg/l;
- Presence of organic acids;
- Low water hardness; and
- Total dissolved solids greater than 1,000 mg/l.

Incrustation is a process that involves the precipitation of salts, minerals and other insoluble substances in the near-well region and on the well screens, as a result of:

- Chemical clogging;
- Biological clogging (biofouling); and
- Mechanical clogging (colmation).

## 5.2.1 Chemical clogging

Over time, minerals and salts are deposited on the well screen and within the near-screen zone under certain hydrochemical conditions and in certain pumping modes. In most cases, the process is not purely chemical but involves biochemical activity. Bacteria are catalytic agents that oxidise certain minerals.

Analyses of the precipitates on the screens and filter packs have shown that the dominant compounds are Fe $(OH)_3$, $CaCO_3$, $Mn(OH)_2$, $Mn(OH)_4$, $SiO_2$ and $H_2O$, with impurities composed of various solid particles.

The most frequent processes of biochemical and chemical clogging can be summarised as follows:

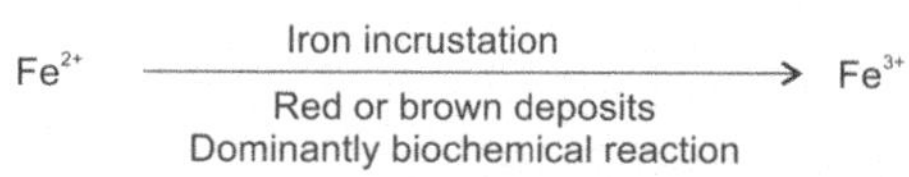


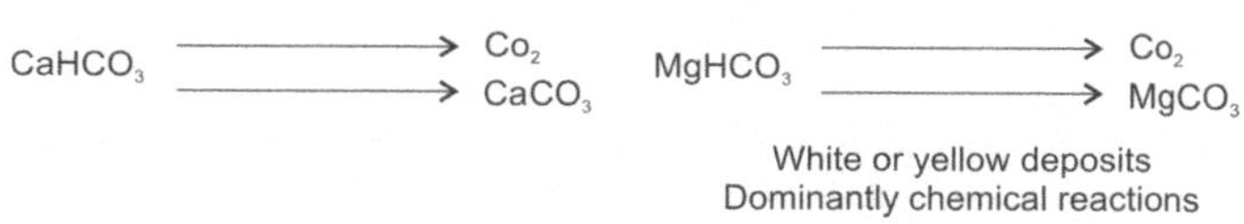


Clearly, clogging occurs due to a disturbed equilibrium between carbonic acid (carbonate equilibrium of the groundwater) and redox processes in the immediate abstraction zone.

The chemical stability of groundwater under natural conditions needs to be determined in order to assess the potential of iron hydrocarbonates to degrade and calcium carbonate to be precipitated. To do so, the results of chemical analyses can be used to calculate the value of pHs that corresponds to the state of equilibrium in the carbonate compound solution. In practice, tables or nomograms are employed to determine (calculate) pHs.

A saturation index was introduced as an indicator of the chemical stability of groundwater, commonly referred to as the Langelier index ($L$), named after Dr. Wilfred F. Langelier:

$$L = \text{pH} - \text{pHs} \tag{5.4}$$

If in a given case the measured pH level is lower than the calculated pHs ($L < 0$), the water contains dissolved $CO_2$ above the equilibrium concentration and the water is able to dissolve carbonate compounds. If not ($L > 0$), the $CO_2$ concentration in the water is below the equilibrium concentration and such water is able to extract and precipitate calcium carbonate.

Ryznar has also conducted extensive research in connection with the stability of carbonate systems in water and he introduced a criterion called the Ryznar number, expressed as

$$R = 2\text{pHs} - \text{pH} \tag{5.5}$$

Based on experimental data, it has generally been observed that at $R < 7$ water tends to precipitate clogging substances. If $R$ is greater than 9, the water is corrosive. However, at $7 < R < 9$ both incrustation and low-intensity corrosion are possible.

After numerous studies, Gavrilko and Alekseev (1985) identified four types of precipitates that lead to the incrustation of well screens:

- Conglomerate type, composed of cemented sand and gravel. Such encrustations are very hard and found at the contact between the filter pack and the water-bearing formation;
- Lightly layered type, composed of fine cemented sand;
- Loose and porous type, characterised by elevated porosity; and
- Soft, paste-like type, which is gelatinous and comprises a considerable amount of organic substances.

According to Johnson, in cases where groundwater is abstracted and the pressure (velocity distribution) and gas regimes are disturbed, the well screen and near-screen zone are incrusted if the groundwater has the following properties:

- Alkaline groundwater, pH >7.5;
- Carbonate hardness greater than 300 mg/l;
- Iron concentration greater than 2 mg/l; and
- Manganese concentration greater than 1 mg/l.

### 5.2.2 Chemical and biochemical clogging (incrustation)

The primary cause of this type of clogging of the screen and near-screen zone (filter pack and original formation) is the presence of iron and manganese bacteria, whose life processes are associated with the iron (or manganese) ion and oxygen in groundwater.

As part of their biological activity, iron bacteria transform and convert bivalent iron into readily precipitable trivalent iron (ferrous to ferric hydroxide), whereby trivalent iron is the primary component to be deposited. Hence, the existence of iron bacteria in groundwater is associated with the presence of dissolved ferrous hydroxide.

After many years of research, Gavrilko & Alekseev (1985) concluded that favourable conditions for the growth of iron and manganese bacteria are connected to the groundwater pH level, bivalent iron and manganese concentrations, and the dissociated hydrogen index (*rH*). This index is not dimensionally homogeneous and is expressed as:

$$rH = Eh/29 + 2\text{pH} \tag{5.6}$$

where *Eh* is the redox potential expressed in mV.

The effect of the microbial factor on the incrustation and biofouling of wells has been studied extensively by Cullimore and associates (Alford & Cullimore, 1999; Cullimore, 1999). Subsequent major contributions came from Houben & Treskatis (2007) and Houben (2003).

Long-term investigations have also been conducted at BGWS and other groundwater sources, particularly from 2005 to 2018 (Dimkić *et al.*, 2008; Dimkić & Pušić, 2008, 2014, 2018; Dimkić *et al.*, 2011b, 2011c, 2012, 2013; Dimkić, 2012; Dimkić & Knežić, 2012; JCI, 2010). These investigations had an overriding influence on the establishment of correlations between biochemical indicators of anoxic groundwater, entrance velocities and increasing hydraulic resistances at well screens. As far as we are aware, these relations were quantified here for the first time on a global scale.

Biofouling is clogging due to the production of organic mass. It largely occurs where conditions are favourable for the development of the biological component, primarily where nutrients are in abundance.

## 5.2.3 Mechanical clogging

Whereas corrosion and incrustation are relatively slow processes and largely associated with geochemical conditions and the presence of iron and manganese bacteria, rapid well ageing is often a result of filtration instability of the near-well region (mechanical clogging).

Mechanical clogging of the well screen and near-screen zone (filter pack) occurs when pores and screen slots become clogged with sand, clay and silt particles washed out from the aquifer matrix or (rarely) the filter pack, or the near-well region has been reached during drilling.

The contact zone between the well casing and filter pack (or original formation) generally becomes clogged by wedged-in sand or gravel particles in the water-receiving section of the well or at the contact between the formation and the filter pack, thus blocking the well entrance and increasing hydraulic losses.

As a rule, the reasons for mechanical clogging of the screen and near-screen zone are associated with inadequate selection of screen and pack parameters (or unskilled or inefficient investigations), as well as inappropriate well construction and development technologies. In places where it occurs, mechanical clogging is relatively rapid and takes only a few hours if the well is not properly developed or is pumped above maximum permissible capacity.

## 5.2.4 Well clogging indicators

Well clogging, which leads to well ageing, occurs where there are relevant physical, chemical and biochemical conditions.

Physical conditions are related to the filtration stability of the soil, groundwater entrance velocities, screen slot characteristics, nature of the filter pack, and similar.

Chemical conditions reflect the chemical stability of the groundwater and any electrochemical reaction between the groundwater and its environment. Certain types of well encrustations are largely chemical (e.g. precipitation of $CaCO_3$).

Biochemical clogging (bioincrustation) is a phenomenon that requires a microbial agent. The best example is $Fe(OH)_3$ precipitation on the well screen, facilitated by iron bacteria.

There can be several types of well clogging acting in synergy. Of course, the size of the screen slots and the pores of the filter pack and near-screen zone affect the clogging process:

$$\text{Large slots} \Rightarrow \text{low entrance velocities} \Rightarrow \text{slower clogging}$$

One type of clogging accelerates another type by reducing the pore or slot size. As the groundwater flow and entrance velocity increases, more well clogging agents become available.

Since well clogging can be mechanical, chemical, biochemical and biological, its indicators are measurable parameters that aid water well designing or can be used to detect the causes of well ageing and select the most appropriate rehabilitation approach.

#### 5.2.4.1 Mechanical soil composition indicators

These indicators are used to determine the following:

- Soil stability to suffosion (permissible entrance velocities, grain size of the filter pack, screen slots); and
- Flow laminarity at the entrance to the well (permissible entrance velocities). Appropriate expressions have been formulated for filtration stability, contact fluidisation, suffosion and maintenance of laminar flow. The relevant parameters are $d_{10}$, $d_{20}$, $d_{40}$, $d_{50}$, $d_{60}$, $D_{50}$, and $K$.

The objective is to design a well ensuring:

- Filtration stability and, at the same time, the largest screen slots and pores of the filter pack; and
- Entrance velocities (and well capacity) in the permissible range, given the need to maintain laminar flow at the entrance to the well.

#### 5.2.4.2 Physicochemical water composition indicators

These indicators are needed to predict the type of well ageing (corrosion, incrustation). If incrustation has already been detected (existing well, groundwater source), the parameters can help select the well rehabilitation approach and consider any further development of the groundwater source as a whole.

To predict well ageing due to the chemical or microbial composition of the groundwater, the knowledge (or prediction) of the quality of groundwater during the course of pumping is more important than the data collected for well design purposes. Johnson and Gavrilko describe the conditions that lead to well screen corrosion. A number of authors discuss the conditions for well incrustation based on the Langelier index and Ryznar number. The physical and chemical parameters, covered by the above criteria, as well as those needed for a general understanding of groundwater quality in connection with well ageing include the following: pH, redox potential, $Ca^{2+}$, $HCO_3^-$, $Fe^{2+}$, $Fe^{3+}$, Mn, $SiO_2$, $Al_2O_3$, temperature, TDS (total dissolved solids), $H_2S$, $SO_4^-$, and $O_2$.

Certain types of chemical incrustation are definitely more pronounced than types of biochemical incrustation, for example $CaCO_3$ or $MgCO_3$ precipitation. However, some are much more intensive as biochemical than chemical processes.

#### 5.2.4.3 Biochemical indicators

Bacteria whose activity leads to the excretion of deposits on the surface or within the well screen, filter pack or the near-screen zone are biochemical agents of well clogging. Bacterial action is required for biochemical incrustation. However, the population size and growth rate are not necessarily proportional to the incrustation rate.

Iron-related and slime-producing bacteria cause iron incrustation. Sulphate-reducing bacteria live within colonies of multiple bacterial species, where conditions are anaerobic. Sulphate-reducing bacteria could be one of the causes of well corrosion.

Biological activity reaction tests (BARTs) are used to determine the presence and activity of certain bacterial species (Cullimore, 1993, 1999).

The well deposits, primarily a mineral crust on the internal surface of the screen, are a very good indicator for identifying the processes that take place within the aquifer. These deposits can help gain insight into the nature of the dominant type of incrustation. The composition of the deposits needs to be cross-analysed with the chemical composition of the water from the pumping well and nearby observation wells, as well as the corresponding biological agent. This analysis is useful in determining the appropriate type of chemical rehabilitation.

#### *5.2.4.4 Indicators and consequences of well screen clogging*

A direct consequence of clogging of the well screen and near-screen zone is an increase in hydraulic losses as the groundwater enters the screen. This results in a decrease of well yield over time.

The main hydraulic parameter that can serve to quantify the extent of clogging of a certain well is the specific drawdown (drawdown per unit well discharge), expressed as the quotient $\Delta S/Q$. The value of $\Delta S/Q$ is referred to as specific hydraulic resistance (loss). $\Delta S$ is the difference between the piezometric head adjacent to the well (in the near-screen zone), or lateral in the case of RCWs, and inside the well or caisson. A change in specific hydraulic resistances over time may or may not be linear, depending on the dominant type and rate of well clogging (Figure 5.1).

It is practical to measure piezometric heads inside the pumping well and a neighbouring observation well that taps the same water-bearing formation.

An increase in hydraulic resistances does not necessarily result in a decrease in well discharge. At the same discharge, the water level will keep dropping until there is no more space left due to the position of the pump or another constraint (Figure 5.2a,b).

The difference between the piezometric head in the observation well ($\Pi_p$) and the pumping well ($\Pi_w$) is:

$$S = \pi_p - \pi_w = S_L - \Delta S \tag{5.7}$$

The hydraulic loss (piezometric head difference) due to laminar flow through the aquifer ($S_L$) does not vary over time (assuming a constant pumping rate):

$$S_L = \pi_p - \pi_d = \text{const.} \tag{5.8}$$

where $\Pi_d$ is the piezometric head on the outer contour of an RCW lateral or the contour of the near-screen zone. Losses at the RCW lateral, screen, filter pack and near-screen zone ($\Delta S$) vary over time and ultimately lead to reduced well discharge:

$$\Delta S = \pi_d - \pi_w = f(t) \tag{5.9}$$

## 5.3 WELL AGEING CAUSED BY IRON

Well ageing is a term that entails processes in wells with multiple consequences. It takes place if there is a favourable combination and interaction of the aquifer, groundwater and well. Ageing leads to the

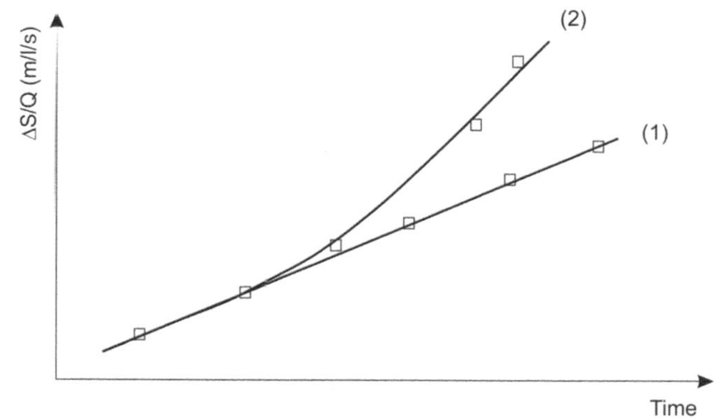


**Figure 5.1** Variation in specific hydraulic resistances over time due to well screen clogging: (1) linear and (2) non-linear (Dimkić & Pušić, 2008).

deterioration of the well's abstraction properties and the quality of construction materials. Well ageing may even cause the cessation of operation.

Well ageing due to clogging with iron occurs where, in conditions of a relatively reduced redox potential, a sufficient quantity of dissolved iron reaches the well screen. Favourable redox conditions in the well need to exist (more oxygen and/or a higher *Eh* value than in the groundwater) to produce the following half-reaction:

$$Fe^{2+} - e^{-} \rightarrow Fe^{3+} \tag{5.10}$$

These reactions occur with the catalytic effect of bacteria, which use energy from the reaction and/or its products for their needs. Iron bacteria are primarily anaerobic, optionally anaerobic and rarely slightly aerobic.

The processes of change in $Fe^{2+}$ into $Fe^{3+}$ hardly ever occur in extremely oxic conditions ($Eh > 300$ mV) and their intensity is low in environments of low oxicity ($150\ \text{mV} < Eh < 300\ \text{mV}$), but they do occur in anoxic environments. Besides the oxic state, the grain-size distribution of an aquifer, screen quality and velocity of groundwater flow towards the well are also relevant for the intensity and kinetics of the process. Several projects have been implemented in recent years, such as a well ageing study on BGWS (JCI, 2010), a scientific project funded by the Ministry of Science of the Republic of Serbia (JCI, 2019), and a project of the Ministry of Agriculture, Forestry and Water Management of the Republic of Serbia (JCI, 2013). These projects involved investigations of approximately 150 water supply wells and drainage wells, which operate in aerobic and anaerobic conditions.

Well ageing of a mechanical nature occurs due to the mobilisation of particles of the porous environment around the screen. Screen slots become plugged (mechanical clogging) and/or fine particles enter the well (mechanical corrosion). This process lasts a relatively short time, from a few hours to several days. Mechanical clogging of the well screen is often combined with biological and chemical clogging. As a result, a cemented conglomerate is formed, which is sometimes impossible to break.

Chemical corrosion ultimately destroys the well screen and casing. The consequence of chemical corrosion is a deposit, which is formed on and around the well screen. Groundwater contact with air and the dissolution of gases, especially oxygen, enhance the deposition.

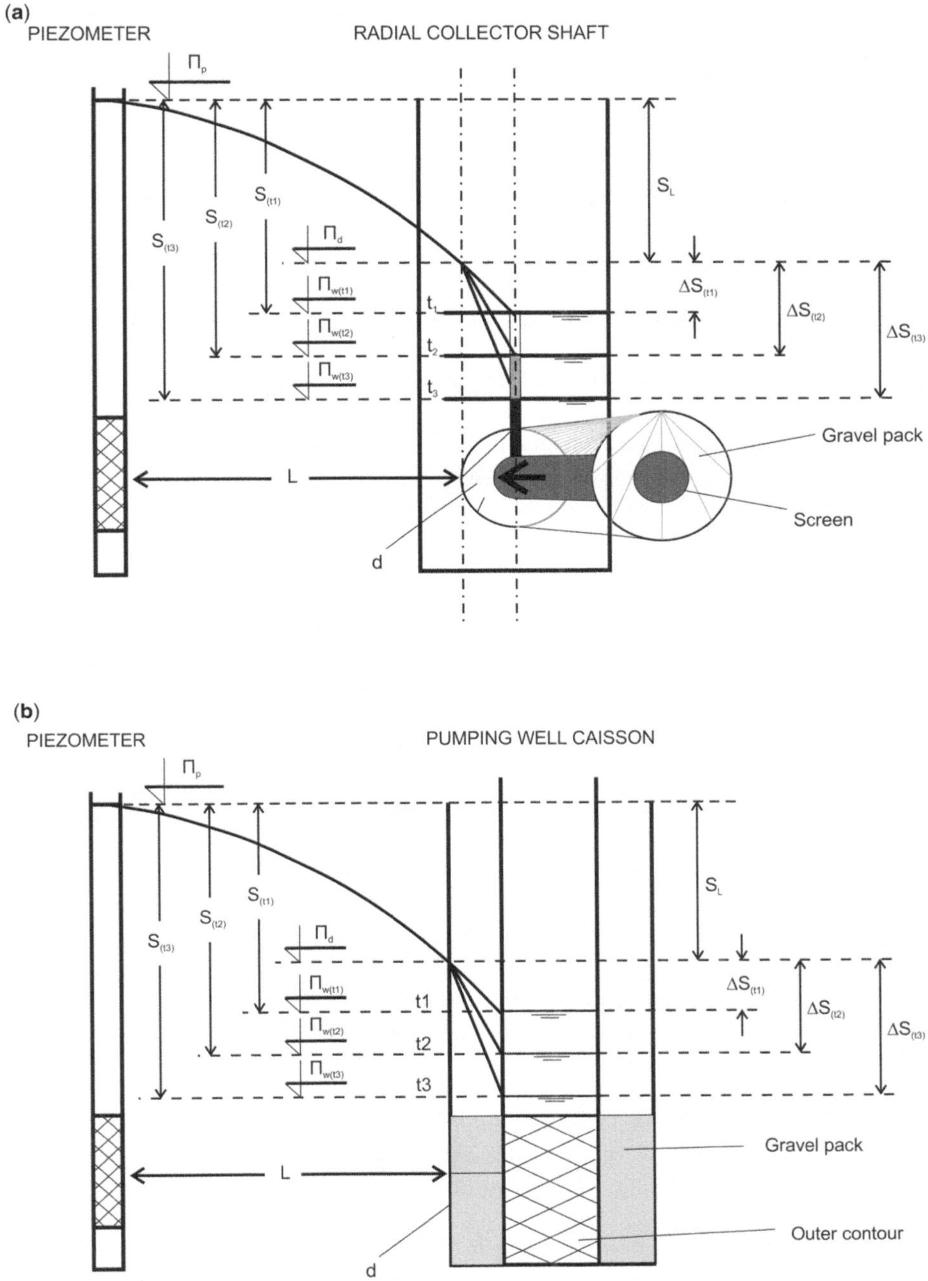


**Figure 5.2** Increase in hydraulic losses: radial collector well lateral (a) and tube well (b) (Dimkić & Pušić, 2008).

In alluvial aquifers, especially along middle and lower courses of large rivers, minerals containing Mn and Fe are usually deposited. In such conditions, large-scale growth of microorganisms occurs in wells, especially of manganese and iron bacteria. The deposits are the products of their metabolism and the bacteria themselves. Apart from clogging, which dominates, bacteria appear in ecological micro-niches in contact with the well casing. They corrode the well material, using it for their life processes (biochemical corrosion).

Above all, these causes of well ageing result in a decrease in well yield and shorten the life-span of the well. Ageing of the well screen is a significant issue in practice. The total quantity of the abstracted water and the capital expenditure, operating expenses and maintenance costs depend on the stability of the well yield and the life-span. In recent years, the introduction of new technologies (materials, design solutions and well construction technologies) have reduced the risk of well ageing, although the problem is still present.

Well ageing has been a research topic worldwide for quite some time. Earlier, more attention was paid to examining well construction criteria (well dimensions, characteristics of the screen, maximum permitted well yield, etc.), focussing on hydraulic parameters as a prerequisite for good quality and long life-span of a well. Among many other researchers on this topic are Sichardt (1928), Abramov (1952), Istomina (1957), Cistin (1965), Alekseev (1968), Gavrilko (1968), Pietraru (1982), Kovacs and Ujfaludi (1983), Gavrilko and Alekseev (1985), and Vuković and Pušić (1992). The standard criteria for defining the permissible groundwater entrance velocity at the well screen are most often related to the grain size distribution of the adjacent material and the hydraulic flow regime (laminar or turbulent).

Over time, it has been observed that in a large number of cases, well ageing occurs due to chemical and biological processes. Many authors have studied these since the 1980s, including Dubinina (1978), Howsam (1990), Leach *et al.*, (1991), Borch *et al.*, (1993), Smith (1995), McLaughlan (1996), Cullimore (1993, 1999), Mansuy (1998), Ellis (2006), and Houben and Treskatis (2007). Chemical clogging with carbonate deposits has been studied to such an extent that LSI (Langelier Saturation Index, Langelier, 1936) and RSI (Ryznar Stability Index, Ryznar, 1944) have been derived for quantifying the tendency of water to cause corrosion or incrustation. These indices have a wider practical significance, apart from their association with wells.

Experience has shown that clogging with iron deposits predominates in wells at urban groundwater sources, located in alluvial aquifers. However, these studies have not made a very important connection and established relationships between the rate of increase in hydraulic resistance at a well screen, the groundwater entrance velocity and biochemical indicators of clogging. Research has now been undertaken in this regard, first at the groundwater source of the city of Belgrade, where there are radial collector wells (Dimkić & Pušić, 2008; Dimkić *et al.*, 2011a, 2011b), and then at other groundwater sources in alluvial environments in Serbia.

Well ageing was monitored by observing changes in the local hydraulic resistance (LHR) of the wells, as a consequence of clogging (Dimkić & Pušić, 2008, Dimkić *et al.*, 2008). This resistance was periodically quantified based on well tests. It was found that the extent and rate of LHR changes are related to certain parameters – indicators of well ageing: in this case to the redox potential, total iron concentration, and microbial population.

Hence, summarising the causes of well ageing in general, it appears that a decrease in well yield is due to shrinkage of the surface through which the groundwater passes, because the pore volume in the near-screen zone and the number and lengths of active RCW laterals are reduced. This is a consequence of three primary causes, two of which are processes (incrustation and corrosion) and the third is corrodible material of the laterals. The reduced number of laterals is a significant consequence of rehabilitation. Laterals are shut off when they are damaged to such an extent that aquifer material penetrates the lateral and plugs it near the caisson, leading to sanding and a dramatic decrease in the flow through the lateral.

Well clogging can also be mechanical, if fine particles carried by the groundwater become blocked in the pore volume of the near-screen zone, plug the pores, and increase the local resistance to the flow. As the entrance velocities decrease, this process becomes less pronounced. There are several criteria in the literature and engineering practice used to select maximum velocities depending on the grain-size distribution and hydraulic conductivity (e.g. those proposed by Abramov, Sichardt and Kovacs).

Apart from corrosion, well clogging can be of a chemical nature, especially where the groundwater is saturated with calcium carbonate. This leads to calcite precipitation and shrinkage of the surface through which the flow passes. Chemical indices, such as the Langelier index and Ryznar number, make it possible to assess the potential for this type of clogging or corrosion.

Where the material of RCW laterals is prone to corrosion and if conditions are favourable, corrosion first increases the roughness of the lateral pipes and creates surfaces suitable for microbial growth and accumulation of their excretions and mineral precipitates. In turn, this creates numerous micro-niches where conditions for corrosion and clogging of the near-screen zone are even more favourable. Ultimately, the degradation of the construction material of the RCW laterals leads to their destruction, especially from mechanical impact during rehabilitation.

If corrosion is not intensive, the flow through the laterals whose mechanical integrity has been preserved is affected by increased local resistances due to a reduced cross-section and clogging of pores by microbial products and biomass.

In order to estimate the level of resistance at the laterals and assess the effectiveness of rehabilitation, the local hydraulic resistance (LHR) at the screen and in the near-screen zone needs to be quantified. LHR is defined as the quotient of the difference between the piezometric heads at the pumping well and a neighbouring observation well, which is within the zone of the laterals ($\Delta S$), and the flow rate through the lateral ($q$). It is possible to use observation wells beyond the zone of the laterals, but in that case a significant part of the difference would be attributable to the flow through the aquifer and LHR would be overestimated. LHR increases over time, except in the event of rehabilitation, during which the change in LHR is practically instantaneous and most often negative. If no rehabilitation is undertaken, in most cases the increase in LHR is linear if there are dramatic variations in well discharge. The rate (kinetics) of local hydraulic resistance variation, abbreviated as KLHR, does not generally change much over time (i.e. it is equal to the slope of the LHR linear regression function).

Consequently, the rate of biochemical clogging could be studied by monitoring RCWs with stainless steel laterals, resistant to corrosion. Because of the small number of wells with new laterals, the focus of study was on laterals assumed to be exposed to these dominant processes as well, for which sufficient LHR data was available.

### 5.3.1 Processes that lead to iron incrustation of wells

At BGWS, iron incrustation of well screens is especially pronounced. The presence of iron in the groundwater is generally the result of dissolved minerals. Dissolution of minerals occurs under variable aerobic/anaerobic conditions, characteristic of the groundwater level fluctuation zone. It is enhanced by the presence of carbon dioxide and reducing agents (organic substances, hydrogen sulphide, nitrites, and similar), whereby the iron found in oxide, sulphide, carbonate and silicate minerals is transformed into a bivalent water-soluble form.

$Fe^{2+}$ to $Fe^{3+}$ transformation agents are bacteria capable of performing that process. They are aerobic or anaerobic bacteria. They accumulate trivalent iron on their surfaces in hydroxide form and use energy released as a result of their life-sustaining activity. The deposits are of different colours and consistencies (slimy, grainy or flaky, commonly known as iron ochre). The sediment is black, brownish-orange, or yellow, with various shades in-between.

The best-known iron bacteria belong to the genuses: *Sphaerotilus, Leptothrix, Crenothrix, Gallionella, Siderocapsa, Siderobacter, Sideromonas, Ferribacterium, Ferrobacillus*, etc.

Other factors, in addition to bivalent iron and manganese, need to be present for the growth and reproduction of iron and manganese bacteria. Some species also require organic substances. There are two general groups of iron bacteria.

The first group, with a relatively small number of species, is made up of obligately acidophilic, true iron bacteria, which use bivalent iron as energy material. Sulphides and sulphuric acid salts serve as a source of iron, whereby pyrites, sulphides, and iron sulphates are oxidised (*Thiobacillus ferroxidans*):

$$4FeS_2 + 15O_2 + 2H_2O \rightarrow 2Fe_2(SO_4)_3 + 2H_2SO_4 \tag{5.11}$$

The subsequent reaction is:

$$Fe_2(SO_4)_3 + 6H_2O \rightarrow 2Fe(OH)_3 + 3H_2SO_4 \tag{5.12}$$

The final product of oxidation is $Fe(OH)_3$, which is separated in the form of scum and deposited around colonies of iron bacteria (Skerman *et al.*, 1957).

The second, larger group is made up of microorganisms that oxidise iron in a neutral or slightly alkaline environment and deposit oxides on cell surfaces. Representatives of this group are common in groundwater and their activity is associated with the formation of iron-manganese concretions (Dubinina, 1978). Bacteria of the genuses *Gallionella* and *Leptothrix* oxidise iron, based on the following reaction:

$$4FeCO_3 + O_2 + 6H_2O \rightarrow 4Fe(OH)_3 + 4CO_2 \tag{5.13}$$

Bacteria of the genus *Crenothrix* oxidise acidic iron carbonate into iron hydroxide:

$$4Fe(HCO_3)_2 + O_2 + 2H_2O \rightarrow 4Fe(OH)_3 + 8CO_2 \tag{5.14}$$

The shape and size of deposits can be substantial and they correlate with the physical and chemical conditions of the medium and the species present. Based on data from various sources, the forms that contribute to rapid well ageing are *Gallionella ferruginea*, *Leptothrix* and *Crenothrix*, while in some cases there can be significant deposits of *Siderocapsaceae*.

The BART method is used at BGWS. A large ecological group of bacteria are tested, which exist in the form of biofilm. In addition to iron-related bacteria (IRB), this method is applied to test the biochemical activity of sulphate-reducing bacteria (SRB), slime-producing (SLYM) bacteria, and denitrifying (DN) bacteria.

## 5.4 INFLOW VELOCITIES AT THE WELL ENTRANCE (WITHIN THE NEAR-SCREEN ZONE)

The criterion for maximum permissible well entrance velocities should be the primary prerequisite for ensuring filtration stability of the near-screen zone and limiting the yield in that regard.

Filtration instability, as a form of deformation of the porous medium caused by groundwater flow, can have a significant impact on the operation and, consequently, the capacity of a well. Filtration instability of the near-screen zone is expressed as suffosion and contact deformation.

Suffosion is a process during which fine particles of a porous formation are washed out, thus increasing porosity.

Suffosion occurs and develops only in soils whose grain-size distribution is non-homogeneous, with a non-uniformity coefficient of $\eta > (10\text{–}15)$, where it is possible to differentiate the skeleton from the fill. As a rule, suffosion in the near-screen zone of a well has a positive effect on well hydraulics, unless it disturbs the stability of the soil skeleton. It increases the hydraulically equivalent radius of the well, decreases parasitic hydraulic losses, and extends the life-span of the well.

This type of (controlled) suffosion needs to be programmed during the course of well development. However, it should be kept in mind that due to radial flow, the flow velocity and rate of washing out of the fill increase with decreasing distance to the well screen.

- Contact deformation is a process during which the porous material deforms and enters the filter pack. A zone of material is formed at the contact of the two formations (original and filter pack), whose mechanical composition is such that the filtration parameters can be poorer than those of both original materials.

Contact deformation can disturb the stability of the porous material skeleton and cause clogging of the screen and filter pack or sanding of the well. From a contact deformation perspective, the permissible critical gradient depends on the ratio of the grain size of the porous material ($d$) to the pack ($D$). The well entrance (near-screen zone) permissible velocity criteria are not uniquely defined. In the past, filtration stability of the near-screen zone (prevention of suffosion and sanding) was the only condition relevant for the selection of well characteristics and pumping mode.

Today, other parameters are considered as well. For example, the longevity of a well is also achieved by ensuring laminar flow in the near-screen zone, defined by the Reynolds number, which is $Re = 6$ for the porous material. Moreover, it should be kept in mind that well ageing depends on a number of hydraulic, chemical and biochemical factors.

A large number of literature sources propose several criteria that define the maximum permissible well entrance velocity. All the criteria have been derived experimentally and are expressed as empirical dependencies of the flow (Darcy's) velocity and the hydraulic conductivity of the porous material (and in special cases a characteristic grain-size).

Commonly used criteria for permissible entrance velocities are those of Gross, Sichardt, Abramov and Kovacs, and laminar flow criteria – Reynolds number.

### Discussion on entrance velocity criteria

Entrance velocity criteria serve to determine the maximum yield of a well, which guarantees the filtration stability of the aquifer and laminar flow. However, this does not solve the well ageing (yield decline) problem.

Well ageing depends on other factors as well (chemical, biochemical, type of material, pumping rate, type and size of screen slots, etc.). Well ageing and yield decline can be pronounced even where the filtration stability criterion has been met and laminar flow ensured.

Figure 5.3 is a plot of $v_p = f(K)$ and $v_p = f(d_{10})$, showing the zone of permissible entrance velocities, which ensures filtration stability of the near-screen zone and laminar flow to the well.

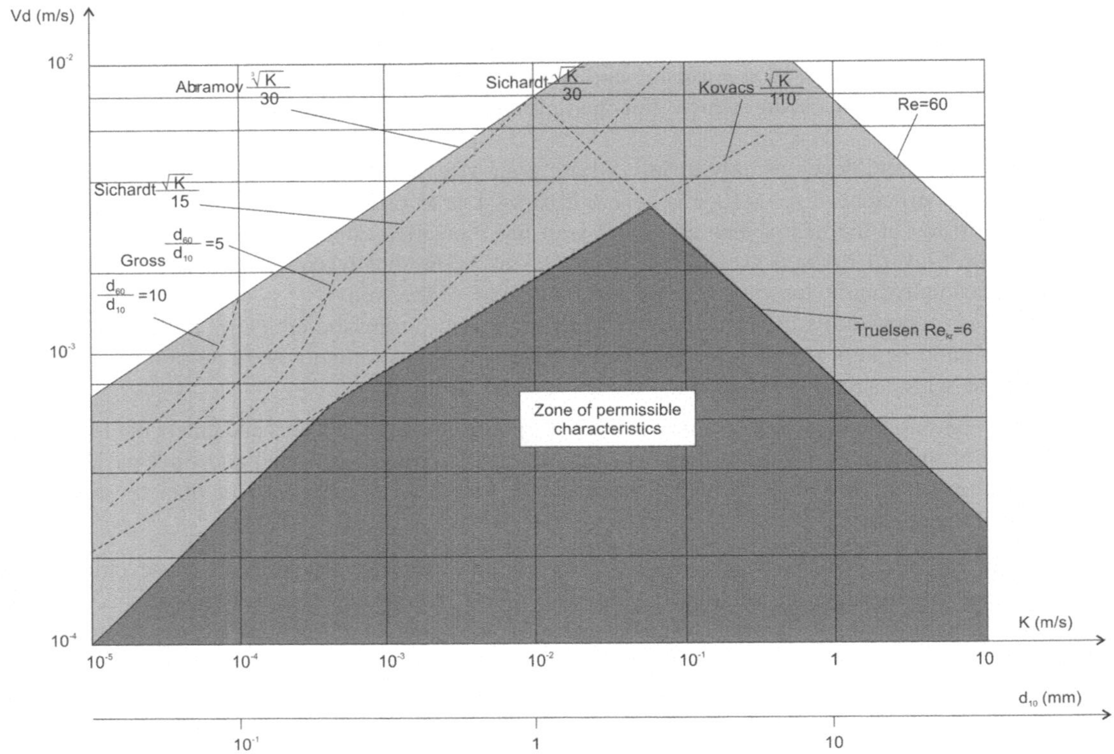


**Figure 5.3** Criteria for permissible well-entrance flow velocity (Babac *et al.*, 1989; Dimkić *et al.*, 2008).

One of the key decisions in connection with designing the installation of new RCW laterals is the number of laterals. More laterals means a larger receiving surface but also lower entrance velocities, which reduces KLHR. The higher the concentration of bivalent iron, the higher the KLHR per lateral at the same discharge, growing exponentially with increasing concentration. A practical way of addressing this in the design is to define the maximum increase in hydraulic losses at the laterals per year. The decision depends on assessing how often rehabilitation will be undertaken and predicting its effectiveness.

## 5.4.1 Local hydraulic losses as a measure of well ageing

The standard parameter used in engineering practice to quantify the performance of a well is its specific capacity, $q$:

$$q = \frac{Q}{S} \tag{5.15}$$

where $Q$ is the capacity of the well (l/s) and $S$ is the well drawdown (m). Drawdown in this case is defined as the difference between the static (undisturbed, natural) groundwater level and the (dynamic) level in the well (assuming that the well is operating).

The inverse, modified quantity is referred to as local hydraulic resistance (loss), or LHR (Dimkić & Pušić, 2008; Dimkić *et al.*, 2012):

$$\mathrm{LHR} = \frac{\Delta S}{v} \tag{5.16}$$

$\Delta S$ (m) is the local drawdown – the difference between the piezometric head of the groundwater on the contour of the near-screen zone (immediately beyond the well) and the head inside the well, and $v$ (m/s) is the average well entrance velocity (in practice, well discharge can be used instead of $v$ or lateral discharge in the case of RCWs). LHR is a parameter that can quantify the extent of well clogging.

In practice, an observation well near the pumping well (i.e. adjacent monitoring well in Figure 5.4) is used to gauge groundwater levels. At the same discharge, the difference in groundwater levels between the observation well and the clogged well zone does not change over time and can be disregarded in ideal conditions. This quantity can be used to estimate LHR with no major errors.

The same assumptions are made in the case of RCWs – the water level inside the well is equal to the water level in the lateral.

An increase in LHR over time does not mean that well discharge will always decrease. At the same discharge, the water level inside the well will decline over time, until there is no more room because of the position of the pump or another constraint.

The rate of clogging, KLHR (kinetics of local hydraulic resistance), is defined as a change in LHR during time interval $\Delta t$:

$$\mathrm{KLHR} = \frac{\Delta(\mathrm{LHR})}{\Delta t} \tag{5.17}$$

Based on the investigations conducted at BGWS, the rate of well clogging depends on several parameters, which can be used as clogging indicators:

$$\mathrm{KLHR} = \mathrm{KLHR}(v, Fe, Eh, B, \Gamma, \ldots) \tag{5.18}$$

where $v$ is the well entrance velocity, $Fe$ is the iron concentration in the well water, $Eh$ is the redox potential of the water, $B$ is the well bacteria growth-rate function, and $\Gamma$ is a function of several parameters associated

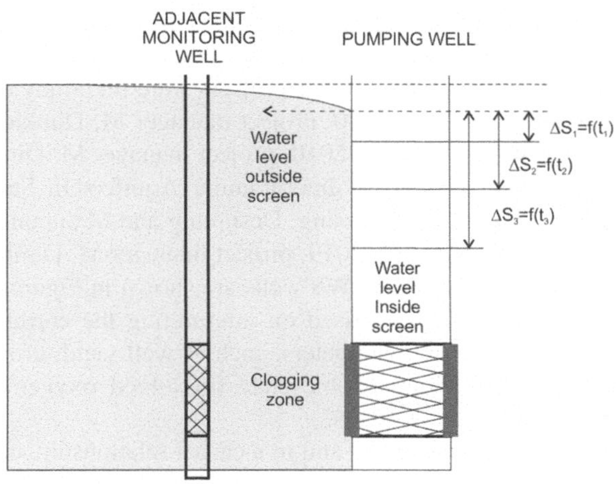


**Figure 5.4** Increase in local hydraulic resistance over time, tube well ($t_3 > t_2 > t_1$) (Dimkić *et al.*, 2008).

with the well construction technology (with or without pack, pack features, type and characteristics of screen slots, etc.), as well as the grain-size distribution of the aquifer.

Moreover, it is clear that KLHR can be viewed as a function of the pumping rate, or the groundwater entrance velocity. Predicting the change in local drawdown is another challenge; in a given time interval, it should not exceed a certain allowable value ($\Delta S_{AV}$). For practical reasons, one year was selected as the time interval $\Delta t$, so $\Delta S_{AV}$ is defined as the maximum permissible drawdown achieved on account of the increase in LHR during the year. It serves as a criterion which, if met, ensures long-term operation of the well.

$$\Delta S_{AV} = v_{perm} \cdot \mathrm{KLHR}_{year}[\mathrm{m}] \tag{5.19}$$

where $\Delta S_{AV}$ is the specified change in LHR during the year, expressed as the increase in well drawdown (m), $v_{perm}$ is the permissible entrance velocity (m/s), and $\mathrm{KLHR}_{year}$ is the variation in LHR during the year. Taking annual variation in LHR as the criterion, permissible entrance velocities (of a tube well or RCW lateral) can be calculated as follows:

$$v_{perm} = \frac{\Delta S_{AV}}{\mathrm{KLHR}_{year}}[\mathrm{m} \cdot \mathrm{s}^{-1}] \tag{5.20}$$

To ensure long-term operation of a well, the above criterion and the filtration stability criterion need to be satisfied. It should be noted that meeting both of these conditions does not guarantee total cessation of well clogging. It is only reduced to a level that allows the annual increase in hydraulic losses between the outside material and the inside of the well to be less than the specified $\Delta S_{AV}$.

In the case of BGWS, as discussed below, $v_{perm}$ of nearly all the wells was found to be much lower than calculated using 'conventional' experimental formulae associated with the filtration stability of the near-screen zone. This was particularly true of anoxic/suboxic groundwater conditions, in this case 70 mV $\leq Eh \leq$ 150–200 mV.

## 5.5 BELGRADE GROUNDWATER SOURCE – AGEING OF RADIAL COLLECTOR WELL LATERALS

Experience gained during a number of large-scale projects helped the authors of this chapter, M. Dimkić and associates, to comprehensively examine and present the processes associated with well ageing, as well as offer their advice and recommendations. Among these projects were the Study on Belgrade Groundwater Source (Jaroslav Černi Water Institute, 2005–2010, project manager M. Dimkić), Study on Well Ageing and Maintenance (Jaroslav Černi Water Institute, 2010, project manager M. Dimkić), Defining the Effect of the Oxic State on the Protection and Use of Intergranular Aquifers in Serbia (2008–2011, project manager M. Dimkić), and Methodology for Assessing, Designing and Maintaining Alluvial Groundwater Sources Depending on the Aerobic State (2011–2019, project manager M. Dimkić).

The location and approximate positions of BGWS wells are shown in Figure 5.5.

The insights detailed in project reports are based on interpreting the correlations established during long-term monitoring of the most important parameters, such as well yield, groundwater flow hydraulics (entrance velocities), grain-size distribution, aerobic state (dissolved oxygen), groundwater pH level, redox potential ($Eh$), iron concentration, and others.

Even though the parameters are interconnected and in a causal relationship, certain practical rules were established in connection with site selection, characteristics of wells in alluvial sediments, and efficient groundwater abstraction in general.

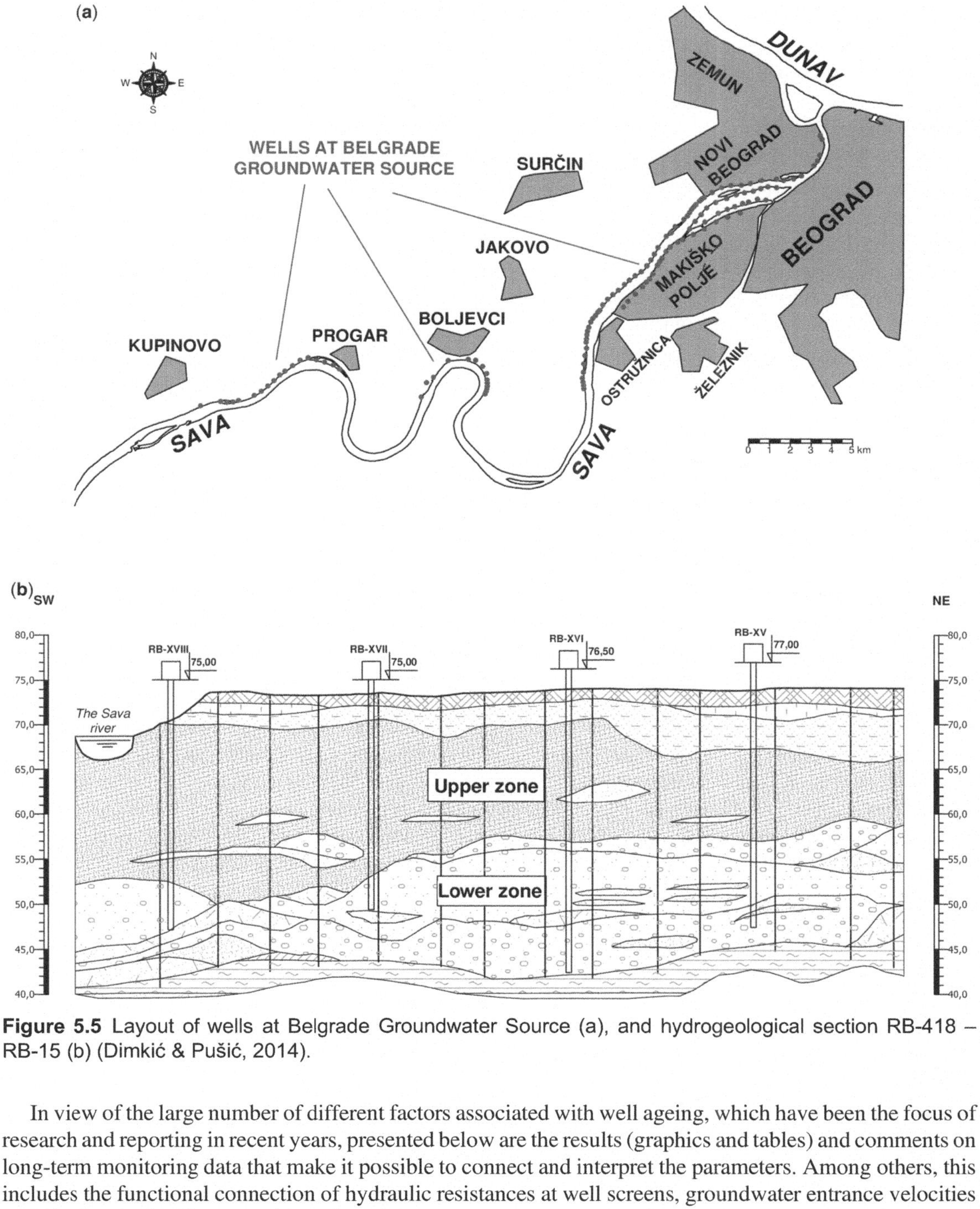


**Figure 5.5** Layout of wells at Belgrade Groundwater Source (a), and hydrogeological section RB-418 – RB-15 (b) (Dimkić & Pušić, 2014).

In view of the large number of different factors associated with well ageing, which have been the focus of research and reporting in recent years, presented below are the results (graphics and tables) and comments on long-term monitoring data that make it possible to connect and interpret the parameters. Among others, this includes the functional connection of hydraulic resistances at well screens, groundwater entrance velocities and biochemical indicators with well ageing.

### 5.5.1 Well yield as impacted by physicochemical parameters

As indicated in the chapter that discussed the aerobic state of aquifers, the discharge capacity of certain BGWS wells remained higher than that of others, even after several decades of pumping. There are many reasons for this, including a combination of local hydrogeological, hydrochemical and microbiological conditions of the aquifer, in light of different oxic states in parts of the aquifer.

The following parameters were considered and analysed:

- Grain-size distribution of the aquifer – natural filtration characteristics of the water-bearing formation;
- Electrochemical properties of the groundwater (redox potential and conductivity) – natural characteristics of the overall groundwater quality;
- Chemical parameters (total iron, total dissolved solids, carbonates) – natural characteristics; and
- Microbial activity based on BART tests of iron-related bacteria (IRB), slime-producing bacteria (SLYM), and sulphate-reducing bacteria (SRB).

A number of wells on the left bank of the Sava River, in New Belgrade, are described for illustration purposes. Figure 5.6 contains three plots showing some of the parameters that might be in a causal relationship with well capacity:

- Hydraulic – RCW discharge and discharge per active lateral;
- Hydrogeological – typical grain size of the formation ($d_{10}$);
- Physicochemical – redox potential of raw (well) water, $Fe^{2+}$ and oxygen concentrations in the aquifer, etc.; and
- Chemical – dissolved oxygen, bivalent iron and total iron concentrations.

A group of high-yield (>100 l/s) wells were identified. These were RCWs RB-21, RB-22 and RB-23. At these wells, the aquifer granulation was pronounced, the redox potential of the well water was high, the concentration of dissolved oxygen was elevated, and the concentrations of bivalent and total iron were relatively low. These parameters clearly show the reasons for the high yield of the wells (zones).

The correlation established between the well yield, iron concentration and redox potential (Figures 5.7 and 5.8) is of special interest. The plots lead to several general conclusions. Well yield is inversely proportional to the iron concentration. The average iron concentration in the water of a well whose measured yield is greater than 100 l/s is 0.85 mg/l. There were three such wells. At a well discharge between 50 and 100 l/s, the average iron concentration was 2 mg/l, and below 50 l/s it was 2.3 mg/l.

High well yield corresponded to a high redox potential. The redox potential (*Eh*) of all three wells in the studied part of BGWS, whose individual yield is above 100 l/s, was greater than 160 mV (250 mV on average).

In the wells whose discharge was between 50 and 100 l/s, the average *Eh* was 140 mV and in the case of less than 50 l/s the average *Eh* was 125 mV. This shows that the redox potential is a good general indicator of well quality. The average discharges for typical iron concentration and redox potential intervals are shown in Table 5.1.

The table shows that if the iron concentration is less than 1 mg/l, the average yield of the well is 75 l/s. At the same time, the average discharge per lateral is 12 l/s. These are the top performing wells at BGWS (there are 13 in the study area).

Where the measured redox potential was greater than 160 mV, the average well discharge was 86 l/s and the discharge per lateral 12.4 l/s.

The results indicate that the discharge of the best BGWS wells is of the order of 80 l/s and the discharge per lateral more than 12 l/s, as well as that the total iron concentration and redox potential are excellent parameters/indicators of well quality.

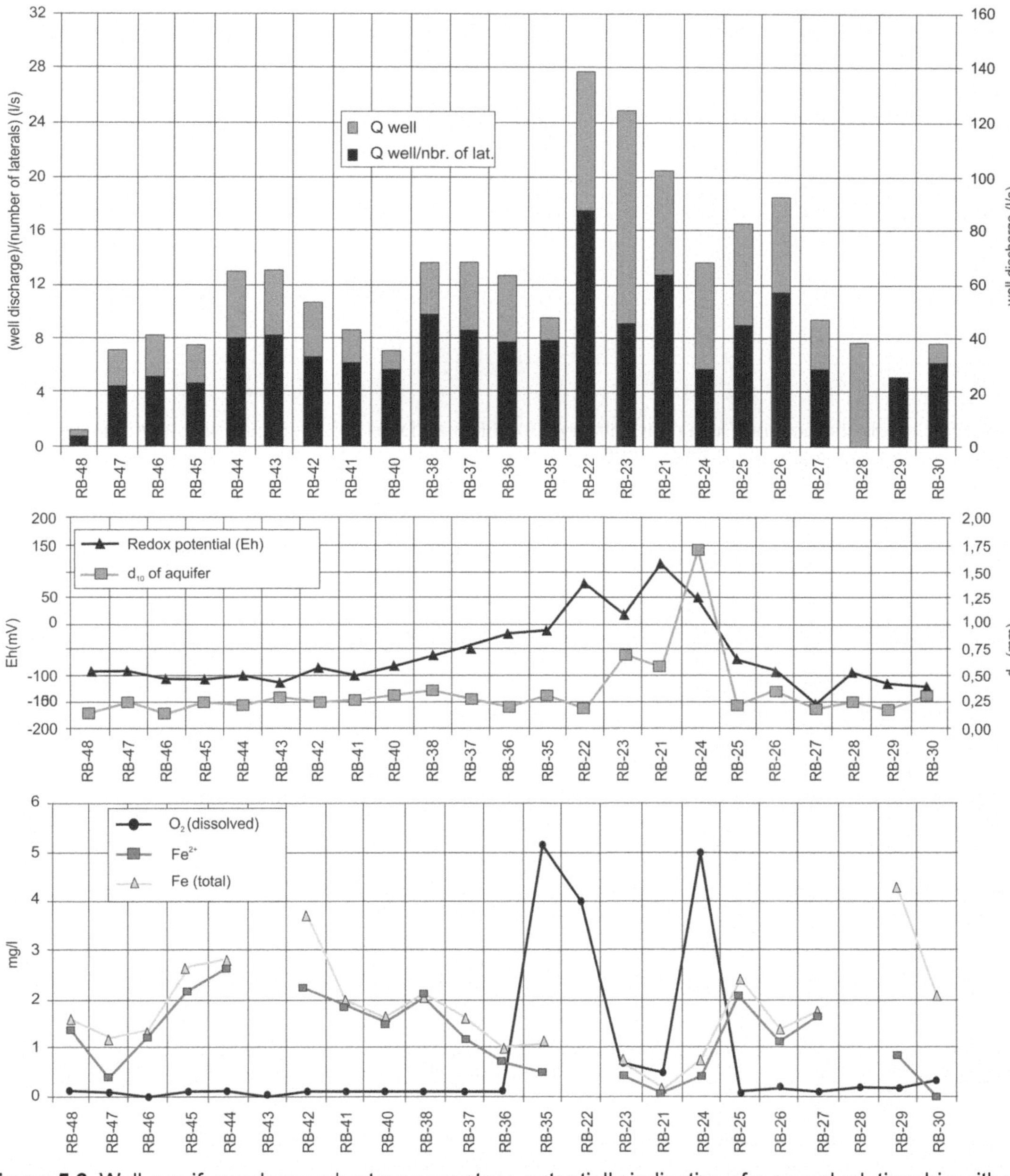


**Figure 5.6** Well, aquifer and groundwater parameters, potentially indicative of a causal relationship with well capacity ($d_{10}$ – grain diameter, 10% of the grain-size distribution curve) (Dimkić & Pušić, 2008).

The correlations among the redox potential, total iron concentration, aquifer grain size and well yield were also analysed. Hence, several well ageing indicators are compared to well capacity based on testing at BGWS from 2005 to 2009.

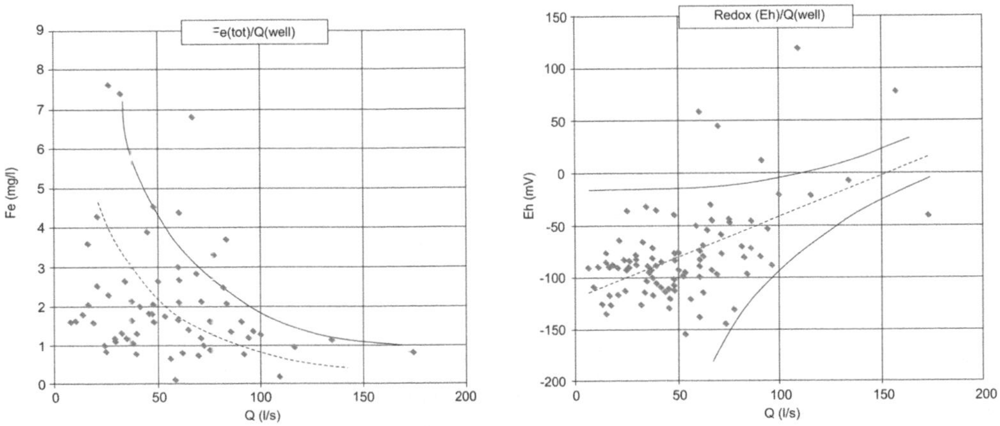


**Figure 5.7** Correlation between iron concentration (total Fe) and well discharge, as well as measured redox potential and discharge (Dimkić & Pušić, 2008).

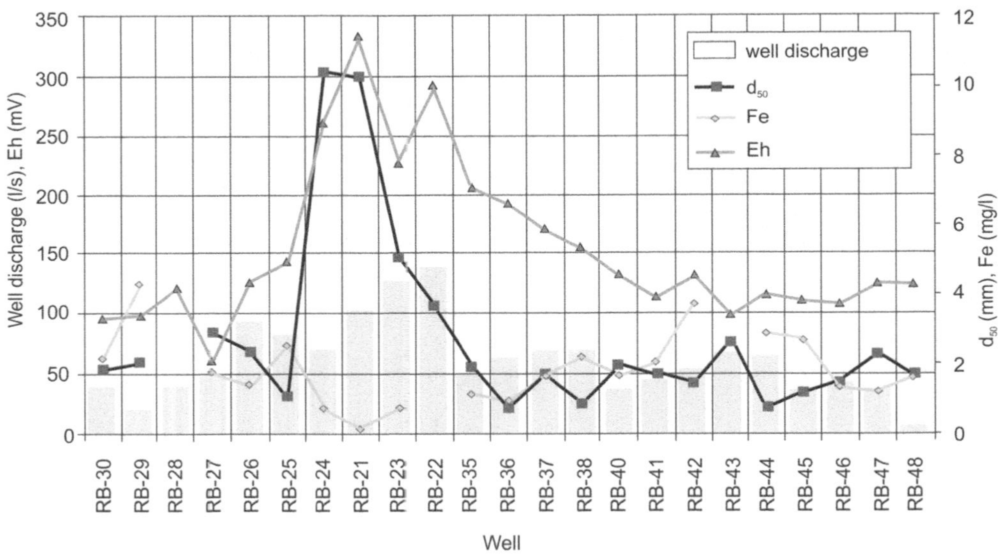


**Figure 5.8** Parameters *Q*, *Eh*, *Fe* and $d_{50}$ for a number of RCWs at BGS along the part of the Sava riverbank (Dimkić & Pušić, 2008).

**Table 5.1** Average well discharge in select iron concentration and redox potential intervals.

| **Fe (mg/l)** | | | | **Redox *Eh* (mV)** | | | |
|---|---|---|---|---|---|---|---|
| Fe (mg/l) | 0–1 | 0–3 | >3 | Redox (mV) | >160 | 110–160 | <110 |
| *Q* (l/s)* | 75 | 53 | 48 | *Q* (l/s)* | 86 | 49 | 45 |
| *Q*/lateral (l/s)* | 12 | 8.5 | 7 | *Q*/lateral (l/s)* | 12.4 | 8.7 | 7.5 |
| * Measured in 2006. | | | | *Measured in 2006. | | | |

The general equation of well (or groundwater source) yield is:

$$Q = Q(G, Eh, Fe, BCH, O_2, M) \tag{5.21}$$

where:

$Q$ is the well yield,
$G$ is the grain size around the laterals,
$Eh$ is the redox potential of the groundwater,
$Fe$ is the total iron concentration in the groundwater,
$BCH$ are the biochemical characteristics,
$O_2$ is the concentration of dissolved oxygen in the groundwater, and
$M$ is the well yield and maintenance regime.

The correlations were not only important for future BGWS management activities, but also interesting from an interpretation perspective in broader scientific circles.

Figure 5.8 shows the corelations between the well capacities and *Eh* and *Fe*.

The above plot is indicative of the correlations among the analysed parameters (high capacity/high redox potential, high capacity/low iron concentration and vice-versa).

Figures 5.9, 5.10 and 5.11 describe the correlations between the iron concentration, redox potential, grain size, and well capacity.

Despite some data scatter, the dependencies of the analysed parameters are obvious and those that coincide can easily be established. For example, the correlation between $Q$ and $Fe$ shows that as $Fe$ increases, $Q$ generally decreases.

The average $Eh$ is about 116 mV. Above that value, the average $Q$ is 64 l/s and for $Eh < 116$ mV, the average well capacity is ~35 l/s. If $Eh > 150$ mV, the average well capacity is ~100 l/s.

In general, as the aquifer grain size increases, so does well yield.

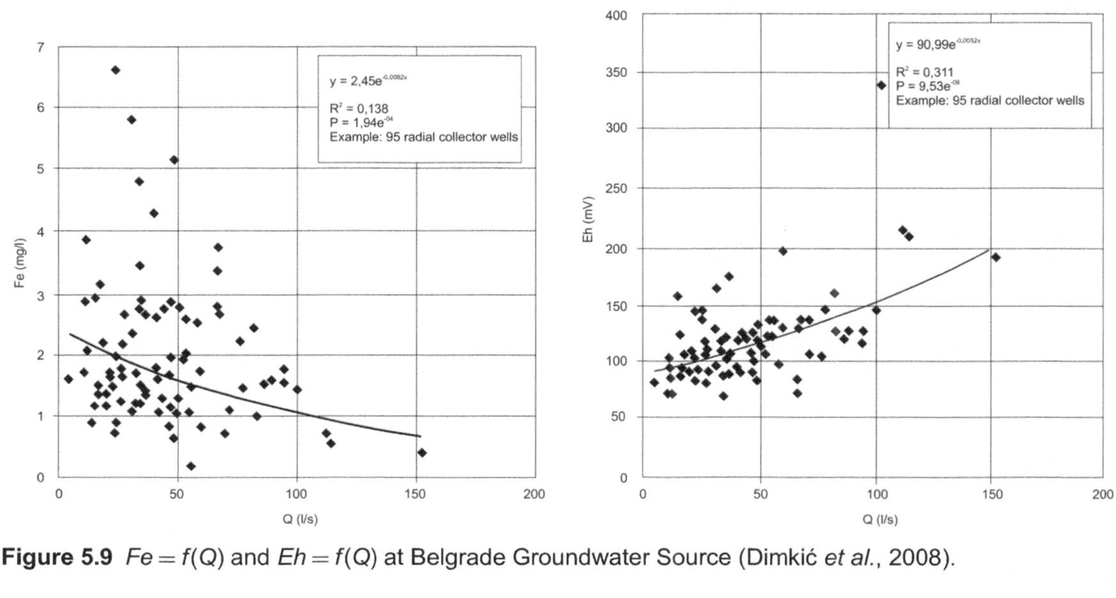


**Figure 5.9** $Fe = f(Q)$ and $Eh = f(Q)$ at Belgrade Groundwater Source (Dimkić *et al.*, 2008).

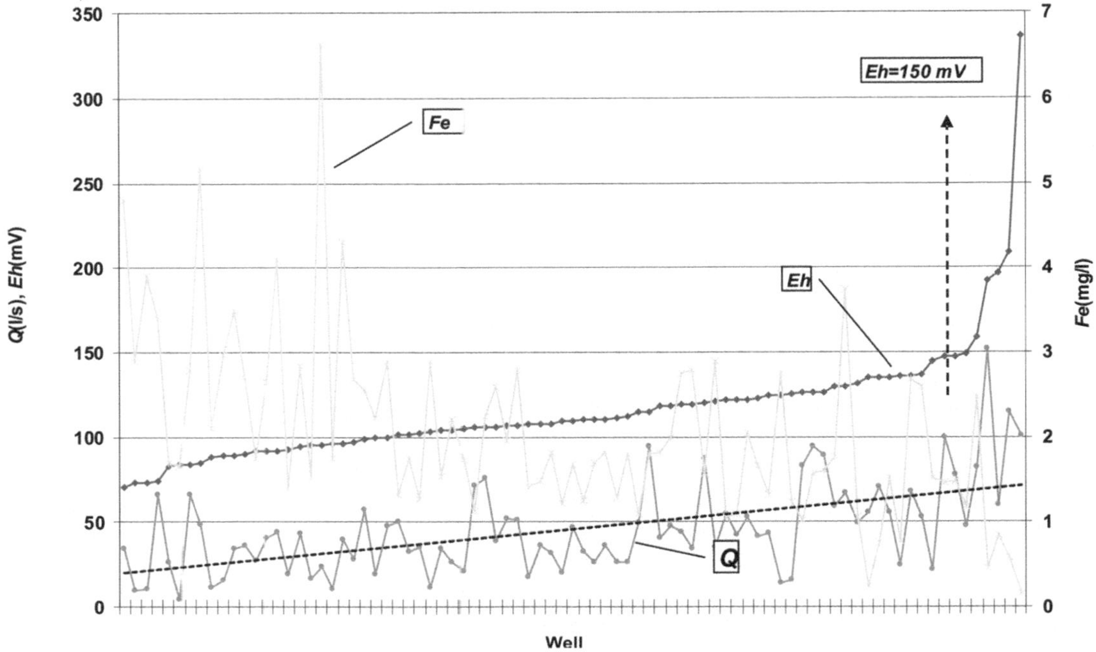


**Figure 5.10** $Fe = f(Q)$ and $Eh = f(Q)$ at Belgrade Groundwater Source (Dimkić *et al.*, 2008).

Figure 5.12 shows that *Fe* decreases with increasing *Eh*, which generally coincides with data collected from identical measurements at the same groundwater source.

Therefore, the capacity of a well depends on several factors, as expressed by the above equation. It is interesting to note that no significant correlation was established between potentially active bacteria (IRB, SRB, SLYM) and well yield.

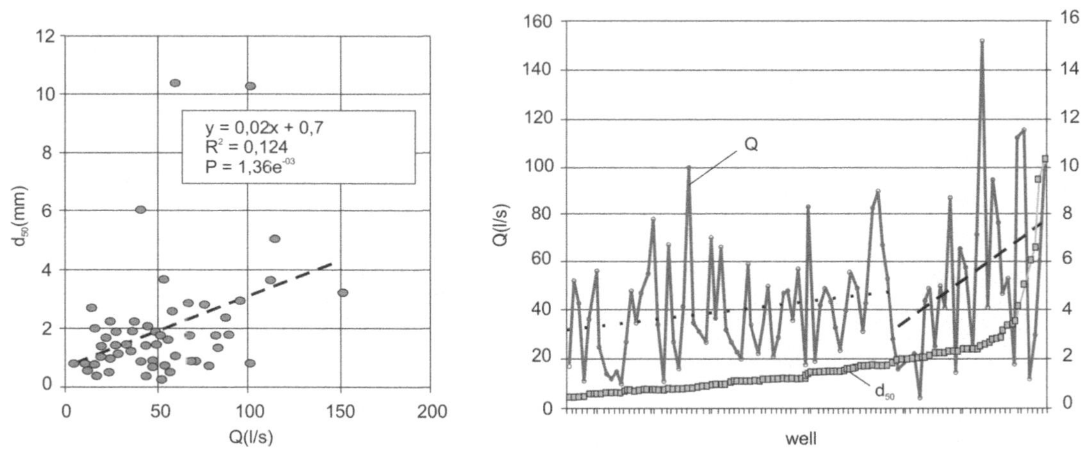


**Figure 5.11** Well yield as a function of aquifer grain size ($d_{50}$) (Dimkić *et al.*, 2008).

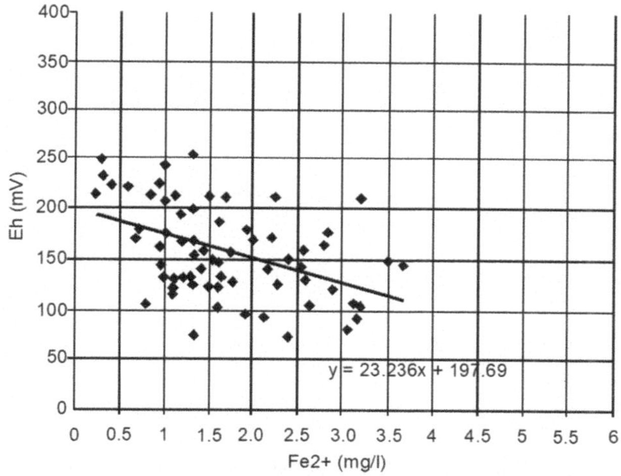


**Figure 5.12** Standard electrode potential as a function of $Fe^{2+}$ concentration at Belgrade Groundwater Source (Petković *et al.*, 2008).

## 5.5.2 Local hydraulic resistances – permissible entrance velocities at Belgrade Groundwater Source

The investigations conducted at BGWS allowed a correlation to be established between KLHR and clogging indicators of RCW laterals ($q$, $Fe$, $Eh$, $B$, $G$). In this context, well clogging also depends on pumping. The practical question is how to estimate LHR, which should not exceed a set value during a given time. The following empirical formula is recommended:

$$\Delta S_{year} = v_{perm} \cdot \mathrm{KLHR}_{year} \leq AV(\mathrm{m}) \quad (5.22)$$

Where $\Delta S_{year}$ is the given, limiting value of the hydraulic resistance of a lateral, generated during a single year (m), and $v_{perm}$ is the permissible entrance velocity of groundwater flow (m/s).

Based on the above, the permissible well (or RCW lateral) entrance velocity can be calculated from the following formula, taking into account the annual variation parameter (e.g. the oxic state) as a criterion:

$$v_{perm} = \frac{\Delta S_{year}}{\mathrm{KLHR}_{year}} \quad (5.23)$$

In practice, the maximum permissible entrance velocity ($v_{perm}$) had previously been associated only with the filtration stability of the near-well region (e.g. Abramov, Sichardt, Kovacs). Now 'permissible clogging or ageing' in the form of KLHR (i.e. permissible drawdown $\Delta S_{year}$) is an additional criterion.

The assumption is that both criteria have to be fulfilled. Of course, if $v_{perm}$ is determined and the well screen sized to meet the condition that the entrance velocity should not exceed the permissible velocity ($v_{en} < v_{perm}$), that does not mean that clogging of the laterals will cease. A correlation was established between $v_{perm}$ and the oxic state of the groundwater, with $Eh$ taken as the indicator.

Figure 5.13 shows both criteria for entrance velocities – the standard criterion that considers only the filtration stability (Abramov, Sichardt, Kovacs) and the proposed approach (KLHR).

BGWS studies show that $v_{perm}$ of nearly all the tested wells was much lower than that derived from 'standard' formulae, based on experience, which are related solely to the filtration stability of the near-well region. This is especially true of anoxic/suboxic groundwater, where 70 mV < $Eh$ < 150–200 mV.

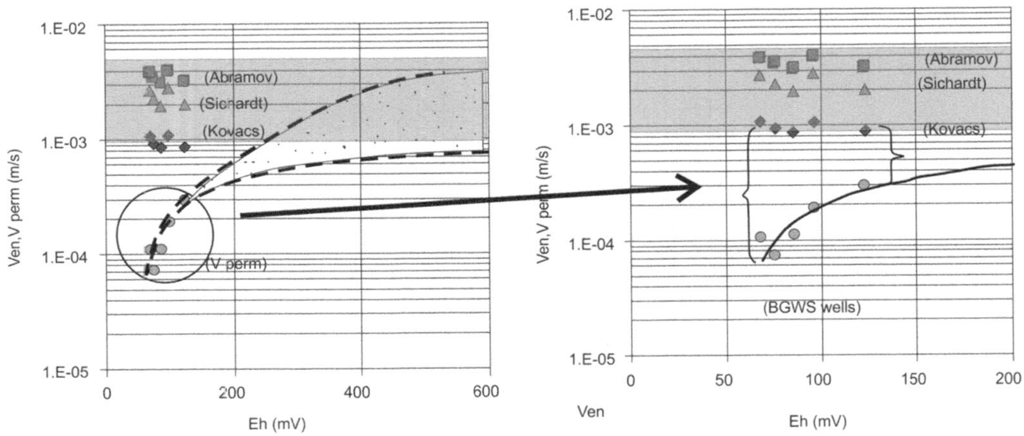


**Figure 5.13** Graphical representation of the correlation between entrance velocities *v* (standard criterion) and $v_{perm}$ with limited permissible permeability decrease (depending on the oxic state), expressed in terms of groundwater *Eh* (Dimkić *et al.*, 2012).

In addition to BGWS, wells from other groundwater sources (Ključ along the Velika Morava River and Mediana along the Nišava River) were tested to examine the effect of high oxic states ($Eh > 200$ mV). The test results generally showed that biochemical processes are extremely important where wells are drilled in anoxic groundwater environments. If the oxic state is high, the relevant criteria are the standard filtration stability and maintenance of laminar flow.

## 5.5.3 Case studies of radial collector wells RB-8 and RB-4

Data collected from two wells are presented below to corroborate that certain parameters affect well ageing.

Wells RB-8 and RB-4 are very different in terms of discharge history and certain biochemical parameters, but the yield of both was initially high. Well RB-8 performed poorly and required numerous and extensive interventions. In contrast, well RB-4 exhibited extraordinary pumping properties and maintained its typically high capacity (except during the latter half of the 1960s). The yield history of the two wells is shown in Figures 5.14 and 5.15.

Figure 5.16 shows internal details of the laterals of well RB-8.

The main characteristics of the RB-8 and RB-4 locations – typical grain-size, redox potential of the pumped water, total and bivalent iron, etc. – are shown in Table 5.2.

Chemical analyses showed that the main difference between RB-4 and RB-8 was that oxidation conditions existed in well RB-4 and reduction conditions in RB-8.

It is interesting that the Langelier index, Ryznar number and bicarbonate concentrations of the two wells were similar, but the iron concentrations differed considerably and were much less favourable at well RB-8.

The conclusion was that well RB-4 has a greater chance of maintaining a high yield in the long term because of its better recharge conditions, higher redox potential and larger grain size.

## 5.5.4 Case studies of radial collector wells RB-5 m, RB-8, RB-15, RB-16 and RB-20

Summary results of analyses of screen clogging of five BGWS RCWs are shown below. These were new wells, tested to examine the causal relationships between groundwater entrance velocities and the primary well ageing indicators: groundwater redox potential and iron concentration.

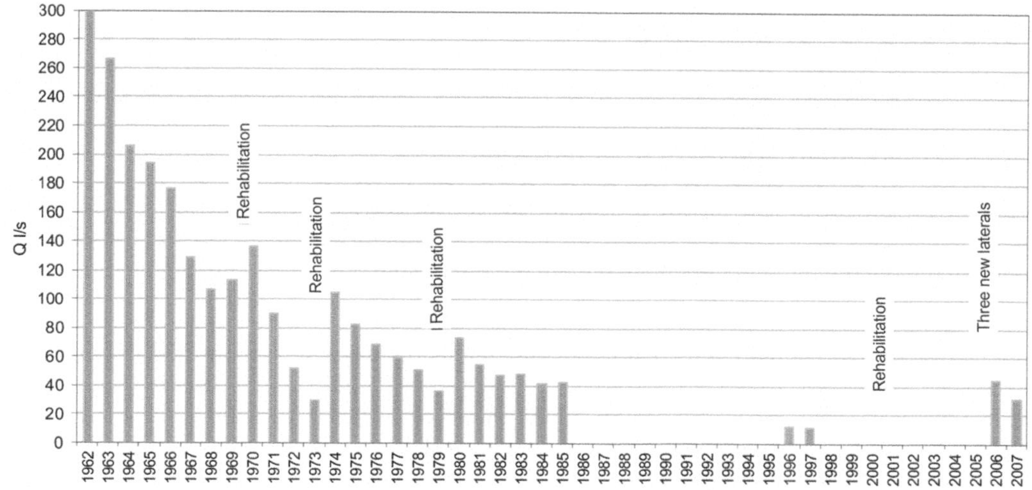


**Figure 5.14** Yield history of well RB-8 and times of rehabilitation (Dimkić *et al.*, 2011b, c).

BGWS is characterised by a relatively low oxygen content and low redox potential. Iron concentrations vary. The groundwater is borderline aerobic/anaerobic, so it is a relatively inert environment for corrosion and mechanical clogging (Table 5.3).

The most efficient investigation of the role of clogging was conducted at five BGWS RCWs, focussing on newly-installed laterals. Table 5.4 shows multiyear averages of select physical and chemical parameters of the sampled well water.

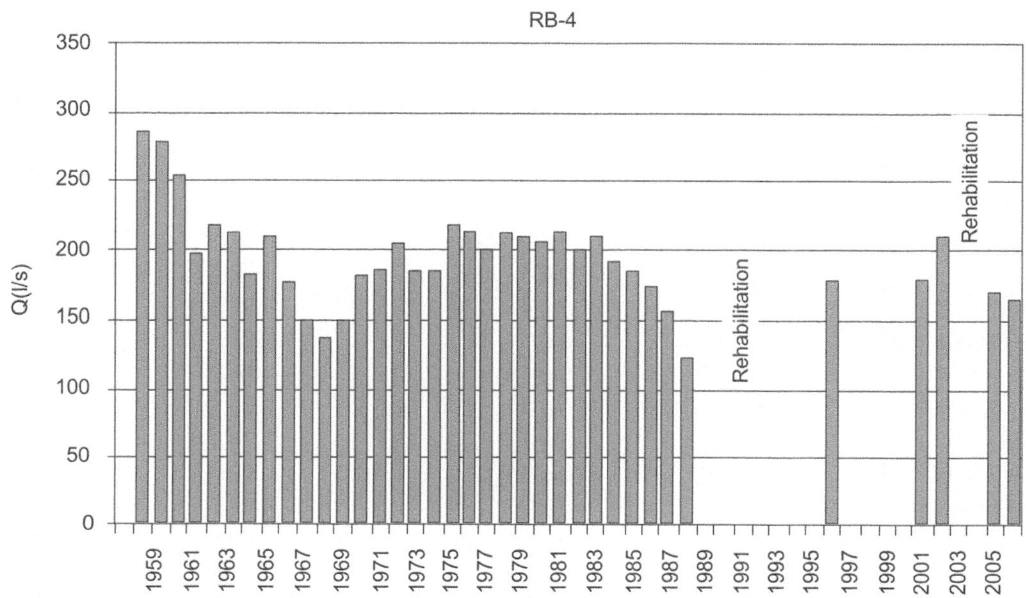


**Figure 5.15** Yield history of well RB-4 and times of rehabilitation (Dimkić *et al.*, 2011b, c).

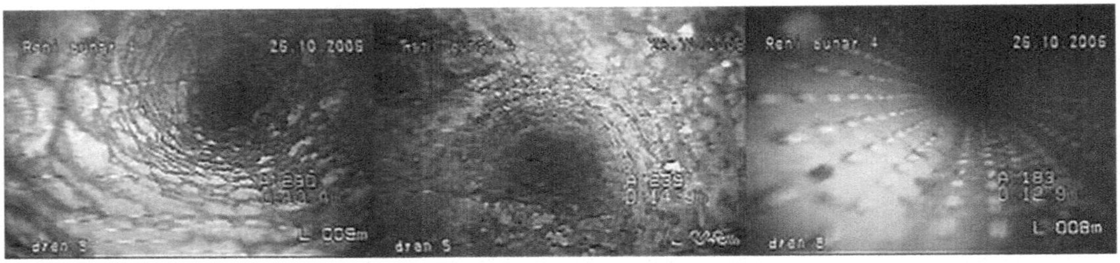

**Figure 5.16** Details of well RB-8 laterals (Dimkić *et al.*, 2011b, c).

Table 5.5 shows the average capacity per well and per lateral, along with the period of operation with new laterals

The fact that it was possible to compare hydraulic entrance resistances to some of the clogging/ageing indicators, such as *Eh*, Fe and bacteria, was of special interest.

Clogging of RCW laterals results in increased local hydraulic losses at the screen entrance. Local piezometric loss is a hydraulic parameter that can be used to quantify the progress of clogging.

### *5.5.4.1 LHR variation at new BGWS wells*

As previously mentioned, the laterals of five BGWS RCWs were replaced between 2005 and 2008. The latest technologies were applied and the screen materials were stainless and nickel-plated steel. Eighteen laterals were installed. Their individual length was 40 or 50 m. The old laterals were shut off. This allowed the

**Table 5.2** Comparison of well RB-8 and RB-4 parameters.

| | | RB-8 | RB-4 |
|---|---|---|---|
| Characteristic aquifer grain-size | $d_{10}$ (mm) | 0.15 | 0.35 |
| | $d_{50}$ (mm) | 0.4 | 4.5 |
| pH | | 6.90 | 7.10 |
| Conductivity at 25°C | Δ*S*/cm | 812 | 745 |
| Measured redox (*Eh*) | mV | 145 | 325 |
| Total hardness | mg$CaCO_3$/l | 417.22 | 336.66 |
| $CO_2$ | mg/l | 176.9 | 162.1 |
| Calcium | mg/l | 131.0 | 83.5 |
| Dry residue at 105°C | mg/l | 545 | 377 |
| Bicarbonates | mg/l | 439.2 | 408.1 |
| $Fe^{2+}$ | mg/l | 3.68 | 0.02 |
| Fe total | mg/l | 5.74 | 0.11 |
| Well discharge | mgN/l | 4.80 | <0.05 |
| Local hydraulic loss | l/s | 44 | 162 |
| Langelier index, *I* | m/l/s | 0.038 | 0.016 |
| Ryznar number, *R* | | −0.27 | −0.6 |
| *RSI* | | 7.43 | 7.39 |

**Table 5.3** Analysed parameters of BGWS radial collector wells (Dimkić *et al.*, 2011b, c).

| Parameter | Minimum Value | Average | Maximum Value | Number of Analyses (approx.) |
|---|---|---|---|---|
| Fe (mg/l) | 0.16 | 2.04 | 6.63 | 350 |
| Eh (mV) | 71 | 118 | 336 | 600 |
| $O_2$ (mg/l) | 0.00 | 0.24 | 2.4 | 600 |
| $d_{50}$ (mm) | 0.37 | 1.77 | 10.3 | 2500 |
| $NH_4^+$ (mg/l) | 0.07 | 0.66 | 2.3 | 350 |
| $NO_3^-$ (mg/l) | <0.05 | 0.46 | 5.02 | 270 |
| pH | 7.1 | 7.3 | 7.6 | 600 |

authors to monitor and correlate the yield of the 'new' wells and certain well ageing indicators, during the same period. These wells, too, exhibited a yield decline after they became operational again. Multiyear testing focussed on the rate of clogging of the new laterals, expressed via the increase in local hydraulic resistances. Monitoring of LHR variation and quantification of KLHR provided a picture of biochemical iron clogging of the five wells. This was achievable for the following reasons:

- Clogging was monitored after the installation of new laterals;
- Iron clogging was dominant (other types of clogging and corrosion were negligible in this case); and
- Observation wells were installed and monitored from the beginning of the process.

Figure 5.17 shows the variation in LHR of the 'new' wells over time. The relative time axis (horizontal axis) represents the period of operation with new laterals. Most of the wells were rehabilitated at least once during the study period (broken lines and denoted by R in the figure). It is clear that in 6–7 years of pumping, LHR increased 20 times on average.

**Table 5.4** Main physical and chemical parameters before and after installation of new laterals.

| Parameter | RB-5 m | | RB-8 | | RCW RB-15 | | RB-16 | | RB-20 | |
|---|---|---|---|---|---|---|---|---|---|---|
| | Before | After | Before | After | Before | After | Before | After | Before | After |
| Conductivity | 527 | 483 | 696 | 670 | 416 | 399 | 413 | 407 | 419 | 424 |
| $Fe^{2+}$ (mg/l) | 2.93 | 2.73 | 4.69 | 4.71 | 0.77 | 0.8 | 0.48 | 0.51 | 1.6 | 1.53 |
| Fe tot (mg/l) | 3.15 | 3.96 | 5.83 | 5.46 | 0.95 | 0.94 | 0.71 | 0.56 | 1.7 | 1.63 |
| Mn (mg/l) | 0.55 | 0.56 | 0.52 | 0.41 | 0.43 | 0.37 | 0.4 | 0.35 | 0.3 | 0.29 |
| Eh (mV) | 99 | 75.5 | 98 | 85.6 | 139 | 96.4 | 120 | 122.7 | 84 | 68.6 |
| $O_2$ (mg/l) | 0.3 | 0.1 | 0.1 | <0.1 | 0.1 | <0.1 | 0.14 | <0.1 | 0.2 | <0.1 |
| $KMnO_4$ (mg/l) | 6.4 | 6.52 | 4.4 | 6.94 | 4.8 | 5.5 | 4.2 | 5.24 | 4.8 | 4.84 |
| $NH_3$ (mg/l) | 0.88 | 0.83 | 0.89 | 0.96 | 0.43 | 0.41 | 0.4 | 0.41 | 0.3 | 0.3 |
| pH | 7.34 | 7.63 | 7.2 | 7.25 | 7.51 | 7.67 | 7.58 | 7.72 | 7.5 | 7.67 |
| IRB | No data | 236 | No data | 1115 | 9000 | 496 | 1162 | 58 | 25 | 53 |
| SRB | 25 | 251 | No data | 83 | No data | 46 | 14 | 53 | No data | 53 |
| SLYM | No data | 19.530 | No data | 2866 | 12500 | 303 | 100 | 3.846 | 500 | 354 |

**Table 5.5** Average pumping capacity per well and per lateral.

| Well | Average Discharge | | Period of Operation with New Laterals (years) | Number of Tested Wells |
|---|---|---|---|---|
| | Per well (l/s) | Per lateral (l/s) | | |
| RB-16 | 64 | 16 | 1.7 | 7 |
| RB-20 | 36 | 9 | 1.4 | 6 |
| RB-15 | 26 | 6.5 | 2.7 | 8 |
| RB-5 m | 32 | 8 | 1.6 | 7 |
| RB-8 | 20 | 10 | 3 | 10 |

The initial LHR values of the five wells were similar (no clogging). LHR increased due to clogging of the laterals and varied among the wells. This finding was extremely important because the wells were practically new. Individual clogging rates of the wells were related to the *Eh* value and Fe concentration.

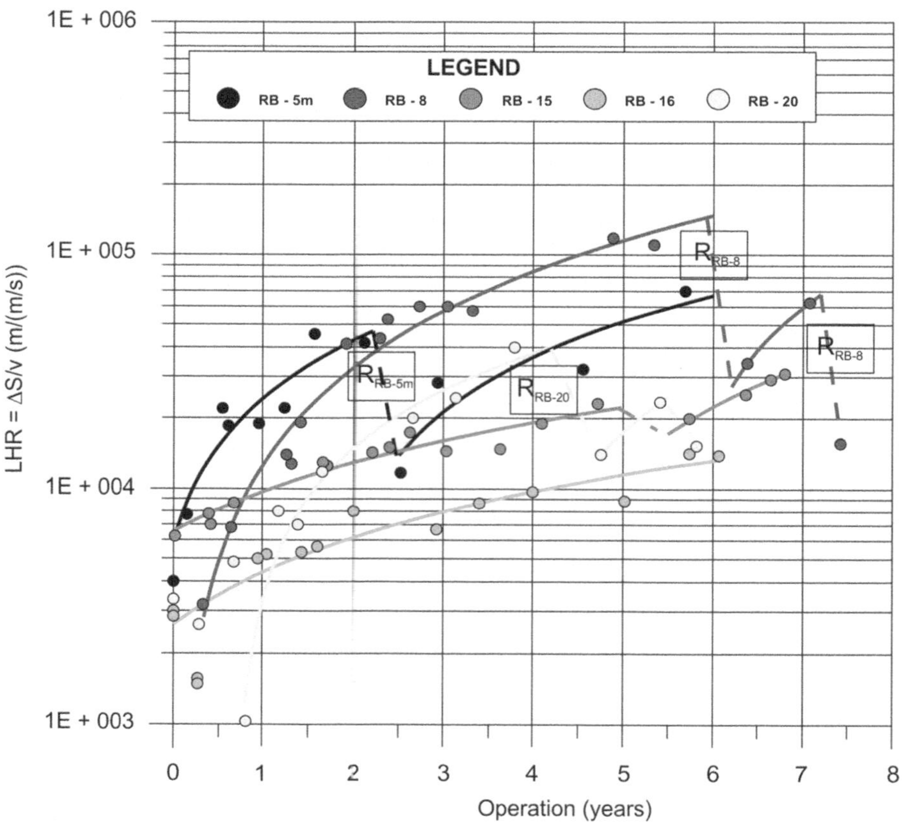


**Figure 5.17** LHR variation after installation of new laterals, R – rehabilitation (Dimkić *et al.*, 2011b, 2012 modified and supplemented).

**Table 5.6** KLHR of five 'new' wells.

| **Well** | **RB-5 m** | **RB-8** | **RB-20** | **RB-15** | **RB-16** |
|---|---|---|---|---|---|
| KLHR(m/l/s)/year | $8.1 \times 10^{-1}$ | $4.4 \times 10^{-1}$ | $1.8 \times 10^{-1}$ | $1.2 \times 10^{-1}$ | $0.7 \times 10^{-1}$ |

Using the plot, it was easy to determine KLHR or the variation in LHR during the study period. KLHR can be calculated as

$$\mathrm{KLHR} = \frac{\mathrm{LHR}(t_1) - \mathrm{LHR}(t_0)}{t_1 - t_0} \tag{5.24}$$

where: $t_1$ is the time of observation of the process and $t_0$ is the time since the installation of laterals. The KLHR values derived for the five wells are shown in Table 5.6.

Since 2005, the public water supply utility has undertaken several rehabilitations of four of the five 'new' wells. Based on insight into the data, KLHR of well RB-5M was calculated for two intervals, pre- and post-rehabilitation. A somewhat slower post-rehabilitation increase in KLHR is attributable to the already created microbial quasi-equilibrium conditions, due to pumping during the previous period.

These case studies of the five BGWS wells indicated a definite correlation between the rate of clogging and the chemical properties of the groundwater, primarily the iron concentration and redox potential. Figure 5.18 shows that the rate of clogging (KLHR) was proportional to the iron concentration and inversely proportional to the redox potential of the groundwater.

The plots should not be considered universal. However, they are indicative of the nature of the relationship KLHR = KLHR (*Fe*, *Eh*.....). It should be noted that the wells were installed in a low anoxic environment.

The fact that each parameter defines KLHR on its own is important, because that value is a result of a set of correlations among inter-connected parameters (Dimkić *et al.*, 2011a, c):

$$\mathrm{KLHR} = f_2(Eh, Fe, B) \tag{5.25}$$

A comparison of KLHR (shown in the above table) to *Fe* and *Eh*, and analysis of Figure 5.18, suggests that rapid clogging corresponds to high Fe concentrations and low *Eh*.

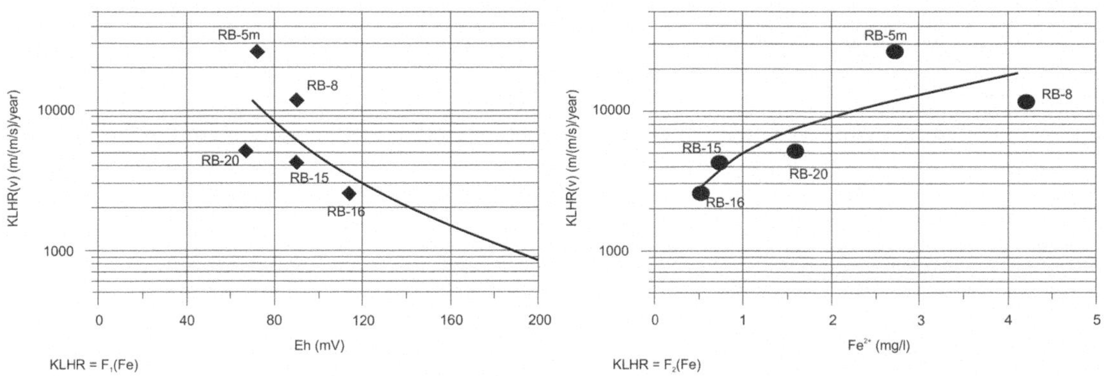


**Figure 5.18** KLHR as a function of Fe concentration and redox potential of pumped water (Dimkić *et al.*, 2011b, c).

The results of testing of these five wells, and similar testing of other BGWS wells (more than 90 RCWs), allowed the authors to identify approximate threshold values; iron concentration $Fe = 1.2$ mg/l and redox potential $Eh = 150$ mV mark the dividing line between 'good' and 'poor' well locations.

If clearly follows from the examples that iron incrustation is an overriding factor of well ageing. It is dominant among the various types of screen clogging.

Taking into account the results of testing of the other BGWS wells, the general conclusion is that rapid well ageing occurs under alluvial anaerobic and semi-anaerobic conditions ($Eh < 150$ mV). A large number of samples were analysed, so the level of confidence is high. On the other hand, the size of BGWS (50 km of riverbank) and different geological and other conditions in some of its parts cause data scatter.

The rate of well clogging is higher under natural anoxic than mildly aerobic conditions. Well ageing estimation is important when locations for future groundwater sources are chosen and well characteristics (e.g. screen type and material) selected. The indicators are also a basis for forecasting future maintenance costs of the wells.

## 5.6 GENERALISED PLOT OF $V_{PERM} = F(EH)$

Past well design practices called for ensuring filtration stability of the water-bearing formation. In other words, aquifer particles should be prevented from mobilising and entering the well. If that happens, the normal operation of a well might be seriously threatened, to the point that it needs to be shut down. Filtration stability is ensured by limiting the pumping capacity and screen entrance velocities. The permissible entrance velocities that guarantee the aquifer's filtration stability in the near-well region have been studied by Istomina (1957), Abramov (1952), Cistin (1965), Johnson (1975), Gavrilko and Alekseev (1985), Truelsen (1988), Vuković and Pušić (1992), Dimkić and Pušić (2008) and others.

Entrance velocities quantified in this manner do not guarantee the absence or at least deceleration of well ageing due to clogging. The increase in LHR, manifested by the so-called parasitic well drawdown, is an important aspect of well maintenance planning.

BGWS keeps records of the pumping history, rehabilitations, number and lengths of RCW laterals, test pumping, groundwater redox potential and bivalent iron concentration. The number of surveys ranges from 1 to more than 10, depending on the well.

A total of 47 wells were selected for analysis, 42 of which were 'old' (no laterals had been replaced) and five were 'new' (old laterals replaced with non-corrodible laterals). Three or more surveys were undertaken.

The objective was to find out whether there was a correlation between *Eh* and LHR increase due to iron incrustation. Their variation over time was analysed based on previously determined local losses. The variation (increase) was monitored via the annual local drawdown ($\Delta S$). The wells were classified as shown in Table 5.7.

**Table 5.7** Classification of BGWS radial collector wells based on annual increase in local drawdown.

| $\Delta S$ (m/year) | Number of Wells |
|---|---|
| $\Delta S < 0.2$ | 9 |
| $0.2 < \Delta S < 0.5$ | 6 |
| $0.5 < \Delta S < 1$ | 8 |
| $1 < \Delta S$ | 27 |

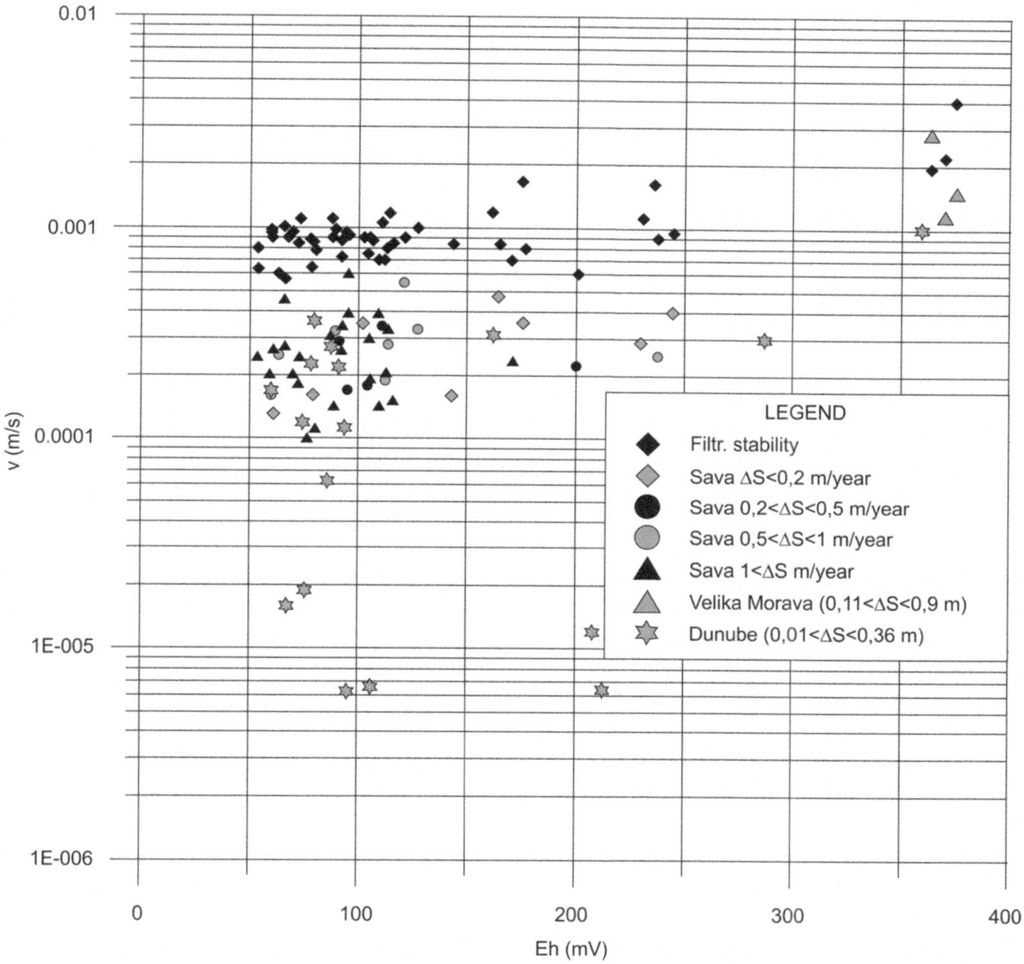


**Figure 5.19** Well entrance velocities as a function of groundwater redox potential (Dimkić & Pušić, 2014).

The value for 15 wells was $\Delta S < 0.5$ m/year and for 35 $\Delta S > 0.5$ m/year. Average entrance velocities ($v$) were calculated for all the wells.

The average entrance velocities were plotted versus the redox potential of the groundwater, $v = f(Eh)$, as shown in Figure 5.19. The figure also includes the criterion for permissible entrance velocities that guarantee filtration stability of the near-well region according to Abramov, Sichardt, Kovacs, and Gross. The Sichardt and Kovacs criteria are the most stringent.

The figures also include other studied locations: Kovin-Dubovac, Knićanin-Čenta and Veliko Gradište along the Danube and Ključ along the Velika Morava River, adding 22 wells to the study.

The correlations among the groundwater redox potential, entrance velocities and the rate of well ageing (i.e. increase in local resistances) served to determine the critical or maximum permissible well entrance velocity. In view of its nature, this concept is both technical and economic. For BGWS, an annual increase in local drawdown of $\Delta S_{AV} = 0.35$ m was adopted for 'good' wells. This is also the limit of permissible well ageing.

**Table 5.8** Wells grouped by representative redox potential (*Eh*) and entrance velocity (*v*).

| *Eh* (mV) | Wells | *Eh* Average (mV) | $v_{(filtr.stab)}$ Average (m/s) | $v(\Delta S = 0.35$ m/y) Average (m/s) |
|---|---|---|---|---|
| 50–100 | RB-15–1, Bp-64, Bp-25, RB-13, RB-49, RB-19-1 | 78 | $8.7 \times 10^{-4}$ | $1.4 \times 10^{-4}$ |
| 100–150 | RB-84, RB-83, RB-8 m, RB-22 | 115 | $8.9 \times 10^{-4}$ | $2.6 \times 10^{-4}$ |
| 150–200 | RB-4, RB-95 | 171 | $8.2 \times 10^{-4}$ | $4.2 \times 10^{-4}$ |
| 200–300 | RB-59, RB-21, Bn-5, RB-23 | 228 | $1.1 \times 10^{-3}$ | $6.2 \times 10^{-4}$ |
| 300–400 | VB-6, VB-2, EB-7 | 370 | $2.7 \times 10^{-3}$ | $1.82 \times 10^{-3}$ |

Wells whose $\Delta S \leq 0.5$ m were used as benchmarks for quantifying the critical entrance velocity.

A table was generated, which also includes wells from groundwater sources along the Velika Morava and the Danube. Fifteen wells were analysed. The data from the wells along the Velika Morava and the Danube were not recalculated to $\Delta S = 0.35$ m/year. Instead, real data was used. The wells along the Velika Morava exhibited virtually no losses, while the pumping rate of the drainage wells along the Danube was rather low, so there could be no significant losses.

Given that the data were not uniform in the considered range, they were redistributed to averages in selected intervals. The results of that interpretation are shown in Table 5.8. The data points were entered into the plot shown in Figure 5.20 and lines were drawn through them (one for $v$ – filtration stability, and the other for $v$ ($\Delta S = 0.35$ m/year)).

Figure 5.20 shows the results and illustrates the correlation between *Eh* and groundwater entrance velocities of wells in certain alluvial aquifers in Serbia. It follows that the velocities that guarantee filtration stability of the near-screen zone do not meet the permissible local hydraulic resistance criterion at low *Eh* (especially if $Eh < 200$ mV). The two criteria are practically equal if *Eh* is greater than 300 mV.

Given that the focus is on well ageing caused by iron incrustation, the correlation between *Eh* and $Fe^{2+}$ was examined in detail. Figure 5.21 shows the averaged results for the studied wells along all three rivers.

There is a definite correlation between the two parameters. At an *Eh* value of 150–200 mV, the average $Fe^{2+}$ concentration is of the order of 0.5 mg/l. As *Eh* decreases, the concentration rapidly increases.

A plot of entrance velocities as a function of $Fe^{2+}$ (Figure 5.22) was constructed based on the established correlation.

It appeared that the variability of local hydraulic resistances had to be taken into account when the maximum permissible entrance velocity was quantified. In nearly all the cases, this quantity was much lower than when estimated using the formula that guarantees filtration stability in the near-screen zone. This is especially true of anoxic groundwater conditions, or in the present case where $Eh \leq 200$ mV and $Fe^{2+} \geq 0.5$ mg/l.

This also sheds light on the phenomenon that has puzzled experts who studied BGWS well ageing. The design criteria pertaining the highest entrance velocities had been stability of the near-screen zone and the screen. However, the effect of biochemical factors (above all redox potential and iron concentration of the groundwater) is introduced here as part of the focus of consideration.

Data from other studied groundwater sources were added to the BGWS data, to supplement the picture of the effect of iron concentration and oxic state on well ageing. Hence, the data pertain to three locations (along the Sava, Danube and Velika Morava) and corroborate the general presence of well clogging processes in alluvial environments.

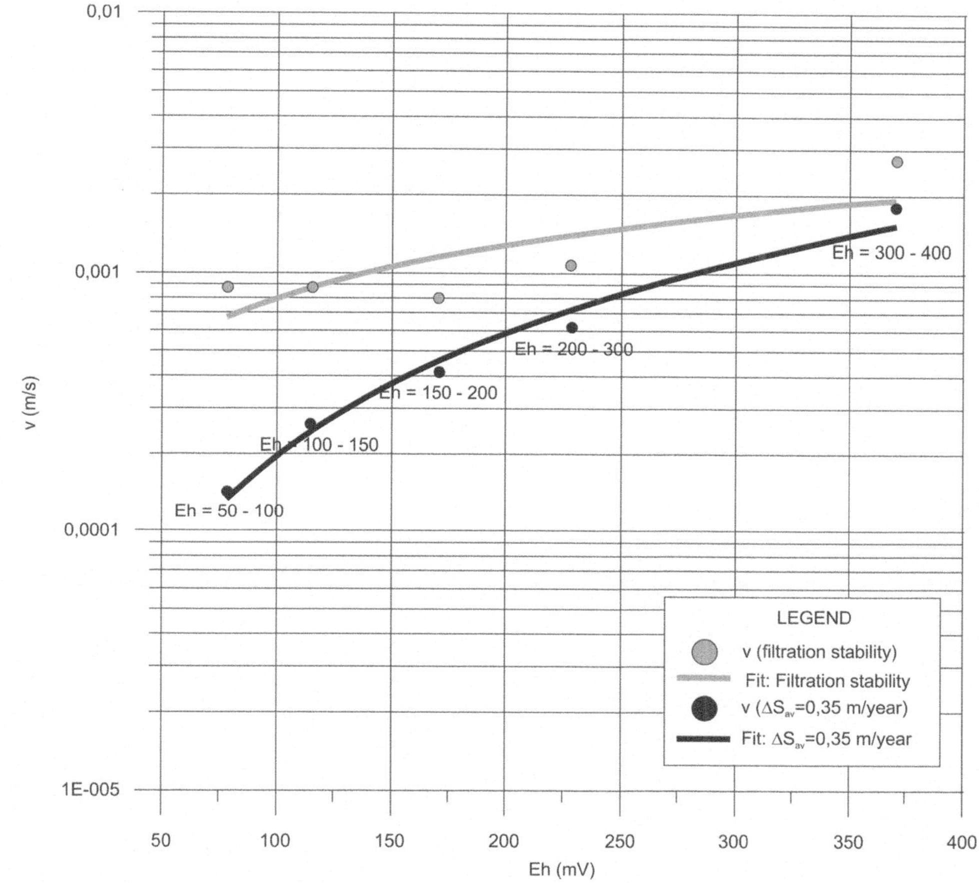


**Figure 5.20** Well entrance velocities as a function of groundwater redox potential, where the annual increase in local drawdown is controlled at $\Delta S = 0.35$ m/year (Dimkić & Pušić, 2014).

For *Eh* greater than 200 mV and $Fe^{2+}$ less than 0.5 mg/l, the criteria derived from the filtration stability condition remain dominant, while the effect of biochemical clogging decreases and ultimately disappears.

## 5.7 A NOVEL APPROACH TO THE DEFINITION OF DESIGN CRITERIA FOR RADIAL COLLECTOR WELLS IN ANOXIC SETTINGS

This approach has already been partially presented as an original scientific work in the Geological Annals of the Balkan Peninsula 2018, 79 (1) 31–45 and Pušić and Dimkić (2017).

### 5.7.1 Case study of radial collector well RB-16

Well RB-16 is part of a wellfield (BGWS) that provides drinking water supply to the city of Belgrade, Serbia. The wells tap an alluvial aquifer along the Sava River. The groundwater is mildly anoxic (*Eh* ~125 mV, $Fe^{2+}$ ~0.46 mg/l, $NO_3^-$ ~0.24 mgN/l, $O_2$ ~0.1 mg/l).

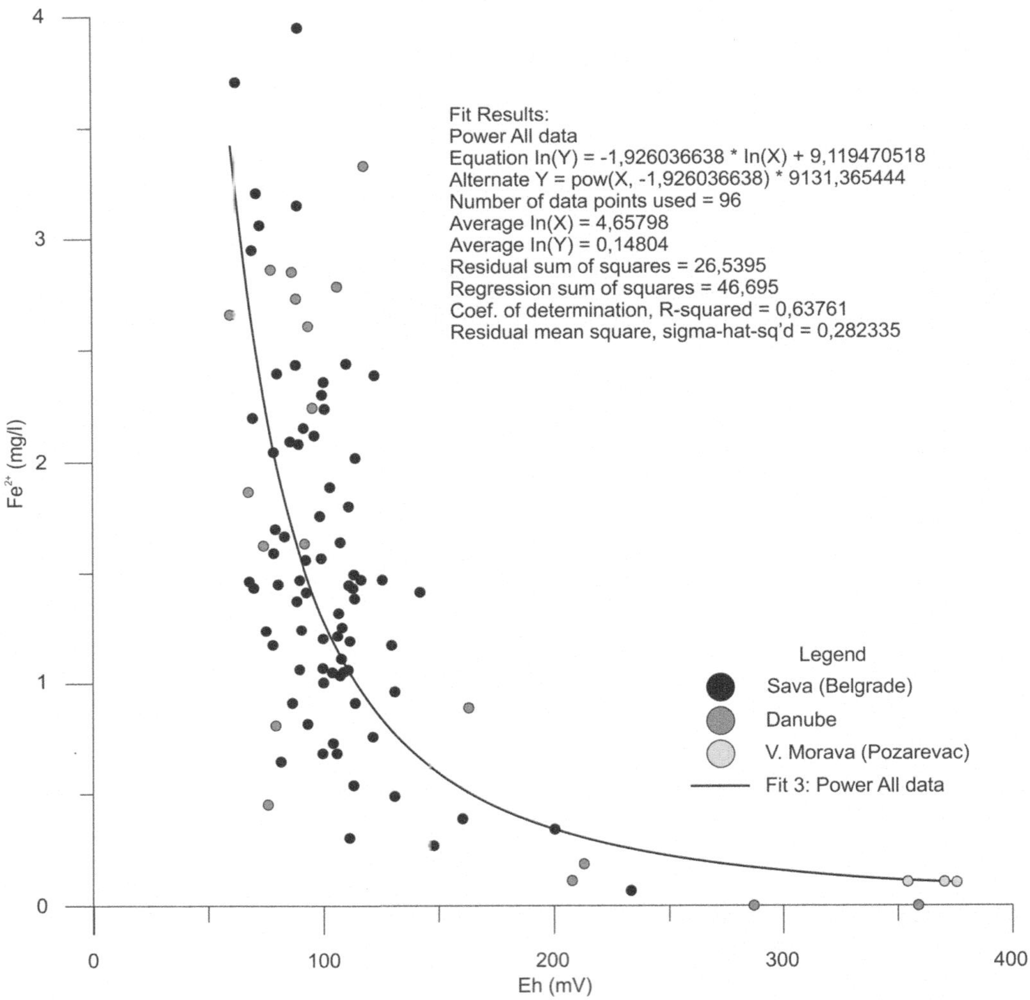


**Figure 5.21** Correlation between *Eh* and $Fe^{2+}$ in the Sava, Danube and Velika Morava alluviums (dots represent averages per well) (Dimkić & Pušić, 2014).

This case is specific because it demonstrates the application of a novel method for defining the propensity of a well to be exposed to iron incrustation under anoxic alluvial conditions. The maximum permissible groundwater flow velocities and the capacities of RCW laterals and the entire well were determined based on the established correlation between the rate of well clogging, redox potential and iron concentration of the groundwater, as well as the entrance velocities at RCW laterals. A condition was imposed that hydraulic resistances due to iron precipitation on the lateral screens should be less than specified. The correlation between the maximum entrance velocities (velocities that still enable an increase in local hydraulic resistances at the lateral to be smaller than specified) and biochemical indicators (*Eh*, $Fe^{2+}$) was taken from previous reports of the authors and their associates. Original

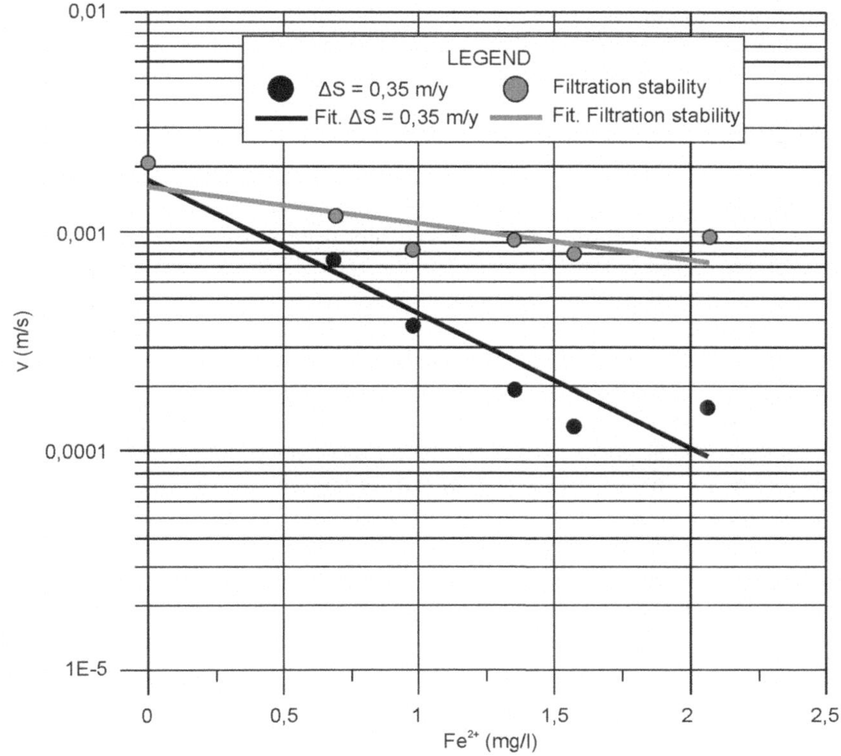


**Figure 5.22** Entrance velocities as a function of $Fe^{2+}$, where the annual increase in local drawdown is controlled at $\Delta S = 0.35$ m/year (Dimkić & Pušić, 2014).

software, which supports 3D analysis, was used to assess the potential yield of the RB-16 location. The boundary conditions were adjusted accordingly.

The authors of this chapter and their associates correlated anoxic aquifer conditions, iron concentrations in groundwater, the intensity and kinetics of screen clogging and the increase in hydraulic resistances at the screens, and established functional relations between them.

As groundwater becomes anoxic, maximum permissible groundwater flow velocities ($v_{perm}$) are increasingly governed by biochemical parameters of the aquifer. The relationships:

$$v_{perm} = f(Fe^{2+}) \tag{5.26}$$

$$v_{perm} = f(Eh) \tag{5.27}$$

have been defined for alluvial sources in Serbia, and this approach has for the first time on a global scale been reported by Dimkić and Pušić (2014) and Dimkić *et al.* (2011a, 2011b, 2011c).

The present chapter describes the complex methodology for determining:

- The groundwater abstraction capacity of a certain location and a particular well; and
- The structural elements of an RCW (number of laterals, entrance velocities),

taking the characteristic biochemical properties of the anoxic aquifer as the starting point. This new design approach has a considerable effect on the longevity and, in general, the operational economics of RCWs. In essence, the approach should also be applicable to other types of wells (e.g. tube wells).

### 5.7.1.1 Setting, analysed processes, and statement of the problem

The alluvial complex tapped by the RCW laterals is characterised by multiple alternations of coarse (gravel) and fine (sand) sediments. The final sequences of the sedimentation cycles are represented by fine, predominantly clay particles, which have hydrogeologically been modelled as semi-permeable interbeds (of which there is locally one or more). The laterals have been installed in the lowest coarse-grain strata, such that the interbeds, along with the colmated riverbed, constitute the primary constraint of the location and well capacities. The greatest losses of energy (hydraulic resistances) during the course of groundwater flow occur at the riverbed and the interbed, as modelled (Figure 5.23). In the immediate vicinity of the laterals of well RB-16, where the concentration of groundwater leakage through the interbed (interlayer) is the greatest, the piezometric head difference between the strata above and below is as large as ten metres. This has been established by a pair of observation wells near the pumping well (Figure 5.24). The filtration characteristics and spread (continuity) of the interlayer directly affect the well discharge capacity and, as such, it is important to characterise the interlayer.

Well RB-16 is located on the right bank of the Sava River, on a river island (Ada Ciganlija) close to downtown Belgrade. It is surrounded on all sides by other wells, roughly at a distance of 400 m. It has four relatively new laterals (installed in 2007); the other laterals have been shut-off and decommissioned.

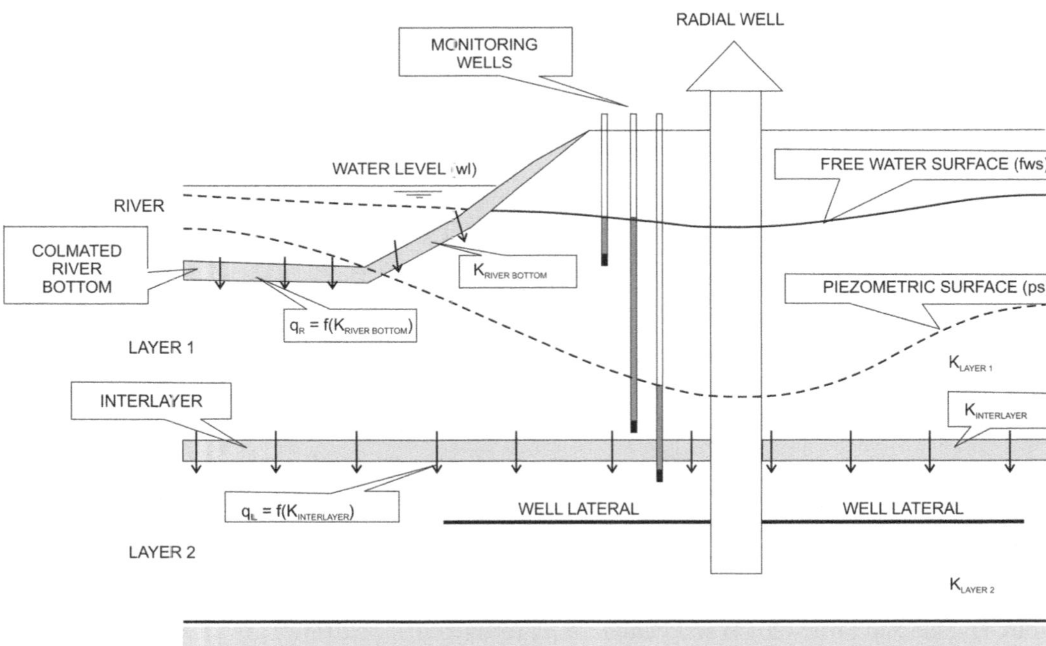


$K_{LAYER\ 1} \sim K_{LAYER\ 2} >> K_{INTERLAYER} \sim K_{RIVER\ BOTTOM}$

**Figure 5.23** Schematic representation of groundwater flow from the river to the radial collector well, where there is a semi-permeable interlayer (Dimkić & Pušić, 2018).

**Figure 5.24** Well RB-16 and its environs; map of a part of Belgrade Groundwater Source (Dimkić & Pušić, 2018).

The screens tap the lowest alluvial water-bearing stratum, at a depth of some 30 m. Given its position in space, the recharge zone of well RB-16 is relatively constricted, primarily by the operation of neighbouring wells. This limits its capacity even further.

The assessment and quantification of the discharge capacity of the well and, in hydraulic terms, the part of the aquifer to which it belongs, are in this case related to the previously-described conditions on the ground. The accuracy of results depends on the number and locations of the required observation wells. In reality, more pairs of observation wells should be put in place, to tap the strata above and below the interbed. Their screens should not be long (about half a metre). Unfortunately, it is not possible to install observation wells where they are most needed: below the riverbed and along the directions of the laterals towards the river.

In addition to the discharge capacity of the well and the location, it is necessary to determine the maximum screen entrance velocities, which in turn requires the quantification of the capacity of individual laterals. The conductivity of a horizontal screen is defined via the coefficient of local hydraulic resistance (LHR), which may vary both along the length of the lateral and over time. Here the maximum (permissible) screen entrance velocities are related to two criteria. The first criterion ensures filtration stability of the porous medium in the near-well region and is determined for site-specific conditions, generally based on experience. The second criterion is much more complex and pertains to well ageing, in the present case study due to screen clogging. The well ageing process is a set of different phenomena that occur simultaneously and whose intensity differ over space and time. They depend on the composition of the formation (grain-size distribution, petrology, mineralogy, geochemistry) and of the groundwater (physical, chemical, biochemical and microbial parameters).

Additionally, well ageing may depend on the operating mode of the well. Well ageing results in increasing LHR along the lateral and decreasing conductivity, until the lateral ultimately ceases to function. Long-term research undertaken at BGWS has revealed a distinct correlation between the above-mentioned drivers and the well discharge capacity (or screen entrance velocities). With regard to two groundwater parameters: redox potential ($Eh$ (mV)) and dissolved iron concentration ($Fe^{2+}$ (mg/l)), identified as well ageing indicators, the correlation with permissible entrance velocities ($v_{perm}$) has been empirically quantified as a function of a given ageing rate (rate of LHR increase) (Dimkić & Pušić, 2014).

$$v_{perm} = f\,(Eh, Fe^{2+}) \tag{5.28}$$

The second criterion is often more stringent and requires lower entrance velocities, resulting in a lower well discharge capacity, but it ensures longevity and requires less frequent rehabilitation of laterals. Groundwater flow is calculated applying the Richards equation, which describes flow through the vadose zone. This equation, as a function of the hydraulic potential, may be written as

$$C(h)\frac{\partial h}{\partial t} - \nabla \cdot (k_r K \nabla h) = q \tag{5.29}$$

where:

$C(h)\frac{\partial h}{\partial t}$ is a specific moisture capacity, $K$ – hydraulic conductivity tensor, $k_r$ – relative conductivity, and $h$ – hydraulic potential (piezometric head).

Three approaches were followed to simulate boundary conditions: Neumann's, Dirichlet's, and Robin's. The Neumann boundary condition applies to the part of the flow boundary at which the rate of discharge is known

$$u \cdot n = gN \tag{5.30}$$

where $u$ is Darcy's velocity, and $n$ is an external unit normal to the boundary. It is used to model the contour that represents the boundary streamline (in plan view, or surface in space, $gN = 0$), the wells ($gN < 0 > gN$), precipitation ($gN > 0$), and evaporation ($gN < 0$). The part of the boundary at which the hydraulic potential is known is modelled applying Dirichlet's boundary condition:

$$h = gD \tag{5.31}$$

Dirichlet's condition is applicable to boundaries that represent water flow (rivers, lakes, channels, etc.), as well as to wells (without LHR) where the piezometric head is known.

Robin's condition is specified in the case of semipermeable boundaries through which groundwater leaks

$$u \cdot n = \Psi(h - gR) \tag{5.32}$$

where: $\Psi = K_K/d$ – leakage coefficient, $K_K$ – hydraulic conductivity of the semi-permeable layer at the boundary, and $d$ – its thickness. A colmated riverbed, an RCW screen, and a tube well screen are typical examples of Robin's condition. Well clogging increases hydraulic resistances due to mechanical, biochemical and microbial processes (Dimkić *et al.*, 2008, 2011b, 2011c; Dimkić & Pušić, 2014).

### *5.7.1.2 Methods*

Spatial data related to well RB-16 were analysed and organised to produce a hydrogeological model of the terrain and alluvial sediments, governed by the principle that a layer is the basic spatial and hydraulic unit (a hydrogeologic entity, a result of sedimentation), whose filtration characteristics are homogenous across the areal extent. Licensed software RockWorks by RockWare, Inc. was used for analysis and display.

The initial values of filtration parameters, as the starting point of hydrodynamic modelling, were determined applying a two-fold approach: based on the grain-size distribution of the material sampled at different borehole depths, and based on the interpretation of multiple well pumping tests.

Well test data were entered and groundwater flow simulation calculations performed using an original solver – WODA (Well Outline and Design Aid), developed at Jaroslav Černi Water Institute, Numerical

Analysis Group, Belgrade. WODA is a simulator of variably saturated well-driven groundwater flow in an anisotropic discontinuous environment with miscible displacements, heat transfer, variable density, sorption, degradation, etc. The mathematical groundwork was prepared and the solver programmed by Vidović *et al.* (2014). WODA does not have its own graphical user interface yet, but it can work with groundwater models constructed using a Lizza interface. Lizza was developed in cooperation with the Bioengineering Research and Development Center – BioIRC, Kragujevac, Serbia. It supports full 3D representation of the results of stationary and non-stationary modelling of groundwater flow within the aquifer and the vadose zone, as well as the results of calculations of mass and heat transport via groundwater. Both software programs are open source and available at:

WODA – http://www.sourceforge.net/projects/wodasolver/
Lizza – http://www.bioirc.ac.rs/index.php/groundwater-flow-software

A special user-friendly feature of the software is ease of specifying the real cross-section of the river and RCW screens in the model (Dimkić *et al.*, 2010, 2011d).

### *5.7.1.3 Analysis and definition of model input data (schematisation)*

The study area was hydrogeologically schematised using available data: drilling data from 139 wells and several hundred grain-size distribution analyses of samples. The schematisation was based on an assessment of descriptions of tapped units, grain-size distributions of samples, and photos of borehole cores. Software was used to generate a 3D hydrogeologic model of the sediments. The first attempt produced eight schematised layers.

Figure 5.25 shows how the layers in the immediate vicinity of well RB-16 were modelled. The vertical dark grey rectangles on the left-hand side of the figure represent the average $d_{10}$ (mm) of the samples collected from boreholes in the near-well region. Light grey horizontal rectangles denote intervals of semi-permeable material, whose grain-size distribution was generally not analysed. The intervals were identified as semi-permeable based on descriptions and relatively few data. As previously mentioned, the profile derived in this manner was compared to the corresponding profiles of neighbouring wells in the

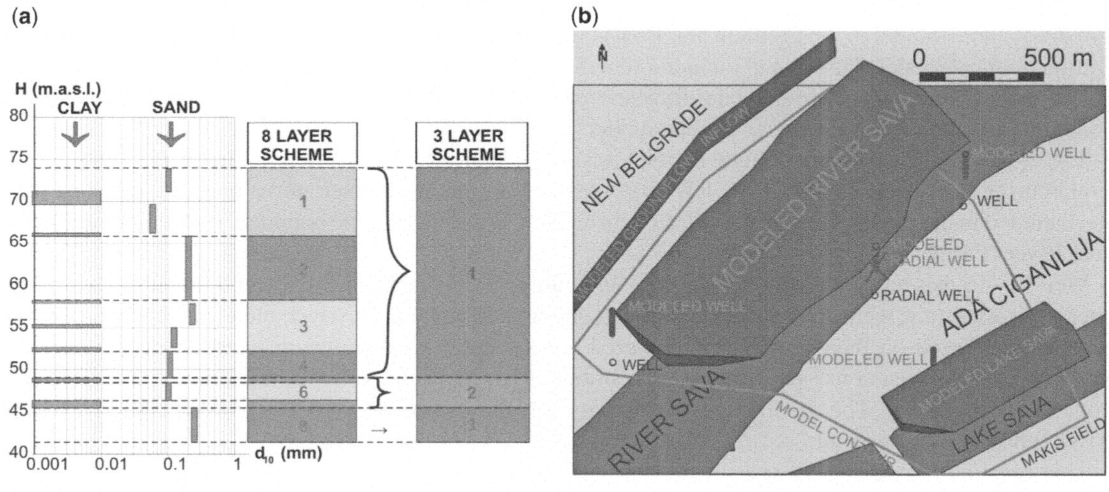


**Figure 5.25** Vertical distribution of grain $d_{10}$ (synthesized well RB-16 location data), switch from 8 to 3 layers (a), and model contours and boundary conditions (b) (Dimkić & Pušić, 2018).

study area, arriving at eight schematised layers. Then, the elevations and lengths of the screens of available observation wells, and the piezometric head recorded while the wells were in operation, were analysed to obtain an indication of the hydraulic relationships of the layers above and below layer packages 5, 6 and 7. In view of the problem addressed and data availability, it was concluded that a simpler hydrogeologic scheme (with fewer layers) could be adopted.

It should be noted that there was a pair of observation wells (RB-16-P-2, RB-16-P-3) with short screens (0.5 m) in the immediate vicinity of the pumping well, a short-screen observation well (RB-16-P-1) in the water-bearing layer, also in close proximity to the pumping well, and two observation wells (P-ut-16–1, P-ut-16-2) with relatively long screens (5 m) in the upper layer package, half way to the neighbouring wells. As concluded, the short screens virtually provided data on a point in space within the water-bearing layer, while the long screens, which covered several layers, provided the resulting piezometric head potential of the individual layers.

It should also be kept in mind that the groundwater flow in the near-well region, around the horizontal screens, is distinctly spatial (3D) and that under such conditions short-screen observation wells provide a more realistic picture. As the distance from the pumping well grew, groundwater flow became increasingly horizontal in nature, such that the screen length ceased to be of overriding importance for the veracity of the measured data.

As a simpler alternative, a hydrogeologic model comprised of three schematised layers was constructed (Figure 5.25). Apart from the knowledge about the study area, the model in this case was a result of extensive local exploration.

Even though it is generally deemed that RCW laterals are horizontal, that does not apply to well RB-16. Two of the four laterals had a vertical bend, especially Lateral 1, whose curvature towards the ground surface was slightly less than 6 m.

Calibration of the model of the extended area of well RB-16 was based on well pumping tests. Datasets from three tests, conducted on 7 October 2007, 24 March 2011, and 22 July 2013 were used in the case study. The first test was carried out upon installation of new laterals (the old laterals were shut-off).

### *5.7.1.4 Groundwater model of radial collector well RB-16*

A hydrodynamic model of the groundwater flow at the location of well RB-16 was constructed based on the hydrogeological model. This model included schematised layers of the alluvial complex of the Sava River. The dominant directions of groundwater flow are from the Sava River, Lake Sava, and New Belgrade. The model output came via the wells within the model.

The model boundaries were defined on the basis of the analysed flow pattern in this part of BGWS. The neighbouring wells (RB-15, RB-16-1, RB-35), a segment of the Sava River, and a part of Lake Sava were also included in the model. Groundwater inflow from New Belgrade was the boundary on the left bank of the Sava River (Figure 5.25).

Well RB-16 was modelled using actual data, including its (impervious) caisson and laterals, whereas the other wells were tube wells, of the corresponding equivalent diameter. The river and the lake were specified in the model based on their previously measured cross-sections. The software interpolated the intervals between the cross-sections, to produce a continuous river channel (whose geometry and filtration properties can vary). The hydraulic conductivity of the colmated riverbed was specified at each cross-section inflection point.

This boundary condition is specific in that the program recognises the river width between banks, depending on the specified river stage. The width of the river was constrained by dykes, or the maximum stage contour.

The existing observation wells were used as control points in groundwater level calculations for model calibration. The pair of observation wells (above and below the semi-permeable interlayer) is also very important, in this specific case for properly defining the well discharge capacity. The pair's piezometric head difference is a function of well discharge and uniquely determines the filtration parameters of the interlayer, which were derived from calibration. It would have been very useful to have observation wells in the water-bearing layer immediately below the riverbed but, unfortunately, that was not the case. Otherwise, it would have been possible to come up with much better inputs for the quantification of riverbed conductivity.

Where there is a semi-permeable interlayer that constitutes a hydraulic barrier for groundwater flow to the well, it is extremely important and useful to have pairs of observation wells, both close and at some distance from the pumping well.

### 5.7.1.5 Model calibration – calculation results

Calibration of the model of the well RB-16 location was a huge and complex task, which involved several steps. Datasets from three pumping tests (2007, 2011, and 2013) were used for calibration. The model was calibrated for the simulation of each test. The results were compared and analysed. Finally, the 2011 test model was selected and, after minor boundary condition adjustments, the calculations were repeated for the other tests. This procedure was followed for both hydrogeological model options: three and eight schematised layers.

At the end, a representative model of the well RB-16 location was adopted. The simulation of well RB-16 tests provided the variation in lateral screen conductivity over time (Figure 5.26a). The conductivity parameter was $K/d$ ($K$ is the hydraulic conductivity of the clogged layer along the edge of the screen (m/s), and $d$ is the adopted representative (unit) thickness, (m)).

Figure 5.26b shows the variation in local hydraulic resistance, LHR, of the lateral, based on information from all the tests, using data on well RB-16 and the so-called 'close' observation well (Dimkić *et al.*, 2011a, 2011b, 2011c; Dimkić & Pušić, 2014). The calculation results from both procedures yielded qualitatively similar results and corroborated well ageing due to biochemical clogging.

Analysis and quantification of the components of groundwater flow to the well are important for assessing the effect of individual boundary conditions (recharge zones) on well discharge capacity and, indirectly, well water quality. Model tests revealed that the discharge capacity of well RB-16 was the

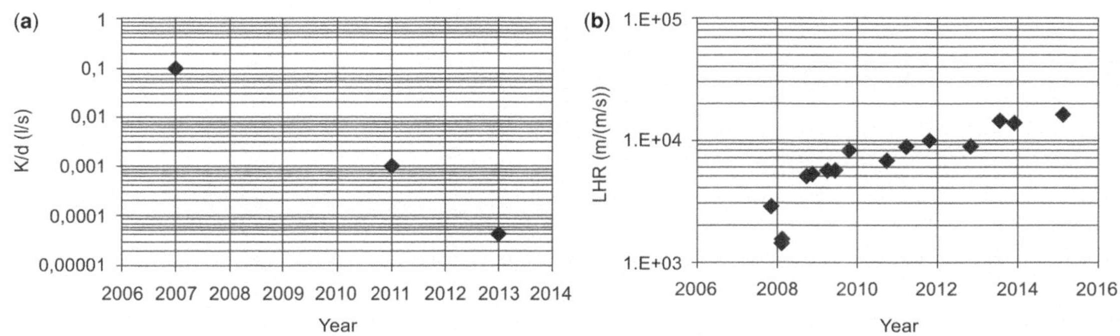


**Figure 5.26** Decrease in conductivity ($K/d$) of radial collector well laterals, based on model simulation of selected pumping tests (a), increase in local hydraulic resistance (LHR) of the laterals, based on analysis of all available pumping test data, via the so-called 'close' observation well (b) (Dimkić & Pušić, 2018).

**Table 5.9** Model calibration results, average values of filtration parameters of schematised layers, riverbed, and lakebed.

| | $K_{horiz(m/s)}$ | $K_{vert(m/s)}$ | $S\ (m^{-1})$ | $S_y\ (-)$ |
|---|---|---|---|---|
| Layer 1 | $3.35 \cdot 10^{-4}$ | $2.37 \cdot 10^{-6}$ | $7.00 \cdot 10^{-5}$ | $1.50 \cdot 10^{-1}$ |
| Layer 2 | $4.46 \cdot 10^{-4}$ | $3.00 \cdot 10^{-7}$ | $7.00 \cdot 10^{-5}$ | $7.00 \cdot 10^{-2}$ |
| Layer 3 | $6.65 \cdot 10^{-4}$ | $6.65 \cdot 10^{-4}$ | $7.00 \cdot 10^{-5}$ | $1.20 \cdot 10^{-1}$ |
| | $K/d\ (s^{-1})$ | | | |
| Sava River | $4.00 \cdot 10^{-7}$ | | | |
| Lake Sava | $3.00 \cdot 10^{-7}$ | | | |

sum of inflows from the Sava River (95%), Lake Sava (4%), and New Belgrade (1%). These proportions resulted from the simulation of the 2011 pumping test (similar results were obtained for different conditions).

Filtration parameters of the schematised layers were another outcome of model calibration. It was obvious that the schematised layers were a result of synthesis of the geometry and filtration characteristics of corresponding real sediments, such that there was certain heterogeneity in plan view and anisotropy of parameters. The representative, average values of filtration parameters of the porous medium and riverbed conductivity in the case of the three-layer model are shown in Table 5.9.

Representative characteristics were also obtained for the Sava River and Lake Sava.

The effect of the interbed on groundwater flow was manifested through the piezometric head difference between the upper and lower water-bearing layers. The conclusion was that it is extremely important for the discharge capacity of the well, and the location as a whole, to determine the hydrogeologic characteristics of the interbed (position, spread, thickness, filtration parameters). Figure 5.27 shows the piezometric head difference between the upper and lower water-bearing layers, as a result of the presence of the schematised continuous interbed, under the operating conditions of well RB-15.

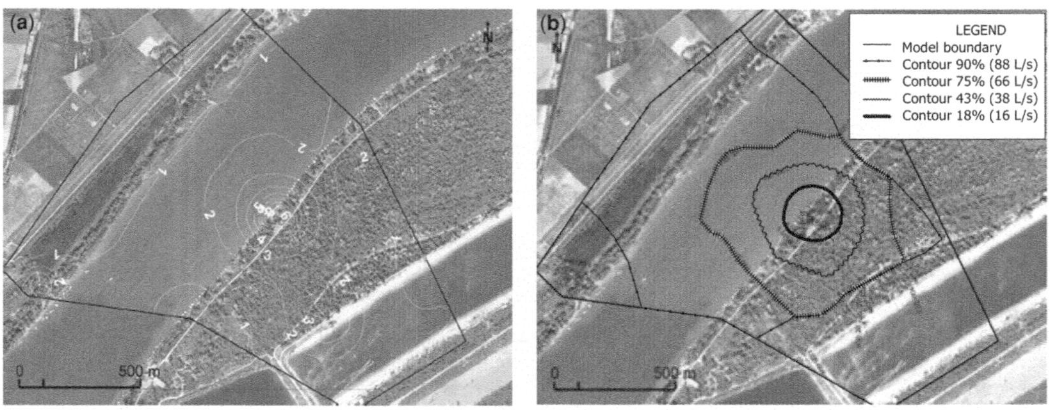


**Figure 5.27** Piezometric head difference between the lower and upper layers (interbed floor and roof) (a), percentage contributions of different zones to well discharge (b) (Dimkić & Pušić, 2018).

### 5.7.1.6 Analysis of the discharge capacity of radial collector well RB-16

In order to predict the achievable discharge capacity of a new or reconstructed well, the design needs to be based on the quantification of discharge under relevant conditions and an analysis of drivers (parameters). Of interest here is the determination of zones with different rates of groundwater flow to the well. Depending on the specific case and as needed, these zones can be treated differently in terms of purpose. Figure 5.27 shows the determined sizes of the zones of well RB-16. The lines delineate areas that contribute a certain percentage of the flow to the well. In the present case study, the basis was the calculation of unit vertical fluxes through the semi-permeable interbed.

The time the groundwater takes to travel through the interbed can be calculated on the basis of the vertical flow velocity distribution and interbed thickness. Figure 5.28 shows the results of such calculations. The first roughly estimated average travel time through the interbed was 12 months (75% contour in Figure 5.27, interbed thickness 4 m). Understandably, the travel time was much shorter within the zone of the laterals (about 100 days for the 18% contour in Figure 5.27, interbed thickness 3.5 m).

Model tests can also provide unit discharge capacities of the laterals, as well as the distribution of entrance velocities. Figure 5.29 shows the results of calculations for a specified water level in the caisson of 55 m above sea level (a.s.l.), corresponding to a discharge of 98.5 l/s. It is apparent that the discharge of Lateral 1 is low; it is bent in the vertical direction (about 6 m). This lateral penetrates the interbed and reaches the water-bearing layer above it. The low discharge is attributable to the fact that this lateral is rather clogged, according to an underwater video.

The total capacity of a well location can be determined based on the calibrated model and set criteria (water level maintained inside and outside the caisson). If the groundwater level is specified in the interval from 52 to 55 m a.s.l. and the water level in the caisson is at 5 m above the laterals, the maximum discharge capacity of well RB-16 will be in the interval from 120 to 150 l/s.

**Figure 5.28** Time zones of groundwater travel through the semi-permeable interbed (in months) (Dimkić & Pušić, 2018).

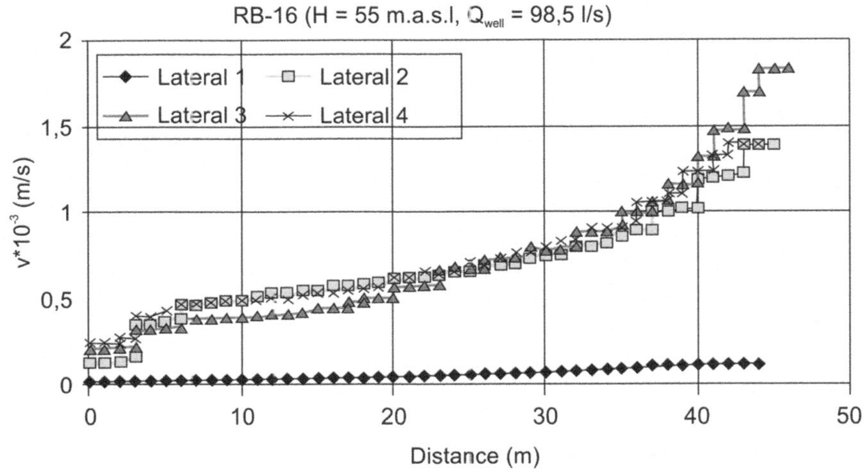


| | l/s |
|---|---|
| Lat. 1 | 2,1 |
| Lat. 2 | 31,5 |
| Lat. 3 | 33,9 |
| Lat. 4 | 31,0 |
| Sum = | 98,5 |

**Figure 5.29** Discharge capacities of individual laterals and distribution of entrance velocities of well RB-16 (water level in the caisson maintained at 55 m above sea level, $Q = 98.5$ l/s) (Dimkić & Pušić, 2018).

### *5.7.1.7 Effect of riverbed conductivity*

Riverbed conductivity is a key driver of the discharge capacity of a well and location. The colmating layer of the river varies over both space and time. It depends on the river discharge regime and the dominant type of groundwater flow. It is especially important where the operation of a bank-filtration type well imposes one-way groundwater flow – constantly from the river to the aquifer.

### *5.7.1.8 Effect of the semi-permeable interbed*

In the case of well RB-16, the semi-permeable interbed causes vertical groundwater filtration from the upper to the lower (tapped) water-bearing layer. Its filtration properties can be so poor as to prevent the flow of water to the well. On the other hand, the interbed is extremely important for the conservation of high well-water quality. Slow and prolonged filtration facilitates complex biochemical processes, which reflect the self-purification potential of the aquifer.

Figure 5.30 shows the piezometric head difference between the interbed floor and roof, when the vertical conductivity is $K_{vert} = 1 \times 10^{-9}$ m/s. The difference in the near-well region can be even greater than 10 m, which is about three times more than under real conditions (Figure 5.26).

The discharge of the well drops to a third of the realistic discharge (34 l/s vs. 98 l/s), at the minimal water level in the caisson ($H = 55$ m a.s.l.). Under these conditions, the entire amount of the water in the well comes from the overbank area.

This is another illustration of the need for more pairs of observation wells, to track hydraulic losses through the semi-permeable interbed.

### *5.7.1.9 Rate of well ageing*

In RCW design, sizing of the discharge capacity is not only a matter of initial permissible groundwater entrance velocities, but also of sustainable pumping, which implies technically and economically viable service over a prolonged period of time.

**Figure 5.30** Piezometric head difference between the lower and upper layers (interbed floor and roof), at reduced interbed conductivity ($K_{vert} = 1 \times 10^{-9}$ m/s) (Dimkić & Pušić, 2018)

The authors of this chapter have been studying the conditions and well ageing processes at BGWS (about 100 horizontal collector wells) for years. A correlation has been established between well ageing and the biochemical properties of the groundwater. Based on maintenance practices and rehabilitation scheduling, the design rate of increase in local hydraulic resistance (LHR) is 0.35 m/year. This criterion determines the maximum permissible groundwater entrance velocities, depending on the redox potential (Eh) and concentration of bivalent iron ($Fe^{2+}$) in the groundwater (Figure 5.31).

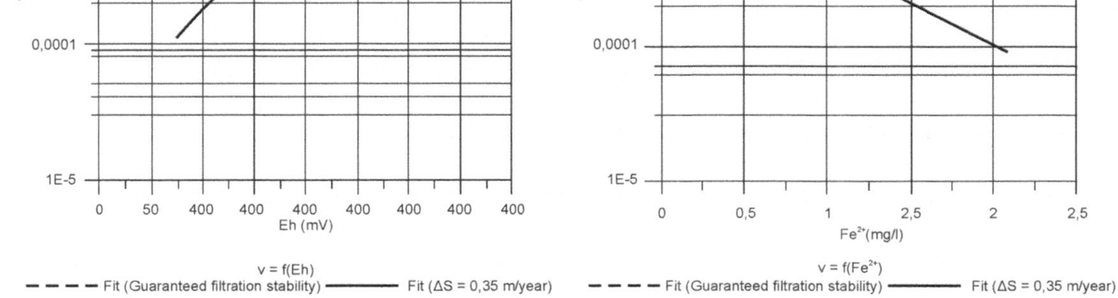


**Figure 5.31** Permissible groundwater entrance velocities as a function of *Eh* and $Fe^{2+}$ (Dimkić & Pušić, 2018).

In the specific case, this means that the well discharge capacity that fulfils these criteria, guarantees a maximum increase in the local drawdown of about 3.5 m over 10 years, which is the preferred interval between two well rehabilitations.

The average groundwater parameters of well RB-16 are: $Eh \sim 137$ mV and $Fe^{2+} = 0.6$ mg/l, which correspond to a maximum permissible entrance velocity of $v = 3.15 \times 10^{-4}$ m/s relative to $Eh$ and $v = 7.5 \times 10^{-4}$ m/s relative to $Fe^{2+}$. Considering the total length of the laterals (four, diameter 0.4 m) of 240 m, the resulting discharge capacities of this well are $Q = 94$ l/s and $Q = 226$ l/s. Obviously, the more stringent criterion needs to be fulfilled. If the well had six laterals, it would be reasonable to expect a sustainable long-term discharge capacity of about 110 l/s.

Taking into account the maximum entrance velocity criterion and the maximum capacity of the location, eight laterals would ensure long-term operation of the well at a capacity of 120–130 l/s. Based on these technical guidelines, the preferred solution becomes a matter of economics.

### *5.7.1.10 Discussion and inferences*

When plans are made to build a well or improve the capacity of an existing well, there is a series of issues that need to be addressed in as precise a manner as possible.

There are several important reasons for using RCWs in alluvial aquifers of relatively large rivers to provide public water supply, generally for big agglomerations. These reasons are:

- Tapping of the most permeable alluvial strata (usually lower sediments);
- Utilisation of the purification potential of semipermeable interbeds; this potential can be harnessed more effectively by RCWs than tube wells; and
- Easier groundwater source management due to high capacities of individual wells, such that fewer wells, pumps, etc. are needed.

Biochemical clogging of wells tends to occur in anoxic alluvial aquifers, where there are sufficient amounts of iron. The correlation between the rate of increase in hydraulic resistance at RCW laterals and biochemical parameters (clogging indicators $Eh$ and $Fe^{2+}$) has already been discussed by the authors (Dimkić *et al.*, 2011a, 2011b, 2011c; Dimkić & Pušić, 2014). The correlation yields a maximum permissible entrance velocity function, depending on the considered indicator

$$v_{perm} = f\,(Fe^{2+});\; v_{perm} = f\,(Eh) \tag{5.33}$$

The procedure presented here generally comprised four stages:

Stage 1: Monitoring of well-water quality and definition of biochemical conditions and rate of LHR increase.

Stage 2: Definition of the capacity of well RB-16 (in terms of its groundwater withdrawal potential), based on monitoring of pumping and changes in piezometric head in observation wells above and below the semi-permeable interbed. The laterals of the considered RCW are below this interbed. The parameters relevant for the determination of the capacity of the location are: permeability of the aquifer strata, permeability and thickness of the semi-permeable interbed, quality of the hydraulic link between the river and the aquifer, and LHR of laterals. A minimal but sufficient network of observation wells was available for achieving this objective.

Original software was used to create a 3D hydraulic model of groundwater flow. Its additional features include excellent modelling of links between rivers and aquifers, as well as RCW laterals and aquifers (Dimkić *et al.*, 2010; BioIRC and Jaroslav Černi Water Institute, 2014; Vidović *et al.*, 2014).

Stage 3: Determination of critical velocities, depending on biochemical parameters. Under anoxic conditions, these velocities are lower or much lower than those derived from the aquifer's filtration characteristics in the region around the laterals. The determination of critical velocities via biochemical parameters allows the capacity and number of laterals to be defined in a way that prolongs their life-span and reduces maintenance costs. The confirmation and/or improvement of the presented relations, and the definition of the roles of individual species of iron bacteria in the well clogging process, are challenging from a scientific perspective.

Stage 4: Monitoring of the operation of the well following rehabilitation or reconstruction (emplacement of new laterals). BGWS is comprised of 99 RCWs and about 50 tube wells. The decline in BGWS capacity, due to iron incrustation, has been substantial – about 200 l/s per year. The proposed method, along with appropriate monitoring, allows considerable well maintenance cost cuts and the selection of best locations for new wells. The method is believed to be a significant contributor to engineering practice and calls for further scientific study.

## Natural setting

Urban agglomerations around rivers generally rely on riparian water supply sources (bank filtration). Well screens are located in the lower part of the water-bearing complex, which is usually the best to be tapped. The hydraulic contact with the river and the filtration parameters of the aquifer are of primary importance. Alluvial sediments, from which groundwater is often abstracted, are characterised by sudden alternations of coarse and fine-grain strata. Semi-permeable interbeds and lenses hinder groundwater flow to a well and limit its capacity.

However, a prolonged groundwater travel time has a positive effect on groundwater quality transformation processes and the natural self-purification potential. Well RB-16, which is the object of the present case study, relies on bank filtration. Within the aquifer system, above the laterals, there are several semi-permeable interbeds.

In order to determine the optimal discharge capacity of a well (from a technoeconomic perspective), the discharge capacity of both the extended location and the well itself needs to be analysed. It should be noted that well RB-16 is surrounded by other BGWS wells. The effect of the extent of riverbed colmation and the filtration parameters of the interbed were assessed, using a hydrodynamic model of the groundwater in the pertinent part of BGWS.

The number and distribution of observation wells from which data were used to calibrate the model governed the schematisation of the aquifer system as a three-layer medium, with a single schematised interbed. This was a result of the presence of a pair of observation wells that registered groundwater levels above and below the series of semi-permeable interbeds. It was noted that a suitable (larger) number of pairs of observation wells would be very useful for this type of analysis.

Based on the calibrated model, the conclusion was that the conductivity of the riverbed was of the order of $1.5 \times 10^{-7}$ m/s. Any decrease in conductivity directly affected the capacity of the location. A decrease by a factor of 100 resulted in total hydraulic separation of the water table from the riverbed. Groundwater flow from the river to the aquifer was solely a function of the river stage and filtration parameters of the aquifer.

The conductivity of the semi-permeable interbed was analogous. In the case of well RB-16, while in service, it is not realistic (but in principle possible) for the piezometric level of the lower water-bearing stratum to drop below its roof and create a vadose zone near the laterals. This would certainly have a considerable impact on the discharge capacity of the well, in particular the biochemical processes that take place in the groundwater.

### Well ageing

Well ageing is a technoeconomic challenge because it affects well longevity and maintenance costs. Groundwater entrance velocities, which determine the discharge capacity of an RCW, also need to guarantee a predefined rate of well ageing (or the rate of discharge capacity decline). The initial water level maintained inside the caisson, which ensures enough (pre-set) time before the next rehabilitation of the well, is defined accordingly.

A maximum permissible entrance velocity of $v = 3 \times 10^{-4}$ m/s was specified for the average ageing indicators of well RB-16 (*Eh* and $Fe^{2+}$); as the current number of laterals is four, the resulting discharge capacity of the well is $Q = 72$ l/s. Given the design and maintenance concept, any increase in discharge capacity needs to be sought in a larger number of laterals. If the well had six laterals, it would be reasonable to expect a sustainable long-term capacity of about 110 l/s.

In closing, the significance of this research lies in the simultaneous determination of the hydraulic capacity of the location of an RCW and the critical groundwater flow velocities to the laterals. The maximum permissible entrance velocities were established on the basis of the relevant biochemical relationship $v_{perm} = f(Fe^{2+}, Eh)$.

The approach outlined above is highly relevant to anoxic alluvial water supply sources, such as BGWS. $v_{perm} = f(Fe^{2+}, Eh)$ was examined for the first time in the research conducted by the authors and their associates.

## 5.8 CONCLUSION RELATING TO WELL CLOGGING

Well ageing due to clogging is a problem that has been evident for a long time. The technical and economic significance is considerable. Mechanical clogging, due to excessive screen-entrance velocities, has been analysed and criteria proposed by renowned scientists in the past (Abramov, 1952; Cistin, 1965; Gavrilko & Alekseev, 1985; Johnson, 1975; Kovacs & Ujfaludi, 1983; Pietraru, 1982; Sichardt, 1928 and others).

Biochemical clogging is a result of processes that take place under anoxic groundwater conditions in the near-well region. A number of authors worldwide have studied anoxic conditions and well ageing processes. However, the extensive investigations described in this chapter were undertaken in order to establish a correlation between some of the primary biochemical parameters (*Eh*, *Fe*) of the groundwater and mechanical parameters (screen entrance velocities and the kinetics or rate of clogging). As far as we are aware, this was among the first attempts to examine the problem based on this approach and is a pioneering work of sorts.

In addition to detailed investigations, the scientific contributions are based on the results of multiyear monitoring of some 150 wells in alluvial environments in Serbia. Testing included several rivers and sites: the Sava River (Belgrade Groundwater Source), the Danube River (riverbank sections from Kovin to Dubovac and Knićanin to Čenta), and the Velika Morava River (Ključ Groundwater Source in the city of Požarevac).

The results showed that a high redox potential (*Eh*) of the groundwater reflected sustainable pumping conditions and slower well ageing. The same was true of low iron concentrations (*Fe*). This suggested that anoxic and sub-anoxic conditions were conducive to well ageing. The following threshold values were determined for $Q = f(Fe, Eh)$, where there are notable quantitative changes: Fe concentration ~12 mg/l and *Eh* between 130 and 150 mV. All the tests showed that there was virtually no biochemical clogging under oxic groundwater conditions (redox potential greater than 250 mV).

Where biochemical processes are dominant, the permissible screen entrance velocity ($v_{perm}$) was proved much lower than derived from 'conventional' experimental formulae that consider only the filtration stability and laminar flow in the new-well region.

Under anoxic conditions (redox potential greater than 50 mV was tested), in addition to the entrance velocity, the redox potential and iron concentration were identified among the main parameters that affect well ageing. The general expression is:

$$\text{KLHR} = \text{KLHR}(v, Fe, Eh, B, \Gamma, \ldots) \tag{5.34}$$

where: $v$ is the groundwater entrance velocity, $Fe$ is the iron concentration in the pumped water, $Eh$ is the redox potential of the water, $B$ is a function of the rate of bacterial growth in the well, and $\Gamma$ is a function of several parameters associated with screen installation (with or without pack, pack properties, type and characteristics of screen slots), as well as the grain-size distribution of the aquifer. Hence, where alluvial aerobic conditions transition into anaerobic conditions, the conventional 'purely hydraulic' criterion relating to the screen entrance velocity becomes insufficient and requires a 'biochemical' criterion as well.

An additional criterion was now introduced – 'permissible well clogging or ageing' in the form of KLHR or permissible drawdown $S_{year}$.

In view of the larger number of parameters and diverse conditions of the studied wells and hydrogeological settings, it was difficult to establish stringent correlations between the rate of clogging and the iron concentration and redox potential of the groundwater. Still, such a relationship is proposed. In addition, permissible entrance velocities are specified. Plots of the $v_{perm} = f(Eh,\ Fe)$ type are recommended for designing both tube wells and RCWs. The same applies to well rehabilitation and reconstruction.

There are several other important aspects of the investigations, both completed and ongoing, such as the origin and nature of deposits, microbial activity, difference in redox potential between different parts of the porous environment, effect of the screen, well rehabilitation method, etc. Some of the investigations remain to be reported and others are pending.

## 5.9 WELL REHABILITATION

### 5.9.1 Introductory remarks

As defined and explained in detail in other sections of this chapter, an 'aged' well is a well whose pumping capacity ($Q_e$) has decreased while the operating drawdown ($S_e$) has increased. Well ageing is most often encountered where the porosity of the water-bearing formation is intergranular. The processes that lead to well ageing depend on many factors, the most important being:

- Aquifer tapping method;
- Groundwater withdrawal method;
- Chemical composition of the groundwater; and
- Chemical and mechanical composition of the water-bearing formation.

Depending on the above factors, well ageing can be caused by mechanical, physicochemical and biochemical processes.

#### *5.9.1.1 Brief description of the processes*

Mechanical processes largely depend on the groundwater withdrawal method, the way the well was constructed, and the hydrogeological properties of the aquifer tapped by the well. Their manifestations are:

- Clogging;
- Suffosion; and in the most difficult case
- Degradation of the water-bearing formation in the near-well region.

When a well is properly constructed, meaning that it has been adapted to the site, there is usually no clogging. Clogging processes largely affect the well screen, steel components, and sand or filter packs that prevent the entry of fine particles into the well. In most cases, the steel components are not adequately protected against corrosion and the pack grains have not been previously separated to ensure their compatibility with the water-bearing formation. As a result, groundwater flow velocities at the contacts between the aquifer and the pack, and the pack and the screen vary. These contacts are critical surfaces where the well is clogged with fine aquifer fractions, or products of degradation of the casing in chemical interaction with the ionic composition of the groundwater and other complex oxidation-reduction processes, or intensified production of phytoplankton and various organogenic-mineral substances.

The opposite process is suffosion (well sanding). Proper well construction allows some suffosion in the well development process, during which a closer connection between the well and its environment is established. However, if suffosion is pronounced, especially after a certain period of pumping, the well might become plugged and the water-bearing formation could degrade in the near-well region. As a rule, such degradation beyond the screen creates three zones:

- Suffosion;
- Caving; and
- Subsidence.

The overall thickness of the three zones is generally 1–1.5 m. Where degradation of the water-bearing formation is substantial (due to excessive or long-term pumping), overhauling or revitalising to improve yield may not be effective.

Physicochemical processes lead to chemical and electrochemical corrosion of the metallic part of the screen, as well as incrustation and precipitation within the well, on the screen and in the near-screen zone. Calcium and magnesium precipitation does not depend solely on electrochemical processes, but also occurs on account of changes in the pressure regime as groundwater from the water-bearing formation enters the screen. Viewed in that context, cation (e.g. calcium as calcium carbonate) precipitation does not depend on the structure of the screen, as is largely the case with electrochemical processes. Incrustation and precipitation of carbonates in the well, screen and the near-screen zone generally depend on the chemical composition of the groundwater and the pumping mode. Groundwater should be viewed as a complex substance, composed of liquid, gaseous and solid phases (colloidal particles). Their state of equilibrium is labile and changes are a result of changes in pressure, temperature and formation. This physicochemical equilibrium of groundwater does not vary by much (except in the aeration zone), and therefore sedimentation and cementation within the water-bearing formation are rare. However, the factors (pressure and velocity) rapidly change when a well is installed and pumped. This alters the equilibrium of the three-phase system. Soluble cation salts are precipitated and incrust the pump pipes, well and screen, and cement the near-screen zone.

Biochemical processes also cause well ageing. Colloidal and organic substances and suspended particles account for a large portion of the chemical composition of groundwater. They cause oxygen ($O_2$) concentrations in groundwater to fluctuate and the complex processes of oxidation, hydrolysis and reduction to take place. The groundwater biocenoses, often including paleo plankton, are subjected to bacterial decomposition, during which oxygen, nitrate, trivalent iron in minerals, sulfate and carbon dioxide are consumed and carbon dioxide, nitrogen gas, hydrogen sulphide and methane are produced, giving groundwater colour, taste and smell. Iron cation $Fe^{2+}$ plays a major role in well screen clogging. Its oxidation and hydrolysis result in brownish-red floccules (limonite). Iron bacteria grow on that substrate, are deposited on the well screen walls and reduce the screen's permeability.

Iron salts occur in groundwater in the form of ferrous bicarbonates or sulphates where bicarbonate concentrations are low. Iron bacteria use ferrous salts as nutrients, converting them into hydroxides. The most common bacterial forms are Crenthrix (filamentous) and callionela (rod-shaped). The size of iron bacteria is in the microns to millimetres range. Temperature, oxygen concentration and pH affect bacterial growth.

The oxygen concentration influences the oxidation of the ferrous ion:

$$Fe^{2} + +5H^{+} + 2O_2 = Fe(OH)_3 + H_2O \tag{5.35}$$

Ferric hydroxide is produced and heat is released, which favours the reproduction of iron bacteria.

Acidic environments (pH <7) are generally suitable for iron bacteria, but they are also found in low alkaline settings. Iron bacteria thrive in groundwaters whose iron concentration is from 0.1 to 0.3 mg/l. The most favourable conditions for the growth of iron bacteria are found in wells drilled in alluvial and moraine sediments. The causes of well ageing discussed in this chapter result in:

- Reduced well yield;
- Clogging of the near screen-zone (colmation, incrustation);
- Sanding (increased entrance velocities); and
- Frequent pump failure.

Deteriorating hydrodynamic parameters due to incrustation, colmation or fluidisation with mineral salts or organogenic-mineral substances in the near-screen zone are addressed by physicochemical-hydraulic well rehabilitation. Before they are rehabilitated, wells need to be tested to examine the condition of the casing, type of material that has penetrated the well during the course of pumping, and current yield.

#### *5.9.1.2 Pre-rehabilitation testing of wells*

Well testing precedes well rehabilitation and its stages are as follows:

- Removal of the submersible pump and visual inspection of the mechanical equipment for any visible damage;
- Airlifting – cleaning of the well casing to examine the type and quantity of material that has entered the well;
- Testing at two or three drawdowns and monitoring of recovery to gain insight into the yield and level of atrophy compared to prior condition; and
- Filming to note any damage or deformation.

### 5.9.2 Well rehabilitation methods

Well rehabilitation is a combination of chemical methods that affect encrustations and cementing materials in the near-screen zone (or products of chemical and biochemical clogging) and physical methods that address mechanical causes of well ageing.

The synergy of chemical and physical methods, based on high air and water pressures, allows chemical action in the near-well region – where the filter pack meets the natural environment.

Well rehabilitation uses chemicals based on hydrochloric acid (HCl) as the reagent and citric acid as an inhibitor, which prevents aggressive action of HCl against the well casing.

#### *5.9.2.1 Mechanical (physical and hydraulic) rehabilitation*

Physical cleaning or ‘loosening’ of the near-screen zone is undertaken in a hydrodynamic fashion, using a suitable piston mounted on the ‘mouth’ of the well casing. This allows sudden pressure changes in the well pipe and hydraulic shock. The generated compression wave propagates almost instantaneously along the well, increasing the pressure and resulting in the detachment and removal of the finest particles around the screen. This material is airlifted. The 'loosening' procedure is repeated until the desired effect is achieved. In brief, physical and hydraulic treatment comprises the following operations:

- Rhythmic surging;
- Airlift and packer combination, to act on the screen by segment;
- Pressure gauging and pressure regulation as needed; and
- Jetting – removal of solution and substances deposited as a result of the physical and chemical treatment. Jetting continues until the water runs clear.

There are several mechanical rehabilitation methods, among which the most common are:

- Jetting;
- Over pumping;
- Flushing by recirculation cell;
- Surging;
- Hydraulic shock; and
- Hydroelevator.

### Jetting

The water must be clean, to avoid chemical or microbial contamination of the aquifer. Special nozzles are used interval by interval. Low-pressure nozzles are shown in Figure 5.32.

Jetting pressures can be as high as 10 bar with flow rates up to 30 l/s. The number of nozzles is usually not greater than 30 and their diameter is up to 10 mm. Nozzle outlet velocities are generally 5 to 20 m/s (Lazić, 2004). Jetting at very high pressures, 50 to 500 bar, is also relatively common. Pressures up to 200 bar are applied to plastic casings and up to 500 bar to steel casings with no major damage due to corrosion.

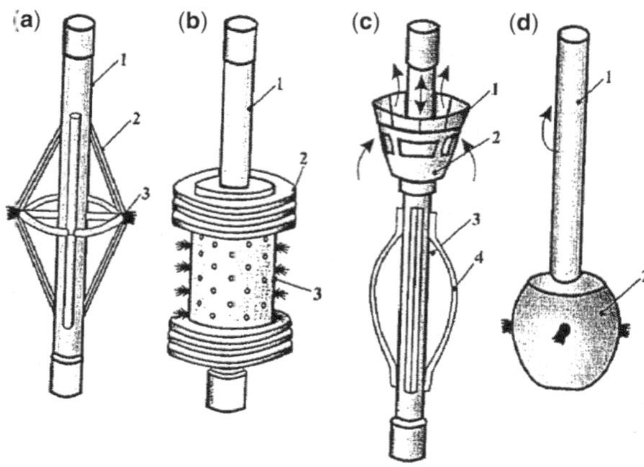


**Figure 5.32** Schematic representation of nozzles used for screen jetting. Legend: (a) simplest jetting device: 1 – jetting pipe, 2 – centraliser, 3 – clean water nozzles; (b) jetting device with two surge blocks: 1 – jetting pipe, 2 – surge block, 3 – nozzle; (c) collar-based device: 1 – leather or impervious cloth collar, 2 – collar body with water openings, 3 – jetting pipe, 4 – centraliser; and (d) circular jetting device: 1 – jetting pipe, 2 – body with nozzles (Lazić, 2004).

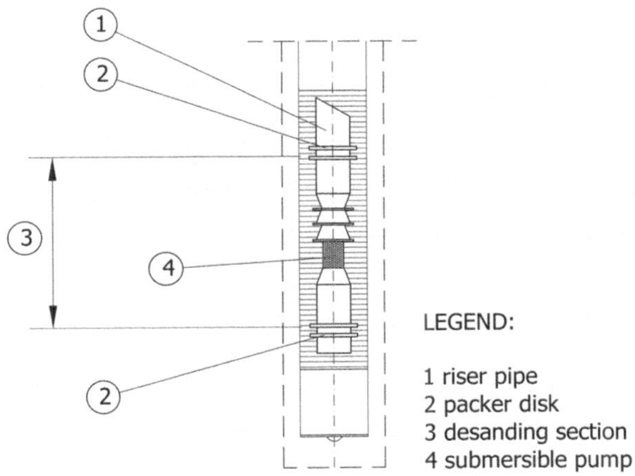


**Figure 5.33** Overpumping method (Houben & Treskatis, 2007).

## Overpumping

This method is based on pumping of water from the part of the casing between two packers (Figure 5.33). The water input velocity needs to be about five times greater than at the maximum rate of regular pumping (Houben & Treskatis, 2007).

## Recirculation cell method

This flushing method involves simultaneous pumping of water into one end and out of the other end of a flushing device, using two pumps. The discharge end of the pump is connected to a flushing tool and the suction end to the source of water (Figure 5.34). The flow rates should be roughly equal or pumping-out slightly more intensive (Houben & Treskatis, 2007).

The water to be pumped-in can be withdrawn from another segment of the casing, another well or any other source of clean water.

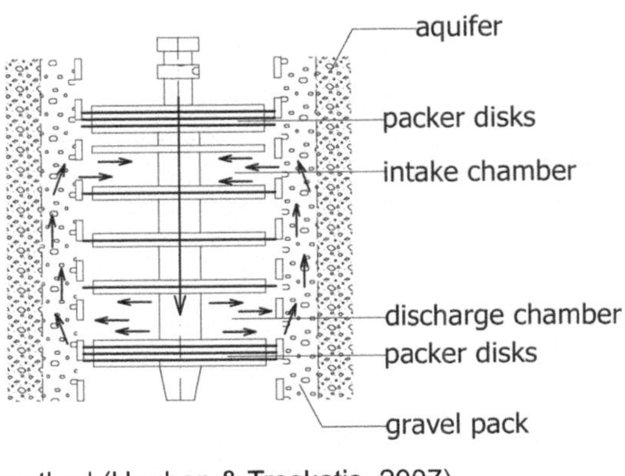


**Figure 5.34** Circulation cell method (Houben & Treskatis, 2007).

### Surging

Surging is a well development and rehabilitation method based on the creation of alternating over- and under-pressures by a special piston, in the zones above the screen (Figure 5.35). The maximum stroke depends on the device used to raise and lower the piston. The stroke length and speed are also selected taking into account the properties of the well casing, to avoid collapse (Lazić, 2004).

The piston is made from hard rawhide or impregnated rubber, in the form of circular plates whose diameter is 3–5 mm smaller than the inner diameter of the casing. A number of rubber or leather plates are packed with wooden or metal mats in-between, whose diameter is 2–3 cm smaller than that of the casing (Figure 5.35, a, b).

The total thickness of the piston should be 15–20 cm. The piston is lowered to a depth of 5–6 m below the water level. Greater depths are not recommended because of high hydrodynamic loads when the piston is pulled. The piston should never be lowered to the screen zone.

### Hydraulic shock

As water is not compressible, a force acting on it is equally transmitted in all directions. This is the basis of one of the well development methods. The impact on the water surface in the well creates pressure that is transmitted to the screen walls and the water-bearing formation. This method is applied to clear clogged screen slots and dislodge fine particles outside the well. The impact is created by a cable tool, whose diameter should be as large as possible. Metal weights are used to make it heavy and more efficient. The weight of a filled cable tool can be 1–1.5 tons. It is allowed to fall freely from a height of 4–5 m, to create a hydraulic shock (Lazić, 2004). Thereafter, the well should be flushed by airlifting or hydroelevating.

### Hydroelevator

Pumping-out by a hydroelevator or upward flow created by special devices is one of the novel and more efficient approaches. Compared to similar methods, it does not require special pumps, pressure conduits, compressors, etc. The essence of well rehabilitation and development by hydroelevating is to use a drilling-rig piston pump and a special device to facilitate the removal of sludge from the well by lifting clean water through a narrow nozzle. It functions as an ejector. The concentration of fine sand and clay particles can be as high as 30%. The piston pump delivers water through drilling rods to a special

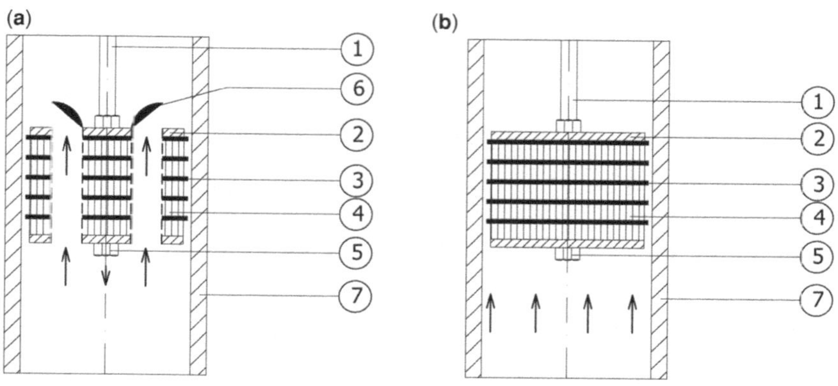


**Figure 5.35** Piston with (a) and without (b) valves. Legend: 1 – water delivery pipe; 2 – valve opening; 3 – metal plate; 4 – rubber or leather rings; 5 – wooden ring; 6 valve cap; 7 – well casing (Lazić, 2004).

device, a so-called hydroelevator which is sealed at the upper end by a packer to ensure suction due to flow velocity and sudden upward jetting. As the water enters the elevator, it changes the circulation route through a 'U' pipe, whose diameter is much smaller than the inner diameter of the drilling rods. This increases the speeds multiple times. A large increase in entrance velocities lowers the hydrostatic pressure, whereby water is drawn from the screen and particles from the filter pack and the water-bearing formation around the well bore where they are entrained. Sudden pumping and a sudden reduction in well drawdown increase entrance velocities and the inflow of water from the formation, and fine particles are removed from the porous medium. Low pump capacities are adequate for efficient hydroelevating.

### 5.9.2.2 Chemical rehabilitation (treatment)

Acid is injected after the well has been tested and chemical piping with valves installed in the wellhead. A chemical pump is used to deliver inhibited hydrochloric acid. The pump is single-stage, with a spiral housing and open rotor. All pump components in contact with the fluid are made from resistant materials. A 7% solution of hydrochloric acid is pumped into the well, to effectively degrade iron carbonates and hydroxides, calcium and magnesium carbonates, and similar.

$$Fe(OH)_3 + 3HCl = FeCl_3 \downarrow + 3H_2O \tag{5.36}$$

$$FeCO_3 + 2HCl = CaCl_2 \downarrow + H_2O + CO_2 \tag{5.37}$$

A 5% solution of citric acid ($C_6H_8O_7$ x $H_2O$) is used as an inhibitor. The sequence of operations is as follows:

- Injection of chemicals based on HCl, as the reagent, and an inhibitor that prevents aggressive action of the reagent on the well casing;
- Recirculation of chemicals by screen segment;
- Pressure gauging and regulation as needed;
- Repeated injection of chemicals, recirculation and flushing; and
- Flushing of the well – removal of the solution and material precipitated as a result of the chemical reaction with the clogging material.

#### Chemical rehabilitation methods

Chemical rehabilitation is the treatment of wells with chemicals that destroy encrustations, converting them into a solution and/or dislodging them from the substrate. A large number of chemical substances can be used for this purpose. They are selected based on the nature and intensity of well clogging, and their concentrations and combinations depend on the type of encrustations. After the solution is injected, the well is gradually pressurised to several bars and maintained at that pressure until the solution reacts (generally 24 hours). Multi-stage treatment (several operations) is more effective than injecting the entire quantity of the acid instantaneously. The volume of the acid should initially correspond to the volume of the well, and should then be increased with each new cycle (Lazić, 2004). The treatment lasts for as long as it is effective. Some of the chemicals used in the process play the role of inhibitors – they prevent adverse effects on the well casing. Some of them, for example sodium tripolyphosphate, are added in low concentrations to prevent precipitation of other types of salts created in the rehabilitation process.

Before a well is rehabilitated, its capacity must be checked by a pumping test, to determine the extent of hydraulic losses due to encrustations, determine the chemical composition, and apply a physical method – usually mechanical descaling, airlifting, surging, etc. All of these are preparatory steps that precede chemical rehabilitation.

Sulphamic acid is often used in well rehabilitation practice. Although it is costlier than hydrochloric acid, it offers many advantages. It is easy to handle because it comes in the form of white grains and the pH level of a water solution is higher, such that corrosive action against steel is weaker. Good results are also achieved with 70% glycolic acid, which is antibacterial and mildly corrosive. The quantity injected is 1/3 of the well volume. In the case of scaling mostly composed of silicon dioxide and alumosilicates, the encrustations are removed by a mixture of HCl (20–25%) and HF (5–8%).

Phosphate compounds facilitate the dissolution of semi-soluble minerals, such as calcium carbonate and iron oxides. Sodium tripolyphosphate ($Na_5P_3O_{10}$ 5-8%) and hexametaphosphate ($(NaPO_3)6$ 5-7%) are commonly used. Sodium dithionite ($Na_2S_2O_4$) is a white powder readily soluble in water; solutions reach a concentration of 6% and their pH is 6 to 7.5.

The chemical mechanisms that dissolve mineral encrustations can generally be classified into three groups: dissolution aided by the hydronium ion (Eqs. (5.38), (5.39), (5.40) and (5.41)), dissolution by reduction (Eq. (5.42)) and dissolution by oxidation (Eqs. (5.43), (5.44) and (5.45)).

$$FeOOH_{(s)} + 3H_3O^+_{(aq)} \rightarrow Fe^{3+}_{(aq)} + 5H_2O_{(l)} \tag{5.38}$$

$$MnOOH_{(s)} + 3H_3O^+_{(aq)} \rightarrow Mn^{3+}_{(aq)} + 5H_2O_{(l)} \tag{5.39}$$

$$MnO_{2\,(s)} + 4H_3O^+_{(aq)} \rightarrow Mn^{4+}_{(aq)} + 6H_2O \tag{5.40}$$

$$FeS_{(s)} + 2H_3O^+_{(aq)} \rightarrow Fe^{2+}_{(aq)} + H_2S_{(g)} + 2H_2O_{(l)} \tag{5.41}$$

Dissolution of oxyhydroxides aided by the hydronium ion requires a pH of at least 1. The presence of calcium carbonate in the aquifer material hinders the dissolution of iron and manganese oxides and oxyhydroxides away from the filter pack, because of its buffer effect (Houben & Treskatis, 2007). Acids can also dissolve iron sulphide (but not iron disulphide).

$$2FeOOH_{(s)} + Na_2S_2^{III}O_{4\,(aq)} + 4H_3O^+_{(aq)} \rightarrow 2Fe^{2+}_{(aq)} + 2Na^+_{(aq)} + 2HS^{VI}O^-_{3\,(aq)} + 2H_2O_{(l)} \tag{5.42}$$

Dissolution by reduction is based on the fact that reducing agents, such as sodium dithionite in chemical equation (5.42), phenols, ascorbic acid, fulvic acid, fructose, etc. transform metals into a lower oxidation state, in which the metal ion is much more soluble (Houben & Treskatis, 2007).

$$4FeS_{(s)} + NaCl^{I}O_{\,(aq)} \rightarrow Fe^{2+}_{(aq)} + 4SO^{2-}_{4\,(aq)} + 4Cl^-_{(aq)} + 4Na^+_{(aq)} \tag{5.43}$$

$$Fe^{2+}_{(aq)} + NaCl^{I}O_{\,(aq)} + 6H_2O_{(l)} \rightarrow Fe^{III}(OH)_3 + Cl^-_{(aq)} + Na^+_{(aq)} + 3H_3O^+_{(aq)} \tag{5.44}$$

$$4FeS_{2\,(s)} + 7NaCl^{I}O_{\,(aq)} + H_2O_{(l)} \rightarrow Fe^{2+}_{(aq)} + 2SO^{2-}_{4\,(aq)} + 7Cl^-_{(aq)} + 7Na^+_{(aq)} + 2H_3O^+_{(aq)} \tag{5.45}$$

Dissolution by oxidation occurs where oxidation agents, like sodium hypochlorite in chemical Eqs. (5.43) and (5.45), oxidise the non-metallic part of minerals, causing metal ions to be dissolved. The bivalent iron ion oxidises in solution, transforms into suspended, amorphous hydroxide particles (Eq. (5.44)), and can be removed to some extent by surging and over pumping. Iron sulphides and disulphides can also be dissolved through oxidation, which is the only method in the case of disulphides. A disadvantage of oxidation treatment, which requires a high concentration of the oxidation agent, is when oxidising a metal ion whose oxidation state is lower in solution (most often bivalent iron), and sorbed to oxyhydroxides that cannot be removed completely from the well and its environs.

Regardless of the type, concentration and quantity of the chemicals used for rehabilitation, the application principle is the same. After the chemicals are injected into the well bore, they are pressurised (5–6 bar) to ensure that they penetrate the filter pack or the water-bearing formation around the screen, and maximise their effectiveness. Pressurisation can be achieved via the wellhead, which must be properly sealed. The wellhead needs to contain openings for a pressure gauge, compressed air, airlifting and gas evacuation. Chemical treatment is followed by physical treatment, to further degrade the products of incrustation and mechanical clogging. Silt and clay particles tend to stick to each other and create a viscous state, which makes it very difficult to remove them from a water-bearing formation composed of gravel and sand. In the case of the wells, where there is silt and clay clogging in addition to chemical incrustation, the most effective treatment is by dispersive and gelatinous components of the polyphosphate family. They are strong enough to separate clay particles. Dispersive agents cause the particles to bounce against each other, sufficiently mobilising them so that they can move when water is being pumped in and out of the well during rehabilitation. Moreover, the calcium, magnesium and iron ions, which adhere to fine particles, become sequestered in the solution. It is thus easier to remove the particles bound to these ions.

Following the chemical reaction, the well is unsealed and airlifted, to remove the products of rehabilitation. The procedure is repeated several times, then the well test-pumped and the water quality analysed before the rehabilitated well is put back into operation.

### *5.9.2.3 Disinfection*

Microorganisms are omnipresent in the underground environment, including aquifers (Cullimore, 1999). Most of them are sessile, attached either to the grains of the aquifer matrix or suspended particles (e.g. in karst systems). Others are of the plankton type, floating freely in groundwater. When a bacterial population finds itself in a setting that is no longer favourable, the bacteria release water, lose their surface electrical charge and become ultramicrobacteria. After they are detached from the surface to which they were fixed, they are carried by the groundwater. If they reach a suitable environment, they revert to their normal state and attach to some surface. The suitable environment can be the region around a pumping well because of the larger influx of nutrients (especially particulate organic substances, including microorganisms and dissolved organic substances, and oxidation agents – oxygen, nitrate ions, sulphate ions), quicker removal of metabolic by-products and, especially in the case of unconfined aquifers, mixing of oxic and anoxic groundwater due to lowered groundwater levels around wells. This means that biofouling bacteria cannot be eliminated permanently, even if they were all removed from the near-well region (which is impossible to achieve). Hence, in the case of biofouling microorganisms, the primary purpose of disinfection is to disturb the physical integrity of the microbial colonies and their extra-cellular deposits, in order to remove as much of the deposited material as possible, rather than remove, even temporarily, the microorganisms from the well and immediate aquifer zone. There are diverse variants of this procedure, which includes mechanical, chemical and disinfection phases.

*BCHT (Blended Chemical Heat Treatment)* comprises three stages:

- The shock stage, including the injection of a heated (30–95°C) chemical solution (based on active chlorine in either acetic or glycolic acid) with non-phosphate polyelectrolyte surfactant into the water-bearing formation (phosphates might facilitate re-growth). This reduces the consumption of chemicals in the next phase – demolition. Biofilms and encrustations are softened and disinfection efficiency improved, enabling a better and more extensive contact between the microorganisms and the disinfectant.
- The demolition stage, which begins after overnight 'soaking', including the injection of a mixture of chemicals that contains another surfactant, again heated to 60°C and 95°C inside the well. The pH level is reduced to the lowest value, to 1 or typically 2. Heating increases the metabolic rate in the borderline zone of influence of the heat, increasing the assimilation of toxic disinfectants.
- The dispersion stage, including physical removal of the destroyed deposits in the captured volume.

*UAB™ (Ultra Acid-Base) procedure* is technically less intensive than BCHT – International School of Well Drilling (ISWD, n.d.) and comprises three stages. Similar to BCHT, the purpose of the first stage is to shock bacterial cells and biofilms, the second stage is to disturb (demolish) the biofilms, and the third stage is to separate biofilms and other materials:

- The UAB shock stage begins after the well is heated with hot water, to increase the temperature inside the well and formation to approximately 65°C. A hot water solution of the disinfectant or acid (such as acetic acid) and a non-phosphate surfactant is injected, the latter to facilitate soaking. The water inside the well and formation is maintained at a temperature of at least 65°C, to improve chemical efficiency. High temperatures increase the chemical reaction rate and reduce the quantity needed for cleansing. The soaking agent facilitates the penetration of hot water and chemicals into the biofilms.
- The demolition stage eliminates bacteria and causes a dramatic shift from a highly acidic to a highly alkaline solution. Even though some bacteria thrive in acidic and others in alkaline environments, none can acclimate to sudden shifts, not even bacteria adapted to circumneutral pH conditions.
- The dispersion stage comprises physical removal of the destroyed deposits from the captured volume.

Microorganisms possess defence mechanisms, such as biofilm dehydration, which increases density, improves mechanical resilience and slows down the diffusion of chemicals, or the buffer effect of biofilms, which mitigates pH changes and protects the microorganisms. Apart from reducing the treatment effectiveness, biofilm shrinkage temporarily reduces flow resistance. The effectiveness can thus be overestimated, if the wells are tested immediately upon cleaning and flushing. Disinfection shortfalls are much less pronounced in the case of injection wells because, after combined aggressive treatment, injection can continue and the dissolved active chlorine can prevent the re-occurrence of microorganisms along the privileged groundwater pathways.

## REFERENCES

Abramov S.K. (1952). Method of Calculation and Selection of Filters for Drilled Wells (in Russian). Works of Hydrogeology and Engineering Geology Section of The Federal Scientific Technical Mining Society, Moscow, Vol. **1**.

Alekseev V.S. (1968). Well-screen Length and Diameter Selection Considering Load Nonuniformity. Power Technology and Engineering (formerly Hydrotechnical Construction), Vol.**2**, No. 8, August 1968 (translated from Gidrotekhnicheskoe stroitel'stvo, No. 8, 46-48, 1968), Springer-Verlag New York, LLC.

Alford G. and Cullimore D. R. (1999). The Application of Heat and Chemicals in the Control of Biofouling Events in Wells CRC Press/Lewis Publishers, Boca Raton, FL, USA.

Babac D., Dimkić M. and Soro A. (1989). General Framework for Hydrodynamic Approaches to Groundwater Source Issues. Hydrodynamic Investigations of Groundwater Sources, Jaroslav Černi Institute, Belgrade, Serbia, Special Editions, Vol. **2**, pp. 7–33 (in Serbian)

BioIrc and Jaroslav Černi Institute for the Development of Water Resources (2014). Lizza – pre and post-processor for modeling of underground water flow with Ranney wells. Retrieved from http://www.bioirc.ac.rs/index.php/groundwater-flow-software (24. 11. 2014).

Borch M.A., Smith S.A. and Noble L.N. (1993). Evaluation, Maintenance, and Restoration of Water Supply Wells. American Water Works Association Research Foundation, Denver, CO, USA. Retrieved from http://www.awwarf.com/research/topicsandprojects/execSum/401.aspx

Cistin J. (1965). Some problems of deformation of mechanical filters (in Slovakian), Vodohospod, Vol. **2**.

Cullimore D.R. (1993) Practical Manual of Groundwater Microbiology. CRC Press LLC, Bocca Raton, Florida, USA. ISBN 978-0873712958

Cullimore D. R. (1999). Microbiology of Well Biofouling. CRC Press LLC, Boca Raton, FL, USA.

Dimkić M. (2012). Processes in alluvial groundwater and their significance (in Serbian). 14th Serbian Symposium on Hydrogeology, 17–20 May 2012, Zlatibor, ISBN 978-86-7352-236-4, pp. 5–10, 2012.

Dimkić M. and Knežić L. (2012). Biochemical Processes, Groundwater management in large river basins. Regional IWA Conference, Belgrade, pp. 169–187, 2007 (First Publishing 2008- London),

Dimkić M. and Pušić M. (2008). Recommendations for Water Well Design Taking into Account Iron Clogging Based on Experience Gained at Belgrade Groundwater Source (in Serbian). *Civil Engineering Calendar* 2008, **40**, 430–496.

Dimkić M. and Pušić M. (2014). Correlation between Entrance Velocities, Increase in Local Hydraulic Resistances and Redox Potential of Alluvial Groundwater Sources. *Water Research and Management*, **4**(4), 3–23, ISSN 2217-5237, Publisher: Water Pollution Control Society of Serbia, Vol. 4, No. 4, pp. 3-23, 2014.

Dimkić M. and Pušić M. (2018). A new approach to the definition of design criteria for radial collector wells in anoxic settings. *Annales Geologiques de la Peninsule Bellqenique*, **79**, 31–45. Doi: 10.2298/6ABP1879031 D

Dimkić M., Brauch H. J. and Kavanaugh M. (2008). Groundwater Management in Large River Basins. IWA Publishing, London, UK.

Dimkić M., Pušić M., Vidović D., Isailović V., Majkić B. and Filipović N. (2010). Numerical Model Assessment of Radial-Well Ageing. *Journal of Computing in Civil Engineering*, **25**(1), 43–49, doi:10.1061/(ASCE)CP.1943-5487.0000063

Dimkić M., Pušić M., Majkić-Dursun B. and Obradović V. (2011a). Certain Implications of Oxic Conditions in Alluvial Groundwater. *Journal of Serbian Water Pollution Control Society "Water Research and Management"*, ISSN 2217-5237, **1**(2), 27–43.

Dimkić M., Pušić M., Vidović D., Petković A. and Boreli-Zdravković Đ. (2011b). Several natural indicators of radial well ageing at the Belgrade Groundwater Source, Part 1. *Water Science & Technology*, **63**(11), 2560–2566.

Dimkić M., Pušić M., Obradović V. and Djurić D. (2011c). Several natural indicators of radial well ageing at Belgrade Groundwater Source, Part 2. In: *Water Science & Technology*, **63**(11), 2567–2574.

Dimkić M., Pušić M., Dursun-Majkić B. and Obradović V. (2011d). Implications of Oxic Conditions in Alluvial Groundwater, "*IWA Specialist Groundwater Conference*", Proceedings, 15–30, September 08-10-2011, Belgrade, ISBN 978-86-82565-31-4.

Dimkić M., Pušić M., Obradović V. and Kovačević S. (2012). The effect of certain biochemical factors on well clogging under suboxic and mildly anoxic conditions. *Water Science & Technology*, **65**(12), 2206–2212, doi: 10.2166/wst.2012.129

Dimkić M., Ranković V., Filipović N., Stojanović B., Isailović V., Pušić M. and Kojić M. (2013). Modeling of radial well lateral screens using 1D finite elements. *Journal of Hydroinformatics*, **15**(2), ISSN 1464-7141, IWA Publishing, 405–415, http://www.wrmjournal.com/images/stories/casopis/No_16/01.pdf

Dotlić M. (2015). Proračun podzemnog toka metodom konačnih zapremina (Calculation of the underground flow using finite volume method). PhD thessis, Faculty of Mathematics, University of Belgrade, Belgrade, Serbia.

Dubinina G. A. (1978). Mechanism of oxidation of divalent iron and manganese by iron bacteria growing in neutral medium. *Mikrobiologiya*, **47**, 591–599.

Ellis D. (2006). Iron Bacteria – Organisms and Their Identification. Merchant Books, USA. ISBN: 978-1933998404.

Gavrilko V.M. (1968). Water Supply, Lowering and Hydrogeological Well Screens (in Russian), Publishing Literature For Mechanical Engineering, Moscow, p. 394.

Gavrilko V.M. and Alekseev V.S. (1985). Water Well Screen. Publishing House "NEDRA", Moscow, pp. 300–304.

Houben G. (2003). Iron Oxide Incrustation in Wells: Parts 1 and 2. Applied Geochemistry, Elsevier Science.

Houben G. and Treskatis C. (2007). Water Well Rehabilitation and Reconstruction. McGraw Hill, New York, USA, ISBN-13: 978-0-07-148651-4, p. 391.

Howsam P. (ed.) (1990). Water Wells Monitoring, Maintenance, and Rehabilitation. Proc. of the International Groundwater Engineering Conference, Cranfield Institute of Technology, UK. E.&F.N. Spon, London.

Institute for the development of water resources "Jaroslav Černi" (JCI) (2010). Studija "Beogradsko izvorište - stanje i pravci razvoja" (Belgrade groundwater source, current state and avenues for development study), Beograd, Srbija.

Institute for the development of water resources "Jaroslav Černi" (JCI) (2013). Studija "Starenje bunara i održavanje objekata" (Well ageing and maintenance of facilities study), Beograd, Srbija.

"Jaroslav Černi" Water Institute (JCI) (2019). Projekat Metodologija ocene, projektovanja i održavanja izvorišta podzemnih voda u aluvijalnim sredinama u zavisnosti od stepena aerobnosti, br. TR37014 (project "Methodology for Assessment, Design and Maintenance of Groundwater Sources in Alluvial Environments Depending on the Aerobic State", No. TR37014), Beograd, Srbija.

Istomina V.S. (1957). Filtration Stability of Soils (in Russian). Gostroizdat, Moscow-Leningrad, USSR.

ISWD. (n.d.). Well Rehabilitation. International School of Well Drilling (ISWD), Lakeland, FL, USA. https://sswm.info/sites/default/files/reference_attachments/ISWD%204000%20Well%20Rehabilitation.pdf (accessed on 16 May 2020)

Johnson E.E. (1975). Groundwater and Wells: A Reference Book for the Water Well Industry. Edward E. Johnson, St. Paul, MN, USA.

Kovacs G. and Ujfaludi L. (1983). Movement of fine grains in the vicinity of well screens. *Hydrological Sciences – Journal – des Sciences Hydrologiques*, **28**(2), 6/1983.

Langelier W.F. (1936). The Analytical Control of Anti-corrosion Water Treatment. *Journal of the American Water Works Association*, **28**, 1500–1521.

Lazić M. (2004). Well Development and Rehabilitation Methods, 2nd edn (in Serbian). University of Belgrade, Faculty of Mining and Geology.

Leach R., Mikell A., Richardson C. and Alford G. (1991). Rehabilitation of Monitoring, Production and Recharge Wells. 15th Annual Army Environmental R&D Symp. 1990 Proceedings. CETHA-TS-CR-91077, U.S. Army Toxic and Hazardous Materials Agency, Aberdeen Proving Grounds, MD, pp. 623–646.

Mansuy N. (1998) Water well rehabilitation: A practical guide to understanding well problems and solutions. CRC Press LLC, Bocca Raton, Florida, USA. ISBN 1-56670-387-5.

McLaughlan R. (1996). Water Well Deterioration, Diagnosis and Control. Technology Transfer Publication 1/96, National Centre for Groundwater Management, University of Technology, Sydney, Australia.

Petković A., Dimkić M. and Boreli-Zdravković Đ. (2008). Merenje Redox potencijala u podzemnim vodama: značaj, koncept i problemi, 37. konferencija o aktuelnim problemima korišćenja i zaštite voda "VODA 2008", 03-06. 06. 2008, Mataruška Banja, str. 221–224, 2008.

Pietraru V. (1982). Rapport general sur les crepines et filtres des puits (in French). Proceedings of The International Symposium on Hydraulic Problems of Ground Water Drainage, Yugoslav association for Hydraulic Research, International Association for Hydraulic Research (Section on Flow Through Porous Medium), Donji Milanovac, Yugoslavia, pp. 39–80.

Pušić M. and Dimkić M. (2017). Case study of the effect of riverbed conductivity and aquifer interbed on groundwater flow to a well (in Serbian). 15th Hydrogeology Symposium, September 2016, Kopaonik, Publisher: University of Belgrade, Faculty of Mining and Geology, ISBN 978-86-7352-316-3, pp. 201–205, 2016.

Ryznar J.W. (1944). A New Index for Determining the Amount of Calcium Carbonate Scale Formed By Water. *Journal of the American Water Works Association*, **36**, 472–484.

Sichardt W. (1928). Method of Stabilisation of Dilled Wells (in German), Springer, Berlin.

Skerman P.J. (1957). Bullamon lucerne (Psoralea eriantha Benth) a Plant Worth Watching. *Journal of the Australian Institute of Agricultural Science*, **23**, 337–339.

Smith S.A. (1995). Monitoring and Remediation Wells: Problem Prevention, Maintenance, and Rehabilitation. CRC Press LLC, Bocca Raton, Florida, USA. ISBN 0-87371-562-4.

Truelsen C. (1988). Groesste Brunnenleistung. Die Wassererschliessung: Grundlagen der Erkundung, Bewirtschaftung und Erschliessung von Grundwasservorkommen. In: H. Schneider (ed.) *Theorie und Praxis*, 3rd ed., Vulkan - Verlag, Essen, W. Germany.

Vidović D. (2014). Groundwater Modeling with WODA, Modeling Course Textbook. Jaroslav Černi Institute, Belgrade, Serbia.

Vuković M. and Pušić M. (1992). Soil Stability And Deformation Due To Seepage, Water Resources Publications, Littleton, Colorado, USA.

Vuković M. and Soro A. (1990). Hydraulics of Water Wells. Water Resources Publications, Littleton, Colorado, USA, 354p.

# Chapter 6

# Pharmaceutical attenuation processes during filtration in groundwater

*Srđan Kovačević and Milan Dimkić*

Indicators of anthropogenic influence which were analysed in the context of the research activities were a selected group of pharmaceuticals and degradation products or metabolites. Awareness of the large number of pharmaceuticals present in aquatic systems, with the tendency for this number to increase due to improvements in analytical technique sensitivity, is a matter of serious concern. However, the site-specific structure of non-organic matter (NOM), which is variable in time and space, and the wide variety of combinations that may occur in treatment (relating to substance mixtures, other constituent mixtures and processes) create a challenging situation both from the point of view of technology optimization and sustainable management.

The discharge of treated effluent from wastewater treatment plants (WWTPs) and non-treated wastewater is a major pathway for the introduction of pharmaceuticals to surface water, and WWTPs should act as a primary barrier against the spread of pharmaceuticals into the environment. It is evident that a large number of pharmaceuticals are not fully removed at the WWTPs and in some countries most of the wastewater is not treated. Current WWTPs are not specifically designed to eliminate pharmaceuticals. Thus, many of these pharmaceuticals are able to pass through wastewater treatment processes by virtue of their persistency and/or their continuous introduction. In addition, monitoring actions for pharmaceuticals have not been well established in most WWTPs.

After drinking water treatment of surface water and groundwater, some pharmaceuticals could still be found in drinking water. The majority of the observed pharmaceuticals could pass through conventional methods of removal, but were mostly removed using advanced methods. However some pharmaceuticals have even been shown to be resilient to advanced methods. Over 100 individual pharmaceutically active compounds have been identified in environmental samples and drinking waters and the number of detected pharmaceuticals is growing. To date, discharge guidelines and standards do not exist for pharmaceuticals.


Editor: Milan Dimkić
doi: 10.2166/9781789060904_0335

Alluvial aquifers are under the direct influence of surface water resources, which are loaded with untreated wastewater. As a result, pharmaceuticals occur in surface water and groundwater.

Riverbank filtration (RBF) can be exploited to remove pharmaceuticals through self-purification processes, among which sorption and biodegradation are the most pronounced. Among others, one important factor for the self-purification capacity of the subsurface is the groundwater residence time (i.e. the travel time of water from the river to a well) which is dependent on the site-specific hydrogeology and can vary from a few days to a few years. A second important factor is the ratio of infiltrated water from the river, because pharmaceuticals originate from anthropogenic activities mainly resulting in emission into surface water. Therefore, the dilution from the hinterland of the well is a very important factor for RBF systems.

The aim of our study was to obtain sufficient information about the occurrence of selected pharmaceuticals in surface water and groundwater, as well as about the removal of these pharmaceuticals through RBF, which is generally the method of choice for drinking water supply in Serbia. This was the first time that a long-term study (7-year survey) of pharmaceutical occurrence in surface water and groundwater was conducted in the Republic of Serbia.

## 6.1 PHARMACEUTICALS: DEFINITION, DIVISION, FLOWS AND PRIORITY ROUTES

The term emerging organic substances (EOSs) refers not only to the substances that have recently been synthesized but also to the existing substances that have recently been detected in the environment, and that have been used for a longer period of time but can have potentially negative effects on the environment (Glassmeyer *et al.*, 2008; NORMAN, 2016). Examples of EOSs from the NORMAN list are: surfactants, flame extinguishers, pharmaceutical products for personal care, gasoline additives and the products of their degradation, biocides, polar pesticides and the products of their degradation, as well as various compounds that are suspected of or have been proven to adversely affect the endocrine system.

Emerging organic substances, even if present in the environment, will not have been considered as potential pollutants until they were detected in the aquatic environment. Pharmaceuticals represent a group of EOSs that are present in the environment and appear constantly at very low concentrations. With the development of new analytical methods for the detection of pharmaceuticals, an ever increasing number of pharmaceuticals can be detected in environmental matrices, and especially so in aquatic environments.

According to the World Health Organization (WHO), a pharmaceutical is any substance or pharmaceutical product used for treating humans and animals which aims to modify or explore physiological systems or pathological conditions for the benefit of the final recipient. Pharmaceuticals is the name used for licensed pharmaceutical products which, based on scientific experiments, reduce symptoms and lead to a favourable outcome of a disease.

### 6.1.1 Division of pharmaceuticals

It is not easy to divide pharmaceuticals. There are several types of divisions in the literature, but the most commonly used is the division of pharmaceuticals by therapeutic class (Farré *et al.*, 2008):

- Analgesics and anti-inflammatory pharmaceuticals;
- Antibiotics;
- Antiepileptics and psychiatric pharmaceuticals;
- Pharmaceuticals for treating tumours;

- Cardiovascular pharmaceuticals;
- Hormones;
- Contrast media for X-ray analysis;
- Prohibited pharmaceuticals – 'Illicit drugs'.

Analgesics and anti-inflammatory pharmaceuticals are intended to relieve pain and they include narcotic analgesics, non-narcotic analgesics, as well as non-steroidal anti-inflammatory drugs (NSAIDs).

The effects of analgesics and anti-inflammatory pharmaceuticals are manifested in different ways by acting on the peripheral and central nervous systems, being used mainly to relieve pain caused by various ailments (Bueno *et al.*, 2012).

Antibiotics are widely used in the treatment of infections that affect the human population or animals. Antibiotics are particularly important because their frequent and widespread use leads to bacterial resistance and to the reduced efficacy of antibiotics for the treatment of infections. Bacteria that become resistant to antibiotics are released into the environment along with human and animal waste which can transmit genes that are responsible for antibiotic resistance (Reinthaler *et al.*, 2003).

Antiepileptics and psychiatric pharmaceuticals are used to control the onset of epilepsy as well as for the antagonism of dopamine receptors. Although many drugs in this group act on other receptors, they function in one of three main ways: they enhance the effect of gamma-butyric acid, inhibit the function of sodium channels, and also inhibit the function of calcium channels.

Cardiovascular pharmaceuticals are used to treat hypertension, and include the following groups of drugs: antihypertensives in the narrow sense of the word (alpha-blockers, direct vasodilators), diuretics, organic nitrates, beta-blockers, calcium channel antagonists, angiotensin converting enzyme inhibitors (ACEs), and angiotensin II inhibitors (ATII inhibitors).

Tumour pharmaceuticals have carcinogenic, mutagenic, teratogenic and ferro-toxic properties and are used in the treatment of tumours or various immune diseases.

Hormones are a group of pharmaceuticals that act as endocrine disruptors and damage the hormonal system of organisms. They include synthetic hormones and contraceptives, as well as other chemicals and their products for human use. Hormones can also cause adverse effects at very low concentrations (ng/l) that are detected in the environment (Ternes, 1998).

Contrast media for X-ray analysis are widely used in human medicine for screening organs or blood vessels when doing diagnostic tests and X-ray imaging. They are metabolically stable and are excreted from the body in a short period of time. Due to their persistence and widespread use, they are often detected in the environment.

Prohibited Pharmaceuticals – 'illicit drugs' are prohibited pharmaceuticals. Most illegal pharmaceuticals are addictive.

Table 6.1 shows a list of four pharmaceuticals that have been analysed in more detail within the chapter. The pharmaceuticals selected for analysis and research belong to different classes, and were chosen based on

**Table 6.1** List of thoroughly analysed pharmaceuticals, abbreviations, and therapeutic classes.

| Pharmaceutical | Abbreviation | Therapeutic Class |
|---|---|---|
| Trimethoprim | TMP | Antibiotics |
| Carbamazepine | CBZ | Antiepileptics and psychiatric pharmaceuticals |
| Diclofenac | DCF | Analgesics and anti-inflammatory pharmaceuticals |
| N-acetyl-4-aminoantipyrine | 4-AAA | |

how much they are used, i.e. their occurrence in: wastewater, surface water, groundwater, as well as after the treatment of drinking water, and their persistence and biodegradability.

The pharmaceuticals analysed in greater detail were: trimethoprim, carbamazepine, diclofenac and the metabolite of metamizole N-acetyl-4-aminoantipyrine (4-AAA). These were selected on the basis of the frequency of their occurrence in surface and groundwater, as well as on the basis of their belonging to different therapeutic classes of pharmaceuticals.

## 6.1.2 Manufacturing and use of pharmaceuticals

Pharmaceutical spending covers the cost of prescription medicines and self-medication, often referred to as over-the-counter products. In some countries, other medical non-durable goods are also included. Pharmaceuticals consumed in hospitals and other health care settings are excluded.

The final spending on pharmaceuticals includes wholesale and retail margins and value-added tax. Total pharmaceutical spending in most countries refers to 'net' spending, i.e. including the adjustment for possible rebates payable by manufacturers, wholesalers or pharmacies (Figure 6.1). This indicator is measured as a share of total health spending, in USD per capita (using economy-wide PPPs – public private partnerships) and as a share of gross domestic product (GDP).

In many countries there has been a gradual decline in the production of pharmaceuticals in recent years (Figure 6.2), but the consumption of certain categories of pharmaceuticals continues to increase.

The amount of resources each country allocates over time for health purposes, both for individual and collective services, depends on a large number of social and economic factors, as well as on the financial and organizational structure of the health system.

Expenditure on health is a measure of the final consumption of health goods products and services. This includes spending by both public and private sources on medical services and goods, and public health and prevention programs and administration, but excludes spending on capital formation (investments).

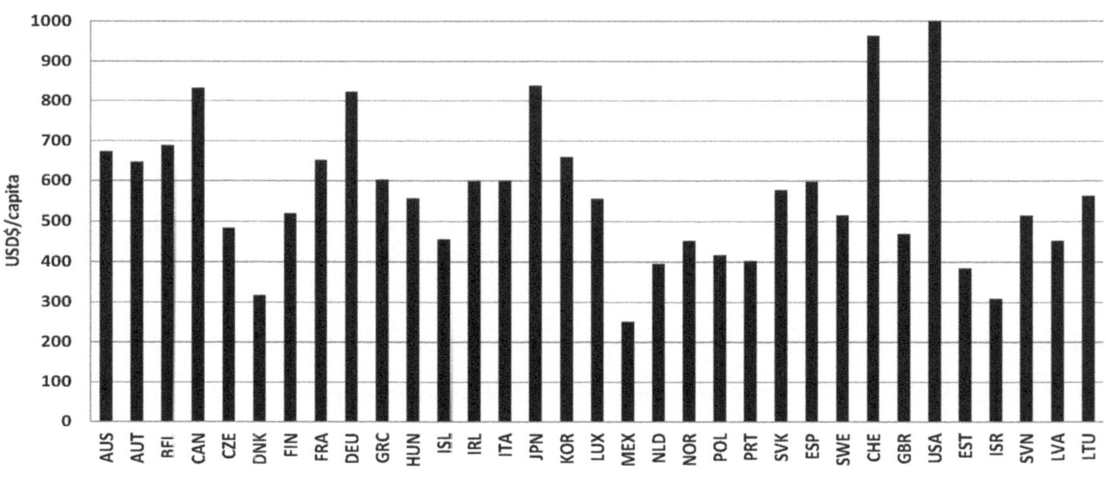


**Figure 6.1** Pharmaceutical spending in USD/capita (Organisation for Economic Co-operation and Development [OECD], 2020).

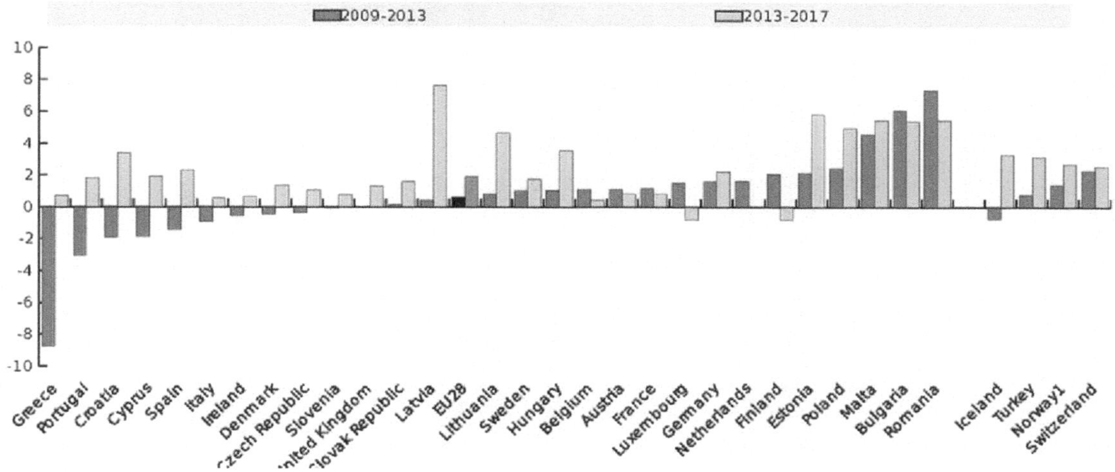


**Figure 6.2** Annual average growth rate in per capita health expenditure, real terms, 2009 to 2017 (OECD Health Statistics 2018, https://doi.org/10.1787/health-data-en; Eurostat Database, Norway 1, stands for Mainland Norway GDP price index used as deflator).

The consumption of some groups of pharmaceuticals is increasing due to an increase in the average age of the population, as well as due to the increasing frequency of chronic diseases.

A study analysed the per capita consumption of three groups of constantly used pharmaceuticals, namely those used to treat chronic diseases: pharmaceuticals for high blood pressure and pharmaceuticals to treat high cholesterol, both belonging to the cardiovascular therapeutic class, as well as antidepressants belonging to the therapeutic group of antiepileptic and psychiatric pharmaceuticals.

The consumption of the mentioned pharmaceuticals is expressed through a defined daily dose (DDD) unit. The defined daily dose is the assumed average daily dosage of the pharmaceutical product for adults. By international expert consensus the DDD is assigned to all the active ingredients in a particular therapeutic class of pharmaceuticals.

High blood pressure is one of the leading public health problems in most countries of the world. It is estimated that one in three adults in the world has problems related to high blood pressure, while 13% of deaths are associated with high blood pressure (WHO, 2012). Between 2000 and 2017, the consumption of pharmaceuticals in the European Union (EU) which are used to reduce blood pressure has increased by an average of 70%, while it has tripled in Estonia and quadrupled in Luxembourg. The highest consumption of pharmaceuticals for hypertension is in Germany, Hungary and the Czech Republic, while the lowest consumption is in Korea and Turkey. Figure 6.3 includes consumption of pharmaceuticals for hypertension in selected EU countries.

The use of cholesterol-lowering pharmaceuticals has tripled in most OECD countries. Increased obesity as well as a higher frequency of monitoring and treatment of increased cholesterol levels may explain the increase in consumption. The United Kingdom, Denmark and Belgium had the highest consumption of cholesterol-lowering pharmaceuticals per capita in 2017, while Chile and Turkey had the lowest consumption. Figure 6.4 includes consumption of cholesterol-lowering drugs in selected EU countries.

Figure 6.5 includes consumption of antidepressant drugs in selected EU countries.

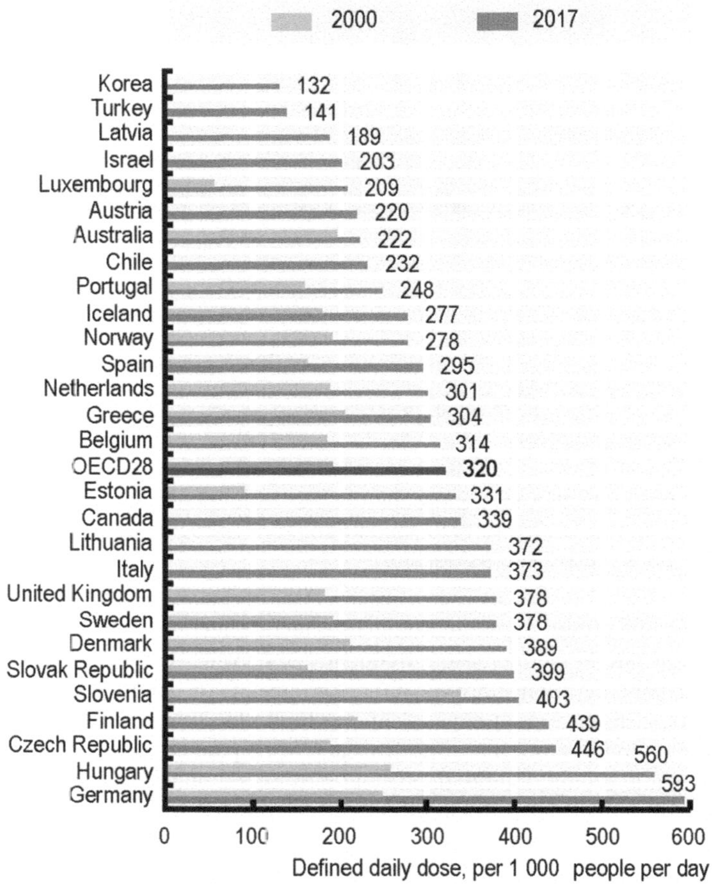


**Figure 6.3** Consumption of pharmaceuticals for hypertension (OECD Health Statistics, 2019).

### 6.1.3 Cycle of pharmaceuticals: waste – surface water – groundwater

The origin of pharmaceuticals in the environment is anthropogenic and is the product of particular human activities that can be divided as follows:

- Pharmaceutical manufacturing industry;
- From households, hospitals or settlements;
- From agriculture (animal husbandry, fish farming).

Intensive industrialization and urbanization have resulted in large quantities of waste substances that reach surface waters as effluents. Over the last two centuries the emission sources of surface and groundwater pollution have changed significantly in line with industrial development.

The pressure on water is ever increasing; consumption is steadily increasing, while water treatment technology (in production, households, etc.) is not sufficiently developed and widespread to purify all the contaminated water. The presence of pharmaceuticals in groundwater is due to the penetration of pharmaceuticals during bank filtration from rivers containing pharmaceuticals. Pollutants can reach an

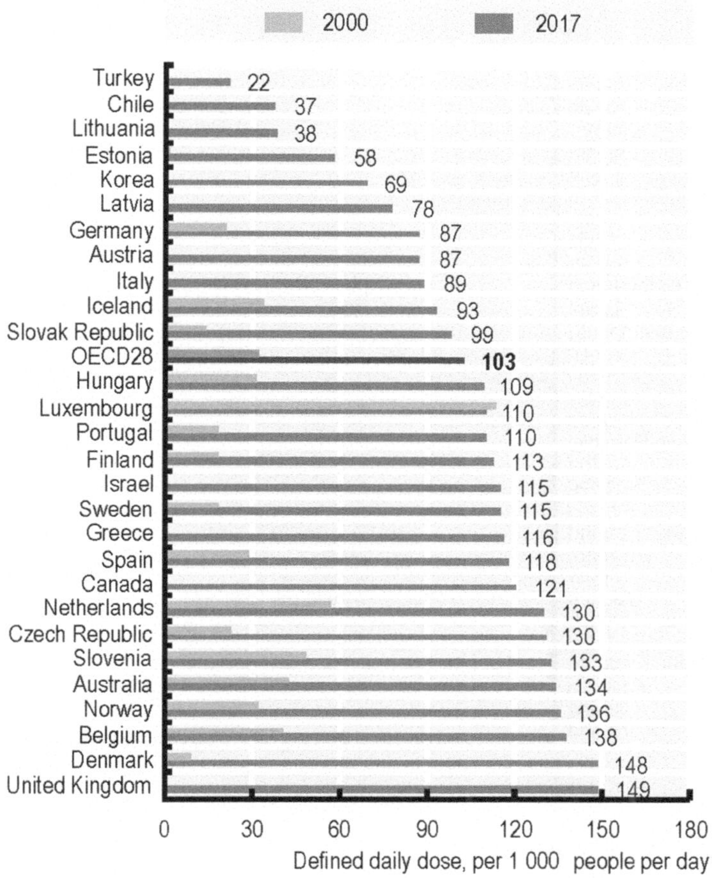


**Figure 6.4** Cholesterol-lowering drugs consumption (OECD Health Statistics, 2019).

aquifer by the infiltration of surface water through soil, sediment and rocks, through the direct infiltration of surface water (rivers, lakes, streams), or as a result of direct inflow through improperly drilled wells, underground reservoirs and similar. In surface waters, pollutants are present due to the inadequate treatment of sewage prior to discharge into recipients. There is also the possibility of a sudden appearance of pharmaceuticals in the aquatic environment due to accidental situations in the manufacture and transport of pharmaceuticals.

There are numerous sources of pollutants in surface water and they most often fall into two groups based on the physical space they occupy:

- Concentrated – Point sources of pollutants;
- Diffuse Sources – Unconcentrated sources of pollutants.

Concentrated – point sources of pollutants – are individual identified sources of pollutants that originate from different locations, but are located on a smaller area, so that the fluxes of pollutant emission into the aquatic environment are more likely to be detected. Industrial wastewater, hospital wastewater,

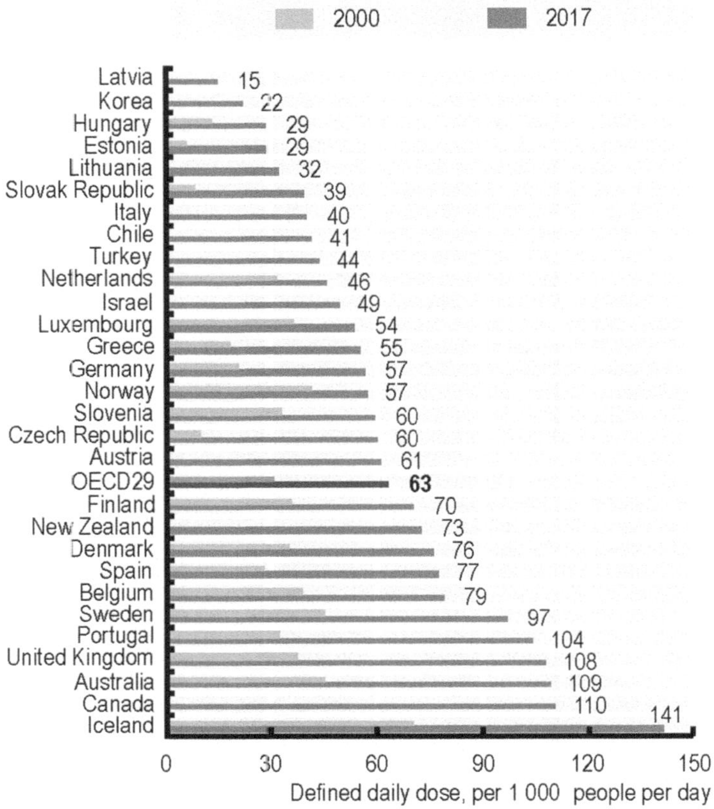


**Figure 6.5** Consumption of antidepressant drugs (OECD Health Statistics, 2019).

municipal wastewater, sanitary landfills and septic tanks represent typical concentrated sources of pollutants that reach the aquatic environment.

Unlike concentrated sources, unconcentrated sources of pollutants are not easy to identify, nor is it easy to calculate their total contribution to the overall surface and groundwater load.

Diffuse sources of pollutants include the excretion from animals, wastewater from animal waste storage, wastewater from illegal dumps, and runoff and leakage in municipal sewage networks, as well as leakage from WWTPs. Compared to concentrated pollutant sources, there is less potential for the natural removal of diffuse pollutant sources in aquatic environments (Murray *et al.*, 2010). Another significant diffuse source of pharmaceuticals in the environment is sludge generated after wastewater treatment.

The main ways of using pharmaceuticals include enteral – oral administration, parenteral administration – injection, and topical – e.g. via the surface of the skin and inhalation. Consumed pharmaceuticals are generally excreted as metabolized or non-metabolized products, and surface-applied substances can be washed off and thus reach the environment. When they reach the wastewater stream, they are combined with municipal wastewater, as well as with process water from certain types of industry. Expired pharmaceuticals can also reach the wastewater directly. Other sources responsible for the presence of pharmaceuticals in wastewater include land where pharmaceuticals may have been disposed of or places where pharmaceuticals are produced, as well as healthcare facilities.

Figure 6.6 illustrates the pathways of pharmaceuticals from a source to various parts of the environment. Five major sources of pharmaceuticals that reach the environment have been identified and they are: landfills, animal waste, hospital waste, industrial waste and municipal waste, whose pathway to the environment has been simplified and summarized on the basis of several literature reviews (Díaz-Cruz & Barceló 2008; Dimkić *et al.*, 2008; Eggen *et al.*, 2012; Heberer, 2002; Lapworth *et al.*, 2012; Pal *et al.*, 2010). The recipients of pharmaceuticals in the environment can be divided into three basic environmental media: soil, surface water and groundwater.

Municipal landfills are sources of pollutants of various origins which can have adverse effects on plants, animals and on human health. Landfills also generate leachate water, which if not treated or if discharged directly into groundwater, as in the case of non-sanitary landfills, represents a potential source of pharmaceuticals in the environment.

Wastewaters originating from various sources represent one of the most important point sources of pharmaceuticals in the aquatic environment. There are numerous studies that have been conducted at different geographical locations that point out the evident occurrence of different pharmaceuticals in the aquatic environment. Wastewater is generated from various sources, such as hospitals, households, industry, septic tanks and animal waste. If it does not reach the environment directly due to the potential leakage of the sewage system, wastewater containing pharmaceuticals is treated in a WWTP. Even after the treatment of wastewater, there is the possibility that certain pharmaceuticals will enter surface and groundwater due to the insufficient degree of disposal in the WWTP. Raw wastewater is distributed to wastewater treatment plants whose main objective is to eliminate pathogenic microorganisms, any

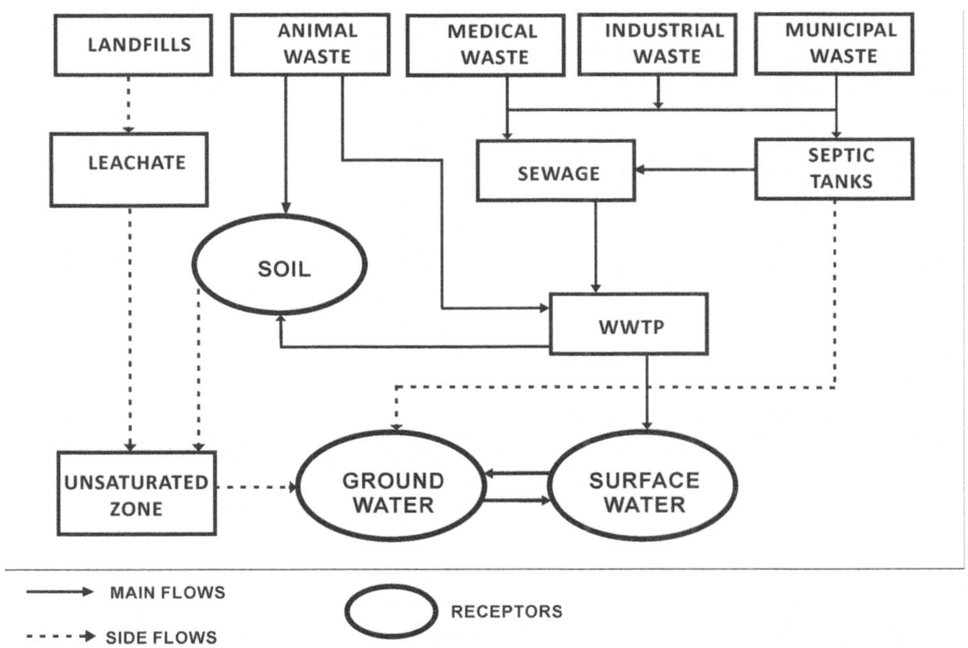


**Figure 6.6** Different pathways of pharmaceuticals into the environment (Díaz-Cruz & Barceló 2008; Dimkić *et al.*, 2008; Eggen *et al.*, 2012; Heberer, 2002; Lapworth *et al.*, 2012; Pal *et al.*, 2010).

biological load of wastewater, and nutrients. There are two basic flows of treated wastewater: treated wastewater and the sludge resulting from wastewater treatment. The treated wastewater can be used for irrigation in agriculture or is discharged into surface water. In coastal regions, treated wastewater is often discharged into seas and oceans. The pharmaceuticals that reach surface waters can reach groundwater as well.

Another way pharmaceuticals get into surface and groundwater is by flushing the soil where pharmaceutical waste and solid waste contaminated by pharmaceutical products have been deposited. The main sources of pharmaceuticals are solid forms of pharmaceutical products from households as well as from industry and agricultural activities that involve the use of pharmaceuticals.

The third possible route for pharmaceuticals to reach surface and groundwater is from the leakage of sewage. Although leakage from a sewage system is occasional and seasonal, its effect on the occurrence of pharmaceuticals in the aquatic environment can be significant.

## 6.2 OCCURRENCE AND TREATMENT OF SELECTED PHARMACEUTICALS IN THE AQUATIC SYSTEM

Based on a literature review of a comprehensive analysis of more than 1,016 original papers and 150 reviewed papers (Aus der Beek *et al.*, 2016), it has been established that pharmaceuticals and products of pharmaceutical transformation have been detected in the environment in 71 countries, including all continents. In total, 631 different pharmaceuticals were detected in the aquatic environment at concentrations above the limit of detection, by applying state of the art analytical methods.

For example, the anti-inflammatory pharmaceutical diclofenac was detected in environmental matrices in 50 countries. On a global level, municipal wastewaters are the dominant pathway for pharmaceuticals to reach the environment, although emissions from industrial production, hospitals, agriculture and aquaculture are important at the local level.

### 6.2.1 Selection of analysed pharmaceuticals

A review is given of the detection of the following four pharmaceuticals in surface and in groundwaters (carbamazepine (CBZ), trimethoprim (TMP), diclofenac (DCF), and metabolite metamizole (4-AAA) N-acetyl-4-aminoantipyrine). Also given is a review of the efficacy of their removal in wastewater treatment plants and water supply plants.

The analysed pharmaceuticals belong to different classes, and were selected on the basis of literature data on the occurrence of pharmaceuticals in surface and groundwater in different regions in the world (Mompelat *et al.*, 2009).

Metamizole metabolites were selected because they belong to the group of analgesics and antipyretics that are used frequently and in large quantities (Moldovan, 2006). Although metamizole is not detected in surface and groundwater, metamizole metabolites are often present in the aquatic environment. These metamizole metabolites are very stable in the aquatic environment and are detected in municipal wastewater and hospital wastewater, as well as in surface and groundwater, in relatively high concentrations (Feldmann *et al.*, 2008; Moldovan, 2006; Wiegel *et al.*, 2004; Zuehlke *et al.*, 2004, 2007). Metamizole metabolites have also been detected in a very high concentration in the effluents from wastewater treatment plants (Martínez Bueno *et al.*, 2007; Qi *et al.*, 2015), indicating a low degree of degradation when using conventional wastewater treatment technologies. Little is known about the behaviour of 4-AAA and 4-FAA in the environment. Gómez *et al.* (2008) investigated the effect of photodegradation on metamizole and metamizole metabolites. 4-AAA and 4-FAA have been found to be resistant to the photodegradation process, so it takes a long time for this to have an effect on the degradation process. These metamizole metabolites can also serve as indicators of the anthropogenic pollution of surface water originating from wastewater (Kahle *et al.*, 2009). Because of the above mentioned facts, the metabolites of the degradation of pharmaceuticals are significant, and it is therefore necessary to monitor their appearance and behaviour in the environment.

### 6.2.2 Removal of pharmaceuticals from wastewater treatment plants

Due to the different compositions and quantities of wastewater, municipal wastewater, industrial wastewater and wastewater from farms and hospitals is often treated separately in developed countries. This is a more efficient solution for controlling the source of pollutants and for precisely monitoring the quality and quantity of wastewater (Terzić *et al.*, 2008). In developing countries and countries in transition, municipal wastewater is most often treated together with wastewater from industry, hospitals and farms (Kaštelan-Macan *et al.*, 2007).

In developing countries, most cities, industrial systems, hospitals and farms discharge untreated water directly into natural or artificial recipients, leading to the continued presence of pharmaceuticals in wastewater, surface water and groundwater (Kovačević *et al.*, 2017a).

On a global level 559 different pharmaceuticals have been detected in wastewater and sludge generated after treatment (Aus der Beek *et al.*, 2016). The large number of pharmaceuticals detected in wastewater is usually correlated with a large number of measurements, for example, in Canada and China. It can be assumed that there is a close link between their occurrence in wastewater and in surface water as wastewater is, after treatment or even without treatment, discharged directly into surface waters such as rivers or lakes (Figure 6.7).

In developed countries, most municipal wastewater, industrial wastewater and wastewater from hospitals and farms is treated. Currently, most WWTPs are not specifically designed for the removal of micropollutants so most of the tested pharmaceuticals are not completely removed by WWTPs.

According to the ICPDR (International Commission for the Protection of the Danube River) since 2013, about 50% of the total amount of wastewater discharged into the Danube basin undergoes tertiary treatment,

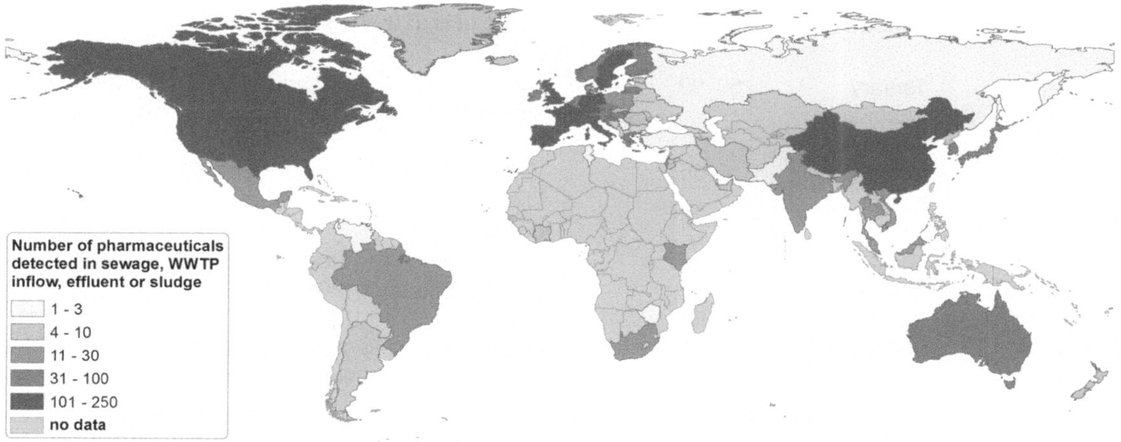


**Figure 6.7** Number of pharmaceuticals detected in sewage, wastewater treatment plants inflow/effluent/sludge by country (Aus der Beek *et al.*, 2016).

about 15% undergoes secondary treatment, and about 1% undergoes primary treatment, while 10% of wastewater is not treated and about 20% of wastewater is not collected or treated. In addition, the countries located in the Danube River Basin upstream of the Republic of Serbia have tertiary treatment (Germany 95%, Austria 100%, Czech Republic 88%, Slovakia 40%, Hungary 63%), while the situation in Croatia and Bosnia and Herzegovina is similar to that in the Republic of Serbia (ICPDR, 2016). Also, about 83 million people live in the Danube River Basin (ICPDR, 2016).

Many pharmaceuticals are persistent and there is a continuous anthropogenic inflow of substances into the environment. In most cases, the monitoring of pharmaceuticals in WWTPs is not carried out in an adequate manner (Bolong *et al.*, 2009), and hence their occurrence and how effectively they are removed in a WWTP cannot be determined precisely. Determining the precision of the removal of pharmaceuticals in WWTPs depends on the composition and quantity of wastewater treated by different technologies in the WWTP, as well as on the analytical method used to analyse the presence of pharmaceuticals in wastewater. Table 6.2 provides a literature review of the efficacy of removing selected pharmaceuticals depending on the degree of treatment in a WWTP.

Trimethoprim is removed relatively well in WWTPs using secondary and tertiary levels of wastewater treatment, while carbamazepine and diclofenac are significantly less removed in WWTPs with secondary levels of wastewater treatment. For the metamizole metabolite 4-AAA, based on the available literature,

**Table 6.2** Effectiveness of the removal of selected pharmaceuticals depending on the degree of treatment in a WWTP (Kovačević, 2017).

| Pharmaceutical | Degree of Treatment | Removal Efficiency (%) | Literature |
|---|---|---|---|
| TMP | Tertiary treatment | 35–99 | Nakada *et al.* (2007), Kima *et al.* (2007), Terzić *et al.* (2008), Miège *et al.* (2009), Radjenovic *et al.* (2009), Tambosi *et al.* (2010), Yang *et al.* (2011), Yang *et al.* (2012), Gracia-Lor *et al.* (2012), Loos *et al.* (2013) |
| | Secondary treatment | 11–70 | Pal *et al.* (2010), Kasprzyk-Hordern *et al.* (2009), Miège *et al.* (2009), Sipma *et al.* (2010), Behera *et al.* (2011) |
| CBZ | Tertiary treatment | 50–99 | Joss *et al.* (2005), Nakada *et al.* (2007), Kima *et al.* (2007), Sui *et al.* (2010), Yang *et al.* (2011), Al-Rifai *et al.* (2011), Yang *et al.* (2012) |
| | Secondary treatment | 0–35 | Vieno *et al.* (2007), Kasprzyk-Hordern *et al.* (2009), Miège *et al.* (2009), Sipma *et al.* (2010), Yu *et al.* (2011) |
| DCF | Tertiary treatment | 55–99 | Clara *et al.* (2005), Joss *et al.* (2005), Suárez *et al.* (2010), Jelić *et al.* (2011) |
| | Secondary treatment | 0–60 | Clara *et al.* (2005), Terzić *et al.* (2008), Radjenovic *et al.* (2009), Kasprzyk-Hordern *et al.* (2009), Behera *et al.* (2011), Gracia-Lor *et al.* (2012), Loos *et al.* (2013) |
| 4-AAA | Secondary treatment | 15–75 | Ternes *et al.* (2001), Zuehlke *et al.* (2007), Feldman *et al.* (2008), Gracia-Lor *et al.* (2012), Gracia-Lor *et al.* (2014) |

analyses showed that 4-AAA was removed relatively well during secondary wastewater treatment in the analysed WWTPs, but there was a high probability of its detection in surface waters due to its incomplete removal in the WWTP.

Based on data on the efficacy of the removal of selected pharmaceuticals, it is evident that the removal of the selected pharmaceuticals after the secondary stage of wastewater treatment is incomplete, while in the tertiary degree of wastewater treatment the complete removal of the selected pharmaceuticals is possible but depends on various factors affecting the operation of WWTP, and hence the selected pharmaceuticals are often not completely removed.

### 6.2.3 Occurrence of selected pharmaceuticals in surface and groundwater

After the incomplete treatment of wastewater or its direct entry into the aquatic environment, pharmaceuticals appear in surface water. The concentrations of pharmaceuticals in surface water are several times smaller than in wastewater due to the effects of dilution, sorption on sediment, the complexation of pharmaceuticals with metals or organic matter, chemical oxidation, photolysis, evaporation and biodegradation. In addition, the accurate quantification of pharmaceuticals, especially in real surface water samples, is a significant analytical challenge, mainly due to the complexity of the pharmaceutical compounds and their occurrence at very low concentrations (ng/l).

The most frequently detected classes of pharmaceuticals in surface waters are: analgesics and anti-inflammatory pharmaceuticals, antibiotics, cardiovascular pharmaceuticals and hormones (Nikolaou *et al.*, 2007). Table 6.3 presents a literature review of the average and maximum concentrations of selected pharmaceuticals, along with the frequency of their occurrence and the number of samples in various regions.

Based on the literature review presented in Table 6.3, it can be concluded that the presence of the pharmaceuticals analysed in more detail in the scope of these studies (TMP, CBZ, DCF and 4-AAA) is evident in surface water. The number of samples is different and varies by several times depending on the conducted research. The detection frequency was relatively high for all the analysed pharmaceuticals so it can be assumed that the inflow of these substances is relatively continuous.

In all the analysed studies the maximum concentration of trimethoprim was the lowest and amounted to 1.4 ng/l in France (Vulliet & Cren-Olivé, 2011), while the highest maximum concentration was observed in the Atibaia River, Brazil, 484 ng/l (Antonio *et al.*, 2011). In the Republic of Serbia, its presence was recorded with a maximum concentration of 8.1 ng/l (Petrović *et al.*, 2014) in the Danube River.

Carbamazepine was detected with a maximum concentration of 2.1 μg/l in Jordan (Zemann *et al.*, 2014), while the lowest maximum concentration was recorded in the Tagus River, Portugal at 17 ng/l (de Jesus Gaffney *et al.*, 2015). A maximum concentration of 35 ng/l of carbamazepine was detected in the Republic of Serbia. (Petrović *et al.*, 2014).

The highest concentration of diclofenac was detected in the Llobregat River, Spain and it equalled 18.74 μg/l (Ginebreda *et al.*, 2010), which is significantly higher than the maximum concentrations detected in other regions. The Llobregat River receives all of the wastewater produced by the city of Barcelona, so there is a realistic explanation for the high concentrations as a large number of residents emit pharmaceuticals into the river with a relatively small flow.

The lowest maximum diclofenac concentration of 11 ng/l was observed in the Tagus River, Portugal (de Jesus Gaffney *et al.*, 2015). In the Republic of Serbia, diclofenac was detected at a maximum concentration of 346 ng/l (Petrović *et al.*, 2014).

**Table 6.3** Review of literature data for the detection of selected pharmaceuticals in surface water (Kovačević, 2017).

| Pharmaceutical | Number of Samples | Detection Frequency (%) | Average Concentration (ng/l) | Maximum Concentration (ng/l) | Region, Sampling Location | Literature |
|---|---|---|---|---|---|---|
| TMP | 6 | 58 | 16 | 36 | Siena River, France | Tamtam *et al.* (2008) |
| | 98 | 64 | 3 | 150 | Brisbane River, Australia | Watkinsona *et al.* (2009) |
| | 10 | 78 | 50 | 484 | Atibaia River, Brazil | Antonio *et al.* (2011) |
| | 70 | 13 | 0.9 | 1.4 | Rhone River, France | Vulliet and Cren-Olivé (2011) |
| | 8 | 100 | 32 | 40 | Elba River, Germany | Wiegel *et al.* (2004) |
| | 10 | 9 | 0.73 | 8.1 | Danube River, Tisa, Begej, Serbia | Petrović *et al.* (2014) |
| | 7 | 100 | 4.8 | 20 | Yangtze River, China | Yan *et al.* (2015) |
| CBZ | 9 | 88 | 230 | 1300 | Tegel Lake, Berlin, Germany | Zuehlke *et al.* (2004) |
| | 26 | 92 | 250 | 1100 | 22 rivers. German | Ternes (1998) |
| | 161 | 57 | 20 | 1238 | Said Ohio, USA | Wu *et al.* (2009) |
| | 14 | 85 | 59 | 121 | Rhine River, The Netherlands | de Jongh *et al.* (2012) |
| | 70 | 61 | 14 | 41.6 | Rhone River, France | Vulliet and Cren-Olivé (2011) |
| | 30 | 100 | 100 | 182 | Fyris River and Lake Mälaren River, Sweden | Daneshvar *et al.* (2010) |
| | 16 | 100 | 65 | 140 | Elba river, Germany | Wiegel *et al.* (2004) |
| | 7 | 100 | 8 | 30 | Yangtze River, China | Yan *et al.* (2015) |
| | 11 | 100 | 5 | 17 | Tagus River, Portugal | de Jesus Gaffney *et al.* (2015) |
| | 12 | 100 | 25 | 35 | Huangpu River, China | Wu *et al.* (2015) |

| | | | | | | |
|---|---|---|---|---|---|---|
| | 95 | 71 | 800 | 2100 | Jordan River, Jordan | Zemann *et al.* (2014) |
| | 10 | 72 | – | 35 | Danube River, Tisa, Begej, Serbia | Petrović *et al.* (2014) |
| DCF | 21 | 100 | 2200 | 18,740 | Llobregat River, Spain | Ginebreda *et al.* (2010) |
| | 7 | 14 | 0.2 | 1.5 | Yangtze River, China | Yan *et al.* (2015) |
| | 11 | 100 | 2 | 11 | Tagus River, Portugal | de Jesus Gaffney *et al.* (2015) |
| | 5 | 60 | 310 | 950 | Hanoi, Vietnam | Tran *et al.* (2014) |
| | 95 | 18 | 78 | 160 | Jordan River, Jordan | Zemann *et al.* (2014) |
| | 10 | 27 | – | 346 | Danube River, Tisa, Begej, Serbia | Petrović *et al.* (2014) |
| 4-AAA | 9 | 77 | 120 | 1000 | Tegel Lake, Berlin, Germany | Zuehlke *et al.* (2004) |
| | 14 | 71 | 76 | 176 | Rhine River, The Netherlands | de Jongh *et al.* (2012) |
| | 16 | 100 | 287 | 939 | Elba River, Germany | Wiegel *et al.* (2004) |

(–) no data available.

The metabolite of metamizole 4-AAA was detected with a maximum concentration of 1,000 ng/l in Lake Tegel, Berlin, Germany (Zuehlke *et al.*, 2004), while the lowest maximum detected concentration of 4a-AAA was recorded in the Rhine, the Netherlands, 176 ng/l (de Jongh *et al.*, 2012). The occurrence of 4-AAA in rivers has not been reported in Serbia so far.

Most review papers dealing with the occurrence of pharmaceuticals in the aquatic environment focus on concentrations in wastewater and surface water where the detected concentrations are generally higher than in groundwater. Since groundwater is an important water resource in many countries and regions, and more than 70% of drinking water is groundwater, it is important to have an insight into the concentrations of pharmaceuticals in groundwater.

Table 6.4 gives an overview of the occurrence of the selected pharmaceuticals (trimethoprim, carbamazepine, diclofenac and 4-AAA) in groundwater in different regions of the world. The table shows the number of samples that were analysed by the study, the frequency of their detection, the average concentration, the maximum concentration of pharmaceuticals, as well as the region and type of aquifer from which the groundwater samples were taken.

Based on the data presented in Table 6.4, it can be concluded that the selected pharmaceuticals in groundwater samples are present in different regions of the world, but the range of average concentrations and frequencies of detection in groundwater is several times smaller than the concentrations and frequencies of detection in surface waters (Table 6.3).

Of the selected pharmaceuticals, trimethoprim appears in the lowest concentration in the range of 0–40 ng/l. The maximum concentration in groundwater has been recorded in China at 40 ng/l (Ma *et al.*, 2015), while the maximum detected concentrations of trimethoprim in other regions are significantly lower and in the range of 0–10 ng/l. The frequency of detection of trimethoprim in groundwater is relatively small 0–20% except in China (80%) where the maximum concentration was registered. In Serbia trimethoprim has not been detected in groundwater to date.

Carbamazepine has been detected at different concentrations ranging between 0–900 ng/l. Based on the literature review, the highest recorded concentration of carbamazepine in groundwater was observed in the USA (420 ng/l), where the highest number of samples $n = 1,231$ was analysed, with a small percentage of positive detection of 1.46% (Fram & Belitz, 2011) and in Germany (900 ng/l) where the detection frequency is slightly higher at 12% in $n = 105$ samples (Sacher *et al.*, 2001).

Based on the literature review, diclofenac was detected in groundwater in the concentration range 0–380 ng/l. The maximum concentration of diclofenac was observed in Spain at 380 ng/l (López-Serna *et al.*, 2013), with a detection rate of 40%. According to the literature data, the other detection frequencies are relatively small 2–6%. In the Republic of Serbia, based on literature data, diclofenac has not been found in groundwater.

Based on available data, a maximum concentration of the metabolite of metamizole 4-AAA was detected in groundwater in Spain (680 ng/l) with a relatively high frequency of detection of 56% (Cabeza *et al.*, 2012). Other literature has reported the presence of 4-AAA in the maximum concentration range of 190–225 ng/l, with a detection frequency of 30–99%.

This points to the metabolite of metamizole 4-AAA as an indicator of the anthropogenic influence in groundwater, but the average concentrations are smaller in groundwater compared to surface water, so it is not completely persistent when filtered into groundwater.

It is important to note that the exact location where the groundwater samples were taken is often not known and neither are the type of aquifer, the hydrogeological conditions or the hydrodynamic conditions of groundwater movement. Therefore, data on the average and maximum concentrations of pharmaceuticals do not represent the actual state in nature, as the infiltration of polluted groundwater can occur, as can the dilution of groundwater from the hinterland.

**Table 6.4** Literature review showing the detection of selected pharmaceuticals in groundwater (Kovačević, 2017c).

| Pharmaceutical | Number of Samples | Detection Frequency (%) | Average Concentration (ng/l) | Maximum Concentration (ng/l) | Region, Aquifer Type | Literature |
|---|---|---|---|---|---|---|
| TMP | 15 | 80 | 9.4 | 40 | China, alluvial type | Ma *et al.* (2015) |
| | 14 | 20 | 0.55 | 5 | Spain, coastal type | López-Serna *et al.* (2013) |
| | 20 | 5 | – | 0.7 | USA, coastal type | Schaider *et al.* (2014) |
| | 26 | 3.8 | – | 10.5 | China, alluvial type | Peng *et al.* (2014) |
| CBZ | 1231 | 1.46 | 40 | 420 | China, alluvial type | Fram and Belitz (2011) |
| | 77 | 24.5 | 70.78 | 118 | Spain, alluvial-coastal type | Teijon *et al.* (2010) |
| | 105 | 12 | – | 900 | Germany, no data | Sacher *et al.* (2001) |
| | 5 | 100 | 27 | 48 | Netherlands, alluvial type | de Jongh *et al.* (2012) |
| | 71 | 70 | 10.4 | 20 | France, no data type | Vulliet and Cren-Olivé (2011) |
| | 14 | 100 | 115 | 136 | Spain, coastal type | López-Serna *et al.* (2013) |
| | 20 | 5 | – | 72 | USA, coastal type | Schaider *et al.* (2014) |
| | 138 | 67 | 0.7 | 9.3 | Singapore, alluvial type | Tran *et al.* (2014) |
| | 121 | 48 | – | 62.4 | Spain, alluvial-coastal type | Cabeza *et al.* (2012) |
| | 51 | 33 | 74 | 500 | Jordan, alluvial type | Wolf *et al.* (2012) |
| | 6 | 69 | – | 4 | Portugal, alluvial type | de Jesus Gaffney *et al.* (2015) |

*(Continued)*

**Table 6.4** Literature review showing the detection of selected pharmaceuticals in groundwater (Kovačević, 2017c) (*Continued*).

| Pharmaceutical | Number of Samples | Detection Frequency (%) | Average Concentration (ng/l) | Maximum Concentration (ng/l) | Region, Aquifer Type | Literature |
|---|---|---|---|---|---|---|
| DCF | 14 | 100 | 225 | 380 | Spain, coastal type | López-Serna *et al.* (2013) |
| | 51 | 2 | 12 | 129 | Germany, alluvial type | Wolf *et al.* (2012) |
| | 138 | 4 | 1.5 | 17 | Singapore, alluvial type | Tran *et al.* (2014) |
| | 16 | 6 | 110 | 480 | Spain, alluvial-coastal type | Cabeza *et al.* (2012) |
| 4-AAA | 99 | 99 | 120 | 190 | Germany, alluvial type | Zuehlke *et al.* (2007) |
| | 77 | 30 | 10 7 | 225 | Spain, alluvial type | Teijon *et al.* (2010) |
| | 16 | 56 | 120 | 680 | Spain, alluvial-coastal type | Cabeza *et al.* (2012) |

(–) no data available.

### 6.2.4 Removal of pharmaceuticals from drinking water treatment plants

The treatment of water intended for drinking purposes involves numerous challenges in order to optimize the technology for the treatment of such water and thus to protect human health. How the treatment of water intended for drinking will be achieved is conditioned by population growth, climate change, the decreased availability of sources of drinking water, by hydrological conditions, and by changes in the quality of raw water (Teodosiu *et al.*, 2018). The demand for drinking water is increasing with population growth but the increase in population and the urbanization of society has resulted in the emergence of pollutants in surface waters, which has led to an increasing risk of using different sources of water for water supply.

Also, with the development of new analytical methods it has become possible to detect new or existing substances that it had not been possible to detect previously. Pharmaceuticals belonging to the group of emergent substances are present in aquatic environments at low concentrations, and their impact on wildlife has not been fully investigated.

Fresh water is the main source of raw water for human consumption, industry, agriculture and energy production. Seawater is used in areas with high water shortages and high economic power due to the high energy and chemical requirements needed for the functioning of such drinking water treatment plants. Pharmaceuticals and numerous emergent substances are constantly being detected in the aquatic environment, and this requires the use of advanced drinking water treatment technologies to achieve water quality suitable for human consumption (Ternes *et al.*, 2015).

Groundwater represents the largest reservoir of the freshwater resource and accounts for more than 97% of freshwater on planet Earth, not counting the glaciers and permanently frozen areas (Gleick, 1996; United Nations Educational, Scientific and Cultural Organization [UNESCO], 2003). Globally, more than 50% of the water used for water supply comes from groundwater resources (UNESCO, 2003). In Europe, the percentage of groundwater use for water supply is slightly higher than 75% (European Commission [EC], 2016). The quality and capacity of groundwater resources are more reliable in comparison to those of surface water and represent a priority in terms of water use, and therefore affect the overall percentage of groundwater used for water supply purposes.

Inadequate wastewater treatment, and the excessive discharge of wastewater from medical facilities are the main sources of pharmaceuticals found in surface water. Surface water is most exposed to effects from industry, agriculture and livestock.

Pharmaceuticals are transported and transformed in an aquatic environment and they are exposed to natural self-purification processes that lead to a decrease from the initial concentration of the source of where pharmaceuticals appear to the raw water that is further treated prior to human use.

In drinking water treatment plants, the transformation of pharmaceuticals can occur, and in some cases, this transformation may result in compounds that may be more toxic, more persistent and less biodegradable than the basic compounds (Farré *et al.*, 2008).

The most significant adverse effects of pharmaceuticals on the environment include bioaccumulation, biomagnification, persistence in the aquatic environment, and ecotoxicity, as well as potential carcinogens, mutagens and teratogenic effects, toxicity and potential endocrine disruption (Guillén *et al.*, 2012).

There are various interactions that occur during the treatment of drinking water, such as: interactions between organic substances themselves (e.g. complexation), between organic substances and other constituents (e.g. complexing with $Ca^{2+}$), and between organic substances and chemicals used during treatment (e.g. disinfectants), as well as interactions between organic substances and other media used for the treatment of drinking water (e.g. sorbents and membranes). It is essential to know the interactions that occur between different substances and how they affect the treatment of drinking water in order to define the best possible way of treating drinking water (Ivancev-Tumbas, 2014).

Conventional drinking water treatment processes are mainly designed and aimed at removing solids of different sizes, and organic matter, as well as microorganisms and pathogens. Conventional processes include grids and sieves, sand removal, pre-oxidation, coagulation, flocculation, settling, filtration on sand filters (fast/slow) or on activated carbon, and disinfection.

The widespread use of emergent substances (ES) has necessitated the use of advanced drinking water treatment techniques, since conventional plants for the treatment of drinking water are not designed to remove emergent substances and therefore pharmaceuticals. The removal of pharmaceuticals and other ES that are detected at very low concentrations in raw water requires a greater use of energy and chemicals during the advanced treatment of drinking water, leading to higher treatment costs. There are relatively few studies addressing the efficacy of removing emergent substances using advanced drinking water treatment, especially considering the number of publications dealing with removal of emerging substances during wastewater treatment (Teodosiu *et al.*, 2018).

Table 6.5 shows a summary of literature pertaining to the removal of pharmaceuticals using the advanced treatment of drinking water in pilot plants and operating facilities.

Based on the literature review, it can be concluded that different purification processes can be used individually or in combination with other processes to achieve the effective removal of pharmaceuticals depending on the matrix of water (structure of substances and properties of substances), as well as depending on the operating conditions of the process (Bui *et al.*, 2016).

**Table 6.5** Effectiveness of the removal of pharmaceuticals using advanced drinking water treatment (Teodosiu *et al.*, 2018).

| Treatment Process | Removal Efficiencies (%) | References |
|---|---|---|
| Ozonization | >98 | Talib and Randhir (2016) |
| UV-photolysis | 30–70 | Pal *et al.* (2010) |
| Nanofiltration | 15–100 | García-Vaquero *et al.* (2014) |
| GAC | 20–50 | Kennedy *et al.* (2015) |
| *Combined processes* | | |
| Deoxychlorination, coagulation/flocculation, sand filtration, ozonation, carbon filtration and final disinfection with chlorine | >98 | Boleda *et al.* (2011) |
| UV pre-disinfection, filtration, nanofiltration/reverse osmosis, remineralization and chlorine disinfection | >85 | Radjenovic *et al.* (2009) |
| Coagulation, ultrafiltration, and GAC filtrate on | 99 | Kim *et al.* (2007) |
| Coagulation, flocculation, sedimentation, rapid sand filtration, ozonation, two-step GAC filtration and UV disinfection | 51–95 | Vieno *et al.* (2007) |
| Advanced oxidation | >99 | Klavarioti *et al.* (2009) |
| Photocatalysis and solar photolysis | 66–82 | Kanakaraju *et al.* (2014) |
| UV/$H_2O_2$integrated into an existing full-scale plant | 67–98 | Kruithof *et al.* (2007) |
| UV-photolysis | 80 | Lekkerkerker-Teunissen *et al.* (2013) |

GAC – granulated active carbon adsorption.

**Table 6.6** Pilot-scale percentage removals of selected pharmaceuticals (Sudhakaran *et al.*, 2013; Wert *et al.*, 2008).

| Treatment | RBF (%) | GAC (%) | NF (%) | RO (%) | OZONE (%) | AOP (%) | UVO (%) |
|---|---|---|---|---|---|---|---|
| Carbamazepine | 13 | 80 | 80 | 99 | 99 | 99 | 88 |
| Diclofenac | 99 | 45 | 80 | 99 | 99 | 99 | 98 |
| Trimethoprim | 99 | 80 | 80 | 99 | 99 | 99 | 94 |

NF – nano-filtration; RO – reverse osmosis; UVO – ultraviolet oxidation.

Advanced drinking water treatment processes, such as AOPs (advanced oxidation processes), adsorption on activated carbon and on other active substances, nanofiltration and reverse osmosis, effectively remove pharmaceuticals from raw water during the treatment of drinking water. Ozonization and reverse osmosis, as a separate process, represent the most efficient technological processes for the removal of pharmaceuticals during the treatment of drinking water, while other technologies, especially in combination, also quite efficiently remove pharmaceuticals during the treatment of drinking water. Newly designed nanomaterials can be used to efficiently remove pharmaceuticals and certain emergent substances as a separate process (sorption), or they can incorporate conventional membrane processes, chemical degradation, and photodegradation processes (Simate *et al.*, 2012), and represent a cheaper technology than reverse osmosis or ozonization (Adeleye *et al.*, 2016).

Based on the available literature, of the analysed pharmaceuticals, trimethoprim, diclofenac and carbamazepine were selected as there is evidence of the degree of their removal during the advanced treatment of drinking water at pilot plants (Table 6.6).

Finally, it can be concluded that the use of a combination of treatment processes that are efficient and cost-effective is the best option given the establishment of effective water quality control and the optimization of operational conditions. It is very important to protect the sources of drinking water in order to reduce the potential emission of pollutants and the threat to human health.

In cases when alluvial sources of groundwater are used, coastal filtration is a very efficient and important form of treatment which can be used for the removal of micropollutants using the self-purification potential of aquifers. Coastal filtration is the first barrier that prevents the inflow of micropollutants and of other polluting substances, but it is necessary to define and adequately set the line of filtration through an alluvial aquifer so that the greatest efficiency of natural of attenuation is achieved and so that it is possible to meet the needs for the appropriate quality and quantity of water at the source. The percentage of elimination of diclofenac and trimethoprim is very high during bank filtration (99%), while it is relatively low (13%) for carbamazepine, and hence despite the efficiency of coastal filtration for the removal of pharmaceuticals, it is still necessary to use advanced water treatment for drinking water to completely eliminate all the micropollutants.

For this reason, it is important to understand the self-purification processes that take place during coastal filtration, the most important of which are sorption, biodegradation and degradation, which will be discussed in the following subsections.

## 6.3 SORPTION OF SELECTED PHARMACEUTICALS

### 6.3.1 Sorption isotherms

The equilibrium distribution of a compound between the solids and the solution depends on the concentration of compounds in the solution. The relationship between the compound concentration in the solution and the total sorbate concentration is referred to as the sorption isotherm. A common approach for fitting experimentally determined sorption is the empirical relationship known as the isotherm.

The most well-known isotherms that are very commonly used in adsorption-desorption analyses are the Linear, Langmuir, Freundlich, Temkin, Brunauer-Emmett-Teller (BET), and Dubinin isotherms.

The first three are widely represented in the interpretation of physical and chemical adsorption, while the BET and Dubinin isotherms are mainly used to describe the adsorption of gases and vapours on porous carbon materials.

The shape of the isotherm depends on the nature of the adsorbent, adsorbate, and solution. Typically, the amount of adsorbed substance per unit mass of adsorbent increases with the concentration in the liquid phase. The increase in the amount of adsorbed substance need not be linear. The simplest mathematical model for describing sorption is a linear model of equilibrium distribution:

$$S = K \cdot C \tag{6.1}$$

wherein the sediment/water distribution coefficient, $K$ ($m^3/kg$), represents the total measure of the substance binding to the sediment and the ratio between the concentration of the substance bound to the solid phase of the sediment, $S$ (kg/kg) and the concentration of the substance in water at an equilibrium, $C$ ($kg/m^3$).

The Langmuir model is based on the assumption that the maximum adsorption corresponds to the saturated layer of solute molecules on the surface of the adsorbent, the adsorption energy is constant and the increase in the amount of adsorbed substance does not have to be linear. Langmuir derived a function that describes the dependence on the degree of surface coverage by molecules of a substance adsorbed from a liquid phase and substance dissolved in liquid phase at a constant temperature.

The Langmuir adsorption model is represented by the equation (Weber, 1972):

$$S = \frac{S_0 \cdot K_L \cdot C}{1 + K_L \cdot C} \tag{6.2}$$

where $C$ ($kg/m^3$) is the equilibrium concentration of adsorbate in solution, $S$ (kg/kg) is the equilibrium adsorption capacity for a given adsorbate, $K_L$ ($m^3/kg$) is Langmuir's equilibrium constant, and $S_0$ (kg/kg) is the theoretical monolayer saturation capacity.

By linearizing Equation (6.2), a linear form is obtained:

$$\frac{C}{S} = \frac{1}{S_0 K_L} + \frac{C}{S_0} \tag{6.3}$$

The graphic representation of $C/S$ in function gives a straight linear dependence with a slope and cross-section of the ordinate. One of the most significant features of the Langmuir adsorption isotherm is the dimensionless constant called the separation factor $R_L$, which is also referred to as the equilibrium parameter in literature and is defined by the relation (Bulut *et al.*, 2008):

$$R_L = \frac{1}{(1 + K_L C_0)} \tag{6.4}$$

The value of the constant $R_L$ determines the shape of the isotherm. Based on its value, the isotherm can be unfavoured ($R_L > 1$), linear ($R_L = 1$), favoured ($0 < R_L < 1$), or irreversible ($R_L = 0$) (Bulut *et al.*, 2008).

The concept of reversible adsorption is, according to Langmuir, mathematically expressed in the form of an equation that is based on the following assumptions:

- There is an exact defined number of localized adsorption centres on the surface of adsorbents;
- Each adsorption centre can bind only one particle;
- The energy state of the adsorbed particles is identical for all the adsorbed particles regardless of their position in space or the state of the adjacent adsorption centre.

The above approach implies that the surface of the adsorbent is flat and homogeneous and that the interactions between the adsorbate particles are negligible. In the experimental study of adsorption, the Langmuir isotherm shows three characteristic forms, namely, the linear part of the dependence of the degree of coverage as a function of the substance pressure in the liquid phase for low pressure conditions, the curvilinear part for moderate pressure conditions, and the asymptotic tendency towards a certain maximum degree of coverage for high pressure conditions. The maximum value corresponds to the situation when the surface is completely covered, so that the degree of coverage is equal to one (Foo & Hameed, 2010).

For the Langmuir sorption model, for small concentrations in solution, where $K_L C \ll 1$, it follows (Dimkic, 2007):

$$S = S_0 \cdot K_L \cdot C \tag{6.5}$$

This fact is very important because in the aquatic environment, especially in groundwater, the studied pharmaceuticals occur in very small concentrations, so the linear model of sorption almost always applies for the sorption of pharmaceuticals in groundwater.

For significantly high concentrations (practically supplemented with the entire layer of adsorbate), where $K_L C \gg 1$, the Langmuir equation approaches the limiting value when:

$$S \rightarrow S_0 \tag{6.6}$$

The Langmuir equation is used for sorption on granular activated carbon. However, it can also be a good basis for a model of sorption in natural systems and even in an aquifer when applying bank filtration and artificial infiltration methods.

The empirical model used to describe the sorption of hydrophobic organic compounds on sediment is the nonlinear Freundlich isotherm (Delle Sita, 2000; Foo & Hameed, 2010):

$$S = K_f \cdot C^{1/n} \tag{6.7}$$

or the logarithmic form:

$$\log S = \log K_f + \frac{1}{n} \log C \tag{6.8}$$

where:

$S$ is the amount of sorbed substance (kg/m$^3$) calculated from the difference in the initial concentration, $C_0$ and equilibrium concentration $C$,
$A = (C_0 - C)$,
$C_0$ is the initial concentration of the substance in the solution (kg/m$^3$),
$C$ is the equilibrium concentration of the substance in the solution (kg/m$^3$),
$K_f$ is Freundlich's constant (units of $K_f$ depend on the units in which $S$ and $C$ are expressed, e.g. kg$^{1-1/n}$ m$^{3\ (1/\mathrm{kg})-1}$),
$1/n$ is Freundlich's exponent.

The Freundlich isotherm is used at low concentrations of aqueous solutions and in situations where the surface of the solution is not energetically homogeneous (Delle Sita, 2000).

The Freundlich exponent, $1/n$, indicates the energy heterogeneity of the surface, that is, the diversity of free energies for the sorption of an organic compound on different components of a heterogeneous system. There are three values of exponent $n$:

- When $n = 1$ ($1/n = 1$), the isotherm is linear and the free energy of sorption is the same at all sorbate concentrations. In cases where $n = 1$, $K_f$ equals $K$, so Equation (6.7) becomes linear. The mentioned hypothesis makes it possible to calculate the sorption coefficient, $K$, which defines the sorption capacity ($C = 1$ ng/ml) (Delle Sita, 2000; Sukul *et al.*, 2008);
- When $n < 1$ ($1/n > 1$), the isotherm is concave and the free sorption energy decreases with increasing sorbate concentration;
- When $n > 1$ ($1/n < 1$), the isotherm is convex, and with an increasing sorbate concentration the free energy for further sorption increases.

The constants $K_f$ and $1/n$ are determined from the linear form of Freundlich's equation, more precisely from the dependence graphs shown by the given equation, where $K_f$ is calculated from the section and the slope represents the ratio $1/n$.

## 6.3.2 Experimental sorption coefficients

In a study with 13 different soil types (Kodešová *et al.*, 2015), which was conducted as a batch experiment, the results showed a *Kd* value of 3.5 ml/g for sediment with a carbon normalized sorption coefficient ($K_{OC} = 9.72$ ml/g). In a batch experiment study (Scheytt *et al.*, 2005), the carbon normalized sorption coefficient for carbamazepine was significantly higher ($K_{OC} = 100.76$ ml/g) with a *Kd* value of 0.13 ml/g. In another batch experiment study (Yu *et al.*, 2013), the carbon normalized sorption coefficient for carbamazepine was $K_{OC} = 3.62$ ml/g, with a *Kd* value of 1.45 ml/g. In a column experiment study (Scheytt *et al.*, 2006), with a high carbon normalized sorption coefficient ($K_{OC} = 715$ ml/g), the obtained *Kd* was 1.43 ml/g.

In a batch experiment study (Scheytt *et al.*, 2005), with a relatively high carbon normalized sorption coefficient ($K_{OC} = 438$ ml/g), the *Kd* value for diclofenac was relatively high (1.87 ml/g). In another batch experiment study (Xu *et al.* 2009), with a lower value of carbon normalized sorption coefficient for diclofenac ($K_{OC} = 1.1$ ml/g), the *Kd* value was 0.55 ml/g. In a column experiment study (Scheytt *et al.*, 2006), with a carbon normalized sorption coefficient of $K_{OC} = 935$ ml/g, the *Kd* value for diclofenac was 0.57 ml/g.

In a study with 14 different sediment types (Kodešová *et al.*, 2015) trimethoprim had a *Kd* value of 3.5 ml/g and $K_{OC} = 110$ ml/g. Also, in a study with activated sludge with a very high organic content, the *Kd* value for trimethoprim was between 3.9–4.2 ml/g.

So far, little information on *Kd* values of 4-AAA can be found in literature sources. A study by Burke *et al.* (2013) reported that 4-AAA was not sorbed on the sediment, $Kd = 0$ ml/g with $f_{OC} = 0.20$. Linear sorption coefficients, carbon normalized sorption coefficients and organic carbon fractions for selected pharmaceuticals are presented in Table 6.7.

Most of the available literature refers to laboratory and column experiments for sediment samples of varying composition, so in practice these data need to be carefully considered and cannot be equally compared.

In the case of sorption tests on aquifer sediments, the conditions in the field differ considerably from those in the laboratory, with the effect of scale significantly affecting the final result. In addition to the effects of scale, the preferred groundwater flows affect the speed of groundwater flow under natural conditions. Furthermore, the diameter of grains of the sediment analysed in the laboratory is much smaller than the

**Table 6.7** Linear sorption coefficients *Kd* (ml/g), carbon normalized sorption coefficients $K_{OC}$ (ml/g) and organic carbon fractions $f_{OC}$ (%) for selected pharmaceuticals.

| Pharmaceutical | *Kd* (ml/g) | $K_{OC}$ (ml/g) | $f_{OC}$ (%) | Literature |
|---|---|---|---|---|
| Trimethoprim | [1]3.5 | [1]9.72 | 0.36 | [1]Kodešová *et al.* (2015) |
| Carbamazepine | [2]0.13 | [2]100.76 | [2]0.002 | [2]Scheytt *et al.* (2005) |
| | [3]1.43 | [3]715 | [3]0.0013 | [3]Scheytt *et al.* (2006) |
| | [4]1.45 | [4]3.62 | [4]0.41 | [4]Yu *et al.* (2013) |
| Diclofenac | [5]0.55 | [5]1.1 | [5]0.5 | [5]Xu *et al.* (2009) |
| | [3]0.57 | [3]438 | [2]0.002 | [6]Burke *et al.* (2013) |
| | [2]1.87 | [2]935 | [3]0.0013 | |
| 4 – AAA | [6]0 | | [6]0.2 | |

average grain diameter of the material in the field, and this significantly affects the sorption effect under laboratory conditions and is not representative of real field conditions.

Fine-grained aquifer sediment samples are dominated by different sorption mechanisms relative to the larger-grained aquifer sediment in nature (groundwater flows through the largest aquifer material), while sorption is much more pronounced on finer aquifer material. The percentage of organic matter ($f_{OC}$) for sediment samples is highly important as the higher the percentage of organic matter, the more pronounced the sorption on organic matter.

Based on a field study (Henzler *et al.*, 2014) and by using a hydrodynamic and transport model, carbamazepine was found to have $Kd = 0.081$ ml/g, which is less than in studies conducted under laboratory experimental conditions (Table 6.7). The same study did not accurately determine the sorption coefficient for diclofenac, but it was found that there was a high tendency for sorption under given conditions.

Based on a literature review of selected pharmaceuticals, the most rapidly transported pharmaceuticals in groundwater are 4-AAA, diclofenac and carbamazepine, while trimethoprim has the highest sorption potential and is slowest in terms of transport in groundwater.

## 6.4 DEGRADATION OF PHARMACEUTICALS IN GROUNDWATER

Intergranular aquifers and filtration through the subsoil can be compared with an efficient and extensive physicochemical and biochemical reactor that contributes to the safety of water supply and the general improvement of groundwater quality. In this case, aquifers are very heterogeneous, and the processes of groundwater filtration, as well as their effects, are numerous (Dimkic & Keckarević, 1990).

The behaviour of organic substances in groundwater depends on the microflora that is present, which later changes, as well as on the properties of the sediment matrix. The sediment matrix itself can modify the structure of the organic compound, and the level of organic and inorganic nutrients for the microflora also changes. The pH value is modified, the microorganisms are retained and the activity of extracellular enzymes is affected.

The biochemical degradation of the sorbed substance does not have to take place by the same mechanisms and dynamics as in solution. Sometimes degradation (on the sorbent) is inhibited. Small molecule degradation is an intracellular enzymatic process which involves the transport of molecules through the cell wall. The activity of enzymes, which changes in the presence of clay materials, is especially important for the extracellular mechanisms of enzymatic degradation.

The change in enzyme activity is conditioned by the presence of inorganic components (metal ions), which may be bound to the substance material by some of the above mechanisms, or by the microenvironment of the enzyme (pH of the medium). The sorbed molecules are used by different microorganisms as a source of carbon, nitrogen, other nutrients or energy, and hence the compounds are transformed by certain kinetics but at a rapid rate of destruction.

A range of protein substances adsorbed on sediments are subject to hydrolytic reactions by microorganisms. In order for the biodegradation of a given compound to occur in nature, certain conditions must be fulfilled (Alexander, 1994):

- In nature, there must be a microbial species that possesses the enzymes necessary for its degradation;
- Microorganisms must be present in the same environment as the given compound;
- The compound (substrate) must be accessible to the microorganism;
- If the decomposition begins under the influence of an extracellular enzyme, the connections that need to be broken must be sufficiently exposed in order for the enzyme to be able to function;
- If the enzyme that catalyses the onset of degradation is intracellular, the substrate molecules must first be transferred to the interior of the cell, to the location at which the enzyme can act;
- As the population of potentially active microorganisms in relation to a given substrate is, as a rule, very small, the environmental conditions must be such that they stimulate the development of the microorganisms.

Two basic approaches are generally used to define the rate of removal of substrates (biodegradable substances). Both are based on relationships that have already been applied in enzymatic fermentation industrial processes. The first approach uses Monod's expression for microbial growth kinetics, while the second approach uses modified chemical kinetics (Eckenfelder) and analytical expressions derived from Michaelis-Menten's empirical relation for the rate of substrate consumption by microorganisms (Appelo & Postma, 2005).

## 6.4.1 Biochemical kinetics and microbial growth kinetics

The Monod equation relates the growth of microorganisms to the concentration of the substrate, i.e. the speed of biomass production:

$$\frac{dS}{dt} = R_{\text{MONOD}} = -k_{\max}\frac{S}{k_{1/2} + S} \tag{6.9}$$

where $S$ is the concentration of the degradable substance (kg/m$^3$); $t$ is time (s); $k_{\max}$ is the maximum degradation level (kg/m$^3$/s); and $k_{1/2}$ is the half-saturation constant (kg/m$^3$).

The degradation level represents half of the maximum degradation level when the concentration of biodegradable substance ($S$) is equal to constant $k_{1/2}$. The Monod equation depends on the concentration of $S$ and on the value of $k_{1/2}$ as shown in Figure 6.8.

When the concentration of $S$ is small, $S < k_{1/2}$, the Monod ratio increases. When the concentrations are higher, $S > k_{1/2}$, the Monod ratio increases less and becomes less dependent on the $S$ concentration. The Monod equation refers to the model of the bacterial metabolism, while $S$ refers to the substrate concentration used by the bacteria for their metabolism.

At low substrate concentrations, bacteria use the substrate in proportion to the concentration, whereas at high substrate concentrations, substrate consumption is limited to the amount that bacteria can consume, since enzymes and other nutrients represent a limiting factor.

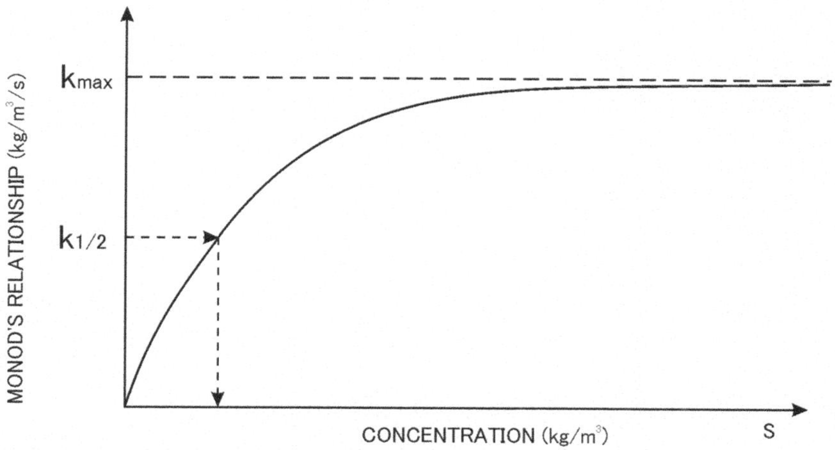


**Figure 6.8** Dependence of the specific biomass growth rate on the concentration of the substrate (Appelo & Postma, 2005).

So when $S$ is negligible with respect to $k_{1/2}$, Equation (6.9) changes to:

$$\frac{dS}{dt} = -\frac{k_{\max}}{k_{1/2}}S \tag{6.10}$$

Monod's relation becomes a relation of the first-order in relation to $S$.

When $k_{1/2}$ is small relative to $S$, Equation (6.9) changes to:

$$\frac{dS}{dt} = -k_{\max} \tag{6.11}$$

Then the Monod equation becomes a zero order equation with respect to $S$, and the concentration of $S$ decreases linearly with time.

The Monod equation represents the influence of enzymes in the cell of microorganisms that bind to the substance and initiate the reaction. Taking enzyme $E$ which binds to $S$, and thus forms complex $ES$, assuming that the reaction depends entirely on the concentration $ES$, we get Equation (6.12):

$$\frac{dS}{dt} = -kES \tag{6.12}$$

If $ES$ is expressed as the total concentration of the enzyme and of the substrate concentration $S$:

$$E + S \ll ES; K_{ES} = \frac{[ES]}{[E][S]} \tag{6.13}$$

While the total enzyme concentration is $E_t$:

$$E_t = \frac{ES}{K_{ES} \times S} + ES = ES\left(\frac{1}{K_{ES} \times S} + 1\right) \tag{6.14}$$

Michaelis-Menten's equation:

$$ES = \frac{E_t \times K_{ES} \times S}{1 + K_{ES} \times S} = \frac{E_t \times S}{(K_{ES})^{-1} + S} \tag{6.15}$$

By combining Equations (6.9) and (6.14), we get:

$$\frac{dS}{dt} = \frac{-k \times E_t \times S}{(K_{ES})^{-1} + S} \tag{6.16}$$

When comparing Equation (6.9) with Equation (6.15) it can be seen that the half-saturation constant $K_{1/2}$, is equal to the inverse association constant $K_{ES}$ in Equation (6.15) (Appelo & Postma, 2005).

### 6.4.2 Half-life periods, $t_{1/2}$, of selected pharmaceuticals

The half-life periods, $t_{1/2}$, obtained from field studies are not sufficiently listed in literature (Henzler *et al.*, 2014). If there are data on periods of half-life, they mainly refer to data obtained in laboratory conditions which are very important, but are often not representative of nature, bearing in mind the complexity and the hydrogeological conditions that exist in natural systems of coastal filtration.

Due to the heterogeneity of bank filtration systems as well as of other systems such as artificial infiltration or infiltration into deep aquifer layers, a deep understanding and an analysis of the processes that occur during the filtration and transport of organic micropollutants in groundwater are necessary. This requires the development of detailed numerical models of the groundwater flow, as well as a multi-component reactive transport model.

The kinetics of the first order degradation gives a rather simplified description of the process of biodegradation (ignoring the effects of physical and chemical conditions and of their changes in time and space), but it is very useful and provides a first estimation of the behaviour and fate of emerging substances, especially of pharmaceuticals in the field.

Based on a review of the available literature, data pertaining to the biodegradation of carbamazepine and diclofenac in field conditions were analysed, whereas only laboratory experimental data were available regarding the degradation kinetics for trimethoprim and 4-AAA.

In a study conducted at the site of Lake Tegel, near Berlin, where the concentration of carbamazepine was monitored during lake bank filtration (Henzler *et al.*, 2014), the half-life, $t_{1/2}$, of carbamazepine was found to be about 66 days. The conditions during the three-year testing of the carbamazepine content in Lake Tegel and in corresponding wells were generally oxic, except in the furthest well (approximately 177 metres) where anoxic conditions were recorded. In another field study that analysed bank filtration on the Rhine River under oxic conditions (Achten *et al.*, 2002), the half-life, $t_{1/2}$, for carbamazepine was found to be about 30 days. In a third study, which was conducted in oxic conditions during artificial infiltration (Stuyfzand *et al.*, 2007) the half-life, $t_{1/2}$, of carbamazepine was found to be about 35 days.

The half-life of diclofenac during coastal filtration in oxic conditions was approximately 36 days (Henzler *et al.*, 2014), whereas in previously conducted studies (Brauch *et al.*, 2000) in oxic conditions the half-life was about 12 days.

For trimethoprim, only the available half-lives based on laboratory experiments were analysed. A half-life, $t_{1/2}$, of 1 hour was obtained based on experiments that simulated bank filtration in a column under oxic conditions (Bertelkamp *et al.*, 2014). The laboratory experiments analysed in a review paper (Greskowiak *et al.*, 2017), mention half-lives equalling from 12 hours (Burke *et al.*, 2016) up to 53 days (Burke *et al.*, 2014a).

For the metamizole 4-AAA metabolite, there is also only literature data obtained from laboratory experiments. Half-life periods, $t_{1/2}$, vary from 1.6 hours (Burke *et al.*, 2014b) to 12 days (Wiese *et al.*, 2011).

The half-lives obtained by applying the QSAR (quantitative structure-activity relationship) model do not correlate and have no connection with data obtained from laboratory and field trials (Greskowiak *et al.*, 2017). It is also not possible to establish any correlation with other data obtained using the QSAR model even if the data are only isolated in relation to temperature change and redox conditions.

The results obtained from laboratory experiments must be considered with some reserve, due to the scale of the experiment, how representative a sample is (particularly granulometric composition of the sediment sample), the preferential groundwater flows, the nutrient and organic matter content and the laboratory conditions, as these can lead to large deviations in the obtained values for the half-life, and thus errors during the calculation of future models of substance transport in groundwater. Certainly, the experimentally obtained data for the half-life have value in that they give a qualitative assessment of the sorption potential of the tested pharmaceuticals.

In addition, even the results obtained from field experimental studies vary considerably. It is therefore necessary, during research, to determine the factors that affect the degradation of the selected pharmaceuticals, among which the temperature, the redox conditions in the aquifer and the chemical and physical heterogeneity of aquifers stand out.

Based on the data on the half-lives for carbamazepine and diclofenac obtained by doing field experiments, it can be concluded that the data are different between field and lab experiments, and mainly depend on:

- Water transport times from the river to the tested wells;
- Distances of wells from surface water;
- Hydrogeological conditions in the aquifer;
- Redox conditions in groundwater;
- Seasonal changes in the temperature of water;
- The mineralogical composition of aquifers;
- The content of organic matter and nutrients in an aquifer.

It is unlikely that models for estimating degradation constants for variable environmental conditions can be developed based on the results that have been analysed. This fact is very important for the creation of future models which serve to evaluate the possible contamination of groundwater sources for drinking water. It is therefore recommended, when designing and developing models of the transport of micropollutants, to use the most favourable available literature data to protect groundwater sources, namely, to in this case, use literature data for maximum half-life periods. This does not guarantee that the results will be completely accurate, but the likelihood of a favourable prognosis is much higher.

## 6.5 EXAMPLES OF RESEARCH CONDUCTED IN THE REPUBLIC OF SERBIA BETWEEN 2009 AND 2015

### 6.5.1 The treatment of municipal wastewater in the Republic of Serbia

In developed countries municipal wastewater, industrial wastewater and wastewater from hospitals and farms are often treated separately because of the different composition and quantity of the wastewater, as it is a more efficient solution for controlling the source of pollutants and for monitoring the quality and quantity of wastewater more accurately (Terzić *et al.*, 2008). In developing countries and those in transition such as the Republic of Serbia, municipal wastewater is most often treated together with wastewater from industry, hospitals and farms (Kaštelan-Macan *et al.*, 2007).

In the Republic of Serbia, the vast majority of cities, industrial systems, hospitals and farms discharge untreated water directly into natural or artificial recipients, leading to the continued emergence of pharmaceuticals in wastewater, and surface and groundwater. This has been noted in several works that have looked at the emergence of pharmaceuticals in the Republic of Serbia over shorter periods of time (Grujić *et al.*, 2009; Kovačević *et al.*, 2013d, 2014; Petrović *et al.*, 2014; Terzić *et al.*, 2008). Furthermore, most wastewater in the Republic of Serbia is discharged into the environment without prior treatment. In the Republic of Serbia, only 10% of the total wastewater is treated, primary treatment is carried out in 16% of the plants, secondary treatment in 71% of plants, and tertiary treatment in 13% of the plants.

In the past few decades Serbia has built just over 50 plants for wastewater treatment in urban settlements with more than 2,000 inhabitants. Of the constructed facilities, 32 are in operation, few of which operate according to the design criteria, while the efficiency of the other facilities is far below the designed efficiency (Water Management Strategy for the Republic of Serbia, 2016). In relation to the total population of the Republic of Serbia, connection to public sewerage systems amounts to about 61% (Figure 6.9).

In the Republic of Serbia, only a few short-term studies have been conducted to monitor the occurrence of pharmaceuticals in municipal and industrial wastewater, for municipal and industrial wastewater in Belgrade (Terzić *et al.*, 2008), as well as for the industrial wastewater in Obrenovac and the municipal wastewater of the city of Novi Sad (Petrović *et al.*, 2014). A large number of tested pharmaceuticals were detected in the municipal and industrial wastewater from the city of Belgrade, while no pharmaceuticals were detected in the industrial wastewater on the territory of the city of Obrenovac, but most of the tested pharmaceuticals were detected in the municipal wastewater from this city. One of the reasons for the emergence of pharmaceuticals in the industrial wastewater of the city of Belgrade is the operation of the factory for the production of pharmaceutical products, 'Galenika'.

In developed countries, most municipal wastewater, industrial wastewater and wastewater from hospitals and farms is treated. Currently, most WWTPs are not specifically designed to remove micropollutants, so most of the tested pharmaceuticals are not removed completely in the WWTPs.

According to data from ICPDR since 2013, about 50% of the total amount of wastewater discharged within the Danube River Basin is treated by tertiary treatment, about 15% undergoes secondary treatment, about 1% undergoes primary treatment, 10% of the waste water is not treated, while about 20% of the wastewater is not collected or treated. In addition, countries located in the Danube basin upstream of the Republic of Serbia mainly have tertiary treatment (Germany 95%, Austria 100%, Czech

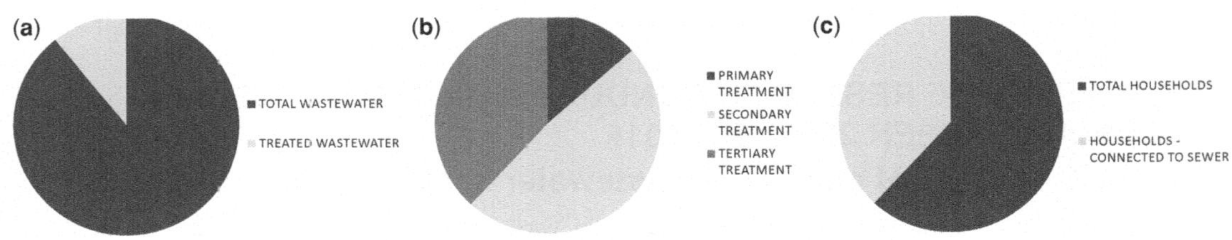


**Figure 6.9** Graphical presentation of wastewater treatment and sewerage in the Republic of Serbia: (a) Ratio of the total amount of wastewater and the amount of treated wastewater; (b) Degree of wastewater treatment; c) The ratio of the number of households connected to the sewerage system and the total number of households (Statistical Office of Republic of Serbia, 2019; Water Management Strategy for the Republic of Serbia, 2016).

Republic 88%, Slovakia 40%, Hungary 63%), while the situation in Croatia and Bosnia and Herzegovina is similar to that in Serbia (ICPDR, 2016). Also, about 83 million people live in the Danube River Basin, more than 42 million of whom are upstream of the Republic of Serbia (ICPDR, 2016).

The JDS III (Joint Danube Survey III) conducted in 2013 looked into the occurrence of carbamazepine within the Danube River Basin and its tributaries (Figure 6.10). Carbamazepine, as a pharmaceutical that persists for a long time in the aquatic environment, can serve as a marker for the possible anthropogenic influence within the Danube River Basin.

Based on research results (ICPDR, 2013) it can be seen that the concentration of carbamazepine in the Danube in the Republic of Serbia (from 845.5 km to 1430.1 km) does not deviate from the concentration of carbamazepine in the Danube basin upstream and downstream of the Republic of Serbia. The mean concentration of carbamazepine in the Danube River and its tributaries is about 30 ng/l.

Within the JDS III program (ICPDR, 2013), a total of 71 surface water samples were analysed. The detection frequency of carbamazepine was 100%, while the registered concentration was in the range of 3.6–68 ng/l, which is in accordance with the research conducted in the Republic of Serbia (Kovacevic, 2017a) where the results show that in Novi Sad (Novi Sad sits 1254 km along the Danube River), of a total of three analysed samples, an average concentration of 30 ng/l with a detection rate of 100% was found, while in Veliko Gradište (Veliko Gradište sits 1059 km along the Danube River), carbamazepine was registered in only one sample with a concentration of 27 ng/l (three samples from the Danube were analysed). Table 6.8 presents the results of the analysis completed to determine the presence of pharmaceuticals in the Danube River in Novi Sad and in Veliko Gradište.

In addition, metamizole metabolites 4-AAA (100–121 ng/l) and 4-FAA (65–71 ng/l) were detected in two samples in the profile of the Danube River near Novi Sad, while in the profile near Veliko Gradište, trimethoprim was detected in two samples (29–90 ng/l), 4-AAA in two samples (75–181 ng/l), and 4-FAA was detected in only one sample with a concentration of 141 ng/l.

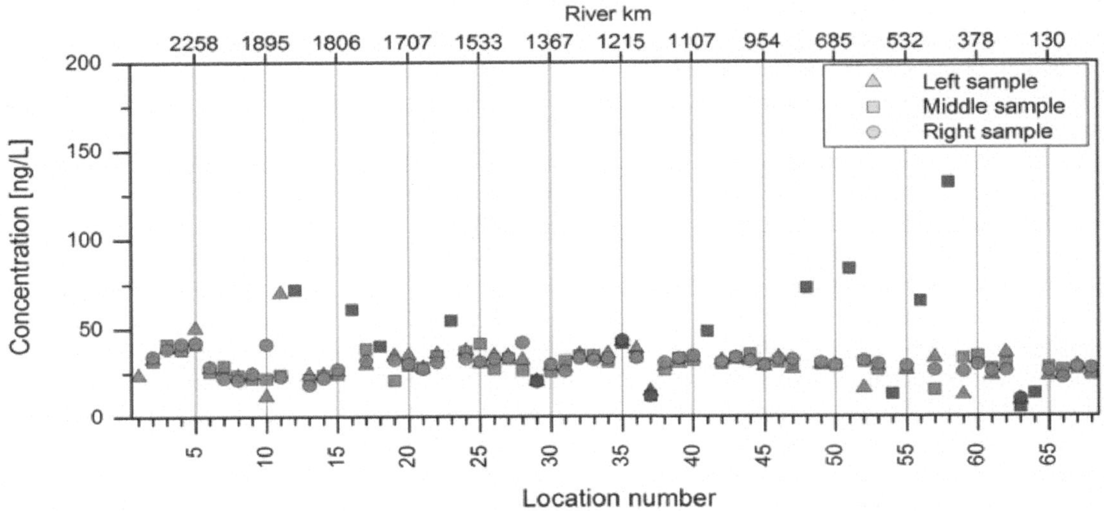


**Figure 6.10** Carbamazepine concentrations along the Danube River and its tributaries (brighter marks indicate the river Danube, the darker marks represent the tributaries of the Danube) (ICPDR, 2013).

**Table 6.8** Results of the analysis completed to determine the presence of pharmaceuticals in the Danube River in Novi Sad and in Veliko Gradište (Kovačević *et al.*, 2013a).

| Detected Pharmaceuticals | Novi Sad Profile, $n = 3$ (ng/l) | Veliko Gradište Profile, $n = 3$ (ng/l) |
|---|---|---|
| Carbamazepine | 27–37 | 27 |
| 4-AAA | 100–121 | 75–181 |
| 4-FAA | 65–71 | 141 |
| Trimethoprim | – | 29–90 |

*n* – number of samples analysed.

The results of the research indicate that, although most of the wastewater is not treated in the Republic of Serbia, there is no increase in the concentration of carbamazepine in the Danube River. This may be due to the greater flow of the river Danube in the Republic of Serbia than in the upstream countries, and hence the dilution of the concentration of the pharmaceuticals that reach the Danube due to anthropogenic activities. Another reason may be the much larger population that inhabits the Danube River Basin upstream of the Republic of Serbia. Consequently they discharge significantly more pollutants into the Danube River which has a smaller flow upstream, so that despite tertiary treatment, carbamazepine enters the river Danube. In countries upstream of the Republic of Serbia, more than 85% of WWTPs apply tertiary treatment, but the tested pharmaceuticals are still present in the Danube River Basin upstream of the Republic of Serbia.

## 6.5.2 Analysed pharmaceuticals and sampling locations

The presence of the analysed pharmaceuticals in the selected larger rivers in the Republic of Serbia, as well as in corresponding wells, was looked into to determine the frequency of their occurrence, as well as their concentration.

Based on the data on the occurrence of the analysed pharmaceuticals, the average concentrations were calculated both in surface and alluvial groundwater. The results obtained for the average concentrations of the detected pharmaceuticals and for average flows in rivers were used to calculate the average pharmaceutical concentration fluxes which were used as input data to define the self-purifying potential of alluvial groundwater at selected locations.

In the 2009–2010 period, water samples for pharmaceutical analyses were prepared to apply solid phase extraction (SPE) and the extracts were analysed using the liquid chromatography tandem mass spectrometry method (LC-MS/MS). Detailed information on the analytical procedures used, the instrumental parameters, SPE recoveries, calibration curves, accuracy, precision, detection limits, and quantification limits has been described in a previously published study (Grujić *et al.*, 2009).

After the year 2010, samples were analysed using slightly modified methods that have been published (Grujić *et al.*, 2009; Jauković *et al.*, 2014) and to which seven cardiovascular pharmaceuticals were added. The method was validated and the detailed results are presented in a study that has been published (Kovačević *et al.*, 2017b).

Between 2009 and 2015 a total of 25 pharmaceuticals and metabolites were analysed from 184 surface and groundwater samples. Ten pharmaceuticals were detected in surface water and eight in groundwater. Carbamazepine and the metabolites of metamizole 4-AAA and 4-FAA have the highest frequency of occurrence in surface water (57.3–68.8%), as well as in groundwater (19.5–43.9%). The maximum concentrations were recorded for the metabolites of metamizole 4-AAA (520 ng/l) and 4-AFAA (248 ng/l) in surface water as well as 4-AAA (128 ng/l) and 4-FAA (150 ng/l) in groundwater (Kovačević *et al.*, 2013b).

No long-term systematic monitoring of pharmaceuticals in the aquatic environment had been previously carried out in Serbia. Hence, the long-term systematic investigation of the occurrence of pharmaceuticals in surface and groundwaters was completed as part of a scientific study reported here for the first time.

The systematic monitoring of the occurrence of pharmaceuticals at selected research sites is aimed at determining the average concentration of the tested pharmaceuticals in surface and alluvial groundwater, as well as at calculating the percentage of removal of the selected pharmaceuticals during bank filtration at all the research sites.

Surface water samples (SW) were collected from the selected sampling locations of larger rivers in the Republic of Serbia. Pharmaceuticals were analysed at five locations on the Danube River, one location on the Velika Morava River, and one location on the Sava River as well as one location on the Tisza River. Figure 6.11 shows the sampling locations while Table 6.9 shows a list of all the surface water samples (SW) and samples of alluvial groundwater (GW).

Surface and groundwater samples were taken periodically, mainly in spring and autumn every year. During one sampling campaign, surface and groundwater samples were taken simultaneously during the investigation period.

Surface water sampling was carried out in the middle of the river at a depth of one metre. Groundwater samples were collected from wells and observation piezometers in the immediate vicinity of the analysed rivers.

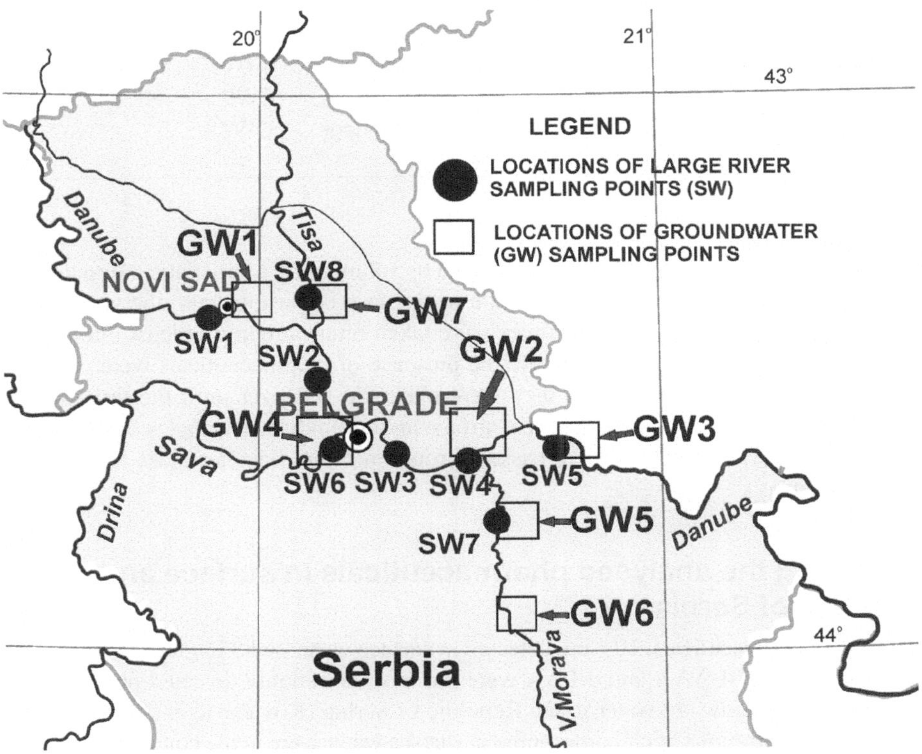


**Figure 6.11** Map of the Republic of Serbia with corresponding sampling locations of surface (SW) and alluvial groundwater (GW) (Kovačević *et al.*, 2017a).

**Table 6.9** Number of samples and labels on samples of surface water (SW) and alluvial groundwater (GW).

| Sample Type | Sample Code | Number of Samples |
|---|---|---|
| *Surface water – the Danube River* | | |
| Novi Sad | SW1 | 4 |
| Stari Banovci | SW2 | 2 |
| Ritopek | SW3 | 2 |
| Kovin | SW4 | 13 |
| Veliko Gradište | SW5 | 3 |
| *Groundwater – an alluvion of the river Danube* | | |
| Water supply source at Novi Sad | GW1 | 3 |
| Kovin-Dubovac drainage system | GW 2 | 39 |
| Water supply source at Veliko Gradište | GW3 | 4 |
| Surface water – the Sava River | SW6 | 14 |
| Groundwater – alluvion of the Sava River – water supply source at Belgrade | GW 4 | 45 |
| Surface water – the Velika Morava River | SW7 | 14 |
| *Groundwater – an alluvion of the Velika Morava River* | | |
| Water supply source at Ključ | GW 5 | 17 |
| Water supply source at Brzan | GW6 | 3 |
| Surface water – the Tisa River | SW8 | 9 |
| Groundwater – alluvion of the Tisa River – Knićanin-Čenta drainage system | GW7 | 12 |

In the case of pump wells, the samples were collected by submerging a peristaltic pump to the level of the filters or drains and abstracting groundwater, while in the case of piezometers, the peristaltic pump was submerged to the level of the filter, and samples were taken after pumping at least three volumes of the piezometer. The water samples taken to analyse the presence of pharmaceuticals were collected in tinted 1 litre glass bottles, and then, unprepared, they were placed in a freezer until the time of analysis (2–3 days after sampling). Prior to the analysis of the surface and groundwater samples for the presence of the tested pharmaceuticals, water samples were passed through a 1-μm diameter glass fibre filter (Whatman GmbH, Dassel, Germany).

### 6.5.3 Monitoring the analysed pharmaceuticals in surface and groundwater in the Republic of Serbia

Of the 25 analysed pharmaceuticals, 10 were detected in surface water and 8 in groundwater in the Republic of Serbia. Carbamazepine, 4-AAA and 4-FAA were the most commonly detected pharmaceuticals in the analysed surface and groundwater water in the Republic of Serbia (Kovačević *et al.*, 2013c).

The registered concentrations of carbamazepine in surface water were in the concentration range of 4–94 ng/l with a detection rate of 57.3%. Carbamazepine was detected in 43.9% of groundwater samples with a concentration range of 2–57 ng/l, which is slightly lower than the surface water concentration.

The high detection rate of carbamazepine in surface and groundwater can be attributed to the relatively low sorption potential and the relatively long persistence of carbamazepine in the aquatic environment (Bertelkamp *et al.*, 2014; Yu *et al.*, 2013), as well as to the relatively small percentage of removal in wastewater treatment plants (Luo *et al.*, 2014; Rivera-Utrilla *et al.*, 2013) and to its widespread and continuous use (as an antiepileptic drug). The concentrations of carbamazepine detected in the larger rivers and alluvial groundwaters of the Republic of Serbia are approximately the same as those observed in a previous study conducted in the Republic of Serbia (Petrović *et al.*, 2014). They are also more or less the same as those recorded in the larger rivers in China (Wu *et al.*, 2015; Yan *et al.*, 2015) and in the larger rivers of France (de Jongh *et al.*, 2012; Tamtam *et al.*, 2008).

Carbamazepine concentrations in the Republic of Serbia are lower than the concentrations recorded in the surface waters of Germany of 250–1100 ng/l (Ternes, 1998; Wiegel *et al.*, 2004; Zuehlke *et al.*, 2004), and those recorded in review papers for the region of Europe (Loos *et al.*, 2013) and the USA (Fram & Belitz, 2011). These carbamazepine concentrations are significantly lower than those observed in Europe (Mendoza *et al.*, 2014). So far, only carbamazepine and lorazepam have been detected in research work carried out in Serbia, which may indicate that the use of antiepileptics and antipsychotics is smaller in Serbia than in Europe. The difference in the occurrence of pharmaceuticals in the surface and groundwater waters of the Republic of Serbia compared to other regions may depend on their use in the area in which the sampling and analyses were conducted.

The metabolite of metamizole 4-AAA was detected in more than half of the surface water samples (59%). The concentration levels of 4-AAA in surface water were in the range of 11–520 ng/l. The registered concentrations of 4-AAA in the Republic of Serbia are generally lower than the concentrations recorded in Germany of 939–1000 ng/l (Wiegel *et al.*, 2004; Zuehlke *et al.*, 2004) and the registered concentrations in Spain of 811–871 ng/l (Gracia-Lor *et al.*, 2012, 2014). The concentrations of 4-AAA registered in the surface waters of the Netherlands are slightly lower at 37–63 ng/l (de Jongh *et al.*, 2012). The frequency of occurrence of 4-AAA in the analysed alluvial groundwater is lower compared to that in surface water, and the concentration range in groundwater is 5–128 ng/l. The 4-AAA concentration levels obtained from the study are close to the results of a study conducted in Germany showing 190 ng/l (Zuehlke *et al.*, 2007) and slightly lower than the concentrations observed in Spain of 362–680 ng/l (Cabeza *et al.*, 2012; Teijon *et al.*, 2010). The slightly lower concentration levels of 4-AAA recorded during the survey may indicate that the load of these pharmaceuticals is lower than in the other analysed regions, even though wastewater in the Republic of Serbia is generally not treated. The lower concentration range of the metamizole metabolite 4-AAA levels may be due to the smaller anthropogenic load (smaller population) relative to the size of the river basin (receiving power of recipients).

During the investigation period, diclofenac was detected in surface waters in the concentration range of 18–53 ng/l, while it was detected with a concentration range of 13–18 ng/l in alluvial groundwater. Similar concentrations of diclofenac were recorded in the surface waters of Portugal at 11 ng/l (de Jesus Gaffney *et al.*, 2015), as well as in the groundwater of Singapore at 17 ng/l (Tran *et al.*, 2014) and Spain at 55 ng/l (Cabeza *et al.*, 2012). Diclofenac has a relatively low frequency of detection in the investigated surface and alluvial groundwaters of the Republic of Serbia which can be attributed to the relatively high photo-degradation of diclofenac in surface water and the high degree of degradation in aerobic groundwater (Stamm *et al.*, 2008).

During the study, trimethoprim was detected only in surface waters with a concentration range of 4–223 ng/l and a relatively low detection rate of 12%, which is in line with data published in a study conducted in the Republic of Serbia (Petrović *et al.*, 2014). Trimethoprim was not detected in the analysed groundwater of the Republic of Serbia, while it was detected in groundwater in the USA with an average concentration of 18 ng/l (Fram & Belitz, 2011), in groundwater in Spain with a concentration range of 5–9 ng/l (López-Serna *et al.*, 2013) and in groundwater in China with an average concentration of 40 ng/l (Ma *et al.*, 2015).

The concentration ranges of all the analysed pharmaceuticals were quite different at different sampling locations. The spatial differences in the concentration levels of the analysed pharmaceuticals in surface waters may be due to different types of water bodies (Wang *et al.*, 2015), different water runoff, and different anthropogenic pressure.

Literature data on the occurrence of pharmaceuticals in groundwater often do not include information on the types of water intake facilities, the hydrogeological conditions, the hydrodynamic conditions of groundwater, or the distance between a river and a well. Literature data on the occurrence of pharmaceuticals in groundwater that do not contain the above-mentioned information should be considered with caution, since that information significantly influences our knowledge of the occurrence of groundwater pharmaceuticals, and has been analysed in more detail as part of this study.

Although there are several studies suggesting seasonal differences in the occurrence of aquatic pharmaceuticals due to changes in temperature (Alidina *et al.*, 2015; Daneshvar *et al.*, 2010; Massmann *et al.*, 2008), our research has found no significant effect of the seasonal changes in temperature on the occurrence of pharmaceuticals in the aquatic environment or on the percentage of their removal during bank filtration. One possible reason for this may be the relatively constant temperature in all the alluvial aquifers that were studied, as well as the similar intervals between when surface and alluvial groundwater samples were taken. According to literature, some of the analysed pharmaceuticals (such as diclofenac) are partially or completely eliminated due to temperature changes (increase of temperature), while carbamazepine and metamizole metabolites are resistant to temperature changes (Aldina *et al.*, 2015; Heberer *et al.*, 2008; Massmann *et al.*, 2008). Due to the inconsistent changes in the concentration of the tested pharmaceuticals under the influence of temperature, it is not possible to predict the trend of occurrence of pharmaceuticals in the aquatic environment as a result of climate change (Sprenger *et al.*, 2011). Fluctuations in surface and groundwater levels have a significantly greater impact on the changes in the concentration of pharmaceuticals during bank filtration, as reported in the literature (Derx *et al.*, 2013).

Based on the obtained results pertaining to the average concentrations of the tested pharmaceuticals in the surface and alluvial groundwaters of the Republic of Serbia, the percentages of removal during bank filtration were calculated for pharmaceuticals with a sufficiently high detection frequency, for all the surface and groundwater samples at all survey sites. The removal efficiency for trimethoprim was 100%, for carbamazepine it was 64%, and for the metabolite of metamizole 4-AAA it was 91.2% (Table 6.10). The detection frequency for diclofenac in surface and alluvial groundwater is too small to be able to quantify the percentage of removal during bank filtration.

Three specific survey sites (where surface and alluvial groundwater sampling frequency was the highest) were analysed in detail as part of the research (Kovačević, 2017) and the percentage of removal during bank filtration was calculated depending on hydrogeological, hydrodynamic and oxic/anoxic groundwater conditions. The samples were taken from the following locations: the Danube River, the Velika Morava River and the Sava River:

- Samples of surface water were taken from the Danube River (SW4, Figure 6.11). Samples of alluvial groundwater were taken from the site of the Kovin-Dubovac drainage system (GW2, Figure 6.11);
- Surface water samples were taken from the Velika Morava River (SW7, Figure 6.11) and alluvial groundwater samples were taken at the Ključ groundwater sources (non-exploitation wells) (AV5);
- Surface water samples were taken from the Sava River (SW6, Figure 6.11) and alluvial groundwater samples were taken at the location of the Belgrade groundwater source (GW4, Figure 6.11).

**Table 6.10** Frequency of occurrence of analysed pharmaceuticals, and minimum and maximum concentrations (Kovačević *et al.*, 2017a).

| Detected Pharmaceuticals Analysed between 2009–2015 | Therapeutic Class | Abbreviation | Surface Water | | Groundwater | | LOD | Removal by RBF |
|---|---|---|---|---|---|---|---|---|
| | | | Frequency of Occurrence ($n = 61$) | Min/Max (ng/l) | Frequency of Occurrence ($n = 123$) | Min/Max (ng/l) | | |
| Sulfametaxazole | Sulfonamide antibiotic | SMZ | 3.2% | 11/23 | nd | Nd | 3.5 | nc |
| Azithromycin | Macrolide antibiotics | AZT | 3.2% | 20/56 | 3.2% | 18/83 | 4.5 | nc |
| Trimethoprim | | TMP | 8.2% | 4/223 | nd | nd | 0.4 | 100% |
| Carbamazepine | Antiepileptic | CBZ | 57.3% | 4/94 | 43.9% | 2/57 | 0.3 | 65.4% |
| Lorazepam | Benzodiazepines | LRZ | 3.2% | 20/34 | 2.4% | 4/30 | 0.6 | nc |
| Diclofenac | Analgoantipyretic | DCF | 3.2% | 18/53 | 1.6% | 13/18 | 4.3 | nc |
| N-Formyl-4-amino-antipyrine (4 – FAA) | Metamizole (analgoantipyretic) metabolites | 4-FAA | 59% | 7/248 | 29.1% | 4/150 | 3.0 | 70% |
| N-acetyl-4-amino-antipyrine (4-AAA) | | 4-AAA | 68.8% | 11/520 | 19.5% | 5/128 | 3.4 | 91.2% |
| Detected cardiovascular pharmaceuticals analysed between 2011–2015 | Therapeutic class | Abbreviation | Frequency of occurrence ($n = 28$) | Min/max (ng/l) | Frequency of occurrence ($n = 99$) | Min/max (ng/l) | | Removal by RBF |
| Metoprolol | Antihypertensives – β-blockers | MTP | 10.7% | 4/35 | 3% | 4/14 | 0.9 | nc |
| Bisoprolol | | BSP | 3.5% | 6 | 3% | 5/6 | 2.2 | nc |

LOD – limit of detection and removal rate by RBF of detected pharmaceuticals in surface water and groundwater; nd – not detected; nc – not calculated.

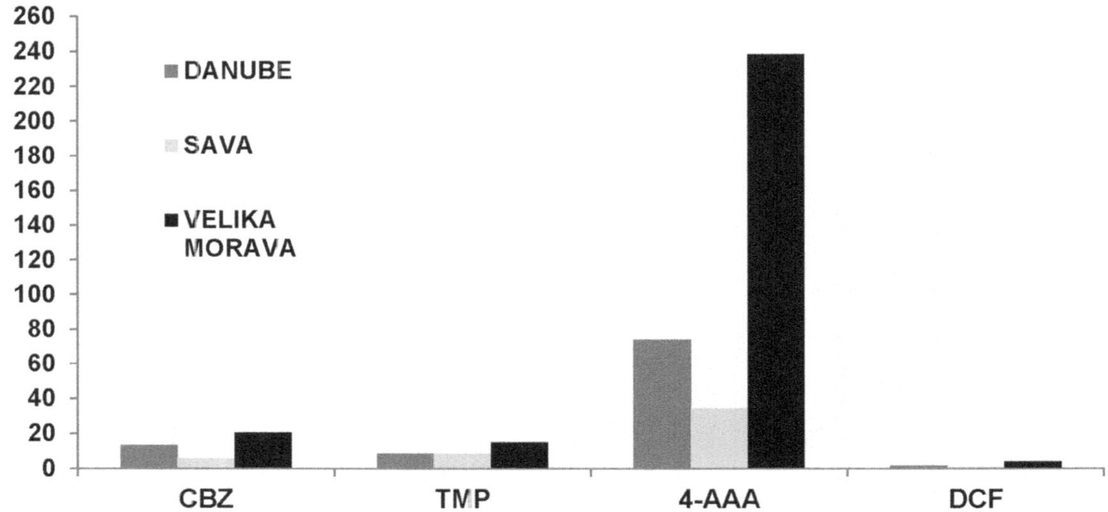


**Figure 6.12** Relationship between the average concentrations of the selected pharmaceuticals for three rivers in the Republic of Serbia (½ LOD – half of the detection limit value was adopted for samples where no selected pharmaceuticals were detected) (Kovačević *et al.*, 2017a).

The concentrations of the selected pharmaceuticals were different at all the three specific locations. Figure 6.12 shows the comparison of the average concentration of the detected pharmaceuticals in surface water.

The results indicate that the average concentrations of the analysed pharmaceuticals vary for different rivers in the Republic of Serbia. The average concentration of the selected pharmaceuticals is the highest in the Velika Morava, followed by the Danube, while the lowest average concentration of the analysed pharmaceuticals was registered in the Sava River. The results indicate that the anthropogenic impact and the emission of pharmaceuticals is the greatest in the Velika Morava River while the river Sava is the least affected. The anthropogenic impact is much higher in the Velika Morava River than in other rivers when looking at the average perennial flow versus the population in the basin (Table 6.11).

**Table 6.11** Average perennial flow in larger rivers in the Republic of Serbia as a function of catchment population (ICDPR, 2016; Water management Strategy for the Republic of Serbia, 2016).

| River | Relative Number of People in the Basin (Million Inhabitants) | Station | Average Annual Flow 1946–2006 ($m^3/s$) | Average Low Flow $Qmin_{95\%}$ ($m^3/s$) | Average High Flow $Qmax_{1\%}$ ($m^3/s$) |
|---|---|---|---|---|---|
| The Danube | 40 | Veliko Gradište | 5413 | 1976 | 7017 |
| The Sava | 8.5 | Sremska Mitrovica | 1535 | 272 | 6379 |
| Tisza | 14 | Novi Bečej | 802 | 134 | 3914 |
| Velika Morava | 4 | Ljubičevski Most | 234 | 35 | 2396 |

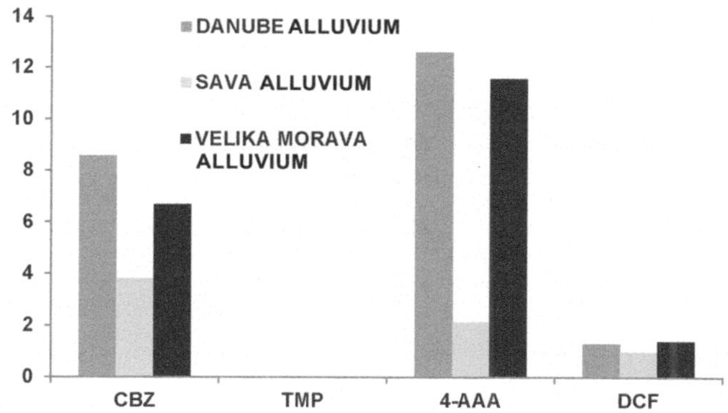


**Figure 6.13** Relationship between the average concentrations of the selected pharmaceuticals for three alluvions of the Republic of Serbia ($C = 0$ ng/l concentration was adopted for samples where the selected pharmaceuticals were not detected) (Kovačević *et al.*, 2017a).

Figure 6.13 shows the results obtained for the average concentrations of the selected pharmaceuticals at the three sampling locations of alluvial groundwater.

Carbamazepine was detected at all three sites, with average concentrations within the 6–8.5 ng/l range. Trimethoprim was not detected at any of the three alluvial groundwater sampling sites. The average concentrations of metamizole 4-AAA metabolites recorded for the Danube alluvion and in the Velika Morava alluvion are several times higher than those in the Sava river alluvion as a result of the anisotropy of the Belgrade aquifer source, the time of water filtration from the river to the wells, and the method of groundwater abstraction (type of wells etc). Diclofenac was detected with a very low average concentration and a low incidence rate at all three study sites.

The type of water intake facility, different hydrogeological and hydrodynamic conditions, as well as oxic/anoxic groundwater conditions can have the greatest influence on the concentration of pharmaceuticals in alluvial groundwater.

### 6.5.4 Characteristic hydrogeological and oxic conditions at selected survey sites

The second objective of the research is to identify the priority factors that have the greatest influence on the removal of pharmaceuticals during bank filtration, such as the infiltration time from a river to a well, the proportion of river water in the tested wells, and the oxic/anoxic status of groundwater.

Hydrogeological conditions affect the efficiency of the aquifer's self-purifying potential as well as the removal of pharmaceuticals during bank filtration. Three alluvions in the Republic of Serbia (the Danube, the Velika Morava and the Sava) were monitored continuously during the study period.

The hydrogeological characteristics as well as the oxic/anoxic environmental conditions were analysed at these locations, as shown in Table 6.12.

Each area that was surveyed represents a different hydrogeological entity. The selected physico-chemical parameters typical of each exploratory alluvion were monitored in situ and laboratory analyses were done of alluvial groundwater samples in the laboratory of the Jaroslav Černi Institute for Water Management. The results of the surveys are presented in Table 6.12. The values refer to the entire source or drainage system and reflect the data for the various wells and piezometers between 2009–2014.

**Table 6.12** Values of the analysed parameters for characteristic locations between 2009–2014 (Dimkić & Pušić, 2014).

| Location | Danube Alluvion (Drainage System – Dubovac) | | | | Alluvion of the Velika Morava River (Ključ Source) | | | | Alluvion of the Sava River (Belgrade Groundwater Source) | | | |
|---|---|---|---|---|---|---|---|---|---|---|---|---|
| **Parameter** | **Min** | **Average** | **Max** | **No. data** | **Min** | **Average** | **Max** | **No. data** | **Min** | **Average** | **Max** | **No. data** |
| *Eh* (mV) | 20 | 96 | 279 | 61 | 312 | 371 | 420 | 15 | 48 | 107 | 261 | 350 |
| $O_2$ (mg/l) | 0 | 0.06 | 0.80 | 79 | 4.4 | 5.77 | 9.1 | 18 | 0.00 | 0.17 | 1.8 | 358 |
| *TER* at 105°C (mg/l) | 223 | 377 | 709 | 87 | 256 | 600 | 996 | 32 | 227 | 320 | 546 | 170 |
| $d_{50}$ (mm) | 0.14 | 4.38 | 13 | 41 | 0.5 | 3 | 10.8 | 15 | 0.37 | 1.75 | 10.3 | 2500 |

*TER* – total evaporated residue.

The oxidation/reduction potential (*Eh*) and the dissolved oxygen concentration ($O_2$) were monitored to define the oxic/anoxic environmental conditions at the three sites that were surveyed. Monitoring of oxic/anoxic environmental conditions is important in order to determine how they affect the removal of pharmaceuticals during bank filtration. In addition to the parameters that determine the oxic/anoxic conditions in the environment, including available data for diameter for the 50% curve of the granulometric composition of the aquifer sediment ($d_{50}$), also computed were the filtration coefficients of alluvial aquifers, as well as the infiltration time of water from the river to the wells and piezometers of the alluvial aquifers. The share of river water at specific locations was determined based on the data on the total evaporated residue.

The thickness of the gravel-sand deposits of the younger quaternary in the alluvial plane of the Kovin-Dubovac drainage system is uneven. The minimum thickness is in the north-western part of the alluvion and equals 1.5–2.0 m. The thickness of the younger quaternary gravel-sand deposits in the coastal area ranges from about 7 m to 20 m. The mean thickness is about 12 m in the eastern part and about 17 m in the western part (Dimkić & Pušić, 2014).

The diameter of the grains in the basic aquifer, $d_{50}$ (the diameter of the sediment grain of fraction $d_{50}$), is generally between 0.5–20 mm, while the grain diameter of the alluvial aquifer sediment fraction $d_{50}$ generally ranges between 0.08 and 0.3 mm. The semi-permeable sediment of the overlying and underlying layer $d_{50}$ is between 0.007 and 0.02 mm.

The coefficient of filtration of the aquifer is on average between $K = 0.5 - 7 \times 10^{-3}$ m/s, while the filtration coefficient of the overlying and underlying layer is $1 \times 10^{-5}$ m/s. Based on the above filtration properties of the aquifer in hydrodynamic terms, the mentioned aquifer of the alluvion of the river Danube is a typical three-layered environment (overlying layer – water-bearing layer – underlying layer) while the filtration coefficients are relatively homogeneous with conductive layers in between. The wells in the first line of drainage and the piezometers are located approximately about 300–400 m from the Danube River together with the inundation, and based on available data, the relative travel time of water from the river to the analysed wells, taking into account the existence the permeable parts of the aquifer, is about 50 days.

The water content of the wells was determined on the basis of data on the total evaporated residue, by comparing the measured average concentrations for the Danube River, the concentration in the observed well and the concentrations in the hinterland. Based on data on the concentration of dry matter in the water (about the same dry matter concentration in the river and in the tested well), it was assumed that there was practically no dilution of the groundwater in the wells for which the occurrence of the tested pharmaceuticals was analysed, that is, it was assumed that the water in the well came mainly from the river water.

In the Kovin-Dubovac drainage system area within the Danube alluvion, the mean value of the redox potential of groundwater is $Eh \approx 96$ mV. The mean dissolved oxygen ($O_2$) in groundwater is about 0.06 mg/l, which means that the groundwater is generally anoxic.

The Ključ groundwater source supplies drinking water to the town of Požarevac and it is located along the alluvion of the Velika Morava River. The aquifers of the Velika Morava River alluvion at the location of the Ključ source are made up of lake/river and alluvial sediments whose total thickness is between 15 and 20 m. The part of the aquifer from which the groundwater is drawn is predominantly composed of sandy gravels. The thickness of the specified gravel layers is from 6 to 11 m, at a distance of 8–12 m from the Velika Morava River. Above the gravel, there is a sand layer that is 2 m to 6 m thick. The explored area of the alluvial aquifer is covered by a semi-permeable overlying layer consisting of loess and clay and powdery materials whose thickness varies from 3 m to 5 m (Dimkić & Pušić, 2014).

The coefficient of filtration of the gravelly part of the aquifer (the water-bearing layer) at the source location of the Ključ is $K = 7 \times 10^{-3}$ m/s, while the sandy part of the water-bearing layer of the aquifer has a filtration coefficient of $K = 10^{-5} - 10^{-4}$ m/s. The filtration coefficient of the semi-permeable overlying layer and of the underlying layer is $K = 10^{-8} - 10^{-6}$ m/s. The hydrogeological characteristics of the wider area of the Ključ groundwater source are relatively uniform (Dimkić & Pušić, 2014).

The tested wells and piezometers are located approximately 40 m from the Velika Morava River. The travel time of water from the river to the well-Vb 2 is ≈ 2–5 days.

More significant is the fact that, on average, about 30% of the water infiltrates from the river to the wells, and this is very important when assessing the removal of the tested pharmaceuticals during bank filtration. The dilution of the concentration of pharmaceuticals directly depends on the river water content in well Vb-2 (Figure 6.14) over time. The percentage of water infiltration from the river to the well at the location of the Ključ source is influenced by the change in water levels.

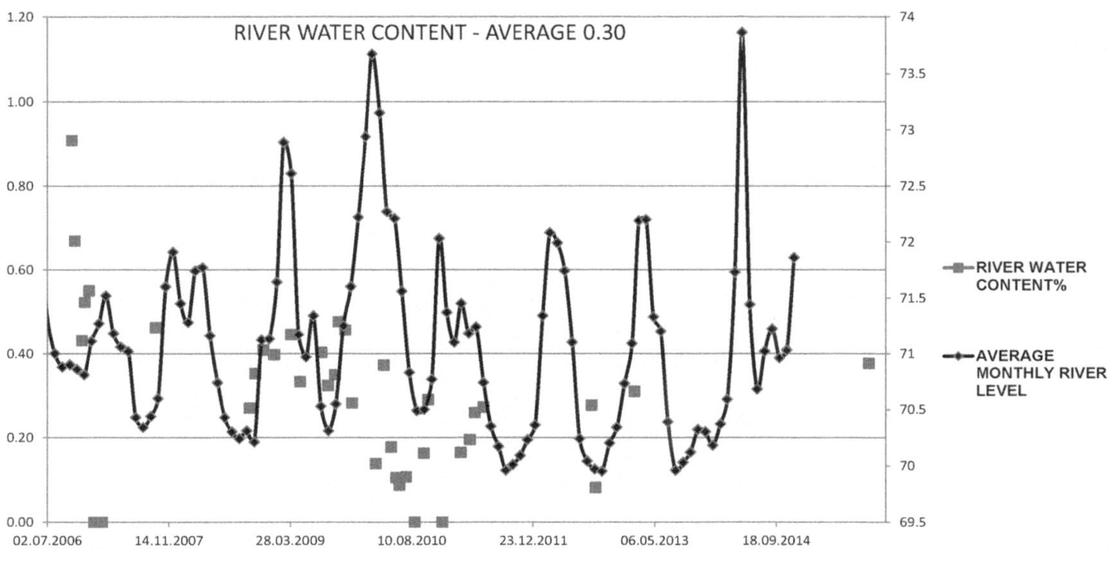


**Figure 6.14** The share of river water in well Vb-2, compared to the average monthly water level between 2006 and 2014 (Kovačević, 2017).

Due to the proximity of the wells to the Velika Morava River, as well as the hydrogeological features of the location of the Ključ source, the river bottom collimation and seasonal changes in water levels play a key role in water infiltration from the Velika Morava River to the Vb-2 well. The average amount of water originating from the Velika Morava River is 30% of the total amount pumped for the Vb-2 well, and hence most of the groundwater comes from the hinterland of the Vb-2 well. In the exploitation wells of the Ključ source there is a constant increase in the concentration of nitrates which originate from settlements and the arable land from the hinterland of the spring. The groundwater infiltrates in the direction of the exploitation wells and there is no influence of it coming from the infiltration basins of the spring. Based on the above, it can be assumed that there is no anthropogenic impact in terms of the emission of pharmaceuticals on the concentration of pharmaceuticals in the groundwater from the hinterland of the Vb-2 well.

Groundwater at the location of the Ključ source is characterized by a high redox potential (*Eh*) of more than 350 mV on average, while the concentration of dissolved oxygen ($O_2$) in the groundwater averages 5.77 mg/l, implying that the groundwater is generally oxic.

The groundwater source for Belgrade is surrounded by the city, and consists of 99 radial wells and about 50 tubular wells located along the banks of the Sava River upstream of the mouth of the Sava River into the Danube. The length of the series of wells of the Belgrade source is about 50 km.

The alluvial aquifer of the Belgrade source was formed in several cycles and sequences of sedimentation and consists of sandy gravel, sands of different granulation, dust and clay sediments. At the location of the Belgrade source, the alluvion of the Sava River, which takes a complex hydrogeological form, there are semi-permeable interlayers within the aquifer.

The thickness of the Quaternary alluvial aquifer is about 25 m. As for the granulometry of the sediment, two zones can be distinguished (Figure 6.15): the water-bearing layer 1 – the lower zone with coarse-grained aquifer material where set lateral drains have been placed. In this layer there are inter-layers of dust, clay and sand, and water-bearing layer 2 – namely the upper zone with a finer granulation of the sediment and with a lower coefficient of filtration.

The first layer of a water-bearing series of aquifers (aquifer 1) consists of sandy gravel with coefficients of filtration of the order of $10^{-3}$ m/s. Covering the mentioned layer is an aquifer with high filtration properties with coefficients of filtration of the order of $5 \times 10^{-4}$ m/s.

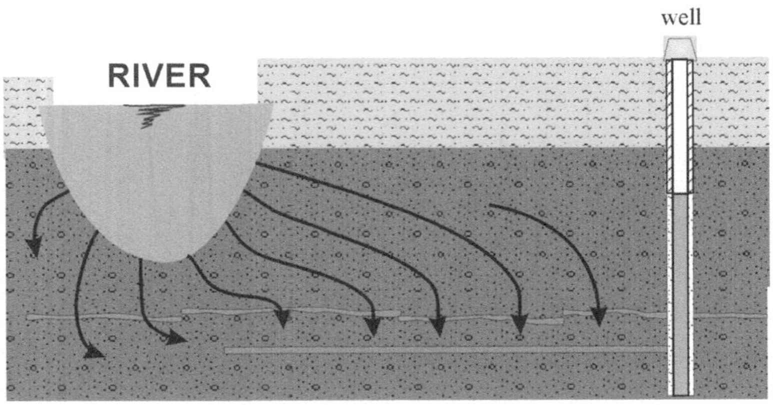


**Figure 6.15** Schematic representation of the influence of low-permeability layers on the flow of water from the river to the well (Kovačević, 2017).

These sediments spread over almost the entire area that was studied. They are covered by highly sandy gravel, and by medium-fine to fine-grained sand. The spread of these layers is almost continuous over the entire analysed space. The filtration coefficients of the complex are of the order of $10^{-4}$ m/s. Semi-permeable sediments (semi-permeable layer) that act as insulators have filtration coefficients of the order of $10^{-6}$ to $10^{-7}$ m/s. Above the semi-permeable layers is a water-bearing layer (aquifer 2) with a filtration coefficient for the fine-grained to medium-grained sands of the order of $5 \times 10^{-5}$ to $1 \times 10^{-4}$ m/s, and for the sandy gravels it is of the order of $1 \times 10^{-4}$ to $5 \times 10^{-4}$ m/s The layer spreads continually over the entire area. The impermeable or poorly permeable overlying layer has a filtration coefficient of $10^{-8}$ to $10^{-7}$ m/s (Dimkić & Pušić, 2014).

The basic characteristics of the alluvion of the Sava River at the Belgrade source are a relatively low mean value of $Eh \approx 107$, generally below 150 mV, and therefore also low values of dissolved oxygen of about 0.17 mg/l. This indicates that the environment is generally anoxic.

Due to the several semi-permeable layers, the water transport time from the Sava River to the Rb-16 well varies and is at least 100 days, but may be over 2 years, depending on the observed trajectory (Pušić & Dimkić, 2017), although the wells are positioned relatively close to the river at ≈ 10–15 m.

Based on the data on the total evaporated residue (TER) in the Sava River (the concentration of TER in the Sava River is approximately the same as the concentration in the tested well), and in the well where the occurrence of pharmaceuticals was monitored, as well as in the wells in the hinterland, it can be assumed that no infiltration of water from the hinterland into the well occurred, and it is therefore assumed that the water in the well mostly comes from the Sava River.

Due to the several low-permeable layers in between, the transport time of water from the Sava River to the exploration wells is considerably long in view of the proximity of the water intake structure on the Sava River. On the other hand, the semi-permeable layers that are in between represent a natural barrier with a pronounced self-purifying potential, where the natural elimination of pharmaceuticals occurs.

The specific filtration characteristics of the studied area of the Belgrade Spring with an inhomogeneous structure of the aquifer must be analysed in detail in the course of the research. This needs to be done to define the hydraulic parameters and explain the removal of the pharmaceuticals.

## 6.5.5 The removal of selected pharmaceuticals during bank filtration

The next phase of the research focused on the following four selected pharmaceuticals (carbamazepine, trimethoprim, diclofenac and metabolite metamizole (4-AAA) N-acetyl-4-aminoantipyrine), whose removal efficiency during the river coast filtration at three trial locations (alluvium of the Danube River, the alluvion of the Great Morava River, the Sava River alluvion) was studied in greater detail. These pharmaceuticals were selected based on the results of research that was done regarding their occurrence in surface and groundwater and on the basis of their belonging to different therapeutic classes of pharmaceuticals. The impact of hydrogeological, hydrodynamic and oxic/anoxic environmental conditions on their removal during bank filtration was analysed for each of the three selected sites separately.

### *6.5.5.1 The removal of the analysed pharmaceuticals in the alluvion of the river Danube*

A significant decrease in the average concentration of the selected pharmaceuticals was observed in the groundwater of the Danube River due to the process of self-purification during bank filtration.

The results obtained after sampling and analysing the surface water of the river Danube and of the corresponding wells of the Kovin-Dubovac drainage system indicate that the average concentration of diclofenac decreases by 30% and of carbamazepine by 35% during bank filtration. Trimethoprim is completely eliminated, while 82% of the metamizole metabolite 4-AAA is removed during bank filtration.

The alluvion of the Danube River at the site of the Kovin-Dubovac drainage system is a predominantly anoxic environment, indicating that the analysed pharmaceuticals are removed to a significant degree under anoxic environmental conditions with a medium time of water infiltration from the river to the wells and without water infiltration from the hinterland to the wells. It is characteristic for the alluvion of the river Danube that the average and the median concentration of pharmaceuticals in the analysed surface water samples is higher than in the analysed samples of groundwater.

#### *6.5.5.2 Removal of pharmaceuticals analysed from the alluvion of the Velika Morava River*

The results of the analysis of surface water samples taken from the Velika Morava River and the groundwater at the location of the Ključ source indicate that the average concentration of carbamazepine and diclofenac decreases by 67%, and that of 4-AAA by 92%, while trimethoprim is completely removed during bank filtration. The decrease in the average concentration of the analysed pharmaceuticals in the alluvial groundwater is mainly due to the dilution of the river water, since on average about 70% of the groundwater in the well comes from the hinterland. It is assumed that the water from the hinterland does not contain pharmaceuticals. The tested wells of the Ključ source have predominantly oxic environmental conditions, and the infiltration time from the river to the wells is relatively short at ≈ 2–5 days, so the dilution of the groundwater in the wells originating from the hinterland (on average 70%) represents the dominant factor affecting the removal of the analysed pharmaceuticals during riverbank filtration.

The average and median concentrations of the analysed pharmaceuticals in the samples taken from the Velika Morava River are several times higher than the concentrations of the analysed pharmaceuticals in alluvial groundwater.

#### *6.5.5.3 Removal of analysed pharmaceuticals in the river Sava alluvion*

Due to the existence of several low-permeability interlayers within the water-bearing layer, the travel time from the river to the well is at least 100 days (Pušić & Dimkić, 2017), although the radial wells are relatively close to the Sava River at ≈ 10–15 m.

Semi-permeable interlayers represent a natural barrier with a marked self-purifying potential where the natural removal of the analysed pharmaceuticals takes place during riverbank filtration. The alluvion of the Sava River at the site of the Belgrade Spring has a relatively small average redox potential value of $Eh =$ 107 mV, generally below 150 mV, as well as a small average value of dissolved oxygen in groundwater of ≈ 0.17 mg/l, so the environment is generally slightly anoxic.

The average concentration decreases by 34% for carbamazepine, and 93% for 4-AAA, while trimethoprim is completely removed. Diclofenac was detected in one sample of surface water and was not detected at all in the groundwater so it is not possible to quantify the percentage of removal during riverbank filtration.

The results indicate that carbamazepine, 4-AAA and trimethoprim are removed under mildly anoxic environmental conditions, but with a long infiltration time from the river to the analysed wells, without the dilution of the water that comes to the well from the hinterland.

The average concentrations of 4-AAA and trimethoprim in the surface water are several times higher than in the groundwater, while the average concentration of carbamazepine in surface water is 1/3 higher than in the groundwater.

### 6.5.6 Factors affecting the removal of analysed pharmaceuticals during riverbank filtration

The results of the study showed that the detected pharmaceuticals are removed during riverbank filtration under different oxic/anoxic, hydrogeological and hydrodynamic environmental conditions.

At the test location of the Belgrade source with slightly anoxic conditions in groundwater, the average concentration of carbamazepine was reduced by 34% over a relatively long period of infiltration of water from the river to the wells (a minimum of 100 days) and without dilution due to the infiltration of groundwater from the hinterland. Because of the relatively small percentage of removal over a relatively long river-to-well path, it can be concluded that carbamazepine is the most persistent of the tested pharmaceuticals, as previously stated in the literature (Hamann *et al.*, 2016; Laws *et al.*, 2011). Based on the available data and research results, the percentage of carbamazepine that is removed amounts to 67% in oxic conditions (alluvion of the Velika Morava River) and 35% in anoxic conditions (alluvion of the Danube River), but with different infiltration times from the river to the well, being approximately 2 days for the oxic environment and approximately 50 days for the anoxic environment. The most important factor influencing the removal of the analysed pharmaceuticals at the location of the Ključ source is the infiltration of groundwater from the hinterland (70% on average). Considering that the water in the well is diluted by the groundwater from the hinterland (it is assumed that the water from the hinterland does not contain pharmaceuticals), and that the infiltration time is relatively small, the removal of the pharmaceuticals must be carefully evaluated so that a mistake is not made if specific parameters and the environmental conditions are considered. The share of groundwater infiltrating from the hinterland, which is on average about 70%, is confirmed by the fact that carbamazepine, which is the most persistent examined pharmaceutical, was removed by about 67% in well Vb-2 for a relatively short groundwater infiltration time from the river to the well which amounts to approximately 2 days.

In the case of the Ključ source (well Vb-2), the greatest influence on the removal of the pharmaceuticals comes from the share of river water in the well, so the pharmaceuticals are mostly attenuated because of the dilution of the water in well Vb-2 by the groundwater from the hinterland. It can be concluded that carbamazepine is the most persistent pharmaceutical at all the analysed locations with different infiltration times of the water from the river to the wells and different oxic/anoxic conditions in the environment.

Research results have shown that the metabolite metamizole 4-AAA is removed efficiently both in oxic and anoxic conditions in groundwater taking into account the impact of the infiltration time from the river to the wells, as well as the effect of dilution due to the groundwater from the hinterland, which is in line with previous literature references (Massmann *et al.*, 2008; Storck *et al.*, 2012; Wiese *et al.*, 2011). At the site of the Belgrade source, the average concentration of 4-AAA was reduced by 93% due to the longer infiltration time from the river to the well. At the site of the Kovin-Dubovac drainage system, the average concentration of 4-AAA decreased by 82% under anoxic conditions in the environment and a medium-long infiltration time from the river to the wells, while at the Ključ source the average concentration of 4-AAA decreased by 92% under oxic conditions in the environment, but with a small contribution of river water (an average of 30%) and a short filtration time from the river to the wells, indicating that 4-AAA decomposes more efficiently under oxic environmental conditions.

Trimethoprim, which is highly susceptible to degradation in the aquatic environment, was completely removed at all the analysed locations, so it was not possible to determine the effect of different hydrogeological, hydrodynamic and oxic/anoxic environmental conditions on the percentage of removal due to the relatively rapid degradation of trimethoprim.

The average concentration of diclofenac decreased by about 30% at the site of the Kovin-Dubovac drainage system, indicating that diclofenac is relatively stable under anoxic environmental conditions with approximately 50 days infiltration time from the river to the wells.

At the Ključ source, where the percentage of dilution from the hinterland is about 70%, 67% of diclofenac was removed, which is similar to that for carbamazepine, suggesting that the oxic/anoxic conditions are not crucial for the removal of diclofenac. In the case of the Belgrade source, diclofenac was not detected in groundwater samples and was detected in only one sample from the river Sava, so it can be assumed that diclofenac is subject to degradation processes. The greatest influence on the removal of diclofenac during riverbank filtration at the site of the Belgrade source comes from the infiltration time from the river to the wells.

### 6.5.7 Field experiment consisting of the injection of tracers and pharmaceuticals

In aquifers of the intergranular type, simulations and the prediction of the groundwater flow as well as of the transport of a substance require detailed knowledge of the basic characteristics and spatial distribution of an aquifer. In most cases, these data are not easily collected and require extensive research and experimentation (Dimkić *et al.*, 2011).

Various research techniques that are applied when drilling and testing the boreholes are used to obtain parameters that are representative of a larger homogeneous area. Generalizing data is often not enough to get real parameters for a given area. Field tracer experiments provide an opportunity to more effectively examine the characteristics of an aquifer and determine relevant environmental parameters. Tracer experiments can be conducted in the laboratory as well as in the field. Field tracer experiments can be performed under natural environmental conditions, without pumping wells, as well as in conditions of an artificially induced hydraulic gradient.

Field tracer experiments can be performed by convergent, divergent and dipole groundwater flow fields. A convergent approach was used within the scope of the study where the tracer, $Cl^-$ (chloride), was injected with a constant flow over a specific period of time into a piezometer located near a well into which a pump was placed and from which water was pumped at a constant rate. The experiment yielded data on the breakthrough curve of the tracer concentration at the well outlet outflow.

In tracer experiments with an artificially caused gradient as well as with defined well pumping conditions, variations in the natural gradient of the environment are reduced, and the duration of the experiment is shortened compared to the tracer experiments in natural conditions, making this method more advantageous from an economic point of view (Sutton *et al.*, 2000).

Pharmaceuticals analysed in greater detail during the study (carbamazepine, diclofenac, trimethoprim and metabolite metamizole 4-AAA) were detected in the Danube River, as well as in the corresponding wells of alluvial groundwater at the location of the Kovin-Dubovac drainage system.

A quantification of the sorption effects during the field experiments consisting of the injection of tracers and of pharmaceuticals has not been conducted so far in the Republic of Serbia. Field tracer experiments are a very useful method for collecting better and more accurate data. Previously in the Republic of Serbia, Dimkić M. conducted tracer experiments that monitored the behaviour of cations of heavy metals and phenol near Kraljevo in 1987 (Žičko Polje and Konarevo). A tracer experiment was also carried out at the Mediana source near Niš (Dimkić, 1997) as shown in the project documentation of the Jaroslav Černi Water Management Institute (Dimkić & Igrutinović, 1987).

The need to conduct field tracer experiments and monitor the behaviour of pharmaceuticals stems from the lack of experimental data on the behaviour of pharmaceuticals.

Tracer experiments are a very effective method for characterizing aquifers. There are numerous ways they can be applied, such as in the conventional testing of aquifers to obtain the effective transport parameters of a given substance (transport rate, dispersion porosity), and thus describe non-reactive (e.g. $Na^+Cl^-$) as well as reactive transport processes (e.g. pharmaceuticals) of a substances in an aquifer. Information on the structure of an aquifer (groundwater priority flows and structural anisotropy of aquifers) can be obtained from field tracer experiments (Käss & Behrens, 1998). Field tracer experiments can also help generate a database (tracer breakthrough curves, parameters pertaining to the transport of substances), which can be used for future predictions of substance transport using deterministic or stochastic models and using inverse stochastic modelling methods (Ptak *et al.*, 2004).

The main objective of conducting a field tracer experiment is to collect sufficient data on the hydrogeological parameters of the environment in order to define the hydrodynamic model of groundwater and the model of transport of the pharmaceuticals in groundwater. The ultimate goal is to quantify the sorption effects of the tested pharmaceuticals based on experimental results obtained in the field. Salt ($Na^+Cl^-$) was used as a tracer for many practical reasons: no toxicity, low natural concentration in groundwater, low detection limit, good water solubility, accessibility, price, negligible influence on the substance transport parameters (density, viscosity, pH), and stability. After the injection of the tracer, the concentration of $Cl^-$ chloride at the outflow of well Bp-2 was monitored and a tracer breakthrough curve was obtained. During the experiment the concentrations of the analysed pharmaceuticals were monitored at the outflow from the well that draws groundwater, and a sampling and analyses program was set up in order to determine the pharmaceutical concentration breakthrough curves. Mass balance calculations were performed for given tracers and analysed pharmaceuticals at the outflow of well Bp-2. The data obtained from the field experiment were used to calculate the theoretical linear coefficient of sorption of the analysed pharmaceuticals. Data were also obtained on the monitoring of groundwater levels, the flow in a well, electrical conductivity, the pH, and the dissolved oxygen in groundwater, as well as data for the oxidation-reduction potential of groundwater.

The relatively short duration of the field experiment (breakthrough of the concentration of the tested pharmaceuticals ranging from several hours to several days) made it possible to precisely observe the effect of sorption. The effects of degradation during the relatively short duration of the experiment could not be accurately quantified, but indications of a potential for degradation were obtained.

The particle size distribution curve was determined by sifting material using a set of sieves according to the JUS L.J9.010 standard. Based on the collected data (Figure 6.16), theoretical filtration coefficients were determined using the Slichter (1899) empirical formula that is most applicable to aquifer material of a diameter $d_{10}$ in the range 0.01–5 mm:

$$K = 7.8 \times n^{3.287} \times (d_{10})^2 \tag{6.17}$$

where:

$K$ = filtration coefficient (m/s);
$d_{10}$ = grain diameter of 10% curve of granulometric composition (m);
$n$ = porosity (–).

The theoretically calculated filtration coefficient based on the empirical formula of Slichter (1899) is in the interval of $K = 1.4 \times 10^{-2} - 3 \times 10^{-4}$ m/s, for different depths of samples, indicating that there are preferential groundwater flows.

The sediment sample was analysed and the content of the total organic carbon fraction $f_{OC} \approx 0.229\%$ was determined. The organic matter content in the sediment sample is relatively small. The dominant

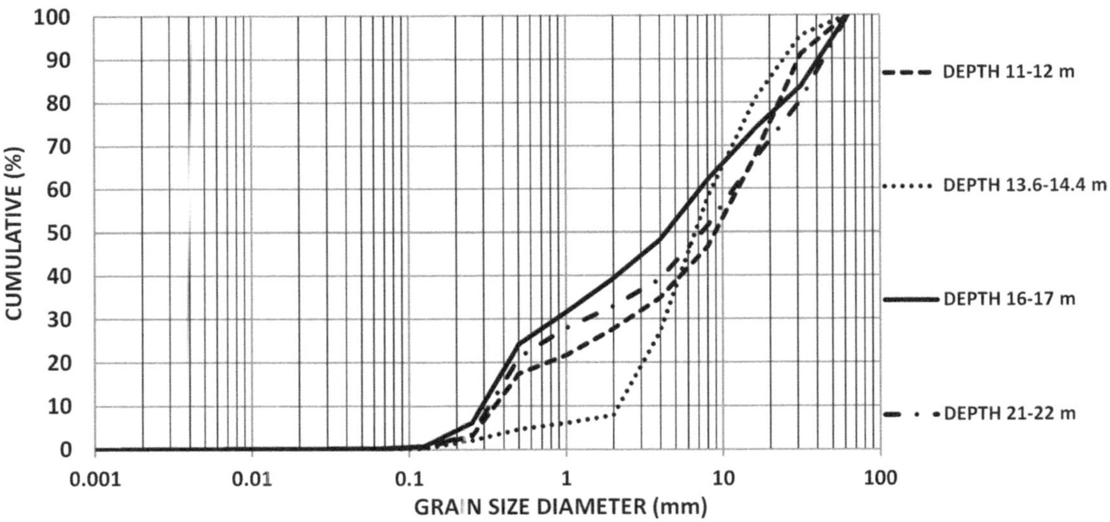


**Figure 6.16** Cumulative distribution curve of granulometric diameter of sediment samples from different depths (Kovačević *et al.*, 2017b).

sediment fraction is 8–16 mm in diameter and contains clay (0%), dust (0.42%), sand (38.33%) and gravel (60.53%).

The percentage of organic matter was determined by analysing the sediment sample from a depth of 10–14 m by taking a sediment sample at each metre and by mixing it into one representative sample. Considering that the sediment sample is 0.2–2 mm in diameter, based on the cumulative distribution curves of the granulometric diameter of the sediment samples (Figure 6.16), it can be assumed that the percentage of organic matter is considerably smaller (approximately 3–5 times) than $f_{OC} \approx 0.229\%$, as 10–20% of the particle size distribution curve includes grains that are less than 2 mm in diameter. The granulometric composition of the sediment that was analysed is not fully representative of the real composition of the sediment in nature due to the effect of ratio and the inhomogeneity of the aquifer.

All the groundwater samples were collected using 1 litre tinted glass bottles. The sampling methods, the chemicals and the chemical methods that were used for the analysis of groundwater samples, as well as the procedures that were used are described in detail in Chapter 10. In situ measurements were carried out by immersing a probe at the well overflow point where there is a discharge from the well itself, as well as by immersing a probe to the level of the piezometer filter, at a depth of about 20 m. The electrical conductivity, the pH, the dissolved oxygen, and the oxidation-reduction potential were monitored continuously throughout the experiment using the HQ40d multiparameter probe, while the titration in the field was carried out using a 'Eppendorf Top Buret M' device. The total organic carbon in sediment samples was analysed using a 'Shimadzu Total Organic Carbon Analyzer (TOC-5050a)'.

The description, verification and optimization of the analytical method for the preparation and analysis of groundwater samples to check for the presence of pharmaceuticals is described in detail in the following papers: Grujić *et al.* (2009), Jauković *et al.* (2014) and Radović *et al.* (2016).

### 6.5.7.1 Experimental setting

The field experiment that is described in detail (Kovačević *et al.*, 2017b) lasted for 16 days and the experimental parameters were monitored continuously. Before the experiment began, baseline quality was monitored to determine the initial conditions. One sample was collected from the observation well before the field experiment. The initial conditions were monitored continuously (every hour) and included the hydraulic head, flow and electrical conductivity. Additionally, the pH, $O_2$ (dissolved oxygen in groundwater) and the oxidation/reduction potential, *Eh*, were monitored periodically during the experiment. The physicochemical characteristics of the investigated compounds, LODs (limit of detection) and LOQs (limit of quantification) are presented in Table 6.13.

During the tracer test, the nonreactive tracer NaCl and reactive pharmaceuticals were injected into the injection piezometer (Figure 6.17) approximately 8.5 m from the observation well. The injection depth was about 15–20 m. A continuous flow peristaltic pump was set up at the observation well at 6 l/s. Also, groundwater levels were monitored with data loggers (divers).

The flow was monitored by a Thompson overflow, the volumetric method and an ultrasonic flow meter 'Nivus PCM Pro'. The NaCl tracer was mixed and dissolved in 1000 litres of water extracted from the observation well. The pharmaceuticals obtained from Sigma Aldrich were dissolved in 1 litre of methanol, then mixed with 100 litres of extracted groundwater and injected into the same piezometer as the NaCl tracer. During the experiment, specific conductivity was monitored as the parameter for calculating the concentration of the NaCl tracer. Samples of the reactive compounds were collected periodically and frozen immediately. The tracer test was conducted in 2014 at one specific observation well and one piezometer (Figure 6.17).

At the beginning of the test, NaCl was dissolved in an injection tank. The amount of pure NaCl that was poured into the tank was 35.7 kg, calculated on the basis of the plots of electrical conductivity as a function of the NaCl concentration. After the dissolution of the NaCl in the 1000-litre injection tank, the NaCl was injected for approximately 52 minutes. The injection piezometer was then flushed with 500 litres of groundwater. The average specific conductivity in the injection tank was 73.250 µS/cm. For a higher reliability and to check the result, electrical conductivity was measured in situ with titration and an electrical conductivity probe.

The pharmaceuticals were injected after the NaCl tracer. The initial concentrations in a 100-litre tank injection were: trimethoprim 2.5 mg/l, carbamazepine 1 mg/l, diclofenac 1 mg/l, and 4-AAA 1 mg/l. The injection was continuous and lasted for approximately 36 minutes. Following the injection, the piezometer was flushed with 200 litres of groundwater previously extracted from the observation well and then sampling was initiated based on an established program.

**Table 6.13** Physicochemical characteristics, LOD and LOQ of the investigated compounds.

| Compounds | LOD (ng/l) | LOQ (ng/l) | Water Solubility (mg/l) (25°C) | *pKa*[1] | $K_{OC}$[2] (ml/g) |
|---|---|---|---|---|---|
| Trimethoprim | 1 | 3 | 2334 | 7.1 | 905 |
| 4-AAA | 1 | 3 | 1590 | 11.9 | 240.7 |
| Carbamazepine | 1 | 3 | 17.7 | 2.4 | 3871 |
| Diclofenac | 2 | 7 | 4.5 | 4.1 | 833.3 |

*pKa* – acid dissociation constant, $K_{OC}$ – with oil organic carbon-water partitioning coefficient.
[1]The e-Chemical Compound Database, http://www.chemspider.com.
[2]The e-Hazardous Substances Data Bank, http://www.toxnet.nlm.nih.gov.

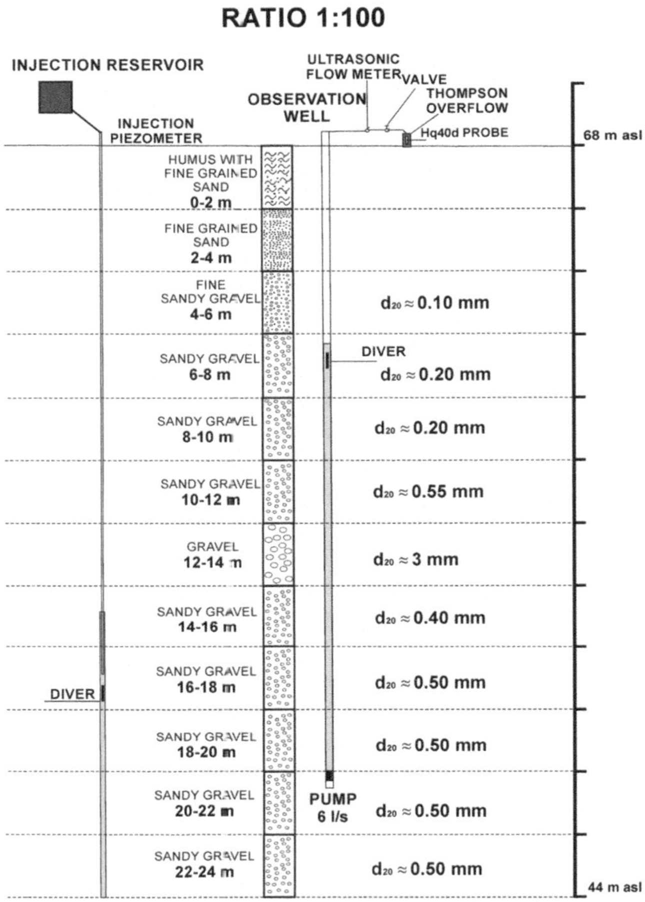


**Figure 6.17** Schematic illustration of the experimental setup and geologic cross-section (asl – above sea level, $d_{20}$ – 20% graduation curve particle size diameter in mm) (Kovačević *et al.*, 2017b).

### *6.5.7.2 Results of a field experiment of tracer filling and pharmacists tested*

On the basis of the calculation, the obtained result showed that the total mass of tracer $Cl^-$ released at the outflow of well Bp-2,($Cl^-$) = 28.8 kg. This is approximately 80% of the initial amount of tracer ($Cl^-$) = 35.7 kg that was dissolved at the beginning of the experiment. This result shows that the tracer loss is about 20% over the duration of the experiment. The sensitivity of the analytical method used to determine the presence of pharmaceuticals in groundwater samples, the short duration of the experiment, and the 20% loss of tracer ($Cl^-$) over the duration of the experiment affect the ability to accurately determine the degree of degradation, so only the potential for the degradation of the analysed pharmaceuticals was looked at in the course of the experiment.

Before the experiment started, the initial conditions were monitored every hour for one day. The results showed that the average values of the monitored parameters were: electrical conductivity ≈ 649 μS/cm, pH ≈ 6.9, $O_2$ (dissolved oxygen in the groundwater) ≈ 0.05 mg/l, oxidation/reduction potential ≈ 91 mV, observation well flow ≈ 6 l/s, injection piezometer hydraulic head ≈ 65.77 m (above sea level), and the hydraulic head in the observation well ≈ 65.61 m (above sea level).

The initial concentrations of the selected pharmaceuticals in the observation well before the experiment were: carbamazepine ≈ 6 ng/l and 4-AAA ≈ 2 ng/l. The thermo-analysed pharmaceuticals (trimethoprim and diclofenac) were not detected before the experiment began.

The results of the initial monitoring of the groundwater showed that it was anoxic and that the hydraulic and flow conditions were quasi-stationary. The results also showed that carbamazepine and diclofenac had background concentrations equal to those in the Danube River.

The highest concentration of the NaCl tracer was established 200 minutes after the start of the test (Figure 6.18). The average calculated hydraulic conductivity based on the water-bearing geometry and tracer (NaCl) velocity was $4 \times 10^{-2}$ m/s. The theoretical hydraulic conductivity that was calculated based on the empirical formula (USBR – United States Bureau of Reclamation) was found to be between $4 \times 10^{-2}$ and $8 \times 10^{-4}$ m/s for different sample depths, which indicates that the preferential flow path has the highest impact on the velocity of groundwater. Hydraulic conductivity could vary locally but the results obtained are consistent with previous research work.

The organic matter is highly variable among different sediment samples. *Kd* can alternatively be expressed as a function of the fraction of organic carbon in the sediment, $f_{OC}$ (%), and of the organic carbon partitioning coefficient, $K_{OC}$ (ml/g), of the compounds according to Equation 6.18:

$$Kd = f_{OC} \cdot K_{OC} \tag{6.18}$$

where: *Kd* is the linear sorption coefficient (ml/g); $f_{OC}$ is the fraction of organic carbon in the sediment (%); and $K_{OC}$ is the carbon normalized sorption coefficient (ml/g).

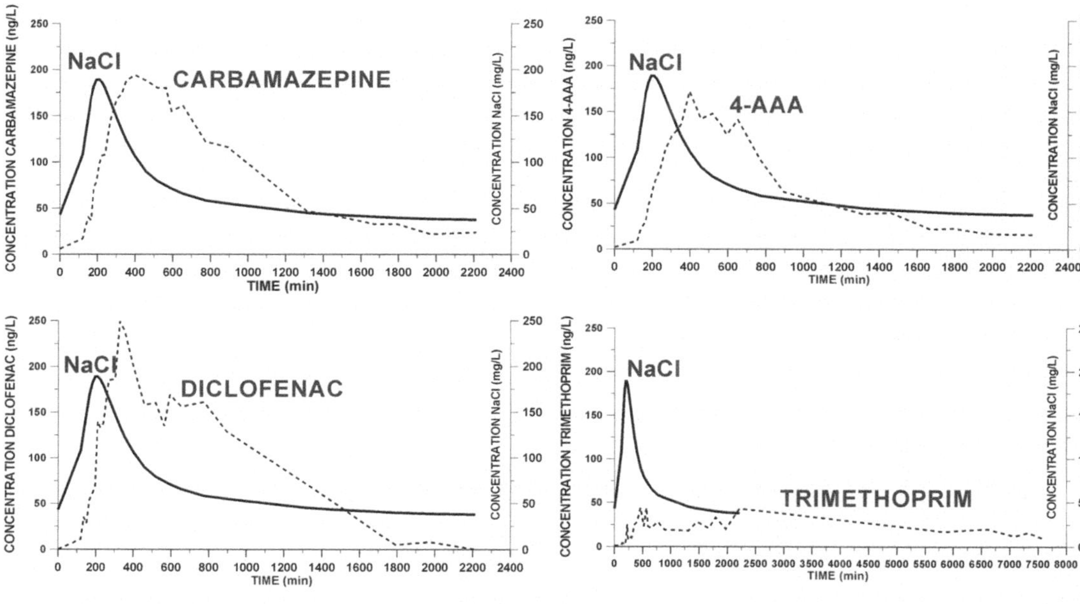


**Figure 6.18** Breakthrough curve of NaCl vs. the breakthrough curve of the investigated pharmaceuticals (Kovačević *et al.*, 2017b).

During the experiment (Kovačević *et al.*, 2017b), the average values of the monitored non-variable parameters were: pH ≈ 7, $O_2$ (dissolved oxygen in the groundwater) ≈ 0.06 mg/l, oxidation/reduction potential ≈ 90 mV, injection hydraulic piezometer head ≈ 65.73 m (asl), and the hydraulic head in the observation well ≈ 65.54 m a.s.l. The results of the monitoring of the non-variable parameters during the experiment showed that the conditions were similar to the initial conditions, the groundwater was anoxic, the hydraulic and flow conditions were quasi-stationary, and a quasi-steady-state radial divergent flow field was established.

After the sampling and analysis, the results of the field experiment revealed a retardation of all the pharmaceuticals relative to the non-reactive tracer NaCl. A breakthrough curve of NaCl vs. that of the investigated pharmaceuticals was obtained (Figure 6.18) based on the results of the tracer test and of the monitoring of the behaviour of the selected pharmaceuticals.

The following equation was used to calculate *Kd* for all the analysed pharmaceuticals:

$$Rd = \frac{v_{\text{NaCl}}}{v_{\text{pharmaceutical}}} = 1 + \frac{\rho_b}{n} Kd \tag{6.19}$$

where: $Rd$ is the retardation coefficient [–]; $v_{\text{NaCl}}$ is the velocity of the NaCl tracer (m/s); $v_{\text{pharmaceutical}}$ is the velocity of the pharmaceutical (m/s); $\rho_b$ is the bulk density of the aquifer material (g/ml); $n$ is the effective porosity (–); and $Kd$ is the linear sorption coefficient (ml/g).

The effective porosity was calculated based on the grain-size distribution curve; in this study the effective porosity was 0.15 and the theoretical soil bulk density of 1.65 g/cm$^3$ was used for the aquifer gravel and sand. The velocities of the NaCl tracer and the pharmaceuticals were calculated from the results of the tracer test and of the monitoring of the behaviour of the selected pharmaceuticals. The linear sorption coefficient *Kd*, the retardation coefficient *Rd* and the carbon normalized sorption coefficient $K_{OC}$ for all the pharmaceuticals were calculated on the basis of the experimental data (Table 6.14).

Based on the results of the tracer test and on other experimental data, the linear sorption coefficients *Kd* for selected pharmaceuticals and the carbon normalized sorption coefficients $K_{OC}$ under field experimental conditions could readily be calculated. It is very important to note that the sorption of the selected pharmaceuticals, with a relatively low linear sorption coefficient compared to the data reported in literature, represents a significant factor that influences the behaviour of pharmaceuticals in groundwater.

The calculated carbon normalized sorption coefficients of $K_{OC}$ are lower than those reported in literature (Kodešová *et al.*, 2015; Scheytt *et al.*, 2005, 2006; Xu *et al.*, 2009; Yu *et al.*, 2013). The obtained *Rd* values are important because they show that there is significant retardation of a selected pharmaceutical although the calculated $K_{OC}$ values in the present study are significantly lower than those reported in literature.

**Table 6.14** Linear sorption coefficients *Kd* (ml/g), carbon normalized sorption coefficients $K_{OC}$ (ml/g), and retardation coefficients *Rd* (–).

| Pharmaceutical | *Rd* Present Study | *Kd* (ml/g) Present Study | $K_{OC}$ (ml/g) Present Study |
|---|---|---|---|
| Trimethoprim | 7.7 | 0.64 | 2.79 |
| Carbamazepine | 2.2 | 0.18 | 0.65 |
| Diclofenac | 1.88 | 0.15 | 0.78 |
| 4-AAA | 1.88 | 0.15 | 0.65 |

The results of the present study clearly indicate that besides the total organic carbon content, another sorption mechanism plays an important role in the total sorption potential of the investigated pharmaceuticals in field experiments.

The results of field experiments carried out to determine *Kd* values, represent relatively unique data because the obtained results are closest to real values for site-specific conditions. In practice, when pharmaceutical transport model calculations are done, the *Kd* values mentioned in literature should be used carefully, because the calculation of results that pertain to transport parameters may deviate significantly from real data.

In this study, trimethoprim had the highest *Rd* and *Kd* values and carbamazepine exhibited higher *Rd* and *Kd* than diclofenac and 4-AAA, which is consistent with data given in literature. According to the results, the fastest moving pharmaceuticals at this experimental site were diclofenac and 4-AAA, while trimethoprim was the slowest. The calculated *Rd*, $K_{OC}$ and *Kd* values were lower than in other studies or in laboratory experiments, except for 4-AAA where the *Rd* was calculated for the first time in a field experiment.

The result of the field experiment showed the delay time pertaining to the breakthrough curve of a pharmaceutical versus the tracer (*Rd*), which was calculated based on the relation between the time required for the outflow concentration breakthrough curve to reach the centre of gravity of the breakthrough curve of the output concentration. The calculated values of *Kd* represent a new and very important result on the basis of which it can be concluded that the investigated pharmaceuticals are sorbed in groundwater, which is confirmed by the delay of the breakthrough curve of the tested pharmaceuticals relative to the breakthrough rate of the tracer concentration. The results confirm that there is a pronounced sorption process. The content of organic matter in groundwater is very low and almost negligible at the site of the field experiment, so it is possible to conclude that in the case of groundwater transport, the physical sorption on aquifer sediment, as well as other sorption mechanisms (chemisorption and ion exchange), have a significant effect on the total sorption potential of the tested pharmaceuticals. A field experiment provided qualitative data regarding the degree of degradation of the tested pharmaceuticals. Carbamazepine did not decompose during the field experiment; while the experiment showed that diclofenac could be degraded, and 4-AAA and trimethoprim did degrade during the field experiment. The precise degree of degradation could not be determined due to the relatively short duration of the experiment, the low concentrations and the sensitivity of the analytical method when analysing the pharmaceuticals in the groundwater samples.

## REFERENCES


Achten Ch., Kolb A. and Püttmann W. (2002). Occurrence of methyl tert-butyl ether (MTBE) in riverbank filtered water and drinking water produced by riverbank filtration. *Environmental Science and Technology*, **36**, 3662–3670.

Adeleye A. S., Conway J. R., Garner K., Huang Y., Su Y. and Keller A. A. (2016). Engineered nanomaterials for water treatment and remediation: costs, benefits, and applicability. *Chemical Engineering Journal*, **286**, 640–662. https://doi.org/10.1016/j.cej.2015.10.105

Alexander M. (1994). Biodegradation and Bioremediation. Academic Press, San Diego, California.

Alidina M., Shewchuk J. and Drewes J. E. (2015). Effect of temperature on removal of trace organic chemicals in managed aquifer recharge systems. *Chemosphere*, **122**, 23–31, doi: 10.1016/j.chemosphere.2014.10.064

Al-Rifai J. H., Khbbaz H. and Schäfer I. A. (2011). Removal of pharmaceuticals and endocrine disrupting compounds in a water recycling process using reverse osmosis systems. *Separation and Purification Technology*, **77**, 60–67, doi: 10.1016/j.seppur.2010.11.020

Antonio M., Locatelli F., Sodré F. F. and Jardim W. F. (2011). Determination of Antibiotics in Brazilian Surface Waters Using Liquid Chromatography–Electrospray Tandem Mass Spectrometry. *Archives of Environmental Contamination and Toxicology*, **60**, 385–393, doi: 10.1007/s00244-010-9550-1

Appelo C. A. J. and Postma D. (2005). Geochemistry, Groundwater and Pollution. A.A. Balkema Publishers, Lieden, Netherland, ISBN: 04-1536-421-3.

Aus der Beek T., Weber F.-A., Bergmann A., Hickmann S., Ebert I., Hein A. and Küster A. (2016). Pharmaceuticals in the environment – Global occurrences and perspectives. *Environmental Toxicology and Chemistry*, **35**, 823–835, doi: 10.1002/etc.3339

Behera S. K., Kim H. W., Oh J. E. and Park H. S. (2011). Occurrence and removal of antibiotics, hormones and several other pharmaceuticals in wastewater treatment plants of the largest industrial city of Korea. *Science of Total Environment*, **409**, 4351–4360, doi: 10.1016/j.scitotenv.2011.07.015

Bertelkamp C., Reungoat J., Cornelissen E. R., Singhal N., Reynisson J., Cabo A. J., van der Hoek J. P. and Verliefde A. R. D. (2014). Sorption and biodegradation of organic micropollutants during river bank filtration: A laboratory column study. *Water Research*, **52**, 231–241.

Boleda M. R., Galceran M. T. and Ventura F. (2011). Behavior of pharmaceuticals and drugs of abuse in a drinking water treatment plant (DWTP) using combined conventional and ultrafiltration and reverse osmosis (UF/RO) treatments. *Environmental Pollution*, **159**, 1584–1591.

Bolong N., Ismail A. F., Salim M. R. and Matsuura T. (2009). A review of the effects of emerging contaminants in wastewater and options for their removal, *Desalination* **239**, 229–246, doi.org/10.1016/j.desal.2008.03.020

Brauch H. -J., Sacher F., Denecke E. and Tacke T. (2000). Wirksamkeit der Uferfiltration für die Entfernung von polaren organischen Spurenstoffen. *gwf Wasser Abwasser*, **141**, 226–234.

Bueno M. J., Gomez M. J., Herrera S., Hernando M. D., Agüera A. and Fernández-Alba A. R. (2012). Occurrence and persistence of organic emerging contaminants and priority pollutants in five sewage treatment plants of Spain: two years pilot survey monitoring. *Environmental Pollution*, **164**, 267–273. https://doi.org/10.1016/j.envpol.2012.01.038

Bui X. T., Vo T. P. T., Ngo H. H., Guo W. S. and Nguyen T. T. (2016). Multicriteria assessment of advanced treatment technologies for micropollutants removal at large-scale applications. *Science of Total Environment*, **563–564**, 1050–1067. https://doi.org/10.1016/j.scitotenv.2016.04.191

Bulut E., Ozacar M. and Sengil I. A. (2008). Adsorption of malachite green onto bentonite: Equilibrium and kinetic studies and process design. *Microporous and Mesoporous Materials*, **115**, 234–246, doi: 10.1016/j.micromeso.2008.01.039

Burke V., Treumann S., Duennbier U., Greskowiak J. and Massmann G. (2013). Sorption behavior of 20 wastewater originated micropollutants in groundwater – Column experiments with pharmaceutical residues and industrial agents. *Journal of Contaminant Hydrology*, **154**, 29–41, doi: 10.1016/j.jconhyd.2013.08.001

Burke V., Richter D., Hass U., Duennbier U., Greskowiak J. and Massmann G. (2014a). Redox-dependent removal of 27 organic trace pollutants: compilation of results from tank aeration experiments. *Environmental Earth Sciences*, **71** (8), 3685–3695, https://doi.org/10.1007/s12665-013-2762-8

Burke V., Greskowiak J., Asmuss T., Bremermann R., Taute T. and Massmann G. (2014b). Temperature dependent redox zonation and attenuation of wastewater derived organic micropollutants in the hyporheic zone. *Science of Total Environment*, **482**, 53–61.

Burke V., Greskowiak J., Grünenbaum N. and Massmann G. (2016). Redox and temperature dependent attenuation of 20 organic micropollutants-a systematic column study. *Water Environment and Research*, **29**, 29.

Cabeza Y., Candela L., Ronen D. and Teijon G. (2012). Monitoring the occurrence of emerging contaminants in treated wastewater and groundwater between 2008 and 2010. The Baix Llobregat (Barcelona, Spain). *Journal of Hazardous Matererial*, **239–240**, 32–39, doi: 10.1016/j.jhazmat.2012.07.032

Clara M., Kreuzinger N., Strenn B., Gans O. and Kroiss H. (2005). The solids retention time – a suitable design parameter to evaluate the capacity of wastewater treatment plants to remove micropollutants. *Water Research*, **39**, 97–106, doi: 10.1016/j.watres.2004.08.036

Daneshvar A., Svanfelt J., Kronberg L., Prévost M. and Weyhenmeyer G. A. (2010). Seasonal variations in the occurrence and fate of basic and neutral pharmaceuticals in a Swedish river–lake system. *Chemosphere*, **80**, 301–309, doi: 10.1016/j.chemosphere.2010.03.060

de Jesus Gaffney V., Almeida C., Rodrigues A., Ferreira E., Benoliel M. J. and Cardoso V. V. (2015). Occurrence of pharmaceuticals in a water supply system and related human health risk assessment. *Water Research*, **72**, 199–208, doi: 10.1016/j.watres.2014.10.027

de Jongh C. M., Kooij P. J. F., de Voogt P. and ter Laak T. (2012). Screening and human health risk assessment of pharmaceuticals and their transformation products in Dutch surface waters and drinking water. *Science of the Total Environment*, **427–428**, 70–77, doi: 10.1016/j.scitotenv.2012.04.010

Delle Sita A. (2000). Factors affecting sorption of organic compounds in natural sorbent/water systems and sorption coefficients for selected pollutants. A Review. *Journal of Physical and Chemical Reference Data*, **30**(1), 187–439, doi: 10.1063/1.1347984

Derx J., Blaschke A. P., Farnleitner A. H., Pang L., Blöschl G. and Schijven J. F. (2013). Effects of fluctuations in river water level on virus removal by bank filtration and aquifer passage — A scenario analysis. *Journal of Contaminant Hydrology*, **147**, 34–44, doi: 10.1016/j.jconhyd.2013.01.001

Díaz-Cruz M. S. and Barceló D. (2008). Trace organic chemicals contamination in ground water recharge. *Chemosphere*, **72**, 333–342, doi: 10.1016/j.chemosphere.2008.02.031

Dimkić M. (1997). *Izveštaj o rezultatima atesta na izvorištu podzemnih voda 'Mediana'*. Institut za vodoprivredu 'Jaroslav Černi', Beograd, Srbija.

Dimkić M. (2007). Samoprečišćavajući efekti filtracije podzemnih voda, Monografija, Izdavač 'Zadužbina Andrejević', Beograd, Srbija, p. 252, ISBN:978 86 7244 629 6

Dimkić M. and Igrutinović D. (1987). Idejno rešenje sanacije uticaja fenola na izvorište podzemne vode u Kraljevu. Institut za vodoprivredu 'Jaroslav Černi', Beograd, Srbija.

Dimkić M. and Keckarević D. (1990). Quantification of selfpurification processes in aquifers as a phishico-biochemical process. X Symposium of Yugoslavian Society for Hidrauluical Research, October 09–13, Sarajevo, Bosna and Herzegovina, pp. 467–479.

Dimkić A. M., Brauch H.-J. and Kavanaugh M. (2008). Groundwater Management in Large river Basins, A. M. Dimkić, H.-J. Brauch and M. Kavanaugh (eds.), IWA Publishing, London, UK, p. 690. Retrieved from www.iwapublishing.com

Dimkić M., Pusić M., Vidovic D., Petković A. and Boreli-Zdravković D. (2011). Several natural indicators of radial well ageing at the Belgrade groundwater source, part 1. *Water Science and Technology*, **63**(11), 2560–2566, doi: 10.2166/wst.2011.590

Dimkić M. and Pušić M. (2014). Correlation Between Entrance Velocities, Increase in Local Hydraulic Resistances and Redox Potential of Alluvial Groundwater Sources. *Water Research and Management*, **4**(1), 3–23.

EC-European Commission, Groundwater Protection in Europe (2016). ISBN 978-92-79-09817-8. https://doi.org/10.1016/j.envpol.2011.02.051

Eggen T., Moeder M. and Arukwe A. (2012). Municipal landfill leachates: a significant source for new and emerging pollutants. *Science of Total Environment*, **408**, 5147–5157, doi: 10.1016/j.scitotenv.2010.07.049

Farré M., Perez S., Kantiani L. and Barcelo D. (2008). Fate and toxicity of emerging pollutants, their metabolites and transformation products in the aquatic environment. *Trends in Analitical Chemistry*, **27**, 991–1007, https://doi.org/10.1016/j.trac.2008.09.010

Feldmann D. F., Zuehlke S. and Heberer. (2008). Occurrence, fate and assessment of polar metamizole (dipyrone) residues in hospital and municipal wastewater. *Chemosphere*, **9**, 1754–1764, doi: 10.1016/j.chemosphere.2007.11.032

Foo K. Y. and Hameed B. H. (2010). Insights into the modeling of adsorption isotherm systems. *Chemical Engineering Journal*, **156**, 2–10, doi: 10.1016/j.cej.2009.09.013

Fram S. M. and Belitz K. (2011). Occurrence and concentrations of pharmaceutical compounds in groundwater used for public drinking-water supply in California. *Science of the Total Environment*, **409**, 3409–3417, doi: 10.1016/j.scitotenv.2011.05.053

García-Vaquero N., Lee E., Jiménez Castáneda R., Cho J. and López-Ramírez J. A. (2014). Comparison of drinking water pollutant removal using a nanofiltration pilot plant powered by renewable energy and a conventional treatment facility. *Desalination*, **347**, 94–102, https://doi.org/10.1016/j.desal.2014.05.036

Ginebreda A. I., Muñoz M. L., de Alda R., Brix J., López D. and Barceló D. (2010). Environmental risk assessment of pharmaceuticals in rivers: Relationships between hazard indexes and aquatic macroinvertebrate diversity indexes in the Llobregat River (NE Spain). *Environment International*, **36**(2), 153–162, doi: 10.1016/j.envint. 2009.10.003

Glassmeyer S. T., Kolpin D. W., Furlong E. T. and Focazio M. J. (2008). Environmental Presence and Persistence of Pharmaceuticals An Overview. In: Fate of Pharmaceuticals in the Environment and in Water Treatment Systems, D. S. Aga (ed.), Taylor & Francis Group, LLC, Boca Raton, FL, 33487–2742 ISBN 978-1-4200-5232-9.

Gleick P. H. (1996). Water resources. In: Encyclopedia of Climate and Weather, S. H. Schneider (ed.), Oxford University Press, New York, Vol. **2**, pp. 817–823.

Gómez M. J., Sirtori C., Mezcua M., Fernández-Alba A. R. and Agüera A. (2008). Photodegradation study of three dipyrone metabolites in various water systems: Identification and toxicity of their photodegradation products. *Water Research*, **42**, 2698–2706, doi: 10.1016/j.watres.2008.01.022

Gracia-Lor E., Sancho J. V., Serrano R. and Hernández F. (2012). Occurrence and removal of pharmaceuticals in wastewater treatment plants at the Spanish Mediterranean area of Valencia. *Chemosphere*, **87**, 453–462, doi: 10.1016/j.chemosphere.2011.12.025

Gracia-Lor E., Ibáñez M., Zamora T., Sancho J. V. and Hernández F. (2014). Investigation of pharmaceutical metabolites in environmental waters by LC-MS/MS. *Environmental Science and Pollution Research*, **21**, 5496–5510, doi: 10.1007/s11356–013-2464-5

Greskowiak J., Hamann E., Burke V. and Massmann G. (2017). The uncertainty of biodegradation rate constants of emerging organic compounds in soil and groundwater e A compilation of literature values for 82 substances. *Water Research*, **126**, 122–133.

Grujić S., Vasiljević T. and Lausević M (2009). Determination of multiple pharmaceutical classes in surface and ground waters by liquid chromatography-ion trap-tandem mass spectrometry. *Journal of Chromatography A*, **1216**, 4989–5000, doi: 10.1016/j.chroma.2009.04.059

Guillén D., Ginebreda A., Farre M., Darbra R. M., Petrovic M., Gros M. and Barcelo D. (2012). Prioritization of chemicals in the aquatic environment based on risk assessment: analytical, modeling and regulatory perspective. *Science of Total Environment*, **440**, 236–252, https://doi.org/10.1016/j.scitotenv.2012.06.064

Hamann E., Stuyfzand P., Greskowiak J., Timmer H. and Massmann G. (2016). The fate of organicmicropollutants during long-term/long-distance river bank filtration. Science of the Total Environment, 545–546, 629–640, doi: 10.1016/j.scitotenv.2015.12.057

Heberer T. (2002). Tracking persistent pharmaceutical residues from municipal sewage to drinking water. *Journal of Hydrology*, **266**, 175–189, doi: 10.1016/S0022-1694(02)00165-8

Heberer T., Massmann G., Fanck B., Taute T. and Dünnbier U. (2008). Behaviour and redox sensitivity of antimicrobial residues during bank filtration. *Chemosphere*, **73**, 451–460, doi: 10.1016/j.chemosphere.2008.06.056

Henzler A. F., Greskowiak J. and Massmann G. (2014). Modeling the fate of organic micropollutants during river bank filtration (Berlin, Germany). *Journal of Contaminant Hydrology*, **156**, 78–92. https://doi.org/10.1016/j.jconhyd. 2013.10.005

International Commission for the Protection of the Danube River (ICPDR). (2016). Danube Watch 1/2016 - Wastewater treatment in the Danube Basin: A successful but still unfinished story. https://www.icpdr.org/

International Commission for the Protection of the Danube River (ICPDR). (2013). Joint Danube Survey 3 (JDS3), Final Scientific Report. http://www.danubesurvey.org/jds3/results

Ivančev-Tumbas I. (2014). The fate and importance of organics in drinking water treatment: a review. *Environmental Science and Pollution Research*, **21**, 11794–11810, doi: 10.1007/s11356-014-2894-8

Jauković Z., Grujić S., Vasiljević T., Petrović S. and Laušević M. (2014). Cardiovascular drugs in environmental waters and wastewaters: method optimization and real sample analysis. *Journal of AOA C International*, **97**(4), 1167–1174, doi: 10.5740/jaoacint.12-121

Jelić A., Gros M., Ginebreda A., Cespedes-Sánchez R., Ventura F., Petrovic M. and Barcelo D. (2010). Occurrence, partition and removal of pharmaceuticals in sewage water and sludge during wastewater treatment. *Water Research*, **45**, 1165–1176, doi: 10.1016/j.watres.2010.11.010

Joss A., Keller E., Alder A. C., Göbel A., McArdell C. S. and Ternes T. A. (2005). Removal of pharmaceuticals and fragrances in biological wastewater treatment. *Water Research*, **39**, 3139–3152, doi: 10.1016/j.watres.2005.05.031

Kahle M., Buerge I. J., Müller M. D. and Poiger T. (2009). Hydrophilic anthropogenic markers for quantification of wastewater contamination in ground- and surface water. *Environmental Toxicology and Chemistry*, **28**, 2528–2536, doi: 10.1897/08-606.1

Kanakaraju D., Motti C. A., Glass B. D. and Oelgemöller M. (2014). Photolysis and $TiO_2$- catalysed degradation of diclofenac in surface and drinking water using circulatingbatch photoreactors. *Environtal Chemistry*, **11**, 51, https://doi.org/10.1071/EN13098

Kasprzyk-Hordern B., Dinsdale R. M. and Guwy A. J. (2009). The removal of pharmaceuticals, personal care products, endocrine disruptors and illicit drugs during wastewater treatment and its impact on the quality of receiving waters. *Water Research*, **43**, 363–380, doi: 10.1016/j.watres.2008.10.047

Käss W. and Behrens H., (1998). Tracing Techniques in Geohydrology. A. A. Balkema, Rotterdam, Brookfield, VT.

Kaštelan-Macan M., Ahel M., Horvat A. J. M., Jabučar D. and Jovančić P. (2007). Water resources and waste water management in Bosnia and Herzegovina, Croatia and the State Union of Serbia and Montenegro. *Water Policy*, **9**, 319–343, doi: 10.2166/wp.2007.003

Kennedy A. M., Reinert A. M., Knappe D. R. U., Ferrer I. and Summers R. S. (2015). Fulland pilot-scale GAC adsorption of organic micropollutants. *Water Research*, **68**, 238–248. https://doi.org/10.1016/j.watres.2014.10.010

Kim S. D., Cho J., Kim I. S., Vanderford B. J. and Snyder S. A. (2007). Occurrence and removal of pharmaceuticals and endocrine disruptors in South Korean surface,drinking, and waste waters. *Water Research*, **41**, 1013–1021, https://doi.org/10.1016/j.watres.2006.06.034

Kima S. D., Choa J., Kima I. S., Vanderfordb J. B. and Snyderb A. S. (2007). Occurrence and removal of pharmaceuticals and endocrine disruptors in South Korean surface, drinking, and waste water. *Water Research*, **41**, 1013–1021, doi: 10.1016/j.watres.2006.06.034

Klavarioti M., Mantzavinos D. and Kassinos D. (2009). Removal of residual pharmaceuticals from aqueous systems by advanced oxidation processes. *Environmental International*, **35**, 402–417. https://doi.org/10.1016/j.envint.2008.07.009

Kodešová R., Grabic R., Kočárek M., Klement A., Golovko O., Fér M., Nikodema A. and Jakšík O. (2015). Pharmaceuticals' sorptions relative to properties of thirteen different soils. *Science of the Total Environment*, **511**, 435–443, doi: 10.1016/j.scitotenv.2014.12.088

Kovačević S. (2017). Analiza samoprečišćavajućeg potencijala podzemnih voda za uklanjanje farmaceutika primenom metode rečne oblaske filtracije (Analysis of Groundwater Self-Purification Potential for Removal of Pharmaceuticals Applying River Bank Filtration Method), PhD thesis, University of Novi Sad, Faculty of Technical Sciences.

Kovačević S., Dimkić M., Vojinović-Miloradov M., Majkić B., Đogo M., Radonić (Jakšić) J. and Turk Sekulić M. (2013a). Occurrence and behavior of selected pharmaceuticals during bank filtration in Danube river – a three year survey in Serbia. International Conference 'Engineering for Environment Protection', 11–13.06.2013, Bratislava, Slovak Republic, pp. 151–155, ISBN: 978-80-227-3955-9.

Kovačević S., Petković A., Pušić M. and Dimkić M. (2013b). Fate of emerging pollutants in alluvial groundwater of big rivers in Serbia. UNESCO Conference on Emerging Pollutants in Water, 09–11.07.2013, Beograd, Republika Srbija, pp. 59–63, ISBN:978-86-82565-39-0.

Kovačević S., Vojinović-Miloradov M., Sremački M., Radonić (Jakšić) J., Petković A. and Dimkić M. (2013c). Natural removal of selected emerging polutants due to the river bank filtration. III International Conference 'Ecology of Urban Areas' 2013, 11.10.2013, Kaštel Ečka-Zrenjanin, Republika Srbija, pp. 238–242, ISBN 978-86-7672-210-5.

Kovačević S., Vojinović-Miloradov M., Petković A., Radonić (Jakšić) J., Sremački M. and Dimkić M. (2013d). Occurance of selected pharmaceuticals in municipial wastewater from city of Novi Sad. 14th European Meeting on Environmental Chemistry EMEC 2013, 04–07.12.2013, Budva, Republika Crna Gora, p. 86, ISBN: 978-9940-9059-1-0.

Kovačević S., Sremački M., Vojinović-Miloradov M. and Dimkić M. (2014). Selected pharmaceuticals load in urban wastewater for the city of Novi Sad. IV International Conference 'Ecology of Urban Areas 2014', 9–10 October 2014, Zrenjanin, Serbia, pp. 311–317. ISBN: 978-86-7672-237-2.

Kovačević S., Radišić M., Laušević M. and Dimkić M. (2017a). Occurrence and behavior of selected pharmaceuticals during riverbank filtration in The Republic of Serbia. *Environmental Science and Pollution Research*, **24**(2), 2075–2088, doi: 10.1007/s11356-016-7959-4

Kovačević S., Radišić M., Mitrinović D., Vojt P. and Živančev N. (2017b). Tracer test and behavior of selected pharmaceuticals. *Water Supply*, **17**(4), 1043–1052. https://doi.org/10.2166/ws.2016.202

Kruithof J. C., Kamp P. C. and Martijn B. J. (2007). UV/$H_2O_2$ treatment: a practical solution for organic contaminant control and primary disinfection, *Ozone: Science & Engineering*, **29**, 273–280. https://doi.org/10.1080/01919510701459311

Lapworth D. J., Baran N., Stuart M. E. and Ward R. S. (2012). Emerging organic contaminants in groundwater: a review of sources, fate and occurrence. *Environmental Pollution*, **163**, 287–303, doi: 10.1016/j.envpol.2011.12.034

Laws B., Dickenson E., Johnson T., Snyderc D. and Drewes J. (2011). Attenuation of contaminants of emerging concern during surface-spreading aquifer recharge. *Science of The Total Environment*, **409**(06), 1087–1094, doi: 10.1016/j.scitotenv.2010.11.021

Lekkerkerker-Teunissen K., Knol A. H., Derks J. G., Heringa M. B., Houtman C. J., Hofman-Caris C. H. M., Beerendonk E. F., Reus A., Verberk J. Q. J. C. and van Dijk J. C. (2013). Pilot plant results with three different types of UV lamps for advanced oxidation. *Ozone: Science & Engineering*, **35**, 38–48. https://doi.org/10.1080/01919512.2013.721317

Loos R., Carvalho R., António D. C., Comero S., Locoro G. and Tavazzi S. (2013). EU-wide monitoring survey on emerging polar organic contaminants inwastewater treatment plant effluents. *Water Research*, **47**, 6475–6487, doi: 10.1016/j.watres.2013.08.024

López-Serna R., Jurado A., Vázquez-Suñé E., Carrera J., Petrović M. and Barceló D. (2013). Occurrence of 95 pharmaceuticals and transformation products in urban groundwaters underlying the metropolis of Barcelona, Spain. *Environmental Pollution*, **174**, 305–315, doi: 10.1016/j.envpol.2012.11.022

Luo Y., Guo W., Hao H., Duc L., Ibney F., Zhang J. and Liang S. (2014). A review on the occurrence of micropollutants in the aquatic environment and their fate and removal during wastewater treatment. *Science of the Total Environment*, **473–474**, 619–641, doi: 10.1016/j.scitotenv.2013.12.065.

Ma Y., Li M., Wu M., Li Z. and Liu X. (2015). Occurrences and regional distributions of 20 antibiotics in water bodies during groundwater recharge. *Science of the Total Environment*, **518–519**, 498–506, doi: 10.1016/j.scitotenv.2015.02.100

Martínez Bueno M. J., Agüera A., Gómez M. J., Hernando M. D., García-Reyes J. F. and Fernández-Alba A. R. (2007). Application of liquid chromatography/quadrupole-linear ion trap mass spectrometry and time-of-flight mass spectrometry to the determination of pharmaceuticals and related contaminants in wastewater. *Analytical Chemistry*, **79**, 9372–9384, doi: 10.1021/ac0715672

Massmann G., Dünnbier U., Heberer T. and Taute T. (2008). Behaviour and redox sensitivity of pharmaceutical residues during bank filtration – Investigation of residues of phenazone-type analgesics. *Chemosphere*, **71**, 1476–1485, doi: 10.1016/j.chemosphere.2007.12.017

Mendoza A., López de Alda M., González-Alonso S., Mastroianni N., Barceló D. and Valcárcel Y. (2014). Occurrence of drugs of abuse and benzodiazepines in river waters from the Madrid Region (Central Spain). *Chemosphere*, **95**, 247–255, doi: 10.1016/j.chemosphere.2013.08.085

Miège M., Choubert J. M., Ribeiro L., Eusèbe M. and Coquery M. (2009). Fate of pharmaceuticals and personal care products in wastewater treatment plants – Conception of a database and first results. *Environmental Pollution*, **157**, 1721–1726, doi: 10.1016/j.envpol.2008.11.045

Moldovan Z. (2006). Occurrences of pharmaceutical and personal care products as micropollutants in rivers from Romania. *Chemosphere*, **64**, 1808–1817, doi: 10.1016/j.chemosphere.2006.02.003

Mompelat S., Le Bot B. and Thomas O. (2009). Occurrence and fate of pharmaceutical products and by-products, from resource to drinking water. *Environment International*, **35**, 803–814, doi: 10.1016/j.envint.2008.10.008

Murray K. E., Thomas S. M. and Bodour A. A. (2010). Prioritizing research for trace pollutants and emerging contaminants in the freshwater environment, *Environmental Pollution*, **158**, 3462–3471, doi: 10.1016/j.envpol.2010.08.009

Nakada N., Shinohara H., Murata A., Kiri K., Managaki S., Sato N. and Takada H. (2007). Removal of selected pharmaceuticals and personal care products (PPCPs) and endocrine-disrupting chemicals (EDCs) during sand filtration and ozonation at a municipal sewage treatment plant. *Water Research*, **41**, 4373–4382.

Nikolaou A., Meric S. and Fatta D. (2007). *Analytical and Bioanalytical Chemistry*, **387**, 1225–1234, doi: 10.1007/s00216-006-1035-8

NORMAN List of Emerging Substances (2016). https://www.norman-network.net/?q=node/81

OECD (2020). Pharmaceutical spending (indicator), doi: 10.1787/998febf6-en (accessed 12 March 2020).

OECD Health Statistics (2018). Eurostat Database. https://doi.org/10.1787/health-data-en

OECD Health Statistics (2019). Eurostat Database. https://doi.org/10.1787/888934018089

Pal A., Gin K. Y., Lin A. Y. and Reinhard M. (2010). Impacts of emerging organic contaminants on freshwater resources: review of recent occurrences, sources, fate and effects. *Science of Total Environment*, **408**, 6062–6069, doi: 10.1016/j.scitotenv.2010.09.026

Peng X., Ou W., Wang C., Wang Z., Huang Q., Jin J. and Tan J. (2014). Occurrence and ecological potential of pharmaceuticals and personal care products in groundwater and reservoirs in the vicinity of municipal landfills in China. *Science of the Total Environment*, **490**, 889–898, doi: 10.1016/j.scitotenv.2014.05.068

Petrović M., Škrbić B., Živančev J., Ferrando-Climent L. and Barcelo D. (2014). Determination of 81 pharmaceutical drugs by high performance liquid chromatography coupled to mass spectrometry with hybrid triple quadruple–linear ion trap in different types of water in Serbia. *Science of the Total Environment*, **468–469**, 415–428, doi: 10.1016/j.scitotenv.2013.08.079

Ptak T., Piepenbrink M. and Martac E. (2004). Tracer tests for the investigation of heterogeneous porous media and stochastic modelling of flow and transport – a review of some recent developments. *Journal of Hydrology*, **294**, 122–163, doi: 10.1016/j.jhydrol.2004.01.020

Pušić M. and Dimkić M. (2017). Comparative assessment of radial collector well elementswith a new approach. *Annales Géologiques De Lapéninsule Balkanique*, **78**, 37–45, doi: 10.2298/GABP1778037P

Qi W., Singer H., Berg M., Müller B., Pernet-Coudrier B., Liu H. and Qu J. (2015). Elimination of polar micropollutants and anthropogenic markers by wastewater treatment in Beijing, China. *Chemosphere*, **119**, 1054–1061, doi: 10.1016/j.chemosphere.2014.09.027

Radjenovic J., Petrovic M. and Barcelo D. (2009). Fate and distribution of pharmaceuticals in wastewater and sewage sludge of conventional activated sludge (CAS) and advanced membrane bioreactor (MBR) treatment. *Water Research*, **43**, 831–841, doi: 10.1016/j.watres.2008.11.043

Radović T., Grujić S., Kovačević S., Laušević M. and Dimkić M. (2016). Sorption of selected pharmaceuticals and pesticides on different river sediments. *Environmental Science and Pollution Research*, **23**(24), 25232–25244, doi: 10.1007/s11356-016-7752-4

Reinthaler F. F., Posch J., Feierl G., Wüst G., Haas D., Ruckenbauer G., Manscher F. and Marth E. (2003). Antibiotic resistance of *E. coli* in sewage and sludge. *Water Research*, **37**, 1685, doi: 10.1016/S0043-1354(02)00569-9

Rivera-Utrilla J., Sánchez-Polo M., Ferro-García M. A., Prados-Joya G. and Ocampo-Pérez R. (2013). Pharmaceuticals as emerging contaminants and their removal from water: A review. *Chemosphere*, **93**, 1268–1287, doi: 10.1016/j.chemosphere.2013.07.059

Sacher F., Lange T. F., Brauch H. J. and Blankenhorn I. (2001). Pharmaceuticals in groundwaters analytical methods and results of a monitoring program in Baden-Württemberg, Germany. *Journal of Chromatography A*, **938**, 199–210, doi: 10.1016/S0021-9673(01)01266-3

Schaider L. A., Rudel R. A., Ackerman J. M., Dunagan S. C. and Brody J. G. (2014). Pharmaceuticals, perfluorosurfactants, and other organic wastewater compounds in public drinking water wells in a shallow sand and gravel aquifer. *Science of Total Environment*, **468–469**, 384–393, doi: 10.1016/j.scitotenv.2013.08.067

Scheytt T., Mersmann P., Lindstädt R. and Heberer T. (2005). Determination of sorption coefficients of pharmaceutically active substances carbamazepine, diclofenac, and ibuprofen, in sandy sediments. *Chemosphere*, **60**, 245–253, doi: 10.1016/j.chemosphere.2004.12.042

Scheytt T., Mersmann P. and Heberer T. (2006). Mobility of pharmaceuticals carbamazepine, diclofenac, ibuprofen, and propyphenazone in miscible-displacement experiments. *Journal of Contaminant Hydrology*, **83**, 53–69, doi: 10.1016/j.jconhyd.2005.11.002

Simate G. S., Iyuke S. E., Ndlovu S., Heydenrych M. and Walubita L. F. (2012). Human health effects of residual carbon nanotubes and traditional water treatment chemicals in drinking water. *Environment International*, **39**, 38–49. https://doi.org/10.1016/j.envint.2011.09.006

Sipma J., Osuna B., Collado N., Monclús H., Ferrero G., Comas J. and Rodriguez-Roda I. (2010). Comparison of removal of pharmaceuticals in MBR and activated sludge systems. *Desalination*, **250**, 653–659, doi: 10.1016/j.desal.2009.06.073

Slichter C. S. (1899). Theoretical Investigation of the Motion of Ground Waters. US Geological Survey, Washington, Reston, Annual report, part 2, 295–384.

Sprenger C., Lorenzen G., Hülshoff I., Grützmacher G., Ronghang M. and Pekdeger A. (2011). Vulnerability of bank filtration systems to climate change. *Science of the Total Environment*, **409**, 655–663, doi: 10.1016/j.scitotenv.2010.11.002

Stamm C., Alder A. C., Fenner K. and Hollender J. (2008). Spatial and temporal patterns of pharmaceuticals in the aquatic environment: A review. *Geography Compass*, **2**(3), 920–955, doi: 10.1111/j.1749-8198.2008.00104.x

Statistical Office of Republic of Serbia. (2019). Statistical Yearbook of the Republic of Serbia, ISSN 0354-4206.

Storck F. R., Schmidt C. K., Lange F. T., Henson J. W. and Hahn K. (2012). Factors controlling micropollutant removal during riverbank filtration. *Journal of American Water Works Association*, **4**(12), 643–652, doi: 10.5942/jawwa.2012.104.0147

Stuyfzand P. J., Segers W. and van Rooijen N. (2007). Behavior of pharmaceuticals and other emerging pollutants in various artificial recharge systems in the Netherlands. ISMAR6 Proceedings.

Suárez S., Lema J. M. and Omil F. (2010). Removal of pharmaceutical and personal care products (PPCPs) under nitrifying and denitrifying conditions. *Water Research*, **44**, 3214–3224, doi: 10.1016/j.watres.2010.02.040

Sudhakaran S., Lattemann S. and Amy G. L. (2013). Appropriate drinking water treatment processes for organic micropollutants removal based on experimental and model studies – A multi-criteria analysis study. *Science of The Total Environment*, **442**, 478–488, https://doi.org/10.1016/j.scitotenv.2012.09.076

Sui Q., Huang J., Deng S., Yu G. and Fan Q. (2010). Occurrence and removal of pharmaceuticals, caffeine and DEET in wastewater treatment plants of Beijing, China. *Water Research*, **44**, 417–426, doi: 10.1016/j.watres.2009.07.010

Sukul P., Lamshöft M., Zühlke S. and Spiteller M. (2008). Sorption of sulfadiazine in soil and soil–manure systems. *Chemosphere*, **73**, 1344–1350, doi: 10.1016/j.chemosphere.2008.06.066.

Sutton D. J., Kabala Z. J., Schaad D. E. and Ruud N. C. (2000). The dipole-flow test with a tracer: a new single-borehole tracer test for aquifer characterization. *Journal of Contaminant Hydrology*, **44**, 71–101, doi: 10.1016/S0169-7722(00)00083-8

Talib A. and Randhir T. O. (2016). Managing emerging contaminants: status, impacts, and watershed-wide strategies. *Exposure and Health*, **8**, 143–158. https://doi.org/10.1007/s12403-015-0192-4

Tambosi J. L., de Sena R. F., Favier M., Gebhardt W., José H. J., Schröder H. F. and de Fátima Peralta Muniz Moreira R. (2010). Removal of pharmaceutical compounds in membrane bioreactors (MBR) applying submerged membranes. *Desalination*, **261**, 148–156, doi: 10.1016/j.desal.2010.05.014

Tamtam F., Mercierb F., Le Botb B., Eurinc J., Tuc Dinh Q., Clément M. and Chevreuil M. (2008). Occurrence and fate of antibiotics in the Seine River in various hydrological conditions. *Science of the Total Environment*, **393**, 84–95, doi: 10.1016/j.scitotenv.2007.12.009

Teijon G., Candela L., Tamoh K., Molina-Díaz A. and Fernández-Alba A. R. (2010). Occurrence of emerging contaminants, priority substances (2008/105/CE) and heavy metals in treated wastewater and groundwater at Depurbaix facility (Barcelona, Spain). *Science of the Total Environment*, **408**, 3584–3595, doi: 10.1016/j.scitotenv.2010.04.041

Teodosiu C., Gilca A. F., Barjoveanu G. and Fiore S. (2018). Emerging pollutants removal through advanced drinking water treatment: A review on processes and environmental performances assessment. *Journal of Cleaner Production*, **197**(1), 1210–1221, https://doi.org/10.1016/j.jclepro.2018.06.247

Ternes T. (1998). Occurrence of drugs in german sewage treatment plants and rivers. *Water Research*, **32**, 3245–3260, doi: 10.1016/S0043-1354(98)00099-2

Ternes T., Bonerz M. and Schmidt T. (2001). Determination of neutral pharmaceuticals in wastewater and rivers by liquid chromatography–electrospray tandem mass spectrometry. *Journal of Chromatography A*, **938**, 175–185, doi: 10.1016/S0021-9673(01)01205-5

Ternes T., Joss A. and Oehlmann J. (2015). Occurrence, fate, removal and assessment of emerging contaminants in water in the water cycle (from wastewater to drinking water). *Water Research*, **72**, 2. https://doi.org/10.1016/j.watres.2015.02.055

Terzić S., Senta I., Ahel M., Gros M., Petrović M., Barcelo D., Müller J., Knepper T., Martí I., Ventura F., Jovančić P. and Jabučar D. (2008). Occurrence and fate of emerging wastewater contaminants in Western Balkan Region. *Science of the Total Environment*, **399**, 66–77, doi: 10.1016/j.scitotenv.2008.03.003

Tran N. H., Urase T. and Thao Ta T. (2014). A Preliminary Study on the Occurrence of Pharmaceutically Active Compounds in Hospital Wastewater and Surface Water in Hanoi, Vietnam. *Clean – Soil, Air, Water*, **42**(3), 267–275, doi: 10.1002/clen.201300021

UNESCO – United Nations Educational, Scientific and Cultural Organization. (2003). Water for People, Water for Life. United Nations World Water Development Report, Part II: A look at the world's freshwater resources.

Vieno N., Tuhkanen T. and Kronberg L. (2007). Elimination of pharmaceuticals in sewage treatment plants in Finland. *Water Research*, **41**, 1001–1012, doi: 10.1016/j.watres.2006.12.017

Vulliet E. and Cren-Olivé C. (2011). Screening of pharmaceuticals and hormones at the regional scale, in surface and groundwaters intended to human consumption. *Environmental Pollution*, **159**, 2929–2934, doi: 10.1016/j.envpol.2011.04.033

Wang Y., Wang Q., Hu L., Lu G. and Li Y. (2015). Occurrence of estrogens in water, sediment and biota and their ecological risk in Northern Taihu Lake in China. *Environmental Geochemistry and Health*, **37**, 147–156, doi: 10.1007/s10653-014-9637-0

Watkinsona A. J., Murbyd E. J., Kolpine D. W. and Costanzof S. D. (2009). The occurrence of antibiotics in an urban watershed: From wastewater to drinking water. *Science of The Total Environment*, **407**, 2711–2723, doi: 10.1016/j.scitotenv.2008.11.059

Water Management Strategy for the Republic of Serbia. (2016). Government of the Republic of Serbia Ministry of Agriculture and Environmental Protection.

Weber W. J. (1972). Physicochemical Processes for Water Quality Control. John Wiley and Sons, Interscience, New York, ISBN: 978-0471924357.

Wert L. C., Snyder H. D., Lei P. and Westerhoff Y. (2008). Removal of EDCs and pharmaceuticals in drinking and reuse treatment processes.

Wiegel S., Aulinger A., Brockmeyer R., Harms H., Löffler J., Reincke H., Schmidt R., Stachel B., von Tümpling W. and Wanke A. (2004). Pharmaceuticals in the river Elbe and its tributaries. *Chemosphere*, **57**, 107–126, doi: 10.1016/j.chemosphere.2004.05.017

Wiese B., Massmann G., Jekel M., Heberer T., Dunnbier U., Orlikowski D. and Grützmacher G. (2011). Removal kinetics of organic compounds and sum parameters under field conditions for managed aquifer recharge. *Water Research*, **45**, 4939–4950.

Wolf L., Zwiener C. and Zemann M. (2012). Tracking artificial sweeteners and pharmaceuticals introduced into urban groundwater by leaking sewer networks. *Science of Total Environment*, **430**, 8–19, doi: 10.1016/j.scitotenv.2012.04.059

World Health Organization – WHO. (2012). High blood pressure - country experiences and effective interventions utilized across the European Region, WHO Regional Office for Europe, UN City, Copenhagen Ø, Denmark.

Wu C., Witter J. D., Spongberg A. L. and Czajkowski K. P. (2009). Occurrence of selected pharmaceuticals in an agricultural landscape, western Lake Erie basin. *Water Research*, **43**, 3407–3416, doi: 10.1016/j.watres.2009.05.014

Wu M., Xiang J., Que C., Chen F. and Xu G. (2015). Occurrence and fate of psychiatric pharmaceuticals in the urban water system of Shanghai, China. *Chemosphere*, **138**, 486–493, doi: 10.1016/j.chemosphere.2015.07.002

Xu J., Wu L. and Chang A. C. (2009). Degradation and adsorption of selected pharmaceuticals and personal care products (PPCPs) in agricultural soils. *Chemosphere*, **77**(10), 1299–305. http://doi.org/10.1016/j.chemosphere.2009.09.063

Yan Q., Zhang Y. X., Kang J., Gan X. M., Xu-Y P., Guo J. S. and Gao X. (2015). A preliminary study on the occurrence of pharmaceutically active compounds in the river basins and their removal in two conventional drinking water treatment plants in Chongqing, China. *Clean – Soil, Air, Water*, **43**(6), 794–803, doi: 10.1002/clen.201400039

Yang X., Flowers R. C., Weinberg H. S. and Singer P. C. (2011). Occurrence and removal of pharmaceuticals and personal care products (PPCPs) in an advanced wastewater reclamation plant. *Water Research*, **45**, 5218–5228, doi: 10.1016/j.watres.2011.07.026

Yang B., Ying G. G., Zhao J. L., Liu S., Zhou L. J. and Chen F. (2012). Removal of selected endocrine disrupting chemicals (EDCs) and pharmaceuticals and personal care products (PPCPs) during ferrate(VI) treatment of secondary wastewater effluents. *Water Research*, **46**, 2194–2204, doi: 10.1016/j.watres.2012.01.047

Yu T.-H., Lin A. Y.-C., Panchangam S. C., Hong P.-K. A., Yang P.-Y. and Lin C.-F. (2011). Biodegradation and bio-sorption of antibiotics and non-steroidal anti-inflammatory drugs using immobilized cell process. *Chemosphere*, **84**(9), 1216–22. http://doi.org/10.1016/j.chemosphere.2011.05.045

Yu Y., Liu Y. and Wu L. (2013). Sorption and degradation of pharmaceuticals and personal care products (PPCPs) in soils. *Environmental Science and Pollution Research*, **20**, 4261–4267, doi: 10.1007/s11356-012-1442-7

Zuehlke S., Duennbier U. and Heberer T. (2004). Determination of polar drug residues in sewage and surface water applying liquid chromatography-Tandem mass spectrometry. *Analytical Chemistry*, **76**, 6548–6554, doi: 10.1021/ac049324m

Zemann M., Wolf L., Pöschko A., Schmidt N., Sawarieh A., Seder N., Tiehm A., Hötzl H. and Goldscheider N. (2014). Sources and processes affecting the spatio-temporal distribution of pharmaceuticals and X-ray contrast media in the water resources of the Lower Jordan Valley, Jordan, *Science of the Total Environment*, **488–489**, 100–114, doi: 10.1016/j.scitotenv.2014.04.063

Zuehlke S., Duennbier U. and Heberer T. (2007). Investigation of the behavior and metabolism of pharmaceutical residues during purification of contaminated ground water used for drinking water supply. *Chemosphere*, **69** (11), 1673–1680, doi: 10.1016/j.chemosphere.2007.06.020

# Chapter 7

# Removal of pesticides from groundwater by filtration

*Nevena Živančev and Milan Dimkić*

## 7.1 PESTICIDES – DEFINITION AND CLASSIFICATION

Pesticides are compounds of chemical or biological origin that are used to prevent direct or indirect undesirable effects on plants or plant products, as well as for suppression or destruction of undesirable organisms. They are most commonly used in agriculture for protection of cultivated crops to increase yields and improve quality, but they are also used in forestry, veterinary medicine and municipal hygiene.

Agriculture has been relying on plant protection products for over 70 years. Before that, unwanted insects and weeds were controlled by means of crop rotation, by planting to avoid periods of the year when pests were most present, by mechanical control of weeds, etc. (Levine, 2007; Matthews, 2015). Modern agricultural production cannot be imagined without the use of pesticides, as it is estimated that without their use, the loss of major crops would be between 26% and 40% worldwide (Matthews, 2015). The positive effect of pesticide use lies in the increased production of crops and food and in the drastic reduction of insect-spread diseases.

Classification of pesticides can be based on different criteria: mode of operation, targeted organism, chemical structure, toxicity to living organisms, production process or periods of their application. Based on mode of operation, organochlorine, organophosphate and carbamate insecticides act as disruptors of the nervous system, while herbicides focus on the pathways of photosynthesis. The way pesticides affect microorganisms does not have to be the same as how they affect targeted organisms. In microorganisms, pesticides affect respiration, photosynthesis, and biosynthetic reactions, as well as cell growth and division and composition of molecules (DeLorenzo *et al.*, 2001).

The most important basic groups based on the type of targeted organisms are herbicides, insecticides and fungicides. Herbicides are used to suppress or destroy harmful weeds and other plant species while fungicides are intended for the suppression or destruction of fungi causing plant diseases, and


Editor: Milan Dimkić
doi: 10.2166/9781789060904_0397

insecticides are intended for the suppression or destruction of harmful insects. In addition to these basic groups, there are also groups of pesticides that affect specific groups of organisms: acaricides (suppression of mites), algicides (suppression of algae), bactericides (suppression of bacteria), desiccants (causing the withering and drying of plants), defoliants (causing premature leaf fall), molluscicides (suppression of mollusks, snails), nematicides (suppression of cylindrical worms), regulators of plant growth, repellents (keep harmful organisms away from cultivated plants), rodenticides (suppression of rodents), etc. (Sabljić, 2009).

Depending on their chemical composition, pesticides can be organic or inorganic. Inorganic pesticides contain arsenic, copper, mercury, sulfur, zinc, etc. (Uneke, 2007). Organic pesticides can be organochlorine, organophosphorus, pyrethroids, carbamates, thiocarbamates, phenoxy compounds, benzimidazoles, triazines etc. (Laws, 2018; Sabljić, 2009). Organochlorine pesticides persist in the environment for long periods of time and are bioaccumulated, which is why they have been replaced by organophosphorus and carbamate pesticides, which remain in the environment for a shorter period of time because they degrade to a greater degree.

Classification of pesticides according to the World Health Organization (WHO, 2019) is done based on how hazardous they are for living organisms. This is expressed in lethal dosages, i.e. the amount of the substance that results in a lethal outcome in 50% of the studied population ($LD_{50}$) in mg of substance per kg of body weight (Table 7.1).

## 7.2 PRODUCTION AND USE IN THE WORLD

Agriculture accounts for the largest share of the total use of pesticides. Approximately 90% of all the pesticides used (Stenersen, 2004) are used in agriculture to achieve higher yields and improve the quality of cultivated crops. Due to population growth worldwide, the need for larger quantities of food is ever increasing, thus increasing the annual consumption of pesticides all over the world.

About 4 million tons of pesticides per year are used worldwide (Ortiz-Hernández *et al.*, 2013; Pretty & Bharuch, 2015; Zhang, 2018). According to the Food and Agriculture Organization (FAO), the largest amount of pesticides is used in Asia – as much as 53% of the total world quantities, about 14% are used in Europe and about 30% in the USA (Figures 7.1 and 7.2) (FAO, 2019).

Figure 7.3 shows the total amount of all pesticides based on FAO data for 2017, while Figure 7.4 shows the 10 countries in the world that use the largest amount of pesticides, based on data for the period from 1990 to 2017 (FAO, 2019).

**Table 7.1** Classification of pesticides by the World Health Organization (WHO, 2019).

| The Class | | $LD_{50}$ for Rats (mg kg$^{-1}$ body weight) | |
|---|---|---|---|
| | | Oral | Dermal |
| Ia | Extremely hazardous | <5 | <50 |
| Ib | Highly hazardous | 5–50 | 50–200 |
| II | Moderately hazardous | 50–2000 | 200–2000 |
| III | Slightly hazardous | >2000 | >2000 |
| U | Unlikely to present acute hazard | ≥5000 | |

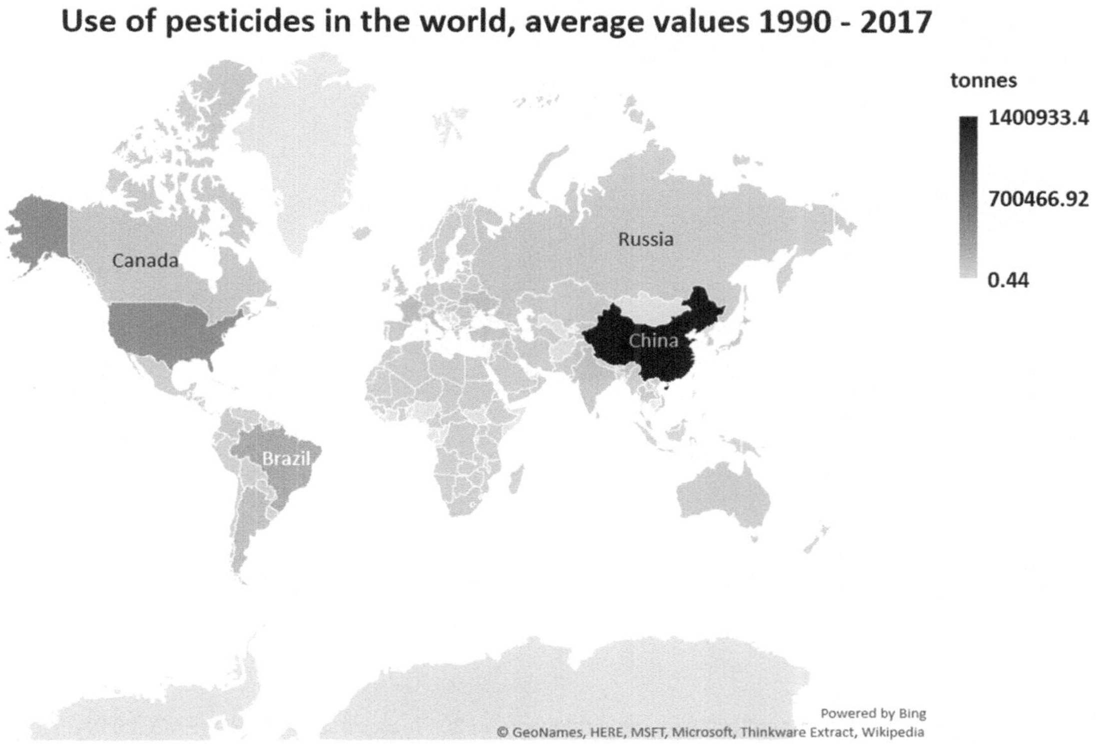


**Figure 7.1** Average annual global use of pesticides between 1990 and 2017, expressed in tons (FAO, 2019).

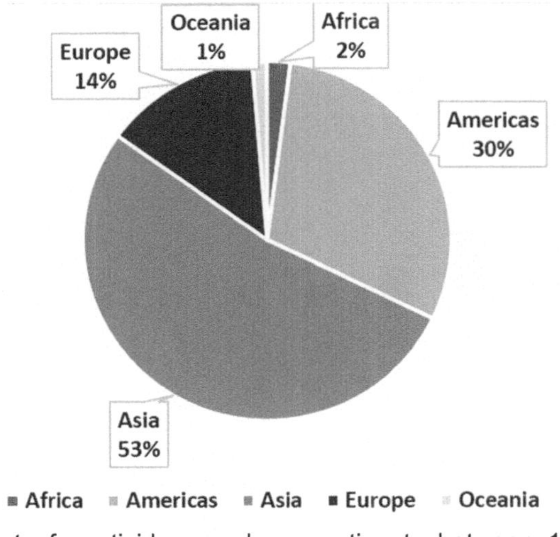


**Figure 7.2** The total amount of pesticides used on continents between 1990–2017, expressed as a percentage (FAO, 2019).

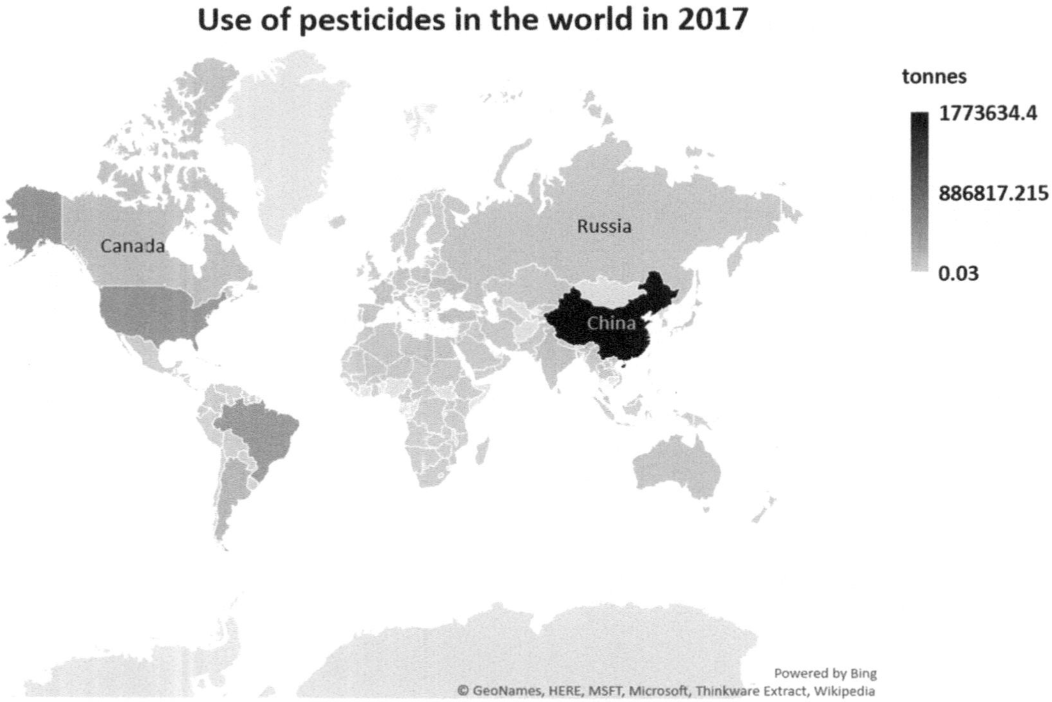


**Figure 7.3** Total quantities of pesticides in use worldwide, in 2017, expressed in tons (FAO, 2019).

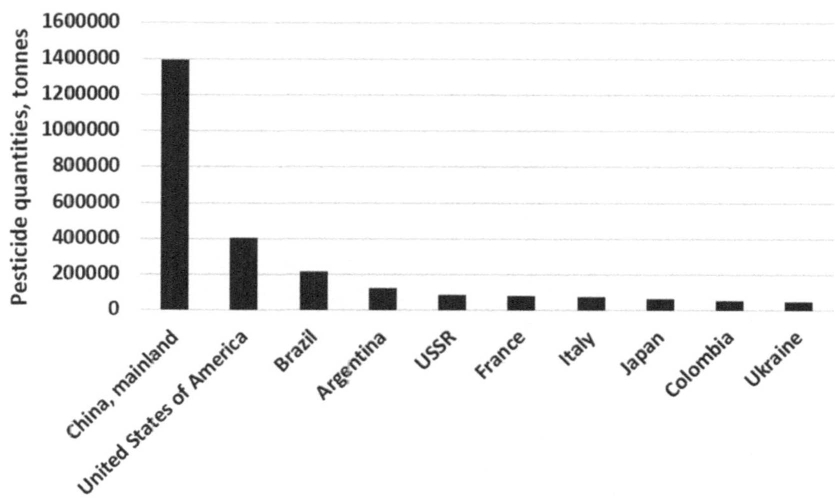


**Figure 7.4** The 10 countries using the largest amount of pesticides in the world, average values for the 1990–2017 period (FAO, 2019).

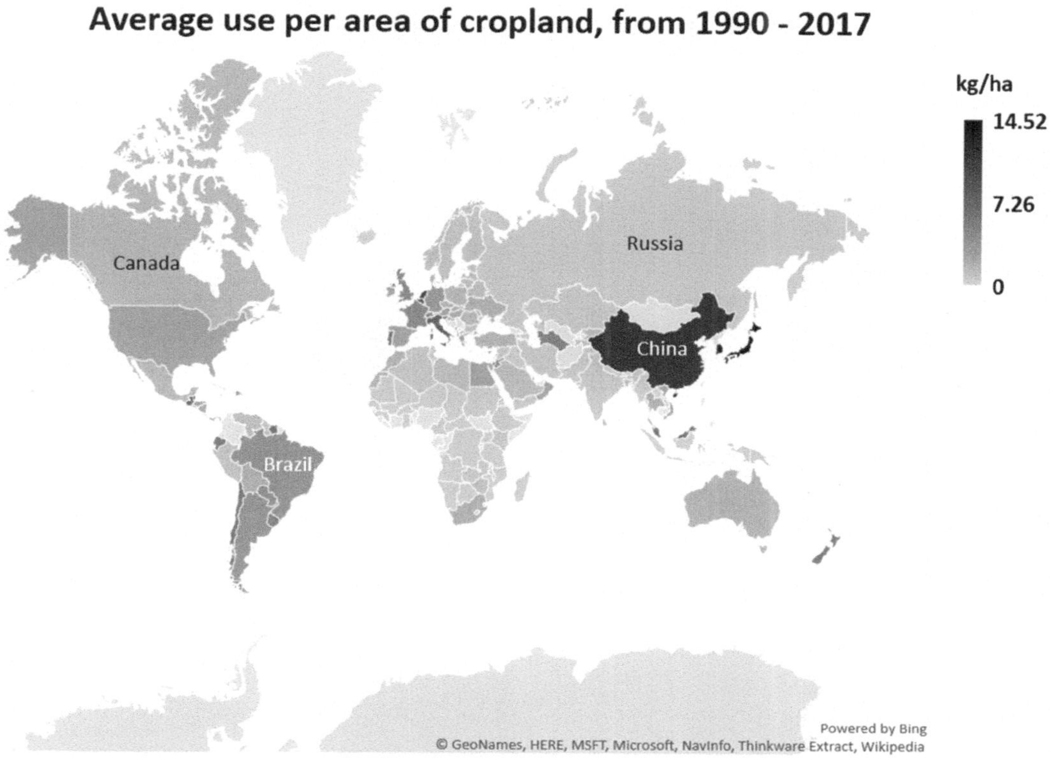


**Figure 7.5** Average use of pesticides per hectare of arable land in the 1990–2017 period (FAO, 2019).

According to FAO (2019), in Figure 7.5 is a map showing the average use of pesticides per hectare of arable land in different countries of the world from 1990 to 2017, while Figure 7.6 shows the data for 2017 only.

Figure 7.7 shows the use of pesticides per hectare of arable land with a focus on European countries. Official data for some countries, such as Serbia, do not exist. The estimated use of pesticides in Serbia is 4 kg/ha (Pavlović *et al.*, 2017).

## 7.2.1 Production and trading

Based on the analysis of data on the quantities of pesticides imported to and exported from individual countries in the world, as well as based on the financial analysis for the 1990 to 2017 period, it can be seen that the largest amount of pesticides were imported by Europe (about 48%), North and South America (25%) and Asia (19%) (Figure 7.8).

Based on FAO data (FAO, 2019) Figure 7.9 shows import of pesticides in different countries of the world in 2017, expressed in thousands of US dollars.

The following countries imported the largest amount of all pesticides in 2017: Brazil (US $ 2,468,937,201), France (US $ 1,976,378,775), Canada (US $ 1,822,916,937), Germany (US $ 1,622,663,501), the US (US $ 1,258,938,110) India (US $ 1,255,574,003), Vietnam (US $ 1,021,278,912), Spain (US $ 947,763,362), the United Kingdom (US $ 935,821,365) (FAO, 2019).

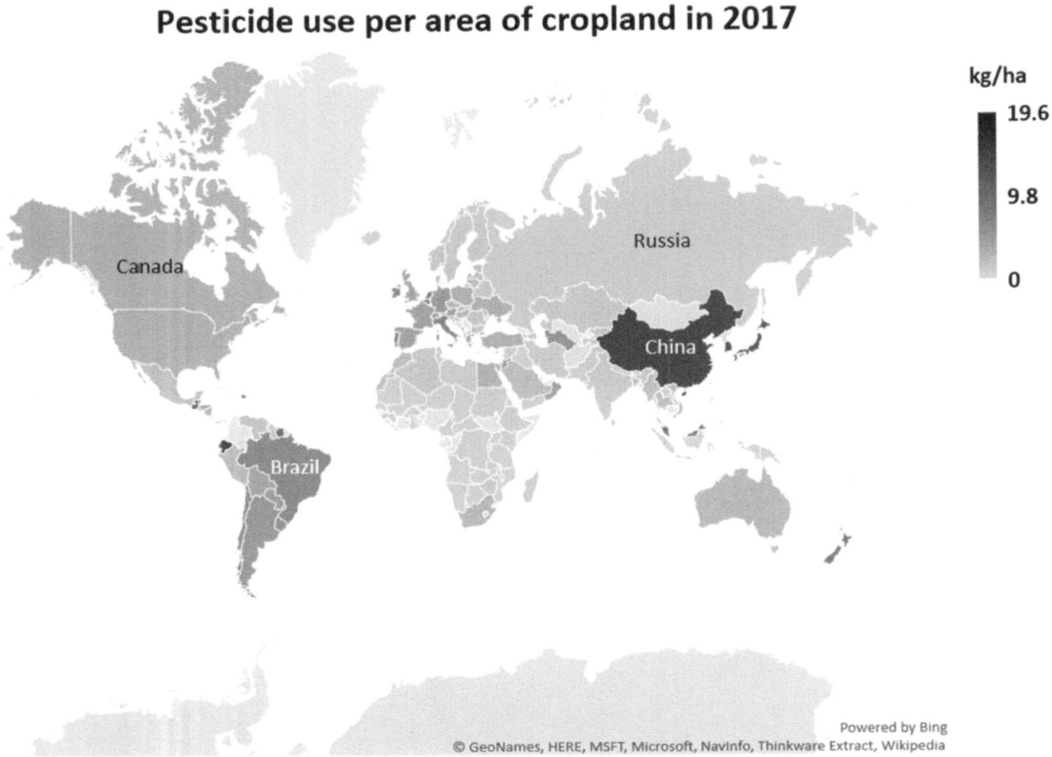


**Figure 7.6** The total amount of pesticides used per hectare of arable land worldwide, data from 2017 (FAO, 2019).

Based on average values for the 1990 to 2017 period, European countries were the largest exporters in the world (58%), followed by countries in Asia (21%) and countries in North and South America (19%) (Figure 7.10).

Figure 7.11 shows the export of pesticides in the world in 2017, expressed in thousands of US dollars. The following countries were the largest exporters of pesticides in the world in 2017: China (US $ 4,762,568,309), Germany (US $ 4,206,702,033), USA (US $ 3,870,910,200), France (US $ 3,592,166,765), India (US $ 2,436,399,062), Belgium (US $ 2,026,466,592), Israel (US $ 1,313,902,000), Spain (US $ 1,300,453,246), the United Kingdom (US $ 1,280,141,485) (FAO, 2019).

## 7.2.2 Health and environmental risks

Pesticides can have certain adverse health effects on humans and other beneficial organisms due to their good solubility in fats and lipids which causes bioaccumulation in organisms that are not the targeted group of a specific pesticide. The dominant sources of pollution come from the direct use for agricultural crops, from any uncontrolled loss during transport or production, from accidental spillage of sometimes large quantities, and from inadequately discarded packaging of the products (Agrawal *et al.*, 2010).

The use of pesticides affects the quality of the environment, but also public health when introduced in the food chain. The effects of pesticides include (Ortiz-Hernández *et al.*, 2013):

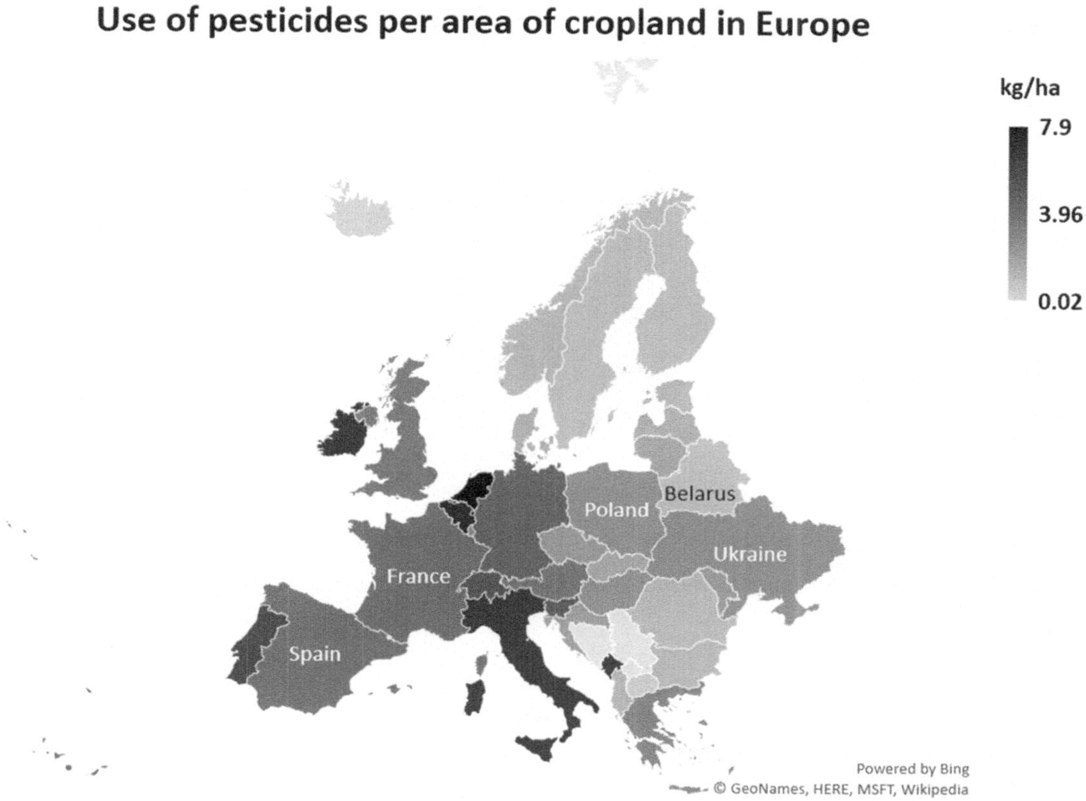


**Figure 7.7** Use of pesticides per hectare of arable land in European countries, 2017 data (FAO, 2019).

- Reduction of the phytoplankton and zooplankton population (which is an indication of water pollution),
- Carcinogenic, neurotoxic and teratogenic effects in invertebrates, fish, amphibians, insects and mammals,
- Development of resistance in the targeted organisms,
- Adverse effect on beneficial populations (e.g. reduction of beneficial insect populations),
- Reduction of macro and microbiota by affecting the biogeochemical cycle,
- Contamination of water bodies due to leaching through soil profile.

Pesticides affect microorganisms in aquatic systems through accidental discharges, runoff or by spray drift. Pesticide toxicity can have a detrimental effect on the structure itself and on the function of microorganism communities. Microorganisms can metabolize or bioaccumulate pesticides. The mechanisms of toxicity vary, depending on the pesticide itself and the type of microorganism exposed to the pesticide. Overall, herbicides are the most toxic to phototrophic microorganisms where photosynthesis is disrupted. Organophosphates and organochlorine insecticides do have a toxic effect on microorganisms, but it is not sufficiently clear what the mechanism of toxicity is in such species that are not targeted by the pesticides. It should be highlighted that just one pesticide can have wide range of toxic effects on different types of microorganisms. When attempting to predict the toxicity of pesticides in aquatic

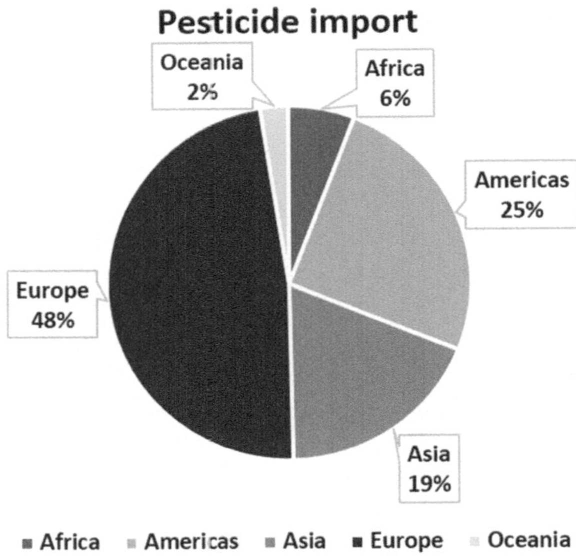


**Figure 7.8** Total quantity of imported pesticides by continent during the 1990 to 2017 period, expressed as a percentage (FAO, 2019).

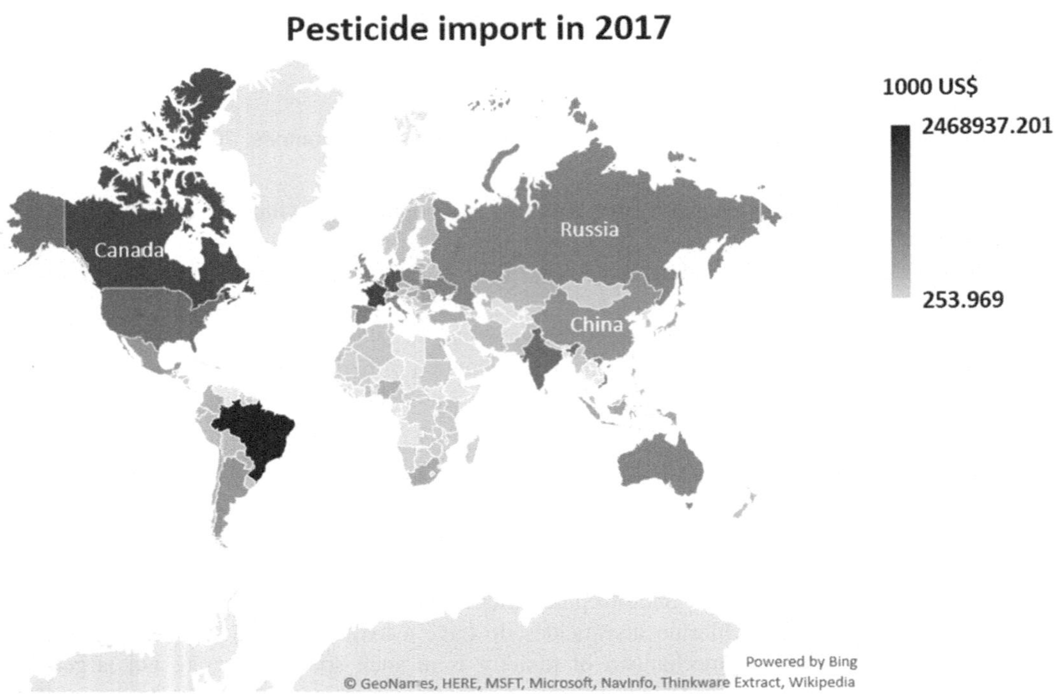


**Figure 7.9** Map showing the pesticides imported in 2017, presented in thousands of US dollars (FAO, 2019).

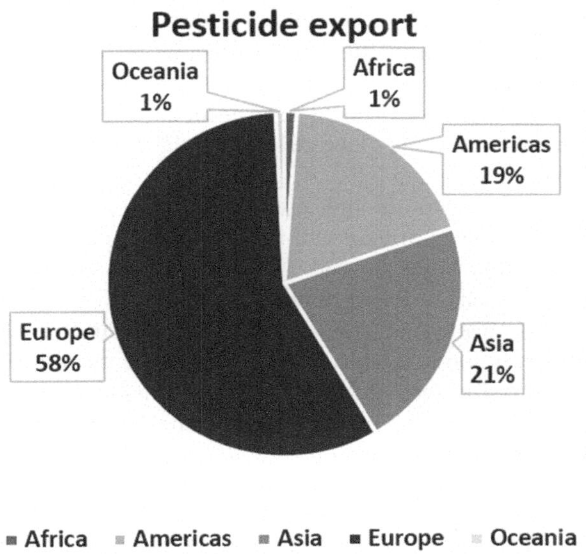


**Figure 7.10** Export of total quantities of pesticides worldwide from 1990 to 2017, expressed as a percentage (FAO, 2019).

**Figure 7.11** Pesticide exports in the world in 2017, expressed in thousands of US dollars (FAO, 2019).

ecosystems, it is also important to consider a mix of pesticides and their interaction with nutrients. The toxicity of pesticides for aquatic microorganisms, and especially for bacteria and protozoa needs further investigation (DeLorenzo *et al.*, 2001).

The effect pesticides have on human health could be acute intoxication (especially as a result of exposure due to occupational use) or indirect exposure of the general population via air, water and food contaminated with trace amounts of pesticide.

Human exposure to pesticides occurs through use in agriculture, use at the workplace, but also in households. One of the possible routes of exposure is also through pesticide contaminated food which implies transport of pesticides from water, air and soil through the food chain to flora and fauna (Kim *et al.*, 2017). Traces of pesticides can be found in our daily food and drink. Pesticides can appear in the food chain by directly reaching a plant or animal, but also indirectly through plants that absorb the pesticides through roots and parts above the ground and then move them to the edible part of the plant (Hamilton & Crossley, 2004). By feeding animals with contaminated plants, it is possible for contaminants to be present in meat or milk. To what extent pesticides are removed from food depends on the treatment preceding consumption (Kaushik *et al.*, 2009; Lehmann *et al.*, 2017; Nicolopoulou-Stamati *et al.*, 2016). Most types of treatment such as washing in water and various chemical solutions, freezing, and peeling or boiling can significantly reduce the amount of pesticide residues in food (Kaushik *et al.*, 2009). The removal of pesticide residues from fruits and vegetables by washing and peeling greatly depends on the type of pesticide, but most pesticides cannot be completely removed (Lehmann *et al.*, 2017; Nicolopoulou-Stamati *et al.*, 2016).

In most cases, the concentrations of pesticides in food and beverages are not above the permissible limits, but the effects of the permissible concentrations are often underestimated because they do not take into account the effects of multiple chemical compounds that have a synergistic effect, which is very common in real conditions (Nicolopoulou-Stamati *et al.*, 2016).

Pesticides enter the human body via the skin, via ingestion, through ocular exposure and inhalation. In the human or animal organism, pesticides can be metabolized, excreted, or bioaccumulated in adipose tissue (Nicolopoulou-Stamati *et al.*, 2016).

Many studies have linked various types of adverse health effects such as dermatological, gastrointestinal, neurological, carcinogenic, respiratory, reproductive, and endocrine effects to pesticide use (Nicolopoulou-Stamati *et al.*, 2016). The risk to human health depends on the toxicity of the active ingredients as well as the level of exposure. For certain categories of people such as children, pregnant women and the elderly, the health risk is higher (Kim *et al.*, 2017). In addition to the active ingredients, most commercial pesticides contain other chemicals that primarily affect the efficacy of the product itself. The impact of such commercial pesticides on the environment and health of living organisms may be due not only to the active ingredients, but also to additives, or to the combined effect of all the substances (Alavanja, 2009).

### 7.2.3 Occurrence of pesticides in the environment

Pesticides are designed to help humans, but in addition to their beneficial effects, adverse environmental effects are also evident. When applied to target organisms, a pesticide may endanger non-target organisms, by reaching different environmental media, threatening the growth and development of plant species or other organisms. Persistent pesticides can last and contaminate for many years after the compound has been used because they accumulate in the soil, and thus can be transmitted through the plant roots to crops for a future season. They can have a potential phytotoxic effect, a negative impact on the organisms in the soil itself and can also affect the quality and fertility of the soil and reach the groundwater (Pavlović *et al.*, 2017). Transport processes, such as runoff from the soil surface, leaching through the soil profile to groundwater, evaporation, and transformation processes (Hamilton & Crossley, 2004) are processes that enable the spread of pesticides and contamination of the wider environment. The most significant processes that influence the fate of pesticides in the environment are processes that take place in the soil: (ad)sorption, degradation and transport. The mobility of pesticides depends to a large extent on the ability of a pesticide to remain as a solute in water as the dominant medium for the transport of pesticides in the environment, as well as on the ability of a pesticide to sorb on solid soil particles (Funari & Vighi, 1995). Evaporation is a process that has a lesser impact on environmental pollution at the site where the pesticide was used, but falls within the broader dispersion mechanism that causes the pollution of remote locations relative to the location where the pesticide was used (Hamilton & Crossley, 2004).

Table 7.2 shows occurrence of the 10 most commonly used classes of pesticides in the environment, with a focus on water bodies.

A review paper by Sousa *et al.* (2018) presents global research done regarding the monitoring of concentrations of priority substances under the EU Directive 2013/39 and the EU Watch List 2015/495 (EU – Decision on water bodies from 2012 to 2017). Pesticides, namely atrazine, simazine, chlorpyrifos, terbutryn, diuron and alachlor, are the most commonly studied priority substances of the Water Framework Directive globally. The highest detected concentrations of pesticides in water bodies were for atrazine, alachlor, trifluralin, imidacloprid, dimethoate and carbofuran (Sousa *et al.*, 2018).

The International Commission for the Protection of the Danube River (ICPDR) carried out research on the presence of pesticides and other pollutants in the river Danube along its entire length through the Joint Danube Survey (JDS). The pesticides that were studied in the river Danube and its tributaries were atrazine, acetamiprid and carbendazim. As part of the first phase of the survey (JDS1) conducted in 2001, atrazine was detected in the Republic of Serbia in the Danube and in its tributaries (Figure 7.12). The concentrations of atrazine in Serbia were higher than those in countries located upstream of Serbia, but also higher than concentrations in countries located downstream. In the second phase of the research (JDS2) conducted in 2007, atrazine was also detected in the Danube River and its tributaries (Figure 7.13). The concentration of atrazine was much lower in the second phase of the study, leading to the conclusion that the ban on atrazine in EU countries in 2004 led to much smaller amounts being detected in the Danube. In the third phase of the research (JDS3) that took place in 2013, atrazine was detected in very low concentrations in the Danube River at 429 river km (Arges, 2 ng $L^{-1}$), at 132 river km (Reni, 2.2 ng $L^{-1}$) and at 31 river km (Sulina, 1.3 ng $L^{-1}$) (ICPDR – Joint Danube Survey report).

During the third phase of the survey (JDS3) in 2013, carbendazim and acetamiprid concentrations were monitored along the entire length of the Danube. Figure 7.14 shows the detected carbendazim concentration in the entire Danube. Detected concentrations of carbendazim in Serbia were below 10 ng $L^{-1}$, which was higher than the concentrations detected in the countries upstream of Serbia, but lower than detected concentrations downstream of Serbia.

Figure 7.15 shows detected concentrations of acetamiprid along the entire course of the river Danube. Acetamiprid was detected in extremely low concentration in the surface water of the Danube, with the concentration in Serbia below 1 ng $L^{-1}$.

A very significant study, also conducted by the ICPDR, called the 'Transnational Monitoring Network' (TNMN), has been conducted since 1996 and monitors the quality of larger rivers in the Danube river basin, based on the concentrations of pollutants in surface water. Figure 7.16 shows the concentrations of atrazine detected in the Danube in the Republic of Serbia in 2016. The maximum detected concentration was 10 ng $L^{-1}$.

As part of the TNMN program, the concentration of atrazine in the river Velika Morava was also monitored, and Figure 7.17 shows the results of the analyses for the 2012 to 2016 period. In 2016, the maximum concentration in August was 9 ng $L^{-1}$, while atrazine concentrations in the other months were generally below 5 ng $L^{-1}$.

**Table 7.2** Occurrence of the most commonly used classes of pesticides in surface and groundwater (Fenner *et al.*, 2013).

| Class of Pesticides | Principal Representative Aactive Substance and Chemical Structure | Pesticide Category by Type | Occurrence in the Environment |
|---|---|---|---|
| Amides | 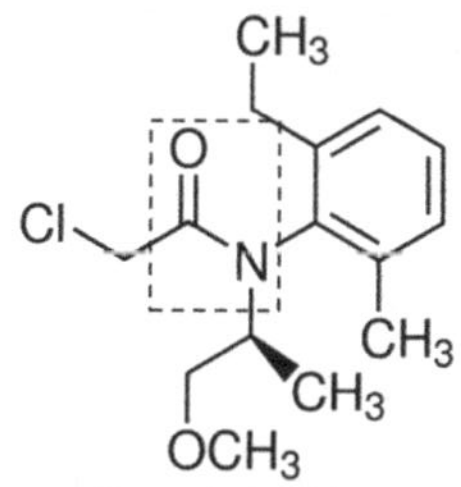 S-metolachlor | Herbicides | Chloroacetanilides and the products of transformation, oxanilic and ethanesulfonic acid, are frequently detected in groundwater; metolachlor and alachlor are detected in distant lake waters |
| Bipyridyls | 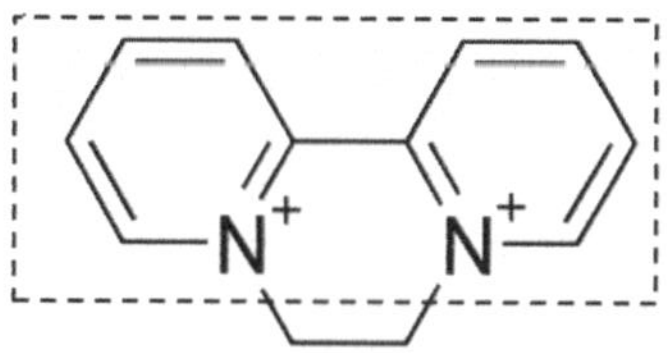 Diquat | Herbicides | Rare occurrence because they are sorbed on sediment material or on soil |
| Carbamates | 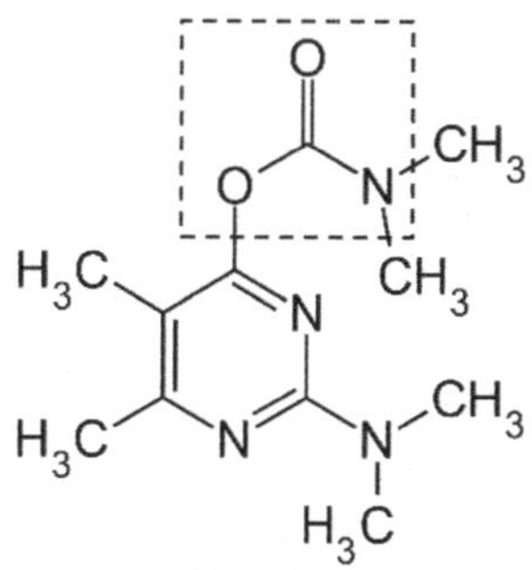 Primicarb | Insecticides/Herbicides | Rare occurrence |

| | | | |
|---|---|---|---|
| Dithiocarbamates | 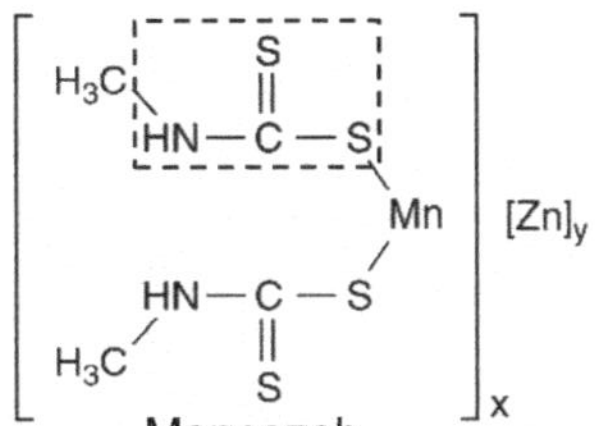 Mancozeb | Fungicides | Rare occurrence |
| Organophosphates | 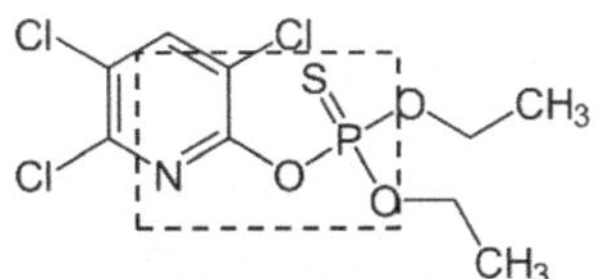 Chlorpyrifos | Insecticides | Glyphosate and aminomethylphosphonic acid are often detected in groundwater; chlorpyrifos, diazinon and disulfuton are detected in rainwater and distant lake waters |
| Phenoxy alkanoic acids | 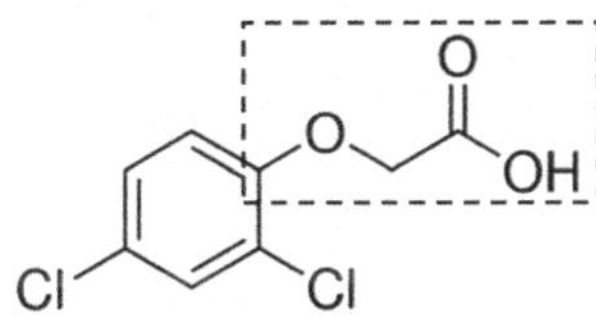 2,4 - D | Herbicides | Parent compounds are often detected in groundwater |
| Pyrethroids | 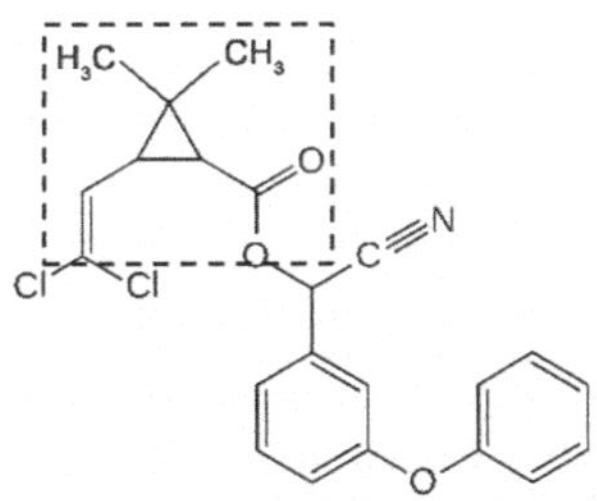 Cypermethrin | Insecticides | Rare occurrence because they are sorbed on sediment or soil materials |

*(Continued)*

**Table 7.2** Occurrence of the most commonly used classes of pesticides in surface and groundwater (Fenner *et al.*, 2013) (*Continued*).

| Class of Pesticides | Principal Representative Aactive Substance and Chemical Structure | Pesticide Category by Type | Occurrence in the Environment |
|---|---|---|---|
| Triazines | 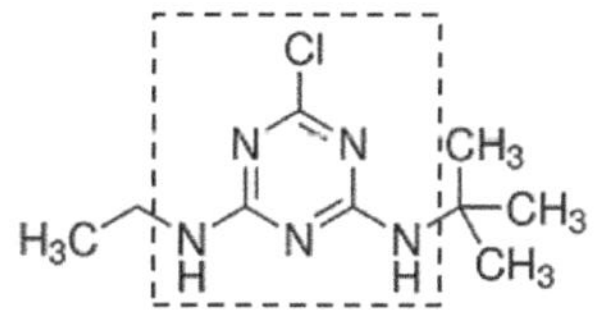 Terbuthylazine | Herbicides | Parent compounds and hydroxy- and dealkylated transformation products are frequently detected in groundwater; atrazine and deethylatrazine are also detected in distant lake waters |
| Triazoles, diazoles | 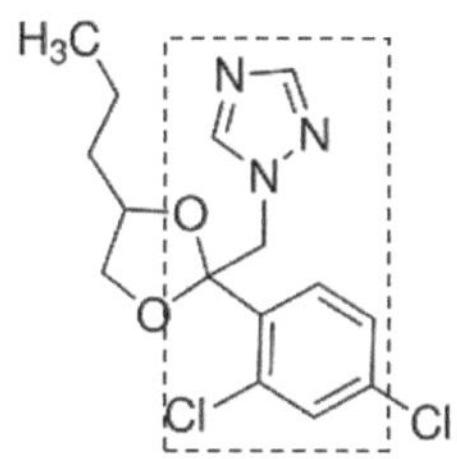 Propiconazole | Fungicides | Flutriafol is detected in remote lake waters |
| Urea derivatives | 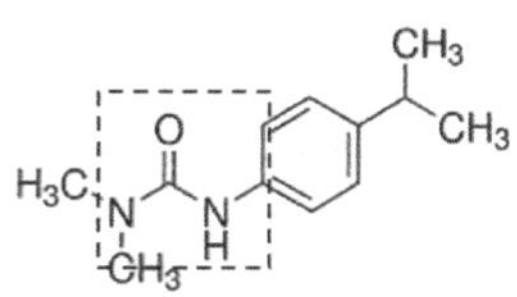 Isoproturon | Herbicides | Parent compounds are frequently detected in groundwater |

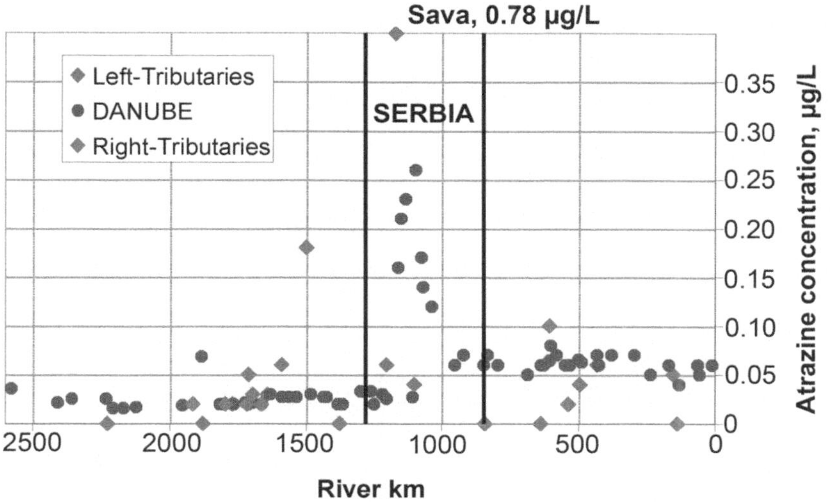


**Figure 7.12** Concentrations of atrazine along the whole course of the river Danube, as part of the first phase of the Joint Danube Survey – JDS1 (ICPDR – Joint Danube Survey report).

## 7.3 MOBILITY OF PESTICIDES IN THE ARABLE GROUND LAYER AND IN AQUIFER LAYERS

In the overall fate of pesticides, the role of soil is to act as a filter, and as a potential degradation zone as pollutants are retained in the soil by active adsorption centers (Tariq *et al.*, 2007). The transport of pesticides below the soil surface takes place both in the unsaturated and saturated zones. The unsaturated

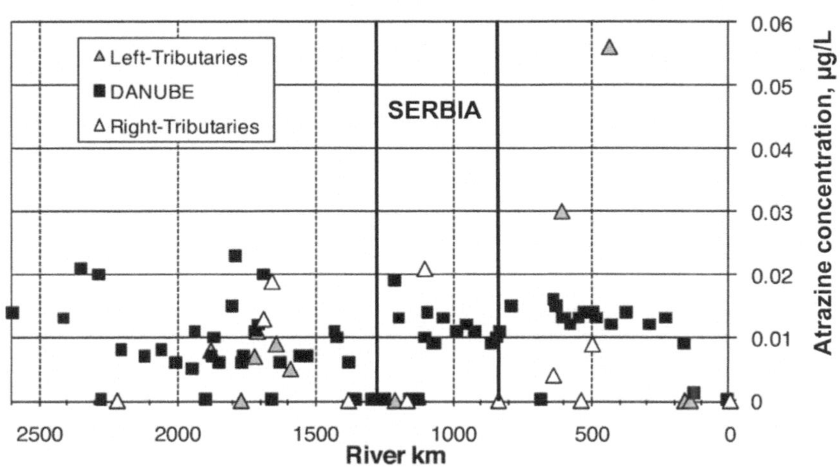


**Figure 7.13** Detected concentrations of atrazine in the river Danube and its tributaries as part of the second phase of the Joint Danube Survey – JDS2 (ICPDR – Joint Danube Survey report).

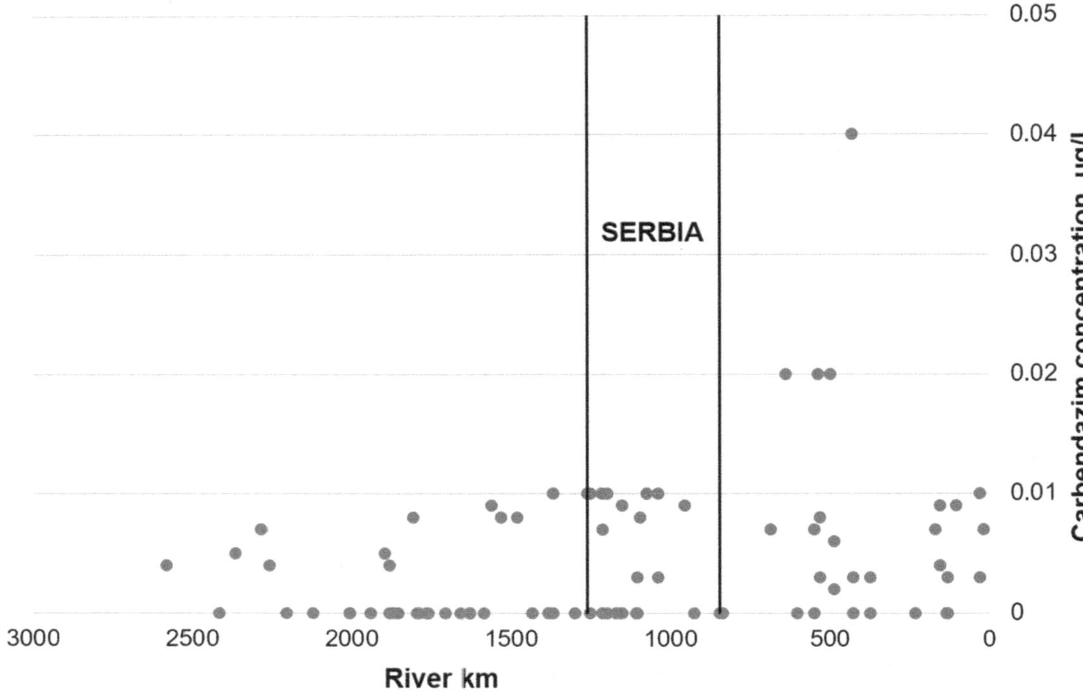


**Figure 7.14** Detected carbendazim concentrations in the Danube River as part of the JDS3 survey (ICPDR – Danube River Water Quality Database).

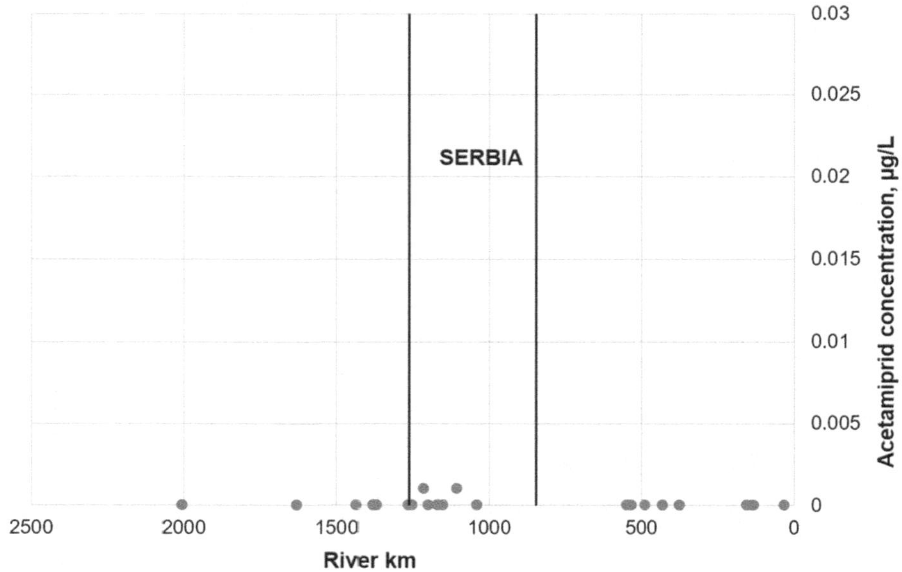


**Figure 7.15** Detected concentrations of acetamiprid in the Danube as part of the third phase of the Joint Danube Survey – JDS3 (ICPDR – Danube River Water Quality Database).

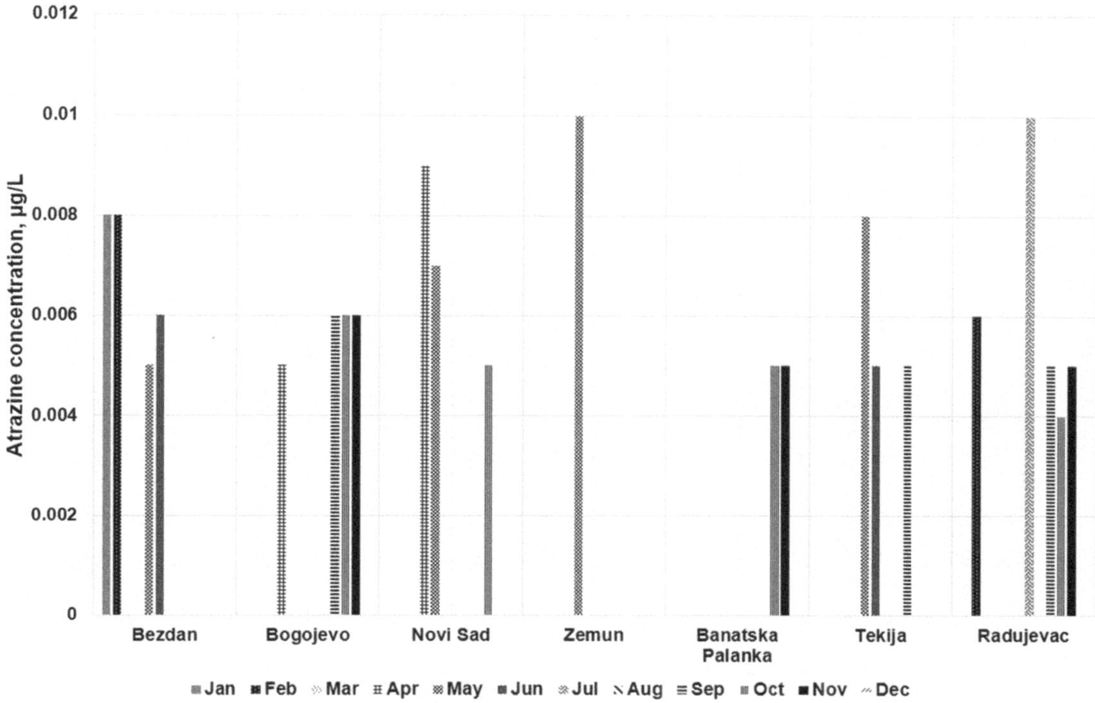


**Figure 7.16** Detected concentrations of atrazine along the entire length of the Danube in Serbia, during 2016 by month (ICPDR – Danube River Water Quality Database).

(vadose) zone represents the layer of soil that lies between the surface of the soil and the area where the pores are completely saturated with water. This zone is a significant part of the land matrix and is studied to determine its influence on the changed quality of the underground environment as a consequence of agricultural, industrial and municipal activities. The unsaturated zone, in which the pores are partly saturated with water and partly with air, is a very significant part of the soil because it serves as a medium for storage, with direct access to the biosphere. The unsaturated zone is important for controlling the transport of pollutants and for the protection of groundwater as this is the zone where various physical and chemical processes which can isolate and slow the exchange of contaminants with other parts of the environment take place (Saifadeen & Gladnyeva, 2012). Agrochemicals used on the soil surface potentially leave the root zone of the plant in the soil and may affect groundwater quality (Figure 7.18).

Above the groundwater level in the unsaturated zone, transport of pesticides predominantly occurs under the influence of three processes:

- The rapid preferential transport through macropores which depends on the soil moisture and retention capacity of the unsaturated zone and the rate of drainage,
- The exchange of pesticides by diffusion between macropores, and
- The slow movement through the soil layer.

The movement rate of pesticides depends very much on the vertical hydraulic gradient.

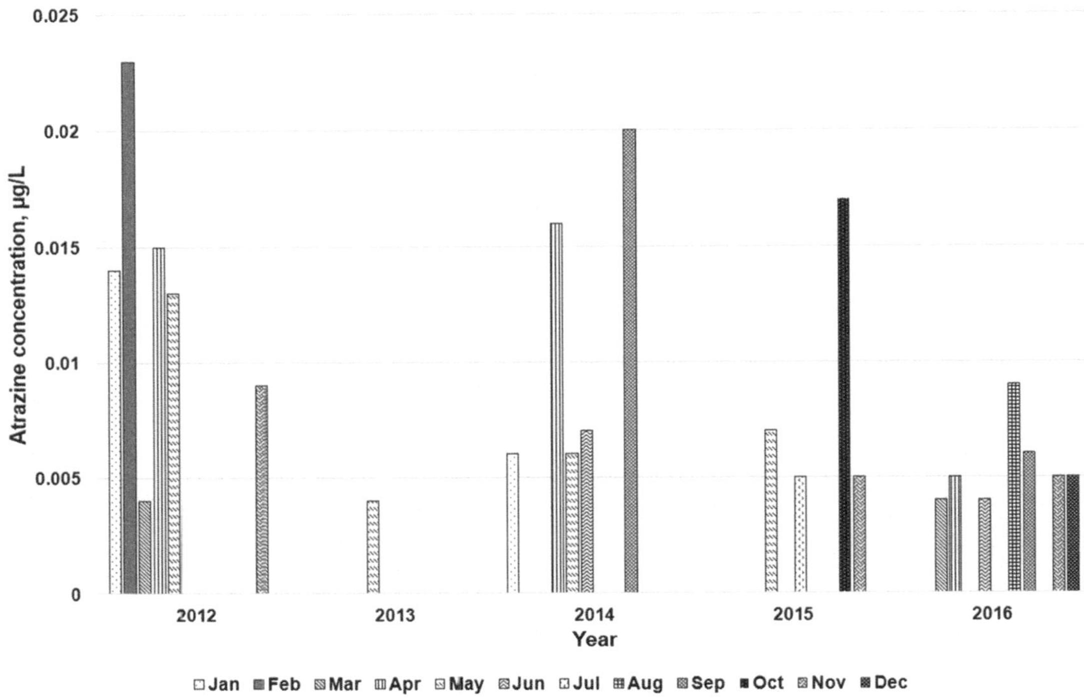


**Figure 7.17** Detected atrazine concentrations in the Velika Morava, at Ljubičevski bridge from 2012 to 2016, by month (ICPDR – Danube River Water Quality Database).

The behavior of pesticides and their fate in groundwater can be observed after a prior simplification of assumptions regarding pesticide sources, transport routes and receptors. The fate and transport of pesticides in groundwater depends on a number of processes, including advection, dilution, dispersion, diffusion, sorption, degradation (microbiological and abiotic), and evaporation, which all depend on the hydrogeological conditions of the environment: the physicochemical nature of the aquifers, the hydrological conditions and saturation with water (Bloomfield *et al.*, 2006).

The advective transport of pesticides in aquifers takes place under the influence of a hydraulic gradient, and the transport rate depends on the filtration coefficient of the aquifer and on porosity. Dispersion reduces the concentration of pesticides by mechanical mixing or by dilution during the movement of groundwater. Diffusion results in the movement of pesticides from areas of higher concentration to areas of lower concentration. It is difficult to separate the processes of dispersion and diffusion and they are viewed together as hydrodynamic dispersion. Pesticide concentrations can be reduced by evaporation in the unsaturated zone, or through degradation and sorption. Evaporation mostly occurs within the first 24 h after application (Hantush *et al.*, 2000).

The water which feeds groundwater can originate from water that gets to the soil surface by natural means in the form of precipitation, but also by means of artificial irrigation, or runoff and leaching from the soil surface or from streams and lakes. Infiltration of precipitation or irrigation water into groundwater is possible if the amount of water is greater than the retention capacity of the soil (the maximum amount of water that the soil can absorb). The infiltrated quantities of water can be stopped by vegetation and

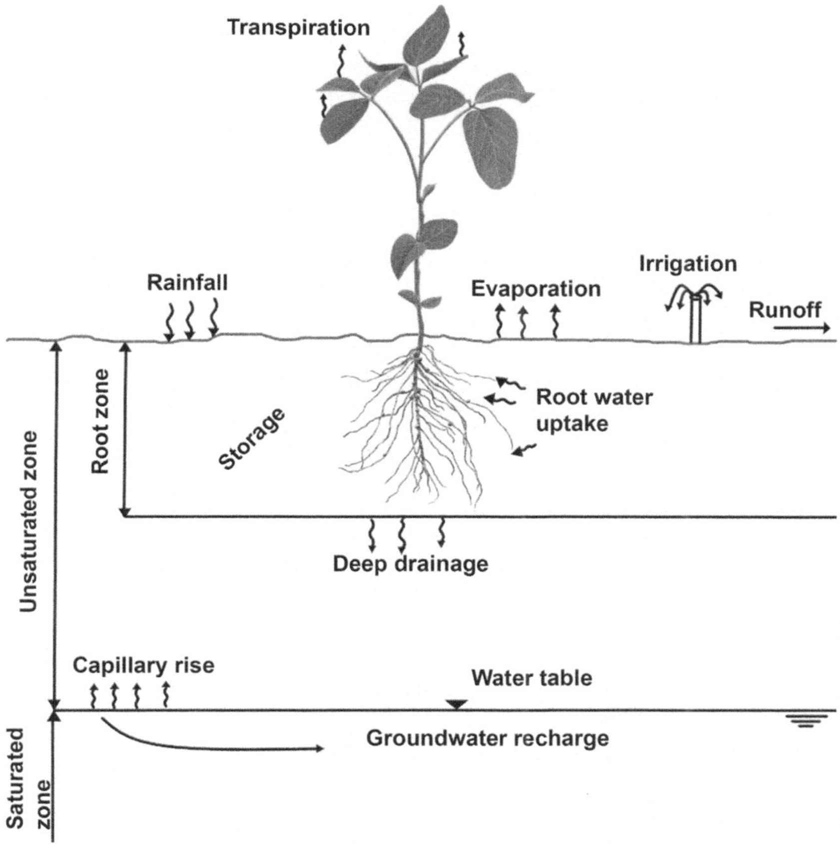


**Figure 7.18** Water fluxes and various hydrological components in the unsaturated zone (Zivancev, 2019).

returned to the atmosphere through evaporation from the surface of leaves, or, if the plant uses the infiltrated water through its roots, it can be returned into the atmosphere by transpiration. The water that is not lost as a result of evapotranspiration can be purified to the deeper unsaturated zone and it can potentially reach groundwater (saturated zone). In the case of shallow groundwater, it is possible for it to move to higher layers of soil, via gas diffusion or capillary lift.

Infiltration is one of the main processes of free groundwater movement, which is influenced by the force of gravity, the hydrostatic gradient and the capillary potential gradient. Infiltration is an extremely complex process which does not depend solely on the hydro-physical properties of soil and on the quantity and duration of precipitation, but also on the initial content of the water, on the surface sealing and compaction, on the vegetation and on the ionic composition of the infiltrated water (Dragišić, 1997).

The process of transporting solutes in an unsaturated zone has been studied for decades, and the conceptual understanding and mathematical description are being perfected. Many analytical and numerical models have been developed to predict the process of water transport and that of dissolved substances in the unsaturated zone. The most commonly used model is Richards equation for a variously saturated flow and the transport equation based on Fick's law for the transport of solute. The deterministic solutions of classical equations are used to predict the movement of water and solutes in an unsaturated zone and to analyze specific laboratory and field experiments regarding unsaturated water flow and the transport of dissolved substances. Transportation models are important for extrapolating information from a limited number of field experiments for different soils, crops and climates, as well as for different water management schemes (Department of Environmental Sciences, 2013).

### 7.3.1 Sorption

Sorption is a process that occurs spontaneously between two phases of a heterogeneous system, whereby the concentration of the components present at the boundary surface of the phases changes. Depending on the kind and type of partition of the substance sorbed on the surface or throughout the sorbent volume this process can be differentiated into adsorption and absorption. Due to the often-simultaneous occurrence of these processes, the common name for the phenomenon is sorption. The component of the system whose presence influences the reduction of the specific surface energy of the system will be absorbed. On the solid surface of the sorbent, the structural components, the molecules, atoms, radicals and ionic spheres do not have a saturated force field, and by sorbing the substance from solution, they reduce the surface free energy.

Sorption includes all the mechanisms for the bonding of sorbates to a solid phase surface, such as ion exchange, precipitation and formation of complex compounds. In a heterogeneous sediment/water system, the aqueous phase component is bound to the solid phase that it comes into contact with.

Sorption processes indirectly affect the concentration of pesticides and therefore the degradation and transport processes. The sorption process can occur through ionic forces, Coulomb forces or attractive forces between the negatively charged soil particles and the positively charged ions of the pesticide molecules. Adsorption dominantly affects the mobility of pesticides in soil because pesticides are retained in the root zone by binding to organo-mineral colloidal particles (Đurović, 2011).

The distribution of pesticides in a two-phase sediment/water system is most often represented by a sorption coefficient, $K_d$ (L kg$^{-1}$), which, under conditions of equilibrium distribution, is equal to the relationship between the pesticide concentration in soil and the concentration in water (Funari & Vighi, 1995). The sorption coefficient depends on the physicochemical properties of the compound and on the soil type, on the content of organic matter and clay, on the ion-exchange capacity of the soil, the pH value and the humidity, which determine the number and reactivity of the active adsorption centers that chemically bind the pesticide molecules. Low values of the $K_d$ coefficient occur in the negatively charged molecules of pesticides because of the repulsion process that occurs due to the negatively charged adsorption centers of the soil, and the positively charged pesticide molecules are strongly sorbed by the Coulomb attractive forces and have high values of $K_d$. The absorbed fraction of the molecules can be transported in the environment by means of erosion and runoff from the area of application. The mentioned fraction can be activated during the desorption processes which predominantly depend on the amount of water that comes into contact with the soil on which the adsorbed pesticide is.

The retention and movement of pesticides in soil is determined by the strength and range of sorption reactions, which take place under the influence of the chemical and physical properties of the soil and pesticides. The interactions that cause the sorption of pesticides to the soil take place on mineral and/or organic components. In soils that contain a higher proportion of organic matter (5% or more) when the interaction with the pesticides occurs the movement of these pollutants is influenced by the content of the total organic matter, with the nature of the organic matter itself having little effect on the sorption process (Spark & Swift, 2002).

The main sorbent in the soil of non-polar and low polarity pesticides, which are not very soluble in water, will be organic matter, due to the dominance of hydrophobic intermolecular interactions (Sekulic *et al.*, 2003). In the absence of organic matter in the soil, adsorption centers of clay minerals, which have a pronounced adsorption power regarding pesticides, can be activated (Sanchez-Martin *et al.*, 2006). When the presence of active adsorption centers of clay minerals is 10 to 30 times greater than the proportion of organic matter, the adsorption on clay minerals becomes dominant (Li *et al.*, 2003). Increasing the content of clay substances leads to a decrease in the mobility of pesticides, with the composition of clays and the presence of major cations in the soil solution being very significant (Spark & Swift, 2002).

The adsorption of polar pesticide molecules takes place, not only on organic matter and clay particles, but also by the alteration of ions, ligands and other adsorption mechanisms.

Based on numerous studies and the fact that most pesticides are non-polar or poorly polar, the widely accepted hypothesis is that organic matter can be approximately regarded as a non-polar surface, and that it is the main sorbent in the upper unsaturated soil layers rich in organic matter.

The adsorption intensity in water-unsaturated soil is largely influenced by the pH value of the medium and the soil moisture. Acidic and/or base properties (ionic and zwitterionic form) of the molecules depend on the acidity of soil, and the sorption that occurs is a combination of the sorption of the ionized and non-ionized fraction of the molecules. The pH of the soil affects the charge of the adsorption centers of the soil that are involved in the sorption.

Soil consists of different structural layers, and the dissolved compound will, in the competitive process, react at different rates and with different intensities with a variety of particles. The heterogeneous two-phase sediment/water system is extremely complex and dynamic, and the equilibrium distribution of molecules of the adsorbate in the observed two-phase system can only be observed conditionally. At least three timescales are usually used to describe the adsorption/desorption kinetics (Wauchope *et al.*, 2002):

(1) *Minutes*: the rapid, reversible diffusion and adsorption of pesticide molecules occurs primarily for available adsorption centers on the soil surface or immediately adjacent to the soil/water interface,
(2) *Hours (up to 48 hours)*: a slightly slower exchange of pesticide molecules occurs between water as the solvent and the adsorption centers of natural soil, and it takes from several hours to one or two days to reach the adsorption equilibrium. The exchange of the pesticides is most likely completely reversible.
(3) *Weeks to a few years*: extremely slow, a completely irreversible adsorption/desorption process, which results in the continuous removal of pesticide molecules from the aqueous phase, with the storage of free pesticide molecules that can subsequently be released, transformed and degraded in other processes.

The contact time between pesticides and the soil is one of the most critical factors which determines the fate of pesticides in soil because it affects the retention of compounds in soil. As a rule, it favors the formation of bound residues through various chemical and physical nonequilibrium processes, whereby the bound fraction does not participate in detoxification processes (mobility, degradation and uptake of pesticides by plants or target organisms).

Research on the sorption behavior of pesticides is most often directed towards testing the soil materials typical for the topsoil layer often used for agricultural purposes so that the conclusions refer to that layer. In the surface, unsaturated soil layer, due to incomplete water saturation, the kinetics of the competitive processes between pesticide molecules and water molecules during sorption is lower, which contributes to the larger amount of pesticide molecules that are sorbed and the smaller amount available for further distribution in the environment.

For the overall fate of pesticides, it is also necessary to observe the sorption behavior in the saturated zone of soil, in which the movement of pesticides along with groundwater occurs. The relatively slow movement of water in the aquifer makes the slowed down movement of pollutants very important, providing an opportunity for the degradation processes to last longer.

Two basic phenomena have been recognized in sorption mechanisms: chemisorption and physisorption. The characteristic physicochemical properties of chemisorption are monolayer sorption, basic chemical bonds and irreversibility. Typical physicochemical properties of physisorption are multilayer sorption, inter-molecular Van der Waals forces and reversibility.

#### *7.3.1.1 Chemisorption*

In chemical sorption, the molecules of the substance (sorbate) bind to the surface of the sorbent by forming primary chemical bonds, such as ionic (Coulomb attraction forces), covalent, covalent with a partial ionic character, short-range coordinatively covalent bonds, and hydrogen bonds from groups of intermolecular long-range connections (Atkins & de Paula, 2010). Chemisorption is a kinetically relatively slow process. Formation of an ionic chemical bond is linked to the ion exchange process and to complexing (Ibanez *et al.*, 2007). The electron transfer energy is high, as local ionic chemical bonds are formed with the active centers on the sorbent, requiring a greater activation energy to prevent the ionized pairs of the chemical substance from moving freely over the sorption surface. Chemisorption is thermodynamically easier to perform at elevated temperatures predominantly in the surface monolayer (Jusoh *et al.*, 2011). Chemisorption is complete when the total surface of the sorbent is saturated with adsorbed molecules, that is, when the monolayer is formed because the total chemical bonding forces between the adsorbate and the adsorbent are strong and short-range. The amount of substance that is chemisorbed depends primarily on the physicochemical and stereo characteristics of the sorbent and sorbate. Chemisorption is generally an irreversible chemical process.

#### *7.3.1.2 Physisorption*

In physical sorption, the physisorption of a substance (sorbate) to a sorbent is based on the formation of intermolecular Van der Waals secondary bonds, including a long-range hydrogen bond, but which individually have low energy bonds. The intermolecular Van der Waals bonds are significantly weaker than the interatomic primary bonds but still sufficient to maintain a stable intermolecular system. All intermolecular bonds are formed primarily due to the attractive electrostatic forces that exist, or they occur, under certain conditions, among the molecules. In addition to the strongest and most commonly occurring hydrogen bond, intermolecular forces include the ion-dipole, dipole-dipole, and the induced dipole-induced dipole forces. Physisorption is a process that is especially expressed at low temperatures and is multilayered. Molecular particles that are physisorbed retain their initial chemical structure and can return to their basic state by desorption, which makes physisorption a reversible process. Through physical sorption, sorbate molecules can receive a certain amount of kinetic energy that allows the molecules to be released from the sorbent surface and returned to the liquid phase. In accordance with the basic principles of reversible dynamic equilibrium processes, desorption increases as a function of time and with the increase of the sorbent surface. The sorption dynamic equilibrium represents the state when the number of sorbate molecules released from the sorbent surface and the number of molecules bound to the sorbent are equalized.

Physical sorption can also be accompanied by chemisorption if there is a chemical similarity between the active sorption centers of the sorbent and the sorbate molecules. In real conditions, both types of sorption occur, and the type of sorption which will be dominant depends on the type of sorbent/sorbate system and physicochemical environmental conditions. Due to the ever-present Van der Waals forces, especially of the hydrogen bonds, physisorption is much more frequent than chemisorption, especially in the pesticide-soil interaction system (Thibodeaux, 1996; Yue *et al.*, 2017).

#### *7.3.1.3 Ion exchange*

The ion exchange within the sorption phenomena is a process in which ions of a single substance concentrate on the surface of a sorbent as a result of the electrostatic Coulomb attraction of charged active centers on the surface. In the intergranular aquifer materials, there are charged colloidal particles, and this enables ion exchange. The sources of ionic changes in the aquifer material may be clay minerals, metal oxides, organic matter, and amorphous minerals and substances. The aquifer material is generally negatively charged and attracts positively charged ions from the solution. By ionic exchange processes, ions are replaced by ions that have a higher affinity for sorption (higher charge and lower ionic mass) (Dimkić *et al.*, 2008).

Certain organic compounds appear in an ionic form in an aqueous solution since they dissolve readily in water in comparison with other molecules that are electroneutral. Contrary to most hydrophobic organic compounds, which are 'dissolved' and retained in the solid organic aquifer material, ions can be sorbed to the solid material of the aquifer via an ion-exchange mechanism.

Based on the summed up electroneutral principle, the sum of all positive charges in a drop of an aqueous solution must be equal to the sum of all the negative charges in the same drop, which is transposed to the solid aquifer materials. Within their basic structure, ion-exchangers (cationic or anionic ion exchangers) have a specific defined capacity of positive or negative charges and they take over the necessary ions from a solution to neutralize the charge. Clays as an inorganic matrix are typical ion-exchangers in aquifer. Clay minerals in the process of isomorphic substitution retrieve ions from water to achieve electroneutrality (Hemond & Fechner, 2014).

Although ion exchange is a seemingly simple ion exchange process, in a phenomenological sense it is actually a very complex one that cannot be fully described by a mathematical equation, as opposed to the idealized sorption of hydrophobic organic compounds that is easily simulated. In the natural environment, for low ionic concentrations of polluting hydrophilic substances in the presence of other ions, the ion exchange process is often regarded approximatively as a linear distribution process, and the use of a simple retardation factor in the transport model is justified. The sorption coefficients for aquifer polluting ions are usually determined experimentally. The most commonly used batch input test, which observes the concentration of ions in the aquifer material and in the pore water coming into contact with the aquifer, and the sorption coefficient is determined as the concentration ratio. The ion exchange is influenced by the concentrations and all other ions in groundwater, and the sorption coefficient can only be used if a constant chemical composition of the surrounding water is assumed.

Based on the data of Sheng *et al.* (2001), the possibility that pure clays, an inorganic matrix, can effectively bind organic molecules, including pesticides, has been documented. Clay plays a significant role in the sorption of organic contaminants and pesticides in soil. It has been observed that in the example of atrazine, when the ratio of clay to organic matter is greater than 62, about 50% of the total absorbed atrazine particles are on clay minerals (Li *et al.*, 2003).

### 7.3.2 Degradation

The purpose of pesticides is to protect plant species from the undesirable impact of weeds, fungi and insects, and it is necessary for the pesticide to be active while the danger to a plant exists. However, once the threat is gone, it is desirable that the pesticide is decomposed to materials that are not harmful (Funari & Vighi, 1995; Levine, 2007).

The transformation of a pesticide is any process in which the molecular structure of the pesticide is changed. Transformation occurs immediately after application, but sometimes even during prior storage. Most pesticides used in the environment are degraded into substances that are universally present such as carbon dioxide, ammonia, water and mineral salts. Different types of chemical compounds are formed before pesticides completely decompose (Somasundaram & Coats, 1991).

Pesticide degradation can be biotic or abiotic. Biotic degradation is when pesticides are broken down by microorganisms or plants, and abiotic degradation occurs as a result of chemical and photochemical reactions. Which transformation process will take place depends on the structural affinity of the pesticide to a specific type of transformation and the environmental conditions that affect the behavior of the pesticide. The redox potential is one of the parameters in soil, sediments or aquifers that determines which biotic or abiotic transformation will take place, while photochemical reactions are only possible in parts of the environment exposed to sunlight (Fenner *et al.*, 2013).

The degree of degradation can be expressed via the half-life of a substance ($DT_{50}$), which is the time it takes for half of the pesticide to be transformed into another form, or until the concentration drops to half the initial concentration (Levine, 2007). Degradation usually results in the formation of non-harmful compounds, while in some cases intermediates may occur that have a greater toxic effect than the parent compound (Aksu, 2005). Transformation processes are very significant as they limit the residence time of pesticides in the environment while at the same time reducing the amount of pesticides (Hamilton & Crossley, 2004).

Most research focuses on the transformation of pesticides in the surface arable land, so little data is available on aquifer materials. The results obtained for the surface layer cannot be directly compared with the results for aquifers since the physicochemical conditions, and in particular the microorganisms, are very different. Laboratory experiments are performed with higher concentrations of pesticides than those found in groundwater, and the results cannot be extrapolated to the low concentrations present in groundwater. Surveys conducted for the surface layer of soil can only be used as an assumption of the fate of pesticides in aquifers (Agertved *et al.*, 1992).

#### *7.3.2.1 Biological degradation*

The microbial degradation of pesticides takes place under the influence of fungi, bacteria and other microorganisms. The organic matter content, soil texture, and properties such as soil moisture, temperature, aeration, and the pH value all influence the microbial degradation in the soil. The activity of microorganisms is greater in warm, moist and well-aerated soils with a neutral pH. Microbial biodegradation occurs to a greater extent in surface aerated soil layers, and especially in parts with a higher amount of organic matter. With increasing depth in the arable layer, degradation decreases, because then the conditions are less favorable for microbial degradation (Kerle *et al.*, 2007).

Degradation occurs by means of intracellular or extracellular enzymes (Parte *et al.*, 2017). Small molecule degradation is an intracellular enzymatic process, which involves the transport of molecules through the cell wall. In extracellular mechanisms of enzymatic degradation, enzyme activity is especially important because it changes in the presence of clay materials. The presence of inorganic components causes a change in enzymatic activity. The sorbed molecules are used by different microorganisms as a source of carbon, nitrogen and other elements, or energy, so that the compounds are transformed by certain kinetics (Dimkić *et al.*, 2008).

From the moment pesticides enter the environment until biodegradation starts, there is a period of adaptation during which the number of microorganisms that break down the pesticide molecules increases to the number necessary to initiate degradation (Aisable & Lloyd-Jones, 1995). In the upper, arable layer, below the soil surface, biological degradation is unlikely, while the physical and biochemical processes that affect the total amount of a pesticide entering groundwater are dominant (Hantush *et al.*, 2000). Degradation in groundwater occurs under the influence of microorganisms, with the possible production of significant amounts of metabolites (Bloomfield *et al.*, 2006; Greskowiak *et al.*, 2017).

Biodegradation may be an effective mechanism for removing low concentrations of pollutants in groundwater, but the results of soil degradation studies using high concentrations of pesticides cannot be extrapolated to oligotrophic aquifer environments where the amount of pesticides is too low to initiate degradation (Klint *et al.*, 1993). At present, there is insufficient knowledge of pesticide biodegradation under these conditions, which may be due to the insufficient monitoring of biodegradation in groundwater over a long period and the ability to isolate relevant microorganisms for degradation in such environments (Fenner *et al.*, 2013).

#### *7.3.2.2 Chemical degradation*

Chemical degradation is the process in which a pesticide reacts with water, oxygen, or with other chemical compounds in soil. In conditions when the pH values of the soil are extremely low or high, microbiological activity is reduced, and chemical degradation is favored in such conditions. Chemical degradation occurs via chemical reactions in the soil, such as hydrolysis, oxidation, and reduction (Levine, 2007; Fenner *et al.*, 2013). The rate and type of chemical reactions that occur, the temperature, humidity, composition and the pH of the soil as well as the binding of molecules of pesticides to the soil through adsorption have the most significant impact on the degree of chemical degradation (Kerle *et al.*, 2007). Chemical reactions can be predicted based on the presence of certain functional groups. Abiotic hydrolysis in aqueous solutions is characteristic for organophosphates, carboxylic acid esters, carbamates, certain halides (methyl bromide, propargyl), and many others. Certain conditions are necessary, such as a high pH or low redox environments in combination with appropriate abiotic catalysts at the site that is studied where the processes are influenced by microorganisms. In this way, the effects of abiotic and biotic transformations are difficult to separate. However, certain chemical transformations can be recognized when studied under appropriate relevant conditions. Examples of the above are triazine hydrolysis under the influence of clay as a catalyst, the transformation of chloroacethylanilyde and nitroaromatic compounds in sulfide environments, or the oxidation of glyphosate by $MnO_2$. Chemical reactions can be the dominant mode of degradation in environments such as groundwater or in lake hypolimnions, which have a hydraulic retention time of several years and where the biomass densities are lower due to the almost complete removal of adaptable organic carbon (Fenner *et al.*, 2013).

#### *7.3.2.3 Photochemical degradation*

Phototransformation can be direct or indirect. By direct transformation, the contaminant absorbs photons, while indirect transformation refers to the process in which reactive species are being formed by absorbing photons through other substances (Fenner *et al.*, 2013). Dissolved natural organic matter is one of the major absorbers of photons, and is a precursor to the excited triplet states, singlet (molecular) oxygen, superoxide anion radicals, and other radicals formed from the dissolved natural organic matter. In such excited states, homolysis, heterolysis, or photoionization can occur (Burrows *et al.*, 2002).

Nitrate and nitrite ions are additional active absorbers that generate hydroxyl radicals under the influence of solar radiation. Indirect phototransformation is a result of the parallel reactions of all the individual reactive species. To evaluate the rate of transformation in the environment, the concentration of all the relevant reactive species as well as the corresponding rate constants of the second order reactions for the selected pesticides must be known. For the hydroxyl radical and the singlet oxygen, an extensive compilation of experimentally determined rate constants for organic compounds is available. In the absence of rate constants, a quantitative structure-activity relationship (QSAR), which can provide an estimate for any substance via the chemical structure, can be used (Fenner *et al.*, 2013).

Indirect photodegradation processes, such as degradation by hydroxyl or carbonate radicals, and by singlet oxygen, are more significant than direct photolysis. Only a small number of pesticides are decomposed by means of direct photochemical transformation, whereas indirect photodegradation processes are much more common (Burrows *et al.*, 2002).

The main transformation pathways of certain classes of pesticides in the environment are shown in Table 7.3.

## 7.4 RESULTS OF SPECIALIZED MONITORING, FIELD AND LABORATORY TESTING

The Jaroslav Černi Institute for the Development of Water Resources from Belgrade, Serbia, conducted monitoring of pesticides in surface and groundwater, from 2009 to 2015. The quality of surface water and of the corresponding groundwater points was monitored in the following four large rivers in Serbia: Danube, Sava, Tisza and Velika Morava. Approximately 170 water samples were analyzed for the presence of pesticides, pharmaceuticals, and for various physical and chemical properties in the aqueous solution. Around 60 analyses were conducted for river water and about 110 for groundwater, and the presence of 15 pesticides was analyzed: monocrotophos, carbendazim, imidacloprid, acetamiprid, dimethoate, monuron, simazine, carbofuran, carbaryl, atrazine, diuron, propazine, linuron, malathion and tebufenozide.

The focus was on testing and analyzing water from four locations:

- Danube and the coastal alluvial anoxic drainage system at the Kovin-Dubovac area,
- Morava and the coastal oxic alluvial aquifer Ključ groundwater source near Požarevac,
- Sava and wells with horizontal drains near the river Sava at the groundwater source with anoxic conditions which serves to supply water to the city of Belgrade,
- Tisza with wells in the anoxic area near the river Tisza.

In addition to the systematization and study of the data obtained from the analyses of river and groundwater pesticides, targeted research of the mechanisms and intensity of sorption was carried out. This research included laboratory and field testing.

### 7.4.1 Results of pesticide monitoring in surface water and groundwater, 2009–2015

The lowest concentrations of pesticides were recorded in the Sava River, followed by the Danube and the Tisza. The highest concentration was observed in the Velika Morava River with a significant presence of carbendazim and atrazine (Table 7.4).

The occurrence of pesticides and the pesticide load depends on several factors:

- The specific use of pesticides (type and amount applied),
- The specific river runoff (influenced by the seasonal fluctuations of the flow and of water levels due to climatic factors),
- Population number of the basin in relation to river flows, etc.

**Table 7.3** Overview of the primary degradation pathways of certain classes of pesticides in the environment (Fenner *et al.*, 2013).

| Class of Pesticides | Principal Representative Active Substance and Chemical Structure | Category of Pesticide Based on Intended Use | Primary Degradation Pathways in the Environment |
|---|---|---|---|
| Amides | 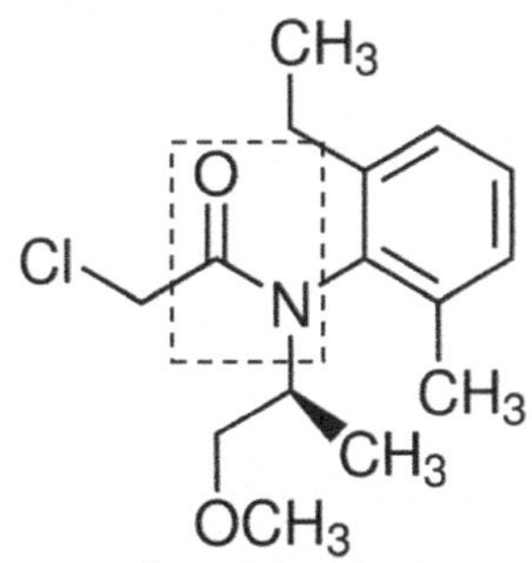 S-metolachlor | Herbicides | Microbial transformation (hydrolysis and glutathione coupling) |
| Bipyridyls | 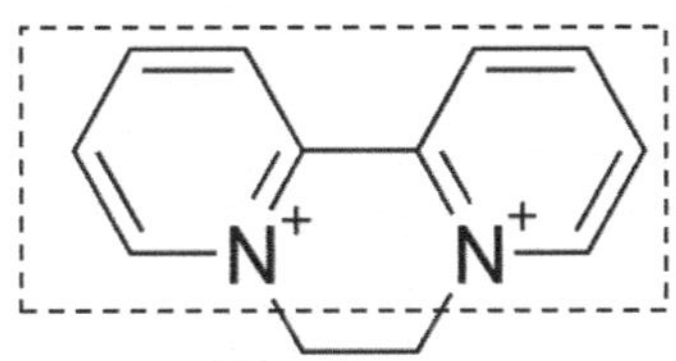 Diquat | Herbicides | Very slow biotransformation due to the strong sorption to soil matrix |
| Carbamates | 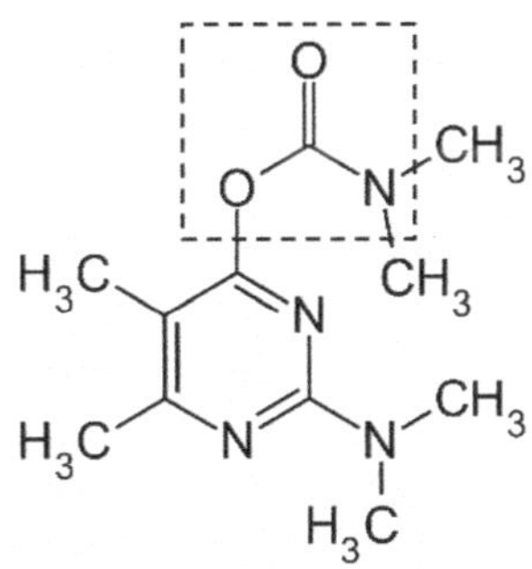 Primicarb | Insecticides/Herbicides | Microbial transformation or catalysis using bases (hydrolysis of ester bond), phototransformation of certain representative substances |

(*Continued*)

**Table 7.3** Overview of the primary degradation pathways of certain classes of pesticides in the environment (Fenner *et al.*, 2013) (*Continued*).

| Class of Pesticides | Principal Representative Active Substance and Chemical Structure | Category of Pesticide Based on Intended Use | Primary Degradation Pathways in the Environment |
|---|---|---|---|
| Dithiocarbamates | 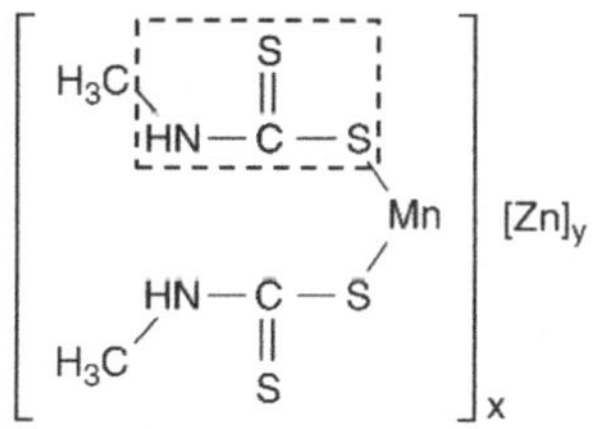 Mancozeb | Fungicides | Acid-catalyzed hydrolysis, formation of potential N-nitrosodimethylamine precursors |
| Organophosphates | 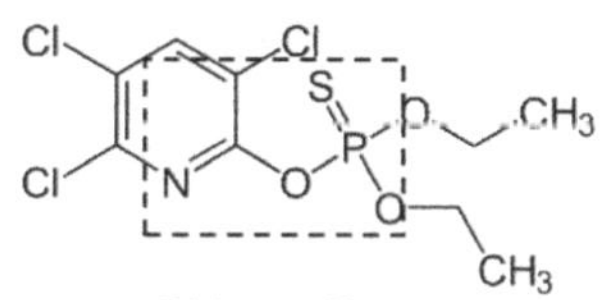 Chlorpyrifos | Insecticides | Microbial transformation (oxidation and hydrolysis) |
| Phenoxy alkanoic acids | 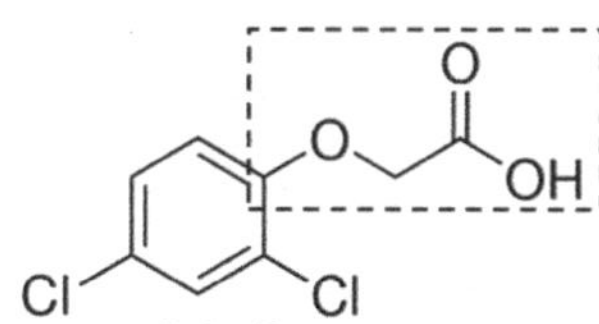 2,4 - D | Herbicides | Microbial transformation (oxidative dealkylation and aromatic ring cleavage) |
| Pyrethroids | 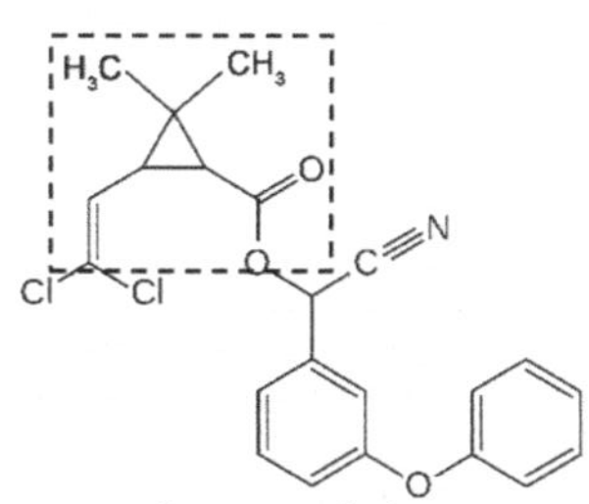 Cypermethrin | Insecticides | Microbiological transformation (hydrolysis, oxidation); phototransformation (direct or indirect) |

| | | | |
|---|---|---|---|
| Triazines | 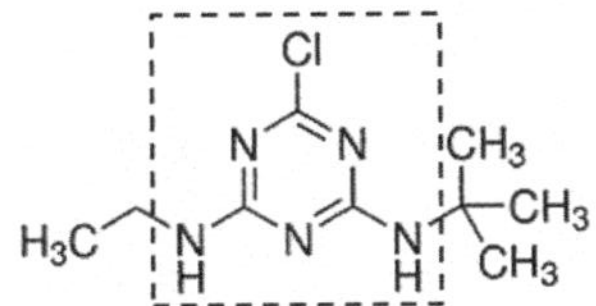<br><br>Terbuthylazine | Herbicides | Microbial transformation (oxidative dealkylation and hydrolysis) |
| Triazoles, diazoles | 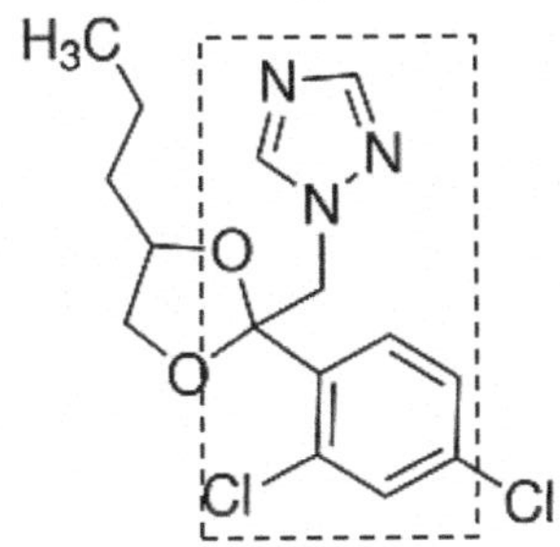<br><br>Propiconazole | Fungicides | Slow microbial transformation (oxidation); photodegradation of specific representative substances |
| Urea derivatives | 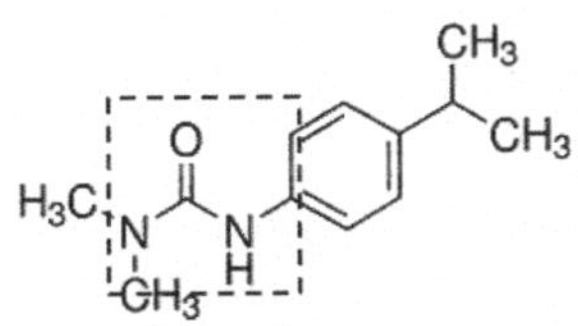<br><br>Isoproturon | Herbicides | Microbiological transformation (oxidative dealkylation and hydrolysis) |

**Table 7.4** Frequency of detected pesticides and the minimum and maximum detected concentrations in surface water.

| Surface Water | Frequency of Detection (%) | Number of Samples | Number of Pesticides Detected | Min/max Detected Concentration (ng $L^{-1}$) |
|---|---|---|---|---|
| Sava | 42% | 12 | 4 | 1/13.8 |
| Danube | 36% | 25 | 3 | 1.5/23 |
| Tisza | 56% | 9 | 4 | 3.7/25 |
| Velika Morava | 47% | 15 | 4 | 3/269 |

The highest concentrations of pesticides were detected in the Velika Morava in May 2010 (atrazine, 165 ng $L^{-1}$) and June 2011 (carbendazim, 269 ng $L^{-1}$). They represent isolated cases resulting from heavy rainfall during the so-called 'spring flush' period when the highest amounts of pesticides are applied after the early rains during the plant growth period.

The higher specific population, which can be expressed as the number of inhabitants per unit of water flow, also exposes the watercourse to greater risks of pollution. In addition to the fact that the Morava Valley is an area with a highly developed agriculture, a high degree of specific population is also evident (Table 7.5). The specific population in the Morava basin is about three times higher than that in the Danube basin, and this also affects the higher concentrations of pesticides in the watercourse (the average concentrations of carbendazim in the Danube are about five times lower than those in the Velika Morava, while atrazine concentrations are seven times smaller in the Danube than in the Velika Morava). In addition to the higher specific density, the way in which watercourse protection measures are implemented also has an impact on the higher concentrations of pesticides in the water.

The results of sampling indicate very low concentrations of the selected pesticides in all four investigated rivers in Serbia and in the corresponding groundwater. In studies of individual study sites, it was found that higher concentrations of the selected pesticides were detected in the alluvion of the Danube River in the area of the Kovin-Dubovac drainage system compared to the concentrations of the same pesticides in the Danube surface water (Table 7.6). The movement of water from the Danube to the wells in the first drainage line from which the groundwater samples were taken, lasts from 30 to 40 days, whereby the selected pesticides would be sorbed on the aquifer material and degradation processes would take place at the

**Table 7.5** Specific population for observed rivers.

| River | Estimated Population Upstream of the Hydrological Station (Expressed in Millions) | Average Value of Minimum Flows from 1950–2017 ($m^3\ s^{-1}$) | Specific Population Density (Population/ $m^3\ s^{-1}$) |
|---|---|---|---|
| Danube (hydrological station at Smederevo) | 50 | 1928 | 25940 |
| Tisza (Senta hydrological station) | 14 | 115 | 122182 |
| Sava (hydrological station Sremska Mitrovica) | 8 | 346 | 23144 |
| Velika Morava (hydrological station Ljubičevski bridge) | 3.5 | 44 | 79500 |

**Table 7.6** Frequency of detected pesticides and the minimum and maximum detected concentrations in groundwater.

| Groundwater | Frequency of Detection (%) | Number of Samples | Number of Detected Pesticides | Min/Max Detected Concentrations (ng $L^{-1}$) |
|---|---|---|---|---|
| Alluvion of the river Sava (Belgrade groundwater source) | 50% | 38 | 3 | 1/24 |
| The Danube alluvion (the Kovin-Dubovac drainage system) | 71% | 34 | 5 | 1/88 |
| The Tisza alluvion (the Knićanin-Čenta drainage system) | 25% | 8 | 2 | 11/47 |
| Alluvion of the Velika Morava (Ključ groundwater source) | 21% | 14 | 2 | 1/11 |

same time. The vertical movement from the surface of the terrain has a dominant influence on the quantities of pesticides in groundwater. The quantities of pesticides in surface water have a lesser impact on the quantities found in groundwater.

The largest number of surface water samples from the Danube and the largest number of groundwater samples from the Danube River alluvion were collected at the Kovin-Dubovac drainage system. Figure 7.19 shows a comparison of the average concentrations of pesticides detected in the surface and groundwater of the Kovin-Dubovac drainage system. It was observed that the average concentration of carbendazim, atrazine, carbofuran and dimethoate was higher in groundwater than in the surface water. Propazine was the only pesticide detected in surface water, while the concentration was below the limit of quantification in groundwater. The maximum determined average concentration of carbendazim in groundwater was 13.95 ng $L^{-1}$, while it was more than two times lower in surface water and equaled 6.13 ng $L^{-1}$. By comparing the average concentration of atrazine in surface and groundwater, it was observed that the maximum determined average concentration of atrazine in groundwater which amounted to 10.25 ng $L^{-1}$ was approximately four times higher than the average concentration in surface water, which was 2.75 ng $L^{-1}$. This confirms the hypothesis that pesticides enter groundwater dominantly from the surface of the soil to which they were applied.

Figure 7.19 shows the comparative average concentrations of pesticides detected in the river Sava and in the alluvion of the river Sava in the area of the Belgrade groundwater source. A total of four pesticides were detected in surface water, while three pesticides were detected in groundwater. The average concentrations of carbendazim were higher in surface water compared to groundwater, while the average concentrations of atrazine and carbofuran were higher in groundwater than in surface water. Dimethoate is a pesticide that was only detected in surface water.

The aquifer of Belgrade's groundwater sources is layered, consisting of polycyclic-formed layers, which were formed by melting glaciers. The drains at the well from which the groundwater samples were taken as part of the sampling are horizontal and are located below the second clay layer, and it takes the water 100 and 200 days to reach the drains (Dimkić *et al.*, 2017). The long-term transport of water and of the dissolved micropollutants in water results in the significant effect of the sorption and degradation of substances, without the detection of selected pollutants or their detection at very low concentrations.

A comparison of the average concentrations of the detected pesticides in the Tisza River and in the alluvion of the Tisza River in the area of the Knićanin-Čenta drainage system are shown in Figure 7.19.

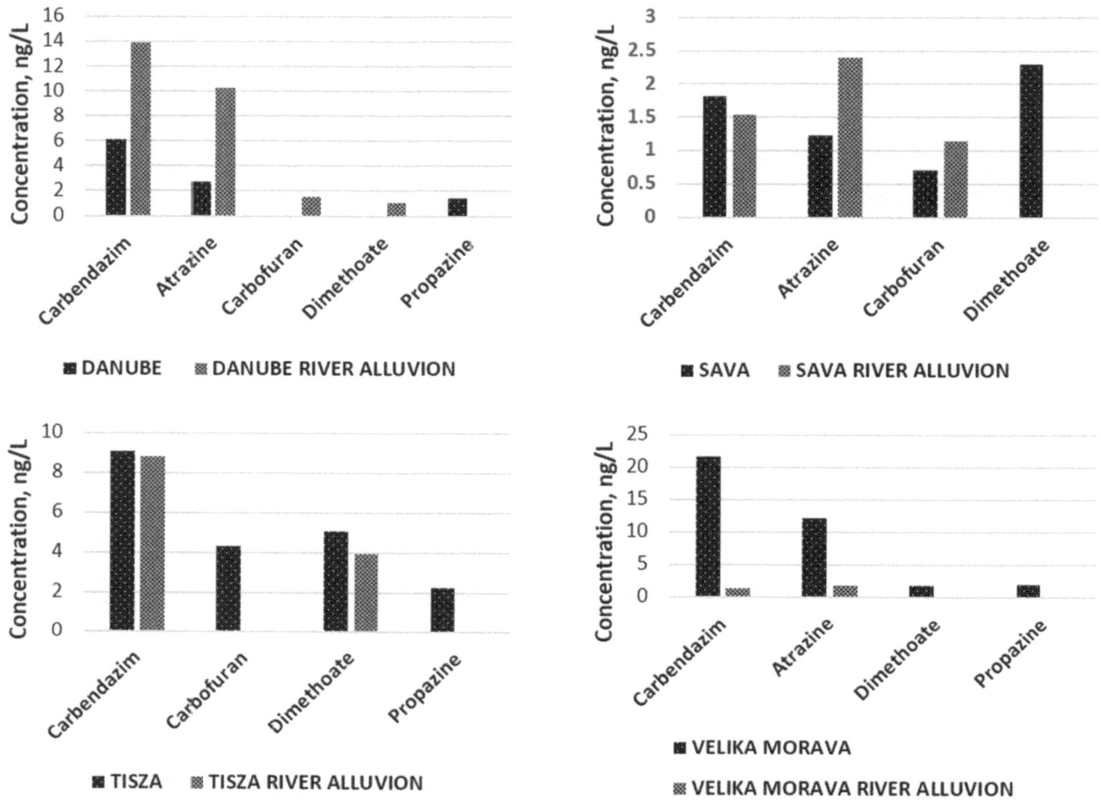


**Figure 7.19** Comparison of average pesticide concentrations in surface and groundwater (Zivancev, 2019).

A total of four pesticides were detected in surface water, two of which were also detected in groundwater. The average concentrations of carbendazim and dimethoate were higher in surface water than in groundwater, while carbofuran and propazine are pesticides that were detected only in surface water. In the study area of the Knićanin-Čenta drainage system, wells draw water from the aquifer that is located below very clayey soil. Since wells pump from a greater depth, pesticide concentrations are diluted, even to the extent that they are not detected in groundwater.

Figure 7.19 shows a comparison of the average concentration of detected pesticides in the Velika Morava river and in the alluvion of the Velika Morava river, in the area of the Ključ source. A total of four pesticides were detected in the surface water, two of which were also detected in groundwater. The average concentrations of carbendazim and atrazine in the surface water were higher than those in the groundwater. Dimethoate and propazine are pesticides that were only detected in surface water.

The research took place after the ban of atrazine in 2008. However, based on the presence of atrazine in 33.03% of the samples of groundwater and 13.11% of surface water samples, as well as based on the period of degradation of atrazine, it can be concluded that there is a constant input of atrazine into the environment. These are very small amounts that may be associated with the use of the active substance terbuthylazine, which in the composition of products may contain atrazine up to 1 g $kg^{-1}$.

Very low concentrations of pesticides were detected in the wells at the Ključ groundwater source and they were significantly lower than the concentrations of the same pesticides in the river itself. The influence of agricultural production on the quality of groundwater in the observed wells was determined based on the movement of groundwater in the entire investigated area, as well as on the occurrence of nitrates in the wells. As the concentrations were lower than those that could have been expected based on the share of river water in the observed wells, as well as on the average concentration in river water, there is a strong implication that the observed pesticides are degraded more rapidly under oxic conditions (Živančev *et al.*, 2020).

An analysis of the concentrations of detected pesticides in the Tisza River and in the alluvion of the Tisza River showed slightly higher concentrations of pesticides in groundwater. In the area of the Knićanin-Čenta drainage system, the alluvial aquifer is anoxic and there is a clear indication of the slower degradation of the observed pesticides under anoxic conditions.

In the Belgrade groundwater source, the average concentrations of atrazine and carbofuran were lower in surface water compared to groundwater while the concentrations of carbendazim and dimethoate were higher in surface water than in groundwater. It should be noted that all the average concentrations of the detected pesticides were very low. Atrazine was detected in more than 50% of the groundwater samples, while it was detected in 25% of the surface water samples. Carbofuran was also detected in a larger number of groundwater samples than surface water samples. The reason for the larger number of detected pesticides in groundwater samples in relation to surface water may be due to the direct inflow of the pesticides into the sampled well, with extremely low concentrations. The movement of water from the Sava River to the wells where groundwater was sampled ranges from 100 days up to more than 2 years, with water being purified as it passes through the aquifer as a result of sorption and degradation processes. The wells draw water from a lower aquifer and the water travels through a semi-permeable layer that reduces the concentration of pesticides. Carbendazim and dimethoate are pesticides whose concentrations were lower in groundwater compared to surface water, precisely because of the sorption and degradation processes that take place.

In the aquifer, water movement can take place in all directions ($v_x$, $v_y$, $v_z$), where the horizontal component ($v_x$, $v_y$) usually dominates. The mineralogical composition of the skeleton depends on the conditions under which the aquifer was formed. Very often, the most permeable aquifer layers have very low or practically no clay minerals and/or organic carbon content. In this case, the physisorption of the organic molecule of the pesticide onto the matter of the inorganic skeleton dominates.

Based on the results obtained by analyzing groundwater movement, nitrate concentrations in piezometers in areas with oxic conditions at the Ključ groundwater source, and the anoxic conditions of the Kovin-Dubovac drainage system, it was possible to confirm that the presence of selected pesticides in groundwater depends on the hydrogeological environmental conditions, and especially on the hydrogeochemical properties of groundwater (oxicity). In highly oxic conditions, the degradation of the observed pesticides is faster and their concentrations are lower than expected, while anoxic conditions affect the persistence and result in slower degradation of the pesticides.

### 7.4.2 Results of field and laboratory tests and tracking of the fate of pesticides

The most common tests for pesticide behavior are, of course, made for the surface soil layer which is subject to arable cultivation, and therefore the conclusions regarding sorption primarily refer to the soil layer that is unsaturated with water. For the overall fate of pesticides in groundwaters the movement of water in the aquifer is also relevant. The relatively slow movement of water in the aquifer is especially important and further slows the movement of pollutants by sorption, allowing more time for the degradation processes that affect the reduction of pesticides.

Preliminary tests performed to define soil sorption that have been reported in literature were mainly done using a composite soil and artificially achieved pH values. Such tests give a certain picture of the sorption of the pesticides that were tested on the soil material, primarily considering the surface, arable soil layers.

The research presented in this chapter was conducted to study and better understand the sorption of pesticides on a material that is more similar to the water-bearing layer of the aquifer. The first activity was carried out under laboratory conditions and sorption on sandy material as a representative of the alluvial aquifer material was investigated. Particular attention was paid to the ratio between the amount of solid material and the solution with pesticides in order to illustrate the real conditions in the aquifer as much as was possible. Another experiment that was carried out in the field consisted of the infiltration of three different pesticides and a non-reactive tracer into the aquifer.

The laboratory experiment pertaining to the sorption of pesticides on sandy material was performed for the three most commonly detected pesticides in the surface and groundwater studies presented earlier and they were atrazine, carbendazim, and carbofuran. After the experiment, sorption isotherms were determined and the linear sorption coefficients shown in Table 7.7 were calculated as a result. The obtained results show that atrazine has the highest affinity for the sandy material.

The results obtained under laboratory conditions for the sorption of atrazine on sandy material are in agreement with the sorption of atrazine in a sandy aquifer in real field conditions reported in literature, where the calculated value of $K_d = 1.01$ L kg$^{-1}$ (Pang & Close, 2001). Based on the obtained result, it was concluded that the way the sorption experiment was set up as well as the results of the experiment were appropriate for further assessment of pesticide transport in groundwater in an alluvial aquifer.

For further research in groundwater, carbendazim was selected, and two more pesticides were included: acetamiprid and imidacloprid.

Figures 7.20–7.22 provide a comparative representation of the change in the concentrations of monitored pesticides and of the non-reactive tracer during the field experiment. The illustrations show a delay of all the three pesticides in relation to the non-reactive tracer, leading to the conclusion that there is a retention of all the three pesticides in the soil layer as a result of sorption processes (Živančev *et al.*, 2019).

The delay in the occurrence of the maximum concentration (peak) of the observed pesticides relative to the occurrence of the maximum concentration of the non-reactive tracer is calculated based on the ratio of the rate of occurrence of the maximum concentration of the non-reactive tracer and the rate of occurrence of half the total mass of the substance that appeared at the observed well. Equation (7.1) is used to calculate the retardation of a pesticide compared to non-reactive tracer:

$$R = \frac{v_{\text{tracer}}}{v_{\text{peticide}}} = \frac{t_{\text{pesticide}}}{t_{\text{tracer}}} = 1 + \frac{\rho_A \cdot K_d}{n_{ef}} \tag{7.1}$$

where:

$v_{\text{tracer}}$ – tracer movement rate (LT$^{-1}$),
$v_{\text{pesticide}}$ – pesticide movement rate (LT$^{-1}$),
$t_{\text{tracer}}$ – the time during which the maximum tracer concentration appears at the observed well (T),
$t_{\text{pesticide}}$ – the time during which half of the total amount of pesticide occurs at the observed well (T),
$\rho_A$ – specific density of aquifer material (ML$^{-3}$),
$K_d$ – linear sorption coefficient (L$^3$ M$^{-1}$),
$n_{ef}$ – effective porosity (-).

**Table 7.7** Linear sorption coefficients for selected pesticides on sandy material.

| **Pesticide** | **Linear Sorption Coefficient $K_d$ (L kg$^{-1}$)** |
|---|---|
| Atrazine | 1.23 |
| Carbendazim | 0.52 |
| Carbofuran | 0.39 |

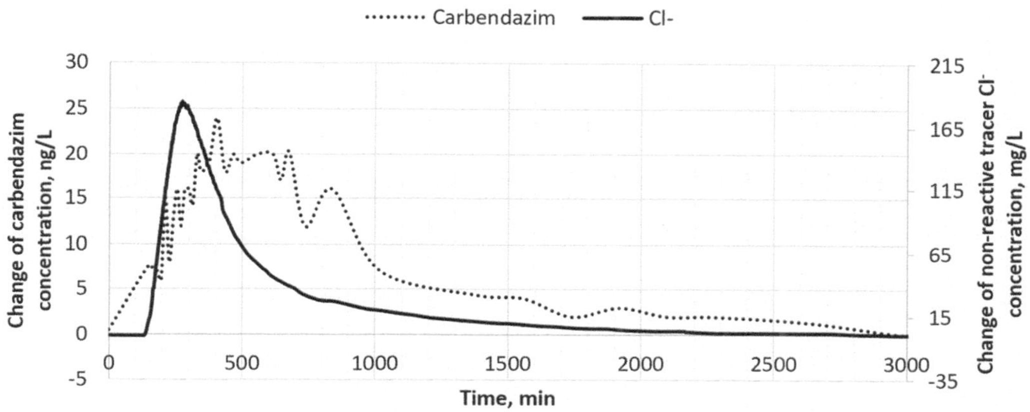


**Figure 7.20** Changes in the concentration of carbendazim and of the non-reactive tracer during the experiment at the observed well (Zivancev, 2019).

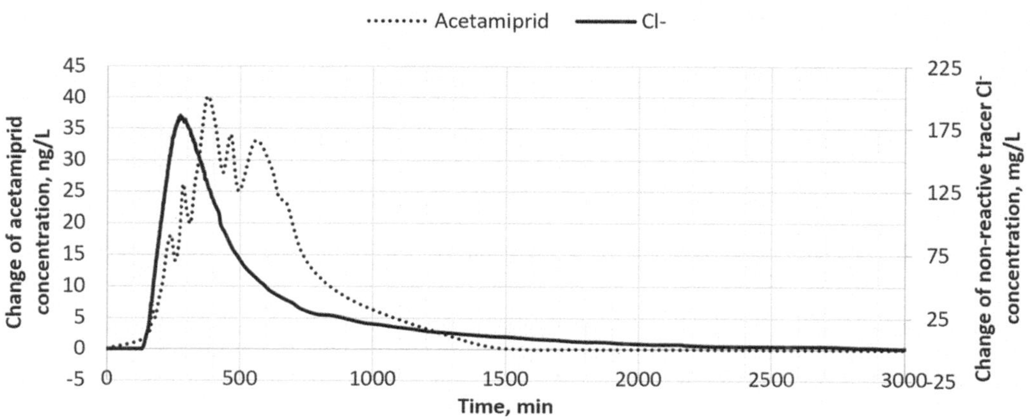


**Figure 7.21** Changes in acetamiprid and non-reactive tracer concentrations during the experiment at the observed well (Zivancev, 2019).

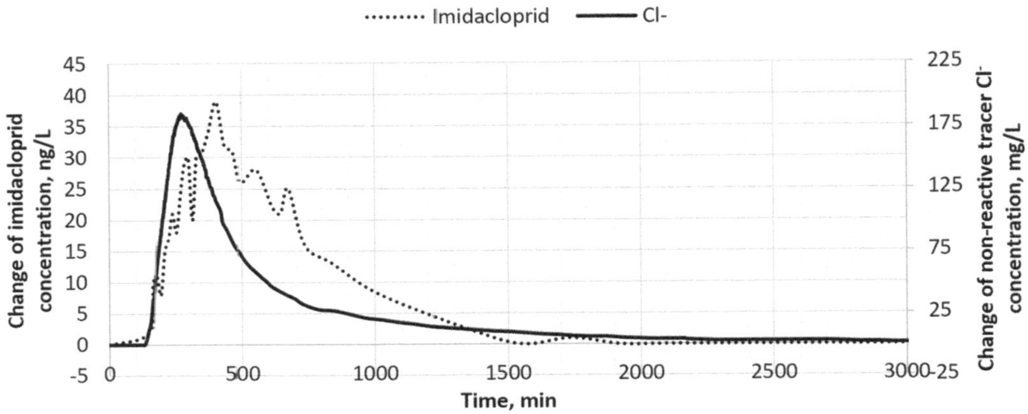


**Figure 7.22** Changes in the concentrations of imidacloprid and of the non-reactive tracer during the experiment at the observed well (Zivancev, 2019).

The delay time of the pesticide at the observed well was determined based on the occurrence of half of the total mass of the pesticide, with half of the total mass of the pesticide calculated using Equation (7.2):

$$\frac{1}{2}m_u = \frac{Q\int C(t)dt}{2} \tag{7.2}$$

where:

$m_u$ – total mass of pesticide that appeared at the observed well (M),
$C$ – pesticide concentration ($ML^{-3}$),
$t$ – sampling time (T),
$Q$ – water flow ($L^3\ T^{-1}$).

Table 7.8 shows the retardation of the pesticide that was calculated using Equation (7.1).

In a field experiment, delays of carbendazim (2.65), acetamiprid (1.88), and imidacloprid (1.88) were observed in relation to the movement of water. Table 7.9 shows the sorption coefficient and the retardation factor of carbendazim in the laboratory and field experiments.

The values in Table 7.9 show that the retardation on sandy material was twice as high in the laboratory experiment compared to that in the field experiment, which is a very new and significant result. When looking at the results of the laboratory and field experiment, it is necessary to consider the difference in the particle size of the material used in the laboratory and the particle size in the most permeable part of

**Table 7.8** Calculated retardation factor of the observed pesticides and the estimated value of the linear sorption coefficient.

| Pesticide | Retardation Factor (*R*) | Linear Sorption Coefficient (L kg$^{-1}$) |
|---|---|---|
| Carbendazim | 2.65 | 0.200 |
| Acetamiprid | 1.88 | 0.107 |
| Imidacloprid | 1.88 | 0.107 |

**Table 7.9** Sorption coefficient and retardation factor for carbendazim obtained from laboratory and field experiments.

| | Laboratory Experiment on Sand Material | Field Experiment |
|---|---|---|
| Sorption coefficient, $K_d$ (L kg$^{-1}$) | 0.52 | 0.20 |
| Retardation factor, $R$ | 5.29 | 2.65 |

an aquifer. The particles of sandy material used in the laboratory are sieved through 2 mm pores, while in the most permeable aquifer layer in the field, the grain diameter with a 20% representation on the granulometric curve is 3 mm. On smaller particles, the specific surface area is larger compared to that of larger particles, as is the sorption capacity of the particles.

## 7.5 SUMMARY COMMENTS

Pesticides in alluvial water mainly come from the surface of a terrain, but the pesticide content in alluvial groundwater is extremely low when compared with the values that are allowed, due to the total self-purification mechanisms of the aquifer. Detected concentrations are lower than the law permitted values for drinking water. Formation of an alluvial aquifer influences the content of micropollutants in groundwater, and the polycyclic genesis of sediments in the area of the Belgrade groundwater source as well as the placement of horizontal drains at greater depths in the aquifer allow for the significantly longer retention and purification of the water before it enters the well filter. Under the oxic groundwater conditions of the source, there are no pesticides in the groundwaters associated with the area of intensive agricultural activities and this is a result of degradation and oxidation processes.

When studying and monitoring pesticides in groundwater, it is necessary to separate the behavior of pesticides in surface, unsaturated, poorly permeable layers from the behavior of pesticides in underlying deeper aquifers. The movement of water in an unsaturated zone with layers containing clay minerals and with layers rich in organic matter can generally be approximated by vertical movement. The slowing down of the movement of pesticides in an unsaturated zone is due to chemisorption, which predominantly occurs on organic carbon or it occurs due to ion exchange on the clay material. Laboratory and field research (Živančev *et al.*, 2019) has shown that there is a physical mechanism of the sorption of pesticides in the sandy and sandy-gravel material of aquifers, and that it slows down the pollutants in an aquifer.

## REFERENCES

Agertved J., Rugge K. and Barker J. F. (1992). Transformation of the Herbicide MCPP and Atrazine under Natural Aquifer Conditions. *Groundwater*, **30**(4), 500–506.

Agrawal A., Pandey R. S. and Sharma B. (2010). Water pollution with special reference to pesticide contamination in India. *Journal of Water Resource and Protection*, **2**, 432–448.

Aisable J. and Lloyd-Jones G. (1995). A review of bacterial degradation of pesticides. *Australian Journal of Soil Research*, **33**(6), 925–942.

Aksu Z. (2005). Application of biosorption for the removal of organic pollutants: a review. *Process Biochemistry*, **40**, 997–1026.

Alavanja M. (2009). Introduction: Pesticides use and exposure, extensive worldwide. *Reviews on Environmental Health*, **24**(4), 303–310.

Atkins P. and de Paula J. (2010). Physical Chemistry. WH Freeman and Company, New York.

Bloomfield J. P., Williams R. J., Gooddy D. C., Cape J. N. and Guha P. (2006). Impacts of climate change on fate and behavior of pesticides in surface and groundwater – a UK perspective. *Science of the Total Environment*, **369**, 163–177.

Burrows H. D., Canle M. L., Santaballa J. A. and Steenken S. (2002). Reaction pathways and mechanisms of photodegradation of pesticides. *Journal of Photochemistry and Photobiology B: Biology*, **67**, 71–108.

DeLorenzo M., Scott G. and Ross P. (2001). Toxicity of pesticides to aquatic microorganisms: a review. *Environmental Toxicology and Chemistry: An International Journal*, **20**(1), 84–98.

Department of Environmental Sciences. (2013). The HYDRUS-1D Software Package for Simulating the One-Dimensional Movement of Water, Heat, and Multiple Solutes in Variably-Saturated Media.

Dimkic M. A., Brauch H.-J. and Kavanaugh M. (2008). Groundwater Management in Large River Basins. IWA Publishing, London, UK.

Dimkic M., Kovacevic S., Pusic M., Dotlic M. and Radisic M. (2017). Modeling of sorption and degradation of selected pharmaceuticals: a case study of Belgrade groundwater source-Well Rb-16. *Geological Annals of the Balkan Peninsula*, **78**(1), 47–59.

Dragišić V. (1997). General Hydrology. Faculty of Mining and Geology, Belgrade, Serbia.

Đurović R. (2011). Processes that determine the fate of pesticides in soil. *Pesticides and Phytomedicine*, **26**(1), 9–22.

FAO. (2019). FAOSTAT. http://www.fao.org/faostat/en/#data/RL (accessed December 2019).

Fenner K., Canonica S., Wackett L. and Elsner M. (2013). Evaluating pesticide degradation in the environment: blind spots and emerging opportunities evaluating pesticide. *Science*, **341**, 752–758.

Funari E. and Vighi M. (1995). Pesticide Risk in Groundwater. Taylor & Francis, London, UK.

Greskowiak J., amann E H., Burke V. and Massmann G. (2017). The uncertainty of biodegradation rate constants of emerging organic compounds in soil and groundwater – a compilation of literature values for 82 substances. *Water Research*, **126**, 122–133.

Hamilton D. and Crossley S. (2004). Pesticide Residues in Food and Drinking Water: Human Exposure and Risks. John Wiley & Sons, Ltd., Chichester, West Sussex, England.

Hantush M. M., Mariño M. A. and Islam M. R. (2000). Models for leaching of pesticides in soils and groundwater. *Journal of Hydrology*, **227**(1–4), 66–83.

Hemond H. and Fechner E. (2014). Chemical Fate and Transport in the Environment. Academic Press, San Diego, CA, USA.

Ibanez J. G., Hernandez-Esparza M., Doria-Serrano C., Fregoso-Infante A. and Singh M. M. (2007). Environmental Chemistry: Fundamentals. Springer Science & Business Media, New York, USA.

ICPDR Danube River Basin Water Quality Database. https://www.icpdr.org/wq-db (accessed December 2019).

ICPDR Joint Danube Survey. http://www.danubesurvey.org (accessed December 2019).

Jusoh A., Hartini W. J. H., Ali N. and Endut A. (2011). A study on the removal of pesticides in agricultural run off by granular activated carbon. *Bioresource Technology*, **102**(9), 5312–5318.

Kaushik G., Satya S. and Naik S. N. (2009). Food processing a tool for pesticide residue dissipation – a review. *Food Research International*, **42**, 26–40.

Kerle E., Jenkins J. and Vogue P. (2007). Understanding pesticide persistence and mobility for groundwater and surface water protection, Oregon State Univ Extension Service EM8561-E.

Kim K.-H., Kabir E. and Jahan S. A. (2017). Exposure to pesticides and associated human health effects. *Science of the Total Environment*, **575**, 525–535.

Klint M., Arvin E. and Jensen B. K. (1993). Degradation of pesticides mecoprop and atrazine in unpolluted sandy aquifers. *Journal of Environmental Quality*, **22**(2), 262–266.

Laws E. A. (2018). Aquatic Pollution: An Introductory Text. John Wiley & Sons, Hoboken, NJ, USA.

Lehmann E., Turrero N., Kolia M., Konaté Y. and de Alencastro L. F. (2017). Dietary risk of pesticides from vegetables and drinking water in gardening areas in Burkina Faso. *Science of the Total Environment*, **601–602**, 1208–1216.

Levine M. J. (2007). Pesticides: A Toxic Time Bomb in Our Midst. Greenwood Publishing Group, Westport, CA, USA.

Li H., Sheng G., Teppen B. J., Johnston C. T. and Boyd S. A. (2003). Sorption and desorption of pesticides by clay minerals and humic acid-clay complexes. *Soil Science Society of America Journal*, **67**, 122–131.

Matthews G. (2015). Pesticides: Health, Safety and the Environment. John Wiley & Sons, Chichester, UK.

Nicolopoulou-Stamati P., Maipas S., Kontampasi C., Stamatis P. and Hens L. (2016). Chemical pesticides and human health: the urgent need for a new concept in agriculture. *Frontiers in Public Health*, **4**, 148.

Ortiz-Hernández M. L., Sánchez-Salinas E., Dantán-González E. and Castrejón-Godínez M. L. (2013). 'Pesticide biodegradation: mechanisms, genetics and strategies to enhance the process' and n Biodegradation-life of science. IntechOpen.

Pang L. and Close M. (2001). A field tracer study of attenuation of atrazine, hexazinone and procymidone in a pumice sand aquifer. *Pest Management Science*, **57**, 1142–1150.

Parte S. G., Mohekar A. D. and Kharat A. S. (2017). Microbial degradation of pesticides: a review. *African Journal of Microbiology Research*, **11**(24), 992–1012.

Pavlovic P., Kostic N., Karadzic B. and Mitrovic M. (2017). The Soils of Serbia. Springer, Dordrecht, The Netherlands.

Pretty J. and Bharuch Z. P. (2015). Integrated pest management for sustainable intensification of agriculture in Asia and Africa. *Insects*, **6**, 152–182.

Sabljić A. (2009). Environmental and Ecological Chemistry – Volume II. EOLSS Publications, Oxford, UK.

Saifadeen A. and Gladnyeva R. (2012). Modeling of solute transport in the unsaturated zone using HYDRUS-1D.

Sanchez-Martin M. J., Rodriguez-Cruz M. S., Andrades M. S. and Sanchez-Camazano M. (2006). Efficiency of different clay minerals modified with a cationic surfactant in adsorptives on pesticides: influence of clay type and pesticide hydrophobicity. *Applied Clay Science*, **31**(3–4), 216–228.

Sekulic P., Kastori R. and Hadzic V. (2003). Soil Protection against Degradation. Scientific Institute of Field and Vegetable Crops, Novi Sad.

Sheng G., Johnston C. T., Teppen B. J. and Boyd S. A. (2001). Potential contribution of smectite clays and organic matter to pesticide retention in soils. *Journal of Agricultural and Food Chemistry*, **49**, 2899–2907.

Somasundaram L. and Coats J. R. (1991). 'Pesticide Transformation Products in the Environment' in Pesticide Transformation Products. American Chemical Society, Washington DC, 2–9.

Sousa J. C. G., Ribeiro A. R., Barbosa M. O., Pereira M. F. R. and Silva A. M. T. (2018). A review on environmental monitoring of water organic pollutants identified by EU guidelines. *Journal of Hazardous Materials*, **344**, 146–162. See http://dx.doi.org/10.1016/j.jhazmat.2017.09.058

Spark K. M. and Swift R. S. (2002). Effect of soil composition and dissolved organic matter on pesticide sorption. *The Science of the Total Environment*, **298**, 147–161.

Stenersen J. (2004). Chemical Pesticides: A Mode of Action and Toxicology. CRC Press, Boca Raton, Florida, USA.

Tariq M., Afzal S., Hussain I. and Sultana N. (2007). Pesticide exposure in Paksitan: a review. *Environment International*, **33**(8), 1107–1122.

Thibodeaux L. J. (1996). Environmental Chemodynamics. John Wiley & Sons, New York, NY, USA.

Uneke C. J. (2007). Integrated Pest Management for Developing Countries: A Systemic Overview. Nova Publishers, New York, USA.

Wauchope R. D., Yeh S., Linders J. B. H. J., Kloskowski R., Tanaka K., Rubin B., Katayama A., Kördel W., Gerstl Z., Lane M. and Unsworth J. B. (2002). Pesticide soil sorption parameters: theory, measurement, uses, limitations and reliability. *Pest Management Science*, **58**(5), 419–445.

WHO. (2019). WHO Recommended Classification of Pesticides by Hazard and Guidelines for Classification.

Yue L., Ge C., Feng D., Yu H., Deng H. and Fu B. (2017). Adsorption-desorption behavior of atrazine on agricultural soils in China. *Journal of Environmental Sciences*, **57**, 180–189.

Zhang W. (2018). 'Global pesticide use: profile, trend, cost/benefit and more'. *Proceedings of the International Academy of Ecology and Environmental Sciences*, **8**(1), 1–27.

Zivancev N. (2019). Analiza korelacije mehanizama rasprostiranja i koncentracionih nivoa pesticida u podzemnoj vodi (Correlation analysis of transport mechanisms and concentration levels of pesticides in groundwater). PhD thesis, Faculty of Technical Sciences, University of Novi Sad, Serbia.

Zivancev N., Kovacevic S., Radovic T., Radisic M. and Dimkic M. (2019). Mobility and sorption assessment of selected pesticides in alluvial aquifer. *Environmental Science and Pollution Research*, **26**(28), 28725–28736.

Zivancev N., Kovacevic S., Perovic M., Calenic A. and Dimkic M. (2020). Influence of oxic and anoxic groundwater conditions on the occurrence of selected agrochemicals. *Water Supply*, **20**(2), 487–498.

# Chapter 8

# Transformation of nitrogen compounds in groundwater

*Marija Perović and Milan Dimkić*

## 8.1 INTRODUCTION

Much of the world's population favours groundwater as the main source of drinking water. About 75% of the European Union (EU) population uses groundwater for water supply. The importance of groundwater in Serbia as a source of water supply is exceptional, as seen from the share of groundwater used for water supply of about 70%. More than 50% of this groundwater comes from alluvial groundwater sources. For groundwater management, it is important to consider the value of the self-purification potential of aquifers, which can be particularly significant in the case of nitrogen compounds. Elevated concentrations of nitrate anions in water ($NO_3^-$) have been observed in many aquifers around the world and since the middle of the 20th century, they have been largely attributed to agricultural activities (Pasten-Zapata *et al.*, 2014).

Global population growth has greatly increased the emission of nitrates from agricultural and other sources. The use of fertilisers in agriculture is one of the most important measures for achieving high yields and quality products, but at the same time inadequate and excessive use can have a direct negative impact and result in high environmental contamination. Leakage of N compounds from agricultural areas, in addition to the negative impact on groundwater quality, also affects the global nitrogen cycle, due to the slow movement of groundwater and the accumulation and biological transformations of N compounds (Nikolenko *et al.*, 2018).

Sources of nitrates such as municipal wastewater which is inappropriately disposed of, and unsanitary and illegal landfills, pose a particular threat to the water quality of shallow, aerobic aquifers with suitable conditions for nitrification processes. It has been found that high concentrations of nitrates in drinking waters can cause *methemoglobinemia*, a haematological disorder, which occurs when red blood cells contain more than 1% of methaemoglobin. The symptoms include cyanosis, bluish discoloration of the mucous membranes, digestive and respiratory problems, headache, drowsiness, fatigue, tachycardia,


Editor: Milan Dimkić
doi: 10.2166/9781789060904_0437

convulsions (methaemoglobin concentrations 1–10%), brain damage and fatal outcome (methaemoglobin concentrations 50–70%) (Pesticide Safety Education Program [PSEP], 2018). In a wider population, nitrate in drinking water is considered to pose a health threat when its concentration is 100–200 mgN/l, but individual negative effects also depend on the amount of nitrate and nitrite ingested (PSEP, 2018).

The US Environmental Protection Agency (EPA) has set the maximum allowable concentration of nitrates in drinking water at 10 mgN/l (nitrate nitrogen). The permissible nitrate concentration prescribed by the EPA is approximately equivalent to the guidelines of the World Health Organization (WHO), which sets a limit of 50 $mgNO_3$/l or 11.3 mgN/l.

The importance of agricultural nitrogen emissions into the environment is reflected in a separate directive adopted by all member states of the European Union, which limits the concentration of nitrate to 50 mg/l in all freshwater bodies (Directive 91/676/EEC, 1991).

The transport and transformation pathways of nitrogen compounds in groundwater, as well as in the environment in general, are diverse and depend on a large number of factors. This section discusses the general conditions for the occurrence and transformation of nitrogen compounds in water and analyses the origin of nitrogen compounds in alluvial groundwater. In particular, the focus is on analytical methods and how the oxygen content contributes and influences the presence and speciation of nitrogen compounds in alluvial groundwaters. The results of extensive research at one aerobic groundwater source (Ključ) in the alluvium of the Great Morava River and one anaerobic source drainage system (Kovin-Dubovac) in the alluvium of the Danube are presented.

## 8.2 ORIGIN OF NITROGEN COMPOUNDS AND NITROGEN TRANSFORMATION CONDITIONS

Sources of nitrogen compounds are organic waste, wastewater, inorganic and organic fertilisers, natural processes in soil (mineralisation), manure, septic tanks, illegal landfills and atmospheric precipitation. Nitrogen is predominantly present in soil in organic form (about 98%), which is not directly available to plants and is not subject to filtration. Reserves of inorganic nitrogen (nitrate and ammonium ions) in the soil are usually very small and quickly depleted, due to uptake by plants and microorganisms. Inadequate irrigation, as well as heavy rainfall, can cause leakage of mobile and persistent forms of nitrogen compounds (nitrates) under the root zone. Nitrogen can be removed from the soil in several ways: by evaporation, filtration, crop removal, denitrification, soil erosion and surface runoff. Soils with a large water permeability coefficient as well as coarse-grained soils are at a higher risk of nitrate leakage.

Certain amounts of nitrogen compounds in water are naturally present and come from the atmosphere, from animals and from the degradation of soil organic matter. Various studies have been conducted to determine the natural concentrations of nitrogen in water. Pano *et al.* (2006) state that the natural concentration of nitrate is in the range from 1.8 to 11.7 $mgNO_3$/l, and in groundwater 0.4–9.3 $mgNO_3$/l. Studies by Cruz and Andrade (2015) report ranges of 8.5–11.7 $mgNO_3$/l for water sources, whereas Kim *et al.* (2015) state that the natural level of nitrogen in the groundwater of alluvial aquifers is 3.0 and 5.5 $mgNO_3$/l. Nitrate nitrogen concentrations in water above 1 or 3 $mgNO_3$-N/l (Dimkić *et al.*, 2008; Kaown *et al.*, 2009; Pastén-Zapata *et al.*, 2014) and ammonium ions above 0.2 mgN/l (WHO, 2004) have been reported as the most likely anthropogenic impact. Ammonium cations ($NH_4^+$) in groundwater can be of autochthonous origin, as a result of anaerobic degradation of organic matter, or artificial, caused by the disposal of organic waste (manure, sewage, septic tanks). Significantly elevated concentrations of $NH_4^+$ in the range of 14–140 mg/l have been observed in aquifers contaminated with landfill leachate or due to wastewater disposal (Bohlke *et al.*, 2006). Inappropriate application of nitrogen fertilisers, inadequate quantities, meteorological conditions, types of fertilisers and soil types, as well as the presence of septic tanks, cause elevated concentrations of $NH_4^+$ (Bohlke *et al.*, 2006). Naturally elevated concentrations of $NH_4^+$ (up to 3 mg/l) have been observed in layers rich in humic substances, iron cations or in aquifers formed below forest ecosystems (Lindebaum, 2012).

The quantity of nitrogen that will be available for infiltration depends on agricultural practices, method of fertiliser application, time of application, weather conditions, and soil and crop types. Area characteristics such as climate, geology, morphology, and hydrology have a significant impact on the fate of nitrogen compounds. Climatic factors also affect the fate of nitrogen compounds, in that elevated temperatures accelerate denitrification and evaporation, elevated temperature and humidity enhance mineralisation, while increased precipitation causes faster transport of chemicals through the unsaturated zone to groundwater. Nitrate mobility has been studied in detail and is most often defined by tracer mobility, due to the stability of nitrate anions in an oxic environment and the inability of negatively charged ions to sorb to sediment and soil particles. Dissolved oxygen, redox, and electron donor availability are considered as geochemical indicators of conditions suitable for denitrification. There are no 'strict' limits on the concentration of dissolved oxygen for the occurrence of the denitrification process. Literature sources most often report that nitrate reduction begins to take place in the presence of organic substances, at an oxygen content of about 0.5 mg/l and *Eh* value lower than 250 mV. Korom (1992), Thayalakumaran *et al.* (2007) and Bottrell *et al.* (2008) state that the denitrification process can take place at dissolved oxygen levels of around 2.0 mg/l. After oxygen is consumed, various ionic and molecular species: nitrates ($NO_3^-$), manganese ($Mn^{4+}$), iron ($Fe^{3+}$), sulphate ($SO_4^{2-}$) and $CO_2$ are suitable electron acceptors.

The circulation of nitrogen within the biosphere through biological and chemical processes constitutes the nitrogen cycle and shows the relationships between different forms of nitrogen compounds and the environmental conditions under which they occur. The main processes that lead to the occurrence of various forms of nitrogen compounds in nature, and thus in water as well, are oxidation-reduction processes of the formation, transformation and transport explained within the nitrogen cycle. Figure 8.1 shows select processes of transformation of nitrogen compounds and transformation products. The main processes of the nitrogen cycle are:

- Fixation (electrochemical, industrial, biological);
- Assimilation (protein synthesis);

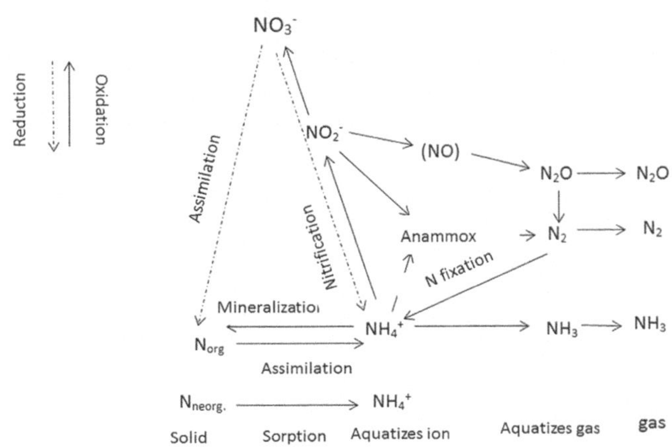


**Figure 8.1** Biogeochemical and physicochemical processes that influence the formation of nitrogen species in aquatic systems (Bohlke *et al.*, 2006).

- Mineralisation (amination, ammonification, nitrification);
- Nitrification;
- Denitrification;
- Dissimilatory nitrate reduction;
- Anaerobic ammonium oxidation – anammox, feammox.

Nitrogen compounds that reach groundwater are subject to complex dynamics of different biogeochemical processes of transformation within the N cycle: denitrification (Den.), dissimilatory (DNRA) and/or assimilatory nitrate reduction to ammonium (ANRA), nitrification, anaerobic ammonium oxidation (anammox, feammox), nitrifying denitrification, sorption, and mineralisation of organic matter (Nikolenko *et al.*, 2018; Weng *et al.*, 2017). Under anoxic/reductive conditions, nitrogen will become the subject of reductive transformations: assimilation into biomass, respiratory denitrification, anammox, feammox and dissimilatory/assimilatory nitrate reduction, usually associated with ammonium cation production (Perović *et al.*, 2017).

## 8.2.1 Nitrogen cycle and isotopic fractionation processes

Nitrogen is found in a wide range of oxidation states (from $-3$ to $+5$) and therefore can participate in various oxidation-reduction reactions, which affect the existence of a large number of different isotopic ratios of nitrogen, present in nature. There are more than 10 known types of nitrogen isotopes. Most are very unstable and radioactive, with the longest half-life being that of $^{13}N$ at 10 minutes. The only two stable forms of nitrogen isotopes are $^{14}N$ and $^{15}N$. The different relative ratio of heavier to lighter isotopes varies due to the fractionation process. Isotopic fractionation is caused by the existence of a stronger chemical bond, by a heavier isotope or due to mass-conditioned processes such as diffusion. Fractionation occurs due to different favourability of a particular isotope during the process that caused the fractionation.

### *8.2.1.1 Nitrogen fixation*

Nitrogen fixation is a reduction process of converting a biounavailable gaseous diatomic nitrogen molecule from air into the form of ammonia. The stable, diatomic molecule $N_2$ is removed from the atmosphere with the help of microorganisms that live in symbiosis with plants (symbiotic fixation), nitrogen-fixing microorganisms (non-symbiotic fixation), electrical discharge in the atmosphere (electrochemical fixation) and industrial processes of nitrogen fixation and synthesis. The nitrogen fixation process requires a certain amount of energy (high temperature, pressure or the presence of a catalyst) in order to break the non-polar triple covalent bond in the inert and stable $N_2$ molecule. Fixation is an extremely important ecological process because all green plants need nitrogen in the form of ammonia or nitrate to synthesise amino acid proteins. Plant proteins are a protein source for animals at the next trophic level. Nitrogen fixation is considered a unique process in terms of interpreting isotope fractionation, because $\delta^{15}N$ values are quantified in a plant or bacteria, regardless of the path of incorporation from $N_{2(g)}$ to organic matter (Sharp, 2007). The value of isotopic fractionation during the fixation process is very small, about 0‰ (Sharp, 2007).

### *8.2.1.2 Nitrogen assimilation*

Assimilation or immobilisation is a process by which $NH_4^+$ or $NO_3^-$ ions are incorporated into biota. The opposite process is degradation of organic matter with the help of heterotrophic bacteria, which release $NH_4^+$ in the process of mineralisation (ammonification). Various studies aimed at quantifying gross soil nitrification have shown that the dominant pathway of transformation and fate of nitrate and ammonium ions in soil is immobilisation by soil organic matter (Bengtsson *et al.*, 2003; Davidson *et al.*, 2003), thus preventing nitrogen loss and transition to groundwater and surface water. Nitrogen can be immobilised by abiotic or biotic processes, and immobilisation of nitrogen into soil organic matter is considered to be predominantly mediated by microorganisms (Bengtsson *et al.*, 2003). Studies have shown that microorganisms in the soil are responsible for 10–50% of immobilised $NH_4^+$ ions, of which one part of $NH_4^+$ becomes very stable and inert due to fixation by minerals of 2:1 structure (illite or vermiculates). The $\delta^{15}$ N values of plants depend on the value of $\delta^{15}N$ in the soil. The fractionation between $N_2$ atmospheric gas and labile organic nitrogen is approximately zero.

### 8.2.1.3 Mineralisation

Mineralisation of organic nitrogen in soil is a set of biochemical and chemical processes, by which the organic nitrogen formed by fixation and assimilation processes is hydrolysed, by proteolytic enzymes, as extracellular products of bacteria, fungi and actinomycetes. Hydrolysis of organic nitrogen produces peptones, amino acids, amines and amides, and inorganic bioavailable ionic forms of nitrogen. Mineralisation of organic nitrogen in the soil takes place through three phases: aminisation, ammonification and nitrification. Aminisation and ammonification are mediated by heterotrophic microorganisms, where organic matter is the energy source. The third phase, nitrification, is facilitated by autotrophic bacteria, which obtain energy by oxidising inorganic compounds and use carbon from atmospheric $CO_2$. Free amino acids, amides and amines, are present in small amounts in the soil because they are very quickly subjected to further changes and ammonification processes. Ammonification is the process of hydrolysis of amino acids, amines and amides in the presence of the enzymes deaminase and deamidase, during which ammonia is generated. This process takes place in anaerobic (hydrolysis and reduction) and aerobic (oxidation) conditions. The ammonia generated can be used by plants directly to produce protein and can be nitrified by autotrophic nitrifying bacteria (Sawyer & McCarty, 1978). The third step of the mineralisation process is nitrification. The nitrification process, and thus the solubility and mobility of the generated nitrates, is conditioned by the concentration of oxygen in the soil and pore water. The average annual rate of nitrogen mineralisation in arable land is in the range from 1 to 7% of total nitrogen (Džamić & Stevanović, 2007).

With regard to mineralisation in the aquifer matrix, it is important to emphasise that mineralisation of organic matter is accompanied by a decrease in pH and often consequent reductive dissolution of $Fe^{3+}$ ions (iron oxide and/or iron oxyhydroxide). Increased acidity of the environment causes the appearance of $Fe^{2+}$ cations and the frequent release of the mobile form of arsenic, the arsenic cation ($As^{3+}$), which is a common constituent of anoxic groundwater in layers rich in arseno-pyrite or iron-arsenic ores (Berg *et al.*, 2001; Weng *et al.*, 2017). If the medium is rich in organic matter, mineralisation can cause the release of mobile ions $Fe^{2+}$, $NH_4^+$ and $As^{3+}$, and the observed linear correlations mainly indicate the *autochthonous* origin of ammonium cations.

The fractionation during the decomposition of soil organic matter to ammonia is small ($0 \pm 1$). Kendall (1998) states that in the interpretation of isotope fractionation in the process of mineralisation, it is most important to determine whether the term mineralisation is considered to be the conversion of organic matter to ammonium ions (slight, small fractionation), or if mineralisation means oxidation to $NO_3^-$. In the latter case, fractionation can be large and variable, but the reason is not mineralisation itself, but nitrification of ammonium ions to nitrates.

### 8.2.1.4 Nitrification

A certain part of ammonium ions obtained by decomposition (mineralisation) in the soil, in oxic ambient conditions, can be transformed into nitrates through a two-stage oxidation process by chemolithotrophic bacteria, which oxidise ammonia and nitrites. This process takes place predominantly in aerated environments rich in oxygen, such as oxic waters and aerated surface layers of soil and sediment (Dimkić *et al.*, 2008). Nitrification is a two-stage process in which the oxidation of ammonium cations to nitrite anions takes place first and is followed by the oxidation of nitrite to nitrate anions. The oxygen needed for nitrification can originate from water or the atmosphere. The energy released during the nitrification process is used by microorganisms.

Nitrification bacteria are chemosynthetic organisms, which synthesise carbohydrates from water and carbon dioxide using the chemical energy released through ammonia oxidation. Ammonia salts are oxidised to nitric acid by bacteria of the genus *Nitrosomonas, Nitrosocystis* and *Nitrospira*, which provide energy for the fixation of $CO_2$ from the combined action of ammonia monooxygenase and hydroxylamine oxidoreductase. In the next phase, autotrophic nitrogen bacteria of the genus *Nitrobacter* oxidise nitrites to nitrates by the action of nitrite oxidoreductase.

Suitable conditions for the nitrification process are soil moisture 60–70% of capillary water, enough oxygen, pH value 6.2–8.2 and temperature 25–30°C.

There are two types of nitrification: (1) autotrophic nitrification and (2) heterotrophic nitrification. Although the two processes use the same substrates and produce the same intermediates and products, they differ in the enzymes they use. Autotrophic nitrification is carried out by two types of microorganisms, under the common name *Nitrobacterriae*: (1) $NH_4^+$ oxidisers, or primary nitrifiers, and (2) $NO_2^-$ oxidisers or secondary nitrifiers. Heterotrophic nitrification is carried out by bacteria: *Paracoccus denitrificans, Thiosphaera pantotropha, Pseudomonas putida* and *Alcakigenes fecalis*, or fungi.

If the soil is acidic, poorly aerated and very moist, the mineralisation process stops at the degree of ammonia formation. In early spring, nitrification slows down due to weakened microbial activity, caused by low soil temperatures and anaerobic conditions as a result of soil water saturation. Therefore, nitrification can be intensified by tillage and fertilisation of the soil as well as calcification, which binds nitrates and encourages ammonia nitrification. Nitrate levels in the soil begin to increase in spring, with the intensification of microbial activity, and reach maximum values in spring and/or summer. The application of organic and mineral fertilisers enriches the soil with nitrogen and other mineral elements and significantly accelerates mineralisation, by introducing organic matter that stimulates the activity of microorganisms (slowing down the decomposition of soil organic matter). Depending on ambient conditions, individual degrees of nitrification can be characterised by different degrees of fractionation. The conversion of $NO_2^-$ to $NO_3^-$ is a quantitative mass process, which means that there will be no isotopic fractionation in this step. The fractionation of ammonia to nitrite and finally nitrate preferentially incorporates a lighter isotope into $NO_3^-$, leading to a decrease in $\delta^{15}N\text{-}NO_3^-$ compared to the ammonium precursor in the range from −18 to −29‰, (Kendall, 1998; Sharp, 2007). According to Kendall and Aravena (2000), the produced $NO_3^-$ will be depleted in $^{15}N$ up to 35‰ (compared to the original $NH_4^+$).

### 8.2.1.5 Denitrification

Denitrification is the process of nitrate and nitrite reduction as terminal electron acceptors, whereby the following gases are formed as reaction products: nitrogen (II) oxide (NO), nitrogen (I) oxide ($N_2O$) and diatomic gas ($N_2$). When the concentration of gaseous nitrogen in groundwater reaches the saturation point, nitrogen tends to leave the saturated zone, which is considered a permanent nitrogen loss process. In addition to nitrogen oxides, $CO_2$ and $H_2$ are released in the denitrification process. Denitrification is an anaerobic process of 'breathing', which is characteristic of low oxygen concentration conditions. Anaerobic and aerobic bacteria and certain types of fungi are mediators of denitrification.

The rate of denitrification is conditioned by the presence and bioavailability of organic carbon in the form of dissolved organic matter and suspended particulate organic carbon, as well as the amount and bioavailability of reducing inorganic species: manganese ($Mn^{2+}$), iron ($Fe^{2+}$) and elemental iron (Fe°) or sulphur (pyrite $FeS_2$, S°), as well as silicate minerals (e.g. pyroxenes, amphibolites), at limited concentration levels of dissolved oxygen (Dimkić *et al.*, 2008; Rivett *et al.*, 2008). There are opinions that nitrate reduction does not take place if the amount of oxygen is above 0.2–0.4 mg/l (Komljenović & Todorović, 1998). The presence of sulphides in the soil will preferably induce dissimilatory nitrate reduction to ammonium (DNRA), compared to denitrification (Hiscock *et al.*, 1991). The presence of sulphur as sulphate and thiosulfate in soil has been shown to be an inhibitory factor of denitrification in soil (Kowalenko, 1979). Denitrification microorganisms are ubiquitous in surface water, groundwater and soil and their presence and activity have been recorded in different types of aquifers and at different depths. Activity of denitrifying microorganisms has been recorded in clayey sands at a depth of up to 289 m, in limestone at a depth of 185 m, and in granite at a depth of 450 m (Francis *et al.*, 1989; Morris *et al.*, 1988; Neilsen *et al.*, 2006). Denitrification takes place by the action of facultative heterotrophic bacteria: *Pseudomonas, Achromobacter, Bacillus* and *Micrococcus*. In addition to coliform bacteria, *Bacillus, Mycobacterium, Pseudomonas, Proteus, Thiobacillus, Clostridium, Alcaligenes, Micrococcus, Nocardia, Leptothrix, Serratia*, and other species are associated with denitrification (Komljenović & Todorović, 1998). Hydroxide ions are released during denitrification, which induces a pH increase. The increase in pH is conditioned by the intensity of denitrification and the buffer capacity of the soil solution. In the denitrification process, by reducing the nitrate concentration, residual $NO_3^-$ is exponentially enriched with the $^{15}N$ isotope, whereby the preferential conversion of the lighter isotope $\delta^{14}N$ to $N_2O$ and $N_2$ occurs (Kendall & Aravena, 2000). During denitrification $\delta^{15}N$ the value of initially produced $NO_3^-$ can be enriched relative to $N_2$ or $N_2O$ by approximately 20–30‰ (Clark, 2015), or 5–40‰ (Kendall, 1998). Laboratory experiments have shown the value of the difference $\delta^{15}N_{N2(g)} - \delta^{15}N_{NO3(aq)}$ to be from −17 to −20 ‰ (Sharp, 2007).

The combined study of $\delta^{15}N$ and $\delta^{18}O$ isotopes in nitrates enabled the assessment of the transformation process in the groundwater environment by microbial denitrification, because the increase in both values in the residual nitrate during the denitrification process is accompanied by a reduction in the nitrate concentration. This effect can be significant, so denitrification of a nitrate mineral fertiliser, which originally had a $\delta^{15}N$ value of 0 ‰, can increase this value in residual $NO_3^-$ from +15 to +30 ‰ $\delta^{15}N$ (Lindebaum, 2012).

### *8.2.1.6 Dissimilatory nitrate reduction to ammonium ions*

The process of dissimilatory nitrate reduction to ammonium ions (DNRA) is similar to denitrification. The conditions that enable DNRA and Den. are the same: anoxia and the presence of an organic substrate. It is considered that the amount of available organic matter determines which anaerobic nitrate reduction process will prevail, Den. or DNRA. Denitrification will become the dominant reduction pathway under carbon-limited conditions, while fermentative DNRA will become the dominant process in nitrate-limited environments rich in labile carbon (Burgin & Hamilton 2007; Kelso *et al.*, 1997; Korom, 1992; Perović *et al.*, 2017; Rivett *et al.*, 2008).

The microbially mediated DNRA process involves dissimilatory nitrate reduction to ammonia, as opposed to the assimilatory process (ANRA), in which nitrogen is incorporated into bacterial cells. Compared to the nitrate anion, the generated ammonium cation is more bioavailable and less mobile than the inorganic N. Although many heterotrophs assimilate nitrate for growth, $NH_4^+$ is the preferred ion. In some environments, there is evidence that the conversion of nitrogen into biomass may be a significant pathway for nitrogen elimination or temporary attenuation. If the conditions are suitable for the DNRA process, the absence of elevated ammonium concentrations in groundwater does not mean that nitrate reduction to ammonium does not take place, but that this process is probably mediated by assimilatory nitrate reduction (ANRA).

There are two known DNRA pathways. The first process is fermentation and the second is sulphur oxidation. Previous research has suggested that DNRA is a fermentative process (Tiedje, 1988), but research in recent years has documented a link between DNRA and the sulphur cycle (Brunet & Garcia-Gil, 1996; Kamp, 2007; Perović *et al.*, 2017).

In fermentative DNRA, electrons released by the oxidation of organic matter reduce nitrates through enzymatic reactions (Kamp, 2007; Megonigal *et al.*, 2005). The microorganisms responsible for this process are *Clostridia, Desulfovibrio, Vibrio* and *Pseudomonas*; these microorganisms can carry out fermentation without the use of nitrates (Tiedje *et al.*, 1984). An environment rich in labile organic carbon favours the metabolism of microorganisms that use electron acceptors with greater efficiency. DNRA transmits $8e^-$ per mole of reduced nitrates, while Den. transmits $5e^-$. In addition to the amount of labile carbon, the DNRA process is affected by the amount of oxygen. Microzones and underground niches of oxygen diffusion will be more suitable for the metabolism of facultative aerobic denitrifiers, in relation to the obligate anaerobic fermentative bacteria.

Another type of the DNRA process is chemolithoautotrophic DNRA, which combines nitrate reduction with oxidation of reduced sulphur, including free sulphide $S^{2-}$, and elemental sulphur. In this process, the dominant fate of reduced nitrates can be determined by the concentration of free sulphides, which are known to inhibit the last two reduction steps of denitrification. Sulphide inhibition of terminal denitrification reactions may induce reduction to ammonia rather than reduction to $N_2O$ and $N_2$. Microorganisms that perform chemolithotrophic DNRA are *Thiobacillus denitrificans, Thioploca, Thiomicrospora,* and *Beggiatoa* (Burgin & Hamilton, 2007; Kamp, 2007). The ecological advantages of DNRA are: (1) DNRA ensures nitrogen conservation in the ecosystem, and (2) microbial assimilation or cation exchange is enabled, therefore this process is considered as a nitrogen conservation process in groundwater.

There is very little data on the DNRA process in groundwater in general, and on the effect of DNRA on isotope fractionation. According to previous modest research, there are indications that the lighter nitrogen form $^{14}N$ is favoured in the DNRA process and therefore the produced $NH_4^+$ is depleted in $^{15}N$ compared to the remaining initial $NO_3^-$, which indicates kinetic fractionation (Lindebaum, 2012; Michener & Lajtha, 2007; Perović, 2019).

Laboratory studies show that $NH_4^+$ sorbed to clays and artificial ion exchange resins, usually become enriched with a heavier isotope $\delta^{15}N$, compared to the $NH_4^+$ remaining in solution, with an isotopic fractionation factor ($\alpha = [\delta^{15}N/\delta^{14}N]_{solid}/[\delta^{15}N/\delta^{14}N]_{dissolved}$) of about 1001 to 1011 (Bohlke *et al.*, 2006). Hübner (1986) showed that the fractionation of ion exchange is usually in the 1–8 ‰ range and that it depends on the concentration and fractionation factor for exchange with clay minerals.

#### *8.2.1.7 Anaerobic ammonium oxidation – Anammox*

The process of anaerobic ammonium oxidation, anammox, takes place in the presence of $NO_2^-$ or $NO_3^-$ as electron acceptors, where $NH_4^+$ is oxidised to $N_2$ and water. Anammox is a chemolithoautotrophic process in which the amount of total nitrogen in the system is reduced. The nitrite that participates in the reaction is most often a product of nitrate reduction. The process is mediated by slow-growing microorganisms, at a temperature of about 12°C. So far, several species of anammox bacteria have been identified: *Brocadia, Kuenenia, Anammoxoglobus, Jettenia* and *Scalindua*.

Anammox requires sufficient amounts of ammonia and nitrite. The inhibitory effect of low molecular weight organic compounds, such as pyruvate, glucose and ethanol, has been shown (Jetten *et al.*, 1998). Anammox can be a significant process in an ecosystem with limited labile organic carbon reserves.

Anammox and respiratory denitrification are considered as significant processes of nitrogen loss in environments depleted in organic carbon.

Research by Robertson *et al.* (2012) and Clark (2015) concludes that anammox leads to moderate enrichment of residual $NH_4^+$, by 4–8 ‰. Low values of the anammox fractionation effect were often observed during field research. Such values may be the result of the presence of a larger amount of $NH_4^+$ sorbed to the solid particles of the aquifer, buffering isotopic enrichment of $\delta^{15}N$ in dissolved $NH_4^+$ (Clark, 2015).

#### *8.2.1.8 Anaerobic ammonium oxidation – Feammox*

Feammox is a process of $NH_4^+$ oxidation where $Fe^{3+}$ is an electron acceptor. Oxidation of ammonium ions coupled with the reduction of $Fe^{3+}$ in the absence of oxygen is a process first observed in coastal wetlands. In this reaction, $Fe^{3+}$ is reduced to $Fe^{2+}$, while $NH_4^+$ is transformed into $NO_2^-$ or $N_2$. $Fe^{3+}$ can be in the form of oxides or oxyhydroxides, where As ions can often be sorbed. $NO_2^-$ produced during this process can be further reduced to $N_2O$, $N_2$ or $NH_4^+$, by dissimilatory $NO_2^-$ reduction or abiotic oxidation of $Fe^{2+}$. Feammox is considered a significant process of nitrogen loss in anoxic and acidic environments rich in iron. Table 8.1 provides an overview of nitrogen transformation processes, the expected changes during a process and the necessary conditions for nitrogen transformation in groundwater.

## 8.3 STATUTORY REGULATIONS

### 8.3.1 Views of the US Environmental Protection Agency and the World Health Organization

The Environmental Protection Agency (EPA) in the United States, based on the Safe Drinking Water Act (SDWA), prescribed the maximum permitted concentration of nitrates in drinking water of 10 mg of N/l (nitrate nitrogen), below which, based on previous scientific findings, no negative impact on health is expected. Besides health effects, potential costs of appropriate technologies for nitrate elimination from water are considered when setting concentration threshold values in practice.

Guidelines of the World Health Organization (WHO) set a limit of 50 $mgNO_3/l$ or 11.3 mgN/l. The maximum allowable concentration has been established with the aim of preventing the occurrence of methemoglobinemia in new-borns; however, other health effects have not been considered, including cancer and adverse reproductive outcomes (Ward *et al.*, 2018).

### 8.3.2 European directives

#### *8.3.2.1 The Nitrates Directive (EEC 1991/676)*

The seriousness of the problem of elevated nitrate concentrations in water bodies in the world can be recognised from the fact that the European Union has adopted a special directive on the protection of waters against nitrate pollution from agricultural sources, clearly confirming that agriculture is one of the main sources of nitrates in surface water and groundwater. The Nitrates Directive (Council Directive

**Table 8.1** Nitrogen transformation processes, expected changes and required conditions.

| The Process | Reaction | Required Conditions and Expected Changes | Fractionation (approximate) |
|---|---|---|---|
| Fixation | $N_{2(g)} + 3H_2O_{(g)} \rightarrow 2NH_3 + 3/2O_{2(g)}$ | Electrical discharges, plant enzymes | 0‰ |
| Mineralisation | $R - CHNH_2 - COOH + O \rightarrow R - CO - COOH + NH_3$ | pH decrease, anoxic conditions until nitrification process | $0 \pm 1$‰ |
| Nitrification | $NH_4^+ + 1.5O_2 \rightarrow NO_2^- + H_2O + 2H^+ NO_2^- + 0.5O_2 \rightarrow NO_3^-$ | pH decrease, oxygen >1 mg/l | Produced $NO_3^-$ will be depleted in $\delta^{15}N$ by −18 to −29‰ and up to −35‰ |
| Denitrification | $5CH_2O + 4NO_3^- \rightarrow 2N_2 + 4HCO_3^- + CO_2 + 3H_2O$ | Dissolved oxygen concentration <2.0 mg/l and *Eh* value <250 mV; pH increase, high TOC, non-sulphidic, low C:N; low TOC, low Fe, high C:N | The remaining nitrate is enriched by 5–40‰ |
| | $10Fe^{2+} + 2NO_3^- + 14H_2O \rightarrow 10FeOOH + N_2 + 18H^+$<br>$5HS^- + 8NO_3^- + 3H^+ \rightarrow 5SO_4^{2-} + 4N_2 + 4H_2O$ | Low TOC, high conc. Fe, Sulphidic waters, FeS, $S^0$ high TOC | |
| Dissimilatory nitrate reduction to ammonium (C >> N) | $2H^+ + NO_3^- + 2CH_2O \rightarrow NH_4^+ + 2CO_2 + H_2O$ | High TOC, non-sulphidic, high C: N-fermentative DNRA | For about + 10‰ the remaining nitrate is enriched, isotopically depleted $NH_4^+$ is generated |
| | $4HS^- + 4NO_3^- + 4H_2O + 4H^+ \rightarrow 4SO_4^{2-} + 4NH_4^+$ | Sulphidic waters, $H_2S$, high TOC, | |
| Anaerobic ammonium oxidation | $NH_4^+ + NO_2^- \rightarrow N_2 + 2H_2O$ | Low TOC, low Fe, low C: N (N > C), respiratory Den. to $NO_2^-$ then anammox | Like denitrification |

TOC – Total Organic Carbon.

91/676/EEC) is one of the most important legal bases of the EU on environmental protection within the framework of the European water policy and represents the basic support to the Water Framework Directive. According to the Nitrates Directive, which was adopted by all member states of the European Union, the concentration of nitrates is limited to 50 mg/l in all freshwater bodies, regardless of their purpose (Directive 91/676/EEC, 1991). Several EU member states (Denmark, Belgium, France, Germany, the Netherlands, Luxembourg, England and Finland) noted and presented the problem of agricultural nitrogen emissions in national reports as part of the implementation of the Nitrates Directive. The objectives of the directive are to reduce water pollution caused or induced by elevated concentrations of nitrate of agricultural origin and to prevent further pollution. In Serbia of all the pollutants in groundwater, limit values have been defined only for nitrates (50 mg/l) and pesticides (Official Gazette of RS No. 55/05, 71/05-corrigendum, 101/07, 65/08 and 16/11). The Nitrates Directive has not yet been transposed into Serbia's legislation. A significant contribution to the determination of nitrate vulnerable areas and the rules of good agricultural practice in Serbia has been made by the team of Prof. P. Marjanović at the Jaroslav Černi Water Institute (JCI) in Belgrade, under the project 'Determination of sensitive areas and vulnerable zones according to the Nitrates Directive and Urban Waste Water Treatment Directive' (NRP, 2016). Through the project 'Analysis of surface water and groundwater quality monitoring network in accordance with the requirements of the Water Framework Directive and accompanying directives' (JCI, 2017) a map of potential vulnerability of groundwater to pollution in Serbia was created (Stevanović & Dokmanović, 2015).

The Nitrates Directive sets forth the following basic requirements:

- Identification of sensitive surface waters intended for water supply as well as sensitive groundwaters that are under the influence/at risk of pollution in case of failure to take protection measures (Article 3.1).
- Determination and delineation of vulnerable zones defined as water catchment areas of the sensitive waters from Article 3.1, which contribute to pollution (Article 3.2).
- Establishment of one or more Rules of good agricultural practice to be applied by agricultural producers (Article 4.1).
- Adoption of an Action Programme for vulnerable zones, whether it is one or more action programmes for individual or groups of specific vulnerable zones (Article 5.1–2).
- Establishment of monitoring programmes to track the efficiency of action programme implementation (Article 5.6).
- Reporting to the European Commission every four years, describing the latest situation in relation to the results of the monitoring programme, the designation of vulnerable zones and summaries of action programmes (Article 10).
- Adoption and implementation of the laws, regulations and administrative procedures necessary to implement in full the provisions of the Nitrates Directive (Article 12).

The criteria for determining and delineating vulnerable zones are defined in Annex I of the Nitrates Directive, whereby member states may decide to implement action programmes throughout their territory (to declare the entire territory as vulnerable) or only within delineated nitrate-vulnerable zones. The criteria for identifying vulnerable zones include:

- Surface water used or intended to be used for water supply in which the nitrate concentration is 50 mg/l (as $NO_3$) or 11.29 mg/l (as N) and water at risk of exceeding these concentrations in the absence of protection measures;

- Groundwater in which nitrate concentrations exceed 50 mg/l (as $NO_3$) or 11.3 mg/l (as N), as well as groundwater at risk of exceeding these concentrations in the absence of protection measures;
- Natural freshwater lakes and other freshwater and coastal and marine waters identified as eutrophic or at risk of becoming eutrophic in the near future;
- Eutrophic waters should be classified as sensitive, even if phosphorus is the nutrient that causes eutrophication, which is often the case in continental (inland) waters.

In vulnerable zones, the Nitrates Directive limits the application of nitrogen originating from manure to a maximum of 170 kgN/ha.

The rules of good agricultural practice defined in Annex II of the Nitrates Directive relate to periods when the application of fertilisers is inappropriate, application of fertilisers on steep terrains, application of fertilisers on water saturated, flooded, frozen or snow-covered land, conditions for land use and construction of manure storage facilities near water courses and include measures to prevent water pollution due to runoff and leaching from manure storage sites and/or drainage from plant material storage such as silage, as well as procedures for fertiliser application (manure and fertilisers), including speed and uniformity of spreading on the soil. The intention of these guidelines and practices is to maintain the loss of nutrients to water at an acceptable level.

According to the Nitrates Directive, action programmes must include:

- Mandatory measures in relation to periods during which the application of certain types of fertilisers is prohibited;
- Minimum manure storage capacity;
- Restrictions on the application of fertilisers on sensitive terrains (terrains under high slopes, wet terrains, flood-prone areas, frozen or snow-covered terrains, terrains near watercourses);
- Other measures defined in the Rules of Good Agricultural Practice.

Mandatory measures in action programmes should be a legal obligation. The Nitrate Directive is binding within Nitrate Vulnerable Zones (NVZs) and requires updates every four years based on current information on the situation in agriculture and the environment. In line with this requirement, member states need to be prepared to review the NVZs as necessary, in particular where an increase in the area of existing NVZs or the designation of new NVZs is justified.

#### *8.3.2.2 The Water Framework Directive (2000/60/EC)*

In 2000, the European Parliament and the Council of the European Union adopted the Water Framework Directive (WFD), which is the most important EU-wide legislative framework in the field of water. The WFD serves as an umbrella directive for all other water-related directives and forms the basis for a coherent and uniform water policy with the aim of successfully realising the concept of integrated environmental management.

The purpose of the WFD is to establish a framework for the protection of all waters: inland surface waters, mixed waters, coastal waters and groundwater, to:

- Prevent further deterioration and protect and enhance the status of aquatic ecosystems, as well as terrestrial and wetland ecosystems, which are directly dependent on the aquatic system;
- Promote sustainable water use based on long-term protection of available water resources;
- Direct and improve the protection of the aquatic environment, through specific measures for progressive reduction of the releases, emissions and losses of priority substances and the cessation or gradual decrease of the releases, emissions and losses of priority hazardous substances;

- Ensure progressive reduction and prevent further pollution of groundwater;
- Contribute to mitigating the effects of floods and droughts.

In so doing, the WFD contributes to:

- The provision of adequate quality surface water and groundwater for the needs of sustainable, balanced and equitable water use;
- Significant reduction of groundwater pollution;
- Protection of inland and marine waters;
- Ensuring that the objectives of important international agreements are achieved, including those aimed at preventing and eliminating marine pollution within the framework of European Union action to eliminate or phase out the discharges, emissions and losses of priority hazardous substances under Article 16, and that the ultimate goal of achieving concentrations of natural substances in seawater, which are close to undisturbed values and close to zero for synthetic substances of anthropogenic origin, is fulfilled.

The main WFD components are:

- Extending the scope of water protection to all waters;
- Achieving good ecological and chemical status of all waters;
- Water management based on river basins;
- 'Combined approach' when it comes to limiting emissions and quality standards;
- Realistic pricing of water services and the polluter pays principle;
- Public participation.

EU member states and candidates are obliged to stop further deterioration of the quality of water bodies and improve and restore the condition of aquatic as well as terrestrial ecosystems and wetlands that are directly dependent on aquatic ecosystems. River basins are defined as basic units for water management, instead of administrative or political boundaries.

Integrated water management is promoted as the most effective method of sustainable water use and it requires coordinated land and water use planning in the river basin. This includes all surface waters, coastal waters and groundwaters, as well as soil. Member states are obliged to transpose all WFD requirements into their national legal systems.

The WFD also contains a number of deadlines, with the aim of achieving a general goal – 'good water status'. In many cases, these requirements alone will not achieve good status, so other laws at national, regional and local levels need to be adapted in accordance with the WFD. Included are the precautionary principle and the polluter pays principle, promoting the sustainable use of water as a resource, based on long-term protection. The economic aspect is also considered, as an important part of water management as well as water policy, and is based on three approaches:

- The polluter pays principle;
- Financial profitability;
- The price of water.

European regulations that establish measures to prevent surface water pollution are implemented at two levels:

- Through the selection of substances relevant to achieving good status of water;
- By defining the measures that each member state must implement at the river basin level to prevent pollution by certain hazardous substances.

## 8.4 ORIGIN OF NITROGEN COMPOUNDS IN WATER BODIES

Global growth in crop production has been achieved through intensive use of agrochemicals, such as pesticides and chemical fertilisers, and the expansion of agricultural land. Compared to 1960, the consumption of mineral fertilisers in the world has increased tenfold (FAO, 2016). North America, Europe and parts of South and East Asia are the regions with the most pronounced impact of elevated concentrations of nutrients on the quality of water bodies (FAO, 2017a).

As agriculture is the main 'consumer' of water, accounting for 70% of the total amount of water used worldwide, the impact of agriculture on water quality is significant (FAO, 2017a). Farms are emitters of agrochemicals, organic substances, drug residues, sludge and excrement into water bodies.

In most high-income countries with developed economies, agricultural pollution is more significant than pollution due to wastewater from urban and industrial areas and represents a major factor of degradation of the quality of inland and coastal waters (eutrophication) (FAO, 2017a). In low-income countries and emerging economies, the greatest impact on the quality of water bodies is still untreated municipal and industrial wastewater (FAO, 2017a). It should be borne in mind that municipal and industrial wastewater is a very significant factor in different countries, which affects the water quality of the hydrographic network.

On average, high-income countries treat about 70% of wastewater generated, while this share decreases to 38% in middle-income countries and to 28% in lower-middle-income countries (United Nations World Water Assessment Program [WWAP], 2017). In low-income countries, only 8% of industrial and municipal wastewater is subjected to some form of treatment (Sato *et al.*, 2013). These estimates support the fact that globally, over 80% of wastewater is discharged into the environment without adequate treatment (WWAP, 2013, 2015).

### 8.4.1 Fertiliser consumption in the world

Nitrates originating from agriculture are the most common chemical contaminants in groundwater worldwide (WWAP, 2013). In the European Union, 38% of water bodies are under high pressure from agricultural pollution (WWAP, 2015). In the United States, agriculture is a major source of river and stream pollution (US EPA, 2016). In China, agriculture is responsible for a large share of surface water pollution and is almost the sole cause of nitrogen pollution of groundwater (FAO, 2013).

Figure 8.2 shows the consumption of fertilisers presented as the amount of fertiliser per hectare of arable land. Fertilisers include nitrogen, potassium and phosphate fertilisers (including mineral phosphate). Manure and 'green manure' fertilisers are not included. Fertiliser consumption is determined by the sum of the quantities of produced and imported fertilisers, from which the quantity of exported fertilisers is deducted. Because certain chemical compounds may have other industrial applications in addition to being used as fertilisers, the values shown may be higher than the actual quantities available to crops. Average annual values are shown. Figures 8.3–8.5 show the average global production of nitrogen fertilisers (Figure 8.3), the application of fertilisers (Figure 8.4) and the application of nitrogen fertilisers in the world (Figure 8.5).

According to Eurostat (2018) data, the nitrogen balance in the countries of the European Union for the period 1990–2015 is graphically represented in Figure 8.6. The nitrogen balance was calculated as the difference between input and total nitrogen loss in kg/ha. The total intake of nitrogen originates from inorganic fertilisers (mineral and complex fertilisers, combined inorganic and organic fertilisers), organic fertilisers (compost, sewage sludge applied to agricultural land), manure, biological nitrogen fixation, nitrogen from atmospheric deposition and other sources (seeds and seed material). Total nitrogen losses represent the sum of the amounts of nitrogen exported by harvest and total fodder production. According

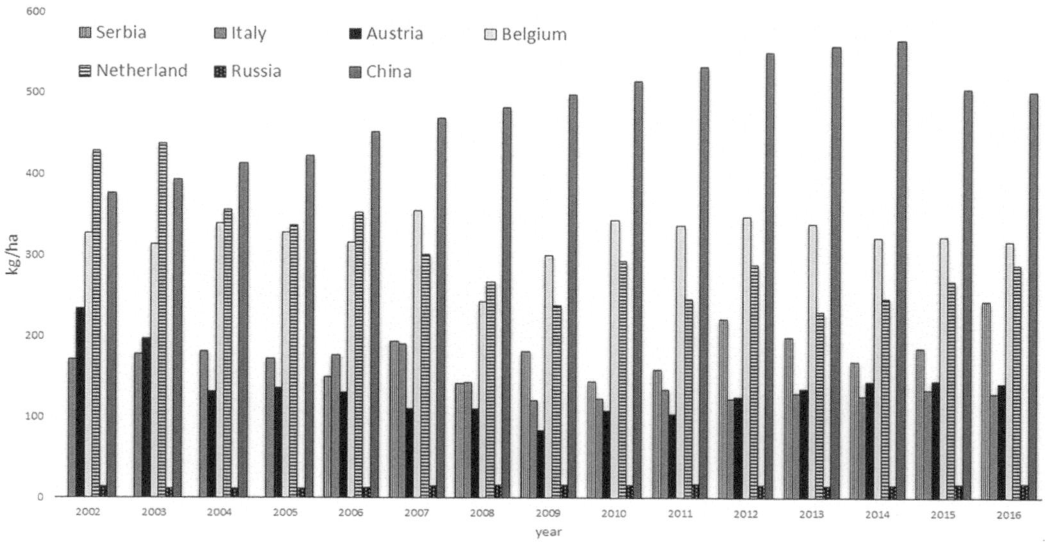


**Figure 8.2** Average annual fertiliser consumption in selected countries (FAO data).

**Figure 8.3** Production of nitrogen fertilisers in the world (FAO, 2017b).

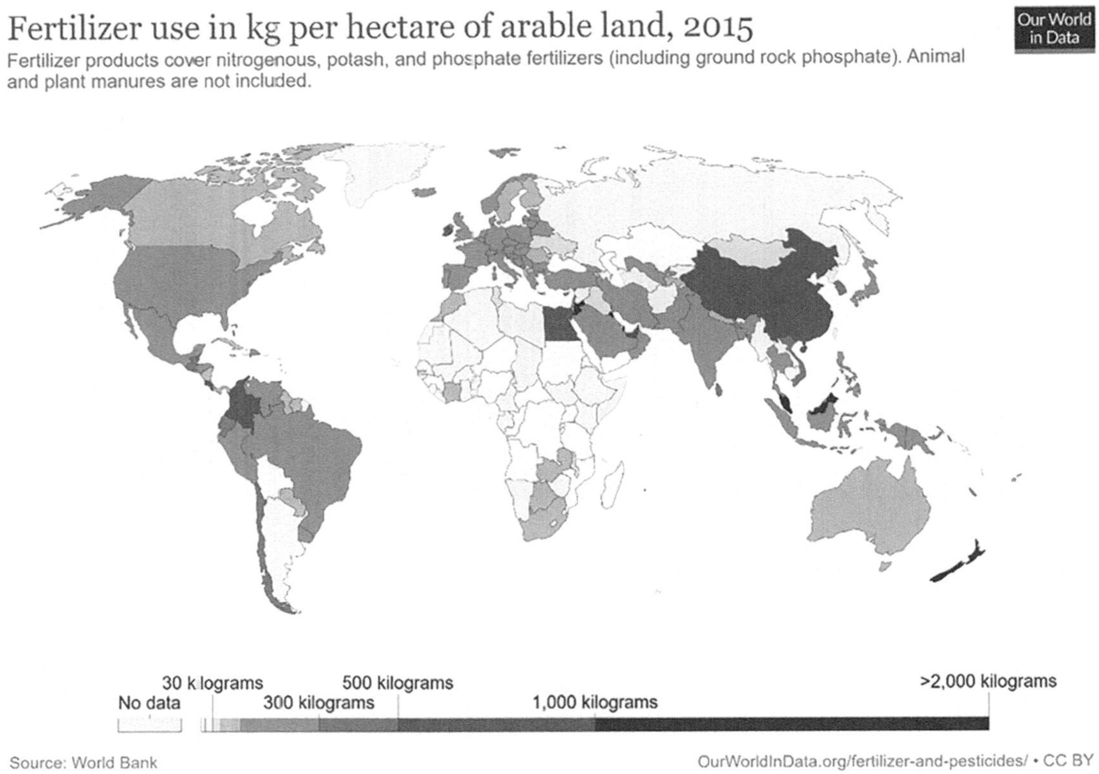


**Figure 8.4** Fertiliser application in the world (FAO, 2017b).

to the above data, a decreasing trend in the N balance is evident in the following developed countries: Austria, Belgium, Denmark, France, the Netherlands, Japan, Germany and Sweden. While less pronounced, a decreasing trend of nitrogen balance is noticeable in Romania, Slovenia, Croatia and Bulgaria (Figure 8.6).

## 8.4.2 Concentrations of nitrogen compounds in European waters

Since 2005, a decreasing trend of nitrate concentration has been observed in groundwaters in Europe, showing that nitrate concentration in 2011 decreased to the levels recorded in 1992. In the rivers of the European Union, the decreasing trend of nitrate concentrations led to an average annual decrease of 0.03 $mgNO_3{-}N/l$, or 0.8% annually for the period 1992–2012 (European Environment Agency [EEA], 2018). The reduction of nitrate nitrogen concentration is the result of measures applied to reduce pressures on water bodies originating from agricultural sources (Nitrates Directive – Directive 91/676/EEC), as well as improvements in wastewater treatment (Urban Wastewater Directive – Directive 91/271/EEC), but also due to the decrease in industrial production in Eastern Europe.

Figure 8.7 shows graphical trends of nitrate and ammonium ion concentrations in surface and ground waters in the European Union (EEA, 2018). The data series shown are the average values of mean annual concentrations in the period 2000–2012. Data for nitrate and ammonium ion concentrations in

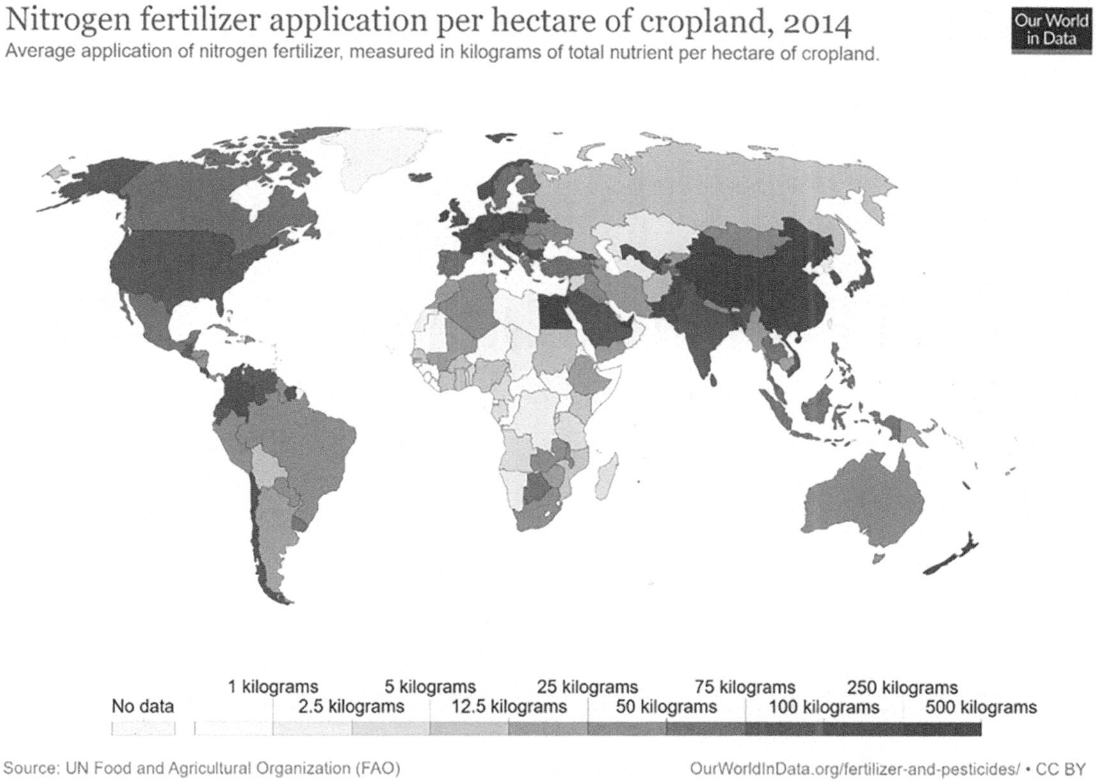


**Figure 8.5** Application of nitrogen fertilisers in the world (FAO, 2017b).

surface waters are presented for: Austria 49 monitoring stations, Bosnia and Herzegovina 23, Belgium 35, Bulgaria 90, Cyprus 8, Denmark 39, Estonia 53, Finland 101, France 246, Germany 136, Ireland 45, Iceland 1, Latvia 19, Liechtenstein 12, Lithuania 28, Luxembourg 3, North Macedonia 19, Norway 10, Poland 60, Romania 118, Slovakia 15, Slovenia 13, Sweden 109, Switzerland 15, and Great Britain 153. The data series for nitrate concentrations in groundwater is presented as mean values of average annual concentrations in groundwater: for Austria at 13, Belgium 24, Bulgaria 24, Cyprus 3, Czech Republic 63, Denmark 107, Estonia 23, Finland 33, France 54, Germany 167, Ireland 86, Italy 5, Liechtenstein 1, Lithuania 3, Netherlands 9, Norway 1, Portugal 10, Slovakia 10, Slovenia 4, Switzerland 1 and England 67 stations.

Between 1992 and 2012, the average concentration of ammonia in European rivers decreased by 231 μgN/l. The average annual decrease was 11.6 μgN/l (−3.5% per year). At 59% of the surface water quality monitoring stations, a trend of significant decrease in ammonia concentrations was observed, with only a few stations reporting an increasing trend in the period 1992–2012. The concentration of ammonium ions in rivers is shown as the mean of annual concentrations. The number of monitored river stations by country was: Austria (49), Belgium (36), Bosnia and Herzegovina (7), Bulgaria (88), Croatia (37), Estonia (53), Finland (100), France (243), Germany (139), Ireland (35), Italy (155), Latvia (19), Liechtenstein (14), Lithuania (28), Luxembourg (3), North Macedonia (19), Norway (10), Romania (117), Serbia (37), Slovenia (14), Sweden (110), and the United Kingdom (12).

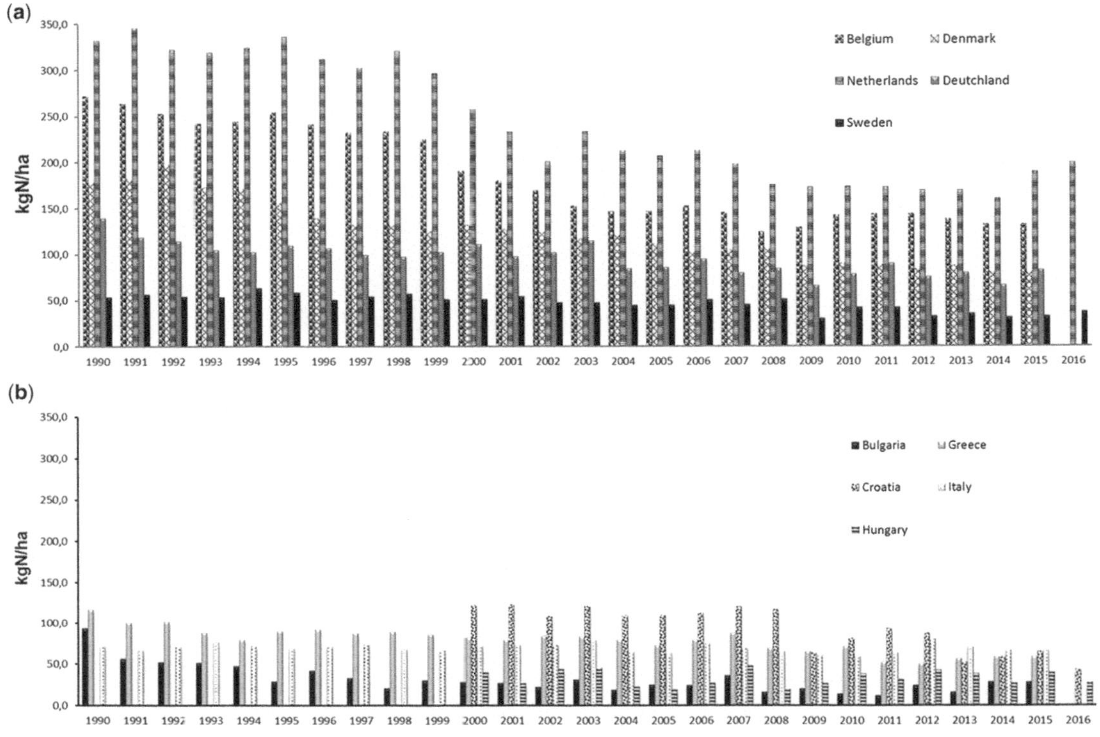


Figure 8.6 N balance sheet trends in selected European countries (Eurostat, 2018 data).

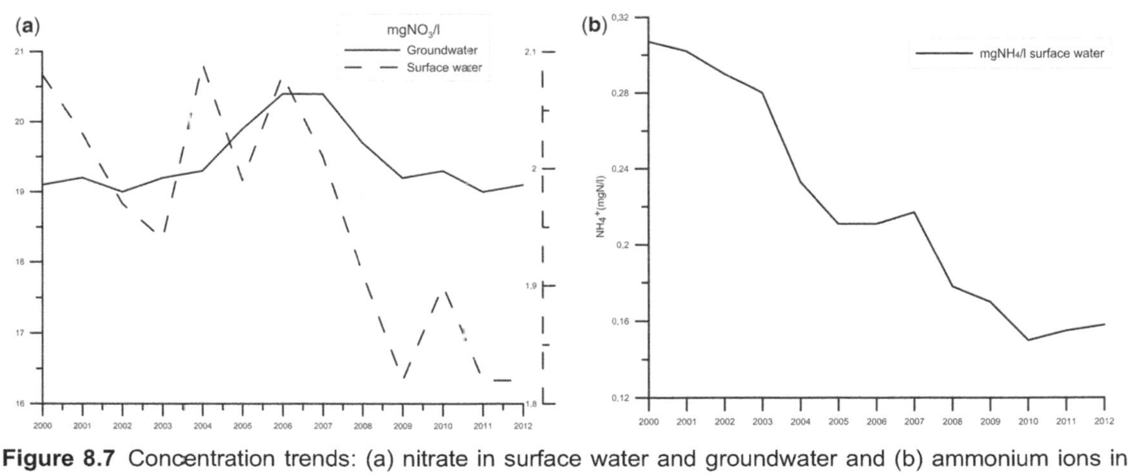


**Figure 8.7** Concentration trends: (a) nitrate in surface water and groundwater and (b) ammonium ions in surface water in the EU in the period 2000–2012 (data source EEA, 2018).

Based on the presented results of the examination of the quality of water bodies, it follows that in Europe there are downward trends of nitrogen in surface water and groundwater. This is certainly due to the implementation of the Nitrates Directive and other water directives. However, it should be emphasised, especially with regard to groundwater, that there are other, very relevant circumstances, on which the nitrogen concentration in groundwater will depend:

- Type of aquifer: shallow (alluvial, etc.) and/or deep, degree of porosity, etc.;
- The oxic state of aquifers, especially alluvial, which also depends on their genesis. In the upper parts of river basins, where sediments are coarse, alluvial aquifers are mostly oxic. In the downstream parts, alluvial aquifers are less oxic, sub-oxic or anoxic. This is, for example, the case with the alluvium of the Danube River. Upstream of Budapest, the springs that capture the Danube alluvium are predominantly oxic. Downstream springs and alluviums are predominantly anoxic. The form and concentration of nitrogen compounds, as well as their transformation in groundwater, are certainly conditioned by the structure of sediments;
- The economic situation of a region/country and use of fertilisers in agriculture. In the Danube River Basin, the rate of nitrogen fertiliser application is greater in upstream regions/countries. Often, the use of fertilisers increased as upstream countries gained economic power, which together with dominant groundwater oxic conditions in the upper basin, resulted in an increased susceptibility of alluvial aquifers to nitrate pollution. A weaker economic situation often results in lower use of nitrogen fertilisers in agriculture. However, in such regions sewage systems and wastewater treatment plants are often underdeveloped. The disposal of municipal waste (liquid and solid) in open pits and many unsanitary landfills also leads to high concentrations of nitrogen compounds in alluvial aquifers.

Elevated nitrate concentrations are the most common cause of poor chemical status of groundwater in Europe (EEA, 2012). It is estimated that 50–80% of the total nitrogen load in European freshwater bodies comes from agriculture (EEA, 2012). For waters to have a good status, they must also achieve a good chemical status of surface water and groundwater, which is defined by the Water Framework Directive and sub-directives. The chemical status of about 25% of groundwater in the EU is poor. Poor chemical status of surface waters does not exceed 10% in Europe (EEA, 2012). The chemical status of 40% of surface waters remains unknown. In addition to intensive agricultural production, population density is considered a major pressure on river quality. River basins with over 40% of arable land and a population density of over 100 inhabitants/km$^2$ have more than two thirds of their water bodies in poor status (EEA, 2012). The implementation of the Urban Wastewater Treatment Directive (Directive 91/271/EEC), along with national legislation, has increased wastewater treatment and resulted in reduced emissions of nutrients and organic pollution from point sources to freshwater recipients.

## 8.5 METHODS FOR DETERMINING THE ORIGIN AND NITROGEN TRANSFORMATION PROCESSES IN WATER BODIES

Modern biogeochemical research focuses on defining the causal relationship between hydrogeological, physicochemical and microbiological parameters of soil and water in order to define the dominant influence on the final transformation process. According to consulted scientific literature, the dominant influences on nitrogen transformation are: oxygen concentration, redox potential, pH value, quantity and quality of available organic carbon, C:N ratio (dissolved organic carbon (DOC) and nitrates ratio – DOC:$NO_3^-$), and concentrations of other electron donors – iron cations, and manganese and sulphide anions (Burgin & Hamilton, 2007; Dimkić *et al.*, 2008; Nikolenko *et al.*, 2018; Rivett *et al.*, 2008).

In order to establish whether there is an anthropogenic impact on water bodies, as well as a potential for conservation or emission of nitrogen compounds, it is necessary to define the prevailing environmental conditions for certain processes and determine the trend of changes in concentrations and correlations between substances involved in oxidation-reduction reactions, as well as chemical species that can be indicators of anthropogenic impact.

As the isotopic signature of nitrogen can change under the influence of various complex biogeochemical mechanisms of transformation: volatilisation, nitrification and denitrification, it cannot be exclusive in the identification of sources (Pasten-Zapata *et al.*, 2014). The simultaneous research of stable nitrogen and oxygen isotopes in nitrates with monitoring of a conservative tracer's movement patterns contribute to more accurate determination of nitrate sources and transformation processes in groundwater.

In order to correctly interpret the observed correlations between the studied groundwater quality parameters, it is necessary to verify the assumed transformation processes and nitrogen sources by simultaneous, comprehensive and complex analyses of physicochemical, hydrogeological, microbiological parameters and isotopic signatures.

Many authors state that if there are anoxic conditions, the amount and bioavailability of organic carbon (TOC) is the next important factor relevant for nitrogen reduction (Burgin & Hamilton, 2007; Rivett *et al.*, 2008).

If the concentration of organic carbon (electron donor) is low, the reduction of nitrate concentration will take place as follows: (1) in the case where a low concentration of organic carbon is accompanied by low concentrations of iron (Fe (II/III)), reduction of nitrate anions will take place by respiratory denitrification to $NO_2^-$ and then anammox to $N_2$, or by respiratory denitrification to $N_2$; (2) if a low concentration of organic carbon is accompanied by high concentrations of iron, denitrification to $N_2$ will take place with oxidation of $Fe^{2+}$. The second case of nitrate reduction occurs in anoxic conditions where there is a high concentration of organic carbon as follows: (1) if a high concentration of organic carbon is accompanied by the presence of hydrogen sulphide, the following processes are expected: dissimilatory $NO_3^-$ reductions to $NH_4^+$, due to the inhibitory effect of $H_2S$ on the denitrification process; and in the case that metal sulphides or zero sulphur are present, denitrification to $N_2$ with sulphide oxidation is expected; (2) if the concentration of TOC is high and the water is not sulphidic, at a low C:N ratio, respiratory denitrification to $N_2$ will take place; and at a high C:N ratio enzymatic DNRA with nitrogen conservation in the form of ammonium ions is expected (Burgin & Hamilton, 2007). This differentiation of the nitrate reduction processes based on different environmental conditions is shown in Figure 8.8 and Table 8.2.

### 8.5.1 Factors that affect nitrate production/consumption and their impact on $\delta^{15}N\text{-}NO_3^-$ and $\delta^{15}N\text{-}NH_4+$ variability

Due to a potentially very significant impact on the isotopic signature of the initial N pollutant, for examination of $\delta^{15}N$ variability in groundwater, it is very important to consider the possibility of mixing waters of different origins and ages and the influence of various environmental factors on the intensity of the transformation processes (Nikolenko *et al.*, 2018).

The significance and rate of the fractionation processes of nitrification, denitrification and anammox depends on the physicochemical conditions of the hydrogeologic system in which these processes take place. Substrate concentration, electron donor availability, dissolved oxygen concentration, temperature, pH, and water retention time have been stated as factors with the greatest influence on fractionation

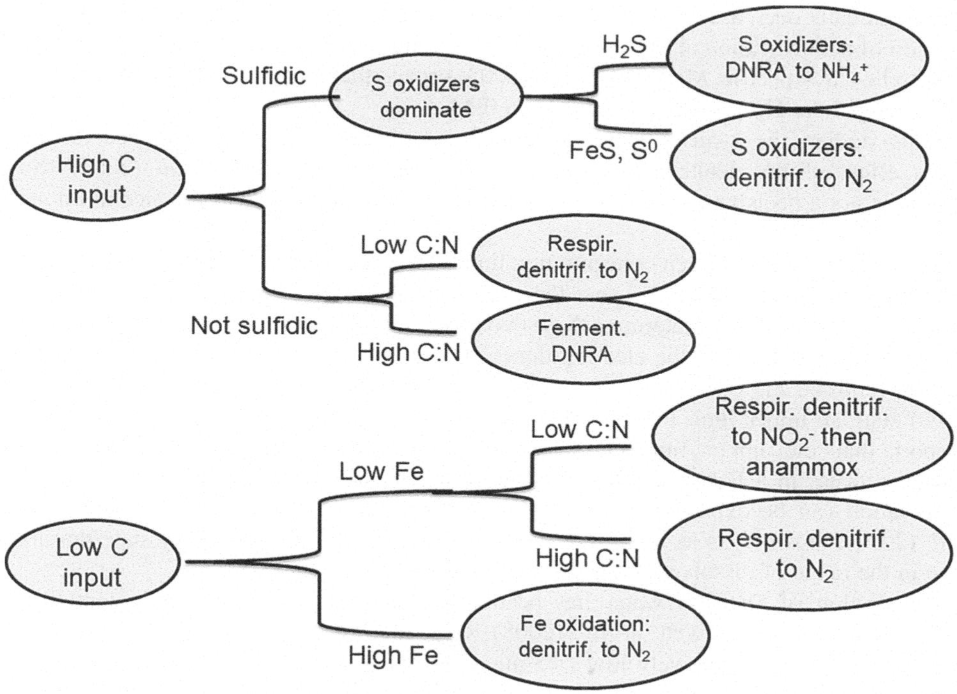


**Figure 8.8** Assumed dominant dissimilatory pathways of nitrate reduction. The figure summarises the conditions under which a specific nitrate reduction pathway was anticipated (C:N – ratio of total organic carbon to nitrate) (Burgin & Hamilton, 2007, modified).

(Nikolenko *et al.*, 2018). Research by Li *et al.* (2007) shows that the amount of substrate determines the degree of fractionation by minimising it in the N-limited system and maximising it in the system with a constant and high inflow of N compounds. Nitrification and fractionation rate will be intensified in the presence of large amounts of $NH_4^+$ (fertiliser application). As $NH_4^+$ is consumed, nitrification

**Table 8.2** Values of $\delta^{15}N\text{-}NO_3^-$ and $\delta^{15}N\text{-}NH_4^+$ for different sources in groundwater (based on data from Nikolenko *et al.*, 2018).

| | Source | Range $\delta^{15}$ N-$NO_3^-$ |
|---|---|---|
| $\delta^{15}$ N-$NO_3^-$ | Organic soil nitrogen | +3 to +8 ‰ |
| | Mineral fertilisers | −8 to +7 ‰ |
| | Organic fertiliser/manure | +5 to +35 ‰ |
| | Municipal wastewater | +3 to +25 ‰ |
| | Rainwater | −12 to +11 ‰ |
| $\delta^{15}$ N-$NH_4^+$ | Organic soil nitrogen | +2.4 to 4.1 ‰ |
| | Rainwater | −13.4 to +2.3 ‰ |
| | Mineral fertilisers | −7.4 to 5.1 ‰ |
| | Municipal wastewater | +5 to +9 ‰ |
| | Organic fertiliser/manure | +8 to +11 ‰ |

fractionation gradually decreases. Rivett *et al.* (2008) showed that $NO_3^-$ in high concentration can induce the completion of denitrification at the level of $N_2O$ generation. The concentration limit for this effect has been shown to be case-specific. Magalhaes *et al.* (2003) proved that increasing the nitrate concentration by 0–4 mgN/l changed the $N_2O$:$N_2$ ratio from 0.11 to 0.34.

In previous studies, the availability of electron donors has largely been discussed in terms of the fractionation effect of the denitrification process. The research by Choi *et al.* (2003) showed that denitrification is not a decisive factor in the increase of $\delta^{15}N$ in $NO_3^-$ where the electron donor content is low.

It is difficult to define the oxygen concentration limit that would strictly separate the conditions suitable for nitrification from denitrification, because there are conditions when both processes can be expected to occur simultaneously. In order to determine these two processes, it is necessary to know the trend patterns of changes in the concentrations of the electron donors and acceptors and changes in pH values. Different authors have identified the occurrence of denitrification at different values of dissolved oxygen. Feast *et al.* (1998) state an upper limit for the occurrence of denitrification of 0.2 mg/l, while Dimkić *et al.* (2008) report that the upper limit of dissolved oxygen concentration for a more pronounced denitrification process in a Belgrade alluvial groundwater source is 0.5 mg/l. Rivett *et al.* (2008) state that denitrification can be expected if the oxygen concentration is below 2 mg/l, while according to Baily *et al.* (2011), this limit is as high as 4 mg/l. Stenstrom and Poduska (1980) state that nitrification is intensive in the range of dissolved oxygen concentrations of 0.3–4 mg/l.

The concentration of DOC in water has been quantified as an indicator of potentially available nitrogen for denitrification. Rivett *et al.* (2008) state that in most aquifers the concentration of dissolved organic carbon is relatively low (<5 mg/l). In the case where the electron donor is reduced sulphur, an increase in the concentration of sulphate ions is expected, with a decrease in the nitrate concentration (in mixed-anoxic conditions). It has been shown that the concentration of DOC decreases with increasing concentration of sulphate ions, which is explained by the reduced solubility of DOC in conditions of increased ionic strength and acidity of water (Clark, 2005). Water temperature controls the microbial activity and the concentration of dissolved oxygen in the water, so seasonal changes can affect $\delta^{15}N$-$NO_3^-$, resulting in higher values of enrichment factors in summer in the aquifer where denitrification occurs, or lower values in groundwater where nitrification occurs (Nikolenko *et al.*, 2018).

It is believed that the optimum pH range for nitrification is 6.5–8 (Buss *et al.*, 2004). For the anammox process, a pH range of 6.5–9.3 was determined with an optimal value of 8 (Tomaszewski *et al.*, 2017).

To determine the nitrogen input to aquifers under farmland, besides the type, amount and method of fertiliser application, it is important to consider the transformation processes of the N cycle (mineralisation and nitrification of soil organic matter, i.e. crop residues) (Savard *et al.*, 2010). The isotopic composition of $NO_3^-$ in groundwater can be significantly influenced by the mineralisation of N fertilisers incorporated into the soil organic matter several decades ago (Sebilo *et al.*, 2013).

The process of volatilisation leads to the enrichment of the remaining $NH_4^+$ with $\delta^{15}N$ values in the 25‰ range, compared to volatilised $NH_3$ (Nikolenko *et al.*, 2018). The intensity of microbiological reactions and the degree of volatilisation are significantly influenced by pH values. If soils are rich in carbonates and alkaline reactions, volatilisation of $NH_3$ will be intensive. In addition to volatilisation, the N mineralisation process also intensifies with increasing pH. Buss *et al.* (2004) state that at pH = 7, $NH_4^+$ is predominantly sorbed to clay surfaces, while at higher pH values it begins to sorb to metal hydroxides and oxyhydroxides (e.g. FeOOH, $MnO_2$). Temperature is directly related to the intensification of $NH_3$ volatilisation, but also to the intensification of bacterial metabolism. The optimal temperature range for mineralisation and nitrification is about 20–40°C and for anammox 30–40°C.

## 8.5.2 Complementary analysis based on examining concentration change patterns of conservative tracers, origin indicators and other stable isotopes

Simultaneous monitoring of changes in the concentrations of $NO_3^-$, $SO_4^{2-}$ and reduced and total arsenic in groundwater ($As^{3+}$ and $As^{3+}$ $As^{5+}$) may indicate the denitrification process by pyrite oxidation (Singh *et al.*, 2018; Verma & Mukherjee, 2018) and the unfolding of the feammox process (Weng *et al.*, 2017). Simultaneous monitoring of changes in the concentrations of boron (Hyer, 2007; Nikolenko *et al.*, 2018; Xue *et al.*, 2009); chloride (Gasser *et al.*, 2014; Hyer, 2007) and sodium (Ellis & Bertrand-Krajewski, 2010) is used to detect the anthropogenic impact (effect of untreated sewage, application of fertilisers/manure).

Detection and quantification of $\delta^{15}N$ values in groundwater are often supplemented with analyses of isotopic enrichment of other isotopes to resolve doubts and support conclusions in the interpretation of $\delta^{15}N$ variations, which relate to overlapping values of $\delta^{15}N$ isotopic signature originating from different sources and due to the simultaneous development of different transformation processes. A review of literature sources showed that stable isotopes of oxygen, boron, carbon, sulphur, and strontium are commonly used for these purposes (Hosono *et al.*, 2014; Widory *et al.*, 2004).

### *8.5.2.1 Boron*

Amphoteric boron is found in nature in the form of boric acid or borate minerals ($H_3BO_3$ or $B(OH)_4^-$) (Lindebaum, 2012; Martinelli *et al.*, 2018). Natural B can originate from mixing of groundwater with seawater, from dissolving sandstones and igneous rocks, or it can be found in certain evaporates such as borax ($Na_2B_4O_5[OH]_4 \cdot 8H_2O$) (Clark, 2015). In these cases, the natural concentration of B in groundwater is typically very low, in the ppb range, but various studies have shown that there are significant variations in the boron concentration in groundwater. Boron is considered a conservative tracer of human impact if elevated concentrations in groundwater are not of geologic origin. Boron can be found in ash and can come from agriculture (application of manure and inorganic fertilisers). Boron is added to mineral fertilisers (Ca-borates, borethanolamine) for use on soils with a low boron content. As B is a plant micronutrient, there is a possibility that B is mineralised together with $NH_4^+$. At the same time, elevated concentrations of B and $NH_4^+$ may indicate their common origin from wastewater (Lindebaum, 2012). One of the most common anthropogenic uses of B compounds is in the production of sodium perborate ($NaBO_3 \cdot nH_2O$), often as auxiliaries such as whitening agents in detergents, soaps and toothpastes. During application and action, these auxiliaries decompose into hydrogen peroxide and an inorganic base, which is most often borate or carbonate. Therefore, boron is a frequent constituent of wastewater (Barth, 1998; Kalagasidis Krušić, 2017; Martinelli *et al.*, 2018).

In addition to boron, the constituent components of the base (filler) in modern detergents are, among others, sodium carbonate and sodium bicarbonate, whose main role is to remove $Ca^{2+}$ and $Mg^{2+}$ from water. As fillers are inactive (for powdered detergents sodium sulphate, for soaps sodium sulphate and sodium chloride, for shampoos sodium and potassium chloride and sodium sulphate) (Kalagasidis Krušić, 2017), the correlation of these compounds and boron in water may indicate anthropogenic impact.

Boron is sorbed to and desorbed from clay minerals and Fe(III)-hydroxide. As boron is considered a nitrate co-migrant and is inert to redox changes during nitrification, denitrification and transport processes, the concentration of B and its isotopic signature are used as indicators of mixing processes in the hydrogeologic system but also as tracers of sewage impact (Kloppmann *et al.*, 2009; Martinelli *et al.*, 2018; Xue *et al.*, 2009). Boron concentrations are significantly elevated in liquid manure and septic tanks (Nikolenko *et al.*, 2018). A high concentration of B followed by a low concentration of $NO_3^-$ is probably an indicator of anthropogenic impact of detergents, whereas a moderate concentration of B along with a moderate concentration of $NO_3^-$ is likely to indicate the effect of excrement from wastewater (Widory *et al.*, 2004). When the pH value is >8, boron is predominantly in the form of $B(OH)_4^-$. At neutral pH values, B is largely in the form of $B(OH)_3$.

### 8.5.2.2 Arsenic

There are several natural sources of arsenic in groundwater. Arsenic in the sediment, due to its pronounced affinity for $Fe^{3+}$, is sorbed to iron-oxyhydroxides and iron oxides ($Fe^{3+}$) and released into water by reductive iron dissolution (Berg *et al.*, 2001). Under anoxic conditions, with a sufficient amount of organic matter, due to a lack of $SO_4^{2-}$, sulphate-reducing bacteria can reduce $Fe^{3+}$ (from oxyhydroxide) and induce the release of $As^{3+}$. If there is a naturally occurring elevated level of organic matter (in river sediments or in the aquifer matrix), its decomposition creates anoxic conditions under which iron oxide dissolves, causing arsenic mobilisation ($As^{3+}$) (Berg *et al.*, 2001; Norrman *et al.*, 2008). If As in groundwater correlates with $NH_4^+$ and $H_2S$, the degradation of organic matter is likely associated with the occurrence of As (Kiurski-Milošević, 2015).

Oxidation of arsenic sulphide minerals may be another natural source of arsenic. Bicarbonate ions can leach arsenic to groundwater, by the carbonation of arsenic sulphide minerals (arseno-pyrite) (Berg *et al.*, 2001). In the presence of sulphide ions, As is rapidly reductively converted and precipitated in the form of insoluble arsenic-sulphides ($As_2S_3$ orpiment and AsS realgar). Under anaerobic conditions, arsenic can form a large number of stable sulphur salts, with different ratios of sulphur to arsenic.

### 8.5.2.3 $\delta^{18}O\text{-}NO_3^-$

A combined analysis of $\delta^{18}O$ and $\delta^{15}N$ from $NO_3^-$ may contribute to the delineation of atmospheric and terrestrial sources of $NO_3^-$, including potential separation of different anthropogenic sources (Xue *et al.*, 2009). Kendall (1998) states that the oxygen isotope ratio $\delta^{18}O\text{-}NO_3$ can indicate whether $N_2O$ originates from nitrification or denitrification.

Quantification of $\delta^{18}O\text{-}NO_3^-$ can help delineate $NO_3^-$ originating from mineral fertilisers or $NO_3^-$ produced by nitrification of $NH_4^+$ or organic N. It has been observed that mineral $NO_3^-$ fertilisers, which are produced from atmospheric $N_2$ have $\delta^{18}O$ values approximately equal to atmospheric values of +23.5‰. Xue *et al.* (2009) state that the isotopic composition of $\delta^{18}O\text{-}NO_3$ mineral fertilisers can vary in the range from +17 to +25‰. Kendall and Aravena (2000) suggest that lower values of $\delta^{18}O$ in $NO_3^-$ originate from other sources (other than the mineral fertilisers) and are a result of nitrification, where one atom of oxygen from air is incorporated in the generated $NO_3^-$ and the remaining two originate from water, lowering the value of $\delta^{18}O$. The isotopic signature of $\delta^{18}O\text{-}NO_3^-$ originating from nitrification is calculated according to Equation (8.1) (Hollocher, 1984):

$$\delta^{18}O_{NO3} = 1/3 \cdot \delta^{18}O_{C2} + 2/3 \cdot \delta^{18}O_{H2O} \tag{8.1}$$

Values of $\delta^{18}O\text{-}NO_3^-$ in the range from 0 to +6‰ indicate a nitrification process (Böhlke *et al.*, 2006; Liu *et al.*, 2006). Oxygen isotopes can be used to monitor the unfolding of denitrification in groundwater, because $\delta^{18}O$ and $\delta^{15}N$ become simultaneously enriched in the residual $NO_3^-$ during bacterial denitrification (Kendall & Aravena, 2000; Sharp, 2007). Several studies have shown constant isotopic enrichment ratios of $^{15}N$ versus $^{18}O$, which indicate the denitrification process: 2:1 (Kendall & Aravena, 2000), 1.5:1 (Baily *et al.*, 2011), and 1.4:1 (Knöller *et al.*, 2011). During denitrification, the isotopic signature of the residual $\delta^{18}O\text{-}NO_3^-$ becomes enriched by 8 to 18‰ compared to the produced $N_2O$ (Clark, 2015; Xue *et al.*, 2009).

## 8.6 NITROGEN IN SHALLOW AQUIFERS IN SERBIA

### 8.6.1 General remarks

Considering the sensitivity to nitrogen pollution, shallow intergranular aquifers are especially interesting. Alluvial aquifers of river terraces and aquifers formed by river and/or lake sedimentation are of primary importance. Particularly significant are shallow aquifers in the alluviums of the Danube and its large tributaries: the Sava, Morava and Tisa, as well as smaller rivers in the Danube River Basin (Figure 8.9).

The importance of shallow aquifers as a natural resource is very high. The state, phenomena and processes of transformation of nitrogen compounds in shallow groundwater in Serbia can be generalised to a certain extent.

This chapter provides basic information about the water quality of the Danube River in Serbia, as well as its major tributaries: right bank tributaries – the Sava (the largest tributary of the Danube) and the Morava, and the largest left tributary – the Tisa River. Official monitoring data on shallow groundwater quality were processed to determine the nitrogen content and potential for nitrification/denitrification to occur depending on hydrochemical conditions and aquifer genesis.

Two relatively large localities were selected (Figure 8.10): the areas of the anoxic aquifer Kovin-Dubovac and oxic aquifer/groundwater source 'Ključ'.

The anoxic aquifer area was examined and defined in hydrological and hydrochemical terms. The quantity and type of nitrogen compounds in the groundwater were determined. In order to examine the nitrogen transport and transformation processes in the anoxic groundwater environment, an experiment

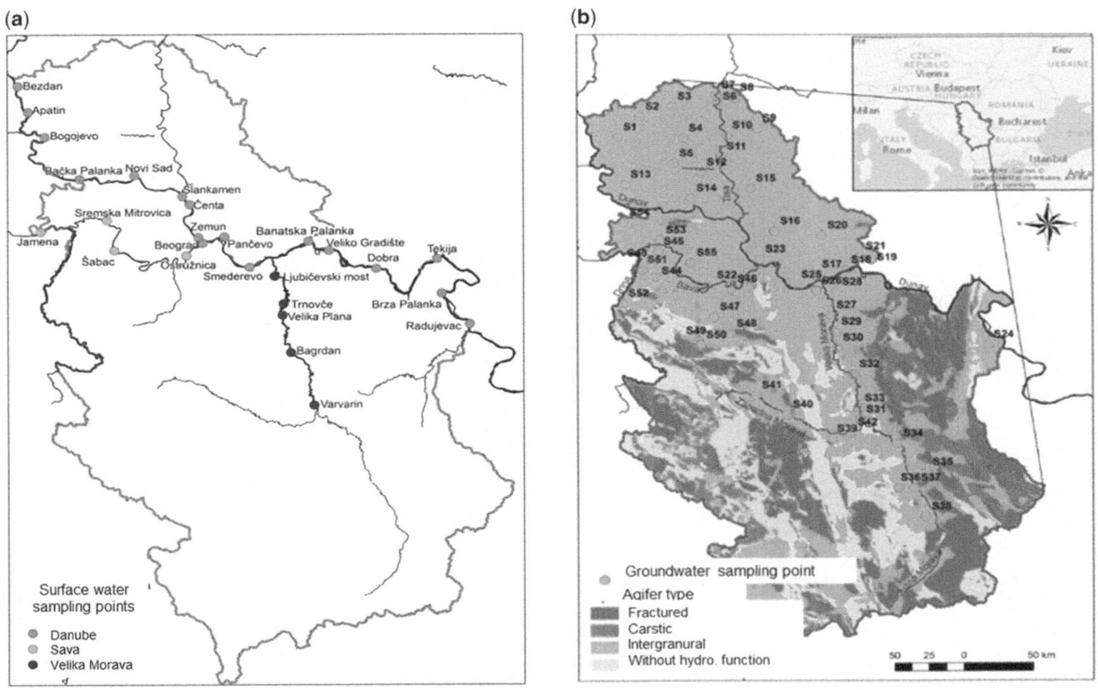


**Figure 8.9** Large rivers in the Danube River Basin in Serbia and surface water and groundwater monitoring stations (Perović, 2019).

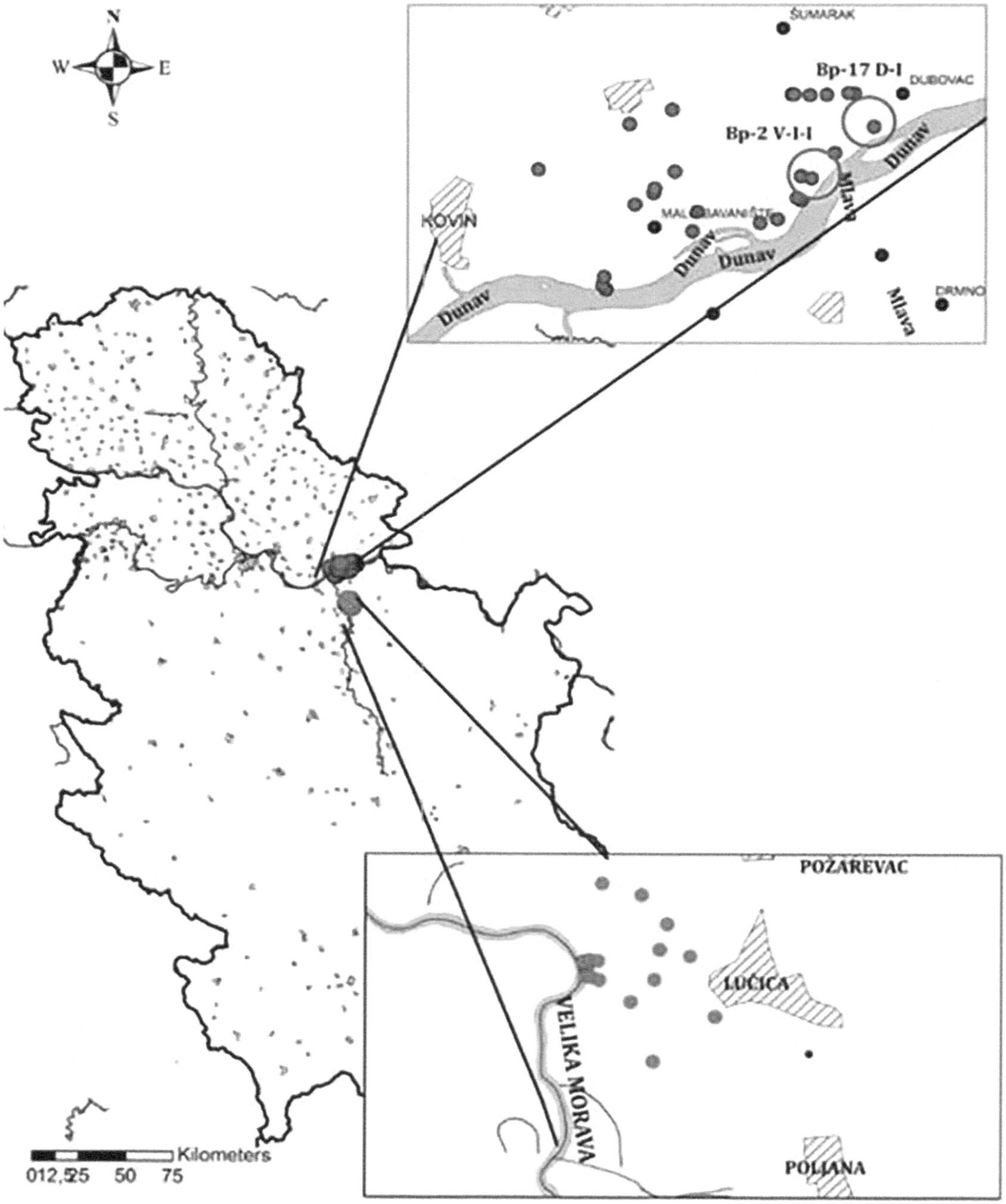


**Figure 8.10** Map cf Serbia and locations of the examined aquifers (Perović, 2019).

with nitrate and tracer salt injection was also conducted. The oxic aquifer Ključ is characterised by elevated nitrate concentrations. Intensive farming and disposal of wastewater in unsanitary septic tanks by rural households are considered as the main sources. A hydrodynamic-hydrogeological analysis of the aquifers was undertaken to determine the origin, transport and transformation of nitrogen in the groundwater. In addition, extensive chemical and biochemical analyses and isotopic signature analysis of nitrates were performed.

Laboratory experiments were conducted in order to examine the transport and transformation of nitrogen compounds in the anoxic aquifer and in the upper soil layers under anaerobic conditions.

### 8.6.2 Large rivers in Serbia

Table 8.3 shows the flow rates at low, medium and high-water levels of the analysed rivers. Upstream from Belgrade, the Danube passes through six countries with some 50 million inhabitants in the river basin.

The water quality of the Danube, Sava, Great Morava and Tisa rivers shows an improvement trend (Figures 8.11 to 8.14).

The selected parameters show a definite water-quality improvement trend of the Danube, Sava, Great Morava and Tisa. This is likely a result of the implementation of European water directives and the construction of many wastewater treatments plants.

The International Commission for the Protection of the Danube River (ICPDR) played an important role in the preparation of the Danube River Water Quality Protection Plan. This commission is the focal point of joint efforts of all countries in the basin to improve and protect the quality of river water.

A decision by the EU Commission adopted by all members of the ICPDR to declare the entire Danube River Basin as a sensitive area is highly significant as it establishes a requirement for tertiary level wastewater treatment in all agglomerations of more than 10,000 population equivalents (PE). Regardless of the fact that this requirement has largely been fulfilled in the upper river basin (Germany, Austria, Slovakia, the Czech Republic and Hungary), the quality of the Danube's water is better downstream. The reasons for this are likely the much higher industrialisation and denser population of the upper part of the basin. By way of illustration, some results of Joint Danube Survey 3 (JDS3) are presented below (Figure 8.15).

Among the studied rivers in Serbia, the highest average concentration of ammonium was detected in the Great Morava (0.19 mgN/l), followed by the Danube (0.10 mgN/l), and the lowest in the Sava River (0.08 mgN/l). The average concentration of nitrate nitrogen was highest in the Danube (1.33 mgN/l),

**Table 8.3** Overview of low, medium and high river discharges in Serbia.

| River | Station Location | Flow Rate ($m^3/s$) | | | Approximate Population in the Basin Upstream |
|---|---|---|---|---|---|
| | | *Qav* | *Qmin* (95%) | *Qmax* (1%) | |
| Danube | V. Gradište (downstream from Belgrade) | 5466 | 1800 | 16114 | ~50 million |
| Tisa | Novi Bečej (confluence with the Danube) | 766 | 123 | 3867 | ~15 million |
| Sava | Sremska Mitrovica (confluence with the Danube) | 1532 | 285 | 6408 | ~15 million |
| Morava | Ljubicevski Most (confluence with the Danube) | 230 | 35 | 2465 | ~4 million |

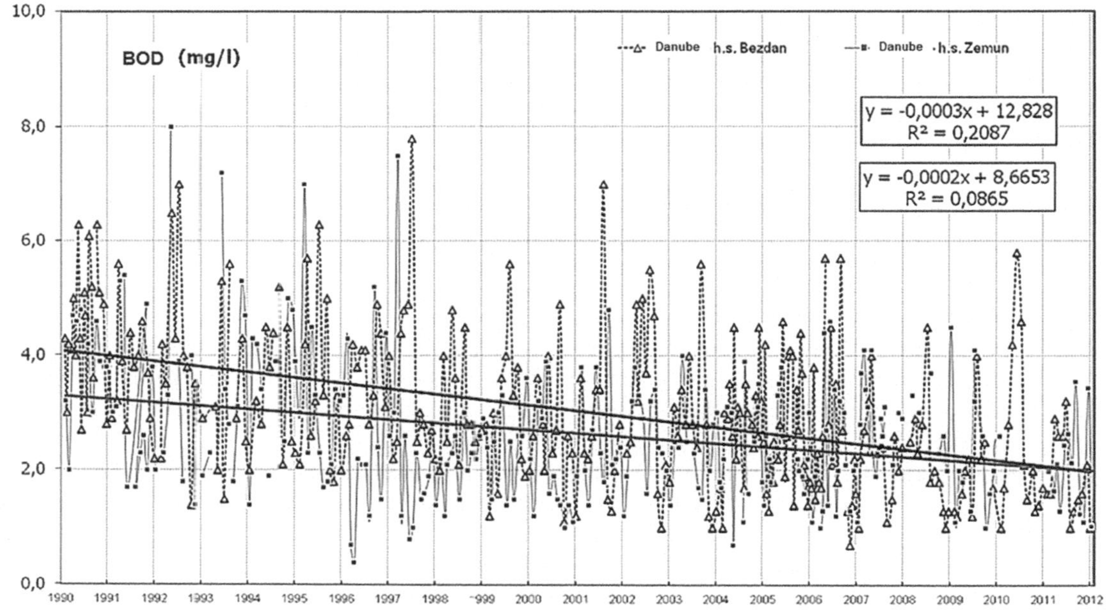


**Figure 8.11** Change in Biological Oxygen Demand in five days ($BOD_5$) concentration (mg/l) in the Danube from 1990 to 2012 (data from the Water Institute 'Jaroslav Černi').

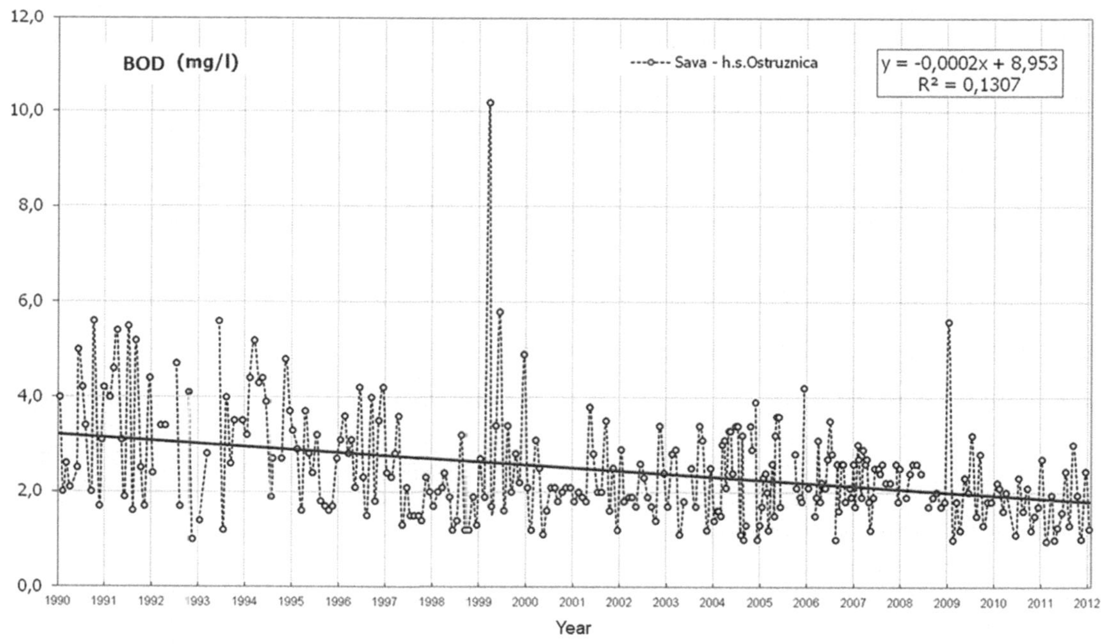


**Figure 8.12** Change in $BOD_5$ concentration (mg/l) in the Sava from 1990 to 2012 (data from the Water Institute 'Jaroslav Černi').

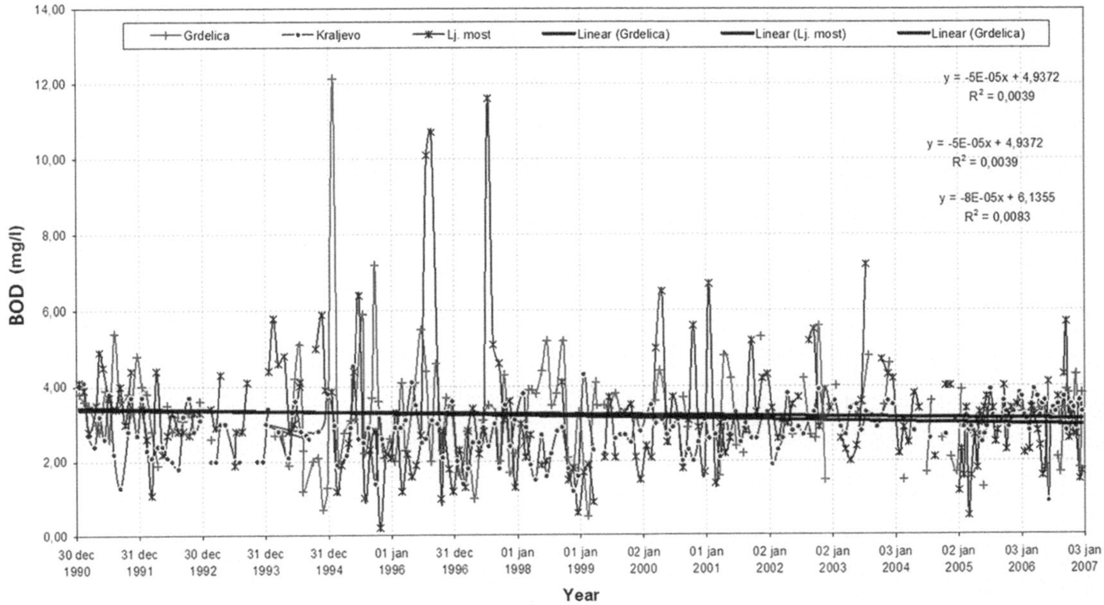


**Figure 8.13** Change in $BOD_5$ concentration (mg/l) in the Great Morava from 1990 to 2007 (data from the Water Institute 'Jaroslav Černi').

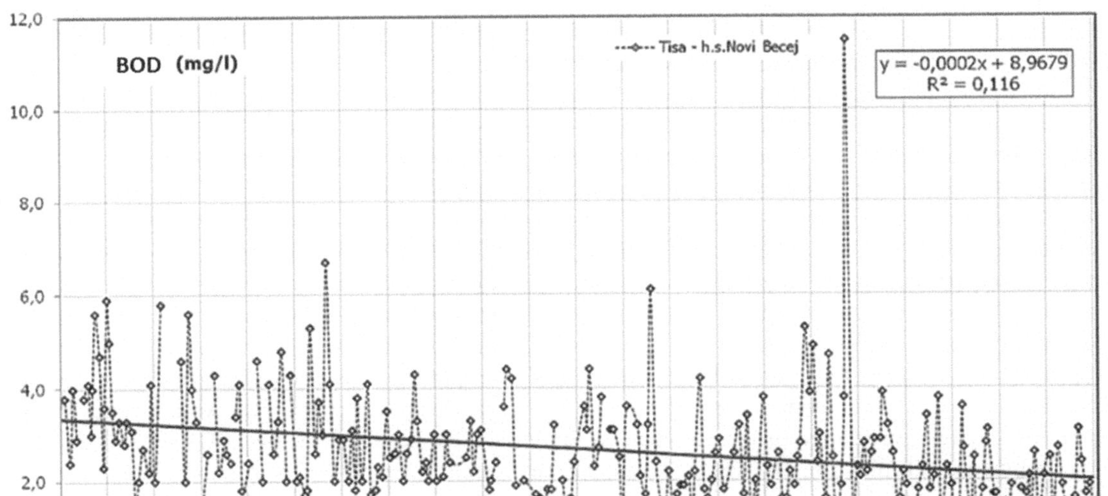


**Figure 8.14** Change in $BOD_5$ concentration (mg/l) in the Tisa from 1990 to 2012 (Data from the Water Institute 'Jaroslav Černi').

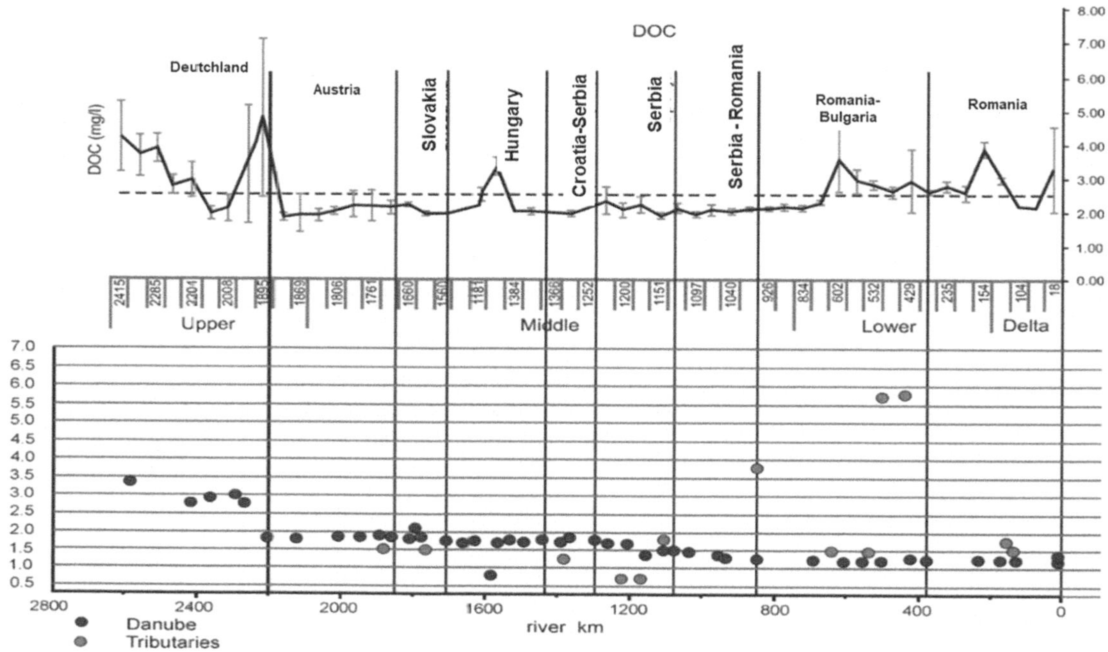


**Figure 8.15** DOC and $NO_3^-$ in the Danube during JDS3 (Joint Danube Survey 3).

then the Great Morova (0.90 mgN/l), and lowest in the Sava (0.62 mgN/l). The most endangered catchment based on the number of inhabitants is that of the Great Morava.

In order to group variables that are directly or indirectly included in the nitrogen cycle, with a similar behaviour pattern, between which it is possible to quantify a significant linear correlation, 13 surface water quality parameters (for monitoring stations on the Sava, Danube and Great Morava) were included in the factor analysis. The following physicochemical parameters were considered: $O_2$, pH, electrical conductivity of water ($\varkappa$), $NH_4^+$, $NO_2^-$, $NO_3^-$, organic nitrogen (ON), total nitrogen (TN), Cl, $SO_4^{2-}$, dissolved iron ($Fe_{diss}$), dissolved manganese ($Mn_{diss}$) and TOC (B was not determined for surface water). Values of the applied Kaiser-Meyer-Olkin test above 0.6 and a significance of the Bartlett sphericity test of $p < 0.01$ indicated that the applied factor analysis can be significant for the processing data sets. The measure of sample adequacy represents the value of the mean correlation between the observed parameters and those in the principal component analysis applied; it is above 0.5. By estimating the factor loadings, the main components (factors) (Principal Components – PCs) were obtained. PCs explain the highest percentage of the parameters' variance. The extracted factors (PC1-PC4) are shown in the columns in Table 8.4. The factors grouped the parameters whose changes in values were in a significant linear correlation. Positive values of Pearson's correlation coefficients indicated that the changes were directly proportional, while a negative sign indicated that the increase in the value of one parameter was accompanied by a linear decrease in the value of the other parameter.

According to this factor analysis, the same pattern of concentration changes and linear relationships between chloride, sulphate and consequently electrical conductivity was observed in all three rivers. In the Danube, which had the highest mean values of nitrate nitrogen, the chloride, sulphate and total

**Table 8.4** Matrix of selected parameters for the Sava, Danube and Great Morava.

| | Sava | | | | Danube | | | | Great Morava | | | |
|---|---|---|---|---|---|---|---|---|---|---|---|---|
| | Extracted Factors | | | | | | | | | | | |
| | PC1 | PC2 | PC3 | PC4 | PC1 | PC2 | PC3 | PC4 | PC1 | PC2 | PC3 | PC4 |
| $O_2$ | | | | 0.898 | 0.737 | | | | | | | |
| pH | | | | | | −0.537 | | | | | 0.854 | |
| $\varkappa$ | 0.838 | | | | 0.826 | | | | 0.729 | | | |
| $NH_4^+$ | | 0.577 | | | | 0.600 | | | 0.581 | | | |
| $NO_2^-$ | | 0.860 | | | | | 0.633 | | | | | −0.689 |
| $NO_3^-$ | | | | | | | | | | 0.852 | | |
| ON | | | | | | | | | | −0.697 | | |
| TN | | 0.731 | | | 0.816 | | | | | | | |
| Cl | 0.876 | | | | 0.828 | | | | 0.535 | | | |
| $SO_4^{2-}$ | 0.775 | | | | 0.768 | | | | 0.717 | | | |
| $Fe_{diss}$ | | | 0.620 | | | | | 0.731 | | | | |
| $Mn_{diss}$ | | | 0.845 | | | | | 0.685 | | | | |
| TOC | | | | −0.529 | | | −0.715 | | | | | −0.817 |

nitrogen increased simultaneously. In the Great Morava, along with chloride and sulphate, a simultaneous increase in the concentration of ammonium ions was evident. In the Danube, the linear inverse dependence with a decrease in pH value and an increase in the ammonium concentration indicated that discharges of organic substances rich in ammonia, their oxidation and consequent decrease in pH value were detectable. It is also characteristic of the Danube that the increase in the concentration of nitrite was accompanied by a decrease in the concentration of total organic carbon, which can be interpreted as a confirmation of the previously drawn conclusion. The observed inverse linear correlation between the concentration of organic nitrogen and nitrates in Great Morava indicates that the nitrates in the river are predominantly of organic nitrogen oxidation origin. Although the Sava River is characterised by the lowest concentrations of ammonium and nitrate ions, a strong correlation was found between the concentrations of total nitrogen, ammonium and nitrite ions.

Considering that only about 10% of wastewater is treated, the present state of surface water quality and concentrations of nitrogen compounds in Serbia is good, but based on the calculated Pearson coefficients of linear correlations, the impacts of untreated water and consequent nitrification of organic nitrogen and ammonium ion are evident.

### 8.6.3 Vulnerability of shallow groundwaters to nitrogen contamination in Serbia

Table 8.5 shows descriptive statistics of select groundwater quality parameters, based on data collected by the Serbian Environmental Protection Agency and the National Hydrometeorological Service of Serbia (2011–2016) from 55 observation wells in shallow aquifers.

The type of aquifer genesis affects the baseline groundwater quality. Shallow aquifers in the alluviums along the lower courses of large lowland rivers, like the Danube, Sava and Tisa, were formed where the energy of the water flow is reduced. Fine-grained material is deposited because of the weakened water flow. This implies higher iron concentrations and less oxygen in the water and an environment predominantly suitable for nitrate reduction processes.

In hilly and mountainous areas south of the Sava and the Danube, the Great Morava River and its short tributaries formed largely shallow aquifers characterised by a coarse-grained structure.

Figure 8.16 presents a map of the overall potential of the examined shallow groundwaters for nitrogen conservation and loss (Perović, 2019). The map was generated based on limiting concentrations of the parameters ($O_2$, TOC, $NO_3^-$, $NH_4^+$) required for certain nitrogen transformation processes. It shows $NH_4^+$ concentrations and sites where there is potential for: (1) $NO_3^-$ conservation through DNRA, (2) nitrogen loss by the Den. process, and (3) nitrogen remaining in the $^{+}5$ state (nitrate vulnerable sites). Locations of low concentrations of dissolved oxygen (<1 mg/l) and a C:N ratio of more than 3.5 were designated as suitable for N conservation (denoted by a star) (Cohen *et al.*, 2012; Darwishe *et al.*, 2017; Heffernan *et al.*, 2010; Henson *et al.*, 2017; Korom 1992; Postma *et al.*, 1991; Rivett *et al.*, 2008). Locations with dissolved oxygen concentrations below 2 mg/l and total organic carbon concentrations above 3 mg/l have been identified as suitable sites for nitrogen loss, by the Den. process (denoted by a flower). All locations where the dissolved oxygen concentration was above 3.5 mg/l were identified as sensitive to nitrate pollution (denoted by the lightning bolt symbol).

## 8.7 ANOXIC ALLUVIAL AQUIFER IN THE KOVIN-DUBOVAC AREA

### 8.7.1 Aquifer characteristics

The Kovin-Dubovac area, located on the left bank along the lower Danube, occupies the southern part of the Banat region and covers an area of 172 km$^2$, between the settlements of Kovin, Dubovac and Gaj. The

**Table 8.5** Statistical data on select parameters of shallow groundwaters in Serbia (2011–2016).

| | | pH | μ | $NH_4^+$ | $NO_2^-$ | $NO_3^-$ | $Fe_{tot}$ | Mn | $Fe_{diss}$ | As | B | TOC |
|---|---|---|---|---|---|---|---|---|---|---|---|---|
| **Unit** | | / | μS/cm | mgN/l | mgN/l | mgN/l | μg/l | μg/l | μg/l | μg/l | μg/l | mg/l |
| No. of samples | | 327 | 327 | 325 | 327 | 327 | 233 | 247 | 254 | 152 | 160 | 214 |
| Aver. value | | 7.45 | 957.65 | 0.35 | 0.01 | 2.29 | 1090.77 | 290.10 | 148.33 | 18.31 | 97.43 | 6.51 |
| Median | | 7.43 | 915.00 | 0.07 | 0.01 | 0.40 | 294.00 | 97.80 | 11.25 | 1.40 | 75.05 | 4.00 |
| Std. deviation | | 0.26 | 356.28 | 0.64 | 0.02 | 4.05 | 2045.36 | 544.05 | 576.82 | 63.38 | 92.10 | 6.99 |
| Range | | 1.70 | 2021.00 | 4.16 | 0.11 | 23.65 | 12,519.50 | 4626.50 | 4700.75 | 520.15 | 594.70 | 36.75 |
| Minimum | | 6.60 | 309.00 | <0.02 | <0.002 | <0.2 | <10 | <10 | <10 | <10 | <10 | <0.5 |
| Maximum | | 8.30 | 2330.00 | 4.16 | 0.11 | 23.65 | 12,520.00 | 4627.00 | 4701.00 | 520.40 | 595.20 | 37.00 |
| Percentiles | 25 | 7.30 | 707.00 | 0.03 | 0.00 | 0.20 | 84.35 | 22.80 | 0.50 | 0.25 | 33.53 | 2.50 |
| | 50 | 7.43 | 915.00 | 0.07 | 0.01 | 0.40 | 294.00 | 97.80 | 11.25 | 1.40 | 75.05 | 4.00 |
| | 75 | 7.60 | 1127.00 | 0.33 | 0.02 | 2.54 | 1115.00 | 307.80 | 90.25 | 3.60 | 126.58 | 7.80 |

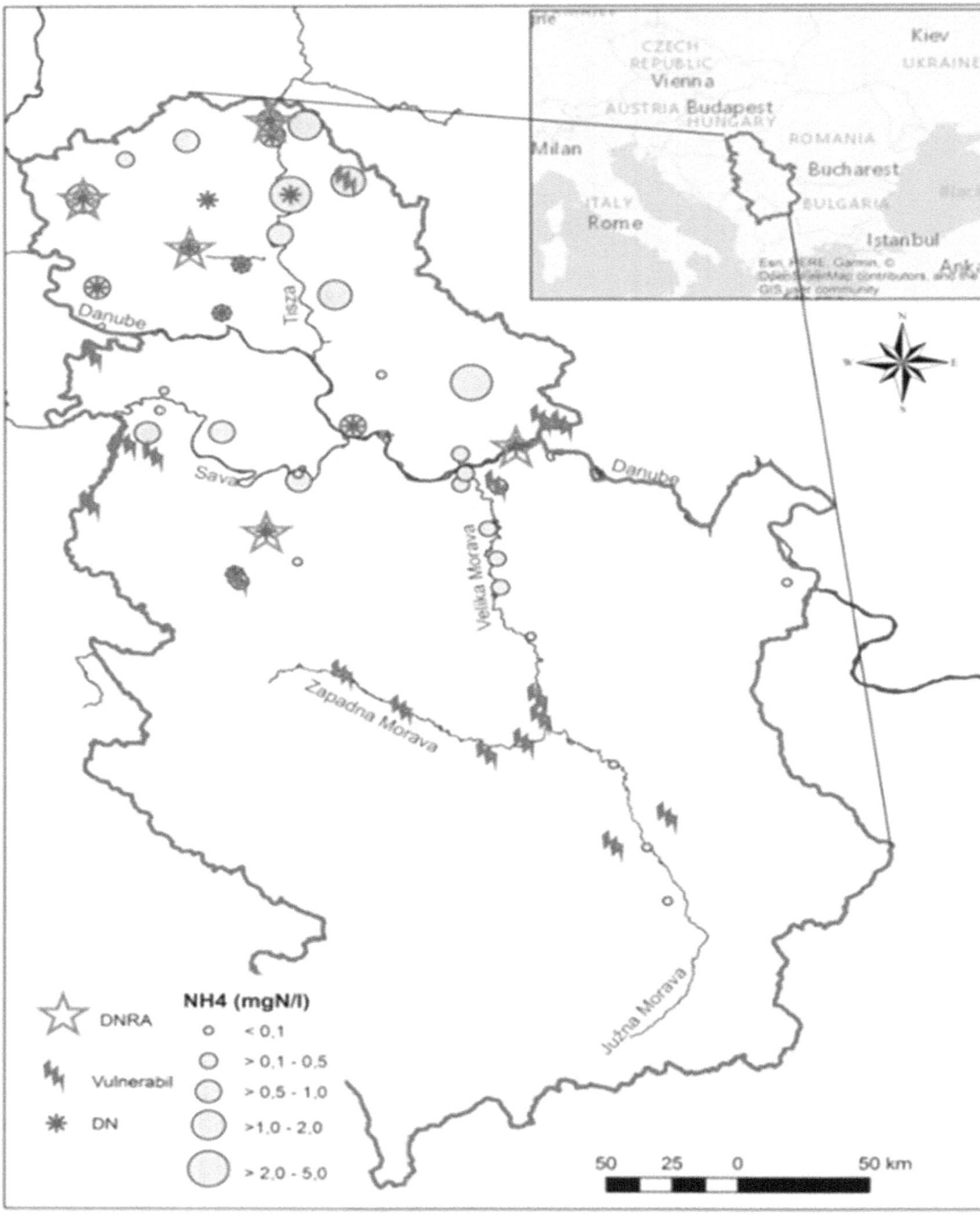


**Figure 8.16** Potential for N transformation and ammonium ion concentrations in shallow intergranular aquifers in Serbia (Perović, 2019).

alluvial plain and aquifer are composed of sand and fine-grained gravel as well as silty and clayey materials. The deposits are overlain by a clay and humus aquitard. The total thickness of the alluvial layers is predominantly in the range from 20 to 30 m.

After the water level of the Danube was permanently raised due to the construction of the dam of the Iron Gate I Hydroelectric Power Station, the agricultural area of Kovin-Dubovac became additionally endangered by high groundwater levels. To protect this agricultural area from flooding, an extensive drainage system of canals and wells was put in place. The drainage wells are free-flowing (without submersible pumps) and accompanied by observation wells, which enabled adequate analyses of groundwater composition in the alluvial aquifer.

In the Kovin-Dubovac area, intensive crop farming takes place with an average amount of applied nitrogen per hectare of about 100 kgN/ha annually (through mineral fertilisers). There are no large settlements or industrial facilities, whose impact on groundwater quality would be significant. Examination of groundwater quality showed that this is an anoxic environment with low concentrations of dissolved oxygen, low redox potential and elevated concentrations of ammonium ions and iron ($Fe^{2+}$ and $\Sigma Fe$) (Table 8.6). The results of Biological Activity Reaction Tests (BART tests) categorise this aquifer as reductive-anoxic, with dominant reactions of fermentation, reduction of $Fe^{3+}$ ions and sporadic denitrification and a very high potential for sulphate reduction at certain locations (Perović *et al.*, 2017). Based on the number of predicted active cells (p.a.c/ml) and the type of reaction in HAB (heterotrophic aerobic bacteria) tests, the area is characterised as rich in organic carbon that is generally mineralised anaerobically.

According to the ratio of $NH_4^+$, $NO_3^-$, $H_2S$ and BART tests, the places suitable for Den. and DNRA are shown in Figure 8.17.

**Table 8.6** Values of select parameters of the Kovin-Dubovac area (2010–2013).

| Parameter | Minimum | Aver. Value | Maximum | No. of Data Points |
|---|---|---|---|---|
| Eh (mV) | 20 | 96 | 279 | 146 |
| $O_2$ (mg/l) | 0 | 0.06 | 1.20 | 153 |
| pH | 6.7 | 7.3 | 8.3 | 156 |
| $NO_3^-$ (mgN/l) | 0.05 | 0.29 | 1.49 | 133 |
| $NH_4^+$ (mgN/l) | 0.02 | 0.95 | 4.70 | 134 |
| $Fe^{2+}$ (mg/l) | 0.10 | 1.92 | 5.41 | 134 |
| $SO_4^{2-}$ (mg/l) | $<2$ | 24.87 | 147.80 | 134 |
| $S^{2-}$ (mg/l) | $<0.02$ | 0.06 | 0.54 | 123 |
| Conductivity (20°C) (μS/cm) | 361 | 629.7 | 1191 | 156 |
| TOC (mg/l) | 0.6 | 2.18 | 5.23 | 119 |
| SRB[a] ($\log_{10}$ p.a.c/ml) | 0.0 | 3.8 | 5.0 | 35 |
| IRB[b] ($\log_{10}$ p.a.c/ml) | 0.9 | 4.6 | 5.1 | 35 |
| DN[c] ($\log_{10}$ p.a.c/ml) | 0.0 | 4.0 | 5.4 | 36 |

[a]Sulphate-reducing bacteria.
[b]Iron-related bacteria.
[c]Denitrifying bacteria.

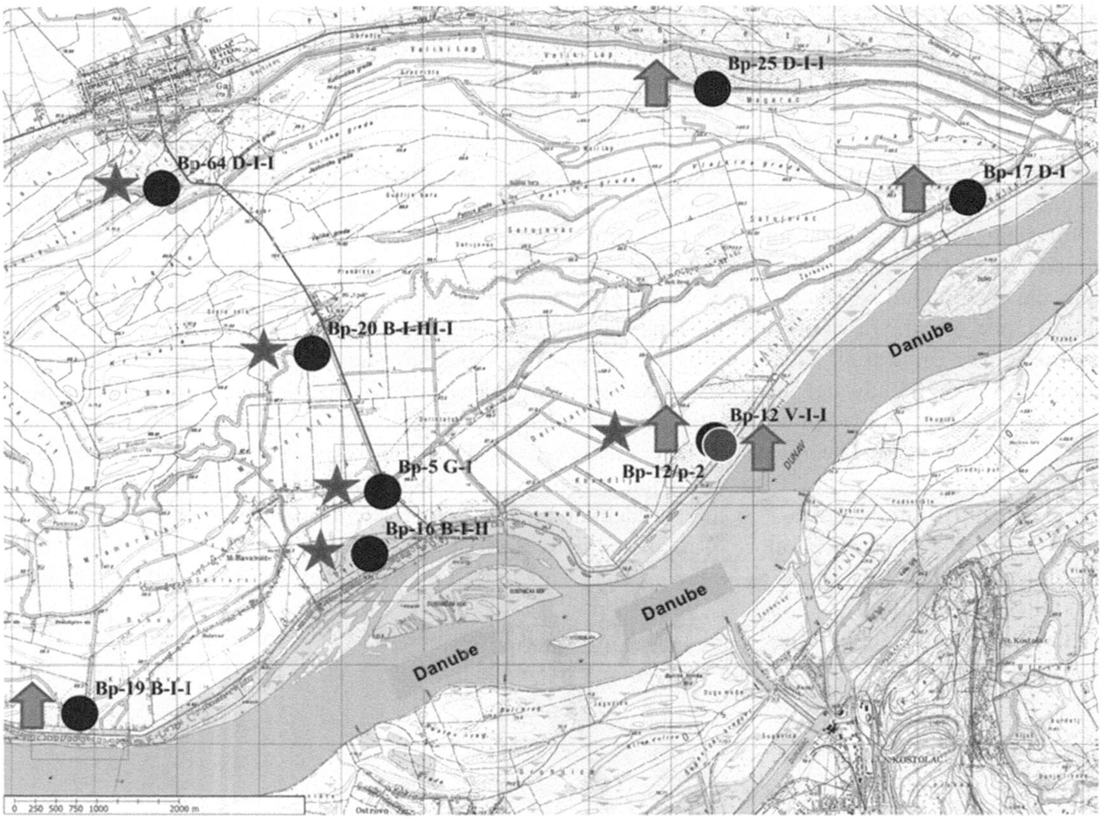


**Figure 8.17** Map of the Kovin-Dubovac drainage system with marked observation points and potential for nitrate removal, based on a comprehensive and complementary analysis of chemical and microbiological parameters in the period 2010–2013 (asterisk indicates potential for DNRA and arrow for respiratory denitrification) (Perović *et al.*, 2017).

## 8.7.2 Tracer and nitrate testing of the Kovin-Dubovac aquifer

*In situ* monitoring of the nitrate transformation pathways and fate was undertaken to determine the sensitivity of the anoxic Kovin-Dubovac aquifer to nitrate pollution and the potential of the anoxic environment for nitrogen conservation. A tracer test (injection of non-reactive salt, sodium chloride, NaCl) and nitrate transport and transformation test (injection of sodium nitrate, $NaNO_3$) were conducted on the first drainage line at Kovin-Dubovac. In order to determine the hydrodynamic characteristics of the studied area, the degree of salt retardation and the corresponding sampling frequency for the nitrate experiment, a tracer test was performed first by injecting sodium chloride into Bp-2/P6 (Kovačević, 2017; Perović *et al.*, 2017). The groundwater level, flow and electrical conductivity were monitored every hour, and parameters such as the pH value, dissolved oxygen and oxidation/reduction potential were observed during the experiment. A detailed description of the tracer test is included in Kovačević (2017) and Perović *et al.* (2017). During the experiment, groundwater conditions were anoxic (standard redox potential approximately 90 mV) and quasi-stationary groundwater flow and a quasi-stationary state of the radial divergent flow were established. The distances between the observation wells and drainage wells were: Bp-2/P6 to Bp-2 = 8 m, Bp-2/P2 to well Bp-2 = 5.30 m and from Bp-2/P1 to Bp-2 = 3.40 m (Figure 8.18). The combination of *in situ* measurements, laboratory analyses of physicochemical parameters and selected microbiological tests enabled the estimation of different nitrate reduction pathways. The non-reactive tracer, NaCl, was injected to a depth of between 15 and 20 m in an injection observation well, located about 8 m from the monitored drainage well. Changes in NaCl concentrations were monitored at the same wells (Bp-2/P2, Bp-2/P1 and Bp-2, respectively) as in the subsequent nitrate experiment. Nitrate-enriched ($NaNO_3$) well water was injected into Bp-2/P6, at a depth of 15 m, at the screen level (Figure 8.18).

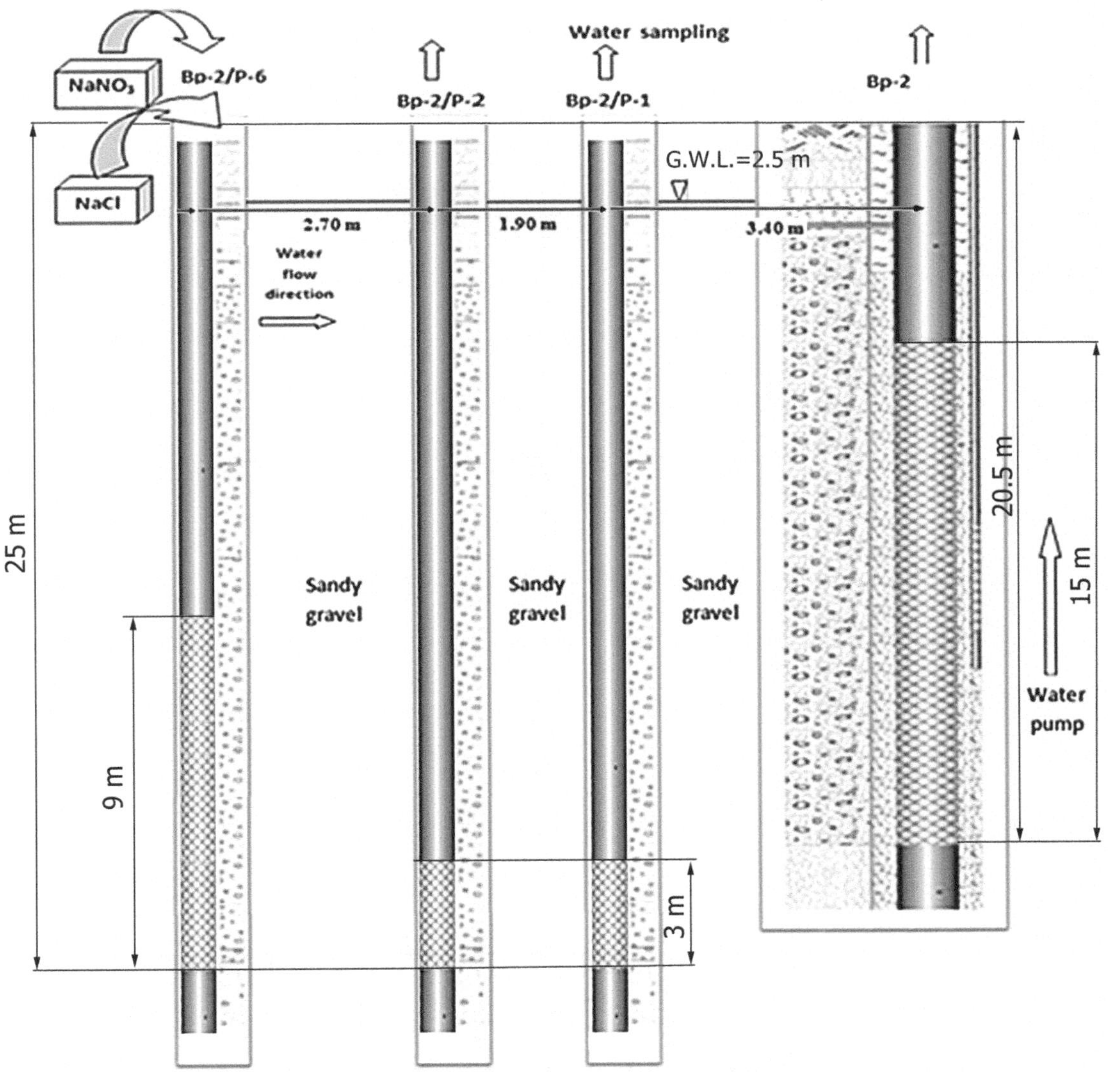


**Figure 8.18** Monitored observation wells and drainage well (Perović *et al.*, 2017).

Ninety-three groundwater samples were collected during the five-day experiment. Nine samples were slated for BART tests. During the experiment, a total of 27 analyses were performed to quantify DN, SRB and IRB.

The results of *in situ* monitoring showed no significant changes in parameters other than electrical conductivity (Figure 8.19). Using the $NO_3^-$ mass balance of drainage well Bp-2, calculations estimated

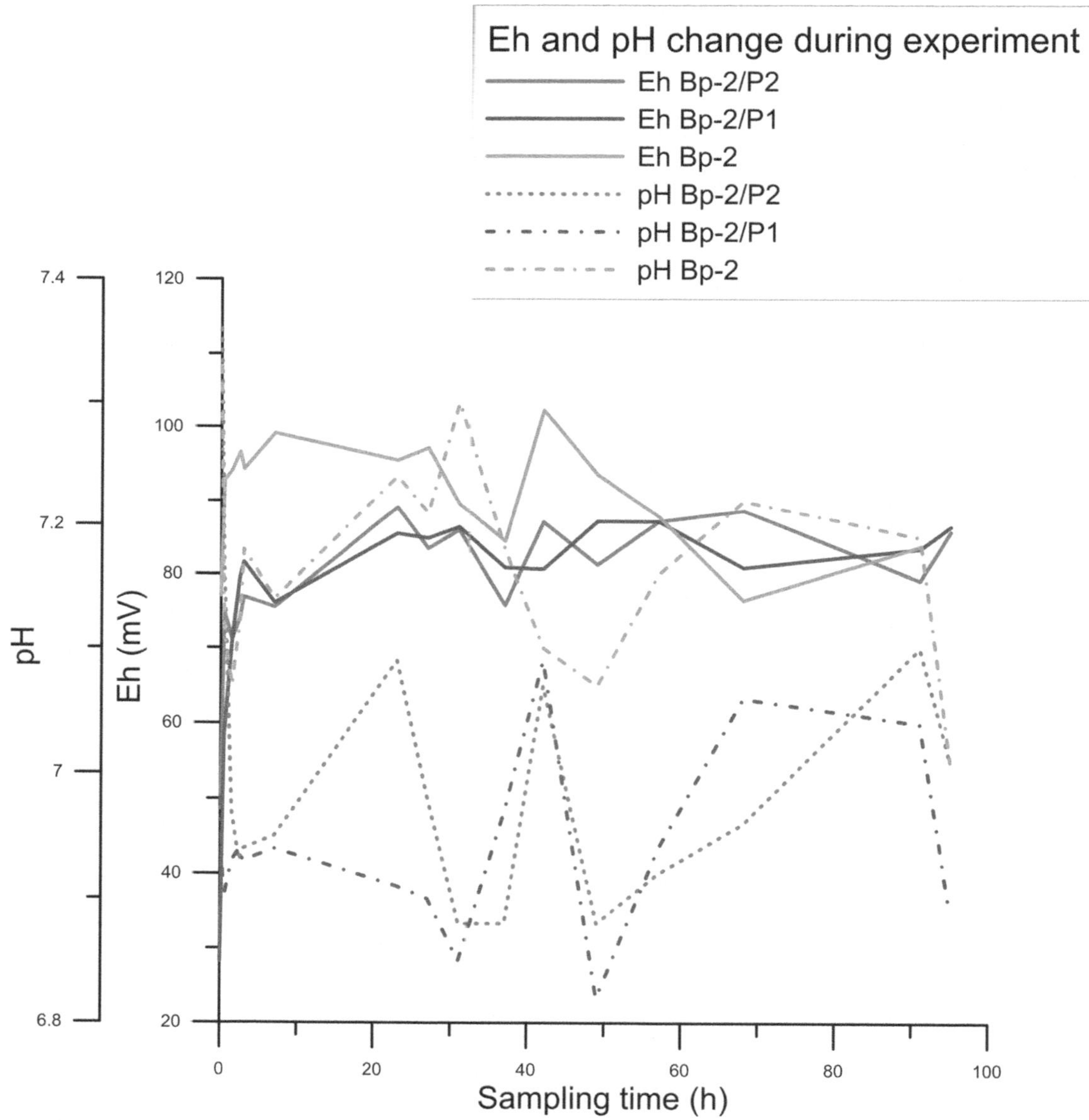


**Figure 8.19** Changes in Eh and pH values during the experiment at wells Bp-2/P2, Bp-2/P1 and Bp-2 (Perović *et al.*, 2017).

that 40.5 g of the injected nitrate nitrogen, $NO_3$-N, came out at the well, which was a quarter of the injected mass.

A comparison of the chemical and microbiological analyses indicated that fermentative DNRA mediated by *Desulfovibrio* was most likely the process observed in the experiment. Perović *et al.* (2017) showed the retardation and delay between the observed ammonium peaks, at various monitored locations. The average concentrations of ammonium ions during the experiment at Bp-2/P2, Bp-2/P1 and Bp-2 were 3.54, 2.83 and 2.21 mgN/l, respectively. The maximum concentrations were 4.97, 3.86 and 3.58 mgN/l, respectively. It was assumed that nitrate reduction took place in accordance with the spatial order of ion arrival with water flow to the observation wells and drainage well (Figure 8.20). Nitrate reduction, dissolution and sorption of positively charged ammonium ions to the aquifer matrix explained the observed differences between the injected volume (400 mgN/l) and the peak concentrations at the monitored sites, as shown on breakthrough curves. The peaks indicated that a certain amount of the produced ammonium ion was carried with the groundwater flow. At Bp-2/P2 from the third day, SRB BART tests (average value 115 000 p.a.c/ml) and a decrease in the total inorganic carbon (TIC) indicated a potential for fermentative DNRA, coupled with the activity of *Desulfovibrio species*. The TIC concentration in Bp-2/P2 was lower than in other wells and a constant decrease during the experiment suggested chemolytoautotrophic uptake of $CO_2$ by autotrophs (e.g., *Gallionela*). It was

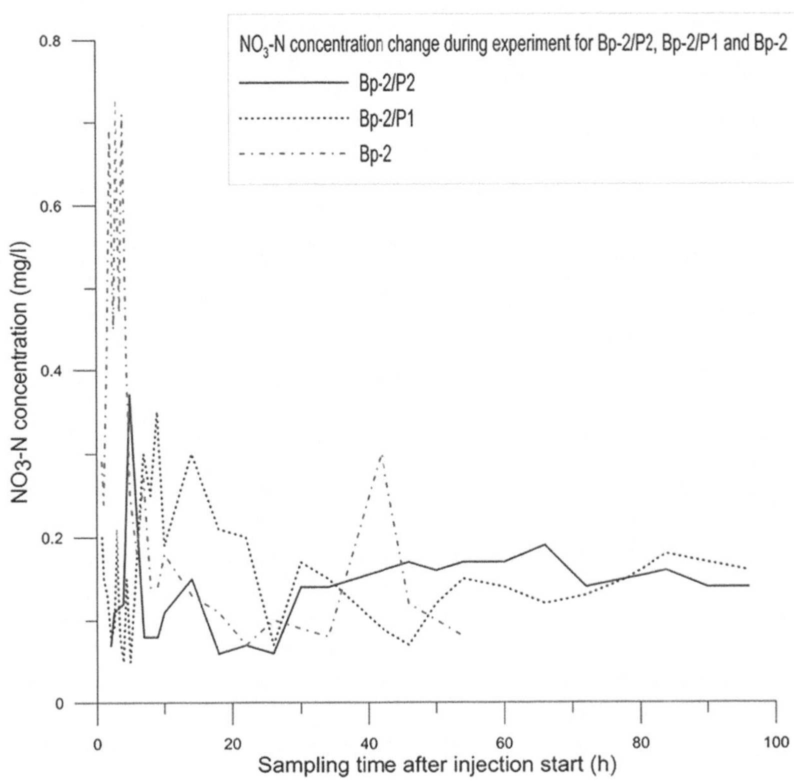


**Figure 8.20** $NO_3^-$ concentration change at Bp-2/P2, Bp-2/P1 and Bp-2 during the experiment (Perović *et al.*, 2017).

hypothesised that in Bp-2/P2 the DNRA process was stimulated by the addition of $NO_3^-$, in the presence of organic matter and reduced sulphur as an inhibitor of respiratory denitrification, or by a high C:N ratio under reductive conditions. In Bp-2/P2, tests showed intensive fermentation (34,800 p.a.c/ml) and high activity of the genus *Desulfovibrio*, sulphate-reducing bacteria (115,000 p.a.c/ml) whose secondary metabolism may be DNRA. In Bp-2/P2, DN BART tests showed the presence of anaerobic bacteria, without denitrification from the third day. In Bp-2/P1, SRB BART tests showed significant activity of sulphate-reducing bacteria (average 85,533 p.a.c/ml). In the same observation well, TIC exhibited a slight increase, indicating the generation of $HCO_3^-/CO_3^{2-}$, the product of complete oxidation of organic matter.

### 8.7.3 Origin and transformation of nitrogen compounds in the anoxic alluvial aquifer in the Kovin-Dubovac area

A very low concentration of dissolved oxygen is the main characteristic of the groundwater in the Kovin-Dubovac area. As the alluvial aquifer is formed in the lower Danube basin, where fine-grained oxidisable material is deposited, anoxic conditions prevail in the aquifer. Certain amounts of organic material originate from ponds and wetlands, largely present before the drainage system was built (1970s).

Decades of intensive agricultural production, nitrate concentrations below or around the detection limit (although groundwater is 2–4 m below land surface), and the fact that the aquifer is slated for regional drinking water supply in the future indicated the importance of examining the origin and fate of nitrogen compounds in this area. The physicochemical parameters of groundwater at 33 wells were analysed (Figure 8.21). Statistical data processing included the results obtained with different sampling frequencies in the period from 2010 to 2018. The average concentration of dissolved oxygen for the entire site was about 0.12 mg/l. The ammonium concentration ranged from 0.01 to 4.70 mgN/l, and the concentration of nitrate nitrogen was generally negligible due to reductive conditions. At Kovin-Dubovac, the concentration of As was below the quantification limit (except at two localities). Boron concentrations had rarely been quantified, so B was not included in the analysis.

Factor analysis/principal component analysis was undertaken to determine whether there were correlations between the selected parameters. Parameters whose variation points to transformation processes or parameters indicative of the origin of N were: pH, Eh, $NH_4^+$, $Cl^-$, $SO_4^{2-}$, $H_2S$, $Fe^{2+}$, $Fe_{tot}$, Na, TOC, and Ca. Based on eigenvalues, four factors (PC1-PC4) were extracted, which explain nearly 70% of the total parameter variance (Table 8.7).

The ammonium isotopic signature was determined in 12 wells: Bp-2, Bp-12, Bp-13 from ditch V-I-I, Bp-16 from B-I-II, Bp-10, Bp-11, Bp-16, Bp- 20, Bp-24 from ditch D-I-I and Bp-2 from B-I-III-II, Bp-5 and Bp-7 from B-I-III-I (Figure 8.22). The groundwater flow direction, water levels in the ditches, water flow at pumping stations, water level of the Danube, $Cl^-$ concentration and $\varkappa$ were considered as indicators of groundwater origin. According to the dominant groundwater origin, it was possible to identify three separate areas. The first area represents the first drainage line, where recharge from the Danube is dominant (wells: Bp-2, Bp-12, Bp-13, Bp-16). The second area is a part where groundwater recharge comes from the direction of Deliblato Sands – north-east (Bp-10, Bp-11, Bp-16, Bp-20, Bp-24 from ditch D-I-I) and the third is where groundwater recharge originates from the direction of the town of Kovin – north-west (Bp-2 B-I-III-II, Bp-5 and Bp-7 B-I-III-I). The second and third drainage lines collect water from the terrain surface (atmospheric water) and groundwater from the hinterland. The values of $\delta^{15}N$-$NH_4^+$ ranged from −0.84 to +6.93‰ in all the examined samples (Perović, 2019). The lowest values of $\delta^{15}N$ (−0.84‰ and −0.33‰) were observed in the groundwater from wells Bp-16 and Bp-24 (D-I-I), where groundwater is recharged from the direction of Deliblato Sands. These values

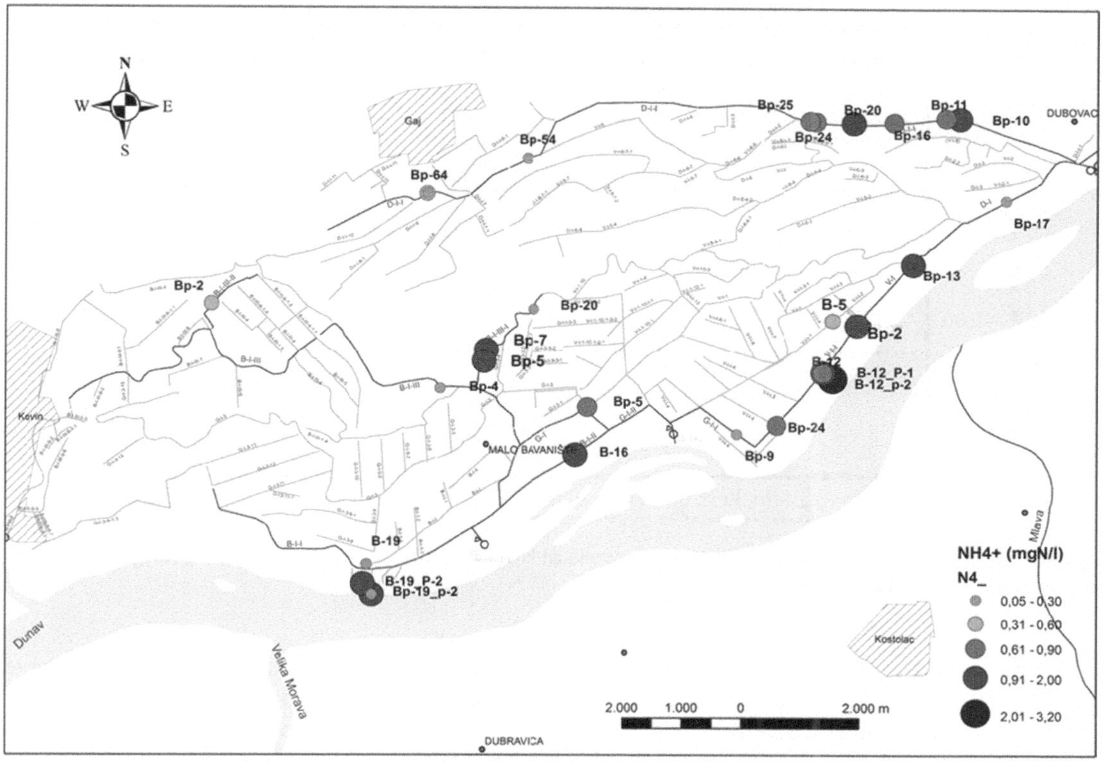


**Figure 8.21** Average ammonium ion concentrations at representative locations in the Kovin-Dubovac area (2010–2018) (Živančev *et al.*, 2020).

overlap with the range for ammonium ion originating from rainwater and from mineral fertilisers. The highest $\delta^{15}N\text{-}NH_4^+$ value of +6.93‰ was observed in Bp-16 (B-I-II) from the first drainage line. Of all the monitored wells, Bp-16 (B-I-II) was the closest to the Danube and its isotopic signature indicated that ammonium probably originated from municipal wastewater. The characteristic range for $NH_4^+$ from the mineralisation of soil organic matter was +2.4 to +4.4‰. It is obvious that most samples were characterised by a high isotopic signature. The $\delta^{15}N$ value of nitrogen originating from the soil organic matter at Kovin (proximity of Bp-12, V-I-I) was determined from subsequent laboratory column experiments. The value of $\delta^{15}N$ nitrogen in the soil from Kovin was +3.65‰. The value of $\delta^{15}N\text{-}NH_4^+$ of the groundwater samples from wells in the first drainage line – Bp-2, Bp-12, Bp-13 (V-I-I) of +4.82

**Table 8.7** Factor analysis of selected parameters in groundwater of the Kovin-Dubovac area.

| | | pH | Eh | $NH_4^+$ | $Cl^-$ | $SO_4^{2-}$ | $H_2S$ | $Fe^{2+}$ | $Fe_{tot}$ | Na | TOC | Ca |
|---|---|---|---|---|---|---|---|---|---|---|---|---|
| Factors | PC1 | | | | 0.709 | | 0.942 | | | 0.792 | | |
| | PC2 | | | | | | | 0.768 | 0.848 | | 0.657 | |
| | PC3 | | | | | −0.864 | | | | | | −0.913 |
| | PC4 | 0.584 | 0.853 | −0.574 | | | | | | | | |

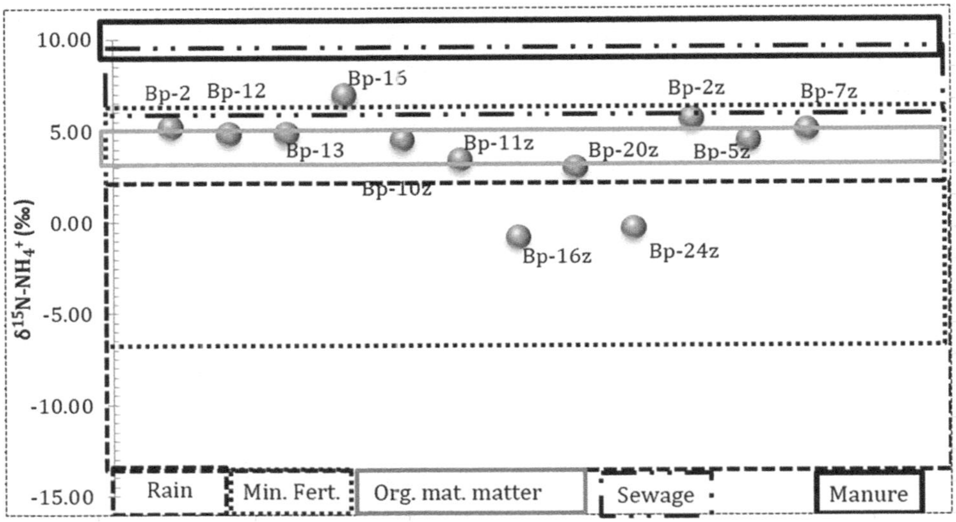


**Figure 8.22** Isotopic signature of $\delta^{15}N\text{-}NH_4^+$ in wells at Kovin-Dubovac and characteristic ranges of different sources (Perović, 2019).

to +5.17‰, indicated that the ammonium ions originated from mineral fertilisers. In the wells recharged from the land surface and from the northwest, the isotopic signature was in the range from +4.45 to +5.63‰, typical of mineral fertilisers. The wells that were recharged from the direction of Deliblato Sands were measured with the lowest values of $\delta^{15}N\text{-}NH_4^+$, from −0.84 to +4.49‰. Elevated $^{15}N$ in Bp-16 was attributable to nitrogen originating from sewage or ammonium nitrification, which resulted in the enrichment of the remaining ammonia with a heavier isotope.

The origin of elevated concentrations of ammonium ions, on the first drainage line, can be explained in two ways. Firstly, the mineralisation of organic matter of *autochthonous* origin was accompanied by nitrification that led to enriched isotopic signature of ammonium, to the range characteristic for mineral fertilisers. This conclusion was corroborated by linear correlations between iron and organic matter on the first drainage line, where elevated ammonium concentrations were also detected. Fluctuating redox conditions, caused by fluctuating stages of the Danube, could have periodically created the conditions necessary for nitrification on the first drainage line. Infiltration of Danube water (first drainage line recharge) rich in oxygen, could have enhanced oxidation of mineralised organic matter, ammonification and nitrification. Due to rapid oxygen consumption (environment rich in organic matter and iron ions), nitrification is probably not very intensive, so it was assumed that enrichment of the remaining $NH_4^+$ was low (1–3‰). Secondly, nitrates from the Danube and from the land surface were reduced by dissimilatory nitrate reduction to ammonium (due to elevated concentrations of Fe and TOC), producing $^{15}N$ depleted ammonium cations, compared to the initial $^{15}N$ originating from nitrates (min. fertilisers $\delta^{15}N\text{-}NO_3^- = -8$ to $+7$‰). Pesticides detected in the wells from the first drainage line at Kovin-Dubovac (Bp-12 V-I-I and Bp-19 B-I-I) also supported this finding (Živančev, 2019; Živančev *et al.*, 2020). Enrichment of ammonium ions with a heavier isotope (observed on the first drainage line, but also in the north-western part, where groundwater flows from the direction of Kovin) may indicate the origin of ammonia from mineral fertilisers or could be caused by nitrification or anammox and/or volatilisation. Depleted ammonium ions (observed in samples of water flowing from the direction of

Deliblato Sands) could have been an indication of another source (fertiliser mineral) or a result of sorption and/or DNRA.

Simultaneous analyses of hydrogeological, physicochemical and microbiological parameters of the groundwater sampled from the drainage and observation wells at Kovin-Dubovac revealed the following:

- Elevated ammonium concentrations were detected mainly on the first drainage line, which recharges predominantly from the Danube and by infiltration from the land surface;
- The reason for the absence of sulphate from the groundwater in the examined area is sulphate reduction in an environment rich in organic matter and $Fe^{2+}$. The sulphides produced in the environment rich in ferrous ions are most likely quickly removed and precipitated in the form of iron sulphides;
- Although the conditions are reducing (average oxygen concentration 0.12 mg/l), high groundwater levels and intensive agricultural production have resulted in elevated concentrations of ammonium ions, derived from mineral fertilisers. There are some ammonium cations (due to elevated organic matter and iron concentration on the first drainage line), which originate from the decomposition of organic matter.

### 8.7.4 Origin and transformation of nitrogen compounds in the oxic alluvial groundwater source 'Ključ'

The oxic alluvial aquifer Ključ is used as a groundwater source for the town of Požarevac. It is located on the right bank of the lower Great Morava River (Figure 8.24). The thickness of the alluvial layers generally ranges from 15 to 25 m. The aquifer mainly consists of gravels and coarse-grained sands (Figure 8.23).

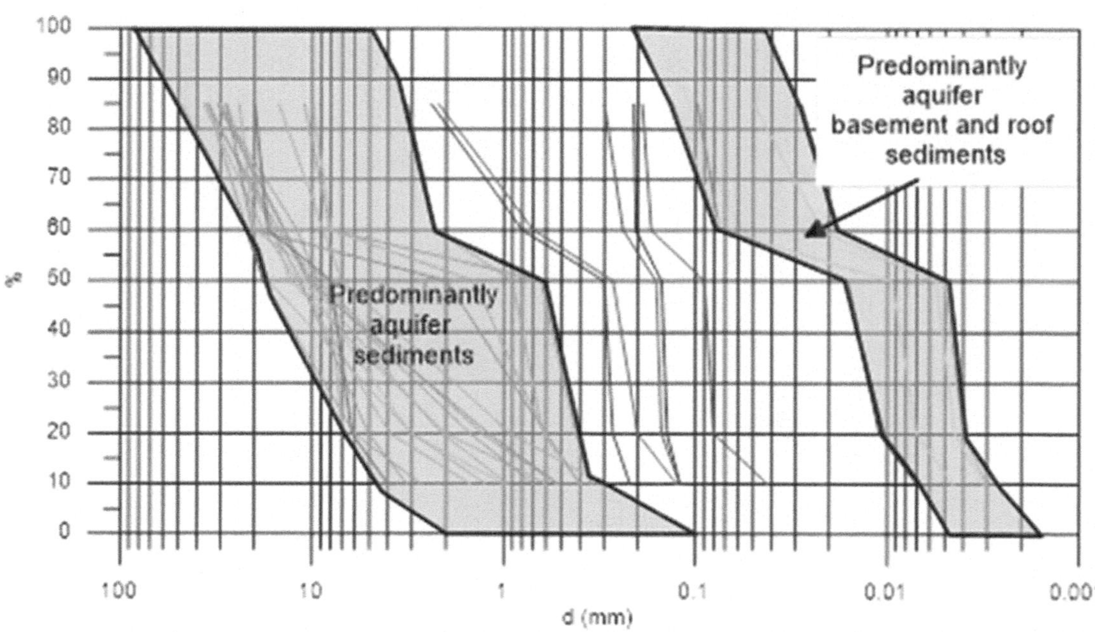


**Figure 8.23** Typical grain size distribution curves of the Ključ aquifer material (Dimkić & Pušić, 2014).

The aquifer cap rock is several metres thick. It consists of sandy clays, silts and topsoil. The aquifer is highly transmissible, and the cap rock is quite permeable.

The Ključ groundwater source relies on bank filtration (marked as 1 in Figure 8.24). The groundwater extracted by wells in the Great Morava alluvium is transported to infiltration basins (2). This approach

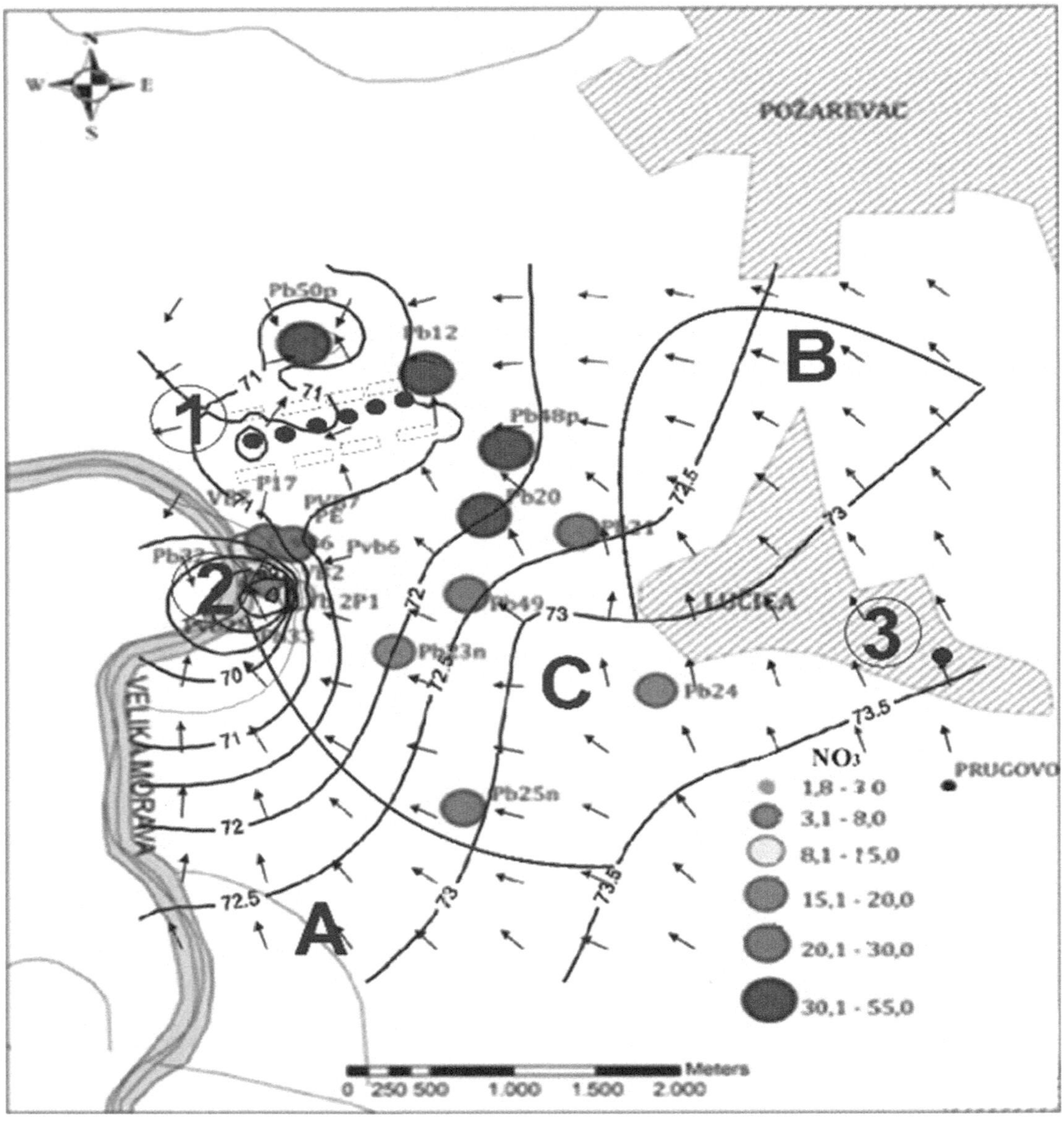


**Figure 8.24** Position of groundwater source Ključ (1), drainage wells (2), Lučica village (3), and predominant groundwater flow direction and average nitrate concentrations in certain parts of the aquifer (Perović, 2019).

(artificial recharge – abstraction) significantly reduces the hydrodynamic influence of the wellfield on the aquifer. The wells on the riverbank capture the water that is transported to the infiltration basins. About 30% of this water originates from the river and the remaining 70% comes from the hinterland. Consequently, the wells on the riverbank (2) have a significant impact on the groundwater flow in the alluvial aquifer. The villages of Lučica and Poljana (Figure 8.25) are evident emitters of nitrogen compounds from municipal wastewater discharged to unsanitary septic tanks.

Considering the high levels of dissolved oxygen in groundwater, processes of nitrogen oxidation and conservation of N in $^{+}5$ state are favoured. Sources of elevated nitrate concentrations include agriculture (manure and mineral fertilisers) and household wastewater. Three areas can be distinguished based on groundwater flow direction data/regime (Figure 8.24).

(A) Area mostly affected by agriculture;
(B) Area with a very pronounced and dominant effect of wastewater from the villages of Lučica and Poljana;
(C) Area with a combined impact of agriculture and domestic wastewater.

The influence, origin and nitrogen transformation pathways were generally determined by complementary hydrogeological, hydrogeochemical, biochemical and isotope analyses.

Apart from the above, the possibility of aerobic denitrification, which leads to nitrate concentration reduction in aerobic conditions (Ji *et al.*, 2015; Patureau *et al.*, 2000; Robertson *et al.*, 1989), is not excluded. Principal component analysis was undertaken to determine any linear correlations between nitrates and chemical species, which might indicate the origin of nitrate in oxic groundwater. The factor

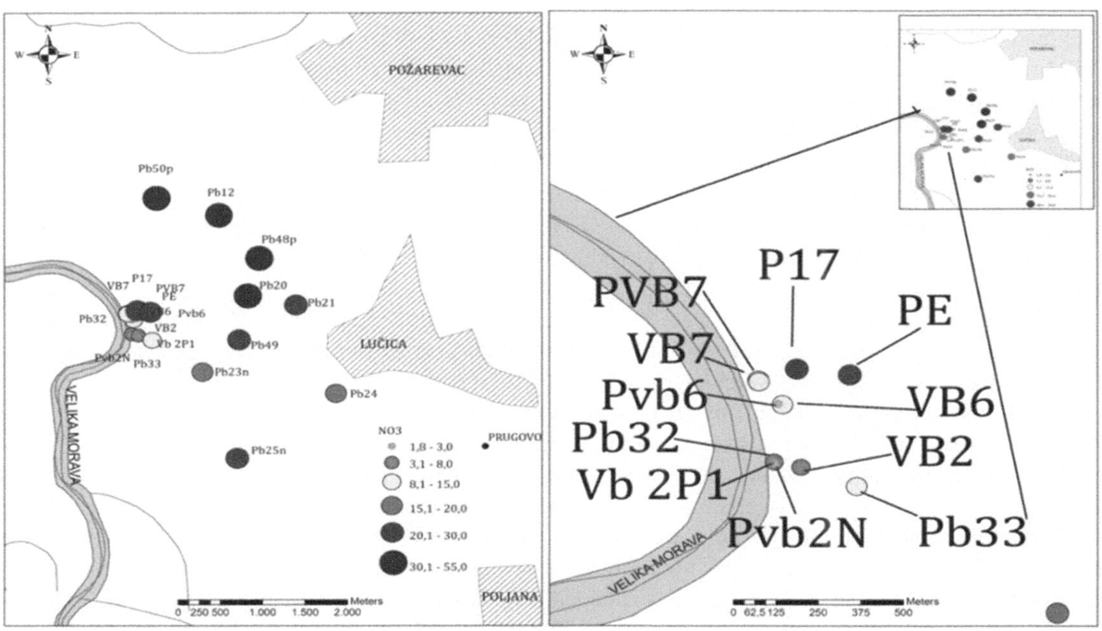


**Figure 8.25** Average nitrate concentrations (mgN/L) from 2010 to 2019 (all monitored wells – left, enlarged part along the river – right) (Živančev *et al.*, 2020).

**Table 8.8** Ključ groundwater source matrix.

| | | pH | $\varkappa$ | Eh | $O_2$ | $NH_4^+$ | $NO_3^-$ | $Cl^-$ | $SO_4^{2-}$ | B | TOC | Na |
|---|---|---|---|---|---|---|---|---|---|---|---|---|
| Factors | PC1 | −0.657 | 0.976 | | 0.767 | | 0.940 | 0.942 | 0.805 | 0.575 | | 0.694 |
| | PC2 | | | | | | | | | 0.633 | 0.917 | |
| | PC3 | 0.706 | | | −0.575 | 0.880 | | | | −0.646 | | |
| | PC4 | | | 0.945 | | | | | | | | |

analysis included 12 physicochemical parameters: pH, $\varkappa$, Eh, $O_2$, $NH_4^+$, $NO_3^-$, Cl, $SO_4^{2-}$, B, Ca, Na, and TOC. The extracted factors are shown in Table 8.8.

The research revealed that the area where groundwater is extracted from the Great Morava alluvium (VB-6, VB-2, VB-7, Vb-2/P-1, Pb-17, Pb-32, Pvb-2N and PVB7) is characterised by slightly elevated pH values and relatively low concentrations of $NO_3^-$, $Cl^-$, $SO_4^{2-}$ and B (compared to the entire area). Water from the hinterland exhibited increased concentrations of dissolved oxygen and lower pH values (indicators of nitrification), and dissolved B, $Cl^-$ and $NO_3^-$. Principal Component Analysis (PCA) and Cluster Analysis (CA) for the third factor showed a simultaneous increase in ammonium and pH values in the wells near the river, in contrast to the hinterland where the concentrations of B and $O_2$ were elevated. This is in a line with the hydrogeological indicators of groundwater origin in the proximity of the river (dominated by river water quality and by infiltration of water from the land surface – rainwater and dissolved substances applied to the soil surface).

In addition to the basic physicochemical and microbiological analyses, the isotopic signatures of sulphate and nitrate were analysed. The isotopic signatures of $\delta^{15}N$ and $\delta^{18}O\text{-}NO_3^-$ were also determined for the Great Morava at Ljubičevski Most. Groundwater from seven observation wells in the hinterland (Pb-12, Pb-20, Pb-21, Pb-24, Pb-48, Pb-49 and Pb-50p) and two near the Great Morava (Pb-23n, Pb-25n) was sampled for isotopic signature analyses. Water from the Great Morava was also sampled. Groundwater values of $\delta^{15}N$ were in the range from +6.70 to +12.90‰ and the oxygen isotope ratio was from +0.28 to +9.70‰. The surface water value of $\delta^{15}N$ was +5.46‰ and $\delta^{18}O$ was +0.48‰ (Figure 8.26). If there is a simultaneous influence of untreated sewage discharge and manure application, it can be difficult to determine the exact origin of the nitrate in groundwater. The absence of a linear relationship of $\delta^{15}N\text{-}NO_3^-$ and/or $\delta^{18}O\text{-}NO_3^-$ with $NO_3^-$ indicates the existence of mixing processes in groundwater and unfolding of several simultaneous biological processes (Wells *et al.*, 2016; Zhang *et al.*, 2013).

The quantified nitrates at Ključ were formed by nitrification. Oxidation of applied manure and anthropogenic excrements, followed by infiltration and transport of nitrates through the permeable and oxic groundwater environment, were determined as the dominant sources. The conclusions are consistent with previous research, reported by Boreli-Zdravković and Radenković (2003), Radenković and Boreli-Zdravković (2003, 2005), Radenković *et al.* (2007) and Miljević *et al.* (2012). Well Pb-23n was characterised by the highest values of $\delta^{15}N$ but not the highest concentrations of $NO_3^-$. The isotopic signature in Pb-23n was in the overlapping range of nitrates formed by ammonium oxidation, originating from human or animal excrements, and according to the groundwater flow direction, it was obvious that Pb-23n was outside the influence of household wastewater. Considering the spatial gradients of the concentrations of the parameters which explained the largest percentage of the variance, and the groundwater flow direction, the conclusion was that 19.46 mgN/l with the highest concentration of ammonium in the area, elevated pH, absence of $Cl^-$ and B with a lower $NO_3^-$ concentration indicated the contribution of fertiliser application.

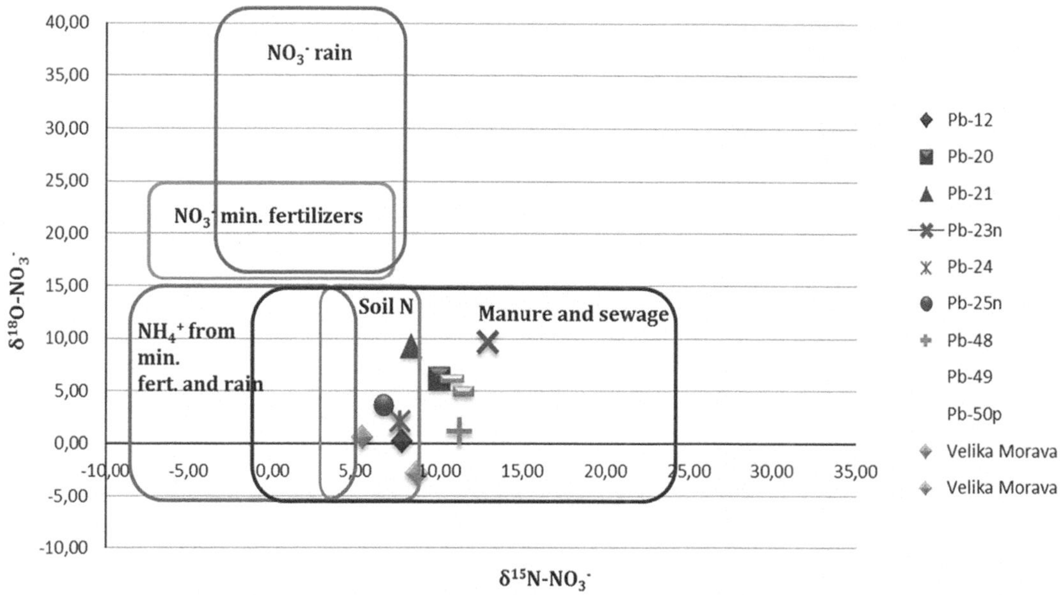


**Figure 8.26** Nitrate isotope ratios with characteristic signatures of different sources (Perović, 2019).

## 8.8 LABORATORY EXPERIMENT

A laboratory experiment was conducted with water-saturated soil columns to determine the dominant nitrate reduction pathway in the soil. The experiment involved *in situ* monitoring of changes in dissolved oxygen concentrations, redox potential, electrical conductivity and pH in the leachate from the water saturated soil columns. The water leached through the soil composite was collected and sampled for chemical analysis from a flow cell attached at the bottom of the column. The experiment lasted for 16 months, during 2017 and 2018. The soil was sampled from a 1.2 m deep excavation pit in the vicinity of Bp-17 D-I at Kovin-Dubovac. The soil was dried at 50°C, sifted through a 2 mm sieve and ground. All soil samples were combined and mixed to obtain a homogenised composite sample. To protect them from sunlight, the columns were placed in non-transparent PVC tubes. The equipment setup is shown in Figure 8.27.

After reaching saturation with distilled water, a $KNO_3$ solution (130 mg/l) and KCl tracer solution (300 mg/l) were applied to the columns. The total leached volume was 32 l or 18 column volumes. The flow through the flow cell was determined by continuously measuring the increase in mass of the leaching bag (in which the solution was collected) and recording the values on the logger. The flow was maintained at a level of up to 2 ml/min. The soil from the columns was sampled from three depths (samples O-II-1 z from 10 cm, O-II-2 z from 20 cm and O-II-3 z from 30 cm), along the entire column height. A dried composite soil sample (used to fill columns O-II-0 z) was also analysed. The ammonium concentration change was monitored during the experiment. After 18 exchanged column pore volumes, the nitrate and ammonium concentrations remained very low, although the concentration of potassium nitrate in the poured water was 130 mg/l. Table 8.9 shows the composition of the soil composite.

According to this laboratory experiment, based on the physicochemical analysis of the effluent and nitrogen residue on the soil matrix, 71% of the nitrogen applied to the columns through the nitrate

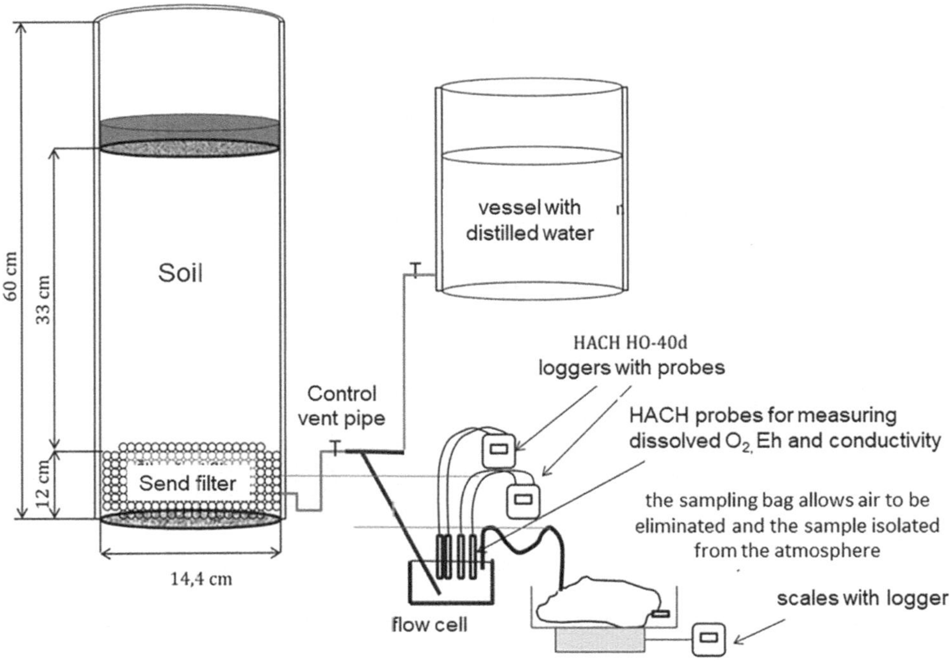


**Figure 8.27** Experiment setup with a water-saturated soil column, loggers, probes, flow cell, scales with loggers and distilled water for rinsing the flow cell (Perović, 2019).

solution was reduced to $NH_4^+$ ions and retained/sorbed to the soil matrix. Of the 12.6% of the nitrogen which leached through the column, 1.6% was in the form of $NO_3^-$ and $NO_2^-$, with the remaining 11% in the form of $NH_4^+$ ions. The conclusion was that in the case of nitrate infiltration into the soil, where

**Table 8.9** Soil quantity and properties.

| | |
|---|---|
| Soil column height [cm] | 33 |
| Mass of soil in column [g] | 9600 |
| Water volume in water-saturated soil [ml] | 1807 |
| Soil porosity [–] | 0.36 |
| Mass ratio of calcite ($CaCO_3$) calculated for dry soil weight [%] | 10. 28 |
| Mass ratio of silicon dioxide ($SiO_2$) calculated for dry soil weight [%] | 20 |
| Mass ratio of dolomite ($MgCO_3$) calculated for dry soil weight [%] | 20 |
| Mass ratio of albite ($NaAlSi_3O_8$) calculated for dry soil weight [%] | 30 |
| Mass ratio of muscovite $(Si_6Al_2)^{IV}(Al_4)^{VI}O_{20}(OH)_4K_2$, chlorite $((X,Y)_{4–6}(Si,Al)_4O_{10}(OH,O)_8)$ and smectite $(Si_{8-x}Al_x)^{IV}(Al_{4-y}Mg_y)^{VI}O_{20}(OH)_4X^+_{x+y} \cdot nH_2O)$ calculated for dry soil weight [%] | 15 |
| Mass ratio of organic carbon calculated for dry soil weight [%] | 2.46 |
| Mass ratio of iron oxides, oxyhydroxides and hydrated oxyhydroxides calculated for dry soil weight [%] | 3 |

anoxic conditions prevail, 16% of the applied nitrates were reduced by denitrification, which resulted in nitrogen loss, while 71% remained bound to the soil in the form of ammonium ions. The laboratory experiment showed that DNRA prevailed over denitrification.

## 8.9 CONCLUSION

Elevated nitrate concentrations in water bodies are a widespread health and environmental hazard, as indicated in numerous reports of European countries. Eutrophication and nitrogen suboxide ($N_2O$) emissions are the most commonly considered adverse environmental impacts. The most documented health disorder caused by elevated levels of nitrate in drinking water is methemoglobinemia. There is scientific evidence that the ingestion of nitrate and nitrite can also have a mutagenic, teratogenic effect and may contribute to the risk of developing different types of cancer.

A certain oxygen concentration is required in a groundwater environment for nitrate to occur and be retained. The oxygen content in an aquifer determines the potential for oxidation-reduction processes. Oxygen in groundwater is predominantly used for the oxidation of organic substances and minerals (lower oxidation state metals, mainly iron), which are sorbed to fine-grained materials. The deposition of fine-grained materials is less pronounced in upper river basins due to a higher flow energy. Therefore, the oxic state of an aquifer is usually higher in the upstream parts of a basin (coarse-grained material), whereas anoxic aquifers are found in lower river basins (Figure 8.28). Aquifers in the alluvial sediments

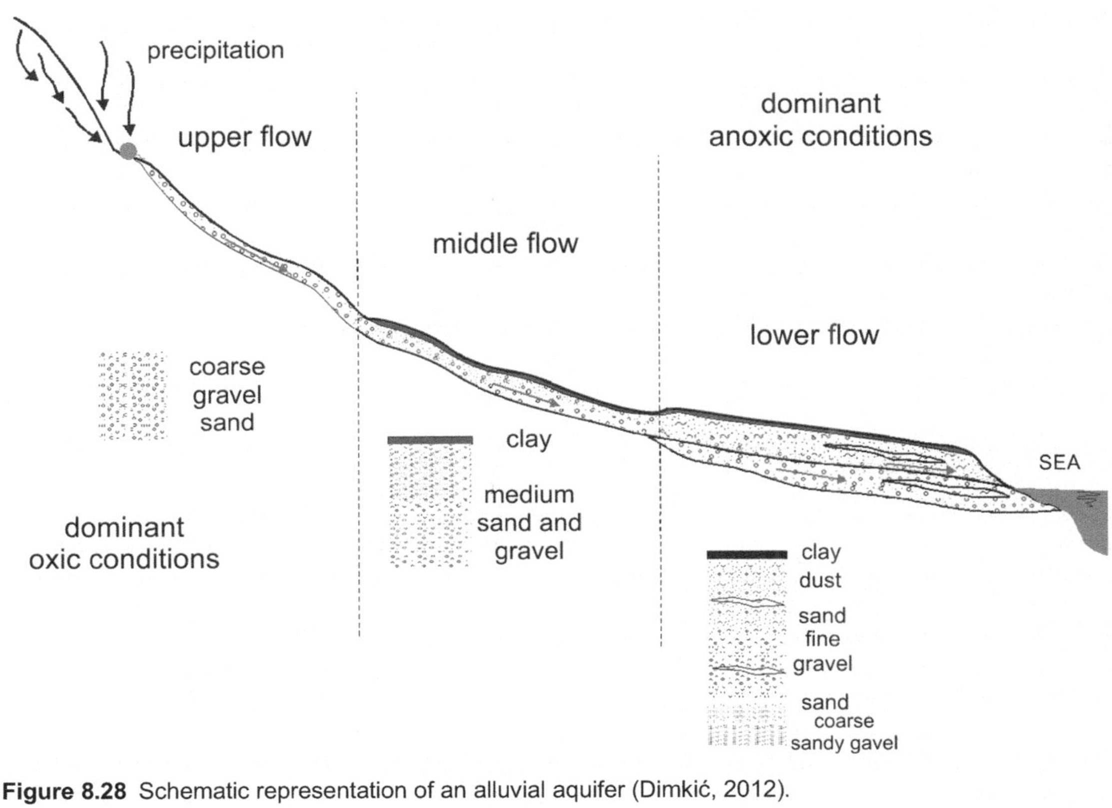


**Figure 8.28** Schematic representation of an alluvial aquifer (Dimkić, 2012).

of the Danube upstream from Budapest are mostly aerobic, so nitrates are not intensively exposed to denitrification and, as a result, the probability of elevated nitrate concentrations is much higher. In order to protect the Black Sea from eutrophication, the entire Danube River Basin has been declared a sensitive area. Hence, tertiary municipal wastewater treatment for the removal of nitrogen and phosphorus is required in the entire basin, as per the Urban Wastewater Treatment Directive.

The Environmental Protection Agency (EPA) of the United States has prescribed a maximum permissible nitrate concentration in drinking water of 10 mgN/l (nitrate nitrogen) under the Safe Drinking Water Act (SDWA). A nitrate limit has been set, below which, based on previous scientific knowledge, no negative health impact is expected. Apart from the health impact, the potential cost and ability of public water supply systems to remove nitrates from water using appropriate technologies were taken into account in setting the concentration limit.

The permissible nitrate concentration prescribed by the EPA is approximately equivalent to the guidelines of the World Health Organization (WHO), which sets a limit of 50 mg$NO_3$/l or 11.3 mgN/l. The maximum allowable concentration was established with the aim of preventing the occurrence of methemoglobinemia in new-borns; however, other health effects, including cancer and unwanted reproductive outcomes, have not been considered.

Europe has adopted a large number of directives related to the protection of the quality of water bodies. For the purposes of understanding the goals and contributions of the research reported herein, the most important are: Directive 2000/60/EC on the establishment of a framework for community action in the field of water policy; Directive 81/271/EEC relating to urban wastewater treatment; Directive 91/676/EEC concerning the protection of waters against pollution caused by nitrates originating from agriculture; and Directive 80/68/EEC on the protection of groundwater against pollution caused by certain hazardous substances. Directive 91/676/EEC stipulates that nitrate concentrations in surface water and groundwater should not exceed 50 mg/l. Preventive measures are also envisaged, which define manure handling (storage and warehousing) technologies, as well as the permitted quantities and manner of application of nitrogen fertilisers. The measures to protect the quality of water bodies listed in the directives are very effective but also a considerable economic burden.

Alluvial aquifers have a characteristic profile: aquifer skeleton composed of clastic material, overlain by a less permeable aquitard. The aquitard may consist of clay-silt and humus.

The rate of transport (load) of nitrogen compounds to the aquifer depends on:

- Precipitation;
- Vegetation cover;
- Types and quantities of applied fertilisers;
- Permeability and thickness of the cap rock;
- Mineralogical-chemical composition of the cap rock.

The aquifer oxygen content is extremely important, as well as its permeability and hydrodynamic features in general. In oxic aquifers, transformation processes move toward nitrogen oxidation and nitrate conservation. There are indications that nitrate reduction may occur to a certain extent even in oxic aquifers (Ji *et al.*, 2015; Patureau *et al.*, 2000; Robertson *et al.*, 1989).

In anoxic aquifers, the processes tend to reduce the introduced nitrates to nitrogen gas ($N_2$), $N_2O$ or $NH_4^+$.

The range of chemical and biochemical transformation processes of nitrogen compounds is very broad. Apart from the dissolved oxygen concentration, the oxic, sub-oxic, anoxic or anaerobic state of the aquifer is also determined by the presence of various oxygen-consuming compounds (iron, sulphur and carbon).

This chapter provided a concise description and overview of the fixation, mineralisation, nitrification, denitrification, dissimilatory nitrate reduction and anaerobic ammonium oxidation processes.

In the Danube River Basin, the occurrence of nitrates in groundwater is more extensive in the upstream part. This is attributable to a more pronounced use of fertilisers and predominantly oxic conditions in alluvial aquifers.

In the lower Danube River Basin, the river water is less exposed to nitrate and phosphate pollution and less loaded with dissolved organic carbon.

In Serbia, shallow aquifers resulting from sedimentation in the downstream parts of large river basins are largely anoxic. As such, there is a tendency towards nitrate reduction processes.

In the mountainous regions of southern Serbia, rivers have deposited coarse-grained sediments. Alluvial aquifers are therefore mostly oxic, with conditions suitable for nitrification and/or nitrate conservation.

A method for determining the origin of elevated nitrates was presented for a shallow oxic alluvial aquifer in which the artificially-recharged groundwater source Ključ is located. The established sources of nitrogen compounds are agriculture (application of manure and fertilisers) and unsanitary household wastewater discharges. The identification of pollution sources and subsequent remediation included the following steps:

- Locating hotspots of nitrate pollution;
- Characterising the hydrogeological composition of the terrain;
- Determining the regime and directions of groundwater flow;
- Defining the distribution of nitrates, as well as other nitrogen compounds in the groundwater;
- Determining the concentrations of oxygen and other relevant chemical substances (for example: B, Na, Cl, Fe, S) in the groundwater;
- Characterising biochemical parameters of the groundwater.

Stable isotope analyses were also conducted to identify nitrogen pollution sources.

The stable isotope (isotopic signature) analysis is a method widely used to determine sources of nitrates in groundwater.

Simultaneous comprehensive and highly complex physicochemical, hydrogeological and microbiological analyses are also required to confirm the assumed transformation processes and nitrogen sources.

The lowest values of $\delta^{15}N\text{-}NO_3^-$ characterise mineral fertilisers, oxidation of soil organic matter, and rainwater, while the highest values are attributable to manure or household wastewater. Their ranges can overlap (Figure 8.29). The combined analysis of $\delta^{18}O$ and $\delta^{15}N$ from $NO_3^-$ may contribute to the

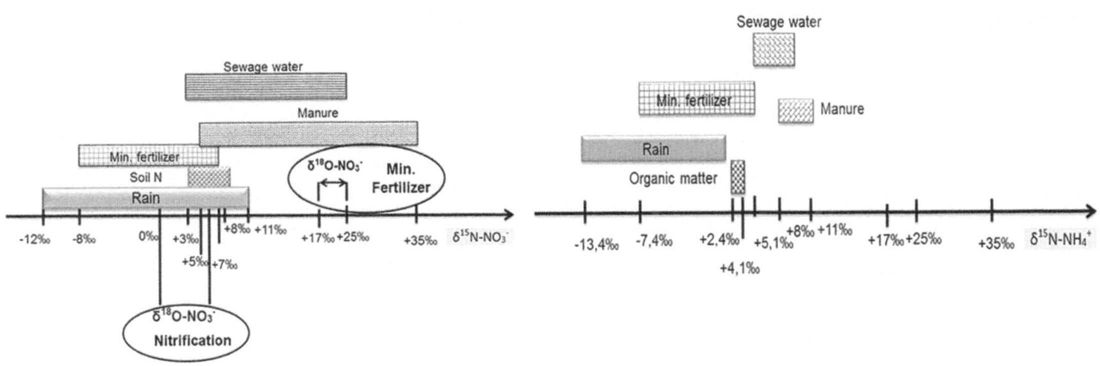


**Figure 8.29** Characteristic ranges of stable N isotopic signatures of different sources (Perović, 2019).

delineation of atmospheric and terrestrial nitrate sources, including potential separation of different anthropogenic sources. The quantification of $\delta^{18}O\text{-}NO_3^-$ can indicate whether $NO_3^-$ originates from mineral fertilisers or if it is a product of nitrification. The lowest values of $\delta^{15}N\text{-}NH_4^+$ were determined in rainwater and the highest in manure and sewage. Organic matter showed slightly higher values of the $\delta^{15}N\text{-}NH_4^+$ isotopic composition, compared to mineral fertilisers and rainwater. However, available experimental evidence suggests that isotopic signatures of different $NH_4^+$ sources may overlap due to specific environmental conditions in certain areas (Figure 8.29).

The research reported herein has shown the theoretical and practical possibilities of detailed determination of nitrogen sources and transport and transformation mechanisms in groundwater by simultaneous and comprehensive hydrogeochemical, isotopic and statistical analyses. The research outcomes contribute to the theoretical understanding of the impact of agricultural practice on groundwater quality in oxic and anoxic environments. They also facilitate assessment of the self-purification potential of groundwater, by defining the conditions suitable for nitrogen conservation ($NH_4^+$) or emission ($N_2$, $N_2O$).

## REFERENCES


Baily A., Rock L., Watson C. J. and Fenton O. (2011). Spatial and temporal variations in groundwater nitrate at an intensive dairy farm in south-east Ireland: insights from stable isotope data. *Agriculture, Ecosystems & Environment*, **144**, 308–318.

Barth S. (1998). Application of boron isotopes for tracing sources of anthropogenic contamination in groundwater. *Water Resources*, **32**, 685–690.

Bengtsson G., Bengtson P. and Mansson F. K. (2003). Gross nitrogen mineralization-, immobilization-, and nitrification rates as a function of soil C/N ratio and microbial activity. *Soil Biology & Biochemistry*, **35**(1), 143–154.

Berg M., Tran H. C., Nguyen T. C., Pham H. V., Schertenleib R. and Giger W. (2001). Arsenic contamination of groundwater and drinking water in Vietnam: a human health threat. *Environmental Science & Technology*, **35**, 2621–2626.

Bohlke J. K., Smith R. L. and Miller D. N. (2006). Ammonium transport and reaction in contaminated groundwater: application of isotope tracers and isotope fractionation studies. *Water Resources Research*, **42**(5), 19, doi: 10.1029/2005WR004349

Boreli-Zdravković Đ. and Radenković Z. (2003). In: Serbian - Nitrati u podzemnim vodama šireg područja izvorišta "Ključ 1" za vodosnabdevanje Požarevca, Zbornik radova 32. Jugoslovenske konferencije "VODA 2003", Jugoslovensko društvo za zaštitu voda i Srpsko društvo za zaštitu voda, Zlatibor 3-6 jun 2003, 357–360.

Bottrell S., Tellam J., Bartlett R. and Hughes A. (2008). Isotopic composition of sulfate as a tracer of natural and anthropogenic influences on groundwater geochemistry in an Urban Sandstone Aquifer, Birmingham, UK. *Applied Geochemistry*, **23**(8), 2382–2394.

Brunet R. C. and Garcia-Gil L. J. (1996). Sulfide-induced dissimilators nitrate reduction to ammonia in anaerobic freshwater sediments. *FEMS Microbiology Ecology*, **21**(2), 131–138, https://doi.org/10.1111/j.1574-6941.1996.tb00340.x

Burgin A. J. and Hamilton S. K. (2007). Have we overemphasized the role of denitrification in aquatic ecosystems? A review of nitrate removal pathways. *Frontiers in Ecology and the Environment*, **5**(2), 89–96.

Buss S. R., Herbert A. W., Morgan P., Thornton S. F. and Smith J. W. N. (2004). A review of ammonium attenuation in soil and groundwater. *Quarterly Journal of Engineering Geology & Hydrogeology*, **37**(4), 347–359.

Choi W. J., Lee S. M. and Ro H. M. (2003). Evaluation of contamination sources of groundwater NO 3− using nitrogen isotope data: A review. *Geosciences Journal*, **7**(1), 81–87.

Clark I. (2015). Contaminant geochemists and isotopes. In: Groundwater Geochemistry and Isotopes. Taylor & Francis Group, LLC, Boca Raton, FL.

Cohen M. J., Heffernan J. B., Albertin A. and Martin J. B. (2012). Inference of riverine nitrogen processing from longitudinal and diel variation in dual nitrate isotopes. *Journal of Geophysical Research*, **117**, G01021, doi: 10.1029/2011JG001715

Cruz J. V. and Andrade C. (2015). Natural background composition in the Azores archipelago (Portugal): a hydrogeochemical study and threshold value determination. *Science of the Total Environment*, **520**, 127–135.

Darwishe H., Khattabi J. E., Chaaban F. *et al.* (2017). Prediction and control of nitrate concentrations in groundwater by implementing a model based on GIS and artificial neural networks (ANN). *Environmental Earth Sciences*, **76**, 649, https://doi.org/10.1007/s12665-017-6990-1

Davidson E. A., Chorover J. and Dail D. B. (2003). A mechanism of abiotic immobilization of nitrate in forest ecosystems: the ferric wheel hypothesis. *Global Change Biology*, **9**(2), 228–236.

Dimkić M. (2012). Importance of the Aerobic State of Alluvial Aquifers for Groundwater use, Nakdong River International Water Week, 7th World Water Forum, September, Andong, Republic of Korea, 69–96. Conference 'Contemporary Issues of Adaptive Water Management', Belgrade, 29–37.

Dimkić M. and Pušić M. (2014). Correlation between entrance velocities, increase in local hydraulic resistances and redox potential of alluvial groundwater sources. *Water Research and Management*, **4**(1), 3–23.

Dimkić M. A., Brauch H.-J. and Kavanaugh M. (2008). Groundwater Management in Large River Basins. IWA Publishing, Alliance House, 12 Caxton Street, London SW1H 0QS, UK.

Dzamic R. and Stevanovic D. (2007). Agrochemistry ISBN: 978-86-7157-353-5. Parthenon, Belgrade.

Ellis B. and Bertrand-Krajewski J. L. (2010). Assessing Infiltration and Exfiltration on the Performance of Urban Sewer Systems, European Water Research Series. IWA Publishing, London, UK, ISBN 184339149X, 9781843391494

European Environment Agency, EEA. (2012). European waters – current status and future challenges. EEA Report No. 9/2012. ISSN 1725-9177.

European Environment Agency, EEA. (2018). Nutrients in freshwater in Europe. https://www.eea.europa.eu/data-and-maps/indicators/nutrients-in-freshwater/nutrients-in-freshwater-assessment-published-6 (accessed March 2019).

Eurostat. (2018). Gross nutrient balance. https://ec.europa.eu/eurostat/data/database (accessed February 2019).

FAO. (2013). Guidelines to control water pollution from agriculture in China, Water Report 40.

FAO. (2016). FAOSTAT. Database. Food and Agriculture Organization of the United Nations (FAO), Rome. Available at http://faostat3.fao.org/browse/R/RP/E (accessed July 2016).

FAO. (2017a). Water Pollution from Agriculture: a Global Review. Food and Agriculture Organization of the United Nations (FAO), Rome. Available at http://www.fao.org/3/a-i7754e.pdf (accessed January 2020).

FAO. (2017b). Tonnes of nitrogen produced per year. UN Food and Agricultural Organization (FAO) (assessed January 2020).

Feast N. A., Hiscock K. M., Dennis P. F. and Andrews J. N. (1998). Nitrogen isotope hydrochemistry and denitrification within the Chalk aquifer system of north Norfolk, UK. *Journal of Hydrology*, **211**, 233–252.

Francis A. J., Slater J. M. and Dodge C. J. (1989). Denitrification in deep sub-surface sediments. *Geomicrobiology Journal*, **7**(1–2), 103–116.

Gasser G., Pankratov I., Elhanany S., Glazman H. and Lev O. (2014). Calculation of wastewater effluent leakage to pristine water sources by the weighted average of multiple tracer approach. *Water Resources Research*, **50**, 4269–4282, doi: 10.1002/2013WR014377

Heffernan J. B., Cohen M. J., Frazer T. K., Thomas R. G., Rayfield T. J., Gulley J., Martin J. B., Delfino J. J. and Graham W.D. (2010). Hydrologic and biotic influences on nitrate removal in a subtropical spring-fed river. *Limnology and Oceanography*, **55**(1), 249–263, doi: 10.4319/lo.2010.55.1.0249

Henson W. R., Huang L., Graham W.D. and Ogram A. (2017). Nitrate reduction mechanisms and rates in an unconfined eogenetic karst aquifer in two sites with different redox potential. *Journal of Geophysical Research: Biogeosciences*, **122**, 1062–1077, doi: 10.1002/2016JG003463

Hiscock K. M., Lloyd J. W. and Lerner D. N. (1991). Review of natural and artificial denitrification of groundwater. *Water Research*, **25**(9), 1099–1111.

Hollocher T. C. (1984). Source of the oxygen atoms of nitrate in the oxidation of nitrite by Nitrobacter agilis and evidence against a PON anhydride mechanism in oxidative phosphorylation. *Archives of Biochemistry and Biophysics*, **233**(2), 721–727.

Hosono T., Tokunaga T., Tsushima A. and Shimada J. (2014). Combined use of δ13C, δ15N, and δ34S tracers to study anaerobic bacterial processes in groundwater flow systems. *Water Research*, **54**, 284–296.

Hübner H. (1986). Isotope effects of nitrogen in the soil and biosphere. In: Handbook of Environmental Isotope Geochemijstry, P. Fritz and J. C. Fontes (eds.), Elsevier, Amsterdam, 361–425.

Hyer K. E. (2007). A multiple-tracer approach for identifying sewage sources to an urban stream system. US Geological Survey Scientific Investigations Report, 2006–5317.

Jaroslav Černi Water Institute. (2017). Analysis of optimization of surface and groundwater quality monitoring network in accordance with the requirements of the Water Framework Directive and accompanying directives, Belgrade.

Jetten M. S. M. and Strous M. van de Pas-Schoonen, Schalk J, van Dongen U.G.J.M., van de Graaf A. A., Logemann S., Muyzer G., van Loosdrecht M.C.M. and Kuenen J. G. (1998). The anaerobic oxidation of ammonium. *FEMS Microbiology Reviews*, **22**(5), 421–437. doi: 10.1111/j.1574-6976.1998.tb00379.x

Ji B., Yang K., Zhu L., Jiang Y., Wang H., Zhou J. and Zhang H. (2015). Aerobic denitrification: a review of important advances of the last 30 years. *Biotechnology and Bioprocess Engineering*, **20**(4), 643–651.

Kalagasidis Krušić M. (2017). Organic Industrial Synthesis. Internal script of the Department of Organic Chemical Technology, Faculty of Technology and Metallurgy in Belgrade, Belgrade.

Kamp A. (2007). Growth and nitrate reduction of Beggiatoa filaments studied in enrichment cultures. Dissertation. Gottfried Wilhelm Leibniz University of Hannover, 81s.

Kaown D., Koh D. C., Mayer B. and Lee K. K. (2009). Identification of nitrate and sulfate sources in groundwater using dual stable isotope approaches for an agricultural area with different land use (Chuncheon, Mid-Eastern Korea). *Agriculture, Ecosystems & Environment*, **132**, 223–231.

Kelso B. H. L., Smith R. V., Laughlin R. J. and Lennox S. D. (1997). Dissimilators nitrate reduction in anaerobic sediments leading to river nitrite accumulation. *Applied and Environmental Microbiology*, **63**(12), 4679–4685.

Kendall C. (1998). Tracing nitrogen sources and cycling in catchments. In: Isotope Tracers in Catchment Hydrology, C. Kendall and J. J. McDonnell (eds.), Elsevier, Amsterdam, 519–576.

Kendall C. and Aravena R. (2000). Nitrate isotopes in groundwater systems. In: Environmental Tracers in Subsurface Hydrology, P. G. Cook and A. L. Herczeg (eds.). Springer, Boston, MA, 261–297. doi: 10.1007/978-1-4615-4557-6_9

Kim K. H., Yun S.-T., Kim H.-K. and Kim J.-W. (2015). Determination of natural backgrounds and thresholds of nitrate in South Korean groundwater using model-based statistical approaches. *Journal of Geochemical Exploration*, **148**, 196–205.

Kiurski-Milošević J. (2015). Groundwater quality assessment model with increased arsenic content using phase logic. Doctoral dissertation, Faculty of Technical Sciences, Novi Sad.

Kloppmann W., Chikurel H., Picot G., Guttman J., Pettenati M., Acharoni A., Guerrot C., Millot R., Gauss I. and Wintgens T. (2009). B and Li isotopes as intrinsic tracers for injection tests in aquifer storage and recovery systems. *Applied Geochemistry*, **24**(7), 1214–1223.

Knöller K., Vogt C., Haupt M., Feisthauer S. and Richnow H. H. (2011). Experimental investigation of nitrogen and oxygen isotope fractionation in nitrate and nitrite during denitrification. *Biogeochemistry*, **103**(1–3), 371–384.

Komljenović I. and Todorović V. (1998). General Farming. Faculty of Agriculture, Belgrade, ISBN 8672620038, 9788672620030.

Korom S. F. (1992). Natural denitrification in the saturated zone: a review. *Water Resources Research*, **28**(6), 1657–1668.

Kovačević S. (2017). Analysis of self-purifying potential of groundwater for removal of pharmaceuticals using the riverbank filtration method. Doctoral dissertation, University of Novi Sad, Faculty of Technical Sciences.

Kowalenko C. G. (1979) The influence of sulfur anions on denitrification. *Canadian Journal of Soil Science*, **59**, 221–223.

Li X. D., Masuda H., Koba K. and Zeng H. A. 2007. Nitrogen isotope study on nitrate-contaminated groundwater in the Sichuan Basin, China. *Water, Air, & Soil Pollution*, **178**, 145–156.

Lindebaum J. (2012). Identification of sources of ammonium in groundwater using stable nitrogen and boron isotopes in Nam Du, Hanoi. Dissertations in Geology at Lund University, No. 300, 38pp. 45.

Liu C. Q., Li S. L., Lang Y. C. and Xiao H. Y. (2006). Using δ15N and δ18O values to identify nitrate sources in karst ground water, Guiyang, Southwest China. *Environmental Science & Technology*, **40**, 6928–6933.

Magalhaes C., Moreira R., Wiebe W. J. and Bordalo A. A. (2003). Salinity and inorganic nitrogen effects on nitrification and denitrification rates in inter-tidal sediments and rocky biofilms: Douro River Estuary, Portugal. Proceedings of the 7th International Specialized IWA Conference on Diffuse Pollution and Basin Management, Dublin, 6–73.

Martinelli G., Dadomo A., De Luca D. A., Mazzolad M., Lasagnac M., Pennisie M., Pillaf G., Sacchi E. and Saccon P. (2018). Nitrate sources, accumulation and reduction in groundwater from Northern Italy: insights provided by a nitrate and boron isotopic database. *Applied Geochemistry*, **91**, 21–35.

Megonigal J. P., Mines M. E. and Visscher P. T. (2005). Linkages to trace gases and anaerobic processes. *Biogeochemistry*, **8**, 350–362.

Michener R. and Lajtha K. (2007). Tracing anthropogenic inputs of nitrogen to ecosystems. In: Stable Isotopes in Ecology and Environmental Science, 2nd edn, C. Kendall, E. M. Elliott and S. D. Wankel (eds.), Blackwell Publishing Ltd., Carlton, Victoria, 375–449.

Miljević N., Boreli-Zdravković Đ., Obradović V., Golobočanin D. and Mayer B. (2012). Evaluation of the origin of nitrate influencing the Kljuc groundwater source, Serbia. *Water Science & Technology*, **66**(3), 472–478, doi: 10.2166/wst.2012.179

Morris J. T., Whiting G. J. and Chapelle F. H. (1988). Potential denitrification rates in deep sediments from the Southeastern Coastal Plain. *Environmental Science & Technology*, **22**(7), 832–836.

Neilsen M. E., Fisk M. R., Istok J. D. and Pedersen K. (2006). Microbial nitrate respiration of lactate at in situ conditions in ground water from a granitic aquifer situated 450 m underground. *Geobiology*, **4**(1), 43–52.

Nikolenko O., Jurado A. V., Borges A. V., Knöller K. and Brouyere S. (2018). Isotopic composition of nitrogen species in groundwater under agricultural areas: a review. *Science of the Total Environment*, **621**, 1415–1432.

Norrman J., Sparrenbom C. J., Berg M., Nhan D. D., Nhan P. Q., Rosqvist H., Jacks G., Sigvardsson E., Baric D. and Moreskog J. (**2008**). Arsenic mobilisation in a new well field for drinking water production along the Red River, *Nam Du, Hanoi: Applied Geochemistry*, **23**, 3127–3142.

NRP. 2016. Nitrate vulnerable areas and Rules of good agricultural practice in Serbia Proposal for determining vulnerable areas according to the EU Nitrates Directive (2016). Ministry of Agriculture and Environment, Republic Water Directorate. NEOKOMERC, doo, Belgrade. ISBN 978-86-919977-0-0. In Serbian Nitratima ranjiva područja i Pravila dobre poljoprivredne prakse u Srbiji Predlog za određivanje ranjivih područja prema EU Nitratnoj Direktivi.

Pano S. V., Kelly W. R., Martinsek A. T. and Hackley K. C. (2006). Estimating background and threshold nitrate concentrations using probability graphs. *Ground Water*, **44**(5), 697–709.

Pasten-Zapata E., Ledesma R., Harter T., Ramirez A. I. and Mahlkencht J. (2014). Assessment of sources and fate of nitrate in shallow groundwater of an agricultural area by using a multi-tracer approach. *Science of the Total Environment*, **470–471**, 855–864.

Patureau D., Zumstein E., Delgenes J. P. and Moletta R. (2000). Aerobic denitrifiers isolated from diverse natural and managed ecosystems. *Microbial Ecology*, **39**(2) 145–152.

Perović M. (2019). Assessment of the influence and regional specificity of hydrogeochemical conditions on the transformation of nitrogen compounds in groundwater. Doctoral dissertation, Faculty of Technical Sciences, Novi Sad, Serbia.

Perović M., Obradović V., Kovačević S., Mitrinović D., Živančev N. and Nenin T. (2017). Indicators of Groundwater Potential for Nitrate Transformation in a Reductive Environment, Water Environment Research, ISSN 1061–4303. *Water Environment Federation*, **89**(1), 4–16.

Pesticide Safety Education Program. (2018). PSEP. Margaret McCasland, Nancy M. Trautmann, and Keith S. Porter, Center for Environmental Research and Robert J. Wagenet Dept. of Agronomy Cornell University. http://psep.cce.cornell.edu/facts-slides-self/facts/nit-heef-grw85.aspx (pose) (accessed June 2018).

Postma D., Boesen C., Kristiansen H. and Larsen F. (1991). Nitrate reduction in an unconfined sandy aquifer: water chemistry, reduction processes, and geochemical modeling. *Water Resources Research*, **27**(8), 2027–2045.

Radenković Z. and Boreli-Zdravković Đ. (2003). Causes of the sudden deterioration of the water quality of the 'Ključ 1' spring. Proceedings of the 32nd Yugoslav Conference 'WATER 2003', Yugoslav Society for Water Protection and Serbian Society for Water Protection 3–6 June, 2003, Zlatibor, 361–366.

Radenković Z. and Boreli-Zdravković Đ. (2005). In: Serbian - Sistem zaštite izvorišta "Ključ" grada Požarevca: faza realizacije objekta, Zbornik radova 34. Jugoslovenske konferencije "VODA 2005", Jugoslovensko društvo za zaštitu voda i Srpsko društvo za zaštitu voda, 7-10 jun 2005. Kopaonik, 449–456.

Radenković Z., Boreli-Zdravković Đ. and Zuber-Radenković V. (2007). Effects of urgent measures implemented to protect the Ključ groundwater source. Proceedings Regional IWA conference on Groundwater management in the Danube river basin and other large river basins, 7–9 June 2007, 'Jaroslav Černi' Institute for the development of water resources, Belgrade, 291–297.

Rivett M. O., Buss S. R., Morgan P., Smith J. W. N. and Bemment C. D. (2008). Nitrate attenuation in groundwater: a review of biogeochemical controlling processes. *Water Research*, **42**, 4215–4232.

Robertson L. A., Cornelisse R., De Vos P., Hadioetomo R. and Kuenen J. G. (1989). Aerobic denitrification in various heterotrophic nitrifiers. *Antonie van Leeuwenhoek*, **56**(4), 289–299.

Robertson W. D., Moore T. A., Spoelstra J., Li L., Elgood R. J., Clark I. D., Schiff S. L., Aravena R. and Neufeld J. D. (2012). Natural attenuation of septic system nitrogen by anammox. *Groundwater*, **50**(4), 541–553.

Sato *et al.* (2013). *Agricultural Water Management*, **130**, 1–13.

Savard M. M., Somers G., Smirnoff A., Paradis D., van Bochove E. and Liao S. (2010). Nitrate isotopes unveil distinct seasonal N-sources and the critical role of crop residues in groundwater contamination. *Journal of Hydrology*, **381**(1), 134–141.

Sawyer C. N. and McCarty P. L. (1978). Chemistry for Environmental Engineering, 3rd edn. McGraw-Hill Book Co, New York.

Sebilo M., Mayer B., Nicolardot B., Pinay G. and Mariotti A. (2013). Long-term fate of nitrate fertilizer in agricultural soils. *Proceedings of the National Academy of Sciences*, **110**(45), 18185–18189.

Sharp Z. (2007). Principles in Stable Isotope Geochemistry. Pearson/Prentice Hall, Upper Saddle River, NJ.

Singh C. K., Kumar A. and Bindal S. (2018). Arsenic contamination in Rapti River Basin, Terai region of India. *Journal of Geochemical Exploration*, **192**, 120–131. https://doi.org/10.1016/j.gexplo.2018.06.010

Stenstrom M. K. and Poduska R. A. (1980). The effect of dissolved oxygen concentration on nitrification. *Water Research*, **14**(6), 643–649.

Stevanović Z. and Dokmanović P. (2015). Project for expansion of the monitoring network for groundwater monitoring in Serbia, RHMZ and RGF Fund, Belgrade.

Thayalakumaran T., Bristow K. L., Charlesworth P. B. and Fass T. (2007). Geochemical conditions in groundwater systems: implications for the attenuation of agricultural nitrate. *Agricultural Water Management*, **95**, 103–115.

Tiedje J. M., Sexstone A. J., Parkin T. B., Revsbech N. P. and Shelton D. R. (1984). Anaerobic processes in soil. *Plant and Soil*, **76**, 197–212.

Tiedje J. M. (1988). Ecology of denitrification and dissimilatory nitrate reduction to ammonium. In: Biology of Anaerobic Microorganisms, A. J. B. Zehnder (ed.), John Wiley and Sons, New York.

Tomaszewski M., Cema G. and Ziembińska-Buczyńska A. (2017). Influence of temperature and pH on the anammox process: a review and meta-analysis. *Chemosphere*, **182**, 203–214.

Toshio S., Manzoor Q., Sadahiro Y., Tsuneyoshi E. and Zahoor A. (2013). Global, regional, and country level need for data on wastewater generation, treatment, and use. *Agricultural Water Management*, **130**, 1–13. doi: 10.1016/j.agwat.2013.08.007

US EPA. 2016. Water Quality Assessment and TMDL Information. United States Environmental Protection Agency (US EPA), Washington, DC. (Available at: https://ofmpub.epa.gov/waters10/attains_index.home).

Verma S. and Mukherjee A. (2018). Groundwater quality, contamination, and processes in brahmaputra river basin aquifers. In: Groundwater of South Asia. Springer Hydrogeology, A. Mukherjee (ed.), Springer, Singapore, Online ISBN 978-981-10-3889-1.

Ward M. H., Jones R. R., Brender J. D., de Kok T. M., Weyer P. J., Nolan B. T., Villanueva C. M. and van Breda S. G. (2018). Drinking water nitrate and human health: an updated review. *International Journal of Environmental Research and Public Health*, **15**(7), 1557, https://doi.org/10.3390/ijerph15071557

Wells N. S., Hakoun V., Brouyère S. and Knöller K. (2016). Multi-species measurements of nitrogen isotopic composition reveal the spatial constraints and biological drivers of ammonium attenuation across a highly contaminated groundwater system. *Water Research*, IWA Publishing, **98**, 363–375, doi: ff10.1016/j.watres. 2016.04.025ff. ffhal-01306976f

Weng T. N., Liu C. W., Kao Y. H. and Hsiao S. S. Y. (2017). Isotopic evidence of nitrogen sources and nitrogen transformation in arsenic-contaminated groundwater. *Science of Total Environment*, **578**, 167–185.

Widory D., Kloppmann W., Chery L., Bonnin J., Rochdi H. and Guinamant J.-L. (2004). Nitrate in groundwater: an isotopic multi-tracer approach. *Journal of Contaminant Hydrology*, **72**, 165–188.

World Heath Organization. (2004). WHO. Guidelines for Drinking Water Quality, 3rd edn. WHO, Geneva.

WWAP. (2013). The United Nations World Water Development Report 2013. United Nations World Water Assessment Program (WWAP). United Nations Educational, Scientific and Cultural Organization, Paris.

WWAP. (2015). The United Nations World Water Development Report 2015: Water for a sustainable world. United Nations World Water Assessment Program (WWAP). United Nations Educational, Scientific and Cultural Organization, Paris.

WWAP. (2017). The United Nations World Water Development Report 2017: Wastewater, the Untapped Resource. United Nations World Water Assessment Program (WWAP). Available at https://unesdoc.unesco.org/ark:/48223/pf0000247553/PDF/247553eng.pdf.multi (accessed January 2020). Paris, United Nations Educational, Scientific and Cultural Organization.

Xue D., Botte J., De Baets B., Accoe F., Nestler A., Taylor P., Van Cleemput O., Berglund M. and Boeckx P. (2009). Present limitations and future prospects of stable isotope methods for nitrate source identification in surface-and groundwater. *Water Research*, **43**(5), 1159–1170.

Zhang Y. C., Prommer H., Broers H., Slomp C. P., Greskowiak J., van der Grift B. and Van Cappellen P. (2013). Model-based integration and analysis of biogeochemical and isotopic dynamics in a nitrate-polluted pyritic aquifer. *Environmental Science and Technology*, **47**(18), 10415–10422.

Živančev N. (2019). Correlation analysis of distribution mechanisms and concentration levels of pesticides in groundwater. Doctoral dissertation, Faculty of Technical Sciences, Novi Sad.

Živančev N., Kovačević S., Perović M., Čalenić A. and Dimkić M. (2020). Influence of oxic and anoxic groundwater conditions on the occurrence of selected agrochemicals. *Water Supply*, doi: 10.2166/ws.2019.178

# Chapter 9

# Methods for trace analysis of pharmaceuticals and pesticides in water and sediment samples

*Ivana Matić Bujagić, Svetlana Grujić, Tatjana Đurkić and Mila Laušević*

## 9.1 INTRODUCTION

This chapter reviews several approaches used in the detection, identification and quantification of pharmaceuticals and pesticides in the environment. The extensive use of pesticides in worldwide agricultural practice and pharmaceuticals in human and veterinary medicine has led to the contamination of water resources. Monitoring trace levels of these contaminants in water bodies is an ongoing challenge and has become possible due to significant improvements in analytical techniques. The determination of trace levels of pesticides and pharmaceutical residues in environmental matrices requires the usage of ultra-sensitive and high-selective analytical methods, such as gas chromatography-mass spectrometry (GC-MS) and liquid chromatography-mass spectrometry (LC-MS). In recent years, the efficiency of chromatographic separations has been significantly improved by the development of ultra-high performance liquid chromatography (UHPLC), while the utilization of high-resolution mass spectrometry (HRMS) facilitated the confirmation of target compounds, as well as the identification of non-target ingredients. Due to the diversity of physicochemical properties among numerous classes of environmental contaminants, many developed analytical methodologies focus only on the most important groups. However, the prerequisite for proper environmental monitoring is the availability of multiresidual analytical methods that would selectively distinguish between a great number of different pollutants in complex mixtures. This section illustrates the latest progress and trends in multiresidual analysis, describing the wide variety of extraction procedures and analytical techniques suitable for the environmental determination of pharmaceuticals and pesticides.


Editor: Milan Dimkić
doi: 10.2166/9781789060904_0495

## 9.2 SAMPLING AND PREPARATION OF ENVIRONMENTAL SAMPLES

### 9.2.1 Water sample preparation

Sampling is the first step in the analysis of pharmaceuticals and pesticides in natural waters. The manner and frequency with which samples are collected can have a significant effect on the quality of the subsequent analytical results and their interpretation (Crt *et al.*, 2010). Conventional water sampling consists of punctual grab sampling, the main objective of which is the collection of representative samples for later water quality analysis. The presence of particulate matter in the water samples can significantly interfere with the extraction and analysis of the obtained extracts. Therefore, the collected samples are usually further processed by filtration through glass microfiber filters (Asghar *et al.*, 2018; Battaglin *et al.*, 2018; Hossain *et al.*, 2018; Kim & Carlson, 2007; Praveena *et al.*, 2018; Sudo *et al.*, 2004). The filtration step can be very significant in the interpretation of the results of the analysis, since a high proportion of pollutants can bind to filterable particulates, which can lead to incorrect conclusions (Petrovic, 2014). The separation of the dissolved phase from the suspended particles is often performed by filtration through $\leq$1.0 µm pore size filters (Asghar *et al.*, 2018; Battaglin *et al.*, 2018; Hossain *et al.*, 2018; Kim & Carlson, 2007; Praveena *et al.*, 2018; Sudo *et al.*, 2004). Seasonal variations in environmental conditions and changes in the weather patterns are also important factors in surface water sampling (Allan *et al.*, 2006). The annual trends in rainfall and the subsequent runoff can be the reason for elevated pesticide levels detected in river water (Antić *et al.*, 2015) while the proximity of wastewater discharge to a sampling site can affect the measured pharmaceutical concentrations (Kasprzyk-Hordern *et al.*, 2008).

Of all the analytical steps in the trace analysis of water samples, sample preparation is one of the most significant and time-consuming processes which greatly influences the precision and accuracy of the overall analysis (Kataoka, 2017). Sample preparation techniques include the extraction, isolation, and pre-concentration of the target compounds, as well as the clean-up of the obtained extracts.

#### *9.2.1.1 Solid-phase extraction*

An ideal sample preparation method should be simple and environmentally friendly, with a low cost and the possibility of automation. However, there is no universal sample preparation technique which is suitable for all samples, since the selection depends on the nature of the analytes and the complexity of the matrices.

As can be seen from literature (Table 9.1), solid-phase extraction (SPE) is unquestionably the preferred technique for the sample preparation of water samples since it represents a convenient, cost-effective alternative to the traditionally used liquid-liquid extraction. The standard SPE procedure involves passing the aqueous phase through a cartridge containing a solid material (sorbent phase) and the subsequent removal of the target analytes from the solid phase by elution with small volumes of an appropriate solvent. The SPE process consists of four consecutive steps: (1) conditioning of the solid-phase material (i.e. activation of sorbent functional groups), (2) sample loading, (3) rinsing to remove the interferences of the sample matrix, and (4) analyte elution by desorption from the sorbent with a suitable solvent. Optimal sorbent and rinsing and elution solvents are usually chosen during the development of the method. One of the major features of the SPE method is the possibility of automation through the coupling of the sample preparation procedure and the separation and detection system. Time consuming sample pre-concentration steps and contact with dirty and hazardous samples have been avoided by employing online SPE (Liang *et al.*, 2019; Meyer *et al.*, 2011; Wang & Gardinali, 2012; Wode *et al.*, 2012).

**Table 9.1** Overview of methods for quantitative determination of pharmaceuticals and pesticides in environmental waters.

| Analytes | Samples | Sample Preparation Technique | Instrumental Technique | Reference |
|---|---|---|---|---|
| Antibiotics | River water | SPE (Oasis HLB cartridges) | HPLC-ESI-IT | Kim and Carlson (2007) |
| Pharmaceuticals Metabolites | River, lake water | SPE (Oasis HLB and LC-18 cartridges) | UHPLC-Q-Orbitrap | Asghar *et al.* (2018) |
| Pharmaceuticals Hormones | River, lake water | SPE (Oasis HLB cartridges-for pharmaceuticals)<br>SPE (octadecylsilyl disk-for hormones) | UHPLC-ESI-MS (pharmaceuticals)<br>GC-MS/MS (hormones) | Battaglin *et al.* (2018) |
| Pharmaceuticals | Surface water, wastewater | SPE (Oasis HLB cartridges) | LC-ESI-IT | Gros and Petrovic (2009) |
| Pharmaceuticals | River water | SPE (Oasis HLB cartridges) | HPLC-ESI-QqQ | Hossain *et al.* (2018) |
| Pharmaceuticals | Tap, lake water and groundwater | Online SPE (Oasis HLB cartridges) | UHPLC-ESI-QqQ | Liang *et al.* (2019) |
| Pharmaceuticals | River water | SPE (Oasis HLB cartridges) | ELISA | Praveena *et al.* (2018) |
| Pharmaceuticals | Drinking water | SPE (C18 cartridges) | LC-ESI-TOF | Reis *et al.* (2019) |
| Pharmaceuticals Pesticides | River water, groundwater | SPE (Oasis HLB cartridges) | LC-ESI-IT | Radović *et al.* (2015) |
| Pharmaceuticals | Surface water, groundwater, wastewater | SPE (Oasis HLB cartridges) | LC-ESI-IT | Grujić *et al.* (2009) |
| Pharmaceuticals | Surface water, wastewater | SPE (Oasis HLB cartridges) | LC-ESI-QqQ | Gros *et al.* (2006) |
| Pharmaceuticals | River water | SPE: (1) Oasis MCX; (2) Lichrolut EN; (3) Bakerbond C18 | LC-ESI-QqQ | Calamari *et al.* (2003) |
| Pharmaceuticals | River water | SPE (Strata-X) | LC-ESI- QqQ | Hilton and Thomas (2003) |
| Pharmaceuticals | River water | SPE (Oasis MCX) | UHPLC-ESI-QqQ | Kasprzyk-Hordern *et al.* (2008) |

(*Continued*)

**Table 9.1** Overview of methods for quantitative determination of pharmaceuticals and pesticides in environmental waters (*Continued*).

| Analytes | Samples | Sample Preparation Technique | Instrumental Technique | Reference |
|---|---|---|---|---|
| Industrial chemicals<br>Pharmaceuticals<br>Flame retardants<br>Pesticides | Drinking and surface water, wastewater | online SPE (C18 cartridges) | UHPLC-HESI-HRMS (Orbitrap) | Wode *et al.* (2012) |
| 5-Nitroimidazoles | Tap and drinking water, surface water, wastewater | Vortex assisted d-SPE (MOFs) | UHPLC-ESI-QqQ | Lu *et al.* (2016) |
| Antibiotics and their metabolites | Aquaculture water | Vortex assisted d-SPE (MOFs) | UHPLC-ESI-QqQ | Cai *et al.* (2017) |
| Pharmaceuticals Metabolites and transformation products | Environmental waters | On-line turbulent flow chromatography | LC-ESI-QqQ | López-Serna *et al.* (2012) |
| Pharmaceuticals | Environmental waters | SPE (MWCNTs) | LC-ESI-IT | Lalović *et al.* (2017) |
| Herbicides<br>Pharmaceuticals | River water | online SPE (Oasis HLB cartridges) | LC-ESI-QqQ | Meyer *et al.* (2011) |
| Antibiotics | Water samples | d-SPE (MWCNTs) | UHPLC-DAD | Herrera-Herrera *et al.* (2013) |
| Pharmaceuticals<br>Personal care products | Environmental waters | SBSE (EG Silicone) | LC-ESI-QqQ | Gilart *et al.* (2013) |
| Pharmaceuticals | Environmental waters | SBSE (MWCNTs) | HPLC-DAD | Abujaber *et al.* (2019) |
| Antibiotics and their metabolites | Surface, tap water, wastewater | HF-LPME | HPLC-DAD/FLD | Payán *et al.* (2011) |
| Pharmaceuticals<br>Emerging contaminants | Drinking water | Passive sampling (POCIS) | LC-ESI-QqQ | Magi *et al.* (2018) |
| Antibiotics | River water | Passive sampling (DGT) | HPLC-PDA | Chen *et al.* (2012) |
| Pharmaceuticals | Environmental water | DLLME ($CCl_4$ as extractant, methanol as dispersive solvent) | GC-FID | Yazdi *et al.* (2008) |
| Estrogenic hormones | Seawater, wastewater | SPE (Strata-X) | GC-EI-Q | Caban *et al.* (2013) |
| NSAID | River water, wastewater, tap water | SPE (Oasis HLB cartridges) | GC-MS<br>LC-MS/MS | Heath *et al.* (2010) |

| | | | | |
|---|---|---|---|---|
| Pesticides | River water | SPE (Oasis HLB cartridges) | HPLC-ESI-IT | Antić *et al.* (2015) |
| Pesticides | Surface water, groundwater | SPE (Oasis HLB cartridges) | LC-ESI-IT | Dujaković *et al.* (2010) |
| Pesticides | Milli-Q water samples | d-SPE (MWCNTs) | nano LC-UV | Asensio-Ramos *et al.* (2011) |
| Herbicides | Environmental water | SPE (Oasis MCX) | HPLC-ESI-QqQ | Zhang *et al.* (2014) |
| Pesticides | River, lake water | SPE (Sep-Pak Plus PS-2 cartridge) | GC-MS | Sudo *et al.* (2004) |
| Organophos-phorus pesticides | Tap, river, pond water | SPE (RP-C18 cartridge) | GC-NPD/FID | Ballesteros and Parrado (2004) |
| Pesticides | Drinking water | SPE (Oasis HLB) | LC-ESI-QqQ | Carvalho *et al.* (2008) |
| Herbicides | Well, tap, canal water | SPE (SWCNTs) | HPLC-PDA | Niu *et al.* (2009) |
| Pesticides | Deionized water | SPE (carbon structured material) | LC-ESI-IT | Vukčević *et al.* (2012) |
| Pesticides | Drinking water | d-SPE (MWCNTs) | LC-ESI-QqQ | Zou *et al.* (2016) |
| Pesticides | Environmental water | magnetic-SPE (MOFs) | HPLC-DAD/FLD | Ma *et al.* (2018) |
| Pharmaceuticals<br>Pesticides<br>Personal care products | Tap water | SPME (polyacrylate coating) | GC-IT | Basaglia and Pietrogrande (2012) |
| Pesticides | River water | SPME (polyacrylate coating) | GC-IT | Jabali *et al.* (2019) |
| Pesticides | Farm water | HF-LPME | GC-ECD | Pan and Ho (2004) |
| Pesticides | Sea water | HF-LPME | GC-MS | Basheer *et al.* (2002) |
| Pesticides | River water | Passive sampling with HF-LPME | GC-EI-Q | Valenzuela *et al.* (2019) |
| Herbicides | Environmental water | DLLME (DCM as extractant, THF as dispersive solvent) | LC-UV | Chou *et al.* (2009) |
| Pesticides | Rainwater, well water, lake water | DLLME ($CCl_3$ as extractant, THF as dispersive solvent) | LC-FLD | Wu *et al.* (2009) |
| Pesticides | River water, well water, farm water | DLLME (chlorobenzene as extractant, acetone as dispersive solvent) | GC-FPD | Berijani *et al.* (2006) |

(*Continued*)

**Table 9.1** Overview of methods for quantitative determination of pharmaceuticals and pesticides in environmental waters (*Continued*).

| Analytes | Samples | Sample Preparation Technique | Instrumental Technique | Reference |
|---|---|---|---|---|
| Pesticides | Drinking water | SPME (PDMS-DVB fiber); HF-LPME | GC–EI-QqQ | Garrido Frenich (2011) |
| Pesticides | Wastewater<br>Suspended particulate matter | SPE (C18 Sep-Pak, Oasis HLB)<br>PLE EtOAc:MeOH (3:1, v/v) GPC clean-up | GC-EI-QqQ<br>UHPLC-ESI-QqQ | Barco-Bonilla *et al.* (2010) |
| Pesticides | Drinking water | SPE (Bond Elut Florisil cartridge) | GC-(EI, NCI)-Q<br>LC-ESI-QqQ | Schwanz *et al.* (2019) |
| Corrosion inhibitors<br>Pesticides Pharmaceuticals | River water, wastewater | SPE (Oasis HLB cartridges) | LC-MS/MS<br>ESI and APCI | Valls-Cantenys *et al.* (2016) |
| Organochlorine pesticides | Surface water | SPE (Strata C-18E) | LC-APCI-QqQ | Chusaksri *et al.* (2006) |
| Organochlorine pesticides | Seawater, river water | SPE (AccuBOND II ODS-C18) | LC- Direct-EI-Q | Famiglini *et al.* (2008) |
| Pesticides<br>Pharmaceuticals Drugs of abuse<br>TPs | Surface water, wastewater | SPE (Oasis HLB cartridges) | LC-ESI-QqQ<br>UHPLC-ESI-QTOF | Masiá *et al.* (2013) |
| Pharmaceuticals Personal care products<br>Steroid hormones<br>Sterols | Wastewater | on-line SPE (Thermo Hypersil aQ) | LC-QqQ<br>HESI,<br>APPI,<br>APCI | Wang and Gardinali (2012) |
| Pesticides | Surface water | SPE (Oasis HLB cartridges) | LC-ESI-Q-Orbitrap | Casado *et al.* (2018) |
| Pesticides<br>TPs | Surface water, wastewater | SPE (Oasis HLB cartridges) | UHPLC-ESI-QTOF | Rousis *et al.* (2017) |

SPE–solid phase extraction; HPLC–high performance liquid chromatography; ESI–electrospray ionization; IT–ion trap detector; UHPLC–ultra-high performance liquid chromatography; Q–quadrupole detector; GC–gas chromatography; LC–liquid chromatography; QqQ–triple quadrupole detector; TOF–time of flight detector; HESI–heated-electrospray ionization; DAD–diode array detector; FLD–fluorescence detector; PDA–photodiode array detector; FID–flame ionization detector; NSAID–non-steroidal anti-inflammatory drugs; UV–ultraviolet detector; d-SPE–dispersive solid phase extraction; MWCNTs–multi-walled carbon nanotubes; SBSE–stir bar sorptive extraction; HF-LPME–hollow fiber-liquid-phase microextraction; DLLME–dispersive liquid-liquid microextraction; SPME–solid phase microextraction; ECD–electron capture detector; EI–electron impact ionization; NCI–negative chemical ionization; APCI–atmospheric pressure chemical ionization; APPI–atmospheric pressure photoionization; QTOF–hybrid quadrupole time of flight mass spectrometry; TPs–transformation products.

#### 9.2.1.2 Sorbent and eluent selection in SPE

Currently, in a conventional solid-phase extraction, a variety of SPE cartridges are commercially available, mainly based on bonded silica phases and polymers. However, as the majority of published methods show, (Table 9.1), Waters Oasis HLB (hydrophilic-lipophilic balanced) cartridge has been mostly used for the preconcentration of both polar and non-polar compounds. Due to its chemical composition (a combination of the lipophilic divinylbenzene and the hydrophilic N-vinylpyrrolidone polymers), acidic, neutral and basic compounds of a wide range of pH values can be extracted using Oasis HLB cartridge. In addition to Oasis HLB, C18 is also a widely used sorbent (Asghar *et al.*, 2018; Ballesteros & Parrado, 2004; Calamari *et al.*, 2003; Reis *et al.*, 2019). However, the use of this sorbent generally requires a pH adjustment of the sample prior to extraction. For example, in order to recover the acid compounds, water samples are acidified, because at neutral pH values the analytes are in their ionized form which is poorly retained by lipophilic sorbents (Gros *et al.*, 2006; Reis *et al.*, 2019). Mixed-mode polymeric sorbents, such as Oasis MCX have also proven to be highly selective in the extraction of acid and basic pharmaceuticals (Calamari *et al.*, 2003; Kasprzyk-Hordern *et al.*, 2008) and herbicides (Zhang *et al.*, 2014). The satisfactory recoveries of 20 pesticides (in the range 74–115%) were achieved using the polymeric sorbent based on styrene divinylbenzene copolymer (Sudo *et al.*, 2004). Hilton and co-workers (Hilton & Thomas, 2003) compared seven different sorbents and found that Strata-X (a polydivinylbenzene resin manufactured by Phenomenex), with properties similar to those of Oasis HLB, can be successfully used to determine the concentrations of a range of human pharmaceuticals in surface and wastewater. For the efficient elution of pharmaceuticals from the solid phase, in most cases, pure methanol is used (Battaglin *et al.*, 2018; Gros *et al.*, 2006; Hossain *et al.*, 2018; Kim & Carlson, 2007; Reis *et al.*, 2019). Pesticides are often eluted with methanol (Carvalho *et al.*, 2008; Meyer *et al.*, 2011), ethyl acetate (Ballesteros & Parrado, 2004), acetone (Sudo *et al.*, 2004) or mixtures of organic solvents (Antić *et al.*, 2015; Dujaković *et al.*, 2010; Radović *et al.*, 2015).

An upgrade of the conventional solid-phase extraction technique has been achieved by using new sorbent materials such as carbon structured and carbon nanostructured materials. Vukčević *et al.* (2012) showed that waste hemp fibers from the textile industry can be carbonized into an efficient pesticide sorbent by the activation at 900°C with KOH. Nanomaterials based on single-walled (SWCNTs) and multi-walled carbon nanotubes (MWCNTs) are also used for the trace analysis of pharmaceuticals and pesticides (see Table 9.1). Carbon nanotubes have been used as a sorbent for many environmental pollutants because of their high affinity and adsorption capacity, good thermal stability, the wide pH range of application, and the possibility of surface chemistry modification (Lalović *et al.*, 2017). One of the major disadvantages of carbon nanotubes is the agglomeration in the SPE cartridge, which can be efficiently overcome by packing CNTs into lab-made disks (Niu *et al.*, 2009).

### 9.2.2 Advancements in water sample preparation

A survey of literature also shows a shift from the traditionally used SPE towards alternative extraction modes such as the dispersive solid-phase extraction and the magnetic solid-phase extraction. In order to reduce time and simplify extraction procedures, a dispersive SPE (d-SPE) has been developed. The dispersive SPE is carried out in a bulk solution, by mixing the dispersive sorbent with the sample. Sorbent material can be used in two ways: for sample clean-up, by retaining matrix interferences onto it or for trapping the target analytes, which are later eluted or desorbed with an appropriate solvent. The advantages of the d-SPE, compared to the traditional SPE, are the minimized clogging of sorbent pores by suspended particles and the possibility of sorbent regeneration and reuse. Asensio-Ramos *et al.* (2011) reported a simple d-SPE method using MWCNTs for pesticide sorption from an aqueous solution. A suspension of water and MWCNTs was shaken for 15 min, and then transferred to a glass tube with a PTFE frit. The elution of the pesticides from the MWCNT sorbent retained on PTFE frit was achieved using ethyl acetate. A d-SPE methodology, based on MWCNTs as stationary phases, was also reported for sulfonamide antibiotic analysis (Herrera-Herrera *et al.*, 2013) and pesticide residue analysis (Zou *et al.*, 2016) in water samples.

Magnetic SPE (MSPE) is another alternative in which a small amount of magnetic adsorbent is dispersed into the sample solution, with the aim of trapping target analytes, and then removed from the solution by using a magnetic field. Analytes are recovered from the adsorbent with a suitable extractant. In MSPE, the most important parameter for high extraction efficiency is the selection of an appropriate magnetic adsorbent. A recent study (Ma *et al.*, 2018) used a combination of water stable and eco-friendly metal organic framework magnetic material (MOF-5) and functionalized $Fe_3O_4$ nanoparticles for the extraction of pesticides from water samples. An efficient extraction of antibiotics, with recoveries ranging from 70% to 109%, was achieved using MOFs as a sorbent for vortex assisted dispersive micro-solid-phase extraction from water samples (Cai *et al.*, 2017; Lu *et al.*, 2016).

In the past few decades, green microextraction techniques involving lower sample volumes (5–100 mL), lower solvent volumes ($\leq$1 mL), simpler equipment and easier handling have emerged. Solid-phase microextraction (SPME) and stir bar sorptive extraction (SBSE) are regarded as solventless extraction techniques. In SPME, extraction is performed by the adsorption of the target analytes on a fiber coated with a suitable stationary phase, which is placed inside a hollow needle in a micro-syringe-like device. Later, the analytes are thermally desorbed directly into the GC column or eluted with the mobile phase in the mode of LC analysis. One of the main advantages of the SPME technique is that it can be performed manually or automatically and is easily coupled to chromatographic systems. A headspace SPME with fibers coated with polyacrylate (PA) was used for the simultaneous extraction of pharmaceuticals and pesticides (Basaglia & Pietrogrande, 2012). The study showed that the optimization of the most significant variables affecting the extraction efficiency (SPME fiber type, extraction temperature and time for extraction, derivatization and desorption) can provide over 80 complete analyses with no apparent fiber degradation (Basaglia & Pietrogrande, 2012). PA fibers were also successfully applied for determining the pesticide contamination of the watershed of Abou Ali River, Lebanon (Jabali *et al.*, 2019).

SBSE, first introduced in 1999, uses a magnetic stirring bar coated with a small amount of sorbent material immobilized on it. During the extraction, the analytes are adsorbed onto a polymeric sorbent material. The desorption of the analytes from the sorbent is then performed by thermal or liquid desorption (Gilart *et al.*, 2013). Nowadays, the commercially available coating based on polydimethylsiloxane (PDMS) used in SBSE is being replaced by new materials, such as EG Silicone (Gilart *et al.*, 2013) and MWNCTs (Abujaber *et al.*, 2019).

Liquid-phase microextraction (LPME) enables the extraction of the target analytes with minimal amounts of the water-immiscible solvent, with high enrichment factors (Farajzadeh *et al.*, 2014). Although LPME can be carried out in different extraction modes, dispersive liquid-liquid microextraction (DLLME) and hollow fiber-LPME (HF-LPME) are the most popular in the analysis of pharmaceuticals and pesticides in environmental samples (see Table 9.1). In the first step of HF-LPME, analytes are extracted from the aqueous sample into a supported liquid membrane sustained in the pores of a hydrophobic porous HF, and later into an acceptor solution placed inside the lumen of the fiber. The microextraction solvent is mechanically protected since it is placed within the lumen of a porous HF and therefore is not in direct contact with the sample solution. Prior to extraction, the organic solvent is immobilized into the pores of the HF by the immersion of the HF into an organic solvent. HF-LPME using toluene as an extraction solvent was used to extract pesticides from farm water (Pan & Ho, 2004) and sea water (Basheer *et al.*,

2002). Excellent preconcentration and clean-up, with enrichment factors up to 1000, was achieved with HF-LPME for the highly sensitive HPLC determination of sulfonamides and their metabolites in environmental waters (Payán *et al.*, 2011).

The DLLME technique is a miniaturized LLE technique that uses μL volumes of the extraction solvent along with a few mL of dispersive solvents. In DLLME the mixture of the extraction and dispersive solvent is rapidly injected into an aqueous sample with a syringe. The vigorous mixing of the sample leads to the formation of a cloudy solution and analyte extraction. In the final step, target analytes are recovered from the cloudy solution by centrifugation. Chlorinated solvents, such as carbon tetrachloride (Yazdi *et al.*, 2008), chloroform (Wu *et al.*, 2009), chlorobenzene (Berijani *et al.*, 2006) and dichloromethane (Chou *et al.*, 2009) have been mostly employed in the DLLME extractions of environmental pollutants. Caldas *et al.* (2016) suggested a modified solvent-based de-emulsification DLLME. For the simultaneous determination of pharmaceuticals, pesticides and personal care products, the authors used a combination of octanol, as a low density extraction solvent, and acetone, as a disperser, thereby omitting the centrifugation step (Caldas *et al.*, 2016). For detailed information on recent developments and modern approaches in microextraction techniques for pharmaceuticals and pesticide determination see Farajzadeh *et al.* (2014), Pérez-Fernández *et al.* (2017) and Primel *et al.* (2012, 2017).

Finally, the use of new passive samplers has received increasing attention throughout the last decade. The evolution of passive sampling has provided low cost and easy transportation devices that reduce sample preparation steps. During passive sampling, both sample collection and preconcentration are performed in one step. Passive samplers contain a stationary phase (receiving phase, i.e. sorbent), which is separated from the aqueous phase by two membranes. The diffusion of water through the passive sampler results in the adsorption of the analytes onto the receiving phase. Passive samplers provide time-weighted-average and equilibrium concentrations over the deployment time, contrary to conventional 'spot' sampling which gives a pollutant's concentrations at the moment of sampling. The passive sampler named Polar Organic Chemical Integrative Sampler (POCIS) has been employed when determining pharmaceuticals and other emerging contaminants in drinking water (Magi *et al.*, 2018). In this study, HLB was used as a sorbent phase and the target analytes were eluted from the sorbent with acetone (Magi *et al.*, 2018). A variation of the passive sampler which uses diffusive gradients in thin films (DGT) was developed for the detection of antibiotics in river water (Chen *et al.*, 2012). A recent study (Valenzuela *et al.*, 2019) reported a new approach combining passive sampling and HF-LPME for the effective determination of pesticides in water sources during prolonged periods. The passive sampling device was built using a plastic bottle containing the liquid phase microextraction system.

### 9.2.3 Sediment sample preparation

Sediment quality is usually assessed by analyzing surface sediments, since the upper 10 cm of sediment are a habitat for benthic organisms. In fact, the major contaminant exposure pathways for most organisms can be determined based on the contaminant concentration in the upper layer of sediments. The collection of surface sediments is performed by using grab samplers, such as the Birge-Ekman and Van Veen sampler. The grab samplers allow so-called non-disruptive sampling, i.e. they protect the sample from disturbance and minimize the washout of fine-grained sediments. The obtained grab samples are then used to determine the horizontal variation of concentrations of pollutants in the sediments. However, in order to determine the depth-related occurrence of contaminants and the vertical variations, as well as to investigate possible historical contamination, it is necessary to collect sediment samples from depths greater than 15 cm. For this purpose, core samplers are used. The collection of core samples is generally quite demanding because core samples should be kept upright and stationary during their transport to the water surface and before sectioning.

Prior to their analysis, sediment samples need to be prepared, and the selection of a proper preparation procedure plays an important role in improving the sensitivity and reducing impurities from complex matrices such as sludge and sediments. Sample pretreatment involves dehydration and the reduction of particle size. Water can be removed by air-drying at room temperature over a period of time (Covaci *et al.*, 2005; Hijosa-Valsero *et al.*, 2016; Matić Bujagić *et al.*, 2016; Pinto *et al.*, 2013), by heating (Lara-Martín *et al.*, 2015) or by the lyophilization process (Carafa *et al.*, 2007; Ccanccapa *et al.*, 2016; da Silva *et al.*, 2011; Jelić *et al.*, 2009; Kalogridi *et al.*, 2014; Lahti & Oikari, 2012; López de Alda *et al.*, 2002; Moreno-González *et al.*, 2015; Vazquez-Roig *et al.*, 2010; Yang *et al.*, 2010). Particle size reduction is done by grinding and sieving through appropriate sieves. The sieving of sediments provides the removal of coarse material (debris, rocks) and information on the granulometric composition of the samples.

The extraction of pharmaceuticals and pesticides from solid samples involves desorption of analytes from the sample matrix and dissolution in a solvent. Various techniques, such as ultrasonic solvent extraction (USE), pressurized liquid extraction (PLE), microwave-assisted extraction (MAE), and QuEChERS (Quick, Easy, Cheap, Effective, Rugged, and Safe), have been applied for the extraction of these contaminants from solid matrices. In order to remove interferences of the samples, extraction is usually followed by an extract clean-up, based on SPE. An overview of the methods used for the quantitative determination of pharmaceuticals and pesticides in sediment samples is presented in Table 9.2.

### 9.2.3.1 *Ultrasonic solvent extraction*

In ultrasonic solvent extraction, ultrasonic vibration erodes the sample matrix and accelerates the matrix contact with the solvent. It is commonly performed in a water bath or with probes and sonoreactors. USE is regarded as a quick and inexpensive method, with equipment which is easy to operate. The extraction procedure is usually performed in two or three cycles, each time with fresh solvents. After USE, the extract is obtained by centrifugation and combined. USE requires a small volume of the solvent (typically 0.1–2 g sample treated with 5–25 mL of solvent) within an extraction time of 10–30 min. The optimization of different USE parameters, such as the type of extraction solvent, irradiation conditions (temperature and amplitude of sonication), sonication time, sample particle size, solid/liquid ratio is usually necessary to maximize the extraction recoveries (Albero *et al.*, 2019). In the extraction done with pure solvents, the most efficient were methanol (Matić Bujagić *et al.*, 2016) and acetonitrile (Pinto *et al.*, 2013; Shi *et al.*, 2014). Since USE usually involves the isolation of multiple classes of pharmaceuticals and/or pesticides, mixtures of methanol/acetone (López de Alda *et al.*, 2002), methanol/dichloromethane (Matić Bujagić *et al.*, 2019; Radović *et al.*, 2015), acetonitrile with a citric buffer (Yang *et al.*, 2010) and a phosphate buffer (Lahti & Oikari, 2012), and acetone/hexane (Sun *et al.*, 2013) are often used. Ultrasonic extraction can also be performed by combining different solvents, i.e. another fresh solvent is added in the next extraction step. For example, in the analysis of river sediment, optimal recoveries of pharmaceuticals were achieved by ultrasonic solvent extraction with pure methanol in the first cycle, followed by an acetone/acetic acid mixture in the second and pure ethyl acetate in the third cycle (Matongo *et al.*, 2015). Pinto *et al.* (2013) found that an increase of the water bath temperature to 80°C had a significant effect on the extraction of pesticides with high octanol-water partitioning coefficients and that it can reduce the number of extraction cycles. However, it is important to bear in mind that an increase in temperature can also lead to the co-extraction of matrix interferences.

**Table 9.2** Overview of methods for quantitative determination of pharmaceuticals and pesticides in sediment samples.

| Analytes | Samples | Sample Preparation Technique | Extraction Solvent | Clean-up Technique | Instrumental Technique | References |
|---|---|---|---|---|---|---|
| Pharmaceuticals | River sediment | PLE | methanol/water mixture (1:2, v/v) | SPE (HLB cartridge) | LC-ESI-QLIT | Jelić *et al.* (2009) |
| Steroid hormones | River sediment | USE | methanol/acetone (1:1, v/v) | SPE (C-18 cartridge) | LC-DAD-Q | López de Alda *et al.* (2002) |
| Steroid hormones Sterols | River sediment | USE | methanol | SPE (silica gel-anhydrous $Na_2SO_4$ cartridge) | LC-ESI-IT | Matić Bujagić *et al.* (2016) |
| Pharmaceuticals | River water and sediment | USE | methanol followed by acetone-acetic acid (20:1 (v/v) and ethyl acetate | SPE (HLB cartridge) | HPLC-DAD-ESI-IT | Matongo *et al.* (2015) |
| Pharmaceuticals | Sea water and sediment | PLE | methanol/water (1:2, v/v) | SPE (HLB cartridge) | UHPLC-ESI-QLIT | Moreno-González *et al.* (2015) |
| Pharmaceuticals Pesticides | River water, sediment | USE | methanol/dichloromethane (1:1, v/v) | – | LC-ESI-IT | Radović *et al.* (2015) |
| Antibiotics | Estuary sediment | USE | acetonitrile | – | UHPLC-ESI-QqQ | Shi *et al.* (2014) |
| Pharmaceuticals | River water, suspended solids, river sediment | PLE | methanol/water (1:2, v/v) | SPE (HLB cartridge) | LC-ESI-QLIT | da Silva *et al.* (2011) |
| Pharmaceuticals | Soil Sediment | PLE | water | SPE (combination of SAX and HLB cartridge) | LC-ESI-QqQ | Vazquez-Roig *et al.* (2010) |
| Antibiotics | River sediment | USE | citric buffer (pH 4) and acetonitrile (50:50, v/v) | SPE (HLB cartridge) | RRLC-ESI-QqQ | Yang *et al.* (2010) |

(*Continued*)

**Table 9.2** Overview of methods for quantitative determination of pharmaceuticals and pesticides in sediment samples (*Continued*).

| Analytes | Samples | Sample Preparation Technique | Extraction Solvent | Clean-up Technique | Instrumental Technique | References |
|---|---|---|---|---|---|---|
| Pharmaceuticals Pesticides Steroids Perfluorinated compounds | River sediment core | USE | methanol/dichloromethane (1:1, v/v) | SPE (silica gel-anhydrous $Na_2SO_4$ cartridge) | LC-ESI-IT | Matić Bujagić *et al.* (2019) |
| Pharmaceuticals | Bay sediment core | PLE | methanol/water (1:2, v/v) | SPE (HLB cartridge) | HPLC-ESI-QqQ-LIT | Lara-Martín *et al.* (2015) |
| Pharmaceuticals | Lake sediment core | USE | acetonitrile and phosphate buffer (pH 2) mixture | SPE (HLB cartridge) | LC-ESI-QqQ | Lahti and Oikari (2012) |
| Pesticides | River sediment core | Soxhlet | hexane/acetone (3:1, v/v) | SPE (acidified silica-anhydrous $Na_2SO_4$-Cu cartridge | GC-ECD | Covaci *et al.* (2005) |
| Pesticides | Lagoon sediment | PLE | 4:1 methanol water mixture | SPE (HLB cartridge) | LC-ESI-QqQ | Carafa *et al.* (2007) |
| Pesticides | River sediment | QuEChERS | 1 g sediment + 7.5 mL $H_2O$ + 10 mL ACN + 6 g $MgSO_4$ + 1.5 g NaCl | – | LC-ESI-QqQ | Ccanccapa *et al.* (2016) |
| Pesticides | River sediment | PLE and QuEChERS | acetonitrile (for PLE) 1 g sediment + 7.5 mL $H_2O$ + 10 mL ACN + 6 g $MgSO_4$ + 1.5 g NaCl | d-SPE (PSA, $MgSO_4$ and C18) | LC-ESI-QqQ | Masiá *et al.* (2015) |
| Pesticides Pharmaceuticals | Sediment | QuEChERS | 1 g sediment + 7.5 mL $H_2O$ + 10 mL ACN + 4 g $MgSO_4$ + 1 g NaCl | d-SPE (PSA, $MgSO_4$ and GCB) | UHPLC-LIT-Orbitrap | Farré *et al.* (2014) |

| | | | | | | |
|---|---|---|---|---|---|---|
| Pesticides | Lake sediment (surface and core samples) | PLE | hexane/acetone (1:1, v/v) | SPE (anhydrous $Na_2SO_4$, alumina, silica gel, anhydrous $Na_2SO_4$ | HPLC-QqQ-LIT | Guo *et al.* (2016) |
| Pesticides | Lake sediment | Solid-liquid extraction | n-hexane/dichloromethane (3:7, v/v) | – | GC-EI-Q | Hijosa-Valsero *et al.* (2016) |
| Pesticides | Water, sediment, fish | Solid-liquid extraction | I step: acetonitrile/water (50:50, v/v)<br>II step: acetate/cyclohexane (75:25, v/v) | – | LC-ESI-QqQ-LIT | Lazartigues *et al.* (2011) |
| Pesticides | River sediment | USE | acetone/n-hexane | – | GC-ECD | Sun *et al.* (2013) |
| Pesticides | Lake sediment | MAE | acetone-hexane (1:1, v/v) | – | LC-ESI-QqQ | Kalogridi *et al.* (2014) |
| Pesticides | Marine sediment | USE | acetonitrile | SBSE | GC-MS | Pinto *et al.* (2013) |

PLE–pressurized liquid extraction; SPE–solid phase extraction; LC–liquid chromatography; ESI–electrospray ionization; QLIT–quadrupole-linear ion trap detector; USE–ultrasonic solvent extraction; DAD–diode array detector; Q–quadrupole detector; IT–ion trap detector; HPLC–high performance liquid chromatography; UHPLC–ultra-high performance liquid chromatography; QqQ–triple quadrupole detector; RRLC–rapid resolution liquid chromatography; LIT–linear ion trap detector; GC–gas chromatography; ECD–electron capture detector; QuEChERS–Quick, Easy, Cheap, Effective, Rugged, and Safe; d-SPE–dispersive solid phase extraction; GCB–graphitized carbon black; EI–electron impact ionization; SBSE–stir bar sorptive extraction.

### *9.2.3.2 Pressurized liquid extraction*

PLE, also known as accelerated solvent extraction, provides good recoveries with a reduced extraction time and small amounts of organic solvents. It takes place at an elevated pressure (1500–2000 psi) and temperature (100–200°C). In these conditions, solvent extraction is more efficient due to the greater solubility and better solvent interaction with the solid matrix. The method is fast (15–20 min) and efficient, with a low solvent consumption (15–40 mL). In order to increase the surface area exposure of the sample, sediment samples are often mixed with inert material, such as Hydromatrix (Jelić *et al.*, 2009; Moreno-González *et al.*, 2015) and sea sand washed with $Na_2$–EDTA (Vazquez-Roig *et al.*, 2010). As can be seen from Table 9.2, PLE has been applied to a wide range of multi-class analyses of pharmaceuticals and pesticides. The high extraction efficiency is derived from the use of both polar and non-polar extraction solvents, as well as their mixtures. For the extraction of pharmaceuticals, pure water (Vazquez-Roig *et al.*, 2010) or a mixture of methanol/water (1:2, v/v) (Jelić *et al.*, 2009; Lara-Martín *et al.*, 2015; Moreno-González *et al.*, 2015) and (4:1, v/v) (Carafa *et al.*, 2007) are commonly applied. Pesticides have been efficiently extracted using acetonitrile (Masiá *et al.*, 2015) and hexane/acetone mixture (1:1, v/v) (Guo *et al.*, 2016). In addition to the extraction solvent, temperature is another parameter that greatly affects the extraction efficiency. In most cases (da Silva *et al.*, 2011; Guo *et al.*, 2016; Jelić *et al.*, 2009; Moreno-González *et al.*, 2015), the extraction temperature was set to 100°C, allowing better penetration of the solvent into the sample matrix, without degradation of the investigated compounds.

### *9.2.3.3 Microwave assisted extraction*

Microwave assisted extraction is performed in a closed vessel system, in which the isolation of the target compounds from solid matrices is achieved by heating the organic solvent using microwaves. The advantages of this method over traditional methods are the shorter extraction time (10–30 min) and low solvent consumption. Microwave extraction efficiency varies depending on the type of organic solvent, the temperature, the extraction time, and the energy of the microwaves applied. Silgoner *et al.* (1998) investigated the influence of different microwave extraction parameters on the determination of the organochlorine pesticide in sediments. Out of the three solvents tested (acetone, acetone/hexane (1:1) and tetrahydrofuran) and four extraction times (5, 10, 20 and 30 min), the best yields were obtained using more polar tetrahydrofuran during 30 min of extraction. Another study (Kalogridi *et al.*, 2014) showed that multiclass pesticide residues can be quantitatively determined using acetone/hexane mixture (1:1, v/v) in just 10 min of extraction.

### *9.2.3.4 QuEChERS*

Quick, Easy, Cheap, Effective, Rugged, and Safe (QuEChERS) is an extraction and clean-up technique originally developed for pesticide multiresidue analyses in food matrices. However, the QuEChERS approach has extended to the determination of environmental contaminants in solid samples, such as soil, sediments and sludge (Bruzzoniti *et al.*, 2014). The method involves sample extraction using acetonitrile (1/1, v/w), followed by manual shaking, and the addition of anhydrous $MgSO_4$ and NaCl to promote water partition from the organic phase and its dehydration. The clean-up of the obtained acetonitrile extract is then performed by dispersive solid-phase extraction (d-SPE). The extract is transferred into a micro-centrifuge vial containing the 'primary secondary amine' (PSA) sorbent, which interacts with matrix interfering compounds, and anhydrous $MgSO_4$. After brief shaking, the extract is centrifuged, and the obtained supernatant is analyzed using an appropriate instrumental technique. Farré *et al.* (2014) reported a modified QuEChERS method for the target and non-target determination of multiple classes of environmental contaminants. Prior to extraction, lyophilized samples were moisturized with 7.5 mL of water and 10 mL of acetonitrile were added to the samples. After vigorous shaking a mixture of 4 g anhydrous $MgSO_4$, 1 g NaCl, 1 g tri-sodium citrate dehydrate, and 0.5 g disodium hydrogen citrate sesquihydrate was added to the extract, which was shaken again and centrifuged. A clean-up of the resulting extracts was done using PSA, anhydrous $MgSO_4$, and GCB (Farré *et al.*, 2014). An assessment of the three different QuEChERS methodologies was performed using 1 g of blank lyophilized sediment and (1) acetate-buffered, (2) citrate-buffered, (3) unbuffered approaches by Masiá *et al.* (2015). The application of sodium citrate resulted in a significant increase in pesticide recoveries. The authors also concluded that the subsequent clean-up should be based on a combination of PSA, $MgSO_4$ and C-18, since the GCB tested retained some of the investigated pesticides (Masiá *et al.*, 2015).

### 9.2.4 Extract clean-up

Due to the complexity of the sediment samples, after extraction, it is necessary to perform a clean-up of the obtained extracts. The components of the matrix that are co-extracted with the analytes can reduce the reproducibility and precision of the analysis. Clean-up is usually carried out using solid-phase extraction (Carafa *et al.*, 2007; Covaci *et al.*, 2005; da Silva *et al.*, 2011; Guo *et al.*, 2016; Jelić *et al.*, 2009; Lahti & Oikari, 2012; Lara-Martín *et al.*, 2015; Matongo *et al.*, 2015; Moreno-González *et al.*, 2015; Vazquez-Roig *et al.*, 2010). For this purpose, Oasis HLB cartridges (Carafa *et al.*, 2007; da Silva *et al.*, 2011; Jelić *et al.*, 2009; Lara-Martín *et al.*, 2015; Matongo *et al.*, 2015; Moreno-González *et al.*, 2015), anion-exchange cartridges (SAX) in tandem with HLB cartridges (Vazquez-Roig *et al.*, 2010) or C-18 cartridges (López de Alda *et al.*, 2002) were used. Silica in combination with anhydrous $Na_2SO_4$ (Matić Bujagić *et al.*, 2016, 2019) and alumina (Guo *et al.*, 2016) is also widely used for extract clean-up. In addition to acidified silica and anhydrous $Na_2SO_4$, Covaci *et al.* (2005), used Cu for the removal of sulfur from the sediment cores.

Additionally, modern solid-phase microextraction techniques, originally developed for the analysis of organic compounds in liquid samples, such as SBSE, have been successfully applied in the purification of the extracts obtained after USE extraction of solid samples (Pinto *et al.*, 2013). Also, as can be seen from the mentioned examples, dispersive SPE has become an integral part of the QuEChERS method (Farré *et al.*, 2014; Masiá *et al.*, 2015).

## 9.3 INSTRUMENTAL TECHNIQUES FOR TRACE ANALYSES OF PHARMACEUTICALS AND PESTICIDES

GC-MS has traditionally been used for the analysis of pesticides, natural and synthetic hormones, and many other organic contaminants (Azzouz *et al.*, 2010; Basaglia & Pietrogrande, 2012; Battaglin *et al.*, 2018; Hijosa-Valsero *et al.*, 2016; Jabali *et al.*, 2019; Pinto *et al.*, 2013; Sudo *et al.*, 2004; Trinh *et al.*, 2011). Despite the many existing applications, GC-MS has certain disadvantages, such as the extensive use of solvents for sample extraction, followed by several evaporation steps, as well as the frequent need for an additional derivatization step, which is used to make polar and non-volatile compounds amenable to GC analysis (Azzouz *et al.*, 2010; Lindholm *et al.*, 2014; Trinh *et al.*, 2011; Wang & Gardinali, 2012). Thus, for new pesticides, which are more polar, liquid chromatography coupled with mass spectrometry (LC-MS/MS) is the preferred analytical technique (Alder *et al.*, 2006).

Liquid chromatography with UV detection, particularly coupled with a photodiode array detector (LC-PDA), still plays a significant role in pharmaceutical analyses (Chadha *et al.*, 2016; Hemdan & Tawakol, 2018). However, due to the LC-PDA lack of sensitivity and selectivity, this method is rarely used for the trace analysis of pharmaceutical residues in complex matrices such as environmental samples. The sensitivity and selectivity of the LC analysis can be significantly improved by the use of mass selective detectors, especially by the use of tandem mass spectrometry (MS/MS), hence LC-MS/MS became the standard method for pharmaceutical analyses in aquatic matrices, as has been revealed in the literature survey presented in Tables 9.1 and 9.2.

Still, LC-MS/MS analyses of environmental samples may show a significant matrix effect (ME) which may lead to a considerable suppression or enhancement of the analyte signal, as well as to occasional false-positive results (Grujić *et al.*, 2009). The ME can be compensated for by an adequate calibration methodology or reduced by using UHPLC. The introduction of HRMS using Orbitrap or time of flight (TOF) mass spectrometers has provided additional structure information needed for the unambiguous identification and confirmation of positive findings. HRMS also enables reliable non-target screening, which is very useful for the identification of metabolites and degradation products.

## 9.3.1 Gas chromatography-mass spectrometry

Due to its high sensitivity and resolution, robustness, as well as reasonable price, GC-MS is widely used for pesticide monitoring in environmental analyses, food safety, and other applications (Botitsi *et al.*, 2011; Hernández *et al.*, 2013; Liang *et al.*, 2015). The most commonly used MS configuration for routine pesticide monitoring contains an electron impact ionization source (EI) and a quadrupole (Q) mass analyzer (Andreu & Picó, 2012; Vasiljević *et al.*, 2012a). The EI is a universal ionization source that provides valuable information on the molecule structure and highly reproducible mass spectra, which can be easily and quickly compared with available commercial spectral libraries (Hernández *et al.*, 2013), while the Q is a low-cost analyzer with good scan speed and sensitivity. For example, GC-EI-Q was successfully used for the analysis of 59 pollutants (polycyclic aromatic hydrocarbons – PAHs, insecticides and herbicides) from lake waters and sediments (Hijosa-Valsero *et al.*, 2016), 29 pesticides belonging to a variety classes from river water (Valenzuela *et al.*, 2019) and 41 multiclass pesticides from drinking water (Schwanz *et al.*, 2019), with obtained limits of quantification (LOQs) lower than 1 $\mu g\ L^{-1}$.

The use of gas chromatography-tandem mass spectrometry (GC-MS/MS) in recent years has improved the sensitivity and selectivity of the analysis of pesticides and many other trace pollutants in environmental matrices (Hernández *et al.*, 2013). As an illustration, Barco-Bonilla with co-authors (Barco-Bonilla *et al.*, 2010) proposed the GC-MS/MS method based on a triple quadrupole analyzer (GC-QqQ) for the analysis of 139 non-polar pesticides from wastewater and obtained LOQs in the range from 0.02 to 0.50 $\mu g\ L^{-1}$. GC-QqQ was also used for determining 77 pesticides in drinking water (Garrido Frenich *et al.*, 2011) and 98 pesticides from soil samples (Vidal *et al.*, 2010), while the GC-MS/MS configuration with an ion trap (GC-IT) was used for the simultaneous analysis of 48 pesticides, belonging to 12 different chemical groups in surface water samples (Jabali *et al.*, 2019).

However, GC is limited to non-polar, volatile, thermally stable compounds. Nowadays, the use of pesticides has shifted towards more polar and easily degradable pesticides, which are not amenable to GC analyses without an additional derivatization step. Thus, in recent years, LC, which is more compatible with polar analytes, is increasingly replacing GC in pesticide analyses. Still, GC-MS and GC-MS/MS remain the best choices for monitoring non-polar, volatile pesticides, such as organochlorines, which are not ionized in the ESI source of LC-MS (He & Aga, 2019; Vasiljević *et al.*, 2012a, 2012b). Furthermore, when pesticides with a wide range of polarities need to be analyzed, the complementary use of GC-MS and LC-MS based methods is advised (Pang *et al.*, 2020; Vidal *et al.*, 2010).

Pharmaceuticals are in general polar, non-volatile compounds, and therefore a derivatization step is usually required to make them suitable for GC analysis (Loos *et al.*, 2016). For example, N-methyl-N-(tert-butyldimethyl-silyl) trifluoroacetamide was used as a derivatizing agent prior to a GC-MS analysis of non-steroidal anti-inflammatory drugs (Heath *et al.*, 2010), while dimethyl (3,3,3-trifluoropropyl) silyldiethylamine was used in the analysis of estrogens (Caban *et al.*, 2013). In order to eliminate the additional derivatization step, an on-fiber SPME derivatization was used for the GC-MS analysis of pharmaceuticals and personal care products in water (Basaglia & Pietrogrande, 2012) and in sewage and sewage sludge (López-Serna *et al.*, 2018). Nevertheless, as can be seen in Tables 9.1 and 9.2, LC-MS methods are used much more often than GC-MS methods for the analysis of pharmaceuticals.

## 9.3.2 Liquid chromatography-mass spectrometry

In the last decade, instrumental techniques for the determination of pharmaceuticals and modern pesticides in the aquatic environment are generally based on LC-MS/MS (Grujić *et al.*, 2009; He & Aga, 2019; Jauković *et al.*, 2014; Lalović *et al.*, 2017; Masiá *et al.*, 2016; Trtić-Petrović *et al.*, 2010; Valls-Cantenys *et al.*, 2016). LC-MS/MS is more compatible with polar organic compounds, which most pharmaceuticals and modern pesticides are, and provides a series of advantages compared to GC-MS, such as the elimination of the derivatization step of non-volatile and thermolabile compounds, an increase in the number of analyzable compounds and a reduction of the total analysis time (Pérez-Fernández *et al.*, 2017).

The most commonly used LC method for pesticide and pharmaceutical analyses is the reverse-phase (RP) chromatography with C18 columns and to a lesser extent C8, C12 or phenyl bonded columns (Botitsi *et al.*, 2011; Kim *et al.*, 2018; Primel *et al.*, 2012). Methanol and acetonitrile are commonly used as organic solvents, while additives such as formic acid, acetic acid, ammonium hydroxide, ammonium acetate, and ammonium formate are added to the mobile phase, in order to enhance the ionization efficiency (Botitsi *et al.*, 2011; Kim *et al.*, 2018; Primel *et al.*, 2012).

As presented in Tables 9.1 and 9.2, most of the published LC-MS/MS methods for pesticide and pharmaceuticals analyses are based on electrospray ionization (ESI), which is the ionization method most suitable for polar analytes. Besides, ESI provides the analysis of a wider range of compounds, in terms of their polarity and molecular mass, than some of the other atmospheric pressure ionization techniques, such as atmospheric pressure chemical ionization (APCI). On the other hand, atmospheric pressure photoionization (APPI) was reported as the preferable ionization method for the ionization of moderate polar analytes such as steroidal hormones (Wang & Gardinali, 2012), while APCI (Chusaksri *et al.*, 2006) or direct-electron ionization (Famiglini *et al.*, 2008) was used for non-polar analytes, such as organochlorinated pesticides. ESI can be used in both the positive (ESI+) or negative (ESI−) mode, and the choice of the mode depends on the proton affinity of the functional group of the analyte. However, for analytes that have more than one functional group, as is true for many pharmaceuticals and pesticides, the selection relies on initial experiments (Valls-Cantenys *et al.*, 2016).

For the LC analysis of trace organic contaminants in environmental matrices, tandem mass spectrometry is routinely used. MS/MS instruments, such as the triple quadrupole (QqQ) (Lahti & Oikari 2012; Vazquez-Roig *et al.*, 2010; Yang *et al.*, 2010), quadrupole – time of flight (QTOF) (Masiá *et al.*, 2013), IT (Matić Bujagić *et al.*, 2016; Radović *et al.*, 2015), or quadrupole – linear ion trap (QLIT) (da Silva *et al.*, 2011; Jelić *et al.*, 2009) are typically used in the so-called selected-reaction-monitoring (SRM) mode. In the SRM mode, a precursor ion (usually a protonated or deprotonated molecular ion of the analyte) is selected in the first analyzer, and the ion is fragmented by introducing a collision gas (usually helium) and electrical energy. One of the most abundant fragment ions, i.e. the product ion, is selected by the second mass analyzer and transferred to the detector for quantitation and confirmation. The SRM mode is a fast and accurate method for the routine quantification of known target analytes. The selection of a mass analyzer depends on the type of research, such as whether a target or non-target analysis is required, and in some cases on the type of compounds that will be analyzed (Loos *et al.*, 2016). QqQ offers a low resolution but provides high sensitivity when working in the SRM mode at a reasonable cost, and it is the most frequently applied mass analyzer in routine pesticide and pharmaceuticals target quantitative analyses. For example, the LC-ESI-QqQ method was applied by Meyer and co-workers (Meyer *et al.*, 2011) for monitoring pharmaceuticals and herbicides in river water and wastewater at ng $L^{-1}$ levels. While QqQ working in SRM mode provides high sensitivity, the number of compounds that can be simultaneously analyzed in a single run is limited and must be previously selected. On the other hand, QTOF provides the selectivity and sensitivity required for wide-scope screening, because it combines high full spectral sensitivity with high mass resolution. Masiá *et al.* (2013) successfully applied a combination of LC-QqQ (for the target analysis of 43 pesticides or degradation products at ng $L^{-1}$ levels) and UHPLC-QTOF (for non-target screening using a home-made database containing more than 1100 organic pollutants), while monitoring pesticides and other contaminants in surface and wastewater samples. LC-IT offers the capability for $MS^n$ analyses, which can be very useful for confirmatory purposes (Grujić *et al.*, 2009).

One notable drawback of the LC-MS quantitative analysis is the matrix effect (ME) which often occurs when analyzing complex matrices, such as environmental samples, and may significantly reduce analysis precision and accuracy (Caban *et al.*, 2013). The prevailing assumption is that the ME is the result of competition between the analyte and co-eluting matrix components during the ionization process (Primel *et al.*, 2012). There are different approaches to overcome the ME, and some rely on an exhaustive sample clean-up, sample dilution, improving chromatographic separation or changing the ion source. More often, the ME is compensated for by the use of an appropriate calibration method, e.g. by using isotopically labeled standards (Gros *et al.*, 2006; Jelić *et al.*, 2009), matrix-match standards (Caldas *et al.*, 2016; Dujaković *et al.*, 2010; Gilart *et al.*, 2013) or the standard-addition method (Van De Steene & Lambert, 2008). A transition to UHPLC may also reduce the ME, due to the better chromatographic resolution (Van De Steene & Lambert, 2008).

### 9.3.3 Recent trends

A survey of literature shows that a significant improvement in a fast separation with an efficient resolution of the analytes in complex matrices was achieved by UHPLC (Behnoush *et al.*, 2015; Churchwell *et al.*, 2005; Leandro *et al.*, 2006; Van De Steene & Lambert, 2008), the use of which has increased exponentially in the last decade (López-Ruiz *et al.*, 2019). UHPLC is an LC technique that uses small diameter particles (less than 2 μm) in the stationary phase, which significantly increases the chromatographic resolution and diminishes its dependence on the mobile phase flow rate. Compared to HPLC, UHPLC offers a higher resolution, lessens the matrix effect (because the co-elution of analyte and matrix components can be avoided), allows a better peak definition and more accurate peak integration (López-Ruiz *et al.*, 2019). In addition, UHPLC decreases the time and cost of analysis and reduces the volume of solvents used (Chen & Kord, 2009), and is therefore considered to be more environmentally friendly than traditional HPLC. UHPLC-MS methods, developed for the multiresidual analysis of pharmaceuticals and pesticides in environmental waters and sediments, had low limits of detection and a short analysis time (Battaglin *et al.*, 2018; Farré *et al.*, 2014; Liang *et al.*, 2019; Moreno-González *et al.*, 2015; Shi *et al.*, 2014; Wode *et al.*, 2012). For example, Liang and co-authors (Liang *et al.*, 2019) developed an automatic SPE-UHPLC-MS/MS method for the trace analysis of 62 pharmaceuticals in water samples, with a complete analysis time of about 14 min per sample, LOQs in the range of 0.00475–2.49 ng $L^{-1}$ and RSDs for the intra-/inter-day precision less than 10.6% and 15.6%, respectively.

Multimode ionization sources such as an ESI and the APCI dual ionization source, which can be operated in ESI only, APCI only and ESI+APCI modes have been introduced recently and were used to simultaneously ionize both polar and non-polar compounds (Cheng *et al.*, 2015). Multimode ionization sources make it possible to analyze a wide range of compounds, in terms of their polarity or thermal stability, in a single run. For example, a dual ionization source was applied in the analysis of benzodiazepines, and the mixed ESI+APCI mode proved to be the most sensitive, generating an increase in the detector signal of an average 35% compared to ESI only and 350% compared to APCI only ionization (Galaon *et al.*, 2013).

Over the last decade, the use of HRMS in environmental analyses has grown substantially. With the help of the high sensitivity full scan acquisition mode, HRMS makes it possible to scan thousands of target and non-target compounds, as well as to separate target and interfering ions with high accuracy (Lin *et al.*, 2015; Valls-Cantenys *et al.*, 2016). There are several types of HRMS instruments, such as time of flight (TOF), sector mass spectrometer (SECTOR), Fourier transform-ion cyclotron resonance-mass spectrometer (FT-ICR-MS), and Orbitrap, amongst which TOF and Orbitrap are most commonly used (Lin *et al.*, 2015). TOF and Orbitrap have been applied in pesticide, pharmaceutical, and many other micropollutants analyses as single-stage mass analyzers (Reis *et al.*, 2019; Wode *et al.*, 2012), or in tandem with high-resolution mass spectrometry (THRMS) configurations (Asghar *et al.*, 2018; Farré *et al.*, 2014). THRMS offers accurate mass measurements and high-resolution MS/MS or $MS^n$ spectra, which are very useful for the confirmation of target compounds, as well as for the reliable identification of unknown compounds (Lin *et al.*, 2015). Some of the most extensively used THRM configurations are QTOF, ITTOF, QOrbitrap and linear ion trap-Orbitrap (LITOrbitrap). The selection of configuration depends on the purpose of analysis. TOF configurations offer a faster acquisition speed and a wider mass range compared to the Orbitrap, while the latter provides a higher resolution and capability of $MS^n$ analysis (Lin *et al.*, 2015). There are different analytical approaches when using LC-THRMS configurations (Casado *et al.*, 2018; Krauss *et al.*, 2010): (1) quantitative targeted analysis of pre-selected substances with reference standards; (2) suspects screening of pre-selected substances with predicted MS/MS fragmentation matching; (3) non-targeted screening of unknowns with peak detection, formula fit and predicted MS/MS fragmentation matching; and (4) post-targeted screening, e.g. searching for additional compounds in already acquired chromatograms. In a recent paper, Pang and co-workers (Pang *et al.*, 2020) proposed a combination of GC-QTOF and LC-QTOF methods, for the simultaneous determination of 733 pesticides in fruits and vegetables. The method is based on accurate mass databases developed for 485 pesticides by GC-QTOF and 525 pesticides by LC-QTOF. Casado *et al.* (2018) developed a highly sensitive (LOQs under 5 ng $L^{-1}$ for most compounds) and selective method using offline SPE and LC-ESI-Q-Orbitrap analyses for determining 252 pesticides in surface water. Compounds were quantified in the full scan acquisition mode, while the data-dependent $MS^2$ analysis was used for unambiguous identification. Nowadays, LC-THRMS is widely used in monitoring and identifying pesticide and pharmaceutical transformation products and metabolites (Asghar *et al.*, 2018; Rousis *et al.*, 2017), which is considered to be of crucial importance for complete environmental risk assessment.

## 9.4 EXAMPLES OF DEVELOPED AND VALIDATED ANALYTICAL METHODS

An overview of the analytical methods for the determination of trace levels of pharmaceuticals and pesticides in surface, groundwater and wastewater, as well as sediments, developed, optimized and validated by the authors has been presented. Some of the most frequently used drugs and pesticides in Serbia were selected for the studies. Surface, groundwater and wastewater samples were prepared for analysis using SPE, while sediments were prepared using USE followed by a clean-up of the sample extract. The resulting extracts were analyzed by several LC-MS methods suitable for the determination of the selected analytes at trace levels. The application of the methods for the trace analyses of the drugs and pesticides set up by the authors in monitoring and risk assessment studies is also described.

### 9.4.1 Determination of pharmaceutical and pesticide residues in environmental samples

Several multiresidual methods have been developed for the identification and quantification of selected pharmaceuticals and pesticides in different water matrices and sediments. The list of analytes investigated using the analytical methods developed by the authors is provided in Table 9.3.

#### *9.4.1.1 Trace analysis of pharmaceuticals and pesticides in water samples*

Methods for the determination and confirmation of 19 drugs from multiple pharmaceutical classes (Grujić *et al.*, 2009) and 14 pesticides from 7 different chemical classes of insecticides, fungicides and herbicides (Dujaković *et al.*, 2010) were optimized and applied in the analysis of surface and groundwater samples. The most commonly used pharmaceuticals, belonging to major drug groups such as antibiotics, sedatives, antiepileptics and analgesics/antipyretics, and agricultural pesticides most frequently used in Serbia, were chosen for the study. The methods were successfully applied in a two-year monitoring study of traces of pharmaceuticals (Radović *et al.*, 2012) and pesticides (Antić *et al.*, 2015) in the surface and groundwater of the Danube River Basin in Serbia. An additional method was also developed (Jauković *et al.*, 2014) for determining eight cardiovascular drugs in surface, groundwater and wastewater. The selected drugs are among the most prescribed medication worldwide and include antihypertensives, antihyperlipemics and anticoagulants. The final method for the simultaneous analysis of the most commonly used pharmaceuticals and pesticides in environmental waters was set up by Radović *et al.* (2015) to be used for monitoring contamination levels in the Danube River Basin.

#### *9.4.1.2 Trace analysis of pharmaceuticals and pesticides in sediment samples*

An initial method for the analysis of sediment samples was developed and validated as part of the ongoing monitoring of drug and pesticide residues in the Danube River Basin (Radović *et al.*, 2015). Subsequently, the method for the joint determination of drug and pesticide residues was successfully applied in investigating the sorption capacity of these analytes on different river sediments (Radović *et al.*, 2016). Finally, the method was slightly modified to be used to investigate the presence and fate of four classes of emerging contaminants (pharmaceuticals, pesticides, steroids and perfluorinated compounds) in the sediment core from the Iron Gate I Reservoir on the river Danube (Matić Bujagić *et al.*, 2019). The analytes selected for the study are generally not covered by monitoring programs in Serbia but are considered as future candidates (Table 9.3). Several pharmaceuticals and pesticides have been proposed for inclusion in future regular monitoring based on their ecotoxicological potential and risk of benthic organisms assessed according to the detected levels.

**Table 9.3** Pharmaceuticals and pesticides investigated using several analytical methods developed by the authors: chemical class, molecular weight (Mw) and MS fragmentation reactions for quantification and confirmation with corresponding collision energies (CE).

| Analyte | Chemical Class | $M_w$ (g mol$^{-1}$) | Precursor Ion ($m/z$) | Quantification Reaction | CE (%) | Confirmation Reaction | CE (%) |
|---|---|---|---|---|---|---|---|
| *Pharmaceuticals* | | | | | | | |
| Ampicillin | Penicillin antibiotic | 349 | 382 $[M+CH_3OH+H]^+$ | 382→223 | 23 | 382→365 | 23 |
| Amoxicillin | Penicillin antibiotic | 365 | 398 $[M+CH_3OH+H]^+$ | 398→381 | 28 | 381→349 | 23 |
| Cephalexin | Cephalosporin antibiotic | 347 | 380 $[M+CH_3OH+H]^+$ | 380→331 | 22 | 380→223 | 22 |
| Sulfamethoxazole | Sulfonamide antibiotic | 253 | 254 $[M+H]^+$ | 254→188 | 34 | 254→156 | 34 |
| Trimethoprim | Used in combination with sulfonamide antibiotics | 290 | 291 $[M+H]^+$ | 291→230 | 44 | 291→123 | 44 |
| Erythromycin | Macrolide antibiotic | 733 | 734 $[M+H]^+$ | 734→576 | 26 | 734→716 | 26 |
| Azithromycin | Macrolide antibiotic | 748 | 749 $[M+H]^+$ | 749→591 | 30 | 591→434 | 28 |
| Doxycycline | Tetracycline antibiotic | 444 | 445 $[M+H]^+$ | 445→428 | 25 | 428→410 | 27 |
| Diazepam | Sedative, benzodiazepine | 284 | 285 $[M+H]^+$ | 285→257 | 40 | 257→228 | 39 |
| Bromazepam | Sedative, benzodiazepine | 315 | 316 $[M+H]^+$ | 316→288 | 36 | 288→261 | 35 |
| Lorazepam | Sedative, benzodiazepine | 320 | 321 $[M+H]^+$ | 321→303 | 32 | 303→275 | 26 |
| Carbamazepine | Antiepileptic | 236 | 237 $[M+H]^+$ | 237→194 | 34 | 237→219 | 34 |
| Phenobarbital | Antiepileptic | 232 | 231 $[M-H]^-$ | 231→188 | 32 | 231→85 | 32 |
| Paracetamol | Analgesic/antipyretic | 151 | 152 $[M+H]^+$ | 152→110 | 34 | – | – |
| Diclofenac | Analgesic/antipyretic | 295 | 296 $[M+H]^+$ | 296→278 | 28 | 278→250 | 22 |
| Ibuprofen | Analgesic/antipyretic | 206 | 205 $[M-H]^-$ | 205→159 | 30 | 159→143 | 31 |
| Metamizole | Analgesic/antipyretic | 310 | 310 $M^-$ | 310→191 | 27 | 191→176 | 31 |
| Flurbiprofen | Analgesic/antipyretic | 244 | 243 $[M-H]^-$ | 243→199 | 30 | – | – |
| Acetylsalicylic acid | Analgesic/antipyretic | 180 | 137 $[M-COCH_3]^-$ | 137→93 | 34 | – | – |

(*Continued*)

**Table 9.3** Pharmaceuticals and pesticides investigated using several analytical methods developed by the authors: chemical class, molecular weight (Mw) and MS fragmentation reactions for quantification and confirmation with corresponding collision energies (CE) (*Continued*).

| Analyte | Chemical Class | $M_w$ (g mol$^{-1}$) | Precursor Ion ($m/z$) | Quantification Reaction | CE (%) | Confirmation Reaction | CE (%) |
|---|---|---|---|---|---|---|---|
| Metoprolol | Antihypertensive, β-blocker | 267 | 268 $[M+H]^+$ | 268→191 | 37 | 268→218 | 37 |
| Bisoprolol | Antihypertensive, β-blocker | 325 | 326 $[M+H]^+$ | 326→116 | 31 | 326→222 | 31 |
| Amlodipine | Antihypertensive, Ca channel blocker | 408 | 409 $[M+H]^+$ | 409→238 | 25 | 409→294 | 25 |
| Enalapril | Antihypertensive, ACE inhibitor | 376 | 377 $[M+H]^+$ | 377→234 | 30 | 377→303 | 30 |
| Cilazapril | Antihypertensive, ACE inhibitor | 417 | 418 $[M+H]^+$ | 418→211 | 25 | 211→183 | 32 |
| Atorvastatin | Antihyperlipemic, statin | 558 | 559 $[M+H]^+$ | 559→466 | 25 | 559→440 | 25 |
| Simvastatin | Antihyperlipemic, statin | 418 | 419 $[M+H]^+$ | 419→285 | 21 | 419→199 | 21 |
| Clopidogrel | Anticoagulant | 321 | 322 $[M+H]^+$ | 322→212 | 28 | 212→184 | 23 |
| *Pesticides* | | | | | | | |
| Monocrotophos | Organophosphate insecticide | 223 | 224 $[M+H]^+$ | 224→193 | 38 | 224→167 | 38 |
| Dimethoate | Organophosphate insecticide | 229 | 230 $[M+H]^+$ | 230→199 | 26 | 199→171 | 22 |
| Malathion | Organophosphate insecticide | 330 | 331 $[M+H]^+$ | 331→285 | 24 | 285→127 | 20 |
| Acephate | Organophosphate insecticide | 183 | 184 $[M+H]^+$ | 184→143 | 40 | 184→113 | 40 |
| Imidacloprid | Neonicotinoid insecticide | 255 | 256 $[M+H]^+$ | 256→210 | 25 | 256→175 | 25 |
| Acetamiprid | Neonicotinoid insecticide | 222 | 223 $[M+H]^+$ | 223→126 | 36 | 223→187 | 36 |
| Carbofuran | Carbamate insecticide | 221 | 222 $[M+H]^+$ | 222→165 | 32 | 165→123 | 27 |
| Carbaryl | Carbamate insecticide | 201 | 202 $[M+H]^+$ | 202→145 | 25 | 145→117 | 31 |
| Tebufenozide | Diacylhydrazine insecticide | 352 | 375 $[M+Na]^+$ | 375→225 | 34 | 375→319 | 34 |
| Carbendazim | Benzimidazole fungicide | 191 | 192 $[M+H]^+$ | 192→160 | 34 | 160→132 | 35 |
| Simazine | Triazine herbicide | 201 | 202 $[M+H]^+$ | 202→124 | 36 | 202→132 | 36 |
| Atrazine | Triazine herbicide | 215 | 216 $[M+H]^+$ | 216→174 | 38 | 174→146 | 35 |
| Propazine | Triazine herbicide | 229 | 230 $[M+H]^+$ | 230→188 | 35 | 230→146 | 35 |
| Linuron | Phenylurea herbicide | 248 | 249 $[M+H]^+$ | 249→160 | 35 | 249→182 | 35 |
| Monuron | Phenylurea herbicide | 198 | 199 $[M+H]^+$ | 199→72 | 30 | – | – |
| Diuron | Phenylurea herbicide | 232 | 233 $[M+H]^+$ | 233→72 | 34 | – | – |

## 9.4.2 LC-MS analysis

When beginning to develop the analytical method, mass spectrometric experiments are performed. For each selected pharmaceutical and pesticide, mass spectra are recorded in the positive and negative ionization modes, in the range from m/z 50 to m/z 1000. The most abundant ions in the mass spectra are subjected to further $MS^n$ fragmentation. The most sensitive transitions are chosen for the identification and quantification of each analyte in the screening method, while additional fragmentation reactions are selected for confirmation purposes (Table 9.3). For most of the investigated analytes, the precursor ion of the quantification reaction is a protonated molecule, $[M+H]^+$. However, there are a few exceptions. The most abundant ion in the mass spectra of ampicillin, amoxicillin and cephalexin is a protonated adduct with methanol, $[M+CH_3OH+H]^+$. The methanol adduct of the three antibiotics is easily formed in a methanol solution and can be effectively used as a precursor ion for their identification and quantification at trace levels (Grujić *et al.*, 2008). In the case of the pesticide tebufenozide, the dominant ion in the mass spectrum is the sodium adduct ($[M+Na]^+$), which is employed as a precursor ion (Dujaković *et al.*, 2010). For several selected pharmaceuticals (phenobarbital, ibuprofen, metamizole, flurbiprofen and acetylsalicylic acid), the precursor ion is determined from the mass spectra recorded in the negative ionization mode. The deprotonated molecule ($[M-H]^-$) is commonly used for the detection of these drugs (Table 9.3). However, in the case of acetylsalicylic acid ($M_w = 180$), due to its rapid degradation to salicylic acid ($M_w = 138$) in solution, deprotonated salicylic acid ($[M-COCH_3-H]^-$) is used for identification and quantification purposes (Grujić *et al.*, 2009).

The confirmation of pharmaceuticals and pesticides detected in real samples is necessary to eliminate false positive results and is performed by using additional $MS^n$ transitions. However, for the drugs paracetamol, acetylsalicylic acid and flurbiprofen, as well as for pesticides monuron and diuron, it was not possible to obtain more than one $MS^2$ transition or stable $MS^3$ fragmentation that could be used in the confirmatory method (Table 9.3). For this reason, the analytes mentioned above were not considered further because positive results could not be confirmed. Several pharmaceuticals and pesticides have been proposed for inclusion in future regular monitoring.

Quantification and confirmation reactions were utilized in setting up several LC-MS methods for determining multiclass pharmaceuticals and pesticides, using the SRM mode. In each developed method, the SRM detection was divided into as many segments as possible. Usually, a maximum of four analytes per segment is acceptable for maintaining the MS detection sensitivity as it decreases with the number of transitions recorded simultaneously.

The LC-MS analysis in several analytical methods developed by the authors was mainly performed using the Surveyor LC system (Thermo Fisher Scientific, US) coupled with the LCQ Advantage quadrupole ion trap mass spectrometer (Thermo Fisher Scientific). However, in recent years, the Dionex UltiMate® 3000 LC system (Thermo Fisher Scientific) coupled with the LTQ XL linear ion trap (Thermo Fisher Scientific) was used. The reverse-phase column (Zorbax Eclipse® XDB-C18; 75 mm × 4.6 mm × 3.5 μm; Agilent Technologies, US) was used for the chromatographic separation of analytes, with a corresponding pre-column placed in front (12.5 mm × 4.6 mm × 5 μm; Agilent Technologies). For the analysis in the positive ionization mode, the gradient mobile phase was a mixture of deionized water, methanol and 10% acetic acid. However, acetic acid was not used in the negative mode as it reduced the anion signal. The flow rate of the mobile phase in both ionization modes was 0.6 mL $min^{-1}$. A volume of 10 μL of the final extract was injected into the LC system and analyzed.

Selected pharmaceuticals and pesticides were ionized using the ESI technique. For both ionization modes, the optimal source parameters were the source voltage (4.5 kV), sheath gas (25 au, i.e. arbitrary units from a scale in the range of 0–100 defined by the LCQ Advantage system) and capillary temperature (290°C).

## 9.4.3 Sample preparation methods for trace analysis

### 9.4.3.1 SPE procedure

Several SPE protocols have been developed for the isolation and preconcentration of selected pharmaceuticals and pesticides from the water matrix. In the initial experiments (Grujić *et al.*, 2009), a number of SPE sorbents were tested for the efficient extraction of chemically diverse compounds, ranging from acidic through neutral to basic drugs, with a high to low polarity. Following the manufacturer's suggestions for the SPE procedure, a variety of cartridges were investigated: with C18, carbon and cationic exchanger packings, as well as with polymeric sorbents such as hydrophobic divinylbenzene polymer, styrene-divinylbenzene copolymer, and HLB copolymer. For most of the 19 selected pharmaceuticals, the best results were obtained using an OASIS HLB cartridge. Since numerous studies have proven the efficiency of this sorbent when wanting to extract acidic, neutral and basic analytes over a wide range of pH, the HLB cartridge was chosen for the extraction of all the investigated drugs and pesticides from all water matrices in all further research.

The pH of the water sample is another important parameter that needs to be optimized in the SPE procedure for isolating analytes from the aqueous matrix. Most pharmaceuticals and pesticides display pH-dependent recoveries. The pH values of 3.0, 4.5, 6.0 or 7.5 (average pH of natural water) are usually tested. In Grujić *et al.* (2009), for instance, a highly acidic environment (pH 3.0) was found to be essential for the successful extraction of azithromycin, doxycycline and acetylsalicylic acid from the water matrix, while the extraction of erythromycin and metamizole was impaired. The results showed that the SPE procedure for the isolation and preconcentration of 19 selected drugs should be performed at pH 3.0 and without a pH adjustment (pH ~7.5). As for the extraction of eight cardiovascular drugs from various aqueous matrices, a pH of 3.0 was shown to be optimal (Jauković *et al.*, 2014). The efficient extraction of 14 most commonly used pesticides from natural waters was achieved at pH 6.0 (Dujaković *et al.*, 2010). In the method used for the joint determination of drugs and pesticides, the highest recoveries for the majority of analytes were obtained at pH 6.0 (Radović *et al.*, 2015).

The next crucial step in the SPE procedure development is the optimization of the extraction solvent. Methanol, dichloromethane, acetonitrile and their mixtures are commonly tested as solvents for the elution of analytes from the SPE sorbent in the multiresidual analysis of pharmaceuticals and pesticides. In general, methanol provided the highest recoveries for most of the investigated drugs as a single extraction solvent (Grujić *et al.*, 2009; Jauković *et al.*, 2014), whereas methanol–dichloromethane mixture (1:1) was found to be optimal for the extraction of few cardiovascular drugs (Jauković *et al.*, 2014) and pesticides (Dujaković *et al.*, 2010). However, for the simultaneous extraction of all the investigated pharmaceuticals and pesticides, the best results were obtained using a methanol–dichloromethane mixture (Radović *et al.*, 2015). Regarding the volume of the elution solvent, although many analytes require a volume as low as 7 mL to completely elute from the column, for several analytes 15 mL is needed (Grujić *et al.*, 2009). Therefore, a volume of 15 mL was determined to be the best choice for eluting selected pharmaceuticals and pesticides, and this parameter was not optimized in other SPE protocols that were later developed. A volume of 15 mL is also the maximum volume that can be retained in the extract collection cuvette.

In the final breakthrough experiments, the water sample volume is optimized. On the one hand, the sample volume should provide the highest possible preconcentration factor; on the other hand, it should not be excessive, as it will cause the loss of the analyte from the column prior to the elution step. It is important to bear in mind that the time of analysis can be significantly shortened by the use of lower sample volumes. The usual tested volumes of a water sample are 50, 100, 250 and 500 mL. For most of the selected pharmaceuticals and pesticides, the optimal sample volumes were reported to be 100 mL (Grujić *et al.*, 2009) or 250 mL (Dujaković *et al.*, 2010; Jauković *et al.*, 2014). The volume of 500 mL was found to be too high, given the analysis time and the loss of each analyte (e.g. up to 60% for doxycycline, Grujić *et al.*, 2009). However, a water sample volume of 250 mL is generally chosen to achieve the higher preconcentration factor required in monitoring the trace levels of analytes. Nevertheless, in the method of combined drug and pesticide analysis (Radović *et al.*, 2015), a volume of 100 mL was selected as optimal for the water sample, as there was a significant loss of analytes (up to 54% for azithromycin) when the volume of 250 mL was preconcentrated.

The finally optimized SPE procedure for the preparation of surface and groundwater samples for the analysis of selected drugs and pesticides is shown in Figure 9.1a. Prior to extraction, the water sample was filtered using a glass fiber filter (<1 μm pore size, Whatman GmbH, Germany). The HLB cartridge (200 mg/6 mL) was preconditioned with 5 mL of methanol, followed by 5 mL of deionized water, and finally 5 mL of deionized water with pH 6.0 (pH of the sample to be subsequently applied to the SPE column). A volume of 100 mL of water sample, with the pH adjusted to 6.0, was loaded dropwise to the column. The sorbent was then dried by vacuum suction for 10 min and the analytes were eluted with 15 mL of a methanol–dichloromethane mixture (1:1). The resulting extract was evaporated to dryness and reconstituted with methanol to a volume of 1 mL. Finally, the obtained extract was vortexed, filtered through a 0.45 μm polyvinylidene difluoride (PVDF) filter (Roth, Germany) into an autosampler vial and analyzed.

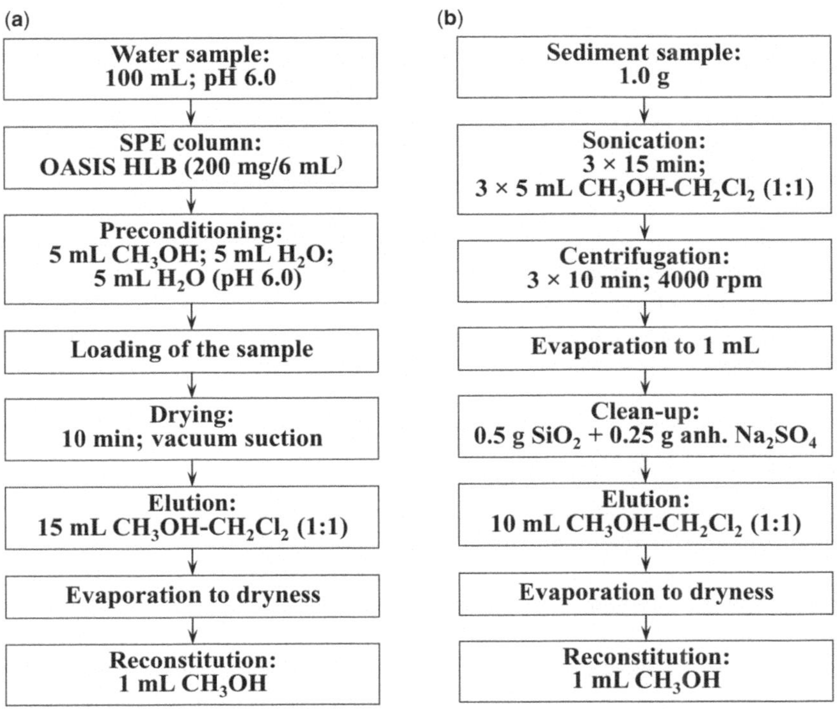


**Figure 9.1** The final procedure for the preparation of: (a) surface and groundwater samples, (b) sediment samples for the analysis of selected pharmaceuticals and pesticides.

### *9.4.3.2 USE procedure*

In a few procedures developed for the extraction of drugs and pesticides from the sediment matrix, the extraction solvent and the extraction time were optimized. The solvents that are usually tested are methanol, dichloromethane, acetonitrile, acetone and ethyl acetate. A methanol–dichloromethane mixture (1:1) was also tested as it was optimal for extracting the same analytes from the aqueous matrix (Radović *et al.*, 2015). The efficient extraction of most selected pharmaceuticals and pesticides from sediment was achieved with a methanol–dichloromethane mixture. However, the β-lactams ampicillin and amoxicillin, and the antibiotic azithromycin, could not be extracted regardless of the solvent used, probably due to irreversible bonding or degradation. For pesticides, high recoveries were generally obtained with all the investigated solvents. In a slightly modified method for the analysis of four different groups of emerging contaminants (Matić Bujagić *et al.*, 2019), the following two extraction solvents were tested: the methanol–dichloromethane mixture, as it was most efficient for extracting pharmaceuticals and pesticides from sediments (Radović *et al.*, 2015), and methanol, which was the optimal solvent for the extraction of sterols and steroid hormones (Matić *et al.*, 2014). The results showed that the methanol–dichloromethane mixture provided higher recoveries for most of the investigated compounds in the multiresidual analysis. However, for as many as eight drugs (azithromycin, doxycycline, lorazepam, metoprolol, bisoprolol, amlodipine, enalapril and cilazapril) very low recoveries were achieved and they were omitted from further investigation. In addition to the extraction solvent, the contact time of the sediment sample and the solvent was also optimized. It was found that, as expected, higher recoveries were achieved when extraction was performed for a longer time (i.e. 45 min vs. 25 min of sonication; Radović *et al.*, 2015).

The final USE procedure for the preparation of sediment samples for the analysis of selected pharmaceuticals and pesticides is presented in Figure 9.1b. The dried and homogenized sediment was sieved through a 500 μm sieve to remove gravel and plant debris. A volume of 5 mL of methanol–dichloromethane mixture (1:1) was added to the sample (1.0 g) and the content was sonicated in an ultrasonic bath for 15 min. The sample was then centrifuged at 4000 rpm for 10 min and the supernatant was separated. The procedure was repeated two more times and the extracts were combined. The resulting extract (15 mL) was evaporated to a volume of 1 mL and transferred onto a clean-up cartridge. The cartridge was prepared by adding 0.5 g of silica gel and 0.25 g of anhydrous sodium sulfate in a tube. It was preconditioned with 5 mL of methanol–dichloromethane mixture. The analyte residues were eluted from the clean-up cartridge with 10 mL of the same extraction solvent. The final extract was evaporated to dryness and reconstituted to a volume of 1 mL with methanol, filtered through a 0.45 μm PVDF filter and analyzed.

## 9.4.4 Calibration and matrix effect

The matrix effect is the suppression or enhancement of the analyte signal in the matrix solution that occurs in the ESI process. It is especially pronounced in the LC-MS analysis of complex matrices, such as wastewater samples (Jauković *et al.*, 2014) or sediments (Radović *et al.*, 2015). To eliminate the matrix effect, an external calibration with the matrix-matched standards was used in the development and optimization of the analytical methods (Dujaković *et al.*, 2010; Grujić *et al.*, 2009; Jauković *et al.*, 2014; Radović *et al.*, 2015). Calibration standards were prepared by adding a working standard solution to the blank extract obtained by the SPE or USE procedure. The peak areas of the analytes detected in the blank samples were subtracted from those obtained for the spiked sample and the matrix-matched standard.

The following equation was used to determine the matrix effect (Matić *et al.*, 2014; Radović *et al.*, 2015):

$$\text{Matrix effect (\%)} = \frac{\text{Area}_{\text{matrix}} - \text{Area}_{\text{blank}}}{\text{Area}_{\text{solvent}}} \times 100 - 100 \qquad (9.1)$$

$\text{Area}_{\text{matrix}}$ is the analyte peak area of the matrix-matched standard; $\text{Area}_{\text{blank}}$ is the peak area of the analyte present in the blank; $\text{Area}_{\text{solvent}}$ is the analyte peak area of the standard solution in methanol. The obtained result represents signal suppression (negative value) or enhancement (positive value) by the matrix components.

For real sample analyses, a standard addition method was used. Each water sample was divided into five to seven aliquots, three to five of which were spiked with a working standard solution at several concentration levels, and two were not spiked and analyzed as blanks. For each sediment sample, equal amounts were spiked at four to six concentration levels and extracted using an optimized procedure.

In general, the matrix effect was not as pronounced when analyzing surface and groundwater, regardless of the analyte selected, since the suppression or enhancement of the signal was less than 20% (Dujaković *et al.*, 2010; Grujić *et al.*, 2009). However, the analyte signal was influenced more by the constituents of the matrix when wastewater (Jauković *et al.*, 2014) or sediments (Radović *et al.*, 2015) were analyzed.

## 9.4.5 Validation of the analytical method

All the developed and optimized methods were validated using real samples spiked at several concentration levels. The performance of each analytical method was evaluated by determining the extraction recoveries and repeatability (expressed as the relative standard deviation, RSD). The experiments were carried out by analyzing three replicate samples, in three consecutive days, as an interday precision study. Analyte recoveries in the range 70–120% and RSDs below 20% are considered acceptable for an efficient analytical method. Of the 19 drugs investigated in Grujić *et al.* (2009), amoxicillin and cephalexin showed low, unacceptable recoveries (55% and 52%, respectively). The method for the analysis of 8 cardiovascular drugs (Jauković *et al.*, 2014) was shown to be unsuitable for the extraction of antihyperlipemic atorvastatin, due to the low recoveries obtained (up to 27%). For these analytes, the methods developed by the authors are not suitable for trace detection in aqueous samples. The validation of the sediment analysis method (Radović *et al.*, 2015) showed that the extraction of ampicillin, amoxicillin and azithromycin cannot be performed. Also, very low recoveries were obtained for the extraction of lorazepam (up to 18%) and diazepam (up to 44%). In addition by using the multiresidual method for the analysis of four classes of emerging contaminants (Matić Bujagić *et al.*, 2019), as many as 8 drugs (azithromycin, doxycycline, lorazepam, metoprolol, bisoprolol, amlodipine, enalapril and cilazapril) were extracted from sediment with recoveries below 21%. Therefore, these drugs cannot be determined in the sediment matrix by the method set up by the authors.

All the investigated pesticides generally showed high extraction recoveries, regardless of the matrix analyzed and the sample preparation procedure employed (Dujaković *et al.*, 2010; Matić Bujagić *et al.*, 2019; Radović *et al.*, 2015).

In all the selected pharmaceuticals and pesticides in water and sediment samples, the RSDs were generally less than 20%, with a few exceptions that were most likely due to a change in the MS sensitivity during the analytical run (Dujaković *et al.*, 2010; Grujić *et al.*, 2009; Radović *et al.*, 2015)

The limits of detection (LODs) and quantification (LOQs) were calculated as concentrations that would produce signal-to-noise (S/N) ratios of 3 and 10, respectively. They were determined using S/N ratios from the analyte's mass chromatograms in a water or sediment matrix obtained in the SRM mode. The low LODs and LOQs achieved for all the selected drugs and pesticides indicate that the developed methods are suitable for trace analyses at ng $L^{-1}$ levels (Dujaković *et al.*, 2010; Grujić *et al.*, 2009; Radović *et al.*, 2015).

The linearity of the developed methods was determined using the results of the method performance evaluation. The analyte peak area of the spiked sample was plotted vs. the determined analyte concentration. As for the linearity of the matrix-matched calibration curves, the curves displayed good linearity over a selected concentration range with high regression correlation coefficients ($R^2$). Regardless of whether the linearity of the method or the calibration curve was determined, the correlation coefficients showed that the methods were linear for all selected analytes (Dujaković *et al.*, 2010; Grujić *et al.*, 2009; Jauković *et al.*, 2014; Matić Bujagić *et al.*, 2019).

## 9.4.6 Application to environmental samples

The analytical methods developed and optimized by the authors have been applied in several surveys on traces of drugs and pesticides in surface water, groundwater, wastewaters and sediments. They have been proven to be suitable for detecting trace levels of selected pharmaceuticals and pesticides in different matrices.

### *9.4.6.1 Application to real water samples*

In the study of selected drug residues in 24 surface and groundwater samples (Grujić *et al.*, 2009), trimethoprim, azithromycin and carbamazepine were detected and confirmed. Carbamazepine was the most commonly detected drug (80% of water samples; in the concentration range 6–130 ng $L^{-1}$), while azithromycin was the most frequently found antibiotic (23%; 25–140 ng $L^{-1}$). In addition, a total of 70 surface and groundwater samples were collected at 38 sampling sites in the Danube River Basin in Serbia and analyzed for traces of pharmaceuticals in Radović *et al.* (2012). As in the previous study, the most frequently detected drug was carbamazepine, detected in the concentration range of 6–94 ng $L^{-1}$. Trimethoprim and lorazepam were also found, up to a level of 223 and 34 ng $L^{-1}$, respectively. Moreover, a study of the traces of cardiovascular drugs in 13 river and groundwater samples showed that river water contained residues of the β-blockers metoprolol and bisoprolol and anticoagulant clopidogrel, whereas the groundwater samples did not contain any of the cardiovascular drugs that were investigated (Jauković *et al.*, 2014). β-Blockers are a widely prescribed group of cardiovascular drugs, and metoprolol is among the most commonly detected drugs in the environment. It was found in the Danube at a level of 92 ng $L^{-1}$. In the same study, the ACE inhibitor enalapril was also detected in influent and effluent samples from two wastewater treatment plants in Serbia, in addition to those cardiovascular drugs found in natural waters. Although present in the influent samples at the highest levels (823 and 456 ng $L^{-1}$), enalapril was efficiently removed with a high elimination rate (up to 82%). A common method for the simultaneous determination of the investigated pharmaceuticals and pesticides was employed in a multi-year monitoring program of the Danube River Basin in Serbia (Radović *et al.*, 2015). As a continuation of the previous study (Radović *et al.*, 2012), 74 surface and groundwater samples were collected over a three-year period, many of which were taken several times at the same sampling site. The results finally showed that the drugs most commonly found in both surface and groundwater are carbamazepine and azithromycin, at the trace levels previously reported. Trimethoprim, lorazepam and sulfamethoxazole were also detected, in just a few surface water samples.

Fourteen samples of surface and groundwater influenced by agricultural practices were analyzed for pesticide traces and were found to contain carbendazim, dimethoate, carbofuran and propazine. As the most frequently detected pesticide, fungicide carbendazim was found in 63% of river water samples (up to a level of 22 ng $L^{-1}$), while insecticides dimethoate and carbofuran, as well as triazine herbicide propazine, were found in one or two water samples at similar levels (Dujaković *et al.*, 2010). In a two-year monitoring survey of the Danube River Basin in Serbia (Antić *et al.*, 2015), the four pesticides previously mentioned, and atrazine were detected. Numerous samples taken in 10 sampling campaigns at 16 sites showed that the highest levels of pesticides were found in May and June, which coincides with agricultural applications. Carbendazim was again the most frequently found pesticide with a very high concentration (up to 269 ng $L^{-1}$). In addition to carbendazim, atrazine was the second pesticide found at a very high level (188 ng $L^{-1}$), although it has been banned in Serbia since 2008. Exceptionally high levels of pesticides were explained by heavy rainfall at the time of the sampling leading to a very pronounced runoff. Both pesticides are considered to be water contaminants due to their persistence and high mobility in the environment. The final results of the previously described monitoring of the Danube River Basin in Serbia using a common analytical method (Radović *et al.*, 2015) showed that the most frequently found pesticides were carbendazim and atrazine, whereas traces of propazine and carbofuran were detected in only one sample.

### *9.4.6.2 Application to real sediment samples*

While conducting the study of drug and pesticide residues in the Danube River Basin in Serbia (Radović *et al.*, 2015), five sediment samples were taken. Using a common method for their analysis, pesticides were detected more often in sediments than in pharmaceuticals. Three samples contained the pesticides dimethoate and atrazine, at levels of 79–1222 ng $g^{-1}$ and 24–392 ng $g^{-1}$, respectively. The insecticide dimethoate was not found in any water sample, probably because it degrades rapidly in water and is thermally unstable. Traces of malathion, carbofuran and propazine were also found in the samples that were tested. As with water samples, the most commonly detected drug in sediments was carbamazepine. Two samples were positive for traces of carbamazepine, at concentrations of up to 214 ng $g^{-1}$. In addition, trace amounts of diazepam and erythromycin were also found.

A risk assessment of the recorded levels of drugs and pesticides in the sediment core from the Iron Gate I Reservoir on the Danube River (Matić Bujagić *et al.*, 2019) showed that several compounds pose a high risk for benthic organisms. The levels of antibiotics sulfamethoxazole and erythromycin in sediment indicate the need for an inclusion in monitoring programs and the necessity to improve wastewater treatment. Based on their ecotoxicological potential, the detected levels of pesticides malathion (approved for use), carbendazim (approved for application in Serbia, but banned in the EU) and linuron (approved for use in Serbia, but recently banned in the EU) can cause adverse effects on sediment-dwelling organisms and should therefore be included in future regular monitoring.

## REFERENCES


Abujaber F., Ahmad S. M., Neng N. R., Martín-Doimeadios R. C. R., Bernardo F. J. G. and Nogueira J. M. F. (2019). Bar adsorptive microextraction coated with multi-walled carbon nanotube phases – application for trace analysis of pharmaceuticals in environmental waters. *Journal of Chromatography A*, **1600**, 17–22.

Albero B., Tadeo J. L. and Pérez R. A. (2019). Ultrasound-assisted extraction of organic contaminants. *Trends in Analytical Chemistry*, **118**, 739–750.

Alder L., Greulich K., Kempe G. and Vieth B. (2006). Residue analysis of 500 high priority pesticides: better by GC–MS or LC–MS/MS? *Mass Spectrometry Reviews*, **25**, 838–865.

Allan I. J., Mills G. A., Vrana B., Knutsson J., Holmberg A., Guigues N., Laschi S., Fouillac A. M. and Greenwood R. (2006). Strategic monitoring for the European Water Framework Directive. *Trends in Analytical Chemistry*, **25**, 704–715.

Andreu V. and Picó Y. (2012). Determination of currently used pesticides in biota. *Analytical and Bioanalytical Chemistry*, **404**, 2659–2681.

Antić N., Radišić M., Radović T., Vasiljević T., Grujić S., Petković A., Dimkić M. and Laušević M. (2015). Pesticide residues in the Danube river basin in Serbia–a survey during 2009–2011. *CLEAN–Soil, Air, Water*, **43**(2), 197–204.

Asensio-Ramos M., D'Orazio G., Hernandez-Borges J., Rocco A. and Fanali S. (2011). Multi-walled carbon nanotubes–dispersive solid-phase extraction combined with nano-liquid chromatography for the analysis of pesticides in water samples. *Analytical and Bioanalytical Chemistry*, **400**, 1113–1123.

Asghar M. A., Zhu Q., Sun S., Peng Y. and Shuai Q. (2018). Suspect screening and target quanti fi cation of human pharmaceutical residues in the surface water of Wuhan, China, using UHPLC-Q-Orbitrap HRMS. *Science of Total Environment*, **635**, 828–837.

Azzouz A., Souhail B. and Ballesteros E. (2010). Continuous solid-phase extraction and gas chromatography–mass spectrometry determination of pharmaceuticals and hormones in water samples. *Journal of Chromatography A*, **1217**, 2956–2963.

Ballesteros E. and Parrado M. J. (2004). Continuous solid-phase extraction and gas chromatographic determination of organophosphorus pesticides in natural and drinking waters. *Journal of Chromatography A*, **1029**, 267–273.

Barco-Bonilla N., Romero-González R., Plaza-Bolaños P., Garrido Frenich A. and Martínez Vidal J. (2010). Analysis and study of the distribution of polar and non-polar pesticides in wastewater effluents from modern and conventional treatments. *Journal of Chromatography A*, **1217**, 7817–7825.

Basaglia G. and Pietrogrande M. C. (2012). Optimization of a SPME/GC/MS method for the simultaneous determination of pharmaceuticals and personal care products in waters. *Chromatographia*, **75**, 361–370.

Basheer C., Lee H. K. and Obbard J. P. (2002). Determination of organochlorine pesticides in seawater using liquid-phase hollow fibre membrane microextraction and gas chromatography – mass spectrometry. *Journal of Chromatography A*, **968**, 191–199.

Battaglin W. A., Bradley P. M., Iwanowicz L., Journey C. A., Walsh H. L. and Blazer V. S. (2018). Pharmaceuticals, hormones, pesticides, and other bioactive contaminants in water, sediment, and tissue from Rocky Mountain National Park, 2012–2013. *Science of Total Environment*, **643**, 651–673.

Behnoush B., Sheikhazadi A., Bazmi E., Fattahi A., Sheikhazadi E. and Saberi Anary S. (2015). Comparison of UHPLC and HPLC in benzodiazepines analysis of postmortem samples. *Medicine*, **94**, 7, article e640.

Berijani S., Assadi Y., Anbia M., Hosseini M.-R. M. and Aghaee E. (2006). Dispersive liquid-liquid microextraction combined with gas chromatography-flame photometric detection Very simple, rapid and sensitive method for the determination of organophosphorus pesticides in water. *Journal of Chromatography A*, **1123**, 1–9.

Botitsi H., Garbis S., Economou A. and Tsipi D. (2011). Current mass spectrometry strategies for the analysis of pesticides and their metabolites in food and water matrices. *Mass Spectrometry Reviews*, **30**, 907–939.

Bruzzoniti M. C., Checchini L., De Carlo R. M., Orlandini S., Rivoira L. and Del Bubba M. (2014). QuEChERS sample preparation for the determination of pesticides and other organic residues in environmental matrices: a critical review. *Analytical and Bioanalytical Chemistry*, **406**, 4089–4116.

Caban M., Czerwicka M., Łukaszewicz P., Migowska N., Stepnowski P., Kwiatkowski M. and Kumirska J. (2013). A new silylation reagent dimethyl(3,3,3–trifluoropropyl)silyldiethylamine for the analysis of estrogenic compounds by gas chromatography–mass spectrometry. *Journal of Chromatography A*, **1301**, 215–224.

Cai Q., Zhang L., Zhao P., Lun X., Li W., Guo Y. and Hou X. (2017). A joint experimental-computational investigation: metal organic framework as a vortex assisted dispersive micro-solid-phase extraction sorbent coupled with UPLC-MS/MS for the simultaneous determination of amphenicols and their metabolite in aquaculture water. *Microchemical Journal*, **130**, 263–270.

Calamari D., Zuccato E., Castiglioni S., Bagnati R. and Fanelli R. (2003). Strategic survey of therapeutic drugs in the Rivers Po and Lambro in Northern Italy. *Environmental Science and Technology*, **37**, 1241–1248.

Caldas S. S., Rombaldi C., de Arias J. L. O., Marube L. C. and Primel E. G. (2016). Multi-residue method for determination of 58 pesticides, pharmaceuticals and personal care products in water using solvent demulsification dispersive liquid–liquid microextraction combined with liquid chromatography-tandem mass spectrometry. *Talanta*, **146**, 676–688.

Carafa R., Wollgast J., Canuti E., Ligthart J., Dueri S., Hanke G., Eisenreich S. J., Viaroli P. and Zaldívar J. M. (2007). Seasonal variations of selected herbicides and related metabolites in water, sediment, seaweed and clams in the Sacca di Goro coastal lagoon (Northern Adriatic). *Chemosphere*, **69**, 1625–1637.

Carvalho J. J., Jerónimo P. C. A., Gonçalves C. and Alpendurada M. F. (2008). Evaluation of a multiresidue method for measuring fourteen chemical groups of pesticides in water by use of LC-MS-MS. *Analytical and Bioanalytical Chemistry*, **392**, 955–968.

Casado J., Santillo D. and Johnston P. (2018). Multi-residue analysis of pesticides in surface water by liquid chromatography quadrupole-Orbitrap high resolution tandem mass spectrometry. *Analytica Chimica Acta*, **1024**, 1–17.

Ccanccapa A., Masiá A., Andreu V. and Picó Y. (2016). Spatio-temporal patterns of pesticide residues in the Turia and Júcar Rivers (Spain). *Science of Total Environment*, **540**, 200–210.

Chadha R., Bali A. and Bansal G. (2016). Characterization of stress degradation products of duloxetine hydrochloride employing LC–UV/PDA and LC–MS/TOF studies. *Journal of Pharmaceutical and Biomedical Analysis*, **121**, 39–55.

Chen C.-E., Zhang H. and Jones K. C. (2012). A novel passive water sampler for in situ sampling of antibiotics. *Journal of Environmental Monitoring*, **14**, 1523–1530.

Chen S. and Kord A. (2009). Theoretical and experimental comparison of mobile phase consumption between ultra-high-performance liquid chromatography and high performance liquid chromatography. *Journal of Chromatography A*, **1216**, 6204–6209.

Cheng S., Jhang S., Huang M. and Shiea J. (2015). Simultaneous detection of polar and nonpolar compounds by ambient mass spectrometry with a dual electrospray and atmospheric pressure chemical ionization source. *Analytical Chemistry*, **87**, 1743–1748.

Chou T.-Y., Lin S.-K. and Fuh M.-R. (2009). Determination of phenylurea herbicides in aqueous samples using partitioned dispersive liquid–liquid microextraction. *Talanta*, **80**, 493–498.

Churchwell M., Twaddle N., Meeker L. and Doerge D. (2005). Improving LC–MS sensitivity through increases in chromatographic performance: Comparisons of UPLC–ES/MS/MS to HPLC–ES/MS/MS. *Journal of Chromatography B*, **825**, 134–143.

Chusaksri S., Sutthivaiyakit S. and Sutthivaiyakit P. (2006). Confirmatory determination of organochlorine pesticides in surface waters using LC/APCI/tandem mass spectrometry◇. *Analytical and Bioanalytical Chemistry*, **384**, 1236–1245.

Covaci A., Gheorghe A., Voorspoels S., Maervoet J., Redeker E. S., Blust R. and Schepens P. (2005). Polybrominated diphenyl ethers, polychlorinated biphenyls and organochlorine pesticides in sediment cores from the Western Scheldt river (Belgium): analytical aspects and depth profiles. *Environment International*, **31**, 367–375.

da Silva B. F., Jelić A., López-Serna R., Mozeto A. A., Petrovic M. and Barceló D. (2011). Occurrence and distribution of pharmaceuticals in surface water, suspended solids and sediments of the Ebro river basin, Spain. *Chemosphere*, **85**, 1331–1339.

Dujaković N., Grujić S., Radišić M., Vasiljević M. and Laušević M. (2010). Determination of pesticides in surface and ground waters by liquid chromatography-electrospray-tandem mass spectrometry. *Analytica Chimica Acta*, **678**, 63–72.

Famiglini G., Palma P., Pierini E., Trufelli H. and Cappiello A. (2008). Organochlorine Pesticides by LC–MS. *Analytical Chemistry*, **80**, 3445–3449.

Farajzadeh M. A., Sorouraddin S. M. and Mogaddam M. R. A. (2014). Liquid phase microextraction of pesticides: a review on current methods. *Microchimica Acta*, **181**, 829–851.

Farré M., Picó Y. and Barceló D. (2014). Application of ultra-high pressure liquid chromatography linear ion-trap orbitrap to qualitative and quantitative assessment of pesticide residues. *Journal of Chromatography A*, **1328**, 66–79.

Galaon T., Vacaresteanu C., Anghel D. and David V. (2013). Simultaneous ESI-APCI(+) ionization and fragmentation pathways for nine benzodiazepines and zolpidem using single quadrupole LC-MS. *Drug Testing and Analysis*, **6**, 439–450.

Garrido Frenich A., Romero-González R., Martínez Vidal J., Martínez Ocaña R. and Baquero Feria P. (2011). Comparison of solid phase microextraction and hollow fiber liquid phase microextraction for the determination of pesticides in aqueous samples by gas chromatography triple quadrupole tandem mass spectrometry. *Analytical and Bioanalytical Chemistry*, **399**, 2043–2059.

Gilart N., Miralles N., Marcé R. M., Borrull F. and Fontanals N. (2013). Novel coatings for stir bar sorptive extraction to determine pharmaceuticals and personal care products in environmental waters by liquid chromatography and tandem mass spectrometry. *Analytica Chimica Acta*, **774**, 51–60.

Gros M., Petrović M. and Barceló D. (2006). Development of a multi-residue analytical methodology based on liquid chromatography-tandem mass spectrometry (LC-MS/MS) for screening and trace level determination of pharmaceuticals in surface and wastewaters. *Talanta*, **70**, 678–690.

Gros M., Petrović M. and Barceló D. (2009). Tracing pharmaceutical residues of different therapeutic classes in environmental waters by using liquid chromatography/quadrupole-linear ion trap mass spectrometry and automated library searching. *Analytical Chemistry*, **81**, 898–912.

Grujić S., Vasiljević T., Laušević M. and Ast T. (2008). Study on the formation of an amoxicillin adduct with methanol using electrospray ion trap tandem mass spectrometry. *Rapid Communications in Mass Spectrometry*, **22**, 67–74.

Grujić S., Vasiljević T. and Laušević M. (2009). Determination of multiple pharmaceutical classes in surface and ground waters by liquid chromatography–ion trap–tandem mass spectrometry. *Journal of Chromatography A*, **1216**, 4989–5000.

Guo J., Li Z., Ranasinghe P., Bonina S., Hosseini S., Corcoran M. B., Smalley C., Kaliappan R., Wu Y., Chen D., Sandy A. L., Wang Y., Rockne K. J., Sturchio N. C., Giesy J. P. and Li A. (2016). Occurrence of atrazine and related compounds in sediments of upper great lakes. *Environmental Science and Technology* **50**, 7335–7343.

He P. and Aga D. (2019). Comparison of GC-MS/MS and LC-MS/MS for the analysis of hormones and pesticides in surface waters: advantages and pitfalls. *Analytical Methods*, **11**, 1436–1448.

Heath E., Kosjek T., Farre M., Quintana J., de Alencastro L., Castiglioni S., Gans O., Langford K., Loos R. and Radjenović J. (2010). Second interlaboratory exercise on non-steroidal anti-inflammatory drug analysis in environmental aqueous samples. *Talanta*, **81**, 1189–1196.

Hemdan A. and Tawakol S. (2018). HPLC–UV chromatographic methods for detection and quantification of undeclared withdrawn synthetic medications in counterfeit herbal medicines with confirmation by HPLC–PDA and mass spectrometry. *Chromatographia*, **81**, 777–783.

Hernández F., Cervera M., Portolés T., Beltrán J. and Pitarch E. (2013). The role of GC-MS/MS with triple quadrupole in pesticide residue analysis in food and the environment. *Analytical Methods*, **5**, 5875–5894.

Herrera-Herrera A. V., Hernández-Borges J., Afonso M. M., Palenzuela J. A. and Rodríguez-Delgado M. Á. (2013). Comparison between magnetic and non magnetic multi-walled carbon nanotubes-dispersive solid-phase extraction combined with ultra-high performance liquid chromatography for the determination of sulfonamide antibiotics in water samples. *Talanta*, **116**, 695–703.

Hijosa-Valsero M., Bécares E., Fernández-Aláez C., Fernández-Aláez M., Mayo R. and Jiménez J. J. (2016). Chemical pollution in inland shallow lakes in the Mediterranean region (NW Spain): PAHs, insecticides and herbicides in water and sediments. *Science of Total Environment*, **544**, 797–810.

Hilton M. J. and Thomas K. V. (2003). Determination of selected human pharmaceutical compounds in effluent and surface water samples by high-performance liquid chromatography-electrospray tandem mass spectrometry. *Journal of Chromatography A*, **1015**, 129–141.

Hossain A., Nakamichi S., Habibullah-Al-Mamun M., Tani K., Masunaga S. and Matsuda H. (2018). Occurrence and ecological risk of pharmaceuticals in river surface water of Bangladesh. *Environmental Research*, **165**, 258–266.

Jabali Y., Millet M. and El-hoz M. (2019). Optimization of a DI-SPME-GC–MS/MS method for multi-residue analysis of pesticides in waters. *Microchemical Journal*, **147**, 83–92.

Jauković Z., Grujić S., Vasiljević T., Petrović S. and Laušević M. (2014). Cardiovascular drugs in environmental waters and wastewaters: method optimization and real sample analysis. *Journal of AOAC International*, **97**, 1167–1174.

Jelić A., Petrović M. and Barceló D. (2009). Multi-residue method for trace level determination of pharmaceuticals in solid samples using pressurized liquid extraction followed by liquid chromatography/quadrupole-linear ion trap mass spectrometry. *Talanta*, **80**, 363–371.

Kalogridi E. C., Christophoridis C., Bizani E., Drimaropoulou G. and Fytianos K. (2014). Part II: temporal and spatial distribution of multiclass pesticide residues in lake sediments of northern Greece: application of an optimized MAE-LC-MS/MS pretreatment and analytical method. *Environmental Science and Pollution Research*, **21**, 7252–7262.

Kasprzyk-Hordern B., Dinsdale R. M. and Guwy A. J. (2008). The occurrence of pharmaceuticals, personal care products, endocrine disruptors and illicit drugs in surface water in South Wales, UK. *Water Research*, **42**, 3498–3518.

Kataoka H. (2017). Sample preparation for liquid chromatography. In: Liquid Chromatography, 2nd edn, S. Fanali, C. F. Poole, P. R. Haddad and M.-R. Riekkola (eds), Elsevier Inc., Amsterdam, Netherlands, 1–37.

Kim S. and Carlson K. (2007). Temporal and Spatial Trends in the Occurrence of Human and Veterinary Antibiotics in Aqueous and River Sediment Matrices. *Environmental Science and Technology*, **41**, 50–57.

Kim C., Ryu H., Chung E., Kim Y. and Lee J. (2018). A review of analytical procedures for the simultaneous determination of medically important veterinary antibiotics in environmental water: Sample preparation, liquid chromatography, and mass spectrometry. *Journal of Environmental Management*, **217**, 629–645.

Krauss M., Singer H. and Hollender J. (2010). LC–high resolution MS in environmental analysis: from target screening to the identification of unknowns. *Analytical and Bioanalytical Chemistry*, **397**, 943–951.

Lahti M. and Oikari A. (2012). Vertical distribution of pharmaceuticals in lake sediments-citalopram as potential chemomarker. *Environmental Toxicology and Chemistry*, **31**, 1738–1744.

Lalović B., Đurkić T., Vukčević M., Janković-Častvan I., Kalijadis A., Laušević Z. and Laušević M. (2017). Solid-phase extraction of multi-class pharmaceuticals from environmental water samples onto modified multi-walled carbon nanotubes followed by LC-MS/MS. *Environmental Science and Pollution Research*, **24**, 20784–20793.

Lara-Martín P.A., Renfro A.A., Cochran J.K. and Brownawell B.J. (2015). Geochronologies of pharmaceuticals in a sewage-impacted estuarine urban setting (Jamaica Bay, New York). *Environmental Scince and Technology*, **49**, 5948–5955.

Lazartigues A., Fratta C., Baudot R., Wiest L., Feidt C., Thomas M. and Cren-Olivé C. (2011). Multiresidue method for the determination of 13 pesticides in three environmental matrices: Water, sediments and fish muscle. *Talanta*, **85**, 1500–1507.

Leandro C., Hancock P., Fussell R. and Keely B. (2006). Comparison of ultra-performance liquid chromatography and high-performance liquid chromatography for the determination of priority pesticides in baby foods by tandem quadrupole mass spectrometry. *Journal of Chromatography A*, **1103**, 94–101.

Liang H. C., Bilon N. and Hay M. (2015). Analytical methods for pesticide residues in the water environment. *Water Environment Research*, **87**, 1923–1937.

Liang Y., Liu J., Zhong Q., Yu D., Yao J., Huang T., Zhu M. and Zhou T. (2019). A fully automatic cross used solid-phase extraction online coupled with ultra-high performance liquid chromatography–tandem mass spectrometry system for the trace analysis of multi-class pharmaceuticals in water samples. *Journal of Pharmaceutical and Biomedical Analysis*, **174**, 330–339.

Lin L., Lin H., Zhang M., Dong X., Yin X., Qu C. and Ni J. (2015). Types, principle, and characteristics of tandem high-resolution mass spectrometry and its applications. *RSC Advances*, **5**, 107623–107636.

Lindholm P. C., Knuutinen J. S., Ahkola H. S. J. and Herve S. H. (2014). Analysis of trace pharmaceuticals and related compounds in municipal wastewaters by preconcentration, chromatography, derivatization, and separation methods. *Bioresources*, **9**, 3688–3732.

Loos G., Van Schepdael A. and Cabooter D. (2016). Quantitative mass spectrometry methods for pharmaceutical analysis. *Philoshopical Transactions of the Royal Society A*, **374**, 25, article 20150366.

López de Alda M. J., Gil A., Paz E. and Barceló D. (2002). Occurrence and analysis of estrogens and progestogens in river sediments by liquid chromatography-electrospray-mass spectrometry. *Analyst*, **127**, 1299–1304.

López-Ruiz R., Romero-González R. and Garrido Frenich A. (2019). Ultrahigh-pressure liquid chromatography-mass spectrometry: an overview of the last decade. *Trends in Analytical Chemistry*, **118**, 170–181.

López-Serna R., Petrovic M. and Barceló D. (2012). Direct analysis of pharmaceuticals, their metabolites and transformation products in environmental waters using on-line TurboFlow$^{TM}$ chromatography–liquid chromatography–tandem mass spectrometry. *Journal of Chromatography A*, **1252**, 115–129.

López-Serna R., Marín-de-Jesús D., Irusta-Mata R., García-Encina P., Lebrero R., Fdez-Polanco M. and Muñoz R. (2018). Multiresidue analytical method for pharmaceuticals and personal care products in sewage and sewage sludge by online direct immersion SPME on-fiber derivatization – GCMS. *Talanta*, **186**, 506–512.

Lu N., Wang T., Zhao P., Zhang L., Lun X., Zhang X. and Hou X. (2016). Experimental and molecular docking investigation on metal-organic framework MIL-101 (Cr) as a sorbent for vortex assisted dispersive micro-solid-phase extraction of trace 5-nitroimidazole residues in environmental water samples prior to UPLC-MS/MS analysis. *Analytical and Bioanalytical Chemistry*, **408**, 8515–8528.

Ma J., Wu G., Li S., Tan W., Wang X., Li J. and Chen L. (2018). Magnetic solid-phase extraction of heterocyclic pesticides in environmental water samples using metal-organic frameworks coupled to high performance liquid chromatography determination. *Journal of Chromatography A*, **1553**, 57–66.

Magi E., Di Carro M., Mirasole C. and Benedetti B. (2018). Combining passive sampling and tandem mass spectrometry for the determination of pharmaceuticals and other emerging pollutants in drinking water. *Microchemical Journal*, **136**, 56–60.

Masiá A., Ibáñez M., Blasco C., Sancho J., Picó Y. and Hernández F. (2013). Combined use of liquid chromatography triple quadrupole mass spectrometry and liquid chromatography quadrupole time-of-flight mass spectrometry in systematic screening of pesticides and other contaminants in water samples. *Analytica Chimica Acta*, **761**, 117–127.

Masiá A., Vásquez K., Campo J. and Picó Y. (2015). Assessment of two extraction methods to determine pesticides in soils, sediments and sludges. Application to the Túria River Basin. *Journal of Chromatography A*, **1378**, 19–31.

Masiá A., Suarez-Varela M., Llopis-Gonzalez A. and Picó Y. (2016). Determination of pesticides and veterinary drug residues in food by liquid chromatography-mass spectrometry: A review. *Analytica Chimica Acta*, **936**, 40–61.

Matić Bujagić I., Grujić S., Jauković Z. and Laušević M. (2016). Sterol ratios as a tool for sewage pollution assessment of river sediments in Serbia. *Environmental Pollution*, **213**, 76–83.

Matić Bujagić I., Grujić S., Laušević M., Hofmann T. and Micić V. (2019). Emerging contaminants in sediment core from the Iron Gate I Reservoir on the Danube River. *Science of Total Environment*, **662**, 77–87.

Matić I., Grujić S., Jauković Z. and Laušević M. (2014). Trace analysis of selected hormones and sterols in river sediments by liquid chromatography-atmospheric pressure chemical ionization-tandem mass spectrometry. *Journal of Chromatography A*, **1364**, 117–127.

Matongo S., Birungi G., Moodley B. and Ndungu P. (2015). Occurrence of selected pharmaceuticals in water and sediment of Umgeni River, KwaZulu-Natal, South Africa. *Environmental Science and Pollution Research*, **22**, 10298–10308.

Meyer B., Pailler J., Guignard C., Hoffmann L. and Krein A. (2011). Concentrations of dissolved herbicides and pharmaceuticals in a small river in Luxembourg. *Environmental Monitoring and Assessment*, **180**, 127–146.

Moreno-González R., Rodriguez-Mozaz S., Gros M., Barceló D. and León V. M. (2015). Seasonal distribution of pharmaceuticals in marine water and sediment from a mediterranean coastal lagoon (SE Spain). *Environmental Research*, **138**, 326–344.

Niu H., Shi Y., Cai Y., Wei F. and Jiang G. (2009). Solid-phase extraction of sulfonylurea herbicides from water samples with single-walled carbon nanotubes disk. *Microchimica Acta*, **164**, 431–438.

Ort C., Lawrence M. G., Rieckermann J. and Joss A. (2010). Sampling for pharmaceuticals and personal care products (PPCPs) and illicit drugs in wastewater systems: are your conclusions valid? A critical review. *Environmental Science and Technology*, **44**, 6024–6035.

Pan H.-J. and Ho W.-H. (2004). Determination of fungicides in water using liquid phase microextraction and gas chromatography with electron capture detection. *Analytica Chimica Acta*, **527**, 61–67.

Pang G., Chang Q., Bai R., Fan C., Zhang Z., Yan H. and Wu X. (2020). Simultaneous Screening of 733 Pesticide Residues in Fruits and Vegetables by a GC/LC-Q-TOFMS Combination Technique. *Engineering*, **6**, 432–441. https://doi.org/10.1016/j.eng.2019.08.008

Payán M. R., López M. Á. B., Fernández-Torres R., Navarro M. V. and Mochón M. C. (2011). Hollow fiber-based liquid phase microextraction (HF-LPME) for a highly sensitive HPLC determination of sulfonamides and their main metabolites. *Journal of Chromatography B*, **879**, 197–204.

Pérez-Fernández V., Rocca L. M., Tomai P., Fanali S. and Gentili A. (2017). Recent advancements and future trends in environmental analysis: Sample preparation, liquid chromatography and mass spectrometry. *Analytica Chimica Acta*, **983**, 9–41.

Petrovic M. (2014). Methodological challenges of multi-residue analysis of pharmaceuticals in environmental samples. Trends environ. *Analytical Chemistry*, **1**, e25–e33.

Pinto M. I., Vale C., Sontag G. and Noronha J. P. (2013). Effects of ultrasonic irradiation and direct heating on extraction of priority pesticides from marine sediments. *International Journal of Environmental Analytical Chemistry*, **93**, 1638–1659.

Praveena M. S., Shaifuddin S. N. M., Syazwani S., Nasir F. A. M., Hanaf Z., Kamarudin N., Ismail T. H. T. and Aris A. Z. (2018). Pharmaceuticals residues in selected tropical surface water bodies from Selangor (Malaysia): occurrence and potential risk assessments. *Science of the Total Environment*, **642**, 230–240.

Primel E. G., Caldas S. S. and Escarrone A. L. V. (2012). Multi-residue analytical methods for the determination of pesticides and PPCPs in water by LC-MS/MS: a review. *Central European Journal of Chemistry*, **10**, 876–899.

Primel E. G., Caldas S. S., Marube L. C. and Escarrone A. L. V. (2017). An overview of advances in dispersive liquid–liquid microextraction for the extraction of pesticides and emerging contaminants from environmental samples. *Trends in Environmental Analytical Chemistry*, **14**, 1–18.

Radović T., Grujić S., Dujaković N., Radišić M., Vasiljević T., Petković A., Boreli-Zdravković Đ., Dimkić M. and Laušević M. (2012). Pharmaceutical residues in the Danube River Basin in Serbia – a two-year survey. *Water Science and Technology*, **66**, 659–665.

Radović T., Grujić S., Petković A., Dimkić M. and Laušević M. (2015). Determination of pharmaceuticals and pesticides in river sediments and corresponding surface and ground water in the Danube River and tributaries in Serbia. *Environmental Monitoring and Assessment*, **187**, 17, article 4092.

Radović T., Grujić T., Kovačević S., Laušević M. and Dimkić M. (2016). Sorption of selected pesticides and pharmaceuticals on different river sediments. *Environmental Science and Pollution Research*, **23**, 25232–25244.

Reis E. O., Foureaux A. F. S., Rodrigues J. S., Moreira V. R., Lebron Y. A. R., Santos L. V. S., Amaral M. C. S. and Lange L. C. (2019). Occurrence, removal and seasonal variation of pharmaceuticals in Brasilian drinking water treatment plants. *Environmental Pollution*, **250**, 773–781.

Rousis N., Bade R., Bijlsma L., Zuccato E., Sancho J., Hernandez F. and Castiglioni S. (2017). Monitoring a large number of pesticides and transformation products in water samples from Spain and Italy. *Environmental Research*, **156**, 31–38.

Schwanz T., Carpilovsky C., Weis G. and Costabeber I. (2019). Validation of a multi-residue method and estimation of measurement uncertainty of pesticides in drinking water using gas chromatography–mass spectrometry and liquid chromatography–tandem mass spectrometry. *Journal of Chromatography A*, **1585**, 10–18.

Shi H., Yang Y., Liu M., Yan C., Yue H. and Zhou J. (2014). Occurrence and distribution of antibiotics in the surface sediments of the Yangtze Estuary and nearby coastal areas. *Marine Pollution Bulletin*, **83**, 317–323.

Silgoner I., Krska R., Lombas E., Gans O., Rosenberg E. and Grasserbauer M. (1998). Microwave assisted extraction of organochlorine pesticides from sediments and its application to contaminated sediment samples. *Fresenius Journal of Analytical Chemistry*, **362**, 120–124.

Sudo M., Kawachi T., Hida Y. and Kunimatsu T. (2004). Spatial distribution and seasonal changes of pesticides in Lake Biwa, Japan. *Limnology*, **5**, 77–86.

Sun X., Zhou Q., Wang Y. and Ren W. (2013). Influence of hydro-geomorphology, land-use and riparian zone characteristics on herbicide occurrence and distribution in sediments in Songhua River Basin, northeastern China. *Geoderma*, **193–194**, 156–164.

Trinh T., Harden N., Coleman H. and Khan S. (2011). Simultaneous determination of estrogenic and androgenic hormones in water by isotope dilution gas chromatography–tandem mass spectrometry. *Journal of Chromatography A*, **1218**, 1668–1676.

Trtić-Petrović T., Đorđević J., Dujaković N., Kumrić K., Vasiljević T. and Laušević M. (2010). Determination of selected pesticides in environmental water by employing liquid-phase microextraction and liquid chromatography–tandem mass spectrometry. *Analytical and Bioanalytical Chemistry*, **397**, 2233–2243.

Valenzuela E. F., Menezes H. C. and Cardeal Z. L. (2019). New passive sampling device for effective monitoring of pesticides in water. *Analytica Chimica Acta*, **1054**, 26–37.

Valls-Cantenys C., Scheurer M., Iglesias M., Sacher F., Brauch H. and Salvadó V. (2016). A sensitive multi-residue method for the determination of 35 micropollutants including pharmaceuticals, iodinated contrast media and pesticides in water. *Analytical and Bioanalytical Chemistry*, **408**, 6189–6200.

Van De Steene J. and Lambert W. (2008). Comparison of matrix effects in HPLC-MS/MS and UPLC-MS/MS analysis of nine basic pharmaceuticals in surface waters. *Journal of The American Society for Mass Spectrometry*, **19**, 713–718.

Vasiljević T., Grujić S., Radišić M., Dujaković N. and Laušević M. (2012a). Pesticide residues in surface water and groundwater. In: Pesticides: Evaluation of Environmental Pollution, H. S. Rathore and L. M. L. Nollet (eds), CRC Press, Taylor & Francis Group, Boca Raton, FL, USA, 259–298.

Vasiljević T., Dujaković N., Radišić M., Grujić S., Laušević M. and Dimkić M. (2012b). Methods for monitoring of pesticide residues in water: current status and recent trends. *Water Science and Technology*, **66**, 965–975.

Vazquez-Roig P., Segarra R., Blasco C., Andreu V. and Picó Y. (2010). Determination of pharmaceuticals in soils and sediments by pressurized liquid extraction and liquid chromatography tandem mass spectrometry. *Journal of Chromatography A*, **1217**, 2471–2483.

Vidal J. L., Sánchez J. A., Plaza-Bolaños P., Frenich A. G. and Romero-González R. (2010). Use of pressurized liquid extraction for the simultaneous analysis of 28 polar and 94 non-polar pesticides in agricultural soils by GC/QqQ-MS/MS and UPLC/QqQ-MS/MS. *Journal of AOAC International*, **93**, 1715–1731.

Vukčević M., Kalijadis A., Radišić M., Pejić B., Kostić M., Laušević Z. and Laušević M. (2012). Application of carbonized hemp fibers as a new solid-phase extraction sorbent for analysis of pesticides in water samples. *Chemical Engineering Journal*, **211–212**, 224–232.

Wang C. and Gardinali P. (2012). Comparison of multiple API techniques for the simultaneous detection of microconstituents in water by on-line SPE-LC-MS/MS. *Journal of Mass Spectrometry*, **47**, 1255–1268.

Wode F., Reilich C., Van Baar P., Dünnbier U., Jekel M. and Reemtsma T. (2012). Multiresidue analytical method for the simultaneous determination of 72 micropollutants in aqueous samples with ultra high performance liquid chromatography–high resolution mass spectrometry. *Journal of Chromatography A*, **1270**, 118–126.

Wu Q., Li Y., Wang C., Liu Z., Zang X., Zhou X. and Wang Z. (2009). Dispersive liquid–liquid microextraction combined with high performance liquid chromatography- fluorescence detection for the determination of carbendazim and thiabendazole in environmental samples. *Analytica Chimica Acta*, **638**, 139–145.

Yang J.-F., Ying G.-G., Zhao J.-L., Tao R., Su H.-C. and Chen F. (2010). Simultaneous determination of four classes of antibiotics in sediments of the Pearl Rivers using RRLC-MS/MS. *Science of Total Environment*, **408**, 3424–3432.

Yazdi A.S., Razavi N. and Yazdinejad S. R. (2008). Separation and determination of amitriptyline and nortriptyline by dispersive liquid–liquid microextraction combined with gas chromatography flame ionization detection. *Talanta*, **75**, 1293–1299.

Zhang P., Bui A., Rose G. and Allinson G. (2014). Mixed-mode solid-phase extraction coupled with liquid chromatography tandem mass spectrometry to determine phenoxy acid, sulfonylurea, triazine and other selected herbicides at nanogram per litre levels in environmental waters. *Journal of Chromatography A*, **1325**, 56–64.

Zou N., Gu K., Liu S., Hou Y., Zhang J., Xu X., Li X. and Pan C. (2016). Rapid analysis of pesticide residues in drinking water samples by dispersive solid-phase extraction based on multiwalled carbon nanotubes and pulse glow discharge ion source ion mobility spectrometry. *Journal of Separation Science*, **39**, 1202–1212.

# Chapter 10

# Artificial aquifer recharge and the compression of layers of a hydrogeological complex due to groundwater abstraction

*Milan Dimkić, Slobodan Vujasinović and Milan Dotlić*

## LIST OF SYMBOLS

| | |
|---|---|
| $\sigma$ | - stress due to ground weight, |
| $\sigma_{EF}$ | - effective stress on the skeleton wall, |
| $P_w$ | - pore water pressure, |
| $m_v$ | - coefficient of volumetric compressibility, |
| $K_c$ | - filtration coefficient of low permeability layer (clay), |
| $K$ | - aquifer filtration coefficient, |
| $E$ | - term corresponding to Young's modulus of elasticity, |
| $\omega$ | - functional dependence for the case of reduction of effective stress (corresponds to increasing groundwater pressure $0 < \omega \leq 1$. In case of continuous non-falling effective stress on the ground skeleton $\omega = 1$, |
| $S_a$ | - aquifer storage parameter - constant, |
| $S_c$ | - water storage parameter - variable size, |
| $S_k$ | - parameter of total storage (water storage) for hydrogeological complex - variable, |
| $D$ | - piezometric level depression in the aquifer, |
| $\Pi$ | - piezometric level in the aquifer, |
| $h_c$ | - piezometric level in low permeability layer (clay), |
| $h_s$ | - the height of the column of water that is released due to the compressible properties of the skeleton when falling D, |
| $h_v$ | - height of the column of water released due to the elastic properties of water when the piezometric level in the aquifer decreases, |
| $h_a$ | - total height of water column released due to elastic properties of aquifer skeleton and water ($h_a$) and water released due to consolidation of aquitard material ($h_s$), |
| $E_{sk}$ | - modulus of elasticity of the skeleton, |


Editor: Milan Dimkić
doi: 10.2166/9781789060904_0531

| | |
|---|---|
| $E_w$ | - modulus of elasticity of water, |
| $M$ | - aquifer thickness, |
| $n$ | - porosity of the material, |
| $S_u$ | - settlement of parameter functions ($\Delta\Pi$, $K$, $m_v$, $t$), |
| $t$ | - time, |
| $Q_c$ | - the amount of water (l/s) flowing into the aquifer from the low-permeability layers, |
| $Q_w$ | - amount of water (l/s) abstracted from the well, |
| $q$ | - vertical inflow of water into the aquifer, |
| $S_q$ | - the amount of water per unit decrease of the piezometric level in the aquifer due to reduction of the pore volume of the clay (settlement), |
| $q_1$ | - inflow from overlaying 'clay' into the aquifer as a result of leakage caused by different conditions at the boundaries of the clay layer, |
| $K(h)$ | - hydraulic conductivity in unsaturated medium, |
| $\boldsymbol{F} = \sqrt{\boldsymbol{m_v} \cdot \boldsymbol{K_c}}$ | - Discharge factor. |

## 10.1 INTRODUCTION

This book is about processes in alluvial aquifers. The consolidation of the lithological layers due to the decrease in the pore pressure of groundwater is more pronounced in deeper, subartesian and artesian aquifers (confined aquifers). Alluvial aquifers are often phreatic (with a free surface), and/or when hydraulically influenced by the river flow, have a much slower propagation of hydraulic disturbances and often less pronounced effects of water pumping. Figure 10.1 shows a schematic diagram of a hydrogeological complex consisting of a deeper aquifer with the water flow under pressure (confined aquifer) and a shallow aquifer.

The porous environment of the artesian hydrogeological complex, which is a typical framework in which this phenomenon occurs, consists of rocks of an intergranular porosity and mainly of incoherent layers of relatively good water permeability and low compressibility (sand-gravel sediments), as well as of coherent layers of relatively poor water permeability and high compressibility (dusty-clay sediments), which are usually intertwined and resting one on top of the other i.e. vertically alternating in a horizontal plane.

This chapter is devoted to the hydrodynamic and geo-mechanical effects of pumping water and to the decrease in the piezometer level in the aquifer. A consolidation of the hydrogeological complex occurs with the consolidation of the coherent plastic (clay) layers. This is particularly significant as it causes the pore space to shrink, expels water from the pores and settles the soil.

Given here are the results which the first author and his collaborators defined back in the 1980s and 1990s. These results have been complemented by a little more knowledge and additional approaches. The mentioned phenomena have been described here by showing the processes in coupled low-permeability soil layers (aquitard) as one-dimensional (vertical direction of the Z axis), while the processes in the aquifer have been regarded as two-dimensional (X, Y). The total treated phenomenon has been formulated by quasi-3D models.

In addition to the quasi-3D approach, a simplified 2D approach is also shown. Here, the water that enters the aquifer due to the consolidation of the aquitard is represented by the coefficient ($S_k$) (storage of the hydrogeological complex). The coefficient ($S_k$) was analysed and formulated in a manner consistent with the given approach.

Today, advanced 3D models are available to describe and quantify the occurrence. However, we believe that it is still useful, and often more rational to address, consider and understand the problems pertaining to

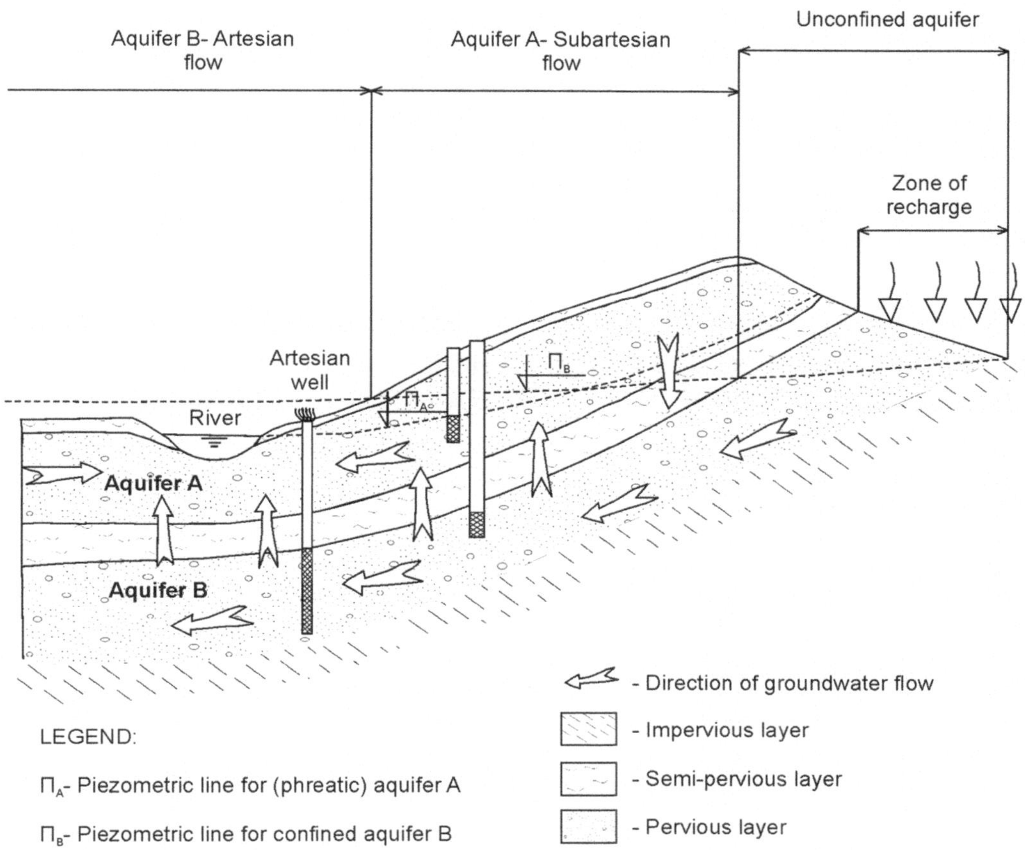


**Figure 10.1** Groundwater flow scheme in artesian and subartesian aquifers (Dimkić & Đukić, 1989).

the consolidation of a hydrogeological complex by using the method described here and that is why we present it.

Today, there are developed approaches to looking at the hydrodynamics of a hydrogeological complex including layer compressibility and the quasi-3D approach. Shown here are:

- A couple of known examples of subsidence due to groundwater pumping;
- Mechanisms of consolidation and expelling of water from an aquitard. Storage analysis of a compressible hydrogeological complex;
- Some specific relationships between the compressibility and water permeability parameters of an aquitard and the amount and kinetics of the water that is released;
- Contributions regarding the compressibility and water permeability of aquifers and aquitards in a hydrogeological complex.

Given also are the significant differences in the calculations of the water flow into the well/source, which may arise as a result of considering or not considering the compressibility of the entire hydrogeological complex.

In addition to the natural, practically unchanging $S_a$ parameter (aquifer storage), given also are the water storage parameter for the aquitard ($S_c$) and the total water storage for the hydrogeological complex ($S_k$).

The very simplified hydrodynamic calculation for one aquifer of a hydrogeological complex is in many cases also the 2D approach. In addition to considering the storage of aquifers, it is also necessary to consider the storage of an aquitard ($S_c$) as well as the storage of the total hydrogeological complex ($S_k$). Unlike $S_a$, which is a constant physical quantity due to the specificity of the parameters in the aquifer, $S_c$ and $S_k$ are time-varying (if water flow from the aquitard is calculated) and also depend on boundary conditions and are therefore discussed as such in this chapter.

## 10.2 A PHYSICAL AND MATHEMATICAL DESCRIPTION OF THE TERRAIN CONSOLIDATION PROCESS AND OF THE SUPPLEMENTARY AQUIFER RECHARGE RESULTING FROM GROUNDWATER ABSTRACTION

The exploitation of water from the sand-gravel layer of an aquifer disturbs the initial state, i.e. the natural equilibrium of a given hydrogeological complex, resulting in a change in pressure and stress both in the liquid (water) and in the solid phase (sediment).

The pumping of water from the aquifer causes a decrease in the piezometer level in it and then gradually in the surrounding poorly permeable layers as well. As the decrease in the piezometer levels occurs, the pore pressures of the water decrease, and the geostatic pressure transfers more and more to the skeletal grains, the effective forces increase resulting in the compression and consolidation of the 'package' of hydrogeological layers. Namely, the geostatic load from the weight of the soil is transferred to the grains of the sediments (effective pressure) and to the water which is in the pore cavities of the skeleton (pore pressure), in an effort to achieve a new stress equilibrium in conditions after the natural equilibrium has been disturbed (see Figure 10.2).

Change of pressure on a particle due to extraction of pore water is described by Equation (10.1) (See detail A in Figure 10.2).

$$\frac{\partial \sigma}{\partial t} = \frac{\partial \sigma_{EF}}{\partial t} + \frac{\partial p_w}{\partial t} \tag{10.1}$$

As there is no change in the load due to the weight of the soil, Equation (10.1) becomes:

$$\frac{\partial \sigma_{EF}}{\partial t} = -\frac{\partial p_w}{\partial t}$$

or

$$\Delta \sigma_{EF} = -\Delta p_w \tag{10.2}$$

The total subsidence of the particle within 'A' will be:

$$dh = m_v d\sigma = -m_v p_w \tag{10.3}$$

The leakage of water from 'A' is the result of the consolidation that takes place as a result of the reducing porosity. This is shown in Figure 10.2.

The difference between the piezometer levels in the aquifer (pumped) and in the less permeable clay layers leads to the consolidation of these layers, the subsidence of the terrain, and to the inflow of water from the less permeable layers to the aquifer layers. This leads to the consolidation of clay more or less

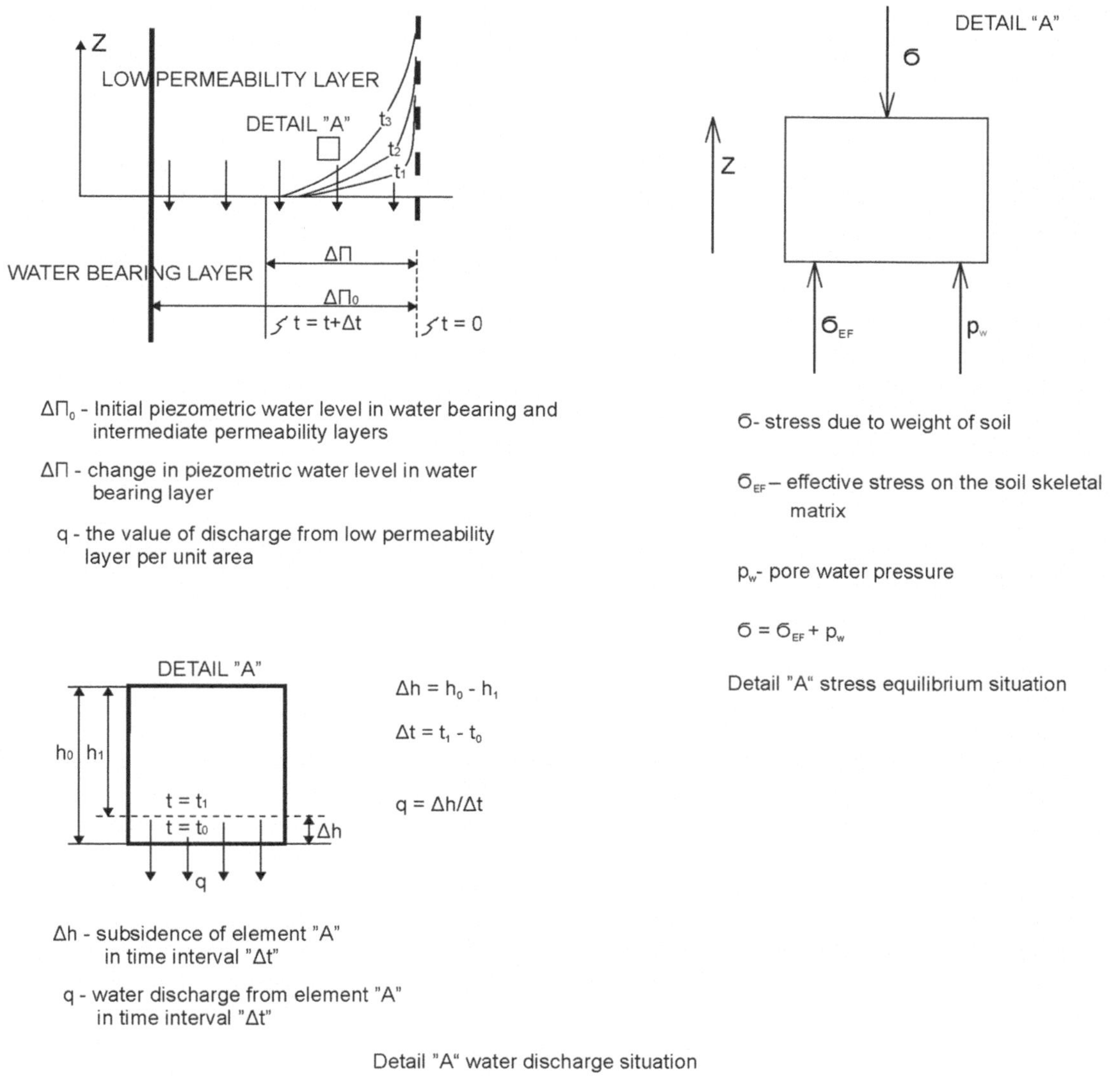


**Figure 10.2** Schematic representation of the leakage of water from poorly permeable layers due to the decrease of piezometer levels in the aquifer (Dimkić & Đukić, 1984a, 1984b, 1984c).

solely at the expense of the reduction in pore volumes. When the clay is completely saturated with water this decrease in pore volume corresponds to the amount of water pushed out from the clay layer.

Figure 10.2 is a schematic presentation of the outflow of water from poorly permeable layers and of their compression.

The described disturbance is manifested through geo-mechanical and hydrodynamic effects that can be measured directly or indirectly. Clearly, a lowering of the groundwater level due to exploitation results in the inflow of water from poorly permeable overlying and underlying layers into the water-bearing layer (this is manifested by a certain increase in the capacity of the aquifer-hydrodynamic effects), followed by a compression of the 'pack' of layers which are affected by the disturbance (on the surface of the terrain this is noted as 'soil subsidence' – geo-mechanical effects).

The inflow and compression processes take place simultaneously and are causally related. Thus, the subsidence of the soil as well as a certain increase in the capacity of the aquifer are a twofold manifestation of a single process. The degree to which the geo-mechanical and hydrodynamic effects are expressed depends on the characteristics of the porous medium, the flow conditions and the intensity of exploitation (water abstraction).

The geo-mechanical and hydrodynamic effects of exploitation are present both in the artesian and in the free phreatic aquifer. Due to the nature of the flow, the lithological constellation, a possible hydraulic connection with surface waters or due to pressure limitations resulting from soil consolidation, the influence on the hydraulics of water in the aquifer are, as a rule, more pronounced in (confined) artesian and subartesian aquifers.

## 10.2.1 Geo-mechanical effects – land subsidence

The geo-mechanical effects of subsidence of the terrain as a result of the intensive pumping of groundwater (as well as oil and gas) and deformation, i.e. layer compression, is well known globally. When looking at a hydrogeological complex, the total subsidence of the terrain above can be very large, especially due to long-term pumping.

However, a much greater role is played by the compressibility of the clay (sometimes several tens of metres or more) when compared with the settling of the aquifer layer (the settling of the aquifer layer is of the order of $(10^{3}-10^{-5})\ \Delta\pi$, where $\Delta\pi$ is the drop in the piezometric level in the aquifer. This means that for the depression of the piezometric level, the subsidence of the incoherent soil of the aquifer is of the order of parts of a millimetre to several centimetres.

However, the consolidation of the coherent layers, due to their high compressibility, can be much more pronounced, causing significant subsidence of the terrain and significant engineering problems. It can result in the deformation, bending and breaking of long structures and of bridges, railways, pipelines, canals, etc.

Table 10.1 shows cases of ground subsidence in groundwater basins worldwide.

The following will illustrate two of the many classic examples of subsidence around the world caused by long-term groundwater abstraction on a wider regional level. The examples mainly refer only to the spatial extent and visualization the subsidence of affected areas.

Example 1. Regional presentation of subsidence in California (USA) due to long-term groundwater pumping (Figure 10.3).

Example 2. Impact of the 25-year groundwater abstraction on subsidence in the Mekong Delta, Vietnam (Figure 10.4).

## 10.2.2 Assumptions and notes on consolidation parameters and the hydrodynamic parameters of hydrogeological layers due to groundwater pumping

The main issue addressed here is the influence of the compressibility of the hydrogeological complex, respecting the accompanying changes, on the groundwater dynamics in the artesian or subartesian aquifers formed in the complex.

The complex flow of water and the consolidation of the lithologic (hydrogeological) coherent and incoherent layers can be looked at using 3D modelling. However, here we adopt some simplifications that we believe can contribute to a more efficient and rational approach to the issue.

**Table 10.1** Known cases of ground subsidence in groundwater basins worldwide (Nelson, 2016).

| | Maximum Subsidence (m) | Area Affected ($km^2$) | Cause |
|---|---|---|---|
| Long Beach, Los Angeles, USA | 9.00 | 50 | Petroleum withdrawal |
| San Joaquin Valley, USA | 8.80 | 13,500 | Groundwater withdrawal |
| Mexico City, Mexico | 8.50 | 225 | Filled lake |
| Tokyo, Japan | 4.50 | 3000 | Groundwater withdrawal |
| Huston, USA | 2.70 | 12,100 | Groundwater withdrawal (possibly oil & gas extraction) |
| Shanghai, China | 2.63 | 121 | Groundwater withdrawal |
| New Orleans, USA | 2.00 | 175 | Groundwater withdrawal, construction dewatering and drainage |
| Bangkok, Thailand | 1.00 | 800 | Groundwater withdrawal |
| Venice, Italy | 0.22 | 150 | Groundwater withdrawal |
| London, England | 0.30 | 295 | Groundwater withdrawal, construction dewatering and drainage |
| Latrobe Valley, Victoria, Austria | 2.00 | Several $km^2$ | Groundwater withdrawal from mining and irrigation |
| Namoi Valley, NSW, Australia | 0.5 | Not available | Groundwater withdrawal |

*Input assumptions – layers of coherent material*

The assumptions adopted here, related to the flow of water in the overlying and underlying layers of bound (clay) material are closely linked to the assumptions made in Terzaghi's theory of one-dimensional consolidation. The basic differences are that the rheological model can, in addition to the viscoelasticity, also take into account plastic deformation, as well as that the filtration coefficient and volume compressibility can also be considered as variables depending on the change in the effective stress on the soil skeleton.

$$\begin{aligned} K_c &= K_c(\sigma_{EF}) \\ m_v &= m_v(\sigma_{EF}) \end{aligned} \tag{10.4}$$

The assumptions are:

- Clay is a visco-elasto-plastic body; the deformation slow movement of clay (creep) is not taken into account (Four models are used for rheological models, essentially derived from Kelvin's viscoelastic body. Models are selected based on the type of problem);
- All pores in the soil are filled with water;
- The water flow in the clay is only vertical;
- The direction of the principal stresses in the clay coincides with the vertical direction;
- In soil consolidation, there is no deformation of the grains of soil and there is no change in the volume of water. The deformation occurs solely at the expense of the change in the volume of the pores in the soil;
- The drainage of water occurs according to Darcy's law;

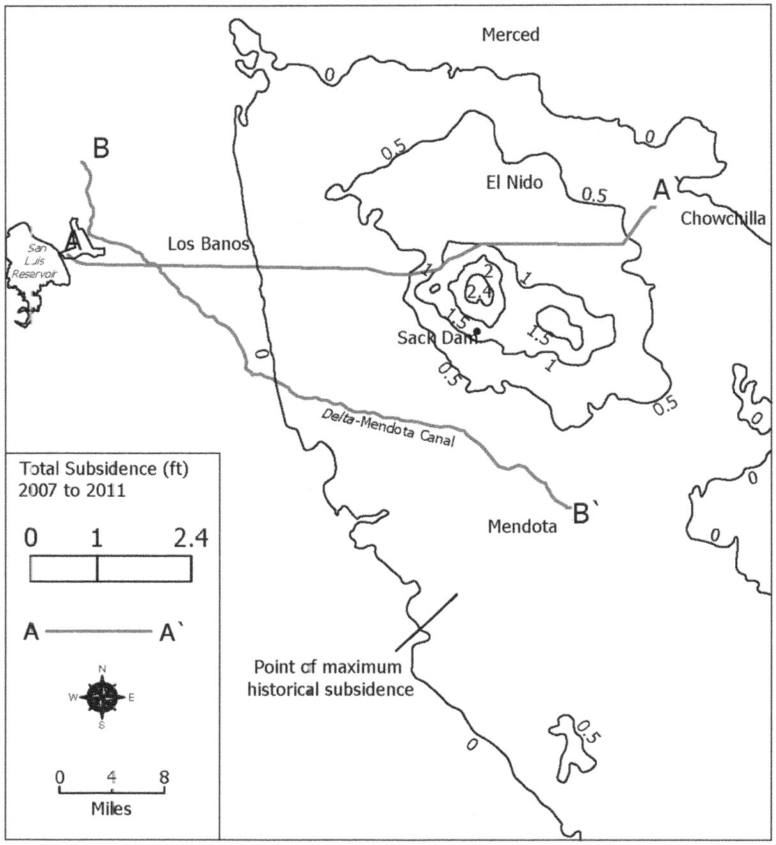


**Figure 10.3** Soil subsidence in the San Joaquin Valley in the period I. 2007 to III. 2011 (Borchers and Carpenter, 2014).

- The vertical soil deformation is small relative to the thickness of the soil layer;
- Changes in the viscosity and in the volume weight of water are not directly taken into account;
- Clay layer is homogeneous and isotropic.

It should be said in relation to the assumption in (i) below, that in his theory of consolidation Terzaghi assumed the homogeneity and isotropy of the clay layer. However, the experiments that were done to determine the $K_c = K_c\ (\sigma_{EF})$ and $m_v = m_v\ (\sigma_{EF})$ relations indicate significant changes in the magnitudes of the filtration coefficients ($K$) and the volume compressibility coefficients ($m_v$) along the depth of the layer of the same bound material and in time (due to the change in the effective stress). In view of what has been said, of all the possible realistic possibilities, two alternatives of the assumption under i) are singled out here (see Figure 10.5).

i-1) The clay layer is homogeneous and isotropic (this assumption is consistent with Terzaghi's theory):

$$\begin{aligned} K_c &= const \\ m_v &= const \end{aligned} \tag{10.5}$$

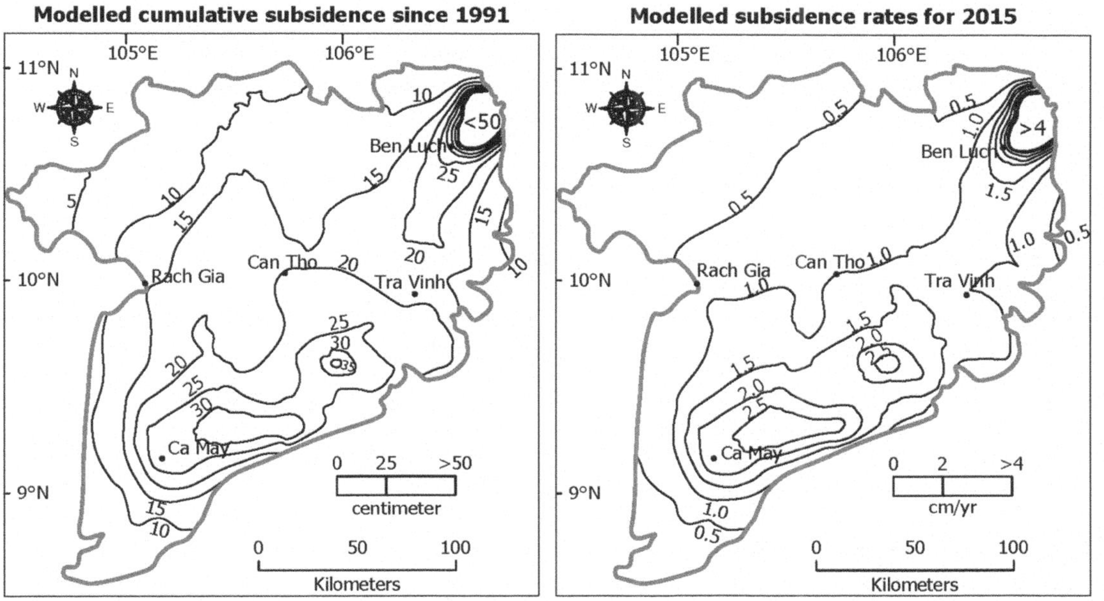


**Figure 10.4** Cumulative subsidence of the studied terrain from 1991 to 2016 and subsidence in 2015 (PSJ Minderhoud etc., IOP Publishing Ltd, Environ. Res. Lett. 12 (2017) 064006).

i-2) The clay layer is of the same material with changes in the characteristics of the filtration coefficients and in the volume compressibility coefficients depending on the changes in the effective stress ($\sigma_{EF}$) in depth and in time:

$$\begin{aligned} K_c &= K_c(\sigma_{EF}) = K_c(Z, T) \\ m_v &= m_v(\sigma_{EF}) = m_v(Z, T) \end{aligned} \tag{10.6}$$

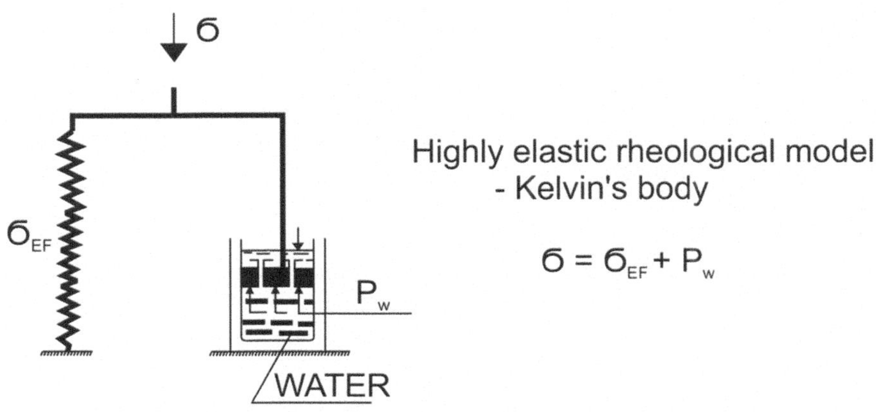


**Figure 10.5** Highly-elastic rheological model – Kelvin's body (Dimkić, 1986).

Assumptions regarding the groundwater flow in an aquifer:

- The skeleton of unbound material in the aquifer represents an elastic – Hooke's – body;
- Equipotential surfaces in the aquifer are vertical, i.e. the water flow is horizontal and the velocity is constant along each vertical (Dupuit's hypothesis)

$$\frac{\partial \Pi}{\partial z} = 0; \quad \frac{\partial V}{\partial z} = 0; \quad v = v(x, y) \tag{10.7}$$

- The vertical deformation in the aquifer is negligible in relation to the thickness of the aquifer (from a hydrodynamic point of view);
- Aquifer parameters – the filtration coefficient ($K$) and the effective porosity due to the compressibility of the soil and water skeleton ($E_{EL}$) do not change over time.

The assumptions defined in this way for the hydrogeological complex of coherent and incoherent hydrogeological layers can be applied to the hydrodynamic calculations of groundwater flows in large aquifers (and relative to the space of relatively thin aquifers) and to overlying, underlying and intercalated layers of coherent (bound) material.

## 10.2.3 Process parameters

*Coherent layer filtration coefficient*

The filtration coefficients of coherent (clay) material are low (for clays they are of the order of less than $10^{-7}–10^{-11}$ m/s). What this parameter is like (size, isotropy, etc.) may depend on a range of factors mentioned earlier. Here, we believe that in coherent soil, the changes in the size of the coefficient of filtration of a certain particle are most often affected by the change in porosity caused by changes in the effective stress on the soil skeleton.

Testing the dependence $K_c = K_c\,(\sigma_{EF})$ involves problems associated with the low values of the filtration coefficient of coherent soil-clay. Figure 10.6 presents the test results (Olsen *et al.*, 1985) from which it is

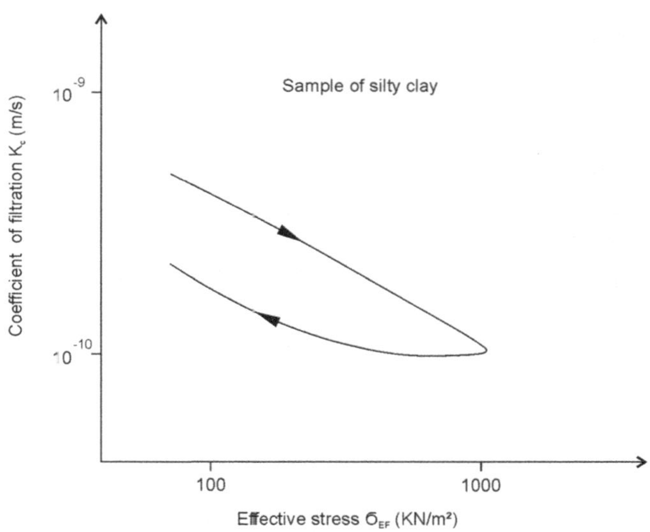


**Figure 10.6** Characteristic dependence of clay $K_c = K_c\,(\sigma_{EF})$ (Dimkić, 1986).

possible to show the change in the clay filtration coefficient depending on the changes in the effective stresses on the soil skeleton. It evident that there is a deviation from the dependence $K = K\ (\sigma_{EF})$ when the stress on the sample is reduced, compared with that registered when there is a continuous increase of the effective stress on the soil skeleton.

*Clay volume compressibility coefficient*

Land subsidence is defined by the change in the effective stress in soil, the thickness of the soil layer and by the coefficient of volume compressibility ($\Delta\sigma_{EF}$, $h$, $m_v$). The $m_v$ coefficient can vary and depends on several factors. Most of the above factors are integrated in Equation (10.8):

$$m_v = m_v(\sigma_{EF}) \tag{10.8}$$

i.e. $\Delta h/h = f\ (\sigma_{EF})$, i.e.: $\Delta h/h = m_v \cdot \Delta\ (\sigma_{EF})$ where $\Delta h\ /h$ is the relative subsidence, and $h$ is the thickness of the sample. It follows that $\mathrm{d}h = m_v \mathrm{d}(\sigma_{EF}) \cdot \mathrm{d}z$ so that $m_v$ is determined by the slope of the tangent on the experimental curve $(\Delta h/h(\sigma_{EF}))$.

Figures 10.7 and 10.8 show $m_v = m_v\ (\sigma_{EF})$ type dependencies for the effective stress rise conditions.

It should be noted that differences in the magnitude of the coefficient of volume compressibility have been registered depending on the mineral composition, structure, granulometric composition, degree of saturation with water, degree of mineralization of pore water, temperature, etc.

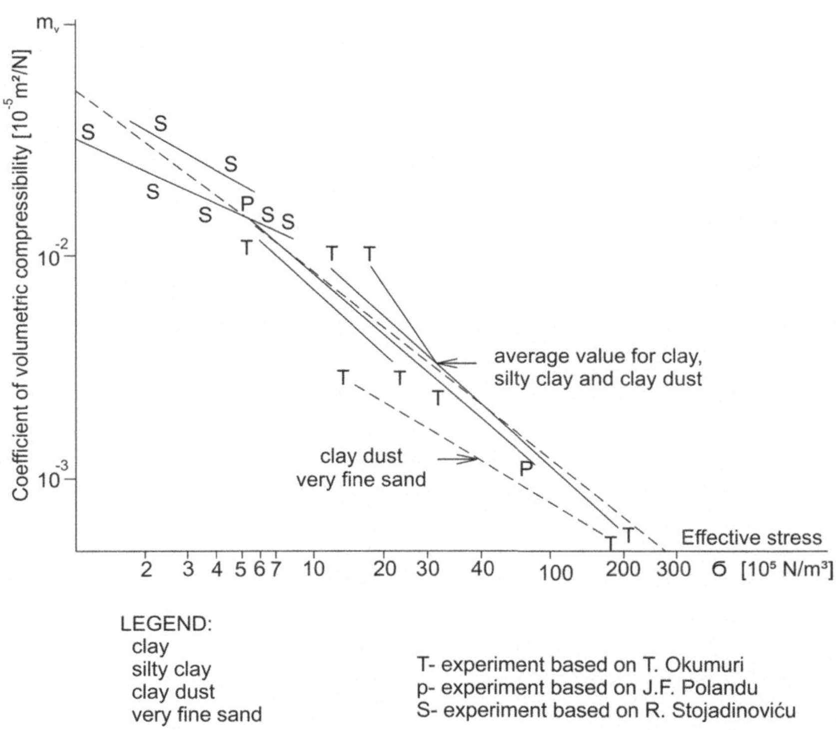


**Figure 10.7** Relationship between the volume compressibility coefficient and effective stress according to various authors (Patankar, 1976).

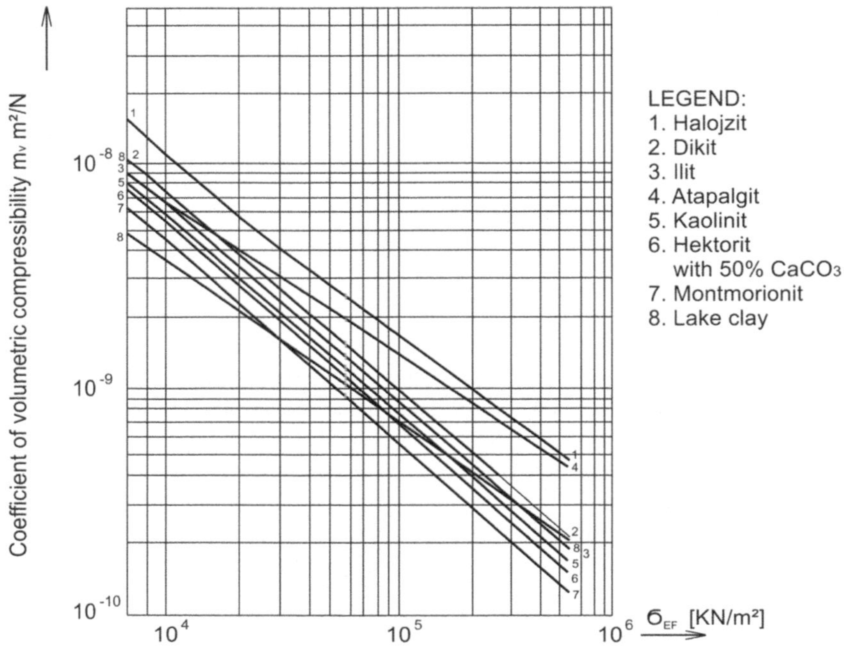


**Figure 10.8** Ratio $m_v = m_v\ (\sigma_{EF})$ according to edometric test results on dry clay soil samples (Chilingar & Rieke, 1976).

In virtually all the experiments conducted to determine the coefficient of volume compressibility, hysteresis loops were detected when reducing the load on samples of coherent soil. This occurs because of the plastic properties of the tested samples. As a rule, the same dependence of $m_v = m_v\ (\sigma_{EF})$ does not apply when there is a reduced stress on the samples as when they are stressed. The obtained values of the volume porosity coefficient when there is reduced stress are slightly lower than those recorded for loaded samples.

*Some rheological models of clay layer bound material*

Here, we schematically list several rheological models of clay layers of bonded material that can be used when considering flow hydrodynamics in the hydrogeological complex of aquifer and aquitard intercalations (Figure 10.9). The first model is that of a visco-elastic Kelvin body. The other three models are basically derived from the Kelvin's visco-elastic body model with some modifications:

## 10.2.4 Rheological model of unbound bedding material (sand, gravel)

It is common for a rheological model of the aquifer skeleton to be regarded as having an elastic (Hooke's) body, although deviations in the rheological behaviour of the material of an aquifer skeleton of intergranular porosity (sand, gravel) from the classical rheological model have also been recorded.

The rheological properties of the aquifer skeleton depend on a number of factors:

- Granulometric composition;
- Porosity;

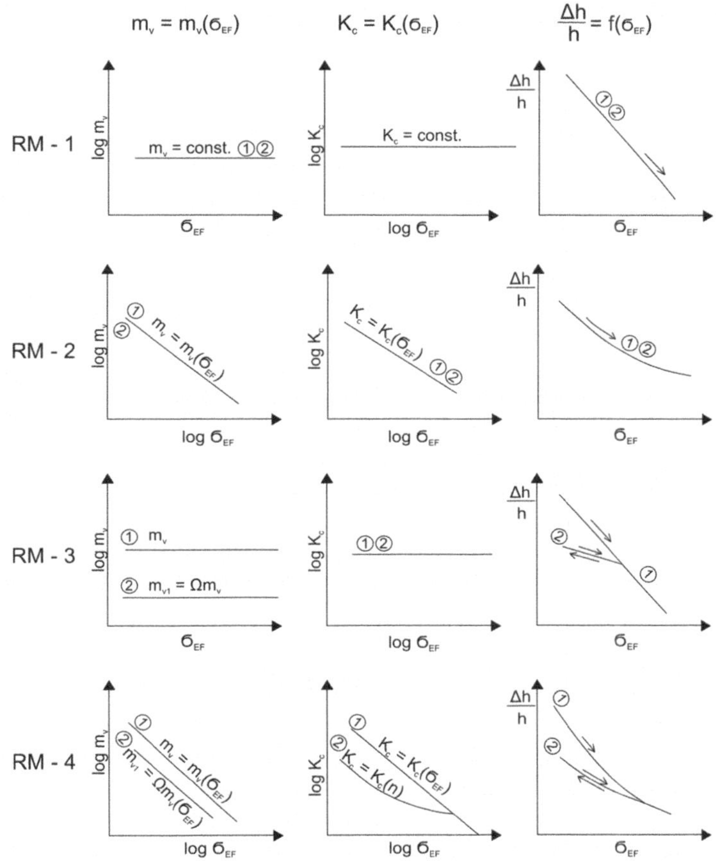


① Continuous increase of effective stress
② Effective stress is smaller than real stress in the preceding period

**Figure 10.9** Schematic representation of the basic parameters of rheological models (Dimkić & Đukić, 1986).

- Mineral or particle composition;
- Pressure exerted by the skeleton;
- Possible previous pressure;
- The shape of the skeleton grain, etc.

Given that the permeability of these materials is relatively good, we could say that there is an 'instantaneous' transfer of stress to the soil skeleton (either due to external loading or due to a drop in the water pressure). For materials of a more uniform granulometric composition, exposed to less pressure and less consolidated, a higher degree of compressibility can be expected. Deformations occur at the expense of soil restructuring – the reduction of pore cavities. In the conditions that have been described, deformations with a marked irreversibility can be expected.

At very great depths in the soil (over several hundred metres), a breakage of grains occurs – first of the larger ones and then of the smaller ones – followed by their restructuring. Such deformations are predominantly irreversible. The rheological behaviour of the skeleton of an intergranular aquifer can be described by the equation:

$$\frac{dh}{h} = \frac{\omega}{E} = d\sigma_{EF} \tag{10.9}$$

where $E$ corresponds to Young's modulus of elasticity, and $\omega$ represents a specific functional dependence in the case of a decrease in effective stress (it corresponds to an increase in groundwater pressure, $0 < \omega \leq 1$. If the effective stress on the soil skeleton continually does not drop, then $\omega = 1$ is true, so the above equation becomes:

$$\frac{dh}{h} = \frac{1}{E} d\sigma_{EF} \tag{10.10}$$

And it is nowadays applied practically when defining the groundwater continuity equation, i.e. in the study of unsteady flows. This approach can be justified to some extent, especially when bearing in mind that the compressibility of gravel and sand sediments (at depths at which groundwater is used for water supply) is generally much lower than the compressibility of poorly permeable aquifer-limiting layers.

The compressibility of sand and gravel can reach the order of compressibility of clays only under very high pressures, when rock grains break (at depths in the soil greater than several hundred metres).

Figure 10.10 shows some results of the quartz sand compressibility experiment in relation to the applied load.

## 10.2.5 Aquifer storage and hydrogeological complex

Aquifer storage ($S_a$) is a physical size that depends on the properties of an aquifer (porosity, skeletal elasticity, etc.) and on water compressibility.

If we want to calculate the linear vertical inflow from bound, weakly permeable, underlying, elevated and intercalated layers (aquitard) using a special equation ($s$) in the balance equation of water flow in an aquifer, aquifer storage as a real physical value will suffice.

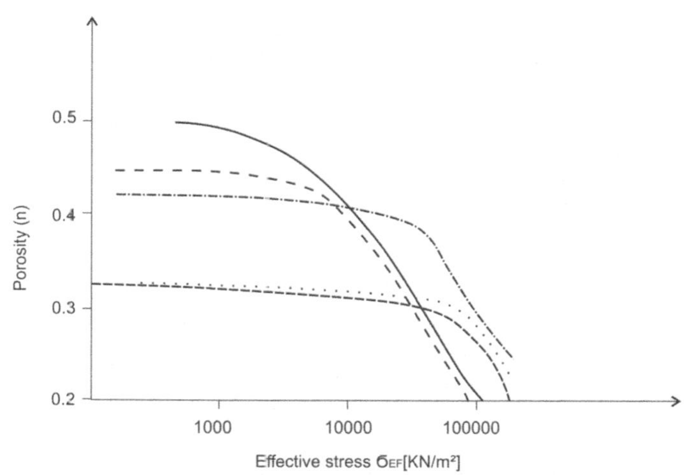


**Figure 10.10** Typical quartz sand compression curves (Roberts, 1976).

However, if we want to simplify the groundwater flow calculations and reduce them to the equation (*s*) of flow in an aquifer (aquifers), the water balance of an aquifer must also include the water released from the aquitard (less permeable bonded layers). We do this by introducing the coefficient of total water storage in a hydrogeological complex ($S_k$) or the coefficient of the momentary effective water storage ($S_{\Delta t}$).

Given here are three different definitions of the storage coefficient which should help us when defining the different approaches to aquifer flow calculations.

*Definition 1: Aquifer storage ($S_a$)*

The aquifer storage is equal to the height of the water column that is released due to the compressibility characteristics of the skeleton of the aquifer and of water, with the unit decrease of the piezometer level in the aquifer.

$$S_a = \frac{h_s + h_v}{D} = \frac{h_a}{D} \tag{10.11}$$

$D$ – depression of the piezometer level in an aquifer, $h_s$ – height of the water column that is released due to the compressibility of the skeleton when the piezometer level decreases in aquifer by an amount, $D$, $h_v$ – height of the water column released due to the elastic properties of the water as the piezometer level drops in the aquifer,

$$h_a = h_s + h_v,$$

$h_a$ – total height of the water column released due to the elastic properties of the aquifer skeleton and water ($h_a$) and the water released due to the consolidation of the aquitard material ($h_s$).

The definition includes the water that is released due to the compression properties of the aquifer and water. This definition does not include the water coming into the aquifer as a result of the consolidation of overlying and underlying layers.

Due to the decrease of the piezometer level in the aquifer (consequence: increase of the effective stress on the soil skeleton), water spreads, as does the compression of the soil skeleton. This results in the release of a certain amount of water, which is defined by storage due to the elastic properties in the aquifer (Equations (10.1)–(10.16)). Thus (Chilingar & Rieke, 1976):

$$S_a = \gamma_v \cdot M \cdot n\left(\frac{1}{E_w} + \frac{\omega}{nE_{sk}}\right) \tag{10.12}$$

If there is no drop in the effective stress on the soil skeleton, parameter $S_a$ (Equation 10.12) becomes:

$$S_a = \gamma_v \cdot M \cdot n\left(\frac{1}{E_w} + \frac{1}{nE_{sk}}\right). \tag{10.13}$$

The elasticity of the water module, when free of the gas globule, is about $E_w = 10^{-7}\ \mathrm{KN/m^2}$. If we adopt the module of elasticity of the skeleton to be $E_{sk} = 10^6\ \mathrm{KN/m^2}$, the porosity $n = 0.3$ and the aquifer thickness $M = 10$ m, we get:

$$S_a = 1 \cdot 10 \cdot 0.3 \cdot (10^{-7} + 3.3 \cdot 10^{-5}) \approx 10^{-4} \tag{10.14}$$

Changes in the magnitude of the effective porosity which includes the elastic properties of aquifers and water due to the decrease in the piezometer level, for depths (pressures) that correspond to those encountered for water supply, can practically be and most often are neglected:

$$\frac{\partial S_a}{\partial t} \approx 0 \tag{10.15}$$

*Definition 2: Total storage of a hydrogeological complex ($S_k$)*

$S_k$ is equal to the quotient of the height of the water column released from the beginning of the drop in the piezometer level until the observed moment and the corresponding depression of the piezometer level. The water released due to the compressibility of the aquifer skeleton and the elasticity of the water, as well as due to the compression of the surrounding hydrogeological layers, is calculated as follows:

$$S_k = \frac{h_0 + h_c}{D}; \quad S_k = \frac{h_0}{D} + \frac{h_c}{D} = S_a + S_C \tag{10.16}$$

$h_c$ – water column coming from clay due to the depression, $D$, of the piezometer level in the aquifer, $S_a$ – storage due to the compressibility of aquifers and water, $S_c$ – storage component due to the inflow of water into the aquifer during the consolidation of the overlying and underlying layers, as well as of all the intercalated clay layers.

Storage defined in this way is not only related to the physical characteristics of the hydrogeological complex, but also depends on other parameters (for example the pumping regimes, time, etc.).

The purpose of introducing the parameter $S_k$ is to make the hydrodynamic calculation of the water flow in the aquifer possible by solving only one differential equation – the water flow equation in the aquifer. The total storage parameter for the hydrogeological complex defined in this way is no longer a one-size-fits-all, as for $S_a$, but also depends on the boundary conditions (for example, the dynamics of aquifer exploitation).

The following remarks are made regarding the storage parameter defined in this way:

- The magnitude of the total effective porosity depends on a number of factors;
- All the factors that go into this dependency are defined $S_k = S_k\,(x, y, t)$;
- When changing the intensity of pumping, a corresponding change in the total storage occurs. These changes can be drastic;
- When looking at the flow in artesian aquifers with superimposed pumping from two points, it is not justified to regard the total storage as solely a function of time, but it should rather be regarded as a variable in space and time. For this superimposed pumping, the change in the storage capacity in the horizontal plane can be very pronounced;
- Changes in the size of the total storage capacity in terms of space can vary significantly depending on the heterogeneity of the overlying or underlying layer in the horizontal plane;
- Dimkić and co-authors introduced the 'outflow factor' parameter $\left(F = \sqrt{m_v \cdot K_c}\right)$, which well defines the intensity of settlement and the outflow of water due to the consolidation of coherent underlying and overlying layers, when pumping from aquifers under pressure;
- When considering the effect of the consolidation of coherent clay layers, the dependence of the filtration coefficient and of the volume compressibility coefficient of the coherent soil on the effective stresses can be taken into account.

*Definition 3: Momentary effective storage ($S_{\Delta t}$)*

The momentary effective storage corresponds to the quotient of the column of water that is released in an infinitely short period of time and the appropriate differential depression of the piezometric level.

This accounts for the water released due to the compressibility of the aquifer skeleton and of the water, as well as for the water entering the aquifer due to consolidation of the aquifer's overlying and underlying layers.

$$S_{\Delta t} = \lim_{\Delta t \to 0}\left(\frac{\Delta(h_a + h_c)}{\Delta D}\right) \tag{10.17}$$

This definition and the use of the current effective storage in some cases has certain advantages when solving equations pertaining to the flow of water in aquifers, in comparison to the previous two.

Definition 1 gives the realistic amount of water that is released both due to the consolidation of the incoherent aquifer skeleton material and the expansion of water when the pore pressure of water in the aquifer decreases.

Definition 2 does not give a one-size-fits-all total storage. Thus:

$$S_k = S_k(\Pi(t, x, y)) \tag{10.18}$$

$\Pi\ (t, x, y)$ – piezometric level in time $t$ and space $(x, y)$.

This means that at a certain moment in time $t_n$, $S_k$ may vary depending on $\Pi\ (t, x, y)$.

Definition 2 ($S_k$) theoretically corresponds to the calculation of the piezometric state in an aquifer in a single instant of time:

$$S_{kn} = S_k(t_n, x, y) \tag{10.19}$$

For some other moment in time, it is theoretically (usually and practically) necessary to correct the value of $S_k$:

$$S_{km} = S_k(t_m, x, y) \tag{10.20}$$

This means that if we want to know the piezometric levels for $n$ different time intervals, we have to perform $n$ special calculations with $n$ corresponding values for $S_k$.

The definition of the momentary effective storage provides the theoretical possibility to have an insight into the piezometric levels for $n$ different time sections by doing one calculation of the flow in an aquifer:

$$\begin{aligned}\Pi(t, x, y) : \Pi_0(t_0, x, y) &\xrightarrow[S\Delta T_1]{} \Pi_1(t_1, x, y) \xrightarrow[S\Delta T_2]{} \Pi_2(t_2, x, y) \xrightarrow[S\Delta T_3]{} \Pi_3(t_3, x, y)\\ &\to \cdots \Pi_n(t_n, x, y) \xrightarrow[S\Delta T_{n+1}]{} \Pi_{n+1}(t_{n+1}, x, y)\end{aligned} \tag{10.21}$$

Of course, hydrodynamic calculations as well as consolidation and soil subsidence calculations of a hydrogeological complex can also be performed using 3D modelling techniques. However, for hydrogeological complexes that extend much further than the value of their thickness and that have aquifers of incoherent skeletal material intercalated with clay layers, we can see the significant advantages of the 2D and quasi-3D approach to calculations. The definitions of storage mentioned here are suitable for such purposes.

### 10.2.6 Aquifer filtration coefficient

Aquifer is loose soil. The aquifer filtration coefficient can be estimated by using the Kozeny formula:

$$K = 8.1 \cdot 10^{-3} \cdot \frac{g \cdot d}{v} \cdot \frac{n^3}{1 - n^2} \tag{10.22}$$

If we consider that the reference (initial) filtration coefficient is $K_o$, and the coefficient of the same material, after a certain decrease in the piezometric level in the aquifer is $K$, then:

$$K = \frac{K}{K_0} \cdot K_0 = \frac{\frac{n^3}{1 - n^2}}{\frac{n_0^3}{1 - n_0^2}} \cdot K_o \tag{10.23}$$

where $n$ is the porosity of the material.

A change in the water pressure in the aquifer will increase the load on the soil skeleton amounting to the value of the reduced water pressure. If we refer to the assumption that all deformations in the soil occur at the expense of pore volume, we can write:

$$n = n_0 - \frac{\Delta M}{M} \tag{10.24}$$

For the values of Jung's model, which is usually valid for the unbound material of the aquifer skeleton, we can consider $\frac{\Delta M}{M = 0}$ and thus Equation (10.23) becomes:

$$K \approx K_0$$

i.e.

$$\frac{\partial K}{\partial t} \approx 0 \tag{10.25}$$

Equation (10.25) applies practically up to the stress area when the fracture of soil grains begins.

## 10.3 CONSIDERATION OF THE INFLUENCE OF COHERENT LAYER PARAMETERS ON ITS CONSOLIDATION AND ON THE RESULTING INFLOW OF WATER TO THE AQUIFER USING THE MATHEMATICAL ANALYSIS METHOD FOR THE DECREASE IN THE PIEZOMETRIC LEVEL IN AN AQUIFER

### 10.3.1 Coherent layer consolidation as a function of its characteristic parameters

For certain initial and boundary conditions it is possible to obtain an analytical solution for the consolidation of a particular layer. Certain relationships regarding the consolidation of a coherent, poorly permeable layer ('clay') will be derived from the size of its parameters (volume compressibility coefficient and filtration coefficient), and the time and drop of the piezometric level in the aquifer. The clay layer is regarded as homogeneous. An instantaneous, stepped drop in the piezometer level in the aquifer will be considered. A layer of clay is considered as a semi-boundary.

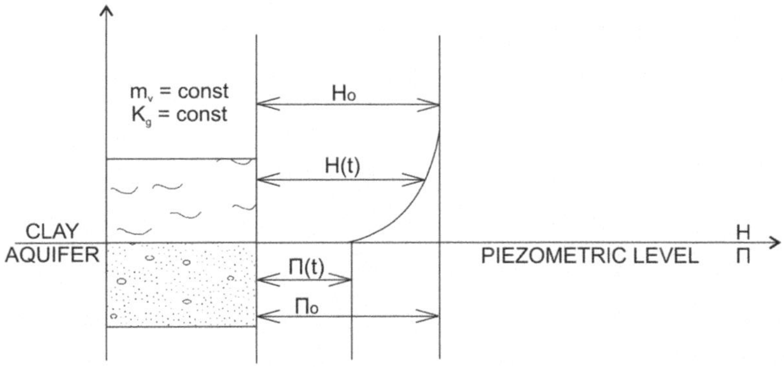


**Figure 10.11** Schematic presentation of the analysed case of clay consolidation (Dimkić, 1986).

By using mathematical analysis methods, this chapter will also look at the values of water leakage to an aquifer due to consolidation from 'clay', as well as at the storage ($S_k$) value for certain boundary conditions (i.e. conditions of changing piezometric levels in an aquifer).

The one-dimensional consolidation equation for a homogeneous layer according to Terzaghi is:

$$\frac{\partial^2 H}{\partial Z^2} = \frac{\gamma_v m_v}{K_c} \cdot \frac{\partial H}{\partial t} \tag{10.26}$$

*Initial and boundary conditions*

The clay layer is considered to be a semi-boundary layer: the upper boundary is infinitely distant (see Figure 10.11).

Initial conditions:

$$H(Z, t = 0) = \Pi_0 \tag{10.27}$$

Boundary conditions:

$$H(t, Z = \text{contact zone}) = \Pi_0 - \Delta\Pi \tag{10.28}$$

Conditions shown in Equations (10.27) and (10.28) assume the same initial piezometric levels in clay ($H$) and in the aquifer ($\Pi$), as well as the same hydrostatic distribution of pore pressure in the clay. The change in the piezometric level in the aquifer occurs in time $t = 0$, while its value further on (during the consolidation process) remains unchanged. For these initial and boundary conditions the solution of Equation (10.26) is:

$$H(z) = \Pi_0 - \Delta\Pi[1 - \phi(\lambda)] \tag{10.29}$$

$$\phi(\lambda) = \frac{2}{\sqrt{\pi}} \int_0^{\lambda} e^{-\lambda^2} \cdot d(\lambda) \tag{10.30}$$

$$\lambda = \frac{z}{2} \cdot \sqrt{\frac{\gamma_v m_v}{K_c \cdot t}} \tag{10.31}$$

Also shown next is the dependence of subsidence of the observed semi-boundary layer on the layer parameters ($m_v$ and $K_c$) as well as on time ($t$) and the subsidence of two layers with parameters ($m_v$ and $k$) with the same initial and boundary conditions (Equations (10.27) and (10.28)).

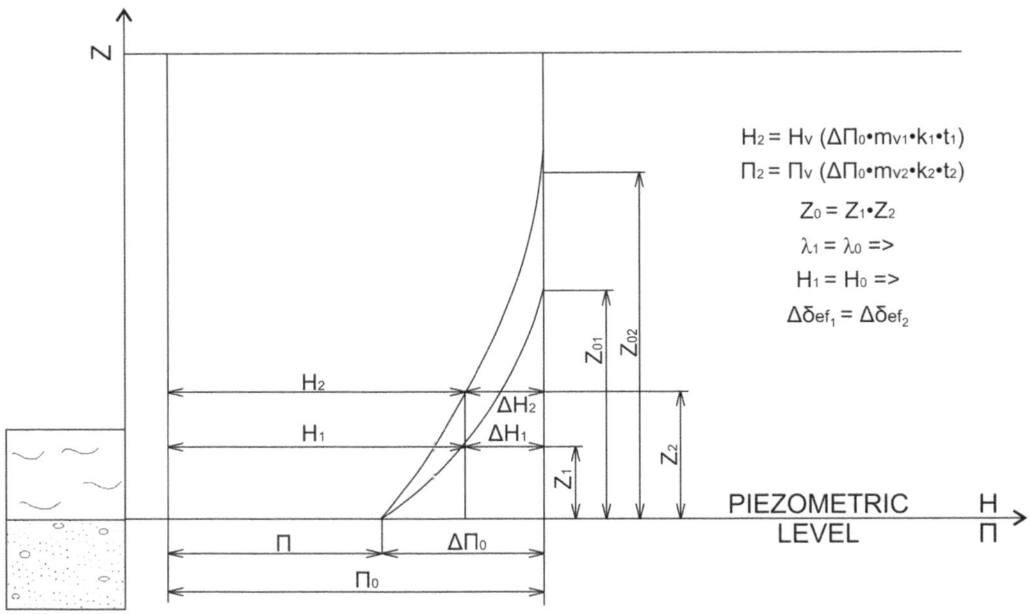


**Figure 10.12** Changes in piezometric levels in the specified layer (clay) for the current fall of the piezometer level in the underlying aquifer (Dimkić, 1986).

We will write expressions for the clay subsidence for two different variants of the characteristics of low-permeability layer: $m_{v1}$, $K_1$ and $m_{v2}$, $K_2$ (according to Figure 10.12), the subsidence expressions are:

$$S_{u1} = \int_0^{z_{01}} m_{v1} \Delta H_1 \cdot \gamma_v \cdot dz_1 \tag{10.32}$$

$$S_{u2} = \int_0^{z_{02}} m_{v2} \Delta H_2 \cdot \gamma_v \cdot dz_2 \tag{10.33}$$

where $Z_{01}$ and $Z_{02}$ are values on the $Z$ axis chosen so that the corresponding $\Delta H(z)$ value is smaller than the pre-set very small value (Figure 10.12).

Geometrically, integrals in Equations (10.32) and (10.33) correspond to the areas limited by:

F1) with line $H_1$ $(z)$, Π- axis and line $H = \Pi_0$,

F2) with line $H_2$ $(z)$, Π- axis and line $H = \Pi_0$.

These surfaces ($F_1$ and $F_2$) multiplied respectively by $m_{v1}$ and $m_{v2}$ give the subsidence values $S_{u1}$ and $S_{u2}$.

The following can be written:

$$S_{u1} = F_1 \cdot m_{v1} \tag{10.34}$$

$$S_{u2} = F_2 \cdot m_{v2} \tag{10.35}$$

The value of subsidence $S_{u1}$ can be given bearing Equation (10.34) in mind, as:

$$S_{u1} = \gamma_v m_{v1} \int_{\Pi_0-\Delta\Pi_0}^{\Pi_0} Z_1(H)dH \tag{10.36}$$

While, bearing in mind Equation (10.35), the value of subsidence $S_{u2}$ can be shown as:

$$S_{u2} = \gamma_v m_{v2} \int_{\Pi_0-\Delta\Pi_0}^{\Pi_0} Z_2(H)dH \tag{10.37}$$

When observing any two ordinates $Z_1$ and $Z_2$ that correspond to one value of the piezometric level $H$, and based on Equation (10.28), it follows that, since the value of $H$ is unique, the following must be true:

$$\lambda_1(Z_1, m_{v1}, K_{c1}, t_1) = \lambda_2(Z_2, m_{v2}, K_{c2}, t_2) \tag{10.38}$$

i.e.

$$\lambda_1 = \frac{Z_1\sqrt{\gamma_v \cdot m_{v1}}}{2\sqrt{K_{c1} \cdot t_1}} = \lambda = \frac{Z_2\sqrt{\gamma_v \cdot m_{v2}}}{2\sqrt{K_{c2} \cdot t_2}} \tag{10.39}$$

so we have that for the unique value of $H$, the following must be satisfied:

$$Z_1 = Z_2 \sqrt{\frac{\frac{K_1}{K_2} \cdot \frac{t_1}{t_2}}{\frac{m_{v1}}{m_{v2}}}} \tag{10.40}$$

If we include Equation (10.40) in Equation (10.36), we get:

$$S_{n1} = m_{v1} \sqrt{\frac{\frac{K_1}{K_2} \cdot \frac{t_1}{t_2}}{\frac{m_{v1}}{m_{v2}}}} \int_{\Pi_0-\Delta\Pi_0}^{\Pi_0} Z_2(H)dH \tag{10.41}$$

By dividing Equation (10.41) by Equation (10.37), we obtain:

$$\frac{S_{n1}}{S_{n2}} = \sqrt{\frac{K_1}{K_2}\frac{m_{v1}}{m_{v2}}\frac{t_1}{t_2}} \tag{10.42}$$

i.e. we can say:

The magnitudes of the two subsidences of the semi-boundary layer of coherent soil in the overlying (underlying) layer of the aquifer, under conditions of an instantaneous drop of the piezometric level in the aquifer, is proportional to the square root of the ratio of the filtration coefficient product, the volume compressibility coefficient and the time elapsed since the piezometric level dropped in the aquifer.

Since the size of the subsidence is a linear function of $\Delta\Pi$ (Equations (10.32) and (10.29)), and taking into account that the subsidence is a function of parameters $S_u = S$ ($\Delta\Pi$, $K$, $m_v$, $t$), as well as Equation (10.41), it follows that:

$$S_u = const\sqrt{K_c \cdot m_v \cdot t} \cdot \Delta\Pi \tag{10.43}$$

The absolute value of the subsidence is proportional to the square root of the product of the filtration coefficient, the volume compressibility coefficient and the consolidation time. The term applies to an unlimited layer of 'clay' i.e. for the period until the effect of (pressure change) reaches the farther boundary of that layer.

### 10.3.2 Dependence of water inflow due to the consolidation of the overlying clay layer of an aquifer and the time under boundary conditions on a sudden step drop in the piezometric level in an aquifer

The function is derived on the assumption that $K = const.$ and $m_v = const.$, for the overlying layer in the case of an instantaneous drop in the piezometric pressure in the aquifer.

As already said in the previous subsection (10.2.4), the following pertains to the amount of subsidence for the conditions that have been mentioned:

$$S_u = const\sqrt{K \cdot m_v \cdot t} \tag{10.44}$$

Considering that subsidence occurs only at the expense of pore reduction (subsection 10.2.2) we have:

$$q_2 = \frac{ds}{dt} = const\sqrt{\frac{K \cdot m_v}{t}} \tag{10.45}$$

This term is valid until the disturbance of pressure in the clay, due to the pressure drop in the aquifer, reaches the upper limit of the water-saturated clay layer.

This means that in the case of an instantaneous decrease in the piezometric level in the aquifer, the leakage of water from the overlying bound (clay) layer in the aquifer is proportional to the square root of the value of the product of the volume compressibility coefficient and the filtration coefficient of these layers, and inversely proportional to the square root of the time elapsed since the level decreased in the aquifer.

### 10.3.3 Total storage of the hydrogeological complex ($S_k$) in the event of an instant fall in the piezometric level in an aquifer

The equation for groundwater flow under pressure is:

$$\nabla^2 \Pi = \frac{S_a}{T} \cdot \frac{\partial \Pi}{\partial t} - \frac{q}{T} \tag{10.46}$$

The important factors in this equation are the storage of an aquifer ($S_a$) and the inflow of water from overlying (underlying) layers ($q$).

If we define $S_k$ porosity as the amount of water released due to the compressible properties of the aquifer and the overlying layer at a unit decrease in the piezometric pressure in the aquifer, this corresponds to the value of total storage $S_k$, which can be applied in the calculations of Equation (10.48):

$$\nabla^2 \Pi = \frac{S_a}{T} \cdot \frac{\partial \Pi}{\partial t} - \frac{q_1}{T} \tag{10.47}$$

Then:

$$S_k = S_a + S_q \tag{10.48}$$

where: $S_a$ – is the total storage due to the elastic properties of the skeleton of aquifers and water, $S_q$ – the amount of water per unit decrease in the piezometric pressure in the aquifer obtained by reducing the pore volume of the clay (subsidence), $q_1$ – inflow from the overlying clay into the aquifer by leakage due to different conditions at the boundaries of the clay layer.

If we assume that the settling in 'clay' is exclusively due to pore compaction and the draining of water, we can write:

$$S_c = \frac{S}{\Delta\Pi} = \frac{A\sqrt{K \cdot m_v \cdot t \cdot \Delta\Pi}}{\Delta\Pi} = A\sqrt{K \cdot m_v \cdot t} \tag{10.49}$$

where $A$ is a constant. The total storage is:

$$S_k = S_a + A\sqrt{K \cdot m_v \cdot t} \tag{10.50}$$

If we know the value of $S_0$ at time $t = t_0$ (for example, we obtain it by calibration of the hydrodynamic model), and assuming ($K = const.$, $m_v = const.$), we can easily derive that:

$$S_k(t) = S_a + (S_{k0} - S_a)\sqrt{\frac{t}{t_0}} \tag{10.51}$$

which is an expression used to describe the sum of the storage coefficient ($S_k$) over time. Again, this term is valid if the following assumptions are fulfilled:

- the drop in piezometric pressure in the aquifer is instantaneous, and its value remains unchanged;
- the disturbance of the piezometric pressures in the clay has not yet reached the limit of the water-saturated layer;
- the clay parameters are constant in time and space ($K_c$ and $m_v$);
- all the initial and boundary conditions given in Equations (10.27) and (10.28) apply;
- the specific gravity of the water is constant;
- the consolidation of the overlying layer is solely due to the change in the soil pore volume.

## 10.4 GROUNDWATER FLOW CALCULATION METHODS FOR PRESSURIZED QUASI-3D AND 2D APPROACHES

With respect to 3D approach calculations, only one overview of the existing 3D approaches for calculating groundwater flow in hydrogeological complexes is given here. Given also is an overview of some of the existing tools for quasi-3D groundwater flow calculations.

The basic presentation in this subsection refers to the quasi-3D and 2D approaches with analyses of the role of the basic parameters of hydrogeological layers. Considered primarily are the differently defined aquifer and aquitard storage coefficients as well as the total storage coefficient of the hydrogeological complex.

### 10.4.1 An overview of 3D and quasi-3D approaches to groundwater flow calculations

The three-dimensional Richards equation is used to mathematically describe groundwater flow (Weill *et al.*, 2009). Software that uses various numerical schemes is available for solving the three-dimensional Richards equation using the finite difference method MODFLOW (Thoms *et al.*, 2006), the finite element method FEFLOW (Diersch, 2013), HydroGeoSphere (Brunner & Simmons, 2012) or the finite volume method

WODA (Vidovic *et al.*, 2014). Since the Richards equation is highly nonlinear, in order to accurately describe the subsurface media, it is necessary to use very small discretization in numerical schemes. When creating regional models, this would result in a computationally demanding process which is usually unusable in practice (Szymkiewicz *et al.*, 2018). Instead, simplified subsurface modelling known as quasi-3D modelling is used (Đukić *et al.*, 1986a, 1986b; Mao *et al.*, 2019; Pikul *et al.*, 1974; Twarakavi *et al.*, 2008), so that the flow in the low-permeability zone is simply viewed as vertical flow, while in the water-bearing strata, it is viewed as two-dimensional horizontal (or three-dimensional).

In the low-permeability zone, the flow is assumed to be one-dimensional (vertical) and described by the one-dimensional Richards equation:

$$C(h)\frac{\partial h}{\partial t} = \frac{\partial}{\partial z}\left[K(h)\frac{\partial h}{\partial z}\right] \tag{10.52}$$

where: $h$ – hydraulic potential (piezometric level), $C(h)$ – specific moisture capacity, $K(h)$ – hydraulic conductivity in an unsaturated environment.

Hydraulic conductivity $K(h)$ is composed of two components: one depends solely on the hydrogeological properties of the formation ($K$) and the other, relative permeability $K_r(h)$, depends on the hydraulic potential $h$. Van Genuchten's model (van Genuchten, 1980), Brooks-Corey's model (Brooks & Corey, 1964) or Gardner's model (Gardner, 1958) is used to determine the relative permeability. A detailed description of how hydraulic conductivity and specific moisture capacity are estimated is available in the doctoral thesis 'Computation of subsurface flow using finite volume method' (Dotlić, 2015).

For flow in a water saturated aquifer, it is assumed that the flow is horizontal and it is described using the conservation equation (known in literature as the Boussinesq equation):

$$\frac{\partial}{\partial x}\left(K_x\frac{\partial h}{\partial x}\right) + \frac{\partial}{\partial y}\left(K_y\frac{\partial h}{\partial y}\right) = \mu_{el}\frac{\partial h}{\partial t} \tag{10.53}$$

where: $K_x$, $K_y$ – the hydraulic conductivity of the saturated aquifer in the $x$, $y$ direction, respectively. The specific yield of the aquifer under pressure is marked as el.

The coupling between the two equations above can be made in three basic ways (Furman, 2008):

- Connection over boundary conditions – calculation points along the interface calculated in the low-permeability zone represent the boundary condition for the equation in the aquifer. While this coupling is easy to implement, the main objection is that such coupling can lead to a large error in terms of balance;
- Full coupling – the calculation points along the interface between two zones should satisfy both equations. The disadvantage of this type of coupling is the cost of the calculation resources;
- Iterative coupling – is similar to coupling across boundary conditions in addition to the fact changing boundary conditions are solved alternately until convergence is achieved.

There are various software solutions that implement quasi-3D modelling by coupling across boundary conditions e.g. LINKFLOW, SVAT-MODFLOW, UZF1-MODFLOW, HYDROS-MODFLOW. There are also software solutions that implement quasi-3D modelling using iterative UBMOD-MODFLOW coupling (Mao *et al.*, 2019). All these codes use MODFLOW for solving the conservation equation in the water-bearing layer, while software solutions that simulate the one-dimensional Richards equation can be used for the unsaturated layer.

The above equations and the methods by which they are solved apply to a broad spectrum of hydrodynamic requirements relating to hydrogeologic formation. However, when groundwater flow needs to be estimated for spacious, deformative hydrogeologic complexes, this might no longer be the case. The compressibility of the hydrogeologic complex then needs to be considered as well.

### 10.4.2 Original groundwater flow calculation method in a compressible hydrogeological complex

#### *10.4.2.1 An innovative approach*

M. Dimkic and his associates (Dimkić & Đukić, 1982, 1983, 1984a, 1984b, 1984c, 1985; Đukić *et al.*, 1986a, 1986b, 1989) set the theoretical approach and developed an appropriate calculation methodology based on equations describing the groundwater flow process through a deformable, compressible, porous medium. By introducing new elements to the classical approach, an area of research has been opened, which significantly influences the general concept of the methodology used for studying groundwater flow under pressure.

As already emphasized, the classical methodology used in the calculations of groundwater flow under pressure generally ignores the water that has entered into an aquifer due to the consolidation of hydrogeological layers. Through selected works, the author and associates have provided possible mathematical approaches that take into account the part of water that is released in this way from the compressible complex and additionally flows into the aquifer. In doing so, the authors critically point to the problems and difficulties of introducing these additional but significant feeds of the aquifer into the calculation.

Three calculation methods are presented here as the author's approach to studying and solving practical problems of groundwater quantification under flow conditions under pressure. At the heart of these methods is the valid hydraulic theory of filtration, the Dupuit hypothesis and Darcy's law on filtration, with appropriate modifications and additions. In each of these methods, the compressibility of the hydrogeological environment is taken into account.

Method I involves the quasi-3D approach: the flow in a deformable aquitard layer is treated as a linear, one-dimensional process in the Terzaghi equation of one-dimensional consolidation. The flow in the aquifer (s) is treated as two-dimensional.

Method II is an approach that, in a very simple way, introduces into the water balance of the aquifer the water introduced due to the compressibility of the overlying, underlying and intercalated aquitards. Recommendations are given as to how to translate this compressibility into a storage coefficient, as a coefficient valid for the entire time period up to the time interval $t_n$ for which the calculation is made.

Method III, also in a simplified manner, considers the water that is released due to the compression of the aquitard. The calculation is 2D and the water is included in the aquifer balance through the coefficients of instant storage ($S_{\Delta t}$).

All three methods take the water that flows into the aquifer due to the consolidation of the overlying and underlying layers into consideration with each method corresponding to one of the stated values of the storage coefficients adapted and defined by the author to the appropriate method.

Methods I and II should find significant application in practice, while Method III is predominantly theoretical.

#### *10.4.2.2 Methods I, II and III*

*Method I*

This method involves modelling, i.e. a mathematical mechanism for describing the flow of water both in the artesian springs as well as in the adjacent, overlying and underlying less permeable layers. This involves the setting and solving of a system of differential flow equations (a), (b) and (c) for all three layers.

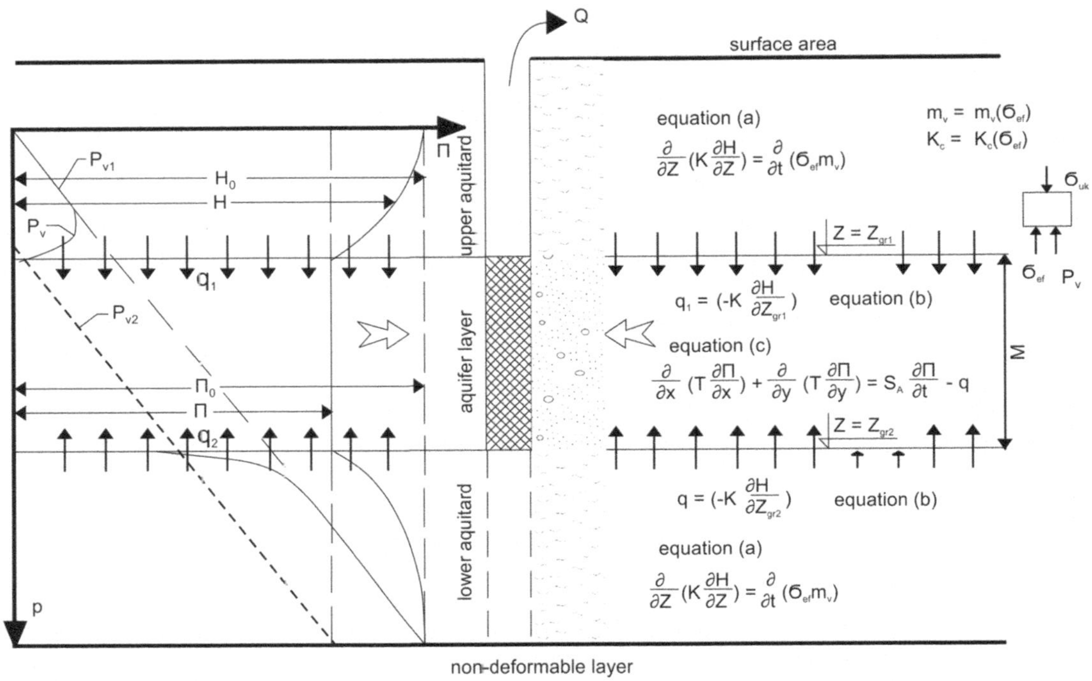


$\Pi$ - Piezometric water level in the aquifer
$\Pi_0$ - Initial piezometric level
$P_{v1}$ - Initial pore pressure
$P_{v2}$ - Final pore pressure
H - Piezometric level in lower permeability layer
M - Thickness of the aquifer
$m_v$ - Module of volumetric compressibility of the aquitard
$K_c$ - Filtration coefficient in the aquitard
T - Aquifer transmissivity T = k • M
K - Filtration coefficient in the aquifer

**Figure 10.13** Schematic presentation and equations of water flow in the subartesian aquifer for a step change of piezometric levels (Dimkić & Đukić, 1989).

To schematically present the underlying-aquifer-overlying complex, the most important assumptions are as follows: The flow in the aquifer is exclusively horizontal while the flow in the overlying and underlying layer is exclusively vertical. The draining of water from the overlying and underlying layers due to their consolidation occurs solely at the expense of a reduction in the porous cavities. The flow throughout the hydrogeological complex occurs in accordance with Darcy's law of filtration.

The basic postulates defined in this way, along with some additional ones mentioned earlier (related primarily to the rheological model and specific yield), lead to the system of equations (a, b, c) – Figure 10.13.

In these equations, the hydrodynamic and compressible characteristics of the layers are represented as parameters:

- For the aquifer (filtration coefficient $K_a$ and aquifer storage coefficient $S_a$);
- For the low permeability layer (filtration coefficient $K_c$ and volume compressibility modulus $m_v$).

The piezometric levels in the aquifer and in the low-permeability layers appear as unknown values for which these equations are solved.

These equations are solved simultaneously. In doing so, equation (a) is solved for the aquifer along coordinates $x$, $y$, $t$ taking into account the vertical inflow of water into the aquifer from the overlying and underlying layers (equation (b)). The flow of water in the aquifer is regarded as horizontal and two-dimensional and the calculation area is appropriately divided into a series of fields, which are assigned values of parameters relevant for defining the aquifer layer.

The solution of equation (a) ($x_1$, $y_1$, $t$) represents the boundary condition for solving equation (c). This equation is solved along the vertical $z_{i,j}$ at an angle of 90 degrees in relation to fields $x_i$, $y_j$, giving, as a result, the arrangement of piezometric levels along the given vertical for the overlying and underlying layer. From here, via equation (b) we get the value of the vertical inflow of water to the aquifer.

This value is included in the further solving of equation (a) for the next iteration (moment) and thus the process of the simulated solution of the system of equations (a), (b) and (c) continues.

By solving the (a, b, c) system of equations, we obtain the hydrodynamic values characteristic of groundwater flow (piezometer levels for the layer package, groundwater flow rates, flows, etc.). Based on these values it is also possible to calculate the values for the subsidence of layers at any time.

The determination of environmental parameters is achieved by the appropriate laboratory testing of undisturbed and disturbed soil samples obtained by exploratory drilling, and by pump testing aquifers as well as by calibrating groundwater flow.

Of particular interest is the parameter that appears in equation (c) and represents the aquifer storage capacity ($S_a$) (aquifer storage, in our country: effective porosity, specific yield). As stated earlier in the book, this is independent of time. Parameters that depend only on the effective stress in the soil are $S_u$, $m_v$ ($\sigma_{EF}$), $K_c$ ($\sigma_{EF}$), while parameter $K$ is constant.

A complex hydrogeological system for multiple aquifers, underlaying, overlaying or intercalated aquitards, can be shown schematically as in Figure 10.14.

Where:

Equation a – equation for the vertical flow of water in the overlying and underlying low permeability layers,

Equation b – vertical flow of water into the aquifer,

Equation c – water flow in the aquifer.

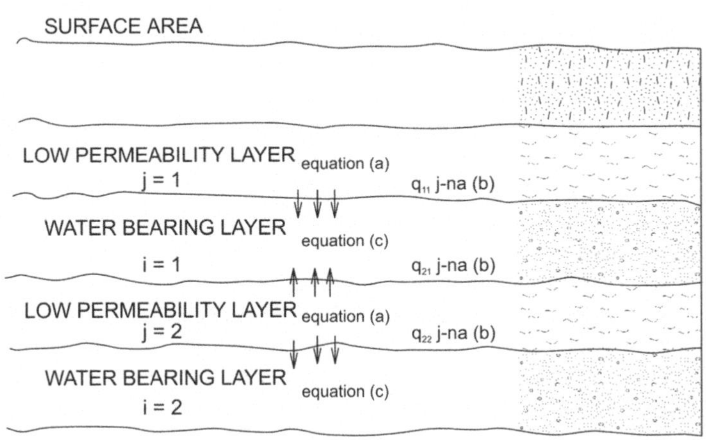


**Figure 10.14** Schematic representation of the flow equation setting in a complex, compressible hydrogeological system by Method I (quasi 3D) (Dimkić, 1986).

*The advantages and disadvantages of Method I*

This method enables the simultaneous calculation of the consolidation of compressible aquitards, of the water that subsequently flows into the aquifer (aquifers of the hydrogeological complex) and of the hydrodynamics in the aquifer (aquifers of the system).

The method requires a slightly simpler numerical mechanism than that using the 3D approach. The method also requires the application of additional exploratory work in relation to the 2D approach.

*Method II*

This considerably simplified hydrodynamic model when compared to Method I involves modelling the groundwater flow only for the aquifer. The method is based on using only one equation for the horizontal flow of water in the aquifer.

$$\frac{\partial}{\partial x}\left(T\frac{\partial n}{\partial x}\right)+\frac{\partial}{\partial y}\left(T\frac{\partial n}{\partial y}\right)=S_k\frac{\partial \pi}{\partial t}-q \tag{10.54}$$

$q$ – inflow into the aquifer, without considering the consolidation of the overlying (underlying) layer, $S_k$– storage of the hydrogeological complex (or total storage). $S_k$ includes water from aquifers and from layers of the aquitard obtained by a unit reduction of the piezometric level in the aquifer (see subsection 10.2.5, definition 2, Equation (10.16)).

According to this definition, the storage is not only related to the physical characteristics of the hydrogeological complex, but also depends on other values (pumping regime, time, space…), i.e.:

$$S_k = S_k(x, y, t, \ldots) \tag{10.55}$$

One way of estimating $S_k$ is given in subsection 10.3.3, see Equations (10.50) and (10.51).

The calculation is performed as many times as necessary to obtain the distribution of piezometric levels in the aquifer for a specific number of times.

$$\pi(x, y, t_i) = \pi(S_i, t_i, \ldots); \quad i = 1, 2, \ldots n \tag{10.56}$$

This method is much more concise than Method I due to the application of a mathematical (numerical) mechanisms. This method indirectly, through the parameter of the total effective porosity (total specific yield), introduces the effect of compressibility of the overlying and underlying layer, and the calculation of water flow is performed using the classical-Boussinesq equation pertaining to water movement in an aquifer. The method assumes the use of $n$ calculations to define the number of the different $n$ states of the piezometric levels in the aquifer, for $n$ time steps.

The above approach is less realistic and precise but much simpler when it comes to implementing the calculations. The groundwater flow equation defined in this way involves solving one equation instead of three. However, introducing the inflow of water into the aquifer due to the consolidation of the overlying and underlying layers, through the value $S_k$ – storage of the hydrogeological complex – pre-supposes the influence of a large number of values (parameters, boundary conditions) on the value of this parameter.

Here $S_k$ represents a parameter that is used to introduce into the calculation not only the water released in the aquifer itself, but also the water obtained as a result of the consolidation of the overlying and underlying layers. This gives a completely different character to the parameter of the specific yield than is true for the same parameter in Method I. This requires a new formulation, i.e. definition of parameter $S_k$, which occupies the same place as parameter $S_a$ in Method I but is different. Definitions and analyses of the parameters of storage complexes ($S_k$) and storage aquifers ($S_a$) are given in this chapter, in subsection 10.2.5.

The values of $S_k$ are therefore variable in time and space. They can be estimated by calibrating the numerical hydrodynamic model, taking into account the aquitard and aquifer parameters in a manner similar to that given in subsection 10.2.5.

*Advantages and disadvantages of Method II*

This simplified method is much simpler, more concise and less deterministic than the previous method, given the applied mathematical model, i.e. numerical mechanisms.

It can also be very useful as a preliminary method, before applying Method I, and in a certain wider range of cases, as a final method that can be used to solve a problem in a satisfactory manner.

On the other hand, when one considers that its application requires certain conditions to be met, as well as that in the calculations the effects of layer compression are expressed indirectly, through the inflow into the aquifer, it is clear that this method is basically less precise than the complex Method I.

*Method III*

Method III corresponds to the parameter defined as instantaneous effective storage ($S_{\Delta t}$) (subsection 10.2.5, definition 3). This parameter corresponds to the solution of the Boussinesq equation:

$$\frac{\partial}{\partial x}\left(T\frac{\partial \pi}{\partial x}\right)+\frac{\partial}{\partial y}\left(T\frac{\partial \pi}{y}\right)=S_{\Delta t}\frac{\partial \pi}{\partial t}-q \tag{10.57}$$

The definition of the instantaneous effective storage provides the theoretical possibility for a single calculation of the flow of water in the aquifer to give us a picture of the state of the piezometric levels for $n$ different times (see subsection 10.2.5, Equation (10.21)).

The method can also be used in practice. However, there may be significant difficulties in determining successive values:

$$S_{\Delta t1} \rightarrow S_{\Delta t2} \rightarrow S_{\Delta t3} \cdots \rightarrow S_{\Delta tn}$$

For a simple insight and for the purpose of comparison, Table 10.2 gives a tabular review of the innovated methods.

## 10.5 CALCULATION OF FLOW TO A SOLITARY WELL IN A COMPRESSIBLE MEDIUM AND UNLIMITED AQUIFER

### 10.5.1 Calculation example

Factors that primarily affect the outflow of water from low-permeability clay layers of the underlaying and overlaying soil and subsoil are: change of piezometric level in the aquifer, time, filtration coefficients and volumetric compressibility of the clay. To consider the influence of the quality of the surface layer on the outflow of water due to the consolidation of clay in the aquifer, a parameter called the 'outflow factor' is suitable.

$$F=\sqrt{m_v \cdot K_c} \tag{10.58}$$

It is not possible to give a general, explicit, dependence of the outflow of water from the clay as a function of all the above factors. It is important to point out that, due to the low filtration coefficient in the clay, as well as due to the significant compressibility of the clay, the process of propagation of water pressure disturbances in the clay is very slow. Due to this, the gradients at the clay-aquifer boundary (and thus the outflow from the clay) are often many times larger than when the distribution of piezometric pressures along the vertical is considered linear:

**Table 10.2** Overview of presented methods used for calculating groundwater flow under pressure (Dimkić & Đukić, 1989)

| Calculation Methods | Equations | Number of Calculations for Defining Piezometric Levels in *n* Time Section Intervals | Application of the Definition of the Specific Yield of an Aquifer |
|---|---|---|---|
| I<br>Direct taking into account the compressible characteristics of a hydrogeological complex | (a) Horizontal flow into an aquifer<br>$\frac{\partial}{\partial x}\left(T\frac{\partial}{\partial x}\pi\right)+\frac{\partial}{\partial y}\left(T\frac{\partial}{\partial y}\pi\right) = Sa\frac{\partial \pi}{\partial t}-q$<br>(b) Inflow into an aquifer from the overlying layer<br>$q = k\frac{\partial H}{\partial Z}/Z = Z_c$<br>$q = k\frac{\partial H}{\partial Z}/Z = Z_g$<br>(c) Vertical flow of water in the overlying (underlying) layer<br>$\frac{\partial^2 H}{\partial Z^2} = \frac{m_v \gamma_v}{K_c}\frac{\partial H}{\partial t}$<br>$\frac{\partial^2 H}{\partial Z^2} = \frac{m_v \gamma_v}{K_g}\frac{\partial H}{\partial t}$ | A single calculation is sufficient to define $\pi(x, y)_k$ for $n$ time intervals $t_k$<br>$K = 1, 2, \ldots, n$ | Def. 1<br>The storage of an aquifer ($S_a$) is equal to the size of the water column which is released per unit decrease of the piezometer level due to the compressibility properties of the aquifer skeleton and of water.<br>$S_a = \gamma_v M n\left(\frac{1}{E_w}+\frac{\omega}{nW_{sk}}\right)$<br>for a decrease of the piezometric level $\omega = 1$ for an increase of the piezometric level $\omega \leq 1$ |

| | | | |
|---|---|---|---|
| II<br>The Boussinesq equation is solved and the water 'released' due to the compressibility of the hydrogeological complex is introduced through the total storage parameter of the hydrogeological complex ($S_k$) | Water flow in the aquifer<br>$\frac{\partial}{\partial x}\left(T\frac{\partial \pi}{\partial x}\right)+\frac{\partial}{\partial y}\left(T\frac{\partial \pi}{\partial y}\right)$<br>$=S_a\frac{\partial \pi}{\partial t}-q$ | ($N$) number of calculations are required to define the piezometric levels at ($n$) moments in time specific yield<br>$\pi(x, y, t_1)S_{k1}$<br>$\pi(x, y, t_2)S_{k2}$<br>.<br>.<br>.<br>.<br>$\pi(x, y, t_n)S_{kn}$ | Def. 2<br>The total storage of the hydrogeological complex ($S_c$) is equal to the quotient of the total water column that is 'released' and has flowed into the aquifer from the hydrogeological complex and to the corresponding decrease in the piezometric level<br>$S_k=\frac{h}{\Delta \pi}$<br>$S_k=S_a+S_c$ |
| III<br>The Boussinesq equation is solved and the water 'released' due to the compressibility of the hydrogeological complex is released as a parameter of the 'current specific yield' ($S_{\Delta t}$) | Water flow in an aquifer<br>$\frac{\partial}{\partial x}\left(T\frac{\partial \pi}{\partial x}\right)+\frac{\partial}{\partial y}\left(T\frac{\partial \pi}{\partial y}\right)$<br>$=S_{\Delta t}\frac{\partial \pi}{\partial t}-q$ | A set of $n$ calculations needs to be done, with the results of the previous calculations used as the 'initial state' for the calculations that follow:<br>$\pi(t_n)=\pi[\mu_1 \cdot \Delta t_1,$<br>$(\mu_1 \cdot \Delta t_2), \ldots . . \mu_n \cdot \Delta t_n)]$ | Def. 3<br>The Instantaneous effective storage for Type III calculations is the quotient of the water column that is 'released' at some time interval and of the decrease in the piezometric level during the same period. Strictly speaking:<br>$S_{\Delta t}=\lim_{\Delta t \to 0}\frac{h(\Delta t)}{\Delta \pi(\Delta t)}$<br>$S_{\Delta t}\approx\frac{h(\Delta t)}{\Delta \pi(\Delta t)}$ |

$$\frac{\partial H}{\delta Z} > \frac{\Delta H}{\Delta Z}(\mathrm{Z} = \text{clay} \; - \; \text{aquifer interface}) \tag{10.59}$$

Clay inflows, which are established during such gradients, can be very important for balancing the flow of water in the aquifer.

The results of the calculation presented here serve to highlight and partially generalize particular conclusions regarding the effect of compressibility and permeability characteristics of a layer with specific hydrodynamic parameters and processes.

Figure 10.15 represents the scheme and equations of the soil compression process due to the fall of the piezometric level in the two-layer medium.

$Q = 18\ \text{l/s}$
$M = 10\ \text{m}$
$m_v = 2 \cdot 10^{-7}\ \text{m/N}$

4 budget variants:

- $K_c = 10^{-8}\ \text{m/s}$;
- $K_c = 10^{-9}\ \text{m/s}$;
- $K_c = 10^{-10}\ \text{m/s}$;
- $K_c = 10^{-11}\ \text{m/s}$.

At the beginning of pumping, the entire amount of water pumped (100%) is compensated from the elastic reserves of the aquifer, and later it is increasingly compensated by water coming from low-permeability layers (so that the percentage of this water in the later period of pumping reaches almost 100%). It is

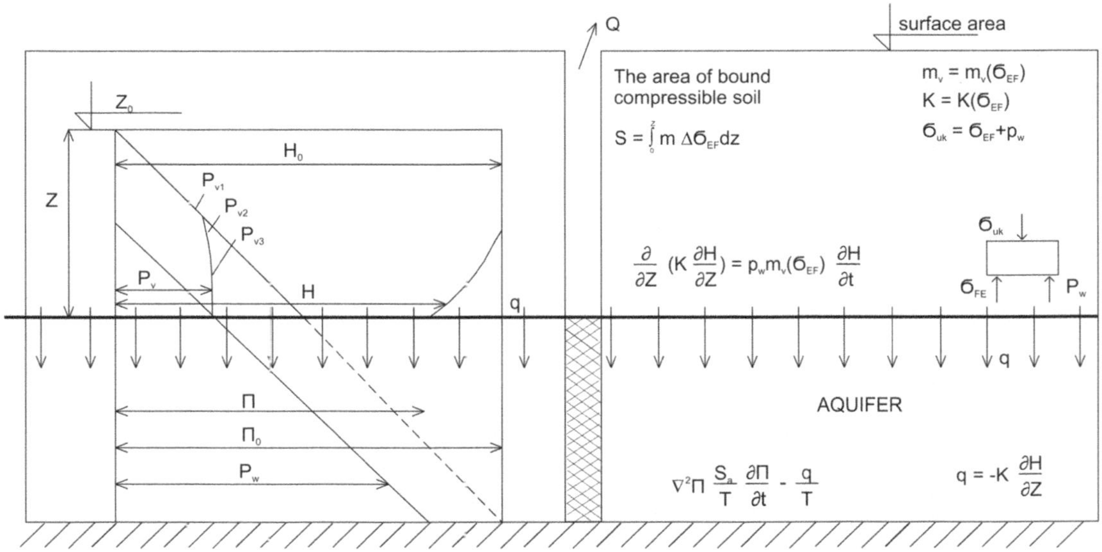


**Figure 10.15** Scheme and equation of the soil compression process due to the fall of the piezometric level in the two-layer medium (Dimkić, 1986).

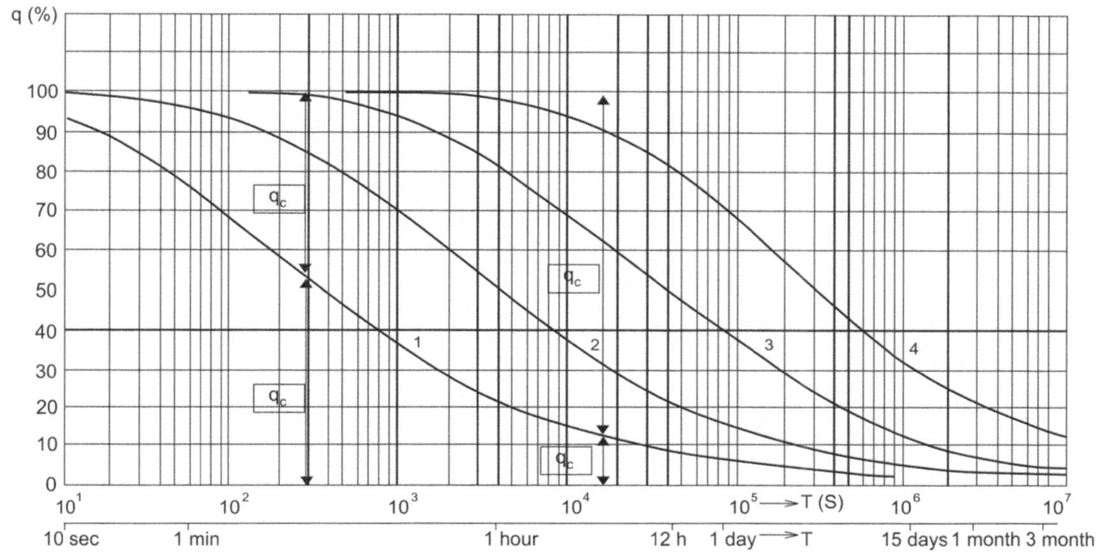


**Figure 10.16** Restoring water unlimited subartesian aquifer depletion in case of single wells. Calculation example (Dimkić, 1986).

interesting to note that, in the example shown in Figure 10.16, after a few days of pumping, most of the pumped water is compensated for by water coming from less permeable layers.

LEGEND:

$Q_{el}$ – amount of water (l/s) formed from elastic reserve released,
$Q_c$ – amount of water (l/s) coming from weakly permeable in the aquifer,
$Q_w$ – amount of water (l/s) pumped from the well.

*Parameters of low-permeability layers

1. $K_c = 1 \cdot 10^{-8}$ m/s;
2. $K_c = 1 \cdot 10^{-9}$ m/s;
3. $K_c = 1 \cdot 10^{-10}$ m/s;
4. $K_c = 1 \cdot 10^{-11}$ m/s.

$$q_{el} = \frac{Q_{el}}{Q_w} \cdot 100 \quad q_c = \frac{Q_c}{Q_w} \cdot 100 \tag{10.60}$$

*Aquifer parameters

$$K_a = 1 \cdot 10^{-4}\,\mathrm{m/s} \quad \varepsilon = 1 \cdot 10^{-4} \tag{10.61}$$

In the initial period of pumping, most of the water that reaches the well originates from the elastic properties of the aquifer. Later, the share of water coming into the aquifer significantly increases, mainly due to the consolidation of low-permeability, compressible layers. With an increasing module of bulk

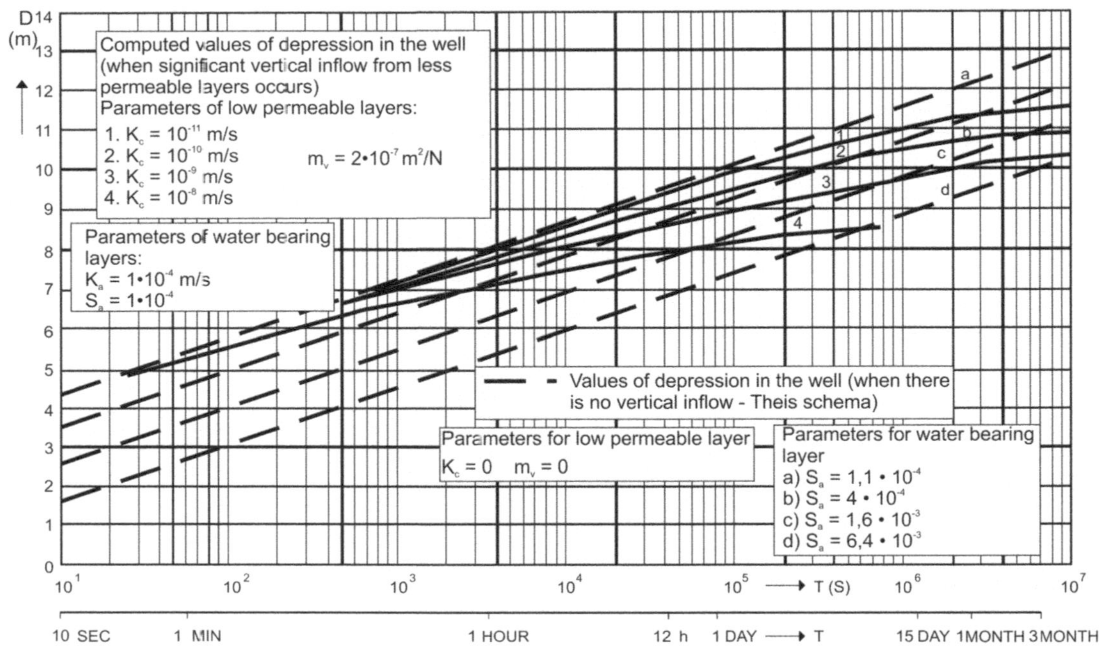


**Figure 10.17** Values of depression in a solitary well as a function of time, for different values of parameters of low-permeability layers – comparison of method I and Theis's solution (Dimkić & Đukić, 1989).

compressibility and the related coefficient of filtration of a low permeability aquitard, the portion of the water that comes from the compressible layers of the aquifer is more pronounced mainly due to the reduction of piezometric levels in the aquifer.

Conversely, by decreasing the values of the mentioned aquitard parameters, we are increasingly approaching the assumptions for which Theis's solution of the Boussinesq equation applies (see Figure 10.17), and the radius (area) of the pumping influence on the piezometric levels in the aquifer is expanding.

This has a significant impact on the resulting piezometric groundwater levels in the confined aquifer, and thus on the explanation of the results of pumping tests, source recharge areas, etc.

Similar to pumping, with the increase of the piezometric level in the aquifer the hydrodynamics of the aquifer are predominantly influenced by the elastic properties of the skeletal aquifer and water. In later periods, the processes of consolidation of compressible aquitards became increasingly important.

## 10.5.2 Relationship between parameters of compressible aquitard ($m_v$, $K_c$), storage aquifer ($S_a$) and storage of hydrogeological complex ($S_k$), based on the calculation results in 10.5.1

Calculations of piezometric levels in the well were performed by the quasi-3D method (Method 1) for four pairs of parameters ($m_v$ and $K_c$), as well as for one value of aquifer storage (see Figure 10.17).

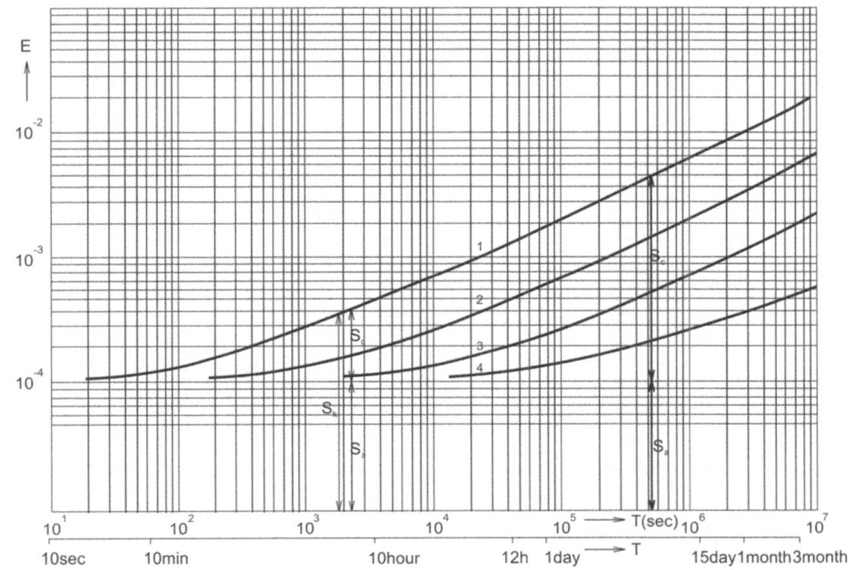


**Figure 10.18** Development of the total storage hydrogeological complex ($S_k$) (Dimkić, 1986).

Calculations by Method II (2D) were performed for four different values of $S_k$ which includes both aquifer storage ($S_a$) and aquitard storage ($S_c$):

$$S_k = S_a + S_c \tag{10.62}$$

The intersection points of the calculated levels indicate an increase in the value $S_k$. So we have a connection (Figure 10.18).

The connection thus established between the characteristic compressibility parameters and the filtration coefficient for the aquitard when using Method I (quasi-3D) and the summary storage hydrogeological complex when using Method II (2D) is valid for the flow to a solitary well. However, for hydrogeological complexes composed of aquifers of intergranular porosity, as well as bound, clayey aquitards, this relationship, between quasi-3D and 2D calculations, is relatively similar. The deviation is that in 2D calculations, summary storage is not a function of time only, but can also be a function of space:

$$S_k = S_k(t, x, y) \tag{10.63}$$

i.e.

$$S_k = S_k(t, x, y, m_v, K_c, S_a, \ldots) \tag{10.64}$$

### 10.5.3 Practical examples of changing the total storage for a hydrogeological complex ($S_k$) in practice – Case of Inđija and Ruma

An estimate of the magnitude of $S_k$ is necessary to forecast the effects of a particular exploitation of a groundwater source. The estimation is based on the extrapolation of an already determined set of values of the parameter $S_k$. Using, as an example, the results of two regional research projects (Dimkić, 1986) which considered the dynamics of groundwater under pressure (spring of Inđija and spring of Ruma), calibration of the change of $S_k$ over time was performed (examples given in the monograph 'The management of groundwater in large river basins' – Chapter 1.5 (Dimkić *et al.*, 2008)).

Both calculations were carried out for regional investigations of groundwater reserves in subartesian aquifers, hydrogeological complexes with subsoil, overflow and interrelations of bound clay aquitards.

Water pumping in both cases led to a decrease in piezometric levels and the release of water from less permeable layers, due to their consolidation. As a consequence of the above, there was an increase in the specific yield in the calculations. It has been shown that the change of this parameter must be taken into account when analysing the effects of source exploitation.

*The source of the city of Inđija*

At the site of the spring in Inđija, two subartesian releases are exploited: the upper one (approximately 50 to 60 m deep) and the lower one (approximately 100 to 190 m deep).

The upper one consists of two interconnected aquifers. The average total thickness of both layers is about 50 m. Piezometric levels of this aquifer are generally below the surface – the aquifer is subartesian.

The lower source consists of a single aquifer. It is also of a subartesian character.

Figure 10.19 shows that the initial value of the specific yield is significantly higher for the first source than for the second one. Also, its change is larger for the first than for the second. The change in specific yield has a very similar shape for the first and second source.

Considering the lithological composition and structure of the materials from which the hydrogeological complex was built in the area of Inđija, such a relationship and a change in the specific yield for the upper and lower sources were expected. Generally speaking, the deeper layers are less compressible and have

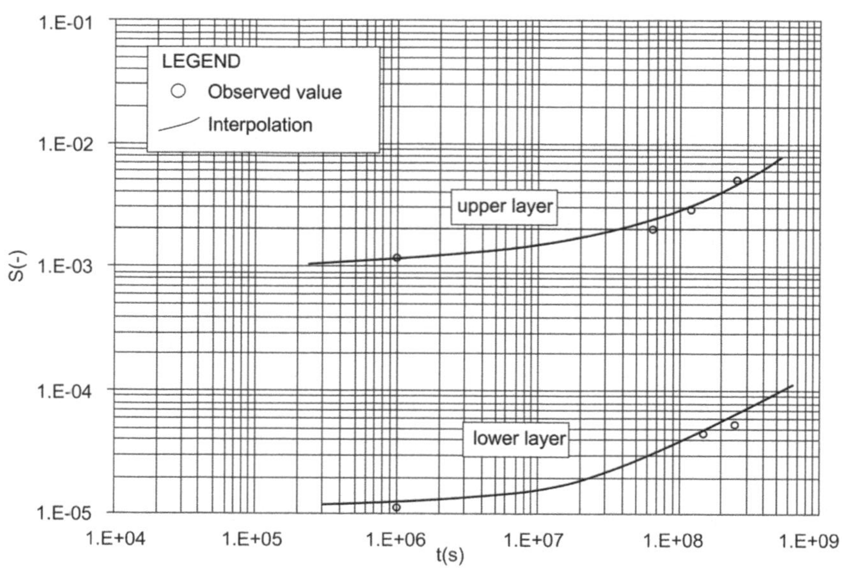


**Figure 10.19** Change of summary storage ($S_k$) for aquifers of the spring of the town of Inđija (Dimkić & Đukić, 1989).

lower filtration characteristics than the shallower layers, because they are more consolidated over time due to the pressure of the rocks above them. This is also confirmed by the composition of the low-permeability layers: the layers of the upper source are made of lesoid and clayey materials, while the layers of the deeper source are mostly made of compacted low-permeability marly clays.

The stated ratio of specific releases for the upper and lower aquifer is one of the main causes of a significantly higher water decrease in the lower aquifer (14.5 m) than in the upper (8.5 m). Calculations have shown that additional replenishment due to the consolidation of the hydrogeological complex is 40–50% of pumped water. Here, taking into account the consolidation of the hydrogeological complex is very important from the point of view of successfully forecasting the effects of future groundwater exploitation.

*The source of the city of Ruma*

For the example of springs in Ruma, the quantification of representative parameters of the natural environment was carried out and the estimation calculation was performed, based on the results of systematic monitoring of long-term groundwater exploitation.

According to the hydraulic mechanism, this source is under pressure, and of subartesian character.

Considering the existing knowledge based on the research carried out for the wider area of the studied source, in the schematization of natural conditions, it is assumed that this source is homogeneous and has isotropic filtration characteristics.

As a result of hydrodynamic calculations of exploitation according to the observed hydrogram, changes in the values of specific yield for the source over time for several calculation schemes were obtained (Figure 10.20).

In both cases (Inđija and Ruma), a constant part of this parameter can be observed, which refers to the elastic properties of the water and subsurface. It represents the initial value, at the beginning of the

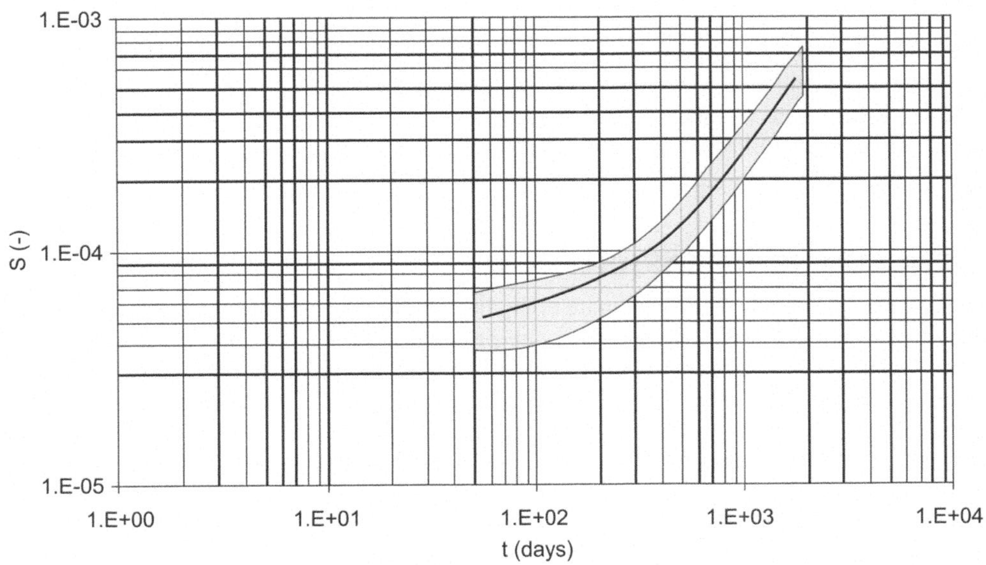


**Figure 10.20** Change in total storage ($S_k$) issued over time, due to groundwater exploitation at the Ruma spring (Dimkić & Đukić, 1989).

exploitation period. The variable part of the parameter, which arises due to the consolidation of the compressible layers released, is not constant and increases over time.

This confirms the set mechanisms of calculation of groundwater flow under pressure. The obtained values of specific yield were determined in accordance with the actual compressibility of certain lithological layers. Of course, in all this, it should be borne in mind that the specific yield can be changed during the calibration of the mathematical model and due to the imperfection of the adopted physical scheme.

### 10.5.4 Relationship between summary storage parameters (application of Method II) and current effective storage

The results of the flow calculation to a solitary well were presented by using Method II (2D) and Method III (also 2D).

As stated in the subsection above, Method II calculates one state of piezometric levels for a given time interval. For each subsequent moment, if calculated consistently, the calculation should be repeated:

$$\begin{aligned} &\text{for} \quad t = t_0 \quad \Pi(x, y) = \Pi(S_{k0}, t_0, x, y) \\ &\text{for} \quad t = t_1 \quad \Pi(x, y) = \Pi(S_{k1}, t_1, x, y) \\ &\text{for} \quad t = t_n \quad \Pi(x, y) = \Pi(S_{kn}, t_n, x, y) \end{aligned} \tag{10.65}$$

In the case of water pumping and a decrease in the piezometric level in the aquifer, there is a need to increase the size of the total storage when calibrating the model:

$$S_0 < S_1 < S_2 < \ldots < S_n \tag{10.66}$$

For the case of calculation by Method III (2D), we perform the calculation with a series of successive computations:

$$t_0 \; S_{\Delta t0} \rightarrow_{\Delta t1} \; S_{\Delta t1} \rightarrow_{\Delta t2} \; S_{\Delta t2} \cdots \rightarrow_{\Delta tn} \; S_{\Delta tn} \tag{10.67}$$

So, we have one calculation in which we use $n$ values of $S_{\Delta t}$ for individual time intervals, and we get matrices $\Pi$ $(x, y, t)$ for each moment of time from $t_0$ to $t_n$.

Here, for the purpose of illustration and comparison of the ratio $S_k$ and $S_{\Delta t}$, the results of the calculation by Method I (quasi-3D), Method II $(2D)^{(+)}$ and Method III $(2D)^{(0)}$ are given (Figure 10.21).

## 10.6 CONCLUDING REMARKS

The increasing use of groundwater leads, more often than not, to the emergence of 'over-exploitation'. The phenomenon is, however, more frequent and more pronounced in deeper and confined aquifers than in shallower, alluvial aquifers. It occurs in conditions when excessive exploitation significantly exceeds the recharge of a certain aquifer (aquifer, hydrogeological complex). We associate the phenomenon of the reaction of vertical, subscale and intercalated, compressible aquitards, to the change (decrease) of piezometric levels in the aquifer. The compressibility of these layers can be many times (up to a couple of orders of magnitude) higher than the compressibility of aquifers. Hence the geo-mechanical effect of subsidence of the terrain, as well as significant hydrodynamic effects that occur due to the inflow of significant amounts of water from compressible aquitards into the aquifer.

If we consider phenomena related to large areas, the dimensions of the affected area can be much larger than the thickness of the affected hydrogeological complex (two, three and more orders of magnitude). In

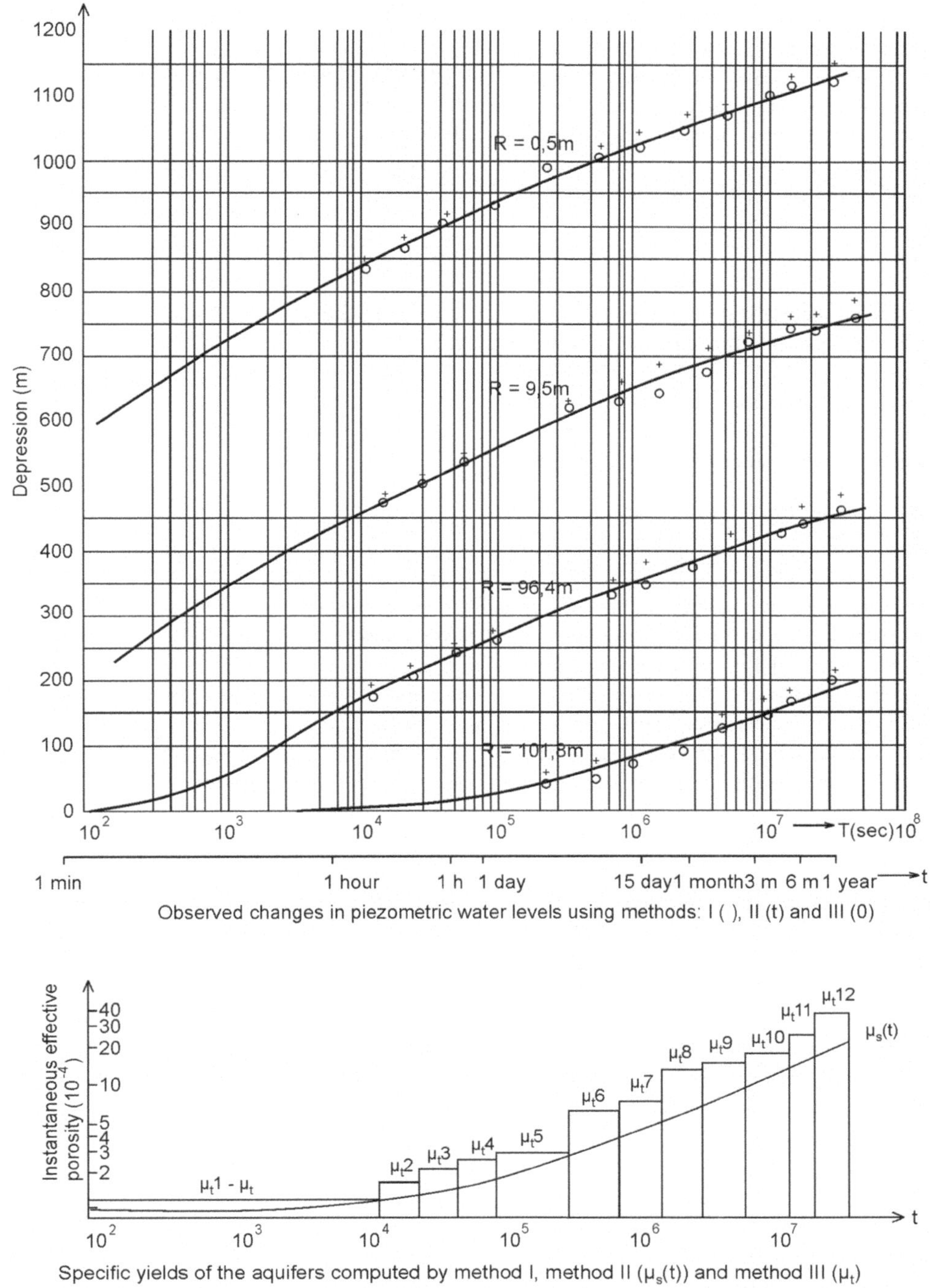


**Figure 10.21** Results of the calculation of the change in the piezometric level during the flow towards a lonely well using methods I, II and III (Dimkić & Đukić, 1989).

this case, it may not be rational to calculate the hydrodynamics and the occurrence of subsidence of the hydrogeological complex using a 3D model.

A method using a quasi-3D approach is proposed and developed here. This method is valid under certain assumptions, also presented here. Among other things, the flow and displacement of water from the aquitard is set as a one-dimensional process in the direction of the vertical axis (z). The flow of water in the aquifer (aquifers) is set as horizontal (x, y axes). As a characteristic parameter for the intensity of subsidence and outflow of water in the aquifer, we define the 'outflow factor':

$$F = \sqrt{K_c \cdot m_v} \tag{10.68}$$

For a sudden, stepwise decrease in the piezometric level in the aquifer ($\Delta\pi$), we derived, under the adopted assumptions, that the subsidence in the aquifer:

$$S_u = const \cdot F \cdot \sqrt{t} \cdot \Delta\pi \tag{10.69}$$

And the outflow of water from the aquitard:

$$q = \frac{dS_u}{dt} = const \frac{F}{\sqrt{t}} \tag{10.70}$$

The dedicated calculation shows that during exploration works (trial pumping), as well as exploitation (long-term pumping), a large part of pumped water is provided from compressible layers of aquitards at the expense of consolidation of lithological layers.

In the quasi-3D approach, aquifer storage is a realistic, unambiguous, natural parameter and depends on the elastic properties of water and soil skeleton (aquifer storage: $S_a$).

Reduction of the budget for computation requires a move towards 2D modelling. The term storage of aquitards ($S_c$) is introduced here, as well as the total storage of the hydrogeological complex ($S_k$):

$$S_k = S_c + S_a \tag{10.71}$$

The storage of the aquifer is constant. For storage of the hydrogeological complex, given the assumptions and a sudden step drop of piezometric levels in the aquifer, the following is derived:

$$S_k = S_a + const \cdot F \cdot \sqrt{t} \tag{10.72}$$

If we know the value of $S_0$ (storage of the total hydrogeological complex) at time $t = t_0$ (for example obtained by calibration of the hydrodynamic model), and assuming $K_c = const.$, $m_v = const.$, it is derived that:

$$S_k(t) = S_a + (S_{k_0} - S_a)\sqrt{\frac{t}{t_0}} \tag{10.73}$$

(see subsection 10.3.3).

Of course, nature can be quite complex, so it is up to the engineers to choose the appropriate research work and the right approach to solve the problems they face. Here we have presented some, original and innovative possibilities.

## REFERENCES

Brooks H. R. and Corey A. T. (1964). Hydraulic properties of porous media. Hydrology papers 3, Collorado state university, Fort Collins, CO, 1964. W.R. Gardner. Some steady-state solutions of the unsaturated moisture flow equation with application to evaporation from a water table. *Soil Science*, **85**(4), 228–232.

Brunner P. and Simmons C. (2012). HydroGeoSphere: a fully integrated, physically based hydrological model. *Groundwater*, **50**, 170–176.

Borchers J. W. and Carpenter M. (2014). Land Subsidence from Groundwater Use in California. Full Report of Findings/April 2014, California Water Foundation.

Chilingar G. V. and Rieke S. C. (1976). Conpressibilities of clays and some means of prodicting and preventing subsidence. Proceedings of Anaheim Symposium of Land Subsidence, Tokyo, Japan, No. 89, pp. 377–39.

Diersch H.-J. G. (2013). FEFLOW: Ffinite Element Modeling of Flow, Mass and Heat Transport in Porous and Fractured Media. Springer Science & Business Media, Berlin, Heidelberg. doi: 10.1007/978-3-642-38739-5

Dimkić M. (1986). A Contribution to the Study of Groundwater Flow in Compressible Subartesian and Artesian Aquifers (in Serbian). Master's thesis, University of Belgrade, School of Civil Engineering, Belgrade.

Dimkić M. and Đukić B. (1982). Several aspects of land subsidence due to groundwater pumping (in Serbian). 8th Hydraulic Research Convention, 02–04 December 1982, Portorož, Yugoslav Hydraulic Research Society, pp. 1–22.

Dimkić M. and Đukić B. (1983). The process and several aspects of land subsidence due to groundwater pumping (in Serbian). Communiques of Jaroslav Černi Water Institute, No. 75–76, pp. 31–46.

Dimkić M. and Đukić B. (1984a). Additional aquifer recharge and compression of hydrogeological strata due to groundwater pumping (in Serbian). Waters of Vojvodina, No. 12, Novi Sad, pp. 249–256.

Dimkić M. and Đukić B. (1984b). Some hydrodynamic aspects of land subsidence due to groundwater extraction (in Serbian). Communiques 77–78 of Jaroslav Černi Water Institute, pp. 27–48.

Dimkić M. and Đukić B. (1984c). Some hydrodynamic aspects of the ground subsidence due to the exploatation of groundwater. 27th International Geological Congress, Moscow, Vol. VII, p. 381.

Dimkić M. and Đukić B. (1985). Comments on the effective porosity parameter and methods for confined groundwater flow analysis in view of the compressibility of hydrogeologic strata (in Serbian). Waters of Vojvodina, No. 13, Novi Sad, pp. 407–424.

Dimkić M. and Đukić B. (1986). Comments on the effective porosity parameter and methods for confined groundwater flow analysis in view of the compressibility of lithologic strata (in Serbian). 9th Convention of the Yugoslav Hydraulic Research Society, 06–09 October 1986, Split, pp. 436–449.

Dimkić M. and Đukić B. (1989). Specific aspects of the modern approach to analyses of groundwater flow in artesian and subartesian aquifers (in Serbian). Hydrodynamic investigations of groundwater sources, Vol. 2, Special editions, Jaroslav Černi Water Institute, Belgrade, pp. 179–257.

Dimkić M., Kavanaugh M. and Brauch H.-J. (2008). Groundwater Management in Large River Basins. IWA Publishing, London, UK, ISBN 1843391902.

Dotlić M. (2015). Computation of subsurface flow using finite volume method. PhD thesis, Mathematical Faculty, Belgrade.

Đukić B., Dimkić M. and Radovanović M. (1986a). Determination of hydrodynamic and compressibility characteristics of the hydrogeologic complex in the extended area of Inđija (in Serbian). Waters of Vojvodina, No. 14, Novi Sad, pp. 113–121.

Đukić B., Dimkić M. and Radovanović M. (1986b). Application of a method for determining hydrodynamic and compressibility characteristics of a hydrogeological complex: case study (in Serbian). 9th Convention of the Yugoslav Hydraulic Research Society, 6–9 October 1986, Split, pp. 450–462.

Đukić B., Dimkić M. and Radovanović M. (1989). Prognosis of hydrodynamic effects of groundwater extraction and estimation of the capacity of M., a subartesian groundwater source in the city of Inđija (in Serbian). Hydrodynamic investigations of groundwater sources - typical examples, Vol. 2, Special editions, Jaroslav Černi Water Institute, Belgrade, pp. 331–349.

Furman A. (2008). Modelling coupled surface-subsurface flow processes: a review. *Vadose Zone Journal*, **7**, 741–756.

Gardner W. R. (1958). Some steady-state solutions of the unsaturated moisture flow equation with application to evaporation from a water table. *Soil Science*, **85**(4), 228–232.

Mao W., Zhu Y., Dai H., Ye M., Yang J. and Wu J. (2019). A comprehensive quasi-3-D model for regional-scale unsaturated-saturated water flow. *Hydrology and Earth System Sciences*, **23**(8), 3481–3502.

Olsen H. W., Nichols R. W. and Rice T. L. (1985). Low gradient Permeability Measurements in a triaxial system. *Geotehnique*, **35**, 145–157.

Patankar S. V. (1976). Numerical Heat Transfer and Fluid Flow. McGraw-Hill Book Company, New York, USA.

Pikul M. F., Street R. L. and Remson I. (1974). A numerical model based on coupled one-dimensional Richards and Boussinesq equations. *Water Resources Research*, **10**(2), 295–302.

Roberts J. E. (1976). Sand compresion as a factor in oil field subsidence. Proceedings of Anaheim Symposium of Land Subsidence, Anaheim, CA, USA.

Nelson S. A. (2016). Subsidence: Dissolution & Human Related Causes. Tulane University, New Orleans, LA, USA.

Szymkiewicz A., Gumuła-Kawęcka A., Šimůnek J., Leterme B., Beegum S., Jaworska-Szulc B., Pruszkowska-Caceres M., Gorczewska-Langner W., Angulo-Jaramillo R. and Jacques D. (2018). Simulations of freshwater lens recharge and salt/freshwater interfaces using the HYDRUS and SWI2 packages for MODFLOW. *Journal of Hydrology and Hydromechanics*, **66**(2), 246–256.

Thoms B., Johnson R. and Healy R. (2006). User's guide to the variably saturated flow (VSF) process to MODFLOW. No. 6-A18.

Twarakavi N. K. C., Šimůnek J. and Seo S. (2008). Evaluating interactions between groundwater and vadose zone using the HYDRUS-based flow package for MODFLOW. *Vadose Zone Journal*, **7**(2), 757–768.

van Genuchten M. T. (1980). A closed form equation for predicting the hydraulic conductivity of unsaturated soils. *Soil Science Society of America Journal*, **44**(5), 892–898.

Vidović D., Dotlić M., Pokorni B., Pušić M. and Dimkić M. (2014). Simulating unsatu- rated flow with a finite volume method. *Water Research and Management*, **4**(1), 23–30.

Weill S., Mouche E. and Patin J. (2009). A generalized Richards equation for surface/subsurface flow modeling. *Journal of Hydrology*, **366**, 9–20.

# Chapter 11

# Drought and alluvial groundwater resources

*Dejan Dimkić, Milan Dimkić and Slobodan Vujasinović*

Alluvial groundwater resources are one of the main sources of water supply and it is therefore very important to study drought and its impact on alluvial groundwater resources. In many countries alluvions are the only source used for the supply of drinking water and in many regions water is abstracted from alluvions for irrigation purposes as well. The amount of water that is being used by people for water supply purposes and for irrigation is on the increase in almost all parts of the world. Population growth has also resulted in a greater demand for water while droughts tend to more and more frequently affect ever wider areas with more severe impacts. In addition, climate change has increased the water scarcity problem in arid and sub-arid areas and has resulted in some climate moderate regions becoming sub-arid (particularly those with a decreasing precipitation trend).

There are numerous examples of water scarcity when referring to human needs in the world and it is impossible to quote them all or combine them in one place. The impact of drought on alluvium is not the same everywhere but differs and depends not only on the climate but on numerous other factors (water use, geological conditions, socio-economic conditions, etc.) as well. The aim of this chapter is to present the basic concepts related to droughts and alluvium, certain important connections between the analysed parameters and to present some examples that are of interest and pertain to the topic of each of the subchapters. At the same time, although there are examples from different parts of the world, the most numerous are those that were most accessible to the authors of this chapter, and they are from Serbia and a somewhat wider area of southern Europe. A reference is given for each of the examples, and depending on the assessed needs, they have been processed by authors.

**The first subchapter** of this chapter describes the different types of drought and the possible damage which can be caused by drought.

**The second subchapter** presents the global changes (largely climatic and hydrological) on a global level and in considerable detail for Serbia. This subchapter also discusses the reliability of any forecast (and


Editor: Milan Dimkić
doi: 10.2166/9781789060904_0573

whether it is possible to increase it) and underlines the importance of the observed data. In this respect, investigations regarding the correlation between climate parameters (temperature before all) versus runoff or the human use of water are very important, and perhaps could help to forecast droughts in some regions.

**The third subchapter** discusses the general elements of the water balance in a basin, as well as the vertical water balance of the overlaying aquitard. Parameters and relationships relevant for infiltration and evapotranspiration, as well as the water capacity of the alluvial aquifer are also been analysed.

In **the last subchapter** some final remarks are given.

## 11.1 GENERAL CONCEPTS ABOUT DROUGHT – TYPES OF DROUGHTS, DAMAGE AND THE IMPORTANCE OF FIGHTING DROUGHTS

### 11.1.1 Types of droughts

Due to its complex behaviour, there is no single acceptable definition of drought (Yevjevich, 1967). One of the widely cited definitions is that drought is a natural disaster that causes enormous damage to human life and/or the environment (Waluyo *et al.*, 2013). It has different characteristics compared to other disasters.

When we think of severe droughts, we usually think of periods of extremely dry weather that persist long enough to cause problems such as crop damage and water shortages. However, the truth is that dry conditions develop for different reasons which is why there is more than one definition of drought. Drought is caused not only by lack of precipitation and high temperatures, but also as a result of overuse and overpopulation, and it can occur in virtually all climates. Consequently, there is a need to compare droughts and categorize them. Drought indicators are variables that describe the magnitude, duration, severity and spatial extent of a drought. Typical indicators are based on meteorological and hydrological variables, such as precipitation, temperature, streamflow, soil moisture, reservoir storage, groundwater levels, etc. Several indicators can be synthesized into a single indicator on a quantitative scale, often called a drought index (Steinemann *et al.*, 2005). A drought index is a prime indicator for assessing the effect of a drought and defining different drought parameters, which include intensity, duration, severity, and spatial extent.

Several types of drought have been identified. Droughts are commonly classified by type as meteorological drought, hydrological drought, agricultural drought, and socio-economic drought (World Meteorological Organization [WMO], 2006). Meteorological drought is related just to weather conditions, so it is easier to define it. However, agricultural, hydrological and socio-economic drought place greater emphasis on the human or social aspects of drought, highlighting the interaction or interplay between the natural characteristics of a meteorological drought and human activities that depend on precipitation to provide adequate water supplies to meet social and environmental demands.

**Meteorological drought** is usually defined by a precipitation deficiency threshold over a predetermined period. The widely accepted meteorological drought index is the Standardized Precipitation Index or SPI. This index has sometimes also been used for other types of drought or generally for drought. The SPI for any location is calculated based on the long-term precipitation record for a desired period. This long-term record is fitted to a probability distribution which is then transformed to a normal distribution so that the mean SPI for the location and desired period is zero.

Some authors have compared the SPI with other indexes. For example, Guttman (1998) compared the Palmer Drought Index (PDI) and SPI. The PDI is used as an indicator of drought severity, and a particular index value is often a signal to begin or discontinue elements of a drought contingency plan. The SPI was developed to quantify a precipitation deficit for different timescales. Historical time series of the PDI have been compared through spectral analysis with time series of the corresponding SPI. Results show that the spectral characteristics of the PDI vary from site to site throughout the US, while those of the SPI do not vary from site to site. They also show that the PDI has a complex structure with an exceptionally long memory, while the SPI is an easily interpreted, simple moving average process.

By identifying the SPI as an index for broad use, the WMO (2012) provided a direction for countries trying to establish early drought warnings. Guttman (1999) created a piece of software which calculates the SPI index at a given location. In 2009, the WMO recommended the SPI as the main meteorological drought index that countries should use to monitor and follow drought conditions (Hayes *et al.*, 2011). Drought events are indicated when the results of the SPI, for whichever timescale is being investigated, become continuously negative and reach a value of −1. The drought event is considered to be ongoing until the SPI reaches a value of 0. McKee *et al.* (1993) stated that drought begins at an SPI of −1 or less (Table 11.1), but there is no standard in place, as some researchers will choose a threshold that is less than 0, but not quite −1, while others will initially classify drought at values less than −1.

Owing to the flexibility of the SPI, calculations can be performed even if data is missing for a specific period for a given location. Even though it works with shorter and incomplete series, it is recommended that series comprise 50–60 or more years, and have more than 90% data (Wu *et al.*, 2005). Most users apply the SPI using monthly datasets, but daily and weekly values are used, as well. The methodology of the SPI does not change based upon whether daily, weekly, or monthly data is used.

The ability to calculate the SPI at various timescales allows for multiple applications, depending on the drought impact in question. SPI values for 3 months or less might be useful for basic drought monitoring, values for 6 months or less for monitoring agricultural impacts, and values for 12 months or longer for hydrological impacts. The SPI can also be calculated on gridded precipitation datasets, which allows for a wider range of users than those just working with station-based data. Using precipitation data only is the greatest strength of the SPI as it makes it very easy to use and calculate. The SPI is applicable in all climate regimes, and SPI values can be compared for very different climates. The fact that the SPI can be computed for short periods of record that contain missing data is also valuable for those regions that may be data-poor or lacking long-term, cohesive datasets. The ability to be calculated over multiple timescales also allows the SPI to have a wide breadth of application.

It is currently used in many countries such as: Argentina, Austria, Belize, Bosnia and Herzegovina, Brazil, Bulgaria, Canada, Chile, Croatia, Cyprus, Dominican Republic, Germany, Greece, Hong Kong, Iran, Israel, Jamaica, Jordan, Kazakhstan, Libya, Lithuania, Macedonia, New Zealand, Pakistan, Peru, Slovenia, Spain, Sri Lanka, Switzerland, Tanzania, Thailand, Trinidad and Tobago, Turkey, Ukraine, and the USA.

With precipitation as the only input, the SPI is deficient when accounting for the temperature component, which is important for the overall water balance and water use of a region. This drawback can make it more

**Table 11.1** Common SPI categories and description.

| | |
|---|---|
| 2.0+ | extremely wet |
| 1.5 to 1.99 | very wet |
| 1.0 to 1.49 | moderately wet |
| −0.99 to 0.99 | near normal |
| −1.0 to −1.49 | moderately dry |
| −1.5 to −1.99 | severely dry |
| −2 and less | extremely dry |

difficult to compare events of similar SPI values but different temperature scenarios. Also, the flexibility of this index which allows it to be calculated for short periods of record, or using data sets that contain many missing values can also lead to the misuse of the output, as the program will provide an output for whatever input is provided. The SPI assumes a prior distribution, which may not be appropriate in all environments, particularly when examining short-duration events or entry into, or exit out of, drought.

Because the SPI is normalized, wetter and drier climates can be represented in the same way; thus, wet periods can also be monitored using the SPI. However, it must be stressed that the SPI is not suitable for climate change analyses because temperature is not an input parameter.

**Agricultural drought** is usually defined by the availability of soil water to support crop and forage growth. Many indexes have been developed, like the Marton aridity index, Ped aridity index, etc. One of the often used indexes is the Hungarian Palfai Drought Index (PDI) which was developed for the numerical characterization of droughts at the beginning of the 1980s. PAI (Palfai's Aridity Index) (Palfai, 1990) is a relative indicator which characterizes the strength of the drought for an agricultural year with one numerical value, and is a good indicator of potential crop failure. It expresses the evaporation (temperature) and precipitation relations (the latter with the time-varying water demand of plants) and considers the groundwater level state. The calculation method of PAI and the associated parameters have been in use since 1988. However, the availability of data for index calculation is often a problem. It requires daily temperature and precipitation values, as well as frequent groundwater level data.

The DMCSEE (Drought Management Centre for South East Europe) project (2009–2012) made it possible to develop an agricultural drought index which is based on easily available meteorological data (only monthly mean air temperature and the sum of precipitation). The calculation is relatively simple and expresses well the impact of meteorological and hydrological conditions that influence droughts and agricultural crops. They called it the Palfai Drought Index (PaDI or PDI) – it is a variant of Palfai's Aridity Index (PAI), and the correlation between them is strong. The comparative results for two cities in Hungary are shown in Figure 11.1.

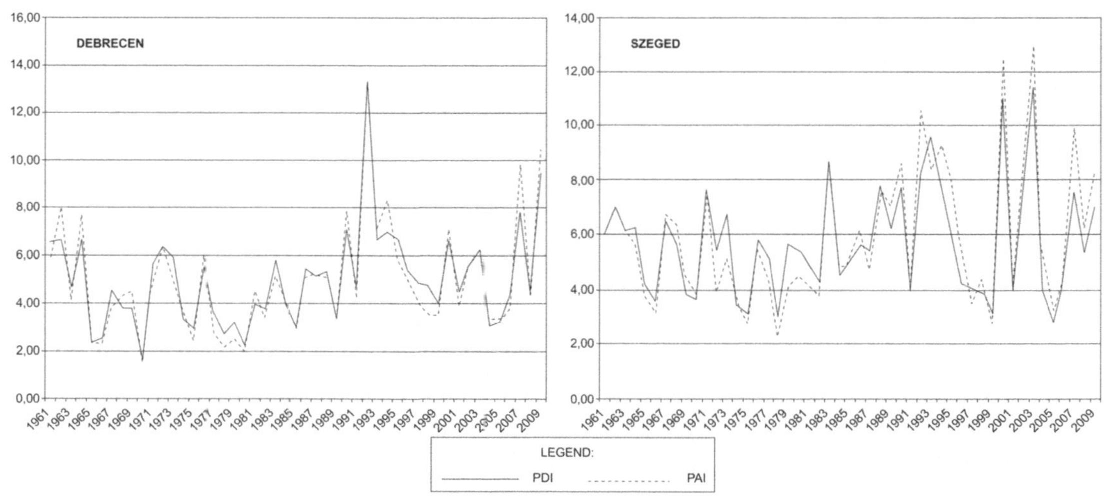


**Figure 11.1** Time series of PAI and PaDI for the 1961–2009 period at the Debrecen and Szeged stations (Hungary) (Deliverable of the International project DMCSEE (2009–2012)).

The calculation procedure is as follows:

$$PaDI = PaDI_0 \times k_1 \times k_2 \times k_3 \tag{11.1}$$

where:

$$PaDI_0 = \frac{\frac{\left[\sum_{i=Apr}^{Avg} T_i\right]}{5*100}}{c + \sum_{i=Oct}^{Sep} (P_i * W_i)} \tag{11.2}$$

where:

$PaDI_0$ – base-value of the drought index (°C/100 mm),
$T_i$ – monthly mean $T$ (Temperature) from April to August (°C),
$P_i$ – monthly sum of $P$ (Precipitation) from October to September (mm),
$w_i$ – weighting factor (0.1 in October; 0.4 in November and December; 0.5 from January to April; 0.8 in May, 1.2 in June; 1.6 in July; 0.9 in August and 0.1 in September),
$c$ – constant value (10 mm).

$$k_1 = \frac{\frac{(T_{Jun} + T_{Jul} + T_{Aug})}{3}}{\frac{(\bar{T}_{Jun} + \bar{T}_{Jul} + \bar{T}_{Aug})}{3}} \tag{11.3}$$

where:

$k_1$ – $T$ correction factor,
$T_{Jun,Jul,Aug}$ – annual mean $T$ for June–August (°C),
$\bar{T}_{Jun,Jul,Aug}$ – multi-annual mean $T$ for June–August (for period 1971–2000) (°C).

$$k_2 = \sqrt[4]{\frac{2*\bar{P}_{summer}^{min}}{MIN(P_{Jun}, P_{Jul}, P_{Aug}) + \bar{P}_{summer}^{min}}} \tag{11.4}$$

where:

$k_2$ – $P$ correction factor,
$\bar{P}_{summer}^{min}$ – the lowest value from multi-annual $P$ sum of three summer months (June, July, August) (mm),
$MIN\ (P_{Jun}, P_{Jul}, P_{Aug})$ – the lowest value from the annual $P$ sum of three summer months (June, July, August) (mm).

$$k_3 = \sqrt[n]{\frac{\bar{P}}{\bar{P}_{36\,m}}} \tag{11.5}$$

where:

$k_3$ – correction factor, which characterizes the precipitation circumstances of the previous period,
$\bar{P}$ – average multi-annual precipitation sum for the period October-September (mm),
$\bar{P}_{36\,m}$ – average precipitation sum for October-September for the previous 3 years (mm),
$n$ – reduction factor, the value is 3.0 on the plain area, on hilly or higher altitudes it is 5.0.

The calculated PaDI results can be categorized as PAI, but considering the climatic circumstances of other countries, the lower and upper limits are extended (Table 11.2).

**Table 11.2** Classification of PaDI and PAI indices.

| PAI (°C/100 mm) | Classification | PaDI (°C/100 mm) | Classification |
|---|---|---|---|
| | | <4 | droughtless year |
| | | 4–6 | mild drought |
| 6–8 | moderate drought | 6–8 | moderate drought |
| 8–10 | medium drought | 8–10 | medium strength drought |
| 10–12 | heavy drought | 10–15 | serious drought |
| | | 15–30 | very serious drought |
| 12< | extremely heavy drought | 30< | extreme drought |

It has been determined that the values of PAI and PaDI are generally the same, the drought classification for the same year is identical, so PaDI substitutes PAI very well in South East Europe. The 10% probability of occurrence of the PaDI index in SEE is shown in Figure 11.2.

This index can be a practical tool to demonstrate the spatial and annual agricultural drought relation changes. Investigations conducted for some other parts of the world indicate that this index (PaDI) could be applied and could also substitute some other agricultural drought indexes which require more, sometimes not available, parameters (for example the self-calibrating Palmer Drought Severity Index (scPDSI) which is based on the soil moisture content). The method using the PaDI index can likely gain a wider currency.

**Hydrological drought** is defined as a period of abnormally dry weather, sufficiently long to significantly reduce water availability in water bodies. The traditional definition of hydrological drought is a lack of water in rivers, lakes, reservoirs, and groundwater due to the decrease in precipitation.

However, in the majority of river basins, water is utilized for public water supply, irrigation, and other purposes. Together with a decrease in rainfall, water utilization could also contribute to hydrological drought. A special case of hydrological drought is water resources drought which gives more attention to

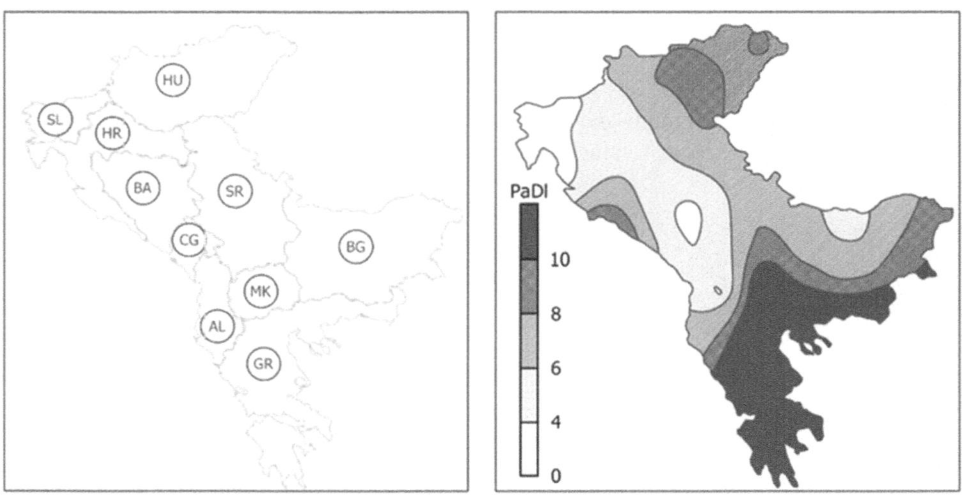


**Figure 11.2** South-East European countries and a 10% probability of occurrence of Palfai DI (PaDI index) in SEE (deliverable of the International project DMCSEE (2009–2012)).

the anthropogenic aspect of drought. It is recommended that the term 'water resources drought' can be applied to hydrological drought in a developed river basin (river basin with a high human impact on water use). Similarly, Arnell *et al.* (2001) and the Intergovernmental Panel on Climate Change (IPCC, 2007) differentiate between hydrological drought and water resources drought: A hydrological drought occurs when river or groundwater levels are low, and a water resources drought occurs when a hydrological drought impacts water use. A water resources drought is defined by Wasserhaushalt (2012) as an occasional condition of the reduction of water availability causing severe water supply deficits during a prolonged period and over a large region. Water resources droughts not only depend on the climatic and hydrological inputs but also on the characteristics of the water resource system and how droughts are managed.

Christierson *et al.* (2009) define hydrological drought as changes in the catchment water balance (precipitation, evaporation and storage) leading to a deficit of runoff, recharge or low groundwater levels over a specific period. On the other hand, a water resources drought is defined as a shortage of water available to meet 'normal' demands (for water supply, industry, or the environment) due to a combination of hydrological drought and socio-economic factors affecting water resources systems. Generally, a water resources drought is more difficult to define.

A schematic approach of links between meteorological, hydrological and water resources droughts is shown in Figure 11.3.

The water exploitation index (WEI) is a water resources drought index often used in Europe. It is defined as a quotient between the water demand and water resource availability in the analysed region (city, municipality, several municipalities, river basin, etc.). Regarding the time scale, it can be calculated on an annual level, but it is particularly used for the water scarce critical part of the year (generally summer and early autumn).

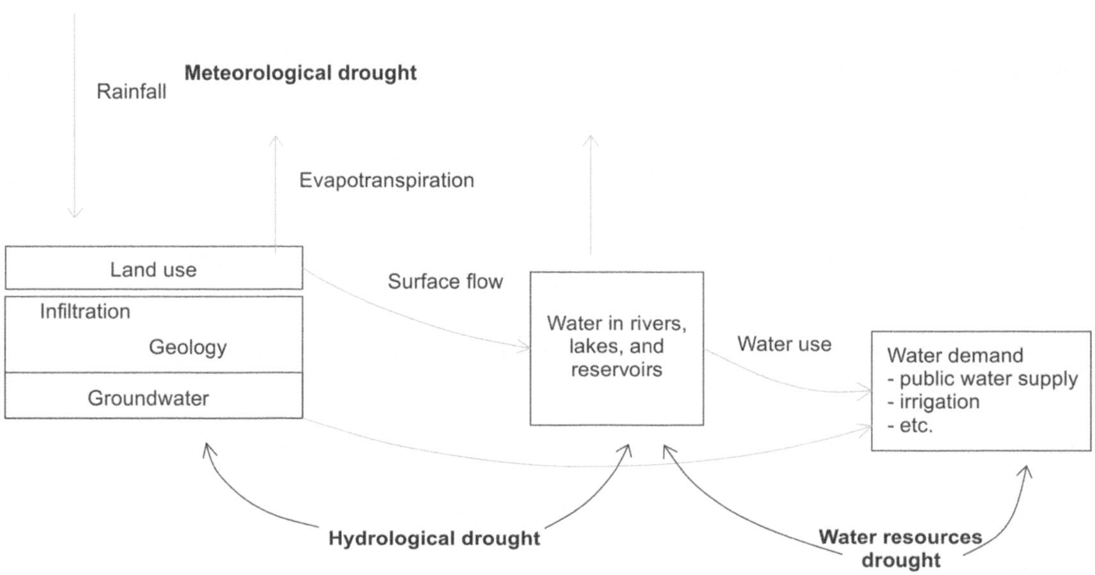


**Figure 11.3** Links between the meteorological, hydrological and water resources droughts-scheme (Waluyo *et al.*, 2013).

Popular hydrological drought indexes in the USA are the Palmer Hydrological Drought Index (PHDI) (Palmer, 1965), used for hydrological drought (not for water resources drought) and the Surface Water Supply Index (SWSI) (Nolan *et al.*, 1991), used for both the hydrological and water resources drought.

**The Socio-economic drought** differs from other types of drought because it reflects the relationship between supply and demand for some commodity or economic good (such as water, livestock forage, or hydroelectric power) dependent on precipitation (United Nations – International Strategy for Disaster Reduction [UN-ISDR], 2009). Wilhite (2005) defines the socio-economic aspect of drought as a deficiency of precipitation from the expected or 'normal' that, extended over a season or longer, is insufficient to meet the demands of human activities and the environment. Drought by itself is not a disaster. Whether it becomes a disaster depends on its impact on local people and the environment. To him, the key to understanding drought is to understand both its natural and social dimensions.

It is not easy to distinguish between a socio-economic drought and a water resources drought. The Environment Agency (2012) of the UK uses the term social drought instead of water resources drought, when an environmental drought (e.g. reduced rainfall) reduces the potential availability of essential water supplies.

### 11.1.2 Damage caused by droughts and the importance of fighting droughts

Damage, as the impact of drought, could be classified as 'direct' or 'indirect'. For example, the dying of farmers' crops would be a 'direct' impact of drought. Dealers, who lose money if they are economically connected with farmers in different ways, would be an 'indirect' impact of drought. Water in rivers may become particularly polluted during times of drought due to a lack of precipitation that would dilute industrial and agricultural chemicals – which could also be classified as an 'indirect' impact.

It is probably better to classify damage in a different way, namely as a type of impact due to drought. Many different drought impacts are often grouped as 'Economic', 'Environmental', and 'Social' impacts (Sandra Collin, FVA Germany).

**Economic** impacts are those impacts of drought that cost people (or businesses) money. For example, farmers can lose money if a drought destroys their crops. If the supply of water to farmers is too low, costs increase for all the many direct and indirect customers (farmers and ranchers, water management companies, power supply companies, marine and other companies and organizations, and individuals).

Drought also affects the **Environment** in many ways. Plants and animals depend on water, just like people. When a drought occurs, their food supply can shrink, and their habitat can be damaged. Sometimes the damage is only temporary and their habitat and food supply return to normal when the drought is over. However, sometimes the impact of a drought on the environment can last a long time, maybe forever. Examples of environmental impacts include: losses or destruction of fish and wildlife habitat, lack of food and drinking water for wild animals, increase in the disease of wild animals due to the reduced amount of food and water supplies, migration of wildlife, increased stress on endangered species or even extinction, lower water levels in reservoirs, lakes, and ponds, loss of wetlands, more wildfires, wind and water erosion of soils and poor soil quality.

**The Social** impacts of drought refer to the ways that drought affects people's health and safety. Social impacts include public safety, health, conflicts between people when there is not enough water to go around, and changes in lifestyle. Examples of social impacts include: anxiety or depression resulting from economic losses caused by drought, health problems related to low water flows and poor water quality, health problems related to dust, the loss of human life, the threat to public safety from an

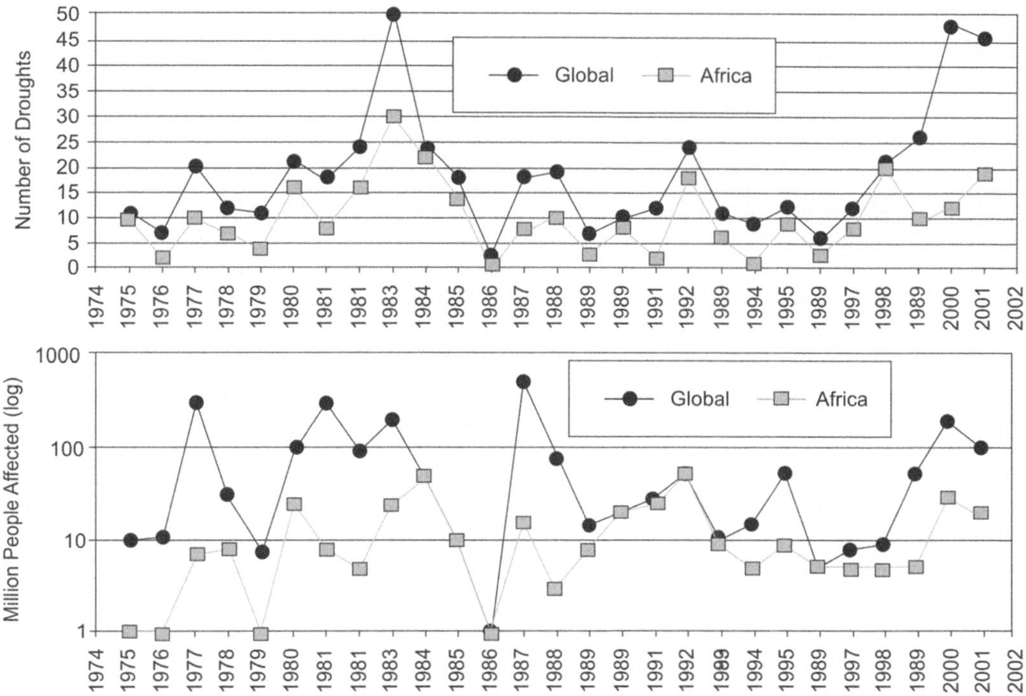


**Figure 11.4** Number of drought events and population affected by drought globally and in Africa between 1975–2002. People affected are defined as population groups requiring immediate assistance to mitigate drought impacts (Dinar *et al.*, 2008).

increased number of forest and range fires, reduced incomes, people having to move from their farms into cities, or from one city to another, and fewer recreational activities.

Droughts have been very present in recent decades, but droughts and their many repercussions, such as extreme famine, have been known to humankind since antiquity, with starvation and malnutrition being some of the tragic outcomes. Of all the continents, Africa (Dinar *et al.*, 2008) is the most vulnerable (Figure 11.4).

Some parts of the world, including sub-Saharan Africa, Spain, Australia, the USA, China and some other regions, have been more frequently affected due to different reasons such as severe climate conditions, population density, high water demand, low economic power, etc. It is possible that many severe droughts in other parts of the world have not even been registered in literature. Several of the best known droughts in history ((List 25 by Teodoras II) https://list25.com/25-most-severe-droughts-ever-recorded/) with severe impacts on different aspects of human life and environment, are given here:

***1° sub-Saharan Africa***

The Sahel droughts were a series of some of the most extreme droughts in recorded history, dating back to at least the 17th century and affecting the Sahel region, a climate zone sandwiched between the African savannah to the south and the Sahara Desert to the north, across West and Central Africa. The famine that followed the severe droughts killed thousands of people and affected the lives of millions for many decades.

Just as the region is experiencing a devastating famine in the 21st century, the Horn of Africa experienced a deadly drought and famine in 1888 that was mainly caused by an extreme lack of rainfall. Over one million reportedly died during that time in Ethiopia, Sudan, and Somalia.

The worst famine to hit Ethiopia in modern history occurred in 1983–1985 and was caused by an extreme drought in the region. It resulted in more than 400 000 deaths.

Another extreme drought in the Sahel, Africa's 'champion' region, induced famine there and in many parts of the neighbouring Senegal River Area from February to August 2010. It was one of many famines that hit the region in recent times due to a drought.

Between July 2011 and mid-2012, a severe drought affected the whole of East Africa. The drought caused a severe food crisis across Somalia, Djibouti, Ethiopia, and Kenya which threatened the lives of more than ten million people.

Since 2012 nearly twenty million people have been facing hunger across eight countries in West Africa including the Sahel region. A combination of failed crops, insect plague, conflict, and extreme drought collectively caused the ensuing famine that is considered as one of the worst on the planet in recent years.

***2° Spain***

The 2014 Spanish Drought. Recently, many parts of Spain have suffered their most intense dry spell in more than a century and a half, with Valencia and Alicante among the worst-affected regions. According to the country's meteorological agency, in the 150 years prior, they had never witnessed such a long and intense drought, which has many experts worrying about the country's water supplies for the near future.

Spatial and temporal drought patterns in the Iberian Peninsula between 1910 and 2000 have been analysed using the standardized precipitation index (Serrano, 2006). Aside from these general patterns, the analyses carried out here show that the temporal evolution of droughts within the Iberian Peninsula can be very diverse, and a drought episode can be restricted to distinct areas without the whole of the Iberian Peninsula being affected. Humid or normal conditions were frequently observed in one region, whilst another suffered drought over the same period. Droughts are sometimes local phenomena and, apart from the most intense episodes, are usually regional. Moreover, even the most widespread drought episodes did not affect the whole of the Iberian Peninsula.

***3° Australia***

The 1829 Major Drought in Western Australia. This drought was so extreme that it nearly destroyed all the agriculture in Western Australia forcing settlers to traverse long distances for water and pastures for their flocks.

The 2000 drought in Australia, also known as the Millennium drought, is said by some to be the worst recorded since settlement. The drought began in 1995 and continued Australia wide until late 2009, with the final areas affected by drought ceasing to be considered as such in early May 2012. When the official end of the drought was declared in 2012, the federal government had provided $ 4.5 billion in drought assistance.

***4° USA***

The Dust Bowl, also known as the Dirty Thirties, was a period of severe dust storms that greatly damaged the ecology and agriculture of the US and Canadian prairies in the 1930s. The drought came in three waves, 1934, 1936, and 1939–40, but some areas of the high plains experienced drought conditions for as many as eight years. The Dust Bowl affected one hundred million acres centred around the panhandles of Texas and Oklahoma and touched the adjacent sections of New Mexico, Colorado, and Kansas, forcing thousands of families to abandon their farms.

The agriculture industry in Texas was hit hard by the drought of 2010–11. According to the media, from November 2010 to August 1, 2011, Texas suffered an estimated $ 5.2 billion in crop and livestock losses, surpassing the previous record loss of $ 4.1 billion in 2006. Due to record heat and drought, violent wildfires destroyed many homes in the area as well.

The 2002 North American Drought. The drought of 2002 was moderately widespread and quite severe in some areas, specifically in the western United States, the Midwest, and the Mountain States. Denver was forced to impose water restrictions for the first time in more than 20 years and the wildfire season was one of the most violent in recent history.

***5° China***

The Chinese Drought of 1941. This is considered by many the worst disaster completely attributed to drought and the lack of rainfall without any other factor such as conflicts, war, or other natural disasters occurring at the same time. The Chinese Drought of 1941 prevented millions from growing and consuming crops. As a result, an estimated three million people died.

The Great Chinese Famine that lasted nearly three years, from 1958 to 1961, occurred for a number of reasons including extreme droughts. It is estimated that over fifteen million people died during these years, while other accounts insist that nearly forty-three million lost their lives in what is considered the worst famine in recorded history.

The 2010 China drought was a series of severe spring droughts that affected a huge swath of the country as well as parts of Southeast Asia, including Vietnam and Thailand. The drought has been referred to as one of the worst droughts in Southwest Asia.

***6° Registered severe droughts in some other parts of the world***

After three consecutive crop failures and continuous droughts that took place in India, the Deccan Famine of 1630–32 was one of the worst in the country. More than two million died during this time.

The Vietnamese drought of 1944 caused the winter-spring harvest to decrease by 20%. This coupled with pests, war and a flood during the harvest season led to what is known as the Vietnamese famine of 1945.

In 2002 there was a famine in Malawi that took about three thousand lives. One of the major factors of this extreme famine was a drought that occurred in the country during the same period.

The 2015 Brazilian drought is an ongoing drought affecting the southeast of Brazil including the metropolitan areas of São Paulo and Rio de Janeiro. It has been described as the worst drought in 80 years. The city of São Paulo appears to be affected the most and by the beginning of February many of its residents were subjected to sporadic water cut offs.

The above examples show that this type of natural disaster very likely has the highest negative impact on human activities and the environment overall, compared to other disasters (except maybe hurricanes). Damage caused by droughts is often underestimated, due to the difficulty in quantifying indirect damage. It is therefore essential to fight them by implementing prevention and adaptation measures.

## 11.2 THE IMPACT OF CLIMATE AND OTHER CHANGES ON DROUGHT

### 11.2.1 Introduction

Many scientific papers, projects and gatherings of experts talk about climate and other changes and their impact on water resources and the occurrence of droughts. While the work from the end of the 20th and the first decade of the 21st century mainly focused on climate change and its impact on water resources, in the last 10 years, more and more people are witnessing and are talking about Global Change. For example, the 5th World Water Council in 2009 looked at global changes defined as a set of pressures on natural resources and ecosystems, and among others included: population growth, migration, urbanization, climate change, changes in land use, water consumption and economic changes. It goes without saying that all these factors are not of the same importance for different regions.

When referring to the types of water resources, the changes most detected were changes in river flow or changes in the flow at karst springs. Most papers considered changes at the mid-year level, including forecasts of some maximum deviations, while intra-year changes were somewhat less represented, mainly at the seasonal (season) level. Monthly changes, in relation to annual and seasonal ones, were analysed much less, but their significance, especially when it comes to water resources and water use, is very significant. The question of the reliability of a forecast has generally not been considered, i.e. has only been considered in a small number of papers, and this issue is very important when it comes to water management planning for the near future (next 20–30 years).

The number of papers analysing other changes (factors), the impact of which is in many cases more significant than climate change, although not negligible, is significantly smaller. As the focus is on water resources, it is much more important to consider the overall impact of all the factors and their changes, than individual ones. The focus of discussion is therefore the impact of global changes (climate and other changes) on water resources.

One of the possible and important indicators (criteria) of the occurrence of a drought is the river flow. Changes in river flows affect the capacities of the associated alluvial springs directly and/or indirectly. Global (GCMs) and regional climate-hydrological models (RCMs) give us a forecast of future flows in an area. Depending on the selected scenarios, applied models and other adopted assumptions, the obtained results can, quite often, for one and the same river differ significantly.

When there is a non-random change in flow in the rivers of an area over a longer period of time (say 70 years or more), then there is a high probability that a similar trend will continue, at least in the near future (several decades). Due to the high degree of uncertainty that accompanies forecasts for the future, the subject of analysis in this subchapter is, unless otherwise indicated, only the near future. We define it as a period of the next 20–30 years.

The observed climatic and hydrological data are crucial for analysing the occurrence of droughts, both in the domain of climatic-hydrological models and regression analyses, and for the establishment of certain correlations between different parameters. The relevance of the periods selected for analysis is very significant. Their detailed analysis can provide new important data related to climatic and hydrological parameters and their most probable values of average change in the near future. Careful analyses can also provide significant correlations between climatic and hydrological parameters. This is shown in the example of Serbia.

## 11.2.2 Climate and some drought parameters trends on the global level

This book does not aim at analysing the issue of climate change in detail as a lot of information on that subject can be found in literature and on the web, but some basic information about the main observed climate changes and trends, important for the occurrence of droughts, should be present here.

Climate change has a significant impact on the frequency of droughts and on their increased intensity. While some regions are likely to get wetter as the world warms, other regions that are already on the dry side are likely to get drier (Climate Hot Map – web, n.d.). Global warming affects evapotranspiration – the movement of water into the atmosphere from land and water surfaces and plants due to evaporation and transpiration – which is expected to lead to:

- **Increased drought in dry areas**. In drier regions, evapotranspiration may produce periods of drought – defined as below-normal levels of rivers, lakes, and groundwater, and the lack of sufficient soil moisture in agricultural areas. Precipitation has been declining in the tropics and subtropics since 1970. Southern Africa, the Sahel region of Africa, southern Asia, the Mediterranean, and the US Southwest, for example, are getting drier. Even areas that remain relatively wet can experience long, dry conditions between extreme precipitation events.

- **Expansion of dry areas**. Scientists expect the amount of land affected by drought to grow by mid-century – and water resources in the affected areas to decline as much as 30%. As jet streams continue to shift to higher latitudes, and storm patterns shift along with them, semi-arid and desert areas are expected to expand.

The Intergovernmental Panel on Climate Change (IPCC) in its AR5 Synthesis Report has shown that each of the last three decades has been successively warmer at the Earth's surface than any preceding decade since 1850. The period from 1983 to 2012 was very likely the warmest 30-year period of the last 800 years in the Northern Hemisphere, where such assessment is possible (high confidence) and likely the warmest 30-year period of the last 1400 years (medium confidence). The globally averaged combined land and ocean surface temperature data as calculated by a linear trend show a warming of 0.85 [0.65 to 1.06]°C during the period 1880 to 2012, for which multiple independently produced datasets exist (Figure 11.5).

The warming of the climate is unequivocal, and since the 1950s, many of the observed changes have been unprecedented for decades to millennia. The atmosphere and oceans have warmed, the amount of snow and ice has diminished and the sea level has risen. Confidence in precipitation change averaged over continental areas since 1901 is low prior to 1951 and medium afterwards (Figure 11.6).

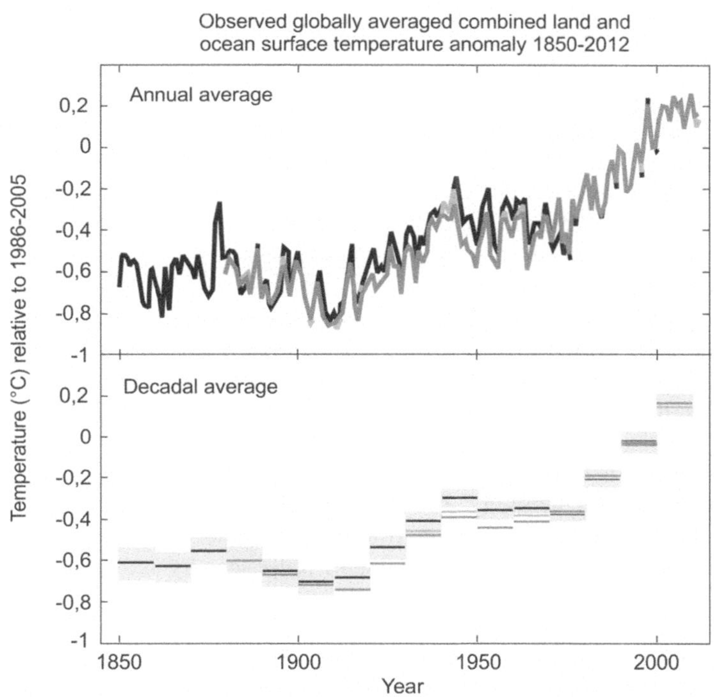


**Figure 11.5** Annually and globally averaged combined land and ocean surface temperature anomalies, with decadal averages, relative to the average over the period 1986 to 2005. Colours indicate different data sets (IPCC, 2014).

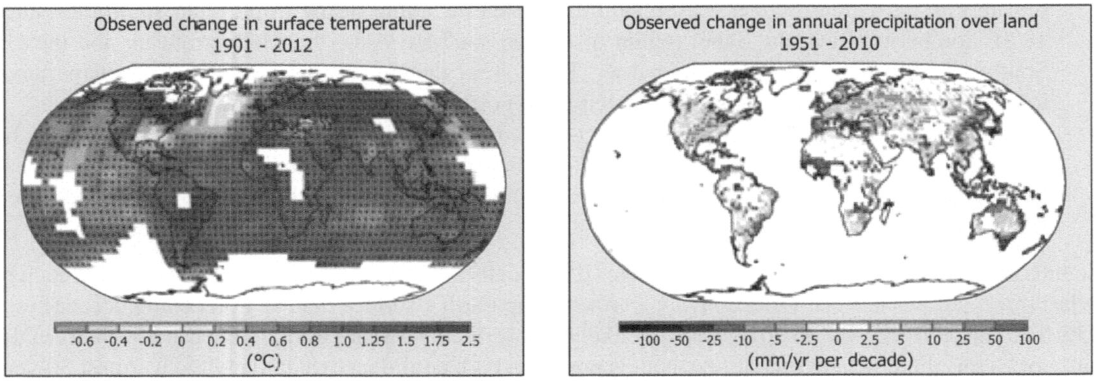


**Figure 11.6** Map of the observed surface temperature change (1901–2012) and map of the observed precipitation change (1951–2012) (IPCC, 2014).

Impacts from recent climate-related extremes, such as heat waves, droughts, floods, cyclones, and wildfires, reveal the significant vulnerability and exposure of some ecosystems and many human systems to current climate variability (very high confidence). Direct and insured losses from weather-related disasters have increased substantially in recent decades, both globally and regionally. The increasing exposure of people and economic assets has been the major cause of long-term increases in economic losses from weather and climate-related disasters (high confidence) (IPCC, 2014).

In many regions of the world there are apparent trends in the increase or the decrease of discharge. The directions of the changes are generally in accordance with the forecasts given in IPCC reports – Figure 11.7, but the magnitudes of these changes may differ. In some regions, the direction of change is uncertain (Arnell, 2003; Ma *et al.*, 2008). However, it is quite certain that an expansion of dry areas and the increased intensity of droughts in dry areas can be expected. By 2025, it is estimated that around 5 billion people, out of a total population of around 8 billion, will be living in countries experiencing water stress (using more than 20% of their available resources) (Arnell, 1999).

The Mediterranean area and the Middle East are among the most vulnerable. In addition to the increase in temperature and, in most Mediterranean countries, the demand for water, this area is probably facing a trend of decreasing rainfall, and an increase in drought can be expected with great certainty. Contrary to the demographic trend in the European part of the Mediterranean, the Middle East has another very important pressure – population growth.

Climate change has, all over the world, also influenced other parameters associated with droughts (surface humidity, soil humidity, groundwater level, different food production, etc). Several regions/countries in the Mediterranean area with an already observed downward availability of water resources, and with the potential for a further increase of water scarcity and drought are given below:

***1° Spain*** (Barajas, 2007; Paneque, 2015)

Within the Mediterranean, designated as an area affected by the adverse impact of climate change on water resources, Spain is one of the countries having the most pronounced problem of insufficient water to meet all its needs. Illegal abstractions of water resources are also on the rise (Figure 11.8). As a consequence of the high water demand, there are significant areas where the overexploitation of groundwater resources occurs (Figure 11.9).

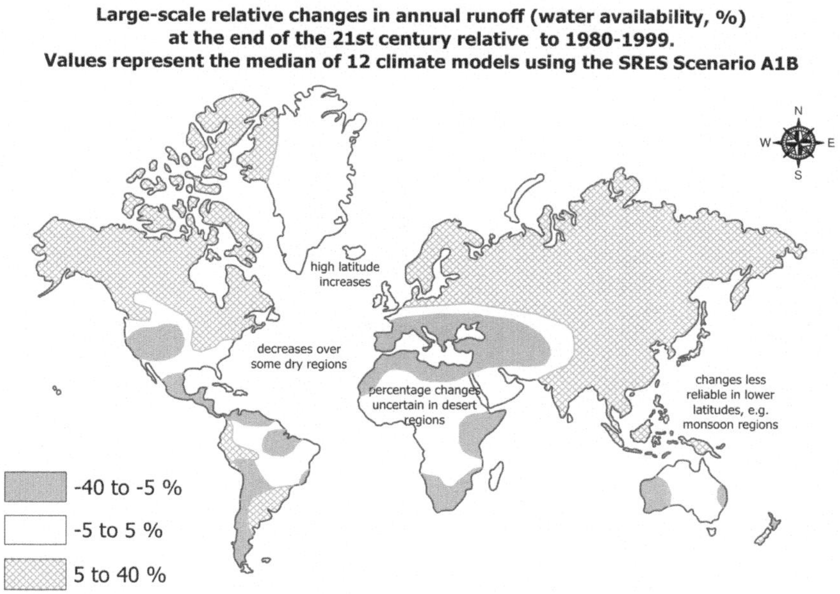


**Figure 11.7** Large-scale relative changes in annual runoff (water availability in %) at the end of the 21st century relative to 1980–1999. Values represent the median of 12 climate models using the SRES Scenario A1B (Synthesis Report of the IPCC (2007, 2014)).

In addition to meeting human needs (water supply, irrigation, industry), ecosystems are also under attack. The Donana National Park, in the south of the country, near Seville, can serve as an example of reducing the capacity of water resources. An analysis of the availability of water resources has been completed by districts (Figure 11.10). It was found that the input to the rivers was reduced by about 15% between

**Figure 11.8** Increase in illegal water extraction in Spain.

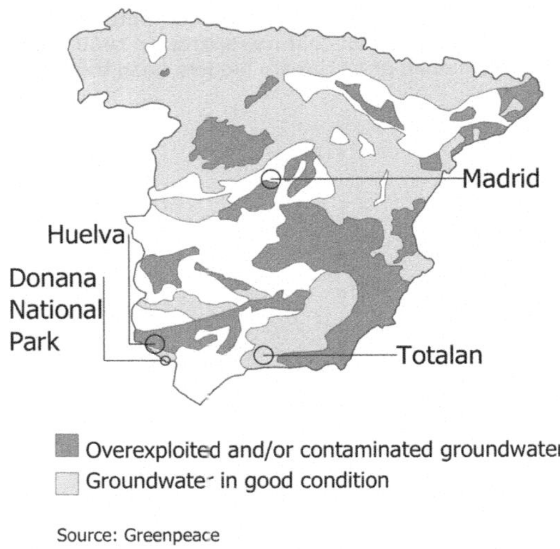


**Figure 11.9** The state of Spanish groundwater sources.

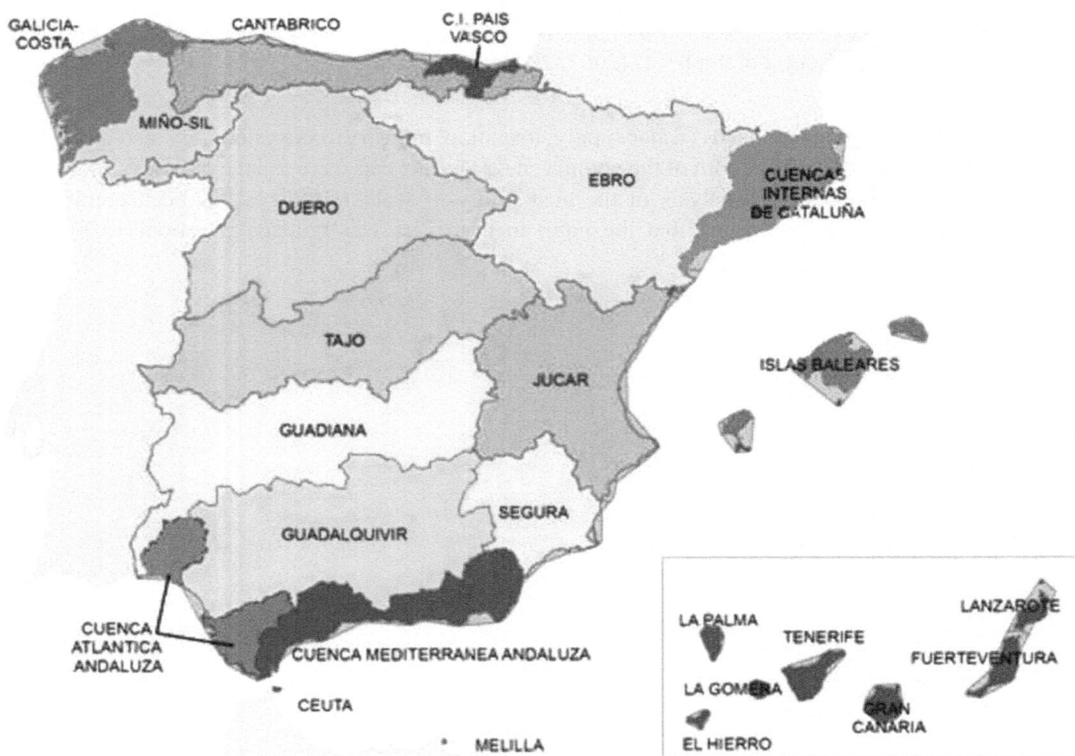


**Figure 11.10** River basin district in Spain (Barajas, 2007).

1996 and 2005 in comparison with the average values obtained during the 1940–1995 period (Table 11.3) (Barajas, 2007). In summary, despite the normal level of precipitation, the input of water resources to the rivers fell by 15% over a period of 10 years (1996–2005).

What are the reasons for this reduction? As mentioned, it could not be precipitation and it is perhaps possible that the cause lies in a combination of other factors. In addition to precipitation, other factors that have been considered are the overexploitation of the aquifers in vast areas of Spain, the expansion of the forestry regions, and the increases in temperature.

The conclusion of the analysis was that the main cause of the dramatic reduction in the water input to the rivers experienced in recent years in Spain was the increase in temperatures. Moreover, given the established tendency, it is foreseeable that the 15% reduction obtained in the 1996–2005 period, in comparison to the 1940–1995 period, will also increase sharply in the coming decades.

Water consumption in Spain continues to increase. The amount of irrigation continues to rise in the majority of basins.

Urban water supply consumption has increased considerably in all the basins, being especially important in the basins in the Mediterranean area due to tourism. Diminishing water resources and water consumption that increases every year point to a situation of greater hydrological and, therefore, environmental unsustainability. Thus, while it was thought in 1990 that the Segura was the only basin with a structural deficit (demands exceeding natural input), it is currently assumed that such a situation exists in one-third of the peninsular surface.

The author of the study (Barajas, 2007) considers that the only solution to this problem is to adopt measures to recover the lost hydrological balance in each basin and sub-basin. This can be achieved by means of increasing the efficiency of water use when possible, and by reducing the demands of those activities that consume the most.

It should be noted that in Spain, there has been a debate for many years about how to prevent and solve the issue of drought. The ongoing debate on water policies in Spain is characterized by a traditional paradigm

**Table 11.3** Mean annual water input ($Hm^3$/year).

| District | Mean Input 1940–1995 | Mean Input 1996–2005 | Reduction of the Water Input in the 1996–2005 Period with Regard to the 1940–1995 Period |
|---|---|---|---|
| North | 43 494 | 38 573 | −11.3% |
| Duero | 13 861 | 11 729 | −15.4% |
| Tajo | 10 533 | 9012 | −14.4% |
| Guadiana | 5464 | 4391 | −19.6% |
| Guadalquivir | 8770 | 8113 | −7.5% |
| Mediterranean-Andalucian Basin | 2446 | 2101 | −14.1% |
| Segura | 817 | 505 | −38.2% |
| Júcar | 3493 | 3057 | −12.5% |
| Ebro | 17 189 | 13 555 | −21.1% |
| Internal Basins of Cataluña | 2742 | 2196 | −19.9% |
| TOTAL | 109 948 | 93 763 | −14.7% |

Source: Spanish Ministry of Environment, Marine and Rural Affairs.

dominated by intervention in hydrological systems through the construction and management of infrastructure. This is progressively being abandoned but is currently still strong in spite of the emergence of new management approaches. In addition, climate change and the Water Framework Directive are the background to increasing challenges to traditional perspectives on drought, and important steps have been taken towards their replacement. This is based on a fundamental conceptual change that places drought in the framework of risk, rather than that of crisis (Paneque, 2015).

***2° Western Turkey***

In the East Mediterranean area, yearly precipitation trends are negative. The effect of climate change on the water resources of western Turkey has been studied by several researchers. For example, the effects of climate change on the water resources of the Büyük Menderes River Basin and its tributaries, the Çine and Akçay (Figure 11.11) rivers in western Turkey has been investigated by Durdu (2009). Fujihara *et al.* (2008) evaluated the impact of climate change on the water resources of the Seyhan River Basin (Figure 11.12) located in the southern part of western Turkey. Both papers concluded that the impact of climate change on the water resources is significantly negative, and that the quantity and way in which the population uses water in the future are crucial for the sustainability of water resources in western Turkey.

The river flow of the Cine and Akçay rivers (Figure 11.11) showed a decreasing trend (Figure 11.13) in the 1963–2007 analysed period. The observed decreasing trend of the flow in the tributaries had a strong correlation with the changes in temperature (increase) and precipitation (decrease). Only the data on the annual water levels was considered.

The potential impacts of climate change on the hydrology and water resources of the Seyhan River Basin (Figure 11.12) were investigated using a dynamic scaling reduction method, called the pseudo-global warming method (PGWM).

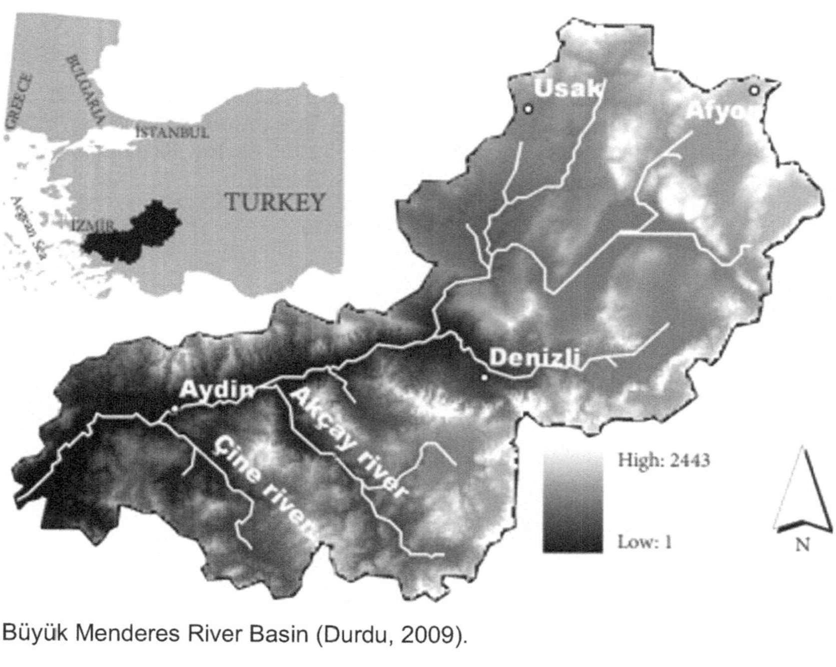


**Figure 11.11** Büyük Menderes River Basin (Durdu, 2009).

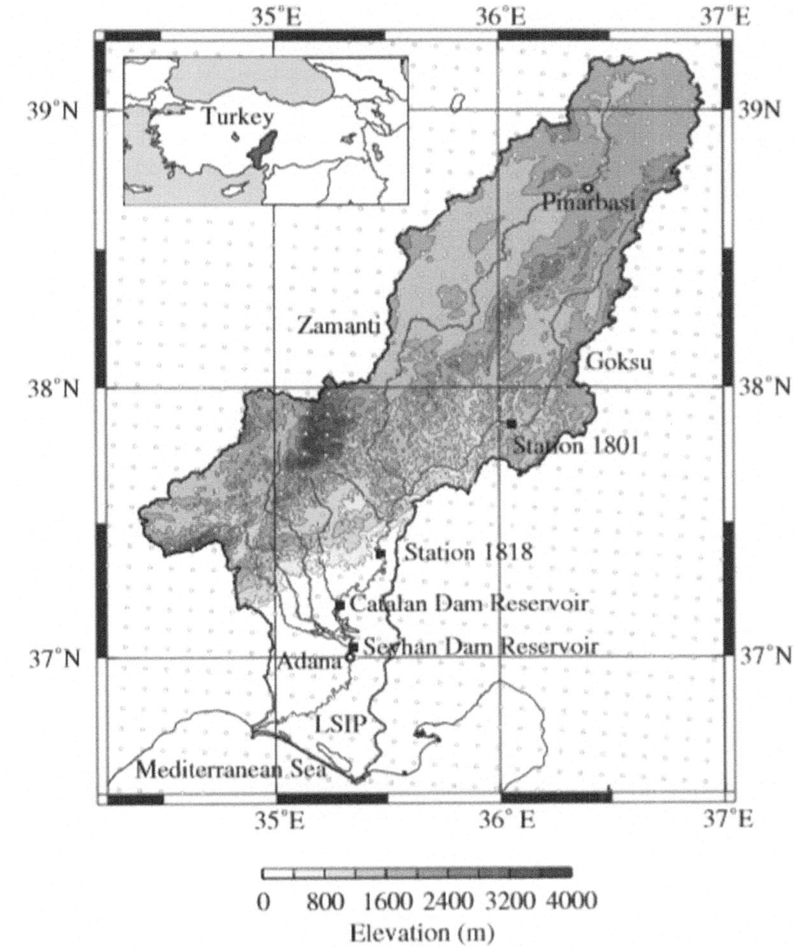


**Figure 11.12** Seyhan River Basin (Fujihara *et al.*, 2008).

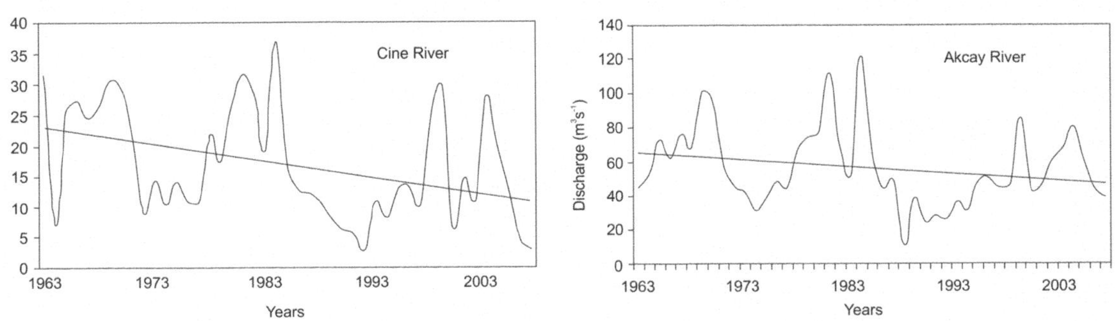


**Figure 11.13** Çine and Akçay rivers streamflow distributions ($m^3/s$) (Durdu, 2009).

An analysis of the water resources systems was conducted by using several scenarios regarding the water demand: from the present state to a demand of double the present state. These analyses indicated that despite the impacts of climate change, water scarcity will not occur in the future if water demand does not increase. However, if the irrigated area is expanded in the future with the current flow expected to remain as it is, water scarcity will occur due to a combination of a decreased inflow and an increased water demand. Thus, in the Seyhan River Basin, water use and management will play more important roles than climate change in controlling future water resource conditions.

***3° Central Greece – Thessaly region***

The impacts of various climate change scenarios on drought impulses were evaluated to provide a framework for sustainable water resources management in the region of Thessaly, central Greece (Loukas *et al.*, 2007). Thessaly was divided into 12 hydrological sub-basins and monthly precipitation data from 50 stations were used for the estimation of sub-basin mean precipitation.

The CGCMa2 model has been used to estimate the precipitation changes for the periods 2020–2050 and 2070–2100 and for three IPCC climate change scenarios (IS92a, SRES A2, SRES B2).

A comparison of the drought characteristics for the historical period (1960–1990) and the future periods indicated that the drought intensity, duration, and severity increased for the three examined climate change scenarios. That conclusion is illustrated in Table 11.4 which shows the Larissa sub-basin results and the number of dry months.

Furthermore, as expected, an analysis of the historical droughts has shown that frequent, extreme and spatially extended droughts should be expected in the future.

***4° The Middle East with a focus on Lebanon*** (Bou-Zeid & Fadel, 2002)

The potential impacts of climate change on water resources were assessed for the Middle East region, with Lebanon as a case study (Figure 11.14).

Despite the significant number of years (17) since the publication of the data presented in this example, this area is one of the most vulnerable in the world in terms of drought both due to the great diversity of some parameters in different countries of this region and due to the present demographic and growing water needs in all of them.

Figure 11.15 shows a fairly good estimate of the forecast and realized population growth in 2017 in Iraq, Lebanon and the Palestinian Authority while the projection was underestimated for Israel and Jordan. For Syria the forecast was overestimated, likely due to the unpredicted war situation and the migration crisis.

The current and projected renewable water resources are presented in Table 11.5 by country. It depicts the baseline conditions, neglecting the effects of potential climate change, and does not account for potential increased capacity through desalination or wastewater reuse. The ranges presented depict varying

**Table 11.4** Number of dry months of 12-month SPI for various drought severity classes, Larissa sub-basin.

| Dry Months | Historical | IS92a | | SRES A2 | | SRES B2 | |
|---|---|---|---|---|---|---|---|
| | 1960–1990 | 2020–2050 | 2070–2100 | 2020–2050 | 2070–2100 | 2020–2050 | 2070–2100 |
| Moderate | 43 | 69 | 69 | 49 | 63 | 65 | 78 |
| Severe | 16 | 38 | 74 | 25 | 88 | 38 | 44 |
| Extreme | 5 | 14 | 51 | 11 | 63 | 20 | 25 |
| TOTAL | 64 | 121 | 194 | 85 | 214 | 123 | 147 |

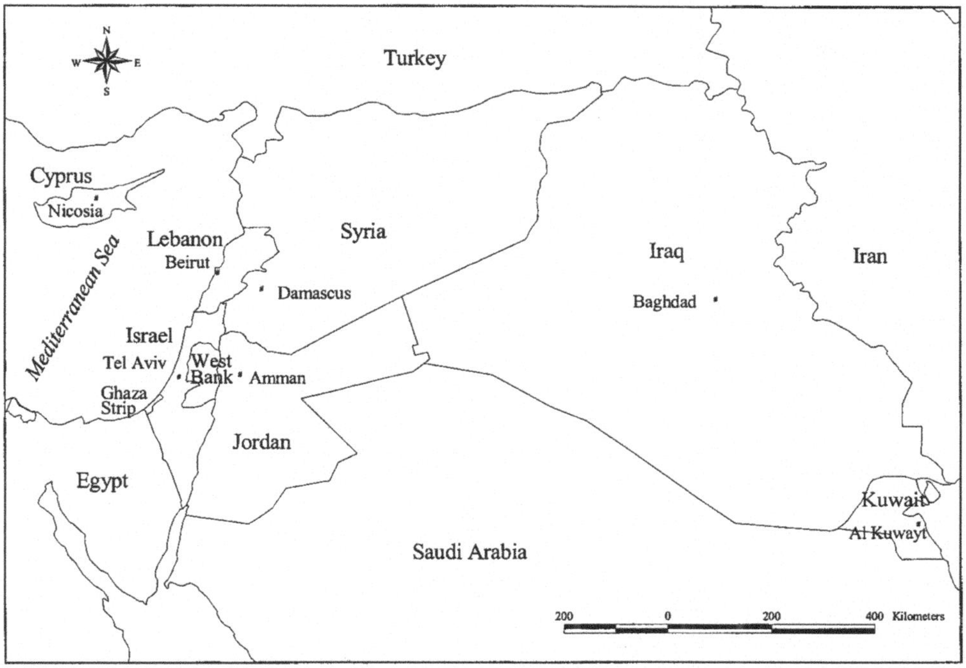


**Figure 11.14** Considered Middle Eastern countries, and rivers and water basins of Lebanon (Bou-Zeid & Fadel, 2002).

assumptions of demographic growth, water use variation, water resources management reform, and likewise. If figures range below 700–1000 $m^3$/capita/year, a country is considered as water poor, and projections indicate water shortage in the future for most countries in the region.

Water shortage is exacerbated by the rapid demographic development in the region (and economic development in some countries as well), and increasing water demand at a relatively high pace. Iraq is the only country in the Middle East that appears to have enough water resources but has the disadvantage of procuring more than two-thirds of them from Turkey (Turkish projects on the Euphrates and Tigris rivers threaten to reduce these significantly).

A water balance model (WATBAL) was used to assess the significance of the analysed changes at two test areas in Lebanon. The first is Ksara, in the Bekaa Valley, and the second is along the Mediterranean coast near Beirut (Figure 11.14). Evaporation increases at the two sites for all scenarios (maximum for 'hot' scenario: 33 mm/year or 6.6% for Beirut and 18 mm/year or 6.1% for Ksara). The soil moisture deficit increases in Beirut, and ranges from 13 mm/year or 1.5% (mild-hot) to 25 mm/year or 3.0% (dry and hot). The impact at Ksara is greater, with a 24–43 mm/year (3–6%) increase. The soil moisture deficit is correlated with the irrigation demand in agricultural areas. Evidently, Ksara, which lies in the agricultural and relatively dry Bekaa Valley (prediction for water demand increases 6% in the 'dry and hot' scenario), is more affected than Beirut, which is located along the west coast. The results also show that surplus water, which ultimately produces runoff or percolates and reaches water bodies, is reduced by 11–34 mm/year (3.5–11.0%) for Ksara and 21–49 mm/year (6.5–15.0%) for Beirut.

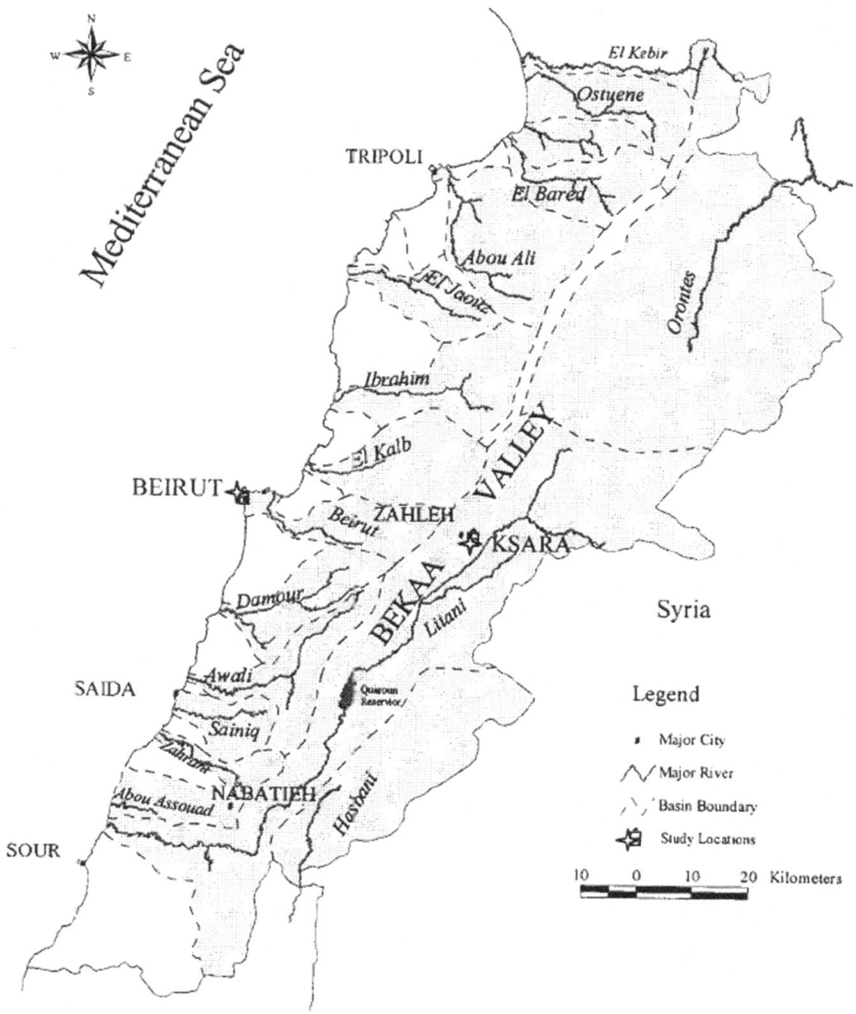


**Figure 11.14** *Continued.*

Since the surplus is the main source of water for surface water bodies and groundwater aquifers, this reduction will directly affect the available water resources in the country. Therefore, the net usable surface water and groundwater resources (about 2000 million $m^3$/year) might be reduced by 5–15% until 2025.

Without important adaptation measures, water shortage is expected in the future in the Middle East countries primarily due to an increase in water demand. Adaptation measures become necessary to a certain level, even without the impact of climate change.

Similar hydrologic impacts might have different socio-economic consequences, depending on the characteristics of a region. In the Middle East, different configurations of water resources might have affected the magnitude of potential adverse effects, such as the reduction of the gross domestic product, population distribution, workforce shifts to alternative economic sectors, etc. (Bou-Zeid & Fadel, 2002).

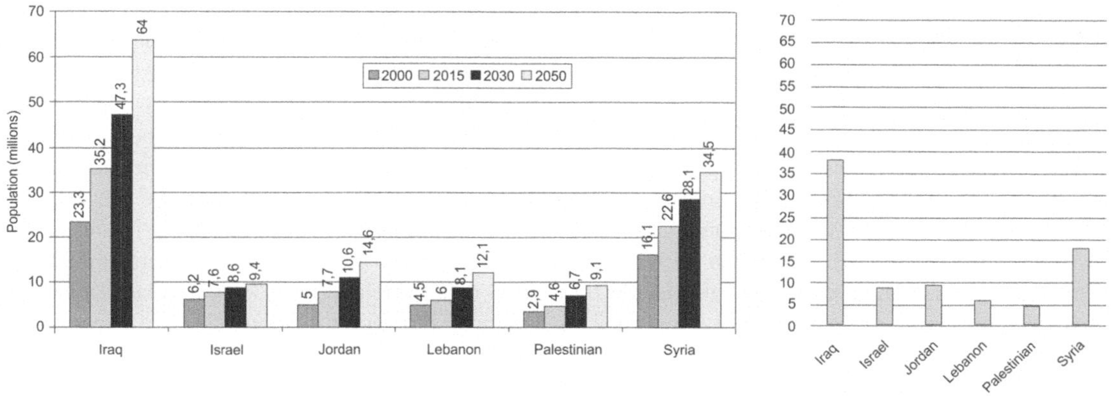


**Figure 11.15** Projection of population (from 2002) and real number (Google Source: World Bank, UN) in 2017.

**Table 11.5** Current and projected renewable water resources in the Middle East countries.

| Country and Access to Safe Drinking Water (% of Population) | Average Rainfall | Area | Total Present[a] Renewable Water Resources | Current and Projected Renewable Water Resources[d] ($m^3$/capita/year) | | |
|---|---|---|---|---|---|---|
| | (mm/year) | ($km^2$) | (billion $m^3$/year) | 2000 | 2015 | 2030 |
| Iraq (77) | 154 | 438320 | 62.85–100.00 | 2697–4292 | 1786–2841 | 1329–2114 |
| Israel (99) | 630 | 20700 | 1.50–2.57 | 242–415 | 197–338 | 174–299 |
| Jordan (98) | 94 | 89210 | 0.75–1.35 | 150–270 | 97–175 | 71–127 |
| Lebanon (94) | 827 | 10452 | 2.00–3.94 | 444–876 | 333–657 | 247–486 |
| Palestine[b] (--[c]) | 350 | 6263 | 0.20–0.22 | 69–76 | 43–48 | 30–33 |
| Syria (88) | 252 | 185180 | 15.00–21.48 | 932–1334 | 664–950 | 534–764 |

[a]2000 year.
[b]Palestinian Authority.
[c]Not available.
[d]The level below which a country is considered water poor varies, and it is around 700–1000 $m^3$/capita/year.

## 11.2.3 Climate and hydrological changes in Serbia

Subchapter 11.2.3 deals with the observed climate and hydrological changes in Serbia. Only temperature and precipitation are considered as parameters for climatic indicators and the flow in rivers for hydrological indicators. Changes are considered primarily at the mid-year level, and given to a lesser extent are both the seasonal and/or monthly registered changes.

### *11.2.3.1 Observed climate change*

All global and regional climate models (RCMs) predict an increase in temperature and, in the majority of scenarios, a small decrease in precipitation in Serbia. The average annual temperature increase is expected to range from 2°C to 6°C at the end of 21st century (from the averages in the period 1971–

2000), largely depending on the selected scenario and to a much lesser extent on the analyst (CC-WaterS, 2009–2012; Giorgi *et al.*, 2004; IPCC, 2007, 2013; SEECOFs, 2008–2010). Annual precipitation predictions are more uncertain and range from current levels (trend ≈ 0) to a small increase/decrease in the near future (of the order of +10%) and to a significant decrease in the distant future (−25%). Each prediction is sensitive to assumption uncertainties and the calculation of imperfections. The quality of a prediction, particularly for the near future, grows with increasing validation based on observed data and recorded trends (Dimkić, 2017). The observed trends for temperature and precipitation are presented on the website of the Ministry for environmental protection of the Republic of Serbia (Figures 11.16 and 11.17).

Another study showed that when different observation periods are chosen, the temperature and precipitation trends change to some extent, but the spatial distribution remains very similar. Table 11.6 shows the observed average monthly, seasonal, and annual temperature trends (°C/100 years), and Table 11.7 shows the precipitation trends (%/100 years) for three analysed periods: 1949–2006 (58 years), 1959–2016 (58 years), and 1949–2016 (68 years).

Yearly T trends differ quite a lot depending on the selected period. This is because of the very high annual values for T in the last 10 years. Compared to the first 10 years of the analysed period (1949–1958), the average T in the last 10 years (2007–2016) is 1.1°C higher (Dimkić, 2019). The observed yearly P trends apparently do not differ as much. They are quite consistent with RCM results (CC-WaterS, 2009–2012; IPCC, 2007, 2013; SEECOFs, 2008–2010). The seasonal and monthly trends are more debatable for both climate parameters.

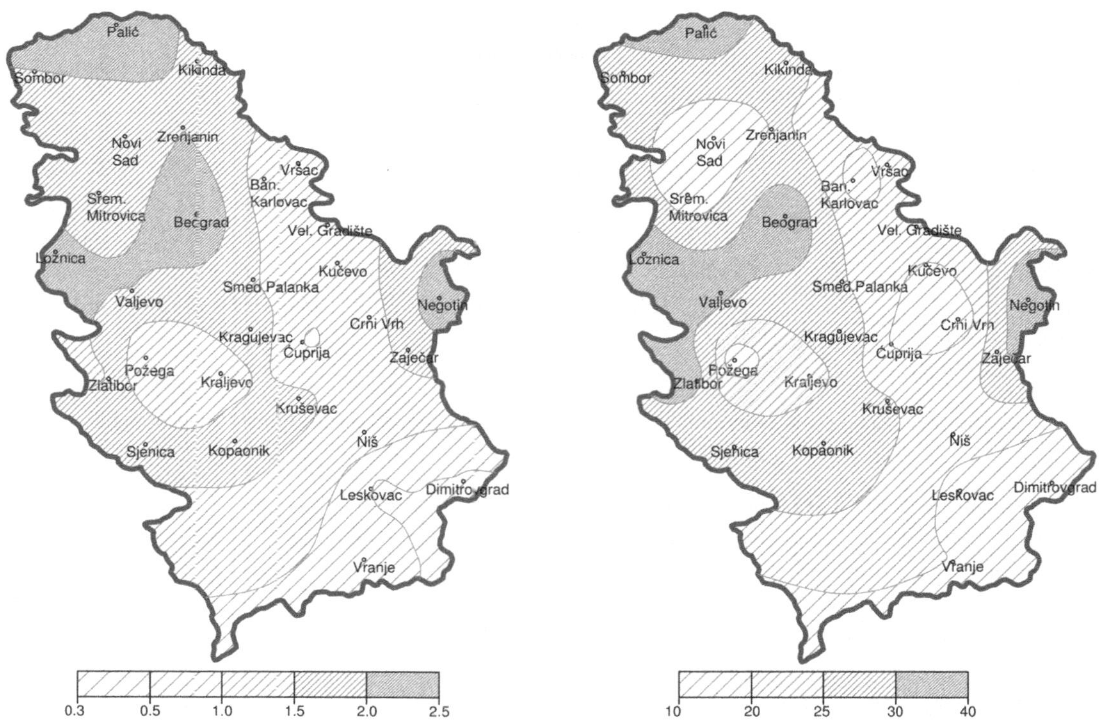


**Figure 11.16** Territorial distribution of trends in air temperatures (°C/100 years), annual (left) and summer (right), in the territory of the Republic of Serbia, period 1951–2013 (Ministry for Environment Protection of Republic of Serbia website, 2019).

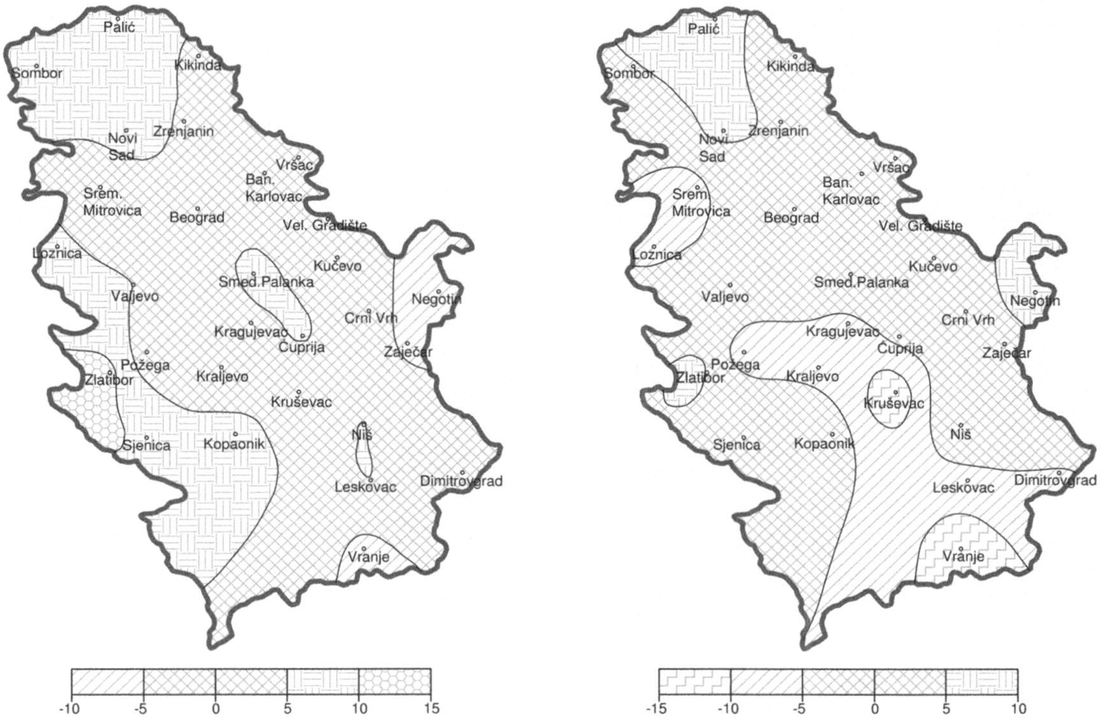


**Figure 11.17** Geographical distribution of the trend in precipitation, annual (left) and summer (right), in the territory of the Republic of Serbia in the period 1951–2014 (in % of N6190/50) (Ministry for Environment Protection of Republic of Serbia website, 2019).

Seasonal T trends for the longest period of available data (1949–2016) are in line with RCMs – the highest increase occurred in summer and, the lowest in autumn. The other two periods have several debatable seasonal T trend values, which could be a consequence of insufficiently long data series (58 years). In addition to July and August, the highest upward monthly temperature trends have been registered in January and March. Negative monthly T trends have been registered in November and December, and there has been no significant change in the month of September. Other months registered increasing trends, in the range of the annual average.

Seasonal P trends show a small decrease in winter and a sizeable increase in autumn. The distribution of certain monthly precipitation trends is particularly questionable: the highest downward trend in the RCMs was almost always predicted for the summer months (often in the order of $-50\%/100$ years), which is inferior to the actual trends in summer. June and July have registered a small decrease in precipitation (in the order of approx. $-10\%/100$years), while August and particularly September registered significant positive trends. For the water sector, regarding the low flow period, it is also important that October exhibits a significant positive trend as well. November and December are the months which exhibit the highest downward P trend. The months from January to May vary. These results underline the importance of detailed analyses of observed data.

Meteorological and hydrological droughts are the result of extreme weather events. In some other studies and projects, changes in temperature and precipitation extremes have also been investigated. For one of them

**Table 11.6** Monthly, seasonal and annual T trends (°C/100years), average of 26 stations.

| Period | T Data Trend | WINTER | | | SPRING | | | SUMMER | | | AUTUMN | | | ANNUAL |
|---|---|---|---|---|---|---|---|---|---|---|---|---|---|---|
| | | DEC | JAN | FEB | MAR | APR | MAY | JUN | JUL | AUG | SEP | OCT | NOV | |
| 1949–2006 | Monthly | −2.2 | 1.9 | 1.3 | 3.2 | −0.1 | 1.7 | 1.1 | 1.1 | 0.8 | −1.3 | 1.2 | −1.9 | **0.6** |
| | Seasonal | | 0.3 | | | 1.6 | | | 1.0 | | | −0.7 | | |
| 1959–2016 | Monthly | 0.5 | 5.3 | 2.9 | 2.6 | 2.1 | 2.4 | 3.7 | 5.1 | 4.7 | 2.0 | 1.0 | 0.1 | **2.7** |
| | Seasonal | | 2.9 | | | 2.4 | | | 4.5 | | | 1.1 | | |
| 1949–2016 | Monthly | −1.0 | 3.3 | 2.6 | 3.6 | 1.5 | 1.8 | 2.1 | 2.6 | 2.5 | 0.1 | 0.9 | 0.2 | **1.7** |
| | Seasonal | | 1.7 | | | 2.3 | | | 2.4 | | | 0.4 | | |
| Average of 3 periods | Monthly | −0.9 | 3.5 | 2.3 | 3.1 | 1.2 | 2.0 | 2.3 | 2.9 | 2.7 | 0.3 | 1.1 | −0.5 | **1.7** |
| | Seasonal | | 1.6 | | | 2.1 | | | 2.6 | | | 0.3 | | |

Table 11.7 Monthly, seasonal and annual P trends (%/100yrs), average of 38 stations.

| Period | P Data Trend | WINTER | | | SPRING | | | SUMMER | | | AUTUMN | | | ANNUAL |
|---|---|---|---|---|---|---|---|---|---|---|---|---|---|---|
| | | DEC | JAN | FEB | MAR | APR | MAY | JUN | JUL | AUG | SEP | OCT | NOV | |
| 1949–2006 | Monthly | −19.8 | −22.0 | −25.5 | −13.8 | 36.0 | −46.8 | −5.9 | 8.9 | 48.6 | 76.3 | 9.7 | −52.7 | −0.6 |
| | Seasonal | | −22.4 | | | −8.2 | | | 17.2 | | | 11.1 | | |
| 1959–2016 | Monthly | −35.9 | 4.3 | 23.4 | 40.6 | −1.0 | 6.6 | −25.4 | −32.6 | 9.1 | 59.3 | 110.9 | −36.7 | 10.2 |
| | Seasonal | | −2.7 | | | 15.4 | | | −16.3 | | | 44.5 | | |
| 1949–2016 | Monthly | −24.1 | 8.8 | 5.5 | 31.4 | 7.7 | 2.3 | −15.1 | −5.3 | 9.9 | 58.9 | 50.9 | −40.9 | 7.5 |
| | Seasonal | | −3.3 | | | 13.8 | | | −3.5 | | | 23.0 | | |
| Average of 3 periods | Monthly | −26.6 | −3.0 | 1.1 | 19.4 | 14.2 | −12.6 | −15.5 | −9.7 | 22.5 | 64.8 | 57.1 | −43.4 | **5.7** |
| | Seasonal | | −9.5 | | | 7.0 | | | −0.9 | | | 26.2 | | |

**Table 11.8** Analysed extreme temperature and precipitation indices.

| Indicator | Abbreviation | Definition |
|---|---|---|
| **Based on temperature (T)** | | |
| Total number of tropical days | Td | Number of days with absolute maximum T >30°C |
| Total number of frost days | Fd | Number of days with absolute minimum T <0°C |
| Maximum number of consecutive frost days | CFD | Maximum number of consecutive days with absolute minimum T <0°C |
| Growing season length (d) | GSL | Number of days between the first occurrence of at least 6 consecutive days with daily mean T >5°C (beginning) and the first occurrence after 1 July of at least 6 consecutive days with daily mean T <5°C (ending) |
| **Based on precipitation (P)** | | |
| Heavy precipitation | R10 | Number of days with P ≥10 mm $d^{-1}$ |
| Maximum number of consecutive dry days | CDD | Maximum number of consecutive days with daily P <1 mm |
| Simple daily intensity index (mm $d^{-1}$) | SDII | Period total/no. of days with P ≥1 mm |

(Krzic *et al.*, 2011), the indices were selected to examine changes in the severity of weather conditions over Serbia under the A1B and A2 scenarios (Table 11.8).

The indices were calculated for the period 1961–1990 and for a future period 2071–2100 for all seasons. Calculations were carried out for the regions of Northern Serbia (NS) and Southern Serbia (SS), using observations at 17 meteorological stations for the reference period and simulations were made using the coupled regional EBU-POM Model. Overall, a simulation of the reference period reproduces the observed climate patterns, although with some bias. Verification was also made in terms of the relative frequencies of temperature and monthly precipitation sums (Figure 11.18). The results show an overall

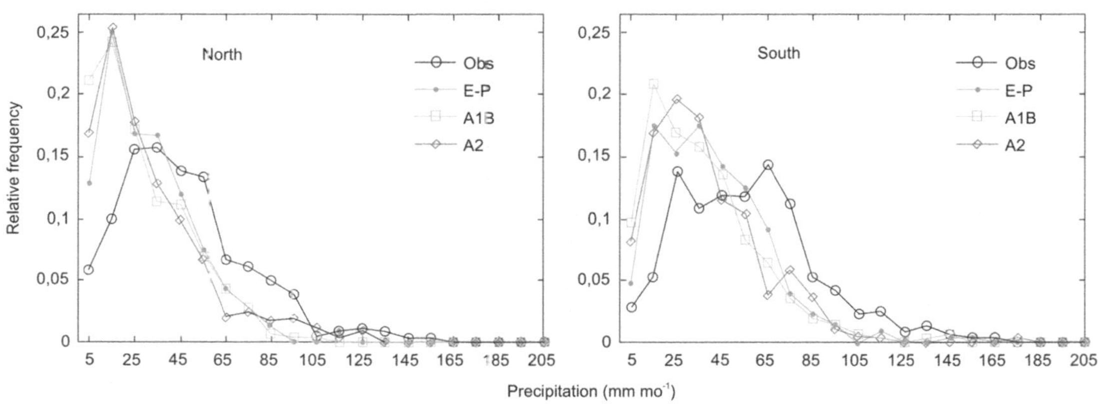


**Figure 11.18** Relative frequency of monthly P sums for Northern and Southern Serbia from observations (Obs), EBU-POM (E-P) simulations for the ref. period 1961–1990, and A1B and A2 scenarios (Krzic *et al.*, 2011).

increase of about 2°C and 4°C in the surface air temperature and a decrease in the seasonal precipitation sums of about 13 and 6 mm for the A1B and A2 scenarios, respectively.

As expected, an increase in the number of tropical days and, at the same time, a decrease in the total number of frost days were predicted, and in line with that, a shorter duration of frost and longer durations of vegetation periods. The increase in Td in the north is greater than that in the south, while the opposite trend is expected for Fd. An increase or decrease in the number of days with heavy rainfall depends on the season and scenario. According to the A2 scenario, dry periods will be longer and the number of days with heavy rain will increase during winter in both regions. For all the indices that were considered, the expected changes according to the A1B scenario are smaller than those expected according to the A2 scenario. All the chosen indices indicate that climate change will have serious implications on the growth of vegetables and fruit, both dominant crops in Serbian agriculture.

### *11.2.3.2 Observed hydrological changes*

The observed hydrological changes have been analysed for the rivers in central Serbia, and the analysed periods are: 1949–2006, 1959–2016 and 1949–2016. Contrary to climate parameters, it is difficult to spatially generalize river discharge trends because several factors affect these trends (Gavrilović *et al.*, 2013; Juckem *et al.*, 2008; Ma *et al.*, 2008; Nachtnebel, 2013; Novotny & Stefan, 2007). They can be grouped in three classes: Climate change (CC), Human use of water (HU), and Land use changes (LU).

Figure 11.19 shows an approximate spatial trend distribution of the downward average annual river flow trends for all three analysed periods. It was compiled based on the trends recorded at 24 selected hydrological stations across central Serbia (Dimkić, 2019). It should be noted that within all river flow trend isolines, there are rivers and monitoring stations that often exhibit significant trend variations (both up and down), as a result of the LU and HU factor (especially if water is transferred upstream from a given station).

Table 11.9 shows the observed average monthly, seasonal and annual river flow trends for all 24 stations and all the three periods (Dimkić, 2019). The yearly trends do not differ much depending on the selected period (−20 to −25%/100 years), and the direction is generally in accordance with the forecasts based on the IPCC reports (2007, 2013) and the observed T and especially P trends. In the eastern and south eastern parts of the country the highest negative trend was exhibited (in the range of −30 to −50%/100 years), while the south-western part exhibited a slightly decreasing trend (somewhere close to zero).

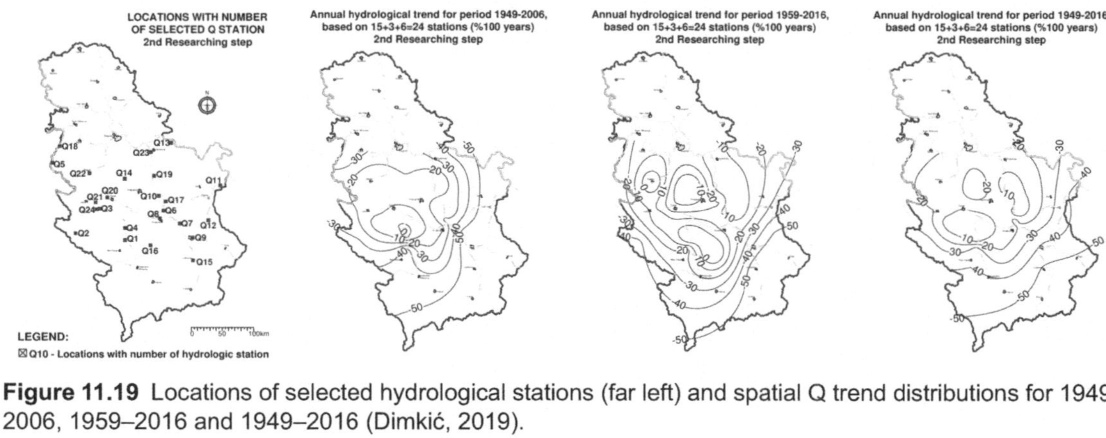


**Figure 11.19** Locations of selected hydrological stations (far left) and spatial Q trend distributions for 1949–2006, 1959–2016 and 1949–2016 (Dimkić, 2019).

**Table 11.9** Monthly, seasonal and annual hydrological trends (%/100years).

| Period | Q Data Trend | WINTER | | | SPRING | | | SUMMER | | | AUTUMN | | | ANNUAL |
|---|---|---|---|---|---|---|---|---|---|---|---|---|---|---|
| | | DEC | JAN | FEB | MAR | APR | MAY | JUN | JUL | AUG | SEP | OCT | NOV | |
| 1949–2006 | Monthly | −53.8 | −0.7 | −45.5 | −13.1 | −14.0 | −74.8 | −21.4 | −31.8 | −7.4 | 30.1 | −7.0 | −55.5 | −29.2 |
| | Seasonal | | −33.3 | | | −34.0 | | | −20.2 | | | −10.8 | | |
| 1959–2016 | Monthly | −2.5 | −4.4 | −55.4 | 24.9 | 9.0 | −39.3 | −73.0 | −88.9 | −17.2 | −10.4 | 16.3 | −6.4 | −18.6 |
| | Seasonal | | −20.8 | | | −1.8 | | | −59.7 | | | −0.2 | | |
| 1949–2016 | Monthly | −45.1 | −1.2 | −36.4 | 16.4 | −14.8 | −34.1 | −40.1 | −48.4 | −24.8 | −0.1 | −23.7 | −36.6 | −21.4 |
| | Seasonal | | −27.5 | | | −10.8 | | | −37.8 | | | −20.1 | | |
| Average of 3 periods | Monthly | −33.8 | −2.1 | −45.7 | 9.4 | −6.6 | −49.4 | −44.8 | −56.4 | −16.5 | 6.5 | −4.8 | −32.8 | **−23.0** |
| | Seasonal | | −27.2 | | | −15.5 | | | −39.2 | | | −10.4 | | |

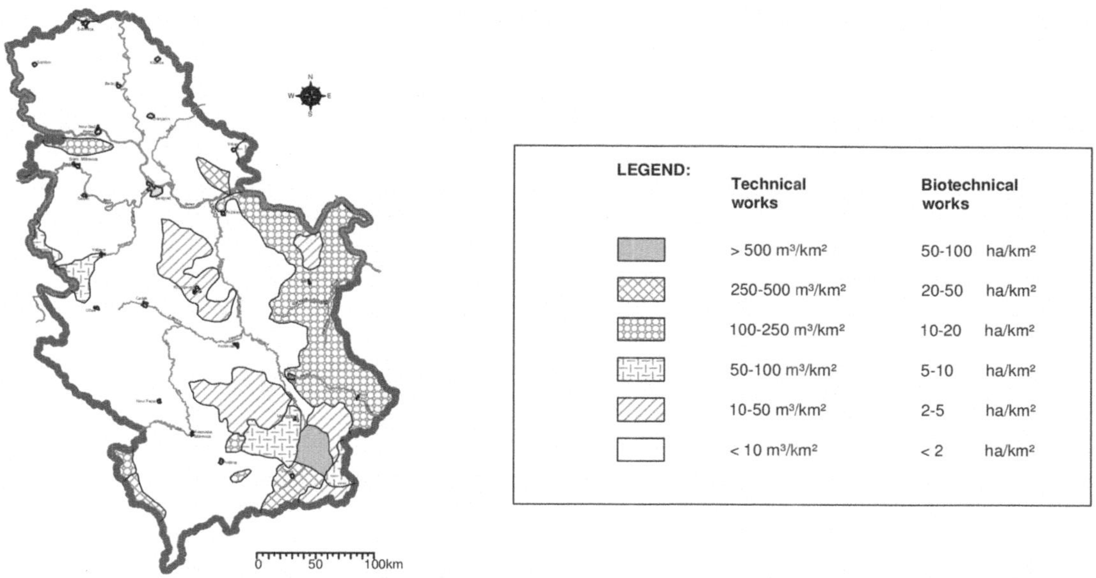


**Figure 11.20** Applied anti-erosion measures in Serbia in the 1970s (erosion barriers and reforestation) (Gavrilović *et al.*, 2013).

Additionally, the LU and HU factors are also present on some rivers. The most important LU changes (reforestation above all) occurred during the 1970s of the 20th century (Figure 11.20) (Gavrilović *et al.*, 2013).

The highest negative seasonal hydrological trend occurs in summer (estimated range −35 to −40%/100 years), followed by winter (−25 to −30%/100 years), and spring and autumn (−10 to −15%/100 years). Two things should be noted regarding the monthly trends:

(1) The highest negative monthly hydrological trend is in May, June, and July (apart from February, where some snow effect is present). Significantly smaller quantities of water in rivers in late spring and early summer represent important information for the water sector, particularly for water use.

(2) The lower negative trend in the low-discharge months (August–October), with a range of −20 to +10%/100 years, is a result of an upward P trend during these months, and often due to the presence of a river reservoir upstream from a given station, which to some extent equalizes the annual discharges. This does not mean, however, that a more significant downward trend will not appear during this period if T continues to rise, particularly at stations where there are no upstream river reservoirs.

Conclusion: The recorded average hydrological trends in Serbia have decreased by about 15% in the last 70 years and depend on numerous factors. Climate change is one of them, which is present at all monitoring stations, but its significance varies. It is generally dominant in the eastern part of the country (along with the LU factor), and in the upper parts of catchments (Gavrilović *et al.*, 2013; Nachtnebel, 2013), but it is often less significant or even minor elsewhere, especially where the human impact is substantial.

## 11.2.4 The importance of observed data and the reliability issue in climate-hydrological projections

In many parts of the world, the observed hydrological changes (subchapter 11.2.2) are significant (Arnell, 2003; CC-WaterS, 2009–2012; Fujihara *et al.*, 2008). The observed hydrological changes for Serbia have shown us that there is a downward average annual river flow trend (subchapter 11.2.3.2). In such cases, the possibility of a drought occurring grows, especially if water demand also grows. The question is: If the global temperature continues to increase, what is to be expected regarding the hydrological trends in a region, and what is the reliability of any forecasting? Regions with an observed downward precipitation trend are especially interesting for the analyses of droughts. Common approaches (RCM models) try to give us an answer using different scenarios. In addition to the common approach, a good way to search for an answer to this question is to analyse what happened in the past with average annual temperature vs. river flow levels, and to establish the correlation between these parameters for one region (Dimkić, 2017).

### *11.2.4.1 Trends and the importance of the observed data for the analysis of a drought*

To understand a problem and get the first picture of a given area, any data is useful, even if discrete (not continuous), which is often the case, especially in countries where there is no long tradition of data monitoring. However, if more serious monitoring of a given parameter or index for droughts over time and the definition of the relevant situation, i.e. the establishment of changes or trends, is desired, it is necessary for the length of the series of observed values to be respectable. If a spatial representation of the value of a parameter or its trend is also desired, then it is necessary that the number of stations registering a given parameter be respectable in the analysed area (at least 10–12), in order to be able to nuance the image and to avoid any excessive influence of unrepresentative arrays (the number of errors that cause unrepresentativeness).

The problem is often the length of the observed data series. The trends (in general) vary to a large extent if the period is shorter than 50 years. Trends become relatively stabilized (much less changeable) where the data series is 70 years long or more. Long data series of parameters regarding climate (above all temperature and precipitation) and hydrological stations are quite numerous in developing countries. But the fact is that long data series in low-income countries and those in transition do not exist or are available from only a few stations. So, in such countries and regions, a shorter period must be found, with trends similar to the long-term trend. The last 100 years can be taken as a relevant period for the long-term trend for each parameter. This is because if there has been a significant trend in the last 100 years, there is a high probability that there will be a similar trend at least in the near future, e.g. in the next 20–30 years.

The trend changes and the importance of the length of data series is illustrated here (Table 11.10, Figure 11.21) using the temperature trend in Belgrade (Dimkić & Despotovic, 2012) as an example (the final year is the latest year in series – 2007 at the time of research, and it is fixed for all periods):

When analysing a trend for periods shorter than approximately 70 years, there is another problem to be solved (minimized): it is necessary to find the beginning and the end of the period (start and end year), which could be approximately representative (similar to the long-term trend, e.g. trend of the last 100 years). A trend is quite sensitive to the effect of data values at the beginning and end of the data series (Table 11.11). This is also illustrated here in the case of the Belgrade T trend: if we fix that period, it must be, for example, 30 or 55 years long, depending on the first and last year, and the (extreme) results will differ as follows:

It is obvious that the extreme trends differ a lot in both periods (30 and 55 years), depending on the first and last years. Even if the last 10 years are excluded (1998–2007), the maximum positive trend will be significantly higher than the maximum negative trend in absolute values.

**Table 11.10** Effect of the length of data series (past 20, 120 years) on the trend line slope – Belgrade TS.

| BELGRADE Temperature Station | OBSERVATION PERIOD | | | | | | | | | | |
|---|---|---|---|---|---|---|---|---|---|---|---|
| | 1888–2007 | 1898–2007 | 1908–2007 | 1918–2007 | 1928–2007 | 1938–2007 | 1948–2007 | 1958–2007 | 1968–2007 | 1978–2007 | 1988–2007 |
| Length of series (years) | 120 | 110 | 100 | 90 | 80 | 70 | 60 | 50 | 40 | 30 | 20 |
| Trend (in °C/100 years) | 1.28 | 1.11 | 1.19 | 1.03 | 1.14 | 1.34 | 1.43 | 2.22 | 3.59 | 5.33 | 3.11 |

The best way to avoid (minimize) this problem is to have a sufficient number of long data series (70 and more years) and a sufficient number of stations. Many different models have been developed in the world to predict future changes in the field of climate or hydrology or extreme events such as drought. Each prediction is sensitive to assumption uncertainties and calculation imperfections. The quality of a prediction, particularly for the near future, grows as the validity of the observed data grows. Moreover, it could be said that the observed data are a crucial source for any relevant analysis. For drought analyses, due to the complexity of this issue, it is maybe even more important.

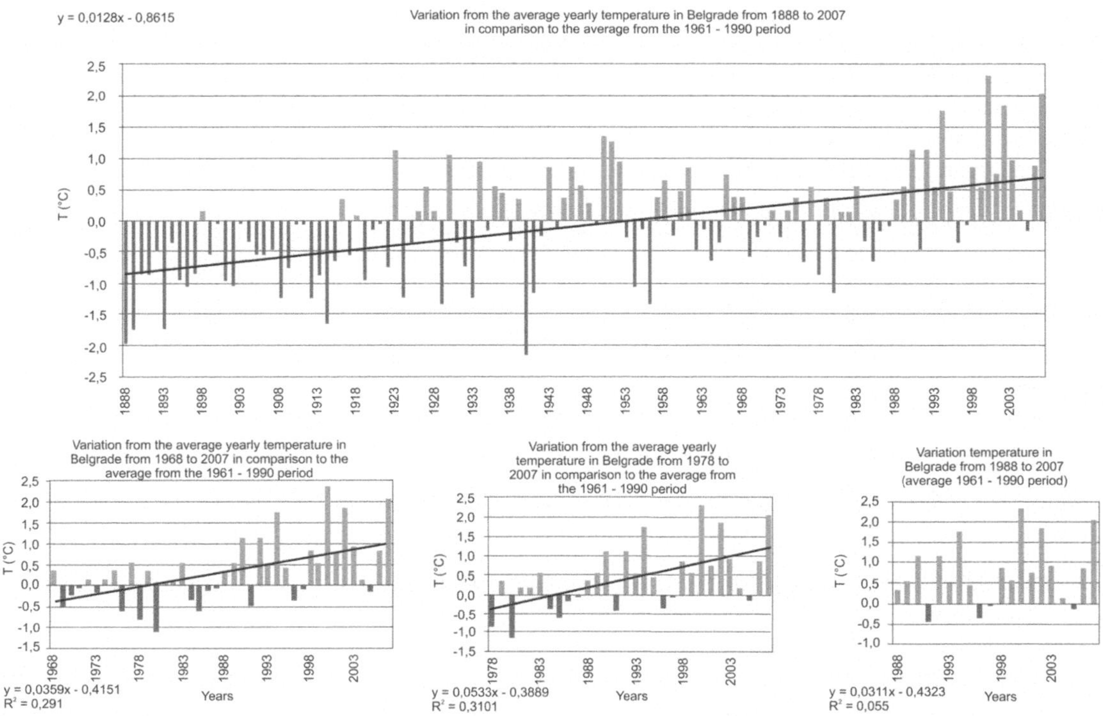


**Figure 11.21** Variation of the temperatures in Belgrade, showing trends for: 1888–2007 (1st graph), 1968–2007 (2nd graph), 1978–2007 (3rd graph) and 1988–2007 (4th graph) (Dimkić & Despotovic, 2012).

**Table 11.11** Variations in Belgrade temperature trends, and the impact of the first and end year of a series.

| | | Maximum (Absolute Value) Negative Trend | Maximum Positive Trend Without Last 10 Years | Maximum Positive Trend for Whole Data Series |
|---|---|---|---|---|
| 30 years | Observation period | 1951–1980 | 1924–1953 | 1978–2007 |
| | Trend (in °C/100 years) | −1.41 | 3.45 | 5.33 |
| 55 years | Observation period | 1926–1980 | 1940–1994 | 1953–2007 |
| | Trend (in °C/100 years) | −0.14 | 0.73 | 2.37 |

### *11.2.4.2 The issue of reliability in projections of hydrological changes*

Many national and international projects and studies address climate change and its impact on water resources. Most recent projects and studies are based on different climate models and different scenarios that provide basic inputs for hydrological models. Apart from climate change (CC), the hydrologic regime of a river is affected by changes in land use (LU) within a catchment area (CA) and by changes in the extent and manner of the human use (HU) of water (Gavrilović *et al.*, 2013; Nachtnebel, 2013; Vasiljević *et al.*, 2013).

There are many uncertainties associated with hydrologic projections. In the case of regional climate-hydrological models (RCMs), the uncertainties regarding the assessment of the impact of climate change on hydrology can, conditionally, be divided into climate prediction uncertainties (including climate change) and uncertainties pertaining to hydrologic regime predictions based on assumed future climate conditions.

The causes of uncertainties regarding climate change predictions should primarily be sought in:

- the natural complexity and variability of the climate itself; and
- insufficient knowledge of the climate system response to climate change.

The causes of unreliable hydrological predictions based on assumed climate conditions include:

- the inability to fully examine the correlation between climate and the hydrology of a given catchment (region);
- the difference in correlations between the present and significantly altered climate conditions; and
- demographic and migration changes in the study area.

In addition to the above uncertainties, apart from those stemming from measurement inaccuracies, the following uncertainties can be associated with regional climate-hydrological models:

- Credibility of the selected climate scenario (assumed GHG increase, assumed economic and social developments, and similar);
- Credibility of the selected hydrological model, non-representative input quantities selected for the given region (e.g. if the base time series reflects 30-year data, significant extremes are often lost, for instance the droughts in England (Fowler *et al.*, 2007) in 1928 and 1995–1996 where the time series included data from 1961–90);
- Behaviour of people depending on future living conditions (perhaps the greatest uncertainty).

Also, some projects and studies based on regional climate-hydrological models consider all three components (CC, LU, HU), while others address only the first type of change (CC). The reliability of the projections usually receives little or no consideration. This subchapter discusses this topic looking at Serbian rivers as a sample and attempts to provide some guidelines.

The majority of Serbia's rivers have recorded a considerable decrease in discharge (subchapter 11.2.3.2.). The discharges of only a small number of rivers have increased, largely due to water transfers from other river basins, which began in the 1970s and 1980s. All three components (CC, LU, HU) are very important, and the degree of significance varies very much from CA to CA. Based on precipitation and temperature trend distributions, the greatest negative trend change was noted in eastern Serbia (Dimkić, 2017, 2019).

However, for stations in some other parts of Serbia, especially where there are significant changes in the human use of water or in land use, the climate change impact is often low or even negligible, compared with the HU and/or LU factor.

Five projects which consider the future hydrology in Serbian rivers have been reviewed, and all the available data on the mean annual level from these projects were analysed. The five projects to analyse were chosen without consideration of their importance. Working on these projects, different groups of experts analysed 10 different rivers at 15 different hydrological stations and considered 54 different cases (different scenarios or climate-hydrology models). All the three impacts on water resources (CC, LU and HU) were analysed in just a few of the 54 cases (Dimkić, 2016). The following main conclusions were reached:

- In all the projects and nearly all the case studies, smaller changes are expected in the near future rather than in the distant future, which seems logical;
- The results vary considerably, even for the same river, especially regarding the distant future.

Let us address the second finding. If hydrologic predictions result in a broad range of possible discharges of the same river – extremes of +20% and −40% have been noted in Serbian rivers – clearly the reliability (probability) of the predictions needs to be examined. Similar or somewhat higher variations exist in hydrological predictions all over the world – for example, predictions for the Nile River in Egypt vary from a 30% increase to a 78% decrease in different models and for the same scenario depicting $CO_2$ conditions (Strzepek & Yates, 1996).

This requires either the selection of certain criteria for assessing the probability of the prediction, which is theoretically possible (e.g. probability of the adopted climate prediction multiplied by the probability of a correctly selected hydrological model), but is rather cumbersome and highly questionable, or the expert judgment of the study (project) analyst, which is certainly debatable.

Instead, let us consider what can, with an important degree of reliability, be expected, on average, for Serbia's internal rivers in the near future. Considering research based on recorded data (Dimkić, 2016), Figure 11.22 shows the relative values of the direct correlation between the average annual temperature and river discharge in 24 study areas. The methodology is described in literature (Dimkić, 2016, 2019), and the good thing is that it can be applied quite quickly after collecting data. An important characteristic of this approach is that it considers all three components: CC, LU and HU. For the purpose of analyses, the available data were split into three different periods: 1949–2006, 1959–2016 and data for the whole period 1949–2016.

It should be noted that the coefficient of determination is high on all three graphs (on the first and third graphs higher than 0.9), leading to the conclusion that a deviation of the average annual T by +1°C has an inversely proportional effect on the average annual river discharge (Q) of about 15%. The results differ from CA to CA, but in most cases this variation is not large. If the linear trends are extrapolated

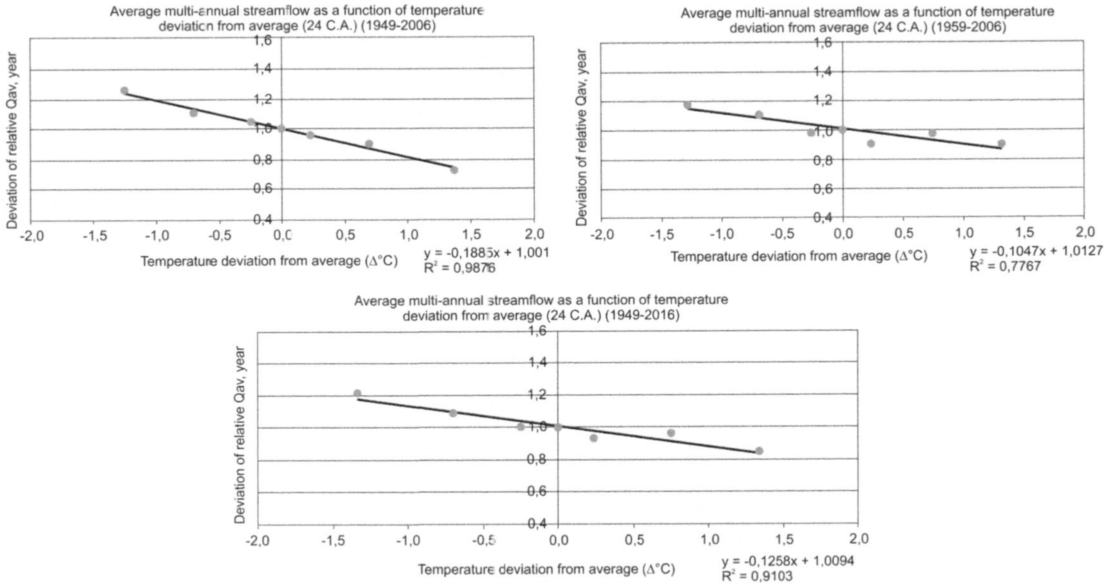


**Figure 11.22** Average relative annual river discharge, as a function of T deviation, along with the linear trend of dependence, for three analysed periods (1949–2006, 1959–2016 and 1949–2016) (Dimkić, 2016, 2019).

to +2°C, the following values are derived for relative river discharges in all 24 catchment areas (Table 11.12).

The results tell us that if the average annual T were to increase by 2°C, based on the correlations established to date between the average annual Q and average annual T, one could expect, as the most probable value, approximately 30% less water in rivers whose catchments largely lie within Serbia.

As already indicated, these correlations refer to annual averages. The distribution of average annual discharges as a function of the average annual temperatures was checked, and it is similar to those given by Gaussian, which corroborates the correlations further. If considerably more water is withdrawn in the future for irrigation, or if adequate climate change adaptation measures are implemented, the above correlation could change, but not by much for the near future.

Considering all the above, it seems that all hydrology projections that pertain to the distant future (50 or more years) should be deemed 'academic' and the focus of professional circles and society as a whole should remain on the near future (about the next 20–30 years), as the average changes in water resource availability are not only probable but to an important extent measurable.

**Table 11.12** Average relative river discharges based on linear trends for different increases in average annual T.

| **Relative River Discharge ($Q_{rel}$)** | **$\Delta T_{av}$ (°C) →** | | **0.5** | **1.0** | **1.5** | **2.0** |
|---|---|---|---|---|---|---|
| 1949–2006 (linear trend) | a = −0.1885 | b = 1.0010 | 0.91 | 0.81 | 0.72 | 0.62 |
| 1959–2016 (linear trend) | a = −0.1047 | b = 1.0127 | 0.96 | 0.91 | 0.86 | 0.80 |
| 1949–2016 (linear trend) | a = −0.1258 | b = 1.0094 | 0.95 | 0.88 | 0.82 | 0.76 |

Perhaps this methodology could help determine which regional climate-hydrological model is appropriate for a certain region for the near future. In order for it to be applied to individual catchments, it might be useful to produce the same RCMs for a number of catchments (at least 10 to 12) and to try to arrive at an average for the analysed region.

## 11.2.5 The impact of climatic parameters on water consumption

When considering the impact which climate factors have on human water consumption, it is important to distinguish between the water that is abstracted for drinking water supply purposes, and the water abstracted for agricultural purposes.

### *11.2.5.1 The impact of climatic parameters on the consumption of drinking water*

This impact is not uniform either temporally (particularly throughout the year) or spatially (different climate regions and countries, different habits, even in different regions of the same country). Locations with cold winters and warm summers are especially sensitive. In such locations two water consumption segments occur in a year: those during the warm part of the year (spring-summer), and those in winter (Weber, 1989). Cold temperatures, freezing precipitation, and snow-covered ground surfaces, tend to restrict water use to indoor activities. On the other hand, spring-summer warm temperatures and snow-free ground surfaces may induce significant increases in water consumption, both indoors and outdoors (Akuoko-Asibey *et al.*, 1993). The main analysed climatic variables are temperature and precipitation. It should be noted that other climatological phenomena such as relative humidity and evaporation also affect spring-summer water use but in most cases are not considered in papers because of the lack of available data.

Produced water (WP) in one water supply system (WSS) – in IWA terminology 'WSS input' – is the amount of water after water treatment. If water treatment plants (WTPs) do not exist in such a WSS, this quantity is equal or close to the abstracted amount. Assuming that all flowmeters in the WSS work properly, and that authorized unbilled consumption is negligible, the consumed amount of water (synonym – water use) in one WSS is the amount of water consumed by category (household, industry, public consumption) or by all categories in total. WP by a WSS is the compound of total water consumption (TWC) – in IWA terminology 'revenue water' – and primary water losses (net amounts), before the consumers' flowmeters 'non-revenue water' – NRW). NRW varies significantly from country to country (and from WSS to WSS in the same country, too), and generally is higher in less- or medium-developed countries. TWC is made up of consumption for household needs including watering of gardens (residential water use, which includes indoor and outdoor activities), consumption by industry (commercial), and public consumption (schools, hospitals, etc.), and varies from less than 50% to more than 90% of WP. Consumption rates depend on the consumer structure and habits, industrial demand, time of year, water pricing, climatic variables, secondary water losses and many other factors.

It would be interesting to investigate the different types of consumption as a function of climatological variables, but the required data are generally not available. It is more important (especially due to climate change) to see how much the increase in temperature or changes in precipitation affect TWC or WP in WSSs. This subchapter presents a few examples of how weather impacts drinking water consumption (or production).

***1° Example of the city of Belgrade (Serbia)*** (Dimkić *et al.*, 2013)

The topic here is the correlation between air temperature and water production. Only daily averages during the warmer part of the year, from May to October, were analysed. Given certain specific features

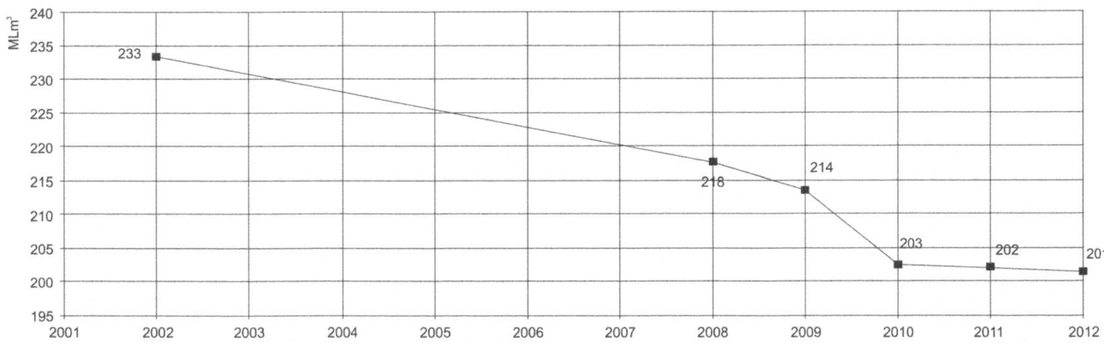


**Figure 11.23** Annual water production of the Belgrade Water Supply System from 2002 to 2012 (Dimkić *et al.*, 2013).

of each of these months (varying use of water for irrigation, summer holidays, etc.), they were grouped into three periods that were analysed and are presented separately: (1) May and June, (2) July and August, and (3) September and October. A linear trend was determined for each plot and the respective equation and coefficient of determination have been graphically represented. Then the percent change in water production for the 1°C difference was computed. It should be said that some water supply restrictions (water demand higher than water resource availability, but not significantly) have existed sporadically.

Public water supply in Belgrade began around 1892 with some 40 L/s withdrawn from springs within the city and its environs, for the then 50 000 inhabitants. The total peak of production, for about 1.6 million inhabitants, occurred at the beginning of the 21st century at an overall annual average rate of 7.5 $m^3$/s.

The efficiency of the system was upgraded considerably in the 2002–2012 period (Figure 11.23), with an annual average production rate of up to 6.5 $m^3$/s. Today, the Belgrade WSS produces about 200 million $m^3$ (originating from groundwater and surface water), which is processed by six water treatment plants. A distribution network that is some 3500 km long delivers water to about 1.5 million inhabitants and numerous industries and businesses.

Figure 11.24 from (a) to (c), shows the correlation between the total water production by the Belgrade WSS and the recorded daily temperatures for May/June, July/August and September/October 2011–2013. The percentage variation in production rates (rate of water production) for 1°C is computed:

In July/Aug the system recorded the highest average rates of water production (6577 L/s) despite the temporarily reduced population. This is attributed to the higher T during this period relative to the other

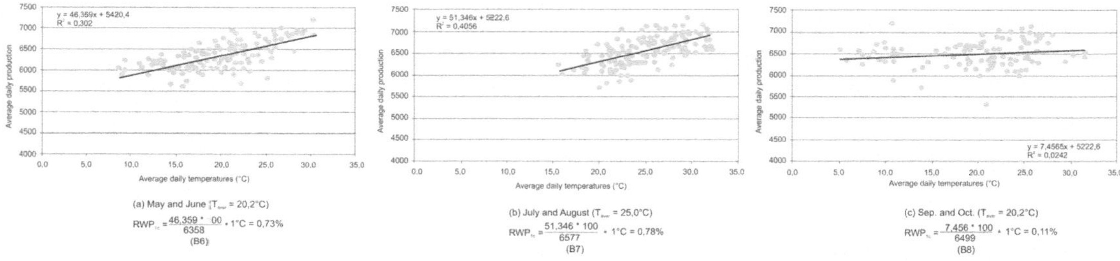


**Figure 11.24** Belgrade WSS: Correlation between the average daily water production and the average daily temperatures (2011–2013) (Dimkić *et al.*, 2013).

two. The Belgrade WSS also reported a higher production in Sept/Oct than in May/June as a result of a higher population count, while the average daily T were the same for the two periods (20.2°C).

The slopes of the trend lines show that the system exhibited a greater impact of T on production during the first two periods, May/June, and July/August. This was very likely due to the watering of gardens and other green surfaces (even though they are not very big) to a much greater extent than in the third period, September–October. During the first two periods it appears that a T increase of 1°C caused an average increase in the rate of water production of approximately 1%.

Results similar to those for Belgrade (T increase of 1°C causes an average increase in production or of the total water consumption of approx. 1%) have been obtained for the city of Niš, Serbia (Dimkić *et al.*, 2013) the city of Seoul, South Korea (Praskievicz & Chang, 2009) and Bahrain (Al-Zubari *et al.*, 2018).

***2° Example related to cities of Calgary and Montreal – Canada***

***The city of Calgary – Alberta (Canada)*** (Akuoko-Asibey *et al.*, 1993)

The city of Calgary (approx. 1.4 million inhabitants) exhibits an impressive climate variability throughout the year. The city has a cold climate, with strong T contrasts between the day and night, as well as between winter and summer. The summers are cool, with a mean daily T of 15°C. Temperatures during the summer have the least variability, and rarely exceed 32°C. The primary objective was to examine the effects of the mean maximum weekly T on the weekly TWC per capita in the spring-summer season. Summer TWC per capita tends to be higher, and in some cases is twice as high as in winter.

Figure 11.25 shows a plot of the mean maximum weekly T and weekly TWC per capita from January 1982 through December 1989. Two distinct segments can be discerned. The weekly TWC per capita seems to be steady when the mean maximum weekly T is below 15°C. In contrast to that, the TWC per capita increases with an increasing mean maximum weekly T when temperatures exceed 15°C. Looking at the trends in Figure 11.25 for temperatures above 15°C, the slope shows that for a T increases of 1°C an average increase in the rate of TWC of approximately 4% occurs:

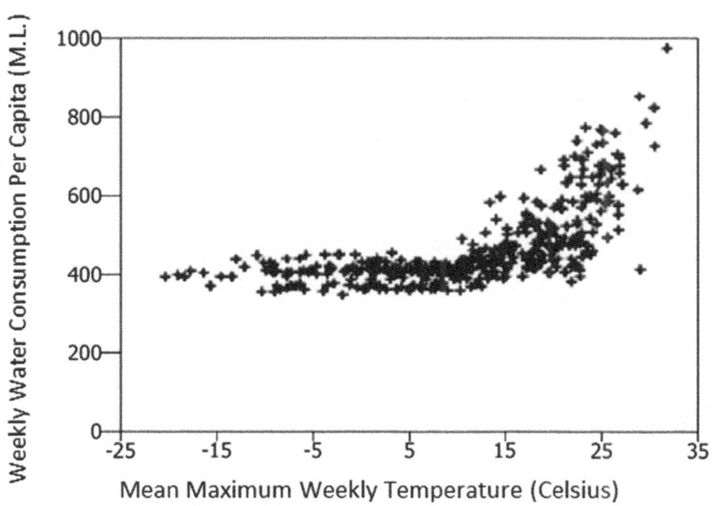


**Figure 11.25** Weekly water consumption per capita and mean maximum T, Jan. 1982–Dec. 1989 for Calgary (Akuoko-Asibey *et al.*, 1993).

For T = 15°C, average weekly consumption per capita is approx. 430 ML;
For T = 30°C, average weekly consumption per capita is approx. 700 ML;

$$RWC_{1c} = \frac{700\,L - 430\,L}{430\,L} \cdot \frac{1^{\circ}\mathrm{C}}{30^{\circ}\mathrm{C} - 15^{\circ}\mathrm{C}} \cdot 100\% = 4.19\%/1^{\circ}\mathrm{C}$$

Figure 11.25 also demonstrates that the weekly TWC per capita during winter is virtually insensitive to changes in the mean maximum weekly T. Consumption sensitivity begins when the temperature exceeds 15°C and this normally begins from the month of May.

***The city of Montreal (Canada)*** (Rasifaghihi *et al.*, 2020)

In the analyses of the observed drinking water consumption of the city of Montreal (approx. 1.8 million inhabitants), a clustering analysis of daily data TWC and T were used. It is proposed to split observations of daily TWC (period 2011–2016, Figure 11.26b) into two main clusters – 0 and 1 (the darkest and brightest spots on Figures 11.26b and 11.26c), and additionally, outliers (anomalous data) in a separate cluster 2 (two groups of spots at beginning of 2011 on Figure 11.26b, and on Figure 11.26c the two uppermost groups of spots). This choice is simple and works well to separate base water use from seasonal water use (Figure 11.26c).

It can be seen that there exists a T threshold (approx. 12°C) for the occurrences of transition from base to seasonal TWC and vice versa. It is acceptable to treat the base water use as independent of T, but the use is subject to weekday/weekend effects. The seasonal water use depends on the daily minimum and maximum T, and on the daily total precipitation (not presented here).

If the trend line were to be added to Figure 11.26c, for T above 12°C, the slope would show that T increases of 1°C cause an average increase in the TWC of approximately 3%:

For T = 12°C, the average daily consumption per capita is approx. 0.35 $m^3$ = 350 L/per capita/per day;
For T = 27°C, the average daily consumption per capita is approx. 0.50 $m^3$ = 500 L/per capita/per day;

$$RWC_{1c} = \frac{500\,L - 350\,L}{350\,L} \cdot \frac{1^{\circ}\mathrm{C}}{27^{\circ}\mathrm{C} - 12^{\circ}\mathrm{C}} \cdot 100\% = 2.86\%/1^{\circ}\mathrm{C}$$

***3° Example (study) related to numerous cities of a large part of the USA*** (Opalinski *et al.*, 2019)

In 2019 researchers in the USA published the following very interesting study related to this topic 'Spatial and Seasonal Response of Municipal Water Use to Weather across the Contiguous US'.

This study used a nationwide approach to characterize municipal water use drivers at both city and regional scales. The analysed monthly municipal water use spanned 229 cities in the contiguous US, had time spans ranging from one to eight years between 2000 and 2007, and was provided by the municipal

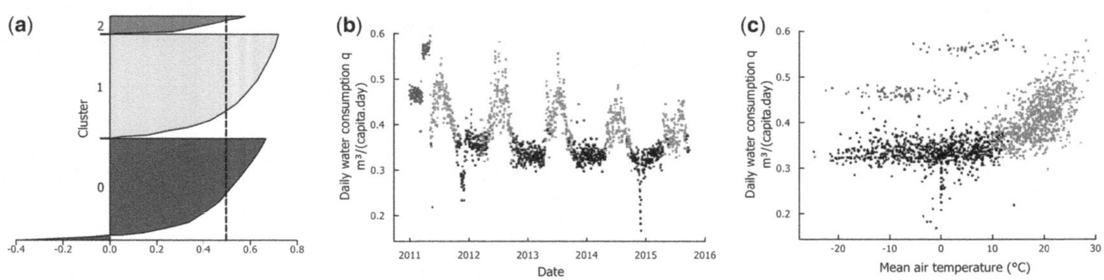


**Figure 11.26** (a) clusters; (b) observations of daily TWC; (c) scatter plot of daily TWC vs. mean T (Rasifaghihi *et al.*, 2020).

utilities with varying accuracy. Despite this, they claim that the data used represent the best existing cross-city picture of water use at a US national scale.

By taking a national approach, it may be possible to evaluate how weather-driven urban water use responses vary across cities and regions. This study used a statistical approach to characterize the short-term relationship between weather and municipal water use for 229 cities in the contiguous US, grouping it into six climate zones (Figure 11.27) aiming to answer the following research questions:

- What portion of the variability in municipal water use is explained by weather, and how does this vary across US cities?
- How does the response of municipal water use to weather compare across seasons and climate regions?

Temperature, precipitation, and the actual evapotranspiration (ET) were considered as explanatory variables to estimate changes in water use corresponding to weather variability. Using monthly municipal water deliveries, and these weather variables, it was possible to demonstrate that variability in water use was generally better explained by weather in dry regions of the western US. In addition, it was possible to estimate the average responses to weather across seasons and climate regions, concluding that water use changes in response to weather are typically higher in summer months in dry climates (Figure 11.28).

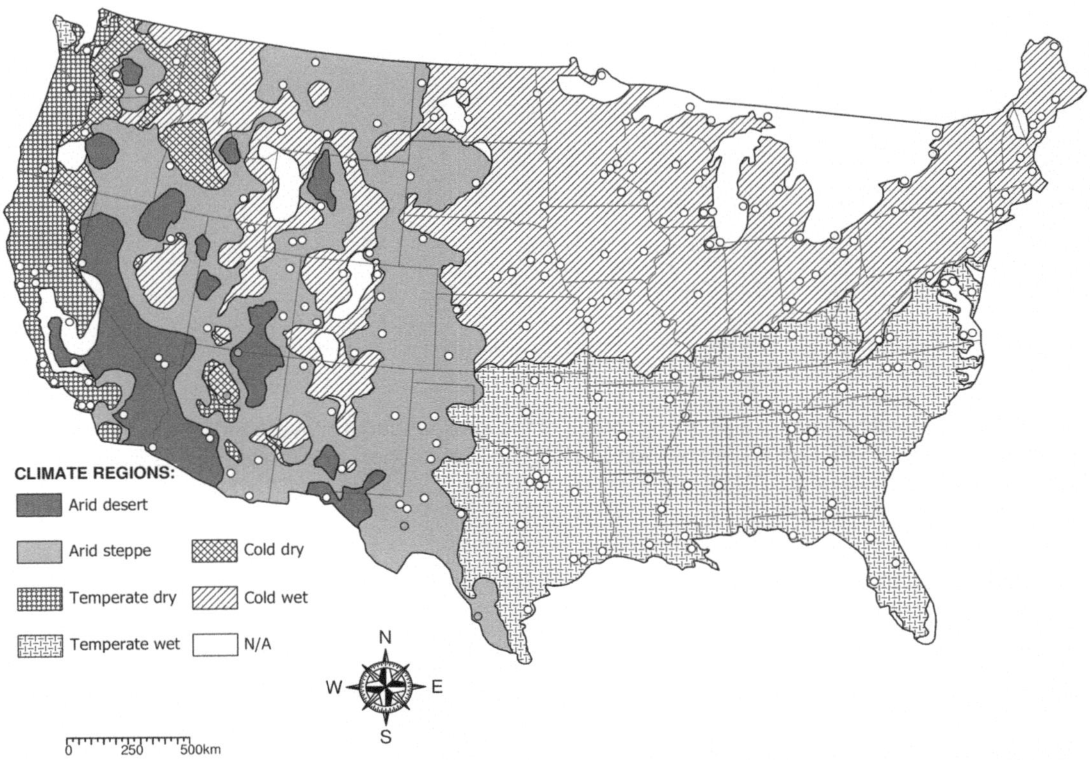


**Figure 11.27** Six climate classification groupings and 229 study cities; N/A denotes regions without study cities (Opalinski *et al.*, 2019).

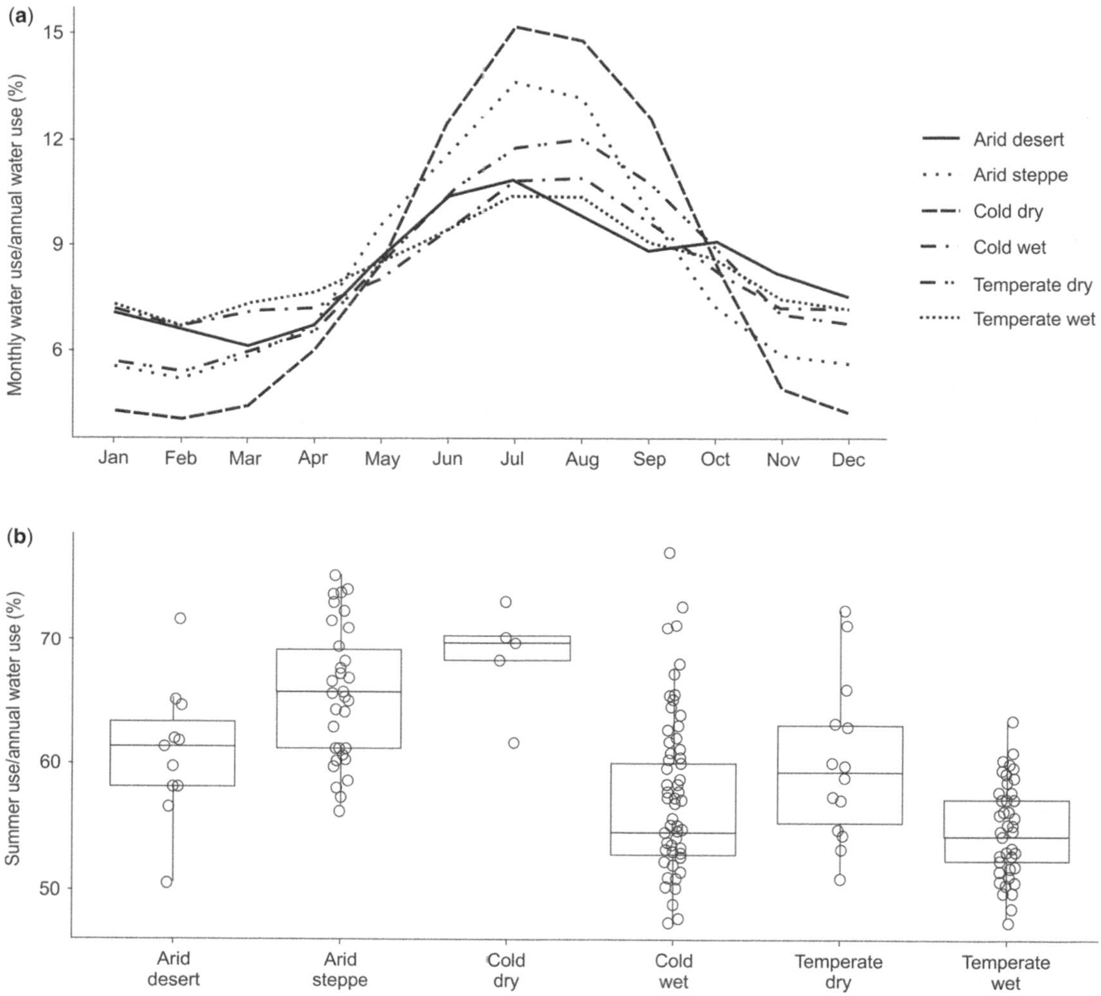


**Figure 11.28** (a) Monthly water use as a percentage of the total annual use, averaged by climate region; (b) Percentage of the total use that is summer water use (April-September) by climate regions, with points for individual cities (Opalinski *et al.*, 2019).

Noteworthy conclusions from this study are as follows:

(1) Seasonality in water use across climate regions was generally related to summertime changes in temperature.

(2) Weather variables alone (temperature, precipitation, and ET) explained most of the variation in monthly municipal water use across the US, with median city specific adjusted coefficient of determination $R^2$ ranging from 63% (temperate wet) to 95% (cold dry).

(3) City-specific regression models suggest that the maximum temperature was highly predictive of water use compared to other weather variables (selected in models for over 80% of cities).

(4) The water use response to temperature and precipitation was found to significantly change across seasons and climate regions of the US.
(5) Across all climate regions, water use increased with the summer maximum temperature and was greater in dry climates compared to wet. On average in dry regions, a 1°C increase in maximum temperature accounted for a 3.2% and 3.9% increase in water use in the winter and summer months, respectively. Comparatively in wet regions, a 1°C increase corresponded to a 1.1% and 3.0% increase in water use in the winter and summer months, respectively.
(6) Water use responses to precipitation were less predictable, especially in winter where increases in water use were sometimes observed with increased precipitation. A 1 mm increase in summer precipitation corresponded to a 0.06% to 0.15% decrease in water use in the wet and dry regions, respectively.
(7) Arid regions of the US were found to be distinct from others in a few ways. First, arid desert regions exhibited less water use seasonality than steppe (semi-arid) regions, and second, desert and steppe regions showed the smallest differences in summer and winter water use responses to maximum temperature. Both results indicate that arid regions have cities where irrigation occurs year-round and therefore the responses to weather in both seasons is similar.

***Conclusion for subchapter 11.2.5.1***

Municipal water use is influenced by a diverse set of climatic, socio-economic, demographic, policy, and landscape factors. Climatic (weather) variables include the temperature (maximum daily, average daily, average monthly, minimum daily), precipitation (amount per time unit, but also frequency and duration), soil humidity, wind (speed intensity, frequency and duration), etc. In addition to climatic, some other most important variables are demographic changes, availability of water resources, income, the price of water, WSS characteristics, habit, and others.

There is no consensus on which weather variables are the best predictors of municipal water use across cities, especially if we consider very different regions in the world. Temperature and precipitation are the most commonly used variables since data are widely available and are highly explanatory, particularly for estimating seasonal or summer-isolated water use. In general, increased water use is related to higher temperatures and lower precipitation (Gutzler & Nims, 2005; Kenney *et al.*, 2008) although some have found precipitation to have minimal effects on water use (Zapata, 2015). One of the reasons for this could be that automated irrigation controllers often do not adjust for rain events (Endter-Wada *et al.*, 2008).

Residential use on average accounts for 60% of the municipal, publicly supplied water in the USA (Dieter *et al.*, 2018) Residential water use could be divided into base use, defined as primarily indoor use with a low dependence on the influence of climate, and seasonal use, which is significantly dependent on climate. The daily base use is sensitive to days of the week and the daily seasonal use exhibits a relation to certain temperature thresholds. Examples from the USA and Canada have shown us that the first threshold value is about 14°C and could vary ±2°C from WSS to WSS. Below this threshold, water use (base use) is almost independent of temperature. Between the first and second threshold (approx. 30°C, also ±2°C from WSS to WSS) changes in water use due to the impact of temperature could be shown sufficiently as a linear relationship. Changes in water consumption at very high average daily temperatures, above the second threshold, could be much more severe (exponential dependency) than below that value (Maidment & Miaou, 1986).

Outdoor water use can account for 22–65% of the annual residential use (De Oreo *et al.*, 2016).

The impact of precipitation and temperature on the consumption of drinking water in the spring-summer months (average daily temperature between 14°C and 30°C) could roughly be expressed numerically:

- A 10 mm increase in summer precipitation corresponds to an approx. 1% decrease in water use, but it could be several times less or more, depending on the frequency and duration of precipitation and whether in a wet or dry region.
- The impact of air temperature on water use also varies: temperature increases of 1°C cause an average increase in the rate of water consumption (or production) from 1% (South Korea, Serbia, Bahrain) to 3–4% (Canada and the USA).

It can be concluded that the rate of the above correlations (between climate parameters and drinking water consumption) depends on water resource availability, water policy, water supply system conditions, and the habits of the people in the various regions.

### *11.2.5.2 The impact of climatic and other parameters on water consumption in agriculture*

Water withdrawals for agriculture account for ~72% of the total water that is withdrawn from surface water and groundwater globally, and for ~90% in developing countries (Cai & Rosegrant, 2002). These numbers tell us that the greatest pressure on water resources, including alluvial sources, comes from the agricultural sector. However, the percentage of abstracted water for the agricultural sector in a particular country or region could be quite different from the global average.

The impact of climatic and other parameters on agriculture can be divided into two parts:

(1) Impact on yield in agriculture; and
(2) Impact on the amount of water abstracted for irrigation.

***The impact of climatic and other parameters on the yield in agriculture***

Many researchers around the world have studied this topic. The results for some climatic parameters could be similar, or quite different due to various factors and the specifics of the analysed region/country. For example, a small change in groundwater levels may not cause significant changes, but if this occurs in a sensitive area, then yields can be significantly reduced (Xu & Su, 2019). Factors and specifics depend on the availability of modern technologies, climatic parameters, the available quantities of water resources, different types of land, biological and biotechnical factors, customs of the people in a given region, etc. Also, a different manifestation of the same climatic parameter, among other factors, has an impact on the income of farmers (for example, if we take the sum of precipitation in the vegetative period as a parameter, the distribution of the given precipitation by months is also important). Finally, the economic development of a given country has a significant role in the degree of impact of climate change. To assess the impact of climate change (predicted climate parameters) on future yields in agriculture, the Ricardian approach (Fleischer *et al.*, 2008; Liu *et al.*, 2004; Seo & Mendelsohn, 2008) is often used. The Ricardian approach is a feasible alternative method used to provide an assessment of the economic impacts of climate change on agriculture, but it often needs to be modified according to the database and characteristics of a country's agriculture.

Some examples of the impact of climatic parameters on the yield in agriculture are given below.

***1 Impact of P and T changes on canola and spring wheat yield in Saskatchewan – Canada*** (Ting Meng, 2017)

The impact of precipitation (P) and temperature (T) changes on the canola and spring wheat yield distributions have been investigated using Environment Canada weather data and crop yield and the planted area data from Statistics Canada for 20 crop districts over the period 1987–2010. Results have shown that the average crop yields are positively associated with the growing season degree-days and pre-growing P season, while they are negatively affected by extremely high T in the growing season.

Higher T tends to decrease wheat yields, confirming earlier Saskatchewan studies. Precipitation in the pre-growing season and P in the early plant growth stages are particularly relevant in providing opportunities to develop new crop varieties and agronomic practices to mitigate climate change impacts.

Higher growing degree-days (GDD) or heat units during the growing season increase the average yield of both canola and spring wheat, with a greater effect on canola yields. Specifically, a 10% increase in GDD enhances the mean yield of canola and spring wheat by 4.3% and 3.4%, respectively. Increasing GDD reduces canola yield variability but with little significant effect on the variance of the spring wheat yield. Our results indicate the temperature stress variable (the number of days with growing season temperatures greater than 30°C) reduces both the canola and spring wheat average yield and increases the spring wheat yield variability. A 20% increase in the number of days with extremely high temperatures (about three additional days) decreases the average yields by 0.6% for canola (8 kg/ha) and spring wheat (11.7 kg/ha).

***2 Impacts of P and T on crop yields in the Pampas – Argentina*** (Verón *et al.*, 2015)

Panel models were used to estimate the contribution of precipitation (P), the average temperature (T) and of the diurnal temperature range (DTR) during the growing season on wheat, maize and soy yield, and yield trends between 1971 and 2012 from 33 counties of the Argentine Pampas. Results have shown a relatively small but significant negative impact of climate trends on the yield which is consistent with the estimated crop and county specific sensitivity of yield to interannual changes in P, T and DTR and their temporal trends. The median yield loss from climate trends for the 1971–2012 period amount to 5.4% of the average yields for maize, 5.1% for wheat, and 2.6% for soy. The crop yield gains for this period could have been 15–20% higher if the climate remained without directional changes in the Pampas. On average, the crop yield responded more to trends in T and DTR than in P. These results indicate that climate trends slow the yield increase.

***3 The relationship between crop yields and climate variations – England*** (Chmielewski & Potts, 1995)

The relationships (correlation analysis and multiple regression) between meteorological variables (average monthly maximum and minimum air temperature, and total monthly precipitation) and the crop yields of a field in southern England have been examined. This investigation has used data related to the grain and straw yields from 1854 to 1967 at Rothamsted, UK.

The obtained results show that the changes in the crop yields at Rothamsted are partially related to climate variations at this site. Precipitation had the biggest influence on the grain yields with negative correlations between the grain yield and rainfall for all the months. The maximum air temperature was positively correlated with the straw yield in all the months except May, June and July. Generally, it is easier to explain the variability of straw yields due to weather variations than is the case with grain yields. With simple multiple regression models, it is possible to explain 33% of the grain and 50% of the straw yield variability due to precipitation and temperature variations. The lowest grain and straw yields were observed in the wet/cold as well as in the wet/warm years whereas the highest grain yields were more frequent in dry or cold years and the highest straw yields in dry or warm years.

***4 Observed trends in phenology and yields of field crops – China*** (Tao *et al.*, 2006)

Chinese researchers synthesized crop and climate daily data from representative stations in central and East China between 1981–2000 to investigate whether there were significant trends in changes of climate variables in different regions, and whether these changes had had a significant impact on the development and production of the staple crops (i.e. rice, wheat, and maize). The obtained results showed that significant warming trends were observed at most of the investigated stations, and the changes in temperature had shifted crop phenology and affected the crop yields during the two decades.

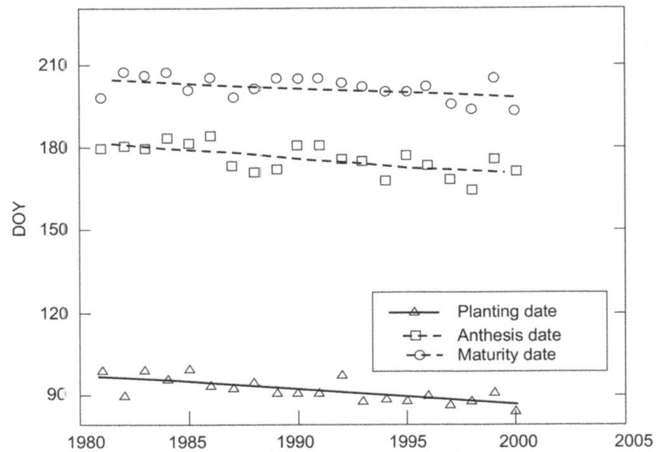


**Figure 11.29** Trends in early rice phenology at Changsha station (Tao *et al.*, 2006).

Some results, in the form of correlation, are shown on Figures 11.29–11.33 (DOY – day of the year).

The sensitivity of crop responses to T change is also influenced by other factors such as changes in other climate related parameters (e.g. precipitation), and management practices, pointing to the importance of adaptation through management practices. This study also highlights the need for further investigations of the combined impacts of temperature and $CO_2$ concentrations on the physiological processes and mechanisms governing crop growth and yield.

***5 Correlation between the annual cereal yield and weather during the growing season – Serbia***

Short analyses were conducted related to the impact of weather (precipitation and temperature) during the growing season (April–August) on the total cereal yield in Serbia. Available data existed for 10 years

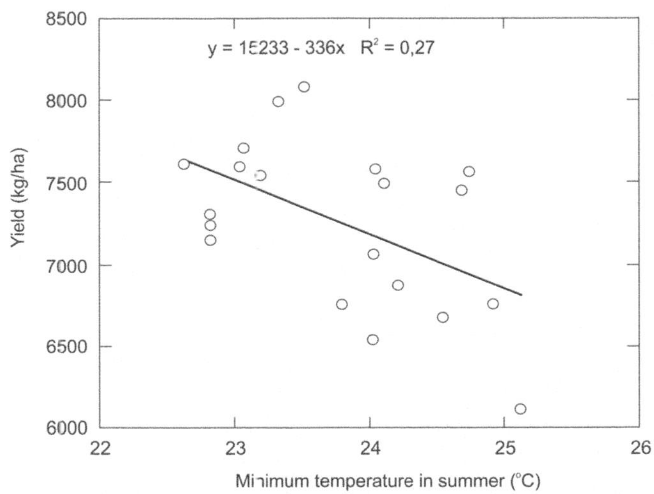


**Figure 11.30** Rice yield – mean minimum T in summer at Hefei station (Tao *et al.*, 2006).

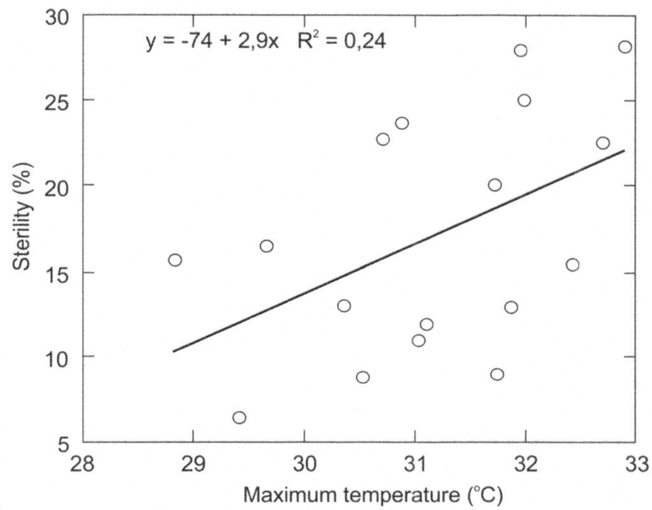


**Figure 11.31** Rice spikelet sterility – mean maximum T during the 20 days before and after anthesis at Hefei station (Tao *et al.*, 2006).

(1997–2006). The relative values were obtained as a quotient (for precipitation and the total yield) and the difference (for temperature) between the value of each parameter (sum for the precipitation and total yield, and the average for temperature) in the growing season of a particular year and as a 10-year average. The correlations are shown in Figure 11.34. As expected, the total yield tends to increase with both a precipitation and a temperature increase. Due to warming and no significant precipitation trend in the growing months (Apr–Aug) in Serbia, one could expect a slightly positive climate impact on the cereal yields. However, extreme weather events and many other factors make such forecasts uncertain.

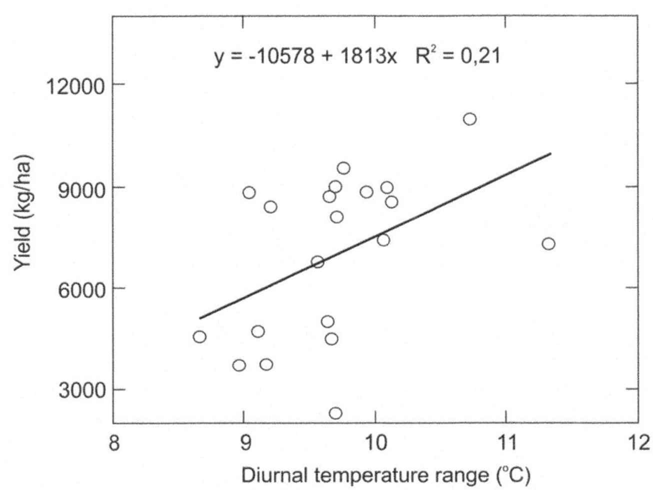


**Figure 11.32** Maize yield – mean diurnal T in summer at Zhengzhou station (Tao *et al.*, 2006).

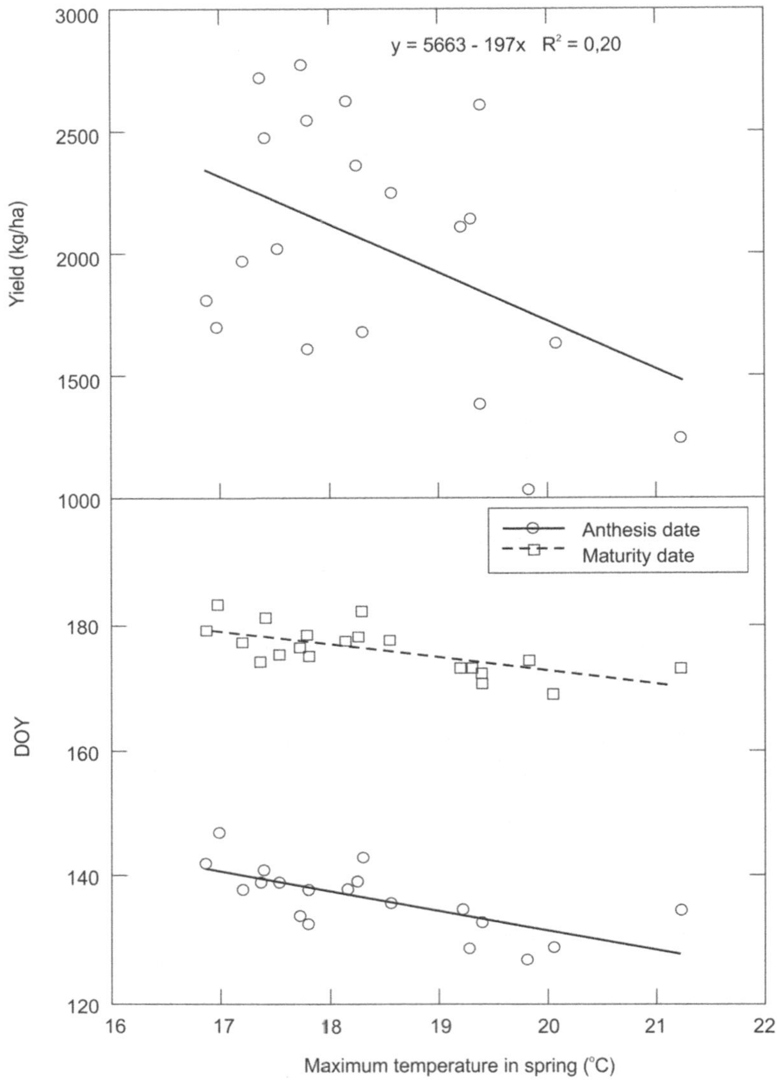


**Figure 11.33** Relations between wheat yield, phenology and mean maximum T in spring at Tianshui station (Tao *et al.*, 2006).

***6 The impact of climate change on African cropland – Africa*** (Kurukulasuriya & Mendelsohn, 2008) Analyses of the present crop net revenues of African farms were performed relying on the Ricardian method in order to investigate the current impact of climate on crop net revenues. Data have been collected from farmers in 11 countries. It was found that farm net revenues are lower in places with higher temperatures. Specifically, the temperature elasticity with respect to the net revenue of African farms is estimated to be −1.3. This means that a 10% increase in temperature will lead to a 13% decline in net revenue. The precipitation elasticity is estimated to be 0.4. African farms are more sensitive to

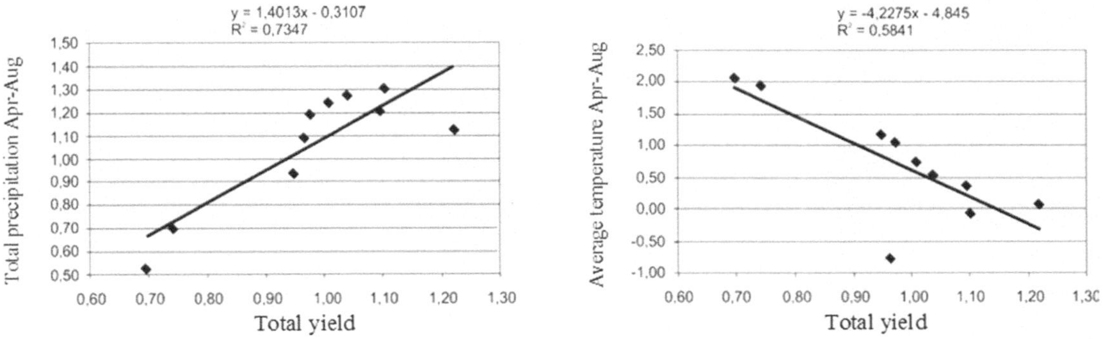


**Figure 11.34** Correlation between the total annual cereal yield and precipitation (a) and temperature (b) in Serbia during the growing season (April-August) for the period 1997–2006 (Serbian Ministry for Agriculture, Directorate for water).

changes in temperature than changes in precipitation. The sensitivity is greatest for the dryland farms with a temperature elasticity of −1.6 and a precipitation elasticity of 0.5. Irrigated farms are much more robust (Figure 11.35) to drought and may increase in value (partly because of their location in temperate regions of Africa) (Kurukulasuriya *et al.*, 2006).

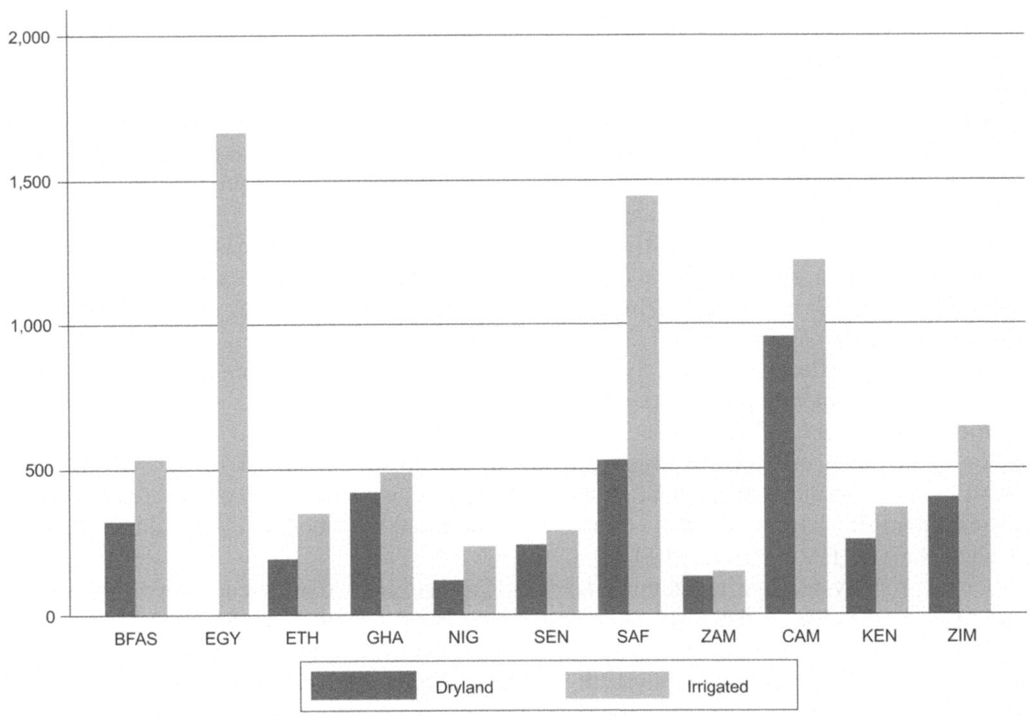


**Figure 11.35** Net revenue per hectare of dryland and irrigated cropland by country ($/ha) (Kurukulasuriya & Mendelsohn, 2008).

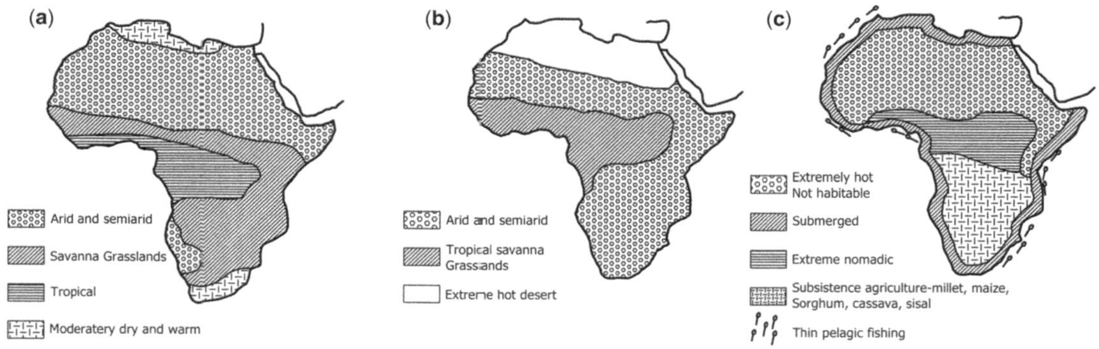


**Figure 11.36** African climatic zones at the beginning of the 21st century (a) and a prediction for 2030 (b). Prediction for African agricultural zones in 2040 (c) (Ngaira, 2007).

Africa is the continent most vulnerable to drought. The conventional approach used to predict the impact of climate is to develop numerous climate models and to single out the average values. African farms are sensitive to climate change in all models. Climate change would likely increase this pressure (Figures 11.36a and 11.36b). The predictions for agricultural zones have been completed for 2040 (Figure 11.36c) (Ngaira, 2007).

***The impact of climatic and other parameters on the amount of water abstracted for irrigation***

Irrigation as a hydrotechnical measure achieves optimal moisture during the vegetation period of plants. The water that is used is from watercourses, artificial lakes or groundwater, while a very small portion is from treated wastewater. Globally, crop yields using a hydrotechnical irrigation measure are 2.5 to 3.0 times higher than crop yields obtained using rainwater alone. With a significant increase in population in the second half of the 20th century, the pressure on food production and increasing yields has increased, and this implies an increase in water consumption in many parts of the world. Unfortunately, water is sometimes used excessively and in an unplanned manner, which can lead to water resource depletion and/or land quality degradation. The area of land equipped with irrigation systems is increasing, having increased from about 170 million ha in 1970 to about 300 million ha in 2010 (Wild *et al.*, 2010).

The impact of population growth and food demand are, for now, a much more important reason for the increase in water consumption for irrigation than climate change.

On the other hand, about 5–7 million ha (about 0.6%) of agricultural land are lost annually due to degradation (partly due to irregular irrigation) and urbanization. In parallel to the reduced areas of agricultural land as a result of the population increase, there is also a reduction of the area of agricultural land per inhabitant (0.4 ha/cap in 1961 versus 0.2 ha/cap in 2005) (Food and Agriculture Organization of the United Nations [FAO], 2011a, 2011b).

In order to properly manage agricultural land and irrigate with a sufficient amount of water, it is necessary to know the characteristics (physical and chemical) of the land to be irrigated. Based on that, it is possible to evaluate the time of water retention in the soil, how much soil moisture there is and at what depth, whether the water is free, physically or chemically bound, as well as how fast the water will infiltrate the deeper layers or evaporate. The quantities of water abstracted for irrigation depend on the needs of the given area for irrigation, the availability of water resources and the economic power of the given country/area to implement the necessary technical measures. The need of a given area for irrigation depends on the crops that are grown and the current water balance in the area. Ideally, irrigation should:

- Provide water in accordance with the needs of a plant at each stage of its development and the adopted agro-technical measures;
- Result in higher yields while not causing erosion, siltation and the salinization of the soil.

For irrigation and drainage, the state of water in the unsaturated environment (aeration zone), which represents the upper layer of the soil from its surface to the groundwater level, is of the greatest importance. The water regime and water balance in this zone is the subject of a complex water balance calculation. Water in an unsaturated environment is closely related to atmospheric precipitation, humidity deficit, etc. By studying the relationship between the balance and the movement of water in this zone with the balance and dynamics of surface and groundwater, it is possible to assess the role of these waters in the supply of the active layer of soil. The needs for land reclamation (Avakumović *et al.*, 2005) define the basic issues relevant for determining:

- Water supply in the aeration zone and its change over time;
- Water balance in the aeration zone and, in general, groundwater movements in a given area.

The land water balance is the sum of all the components of the inputs ($Pr$) and outputs ($R$) of water in a given time interval:

$$\Delta V = \mathrm{Pr} - R \tag{11.6}$$

The plant is the main consumer of water. Water inputs on the observed land in the same time interval $t$ originate from precipitation and the water runoff and seepage flow from the surrounding areas and from the underground. Outputs are related to evapotranspiration, plant mass formation, and surface and underground runoff into deeper layers and outside of the observed area, as well as to the soil microflora and microfauna.

Based on the water balance equation, three cases are possible (Figure 11.37) regardless of the duration of the interval in which the inputs and outputs are calculated: 1. $\Delta V = 0$; 2. $\Delta V < 0$; 3. $\Delta V > 0$.

In the first case, reclamation hydrotechnical measures are unnecessary, because the inputs and outputs are equal: $Pr \approx R$ (Figure 11.37a). This case is only temporary, because as soon as a demand is made to increase the harvest yields, this equality is violated.

In the second case, the inputs are constantly lower than the outputs: $Pr < R$ (Figure 11.37b) so there is a shortage of water, which should be compensated for by irrigation.

In the third case, the inputs are consistently higher than the outputs: $Pr > R$ (Figure 11.37c) so there is excess water, which requires drainage.

In cases 2 and 3, there is no rational agricultural production without water supply or drainage, so irrigation or drainage are necessary.

In temperate climates, in sub-humid regions, the most complex cases occur because the inputs and outputs are variable and hence $Pr < R$ and $Pr > R$ (Figure 11.37d) can happen, depending on the duration of the interval in which they are summarized. The water regime is unstable, so drainage is needed in one part of the year and irrigation in another. Here, irrigation and drainage are not imperative measures, as for the second and third case, but they are a supplementary tool, the usefulness of which is not always immediately apparent, and may occur on a case-by-case basis.

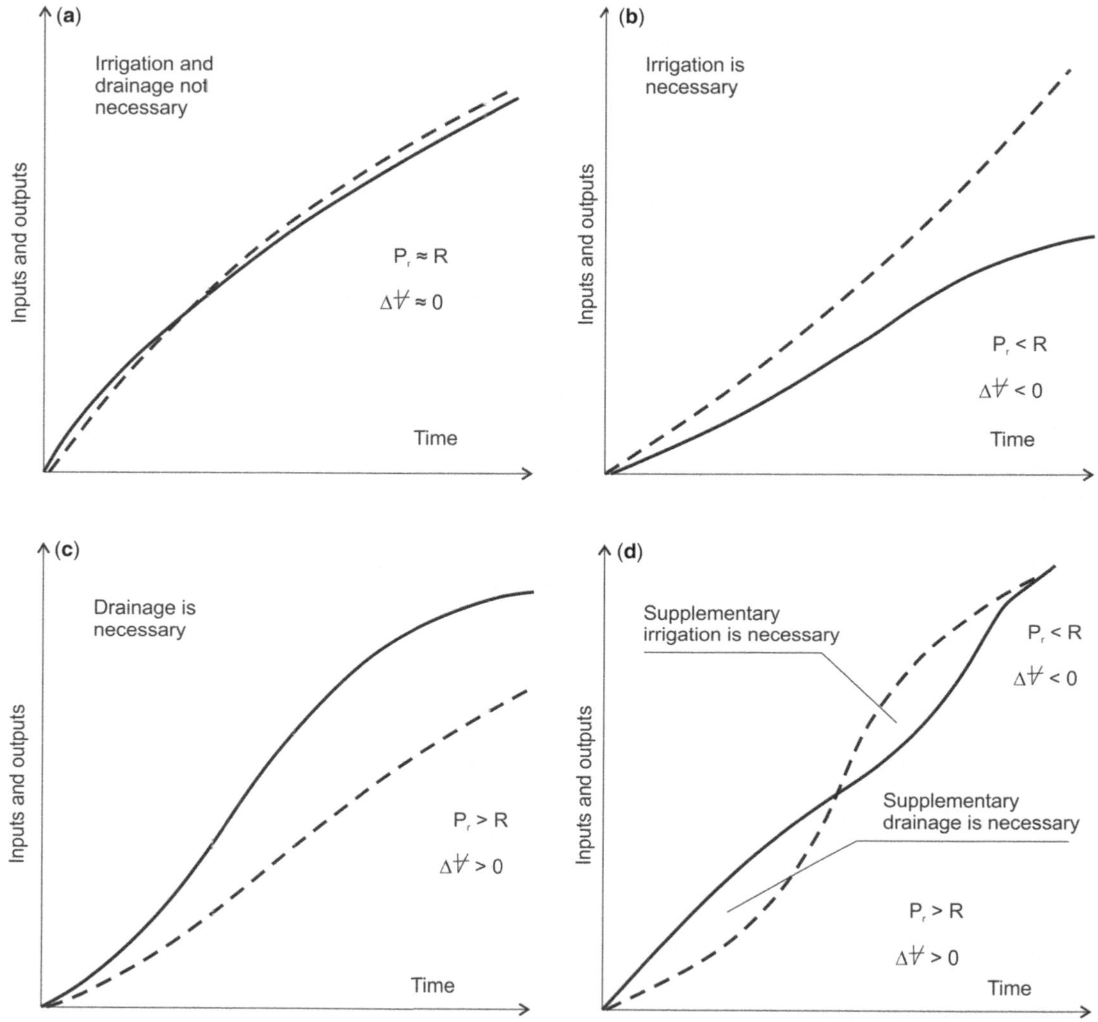


**Figure 11.37** Variation of land water balance input and output curves (Avakumović *et al.*, 2005).

The required amounts of water are often expressed through the term irrigation norms. Water consumption for evapotranspiration is covered in three possible ways: precipitation, irrigation and the available groundwater from the soil (water in the active layer plays the most important role). Depending on this, the irrigation norm can be defined as the amount of water supplied by the irrigation system to one hectare of area sown for a crop for the entire irrigation period in one season, in order to obtain the projected yields. There is a difference between net and gross irrigation rates. The net norm of irrigation is the amount of water used by a plant and the soil for evapotranspiration (Avakumović *et al.*, 2005).

The net irrigation rate is determined from the equation of the water balance of land:

$$\Delta V = ET - (Pe + W + Wk) \tag{11.7}$$

where:

$ET$ – amount of water consumed by the land during the growing season. It is usually taken to be equal to evapotranspiration,

$Pe$ – effective precipitation,

$W$ – water supply from the active layer of soil, which plants can use,

$Wk$ – groundwater, which plants can use by capillary uptake.

The task (norms) of irrigation is to meet the water balance deficit, i.e. the difference between evapotranspiration and the natural water reserves in the soil.

The gross rate of irrigation is the amount of water to be taken from an irrigation source and is determined by the expression:

$$V_b = \frac{V_n}{\eta} \tag{11.8}$$

where:

$\eta$ – efficiency of the system,

The irrigation rate is expressed in mm or $m^3/ha$.

Depending on the climate, soil conditions and cultivated crops, the components of the irrigation norm can have various numerical values.

Effective precipitation is a function of many parameters, so its more accurate determination is quite complex. Therefore, more practical calculations of effective precipitation are acceptable. By studying a number of meteorological stations and irrigation systems, the U.S. Department of Agriculture (USDA) has made certain recommendations for calculating effective rainfall (USDA, 1970). According to Doorenbos and Pruitt (1975) and USDA (1970), effective precipitation can be determined as a function of actual precipitation, evapotranspiration, and irrigation norms. Evapotranspiration is characterized by the intensity of water consumption, and at the same time the possibility for precipitation to be accepted and used, to become effective. The watering norm, depending on the crop and soil, characterizes the depth of the active layer and its capacity, i.e. the readiness to accept precipitation and to use it. Figure 11.38 gives the nomogram for determining the effective rainfall depending on the mean monthly rainfall and the mean monthly evapotranspiration. Depending on the watering rate, the given value should be multiplied by coefficient $\sigma$ (Table 11.13).

For example: The height of the average monthly precipitation is 100 mm; $ET$ (culture) is 150 mm and $Dn$ is 50 mm. The average monthly effective precipitation is: $0.93 \times 74 = 69$ mm.

However, this method of determining effective precipitation did not take into account either the intensity of individual precipitation events or the intensity of absorption into the soil. If the intensity of absorption is low (heavy soils or soil that is very wet), and the intensity of rain is high, most of the precipitation can be lost by swelling. In such cases, the effective precipitation, determined in this way, is experientially reduced, having in mind the slope of the terrain, the size of the system etc.

The water supply in the active layer of the soil is determined by the following expression:

$$W = \frac{wh_a}{100}[mm] \tag{11.9}$$

where:

$h_a$ – depth of the active layer of soil for a particular plant in mm,

$w$ – volume moisture of the soil in%.

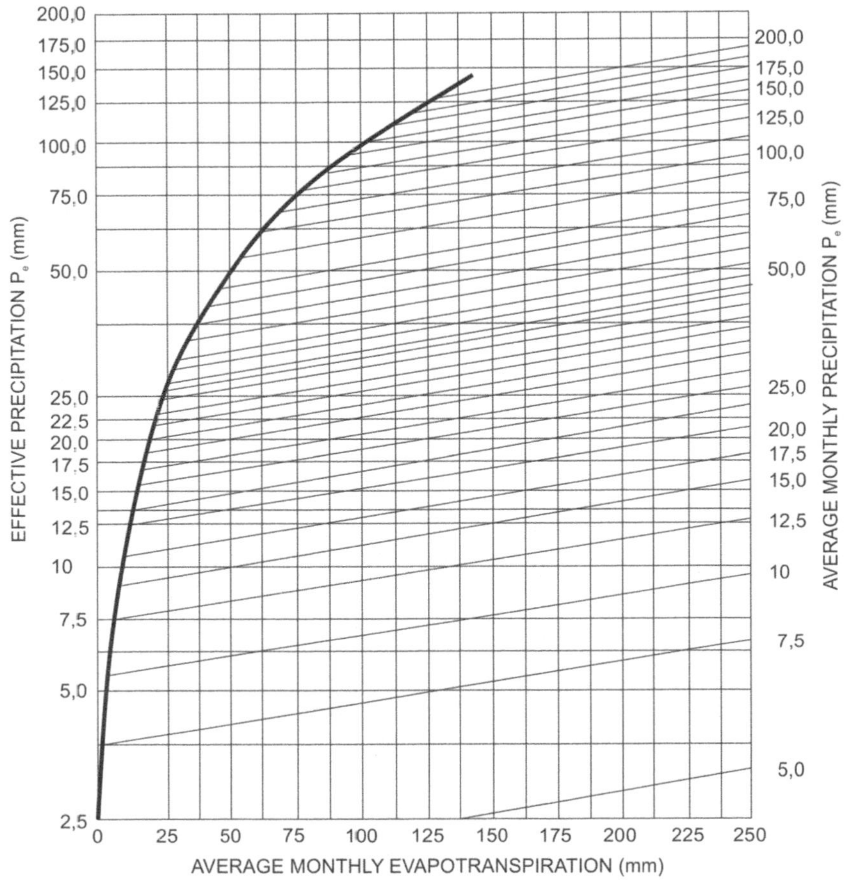


**Figure 11.38** Nomogram for determining effective precipitation depending on the mean monthly rainfall and the mean monthly evapotranspiration for watering rate *Dn* 75 ($\sigma = 1.00$) (Avakumović *et al.*, 2005).

**Table 11.13** Dependence of coefficient $\sigma$ on the watering norm *Dn*.

| **Watering Rate *Dn* (mm)** | **20** | **25** | **37.5** | **50** | **62.5** | **75** | **100** | **125** | **150** | **175** | **200** |
|---|---|---|---|---|---|---|---|---|---|---|---|
| Coefficient $\sigma$ | 0.73 | 0.77 | 0.86 | 0.93 | 0.97 | 1.00 | 1.02 | 1.04 | 1.06 | 1.07 | 1.08 |

If the groundwater level is shallow, then the water reserves in the active layer of the soil are constantly replenished by capillary rise from the groundwater level to the root system of plants.

The amount of groundwater that plants can use by capillary uptake, *Wk*, depends on many factors, the most important of which are the depth of the groundwater levels, the capillary properties of the soil, climatic conditions and the type of crop. Groundwater in such cases is very mobile and is of great importance for providing plants with water as well as for the physical and chemical processes in the soil. In the moderately moist and wetter regions, the amount of capillary water is generally negligible or small in relation to the amount of water in the active layer, but in drier climates it is the only reserve for the plant. Capillary water moves under the influence of capillary forces, i.e. the surface forces of soil particles that occur in pores with a diameter of 2–10 microns. Capillary water is available to plants and is the most important part of water for them. The amount of capillary water in a soil is referred to as the field moisture capacity or the capillary capacity of the soil. Table 11.14 gives the approximate values of water needs of different types of agricultural plants.

**Table 11.14** Average water demand of crops and crop water footprint (Hoekstra & Mekonnen, 2011; Hoekstra *et al.*, 2011; Mekonnen & Hoekstra, 2011, 2014).

| Crop | Water Demand ($m^3$/t) | Crop | Water Demand ($m^3$/t) |
|---|---|---|---|
| Wheat | 171 | Spinach | 7 |
| Wheat flour | 173 | Tomatoes | 31 |
| Wheat bread | 150 | Tomato paste | 126 |
| Dry pasta | 173 | Tomato ketchup | 79 |
| Rice, paddy | 170 | Cauliflowers and broccoli | 10 |
| Rice flour | 267 | Pumpkins and squash | 12 |
| Maize (corn) | 40 | Chilies and peppers, green | 21 |
| Maize (corn) flour | 41 | Garlic | 40 |
| Maize (corn) oil | 85 | Garlic powder | 156 |
| Rye | 12 | Beans, green | 27 |
| Rye flour | 16 | Peas, green | 31 |
| Potatoes | 16 | Bananas | 48 |
| Sugar cane | 28 | Oranges | 55 |
| Refined sugar | 243 | Orange juice | 99 |
| Beans, dry | 62 | Lemons and limes | 76 |
| Peas, dry | 16 | Grapefruit | 42 |
| Chickpeas | 112 | Apples, fresh | 66 |
| Cashew nuts | 460 | Strawberries | 54 |
| Chestnuts | 87 | Raspberries | 26 |
| Almonds, with shell | 954 | Blueberries | 167 |
| Walnuts, with shell | 649 | Grapes | 48 |
| Pistachios | 3801 | Pineapples | 4 |
| Hazelnuts, with shell | 545 | Coffee, green | 58 |
| Soya beans | 35 | Coffee, roasted | 69 |
| Coconuts | 1 | Cocoa beans | 2 |
| Olives | 249 | Green and black tea | 449 |
| Sunflower seed oil, refined | 149 | Vanilla beans | 19,524 |
| Cabbages and other brassicas | 13 | Cinnamon | 20 |
| Ginger | 20 | Watermelons | 12 |

In the most common modern method of determining the net rate of irrigation, the crop water consumption in the crop rotation (inputs) should be calculated using the Penman-Monteith method and the effective precipitation (outputs) as mentioned in the USDA method.

The Pennman-Monteith method is the only method recognized by the FAO for calculating reference evapotranspiration ($ET_0$) values. The reference evapotranspiration is a climatic parameter that expresses the atmospheric power to evaporate from the reference surface. The reference area is a hypothetical grass crop with a height of 0.12 m, a fixed surface resistance of 70 s/m and an albedo of 0.23. The reference area is very similar to extensive green, well-irrigated grass of uniform height during active growth and a complete land cover. Fixed surface resistance essentially means a medium-dry land area caused by a weekly watering pattern. Reference evapotranspiration is calculated based on the expression:

$$ET_o = \frac{0,408 \cdot \Delta \cdot (R_n - G) + \gamma \cdot (\frac{900}{T+273}) \cdot u_2 \cdot (e_s - e_a)}{\Delta + \gamma \cdot (1 + 0,34 \cdot u_2)} \tag{11.10}$$

where:

$ET_0$ – reference evapotranspiration (mm/day),
$R_n$ – net radiation from the crop surface (MJ/m$^2$/day),
$G$ – energy used to heat the land (MJ/m$^2$ day),
$(e_s - e_a)$ – water vapour voltage deficit (kPa),
$e_s$ – saturated voltage vp (kPa),
$e_a$ – actual voltage vp (kPa),
$\Delta$ – slope of the water vapour voltage curve (kPa/°C),
$\gamma$ – psychrometric constant (kPa/°C),
$T$ – air temperature measured at a 2 m height (in °C),
$u_2$ – wind speed measured at a 2 m height (m/s).

Crop evapotranspiration (expenditure) is a product of the crop coefficient, $k_c$, which contains the characteristics of the crop and the reference evapotranspiration ($ET_0$) in which different climatic conditions are taken into account. Crop evapotranspiration means that the crop is grown under standard conditions, i.e. that the crop is of an adequate sowing density, well provided with water and nutrients, that it is not under stress due to salinity and that it is free of diseases, pests and weeds. The evapotranspiration of a given crop is therefore obtained from the relationship:

$$ETc = ETo \cdot k_c \tag{11.11}$$

The values of the crop coefficients ($k_c$) for a given climate have been taken from literature.

When designing an irrigation system, the irrigation norm for typical crop rotation in an irrigation system is determined. It is most economical to determine the amount of water that will meet the water needs of crops in 4 out of 5 years or in 3 out of 4 years during the peak consumption period. In temperate-continental climatic conditions, such as in Serbia, the months are usually July and/or August. By considering the probability of drought individually by months and for the critical month during the entire vegetation season (July), it was determined that the most appropriate irrigation norm for Serbia is the one with a frequency of 23%, which means that a 100 year drought, whenever it may occur, will be mitigated by irrigation 77 times.

***Conclusion for subchapter 11.2.5.2***

Freshwater consumption is constantly increasing worldwide and it is happening faster than population growth. Arid and sub-arid regions are the most vulnerable to drought, but other climate areas are also affected. The amount of water used for irrigation varies very much from country to country and from region to region, depending on the irrigation demand, water resources availability and the economic power of a given country. The increase in population and the demand for food and a better life have, for now, had a more significant impact on the increase in freshwater consumption than has climate change. However, climate change further accelerates the increase in water demand.

Agriculture is expected to be one of the economic activities most impacted by climate change because weather is an essential input in agricultural production. The overall impact of climate change on human wellbeing will depend on a combination of the natural resilience of ecosystems and on the adaptation measures taken by farmers and other stakeholders.

## 11.3 IMPACT OF DROUGHT ON THE ABUNDANCE OF ALLUVIAL GROUNDWATER RESOURCES

### 11.3.1 General remarks

The classical alluvial environment, characteristic of areas of medium latitudes, consists of deposits of eroded gravelly, sandy siltstone and clay material. These environments are characterized by a narrower alluvium, less rich in material in the upper parts of the basin. The material of the alluvial aquifer is predominantly of large granulometric composition and predominantly high groundwater oxicity.

In the middle and lower part of the basin, the alluvial plain widens and the appearance of river terraces is characteristic (Figure 11.39). Alluvial aquifers become more abundant, and the oxicity of water in the aquifers decreases going downstream. In the lower parts of the basins of larger rivers, the material that makes up the matrix (skeleton) of the alluvial aquifer is increasingly fine-grained, and the groundwater is often anoxic.

In the previous part of the chapter, examples were given of meteorological, hydrological, agricultural and sociological droughts.

Meteorological droughts are a consequence of a lack of precipitation. They can be a phenomenon that usually occurs in one climate system. Climate change is happening and is expected, and will lead to changes in the regime of precipitation and runoff on the planet (floods and droughts frequency and duration). For Serbia, it has been estimated that in the case of a global increase in the average temperature of about 2°C, the flow in rivers may decrease by up to 30%. Changing runoff can therefore lead to different types of drought in different parts of the world.

Agricultural droughts occur primarily due to the 'escape' of water to deeper soil layers, out of the plants' root reach, resulting in reduced agricultural yields, and even to total loss of yield. With a significant decrease

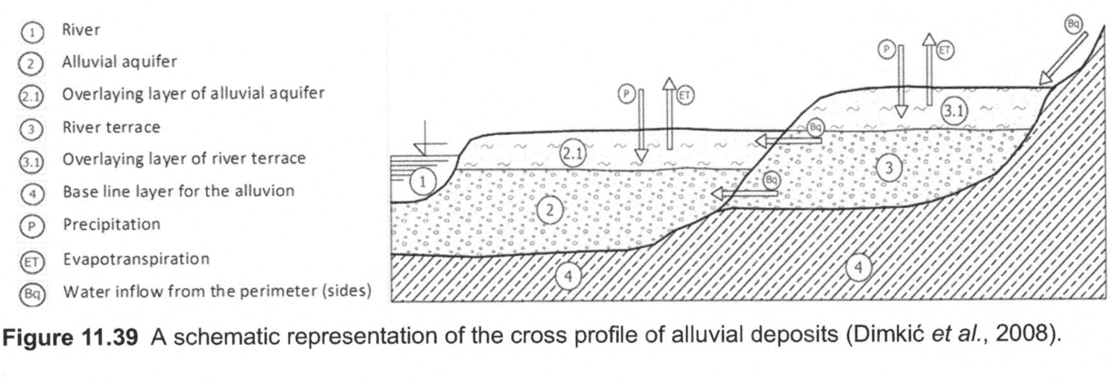


**Figure 11.39** A schematic representation of the cross profile of alluvial deposits (Dimkić *et al.*, 2008).

in precipitation and an increase in air temperature, there is an imbalance of moisture in the soil. The evaporation and transpiration of plants exceeds the income of moisture and groundwater is withdrawn to lower levels here and thus reduces evaporation. Plants therefore have to collect water from lower levels in the soil, reduce vegetative activity, and ultimately dry out. Plants that inhabit sub-arid and arid regions have a deeper root than average, and some fight moisture loss by shaping their leaves in a way that reduces evapotranspiration (Gunduz and Simsek, 2011).

Social droughts occur due to the increase in the number of inhabitants on the planet, or in certain areas. World water demands are estimated at 3000–4000 $km^3$/year, and it is already difficult to maintain the water resources management system. The continued trend of population change indicates an increase in consumption by 2050 of 30–50%. This would lead to a demand for water of about 5000 $km^3$ per year. The further concentration of inhabitants in the already densely populated areas of alluvial plains, leads to 'conflicts' between the expansion of agglomerations and the maintenance of natural resources in the alluvium.

In the end, the factors that influence the occurrence, types and effects of droughts include the climate and climate change, water consumption, the lithological soil structure, the hydrological balance of rivers, the degree of afforestation, urbanization and population density, the degree of land covered with forests, crops and other plants, irrigation, etc. This indicates a large number of drought factors and the need to respect the expressed specifics of individual areas/problems.

## 11.3.2 Basics of the water basin balance

Given below is one way of writing the equation of the water balance for a river basin for a certain period of time (Lebedev, 1967):

$$P = Q_1 + Q_2 + ET + q_1 - q_2 + S_{T1} + S_{T2} + D \tag{11.12}$$

where:

$P$ – average for the whole basin, the amount of atmospheric precipitation,
$Q_1$ – water runoff in the final profile of the river,
$Q_2$ – water runoff in the final profile of the alluvial aquifer,
$ET$ – average amount of total evaporation and transpiration,
$q_1$ – water intake from the river and aquifer,
$q_2$ – used return water,
$S_{T1}$ – average for the basin, change of groundwater supply in the aquifer (aquifers),
$S_{T2}$ – average for the pool, change of moisture supply in the vadose zone,
$D$ – deviation in the water balance, including exchange in other basins and errors in the assessment of individual quantities.

$S_{T1}$ is calculated from:

$$S_{T1} = \mu \Delta H \tag{11.13}$$

where:

$\mu$ – specific effective porosity (effective storage),
$\Delta H$ – the change in groundwater piezometric water level.

Equation (11.13) is given very roughly, practically in symbolic form. It can be broken down and reshaped in various ways for the needs of balancing water in the basin, alluvial aquifer or overlying layer, for the purposes of calculating the required amount of water for irrigation and similar.

The amount of precipitation, water consumption and the amount of discharged/returned wastewater to the basin/river can be clearly set in a relatively well-managed water basin and is not the topic of this chapter.

The subsurface runoff and the calculation of changes in the groundwater reserves as well as the moisture in the vadose zone is a more complicated task.

The same is true for the calculation of evapotranspiration which can be separated into evaporation ($E$) and the transpiration of plants ($T$):

$$ET = E + T \tag{11.14}$$

In this case, evaporation can occur from the surface of the terrain as well as from the somewhat deeper layers of the soil. With the increasing depth of groundwater, evaporation from its surface decreases.

The capillary action climbing of groundwater moisture is very important for the water content humidity in the vadose above-ground zone. Humidity in the above-ground zone is very important for the successful production of agricultural crops as well as other plant vegetation.

By calculating the water runoff, an important parameter of the water balance is obtained:

$$\text{runoff}(O) = \frac{Q_1 \times t}{F} \tag{11.15}$$

If we divide the average precipitation by the runoff, we get the surface runoff for a basin. This is a very important parameter of the water balance and water management in the basin and, as was shown in previous subchapters, it depends on the relief, climate and climate change, and on the plant and pedological cover, etc.

Figure 11.39 shows a characteristic cross-section of the mid-reach alluvium and downstream sections, for medium-sized and larger rivers. We see that for a significant part of the basin and the balancing of water in such basins, the hydrogeological profile can be viewed as a two-layered environment plus the terrace sediments. The two-layer medium is represented by a well-permeable alluvial aquifer and a less permeable roof layer confining bed (overlaying aquitard). The difference in water permeability between these two layers, as a rule, is several orders of magnitude. The flow of water in the upstream layer is predominantly in the vertical direction. The flow of water in the aquifer can usually be viewed as horizontal. The flow in the low-permeability roof layer upper confining bed can be the flow in a water-saturated and/or water-unsaturated medium.

The impact of drought on the alluvial space can be observed from several aspects of which we emphasize two important ones:

(1) the plant water demand and the damage that can occur in alluvial plains due to drought;
(2) the impact of drought on the water content of alluvial aquifers as a water resource.

For both aspects, the soil moisture in the topsoil is important. This is especially true for the issue under point 2, where the vertical water balance through the less permeable layers dominates. In view of this, in subchapter 11.3 we will focus on:

- Some basic concepts and equations relevant to the groundwater flow in unsaturated environments;
- Water capillary action (rise) climbing of water and the formation of an equilibrium moisture profile in the soil;
- The evaporation of water from the free surface, from plants and from shallower soil layers and transpiration;
- The coefficients of water permeability and the air permeability of soil in unsaturated environments and the precipitation infiltration;

- The dependence of the humidity profile in relation to points 1) and 2) above on the hydrogeological characteristics of the terrain;
- The relationship between the depth of the roots of plants and the soil moisture profile.

## 11.3.3 Vertical water balance in the alluvial outcrop

Intensive groundwater exchange with the atmosphere takes place through the aeration (vadose) zone. Here, groundwater is lost by evaporation as well as transpiration. Groundwater inputs can be from precipitation, by contact from a river into the alluvial aquifer, and possibly from irrigation, artificial infiltration or the discharge of used water. Groundwater use can also be due to abstraction withdrawals by people to fulfil their own needs.

The vertical water balance therefore depends on many hydrological and climatic factors, the structure of the geological/pedological environment, plant cover, etc.

The hydrological regime of the river flow is important here. The river represents the boundary of the groundwater flow whether it drains or replenishes the alluvial outcrop aquifer. The closer we are to the river, the more important is the water balance in the aquifer and the upstream layer of the river. As you move away from the river, the vertical water balance becomes more and more dominant.

The rational theory of groundwater infiltration and evaporation is covered by the laws of water and air flow in an unsaturated porous medium. Equations pertaining to the water flow in unsaturated media also define the effective porosity (specific yield and lack of saturation). Knowing the exact value of the effective porosity is of great importance for all calculations regarding groundwater oscillations. As will be shown, effective porosity often does not have a static character but depends on the flow regime and is therefore defined by the equations pertaining to the water flow in unsaturated media. In the case of slow, controlled infiltration, air pressure must be taken into account because the increase in air pressure is the main reason for the decrease in infiltration. The hydraulics of the water flow in unsaturated environments has led to the need for a more precise definition of a number of terms that have been used, and are still used today, in a number of sciences that are interested in groundwater problems in the aeration zone.

### *11.3.3.1 Basic concepts and equations relevant to the groundwater flow in unsaturated environments*

All changes in humidity in the so-called aerated (vadose) zone, i.e. the unsaturated porous environment, are accompanied by water and air flow, whereby during drainage the liquid phase (water) is replaced by the gaseous one, while in the case of infiltration and humidification, the process is reversed. It follows from this that the flow of water and air in the aerated zone should be treated with equations relating to two-phase fluid flows in porous media.

The following equations assume that the bulk density of the porous medium is constant, as is the specific gravity of water and air. The solubility of air in water is neglected. It is assumed that the process is isothermal and that there is no ion exchange between water and the solid phase.

The skeleton of the porous medium matrix is characterized by porosity $n$ which represents the ratio of the pore volume to the total volume. The pore volume is filled with water or air. The degree of water saturation $S$ is defined by the ratio of the volume of water to the total volume of the cavities; the degree of air saturation is $1/S$. The moisture content in the soil ($\theta$) is:

$$\theta = n \cdot S \tag{11.16}$$

It is assumed that the phases are continuous, which means that it is possible to move from one point of one phase to any other point of that same phase without intersecting another phase.

The first equation we introduce relates air pressure to water pressure:

$$P_w + P_a = P_c \tag{11.17}$$

where:

$P_c$ – capillary pressure,
$P_w$ – pressure in the liquid phase (water),
$P_a$ – air pressure.

The difference in pressure between the water and air appears due to the existence of the curvature of the meniscus. Capillary pressure is always negative and depends on the degree of soil saturation with water (for soil with given characteristics).

The mass balance equation for water, for the made assumptions reads:

$$n\frac{\partial S}{\partial t} + divV_w = 0 \tag{11.18}$$

i.e.

$$n\frac{\partial S}{\partial t} = \frac{\partial \theta}{\partial t} = -\left[\frac{\partial u}{\partial x} + \frac{\partial v}{\partial y} + \frac{\partial w}{\partial z}\right] \tag{11.19}$$

where:

$V_w$ – fictitious Darcy's filtration velocity whose components are $u$, $v$ and $w$.

The mass balance equation for air has the form:

$$n\frac{\partial(1-S)}{\partial t} + divV_a = 0 \tag{11.20}$$

Finally, it is assumed that the relationship between the velocity $V_w$ and the piezometric height $h$ is one equation – the so-called generalized Darcy equation – analogous to the Darcy equation for saturated media, both for air and water. For water it is:

$$V_w = K_w \cdot grad\ h_w \tag{11.21}$$

where $K_w$ is the Darcy's filtration coefficient, which is often called conductivity in world literature.

For air, the generalized Darcy's equation is:

$$q_{\mathrm{a}} = K_a \cdot grad\ h_{\mathrm{a}} \tag{11.22}$$

Piezometric heights $h_w$ and $h_{\mathrm{a}}$ are given by the expressions:

$$h_w = \frac{P_w}{\gamma_w} + z \quad h_a = \frac{P_a}{\gamma_a} + z \tag{11.23}$$

Since $\gamma_a$ is small relative to $P_a/\gamma_a$ the term $z$ in Equation (11.23) can be neglected.

The quantities $K_w$ and $K_a$ represent the permeability coefficient in the Darcy's sense, so they have a velocity dimension. On the one hand, they are a function of the degree of saturation, and on the other, a characteristic of the fluid. Therefore, the concept of geometric permeability $C$, is defined by the expression:

$$K = C\frac{g}{v} \tag{11.24}$$

The $C$ values for water as well as for air for a particular undeformed soil depend only on the degree of saturation. $C$ has a dimension of $cm^2$, which is why it is called geometric permeability. From Equation (11.24) we would get that the permeability coefficient $K = 1$ cm/s corresponds to:

$$T = 12^{\circ}(v = 0.124\,\mathrm{cm}^2/\,\mathrm{sec});\, C = 1.25 \cdot 10^{-5}\mathrm{cm}^2 \tag{11.25}$$

In the following text, the equations with the Darcy's filtration coefficient will be used.

In all, we obtained total of five equations: scalar Equations (11.17, 11.19, 11.22) and vector Equations (11.21, 11.22) with nine unknowns, as follows:

$P_a$, $P_w$ and $S$, three components of velocity speed $V_w$ for water and three components of velocity speed $V_a$ for air. In general, the given equations cannot be solved explicitly.

#### *11.3.3.2 Moisture balance profile during drainage*

Consider a soil column that is in contact via its base with a reservoir whose level corresponds to the surface of the soil. Suppose the soil is completely saturated with water; the pressure distribution in the water is then hydrostatic.

Lower the tank to near the lower level of the column. Due to the weight of the water, water begins to evacuate from the pores, but capillary forces limit this evacuation. After some time, an equilibrium profile will be established with the saturation level of water in the soil, which is shown in the form of a diagram as follows (Figure 11.40).

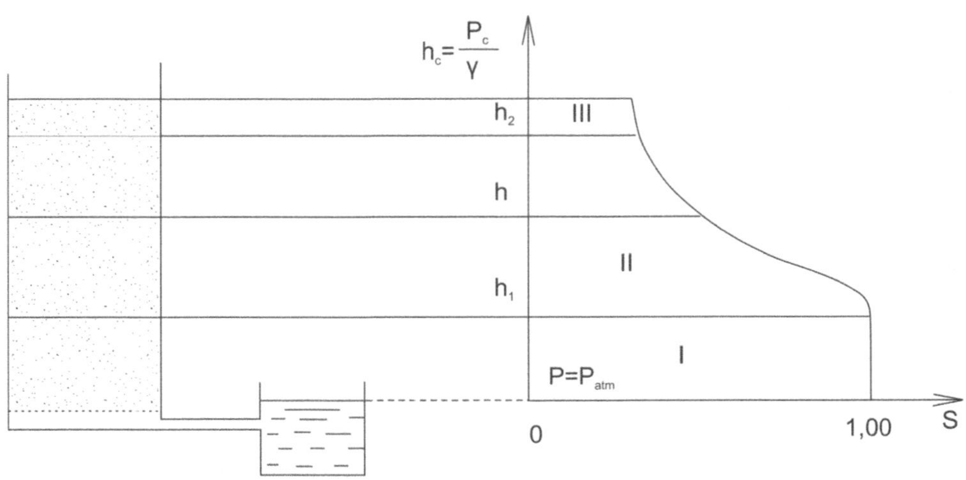


**Figure 11.40** Schematic presentation of the dependence of capillary pressure on the degree of saturation in the drainage cycle: Zones: I – zone of complete saturation, II – zone of the continuity of the liquid and gaseous phase, III – zone of the discontinuity of the liquid phase (Boreli, 1967).

In the zone where the water is in a continuous phase, the hydrostatic law of pressure prevails, so for water we have:

$$P + \gamma_w \cdot h = const. \tag{11.26}$$

Since the air pressure is equal to the atmospheric pressure, the elevation relative to the water level in the tank corresponds to the capillary pressure:

$$P_e = -\gamma_w \cdot h \tag{11.27}$$

where height $h$ in the considered state represents the height of the capillary pressure for the moisture content, i.e. the saturation that occurs at that height.

In the case of an equilibrium profile we have:

$$h = h_c = \frac{P_c}{\gamma_w} \tag{11.28}$$

Using the known equation that relates the capillary pressure and the capillary radius, it can be said that the following relationship exists between capillary pressure and the mean radius of the curvature of different menisci:

$$P_c = \frac{2\sigma}{r} \tag{11.29}$$

where:

$\sigma$ – capillary stress ($\sigma \approx 0.075$ g/cm).

There will be three zones on the equilibrium humidity profile.

The first zone, in which the soil is completely saturated with water, is in the range $0 < h < h_l$. This zone occurs because in the observed soil there can be no meniscus whose radius is greater than:

$$r_1 = \frac{2\sigma}{h_1} \cdot \frac{1}{\gamma_w} \tag{11.30}$$

The value of $r_1$ directly defines the value of the largest pores of the soil and depends on the dimensions of the particles, the porosity of the soil, the size of any cracks, etc.

This occurs in the soil and air (zone II) only for values of $h$ that are greater than $h_1$; and the largest menisci appear. This leads to a decrease in saturation. If $h$ grows, the finer menisci meet, and the pores gradually empty. While this is happening, the water is still in a continuous phase.

Starting from some height $h_2$, the menisci become too small, the water is no longer in a continuous phase; either isolated droplets or a film of absorbed water covering the grains occur (zone III).

In the zone in which water appears in the form of a discontinuous phase, the exchange takes place by means of water vapour; and pressure obeys Kelvin's law.

Since the minimum value of the radius of the curvature of the meniscus is very small, the molecular dimensions $h_c$ and $P_c$ can have a very large value. Therefore, the logarithmic exchange of $PF$ is introduced, so that:

$$PF = \log\left(\frac{-P_c}{\gamma}\right) = \log_{10} h_c \tag{11.31}$$

where $h_c$ is expressed in cm.

**Table 11.15** Capillary rise calculations, partially based on Adamski *et al.* (2005).

| Sediment | Average Grain Diameter (cm) | Representative $d_{10}$ (cm) | Pore Radius $=d_{10}/d_5$ (cm) | Capillary Rise (cm) |
|---|---|---|---|---|
| Fine silt | 0.0008 | 0.0002 | 0.00004 | 3750 |
| Coarse silt | 0.0025 | 0.00063 | 0.00013 | 1154 |
| Very fine sand | 0.0075 | 0.0019 | 0.00037 | 405 |
| Fine sand | 0.0150 | 0.00375 | 0.00075 | 200 |
| Medium sand | 0.03 | 0.0075 | 0.0015 | 100 |
| Coarse sand | 0.05 | 0.0125 | 0.0025 | 60 |
| Very coarse sand | 60 | 0.050 | 0.010 | 15 |
| Fine gravel | 0.50 | 0.125 | 0.025 | 6 |

Capillarity action in rocks is the phenomenon whereby water or some other liquid moves and rises in rocks that contain capillary pores and cracks, due to the action of adhesion and cohesion forces (Table 11.15). Capillary pores are all those pores in a rock of 0.5–0.0002 mm in size, and capillary cracks are all those cracks of 0.25–0.0001 mm in size.

Pores and cracks larger than these are known as the super-capillary porosity of the rocks, while those that are smaller are known as the sub-capillary porosity of the rocks, mm (at 0°C).

$$H_K = \frac{2a^2}{d} \tag{11.32}$$

where:
$H_K$ – capillarity rise (m),
$d$ – pore diameter (m),
$a$ – capillarity constant.

Each pore represents an equivalent for a different capillary tube which means that the value of the capillary rise is different in laboratory conditions compared to field conditions. It is difficult, if not impossible, to even out the pores, even in controlled laboratory experimental conditions. Due to the complexity of the soil and the distribution of pores that appear unevenly, it is very difficult to accurately predict the capillary rise in the soil under natural conditions.

It is known that in nature, capillary rise for some clays can exceed the above-mentioned values.

#### *11.3.3.3 Relative permeability coefficients and infiltrations in unsaturated media*

As a rule, the permeability coefficients of unsaturated soil are expressed in relation to the permeability coefficient when the soil is completely saturated with the appropriate phase. This gives a quotient called the relative permeability coefficient (index $r$).

For water it is:

$$K_{rw} = \frac{K_w}{K_{0w}} \tag{11.33}$$

where $K_{0w}$ is the coefficient of permeability for water for $S = 1$.

i.e. for $\theta = \theta = n$

For air it is:

$$K_{ra} = \frac{K_a}{K_{0a}} \tag{11.34}$$

where $K_{0a}$ is the permeability coefficient for air for $S = 0$.

When the soil is 100% saturated, the water permeability is maximum ($K_w = K_{0w}$), while the air permeability $K_a = 0$. When the degree of saturation decreases, the water permeability coefficient decreases and becomes equal to zero for some saturation $S_2$, starting from which the liquid phase of water is no longer continuous.

When the degree of saturation reaches the value of $S_1$ during the decrease of humidity, the air phase becomes continuous. The value of $K_{ra}$ increases, while $S$ decreases, and reaches the maximum value ($K_{ra} = 1$), with $S = 0$.

Typical curves $K_{ra}$, $K_{rw}$, for the drainage cycle are shown in Figure 11.41.

It can be seen that for air $K_{ra} = 1$, for $S \approx 10\%$; the very fine menisci that still exist at that saturation and which are in contact with the grains no longer provide any measurable resistance to air movement. For water, $K_{rw}$ does not differ significantly in the drainage and irrigation cycle.

This means that:

- A higher intensity of infiltration is achieved via the precipitation system – infiltration or water supply through the infiltration basin – while achieving a greater saturation of the soil with water;
- A higher infiltration intensity is achieved for soil with a higher absolute filtration coefficient.

An interesting phenomenon is the so-called inhibited infiltration.

During infiltration that follows intense precipitation the soil can be completely saturated with water if the soil is of a heavier composition in lowland terrains that occupy a larger area. The wet front pushes the air in front of it, compressing it. A significant increase in the piezometric height can often be observed after precipitation even before the wet front could merge with groundwater. In addition, the increase in pressure may be disproportionate to the amount of infiltrated water.

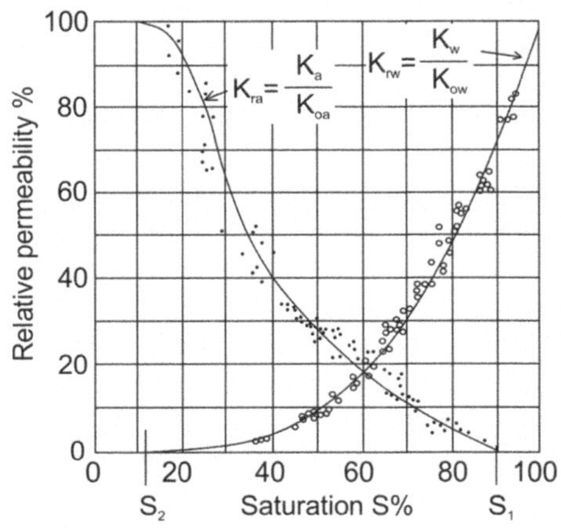


**Figure 11.41** Relative permeability for air ($K_{ra}$) and water ($K_{rw}$) according to Scheidegger (1957).

### *11.3.3.4 Evaporation*

Evaporation can be potential as well as actually realized. Potential evaporation is evaporation from the free surface of water. However, real evaporation is limited by:

- Rainfall;
- Infiltration into the soil;
- Transpiration of plants.

Infiltration and transpiration are often viewed together as a single quantity and are called evapotranspiration.

Three factors are dominant for the correction of evaporation from the free surface of water:

(a) Temperature;
(b) The density of water molecules in the gas layer above the water surface, described by Fick's law;
(c) Vortex diffusion – depends on the density of water molecules above the free surface of water, as well as the vertical component of the air flow rate.

Water evaporation takes place from the surface of water (surface water – sea, lakes, rivers, etc.) as well as from the surface of the land. It can also evaporate from the subsurface layers of the soil (Assouline *et al.*, 2013). With depth, evaporation decreases. If the part of precipitation that infiltrates into the upper layers of the soil (aquifer overlaying strata) is less than the evaporation of water from the soil, then:

- Capillary forces extract water from the underlying alluvial aquifer and compensate water deficit; and/or
- Groundwater table is lowered deeper and in that way the precipitation balance is re-established by evaporation.

It is often difficult to separate the intensity of evaporation from the transpiration of plants, so that evaporation and transpiration are observed as an integral manifestation and are referred to by one name – evapotranspiration.

Also, as is the case with surface waters, evaporation from the groundwater level or the vadose belt depends on a number of factors, one of which is certainly the composition and properties of the overlying layer.

During the construction of the first of the two large dams on the Danube, Iron Gate I and Iron Gate II, the river water levels rose on average by about 20 m. A rough disposition of the system is given in Figure 11.42.

The hydraulic system Djerdap/Iron Gate consisting of the Iron Gate I and Iron Gate II dams has an impact on about 200,000 ha of land.

The larger dam, Iron Gate I and most of the protection works for affected land areas were completed in 1972, and the Iron Gate II dam and the system as a whole were completed in the early 1980s.

Measures to protect the affected areas are outlined in the book Groundwater Management in Large River Basins (Dimkić *et al.*, 2008, pages 109–116).

In order to appropriately design the protection system, it was necessary to set up a methodology to calculate the groundwater levels and the effects of the system of protection.

Based on monitoring results covering many years, envelope diagrams of the evapotranspiration from the groundwater surface were synthesized, as well as for the relation dependence between infiltration and precipitation for characteristic regions of the protected Danube banks (Figures 11.43 and 11.44).

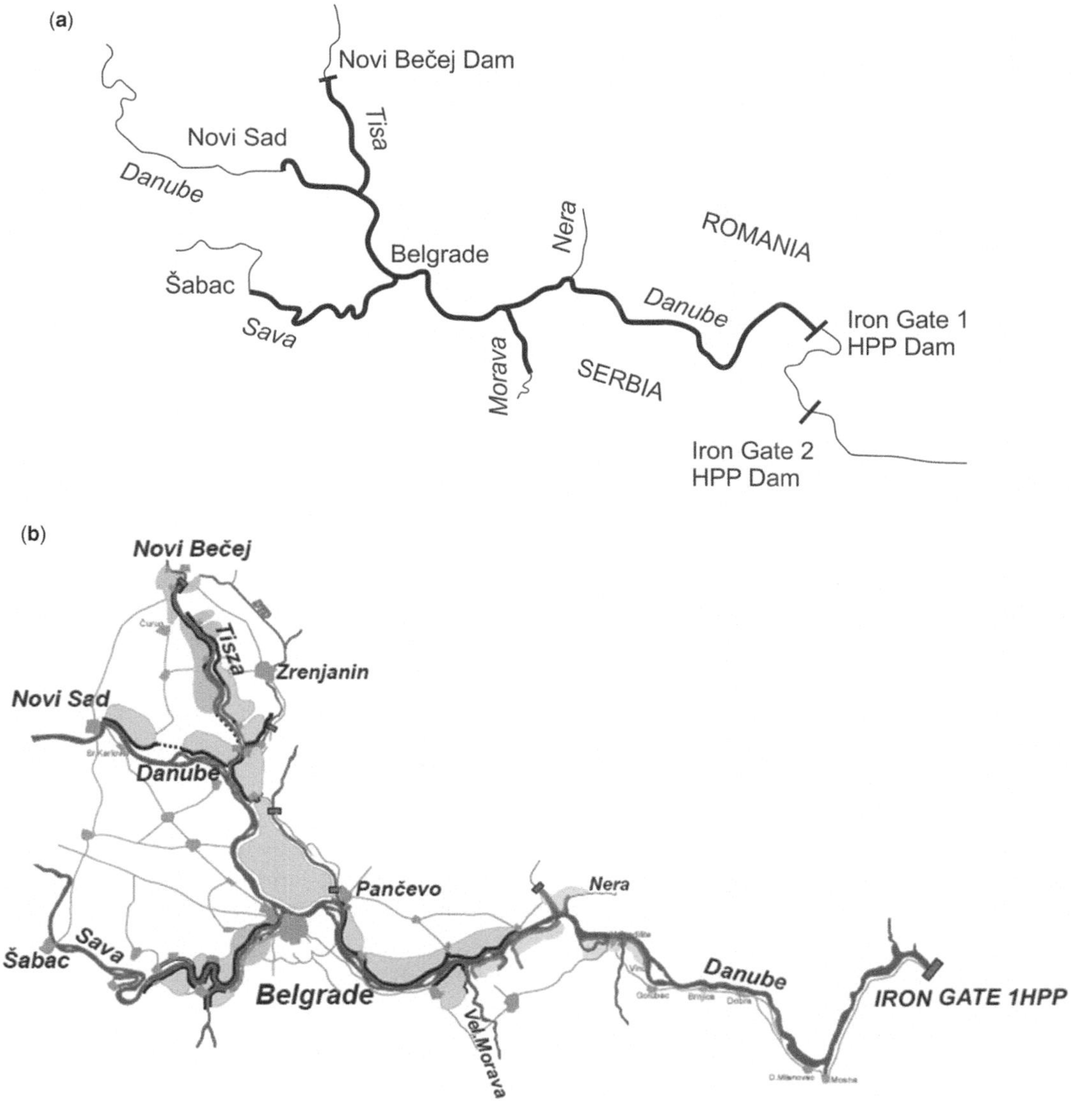


**Figure 11.42** (a) The Iron Gate HPP (hydro power plant) disposition (Dimkić *et al.*, 2008). (b) Levels and protected riparian lands from the Iron Gate HPNS (Dimkić *et al.*, 2008).

### *11.3.3.5 The Iron Gate hydroelectric system – solving the problem of drainage and drought control*

The protected 200,000 ha of land can be hydrogeologically represented as a two-layered environment:

- The lower layer – alluvial aquifer that rests on a largely impermeable/poorly permeable underlying layer;
- The top layer – semi-permeable/low-permeability overlaying aquifer.

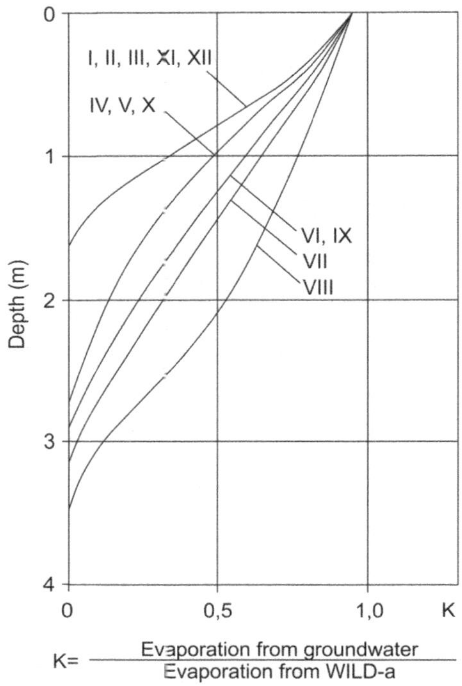


**Figure 11.43** Relationship between evapotranspiration from the groundwater level and evaporation from the free water surface as a function of the depth of groundwater (Vuković, 1967).

In order to have undisturbed, good quality agricultural production, it is necessary for the piezometric level of groundwater in the aquifer to be 0.8–1.2 m below the surface of the ground during the vegetative period. For smaller settlements, the piezometric level of the groundwater in the aquifer should be deeper than 2 m below

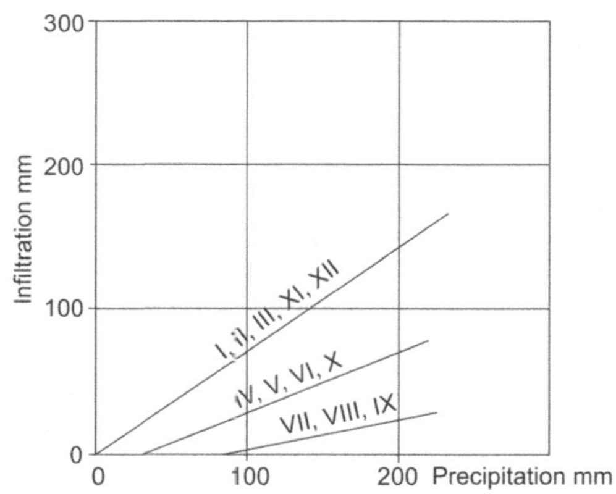


**Figure 11.44** Relationship between infiltration and precipitation (Vuković, 1967).

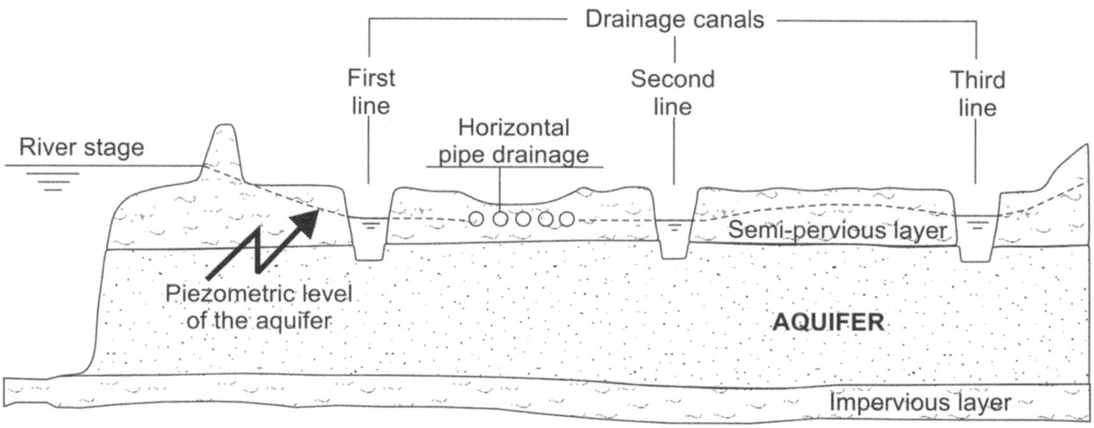


**Figure 11.45** Basic types of protection measures and farmland development (Dimkić *et al.*, 2008).

the ground surface of the terrain. For larger cities, the piezometer levels in the aquifer are required to be at least 3 m below the ground surface of the terrain.

Here, large reclamation systems (Figures 11.45 and 11.46) make it possible to overcome the adverse effects of the wetting of agricultural land, as well as prolonged dry periods.

The influence of the river level on the banks is pronounced up to the nearest drainage line (canals, wells). However, the system is designed so that the desired water regime is achieved in the entire protected area.

During dry periods, the water supply of agricultural crops is achieved by the capillary raising of water levels to the roots of plants. If necessary, the water level in the drainage canals is adjusted, at which point they change their role and become water irrigation canals.

If necessary, some of the conventional methods of watering crops are applied at some sections.

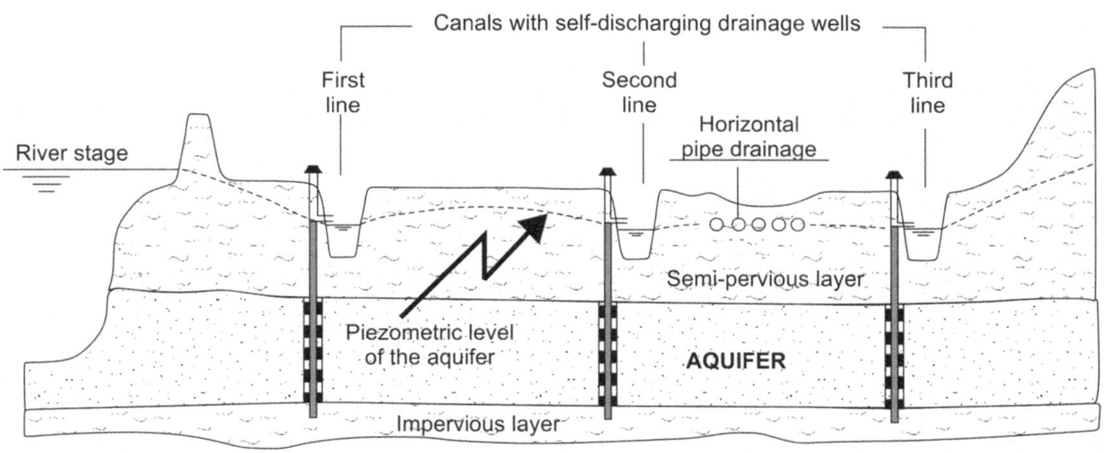


**Figure 11.46** Basic types of protection measures and farmland development (Dimkić *et al.*, 2008).

### 11.3.3.6 System of equations used to calculate the water regime in the two-layer environment of the Iron Gate hydropower system

The scheme according to which the equations and methodology for calculating the groundwater flow in the alluvial aquifer and the semi-permeable layer (aquitard) were developed is adapted to the hydrogeological and hydrodynamic features of the protected area of the Danube riverbank.

The term 'two-layer porous medium' means that there is an appropriate hydrogeological medium consisting of two layers of different filtration characteristics which lie on a water-permeable substrate (bottom). The lower layer, which is, as a rule, of greater capacity, is an aquifer, while the layer with weaker filtration characteristics spreads on the surface. The flow in the porous medium thus defined has a quasi-artesian character; in the lower aquifer the flow is under pressure, and in the surface less-permeable layer, a free surface of groundwater is formed.

In order to properly set up the program to explore the hydrogeological conditions which can be schematized by a two-layer porous medium, and to appropriately interpret the obtained results, it is necessary, first of all, to understand the specifics of a flow under the mentioned conditions, and this is presented below:

(1) The notion of pressure in the lower aquifer must be clearly distinguished from the notion of the groundwater level (free release) which is formed in the surface less-permeable layer;

(2) The equipotential lines (lines of the same pressures) in the lower aquifer are quasi-vertical. In contrast, in the surface layer, where the flow takes place under the effect of the pressure difference $H$ and the groundwater level $h$ (positive value of the difference $H$-$h$ means the flow is directed upwards, and a negative difference means that the flow is in the downwards direction), the equipotential lines are quasi-horizontal.

    Thus, the observation results do not depend on the position of the filter of the vertical piezometer which registers the pressure in the lower aquifer. However, the position of the piezometer filter in the surface layer (so-called shallow piezometers) must be taken into account;

(3) The water level registered in shallow piezometers does not represent the groundwater level, but the position of the groundwater level must be calculated based on the results of observations on the battery of the piezometer.

For the above reasons, the shallow piezometer filter must have a minimum length, since an undefined averaged value is otherwise obtained (the piezometer is used as a water conduit).

Figure 11.47 gives a schematic representation of the flow in conditions of a two-layer porous medium, which represents the basic scheme for the appropriate mathematical calculation.

To set the differential equations that describe the flow in a two-layer porous medium, the scheme given in Figure 11.41 is used, with the introduction of the following assumptions:

(a) the flow is slow enough, so the inertial term $\frac{\partial v}{\partial t}$ can be neglected;

(b) the coefficient of the effective porosity of the surface layer $\varepsilon$ is considered to be a constant value and does not depend on the flow regime, gradients and time; and

(c) the oscillations of groundwater levels in the surface layer take place exclusively within this layer, while the flow in the lower aquifer is under pressure.

Considering the set scheme, the following two partial differential equations are obtained, which define the nonstationary flow in a two-layer porous medium in conditions of a plane-two-dimensional flow

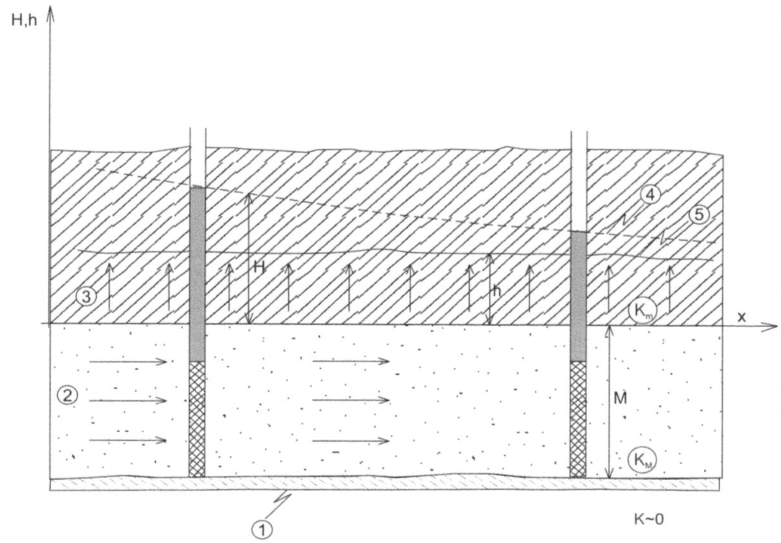


**Figure 11.47** Schematic representation of flow in a two-layer porous medium: 1 – watertight layer, 2 – aquifer, 3 – surface less-permeable layer, 4 – piezometric level in the lower aquifer, 5 – free water surface in the surface layer (Vuković, 1967).

(Vuković, 1967):

$$\frac{\partial^2 H}{\partial x^2} = \frac{K_m}{K_M \cdot M}\left(\frac{H-h}{h}\right)$$
$$\frac{\partial h}{\partial t} = \frac{K_m}{\varepsilon}\left(\frac{H-h}{h}\right) + \frac{W}{\varepsilon} \quad (11.35)$$

where:

$W$ – amount of infiltration water (a positive sign of this term shows the dominant influence of infiltration from the surface of the terrain – filling the underground, and a negative sign shows the dominant influence of evapotranspiration – emptying the underground),
$H$ – piezometric level in the aquifer,
$h$ – groundwater level,
$x$ – coordinates in the direction of the groundwater flow,
$t$ – time,
$K_m$ – filtration coefficient of the overburden layer,
$K_M$ – aquifer filtration coefficient,
$\varepsilon$ – effective porosity of the top layer.

This presupposes that the flow in the upper aquitard is exclusively vertical. The flow in the aquifer is exclusively horizontal. This is justified if the permeability of the aquifer is sufficiently greater than the permeability of the top layer.

We apply the quasi-2D approach if we set the computation scheme for the aquifer as one-dimensional. Then the coordinates along path $x$ coincide with the flow stream in the aquifer.

We apply the quasi-3D approach when we observe the current in the alluvial aquifer as horizontal, in the $x$, $y$ directions, and the flow in the aquifer as exclusively vertical.

### 11.3.3.7 Some global indicators of evaporation and precipitation

***Aridity zones defined across the $P/E_{pot}$ relationship***

Evaporation, if viewed as an element of the water balance and climate in general, and above all on a global scale, depends mostly on climatic zones. Table 11.16 gives the basic climate zones on earth.

Table 11.16 shows that deserts and their semi-arid margins occupy 37.5% of the total land area. If dry sub-humid regions are included here, where the average annual precipitation ranges between 750–1500 mm, this amounts to 47.5% of the total land area. It should be noted that every fifth person on the planet lives in these dry areas, which indicates the great importance of knowledge of evaporation as an element of climate, life and the future climate change in these countries.

***Overview of the basic climatic regions in the world and annual precipitation in them***

Figure 11.48 shows the average annual precipitation in the described climatic regions.

***Ratios of precipitation (P) evaporation (E), runoff (P-E), and interception (I) in the 2003–2007 period at a global level***

Table 11.17 and Figure 11.49 provide basic climatic data related to $P$, $E$, $P$-$E$, $I$, by biome, continent, and latitude.

***Evaporation from the ocean surface depending on latitude***

Table 11.18 shows the total annual evaporation from the ocean surface in mm.

***Evaporation in deserts***

The desert is a region that receives very little rainfall – less than 250 mm per year (about 20 cm). Approximately one-third of the Earth's surface is desert. There are different types of deserts based on their geographical situation: tropical deserts, subtropical, polar, cold winter deserts, cold coastal deserts, etc. So, deserts occur on all continents (Figure 11.50).

The average temperatures in tropical warm deserts are usually between 29 and 35°C, and at noon the readings are 43–46°C (a record of 56.7°C in Death Valley). On the other hand, the Arctic and Antarctic desert regions have a polar climate because they have average summer temperatures below 10°C, and in

**Table 11.16** Aridity zones defined by $P/E_{pot}$ ratios.

| Climate zone | $P/E_{pot}$ ratio | % of the World Covered |
|---|---|---|
| Hyper-arid | <0.05 | 7.5 |
| Arid | 0.05–0.20 | 12.5 |
| Semi-arid | 0.21–0.50 | 17.5 |
| Dry subhumid | 0.51–0.65 | 9.9 |
| Humid | >0.65 | 39.1 |
| Cold | >0.65 | 13.5 |

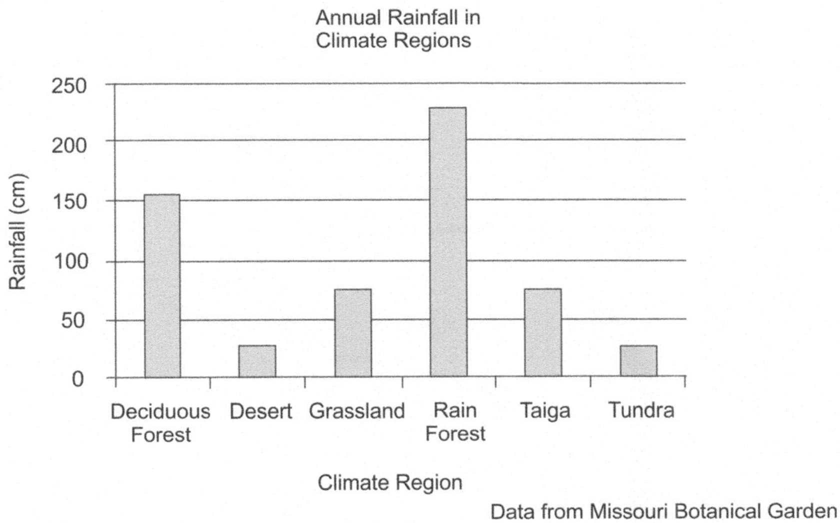


**Figure 11.48** Overview of the basic climatic regions in the world and the annual precipitation in them (Data from Missouri Botanical Garden).

winter up to record levels of −98.6°C. Table 11.19 shows average annual precipitation and evapotranspiration for some deserts.

***Precipitation and evaporation for different latitudes (P-E)***

At the end of this brief overview of some global indicators of evaporation and precipitation, a generally known approximate illustration of precipitation and evaporation from the water surface for different latitudes is given in Figure 11.51.

**Table 11.17** Annual precipitations (*P*), evaporation (*E*), water available for runoff (*P-E*) and rainfall interception loss (*I*) per biome type for 2003–2007.

| Biome | *P* | *E* | | | *P-E* | | | *I* | | |
|---|---|---|---|---|---|---|---|---|---|---|
| | mm | mm | $10^3Km^3$ | %*P* | mm | $10^3Km^3$ | %*P* | mm | $10^3Km^3$ | %*P* |
| Tropical forest | 2250 | 1182 | 19.6 | 53 | 1068 | 17.7 | 47 | 232 | 3.8 | 10 |
| Temperate forest | 718 | 512 | 4.7 | 71 | 207 | 1.9 | 29 | 95 | 0.9 | 13 |
| Boreal forest | 594 | 372 | 2.9 | 63 | 222 | 1.7 | 37 | 79 | 0.6 | 13 |
| Shrubland | 503 | 315 | 8.2 | 63 | 187 | 4.9 | 37 | 9 | 0.2 | 2 |
| Savanna | 1339 | 806 | 14.7 | 60 | 533 | 9.7 | 40 | 51 | 0.9 | 4 |
| Grassland | 689 | 462 | 4.2 | 67 | 227 | 2.1 | 33 | 15 | 0.1 | 2 |
| Cropland | 878 | 542 | 10.7 | 62 | 336 | 6.6 | 38 | 31 | 0.6 | 4 |
| Permanent snow | 225 | 27 | 0.4 | 12 | 198 | 3.3 | 88 | 0 | 0.0 | 0 |
| Desert | 167 | 112 | 2.6 | 67 | 54 | 1.2 | 33 | 0 | 0.0 | 0 |
| **Total** | **799** | **463** | **67.9** | **58** | **336** | **49.0** | **42** | **50** | **7.2** | **6** |

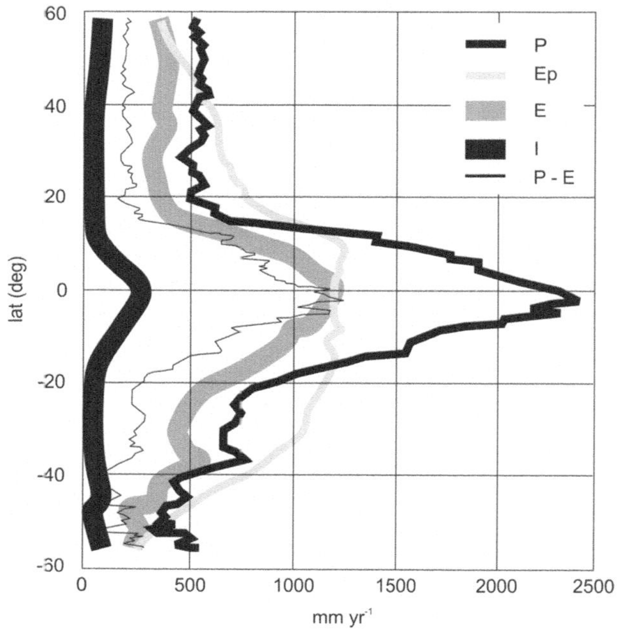


**Figure 11.49** Magnitude of the different hydrological fluxes as an average along the latitudinal bands (modified from Fleischer *et al.*, 2008). The results correspond to the application of GLEAM for the 2003–2007 period. $E_p$ refers to potential evaporation.

**Table 11.18** Total annual evaporation from the surface of the oceans (mm) (Babkin, 2008).

| Latitude, in Degrees | Arctic Ocean | Atlantic Ocean | Indian Ocean | Pacific Ocean | World Ocean |
|---|---|---|---|---|---|
| 90–80NL | 75 | | | | 75 |
| 80–70 | 205 | | | | 205 |
| 70–60 | 405 | 798 | | 308 | 520 |
| 60–50 | 390 | 1100 | | 565 | 795 |
| 50–40 | | 1420 | | 953 | 1150 |
| 40–30 | | 1720 | | 1560 | 1630 |
| 30–20 | | 1840 | 1920 | 1810 | 1830 |
| 20–10 | | 1950 | 2000 | 2020 | 2000 |
| 10–0 | | 1720 | 1940 | 1810 | 1820 |
| 0–10SL | | 1640 | 1810 | 1630 | 1680 |
| 10–20 | | 1700 | 2060 | 1900 | 1910 |
| 20–30 | | 1560 | 1960 | 1890 | 1830 |
| 30–40 | | 1350 | 1380 | 1530 | 1440 |
| 40–50 | | 672 | 773 | 1100 | 876 |
| 50–60 | | 409 | 569 | 735 | 598 |
| 60–70 | | 239 | 372 | 409 | 355 |
| Mean from the ocean | 220 | 1360 | 1420 | 1510 | 1400 |

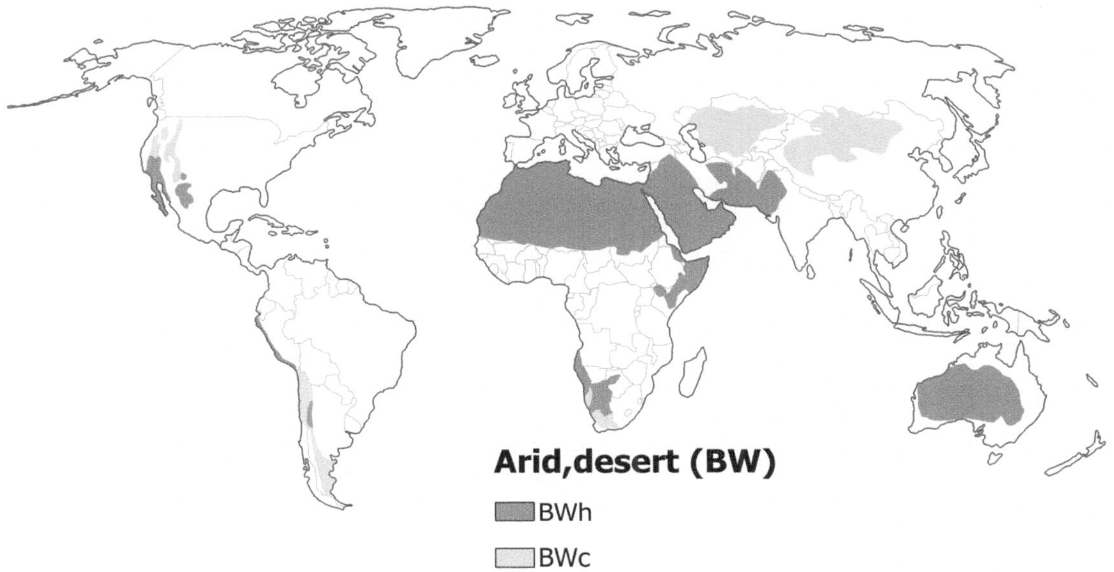


**Figure 11.50** Regions with desert climates BWh (hot desert climates) BWc (cold desert climates) (www.bing.com).

**Table 11.19** Data on the average annual precipitation and evaporation for some deserts.

| Desert | *P* mm/y | *E* mm/y |
|---|---|---|
| Namib Desert | 5–85 | Ekstremna |
| Sahara Desert | 100–250 | 2500–6000 |
| Kalahari Desert | 110–500 | 3000 |

### *11.3.3.8 Soil moisture profile and depth of plant rooting*

There are, in principle, very good reasons why the vertical flow and water balance in the overlying, surface (roof) layer of the alluvial aquifer are important. One is associated with the provision of water for the maintenance and development of flora, both natural and those grown for agricultural purposes. The second is related to the capacity of groundwater aquifers – in our case – alluvial aquifers.

The vertical movement of water through the roof (overlaying) aquitard depends on the amount and regime of precipitation, evaporation, transpiration of plants and the retention of precipitation on vegetation (interception loss). The following hydrogeological and hydrodynamic conditions are important – the composition and interrelation of the alluvial aquifer and the overlying, less-permeable layer, as well as the possibility of the capillary water uplift, and water and air flow in a water unsaturated environment.

Physical evaporation from the surface of a water table is often taken as the reference evaporation. Groundwater evaporation decreases as the depth of groundwater increases. One practical example of taking this evaporation into account is given in subchapter 11.3.3.4, Figures 11.43 and 11.44.

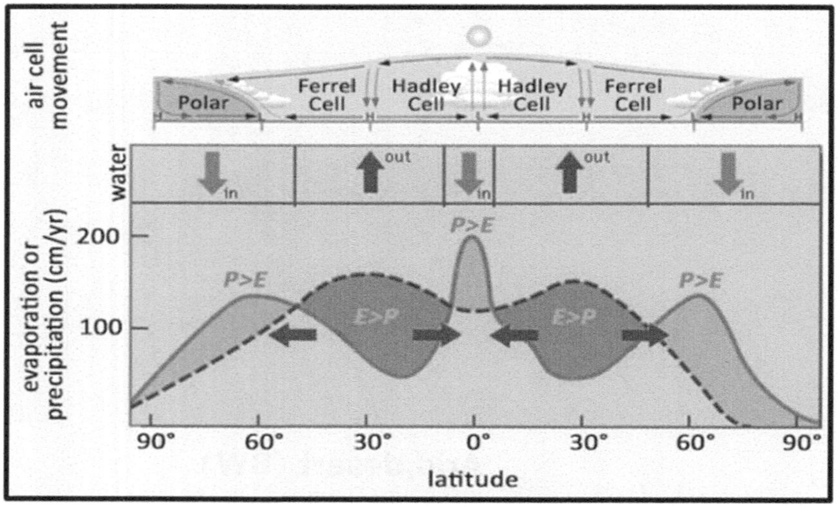


**Figure 11.51** Indicative illustration of precipitation and evaporation from the free water surface area in cm/year depending on the latitude (www.bing.com).

Trees and plants generally use or lose water in two separate processes. First, roots take water from the soil and water evaporates through pores or stomata on the surfaces. This is called transpiration and is a physiological process that reacts to soil and atmospheric factors. The second procedure is retention – interception of water by the surface of leaves, branches and trunks during rainfall, and its later evaporation. Interception losses are amplified by high atmospheric turbulence. Taken together, these two processes are often called evapotranspiration. Both are strongly influenced by the amount of sunlight, temperature and humidity, and wind speed.

The annual evaporation loss for different land coverings is shown in Table 11.20.

Conventionally, potential evapotranspiration ($PE_t$) is defined as the total loss of water by evaporation from an actively growing, short, green crop (grass) that never lacks water.

Rooting depth varies from very shallow ($<0.01$ m) to very deep ($>70$ m). The depth of rooting depends on the infiltration of precipitation, the depth of groundwater, soil characteristics, latitude, etc. However, the type of plant is the primary factor in terms of the depth of rooting. The roots of some species remain shallow,

**Table 11.20** Typical range of annual evaporation losses (mm) for different land covers receiving 1000 mm annual rainfall (Nisbet, 2005).

| **Typical Range of Annual Evaporation Losses (mm) for Different Land Covers Receiving 1000 mm Annual Rainfall** | | | |
|---|---|---|---|
| Conifers | 300–350 | 250–450 | 550–800 |
| Broadleaves | 300–390 | 100–250 | 400–640 |
| Grass | 400–600 | – | 400–600 |
| Heather | 200–420 | 160–190 | 360–610 |
| Bracken | 400–600 | 200 | 600–800 |
| Arable | 370–430 | – | 370–430 |

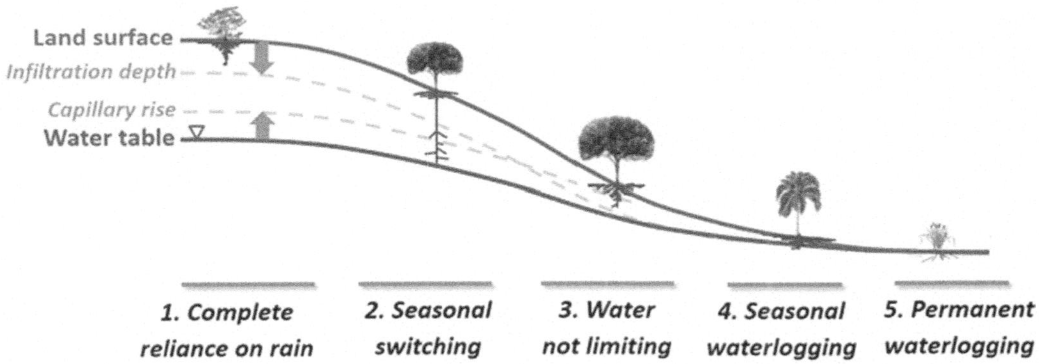


**Figure 11.52** Schematic profile of groundwater showing systematic changes in the rooting depth (Ying & Gonzalo, 2017).

to avoid oxygen stress below groundwater levels. On the other hand, drought can cause the roots to descend many metres lower to the capillary belt (Figures 11.52 and 11.53).

Moisture profile issues are very important both from the point of view of the maintenance of flora and from the point of view of feeding and the total capacity of the alluvial aquifer itself. The larger material in the alluvial aquifer cover allows for a higher percentage of the infiltrated water than precipitation

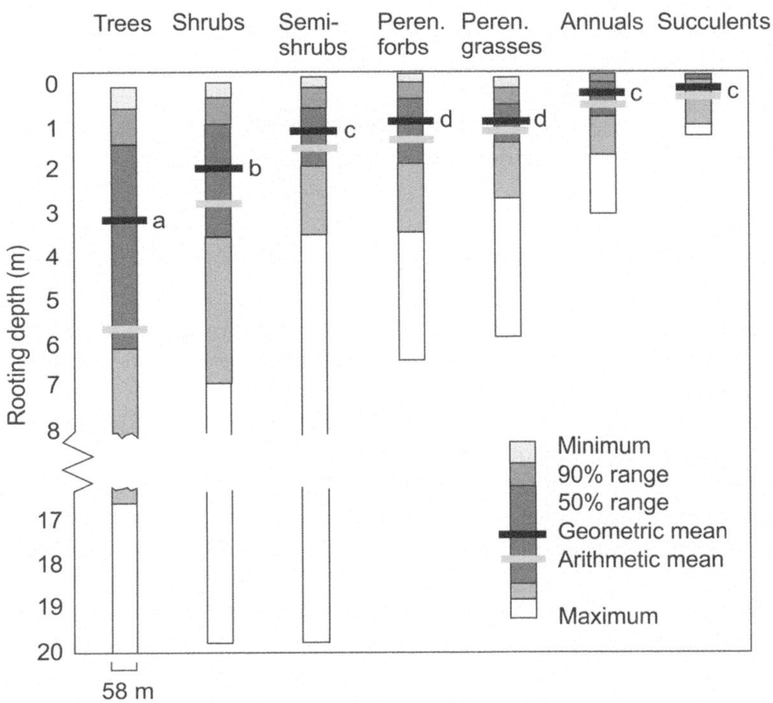


**Figure 11.53** Maximum rooting depths of plant growth forms (Schenk & Jackson, 2002).

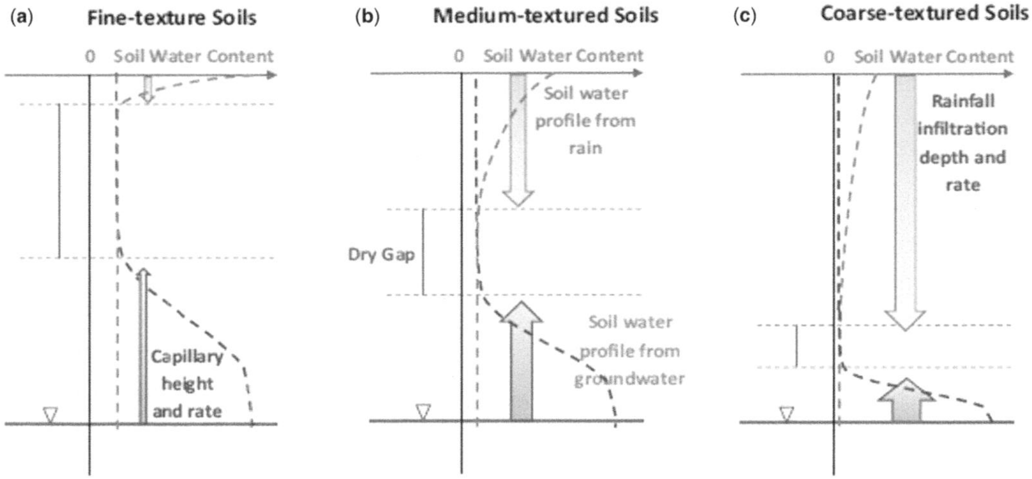


**Figure 11.54** Schematic representation of two opposite water fluxes in soil in different soil types (Ying & Gonzalo, 2017).

provides. However, the drying rate of the surface layer of the soil is also increased if precipitation ceases. Heavier soils retain moisture for a slightly longer time, but at the same time reduce infiltration from precipitation into the aquifer.

The larger material in the upstream layer reduces the capillary forces and can significantly limit capillary rise. The finer material of the topsoil can allow better transfer of the water from the aquifer to the topsoil and the roots of plants. Lowering the groundwater level in the aquifer below the level of the boundary between the aquifer and the aquitard can reduce and cut off the inflow of water from the alluvial aquifer into the aquitard.

Figure 11.54 gives a schematic representation of how soil texture (clay, silt, sand) regulates two water fluxes – the infiltration flux from the top down and the capillary climbing flux from the bottom up above the considered level.

### *11.3.3.9 Influence of climate and other changes on the capacities of alluvial aquifers – examples*

It is well known that on a global scale, at least 100 times more water is stored in aquifers than in rivers and lakes combined. This makes groundwater systems comparatively resilient to the short-term, most often seasonal, or even longer-term shortage of rainfall.

Groundwater is often, especially in arid regions, the most reliable source of water – if not the only one – for domestic water supply and irrigation of crops. Climate change is, nevertheless, likely to impact upon the quantity and quality of groundwater resources and alluvial aquifers in arid and sub-arid regions which rank as the most vulnerable to climate change (Van der Gun, 2009). In some moderate climate regions (particularly those with a decreasing precipitation trend), climate change is pushing them to become sub-arid.

Alluvial groundwater systems are extremely vulnerable to changes in the amount of precipitation as they are directly and/or indirectly recharged from rivers. Temperature is also important in the alluvial aquifer water balance due to higher evapotranspiration which is linked to increasing temperature. In addition to the impact of climate parameters, some other changes, like land use or vegetation changes also influence the capacity of an alluvial source. However, the greatest pressure on alluvions is usually due to the increasing water demand and potential pollution. So, in the studies where the goal is to estimate only the impact of climate change on the future alluvial capacity, the assumption must be that the future water demand stays the same, and that potential pollution does not compromise the source.

The impact of climate and other changes on the capacities of alluvial groundwater resources is evaluated through the following two examples.

***1° Example of the Pek River alluvial sources (Serbia)*** (Dimkić *et al.*, 2018)

Decreasing trends in river discharge regimes have been registered in Serbia (subchapter 11.2.3.2). Within the scope of the international project CC-WaterS, (2009–2012), the Pek River and two of its alluvial water sources (ALWSs) Jelak (providing drinking water supply for the City of Veliko Gradište) and Mlaka (servicing the City of Kučevo) represented a pilot study area and the estimated impact of climate change on the water resources was analysed. These two ALWSs were selected because they face similar water quantity issues in terms of the river discharge, but with possibly different impacts on the alluvial source capacities.

The Pek River is situated in central-eastern Serbia (Figure 11.55a) and is a tributary of the Danube River. Their confluence is at the city of V. Gradište. The catchment area (CA) of the Pek River is 1230 $km^2$ (Figure 11.55b), with a population of 45,000. The majority of the population lives in three cities: Majdanpek (the most upstream city), Kučevo, and V. Gradište at the mouth of the river. The total drinking water demand in the Pek watershed is approximately 5.3 $Mm^3/y$ (170 L/s). The irrigation demand is much lower – less than 10% of the drinking water demand.

The climate is temperate continental with some steppe characteristics. The mean annual temperature (T) and precipitation (P) are about 11°C and 660 mm/year in the CA. Figure 11.55c shows the altitudes of three cities, and two temperature and four precipitation stations.

The water scarcity problem is growing since this region, in addition to increases in T (and evapotranspiration), is also facing a slightly decreasing P trend. As a result, the Pek River is already exhibiting a decreasing discharge trend. There are three hydrological stations on the Pek River: an upstream station at Debeli Lug (near Majdanpek), a central station at Kučevo, and a downstream station at Kusiće (near Veliko Gradište). Figure 11.56 shows their observed annual discharges. The registered

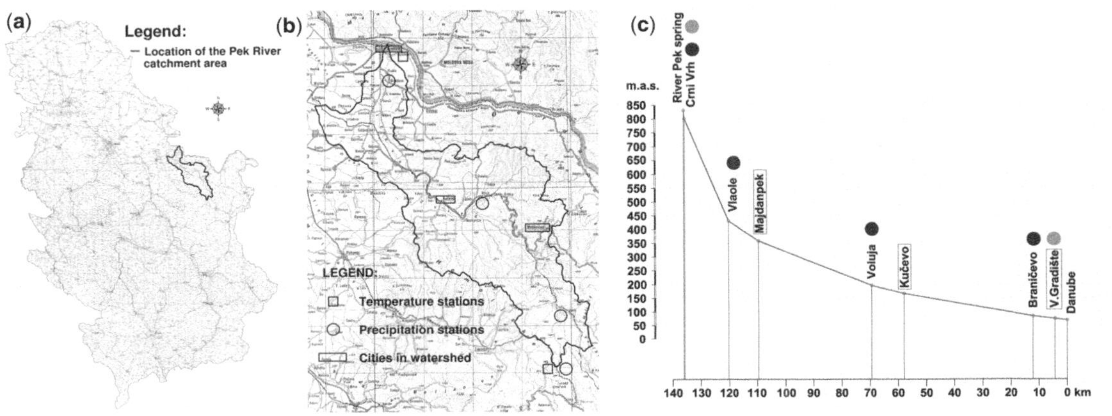


**Figure 11.55** (a) Location of the Pek River CA in Serbia, (b) positions of the cities, T and P stations in the CA, and (c) altitudes of the cities, and temperature and precipitation stations (Dimkić *et al.*, 2018).

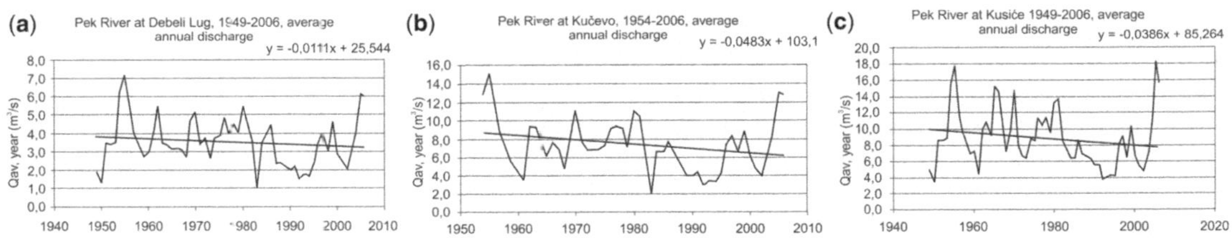


**Figure 11.56** Observed annual trends at the three hydrological stations: (a) the most upstream, (b) in the middle, and (c) close to the mouth of the Pek River (Dimkić *et al.*, 2018).

long-term trend for the Pek River is about (−3 to −4%)/10 years. The greatest impact factor is attributed to CC, after that to LU (subchapter 11.2.3.2).

The geology is characterized by alluvial sediments, limestone and dolomites. The flow at the Jelak source occurs in two-layer media, and is partly unconfined and partly confined. The flow in the Mlaka source occurs in two-layer media and is unconfined. The interaction between the alluvial aquifers and the Pek River is direct. Initial studies included the preparation and validation of the hydrodynamic models for two ALWSs: Jelak for the period Sep–Dec 2010, and Mlaka for the period Mar–Jun 1997 (Figure 11.57).

The methodology applied for estimating future source capacities exposed to climate and other changes were quite complex (Dimkić *et al.*, 2018). Climate predictions were made through three RCM models, using

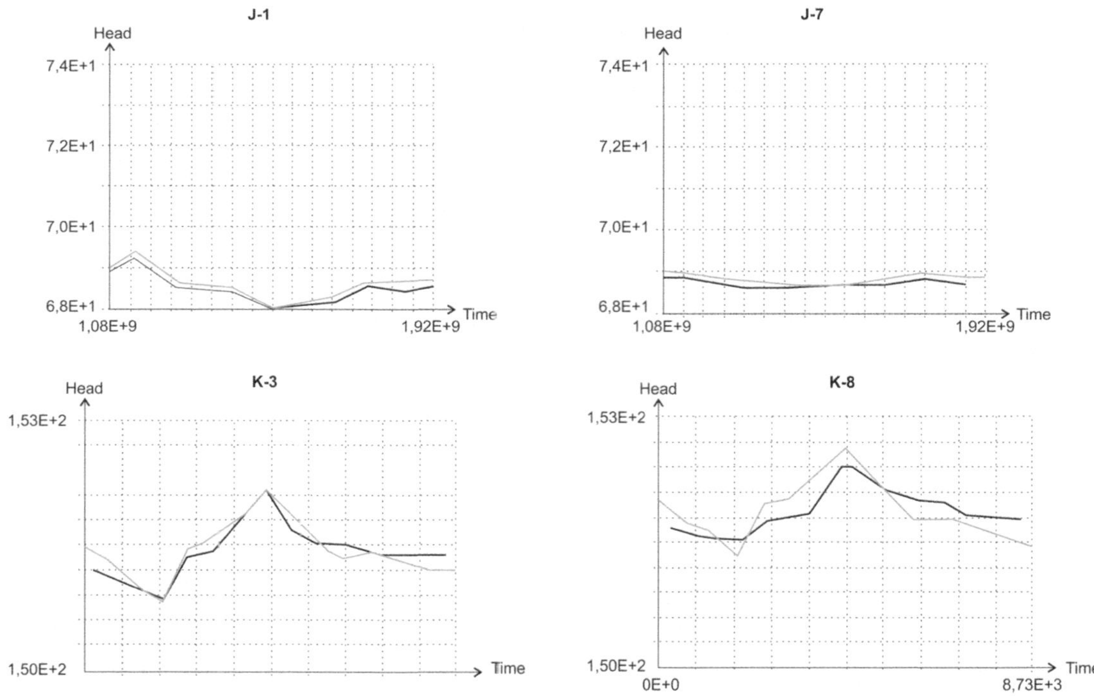


**Figure 11.57** Hydrodynamic model calibration results for two piezometers at both alluvial sources. Note: The ordinate shows the piezometric head in m a.s.l., and the abscissa shows the simulation time in seconds (in exponential notation) (Dimkić *et al.*, 2018).

scenario A1B. In brief, the basic assumption of the methodology was that in comparison with the present, only the following will change in the future: the temperature, precipitation, the upstream drinking water and irrigation water demand, and, as a consequence of climate change: recharge, evapotranspiration, river discharge (water level) and the upland boundary conditions.

Results indicate that, for the near future (until 2050), we can expect that the capacities will decline by about 10–15% on both alluvial water sources, while this percentage is higher for the distant future (until 2100): about 30% for Jelak and about 45% for the Mlaka alluvial water source.

When the results derived for the two ALWSs are compared, it is apparent that Mlaka (Kučevo) is more sensitive to climate change than Jelak (Veliko Gradište). The explanation for this may be sought in the lower permeability of Mlaka, as well as in the proximity of Jelak to the river Danube.

*2° **Example of the alluvial aquifers of Yemen*** (Van der Gun, 2009)

Yemen with its 30 million inhabitants (about one-quarter in the capital Sana'a) is regarded as a poor country in terms of water resources availability. The long-term mean rate of the renewal of surface water and groundwater combined is less than 200 $m^3$ per capita per year. A significant part of the country's agriculture is dependent upon alluvial aquifers, which underlines the need to slow down groundwater depletion in these aquifers and to implement adequate adaptation measures.

Arid regions have low annual rainfall (Yemen's average is about 130 mm/year) characterized by irregular and infrequent events. Occasional surface water runoff tends to be discharged by natural drainage channels called wadis. Wadis have intermittent or seasonal flows and have no significant base flow during the long dry periods.

Wadi channels are usually accompanied by shallow, narrow and elongated alluvial aquifers. They consist of sediments carried and deposited by the wadis, and the groundwater they contain is recharged mainly by wadi flows. Wadi channels and alluvial aquifers tend to be so inseparably linked, that the word 'wadi' is often used to refer to both the wadi channel and the related alluvial aquifer (for this reason it is also called a 'wadi aquifer'). In general, two types and five subtypes can be recognized (Table 11.21).

The first type is an alluvial strip aquifer, and it can be split into three subtypes depending on the bottom of the aquifer: (a) embedded in impervious rocks; (b) hydraulically connected to other permeable rock units; and (c) underlain by a regional aquifer.

The second type is the wadi plain aquifer, and it can be split into two subtypes depending on the catchment part: (a) mountain plain aquifer; and (b) lowland plain aquifer.

Their locations are shown in Figure 11.58. Due to tectonic setting, most of the wadi plain aquifers (type 2) have a much larger surface area, but they are also much thicker than alluvial strip aquifers (type 1). For an equal size of the catchments recharging them, wadi plain aquifers have a much larger storage capacity than strip aquifers, and the residence time of water in them is much longer (often in the order of hundreds of years). The main characteristics of the five aquifer type examples are presented in Table 11.22.

An increasing temperature trend has been observed, in the range of 1°C for the 20th century. There is no conclusive evidence of either an increase or decrease in precipitation in Yemen. Global climate change projections for Yemen (Synthesis Report of the IPCC, 2007, 2013) indicate that the temperature will increase in the 21st century by between 2.5°C (B1 scenario) and 4.0°C (A2 scenario). The GCMs projections regarding rainfall are very uncertain.

Natural groundwater recharge in arid regions is not only generally low, it is also very sensitive to changes in rainfall and temperature regimes. Among the five selected Yemen aquifers, Wadi Hadramawt (1b) is most likely to experience more prominent changes in natural groundwater recharge than any of the other four aquifers, due to the extremely low rainfall in the zone.

**Table 11.21** Different types of wadi aquifers in Yemen with examples.

| Type and Subtype | Example of Aquifer Type | Remarks | Scheme |
|---|---|---|---|
| 1 Alluvial strip aquifers | | | 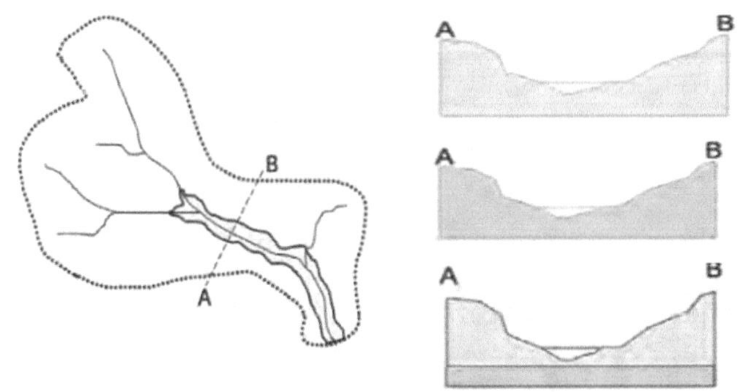 |
| a) Embedded in impervious rocks | Wadi Fallah alluvial aquifer | Near Sa'dah Plain, North-west part | |
| b) Hydraulically connected to other permeable rock units | Wadi Hadramawt alluvial aquifer[4] | Shibam-Sayun zone, Yemen central part | |
| c) Underlain by a regional aquifer[1] | Wadi As Sirr alluvial aquifer | North-east of Sana'a Plain | |
| 2 Wadi plain aquifers[2] | | | 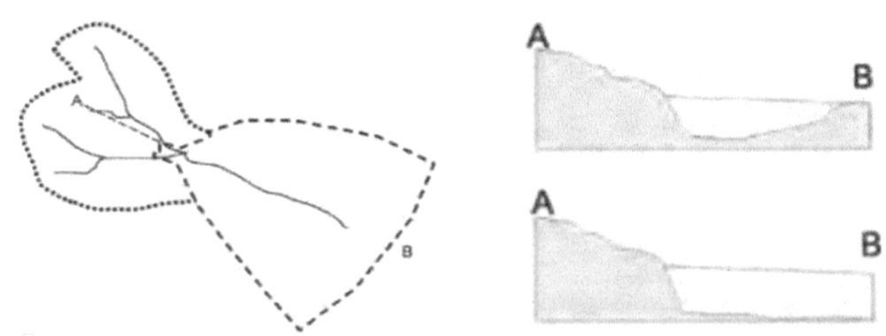 |
| a) Mountain plain aquifer[3] | Sana'a plain alluvial aquifer | In the central highlands of Yemen | |
| b) Lowland plain aquifer | Tihama (Siham) plain alluvial aquifer | Bordering the Red Sea | |

[1]Embedded in poorly permeable rocks, but sufficiently close to an underlying regional aquifer to allow an artificial hydraulic connection by means of wells penetrating the separating lithological unit. Under virgin conditions, this aquifer subtype is similar to subtype 1a; but after drilling wells and significant connection with the underlying regional aquifer, its behaviour may become more like aquifer subtype 1b.
[2]Where topographic or tectonic conditions force the wadi gradients to be suddenly reduced, alluvial plains tend to develop by deposition of the sediments carried by the wadis during floods.
[3]The mountain plain aquifer is limited downstream by impermeable rock.
[4]Very large, the Mukalla Sandstone groundwater system recharges Wadi Hadramawt alluvial aquifer.

**Figure 11.58** Map of Yemen and locations of the analysed aquifers (Van der Gun, 2009).

Natural groundwater discharge may occur through different routes: springs, seepage to streams (baseflow), evapotranspiration in shallow-water-table zones, and subsurface outflow. The latter two are active just in the Tihama aquifer (2b). A sea level rise in the order of decimetres is expected to have a minor influence on the submarine fresh-groundwater outflow (around 0.1% reduction).

The expected temperature rise will impact groundwater abstraction: for example, an increase of 3°C would increase the potential evapotranspiration of crops by about 75 mm/year. This means that irrigated

**Table 11.22** The main characteristics of the five aquifer type examples.

| Characteristic | 1a Wadi Fallah | 1b Wadi Hadramawt | 1c Wadi As Sirr | 2a Sana'a Plain | 2b Tihama Plain |
|---|---|---|---|---|---|
| Elevation (m a.s.l.) | 2500–1900 | 800–500 | 2600–2200 | Approx. 2200 | 700–100 |
| Catchment ($km^2$) | 100 | 22 500 | 200 | 300[1] | Huge, no clear border |
| Rainfall (mm/year) | 150 | less than 100 | 100–150 | 230 | East: 300<br>West: 50 |
| Length (km) | 15–20 | 90 (Dir. WSW – ENE) | All three (L, W, T) characteristics slightly larger than in case 1a | | 50 (Direc. E – W) |
| Width (m) | 20–150 | 1500–20 000 | | | 2500–60 000 |
| Thickness (m) | 10–20 | 100–150 | | 20–200 | 200–300 |
| Stored volume (mil. $m^3$) | 1–2 | 5000–10,000 | 3–8 | More than 1000 (50% in alluvial) | at least 250 000[2] |
| Abstracted water (mil. $m^3$/year) | 0.5 | 250 | 41 | 1972: 16<br>1993: 182 increasing further | 1994: 810 increasing further |
| Note | Groundwater levels decline significantly throughout the dry season | Observed decline more than 0.5 m/year | Hydraulic links through wells with 2nd regional aquifer: Cretaceous Tawilah Sandstone | GW levels in the underlying sandstones are falling by a rate of several metres a year. | Abstraction exceeds the annual recharge (about 550 mil. $m^3$) |

[1]Wadis surrounding the plain in a radial pattern bring their occasional flood waters into the plain to recharge the alluvial aquifer and underlying aquifers. Their catchment areas combined with the Sana'a plain constitute the Sana'a basin of app. 3200 $km^2$.
[2]Although this huge stored volume contributes to the aquifer's resilience to stress, it does not prevent declining groundwater levels – especially near the foothills – and steadily increasing groundwater salinity problems in the coastal area.

lands will require about 10% more irrigation water than at present. Water demand in the domestic water use sector will also increase, but this impact is less significant. However, one may expect that the impact of climate change alone will lead to at least a 5–10% higher water abstraction, i.e. 200–400 million $m^3$ per year. This corresponds to an 8–15% increase in the current groundwater abstraction rate, which is already unsustainable.

Climate change may lead to adaptive measures related to land use. Farmers may respond by intensifying irrigation water applications. But, if water is a really limiting factor, they may decide to reduce their irrigation areas – that is likely to occur in aquifer systems such as those of Wadi Fallah.

Based on all factors considered, the following tentative assessment can be made for aquifer storage conditions by the end of the 21st century:

(1a) The Wadi Fallah groundwater storage will be very significantly reduced, particularly during the dry season;
(1b) The Wadi Hadramawt groundwater storage will be significantly reduced in the axial zones of the wadi. Near the canyon walls, however, the depletion process may be slowed down considerably by recharging groundwater flows from the huge Mukalla Sandstone into the alluvial aquifer beds, which are expected to react extremely slowly to climate change, due to their storage;
(1c) The Wadi As Sirr groundwater storage may remain relatively stable as long as conditions in the regional Tawilah Sandstone aquifer (the main pressure is due to the intensive exploitation in the Sana'a Plain) allow local irrigation to continue at its present intensity;
(2a) Parts of the Sana'a Plain alluvial aquifer are dry already, and other parts will gradually become dry as well due to intensive groundwater pumping, and also accelerated by climate change. The saturated part of the Sana'a Plain alluvial aquifer will probably have shrunk to only a few small zones;
(2b) The Tihama Plain groundwater storage is already being depleted by abstraction. It is most pronounced in the eastern zone of Tihama, near the foothills. The present rate of storage depletion per annum is not more than about 0.1% of the total fresh groundwater storage, but problems could increase related to water quality and the economy, triggered by declining groundwater levels (e.g. increasing groundwater salinity, higher pumping cost).

The rate of the expected impact on the capacities of the alluvial sources in the 21st century varies from low or negligible to more than 50%, depending on the geology and hydrogeology of the water source, on the amount of rainfall and on the future human water demand, above all other factors.

The increasing water demand represents the main pressure on groundwater resources with the related set of complex problems escalated by climate change. The aquifers and their stakeholders become more vulnerable if the ratio of the stored groundwater volume to groundwater recharge is smaller, and if there are locally no other water resources available. People at isolated alluvial aquifers are likely to be among the most seriously and directly affected victims of climate change.

## 11.4 SOME REMARKS RELATED TO THE OCCURRENCE AND IMPACT OF DROUGHTS ON ALLUVIAL RELEASES

Two important characteristics of the time in which we live are the increase in the number of people on the planet and their activities, and certain climate change.

The human population is expected to increase from the current level of over 7 billion to about 9 billion inhabitants by 2050. Consequently, the issue of resources is becoming more acute. It is estimated that the world's water needs could increase from the current 3000–4000 $km^3$/year, to about 5000 $km^3$/year by 2050.

Climate change on the planet is a phenomenon that occurs periodically and lasts for longer intervals. In the Pleistocene and Holocene, these were ice age changes and interglacials. However, in recent decades, much faster systematic changes have been recorded. The Intergovernmental Panel on Climate Change (IPCC) in the synthesis of AR5 (IPCC, 2014) showed that each of the last three decades on the Earth's surface has been warmer than any previous decade since 1850. The considered period (1983–2012) was estimated to be probably the warmest 30-year period in the last 800 years in the northern hemisphere. The average annual temperature on land and sea increased in the range (0.65–1.06°C) from 1880 to 2012, with the trend being higher in the last few decades. It is estimated that the increasing temperature trend is likely to continue.

It is estimated that with further warming some wet regions will be even wetter, while some now relatively dry regions will be even drier (IPCC, 2014). The following are expected:

- increased droughts in dry areas (intensity and/or duration);
- expansion of dry areas;
- changes in surface runoff.

With an increase in average temperatures, the expansion of water-deficient areas is expected in arid and sub-arid areas, and areas poorer in precipitation.

For significant areas with a temperate climate, when the average temperature increases, a decrease in the surface runoff is expected and consequently a certain decrease in the average runoff. In addition, changes are expected in the regime of the vertical water flow in the return low-permeability layer (aquitard).

For the area of Serbia, our own analysis showed that, in the near future, for an increase of the average annual temperature by 2°C, the most probable reduction of the mean annual flow in rivers would be about 30%, and that the distribution is similar to those given by Gaussian.

Factors that cause the occurrence and the types and height of drought effects can include the climate and climate change, water consumption, the lithological soil structure, the hydrological balance of rivers, the degree of afforestation, urbanization and population density, the degree of forest cover, crops and other plants, irrigation etc.. This indicates a large number of factors that affect droughts and the need to respect the pronounced specifics of individual areas/problems.

Problems related to meeting the needs of the water-food-energy chain are expected for the entire planet, but they are more pronounced for some of its regions. Achieving sustainability in resource management remains one of the important goals.

Alluvial plains can be considered in terms of drought in several ways:

(1) Alluvial plains where droughts damage human needs and activities, as well as natural habitats;
(2) Climate change and drought which, among other factors, affect the state of water resources and the capacity of alluvial aquifers.

We touched on some examples and ways of indexing meteorological, hydrological, agricultural and sociological droughts. We paid some attention to factors regarding the vertical water balance in the surface – roof (above the aquifer) layers. These are precipitation, evaporation and transpiration at the depths of the root canal, and the capillary rise of water. Many of these factors make it difficult to accurately generalize drought. We would say that the occurrence of a drought is very much related to the region and locality and that punctual consideration is often expedient.

From the point of view of preserving the production of agricultural crops and natural vegetation in two-layered alluvial environments, it is of great importance to preserve the hydrological connection between the alluvial aquifer and the adjacent low-permeability aquifer. In the case of longer periods without rain, evapotranspiration is compensated for by capillary water from the alluvial aquifer. The withdrawal of groundwater to a greater depth in the soil reduces evaporation and, to a certain extent, enables the maintenance of a vertical water balance in the soil. Lowering groundwater levels below the aquifer-surface aquitard limit leads to a reduction or the interrupting of the capillary inflow of water to the area used by plant roots with consequences that can be very unfavourable.

The level and amount of water in an alluvial aquifer mainly depends on:

- The vertical precipitation water balance – evapotranspiration;
- The use of water from the aquifer, as well as the return of used water to the river/aquifer;
- The hydrological regime of the river;
- The inflow from the hinterland.

Reducing the amount of water in rivers can exacerbate the problem of water quality. This also applies to the irrigation process. Irrigation without appropriate accompanying measures (e.g. drainage) can lead to increased soil mineralization. It is also very important to determine the impact of agrotechnical measures on the quality of alluvial groundwater (fertilizers, pesticides, etc.)

The expansion of human agglomerations and infrastructure which are in opposition to the preservation of (alluvial) resources is also a very important issue.

The approach to the evaluation of the hydrological and hydrodynamic situation in the alluvial aquifer, the aquitard and the arable layer can be deterministic, stochastic and mixed. We have presented some basic equations of an approach for unsaturated and groundwater saturated porous media. We also presented some of the stochastic approaches, as well as indexing different types of drought.

We considered the question of water shortages more by looking at how to meet demands and use and maintain resources on the planet. Here it is important to consider the water-food-energy nexus in an appropriate way.

The questions that are asked are practical, but at the same time philosophical, political, etc. The use and management of resources must certainly be local, regional and global. Questions pertaining to responsibility and the methods of management, and to social and climate change in the world are questions which continue to be very topical and largely unanswered.

The issue of sustainability and of the conservation of water resources is very complex. In addition to climate, the sustainability of water resources management is influenced by economic, social, political and other factors. The planet's water resources are not homogeneously distributed. In addition to large climate differences, there is an evident difference in terms of the type, direction and intensity of the expected climate change. Social and economic opportunities and future shifts are factors in the sustainability of the use and protection of water resources.

The issues of response to various challenges are certainly very complex. Answers should be sought primarily based on very good knowledge of natural and social processes, as well as of the state of resources. It is necessary to have excellent knowledge of the possibilities of mastering technical solutions. In this regard, sufficient knowledge of the sector and of various issues in the field of water is required. It is necessary for human society to be aware of the necessary scientific, professional, institutional and operational capacities. Determining the political approach and its efficiency, as well as the possibility of its application, is also a very important factor.

The dispersion of the use of water resources can help increase adaptability to processes of change. This includes the use of groundwater and surface water, seawater desalination processes, wastewater recycling and stockpiling of reserves where geologic and other conditions permit it, primarily in groundwater aquifers, and then in surface reservoirs.

The issue of agricultural production indicates the need to focus on those crops whose cultivation is consistent with the capabilities of a particular community in terms of water. Climate change will lead to improvements in some areas (for example, two harvests a year), and to worsening agricultural conditions elsewhere. In some places, it will be necessary to switch to new agricultural crops. With the application of irrigation, it will be increasingly necessary to implement a system that will include the maximum rationalization of water consumption.

Any potential threat to alluvial plains and aquifers requires very careful planning and the systematic implementation of protection measures. It is not enough to just adapt to certain climatic and other challenges. It is certainly necessary to monitor the situation along with climate change, the trends in society, economic trends, etc.

The earth climate system and climate change processes are huge issues today and go beyond what is covered in this book.

## REFERENCES

Adamski M., Kremesec V., Kolhatkar R., Pearson C. and Rowan B. (2005). LNAPL in fine-grained soils: Conceptualization of saturation, distribution, recovery, and their modeling. *Ground Water Monitoring & Remediation*, **25**(1), 100–112.

Akuoko-Asibey A., Nkemdirim L. C. and Draper D. L. (1993). The impacts of climatic variables on seasonal water consumption in Calgary, Alberta. *Canadian Water Resources Journal*, **18**(2), 107–116, doi: 10.4296/cwrj1802107, https://doi.org/10.4296/cwrj1802107

Al-Zubari W. K., El-Sadek A. A., Al-Aradi M. J. and Al-Mahal H. A. (2018). Impacts of climate change on the municipal water management system in the Kingdom of Bahrain: vulnerability assessment and adaptation options. Available at http://www.researchgate.net/publication/322980266_Impacts_of_climate_change_on_the_municipal_water_management_system_in_the_Kingdom_of_Bahrain_Vulnerability_assessment_and_adaptation_options.

Arnell N. W. (1999). Climate change and global water resources. *Global Environmental Change*, **9**, S31–S49.

Arnell N. W. (2003). Effects of IPCC SRES emissions scenarios on river runoff: a global perspective. *Hydrology and Earth System Sciences*, **7**, 619–641.

Arnell N., Liu C. and Compagnucci R. (2001). 'Hydrology and Water Resources'. In: Climate Change 2001: Impacts, Adaptation, and Vulnerability, A. Becker (Germany) and Jianyun Zhang (China) (eds.), The Press Syndicate of the University of Cambridge, Cambridge, UK, p. 191234.

Assouline S., Tyler S. W., Selker J. S., Lunati I., Higgins C. W. and Parlange M. B. (2013). Evaporation from a shallow water table: diurnal dynamics of water and heat at the surface of drying sand. *Water Resources Research*, **49**, 4022–4034, doi: 10.1002/wrcr.20293

Avakumović D., Stričević R., Urović N., Stanić M., Dašić T. and Ukić V. (2005). Savremena analiza potrebnih količina vode za navodnjavanje, VODOPRIVREDA 0350-0519, 37(2005) 213–215, pp. 11–20.

Babkin V. I. (2008). Evaporation from the Surface of the Globe. State Hidrological Institute, St. Petesburg, Russia.

Barajas S. M. (2007). Ecologistac en accion, The Effects of Climate Change on Water Resources in Spain, 2007. Available at https://spip.ecologistasenaccion.org/IMG/pdf_CC_W_in_Spain.pdf

Boreli M. (1967). Podzemna voda u zoni aeracije tretirana jednačinama strujanja vode u nezasićenim sredinama, Seminar Bilans podzemnih voda, Mladen Boreli, Beograd 1967.

Bou-Zeid E. and El Fadel M. (2002). Climate change and water resources in Lebanon and the Middle East. *Journal of Water Resources Planning and Management*, **128**: 343–355.

Cai X. and Rosegrant M. W. (2002). Global water demand and supply projections. Part **1**. A modelling approach. *Water International*, **27**(2), 159–169.

Chmielewski F.-M. and Potts J. M. (1995). The relationship between crop yields from an experiment in southern England and long-term climate variations. *Agricultural and Forest Meteorology*, **73**(1–2), 43–66.

von Christierson B., Hannaford J., Lonsdale K. and Parry S. (2009). Impacts of Long Droughts on Water Resources. Environment Agency, Bristol, UK.

Climate Change and Impacts on Water Supply (CC-WaterS). (2009–2012). International Project for SE Europe, 18 Institutions from SE Europe, May 2009–May 2012, [Available online at http://www.ccwaters.eu/index.php?option=com_content&view=article&id=48&Itemid=54&56b00064c3e6beb26da3b96d1578b92a=caac98a9a4248cd115a194c70c97a142].

Climate Hot Map. Global warming effects around the world; Union of Concerned Scientists, Available at: https://www.climatehotmap.org/global-warming-effects/drought.html

Deliverable of the International project DMCSEE. (2009–2012). Palfai Drought Index (PaDI) - Expansion of applicability of Hungarian PAI for Southeast Europe (SEE) region, Available at http://www.dmcsee.org/uploads/file/306_wp322_padi_expansionofapplicability.pdf

DeOreo W. B., Mayer P. W., Dziegielewski B., Kiefer J. and Water Research Foundation. (2016). Residential End Uses of Water, Version 2. Water Research Foundation, Denver, CO.

Dieter C. A., Maupin M. A., Caldwell R. R., Harris M. A., Ivahnenko T. I., Lovelace J. K., Barber N. L. and Linsey K. S. (2018). Estimated Use of Water in the United States in 2015. USGS Numbered Series. U.S. Geological Survey, Reston, VA.

Dimkić D. (2016). Reliability Issue in Projections of the Impacts of Climate Change and Other Changes on Water Resources. *Journal of Serbian Water Pollution Control Society*, **6**(1), 19–24, 'Water Research and Management', ISSN 2217-5237, Publisher: Serbian assotiation for the water protection. Available at http://www.wrmjournal.com/images/stories/casopis/No_21/03.pdf

Dimkić D. (2017). Present and future mean hydrological trends in Serbia as a function of climate trends. *Desalination and Water Treatment*, **99**, 10–17. ISSN 1944-3994, Publisher: Desalination Dimkić D., Characteristics of annual, seasonal and monthly observed climate and hydrologic changes in Serbia in the last seventy years, Desalination and Water Treatment, Publisher: Desalination Publications, 167, 359–369, November 2019. https://doi.org/10.5004/dwt.2019.24575Publications.

Dimkić D. (2019). Characteristics of annual, seasonal and monthly observed climate and hydrologic changes in Serbia in the last seventy years. *Desalination and Water Treatment*, **167**, 359–369, Publisher: Desalination Publications, https://doi.org/10.5004/dwt.2019.24575

Dimkić D. and Despotović J. (2012). Analysis of the Changes of the Streamflows in Serbia Due to Climate Changes. In: 'Climate Change - Inferences from Paleoclimate and Regional Aspects', A. Berger, F. Mesinger and D. Šijački (eds.), Springer, Wurzburg, pp. 167–177, ISBN 978-3-7091-0972-4.

Dimkić M., Brauch H. J. and Kavanaugh M. (2008). Water Management in Large River Basins. IWA Publishing, London.

Dimkić D., Slimak T. and Radenković Z. (2013). Correlation Between Average Daily Temperatures and Production of Water in Water Supply Systems. International Conference Climate Change Impacts on Water Resources, 17–18 October 2013, Belgrade, Serbia, pp. 153–160, Publisher: Jaroslav Černi Institute for the Development of Water Resources, ISBN 978-86-82565-41-3.

Dimkić D., Prohaska S., Stanković B. and Pajić P. (2018). Alluvial Water Source Capacity under the Climate Change and Other Impacts – Case Study of the Pek River in Serbia. Proceedings of EWAS3 Conference, 2018, Lefkada, paper published by Elsevier in Procedia Engineering, http://www.mdpi.com/2504-3900/2/11/596/pdf

Dinar A., Hassan R., Mendelsohn R. and Benhin J. (2008). Climate Change and Agriculture in Africa: Impact Assessment and Adaptation Strategies. EarthScan, London.

Doorenbos J. and Pruitt W. D. (1975). Les besoins en eau des cultures, Bulltin d'Irr. et de drain. No 24, OUN – FAO, Rome.

Durdu Ö. F. (2009). Effects of climate change on water resources of the Büyük Menderes river basin, western Turkey, 2009. Available at https://dergipark.org.tr/en/download/article-file/119648

Endter-Wada J., Kurtzman J., Keenan S. P., Kjelgren R. K. and Neale C. M. U. (2008). 'Situational Waste in Landscape Watering: Residential and Business Water Use in an Urban Utah Community'. *Journal of the American Water Resources Association*, **44**, 902–920.

Environment Agency Horizon House. (2012). South West Drought Plan. Deanery Road Bristol, UK.

FAO (Food and Agriculture Organization of the United Nations). (2011a). The State of the World's Land and Water Resources: Managing Systems at Risk. Earthscan, London.

FAO (Food and Agriculture Organization of the United Nations). (2011b). AQUASTAT Online Database. FAO, Rome. http://www.fao.org/nr/water/aquastat/data/query/index.html

Fleischer A., Lichtman I. and Mendelsohn R. (2008). Climate change, irrigation, and Israeli agriculture: will warming be harmful? *Ecological Economics*, **67**, 109–116.

Fowler H. J., Kilsby C. G. and Stunell J. (2007). Modelling the impacts of projected future climate change on water resources in north-west England. *Hydrology and Earth System Sciences*, **11**(3), 1115–1126.

Fujihara Y., Tanaka K., Watanabe T., Nagano T. and Kojiri T. (2008). Assessing the impacts of climate change on the water resources of the Seyhan River Basin in Turkey: use of dynamically downscaled data for hydrologic simulations. *Journal of Hydrology*, **353**, 33–48, http://www.sciencedirect.com/science/article/pii/ S0022169 408000346

Gavrilović Z., Dimkić D. and Stefanović M. (2013). The effects of Forest on Runoff Regimes in Serbia. International Conference, 17–18 October, pp. 203–210. Climate Change Impacts on Water Resources; ISBN 978-86-82565-41-3

Giorgi F., Bi X. and Pal J. S. (2004). Mean, interannual variability and trends in a regional climate change experiment over Europe. II. Climate change scenarios (2071–2100). *Climate Dynamics*, **23**, 839–858.

Gunduz O. and Simsek C. (2011). Influence of Climate Change on Shallow Groundwater Resources: The Link Between Precipitation and Groundwater Levels in Alluvial Systems, Available at: https://link.springer.com/chapter/10.1007/978-94-007-1143-3_25b

Guttman N. B. (1998). Comparing the Palmer Drought Index and the Standardized Precipitation Index. *Journal of the American Water Resources Association*, **34**, 113–121, doi: 10.1111/j.1752-1688.1998.tb05964.x. (For more information on this paper, please contact the IDMP HelpDesk).

Guttman N. B. (1999). Accepting the Standardized Precipitation Index: A Calculation Algorithm. *Journal of the American Water Resources Association*, **35**, 311–322, doi: 10.1111/j.1752-1688.1999.tb03592.x. (For more information on this paper, please contact the IDMP HelpDesk).

Gutzler D. S. and Nims J. S. (2005). Interannual Variability of Water Demand and Summer Climate in Albuquerque, New Mexico. *Journal of Applied Meteorology and Climatology*, **44**(12), 1777–1787, https://doi.org/10.1175/JAM2298.1

Hayes M., Svoboda M., Wall N. and Widhalm M. (2011). The Lincoln Declaration on Drought Indices: Universal Meteorological Drought Index Recommended. *Bulletin of the American Meteorological Society*, **92**(4), 485–488, doi: 10.1175/2010BAMS3103.1

Hoekstra A. Y., Chapagain A. K., Aldaya M. M. and Mekonnen M. M. (2011). The Water Footprint Assessment Manual: Setting the Global Standard. Earthscan, London and Washington, DC. ISBN: 978-1-84971-279-8.

Hoekstra Y. and Mekonnen M. M. (2011). The water footprint of humanity. *PNAS Direct Submission*, **109**(9), 3232–3237.

IPCC. (2007). Assessment Reports Working Group II: Impacts, Adaptation and Vulnerability.

Intergovernmental Panel on Climate Change (IPCC). (2014). AR5 Synthesis Report, Available at: https://www.ipcc.ch/site/assets/uploads/2018/05/SYR_AR5_FINAL_full_wcover.pdf

Juckem P. F., Hunt R. J., Anderson M. P. and Robertson D. M. (2008). Effects of climate and land management change on stream flow in the driftless area of Wisconsin. *Journal of Hydrology*, **355**, 123–130.

Kenney D. S., Goemans C., Klein R., Lowrey J. and Reidy K. (2008). 'Residential Water Demand Management: Lessons from Aurora, Colorado'. *Journal of the American Water Resources Association*, **44**, 192–207.

Krzic A., Tosic I., Djurdjevic V., Veljovic K. and Rajkovic B. (2011). Changes in some indices over Serbia according to the SRES A1B and A2 scenarios. *Climate Research*, **49**, 73–86. doi: 10.3354/cr01008. Available at https://www.int-res.com/articles/cr2012/49/c049p073.pdf

Kurukulasuriya P. and Mendelsohn R. (2008). A Ricardian analysis of the impact of climate change on African cropland. *African Journal of Agricultural and Resource Economics*, **2**, 1–23.

Kurukulasuriya P., Mendelsohn R., Hassan R., Benhin J., Diop M., Eid H. M., Fosu K. Y., Gbetibouo G., Jain S., Mahamadou A., El-Marsafawy S., Ouda S., Ouedraogo M., Sene I., Seo N., Maddison D. and Dinar A. (2006). Will African agriculture survive climate change? *World Bank Economic Review*, **20**, 367–388.

Lebedev. (1967). Izrada bilansa podzemnih voda za rečni sliv, Seminar Bilans podzemnih voda, Mladen Boreli, Beograd 1967.

List 25 by Teodoras II. https://list25.com/25-most-severe-droughts-ever-recorded/

Liu H., Li X., Fischer G. and Sun L. (2004). Study on the impacts of climate change on China's agriculture. *Climate Change*, **65**(1–2), 125–148.

Loukas A., Vasiliades L. and Tzabiras J. (2007). Evaluation of climate change on drought impulses in Thessaly, *Greece.* European Water, **17/18**, 17–28.

Ma Z., Kang S., Zhang L., Tong L. and Su X. (2008). Analysis of impacts of climate variability and human activity on stream flow for a river basin in arid region of northwest China. *Journal of Hydrology*, **352**, 239–249.

Maidment D. R. and Miaou S.-P. (1986). 'Daily Water Use in Nine Cities'. *Water Resources Research*, **22**, 845–851.

McKee T. B., Doesken N. J. and Kleist J. (1993). The Relationship of Drought Frequency andDuration to Time Scales. Proceedings of the 8th Conference on Applied Climatology, 17–22 January 1993, Anaheim, CA, Boston, MA, American Meteorological Society.

Mekonnen M. M. and Hoekstra A. Y. (2011). The green, blue and grey water footprint of crops and derived crop products. *Hydrology and Earth System Science*, **15**, 1577–1600, doi: 10.5194

Ministry for Environment Protection of Republic of Serbia. (2019). Available at http://indicator.sepa.gov.rs/pretrazivanje-indikatora/indikatorilat/allfindu/2541db7873af4a7c89a8af252092ce4f

Ministry for Environment Protection of Republic of Serbia. (2019). Available at http://indicator.sepa.gov.rs/pretrazivanje-indikatora/indikatorilat/allfindu/448a2a2911e04de49d29a280dd496ddb

Nachtnebel H. P. (2013). Water Resources Under Pressure: Direct Human Interventions and Climate Change. International Conference, 17–18 October: Climate Change Impacts on Water Resources, p. 31, ISBN 978-86-82565-41-3.

Ngaira J. K. W. (2007). Impact of climate change on agriculture in Africa by 2030. *Scientific Research and Essays*, **2**(7), 238–243, Available at http://www.academicjournals.org/SRE

Nisbet T. (2005). Water use by trees. Forestry Commission Information Note 65. Forestry Commission Edinburgh, UK.

Nolan J. D., McKee T. B. and Kleist J. (1991). Development of a surface water supply index for the western United States, Final report, October 1991. http://www.droughtmanagement.info/literature/CSU_Development_SWSI_Western_United_States_1991.pdf

Novotny E. V. and Stefan H. G. (2007). Stream flow in Minnesota: Indicator of climate change. *Journal of Hydrology*, **334**, 319–333.

Opalinski N. F., Bhaskar A. S. and Manning D. T. (2019). Spatial and Seasonal Response of Municipal Water Use to Weather across the Contiguous US. *Journal of the American Water Resources Association*, **2019**, 1–14, Paper No. JAWRA-18-0117-P.

Palfai I. (1990). Description and forecasting of droughts in Hungary. Proceedings of 14th Congress on Irrigation and Drainage (ICID), 1990, Rio de Janerio, Vol. 1-C, pp. 151–158.

Palmer W. C. (1965). Meteorological Drought. Research Paper No. 45. United States Weather Bureau, Washington, DC. http://www.droughtmanagement.info/palmer-hydrological-drought-index-phdi/

Paneque P. (2015). Drought Management Strategies in Spain, 2015, Available at https://www.researchgate.net/publication/284721751_Drought_Management_Strategies_in_Span

Praskievicz S. and Chang H. (2009). Identifying the Relationships Between Urban Water Consumption and Weather Variables in Seoul, Korea. *Physical Geography*, **30**(4), 324–337, doi: 10.2747/0272-3646.30.4.324

Rasifaghihi N., Li S. S. and Haghighat F. (2020). Forecast of urban water consumption under the impact of climate change. *Sustainable Cities and Society*, **52**, Elsevier, https://doi.org/10.1016/j.scs.2019.101848

Sandra Collin. Editorial office: FVA, Germany, Available at: https://www.waldwissen.net/waldwirtschaft/schaden/trockenheit/fva_trockenheit__1/index_EN

Scheidegger A.E. (1957). The Physics of Flow through Porous Media. New York, Macmillan.

Schenk J. H. and Jackson R. B. (2002). Rooting depths, lateral root spreads and below-ground/above-ground allometries of plants in water-limited ecosystems. *Journal of Ecology*, **90**, 480–494. Department of Biology and Nicholas School of the Environment and Earth Sciences, Duke University, Durham, North Carolina 27708, USA.

Seo S. N. and Mendelsohn R. (2008). A Ricardian analysis of the impact of climate change on South American farms. *Chilean Journal of Agricultural Research*, **68**(1), 69–79.

Steinemann A., Hayes M. J. and Cavalcanti L. (2005). Drought Indicators and Triggers. Drought and Water Crises: Science, Technology, and Management Issues: 1–15.

Strzepek K. M. and Yates D. N. (1996). Economic and social adaptations to climate change impacts on water resources: a Case study of Egypt. *Water Resources Development*, **12**, 141–150.

Synthesis Report of the IPCC. (2007, 2013). www.ipcc.ch South East Europe Climate Outlook Forums (SEECOFs), 2008–2010. International Cooperation, Available online at http://www.google.rs/search?hl=sr&source=hp&q=climate+change+seecof&btnG=Google+%D0%BF%D1%80%D0%B5%D1%82%D1%80%D0%B0%D0%B3%D0%B0&meta=&gbv=2&oq=climate+change+seecof&aq=f&aqi=&aql=&gs_sm=s&gs_upl=1280l1323l10l1733l4l23l23l1l15l3l0l297l346l0.9.8l17l0

Tao F., Yokozawa M., Xu Y., Hayashi Y. and Zhang Z. (2006). Climate change and trends in phenology and yields of field crops in China, 1981–2000. *Agricultural and Forest Meteorology*, **138**, 82–92.

Ting Meng. (2017). Analyzing Temperature and Precipitation Influences on Yield Distributions of Canola and Spring Wheat in Saskatchewan, JAMC April 2017, Available at: https://journals.ametsoc.org/doi/full/10.1175/JAMC-D-16-0258.1

UNISDR. (2009). Drought Risk Reduction Framework and Practices. United Nations Secretariat of the International Strategy for Disaster Reduction, Geneva, Switzerland.

USDA Irrigation Water Requirements. (1970). Technical Release No 21 2nd Edition, Washington.

Van der Gun J. A. M. (2009). Climate Change and Alluvial Aquifers in Arid Regions: Example from Yemen, Available at: https://books.google.rs/books?id=fOQ-BUNIjJ0C&pg=PA159&lpg=PA159&dq=climate+change+and+alluvial+sources&source=bl&ots=PO9Ck6h7_q&sig=ACfU3U1zMgh5-_-VYpN-ksuYu5RNa6b0VA&hl=en&sa=X&ved=2ahUKEwjp4vSOjbXmAhWumIsKHUqrATk4ChDoATAAegQIChAB#v=onepage&q=climate%20change%20and%20alluvial%20sources&f=false

Vasiljević B., Dimkić D. and Đurđević V. (2013). Historical Overview and Different Methodologies Applied in Climate Change Studies. International Conference: Climate Change Impacts on Water Resources, 17–18 October, pp. 183–194, ISBN 978-86-82565-41-3.

Verón S. R., de Abelleyra D. and Lobell D. B. (2015). Impacts of precipitation and temperature on crop yields in the Pampas. *Climatic Change*, **130**, 235–245, https://doi.org/10.1007/s10584-015-1350-1

Vicente-Serrano S. M. (2006). Spatial and temporal analysis of droughts in the Iberian Peninsula (1910–2000). *Hydrological Sciences Journal*, **51**(1), 83–97, doi: 10.1623/hysj.51.1.83

Vuković M. (1967). Određivanje parametara vodonosne sredine i faktora bilansa u dvoslojevitoj i višeslojevitoj sredini iz pijezometarskih podataka, Seminar Bilans podzemnih voda, Mladen Boreli, Beograd 1967.

Waluyo H. R., Triweko W. and Hadihardaja I. K. (2013). Water resources drought: Hydrological drought in developed river basins. 4th International Seminar of HATHI, 6–8 September 2013, Yogyakarta; https://www.researchgate.net/publication/275341859_Water_Resources_Drought_Hydrological_Drought_in_Developed_River_Basins

Wasserhaushalt. (2012). 'How to Define a Drought'. ETH Zurich. Wilhite, DA, and MH Glantz. 1985. 'Understanding: The Drought Phenomenon: The Role of Definitions'. Water International (May 2013): 37–41.

Weber J. A. (1989). 'Forecasting Demand and Measuring Price Elasticity'. *American Water Works Association Journal*, **81**(5), 57–65.

Wild D., Francke C.-J., Menzli P. and Schön U. (2010). Water: A Market of the Future. A Sustainable Asset Management (SAM) Study. SAM, Switzerland.

Wilhite D. A. (2005). Drought and Water Crises, Science, Technology and Management Issues. Taylor and Francis Group, Boca Raton.

WMO. (2006). World Meteorological Organization CAgM Report No. 101 Impacts of Desertification and Drought. Geneva, Switzerland.

World Meteorological Organization. (2012). Standardized Precipitation Index User Guide, (WMO-No. 1090). World Meteorological Organization, Geneva, Switzerland.

World Water Council. (2009). Istanbul Water Consensus For Local and Regional Authorities. Proceedings of the 5th World Water Forum, Istanbul, Turkey. http://www.worldwatercouncil.org/fileadmin/wwc/World_Water_Forum/WWF5/IstanbulWaterGuide_Final_05-03-09.pdf

Wu H., Hayes M. J., Wilhite D. A. and Svoboda M. D. (2005). The Effect of the Length of Record on the Standardized Precipitation Index Calculation. *International Journal of Climatology*, **25**(4), 505–520, doi: 10.1002/joc.1142

Xu W. and Su X. (2019). Challenges and impacts of climate change and human activities on groundwater-dependent ecosystems in arid areas – a case study of the Nalenggele alluvial fan in NW China. *Journal of Hydrology*, **571**, 376–385.

Yevjevich V. (1967). An Objective Approach to Definitions and Investigations of Continental Hydrologic Droughts. Hydrology Papers Colorado State University Fort Collins, Colorado.

Ying F. and Gonzalo M.-M. (2017). Hydrologic regulation of plant rooting depth, 2017.

Zapata O. (2015). More Water Please, It's Getting Hot! The Effect of Climate on Residential Water Demand. *Water Economics and Policy*, **01**, 1550007, https://doi.org/10.1142/S2382624X15500071

# Appendix 1

# Surface tension, capillarity and viscosity

*Srđan Kovačević, Slobodan Vujasinović and Milan Dimkić*

## A1.1 SURFACE TENSION

The free level of still fluid does not remain horizontal at the point where it contacts a solid wall. The deformation of the surface is caused by the surface potential. The boundaries between two liquids of different densities that do not mix are internal boundary surfaces. The outer or free surface is formed at places where liquids and gas meet (liquid-gas phase boundary layer).

Inside the fluid, the particles are located in the middle of intermolecular attracting forces, and the axial field is spherical (Figure A1.1). The resultant force on the particles is equal to zero. The particles located directly on the free surface are not affected by the attraction forces of the gas (negligibly small relative to the liquid), so that the resultant force $F$ is directed towards the interior of the liquid. The thickness of the free surface layer is comparable to the area where intermolecular forces are active (about 10 molecule diameters, i.e. about $10^{-9}$ m = 1 nm).

Due to the effect of the resultant force $R$, the particles on the surface exert pressure on the particles inside the liquid, which makes the inner particles tend to stretch the free surface and is why the particles on the free surface have more energy than the particles inside. Each free surface contains extra energy. Because of this extra energy, nature has made it possible to build a free surface with a minimum of particles. Thus a minimal free surface was created.

Fluid that is in equilibrium has minimal potential energy. Therefore, any liquid body affected by external forces must have a shape at which the surface of its envelope will be the smallest. For an incompressible fluid, the minimum quotient of the envelope surface and of the corresponding volume is the necessary condition for a stationary equilibrium. Spherical bodies satisfy this condition.

Surface tension is responsible for the capillary properties that allow water (and the dissolved substances in it) to move through the roots of plants and through blood vessels in the bodies of animals The layer has the


Editor: Milan Dimkić
doi: 10.2166/9781789060904_0667

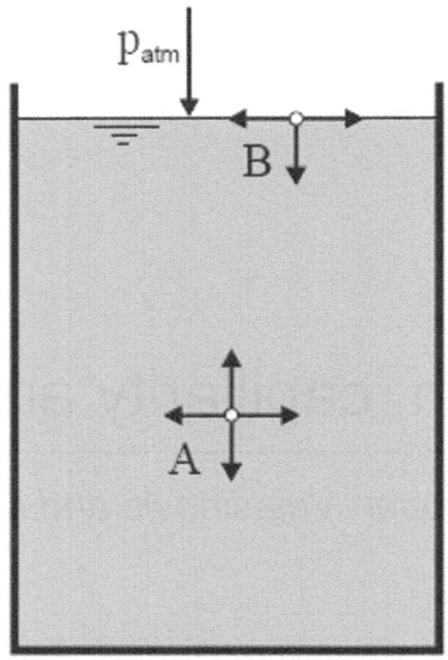


**Figure A1.1** Balance of forces when fluid is still (Pušić, 2012).

properties of a 'taut' (tightened) surface. The equal attraction between the liquid and gas molecules is impaired at the surface. The molecules at the surface are subject to more intense attraction of internal molecules. The result is a reduction in the surface area and the formation of a tightened 'membrane' [N/m].

Liquid is characterized by surface tension, which is, in addition to vapour pressure, another measure of the strength of intermolecular forces. This is why a very small drop of liquid has the shape of a sphere, because in this geometric form the relationship between the surface and the volume of the liquid is the smallest. Consequently, in order to change the shape of the fluid droplets, action is necessary, and how big the action will be depends on the strength of the intermolecular forces in the liquid. The work required to spread a layer of fluid per unit area, i.e. for the liquid to 'get a surface wet', represents the surface tension.

$$\gamma = \frac{F}{l} \tag{A1.1}$$

where $\gamma$ – the surface tension [N/m], $F$ – the total force on the interface [N], $l$ – the length of the interface [m].

The surface tension is due to the interaction of the liquid molecules. If the fluid is in a state of equilibrium (stationary), then all the forces acting at point A are cancelled by equal and opposite forces from neighbouring fluid particles (Figure A1.1).

According to the same principle, only the horizontal forces at point B on the surface of the fluid are (cancelled out as equal and opposite) in equilibrium. These horizontal forces act only along the surface and are called surface tension forces.

In addition, a vertical force acts at point B, along with the force of gravity. This force is called the force of the internal fluid pressure, or molecular pressure.

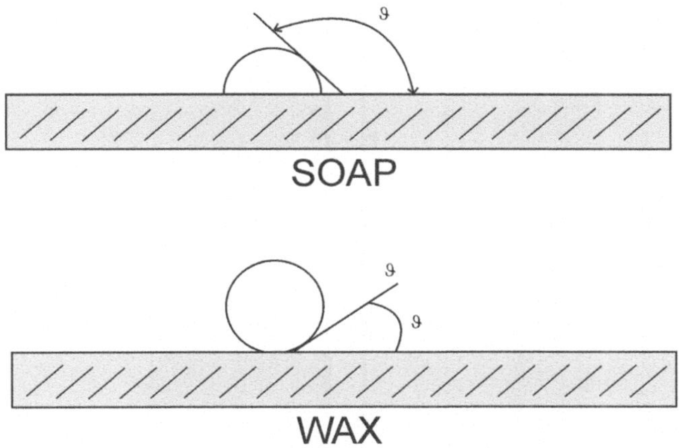


**Figure A1.2** Leaving a surface wet depending on the angle of contact (Pušić, 2012).

The shape of the surface of the liquid that is in contact with a solid body is determined by the relative value of the surface energy. Liquid-solid combinations in this field are classified as wetting or non-wetting, depending on the contact angle between the liquid and the solid surface. Figure A1.2 shows water droplets on soap and on a freshly waxed surface. If the contact angle is $>90°$, the liquid will leave the surface wet and for $<90°$ the liquid will not leave the surface wet.

## A1.2 CAPILLARITY

Due to the different intensities of adhesion forces (forces exerted by a solid surface on fluid particles) and on cohesion (intermolecular fluid forces), the level of fluid in relatively narrow tubes may be higher or lower than is consistent with the law of coupled vessels. The capillary rise or fall of the fluid level is brought about by the surface tensions. The liquid rises or falls in the capillary tube (relative to the free surface level) depending on the shape of the dividing liquid surface, that is, on the type of fluid and on the capillary material.

If a glass tube of a small diameter (capillary) is immersed in water, the water will rise in it and a concave meniscus will be created, the curved surface of the fluid (Figure A1.3).

The height of the rise of water in the pipe, *h,* is expressed by the equation:

$$h = \frac{2\gamma \cos \vartheta}{\rho g r} \tag{A1.2}$$

where:

$\gamma$ – surface tension of water, [N/m$^2$];
$\vartheta$ – the angle of curvature of the meniscus, for water $\vartheta = 0$;
$\rho$ – density, [kg/m$^3$];
$g$ – gravity, [m/s$^2$];
$r$ – pipe radius, [m].

Figure A1.4 shows the height of water rising in a capillary depending on its diameter (logarithmic scale).

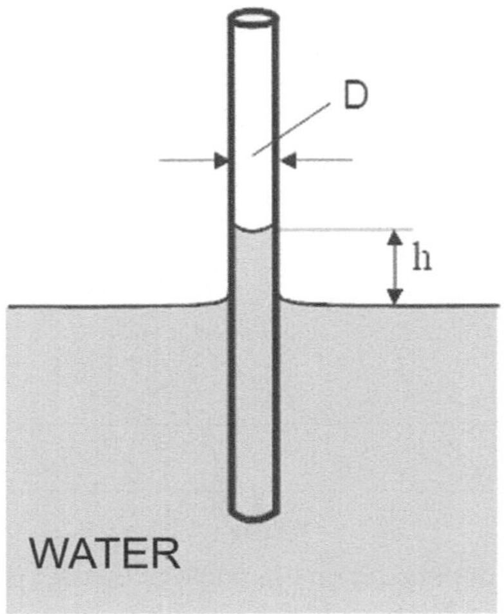


**Figure A1.3** Water capillarity in a glass tube (Pušić, 2012).

The height of water rising in a capillary depends on the:

- Temperature (temp ↑, γ ↓);
- Dissolved substance concentration;
- Inorganic salts, γ ↑;
- Alcohols, γ ↓;
- Surfactants (surfactants), γ ↓.

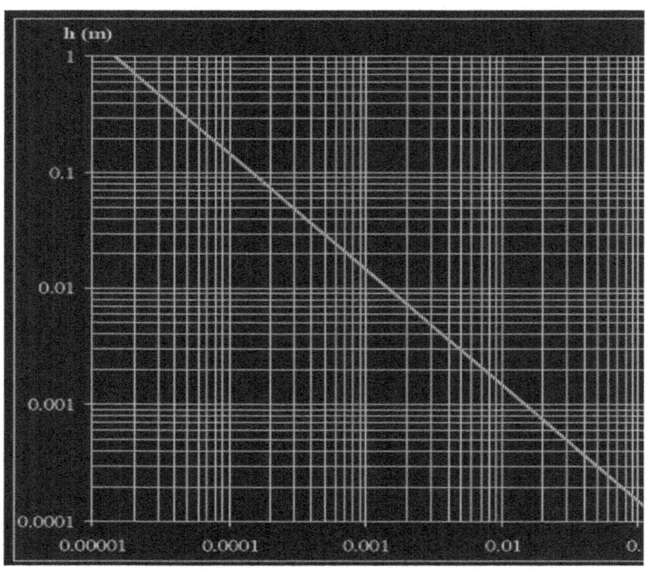


**Figure A1.4** The height of water rising in a capi lary depending on its diameter (Pušić, 2012).

Capillarity in rocks is a phenomenon that consists of water or some other fluid moving and rising in rocks containing capillary pores and cracks, under the influence of adhesion and cohesion forces. Capillary pores are all those pores in a rock that range in size between 0.5–0.0002 mm, and capillary cracks are all cracks whose size ranges between 0.25–0.0001 mm.

Pores and cracks larger than those mentioned belong to rocks of super capillary porosity, and less so to rocks of sub-capillary porosity. The height of the capillary rise ($H_k$) depends on the diameter of the capillary pores (Figure A1.5) or cracks ($d$) and on size ($a$) which, for example, for water has a constant value of 15.4 mm (at 0°C).

$$H_K = \frac{2a^2}{d} \tag{A1.3}$$

An understanding of the value of the capillary rise of groundwater influences different areas of human activity such as: remediation and environmental protection, and construction work, as well as agricultural production (Salim, 2016). The capillary uplift zone is important for studying the biodegradation processes and chemical reactions that take place in groundwater at the boundaries of three environmental matrices: water, air, and soil.

The processes of the biodegradation and chemical reactions which take place at the border of the three phases are particularly important to the field of remediation and groundwater protection as well as for environmental protection, especially when at a particular location, contamination occurs in the saturated zone (below the top of the capillary edge) and in the unsaturated zone (above the capillary edge).

Due to different conditions, the capillary edge should be viewed as a separate zone, and as such the capillary edge must be defined for a test area. In order to define the area of the capillary edge, the capillary rise of water or other fluids in the soil that is tested must be determined.

The capillary rise of water is also very important in the construction industry because the foundations of a building could be above the groundwater level but are in the capillary rise zone of water, which leads to an increase in moisture in the foundation of the building. It is also necessary to be familiar with the capillary lifting for various building materials.

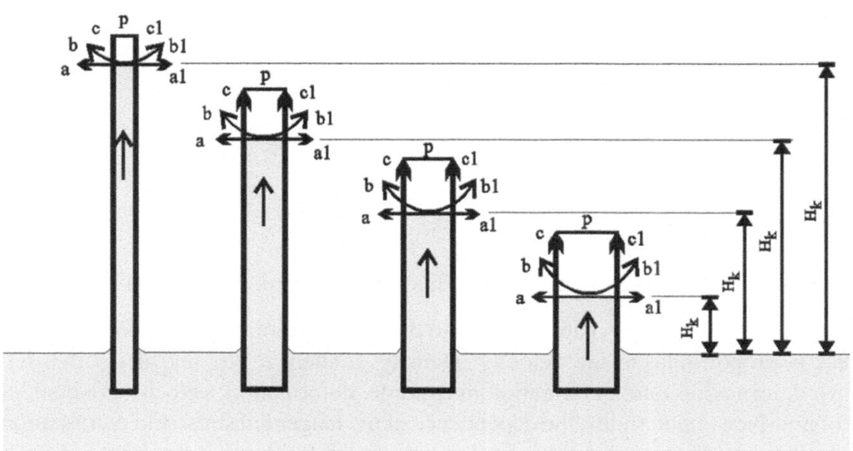


**Figure A1.5** Schematic illustration of the rising of water in pipes of different diameters (Dragišić, 1997).

**Table A1.1** Capillary rise partially based on Acamski *et al.*'s (2005) calculations.

| Sediment | Average Grain Diameter (cm) | Representative $d_{10}$ (cm) | Pore Radius = $d_{10}/d_5$ (cm) | Capillary Rise (cm) |
|---|---|---|---|---|
| Fine silt | 0.0008 | 0.0002 | 0.00004 | 3750 |
| Coarse silt | 0.0025 | 0.00063 | 0.00013 | 1154 |
| Very fine sand | 0.0075 | 0.0019 | 0.00037 | 405 |
| Fine sand | 0.0150 | 0.00375 | 0.00075 | 200 |
| Medium-sized sand | 0.03 | 0.0075 | 0.0015 | 100 |
| Coarse sand | 0.05 | 0.0125 | 0.0025 | 60 |
| Very coarse sand | 60 | 0.050 | 0.010 | 15 |
| Fine gravel | 0.50 | 0.125 | 0.025 | 6 |

Knowing and determining the value of the capillary rising of water for different soil types is particularly important when looking at the irrigation of farmland. Capillary lifting can also affect the salinity of soil used for agricultural production. If groundwater is salinized, the capillary rise of water can lead to the formation of salt deposits due to evaporation as a result of the capillary lifting of water. Table A1.1 shows the experimentally obtained values pertaining to the capillary rise of water for different types of soil with different granulometric compositions.

Each pore represents the equivalent for a different capillary tube, which means that the value of the capillary rise is different under laboratory conditions compared to that in field conditions. It is difficult, if not impossible, to equalise the pores, even in controlled laboratory experimental conditions. Due to the complexity of the soil and the distribution of pores that appear uneven and lack regularity, it is very difficult to accurately predict the capillary rise in the soil in natural conditions.

## A1.3 VISCOSITY

Viscosity is a natural property of a fluid that produces a shear stress (tangential stress) at the contact surface of two fluid layers. According to Newton, the shear stresses in the contact layers for a uniform plane current are equal to:

$$\tau = \eta \frac{dv}{dy} \tag{A1.4}$$

where

- $\tau$ [Pa] – tangential voltage;
- $\eta$ [Pa · s] – dynamic viscosity;
- $dv/dy$ $[(m/s)/s]$ – change of velocity $v$ of a fluid in a direction normal to a current flow ($v = f(y)$).

Fundamental regularity $\tau = \eta(dv/dy)$ applies only to the laminar motion of Newtonian fluids in which the tangential stress is proportional to the velocity intensity gradient $dv/dy$ and it is often replaced by the following terms: deformation rate, deformation magnitude, deformation, velocity gradient, etc.

In the case of non-Newtonian fluids, the dependence of the tangential stress and deformation is given by a whole series of relations, depending on the rheological type of fluid, the instrument used to determine the dependence and the level of knowledge pertaining to the properties of fluids.

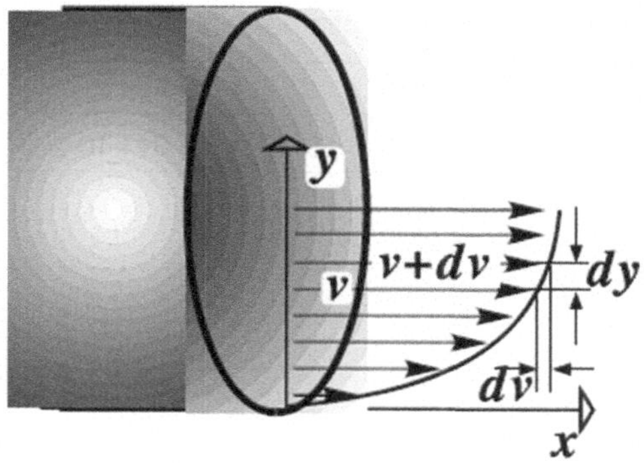


**Figure A1.6** Velocity distribution along the cross-section of a fluid flow (Pušić, 2012).

Dynamic viscosity (in Newtonian fluids) divided by fluid density is called kinematic viscosity:

$$\nu = \frac{\eta}{\rho}\left[\frac{m^2}{s}\right] \tag{A1.5}$$

Viscosity is the property of a fluid that resists the relative movement of its particles, that is, a condition that causes the occurrence of tangential stresses ('internal friction') when the fluid flows. When a fluid moves there is friction between its adjacent layers. It also causes friction between the fluid and the surface where the fluid moves (riverbed and the sides of a river, pipe, etc.). As a result of the resistance that occurs during sliding of one layer of fluid over another (the layer that moves more rapidly 'pulls' the adjacent layer that moves more slowly), a force of internal friction occurs and it is called the viscosity of a fluid, Figure A1.6. In this movement, mechanical movement is converted into thermal energy. The word viscosity comes from the English word viscosity meaning stickiness, and it was introduced by W. Thomson (1865).

It is impossible to distinguish individual layers in a liquid and that is why the term infinitely thin layer has been introduced. This term was introduced by Isaac Newton, who defined the law of the internal friction of a fluid (1687). According to him, the force of internal friction (with the so-called Newtonian fluids) is:

- Proportional to the velocity gradient perpendicular to the direction of travel;
- Proportional to the contact surface;
- Not dependent on pressure;
- Dependent on the physical properties of a liquid (its type and its temperature).

The tangential stress or 'internal friction' of a real fluid is often high and should not be neglected in many cases. Some specific fluids (heavy oils, rubber solutions, honey, blood clots, etc.) do not behave according to Newton's law, i.e. the relationship between the force of internal friction and the velocity gradient is not linear. These fluids are also called non-Newtonian fluids.

In practice, there are also cases where the friction forces caused by the viscosity occur as negligible magnitudes with respect to other forces acting on the fluid. In these special cases, it is possible to neglect the viscosity and assume that there is no tangential stress in the moving fluid. The viscosity of the fluid varies with temperature. As the temperature increases, the viscosity of the liquid decreases. Table A1.2 and Figure A1.7 give the values of the kinematic coefficient of water viscosity as a function of temperature.

**Table A1.2** Change in the kinematic coefficient of water viscosity as a function of temperature (Pušić, 2012).

| *T* (°C) | 0 | 5 | 10 | 20 | 50 | 100 |
|---|---|---|---|---|---|---|
| $v\ (m^2/s)$ | $1.78\times10^{-6}$ | $1.52\times10^{-6}$ | $1.31\times10^{-6}$ | $1.01\times10^{-6}$ | $0.55\times10^{-6}$ | $0.28\times10^{-6}$ |

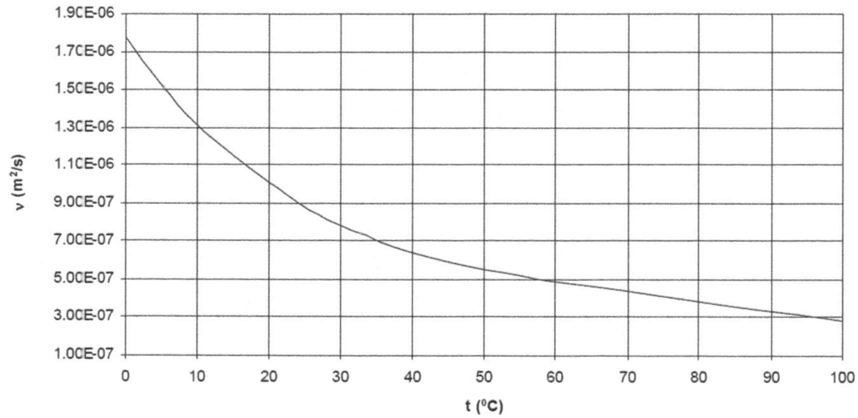


**Figure A1.7** Change in the kinematic coefficient of water viscosity as a function of temperature (Pušić, 2012).

## REFERENCES


Adamski M., Kremesec V., Kolhatkar R., Pearson C. and Rowan B. (2005) LNAPL in fine-grained soils: Conceptualization of saturation, distribution, recovery, and their modeling. *Ground Water Monitoring & Remediation*, **25**(1), pp. 100–112.

Dragišić V. (1997) General Hydrogeology. Faculty of Mining and Geology, Belgrade.

Pušić M. (2012) Groundwater Hydrogeology. Faculty of Mining and Geology, Belgrade.

Salim R. L. (2016) Extent of Capillary Rise in Sands and Silts, Master's Theses, Western Michigan University, 688.

# Appendix 2

# Determining the coefficient of filtration based on data on the granulometric composition of soil

*Srđan Kovačević, Slobodan Vujasinović and Milan Dimkić*

This appendix looks at the use of the 10 most commonly used empirical equations to determine the coefficient of filtration on the basis of data on granulometric composition with concretely given elements which are a component part of each formula. The possible area of use has also been indicated wherever appropriate data were obtained.

To simplify the application and comparison, all the considered empirical formulas are reduced to the same, dimensionally homogeneous form:

$$K = \frac{g}{\nu} C \cdot \varphi(n) \cdot d_e^2 \tag{A2.1}$$

where: $C$ – dimensionless coefficient; $\varphi(n)$ – porosity function; $d_e$ – effective grain diameter and $n$ – porosity.

A review of all the considered ways of determining the filtration coefficient on the basis of data on granulometric composition is provided in a homogeneous form in Table A2.1.

However, when determining the effective grain diameter ($d_e$) some authors, in a somewhat modified form, use the entire particle size distribution in an area (the specific surface area of a grain).

Figure A2.1 gives the generalized results of the calculation of the effective grain diameter for a larger number of natural and synthetic granulometric curves in the form of a function:

In addition to the results of the calculation, the same figure shows the effective diameters of grains in a porous material as recommended by other authors ($d_{10}$, $d_{17}$, $d_{20}$).

In a general case, the dimensionless coefficient $C$ has a constant value and includes parameters that affect the filtration coefficient (grain shape, content of the clay fraction, etc.).

Authors who do not express the filtration coefficient via porosity take coefficient $C$ as a variable value dependent on the uniformity coefficient or the effective grain diameter.


Editor: Milan Dimkić
doi: 10.2166/9781789060904_0675

**Table A2.1** The most frequently used empirical formulas for calculating the filtration coefficient (Dimkić *et al.*, 2008).

| Nr. | The Author | Formula (dimensionless, homogeneous) | C | Effective Grain Diameter $d_e$ | Application Domain |
|---|---|---|---|---|---|
| 1 | Hazen | $K = \frac{g}{v} C \cdot [1 + 10(n - 0.26)] \cdot d_e^2$ | $6 \cdot 10^{-4}$ | $d_e = d_{10}$ | $0.1 \text{ mm} < d_e < 3 \text{ mm}, \eta < 5$ |
| 2 | Slichter | $K = \frac{g}{v} C \cdot n^{3.287} \cdot d_e^2$ | $1 \cdot 10^{-2}$ | $d_e = d_{10}$ | $0.01 \text{ mm} < d_e < 5 \text{ mm}$ |
| 3 | Terzaghi | $K = \frac{g}{v} C \cdot \left(\frac{n - 0.13}{\sqrt[3]{1 - n}}\right)^2 \cdot d_e^2$ | $10.7 \cdot 10^{-3} > C > 6.1 \cdot 10^{-3}$ | $d_e = d_{10}$ | coarse-grained sand |
| 4 | Beyer | $K = \frac{g}{v} C \cdot d_e^2$ | | $d_e = d_{10}$ | $0.06 \text{ mm} < d_e < 0.6 \text{ mm}$<br>$1 < \eta < 20$ |
| 5 | Zauerbrei | $K = \frac{g}{v} C \cdot \frac{n^3}{(1 - n)^2} d_e^2$ | $3.75 \cdot 10^{-3}$ | $d_e = d_{17}$ | sandy and sand-clay materials, $d < 0.5$ mm. |
| 6 | Kruger | $K = \frac{g}{v} C \cdot \frac{n}{(1 - n)^2} d_e^2$ | $4.35 \cdot 10^{-5}$ | $\frac{1}{d_e} = \sum_{i=1}^{i=n} \left(\Delta g_i \frac{2}{d_i^g + d_i^d}\right)$ | medium-grained sands, $\eta > 5$ |
| 7 | Kozeny | $K = \frac{g}{v} C \cdot \frac{n^3}{(1 - n)^2} d_e^2$ | $8.3 \cdot 10^{-3}$ | $\frac{1}{d_e} = \frac{3\Delta g_i}{2d_i} + \sum_{i=2}^{i=n} \left(\Delta g_i \frac{d_i^g + d_i^d}{2d_i^g + d_i^d}\right)$ | coarse-grained sands |
| 8 | Tsunker | $K = \frac{g}{v} C \cdot \left(\frac{n}{1 - n}\right)^2 \cdot d_e^2$ | $0.7 \cdot 10^{-3} > C > 2.4 \cdot 10^{-3}$ | $\frac{1}{d_e} = \frac{3\Delta g_i}{2d_i} + \sum_{i=2}^{i=n} \left(\Delta g_j \frac{d_i^g - d_i^d}{d_i^g \cdot d_i^d (\ln d_i^g - \ln d_i^d)}\right)$ | Fine-grained and medium-grained sand |
| 9 | Zamarin | $K = \frac{g}{v} C \cdot \frac{n^3}{(1 - n)^2}$<br>$(1.275 - 1.5n)^2 \cdot d_e^2 \cdot 8.2 \cdot 10^3$ | $8.3 \cdot 10^{-3}$ | $\frac{1}{d_e} = \frac{3\Delta g_i}{2d_i} + \sum_{i=2}^{i=n} \left(\Delta g_i \frac{\ln d_i^g - \ln d_i^d}{d_i^g - d_i^d}\right)$ | coarse-grained sands |
| 10 | USBR | $K = \frac{g}{v} C \cdot d_e^2$ | | $d_e = d_{20}$ | medium-grained sand, $\eta < 5$ |

$v$ – kinematic viscosity coefficient [m$^2$/s], $g$ – gravitational acceleration [m/s$^2$], $g_i$ – the fraction (by weight) of the i-th fraction in the total sample weight, $d_i$ – the average grain diameter of the i-th fraction, $d_{ig}$ – the maximum grain diameter of the i-th fraction, $d_{id}$ – the minimum grain diameter of the i-th fraction.

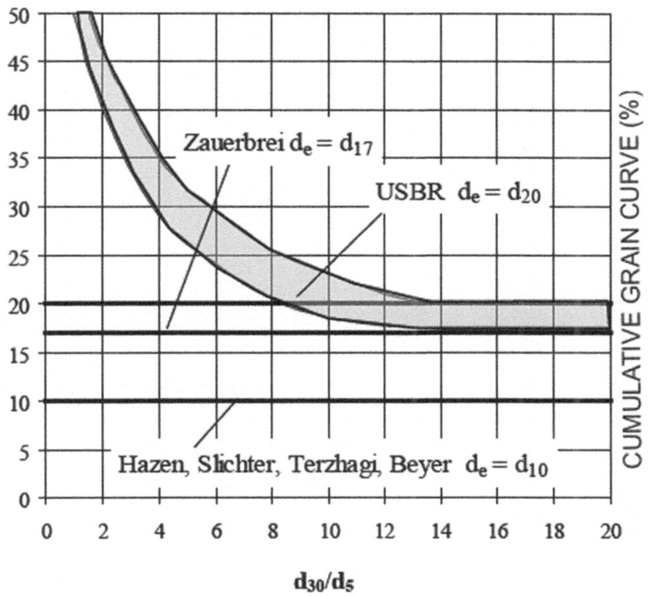


**Figure A2.1** Dependence of the effective grain diameter ($d_e$) on the uniformity coefficient of the material, expressed via the $d_{30}/d_5$ ratio; USBR – United States Bureau of Reclamation (Dimkić *et al.*, 2008).

The analyses and experiments presented here indicate that it is most realistic to express the porosity function as:

$$\varphi(n) = \frac{n^3}{(1-n)^2} \tag{A2.2}$$

Kozeny and Zauerbrei use this functional relationship in their formulas. Some authors use the porosity ($n$) function in a somewhat modified form, while it is omitted in the Bayer and USBR formulas.

A theoretical analysis based on the Darcy-Weissbachlaw formula shows that the filtration coefficient is directly proportional to the square of the effective grain diameter of the porous medium that is being considered. Most authors suggest that the size of a grain with a specific weight share in the total mass (10% – Hazen, Slichter, Terzaghi and Beyer, 17% – Zamarin and 20% – USBR) should be considered as the effective diameter.

In order to analyse the empirical formulas used in the calculation of the coefficient of filtration based on data on the granulometric composition, a comparison was made of the values of that parameter calculated according to certain authors for material (porous medium) where the mechanical composition was uniform (coefficient of uniformity of $\eta = 1$). For such (idealized) material of a specific grain structure, the effective diameter of the grain that is used in all empirical equations has the same value ($d_e = d$).

A comparative analysis was carried out (Figure A2.2) for three different values of the porosity of the material (the maximum density of a uniform material with spherical grains – $n = 0.26$, the minimum compaction – $n = 0.476$; the average compaction – $n = 0.35$).

The results indicate that even if we have a material whose geometrical characteristics are homogeneous (uniform grain diameter), there are significant differences between the values of the coefficient of filtration

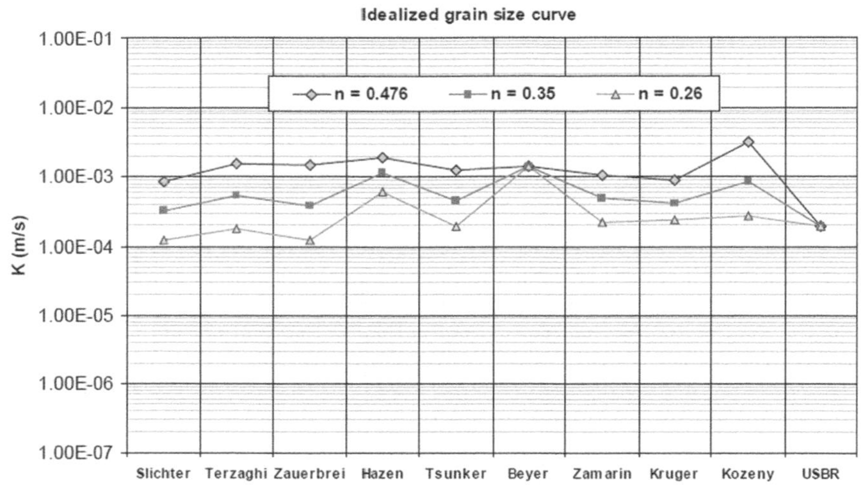


**Figure A2.2** Filtration coefficient of a mechanically uniform material – porous medium – uniformity coefficient $\eta = 1$ (Dimkić *et al.*, 2008).

calculated based on the empirical formulas proposed by some authors. The values of the filtration coefficient calculated by using these formulas can differ by as many as 30 times or more.

The results shown here indicate that the use of different empirical formulas can lead to significantly different results for the coefficient of filtration if an inappropriate value for porosity is adopted. However, some authors believe that the difference is insignificant or even non-existent.

In actual practice, the comparison of empirical formulas using the calculated values of the filtration coefficient of materials with a natural granulometric composition is especially interesting.

The materials of an intergranular porous medium, which Vuković and Soro (1992) mostly encountered in their practice, were synthesized by five types of granulometric curves (Figure A2.3). In addition, the figure shows where the corresponding empirical formulas could be used as suggested by the authors, based on our free interpretation, i.e. it shows the domains in which the experiments were conducted.

The filtration coefficient calculations were performed for the adopted types of granulometric curves of natural material using all 10 considered empirical approaches. The following characteristics and assumptions were adopted in the calculations:

- The obtained porosity values are based on the Istomina diagram (Figure A2.3) and are expressed depending on the coefficient of the uniformity of the material;
- Each considered granulometric curve is represented by 20 fractions (the separation of the curves was done at intervals of 5% of the weight share);
- The water temperature is $t = 15°C$.

The results of the filtration coefficient calculations performed using all of the 10 considered empirical formulas are given in Figure A2.4. An analysis of the calculations of the filtration coefficient in relation to the five types of curves of natural mechanical composition of the material, as well as their comparison indicate (in the case of uniform material) a relatively large dispersion of the results of calculation.

The analysis shows that the values of the filtration coefficient, depending on the empirical formula that is applied, can differ by as many as 10 times and under special conditions more than 20 times. As confirmed by

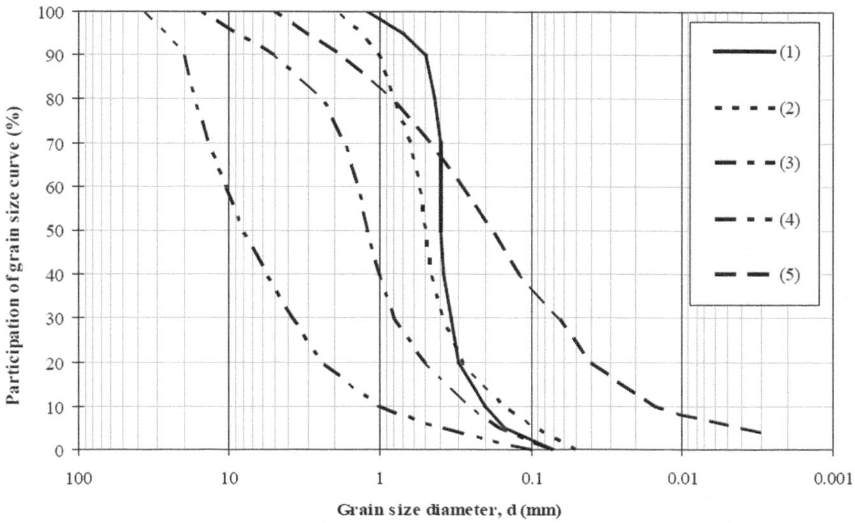


**Figure A2.3** Five types of granulometric curves for which comparative calculations were performed (Dimkić *et al.*, 2008).

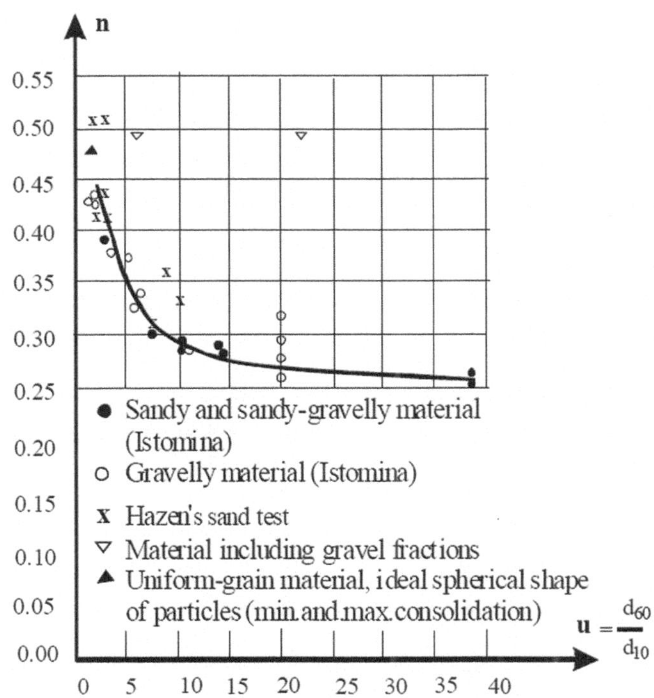


**Figure A2.4** Porosity variation as a function of the coefficient of uniformity – $\eta$ (Istomina, 1957).

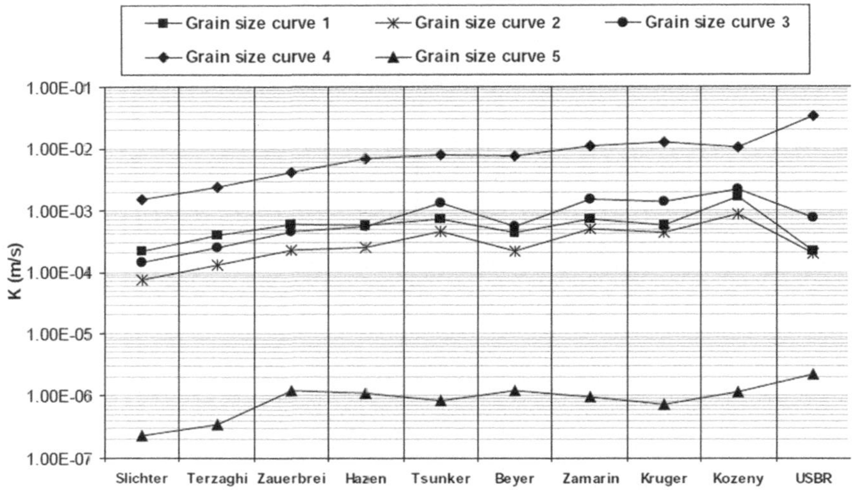


**Figure A2.5** Filtration coefficient for five types of curves with respect to the natural mechanical composition of the material (Dimkić *et al.*, 2008).

the analysis of the calculated values of the filtration coefficient of mechanically uniform materials ($\eta = 1$), of all the discussed formulas the Slichter formula gives the lowest value while Terzaghi's formula also gives a low value of the filtration coefficient (Figure A2.5). This means that when we apply these empirical formulas, we do not need to get a real (or representative) value of the filtration coefficient.

The Kozeny, Zamarin, and Tsunker formulas gave the highest filtration coefficient values for materials with a low uniformity coefficient (Carrier, 2003; Ishaku *et al.*, 2011; Odong, 2007; Petalas & Pilakas, 2011; Salarashayeri & Siosemarde, 2012). However, for materials with a high uniformity coefficient (sandy materials with significant clay content), the values of the filtration coefficient obtained using the USBR formula are by far the highest. It is interesting to note that the empirical formulas for calculating the effective grain diameter take into account the whole granulometric curve (Kruger, Kozeny, Tsunker, and Zamarin) and obtain significantly higher values for materials with a high coefficient of uniformity than in the case of other formulas (although these need not be maximum values).

It is obvious that the results of this analysis and the conclusions refer to the cases where the porosity is determined in the manner that has been described (via $n = f(\eta)$). Any other method will give other absolute values of the filtration coefficient, as well as other relations of these values.

A comparison of the empirical formulas that were considered showed that applying different formulas for the same material of a porous medium (with the same granulometric curve) yielded calculated values of filtration coefficients that differed by 10 to 20 times or more.

## REFERENCES

Beyer W. (1966). Hydrogeological investigations in the deposition of water pollutants. *Journal of Applied Geology*, **12.1**, 599–606.

Carrier D. (2003). Goodbye, hazen; hello, kozeny-carman. *Journal of Geotechnical and Geoenvironmental Engineering, ASCE*, **131**(8), 1057–1058, http://www.engr.uconn.edu/~lanbo/Carrier2003.pdf., doi: 10.1061/(ASCE)1090-0241(2005)131:8(1057)

Dimkic A. M., Brauch H. J. and Kavanaugh M. (2008). Groundwater management in large river basins. In: A. M. Dimkic, H. J. Brauch and M. Kavanaugh (eds.), IWA Publishing, Alliance House, 12 Caxton Street, London SW1H 0QS, UK. Retrieved from www.iwapublishing.com

Hazen A. (1892). Some physical properties of sands and gravels. Massachusetts State Board of Health, Annual Report, 539–556.

Ishaku E., Gadzama J. and Kaigama U. (2011). Evaluation of empirical formulas for determination of hydraulic conductivity based on grain-size analysis. *Journal of Geology and Mining Research*, **34**, 105–113.

Istomina V. S. (1957). Filtration stability of soils.

Odong J. (2007). Evaluation of empirical al formulas for determination of hydraulic conductivity based on grain size analysis. *Journal of American Science*, **3**(3), 54–60, http://www.sciencepub.net/american/0303/10-0284-Odong-Evaluation-am.pdf

Petalas P. and Pilakas F. (2011). Determination of hydraulic conductivity of unconsolidated river alluvium from permeameter tests, empirical formulas, and statistical parameters effect analysis. *Water Resources Management*, **25**, 2877–2899.

Salarashayeri A. F. and Siosemarde M. (2012). Prediction of soil hydraulic conductivity from particle-size distribution. World Academy of Science, Engineering and Technology, **6**(1), 16–20, http://www.waset.org/journals/ waset/v61/v61-81.pdf

Slichter C. S. (1898). Theoretical investigations of groundwater motions. *USGS*, **19**, 295–384, http://pubs.er.usgs.gov/publication/ar19_2

USBR. (1978). Drainage Manual: A Water Resources Technical Publication. http://www.usbr.gov/pmts/wquality_land/DrainMan.pdfg

Vukovic M. and Soro A. (1992). Determination of Hydraulic Conductivity of Porous Media from Grain-Size Composition, Water Resources Publications, LLC Highlands Ranch, Clorado, USA.

Zunker F. (1932). Fertilization and soil science. *Plant Nutrition Journal*, **A25**, Bd S. 8.

# Appendix 3

# Chemical structure and selected properties that influence the fate of pesticides in the environment

*Nevena Živančev*

The fate of pesticides in the environment is governed by the physico-chemical properties of these compounds. Solubility in water is one of the physico-chemical properties which can be used in determination of the distribution of that chemical in the environment and its degradation pathways; e.g. if a chemical has high solubility in water, it will remain in the water and tend not to sorb onto soil or living organisms. Henry's law constant of a chemical is an important property for determining if a pesticide will volatilize into the air or if it will remain in the water. If a pesticide has a high value of Henry's law constant it will volatilize from water into the air and be distributed over a large area, and if the value is low, the chemical will persist in water and depending on other physico-chemical properties, it may be sorbed onto soil. The half-life of pesticides in the soil is defined as the time required for the amount of a compound to be reduced by half, usually through degradation, and it is represented as $DT_{50}$ (DT – degradation time). These values are often used to indicate stability of a chemical in soil. Groundwater Ubiquity Score (GUS) is an experimentally calculated value used for assessing the potential of a pesticide to leach to groundwater. If a pesticide has extremly low or low potential for movement toward groundwater, the GUS value is below 0 and 1.8, respectively. If the GUS value is higher than 2.8, the potential for the pesticide leaching to groundwater is high.


Editor: Milan Dimkić
doi: 10.2166/9781789060904_0683

### 2,4-D

Identification of the substance

| | |
|---|---|
| CAS-No.: | 94-75-7 |
| IUPAC name: | (2,4-dichlorophenoxy)acetic acid |
| Molecular mass (g mol$^{-1}$): | 221.04 |
| Molecular formula: | $C_8H_6Cl_2O_3$ |
| Structural formula: | 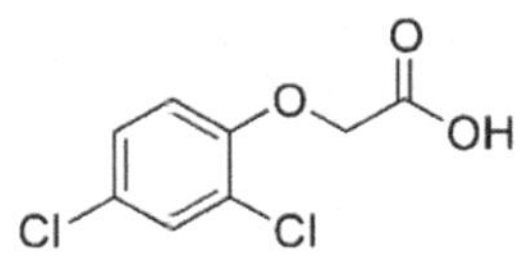 |
| Physico-chemical properties | |
| Physical state: | White powdery solid [1] |
| Water solubility: | 24,300 mg L$^{-1}$ at 20°C [1] |
| Soil degradation, $DT_{50}$ (days): | 4.4 (lab at 20°C), 28.8 (field) [1] |
| Henry's law constant: | $4.0 \times 10^{-6}$ Pa m$^3$ mol$^{-1}$ at 25°C [1] |
| GUS leaching potential index: | 3.82 (high leachability) [1] |

### ACEPHATE

Identification of the substance

| | |
|---|---|
| CAS-No.: | 30560-19-1 |
| IUPAC name: | (*RS*)-*N*-[methoxy(methylthio)phosphinoyl]acetamide |
| Molecular mass (g mol$^{-1}$): | 183.17 |
| Molecular formula: | $C_4H_{10}NO_3PS$ |
| Structural formula: | 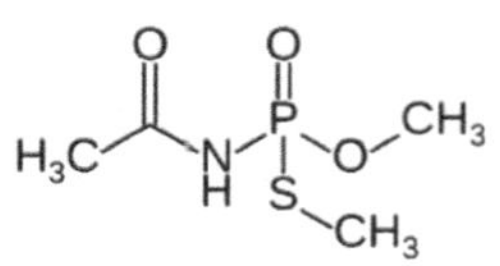 |
| Physico-chemical properties | |
| Physical state: | Colourless/white solid [1] |
| Water solubility: | 790,000 mg L$^{-1}$ at 20°C [1] |
| Soil degradation, $DT_{50}$ (days): | 3 (field) [1] |
| Henry's law constant: | $5.15 \times 10^{-8}$ Pa m$^3$ mol$^{-1}$ at 25°C [1] |
| GUS leaching potential index: | 1.14 (low leachability) [1] |

## ACETOCHLOR

Identification of the substance

| | |
|---|---|
| CAS-No.: | 34256-82-1 |
| IUPAC name: | 2-chloro-*N*-ethoxymethyl-6′-ethylacet-o-toluidide |
| Molecular mass (g $mol^{-1}$): | 269.77 |
| Molecular formula: | $C_{14}H_{20}ClNO_2$ |
| Structural formula: | 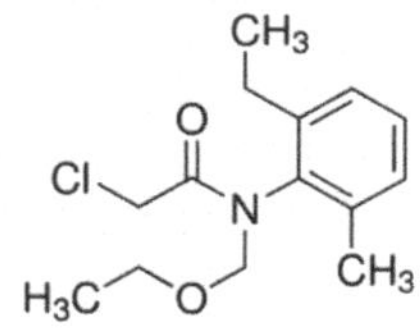 |
| Physico-chemical properties | |
| Physical state: | Pale yellow liquid [1] |
| Water solubility: | 282 mg $L^{-1}$ at 20°C [1] |
| Soil degradation, $DT_{50}$ (days): | 14 (typical); 10.6 (lab at 20°C); 12.1 (field) [1] |
| Henry's law constant: | $2.1 \times 10^{-3}$ Pa $m^3$ $mol^{-1}$ at 25°C [1] |
| GUS leaching potential index: | 1.67 (low leachability) [1] |

## ACIFLUORFEN

Identification of the substance

| | |
|---|---|
| CAS-No.: | 50594-66-6 |
| IUPAC name: | 5-(2-chloro-α,α,α-trifluoro-p-tolyloxy)-2-nitrobenzoic acid |
| Molecular mass (g $mol^{-1}$): | 361.66 |
| Molecular formula: | $C_{14}H_7ClF_3NO_5$ |
| Structural formula: | 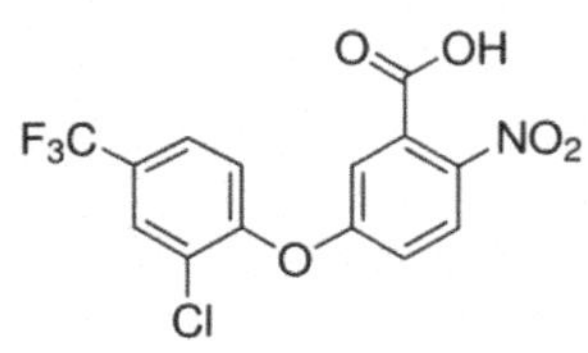 |
| Physico-chemical properties | |
| Physical state: | Off-white coloured solid [1] |
| Water solubility: | 250,000 mg $L^{-1}$ at 20°C [1] |
| Soil degradation, $DT_{50}$ (days): | 54 (typical) [1] |
| Henry's law constant: | – |
| GUS leaching potential index: | 3.11 (high leachability) [1] |

## ATRAZINE

Identification of the substance

| | |
|---|---|
| CAS-No.: | 1912-24-9 |
| IUPAC name: | 6-chloro-*N*2-ethyl-*N*4-isopropyl-1,3,5-triazine-2,4-diamine |
| Molecular mass (g $mol^{-1}$): | 215.68 |
| Molecular formula: | $C_8H_{14}ClN_5$ |
| Structural formula: | 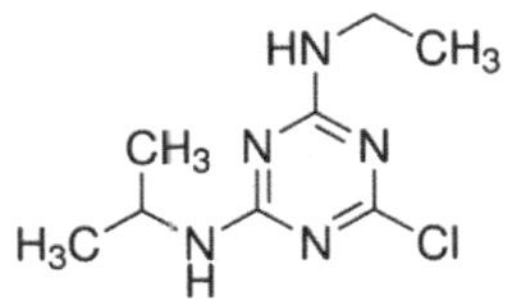 |

| | |
|---|---|
| Physico-chemical properties | |
| Physical state: | White crystals [1] |
| Water solubility: | 35 mg $L^{-1}$ at 20°C [1] |
| Soil degradation, $DT_{50}$ (days): | 75 (typical); 66 (lab at 20°C); 29 (field) [1] |
| Henry's law constant: | $1.5 \times 10^{-4}$ Pa $m^3$ $mol^{-1}$ at 25°C [1] |
| GUS leaching potential index: | 2.57 (transition state) [1] |

## AZOXYSTROBIN

Identification of the substance

| | |
|---|---|
| CAS-No.: | 131860-33-8 |
| IUPAC name: | methyl (*E*)-2-{2-[6-(2-cyanophenoxy)pyrimidin-4-yloxy]phenyl}-3-methoxyacrylate |
| Molecular mass (g $mol^{-1}$): | 403.4 |
| Molecular formula: | $C_{22}H_{17}N_3O_5$ |
| Structural formula: | 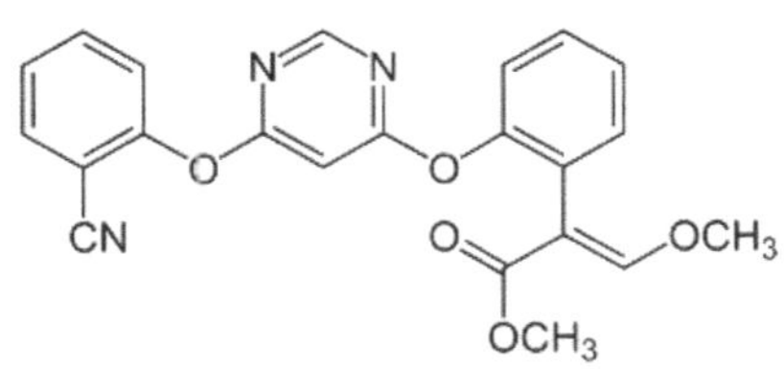 |

| | |
|---|---|
| Physico-chemical properties | |
| Physical state: | White crystalline solid [1] |
| Water solubility: | 6.7 mg $L^{-1}$ at 20°C [1] |
| Soil degradation, $DT_{50}$ (days): | 78 (typical); 84.5 (lab at 20°C); 180.7 (field) [1] |
| Henry's law constant: | $7.4 \times 10^{-9}$ Pa $m^3$ $mol^{-1}$ at 25°C [1] |
| GUS leaching potential index: | 3.10 (high leachability) [1] |

**BENSULIDE**

Identification of the substance

| | |
|---|---|
| CAS-No.: | 741-58-2 |
| IUPAC name: | *O*, *O*-diisopropyl *S*-2-phenylsulfonylaminoethyl phosphorodithioate |
| Molecular mass (g $mol^{-1}$): | 397.51 |
| Molecular formula: | $C_{14}H_{24}NO_4PS_3$ |
| Structural formula: | 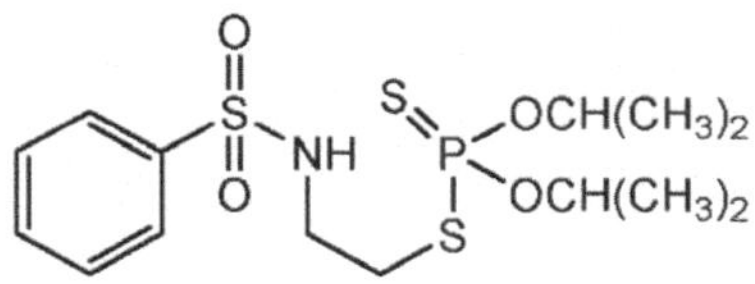 |
| Physico-chemical properties | |
| Physical state: | Colourless to yellow viscous liquid [1] |
| Water solubility: | 25 mg $L^{-1}$ at 20°C [1] |
| Soil degradation, $DT_{50}$ (days): | 90 (typical); 120 (field) [1] |
| Henry's law constant: | $2.11 \times 10^{-3}$ Pa $m^3$ $mol^{-1}$ at 25°C [1] |
| GUS leaching potential index: | 1.74 (low leachability) [1] |

**BENTAZONE**

Identification of the substance

| | |
|---|---|
| CAS-No.: | 25057-89-0 |
| IUPAC name: | 3-isopropyl-1*H*-2,1,3-benzothiadiazin-4(3*H*)-one 2,2-dioxide |
| Molecular mass (g $mol^{-1}$): | 240.3 |
| Molecular formula: | $C_{10}H_{12}N_2O_3S$ |
| Structural formula: | 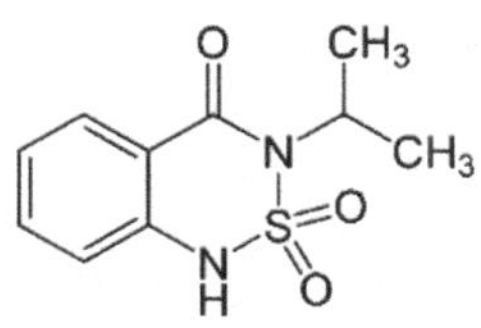 |
| Physico-chemical properties | |
| Physical state: | White to yellow coloured crystals depending upon purity [1] |
| Water solubility: | 7,112 mg $L^{-1}$ at 20°C [1] |
| Soil degradation, $DT_{50}$ (days): | 20 (typical); 20 (lab at 20°C); 7.5 (field) [1] |
| Henry's law constant: | $7.20 \times 10^{-5}$ Pa $m^3$ $mol^{-1}$ at 25°C [1] |
| GUS leaching potential index: | 1.95 (transition state) [1] |

## BROMOXYNIL

Identification of the substance

| | |
|---|---|
| CAS-No.: | 1689-84-5 |
| IUPAC name: | 3,5-dibromo-4-hydroxybenzonitrile |
| Molecular mass (g $mol^{-1}$): | 276.9 |
| Molecular formula: | $C_7H_3Br_2NO$ |
| Structural formula: | 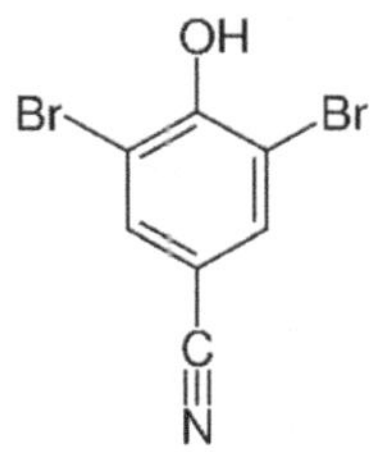 |



| | |
|---|---|
| Physico-chemical properties | |
| Physical state: | Clear to white crystalline solid [1] |
| Water solubility: | 38,000 mg $L^{-1}$ at 20°C [1] |
| Soil degradation, $DT_{50}$ (days): | 1.04 (typical); 1.04 (lab at 20°C); 8 (field) [1] |
| Henry's law constant: | $8.7 \times 10^{-7}$ Pa $m^3$ $mol^{-1}$ at 25°C [1] |
| GUS leaching potential index: | 1.71 (low leachability) [1] |

## CAPTAN

Identification of the substance

| | |
|---|---|
| CAS-No.: | 133-06-2 |
| IUPAC name: | *N*-(trichloromethylthio)cyclohex-4-ene-1,2-dicarboximide |
| Molecular mass (g $mol^{-1}$): | 300.61 |
| Molecular formula: | $C_9H_8Cl_3NO_2S$ |
| Structural formula: | 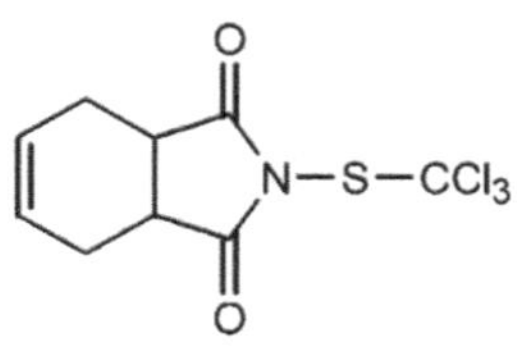 |



| | |
|---|---|
| Physico-chemical properties | |
| Physical state: | Dirty white powder [1] |
| Water solubility: | 5.2 mg $L^{-1}$ at 20°C [1] |
| Soil degradation, $DT_{50}$ (days): | 0.8 (typical); 0.8 (lab at 20°C); 3.7 (field) [1] |
| Henry's law constant: | $3.00 \times 10^{-4}$ Pa $m^3$ $mol^{-1}$ at 25°C [1] |
| GUS leaching potential index: | 0.97 (low leachability) [1] |

**CHLOROTHALONIL**

Identification of the substance

| | |
|---|---|
| CAS-No.: | 1897-45-6 |
| IUPAC name: | tetrachloroisophthalonitrile |
| Molecular mass (g $mol^{-1}$): | 265.91 |
| Molecular formula: | $C_8Cl_4N_2$ |
| Structural formula: | 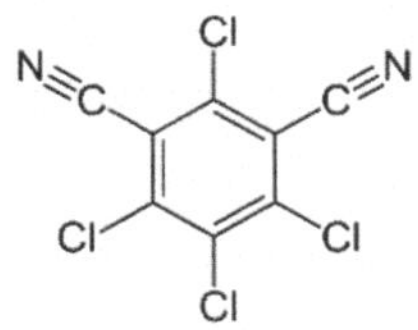 |
| Physico-chemical properties | |
| Physical state: | White crystalline solid at 99.6% purity [1] |
| Water solubility: | 0.81 mg $L^{-1}$ at 20°C [1] |
| Soil degradation, $DT_{50}$ (days): | 3.53 (typical); 3.53 (lab at 20°C); 17.9 (field) [1] |
| Henry's law constant: | $2.50 \times 10^{-2}$ Pa $m^3$ $mol^{-1}$ at 25°C [1] |
| GUS leaching potential index: | 1.12 (low leachability) [1] |

**CHLOROPICRIN**

Identification of the substance

| | |
|---|---|
| CAS-No.: | 76-06-2 |
| IUPAC name: | trichloronitromethane |
| Molecular mass (g $mol^{-1}$): | 164.37 |
| Molecular formula: | $CCl_3NO_2$ |
| Structural formula: | 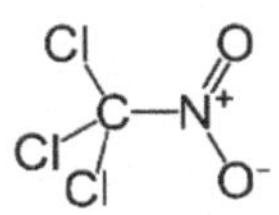 |
| Physico-chemical properties | |
| Physical state: | Colourless liquid [1] |
| Water solubility: | 10,000 mg $L^{-1}$ at 20°C [1] |
| Soil degradation, $DT_{50}$ (days): | 4.2 (typical); 4.2 (lab at 20°C); 3.0 (field) [1] |
| Henry's law constant: | $3.44 \times 10^2$ Pa $m^3$ $mol^{-1}$ at 25°C [1] |
| GUS leaching potential index: | 1.06 (low leachability) [1] |

## CHLORPYRIFOS

Identification of the substance

| | |
|---|---|
| CAS-No.: | 2921-88-2 |
| IUPAC name: | O,O-diethyl O-3,5,6-trichloro-2-pyridyl phosphorothioate |
| Molecular mass (g $mol^{-1}$): | 350.58 |
| Molecular formula: | $C_9H_{11}Cl_3NO_3PS$ |
| Structural formula: | 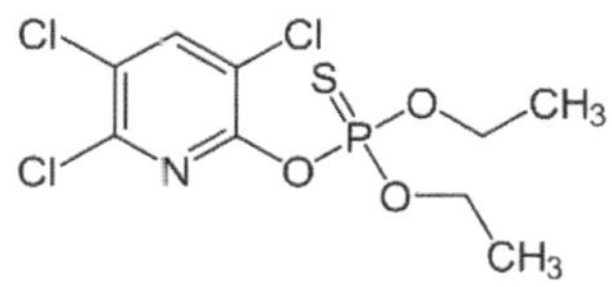 |
| Physico-chemical properties | |
| Physical state: | White to light brown coloured crystals [1] |
| Water solubility: | 1.05 mg $L^{-1}$ at 20°C [1] |
| Soil degradation, $DT_{50}$ (days): | 386 (typical); 386 (lab at 20°C); 27.6 (field) [1] |
| Henry's law constant: | 0.478 Pa $m^3$ $mol^{-1}$ at 25°C [1] |
| GUS leaching potential index: | 0.58 (low leachability) [1] |

## CLOMAZONE

Identification of the substance

| | |
|---|---|
| CAS-No.: | 81777-89-1 |
| IUPAC name: | 2-(2-chlorobenzyl)-4,4-dimethyl-1,2-oxazolidin-3-one |
| Molecular mass (g $mol^{-1}$): | 239.7 |
| Molecular formula: | $C_{12}H_{14}ClNO_2$ |
| Structural formula: | 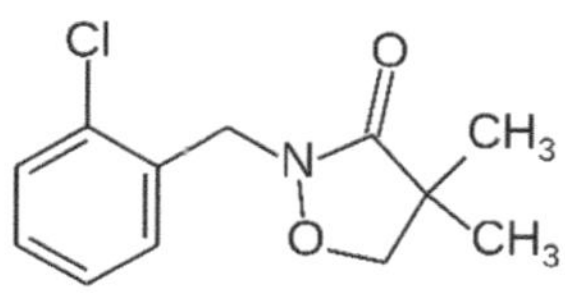 |
| Physico-chemical properties | |
| Physical state: | Pale straw coloured liquid [1] |
| Water solubility: | 1,212 mg $L^{-1}$ at 20°C [1] |
| Soil degradation, $DT_{50}$ (days): | 22.6 (typical); 22.6 (lab at 20°C); 27.3 (field) [1] |
| Henry's law constant: | $5.9 \times 10^{-3}$ Pa $m^3$ $mol^{-1}$ at 25°C [1] |
| GUS leaching potential index: | 2.72 (transition state) [1] |

## CYHALOFOP

Identification of the substance

| | |
|---|---|
| CAS-No.: | 122008-78-0 |
| IUPAC name: | (*R*)-2-[4-(4-cyano-2-fluorophenoxy)phenoxy]propionic acid |
| Molecular mass (g $mol^{-1}$): | 357.38 |
| Molecular formula: | $C_{16}H_{12}FNO_4$ |
| Structural formula: | 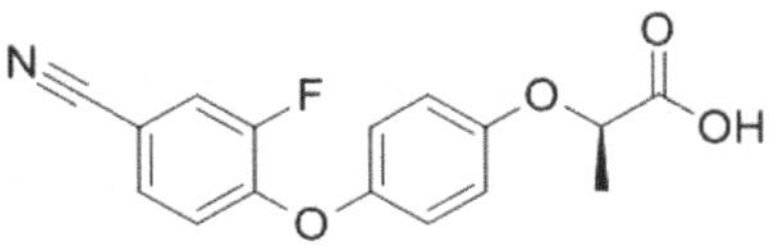 |
| Physico-chemical properties | |
| Physical state: | Odourless colourless-to-cream-coloured crystalline solid [2] |
| Water solubility: | 180 mg $L^{-1}$ at 20°C [1] |
| Soil degradation, $DT_{50}$ (days): | 0.71 (typical); 0.71 (lab at 20°C) [1] |
| Henry's law constant: | $2.41 \times 10^{-9}$ Pa $m^3$ $mol^{-1}$ at 25°C [1] |
| GUS leaching potential index: | −0.39 (low leachability) [1] |

## CYHALOTHRIN-LAMBDA

Identification of the substance

| | |
|---|---|
| CAS-No.: | 76703-65-6 |
| IUPAC name: | [(R)-cyano-(3-phenoxyphenyl)methyl] (1S,3S)-3-[(Z)-2-chloro-3,3,3-trifluoroprop-1-enyl]-2,2-dimethylcyclopropane-1-carboxylate |
| Molecular mass (g $mol^{-1}$): | 449.85 |
| Molecular formula: | $C_{23}H_{19}ClF_3NO_3$ |
| Structural formula: | 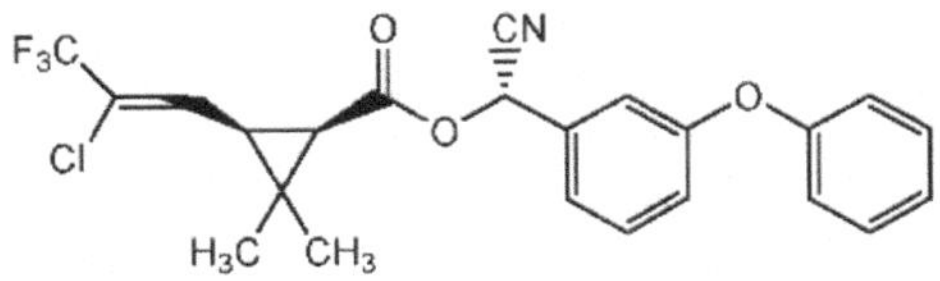 |
| Physico-chemical properties | |
| Physical state: | Colourless-to-beige solid (technical grade) [2]<br>White solid [3] |
| Water solubility: | 0.004 mg $L^{-1}$ at 20°C, pH 5 & pH 9.2 [3]<br>0.005 mg $L^{-1}$ at 20°C, pH 6.5 [3] |
| Soil degradation, $DT_{50}$ (days): | 22 to 83 [3] |
| Henry's law constant: | $2 \times 10^{-2}$ Pa $m^3$ $mol^{-1}$ at 25°C [3] |
| GUS leaching potential index: | −3.28 (low leachability)[4] |

**DICAMBA**

Identification of the substance

| | |
|---|---|
| CAS-No.: | 1918-00-9 |
| IUPAC name: | 3,6-dichloro-o-anisic acid |
| Molecular mass (g $mol^{-1}$): | 221.04 |
| Molecular formula: | $C_8H_6Cl_2O_3$ |
| Structural formula: | 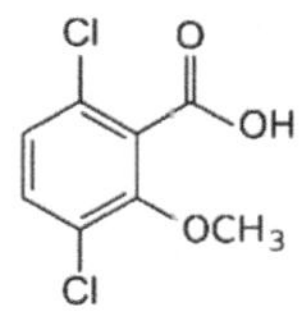 |
| Physico-chemical properties | |
| Physical state: | White granules [1] |
| Water solubility: | 250,000 mg $L^{-1}$ at 20°C [1] |
| Soil degradation, $DT_{50}$ (days): | 4.0 (typical); 4.0 (lab at 20°C); 3.9 (field) [1] |
| Henry's law constant: | $1.0 \times 10^{-4}$ Pa $m^3$ $mol^{-1}$ at 25°C [1] |
| GUS leaching potential index: | 1.72 (low leachability) [1] |

**1,3 – DICHLOROPROPENE**

Identification of the substance

| | |
|---|---|
| CAS-No.: | 542-75-6 |
| IUPAC name: | (*EZ*)-1,3-dichloropropene |
| Molecular mass (g $mol^{-1}$): | 110.97 |
| Molecular formula: | $C_3H_4Cl_2$ |
| Structural formula: | 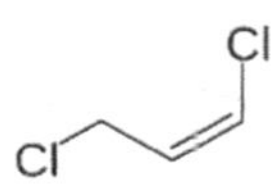 |
| Physico-chemical properties | |
| Physical state: | Colourless to pale straw-coloured, volatile liquid [1] |
| Water solubility: | 2,450/2,520 mg $L^{-1}$ at 20°C [1] |
| Soil degradation, $DT_{50}$ (days): | 10.3 (typical); 10.3 (lab at 20°C) [1] |
| Henry's law constant: | 170/101 Pa $m^3$ $mol^{-1}$ at 25°C [1] |
| GUS leaching potential index: | 2.50 (transition state) [1] |

**DICROTOPHOS**

Identification of the substance

| | |
|---|---|
| CAS-No.: | 141-66-2 |
| IUPAC name: | (*E*)-2-dimethylcarbamoyl-1-methylvinyl dimethyl phosphate |
| Molecular mass (g $mol^{-1}$): | 237.19 |
| Molecular formula: | $C_8H_{16}NO_5P$ |
| Structural formula: | 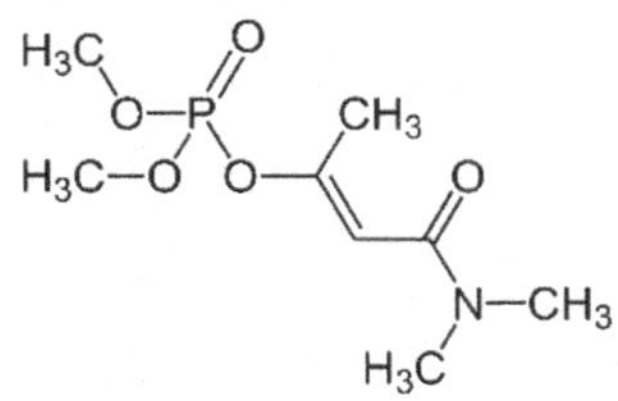 |
| Physico-chemical properties | |
| Physical state: | Yellow brown liquid [1] |
| Water solubility: | 1,000,000 mg $L^{-1}$ at 20°C [1] |
| Soil degradation, $DT_{50}$ (days): | 28 (typical) [1] |
| Henry's law constant: | $5.10 \times 10^{-6}$ Pa $m^3$ $mol^{-1}$ at 25°C [1] |
| GUS leaching potential index: | 3.08 (high leachability) [1] |

**DIMETHENAMID(-P)**

Identification of the substance

| | |
|---|---|
| CAS-No.: | 87674-68-8 |
| IUPAC name: | (*RS*)-2-chloro-*N*-(2,4-dimethyl-3-thienyl)-*N*-(2-methoxy-1-methylethyl)acetamide |
| Molecular mass (g $mol^{-1}$): | 275.79 |
| Molecular formula: | $C_{12}H_{18}ClNO_2S$ |
| Structural formula: | 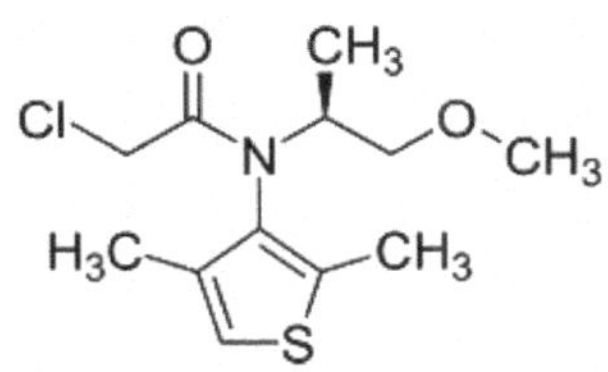 |
| Physico-chemical properties | |
| Physical state: | Yellow-brown viscous liquid [1] |
| Water solubility: | 1,200 mg $L^{-1}$ at 20°C [1] |
| Soil degradation, $DT_{50}$ (days): | 11 (typical); 23 (lab at 20°C); 7 (field) [1] |
| Henry's law constant: | $8.60 \times 10^{-3}$ Pa $m^3$ $mol^{-1}$ at 25°C [1] |
| GUS leaching potential index: | 2.41 (transition state) [1] |

**DIURON**

Identification of the substance

| | |
|---|---|
| CAS-No.: | 330-54-1 |
| IUPAC name: | 3-(3,4-dichlorophenyl)-1,1-dimethylurea |
| Molecular mass (g $mol^{-1}$): | 233.09 |
| Molecular formula: | $C_9H_{10}Cl_2N_2O$ |
| Structural formula: | 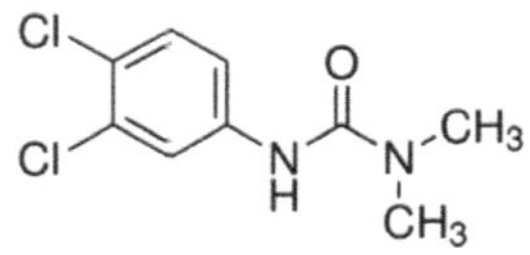 |
| Physico-chemical properties | |
| Physical state: | Off-white powder [1] |
| Water solubility: | 35.6 mg $L^{-1}$ at 20°C [1] |
| Soil degradation, $DT_{50}$ (days): | 146.6 (typical); 146.6 (lab at 20°C); 229 (field) [1] |
| Henry's law constant: | $2.00 \times 10^{-6}$ Pa $m^3$ $mol^{-1}$ at 25°C [1] |
| GUS leaching potential index: | 2.65 (transition state) [1] |

**EPTC**

Identification of the substance

| | |
|---|---|
| CAS-No.: | 759-94-4 |
| IUPAC name: | *S*-ethyl dipropyl(thiocarbamate) |
| Molecular mass (g $mol^{-1}$): | 189.3 |
| Molecular formula: | $C_9H_{19}NOS$ |
| Structural formula: | 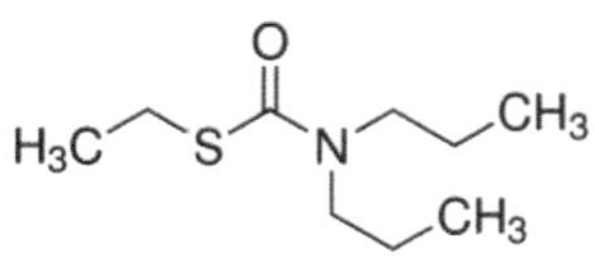 |
| Physico-chemical properties | |
| Physical state: | Pale yellow liquid [1] |
| Water solubility: | 370 mg $L^{-1}$ at 20°C [1] |
| Soil degradation, $DT_{50}$ (days): | 6 (typical); 18 (field) [1] |
| Henry's law constant: | $5.00 \times 10^{-6}$ Pa $m^3$ $mol^{-1}$ at 25°C [1] |
| GUS leaching potential index: | 2.17 (transition state) [1] |

## ETHALFLURALIN

Identification of the substance

| | |
|---|---|
| CAS-No.: | 55283-68-6 |
| IUPAC name: | *N*-ethyl-α,α,α-trifluoro-*N*-(2-methylallyl)-2,6-dinitro-p-toluidine |
| Molecular mass (g $mol^{-1}$): | 333.3 |
| Molecular formula: | $C_{13}H_{14}F_3N_3O_4$ |
| Structural formula: | 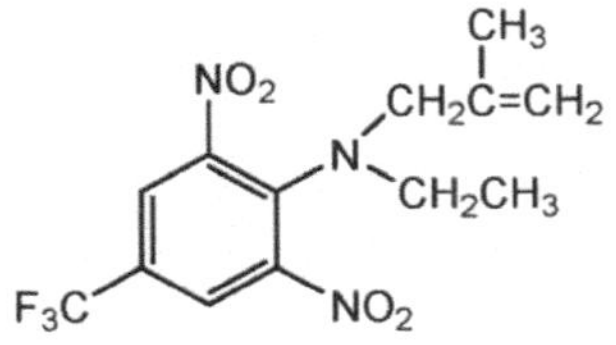 |

| | |
|---|---|
| Physico-chemical properties | |
| Physical state: | Yellow to orange coloured crystals [1] |
| Water solubility: | 0.01 mg $L^{-1}$ at 20°C [1] |
| Soil degradation, $DT_{50}$ (days): | 45 (typical); 39.8 (lab at 20°C); 53 (field) [1] |
| Henry's law constant: | 18.0 Pa $m^3$ $mol^{-1}$ at 25°C [1] |
| GUS leaching potential index: | 0.47 (low leachability) [1] |

## FLUMIOXAZIN

Identification of the substance

| | |
|---|---|
| CAS-No.: | 103361-09-7 |
| IUPAC name: | *N*-(7-fluoro-3,4-dihydro-3-oxo-4-prop-2-ynyl-2*H*-1,4-benzoxazin-6-yl)cyclohex-1-ene-1,2-dicarboxamide |
| Molecular mass (g $mol^{-1}$): | 354.33 |
| Molecular formula: | $C_{19}H_{15}FN_2O_4$ |
| Structural formula: | 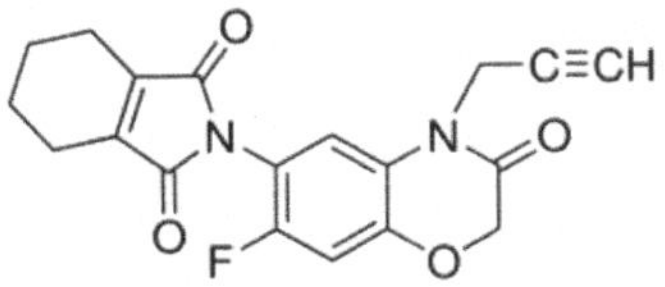 |

| | |
|---|---|
| Physico-chemical properties | |
| Physical state: | White when pure (99.6%), yellow-brown powder (97.65%) [1] |
| Water solubility: | 0.786 mg $L^{-1}$ at 20°C [1] |
| Soil degradation, $DT_{50}$ (days): | 21.9 (typical); 21.9 (lab at 20°C); 17.6 (field) [1] |
| Henry's law constant: | 0.145 Pa $m^3$ $mol^{-1}$ at 25°C [1] |
| GUS leaching potential index: | 1.31 (low leachability) [1] |

**FOMESAFEN**

Identification of the substance

| | |
|---|---|
| CAS-No.: | 72178-02-0 |
| IUPAC name: | 5-(2-chloro-α,α,α-trifluoro-p-tolyloxy)-*N*-mesyl-2-nitrobenzamide |
| Molecular mass (g mol$^{-1}$): | 438.76 |
| Molecular formula: | $C_{15}H_{10}ClF_3N_2O_6S$ |
| Structural formula: | 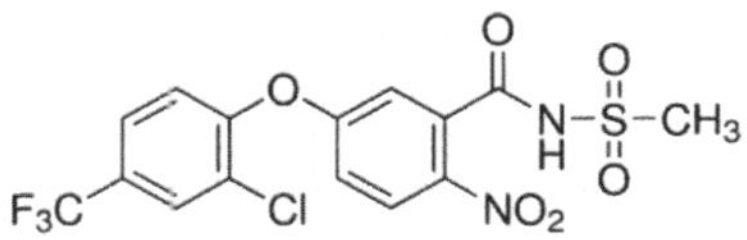 |
| Physico-chemical properties | |
| Physical state: | White crystals [1] |
| Water solubility: | 50 mg L$^{-1}$ at 20°C [1] |
| Soil degradation, $DT_{50}$ (days): | 86 (typical); 86 (field) [1] |
| Henry's law constant: | $2.00 \times 10^{-7}$ Pa m$^3$ mol$^{-1}$ at 25°C [1] |
| GUS leaching potential index: | 3.18 (high leachability) [1] |

**GLUFOSINATE**

Identification of the substance

| | |
|---|---|
| CAS-No.: | 51276-47-2 |
| IUPAC name: | (2*RS*)-2-amino-4-[hydroxy(methyl)phosphinoyl]butyric acid |
| Molecular mass (g mol$^{-1}$): | 181.13 |
| Molecular formula: | $C_5H_{12}NO_4P$ |
| Structural formula: | 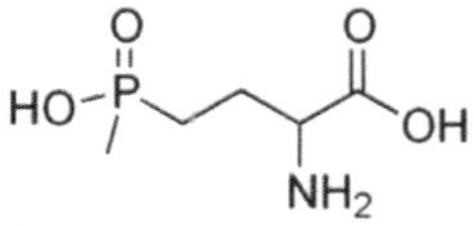 |
| Physico-chemical properties | |
| Physical state: | Solid [1]<br>White to light yellow crystalline powder [2] |
| Water solubility: | 1,370 g L$^{-1}$ at 22°C [2] |
| Soil degradation, $DT_{50}$ (days): | 6 to 10 (lab at 20°C) [6] |
| Henry's law constant: | $4.48 \times 10^{-9}$ Pa m$^3$ mol$^{-1}$ at 25°C [6] |
| GUS leaching potential index: | – |

## GLYPHOSATE

Identification of the substance

| | |
|---|---|
| CAS-No.: | 1071-83-6 |
| IUPAC name: | N-(phosphonomethyl)glycine |
| Molecular mass (g $mol^{-1}$): | 169.1 |
| Molecular formula: | $C_3H_8NO_5P$ |
| Structural formula: | 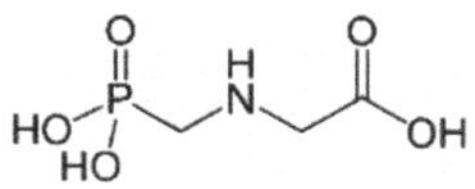 |
| Physico-chemical properties | |
| Physical state: | Colourless crystals [1] |
| Water solubility: | 10,500 mg $L^{-1}$ at 20°C [1] |
| Soil degradation, $DT_{50}$ (days): | 15.0 (typical); 15.0 (lab at 20°C); 23.79 (field) [1] |
| Henry's law constant: | $2.1 \times 10^{-7}$ Pa $m^3$ $mol^{-1}$ at 25°C [1] |
| GUS leaching potential index: | −0.29 (low leachability) [1] |

## HEXAZINONE

Identification of the substance

| | |
|---|---|
| CAS-No.: | 51235-04-2 |
| IUPAC name: | 3-cyclohexyl-6-dimethylamino-1-methyl-1,3,5-triazine-2,4(1*H*,3*H*)-dione |
| Molecular mass (g $mol^{-1}$): | 252.31 |
| Molecular formula: | $C_{12}H_{20}N_4O_2$ |
| Structural formula: | 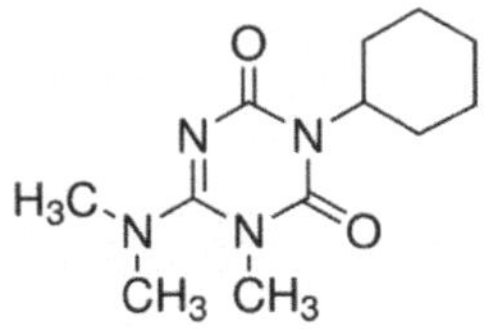 |
| Physico-chemical properties | |
| Physical state: | Colourless crystals [1] |
| Water solubility: | 33,000 mg $L^{-1}$ at 20°C [1] |
| Soil degradation, $DT_{50}$ (days): | 105 (typical); 90 (lab at 20°C) [1] |
| Henry's law constant: | $1.10 \times 10^{-7}$ Pa $m^3$ $mol^{-1}$ at 25°C [1] |
| GUS leaching potential index: | 4.43 (high leachability) [1] |

**IMAZETHAPYR**

Identification of the substance

| | |
|---|---|
| CAS-No.: | 81335-77-5 |
| IUPAC name: | 5-ethyl-2-[(*RS*)-4-isopropyl-4-methyl-5-oxo-2-imidazolin-2-yl]nicotinic acid |
| Molecular mass (g $mol^{-1}$): | 289.33 |
| Molecular formula: | $C_{15}H_{19}N_3O_3$ |
| Structural formula: | 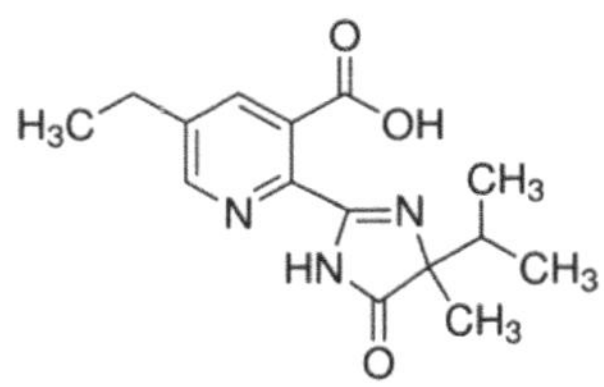 |
| Physico-chemical properties | |
| Physical state: | Colourless crystals [1] |
| Water solubility: | 1,400 mg $L^{-1}$ at 20°C [1] |
| Soil degradation, $DT_{50}$ (days): | 90 (typical); 513 (lab at 20°C); 51 (field) [1] |
| Henry's law constant: | $1.30 \times 10^{-2}$ Pa $m^3$ $mol^{-1}$ at 25°C [1] |
| GUS leaching potential index: | 3.90 (high leachability) [1] |

**MALATHION**

Identification of the substance

| | |
|---|---|
| CAS-No.: | 121-75-5 |
| IUPAC name: | diethyl (dimethoxyphosphinothioylthio)succinate |
| Molecular mass (g $mol^{-1}$): | 330.36 |
| Molecular formula: | $C_{10}H_{19}O_6PS_2$ |
| Structural formula: | 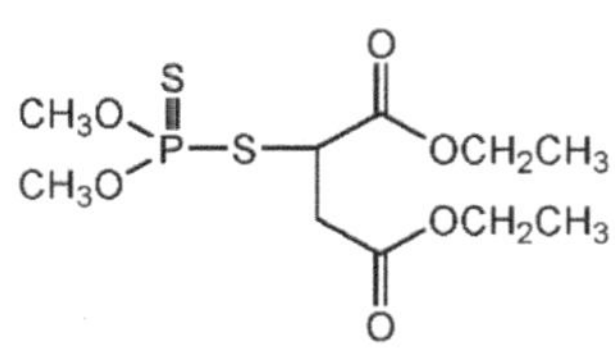 |
| Physico-chemical properties | |
| Physical state: | Clear to amber liquid [1] |
| Water solubility: | 148 mg $L^{-1}$ at 20°C [1] |
| Soil degradation, $DT_{50}$ (days): | 0.17 (typical); 0.17 (lab at 20°C); 1 (field) [1] |
| Henry's law constant: | $1.00 \times 10^{-3}$ Pa $m^3$ $mol^{-1}$ at 25°C [1] |
| GUS leaching potential index: | 0.00 (low leachability) [1] |

## MANCOZEB

Identification of the substance

| | |
|---|---|
| CAS-No.: | 8018-01-7 |
| IUPAC name: | manganese ethylenebis(dithiocarbamate) (polymeric) complex with zinc salt |
| Molecular mass (g $mol^{-1}$): | 271.3 |
| Molecular formula: | $(C_4H_6MnN_2S_4)_n(Zn)_m$ |
| Structural formula: | 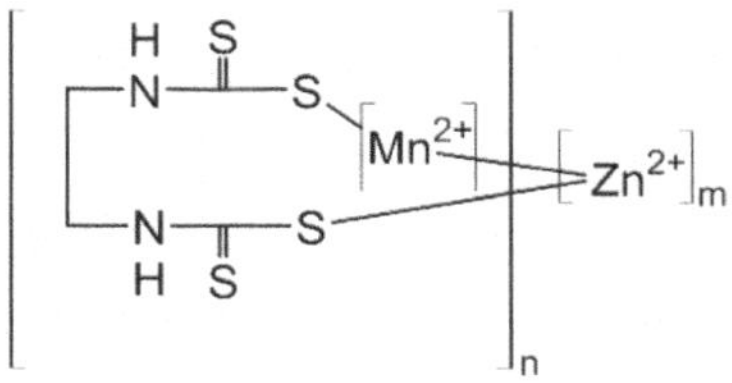 |
| Physico-chemical properties | |
| Physical state: | Grey to yellow coloured powder [1] |
| Water solubility: | 6.2 mg $L^{-1}$ at 20°C [1] |
| Soil degradation, $DT_{50}$ (days): | 0.05 (typical); 0.05 (lab at 20°C) [1] |
| Henry's law constant: | $6.17 \times 10^{-2}$ Pa $m^3$ $mol^{-1}$ at 25°C [1] |
| GUS leaching potential index: | −1.45 (low leachability) [1] |

## MCPA

Identification of the substance

| | |
|---|---|
| CAS-No.: | 94-74-6 |
| IUPAC name: | 4-chloro-o-tolyloxyacetic acid |
| Molecular mass (g $mol^{-1}$): | 200.62 |
| Molecular formula: | $C_9H_9ClO_3$ |
| Structural formula: | 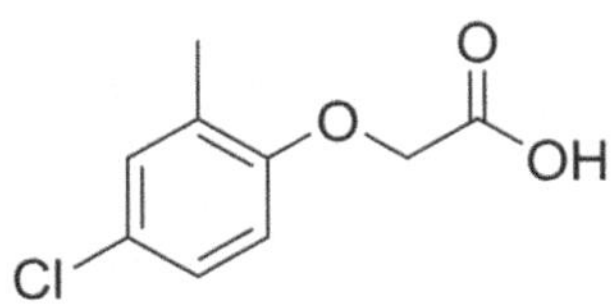 |
| Physico-chemical properties | |
| Physical state: | White crystalline powder [1] |
| Water solubility: | 29,390 mg $L^{-1}$ at 20°C [1] |
| Soil degradation, $DT_{50}$ (days): | 24 (typical); 24 (lab at 20°C); 25 (field) [1] |
| Henry's law constant: | $5.50 \times 10^{-5}$ Pa $m^3$ $mol^{-1}$ at 25°C [1] |
| GUS leaching potential index: | 2.98 (high leachability) [1] |

## MESOTRIONE

Identification of the substance

| | |
|---|---|
| CAS-No.: | 104206-82-8 |
| IUPAC name: | 2-(4-mesyl-2-nitrobenzoyl)cyclohexane-1,3-dione |
| Molecular mass (g $mol^{-1}$): | 339.32 |
| Molecular formula: | $C_{14}H_{13}NO_7S$ |
| Structural formula: | 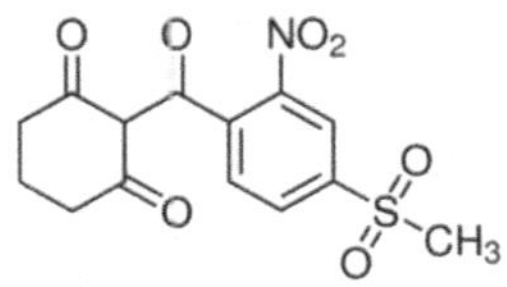 |
| Physico-chemical properties | |
| Physical state: | Yellow to tan coloured solid [1] |
| Water solubility: | 1,500 mg $L^{-1}$ at 20°C [1] |
| Soil degradation, $DT_{50}$ (days): | 19.6 (typical); 19.6 (lab at 20°C); 5 (field) [1] |
| Henry's law constant: | $5.10 \times 10^{-7}$ Pa $m^3$ $mol^{-1}$ at 25°C [1] |
| GUS leaching potential index: | 1.45 (low leachability) [1] |

## METAM POTASSIUM

Identification of the substance

| | |
|---|---|
| CAS-No.: | 137-41-7 |
| IUPAC name: | potassium methylcarbamodithioate |
| Molecular mass (g $mol^{-1}$): | 145.28 |
| Molecular formula: | $C_2H_4KNS_2$ |
| Structural formula: | 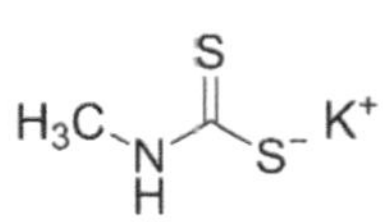 |
| Physico-chemical properties | |
| Physical state: | Colourless crystalline solid [1] |
| Water solubility: | 722,000 mg $L^{-1}$ at 20°C [1] |
| Soil degradation, $DT_{50}$ (days): | – |
| Henry's law constant: | – |
| GUS leaching potential index: | – |

## METOLACHLOR -S

Identification of the substance

| | |
|---|---|
| CAS-No.: | 87392-12-9 |
| IUPAC name: | 2-chloro-*N*-(2-ethyl-6-methylphenyl)-*N*-[(2*S*)-1-methoxypropan-2-yl]acetamide |
| Molecular mass (g $mol^{-1}$): | 283.79 |
| Molecular formula: | $C_{15}H_{22}ClNO_2$ |
| Structural formula: | 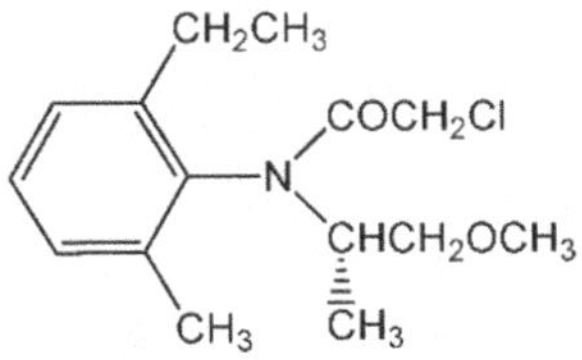 |



| Physico-chemical properties | |
|---|---|
| Physical state: | Clear straw-coloured liquid [1] |
| Water solubility: | 480 mg $L^{-1}$ at 20°C [1] |
| Soil degradation, $DT_{50}$ (days): | 51.8 (typical); 51.8 (lab at 20°C); 23.17 (field) [1] |
| Henry's law constant: | $2.20 \times 10^{-3}$ Pa $m^3$ $mol^{-1}$ at 25°C [1] |
| GUS leaching potential index: | 2.32 (transition state) [1] |

## METRIBUZIN

Identification of the substance

| | |
|---|---|
| CAS-No.: | 21087-64-9 |
| IUPAC name: | 4-amino-6-tert-butyl-4,5-dihydro-3-methylthio-1,2,4-triazin-5-one |
| Molecular mass (g $mol^{-1}$): | 214.29 |
| Molecular formula: | $C_8H_{14}N_4OS$ |
| Structural formula: | 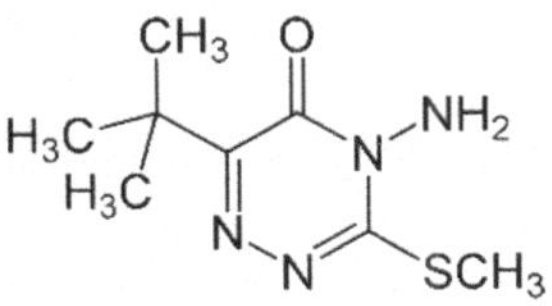 |



| Physico-chemical properties | |
|---|---|
| Physical state: | White crystalline solid [1] |
| Water solubility: | 10,700 mg $L^{-1}$ at 20°C [1] |
| Soil degradation, $DT_{50}$ (days): | 7.03 (typical); 7.03 (lab at 20°C); 19 (field) [1] |
| Henry's law constant: | $0.25 \times 10^{-4}$ Pa $m^3$ $mol^{-1}$ at 25°C [1] |
| GUS leaching potential index: | 2.96 (high leachability) [1] |

**MSMA**

Identification of the substance

| | |
|---|---|
| CAS-No.: | 2163-80-6 |
| IUPAC name: | sodium hydrogen methylarsonate |
| Molecular mass (g $mol^{-1}$): | 161.95 |
| Molecular formula: | $CH_4AsNaO_3$ |
| Structural formula: | 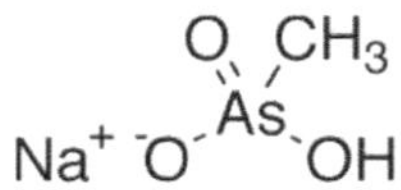 |
| Physico-chemical properties | |
| Physical state: | Technical form is a pale yellow liquid [1] |
| Water solubility: | 580,000 mg $L^{-1}$ at 20°C [1] |
| Soil degradation, $DT_{50}$ (days): | 200 (typical) [1] |
| Henry's law constant: | – |
| GUS leaching potential index: | – |

**OXYFLUORFEN**

Identification of the substance

| | |
|---|---|
| CAS-No.: | 42874-03-3 |
| IUPAC name: | 2-chloro-α,α,α-trifluoro-p-tolyl 3-ethoxy-4-nitrophenyl ether |
| Molecular mass (g $mol^{-1}$): | 361.7 |
| Molecular formula: | $C_{15}H_{11}ClF_3NO_4$ |
| Structural formula: | 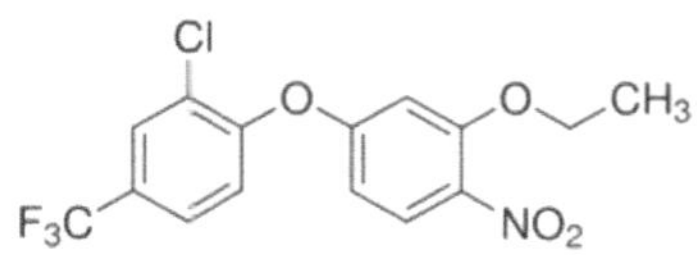 |
| Physico-chemical properties | |
| Physical state: | Orange to brown crystalline powder [1] |
| Water solubility: | 0.116 mg $L^{-1}$ at 20°C [1] |
| Soil degradation, $DT_{50}$ (days): | 35 (typical); 138 (lab at 20°C); 73 (field) [1] |
| Henry's law constant: | 0.02382 Pa $m^3$ $mol^{-1}$ at 25°C [1] |
| GUS leaching potential index: | 0.23 (low leachability) [1] |

**PARAQUAT**

Identification of the substance

| | |
|---|---|
| CAS-No.: | 4685-14-7 |
| IUPAC name: | 1,1′-dimethyl-4,4′-bipyridinium |
| Molecular mass (g mol$^{-1}$): | 186.25 |
| Molecular formula: | $C_{12}H_{14}N_2$ |
| Structural formula: | 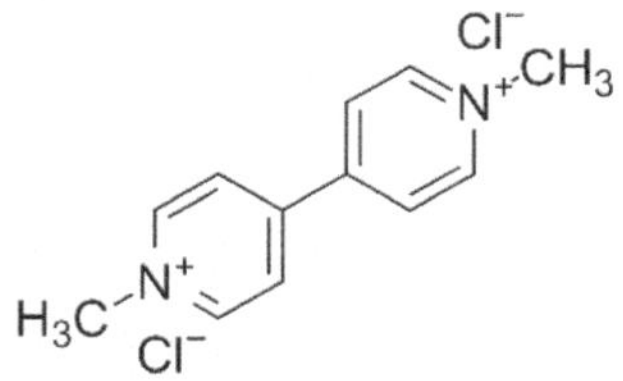 |
| Physico-chemical properties | |
| Physical state: | Colourless to pale yellow crystalline solid [1] |
| Water solubility: | 620,000 mg L$^{-1}$ at 20°C [1] |
| Soil degradation, $DT_{50}$ (days): | 3000 (typical); 5000 (lab at 20°C); 2800 (field) [1] |
| Henry's law constant: | $4.00 \times 10^{-12}$ Pa m$^3$ mol$^{-1}$ at 25°C [1] |
| GUS leaching potential index: | −6.89 (low leachability) [1] |

**PENDIMETHALIN**

Identification of the substance

| | |
|---|---|
| CAS-No.: | 40487-42-1 |
| IUPAC name: | *N*-(1-ethylpropyl)-2,6-dinitro-3,4-xylidine |
| Molecular mass (g mol$^{-1}$): | 281.31 |
| Molecular formula: | $C_{13}H_{19}N_3O_4$ |
| Structural formula: | 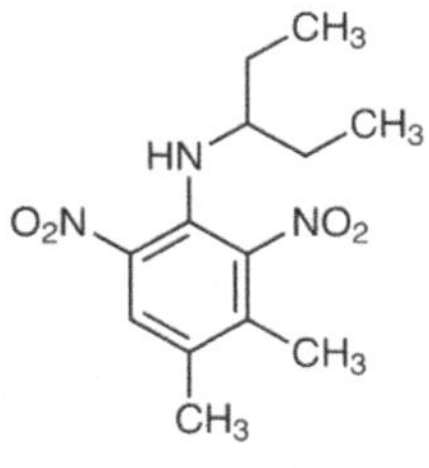 |
| Physico-chemical properties | |
| Physical state: | Orange yellow crystalline solid [1] |
| Water solubility: | 0.33 mg L$^{-1}$ at 20°C [1] |
| Soil degradation, $DT_{50}$ (days): | 182.3 (typical); 182.3 (lab at 20°C); 100.6 (field) [1] |
| Henry's law constant: | 1.27 Pa m$^3$ mol$^{-1}$ at 25°C [1] |
| GUS leaching potential index: | −0.28 (low leachability) [1] |

**PICLORAM**

Identification of the substance

| | |
|---|---|
| CAS-No.: | 1918-02-1 |
| IUPAC name: | 4-amino-3,5,6-trichloropyridine-2-carboxylic acid |
| Molecular mass (g $mol^{-1}$): | 241.46 |
| Molecular formula: | $C_6H_3Cl_3N_2O_2$ |
| Structural formula: | 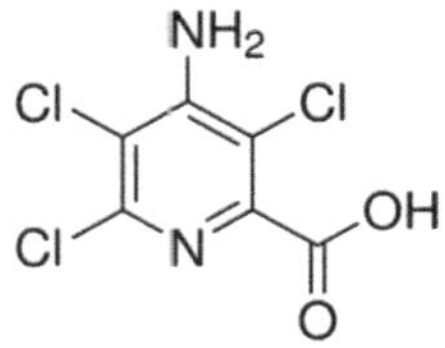 |
| Physico-chemical properties | |
| Physical state: | Light brown to tan coloured solid [1] |
| Water solubility: | 560 mg $L^{-1}$ at 20°C [1] |
| Soil degradation, $DT_{50}$ (days): | 82.8 (typical); 82.8 (lab at 20°C); 36 (field) [1] |
| Henry's law constant: | $3.0 \times 10^{-7}$ Pa $m^3$ $mol^{-1}$ at 25°C [1] |
| GUS leaching potential index: | 4.89 (high leachability) [1] |

**PROMETRYN**

Identification of the substance

| | |
|---|---|
| CAS-No.: | 7287-19-6 |
| IUPAC name: | *N*2,*N*4-diisopropyl-6-methylthio-1,3,5-triazine-2,4-diamine |
| Molecular mass (g $mol^{-1}$): | 241.36 |
| Molecular formula: | $C_{10}H_{19}N_5S$ |
| Structural formula: | 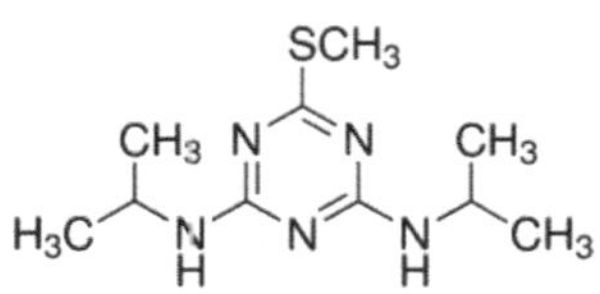 |
| Physico-chemical properties | |
| Physical state: | Colourless crystalline solid [1] |
| Water solubility: | 33 mg $L^{-1}$ at 20°C [1] |
| Soil degradation, $DT_{50}$ (days): | 41 (typical); 41 (lab at 20°C) [1] |
| Henry's law constant: | $1.20 \times 10^{-3}$ Pa $m^3$ $mol^{-1}$ at 25°C [1] |
| GUS leaching potential index: | 0.59 (low leachability) [1] |

**PROPANIL**

Identification of the substance

| | |
|---|---|
| CAS-No.: | 709-98-8 |
| IUPAC name: | 3′,4′-dichloropropionanilide |
| Molecular mass (g $mol^{-1}$): | 218.08 |
| Molecular formula: | $C_9H_9Cl_2NO$ |
| Structural formula: | 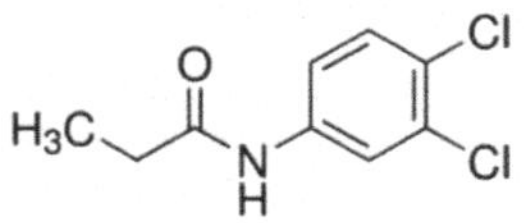 |
| Physico-chemical properties | |
| Physical state: | Brown crystalline solid [1] |
| Water solubility: | 95.0 mg $L^{-1}$ at 20°C [1] |
| Soil degradation, $DT_{50}$ (days): | 0.4 (typical); 0.4 (lab at 20°C) [1] |
| Henry's law constant: | $4.4 \times 10^{-4}$ Pa $m^3$ $mol^{-1}$ at 25°C [1] |
| GUS leaching potential index: | −0.51 (low leachability) [1] |

**PROPARGITE**

Identification of the substance

| | |
|---|---|
| CAS-No.: | 2312-35-8 |
| IUPAC name: | 2-(4-tert-butylphenoxy)cyclohexyl prop-2-ynyl sulphite |
| Molecular mass (g $mol^{-1}$): | 350.47 |
| Molecular formula: | $C_{19}H_{26}O_4S$ |
| Structural formula: | 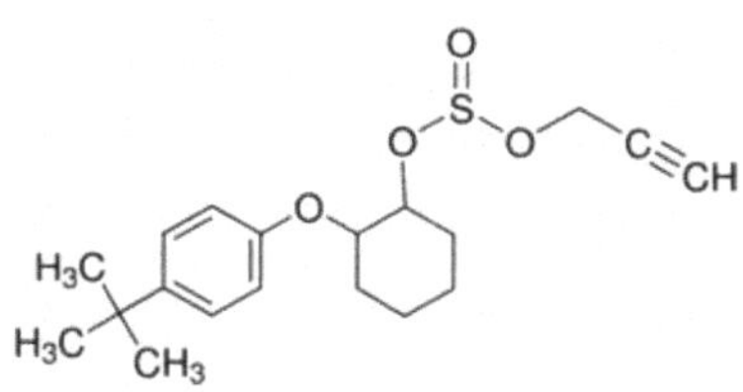 |
| Physico-chemical properties | |
| Physical state: | Dark yellow-brown viscous liquid [1] |
| Water solubility: | 0.215 mg $L^{-1}$ at 20°C [1] |
| Soil degradation, $DT_{50}$ (days): | 56 (typical); 72.2 (lab at 20°C); 16.7 (field) [1] |
| Henry's law constant: | $6.4 \times 10^{-2}$ Pa $m^3$ $mol^{-1}$ at 25°C [1] |
| GUS leaching potential index: | −0.86 (low leachability) [1] |

**PROPICONAZOLE**

Identification of the substance

| | |
|---|---|
| CAS-No.: | 60207-90-1 |
| IUPAC name: | (2*RS*,4*RS*;2*RS*,4*SR*)-1-[2-(2,4-dichlorophenyl)-4-propyl-1,3-dioxolan-2-ylmethyl]-1*H*-1,2,4-triazole |
| Molecular mass (g mol$^{-1}$): | 342.22 |
| Molecular formula: | $C_{15}H_{17}Cl_2N_3O_2$ |
| Structural formula: | 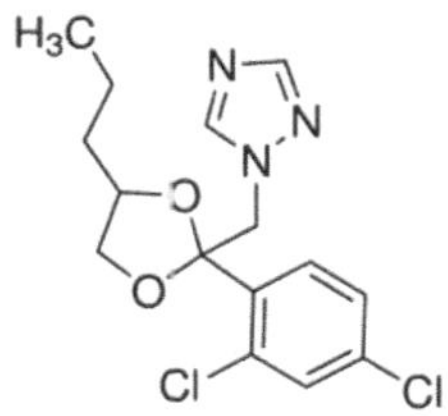 |
| Physico-chemical properties | |
| Physical state: | Clear to pale yellow viscous liquid depending on the ratio of isomers |
| Water solubility: | 150 mg L$^{-1}$ at 20°C [1] |
| Soil degradation, $DT_{50}$ (days): | 71.8 (typical); 71.8 (lab at 20°C); 35.2 (field) [1] |
| Henry's law constant: | $9.20 \times 10^{-5}$ Pa m$^3$ mol$^{-1}$ at 25°C [1] |
| GUS leaching potential index: | 1.58 (low leachability) [1] |

**QUINCLORAC**

Identification of the substance

| | |
|---|---|
| CAS-No.: | 84087-01-4 |
| IUPAC name: | 3,7-dichloroquinoline-8-carboxylic acid |
| Molecular mass (g mol$^{-1}$): | 242.06 |
| Molecular formula: | $C_{10}H_5Cl_2NO_2$ |
| Structural formula: | 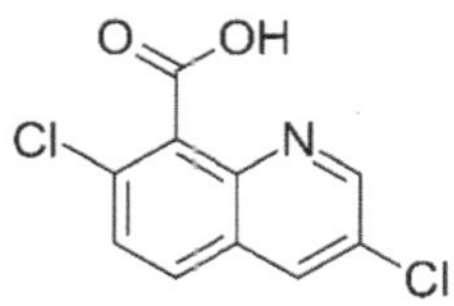 |
| Physico-chemical properties | |
| Physical state: | Colourless crystals [1] |
| Water solubility: | 0.065 mg L$^{-1}$ at 20°C [1] |
| Soil degradation, $DT_{50}$ (days): | 450 (typical); 540.5 (lab at 20°C) [1] |
| Henry's law constant: | $3.72 \times 10^{-2}$ Pa m$^3$ mol$^{-1}$ at 25°C [1] |
| GUS leaching potential index: | 6.29 (high leachability) [1] |

**SIMAZINE**

Identification of the substance

| | |
|---|---|
| CAS-No.: | 122-34-9 |
| IUPAC name: | 6-chloro-*N*2,*N*4-diethyl-1,3,5-triazine-2,4-diamine |
| Molecular mass (g $mol^{-1}$): | 201.66 |
| Molecular formula: | $C_7H_{12}ClN_5$ |
| Structural formula: | 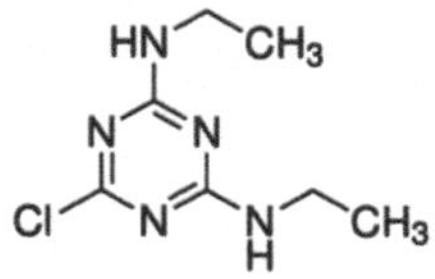 |
| Physico-chemical properties | |
| Physical state: | Colourless to white solid [1] |
| Water solubility: | 5 mg $L^{-1}$ at 20°C [1] |
| Soil degradation, $DT_{50}$ (days): | 60 (typical); 90 (field) [1] |
| Henry's law constant: | $5.60 \times 10^{-5}$ Pa $m^3$ $mol^{-1}$ at 25°C [1] |
| GUS leaching potential index: | 2.20 (transition state) [1] |

**SULFENTRAZONE**

Identification of the substance

| | |
|---|---|
| CAS-No.: | 122836-35-5 |
| IUPAC name: | 2′,4′-dichloro-5′-(4-difluoromethyl-4,5-dihydro-3-methyl-5-oxo-1*H*-1,2,4-triazol-1-yl)methanesulfonanilide |
| Molecular mass (g $mol^{-1}$): | 387.19 |
| Molecular formula: | $C_{11}H_{10}Cl_2F_2N_4O_3S$ |
| Structural formula: | 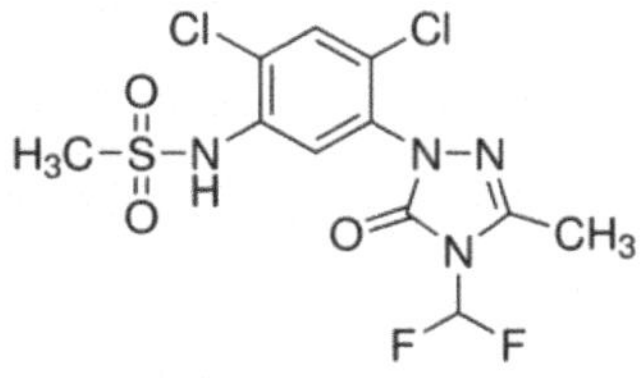 |
| Physico-chemical properties | |
| Physical state: | Light tan coloured solid formulated as an off-white liquid with faint sulfur-like odour [1] |
| Water solubility: | 780 mg $L^{-1}$ at 20°C [1][2] |
| Soil degradation, $DT_{50}$ (days): | 541 (typical); 400 (lab at 20°C) [1]; 110 to 280 (field) [2] |
| Henry's law constant: | $6.37 \times 10^{-13}$ atm-$m^3$ $mol^{-1}$ [2] |
| GUS leaching potential index: | 6.16 (high leachability) [1] |

## THIOBENCARB

Identification of the substance

| | |
|---|---|
| CAS-No.: | 28249-77-6 |
| IUPAC name: | *S*-4-chlorobenzyl diethyl(thiocarbamate) |
| Molecular mass (g $mol^{-1}$): | 257.8 |
| Molecular formula: | $C_{12}H_{16}ClNOS$ |
| Structural formula: | 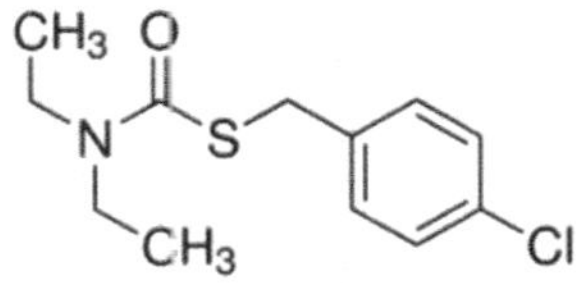 |
| Physico-chemical properties | |
| Physical state: | Colourless viscous liquid [1] |
| Water solubility: | 16.7 mg $L^{-1}$ at 20°C [1] |
| Soil degradation, $DT_{50}$ (days): | 21 (typical); 79 (lab at 20°C); 4 (field) [1] |
| Henry's law constant: | $3.68 \times 10^{-2}$ Pa $m^3$ $mol^{-1}$ at 25°C [1] |
| GUS leaching potential index: | 0.59 (low leachability) [1] |

## THIOPHANATE-METHYL

Identification of the substance

| | |
|---|---|
| CAS-No.: | 23564-05-8 |
| IUPAC name: | dimethyl 4,4′-(o-phenylene)bis(3-thioallophanate) |
| Molecular mass (g $mol^{-1}$): | 342.39 |
| Molecular formula: | $C_{12}H_{14}N_4O_4S_2$ |
| Structural formula: | 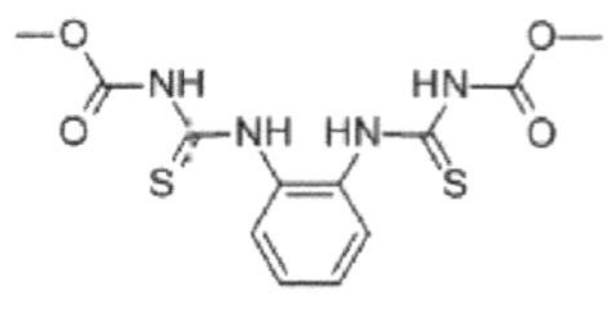 |
| Physico-chemical properties | |
| Physical state: | Colourless to light brown crystalline powder [1] |
| Water solubility: | 18.5 mg $L^{-1}$ at 20°C [1] |
| Soil degradation, $DT_{50}$ (days): | 0.5 (lab at 20°C); 2.0 (field) [1] |
| Henry's law constant: | $1.67 \times 10^{-4}$ Pa $m^3$ $mol^{-1}$ at 25°C [1] |
| GUS leaching potential index: | 0.50 (low leachability) [1] |

## TRICLOPYR

Identification of the substance

| | |
|---|---|
| CAS-No.: | 55335-06-3 |
| IUPAC name: | 3,5,6-trichloro-2-pyridyloxyacetic acid |
| Molecular mass (g mol$^{-1}$): | 256.47 |
| Molecular formula: | $C_7H_4Cl_3NO_3$ |
| Structural formula: | 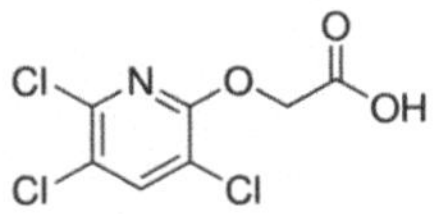 |
| Physico-chemical properties | |
| Physical state: | White solid [1] |
| Water solubility: | 8,100 mg L$^{-1}$ at 20°C [1] |
| Soil degradation, $DT_{50}$ (days): | 18.81 (lab at 20°C); 30 (field) [1] |
| Henry's law constant: | $2.90 \times 10^{-3}$ Pa m$^3$ mol$^{-1}$ at 25°C [1] |
| GUS leaching potential index: | 3.30 (high leachability) [1] |

## TRIFLURALIN

Identification of the substance

| | |
|---|---|
| CAS-No.: | 1582-09-8 |
| IUPAC name: | α,α,α-trifluoro-2,6-dinitro-*N*, *N*-dipropyl-p-toluidine |
| Molecular mass (g mol$^{-1}$): | 335.28 |
| Molecular formula: | $C_{13}H_{16}F_3N_3O_4$ |
| Structural formula: | 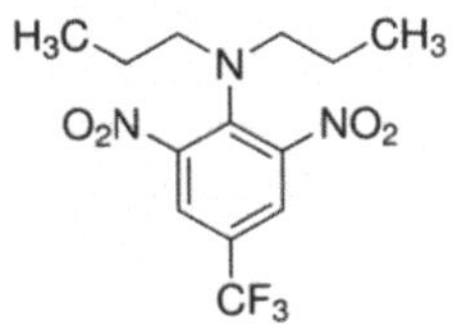 |
| Physico-chemical properties | |
| Physical state: | Orange-yellow crystalline solid [1] |
| Water solubility: | 0.221 mg L$^{-1}$ at 20°C [1] |
| Soil degradation, $DT_{50}$ (days): | 181 (lab at 20°C); 170 (field) [1] |
| Henry's law constant: | 10.2 Pa m$^3$ mol$^{-1}$ at 25°C [1] |
| GUS leaching potential index: | 0.13 (low leachability) [1] |

## ZIRAM

Identification of the substance

| | |
|---|---|
| CAS-No.: | 137-30-4 |
| IUPAC name: | zinc bis(dimethyldithiocarbamate) |
| Molecular mass (g $mol^{-1}$): | 305.84 |
| Molecular formula: | $C_6H_{12}N_2S_4Zn$ |
| Structural formula: | 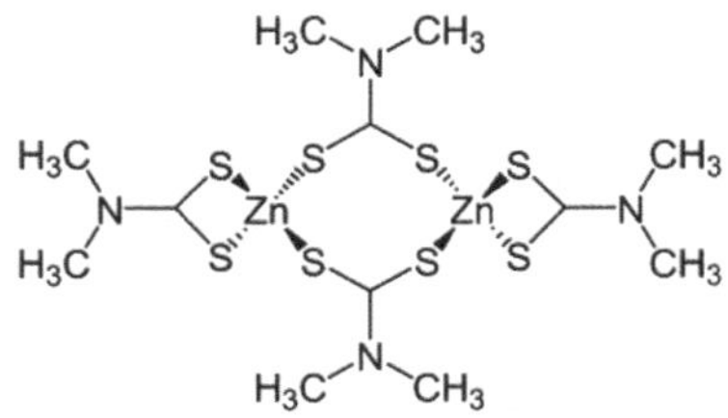 |
| Physico-chemical properties | |
| Physical state: | White powder [1] |
| Water solubility: | 0.967 mg $L^{-1}$ at 20°C [1] |
| Soil degradation, $DT_{50}$ (days): | 1.62 (lab at 20°C); 6.3 (field) [1] |
| Henry's law constant: | $2.3 \times 10^{-6}$ Pa $m^3$ $mol^{-1}$ at 25°C [1] |
| GUS leaching potential index: | 0.07 (low leachability) [1] |

# REFERENCES


1. https://sitem.herts.ac.uk/aeru/ppdb/en/index.htm
2. https://pubchem.ncbi.nlm.nih.gov/
3. http://www.fao.org/fileadmin/templates/agphome/documents/Pests_Pesticides/JMPR/Evaluation08/Lambda-cyhalotrhin.pdf
4. https://www.intechopen.com/books/groundwater-resource-characterisation-and-management-aspects/development-of-a-method-for-prediction-of-risk-of-surface-and-groundwater-contamination-with-pestici
5. http://www.thegoodscentscompany.com/data/rw1077161.html
6. EFSA Scientific Report (2005) 27, 1-81, Conclusion on the peer review of glufosinate https://efsa.onlinelibrary.wiley.com/doi/pdf/10.2903/j.efsa.2005.27r

# Appendix 4

# The chronology of the quaternary period

*David Mitrinović and Jelena Zarić*

The Quaternary chronology was originally developed by observing and studying the succession of glaciations through sedimentary deposits and geomorphological forms – soils that developed in warm climates positioned between glacial sediments deposited during glacials, marine deposits between fluvio-glacial sediments, or river terraces in the Alps. The chronology developed in this way suggested that the Pleistocene consisted of four or five major glacials separated by interglacials with a climate that was largely similar to the contemporary one (Hilton Johnson, 1991).

The two main classical systems of terrestrial Pleistocene chronology in Europe are Alpine (Eberl, 1930; Penck & Brückner, 1909; Schaefer, 1953) and northern European (Dubois, 1925; Keilhack, 1926; Madsen, 1928; Penck, 1922), on whose examples the principles of Quaternary chronology forming will be shown. The youngest four glacials and three interglacials, recognised by the 'classical' Alpine and northern European chronologies, correlated with the continuous records of the share of oxygen isotopes $^{18}O$ in the fossils of marine organisms (foraminifera) at the ocean floor, using loess sections and river terraces as a link. It has been established that alpine 'ice' stages consist of sediments formed during both glacials and interglacials, that classical alpine interglacials do not represent just the episodes of warmer climates, but probably intervals of intensified, accelerated tectonic movements, and that the physical evidence on which the northern European classical division is based is in part wrongly interpreted due to the long hiatuses in the sedimentary record (Kukla, 1978).

Due to the above, it was recommended back in the 1970s to discontinue the use of classical terminology in all interregional correlations and to base the chronostratigraphic division of the Pleistocene on the share of $^{18}0$ isotope in deep-sea sediments (Kukla, 1978).

In the system of Alpine chronology, glacials in the first, classical variant were named after the tributaries of the Danube and the Isar in Bavaria: Günz, Mindel, Riss and Würm (Penck & Brückner, 1909), and later Biber and Donau were added (Eberl, 1930; Schaefer, 1953). A typical site with all the geomorphological


Editor: Milan Dimkić
doi: 10.2166/9781789060904_0711

**Table A4.1** Geomorphological forms and units of specific lithogenesis that served for the chronological reconstruction of the glaciation sequence – classical Alpine and northern European Quaternary chronology (according to Kukla (1978)).

| Glacial or Interglacial | Geomorphological Unit | Glacial or Interglacial | Geomorphological or Units of Specific Lithogenesis |
|---|---|---|---|
| Biber | Highland gravel | | |
| Biber-Donau | | | |
| Donau | 'Donau' gravel | | |
| Donau-Günz | | | |
| Günz | Older cover gravel | | |
| Günz-Mindel | | Cromer | Alluvial sand, flood plain clays, with an abundance and a great variety of faunal fossils (Great Britain – East Anglia) |
| Mindel | Younger cover gravel | Elster | Glacial-lacustrine sediments (central Germany) |
| Mindel-Riss | | Holstein | Marine deposits over the glacio-fluvial Elster deposits (Germany – Schleswig-Holstein) |
| Riss | Higher terrace | Saale (Warthe) | Terminal moraines (central Germany) |
| Riss-Würm | | Eem | Marine clays and brackish water sediments (The Netherlands – Amersfoort) |
| Würm | Lower terrace | Weichsel | Terminal moraines (northern Germany and Poland) |
| Holocene | Recent alluvium | Flanders | Marine clays and brackish water sediments (The Netherlands) |

forms that served for the chronological reconstruction of the glaciation sequence is the Donau-Iller-Lech plateau. Geomorphological forms are given in Table A4.1. The lower terrace is associated with morphologically preserved moraines that do not look old, while the upper terrace is connected with much less preserved, so-called old moraines. The surface of the younger and older 'cover' gravel is decomposed and forms the red clay called 'Ferretto'. According to the theory that is the basis of this chronology, the interglacials left a trace in the form of erosion of the mentioned terraces and 'cover' gravel. The hypsometric difference between the younger 'cover' gravel and the higher terrace is particularly large in several locations, which was initially interpreted as a consequence of the long duration of the Mindel-Riss interglacial (Kukla, 1978).

The classical glacials of the northern European chronology of the Quaternary are in an area that was covered by a Scandinavian continental ice sheet during the Pleistocene and are named (in the order of increasing age) Weichsel, Warthe, Saale, and Elster. They were singled out on the basis of a system of terminal moraines, which stretches across Denmark, the Netherlands, northern Germany and Poland, and were named after the rivers from this area (Kukla, 1978). The morphology of the moraines from Weichsel and Warthe is preserved, the moraines of Saale are interrupted in many places and are mostly lowered and levelled, while the moraines of Elster have completely lost their original morphology, so the ice plate extent in that glacial is, in the most part, reconstructed on the basis of the position of erratic blocks and identification of redeposited moraine material. The older moraines extend further south and

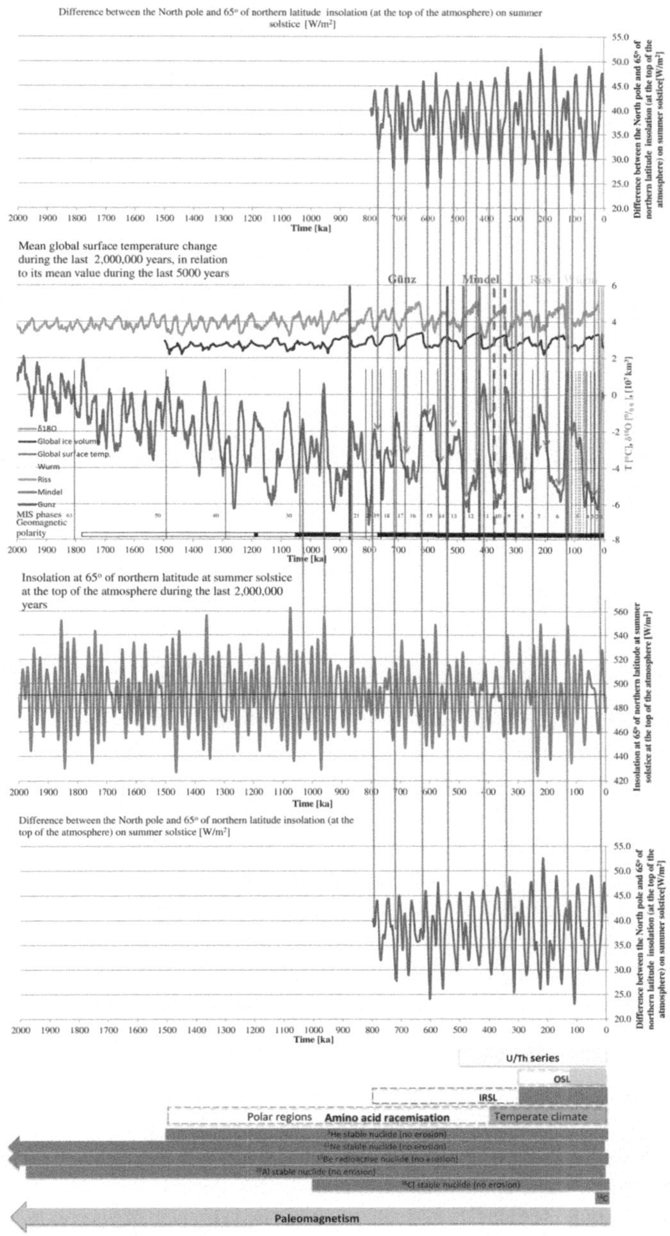


**Figure A4.1** Influence of astronomical factors (Milankovitch Orbital Data Viewer [MODW], 2020) on the mean global surface temperature (Snyder, 2016) and global ice volume (Pollard & DeConto, 2009) during the last 2,000,000 years, MIS phases and $^{18}O$ isotope content (Snyder, 2016), Alpine chronology (Deutsche Stratigraphische Kommission, 2016), geomagnetic scale with chrones and subchrones, and reach of dating methods (IRSL – Infra Red Stimulated Luminescence, OSL – Optically Stimulated Luminescence, cosmogenic nuclides, paleomagnetism, amino acid racemisation, U/Th series).

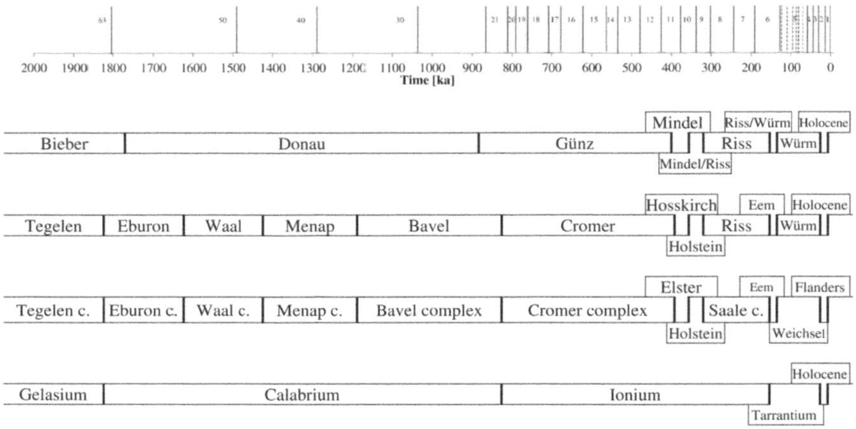


**Figure A4.1** (*Continued*). Chosen Quaternary chronology systems (Deutsche Stratigraphische Kommission, 2016).

are therefore more or less preserved. The classical interglacials of northern Europe are, in order of increasing age, Eem, Holstein and Cromer. They have been identified on the basis of marine ingress deposits during periods of higher sea levels, ice disappearance and warmer climates, and peat whose pollen content documents the presence of deciduous tree forests in north-western Europe. The stratigraphic position of the Eem interglacial and transgression was determined by its sediments being located between the redeposited moraine material of the Saale and Warthe glacials and the Weichsel moraines (Kukla, 1978). Similarly, marine deposits from the Holstein interglacial are located above the glacio-fluvial sediments from the Elster glacial, and below the Saale glacial deposits. The deposits of the so-called Cromer 'Forest' layers (rich in various faunal fossils) represented by clays, silt and sands of alluvial origin are a typical unit of the Cromer interglacial.

According to a more recent version of northern European chronology, the Saale complex (Figure A4.1) includes, in order of decreasing age, Fuhne, Drenthe (Saale in the stricter sense, the glacial with the greatest extent during the Saale complex) and Warthe glacials (Sirocko *et al.*, 2007). The Cromer complex (Figure A4.1) includes, in order of decreasing age, interglacial I (Cromer I), glacial A, interglacial II (Cromer II), glacial B, interglacial III (Cromer III), glacial C, and interglacial IV (Cromer IV or Volgstedt) (Ehlers *et al.*, 2011; Deutsche Stratigraphische Kommission, 2016).

## REFERENCES


Deutsche Stratigraphische Kommission (2016). Stratigraphische Tabelle von Deutschland 2016 (Stratigraphic table for Germany 2016). http://www.stratigraphie.de/std/Bilder/5_2.pdf (accessed 25 March 2020)

Dubois G. (1925). Recherches sur les terrains quaternaires du Nord de la France. *Mémoires de la Société Géologique du Nord*, **8**, 1–356.

Eberl B. (1930). Die Eiszeitenfolge im nordlichen Alpenvorland. Benno Filser Verlag, Augsburg.

Ehlers J., Gibbard P. L. and Hughes P. D. (eds.) (2011). Quaternary Glaciations – Extent and Chronology, 1st edn, Developments in Quaternary Science. Elsevier Science, Vol. **15**, ISBN: 9780444534477

Hilton Johnson W. (1991). Quaternary Period and Pleistocene Epoch, 15th edn. Encyclopedia Britannica, Encyclopædia Britannica, Inc., Chicago, IL, USA, Vol. **19**, pp. 858–867.

Keilhack K. (1926). Das Quartär in Grundzüge der Geologie. W. Salomon, ed., Stuttgart, Vol. **2**.

Kukla G. (1978). The Classical European Glacial Stages: Correlation with Deep-Sea Sediments. Transactions of the Nebraska Academy of Sciences, Nebraska Academy of Sciences, Lincoln, NE, USA, Vol. **VI**, pp. 57–93.

Madsen V. (1928). Übersicht über die Geologie von Danmark. Danmarks Geologiske Undersögelse, Copenhagen, Vol. **5**, p. 4.

Milankovitch Orbital Data Viewer (MODW) (2020). Milankovitch orbital parameters over time based on the calculations of: J. Laskar et al. (2004). A long term numerical solution for the insolation quantities of the Earth. *Astronomy and Astrophysics* **428**, 261–285, doi: 10.1051/0004-6361:20041335. https://biocycle.atmos.colostate.edu/shiny/Milankovitch/ (accessed 22 May 2020).

Penck A. (1922). Die Terrassen des Isartales in den Alpen. Sonderberichte d. preussischen Akadernie d. Wissenschaften. Phys.-math. KI.

Penck A. and Bruckner E. (1909). Die Alpen im Eiszeitalter. Chr.-Herm. Tauchnitz., Leipzig, Bd. **1–3**

Pollard D. and DeConto R. M. (2009). Modelling West Antarctic ice sheet growth and collapse through the past five million years. *Nature*, **458**, 329–332.

Schaefer I. (1953). Die donaueiszeitlichen Ablagerungen an Lech und Wertach. *Geologia Bavarica*, **19**, 13–64

Sirocko F., Claussen M., Litt T. and Sanchez-Goni M. F. (eds.) (2006). The Climate of Past Interglacials, 1st edn, Developments in Quaternary Science. Elsevier Science, Vol. **7**, ISBN: 9780444529558

Snyder C. W. (2016). Evolution of global temperature over the past two million years. Nature, **538**, 226–228.

# Index

## W

IWA Publishing's authorised EU representative for General Product Safety Regulations is Diane D'Arras, 15 rue Duret, 75116 Paris, France, e-mail: safety@iwap.co.uk.

Printed and bound by CPI Group (UK) Ltd, Croydon, CR0 4YY
12/05/2026
02108866-0016